Selected Keys of the Graphing Calculator

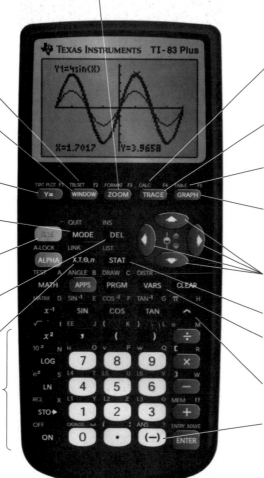

Magnifies or reduces a portion of the curve being viewed and can "square" the graph to reduce distortion.

Controls the values that are used when creating a table.

Determines the portion of the curve(s) shown and the scale of the graph.

Used to enter the equation(s) that is to be graphed.

Controls whether graphs are drawn sequentially or simultaneously and if the window is split.

Activates the secondary functions printed above many keys.

Used to delete previously entered characters.

Used to write the variable, x.

These keys are similar to those found on a scientific calculator.

Used to determine certain important values associated with a graph.

Used to display the coordinates of points on a curve.

Used to display x- and y-values in a table.

Used to graph equations that were entered using the Y= key.

Used to move the cursor and adjust contrast.

Used to fit curves to data.

Used to access a previously named function or equation.

Used to raise a base to a power.

Used as a negative sign.

Elementary and Intermediate Algebra: Graphs and Models

Elementary and Intermediate Algebra: Graphs and Models

THIRD EDITION

Marvin L. Bittinger

Indiana University Purdue University Indianapolis

David J. Ellenbogen

Community College of Vermont

Barbara L. Johnson

Indiana University Purdue University Indianapolis

PEARSON

Addison
Wesley

Boston   San Francisco   New York
London   Toronto   Sydney   Tokyo   Singapore   Madrid
Mexico City   Munich   Paris   Cape Town   Hong Kong   Montreal

Publisher	Greg Tobin
Editor in Chief	Maureen O'Connor
Acquisitions Editor	Randy Welch
Executive Project Manager	Kari Heen
Project Editor	Katie Nopper
Editorial Assistant	Antonio Arvelo
Production Manager	Ron Hampton
Editorial and Production Services	Martha K. Morong/Quadrata, Inc.
Art Editor and Photo Researcher	The Davis Group, Inc.
Compositor	BeaconPMG
Senior Media Producer	Ceci Fleming
Software Development	Mary Dougherty and Marty Wright
Marketing Manager	Jay Jenkins
Marketing Coordinator	Alexandra Waibel
Prepress Supervisor	Caroline Fell
Manufacturing Manager	Evelyn Beaton
Senior Media Buyer	Ginny Michaud
Text Designer	The Davis Group, Inc.
Cover Designer	Leslie Haimes
Cover Photograph	© Getty Images/Jeremy Walker

Photo credits appear on page G-8.

Library of Congress Cataloging-in-Publication Data

Bittinger, Marvin L.

　Elementary and intermediate algebra: graphs and models / Marvin L. Bittinger, David J. Ellenbogen, Barbara L. Johnson.—3rd ed.

　　p. cm.

　ISBN-13: 978-0-321-42240-8/ISBN-10: 0-321-42240-6

　　1. Algebra.　I. Ellenbogen, David.　II. Johnson, Barbara L.
　　(Barbara Loreen), 1962–III.

　Title.

QA152.3.B55　2007

512.9—dc22　　　　　　　　　　　　　　　　　　　　　　　2006049397

4 5 6 7 8 9 10—DOW—10 09

Contents

3   Introduction to Graphing and Functions   159

4   Systems of Equations in Two Variables   285

10   Exponents and Radical Functions   723

11   Quadratic Functions and Equations   803

*This chapter is found on the CD included in the back of this book.

*This chapter is found on the CD included in the back of this book.

Preface

Appropriate for two consecutive courses or a course combining the study of elementary and intermediate algebra, *Elementary and Intermediate Algebra: Graphs and Models*, Third Edition, covers both elementary and intermediate algebra topics without the repetition necessary in two separate texts. This text is more interactive than most other elementary and intermediate algebra texts. Our goal is to enhance the learning process by encouraging students to visualize the mathematics and by providing as much support as possible to help students in their study of algebra. This text is part of a series that includes the following texts:

- *Elementary Algebra: Graphs and Models*
- *Intermediate Algebra: Graphs and Models,* Third Edition

Content Features

Problem Solving

One distinguishing feature of our approach is our treatment of and emphasis on problem solving. We use problem solving and applications to motivate the material wherever possible, and we include real-life applications and problem-solving techniques throughout the text. Problem solving not only encourages students to think about how mathematics can be used, it helps to prepare them for more advanced material in future courses.

In Chapter 2, we introduce the five-step process for solving problems: (1) *Familiarize*, (2) *Translate*, (3) *Carry out*, (4) *Check*, and (5) *State* the answer. These steps are then used consistently throughout the text whenever we encounter a problem-solving situation. Repeated use of this problem-solving strategy gives students a sense that they have a starting point for any type of problem they encounter, and frees them to focus on the mathematics necessary to successfully translate the problem situation. We often use estimation and carefully checked guesses to help with the *Familiarize* and *Check* steps (see pp. 118, 123, and 324–325).

Algebraic/Graphical Side-by-Sides

Algebraic/graphical side-by-sides give students a direct comparison between these two problem-solving approaches. They show the connection between algebraic and graphical or visual solutions and demonstrate that there is more than one way to obtain a result. This feature also illustrates the comparative efficiency and accuracy of the two methods. (See pp. 318, 577, and 639.) Instructors using this text have found that it works superbly both in courses where the graphing calculator is required and in courses where it is optional.

Applications

Interesting applications of mathematics help motivate both students and instructors. Solving applied problems gives students the opportunity to see their conceptual understanding put to use in a real way. In the third edition of *Elementary and Intermediate Algebra: Graphs and Models*, the number of applications and source lines has been increased, and effort has been made to present the most current and relevant applications. As in the past, art is integrated into the applications and exercises to aid the student in visualizing the mathematics. (See pp. 130, 218, and 509.)

Interactive Discoveries

Interactive Discoveries invite students to develop analytical and reasoning skills while taking an active role in the learning process. These discoveries can be used as lecture launchers to introduce new topics at the beginning of a class and quickly guide students through a concept, or as out-of-class concept discoveries. (See pp. 222, 378, and 451.)

Pedagogical Features

New! **Concept Reinforcement Exercises.** This feature is designed to help students build their confidence and comprehension through true/false, matching, and fill-in-the-blank exercises at the beginning of most exercise sets. Whenever possible, special attention is devoted to increasing student understanding of the new vocabulary and notation developed in that section. (See pp. 90, 213, and 296.)

New! **Visualizing the Graph.** These matching exercises provide students with an opportunity to match an equation with its graph by focusing on the characteristics of the equation and the corresponding attributes of the graph. This feature occurs once in each chapter at the end of a related section. (See pp. 295, 381, and 511.)

New! **Student Notes.** These comments, strategically located in the margin within each section, are specific to the mathematics appearing on that page. Remarks are often more casual in format than the typical exposition and range from suggestions on how to avoid common mistakes to how to best read new mathematical notation. (See pp. 235, 287, and 310.)

New! **Tabbing for Success.** The new Tabbing for Success page, found on page xxv, provides students with forty color-coded, reusable tabs to quickly locate key examples, review important summaries, flag text topics, and highlight areas where they need help. Together, these features will help students better use their time and their textbooks to succeed in the course.

Connecting the Concepts. To help students understand the big picture, Connecting the Concepts subsections relate the concept at hand to previously learned and upcoming concepts. Because students occasionally lose sight of the forest because of the trees, this feature helps students keep their bearings as they encounter new material. (See pp. 169, 316, and 656.)

Study Tips. These remarks, located in the margin near the beginning of each section, provide suggestions for successful study habits that can be applied to both this and other college courses. Ranging from ideas for better time management to suggestions for test preparation, these comments can be useful even to experienced college students. (See pp. 9, 110, and 474.)

Skill Maintenance Exercises. Retention of skills is critical to a student's success in this and future courses. Thus, beginning in Section 1.2, every exercise set includes Skill Maintenance exercises that review skills and concepts from preceding sections of the text. Often, these exercises provide practice with specific skills needed for the next section of the text. **New to this edition,** some Skill Maintenance sections are titled **Focused Review.** These sections provide a set of mixed exercises **focused on reviewing connected skills,** such as factoring or solving equations, covered in separate sections or chapters. (See pp. 143, 306, and 359.)

Synthesis Exercises. Following the Skill Maintenance section, every exercise set ends with a group of Synthesis exercises that offers opportunities for students to synthesize skills and concepts from earlier sections with the present material, and often provide students with deeper insights into the current topic. Synthesis exercises are generally more challenging than those in the main body of the exercise set and occasionally include **Aha!** exercises (exercises that can be solved more quickly by intuition than by computation). (See pp. 117, 220, and 412.)

Thinking and Writing Exercises. Writing exercises have been found to aid in student comprehension, critical thinking, and conceptualization. Thus every set of exercises includes at least four Thinking and Writing exercises. Two of these appear just before the Skill Maintenance exercises. The others are more challenging and appear as Synthesis exercises. All are marked with **TW** and require answers that are one or more complete sentences. Because some instructors may collect answers to writing exercises, and because more than one answer may be correct, answers to the Thinking and Writing exercises are listed at the back of the text only when they are within review exercises. (See pp. 108, 173, and 371.)

Collaborative Corners. Studies have shown that students who work together generally outperform those who do not. Throughout the text, we provide optional Collaborative Corner features that require students to work in groups to explore and solve problems. There is at least one Collaborative Corner per chapter, each one appearing after the appropriate exercise set. (See pp. 100, 174, and 334.)

Chapter Summary. Each chapter summary contains a list of key terms from the chapter, with definitions, as well as formulas from the chapter. The second part of the summary is a two-column table: An important concept is shown in the first column, with an example explaining that concept appearing in the second column. Page numbers or section references are provided so students can reference the corresponding exposition in the chapter. The Summary provides a terrific point from which to begin reviewing for a chapter test. (See pp. 154, 273, and 346.)

What's New in the Third Edition?

We have rewritten many key topics in response to user and reviewer feedback and have made significant improvements in design, art, pedagogy, and an expanded supplements package. Information about the content changes is available in the form of a conversion guide in the *Instructor and Adjunct Support Manual*. Following is a list of the major changes in this edition.

New Design

While incorporating a new layout, a fresh palette of colors, and new features, we have maintained the larger page dimension for an open look and a typeface that is easy to read. As always, it is our goal to make the text look mature without being intimidating. In addition, we continue to pay close attention to the pedagogical use of color to make sure that it is used to present concepts in the clearest possible manner.

Content Changes

A variety of content changes have been made throughout the text. Some of the more significant changes are listed below.

- All of the topics related to graphing linear equations are now located in Chapter 3.
- Functions are introduced, as before, in Chapter 3. The algebra of functions, however, is moved to Chapter 5, where it is applied using polynomial operations.
- Solving systems of equations in two variables is moved to Chapter 4, immediately following the chapter covering linear equations. Systems of equations in three variables are discussed in Chapter 9, along with solving using matrices and Cramer's rule.
- Solving equations by graphing is discussed after systems of equations are introduced.
- Negative exponents are covered in Section 5.2, immediately following the discussion of properties of whole-number exponents.
- Chapter 5 now includes an improved discussion of the use of scientific notation and significant digits.
- Chapter 7 now places greater emphasis on identifying the domain of a rational function.
- Topics involving inequalities (other than linear inequalities, which remain in Chapter 2) are discussed in Chapter 8.
- Chapters 13 and 14 are now included on a CD, thus reducing the length of the text while retaining full coverage of conics and sequences.
- An appendix covering unit conversion and dimension analysis is added for those schools wishing to include this topic or for students who need a review of this topic.
- This edition includes an appendix covering the distance and midpoint formulas and equations of circles, providing flexibility in covering these topics at various points in a course syllabus.
- Throughout the text, there is an increased emphasis on students learning how to distinguish between equivalent expressions and equivalent equations.

Ancillaries

The following ancillaries are available to help both instructors and students use this text more effectively.

STUDENT SUPPLEMENTS

Student's Solutions Manual
(ISBN-13 978-0-321-42906-3)
(ISBN-10 0-321-42906-0)
- By James J. Ball and Rhea Meyerholtz, *Indiana State University*
- Contains completely worked-out solutions for all the odd-numbered exercises in the text, with the exception of the Thinking and Writing exercises, as well as completely worked-out solutions to all the exercises in the Chapter Reviews, Chapter Tests, and Cumulative Reviews.

Graphing Calculator Manual
(ISBN-13 978-0-321-42613-0)
(ISBN-10 0-321-42613-4)
- By Judith A. Penna, *Indiana University Purdue University Indianapolis*
- Uses actual examples and exercises from the text to help teach students to use the graphing calculator.
- Order of topics mirrors order in the text, providing a just-in-time mode of instruction.

Video Lectures on CD
(ISBN-13 978-0-321-42497-6)
(ISBN-10 0-321-42497-2)
- Complete set of digitized videos on CD-ROMs for student use at home or on campus.
- Presents a series of lectures correlated directly to the content of each section of the text.
- Features an engaging team of instructors including authors Barbara Johnson and David Ellenbogen who present material in a format that stresses student interaction, often using examples and exercises from the text.
- Ideal for distance learning or supplemental instruction.
- Includes an expandable window that shows text captioning. Captions can be turned on or off.

INSTRUCTOR SUPPLEMENTS

Annotated Instructor's Edition
(ISBN-13 978-0-321-42293-4)
(ISBN-10 0-321-42293-7)
- Includes answers to all exercises printed in blue on the same page as those exercises.

Instructor's Solutions Manual
(ISBN-13 978-0-321-43044-1)
(ISBN-10 0-321-43044-1)
- By James J. Ball and Rhea Meyerholtz, *Indiana State University*
- Contains full, worked-out solutions to all the exercises in the exercise sets, including the Thinking and Writing exercises, and worked-out solutions to all the exercises in the Chapter Reviews, Chapter Tests, and Cumulative Reviews.

Printed Test Bank
(ISBN-13 978-0-321-42905-6)
(ISBN-10 0-321-42905-2)
- By Carrie Green
- Provides 8 revised test forms for every chapter and 8 revised test forms for the final exam.
- For the chapter tests, test forms are organized by topic order following the chapter tests in the text, and 2 test forms are multiple choice.

New! Instructor and Adjunct Support Manual
(ISBN-13 978-0-321-43043-4)
(ISBN-10 0-321-43043-3)
- Features resources and teaching tips designed to help both new and adjunct faculty with course preparation and classroom management.
- Resources include extra practice sheets, conversion guide, video index, and transparency masters.
- Also available electronically so course/adjunct coordinators can customize material specific to their schools.

STUDENT SUPPLEMENTS

Addison-Wesley Math Tutor Center
www.aw-bc.com/tutorcenter

- The Addison-Wesley Math Tutor Center is staffed by qualified mathematics instructors who provide students with tutoring on examples and odd-numbered exercises from the textbook. Tutoring is available via toll-free telephone, toll-free fax, e-mail, or the Internet. White Board technology allows tutors and students to actually see problems worked while they "talk" in real time over the Internet during tutoring sessions.

MathXL® Tutorials on CD
(ISBN-13 978-0-321-42785-4)
(ISBN-10 0-321-42785-8)

- Provides algorithmically generated practice exercises that correlate at the objective level to the content of the text.
- Includes an example and a guided solution to accompany every exercise and video clips for selected exercises.
- Recognizes student errors and provides feedback; generates printed summaries of students' progress.

INSTRUCTOR SUPPLEMENTS

TestGen with Quizmaster
(ISBN-13 978-0-321-42495-2)
(ISBN-10 0-321-42495-6)

- Enables instructors to build, edit, print, and administer tests.
- Features a computerized bank of questions developed to cover all text objectives.
- Algorithmically based content allows instructors to create multiple but equivalent versions of the same question or test with a click of a button.
- Instructors can also modify test-bank questions or add new questions by using the built-in question editor, which allows users to create graphs, input graphics, and insert math notation, variable numbers, or text.
- Tests can be printed or administered online via the Internet or another network. Quizmaster allows students to take tests on a local area network.
- Available on a dual-platform Windows/Macintosh CD-ROM.

MathXL® www.mathxl.com   MathXL is a powerful online homework, tutorial, and assessment system that accompanies Addison-Wesley textbooks in mathematics or statistics. With MathXL, instructors can create, edit, and assign online homework and tests using algorithmically generated exercises correlated at the objective level to the textbook. They can also create and assign their own online exercises and import TestGen tests for added flexibility. All student work is tracked in MathXL's online gradebook. Students can take chapter tests in MathXL and receive personalized study plans based on their test results. The study plan diagnoses weaknesses and links students directly to tutorial exercises for the objectives they need to study and retest. Students can also access supplemental animations and video clips directly from selected exercises. MathXL is available to qualified adopters. For more information, visit our Web site at www.mathxl.com or contact your Addison-Wesley representative.

MyMathLab® www.mymathlab.com   MyMathLab is a series of text-specific, easily customizable online courses for Addison-Wesley textbooks in mathematics and statistics. Powered by CourseCompass™ (Pearson Education's online teaching and learning environment) and MathXL® (our online homework, tutorial, and assessment system), MyMathLab gives instructors the tools they need to deliver all or a portion of their course online, whether students are in a lab setting or working from home. MyMathLab provides a rich and flexible set of course materials, featuring free-response exercises that are algorithmically generated for unlimited practice and mastery. Students can also use online tools, such as video lectures, animations, and a multimedia textbook, to independently improve their understanding and performance. Instructors can use MyMathLab's homework and test managers to select and assign online exercises correlated directly to the textbook, and they can also create and assign their own online exercises and import TestGen tests for added flexibility. MyMathLab's online gradebook—designed specifically for mathematics and statistics—automatically tracks students' homework and test results and gives the instructor control over how to calculate final grades. Instructors can also add offline (paper-and-pencil) grades to the gradebook. MyMathLab is available to qualified adopters. For more information, visit our Web site at www.mymathlab.com or contact your Addison-Wesley representative.

InterAct Math® Tutorial Web site www.interactmath.com   Get practice and tutorial help online! This interactive tutorial Web site provides algorithmically generated practice exercises that correlate directly to the exercises in the textbook. Students can retry an exercise as many times as they like with new values each time for unlimited practice and mastery. Every exercise is accompanied by an interactive guided solution that provides helpful feedback for incorrect answers, and students can also view a worked-out sample problem that steps them through an exercise similar to the one they're working on.

Addison-Wesley Math Adjunct Support Center   The Addison-Wesley Math Adjunct Support Center is staffed by qualified mathematics instructors with over 50 years of combined experience at both the community college and university level. Assistance is provided for faculty in the following areas:

- Suggested syllabus consultation
- Tips on using materials packed with your book
- Book-specific content assistance
- Teaching suggestions including advice on classroom strategies

For more information, visit www.aw-bc.com/tutorcenter/math-adjunct.html

Acknowledgments

No book can be produced without a team of professionals who take pride in their work and are willing to put in long hours. Thanks to James J. Ball and Rhea Meyerholtz, for authoring the *Instructor's Solutions Manual* and the *Student's Solutions Manual,* and to Carrie Green, for authoring the *Printed Test Bank.* Sybil MacBeth, Perian Herring, Holly Martinez, and Sharon O'Donnell provided enormous help, often in the face of great time pressure, as accuracy checkers. We are also indebted to Allegra Atkinson for her help with applications research.

Martha Morong, of Quadrata, Inc., provided editorial and production services of the highest quality imaginable—she is simply a joy to work with. Geri Davis, of the Davis Group, Inc., performed superb work as designer, art editor, and photo researcher, and always with a disposition that can brighten an otherwise gray day. Network Graphics generated the graphs, charts, and many of the illustrations. Not only are the people at Network reliable, but they clearly take pride in their work. The many representational illustrations appear thanks to Bill Melvin, a gifted artist with true mathematical sensibilities.

Our team at Addison-Wesley deserves special thanks. Editorial Assistant Antonio Arvelo managed many of the day-to-day details—always in a pleasant and reliable manner. Project Editor Katie Nopper expertly provided information and a steadying influence along with gentle prodding at just the right moments. Acquisitions Editor Randy Welch provided many fine suggestions along with unflagging support. Production Manager Ron Hampton exhibited careful supervision and an eye for detail throughout production. Marketing Manager Jay Jenkins and Marketing Coordinator Alexandra Waibel skillfully kept us in touch with the needs of faculty. Senior Media Producer Ceci Fleming provided us with the technological guidance so necessary for our many supplements and our fine video series. To all of these people we owe a real debt of gratitude.

Reviewers

Renee Aschbrenner, *Hawkeye Community College*
Alan Bass, *Mesa College*
Peggy Clifton, *Redlands Community College*
Andrea Hendricks, *Georgia Perimeter College–Clarkston*
Bonnie Hodge, *Austin Peay State University*
Kathy Kopelousos, *Lewis & Clark Community College*
Kim Nunn, *Northeast State Technical Community College*
Sandra Pierce, *Maryland Community College*
Barbara Sehr, *Indiana University–Kokomo*
Scott Owen Thompson, *Sheridan College*
Julie Turnbow, *Collin County Community College*
Kevin Yokoyama, *College of the Redwoods*

M.L.B.
D.J.E.
B.L.J.

Feature Walkthrough

What's New

New!

TABBING FOR SUCCESS

The new Tabbing for Success page, found on page xxv, provides students with forty color-coded, reusable tabs to quickly locate key examples, review important summaries, flag text topics, and highlight areas where they need help. Together, these features will help students better use their time and their textbooks to succeed in the course.

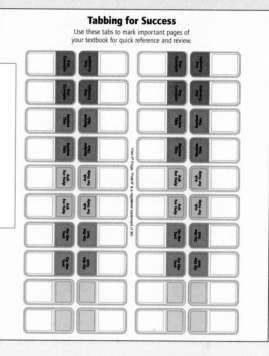

Tabbing for Success

Use these tabs to mark important pages of your textbook for quick reference and review.

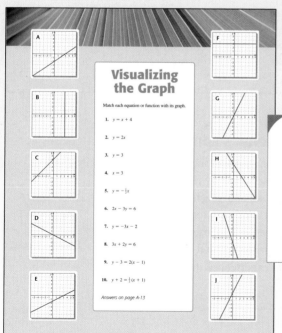

Visualizing the Graph

Match each equation or function with its graph.

1. $y = x + 4$
2. $y = 2x$
3. $y = 3$
4. $x = 3$
5. $y = -\frac{1}{2}x$
6. $2x - 3y = 6$
7. $y = -3x - 2$
8. $3x + 2y = 6$
9. $y - 3 = 2(x - 1)$
10. $y + 2 = \frac{1}{2}(x + 1)$

Answers on page A-13

New!

VISUALIZING THE GRAPH

Occurring once in each chapter, Visualizing the Graph matching exercises provide students with an opportunity to match an equation with its graph by focusing on the characteristics of the equation and the corresponding attributes of the graph.

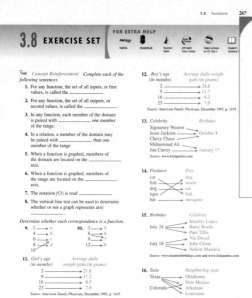

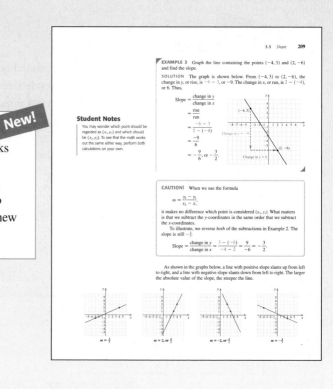

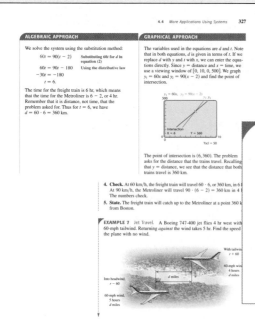

CONCEPT REINFORCEMENT EXERCISES — New!

Concept Reinforcement Exercises are designed to help students build their confidence and comprehension through true/false, matching, and fill-in-the-blank exercises at the beginning of most exercise sets. Whenever possible, special attention is devoted to increasing student understanding of the new vocabulary and notation developed in that section.

STUDENT NOTES — New!

Student Notes are strategically located remarks in the margin within each section, and are specific to the mathematics appearing on that page. They range from suggestions on how to avoid common mistakes to how to best read new mathematical notation.

ANNOTATED INSTRUCTOR'S EDITION

The Annotated Instructor's Edition includes all the answers to the exercise sets, usually right on the page where the exercises appear, and Teaching Tips in the margins that give insights and classroom discussion suggestions that will be especially useful for new instructors. These handy answers and Teaching Tips will help both new and experienced instructors save preparation time.

CHAPTER OPENERS

Each chapter opens with a real-data application, including a data table (numerical representation) and a graphical representation of the data. Data tables and graphs are used frequently throughout the body of the text to show the relevance of the material as well as to appeal to the visual learners.

CALCULATOR REFERENCES

Throughout each chapter, calculator references act as an in-text manual to introduce and explain various graphing-calculator techniques. The references are placed so that the student can learn the technique just prior to practicing it.

STUDY TIPS

To help students develop good study habits throughout this course, Study Tips are placed near the beginning of each section in the margins. These tips range from how to approach assignments, to reminders of the various study aids that are available, to strategies for preparing for a final exam.

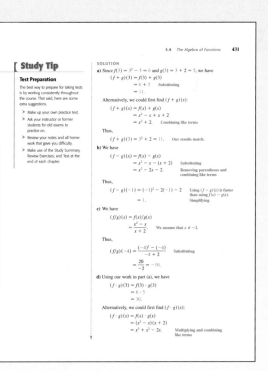

Examples

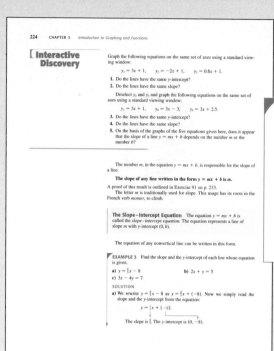

[Interactive Discovery

Graph the following equations on the same set of axes using a standard viewing window:

$$y_1 = 3x + 1, \qquad y_2 = -2x + 1, \qquad y_3 = 0.6x + 1.$$

1. Do the lines have the same y-intercept?

2. Do the lines have the same slope?

Deselect y_2 and y_3 and graph the following equations on the same set of axes using a standard viewing window:

$$y_1 = 3x + 1, \qquad y_4 = 3x - 3, \qquad y_5 = 3x + 2.5.$$

3. Do the lines have the same y-intercept?

4. Do the lines have the same slope?

5. On the basis of the graphs of the five equations given here, does it appear that the slope of a line $y = mx + b$ depends on the number m or the number b?

The number m, in the equation $y = mx + b$, is responsible for the slope of a line.

The slope of any line written in the form $y = mx + b$ is m.

A proof of this result is outlined in Exercise 91 on p. 233.

The letter m is traditionally used for slope. This usage has its roots in the French verb *monter*, to climb.

> **The Slope–Intercept Equation**   The equation $y = mx + b$ is called the *slope–intercept equation*. The equation represents a line of slope m with y-intercept $(0, b)$.

The equation of any nonvertical line can be written in this form.

EXAMPLE 3   Find the slope and the y-intercept of each line whose equation is given.

a) $y = \frac{4}{5}x - 8$ **b)** $2x + y = 5$
c) $3x - 4y = 7$

SOLUTION

a) We rewrite $y = \frac{4}{5}x - 8$ as $y = \frac{4}{5}x + (-8)$. Now we simply read the slope and the y-intercept from the equation:

$$y = \tfrac{4}{5}x + (-8).$$

The slope is $\frac{4}{5}$. The y-intercept is $(0, -8)$.

INTERACTIVE DISCOVERIES

Interactive Discoveries invite students to develop analytical and reasoning skills while taking an active role in the learning process. These discoveries can be used as lecture launchers to introduce new topics at the start of a class and quickly guide students through a concept, or as out-of-class concept discoveries.

ANNOTATED EXAMPLES

Learning is carefully guided with numerous color-coded art pieces and step-by-step annotations alongside examples. Substitutions and annotations are highlighted in red so students can see exactly what is happening in each step.

Dependency, Inconsistency, and Geometric Considerations

Each equation in Examples 2, 3, and 4 has a graph that is a plane in three dimensions. The solutions are points common to the planes of each system. Since three planes can have an infinite number of points in common or no points at all in common, we need to generalize the concept of *consistency*.

Planes intersect at one point. System is consistent and has one solution.

Planes intersect along a common line. System is consistent and has an infinite number of solutions.

Three parallel planes. System is *inconsistent*; it has no solution.

Planes intersect two at a time, with no point common to all three. System is *inconsistent*; it has no solution.

> **Consistency**
> A system of equations that has at least one solution is said to be **consistent**.
> A system of equations that has no solution is said to be **inconsistent**.

EXAMPLE 5   Solve:

$$
\begin{aligned}
y + 3z &= 4, & (1)\\
-x - \ y + 2z &= 0, & (2)\\
x + 2y + \ z &= 1. & (3)
\end{aligned}
$$

SOLUTION   The variable x is missing in equation (1). By adding equations (2) and (3), we can find a second equation in which x is missing:

$$
\begin{aligned}
-x - \ y + 2z &= 0 & (2)\\
x + 2y + \ z &= 1 & (3)\\
\hline
y + 3z &= 1. & (4) \quad \text{Adding}
\end{aligned}
$$

Equations (1) and (4) form a system in y and z. We solve as before:

$$
\begin{aligned}
y + 3z &= 4, && \text{Multiplying both sides} && -y - 3z = -4 \\
y + 3z &= 1 && \text{of equation (1) by } -1 && \underline{\ \ y + 3z = \ \ 1} \\
& && && \text{This is a contradiction.} \longrightarrow 0 = -3. \quad \text{Adding}
\end{aligned}
$$

Since we end up with a *false* equation, or contradiction, we know that the system has no solution. It is *inconsistent*.

The notion of *dependency* from Section 4.1 can also be extended.

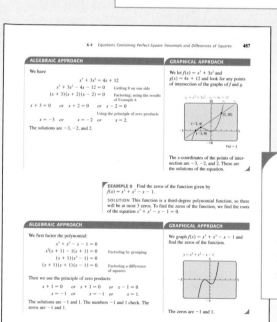

ALGEBRAIC APPROACH

We have

$$
\begin{aligned}
x^3 + 3x^2 &= 4x + 12 \\
x^3 + 3x^2 - 4x - 12 &= 0 && \text{Getting 0 on one side} \\
(x + 3)(x + 2)(x - 2) &= 0 && \text{Factoring; using the results of Example 6} \\
x + 3 = 0 \quad or \quad x + 2 = 0 & \quad or \quad x - 2 = 0 && \text{Using the principle of zero products} \\
x = -3 \quad or \quad x = -2 & \quad or \quad x = 2.
\end{aligned}
$$

The solutions are -3, -2, and 2.

GRAPHICAL APPROACH

We let $f(x) = x^3 + 3x^2$ and $g(x) = 4x + 12$ and look for any points of intersection of the graphs of f and g.

The x-coordinates of the points of intersection are -3, -2, and 2. These are the solutions of the equation.

EXAMPLE 9   Find the zeros of the function given by $f(x) = x^3 + x^2 - x - 1$.

SOLUTION   This function is a third-degree polynomial function, so there will be at most 3 zeros. To find the zeros of the function, we find the roots of the equation $x^3 + x^2 - x - 1 = 0$.

ALGEBRAIC APPROACH

We first factor the polynomial:

$$
\begin{aligned}
x^3 + x^2 - x - 1 &= 0 \\
x^2(x + 1) - 1(x + 1) &= 0 && \text{Factoring by grouping} \\
(x + 1)(x^2 - 1) &= 0 \\
(x + 1)(x + 1)(x - 1) &= 0. && \text{Factoring a difference of squares}
\end{aligned}
$$

Then we use the principle of zero products:

$$
\begin{aligned}
x + 1 = 0 \quad or \quad x + 1 = 0 & \quad or \quad x - 1 = 0 \\
x = -1 \quad or \quad x = -1 & \quad or \quad x = 1.
\end{aligned}
$$

The solutions are -1 and 1. The numbers 1 and 1 check. The zeros are -1 and 1.

GRAPHICAL APPROACH

We graph $f(x) = x^3 + x^2 - x - 1$ and find the zeros of the function.

$$y = x^3 + x^2 - x - 1$$

The zeros are -1 and 1.

ALGEBRAIC/GRAPHICAL SIDE-BY-SIDES

Algebraic/Graphical Side-by-Sides give students a direct comparison between these two problem-solving approaches. They show the connection between algebraic and graphical or visual solutions and demonstrate that there is more than one way to obtain a result. This feature also illustrates the comparative efficiency and accuracy of the two methods.

Student Guidance

FIVE-STEP PROBLEM-SOLVING PROCESS

Bittinger's five-step problem-solving process (Familiarize, Translate, Carry Out, Check, State) is introduced early in the text and used consistently throughout the text whenever students encounter an application problem. This provides students with a reliable framework for solving application problems.

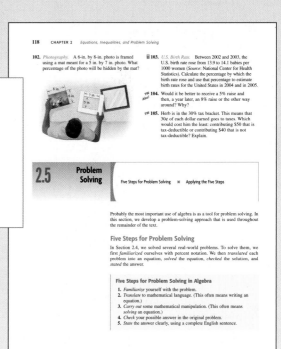

Using the Tabbing for Success

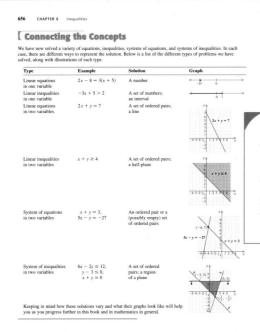

CONNECTING THE CONCEPTS

To help students understand the big picture, Connecting the Concepts subsections relate the concept at hand to previously learned and upcoming concepts. Because students occasionally lose sight of the forest because of the trees, this feature helps students keep their bearings as they encounter new material.

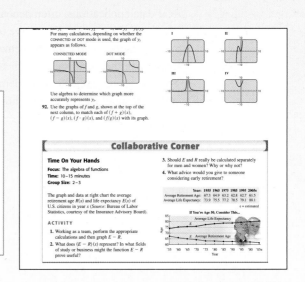

COLLABORATIVE CORNERS

Collaborative Corner exercises appear one to three times per chapter, after the section exercise set. Designed for group work, these optional activities provide instructors and students with the opportunity to incorporate group learning and/or class discussions.

Exercises

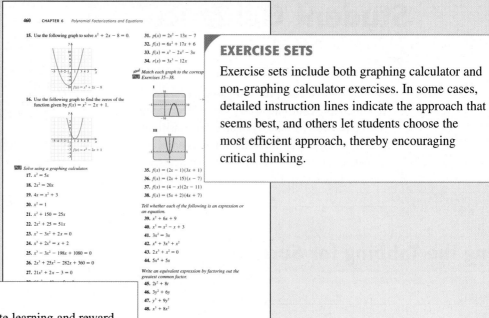

15. Use the following graph to solve $x^2 + 2x - 8 = 0$.

$f(x) = x^2 + 2x - 8$

16. Use the following graph to find the zeros of the function given by $f(x) = x^2 - 2x + 1$.

$f(x) = x^2 - 2x + 1$

Solve using a graphing calculator.

17. $x^2 = 5x$

18. $2x^2 = 20x$

19. $4x = x^2 + 3$

20. $x^2 = 1$

21. $x^2 + 150 = 25x$

22. $2x^2 + 25 = 51x$

23. $x^3 - 3x^2 + 2x = 0$

24. $x^3 + 2x^2 = x + 2$

25. $x^3 - 3x^2 - 198x + 1080 = 0$

26. $2x^3 + 25x^2 - 282x + 360 = 0$

27. $21x^2 + 2x - 3 = 0$

31. $p(x) = 2x^2 - 13x - 7$

32. $f(x) = 6x^2 + 17x + 6$

33. $f(x) = x^3 - 2x^2 - 3x$

34. $r(x) = 3x^3 - 12x$

Match each graph to the corresp... Exercises 35–38.

I

III

35. $f(x) = (2x - 1)(3x + 1)$

36. $f(x) = (2x + 15)(x - 7)$

37. $f(x) = (4 - x)(2x - 11)$

38. $f(x) = (5x + 2)(4x + 7)$

Tell whether each of the following is an expression or an equation.

39. $x^2 + 6x + 9$

40. $x^3 = x^2 - x + 3$

41. $3x^2 = 3x$

42. $x^4 + 3x^3 + x^2$

43. $2x^3 + x^2 = 0$

44. $5x^4 + 5x$

Write an equivalent expression by factoring out the greatest common factor.

45. $2t^2 + 8t$

46. $3y^2 + 6y$

47. $y^3 + 9y^2$

48. $x^3 + 8x^2$

EXERCISE SETS

Exercise sets include both graphing calculator and non-graphing calculator exercises. In some cases, detailed instruction lines indicate the approach that seems best, and others let students choose the most efficient approach, thereby encouraging critical thinking.

AHA! EXERCISES

In an effort to discourage rote learning and reward students who think a problem through before solving it, we include exercises that can be solved quickly if the student has the proper insight. The *Aha!* designation is used the first time a new insight can be used on a particular type of exercise and indicates to the student that there is a way to complete the exercise with more concise computation. It's then up to the student to find the simpler approach and, in subsequent exercises, to determine if and when that particular insight might be used again.

REAL-DATA APPLICATIONS

Real-data applications, often denoted in examples and exercises with a source line, serve to motivate students by showing them how the mathematics they are studying in their text relates to their everyday lives. The number of application problems has been increased significantly in this edition.

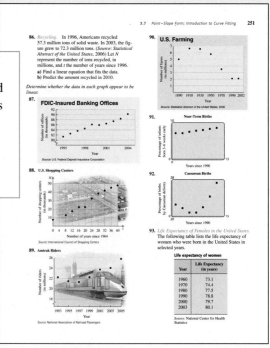

86. *Recycling.* In 1996, Americans recycled 57.3 million tons of solid waste. In 2003, the figure grew to 72.3 million tons. (*Source: Statistical Abstract of the United States*, 2006) Let N represent the number of tons recycled, in millions, and t the number of years since 1996.
 a) Find a linear equation that fits the data.
 b) Predict the amount recycled in 2010.

Determine whether the data in each graph appear to be linear.

87. **FDIC-Insured Banking Offices**

88. **U.S. Shopping Centers**

89. **Amtrak Riders**

90. **U.S. Farming**

91. **Near-Term Births**

92. **Caesarean Births**

93. *Life Expectancy of Females in the United States.* The following table lists the life expectancy of women who were born in the United States in selected years.

Life expectancy of women

Year	Life Expectancy (in years)
1960	73.1
1970	74.4
1980	77.5
1990	78.8
2000	79.7
2003	80.1

Source: National Center for Health Statistics

109. *Trade-In Value.* The trade-in value of a Homelite snowblower can be determined using the function given by $v(n) = -150n + 900$. Here $v(n)$ is the trade-in value, in dollars, after n winters of use.

_____nify?

_____/blower

_____hn

_____d using

_____0.

_____fter x

_____gnify?

_____200?

111. Explain why the domain of the function given by $f(x) = \dfrac{x+3}{2}$ is $\mathbb{R}$, but the domain of the function given by $g(x) = \dfrac{2}{x+3}$ is not $\mathbb{R}$.

112. Alayna asserts that for a function described by a set of ordered pairs, the range of the function will always have the same number of elements as there are ordered pairs. Is she correct? Why or why not?

Focused Review

Graph by hand.

113. $y = 2x + 3$ [3.6]

114. $y - 3 = -(x + 4$ [3.7]

115. $y = -1$ [3.3]

116. $f(x) = \frac{1}{2}x - 2$ [3.6], [3.8]

Graph using a graphing calculator. [3.2]

117. $y = x^2$

118. $y = 0.1x - 5.7$

119. $y = |x|$

120. $y = \frac{1}{3}x - 15$

Synthesis

121. For the function given by $n(z) = ab + wz$, what is the independent variable? How can you tell?

122. Explain in your own words why every function is a relation, but not every relation is a function.

For Exercises 123 and 124, let $f(x) = 3x^2 - 1$ and $g(x) = 2x + 5$.

123. Find $f(g(-4))$ and $g(f(-4))$.

124. Find $f(g(-1))$ and $g(f(-1))$.

125. If f represents the function in Exercise 14, find $f(f(f(f(tiger))))$.

Pregnancy. For Exercises 126–129, use the following graph of a woman's "stress test." This graph shows the size of a pregnant woman's contractions as a function of time.

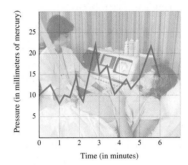

126. How large is the largest contraction that occurred during the test?

127. At what time during the test did the largest contraction occur?

128. On the basis of the information provided, how large a contraction would you expect 60 sec after the end of the test? Why?

131. Suppose that a function g is such that $g(-1) = -7$ and $g(3) = 8$. Find a formula for g if $g(x)$ is of the form $g(x) = mx + b$, where m and b are constants.

Given that $f(x) = mx + b$, classify each of the following as true or false.

132. $f(c + d) = f(c) + f(d)$

133. $f(cd) = f(c)f(d)$

134. $f(kx) = kf(x)$

135. $f(c - d) = f(c) - f(d)$

Summary

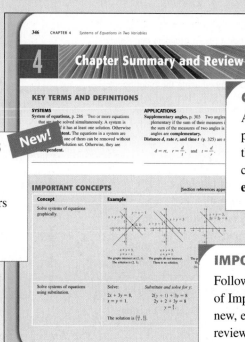

KEY TERMS AND DEFINITIONS

New!

A list of Key Terms and Definitions from the chapter is provided along with the corresponding page numbers to serve as a reference point for students.

CHAPTER SUMMARY AND REVIEW

At the end of each chapter, students can practice all they have learned as well as tie the current chapter material to material covered in earlier chapters in the **enhanced** chapter summary and review.

IMPORTANT CONCEPTS

Following the Key Terms and Definitions, a list of Important Concepts with new examples in a new, enhanced format offers students a quick review before beginning the review exercises.

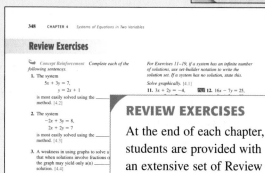

REVIEW EXERCISES

At the end of each chapter, students are provided with an extensive set of Review Exercises that provides practice for the key chapter material. These exercises include a wide variety of exercises, including concept reinforcement and synthesis exercises.

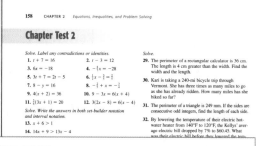

CHAPTER TEST

Following the Review Exercises, a sample Chapter Test allows students to review and test comprehension of chapter skills prior to taking an instructor's exam.

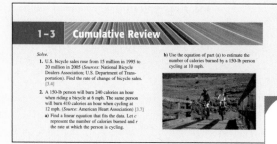

CUMULATIVE REVIEW

Following every three chapters, students will find a cumulative review, which covers skills and concepts from all preceding chapters of the text.

1

Introduction to Algebraic Expressions

P roblem solving is the focus of this text. Chapter 1 presents some preliminaries that are needed for the problem-solving approach that is developed in Chapter 2 and used throughout the rest of the book. These preliminaries include a review of arithmetic, a discussion of real numbers and their properties, and an examination of how real numbers are added, subtracted, multiplied, divided, and raised to powers. The graphing calculator is also introduced as a problem-solving tool.

APPLICATION   *Number of Drivers.*

The following table lists the number of vehicle miles traveled annually per household by the number of drivers in the household. Find an equation for the number v of vehicle miles driven in a household with d drivers.

NUMBER OF DRIVERS	NUMBER OF VEHICLE MILES TRAVELED
1	10,000
2	20,000
3	30,000
4	40,000

Source: Energy Information Administration

This problem appears as Exercise 83 in Exercise Set 1.1.

1.1   Introduction to Algebra

Algebraic Expressions ◼ Translating to Algebraic Expressions ◼
Translating to Equations ◼ Models

This section introduces some basic concepts and expressions used in algebra. Solving real-world problems is an important part of algebra, so we will focus on the wordings and mathematical expressions that often arise in applications.

Algebraic Expressions

Probably the greatest difference between arithmetic and algebra is the use of *variables* in algebra. When a letter can represent any one of a set of numbers, that letter is a **variable.** For example, if n represents the number of tickets sold for a Mariah Carey concert, then n will vary, depending on factors like price and day of the week. Thus the number n is a variable. If every ticket costs \$45, then a total of $45 \cdot n$ dollars will be paid for tickets. Note that $45 \cdot n$ means 45 *times n.* The number 45 is an example of a **constant** because it does not change.

The expression $45 \cdot n$ is a **variable expression** because its value varies with the replacement for n. In the following chart, we replace n with a variety of values and compute the total amount collected. In doing so, we are **evaluating the expression** $45 \cdot n$.

Cost per Ticket (in dollars), 45	Number of Tickets Sold, n	Total Collected (in dollars), $45 \cdot n$
45	400	\$18,000
45	500	22,500
45	600	27,000

Variable expressions are examples of *algebraic expressions.* An **algebraic expression** consists of variables and/or numerals, often with operation signs and grouping symbols. Examples are

$$t + 97, \quad 5 \cdot n, \quad 3a - b, \quad 18 \div y, \quad \frac{9}{7}, \quad \text{and} \quad 4r(s + t).$$

Recall that a fraction bar is a division symbol: $\frac{9}{7}$, or 9/7, means $9 \div 7$. Similarly, multiplication can be written in several ways. For example, "5 times x" can be written as $5 \cdot x$, $5 \times x$, $5(x)$, $5 * x$, or simply $5x$.

To evaluate an algebraic expression, we **substitute** a number for each variable in the expression and calculate the result.

EXAMPLE 1   Evaluate each expression for the given values.

a) $x + y$ for $x = 37$ and $y = 28$ **b)** $5ab$ for $a = 2$ and $b = 3$

SOLUTION

a) We substitute 37 for x and 28 for y and carry out the addition:

$$x + y = 37 + 28 = 65.$$

The number 65 is called the **value** of the expression.

b) We substitute 2 for a and 3 for b and multiply:

$$5ab = 5 \cdot 2 \cdot 3 = 10 \cdot 3 = 30. \qquad \textbf{5}ab \textbf{ means 5 times } a \textbf{ times } b.$$

EXAMPLE 2   The area A of a rectangle of length l and width w is given by the formula $A = lw$. Find the area when l is 17 in. and w is 10 in.

SOLUTION   We evaluate, substituting 17 in. for l and 10 in. for w and carrying out the multiplication:

$$
\begin{aligned}
A = lw &= (17 \text{ in.})(10 \text{ in.}) \\
&= (17)(10)(\text{in.})(\text{in.}) \\
&= 170 \text{ in}^2, \text{ or } 170 \text{ square inches.}
\end{aligned}
$$

Note that we always use square units for area and $(\text{in.})(\text{in.}) = \text{in}^2$. Exponents like the 2 in the expression in^2 are discussed further in Section 1.8.

EXAMPLE 3   The area of a triangle with a base of length b and a height of length h is given by the formula $A = \frac{1}{2}bh$. Find the area when b is 8 m and h is 6.4 m.

SOLUTION   We substitute 8 m for b and 6.4 m for h and then multiply:

$$
\begin{aligned}
A = \tfrac{1}{2}bh &= \tfrac{1}{2}(8 \text{ m})(6.4 \text{ m}) \\
&= \tfrac{1}{2}(8)(6.4)(\text{m})(\text{m}) \\
&= 4(6.4) \text{ m}^2 \\
&= 25.6 \text{ m}^2, \text{ or } 25.6 \text{ square meters.}
\end{aligned}
$$

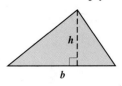

Student Notes

Using a graphing calculator is not a substitute for understanding mathematical procedures. Be sure that you understand a new procedure before you rely on a graphing calculator to do the operations. A calculator can be useful to check your answers.

Introduction to the Graphing Calculator

Graphing calculators and computers equipped with graphing software can be valuable aids in understanding and applying algebra. Features and keystrokes vary among the many brands and models of graphing calculators available. In this text, we use features that are common to most graphing calculators. Specific keystrokes and instructions for certain calculators are included in the Graphing Calculator Manual that accompanies this book. For other procedures, you should consult your instructor or the user's manual for your particular calculator.

There are two important things to keep in mind as you proceed through this text:

1. You will not learn to use a graphing calculator by simply reading about it; you must in fact *use* the calculator. Press the keys on your own

(*continued*)

calculator as you read the text, do the calculator exercises in the exercise set, and experiment as you learn new procedures.

2. Your user's manual contains more information about your calculator than appears in this text. If you need additional explanation and examples, be sure to consult the manual.

Keypad   A diagram of the keypad of a graphing calculator appears at the front of this text. The organization and labeling of the keys differ for different calculators. Note that there are options written above keys as well as on the keys. To access the options shown above the keys, press **2ND** or **ALPHA**, depending on the color of the desired option, and then the key below the desired option.

Screen   After you have turned the calculator on, you should see a blinking rectangle, or **cursor,** at the top left corner of the screen. If you do not see anything, try adjusting the **contrast.** On many calculators, this is done by pressing **2ND** and the up or down arrow keys. The up key darkens the screen; the down key lightens the screen. To perform computations, you should be in the **home screen.** Pressing (QUIT) (often the 2nd option associated with the MODE key) will return you to the home screen.

Performing Operations   To perform addition, subtraction, multiplication, and division using a graphing calculator, type in the expression as it is written. The entire expression will appear on the screen, and you can check your typing. A multiplication symbol will appear on the screen as *, and division is usually shown by the symbol /. Grouping symbols such as brackets or braces are entered as parentheses. If the expression appears to be correct, press **ENTER**. At that time, the calculator will evaluate the expression and display the result on the screen.

Catalog   A graphing calculator's catalog lists all the functions of the calculator in alphabetical order. On many calculators, CATALOG is the 2nd option associated with the (0) key. To copy an item from the catalog to the screen, press (CATALOG), and then scroll through the list using the up and down arrow keys until the desired item is indicated. To move through the list more quickly, press the key associated with the first letter of the item. The indicator will move to the first item beginning with that letter. When the desired item is indicated, press **ENTER**.

Error Messages   When a calculator cannot complete an instruction, a message similar to the one shown below appears on the screen.

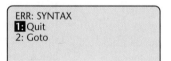

Press (1) to return to the home screen, and press (2) to go to the instruction that caused the error. Not all errors are operator errors; ERR:OVERFLOW indicates a result too large for the calculator to handle.

We can use a graphing calculator to evaluate algebraic expressions.

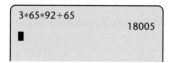

EXAMPLE 4   Use a graphing calculator to evaluate $3xy + x$ for $x = 65$ and $y = 92$.

SOLUTION   We enter the expression in the graphing calculator as it is written, replacing x with 65 and y with 92. Note that since $3xy$ means $3 \cdot x \cdot y$, we must supply a multiplication symbol ($*$) between 3, 65, and 92. We then press **ENTER** after the expression is complete. The value of the expression appears at the right on the screen shown at left.
   We see that $3xy + x = 18{,}005$ for $x = 65$ and $y = 92$.

Translating to Algebraic Expressions

Before attempting to translate problems to equations, we need to be able to translate certain phrases to algebraic expressions.

Important Words	Sample Phrase or Sentence	Translation
Addition (+)		
added to	700 lb was added to the car's weight.	$w + 700$
sum of	The sum of a number and 12	$n + 12$
plus	53 plus some number	$53 + x$
more than	8 more than Biloxi's population	$p + 8$
increased by	Jake's original guess, increased by 4	$n + 4$
Subtraction (−)		
subtracted from	2 oz was subtracted from the bag's weight.	$w - 2$
difference of	The difference of two scores	$m - n$
minus	A team of size s, minus 2 injured players	$s - 2$
less than	9 less than the given population	$p - 9$
decreased by	The car's speed, decreased by 8 mph	$s - 8$
Multiplication (·)		
multiplied by	The number of reservations, multiplied by 3	$r \cdot 3$
product of	The product of two numbers	$m \cdot n$
times	5 times the dog's weight	$5w$
twice	Twice Jackie's age	$2a$
of	$\frac{1}{2}$ of Amelia's salary	$\frac{1}{2}s$
Division (÷)		
divided by	A 2-lb coffee cake, divided by 3	$2 \div 3$
quotient of	The quotient of 14 and 7	$14 \div 7$
divided into	4 divided into the delivery fee	$f \div 4$
ratio of	The ratio of \$500 to the cost of a new car	$500/n$
per	There were 18 models per class of size s.	$18/s$

EXAMPLE 5　Translate each phrase to an algebraic expression.

a) Eighteen more than a number

b) Four less than Jean's height, in inches

c) A day's pay, in dollars, divided by eight

SOLUTION　To help think through a translation, we sometimes begin with a specific number in place of a variable.

a) If we knew the number to be 10, the translation would be $10 + 18$, or $18 + 10$. If we use t to represent "a number," the translation of "Eighteen more than a number" is $t + 18$, or $18 + t$.

b) If the height were 60, then 4 less than 60 would mean $60 - 4$. If we use h to represent "Jean's height, in inches," the translation of "Four less than Jean's height, in inches" is $h - 4$.

c) We let d represent "a day's pay, in dollars." If the pay were \$78, the translation would be $78 \div 8$, or $\frac{78}{8}$. Thus our translation of "a day's pay, in dollars, divided by eight" is $d \div 8$, or $\frac{d}{8}$.

> **CAUTION!**　The order in which we subtract and divide affects the answer! Answering $4 - h$ or $8 \div d$ in Examples 5(b) and 5(c) is incorrect.

Student Notes

Try looking for "than" or "from" in a phrase and writing what follows it first. Then add or subtract the necessary quantity. (See Examples 6c and 6d.)

EXAMPLE 6　Translate each of the following.

a) Kate's age increased by five

b) Half of some number

c) Three more than twice a number

d) Six less than the product of two numbers

e) Seventy-six percent of the town's population

SOLUTION

Phrase	Algebraic Expression
a) Kate's age increased by five	$a + 5$, or $5 + a$
b) Half of some number	$\frac{1}{2}t$, or $\frac{t}{2}$, or $t \div 2$
c) Three more than twice a number	$2x + 3$, or $3 + 2x$
d) Six less than the product of two numbers	$mn - 6$
e) Seventy-six percent of the town's population	76% of p, or $0.76p$

Translating to Equations

The symbol $=$ ("equals") indicates that the expressions on either side of the equals sign represent the same number. An **equation** is a number sentence with the verb $=$. Equations may be true, false, or neither true nor false.

EXAMPLE 7   Determine whether each equation is true, false, or neither.

a) $8 \cdot 4 = 32$ **b)** $7 - 2 = 4$ **c)** $x + 6 = 13$

SOLUTION

a) $8 \cdot 4 = 32$ The equation is *true*.

b) $7 - 2 = 4$ The equation is *false*.

c) $x + 6 = 13$ The equation is *neither* true nor false, because we do not know what number x represents.

Solution   A replacement or substitution that makes an equation true is called a *solution*. Some equations have more than one solution, and some have no solution. When all solutions have been found, we have *solved* the equation.

To determine whether a number is a solution, we evaluate all expressions in the equation. If the values on both sides of the equation are the same, the number is a solution.

EXAMPLE 8   Determine whether 7 is a solution of $x + 6 = 13$.

SOLUTION

$$\begin{array}{ll} x + 6 = 13 & \text{Writing the equation} \\ \overline{7 + 6 \ \big| \ 13} & \text{Substituting 7 for } x \\ \quad 13 \overset{?}{=} 13 & 13 = 13 \text{ is TRUE.} \end{array}$$

Since the left-hand and the right-hand sides are the same, 7 is a solution.

Although we do not study solving equations until Chapter 2, we can translate certain problem situations to equations now. The words "is the same as," "equal," "is," and "are" translate to "=."

Words indicating equality, = : "is the same as," "equal," "is," "are"

EXAMPLE 9   Translate the following problem to an equation.

What number plus 478 is 1019?

SOLUTION   We let y represent the unknown number. The translation then comes almost directly from the English sentence.

$$\begin{array}{ccccc} \text{What number} & \text{plus} & 478 & \text{is} & 1019? \\ \downarrow & \downarrow & \downarrow & \downarrow & \downarrow \\ y & + & 478 & = & 1019 \end{array}$$

Note that "plus" translates to "+" and "is" translates to "=."

Sometimes it helps to reword a problem before translating.

EXAMPLE 10   Translate the following problem to an equation.

The Petronas Twin Towers in Kuala Lumpur are the world's tallest buildings. At 1483 ft, they are 33 ft taller than the Sears Tower. (*Source: New York Times*) How tall is the Sears Tower?

SOLUTION   We let h represent the height, in feet, of the Sears Tower. A rewording and translation follow:

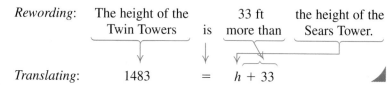

Rewording: The height of the Twin Towers   is   33 ft more than   the height of the Sears Tower.

Translating: 1483 $=$ $h + 33$

Models

When we translate a problem into mathematical language, we say that we **model** the problem. A **mathematical model** is a representation, using mathematics, of a real-world situation. An equation can be a model.

Information about a problem is often given in a set of numbers, called **data.** Sometimes data follow a pattern that can be modeled using an equation.

EXAMPLE 11   Postal Rates.   The following table lists the costs for shipping books using Basic Media Mail or 5-Digit Media Mail for various weights. Find an equation giving the cost d for 5-Digit shipping when the cost of Basic shipping is b.

Weight (in pounds)	Basic Media Mail	5-Digit Media Mail
1	$1.26	$0.90
2	1.74	1.38
3	2.22	1.86
4	2.70	2.34
5	3.18	2.82

Source: United States Postal Service, January 8, 2006

SOLUTION   We look for a pattern in the data. We note that the cost of shipping using 5-Digit Media Mail is less than the cost of shipping using Basic Media Mail and that both numbers increase as the number of pounds increases. If we subtract, we find that the difference is $0.36 or 36 cents for every pound increase.

For 1 pound: $1.26 - 0.90 = 0.36$;

For 2 pounds: $1.74 - 1.38 = 0.36$;

For 3 pounds: $2.22 - 1.86 = 0.36$;

For 4 pounds: $2.70 - 2.34 = 0.36$;

For 5 pounds: $3.18 - 2.82 = 0.36$.

We reword and translate as follows:

Rewording: The cost of   0.36   the cost of
                5–Digit shipping   is   less than   Basic shipping.

Translating: d $=$ b $-$ 0.36

Instructor:

Name _____

Office hours and location _____

Phone number _____

Fax number _____

E-mail address _____

Find the names of two students whom you could contact for information or study questions:

1. Name _____

   Phone number _____

   E-mail address _____

2. Name _____

   Phone number _____

   E-mail address _____

Math lab on campus:

Location _____

Hours _____

Phone _____

Tutoring:

Campus location _____

Hours _____

Addison-Wesley Tutor Center _____

To order, call _____

(See the preface for important information concerning this tutoring.)

Important supplements:
(See the preface for a complete list of available supplements.)

Supplements recommended by the instructor

_____

_____

1.1 EXERCISE SET

FOR EXTRA HELP

  MathXL

  MyMathLab

  InterAct Math   Tutor Center   AW Math Tutor Center   Video Lectures on CD: Disc 1   Student's Solutions Manual

🖐 *Concept Reinforcement*   *Classify each of the following as either an expression or an equation.*

1. $4x + 7$

2. $3x = 21$

3. $2x - 5 = 9$

4. $8x - 3$

5. $38 = 2t$

6. $45 = a - 1$

7. $4a - 5b$

8. $3t + 4 = 19$

9. $2x - 3y = 8$

10. $12 - 4xy$

11. $7 - 4rt$

12. $9a + b$

Evaluate.

13. $3a$, for $a = 9$

14. $8x$, for $x = 7$

15. $t + 6$, for $t = 2$

16. $13 - r$, for $r = 9$

17. $\dfrac{x + y}{4}$, for $x = 2$ and $y = 14$

(*Hint:* Add $x + y$ before dividing by 4.)

18. $\dfrac{p + q}{7}$, for $p = 15$ and $q = 20$

19. $\dfrac{m - n}{2}$, for $m = 20$ and $n = 6$

20. $\dfrac{x - y}{6}$, for $x = 23$ and $y = 5$

21. $\dfrac{9m}{q}$, for $m = 6$ and $q = 18$

22. $\dfrac{5z}{y}$, for $z = 9$ and $y = 15$

To the student and instructor: *The calculator symbol,* 🖩*, is used to indicate those exercises designed to be solved with a calculator.*

🖩 *Evaluate using a calculator.*

23. $27a - 18b$, for $a = 136$ and $b = 13$

24. $19xy - 9x + 13y$, for $x = 87$ and $y = 29$

Substitute to find the value of each expression.

25. *Hockey.*   The area of a rectangle with base b and height h is bh. A regulation hockey goal is 6 ft wide and 4 ft high. Find the area of the opening.

26. *Travel Time.*   The length of a flight from Seattle, Washington, to St. Paul, Minnesota, is approximately 1400 mi. The time, in hours, for the flight is

$$\frac{1400}{v},$$

where v is the velocity, in miles per hour. How long will a flight take at a velocity of 400 mph?

27. *Zoology.*   A great white shark has triangular teeth. Each tooth measures about 5 cm across the base and has a height of 6 cm. Find the surface area of the front side of one such tooth. (See Example 3.)

28. *Work Time.*   Enrico takes five times as long to do a job as Rosa does. Suppose t represents the time it takes Rosa to do the job. Then $5t$ represents the time it takes Enrico. How long does it take Enrico if Rosa takes **(a)** 30 sec? **(b)** 90 sec? **(c)** 2 min?

29. *Olympic Softball.*   A softball player's batting average is h/a, where h is the number of hits and a is the number of "at bats." In the 2004 Summer Olympics, Laura Berg had 7 hits in 19 at bats. What was her batting average? Round to the nearest thousandth.

30. *Area of a Parallelogram.*   The area of a parallelogram with base b and height h is bh. Find the area of the parallelogram when the height is 6 cm (centimeters) and the base is 7.5 cm.

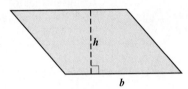

Translate to an algebraic expression.

31. 5 more than Ron's age

32. The product of 4 and a

33. 6 more than b

34. 7 more than Lou's weight

35. 9 less than c

36. 4 less than d

37. 6 increased by q

38. 11 increased by z

39. 9 times Phil's speed

40. c more than d

41. x less than y

42. 2 less than Lorrie's age

43. x divided by w

44. The quotient of two numbers

45. m subtracted from n

46. p subtracted from q

47. The sum of the box's length and height

48. The sum of d and f

49. The product of 9 and twice m

50. Penny's speed minus twice the wind speed

51. One quarter of some number

52. One third of the sum of two numbers

53. 64% of the women attending

54. 38% of a number

Determine whether the given number is a solution of the given equation.

55. 15; $x + 17 = 32$

56. 75; $y + 28 = 93$

57. 93; $a - 28 = 75$

58. 12; $8t = 96$

59. 63; $\dfrac{t}{7} = 9$

60. 52; $\dfrac{x}{8} = 6$

61. 3; $\dfrac{108}{x} = 36$

62. 7; $\dfrac{94}{y} = 12$

Translate each problem to an equation. Do not solve.

63. What number added to 73 is 201?

64. Seven times what number is 1596?

65. When 42 is multiplied by a number, the result is 2352. Find the number.

66. When 345 is added to a number, the result is 987. Find the number.

67. *Chess.*   A chess board has 64 squares. If pieces occupy 19 squares, how many squares are unoccupied?

68. *Hours Worked.*   A carpenter charges $25 an hour. How many hours did she work if she billed a total of $53,400?

69. *Recycling.*   Currently, Americans recycle or compost 28.5% of all municipal solid waste (*Source*: Biocycle/EEC Survey). This is the same as recycling or composting 110.4 million tons. What is the total amount of waste generated?

70. *Travel to Work.*   In Maryland, the average commute to work is 30.2 min. The average commuting time in California is 3.7 min less. (*Source*: U.S. Bureau of the Census, 2003 American Community Survey) How long is the average commute in California?

In each of Exercises 71–78, match the phrase or sentence with the appropriate expression or equation from the column on the right.

71. _____ Twice the sum of two numbers

72. _____ Five less than a number is nine.

73. _____ Two more than a number is five.

74. _____ Half of the product of two numbers

75. _____ Three times the sum of a number and five

a) $\dfrac{x}{y} + 6$

b) $2(x + y) = 48$

c) $\dfrac{1}{2} \cdot a \cdot b$

d) $t + 2 = 5$

e) $ab - 1 = 49$

f) $2(m + n)$

g) $3(t + 5)$

h) $x - 5 = 9$

76. _____ Twice the sum of two numbers is 48.

77. _____ One less than the product of two numbers is 49.

78. _____ Six more than the quotient of two numbers

79. *Gasoline Prices.* The following table lists the price per gallon of unleaded premium and unleaded regular gasoline at four different gas stations. Find an equation for the price p of unleaded premium gas when the price of unleaded regular gas is r cents per gallon.

Price, r, of Unleaded Regular Gas (in cents per gallon)	Price, p, of Unleaded Premium Gas (in cents per gallon)
182.9	201.9
206.9	225.9
291.9	310.9
228.9	247.9

80. *Tuition.* The following table lists the tuition costs for students taking various numbers of hours of classes. Find an equation for the cost c of tuition for a student taking h hours of classes.

Number of Class Hours, h	Tuition, c
12	$1200
15	1500
18	1800
21	2100

81. *Nutrition.* The number of grams of dietary fiber recommended daily for children depends on the age of the child, as shown in the following table. Find an equation for the number of grams of fiber f that is recommended daily for a child of age a.

Age of Child, a (in years)	Grams of Dietary Fiber Recommended Daily, f
3	8
4	9
5	10
6	11
7	12
8	13

Source: The American Health Foundation

82. *Shipping Costs.* The following table lists the total cost, including postage and handling, for various items ordered from a catalog. Find an equation for the total cost c for an item of price p.

Price of Item, p	Total Cost, c
$4.59	$ 7.59
2.17	5.17
8.79	11.79
6.53	9.53

83. *Number of Drivers.* The following table lists the number of vehicle miles traveled annually per household by the number of drivers in the household. Find an equation for the number v of vehicle miles driven in a household with d drivers.

Number of Drivers, d	Number of Vehicle Miles Traveled, v
1	10,000
2	20,000
3	30,000
4	40,000

Source: Energy Information Administration

84. *Meteorology.* The following table lists the number of centimeters of water to which various amounts of snow will melt under certain conditions. Find an equation for the number w of centimeters of water to which s centimeters of snow will melt.

Depth of Snow, s (in centimeters)	Depth of Water, w (in centimeters)
120	12
135	13.5
160	16
90	9

To the student and the instructor: Thinking and writing exercises, denoted by TW, *are meant to be answered using one or more English sentences. Because answers to many writing exercises will vary, solutions are not listed in the answers at the back of the book.*

TW **85.** What is the difference between a variable, a variable expression, and an equation?

TW **86.** What does it mean to evaluate an algebraic expression?

Synthesis

To the student and the instructor: Synthesis exercises are designed to challenge students to extend the concepts or skills studied in each section. Many synthesis exercises require the assimilation of skills and concepts from several sections.

TW **87.** If the lengths of the sides of a square are doubled, is the area doubled? Why or why not?

TW **88.** Write a problem that translates to $1998 + t = 2006$.

89. Signs of Distinction charges \$90 per square foot for handpainted signs. The town of Belmar commissioned a triangular sign with a base of 3.0 ft and a height of 2.5 ft. How much will the sign cost?

90. Find the area that is shaded.

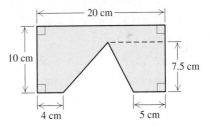

91. Evaluate $\dfrac{x - y}{3}$ when x is twice y and $x = 12$.

92. Evaluate $\dfrac{x + y}{2}$ when y is twice x and $x = 6$.

93. Evaluate $\dfrac{a + b}{4}$ when a is twice b and $a = 16$.

94. Evaluate $\dfrac{a - b}{3}$ when a is three times b and $a = 18$.

Answer each question with an algebraic expression.

95. If $w + 3$ is a whole number, what is the next whole number after it?

96. If $d + 2$ is an odd number, what is the preceding odd number?

Translate to an algebraic expression.

97. The perimeter of a rectangle with length l and width w (perimeter means distance around)

98. The perimeter of a square with side s (perimeter means distance around)

99. Ella's race time, assuming she took 5 sec longer than Kyle and Kyle took 3 sec longer than Ava. Assume that Ava's time was t seconds.

100. Ray's age 7 yr from now if he is 2 yr older than Monique and Monique is a years old

TW 101. If the height of a triangle is doubled, is its area also doubled? Why or why not?

Collaborative Corner

Teamwork

Focus: Group problem solving; working collaboratively
Time: 15 minutes
Group Size: 2

Working and studying as a team often enables students to solve problems that are difficult to solve alone.

ACTIVITY

1. The left-hand column below contains the names of 12 colleges. A scrambled list of the names of their sports teams is on the right. As a group, match the names of the colleges to the teams.

1. University of Texas	**a.** Antelopes
2. Western State College of Colorado	**b.** Fighting Banana Slugs
3. University of North Carolina	**c.** Sea Warriors
4. University of Massachusetts	**d.** Gators
5. Hawaii Pacific University	**e.** Mountaineers
6. University of Nebraska	**f.** Sailfish
7. University of California, Santa Cruz	**g.** Longhorns
8. University of Louisiana at Lafayette	**h.** Tar Heels
9. Grand Canyon University	**i.** Seawolves
10. Palm Beach Atlantic College	**j.** Ragin' Cajuns
11. University of Alaska, Anchorage	**k.** Cornhuskers
12. University of Florida	**l.** Minutemen

2. After working for 5 min, confer with another group and reach mutual agreement.

3. Does the class agree on all 12 pairs?

4. Do you agree that group collaboration enhances our ability to solve problems?

1.2 The Commutative, Associative, and Distributive Laws

Equivalent Expressions ◼   The Commutative Laws ◼
The Associative Laws ◼   The Distributive Law ◼
The Distributive Law and Factoring

In order to solve equations, we must be able to manipulate algebraic expressions. The commutative, associative, and distributive laws discussed in this section enable us to write *equivalent expressions* that will simplify our work. Indeed, much of this text is devoted to finding equivalent expressions.

Equivalent Expressions

The expressions $4 + 4 + 4$, $3 \cdot 4$, and $4 \cdot 3$ all represent the same number, 12. Expressions that represent the same number are said to be **equivalent.** The equivalent expressions $t + 18$ and $18 + t$ were used on p. 6 when we translated "eighteen more than a number." To check that these expressions are equivalent, we make some choices for t:

$$\text{When } t = 3, \quad t + 18 = 3 + 18 \quad \text{and} \quad 18 + t = 18 + 3$$
$$= 21 \qquad\qquad\qquad = 21.$$

$$\text{When } t = 40, \quad t + 18 = 40 + 18 \quad \text{and} \quad 18 + t = 18 + 40$$
$$= 58 \qquad\qquad\qquad = 58.$$

The Commutative Laws

Recall that changing the order in addition or multiplication does not change the result. Equations like $3 + 78 = 78 + 3$ and $5 \cdot 14 = 14 \cdot 5$ illustrate this idea and show that addition and multiplication are **commutative.**

> **The Commutative Laws**   *For Addition.* For any numbers a and b,
>
> $$a + b = b + a.$$
>
> (Changing the order of addition does not affect the answer.)
>
> *For Multiplication.* For any numbers a and b,
>
> $$ab = ba.$$
>
> (Changing the order of multiplication does not affect the answer.)

EXAMPLE 1   Use the commutative laws to write an expression equivalent to each of the following: **(a)** $y + 5$; **(b)** $9x$; **(c)** $7 + ab$.

SOLUTION

a) $y + 5$ is equivalent to $5 + y$ by the commutative law of addition.

b) $9x$ is equivalent to $x \cdot 9$ by the commutative law of multiplication.

c) $7 + ab$ is equivalent to $ab + 7$ by the commutative law of *addition*.

$7 + ab$ is also equivalent to $7 + ba$ by the commutative law of *multiplication*.

$7 + ab$ is also equivalent to $ba + 7$ by the two commutative laws, used together.

The Associative Laws

Parentheses are used to indicate groupings. We normally simplify within the parentheses first. For example,

$$3 + (8 + 4) = 3 + 12 \quad \text{and} \quad (3 + 8) + 4 = 11 + 4$$
$$= 15 \qquad\qquad\qquad\qquad = 15.$$

Similarly,

$$4 \cdot (2 \cdot 3) = 4 \cdot 6 \quad \text{and} \quad (4 \cdot 2) \cdot 3 = 8 \cdot 3$$
$$= 24 \qquad\qquad\qquad\qquad = 24.$$

Note that, so long as only addition or only multiplication appears in an expression, changing the grouping does not change the result. Equations such as $3 + (7 + 5) = (3 + 7) + 5$ and $4(5 \cdot 3) = (4 \cdot 5)3$ illustrate that addition and multiplication are **associative.**

Student Notes

Examine and compare the statements of the commutative and associative laws. Note that the order of the variables changes in the commutative laws. In the associative laws, the order does not change, but the grouping does.

The Associative Laws   *For Addition.* For any numbers a, b, and c,

$$a + (b + c) = (a + b) + c.$$

(Numbers can be grouped in any manner for addition.)

For Multiplication. For any numbers a, b, and c,

$$a \cdot (b \cdot c) = (a \cdot b) \cdot c.$$

(Numbers can be grouped in any manner for multiplication.)

EXAMPLE 2   Use an associative law to write an expression equivalent to each of the following: **(a)** $y + (z + 3)$; **(b)** $(8x)y$.

SOLUTION

a) $y + (z + 3)$ is equivalent to $(y + z) + 3$ by the associative law of addition.

b) $(8x)y$ is equivalent to $8(xy)$ by the associative law of multiplication.

When only addition or only multiplication is involved, parentheses do not change the result. For that reason, we sometimes omit them altogether. Thus,

$$x + (y + 7) = x + y + 7, \quad \text{and} \quad l(wh) = lwh.$$

A sum such as $(5 + 1) + (3 + 5) + 9$ can be simplified by pairing numbers that add to 10. The associative and commutative laws allow us to do this:

$$(5 + 1) + (3 + 5) + 9 = 5 + 5 + 9 + 1 + 3$$
$$= 10 + 10 + 3 = 23.$$

EXAMPLE 3   Use the commutative and/or associative laws of addition to write two expressions equivalent to $(7 + x) + 3$. Then simplify.

SOLUTION

$(7 + x) + 3 = (x + 7) + 3$	Using the commutative law; $(x + 7) + 3$ is one equivalent expression.
$\qquad = x + (7 + 3)$	Using the associative law; $x + (7 + 3)$ is another equivalent expression.
$\qquad = x + 10$	Simplifying

EXAMPLE 4   Use the commutative and/or associative laws of multiplication to write two expressions equivalent to $2(x \cdot 3)$. Then simplify.

SOLUTION

$2(x \cdot 3) = 2(3x)$	Using the commutative law; $2(3x)$ is one equivalent expression.
$\qquad = (2 \cdot 3)x$	Using the associative law; $(2 \cdot 3)x$ is another equivalent expression.
$\qquad = 6x$	Simplifying

Student Notes

To remember the names *commutative*, *associative*, and *distributive*, first understand the concept. Next, use everyday life to link the word to the concept. For example, think of commuting to and from work as changing the order of appearance.

The Distributive Law

The *distributive law* is probably the single most important law for manipulating algebraic expressions. Unlike the commutative and associative laws, the distributive law uses multiplication together with addition.

You have already used the distributive law although you may not have realized it at the time. To illustrate, try to multiply $3 \cdot 21$ mentally. Many people find the product, 63, by thinking of 21 as $20 + 1$ and then multiplying 20 by 3 and 1 by 3. The sum of the two products, $60 + 3$, is 63. Note that if the 3 does not multiply both 20 and 1, the result will not be correct.

EXAMPLE 5   Compute in two ways: $4(7 + 2)$.

SOLUTION

a) As in the discussion of $3(20 + 1)$ above, to compute $4(7 + 2)$, we can multiply both 7 and 2 by 4 and add the results:

$4(7 + 2) = 4 \cdot 7 + 4 \cdot 2$	Multiplying both 7 and 2 by 4
$\qquad = 28 + 8 = 36.$	Adding

b) By first adding inside the parentheses, we get the same result in a different way:

$$4(7 + 2) = 4(9) \quad \text{Adding; } 7 + 2 = 9$$
$$= 36. \quad \text{Multiplying}$$

The Distributive Law   For any numbers a, b, and c,

$$a(b + c) = ab + ac.$$

(The product of a number and a sum can be written as the sum of two products.)

EXAMPLE 6   Multiply: $3(x + 2)$.

SOLUTION   Since $x + 2$ cannot be simplified unless a value for x is given, we use the distributive law:

$$3(x + 2) = 3 \cdot x + 3 \cdot 2 \quad \text{Using the distributive law}$$
$$= 3x + 6. \quad \text{Note that } 3 \cdot x \text{ is the same as } 3x.$$

The expression $3x + 6$ has two *terms*, $3x$ and 6. In general, a **term** is a number, a variable, or a product or a quotient of numbers and/or variables. Thus, t, 29, $5ab$, and $\dfrac{2x}{y}$ are terms in $t + 29 + 5ab + \dfrac{2x}{y}$. Note that terms are separated by plus signs.

EXAMPLE 7   List the terms in $7s + st + \dfrac{3}{t}$.

SOLUTION   Terms are separated by plus signs, so the terms in $7s + st + \dfrac{3}{t}$ are $7s$, st, and $\dfrac{3}{t}$.

The distributive law can also be used when more than two terms are inside the parentheses.

EXAMPLE 8   Multiply: $6(s + 2 + 5w)$.

SOLUTION

$$6(s + 2 + 5w) = 6 \cdot s + 6 \cdot 2 + 6 \cdot 5w \quad \text{Using the distributive law}$$
$$= 6s + 12 + (6 \cdot 5)w \quad \text{Using the associative law for multiplication}$$
$$= 6s + 12 + 30w$$

Because of the commutative law of multiplication, the distributive law can be used on the "right": $(b + c)a = ba + ca$.

EXAMPLE 9   Multiply: $(c + 4)5$.

SOLUTION

$$(c + 4)5 = c \cdot 5 + 4 \cdot 5 \qquad \text{Using the distributive law on the right}$$
$$= 5c + 20$$

> **CAUTION!**   To use the distributive law for removing parentheses, be sure to multiply *each* term inside the parentheses by the multiplier outside.

The Distributive Law and Factoring

If we use the distributive law in reverse, we have the basis of a process called **factoring:** $ab + ac = a(b + c)$. Factoring involves multiplication: To **factor** an expression means to write an equivalent expression that is a *product*. The parts of the product are called **factors.** Note that "factor" can be used as either a verb or a noun. Thus in the expression $5t$, the factors are 5 and t. In the expression $4(m + n)$, the factors are 4 and $(m + n)$.

EXAMPLE 10   Use the distributive law to factor each of the following.

a) $3x + 3y$ **b)** $7x + 21y + 7$

SOLUTION

a) By the distributive law,

$$3x + 3y = 3(x + y). \qquad \text{The } common \ factor \text{ for } 3x \text{ and } 3y \text{ is 3.}$$

b) $7x + 21y + 7 = 7 \cdot x + 7 \cdot 3y + 7 \cdot 1$ The common factor is 7.

$$= 7(x + 3y + 1) \qquad \text{Using the distributive law}$$

Be sure to include both the 1 and the common factor, 7.

To check our factoring, we multiply to see if the original expression is obtained. For example, to check the **factorization** in Example 10(b), note that

$$7(x + 3y + 1) = 7x + 7 \cdot 3y + 7 \cdot 1$$
$$= 7x + 21y + 7.$$

Since $7x + 21y + 7$ is what we began with in Example 10(b), we have a check.

1.2 EXERCISE SET

FOR EXTRA HELP

MathXL   MyMathLab   InterAct Math   AW Math Tutor Center   Video Lectures on CD: Disc 1   Student's Solutions Manual

Concept Reinforcement   *Complete each sentence using one of these terms:* commutative, associative, *or* distributive.

1. $8 + t$ is equivalent to $t + 8$ by the _____ law for addition.

2. $3(xy)$ is equivalent to $(3x)y$ by the _____ law for multiplication.

3. $(5b)c$ is equivalent to $5(bc)$ by the _____ law for multiplication.

4. mn is equivalent to nm by the _____ law for multiplication.

5. $x(y + z)$ is equivalent to $xy + xz$ by the _____ law.

6. $(9 + a) + b$ is equivalent to $9 + (a + b)$ by the _____ law for addition.

7. $a + (6 + d)$ is equivalent to $(a + 6) + d$ by the _____ law for addition.

8. $t + 4$ is equivalent to $4 + t$ by the _____ law for addition.

9. $x \cdot 7$ is equivalent to $7 \cdot x$ by the _____ law for multiplication.

10. $2(a + b)$ is equivalent to $2 \cdot a + 2 \cdot b$ by the _____ law.

Use the commutative law of addition to write an equivalent expression.

11. $7 + x$ **12.** $a + 2$

13. $ab + c$ **14.** $x + 3y$

15. $9x + 3y$ **16.** $3a + 7b$

17. $5(a + 1)$ **18.** $9(x + 5)$

Use the commutative law of multiplication to write an equivalent expression.

19. $2 \cdot a$ **20.** xy

21. st **22.** $4x$

23. $5 + ab$ **24.** $x + 3y$

25. $5(a + 1)$ **26.** $9(x + 5)$

Use the associative law of addition to write an equivalent expression.

27. $(a + 5) + b$ **28.** $(5 + m) + r$

29. $r + (t + 7)$ **30.** $x + (2 + y)$

31. $(ab + c) + d$ **32.** $(m + np) + r$

Use the associative law of multiplication to write an equivalent expression.

33. $(8x)y$ **34.** $(9a)b$

35. $2(ab)$ **36.** $9(rp)$

37. $3[2(a + b)]$ **38.** $5[x(2 + y)]$

Use the commutative and/or associative laws to write two equivalent expressions. Answers may vary.

39. $r + (t + 6)$ **40.** $5 + (v + w)$

41. $(17a)b$ **42.** $x(3y)$

Use the commutative and/or associative laws to show why the expression on the left is equivalent to the expression on the right. Write a series of steps with labels, as in Example 4.

43. $(5 + x) + 2$ is equivalent to $x + 7$

44. $(2a)4$ is equivalent to $8a$

45. $(m \cdot 3)7$ is equivalent to $21m$

46. $4 + (9 + x)$ is equivalent to $x + 13$

Multiply.

47. $4(a + 3)$ **48.** $3(x + 5)$

49. $6(1 + x)$ **50.** $6(v + 4)$

51. $3(x + 1)$ **52.** $9(x + 3)$

53. $8(3 + y)$ **54.** $7(s + 5)$

55. $9(2x + 6)$ **56.** $9(6m + 7)$

57. $5(r + 2 + 3t)$ **58.** $4(5x + 8 + 3p)$

59. $(a + b)2$ **60.** $(x + 2)7$

61. $(x + y + 2)5$ **62.** $(2 + a + b)6$

List the terms in each expression.

63. $x + xyz + 19$ **64.** $9 + 17a + abc$

65. $2a + \dfrac{a}{b} + 5b$

66. $3xy + 20 + \dfrac{4a}{b}$

Use the distributive law to factor each of the following. Check by multiplying.

67. $2a + 2b$ **68.** $5y + 5z$

69. $7 + 7y$ **70.** $13 + 13x$

71. $18x + 3$ **72.** $20a + 5$

73. $5x + 10 + 15y$ **74.** $3 + 27b + 6c$

75. $12x + 9$ **76.** $6x + 6$

77. $3a + 9b$ **78.** $5a + 15b$

79. $44x + 11y + 22z$ **80.** $14a + 56b + 7$

List the factors in each expression.

81. st **82.** $5x$

83. $3(x + y)$ **84.** $(a + b)6$

85. $7 \cdot a$ **86.** $m \cdot 2$

87. $(a - b)(x - y)$ **88.** $(3 - a)(b + c)$

TW 89. Is subtraction commutative? Why or why not?

TW 90. Is division associative? Why or why not?

Skill Maintenance

To the student and the instructor: Exercises included for Skill Maintenance review skills previously studied in the text. Often these exercises provide preparation for the next section of the text. The numbers in brackets immediately following the directions or exercise indicate the section in which the skill was introduced. The answers to all Skill Maintenance exercises appear at the back of the book. If a Skill Maintenance exercise gives you difficulty, review the material in the indicated section of the text.

Translate to an algebraic expression. [1.1]

91. Twice Kara's salary   **92.** Half of m

Synthesis

TW 93. Are terms and factors the same thing? Why or why not?

TW 94. Explain how the distributive, commutative, and associative laws can be used to show that $2(3x + 4y)$ is equivalent to $6x + 8y$.

Tell whether the expressions in each pairing are equivalent. Then explain why or why not.

95. $8 + 4(a + b)$ and $4(2 + a + b)$

96. $7 \div 3m$ and $m \cdot 3 \div 7$

97. $(rt + st)5$ and $5t(r + s)$

98. $yax + ax$ and $xa(1 + y)$

99. $30y + x15$ and $5[2(x + 3y)]$

100. $[c(2 + 3b)]5$ and $10c + 15bc$

TW 101. Evaluate the expressions $3(2 + x)$ and $6 + x$ for $x = 0$. Do your results indicate that $3(2 + x)$ and $6 + x$ are equivalent? Why or why not?

TW 102. Factor $15x + 40$. Then evaluate both $15x + 40$ and the factorization for $x = 4$. Do your results *guarantee* that the factorization is correct? Why or why not? (*Hint:* See Exercise 101.)

1.3 Fraction Notation

Factors and Prime Factorizations ◼ Fraction Notation ◼ Multiplication, Division, and Simplification ◼ More Simplifying ◼ Addition and Subtraction

This section covers multiplication, addition, subtraction, and division with fractions. Although much of this may be review, note that fraction expressions that contain variables are also included.

Factors and Prime Factorizations

In preparation for work with fraction notation, we first review how *natural numbers* are factored. **Natural numbers** can be thought of as the counting numbers:

1, 2, 3, 4, 5, ...*

(The dots indicate that the established pattern continues without ending.)

To factor a number, we simply express it as a product of two or more numbers.

EXAMPLE 1 Write several factorizations of 12. Then list all factors of 12.

SOLUTION The number 12 can be factored in several ways:

$$1 \cdot 12, \quad 2 \cdot 6, \quad 3 \cdot 4, \quad 2 \cdot 2 \cdot 3.$$

The factors of 12 are 1, 2, 3, 4, 6, and 12.

Some numbers have only two factors, the number itself and 1. Such numbers are called **prime.**

Prime Number A *prime number* is a natural number that has exactly two different factors: the number itself and 1. The first several primes are 2, 3, 5, 7, 11, 13, 17, 19, and 23.

EXAMPLE 2 Which of these numbers are prime? 29, 4, 1

SOLUTION

29 is prime. It has exactly two different factors, 29 and 1.

4 is not prime. It has three different factors, 1, 2, and 4.

1 is not prime. It does not have two *different* factors.

---

*A similar collection of numbers, the **whole numbers,** includes 0: 0, 1, 2, 3, ...

If a natural number, other than 1, is not prime, we call it **composite.** Every composite number can be factored into a product of prime numbers. Such a factorization is called the **prime factorization** of that composite number.

Student Notes

When writing a factorization, you are writing an equivalent expression for the original number. Some students do this with a tree diagram:

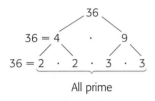

All prime

▸ **EXAMPLE 3**   Find the prime factorization of 36.

SOLUTION   We first factor 36 in any way that we can. One way is like this:

$$36 = 4 \cdot 9.$$

The factors 4 and 9 are not prime, so we factor them:

$$36 = 4 \cdot 9$$
$$= 2 \cdot 2 \cdot 3 \cdot 3. \qquad \textbf{2 and 3 are both prime.}$$

The prime factorization of 36 is $2 \cdot 2 \cdot 3 \cdot 3$.   ◣

Fraction Notation

An example of **fraction notation** for a number is

$$\frac{2}{3}. \quad \begin{array}{l} \longleftarrow \textbf{Numerator} \\ \longleftarrow \textbf{Denominator} \end{array}$$

The top number is called the **numerator,** and the bottom number is called the **denominator.** When the numerator and the denominator are the same nonzero number, we have fraction notation for the number 1.

Student Notes

Fraction notation for 1 appears in many contexts in algebra. Keep in mind that when the (nonzero) numerator of a fraction is the same as the denominator, the fraction is equivalent to 1.

Fraction Notation for 1   For any number a, except 0,

$$\frac{a}{a} = 1.$$

(Any nonzero number divided by itself is 1.)

Multiplication, Division, and Simplification

Recall from arithmetic that fractions are multiplied as follows.

Multiplication of Fractions   For any two fractions $\dfrac{a}{b}$ and $\dfrac{c}{d}$,

$$\frac{a}{b} \cdot \frac{c}{d} = \frac{ac}{bd}.$$

(The numerator of the product is the product of the two numerators. The denominator of the product is the product of the two denominators.)

▶ **EXAMPLE 4** Multiply: **(a)** $\dfrac{2}{3} \cdot \dfrac{7}{5}$; **(b)** $\dfrac{4}{x} \cdot \dfrac{8}{y}$.

SOLUTION We multiply numerators as well as denominators.

a) $\dfrac{2}{3} \cdot \dfrac{7}{5} = \dfrac{2 \cdot 7}{3 \cdot 5} = \dfrac{14}{15}$ **b)** $\dfrac{4}{x} \cdot \dfrac{8}{y} = \dfrac{4 \cdot 8}{x \cdot y} = \dfrac{32}{xy}$ ◢

Two numbers whose product is 1 are **reciprocals,** or **multiplicative inverses,** of each other. All numbers, except zero, have reciprocals. For example,

the reciprocal of $\frac{2}{3}$ is $\frac{3}{2}$ because $\frac{2}{3} \cdot \frac{3}{2} = \frac{6}{6} = 1$;

the reciprocal of 9 is $\frac{1}{9}$ because $9 \cdot \frac{1}{9} = \frac{9}{9} = 1$; and

the reciprocal of $\frac{1}{4}$ is 4 because $\frac{1}{4} \cdot 4 = 1$.

Reciprocals are used to rewrite division as multiplication.

Division of Fractions To divide two fractions, multiply by the reciprocal of the divisor:

$$\frac{a}{b} \div \frac{c}{d} = \frac{a}{b} \cdot \frac{d}{c}.$$

▶ **EXAMPLE 5** Divide: $\dfrac{1}{2} \div \dfrac{3}{5}$.

SOLUTION Note that the *divisor* is $\dfrac{3}{5}$:

$$\frac{1}{2} \div \frac{3}{5} = \frac{1}{2} \cdot \frac{5}{3} \qquad \text{$\frac{5}{3}$ is the reciprocal of $\frac{3}{5}$}$$

$$= \frac{5}{6}.$$

◢

When one of the fractions being multiplied is 1, multiplying yields an equivalent expression because of the *identity property of* 1.

The Identity Property of 1 For any number a,

$$a \cdot 1 = a.$$

(Multiplying a number by 1 gives that same number.) The number 1 is called the *multiplicative identity*.

▶ **EXAMPLE 6** Multiply $\dfrac{4}{5} \cdot \dfrac{6}{6}$ to find an expression equivalent to $\dfrac{4}{5}$.

SOLUTION We have

$$\frac{4}{5} \cdot \frac{6}{6} = \frac{4 \cdot 6}{5 \cdot 6} = \frac{24}{30}.$$

Since $\frac{6}{6} = 1$, the expression $\frac{4}{5} \cdot \frac{6}{6}$ is equivalent to $\frac{4}{5} \cdot 1$, or simply $\frac{4}{5}$. Thus, $\frac{24}{30}$ is equivalent to $\frac{4}{5}$.

The steps of Example 6 are reversed by "removing a factor equal to 1"— in this case, $\frac{6}{6}$. By removing a factor that equals 1, we can *simplify* an expression like $\frac{24}{30}$ to an equivalent expression like $\frac{4}{5}$.

To simplify, we factor the numerator and the denominator, looking for the largest factor common to both. This is sometimes made easier by writing prime factorizations. After identifying common factors, we can express the fraction as a product of two fractions, one of which is in the form $\dfrac{a}{a}$.

![triangle] **EXAMPLE 7**   Simplify: (a) $\dfrac{15}{40}$; (b) $\dfrac{36}{24}$.

SOLUTION

a) Note that 5 is a factor of both 15 and 40:

$$\frac{15}{40} = \frac{3 \cdot 5}{8 \cdot 5} \qquad \text{Factoring the numerator and the denominator, using the common factor, 5}$$

$$= \frac{3}{8} \cdot \frac{5}{5} \qquad \text{Rewriting as a product of two fractions; } \tfrac{5}{5} = 1$$

$$= \frac{3}{8} \cdot 1 = \frac{3}{8}. \qquad \text{Using the identity property of 1 (removing a factor equal to 1)}$$

b) $\dfrac{36}{24} = \dfrac{2 \cdot 2 \cdot 3 \cdot 3}{2 \cdot 2 \cdot 2 \cdot 3}$ Writing the prime factorizations and identifying common factors; 12/12 could also be used.

$$= \frac{3}{2} \cdot \frac{2 \cdot 2 \cdot 3}{2 \cdot 2 \cdot 3} \qquad \text{Rewriting as a product of two fractions; } \tfrac{2 \cdot 2 \cdot 3}{2 \cdot 2 \cdot 3} = 1$$

$$= \frac{3}{2} \cdot 1 = \frac{3}{2} \qquad \text{Using the identity property of 1}$$

It is always wise to check your result to see if any common factors of the numerator and the denominator remain. (This will never happen if prime factorizations are used correctly.) If common factors remain, repeat the process by removing another factor equal to 1 to simplify your result.

More Simplifying

You may have used a shortcut called "canceling" to remove a factor equal to 1 when working with fraction notation. With *great* concern, we mention it as a possible way to speed up your work. Canceling can be used only when removing common factors in numerators and denominators. Canceling *cannot* be used in sums or differences. Our concern is that "canceling" be used with understanding. Example 7(b) might have been done faster as follows:

$$\frac{36}{24} = \frac{\cancel{2} \cdot \cancel{2} \cdot 3 \cdot \cancel{3}}{\cancel{2} \cdot \cancel{2} \cdot 2 \cdot \cancel{3}} = \frac{3}{2}, \quad \text{or} \quad \frac{36}{24} = \frac{3 \cdot \cancel{12}}{2 \cdot \cancel{12}} = \frac{3}{2}, \quad \text{or} \quad \frac{\overset{3}{\cancel{\overset{18}{\cancel{36}}}}}{\underset{2}{\cancel{\underset{12}{\cancel{24}}}}} = \frac{3}{2}.$$

Student Notes

Canceling is the same as removing a factor equal to 1; each pair of slashes indicates a factor of 1. Thus numerators and denominators must be factored, or written as products, before canceling is used.

CAUTION!   Unfortunately, canceling is often performed incorrectly:

$$\frac{\cancel{2} + 3}{\cancel{2}} = 3, \qquad \frac{\cancel{4} - 1}{\cancel{4} - 2} = \frac{1}{2}, \qquad \frac{1\cancel{5}}{\cancel{5}4} = \frac{1}{4}.$$

The above cancellations are incorrect because the expressions canceled are *not* factors. Correct simplifications are as follows:

$$\frac{2 + 3}{2} = \frac{5}{2}, \qquad \frac{4 - 1}{4 - 2} = \frac{3}{2}, \qquad \frac{15}{54} = \frac{5 \cdot \cancel{3}}{18 \cdot \cancel{3}} = \frac{5}{18}.$$

Remember: **If you can't factor, you can't cancel! If in doubt, don't cancel!**

Sometimes it is helpful to use 1 as a factor in the numerator or the denominator when simplifying.

EXAMPLE 8   Simplify: $\dfrac{9}{72}$.

SOLUTION

$$\frac{9}{72} = \frac{1 \cdot 9}{8 \cdot 9} \qquad \text{Factoring and using the identity property of 1 to write 9 as } 1 \cdot 9$$

$$= \frac{1 \cdot \cancel{9}}{8 \cdot \cancel{9}} = \frac{1}{8} \qquad \text{Simplifying by removing a factor equal to 1: } \frac{9}{9} = 1$$

Addition and Subtraction

When denominators are the same, fractions are added or subtracted by adding or subtracting numerators and keeping the same denominator.

Addition and Subtraction of Fractions   For any two fractions $\dfrac{a}{d}$ and $\dfrac{b}{d}$,

$$\frac{a}{d} + \frac{b}{d} = \frac{a + b}{d} \quad \text{and} \quad \frac{a}{d} - \frac{b}{d} = \frac{a - b}{d}.$$

EXAMPLE 9   Add and simplify: $\dfrac{4}{8} + \dfrac{5}{8}$.

SOLUTION   We add the numerators and keep the common denominator:

$$\frac{4}{8} + \frac{5}{8} = \frac{4 + 5}{8} = \frac{9}{8}. \qquad \text{You can think of this as 4 eighths + 5 eighths = 9 eighths, or } \frac{9}{8}.$$

In arithmetic, we often write $1\frac{1}{8}$ rather than the "improper" fraction $\frac{9}{8}$. In algebra, $\frac{9}{8}$ is generally more useful and is quite "proper" for our purposes.

When denominators are different, we use the identity property of 1 and multiply to obtain a common denominator. Then we add, as in Example 9.

▶ **EXAMPLE 10**   Add or subtract as indicated: **(a)** $\dfrac{7}{8} + \dfrac{5}{12}$; **(b)** $\dfrac{9}{8} - \dfrac{4}{5}$.

SOLUTION

a) The number 24 is divisible by both 8 and 12. We multiply both $\frac{7}{8}$ and $\frac{5}{12}$ by suitable forms of 1 to obtain two fractions with denominators of 24:

$$\frac{7}{8} + \frac{5}{12} = \frac{7}{8} \cdot \frac{3}{3} + \frac{5}{12} \cdot \frac{2}{2} \qquad \begin{array}{l}\text{Multiplying by 1.}\\ \text{Since } 8 \cdot 3 = 24, \text{ we multiply } \frac{7}{8} \text{ by } \frac{3}{3}.\\ \text{Since } 12 \cdot 2 = 24, \text{ we multiply } \frac{5}{12} \text{ by } \frac{2}{2}.\end{array}$$

$$= \frac{21}{24} + \frac{10}{24} = \frac{31}{24}. \qquad \text{Adding fractions}$$

b) $\dfrac{9}{8} - \dfrac{4}{5} = \dfrac{9}{8} \cdot \dfrac{5}{5} - \dfrac{4}{5} \cdot \dfrac{8}{8}$ Using 40 as a common denominator

$$= \frac{45}{40} - \frac{32}{40} = \frac{13}{40} \qquad \text{Subtracting fractions}$$ ◀

After adding, subtracting, multiplying, or dividing, we may still need to simplify the answer.

▶ **EXAMPLE 11**   Perform the indicated operation and, if possible, simplify.

a) $\dfrac{7}{10} - \dfrac{1}{5}$ 　　　　**b)** $8 \cdot \dfrac{5}{12}$ 　　　　**c)** $\dfrac{\frac{5}{6}}{\frac{25}{9}}$

SOLUTION

a) $\dfrac{7}{10} - \dfrac{1}{5} = \dfrac{7}{10} - \dfrac{1}{5} \cdot \dfrac{2}{2}$ Using 10 as the common denominator

$$= \frac{7}{10} - \frac{2}{10}$$

$$= \frac{5}{10} = \frac{1 \cdot \cancel{5}}{2 \cdot \cancel{5}} = \frac{1}{2} \qquad \text{Removing a factor equal to 1: } \tfrac{5}{5} = 1$$

b) $8 \cdot \dfrac{5}{12} = \dfrac{8 \cdot 5}{12}$ Multiplying numerators and denominators. Think of 8 as $\frac{8}{1}$.

$$= \frac{2 \cdot 2 \cdot 2 \cdot 5}{2 \cdot 2 \cdot 3} \qquad \text{Factoring; } \frac{4 \cdot 2 \cdot 5}{4 \cdot 3} \text{ can also be used.}$$

$$= \frac{\cancel{2} \cdot \cancel{2} \cdot 2 \cdot 5}{\cancel{2} \cdot \cancel{2} \cdot 3} \qquad \text{Removing a factor equal to 1: } \frac{2 \cdot 2}{2 \cdot 2} = 1$$

$$= \frac{10}{3} \qquad \text{Simplifying}$$

➤ Usually an instructor assigns some odd-numbered exercises. When you complete these, you can check your answers at the back of the book. If you miss any, closely examine your work, and if necessary, consult the *Student's Solutions Manual* or your instructor for guidance.

➤ Whether or not your instructor assigns the even-numbered exercises, try to do some on your own. There are no answers given for them, so you will gain practice doing exercises that are similar to quiz or test problems. Check your answers later with a friend or your instructor.

c) $\dfrac{\dfrac{5}{6}}{\dfrac{25}{9}} = \dfrac{5}{6} \div \dfrac{25}{9}$ Rewriting horizontally. Remember that a fraction bar indicates division.

$\qquad = \dfrac{5}{6} \cdot \dfrac{9}{25}$ Multiplying by the reciprocal of $\frac{25}{9}$

$\qquad = \dfrac{5 \cdot 3 \cdot 3}{2 \cdot 3 \cdot 5 \cdot 5}$ Writing as one fraction and factoring

$\qquad = \dfrac{\cancel{5} \cdot \cancel{3} \cdot 3}{2 \cdot \cancel{3} \cdot \cancel{5} \cdot 5}$ Removing a factor equal to 1: $\dfrac{5 \cdot 3}{3 \cdot 5} = 1$

$\qquad = \dfrac{3}{10}$ Simplifying

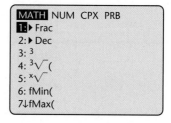

Fraction Notation and Menus

Some graphing calculators can perform operations using fraction notation. Others can convert answers to fraction notation. Often this conversion is performed using a command found in a **menu.**

Menus   A menu is a list of options that appears when a key is pressed. To select an item from the menu, highlight its number using the up or down arrow keys and press **ENTER** or simply press the number of the item. For example, pressing **MATH** results in a screen like the following. Four menu titles, or **submenus,** are listed across the top of the screen. Use the left and right arrow keys to highlight the desired submenu.

```
MATH NUM CPX PRB
1:▶Frac
2:▶Dec
3: 3
4: 3√(
5: x√
6: fMin(
7↓fMax(
```

In the screen shown above, the MATH submenu is highlighted. The options in that menu appear on the screen. We will refer to this submenu as MATH MATH, meaning that we first press **MATH** and then highlight the MATH submenu. Note that the submenu contains more options than can fit on a screen, as indicated by the arrow in entry 7. The remaining options will appear as the down arrow is pressed.

Fraction Notation   To convert a number to fraction notation, enter the number on the home screen and choose the Frac option from the MATH MATH submenu. After the notation ▶Frac is copied on the home screen, press **ENTER**.

Reciprocals   To find the reciprocal of a number, enter the number, press **x⁻¹**, and then press **ENTER**. The reciprocal will be given in decimal notation and can be converted to fraction notation using the Frac option.

EXAMPLE 12 Use a graphing calculator to find fraction notation for $\frac{2}{15} + \frac{7}{12}$.

SOLUTION A fraction bar indicates division, so we enter $\frac{2}{15}$ as 2/15 and $\frac{7}{12}$ as 7/12. The answer is given in decimal notation. We then convert to fraction notation by pressing **MATH** and selecting the Frac option. The Ans notation on the screen indicates the result, or answer, of the most recent operation. In this case, the notation Ans ▶Frac indicates that the calculator will convert 0.7166666667 to fraction notation.

```
2/15+7/12
              .7166666667
Ans▶Frac
                    43/60
```

This procedure can be done in one step using the keystrokes

 **MATH** (1) **ENTER**.

```
2/15+7/12▶Frac
                    43/60
```

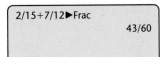

1.3 EXERCISE SET

*To the student and the instructor: **Beginning in this section, selected exercises are marked with the symbol** Aha! **Students who pause to inspect an Aha! exercise should find the answer more readily than those who proceed mechanically. This is done to discourage rote memorization. Some later "Aha!" exercises in this exercise set are unmarked, to encourage students to always pause before working a problem.***

Concept Reinforcement Label each of the following numbers as prime, composite, or neither.

1. 9   **2.** 15   **3.** 31   **4.** 35

5. 25   **6.** 37   **7.** 2   **8.** 1

9. 0   **10.** 4

Concept Reinforcement *In each of Exercises 11–14, match the description with a number from the list on the right.*

11. ___ A factor of 35   **a)** 2

12. ___ A number that has 3 as a factor   **b)** 7

c) 60

13. ___ An odd composite number   **d)** 65

14. ___ The only even prime number

Write at least two factorizations of each number. Then list all the factors of the number.

15. 50   **16.** 70   **17.** 42   **18.** 60

Find the prime factorization of each number. If the number is prime, state this.

19. 26   **20.** 15   **21.** 30

22. 55   **23.** 27   **24.** 98

25. 18   **26.** 54   **27.** 40

28. 56   **29.** 43   **30.** 120

31. 210   **32.** 79   **33.** 115

34. 143

Simplify.

35. $\dfrac{14}{21}$ **36.** $\dfrac{10}{14}$ **37.** $\dfrac{16}{56}$

38. $\dfrac{72}{27}$ **39.** $\dfrac{6}{48}$ **40.** $\dfrac{12}{70}$

41. $\dfrac{49}{7}$ **42.** $\dfrac{132}{11}$ **43.** $\dfrac{19}{76}$

44. $\dfrac{17}{51}$ **45.** $\dfrac{150}{25}$ **46.** $\dfrac{170}{34}$

47. $\dfrac{42}{50}$ **48.** $\dfrac{75}{80}$ **49.** $\dfrac{120}{82}$

50. $\dfrac{75}{45}$ **51.** $\dfrac{210}{98}$ **52.** $\dfrac{140}{350}$

Perform the indicated operation and, if possible, simplify. Check using a calculator.

53. $\dfrac{1}{2} \cdot \dfrac{3}{7}$ **54.** $\dfrac{11}{10} \cdot \dfrac{8}{5}$ **55.** $\dfrac{9}{2} \cdot \dfrac{3}{4}$

Aha! **56.** $\dfrac{11}{12} \cdot \dfrac{12}{11}$ **57.** $\dfrac{1}{8} + \dfrac{3}{8}$ **58.** $\dfrac{1}{2} + \dfrac{1}{8}$

59. $\dfrac{4}{9} + \dfrac{13}{18}$ **60.** $\dfrac{4}{5} + \dfrac{8}{15}$ **61.** $\dfrac{3}{a} \cdot \dfrac{b}{7}$

62. $\dfrac{x}{5} \cdot \dfrac{y}{z}$ **63.** $\dfrac{4}{a} + \dfrac{3}{a}$ **64.** $\dfrac{7}{a} - \dfrac{5}{a}$

65. $\dfrac{3}{10} + \dfrac{8}{15}$ **66.** $\dfrac{7}{8} + \dfrac{5}{12}$ **67.** $\dfrac{9}{7} - \dfrac{2}{7}$

68. $\dfrac{12}{5} - \dfrac{2}{5}$ **69.** $\dfrac{13}{18} - \dfrac{4}{9}$ **70.** $\dfrac{13}{15} - \dfrac{8}{45}$

Aha! **71.** $\dfrac{20}{30} - \dfrac{2}{3}$ **72.** $\dfrac{5}{7} - \dfrac{5}{21}$ **73.** $\dfrac{7}{6} \div \dfrac{3}{5}$

74. $\dfrac{7}{5} \div \dfrac{3}{4}$ **75.** $\dfrac{8}{9} \div \dfrac{4}{15}$ **76.** $\dfrac{9}{4} \div 9$

77. $12 \div \dfrac{3}{7}$ **78.** $\dfrac{1}{10} \div \dfrac{1}{5}$ *Aha!* **79.** $\dfrac{7}{13} \div \dfrac{7}{13}$

80. $\dfrac{17}{8} \div \dfrac{5}{6}$ **81.** $\dfrac{\frac{2}{7}}{\frac{5}{3}}$ **82.** $\dfrac{\frac{3}{8}}{\frac{1}{5}}$

83. $\dfrac{9}{\frac{1}{2}}$ **84.** $\dfrac{\frac{7}{3}}{5}$

TW 85. Under what circumstances would the sum of two fractions be easier to compute than the product of the same two fractions?

TW 86. Under what circumstances would the product of two fractions be easier to compute than the sum of the same two fractions?

Skill Maintenance

Use a commutative law to write an equivalent expression. There can be more than one correct answer. [1.2]

87. $5(x + 3)$ **88.** $7 + (a + b)$

Synthesis

TW 89. Bryce insists that $\dfrac{2 + x}{8}$ is equivalent to $\dfrac{1 + x}{4}$.
What mistake do you think is being made and how could you demonstrate to Bryce that the two expressions are not equivalent?

TW 90. Why are 0 and 1 considered neither prime nor composite?

91. In the following table, the top number can be factored in such a way that the sum of the factors is the bottom number. For example, in the first column, 56 is factored as $7 \cdot 8$, since $7 + 8 = 15$, the bottom number. Find the missing numbers in each column.

Product	56	63	36	72	140	96	168
Factor	7						
Factor	8						
Sum	15	16	20	38	24	20	29

92. *Packaging.* Tritan Candies uses two sizes of boxes, 6 in. long and 8 in. long. These are packed end to end in bigger cartons to be shipped. What is the shortest-length carton that will accommodate boxes of either size without any room left over? (Each carton must contain boxes of only one size; no mixing is allowed.)

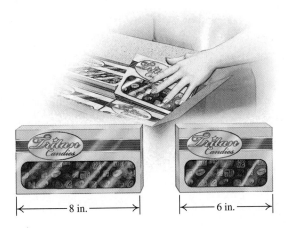

Simplify.

93. $\dfrac{16 \cdot 9 \cdot 4}{15 \cdot 8 \cdot 12}$

94. $\dfrac{9 \cdot 8xy}{2xy \cdot 36}$

95. $\dfrac{27pqrs}{9prst}$

96. $\dfrac{512}{192}$

97. $\dfrac{15 \cdot 4xy \cdot 9}{6 \cdot 25x \cdot 15y}$

98. $\dfrac{10x \cdot 12 \cdot 25y}{2 \cdot 30x \cdot 20y}$

99. $\dfrac{\frac{27ab}{15mn}}{\frac{18bc}{25np}}$

100. $\dfrac{\frac{45xyz}{24ab}}{\frac{30xz}{32ac}}$

101. $\dfrac{5\frac{3}{4}\,rs}{4\frac{1}{2}\,st}$

102. $\dfrac{3\frac{5}{7}\,mn}{2\frac{4}{5}\,np}$

Find the area of each figure.

103.

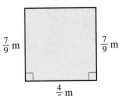

104.

105. Find the perimeter of a square with sides of length $3\frac{5}{9}$ m.

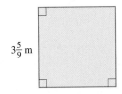

106. Find the perimeter of the rectangle in Exercise 103.

107. Find the total length of the edges of a cube with sides of length $2\frac{3}{10}$ cm.

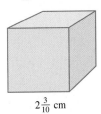

1.4 Positive and Negative Real Numbers

The Integers ■ The Rational Numbers ■ Real Numbers
and Order ■ Absolute Value

A **set** is a collection of objects. The set containing 1, 3, and 7 is usually written $\{1, 3, 7\}$. In this section, we examine some important sets of numbers.

The Integers

Two sets of numbers were mentioned in Section 1.3. We represent these sets using dots on the number line.

Natural numbers = {1, 2, 3, ...}

0 1 2 3 4 5 6 7

Whole numbers = {0, 1, 2, 3, ...}

To create the set of *integers*, we include all whole numbers, along with their *opposites*. To find the opposite of a number, we locate the number that is the same distance from 0 but on the other side of the number line. For example,

the opposite of 1 is negative 1, written -1;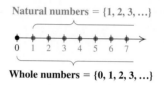

and

the opposite of 3 is negative 3, written -3.

The **integers** consist of all whole numbers and their opposites.

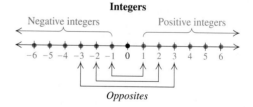

Opposites are discussed in more detail in Section 1.6. Note that, except for 0, opposites occur in pairs. Thus, 5 is the opposite of -5, just as -5 is the opposite of 5. Note that 0 acts as its own opposite.

Set of Integers

The set of integers = $\{\ldots, -4, -3, -2, -1, 0, 1, 2, 3, 4, \ldots\}$.

Integers are associated with many real-world problems and situations.

▸ **EXAMPLE 1**   State which integer(s) corresponds to each situation.

a) In 2003, the Colonial Pipeline Company was fined a record $34 million for pollution crimes (*Source*: GreenConsumerGuide.com).

b) Death Valley is 280 ft below sea level.

c) To lose one pound of fat, it is necessary for most people to create a 3500-calorie deficit (*Source*: World Health Organization).

SOLUTION

a) The integer $-34,000,000$ corresponds to a fine of $34 million.

b) The integer -280 corresponds to 280 ft below sea level. The elevation is -280 ft.

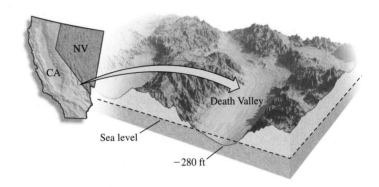

c) The integer -3500 corresponds to a deficit of 3500 calories.

The Rational Numbers

Although numbers like $\frac{5}{9}$ are built out of integers, these numbers are not themselves integers. Another set, the **rational numbers,** contains fractions and decimals that repeat or terminate, as well as the integers. Some examples of rational numbers are

$$\frac{5}{9}, \quad -\frac{4}{7}, \quad 95, \quad -16, \quad 0, \quad \frac{-35}{8}, \quad 2.4, \quad -0.31.$$

In Section 1.7, we show that $-\frac{4}{7}$ can be written as $\frac{-4}{7}$ or $\frac{4}{-7}$. Indeed, every number listed above can be written as an integer over an integer. For example, 95 can be written as $\frac{95}{1}$ and 2.4 can be written as $\frac{24}{10}$. In this manner, any *ratio*nal number can be expressed as the *ratio* of two integers. Rather than attempt to list all rational numbers, we use this idea of ratio to describe the set as follows.

Set of Rational Numbers

The set of rational numbers $= \left\{ \dfrac{a}{b} \,\middle|\, a \text{ and } b \text{ are integers and } b \neq 0 \right\}.$

This is read "the set of all numbers $\dfrac{a}{b}$, where a and b are integers and b does not equal zero."

To *graph* a number is to mark its location on the number line.

EXAMPLE 2   Graph each of the following rational numbers.

a) $\frac{5}{2}$ **b)** -3.2 **c)** $\frac{11}{8}$

SOLUTION

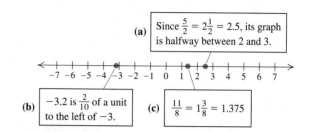

(a) Since $\frac{5}{2} = 2\frac{1}{2} = 2.5$, its graph is halfway between 2 and 3.

(b) -3.2 is $\frac{2}{10}$ of a unit to the left of -3.

(c) $\frac{11}{8} = 1\frac{3}{8} = 1.375$

Every rational number can be written as a fraction or a decimal.

EXAMPLE 3   Convert to decimal notation: $-\frac{5}{8}$.

SOLUTION   We first find decimal notation for $\frac{5}{8}$. Since $\frac{5}{8}$ means $5 \div 8$, we divide.

```
      0.6 2 5
  8)5.0 0 0
    4 8 0 0
      2 0 0
      1 6 0
        4 0
        4 0
          0  ←—— The remainder is 0.
```

Thus, $\frac{5}{8} = 0.625$, so $-\frac{5}{8} = -0.625$.

Because the division in Example 3 ends with the remainder 0, we consider -0.625 a **terminating decimal.** If we are "bringing down" zeros and a remainder reappears, we have a **repeating decimal,** as shown in the next example.

EXAMPLE 4   Convert to decimal notation: $\frac{7}{11}$.

SOLUTION   We divide:

```
      0.6 3 6 3...
  1 1)7.0 0 0 0
      6 6
        4 0  ←
        3 3
          7 0
          6 6
            4 0 ←
```

4 reappears as a remainder, so the pattern of 6's and 3's in the quotient will continue.

We abbreviate repeating decimals by writing a bar over the repeating part—in this case, $0.\overline{63}$. Thus, $\frac{7}{11} = 0.\overline{63}$. ◢

Although we do not prove it here, every rational number can be expressed as either a terminating or a repeating decimal, and every terminating or repeating decimal can be expressed as a ratio of two integers.

Real Numbers and Order

Some numbers, when written in decimal form, neither terminate nor repeat. Such numbers are called **irrational numbers.**

What sort of numbers are irrational? One example is π (the Greek letter *pi*, read "pie"), which is used to find the area and the circumference of a circle: $A = \pi r^2$ and $C = 2\pi r$.

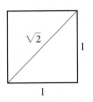

Another irrational number, $\sqrt{2}$ (read "the square root of 2"), is the length of the diagonal of a square with sides of length 1 (see the figure at left). It is also the number that, when multiplied by itself, gives 2. No rational number can be multiplied by itself to get 2, although some approximations come close:

1.4 is an *approximation* of $\sqrt{2}$ because $(1.4)(1.4) = 1.96$;

1.41 is a better approximation because $(1.41)(1.41) = 1.9881$;

1.4142 is an even better approximation because $(1.4142)(1.4142) = 1.99996164$.

Square roots are discussed in Chapter 10.

To approximate $\sqrt{2}$ on most graphing calculators, we press 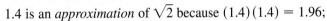 and then enter 2 enclosed by parentheses. Some calculators will supply the left parenthesis automatically when 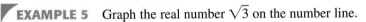 is pressed.

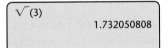

▶ **EXAMPLE 5** Graph the real number $\sqrt{3}$ on the number line.

SOLUTION We use a calculator as shown at left and approximate: $\sqrt{3} \approx 1.732$ ("$\approx$" means "approximately equals"). Then we locate this number on the number line.

◢

The rational numbers and the irrational numbers together correspond to all the points on the number line and make up what is called the **real-number system.**

Set of Real Numbers

The set of real numbers = The set of all numbers corresponding to points on the number line.

The following figure shows the relationships among various kinds of numbers.

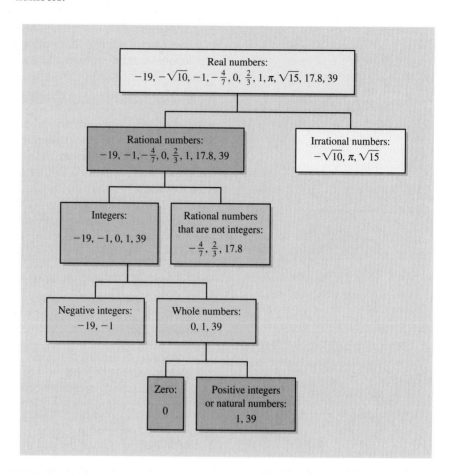

EXAMPLE 6   Which numbers in the following list are **(a)** whole numbers? **(b)** integers? **(c)** rational numbers? **(d)** irrational numbers? **(e)** real numbers?

$$-38, \quad -\frac{8}{5}, \quad 0, \quad 3, \quad 4.5, \quad \sqrt{30}, \quad 52$$

SOLUTION

a) 0, 3, and 52 are whole numbers.

b) $-38, 0, 3,$ and 52 are integers.

c) $-38, -\frac{8}{5}, 0, 3, 4.5,$ and 52 are rational numbers.

d) $\sqrt{30}$ is an irrational number.

e) $-38, -\frac{8}{5}, 0, 3, 4.5, \sqrt{30},$ and 52 are real numbers.

Real numbers are named in order on the number line, with larger numbers further to the right. For any two numbers, the one to the left is less than the one to the right. We use the symbol $<$ to mean "**is less than.**" The sentence $-8 < 6$

means "−8 is less than 6." The symbol **>** means "**is greater than.**" The sentence −3 > −7 means "−3 is greater than −7."

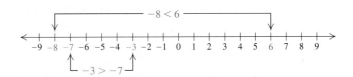

EXAMPLE 7 Use either < or > for ▦ to write a true sentence.

a) 2 ▦ 9   **b)** −3.45 ▦ 1.32   **c)** 6 ▦ −12

d) −18 ▦ −5   **e)** $\frac{7}{11}$ ▦ $\frac{5}{8}$

SOLUTION

a) Since 2 is to the left of 9 on the number line, we know that 2 is less than 9, so 2 < 9.

b) Since −3.45 is to the left of 1.32, we have −3.45 < 1.32.

c) Since 6 is to the right of −12, we have 6 > −12.

d) Since −18 is to the left of −5, we have −18 < −5.

e) We convert to decimal notation: $\frac{7}{11} = 0.\overline{63}$ and $\frac{5}{8} = 0.625$. Thus, $\frac{7}{11} > \frac{5}{8}$.
   We also could have used a common denominator: $\frac{7}{11} = \frac{56}{88} > \frac{55}{88} = \frac{5}{8}$.

Sentences like "$a < -5$" and "$-3 > -8$" are **inequalities.** It is useful to remember that every inequality can be written in two ways. For example,

$$-3 > -8 \quad \text{has the same meaning as} \quad -8 < -3.$$

It may be helpful to think of an inequality sign as an "arrow" with the smaller side pointing to the smaller number.

Note that $a > 0$ means that a represents a positive real number and $a < 0$ means that a represents a negative real number.

Statements like $a \le b$ and $a \ge b$ are also inequalities. We read $a \le b$ as "**a is less than or equal to b**" and $a \ge b$ as "**a is greater than or equal to b.**"

EXAMPLE 8 Classify each inequality as true or false.

a) $-3 \le 5$ **b)** $-3 \le -3$ **c)** $-5 \ge 4$

SOLUTION

Student Notes

It is important to remember that just because an equation or inequality is written or printed, it is not necessarily *true*. For instance, 6 = 7 is an equation and 2 > 5 is an inequality. Of course, both statements are *false*.

a) $-3 \le 5$ is *true* because $-3 < 5$ is true.

b) $-3 \le -3$ is *true* because $-3 = -3$ is true.

c) $-5 \ge 4$ is *false* since neither $-5 > 4$ nor $-5 = 4$ is true.

Absolute Value

There is a convenient terminology and notation for the distance a number is from 0 on the number line. It is called the **absolute value** of the number.

> **Absolute Value**   We write $|a|$, read "the absolute value of a," to represent the number of units that a is from zero.

EXAMPLE 9   Find each absolute value: **(a)** $|-3|$; **(b)** $|7.2|$; **(c)** $|0|$.

SOLUTION

a) $|-3| = 3$ since -3 is 3 units from 0.

b) $|7.2| = 7.2$ since 7.2 is 7.2 units from 0.

c) $|0| = 0$ since 0 is 0 units from itself.

Distance is never negative, so numbers that are opposites have the same absolute value. If a number is nonnegative, its absolute value is the number itself. If a number is negative, its absolute value is its opposite.

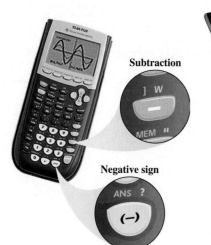

Subtraction

Negative sign

Negative Numbers and Absolute Value

Graphing calculators have different keys for writing negatives and subtracting. The key labeled (-) is used to create a negative sign whereas the one labeled − is used for subtraction. Using the wrong key may result in an ERR:SYNTAX message.

Many graphing calculators use the notation "abs" with parentheses to indicate absolute value. Thus, abs(2) = $|2|$ = 2. The abs notation is often found in the MATH NUM submenu. Some calculators automatically supply the left parenthesis.

To check Example 9(a) using a calculator, we press **MATH** ▷ ① (-) ③) **ENTER**. We see that $|-3| = 3$.

abs(−3)
                    3

1.4 EXERCISE SET

Concept Reinforcement   *In each of Exercises 1–8, fill in the blank using one of the following words:* natural number, whole number, integer, rational number, terminating, repeating, irrational number, absolute value.

1. Division can be used to show that $\frac{4}{7}$ can be written as a(n) _____ decimal.

2. Division can be used to show that $\frac{3}{20}$ can be written as a(n) _____ decimal.

3. If a number is a(n) _____, it is either a whole number or the opposite of a whole number.

4. 0 is the only _____ that is not a natural number.

5. Any number of the form a/b, where a and b are integers, with $b \neq 0$, is an example of a(n) _____.

6. A number like $\sqrt{5}$, which cannot be written precisely in fraction or decimal notation, is an example of a(n) _____.

7. If a number is a(n) _____, then it can be thought of as a counting number.

8. When two numbers are opposites, they have the same _____.

State which real number(s) correspond to each situation.

9. The Dead Sea is 1395 feet below sea level, whereas Mt. Everest is 29,035 feet above sea level.

10. Using a NordicTrack exercise machine, Kit burned 150 calories. She then drank an isotonic drink containing 65 calories.

11. The highest temperature ever reached on the earth was 950 million degrees Fahrenheit (F). The lowest temperature ever reached on the earth was approximately 460°F below zero. (*Source: Guinness World Records* 2001)

12. In Burlington, Vermont, the record low temperature for Washington's Birthday is 19°F below zero. The record high for the date is 59°F above zero.

13. In the 2003 Masters Tournament, golfer Tiger Woods finished two over par. In the 2005 Masters Tournament, he finished twelve under par.

14. Ignition occurs 10 seconds before liftoff. A spent fuel tank is detached 235 seconds after liftoff.

15. Janice deposited $750 in a savings account. Two weeks later, she withdrew $125.

16. *Birth and Death Rates.*   Recently, the world birth rate was 21 per thousand. The death rate was 9 per thousand. (*Source:* PRB 2005 World Population Data Sheet)

17. In bowling, the Jets are 34 pins behind the Strikers after one game. Describe the situation from the viewpoint of each team.

18. The halfback gained 8 yd on the first play. The quarterback was tackled for a 5-yd loss on the second play.

Graph each rational number on the number line.

19. -2

20. 5

21. -4.3

22. 3.87

23. $\frac{10}{3}$

24. $-\frac{17}{5}$

Write decimal notation for each number.

25. $\frac{7}{8}$

26. $-\frac{1}{8}$

27. $-\frac{3}{4}$

28. $\frac{5}{6}$

29. $\frac{7}{6}$

30. $\frac{5}{12}$

31. $\frac{2}{3}$

32. $\frac{1}{4}$

33. $-\frac{1}{2}$

34. $-\frac{3}{8}$ *Aha!* **35.** $\frac{13}{100}$

36. $-\frac{7}{20}$

Write a true sentence using either $<$ or $>$.

37. $-9 \;\blacksquare\; 4$

38. $9 \;\blacksquare\; 0$

39. $7 \;\blacksquare\; 0$

40. $8 \;\blacksquare\; -8$

41. $-6 \;\blacksquare\; 6$

42. $0 \;\blacksquare\; -7$

43. $-8 \;\blacksquare\; -5$

44. $-4 \;\blacksquare\; -3$

45. $-5 \;\blacksquare\; -11$

46. $-3 \;\blacksquare\; -4$

47. $-12.5 \;\blacksquare\; -9.4$

48. $-10.3 \;\blacksquare\; -14.5$

49. $\frac{5}{12} \;\blacksquare\; \frac{11}{25}$

50. $-\frac{14}{17} \;\blacksquare\; -\frac{27}{35}$

For each of the following, write a second inequality with the same meaning.

51. $-7 > x$

52. $a > 9$

53. $-10 \leq y$

54. $12 \geq t$

Classify each inequality as either true or false.

55. $-3 \geq -11$

56. $5 \leq -5$

57. $0 \geq 8$

58. $-5 \leq 7$

59. $-8 \leq -8$

60. $8 \geq 8$

Find each absolute value.

61. $|-58|$

62. $|-47|$

63. $|17|$

64. $|3.1|$

65. $|5.6|$

66. $\left|-\frac{2}{5}\right|$

67. $|329|$

68. $|-456|$

69. $\left|-\frac{9}{7}\right|$

70. $|8.02|$

71. $|0|$

72. $|-1.07|$

73. $|x|$, for $x = -8$

74. $|a|$, for $a = -5$

For Exercises 75–80, consider the following list:

$$-83, \quad -4.7, \quad 0, \quad \tfrac{5}{9}, \quad \pi, \quad \sqrt{17}, \quad 8.31, \quad 62.$$

75. List all rational numbers.

76. List all natural numbers.

77. List all integers.

78. List all irrational numbers.

79. List all real numbers.

80. List all nonnegative integers.

TW 81. Is every integer a rational number? Why or why not?

TW 82. Is every integer a natural number? Why or why not?

Skill Maintenance

83. Evaluate $3xy$ for $x = 2$ and $y = 7$. [1.1]

84. Use a commutative law to write an expression equivalent to $ab + 5$. [1.2]

Synthesis

TW 85. Is the absolute value of a number always positive? Why or why not?

TW 86. How many rational numbers are there between 0 and 1? Justify your answer.

TW 87. Does "nonnegative" mean the same thing as "positive"? Why or why not?

List in order from least to greatest.

88. $13, -12, 5, -17$

89. $-23, 4, 0, -17$

90. $-\frac{2}{3}, \frac{1}{2}, -\frac{3}{4}, -\frac{5}{6}, \frac{3}{8}, \frac{1}{6}$

91. $\frac{4}{5}, \frac{4}{3}, \frac{4}{8}, \frac{4}{6}, \frac{4}{9}, \frac{4}{2}, -\frac{4}{3}$

Write a true sentence using either $<$, $>$, or $=$.

92. $|-5| \;\blacksquare\; |-2|$

93. $|4| \;\blacksquare\; |-7|$

94. $|-8| \;\blacksquare\; |8|$

95. $|23| \;\blacksquare\; |-23|$

96. $|-3| \;\blacksquare\; |5|$

Solve. Consider only integer replacements.

Aha! **97.** $|x| = 7$

98. $|x| < 3$

99. $2 < |x| < 5$

Given that $0.\overline{33} = \frac{1}{3}$ *and* $0.\overline{66} = \frac{2}{3}$, *express each of the following as a ratio of two integers.*

100. $0.1\overline{1}$ **101.** $0.9\overline{9}$

102. $5.5\overline{5}$ **103.** $7.7\overline{7}$

104. When Helga's calculator gives a decimal value for $\sqrt{2}$ and that value is promptly squared, the result

is 2. Yet when that same decimal approximation is entered by hand and then squared, the result is not exactly 2. Why do you suppose this is?

105. Is the following statement true? Why or why not?
$$\sqrt{a^2} = |a| \quad \text{for any real number } a.$$

1.5 Addition of Real Numbers

Adding with the Number Line ■ Adding without the Number Line ■ Problem Solving ■ Combining Like Terms

We now consider addition of real numbers. To gain understanding, we will use the number line first. After observing the principles involved, we will develop rules that allow us to work more quickly without the number line.

Adding with the Number Line

To add $a + b$ on the number line, we start at a and move according to b.

a) If b is positive, we move to the right (the positive direction).

b) If b is negative, we move to the left (the negative direction).

c) If b is 0, we stay at a.

EXAMPLE 1   Add: $-4 + 9$.

SOLUTION   To add on the number line, we locate the first number, -4, and then move 9 units to the right. Note that it requires 4 units to reach 0. The difference between 9 and 4 is where we finish.

$$-4 + 9 = 5$$

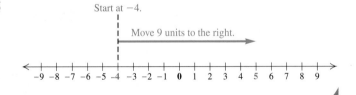

EXAMPLE 2   Add: $3 + (-5)$.

SOLUTION   We locate the first number, 3, and then move 5 units to the left. Note that it requires 3 units to reach 0. The difference between 5 and 3 is 2, so we finish 2 units to the left of 0.

Student Notes

Parentheses are used when a negative sign follows an operation. Just as we would never write $8 \div \times 2$, it is improper to write $3 + -5$.

$3 + (-5) = -2$

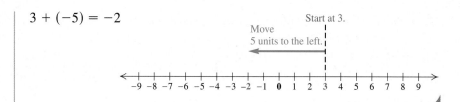

EXAMPLE 3   Add: $-4 + (-3)$.

SOLUTION   After locating -4, we move 3 units to the left. We finish a total of 7 units to the left of 0.

$-4 + (-3) = -7$

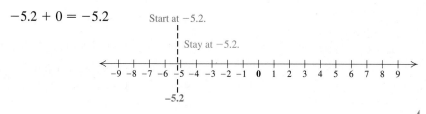

EXAMPLE 4   Add: $-5.2 + 0$.

SOLUTION   We locate -5.2 and move 0 units. Thus we finish where we started, at -5.2.

$-5.2 + 0 = -5.2$

Start at -5.2.

Stay at -5.2.

$$\begin{array}{c}
\longleftarrow\!\!+\!\!+\!\!+\!\!+\!\!+\!\!+\!\!+\!\!+\!\!+\!\!+\!\!+\!\!+\!\!+\!\!+\!\!+\!\!+\!\!+\!\!+\!\!\longrightarrow \\
\!-9\ -8\ -7\ -6\ -5\ -4\ -3\ -2\ -1\ \ 0\ \ 1\ \ 2\ \ 3\ \ 4\ \ 5\ \ 6\ \ 7\ \ 8\ \ 9
\end{array}$$

-5.2

From Examples 1–4, we observe the following rules.

Rules for Addition of Real Numbers

1. *Positive numbers*: Add as usual. The answer is positive.
2. *Negative numbers*: Add absolute values and make the answer negative (see Example 3).
3. *A positive number and a negative number*: Subtract the smaller absolute value from the greater absolute value. Then:

   a) If the positive number has the greater absolute value, the answer is positive (see Example 1).
   b) If the negative number has the greater absolute value, the answer is negative (see Example 2).
   c) If the numbers have the same absolute value, the answer is 0.

4. *One number is zero*: The sum is the other number (see Example 4).

Rule 4 is known as the **identity property of 0.**

> **The Identity Property of 0**   For any real number a,
>
> $$a + 0 = a.$$
>
> (Adding 0 to a number gives that same number.) The number 0 is called the *additive identity*.

Adding without the Number Line

The rules listed above can be used without drawing the number line.

EXAMPLE 5   Add without using the number line.

a) $-12 + (-7)$
b) $-1.4 + 8.5$
c) $-36 + 21$
d) $1.5 + (-1.5)$
e) $-\frac{7}{8} + 0$
f) $\frac{2}{3} + \left(-\frac{5}{8}\right)$

SOLUTION

a) $-12 + (-7) = -19$ **Two negatives.** *Think*: Add the absolute values, 12 and 7, to get 19. Make the answer *negative*, -19.

b) $-1.4 + 8.5 = 7.1$ **A negative and a positive.** *Think*: The difference of absolute values is $8.5 - 1.4$, or 7.1. The positive number has the larger absolute value, so the answer is *positive*, 7.1.

c) $-36 + 21 = -15$ **A negative and a positive.** *Think*: The difference of absolute values is $36 - 21$, or 15. The negative number has the larger absolute value, so the answer is *negative*, -15.

d) $1.5 + (-1.5) = 0$ **A negative and a positive.** *Think*: Since the numbers are opposites, they have the same absolute value and the answer is 0.

e) $-\dfrac{7}{8} + 0 = -\dfrac{7}{8}$ **One number is zero.** The sum is the other number, $-\frac{7}{8}$.

f) $\dfrac{2}{3} + \left(-\dfrac{5}{8}\right) = \dfrac{16}{24} + \left(-\dfrac{15}{24}\right)$ This is similar to part (b) above. We find a common denominator and then add.

$$= \dfrac{1}{24}$$

If we are adding several numbers, some positive and some negative, the commutative and associative laws allow us to add all the positive numbers, then add all the negative numbers, and then add the results. Of course, we can also add from left to right, if we prefer.

Study Tip

Two (or More) Heads Are Better Than One

Consider forming a study group with some of your fellow students. Exchange telephone numbers, schedules, and any e-mail addresses so that you can coordinate study time for homework and tests.

EXAMPLE 6 Add: $15 + (-2) + 7 + 14 + (-5) + (-12)$.

SOLUTION

$15 + (-2) + 7 + 14 + (-5) + (-12)$

$= 15 + 7 + 14 + (-2) + (-5) + (-12)$ Using the commutative law of addition

$= (15 + 7 + 14) + [(-2) + (-5) + (-12)]$ Using the associative law of addition

$= 36 + (-19)$ Adding the positives; adding the negatives

$= 17$ Adding a positive and a negative

A calculator can be helpful when we are adding real numbers. However, it is not a substitute for knowledge of the rules for addition: When the rules are new to you, a calculator should be used only when checking your work.

Problem Solving

Addition of real numbers occurs in many real-world applications.

EXAMPLE 7 Lake Level. From the spring of 2004 through the spring of 2006, the water-surface elevation of the Great Salt Lake at Boat Harbor rose $\frac{9}{10}$ ft, dropped $2\frac{2}{5}$ ft, and rose $3\frac{1}{2}$ ft (*Source*: U.S. Geological Survey). By how much did the level change over the 2-yr period?

SOLUTION The problem translates to a sum:

Rewording: The 1st change plus the 2nd change plus the 3rd change is the total change.

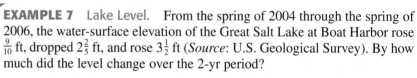

Translating: $\frac{9}{10}$ $+$ $\left(-2\frac{2}{5}\right)$ $+$ $3\frac{1}{2}$ $=$ Total change.

Writing with common denominators and adding from left to right, we have

$$\frac{9}{10} + \left(-2\frac{4}{10}\right) + 3\frac{5}{10} = \frac{9}{10} + \left(-\frac{24}{10}\right) + \frac{35}{10}$$

$$= -\frac{15}{10} + \frac{35}{10} = \frac{20}{10} = 2.$$

The lake level rose 2 ft between the spring of 2004 and the spring of 2006.

Combining Like Terms

When two terms have variable factors that are exactly the same, like $5ab$ and $7ab$, the terms are called **like,** or **similar, terms.*** The distributive law enables us to **combine,** or **collect, like terms.** The above rules for addition will again apply.

---

*Like terms are discussed in greater detail in Section 1.8.

EXAMPLE 8 Combine like terms.

a) $-7x + 9x$ b) $2a + (-3b) + (-5a) + 9b$

c) $6 + y + (-3.5y) + 2$

SOLUTION

a) $-7x + 9x = (-7 + 9)x$ Using the distributive law

$\qquad\qquad\quad = 2x$ Adding -7 and 9

b) $2a + (-3b) + (-5a) + 9b$

$\quad = 2a + (-5a) + (-3b) + 9b$ Using the commutative law of addition

$\quad = (2 + (-5))a + (-3 + 9)b$ Using the distributive law

$\quad = -3a + 6b$ Adding

c) $6 + y + (-3.5y) + 2 = y + (-3.5y) + 6 + 2$ Using the commutative law of addition

$\qquad\qquad\qquad\qquad = (1 + (-3.5))y + 6 + 2$ Using the distributive law

$\qquad\qquad\qquad\qquad = -2.5y + 8$ Adding

With practice we can leave out some steps, combining like terms mentally. Note that numbers like 6 and 2 in the expression $6 + y + (-3.5y) + 2$ are constants and are also considered to be like terms.

1.5 EXERCISE SET

FOR EXTRA HELP

MathXL   MyMathLab   InterAct Math   Tutor Center   AW Math Tutor Center   Video Lectures on CD: Disc 1   Student's Solutions Manual

Concept Reinforcement   *In each of Exercises 1–6, match the term with a like term from the column on the right.*

1. ____ $8n$

2. ____ $7m$

3. ____ 43

4. ____ $28z$

5. ____ $-2x$

6. ____ $-9t$

a) $-3z$

b) $5x$

c) $2t$

d) $-4m$

e) 9

f) $-3n$

Add using the number line.

7. $5 + (-8)$ 8. $2 + (-5)$

9. $-5 + 9$ 10. $-3 + 8$

11. $8 + (-8)$ 12. $6 + (-6)$

13. $-3 + (-5)$ 14. $-4 + (-6)$

Add. Do not use the number line except as a check.

15. $-27 + 0$ 16. $-6 + 0$

17. $0 + (-8)$ 18. $0 + (-2)$

19. $12 + (-12)$

20. $17 + (-17)$

21. $-24 + (-17)$

22. $-17 + (-25)$

23. $-13 + 13$

24. $-18 + 18$

25. $18 + (-11)$

26. $8 + (-5)$

27. $10 + (-12)$

28. $9 + (-13)$

29. $-3 + 14$

30. $13 + (-6)$

31. $-14 + (-19)$

32. $11 + (-9)$

33. $19 + (-19)$

34. $-20 + (-6)$

35. $23 + (-5)$

36. $-15 + (-7)$

37. $-31 + (-14)$

38. $40 + (-8)$

39. $40 + (-40)$

40. $-25 + 25$

41. $85 + (-65)$

42. $63 + (-18)$

43. $-3.6 + 1.9$

44. $-6.5 + 4.7$

45. $-5.4 + (-3.7)$

46. $-3.8 + (-9.4)$

47. $\frac{-3}{5} + \frac{4}{5}$

48. $\frac{-2}{7} + \frac{3}{7}$

49. $\frac{-4}{7} + \frac{-2}{7}$

50. $\frac{-5}{9} + \frac{-2}{9}$

51. $-\frac{2}{5} + \frac{1}{3}$

52. $-\frac{4}{13} + \frac{1}{2}$

53. $\frac{-4}{9} + \frac{2}{3}$

54. $\frac{-1}{6} + \frac{1}{3}$

55. $35 + (-14) + (-19) + (-5)$

56. $28 + (-44) + 17 + 31 + (-94)$

Aha! **57.** $-4.9 + 8.5 + 4.9 + (-8.5)$

58. $24 + 3.1 + (-44) + (-8.2) + 63$

Solve. Write your answer as a complete sentence.

59. *Gas Prices.* In a recent year, the price of a gallon of 87-octane gasoline was \$1.69. The price then rose 5¢, dropped 3¢, and then rose 7¢. By how much did the price change during that period?

60. *Oil Prices.* In a recent year, the price of a gallon of home heating oil was \$1.28. The price then rose 6¢, dropped 2¢, and then rose 5¢. By how much did the price change during that period?

61. *Telephone Bills.* Maya's cell-phone bill for July was \$82. She sent a check for \$50 and then ran up \$63 in charges for August. What was her new balance?

62. *Profits and Losses.* The following table shows the profits and losses of Fitness City over a 3-yr period. Find the profit or loss after this period of time.

Year	Profit or loss
2002	−\$26,500
2003	−\$10,200
2004	+\$32,400

63. *Yardage Gained.* In an intramural football game, the quarterback attempted passes with the following results.

First try	13-yd gain
Second try	12-yd loss
Third try	21-yd gain

Find the total gain (or loss).

64. *Account Balance.* Leah has \$350 in a checking account. She writes a check for \$530, makes a deposit of \$75, and then writes a check for \$90. What is the balance in the account?

65. *Credit-Card Bills.* Lyle's credit-card bill indicates that he owes \$470. He sends a check to the credit-card company for \$45, charges another \$160 in merchandise, and then pays off another \$500 of his bill. What is Lyle's new balance?

66. *Stock Growth.* In the course of one day, the value of a share of Google rose \$0.82, fell \$0.90, and then rose \$0.25. How much had the stock's value risen or fallen at the end of the day?

67. *Peak Elevation.* The tallest mountain in the world, as measured from base to peak, is Mauna Kea in Hawaii. From a base 19,684 ft below sea level, it rises 33,480 ft. (*Source: The Guinness Book of Records*, 1999) What is the elevation of its peak?

68. *Class Size.* During the first two weeks of the semester, 5 students withdrew from Elisa's algebra class, 8 students were added to the class, and 6 students were dropped as "no-shows." By how many students did the original class size change?

Combine like terms.

69. $8a + 6a$

70. $3x + 8x$

71. $-3x + 12x$

72. $2m + (-7m)$

73. $5t + 8t$

74. $5a + 8a$

75. $7m + (-9m)$

76. $-4x + 4x$

77. $-5a + (-2a)$

78. $10n + (-17n)$

79. $-3 + 8x + 4 + (-10x)$

80. $8a + 5 + (-a) + (-3)$

Find the perimeter of each figure.

81.

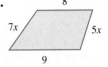

82.

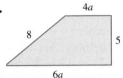

83.

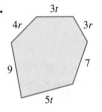

84.

85.

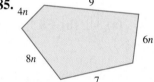

86.

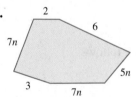

TW 87. Explain in your own words why the sum of two negative numbers is negative.

TW 88. Without performing the actual addition, explain why the sum of all integers from -10 to 10 is 0.

Skill Maintenance

89. Multiply: $7(3z + y + 2)$. [1.2]

90. Divide and simplify: $\frac{7}{2} \div \frac{3}{8}$. [1.3]

Synthesis

TW 91. Under what circumstances will the sum of one positive number and several negative numbers be positive?

TW 92. Is it possible to add real numbers without knowing how to calculate $a - b$ with a and b both non-negative and $a \geq b$? Why or why not?

93. *Stock Prices.* The value of EKB stock rose $2.38 and then dropped $3.25 before finishing at $64.38. What was the stock's original value?

94. *Sports-Card Values.* The value of a sports card dropped $12 and then rose $17.50 before settling at $61. What was the original value of the card?

Find the missing term or terms.

95. $4x + \underline{\hspace{0.5in}} + (-9x) + (-2y) = -5x - 7y$

96. $-3a + 9b + \underline{\hspace{0.5in}} + 5a = 2a - 6b$

97. $3m + 2n + \underline{\hspace{0.5in}} + (-2m) = 2n + (-6m)$

98. $\underline{\hspace{0.5in}} + 9x + (-4y) + x = 10x - 7y$

Aha! **99.** $7t + 23 + \underline{\hspace{0.5in}} + \underline{\hspace{0.5in}} = 0$

100. *Geometry.* The perimeter of a rectangle is $7x + 10$. If the length of the rectangle is 5, express the width in terms of x.

101. *Golfing.* After five rounds of golf, a golf pro was 3 under par twice, 2 over par once, 2 under par once, and 1 over par once. On average, how far above or below par was the golfer?

1.6 Subtraction of Real Numbers

Opposites and Additive Inverses ◼ Subtraction ◼ Problem Solving

In arithmetic, when a number b is subtracted from another number a, the difference, $a - b$, is the number that when added to b gives a. For example, $45 - 17 = 28$ because $28 + 17 = 45$. We will use this approach to develop an efficient way of finding the value of $a - b$ for any real numbers a and b. Before doing so, however, we must develop some terminology.

Opposites and Additive Inverses

Numbers such as 6 and -6 are *opposites*, or *additive inverses*, of each other. Whenever opposites are added, the result is 0; and whenever two numbers add to 0, those numbers are opposites.

EXAMPLE 1   Find the opposite of each number: **(a)** 34; **(b)** -8.3; **(c)** 0.

SOLUTION

a) The opposite of 34 is -34: $34 + (-34) = 0$.
b) The opposite of -8.3 is 8.3: $-8.3 + 8.3 = 0$.
c) The opposite of 0 is 0: $0 + 0 = 0$.

To write the opposite, we use the symbol $-$, as follows.

Opposite   The *opposite*, or *additive inverse*, of a number a is written $-a$ (read "the opposite of a" or "the additive inverse of a").

Note that if we take a number, say 8, and find its opposite, -8, and then find the opposite of the result, we will have the original number, 8, again.

The Opposite of an Opposite
For any real number a,

$$-(-a) = a.$$

(The opposite of the opposite of a is a.)

> **EXAMPLE 2**   Find $-x$ and $-(-x)$ when $x = 16$.
>
> **SOLUTION**
>
> If $x = 16$, then $-x = -16$.   The opposite of 16 is -16.
>
> If $x = 16$, then $-(-x) = -(-16) = 16$.   **The opposite of the opposite of 16 is 16.**

> **EXAMPLE 3**   Find $-x$ and $-(-x)$ when $x = -3$.
>
> **SOLUTION**
>
> If $x = -3$, then $-x = -(-3) = 3$.   The opposite of -3 is 3.
>
> If $x = -3$, then $-(-x) = -(-(-3)) = -(\ 3\ ) = -3$.

Student Notes

As you read mathematics, it is important to verbalize correctly the words and symbols to yourself. Consistently reading the expression $-x$ as "the opposite of x" is a good step in this direction.

Note in Example 3 that an extra set of parentheses is used to show that we are substituting the negative number -3 for x. The notation $- -x$ is not used.

A symbol such as -8 is usually read "negative 8." It could be read "the additive inverse of 8," because the additive inverse of 8 is negative 8. It could also be read "the opposite of 8," because the opposite of 8 is -8.

A symbol like $-x$, which has a variable, should be read "the opposite of x" or "the additive inverse of x" and *not* "negative x," since to do so suggests that $-x$ represents a negative number.

The symbol "$-$" is read differently depending on where it appears. For example, $-5 - (-x)$ is read "negative five minus the opposite of x."

> **EXAMPLE 4**   Write each of the following in words.
>
> **a)** $2 - 8$ **b)** $5 - (-4)$ **c)** $6 - (-x)$
>
> **SOLUTION**
>
> **a)** $2 - 8$ is read "two minus eight."
>
> **b)** $5 - (-4)$ is read "five minus negative four."
>
> **c)** $6 - (-x)$ is read "six minus the opposite of x."

As we saw in Example 3, $-x$ can represent a positive number. This notation can be used to restate a result from Section 1.5 as *the law of opposites.*

> **The Law of Opposites**   For any two numbers a and $-a$,
>
> $$a + (-a) = 0.$$
>
> (When opposites are added, their sum is 0.)

A negative number is said to have a "negative *sign*." A positive number is said to have a "positive *sign*." If we change a number to its opposite, or additive inverse, we say that we have "changed or reversed its sign."

EXAMPLE 5 Change the sign (find the opposite) of each number: **(a)** -3; **(b)** -10; **(c)** 14.

SOLUTION

a) When we change the sign of -3, we obtain 3.

b) When we change the sign of -10, we obtain 10.

c) When we change the sign of 14, we obtain -14.

Subtraction

Opposites are helpful when subtraction involves negative numbers. To see why, look for a pattern in the following:

Subtracting		Adding the Opposite
$9 - 5 = 4$	since $4 + 5 = 9$	$9 + (-5) = 4$
$5 - 8 = -3$	since $-3 + 8 = 5$	$5 + (-8) = -3$
$-6 - 4 = -10$	since $-10 + 4 = -6$	$-6 + (-4) = -10$
$-7 - (-10) = 3$	since $3 + (-10) = -7$	$-7 + 10 = 3$
$-7 - (-2) = -5$	since $-5 + (-2) = -7$	$-7 + 2 = -5$

The matching results suggest that we can subtract by adding the opposite of the number being subtracted. This can always be done and often provides the easiest way to subtract real numbers.

> **Subtraction of Real Numbers** For any real numbers a and b,
> $$a - b = a + (-b).$$
> (To subtract, add the opposite, or additive inverse, of the number being subtracted.)

EXAMPLE 6 Subtract each of the following and then check with addition.

a) $2 - 6$ **b)** $4 - (-9)$ **c)** $-4.2 - (-3.6)$

d) $-1.8 - (-7.5)$ **e)** $\frac{1}{5} - \left(-\frac{3}{5}\right)$

SOLUTION

a) $2 - 6 = 2 + (-6) = -4$ The opposite of 6 is -6. We change the subtraction to addition and add the opposite. *Check*: $-4 + 6 = 2$.

b) $4 - (-9) = 4 + 9 = 13$ The opposite of -9 is 9. We change the subtraction to addition and add the opposite. *Check*: $13 + (-9) = 4$.

c) $-4.2 - (-3.6) = -4.2 + 3.6$ Adding the opposite of -3.6. *Check*: $-0.6 + (-3.6) = -4.2$.

$= -0.6$

d) $-1.8 - (-7.5) = -1.8 + 7.5$ Adding the opposite

$\qquad\qquad\qquad\quad = 5.7$ *Check:* $5.7 + (-7.5) = -1.8$.

e) $\dfrac{1}{5} - \left(-\dfrac{3}{5}\right) = \dfrac{1}{5} + \dfrac{3}{5}$ Adding the opposite

$\qquad\qquad\qquad = \dfrac{1 + 3}{5}$ A common denominator exists so we add in the numerator.

$\qquad\qquad\qquad = \dfrac{4}{5}$

Check: $\dfrac{4}{5} + \left(-\dfrac{3}{5}\right) = \dfrac{4}{5} + \dfrac{-3}{5} = \dfrac{4 + (-3)}{5} = \dfrac{1}{5}$.

EXAMPLE 7   Simplify: $8 - (-4) - 2 - (-5) + 3$.

SOLUTION

$8 - (-4) - 2 - (-5) + 3 = 8 + 4 + (-2) + 5 + 3$ To subtract, we add the opposite.

$\qquad\qquad\qquad\qquad\qquad\quad = 18$

   Recall from Section 1.2 that the terms of an algebraic expression are separated by plus signs. This means that the terms of $5x - 7y - 9$ are $5x$, $-7y$, and -9, since $5x - 7y - 9 = 5x + (-7y) + (-9)$.

EXAMPLE 8   Identify the terms of $4 - 2ab + 7a - 9$.

SOLUTION   We have

$4 - 2ab + 7a - 9 = 4 + (-2ab) + 7a + (-9)$,   Rewriting as addition

so the terms are 4, $-2ab$, $7a$, and -9.

EXAMPLE 9   Combine like terms.

a) $1 + 3x - 7x$ **b)** $-5a - 7b - 4a + 10b$

c) $4 - 3m - 9 + 7m$

SOLUTION

a) $1 + 3x - 7x = 1 + 3x + (-7x)$ Adding the opposite

$\qquad\qquad\quad = 1 + (3 + (-7))x$ Using the distributive law. Try to do this mentally.

$\qquad\qquad\quad = 1 + (-4)x$

$\qquad\qquad\quad = 1 - 4x$ Rewriting as subtraction to be more concise

b) $-5a - 7b - 4a + 10b = -5a + (-7b) + (-4a) + 10b$ Adding the opposite

$$= -5a + (-4a) + (-7b) + 10b$$ Using the commutative law of addition

$$= -9a + 3b$$ Combining like terms mentally

c) $4 - 3m - 9 + 7m = 4 + (-3m) + (-9) + 7m$ Rewriting as addition

$$= 4 + (-9) + (-3m) + 7m$$ Using the commutative law of addition

$$= -5 + 4m$$

Problem Solving

We use subtraction to solve problems involving differences. These include problems that ask "How much more?" or "How much higher?"

EXAMPLE 10 Temperature Extremes. The highest temperature ever recorded in the United States is 134°F in Greenland Ranch, California, on July 10, 1913. The lowest temperature ever recorded is −80°F in Prospect Creek, Alaska, on January 23, 1971. (*Source*: National Oceanographic and Atmospheric Administration) How much higher was the temperature in Greenland Ranch than that in Prospect Creek?

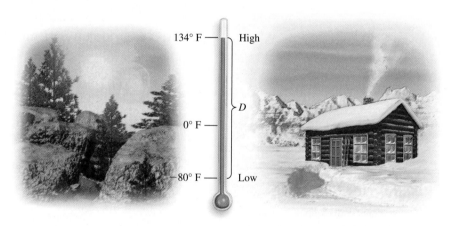

SOLUTION To find the difference between two temperatures, we always subtract the lower temperature from the higher temperature:

Higher temperature − Lower temperature

$$134 \qquad - \qquad (-80)$$

$$= 134 + 80$$

$$= 214.$$

The temperature in Greenland Ranch was 214°F higher than the temperature in Prospect Creek.

1.6 EXERCISE SET

FOR EXTRA HELP

 MathXL

 MyMathLab

InterAct Math

 AW Math Tutor Center

Video Lectures on CD: Disc 1

 Student's Solutions Manual

Concept Reinforcement In each of Exercises 1–8, match the expression with the appropriate wording from the column on the right.

1. ____ $-x$

2. ____ $12 - x$

3. ____ $12 - (-x)$

4. ____ $x - 12$

5. ____ $x - (-12)$

6. ____ $-x - 12$

7. ____ $-x - x$

8. ____ $-x - (-12)$

a) x minus negative twelve

b) The opposite of x minus x

c) The opposite of x minus twelve

d) The opposite of x

e) The opposite of x minus negative twelve

f) Twelve minus the opposite of x

g) Twelve minus x

h) x minus twelve

Write each of the following in words.

9. $4 - 10$

10. $5 - 13$

11. $2 - (-9)$

12. $4 - (-1)$

13. $9 - (-t)$

14. $8 - (-m)$

15. $-x - y$

16. $-a - b$

17. $-3 - (-n)$

18. $-7 - (-m)$

Find the opposite, or additive inverse.

19. 39

20. -17

21. -9

22. $\frac{7}{2}$

23. -3.14

24. 48.2

Find $-x$ when x is each of the following.

25. 23

26. -26

27. $-\frac{14}{3}$

28. $\frac{1}{328}$

29. 0.101

30. 0

Find $-(-x)$ when x is each of the following.

31. 72

32. 29

33. $-\frac{2}{5}$

34. -9.1

Change the sign. (Find the opposite.)

35. -1

36. -7

37. 7

38. 10

Subtract.

39. $6 - 8$

40. $4 - 13$

41. $0 - 5$

42. $0 - 8$

43. $3 - 9$

44. $3 - 13$

45. $0 - 10$

46. $0 - 7$

47. $-9 - (-3)$

48. $-9 - (-5)$

Aha! **49.** $-8 - (-8)$

50. $-10 - (-10)$

51. $14 - 19$

52. $12 - 16$

53. $30 - 40$

54. $20 - 27$

55. $-7 - (-9)$

56. $-8 - (-3)$

57. $-9 - (-9)$

58. $-40 - (-40)$

59. $5 - 5$

60. $7 - 7$

61. $4 - (-4)$

62. $6 - (-6)$

63. $-7 - 4$

64. $-6 - 8$

65. $6 - (-10)$

66. $3 - (-12)$

67. $-4 - 15$

68. $-14 - 2$

69. $-6 - (-5)$

70. $-4 - (-3)$

71. $5 - (-12)$

72. $5 - (-6)$

73. $0 - 5$

74. $0 - 6$

75. $-5 - (-2)$

76. $-3 - (-1)$

77. $-7 - 14$

78. $-9 - 16$

79. $0 - (-5)$

80. $0 - (-1)$

81. $-8 - 0$

82. $-9 - 0$

83. $3 - (-7)$

84. $12 - (-5)$

85. $2 - 25$

86. $18 - 63$

87. $-42 - 26$

88. $-18 - 63$

89. $-51 - 7$

90. $-45 - 4$

91. $3.2 - 8.7$

92. $1.5 - 9.4$

93. $0.072 - 1$

94. $0.825 - 1$

95. $\frac{2}{11} - \frac{9}{11}$

96. $\frac{3}{7} - \frac{5}{7}$

97. $\frac{-1}{5} - \frac{3}{5}$

98. $\frac{-2}{9} - \frac{5}{9}$

99. $-\frac{4}{17} - \left(-\frac{9}{17}\right)$

100. $-\frac{2}{13} - \left(-\frac{5}{13}\right)$

In each of Exercises 101–104, translate the phrase to mathematical language and simplify. See Example 10.

101. The difference between 3.8 and −5.2

102. The difference between −2.1 and −5.9

103. The difference between 114 and −79

104. The difference between 23 and −17

105. Subtract 37 from −21.

106. Subtract 19 from −7.

107. Subtract −25 from 9.

108. Subtract −31 from −5.

Simplify.

109. $25 - (-12) - 7 - (-2) + 9$

110. $22 - (-18) + 7 + (-42) - 27$

111. $-31 + (-28) - (-14) - 17$

112. $-43 - (-19) - (-21) + 25$

113. $-34 - 28 + (-33) - 44$

114. $39 + (-88) - 29 - (-83)$

Aha! **115.** $-93 + (-84) - (-93) - (-84)$

116. $84 + (-99) + 44 - (-18) - 43$

Identify the terms in each expression.

117. $-7x - 4y$

118. $7a - 9b$

119. $9 - 5t - 3st$

120. $-4 - 3x + 2xy$

Combine like terms.

121. $4x - 7x$

122. $3a - 14a$

123. $7a - 12a + 4$

124. $-9x - 13x + 7$

125. $-8n - 9 + n$

126. $-7 + 9n - 8$

127. $3x + 5 - 9x$

128. $2 + 3a - 7$

129. $2 - 6t - 9 - 2t$

130. $-5 + 3b - 7 - 5b$

131. $5y + (-3x) - 9x + 1 - 2y + 8$

132. $14 - (-5x) + 2z - (-32) + 4z - 2x$

133. $13x - (-2x) + 45 - (-21) - 7x$

134. $8x - (-2x) - 14 - (-5x) + 53 - 9x$

Solve.

135. *Basketball.* A team's scoring differential is the difference between points scored and points allowed. The Phoenix Suns improved their scoring differential from −3.70 in 2003–2004 to +5.54 in 2005–2006 (*Source: The Wall Street Journal*, May 26, 2006). By how many points did they improve?

136. *Record Temperature Drop.* The greatest recorded temperature change in one day occurred in Browning, Montana, when the temperature fell from 44°F to −56°F (*Source: The Guinness Book of Records*, 1999). How much did the temperature drop?

137. *Elevation Extremes.* The lowest elevation in Asia, the Dead Sea, is 1310 ft below sea level. The highest elevation in Asia, Mount Everest, is 29,035 ft. (*Source: 2005 World Book Encyclopedia*) Find the difference in elevation.

138. *Elevation Extremes.* The elevation of Mount Whitney, the highest peak in California, is 14,777 ft more than the elevation of Death Valley, California (*Source: 2005 World Book Encyclopedia*). If Death Valley is 282 ft below sea level, find the elevation of Mount Whitney.

139. *Changes in Elevation.* The lowest point in Africa is Lake Assal, which is 156 m below sea level. The lowest point in South America is the Valdes Peninsula, which is 40 m below sea level. How much lower is Lake Assal than the Valdes Peninsula?

140. *Underwater Elevation.* The deepest point in the Pacific Ocean is the Marianas Trench, with a depth of 10,415 m. The deepest point in the Atlantic Ocean is the Puerto Rico Trench, with a depth of 8648 m. What is the difference in elevation of the two trenches?

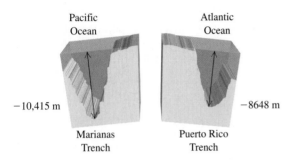

141. Jeremy insists that if you can *add* real numbers, then you can also *subtract* real numbers. Do you agree? Why or why not?

142. Are the expressions $-a + b$ and $a + (-b)$ opposites of each other? Why or why not?

Skill Maintenance

143. Find the area of a rectangle when the length is 36 ft and the width is 12 ft. [1.1]

144. Find the prime factorization of 864. [1.3]

Synthesis

145. Why might it be advantageous to rewrite a long series of additions and subtractions as all additions?

146. If a and b are both negative, under what circumstances will $a - b$ be negative?

147. *Power Outages.* During the Northeast's electrical blackout of August 14, 2003, residents of Bloomfield, New Jersey, lost power at 4:00 P.M. One resident returned from vacation at 3:00 P.M. the following day to find the clocks in her apartment reading 8:00 A.M. At what time, and on what day, was power restored?

Tell whether each statement is true or false for all real numbers m and n. Use various replacements for m and n to support your answer.

148. If $m > n$, then $m - n > 0$.

149. If $m > n$, then $m + n > 0$.

150. If m and n are opposites, then $m - n = 0$.

151. If $m = -n$, then $m + n = 0$.

152. A gambler loses a wager and then loses "double or nothing" (meaning the gambler owes twice as much) twice more. After the three losses, the gambler's assets are $-\$20$. Explain how much the gambler originally bet and how the \$20 debt occurred.

153. List the keystrokes needed to compute $-9 - (-7)$.

154. If n is positive and m is negative, what is the sign of $n + (-m)$? Why?

1.7 Multiplication and Division of Real Numbers

Multiplication ▪ Division

We now develop rules for multiplication and division of real numbers. Because multiplication and division are closely related, the rules are quite similar.

Multiplication

We already know how to multiply two nonnegative numbers. To see how to multiply a positive number and a negative number, consider the following pattern in which multiplication is regarded as repeated addition:

This number ⟶ decreases by 1 each time.

$$4(-5) = (-5) + (-5) + (-5) + (-5) = -20$$
$$3(-5) = \qquad\qquad (-5) + (-5) + (-5) = -15$$
$$2(-5) = \qquad\qquad\qquad\qquad (-5) + (-5) = -10$$
$$1(-5) = \qquad\qquad\qquad\qquad\qquad\qquad (-5) = -5$$
$$0(-5) = \qquad\qquad\qquad\qquad\qquad\qquad\qquad\ 0 = 0$$

⟵ This number increases by 5 each time.

This pattern illustrates that the product of a negative number and a positive number is negative.

The Product of a Negative Number and a Positive Number
To multiply a positive number and a negative number, multiply their absolute values. The answer is negative.

EXAMPLE 1 Multiply: **(a)** $8(-5)$; **(b)** $-\frac{1}{3} \cdot \frac{5}{7}$.

SOLUTION The product of a negative number and a positive number is negative.

a) $8(-5) = -40$ \qquad *Think*: $8 \cdot 5 = 40$; make the answer negative.

b) $-\frac{1}{3} \cdot \frac{5}{7} = -\frac{5}{21}$ \qquad *Think*: $\frac{1}{3} \cdot \frac{5}{7} = \frac{5}{21}$; make the answer negative.

The pattern developed above includes not just products of positive and negative numbers, but a product involving zero as well.

The Multiplicative Property of Zero   For any real number a,

$$0 \cdot a = a \cdot 0 = 0.$$

(The product of 0 and any real number is 0.)

EXAMPLE 2   Multiply: $173(-452)0$.

SOLUTION   We have

$$173(-452)0 = 173[(-452)0]$$ Using the associative law of multiplication

$$= 173[0]$$ Using the multiplicative property of zero

$$= 0.$$ Using the multiplicative property of zero again

Note that whenever 0 appears as a factor, the product will be 0.

We can extend the above pattern still further to examine the product of two negative numbers.

This number→ $2(-5) =$ $(-5) + (-5) = -10$ ← This number
decreases by $1(-5) =$ $(-5) = -5$ increases by
1 each time. $0(-5) =$ $0 = 0$ 5 each time.
                 $-1(-5) =$ $-(-5) = 5$
                 $-2(-5) = -(-5) - (-5) = 10$

According to the pattern, the product of two negative numbers is positive.

> **The Product of Two Negative Numbers**   To multiply two negative numbers, multiply their absolute values. The answer is positive.

EXAMPLE 3   Multiply: **(a)** $(-6)(-8)$; **(b)** $(-1.2)(-3)$.

SOLUTION   The product of two negative numbers is positive.

a) The absolute value of -6 is 6 and the absolute value of -8 is 8. Thus,

$$(-6)(-8) = 6 \cdot 8$$ Multiplying absolute values. The answer is positive.

$$= 48.$$

b) $(-1.2)(-3) = (1.2)(3)$ Multiplying absolute values. The answer is positive.

$$= 3.6$$ Try to go directly to this step.

When three or more numbers are multiplied, we can order and group the numbers as we please because of the commutative and associative laws.

EXAMPLE 4   Multiply: **(a)** $-3(-2)(-5)$; **(b)** $-4(-6)(-1)(-2)$.

SOLUTION

a) $-3(-2)(-5) = 6(-5)$ Multiplying the first two numbers. The product of two negatives is positive.

$$= -30$$ The product of a positive and a negative is negative.

b) $-4(-6)(-1)(-2) = 24 \cdot 2$ Multiplying the first two numbers and the last two numbers

$$= 48$$

We can see the following pattern in the results of Example 4.

The product of an even number of negative numbers is positive.

The product of an odd number of negative numbers is negative.

Division

Recall that $a \div b$, or $\frac{a}{b}$, is the number, if one exists, that when multiplied by b gives a. For example, to show that $10 \div 2$ is 5, we need only note that $5 \cdot 2 = 10$. Thus division can always be checked with multiplication.

Student Notes

Try to regard "undefined" as a mathematical way of saying "we do not give any meaning to this expression."

EXAMPLE 5   Divide, if possible, and check your answer.

a) $14 \div (-7)$ **b)** $\dfrac{-32}{-4}$ **c)** $\dfrac{-10}{9}$ **d)** $\dfrac{-17}{0}$

SOLUTION

a) $14 \div (-7) = -2$ We look for a number that when multiplied by -7 gives 14. That number is -2. *Check*: $(-2)(-7) = 14$.

b) $\dfrac{-32}{-4} = 8$ We look for a number that when multiplied by -4 gives -32. That number is 8. *Check*: $8(-4) = -32$.

c) $\dfrac{-10}{9} = -\dfrac{10}{9}$ We look for a number that when multiplied by 9 gives -10. That number is $-\frac{10}{9}$. *Check*: $-\frac{10}{9} \cdot 9 = -10$.

d) $\dfrac{-17}{0}$ is **undefined.**   We look for a number that when multiplied by 0 gives -17. There is no such number because if 0 is a factor, then the product is 0, not -17.

The sign rules for division are the same as those for multiplication: The quotient of a positive number and a negative number is negative; the quotient of two negative numbers is positive.

> **Rules for Multiplication and Division**   To multiply or divide two nonzero real numbers:
>
> **1.** Using the absolute values, multiply or divide, as indicated.
> **2.** If the signs are the same, the answer is positive.
> **3.** If the signs are different, the answer is negative.

Had Example 5(a) been written as $-14 \div 7$ or $-\frac{14}{7}$, rather than $14 \div (-7)$, the result would still have been -2. Thus from Examples 5(a)–5(c), we have the following:

$$\frac{-a}{b} = \frac{a}{-b} = -\frac{a}{b} \quad \text{and} \quad \frac{-a}{-b} = \frac{a}{b}.$$

EXAMPLE 6 Rewrite each of the following in two equivalent forms: **(a)** $\frac{5}{-2}$; **(b)** $-\frac{3}{10}$.

SOLUTION   We use one of the properties just listed.

a) $\dfrac{5}{-2} = \dfrac{-5}{2}$ and $\dfrac{5}{-2} = -\dfrac{5}{2}$

b) $-\dfrac{3}{10} = \dfrac{-3}{10}$ and $-\dfrac{3}{10} = \dfrac{3}{-10}$

Since $\dfrac{-a}{b} = \dfrac{a}{-b} = -\dfrac{a}{b}$

When a fraction contains a negative sign, it may be helpful to rewrite (or simply visualize) the fraction in an equivalent form.

EXAMPLE 7 Perform the indicated operation.

a) $\left(-\frac{4}{5}\right)\left(\frac{-7}{3}\right)$ b) $-\frac{2}{7} + \frac{9}{-7}$

SOLUTION

a) $\left(-\dfrac{4}{5}\right)\left(\dfrac{-7}{3}\right) = \left(-\dfrac{4}{5}\right)\left(-\dfrac{7}{3}\right)$ Rewriting $\dfrac{-7}{3}$ as $-\dfrac{7}{3}$

$= \dfrac{28}{15}$ Try to go directly to this step.

b) Given a choice, we generally choose a positive denominator:

$-\dfrac{2}{7} + \dfrac{9}{-7} = \dfrac{-2}{7} + \dfrac{-9}{7}$ Rewriting both fractions with a common denominator of 7

$= \dfrac{-11}{7}$, or $-\dfrac{11}{7}$.

To divide with fraction notation, it is usually easiest to find a reciprocal and then multiply.

EXAMPLE 8 Find the reciprocal of each number, if it exists.

a) -27 b) $\frac{-3}{4}$ c) $-\frac{1}{5}$ d) 0

SOLUTION   Recall from Section 1.3 that we can check that two numbers are reciprocals of each other by confirming that their product is 1.

a) The reciprocal of -27 is $\frac{1}{-27}$. More often, this number is written as $-\frac{1}{27}$.
   Check: $(-27)\left(-\frac{1}{27}\right) = \frac{27}{27} = 1.$

b) The reciprocal of $\frac{-3}{4}$ is $\frac{4}{-3}$, or, equivalently, $-\frac{4}{3}$.
   Check: $\frac{-3}{4} \cdot \frac{4}{-3} = \frac{-12}{-12} = 1.$

c) The reciprocal of $-\frac{1}{5}$ is $-\frac{5}{1}$, or -5.   *Check*: $-\frac{1}{5}(-5) = \frac{5}{5} = 1.$

d) The reciprocal of 0 does not exist. To see this, recall that there is no number r for which $0 \cdot r = 1.$

EXAMPLE 9   Divide: **(a)** $-\frac{2}{3} \div \left(-\frac{5}{4}\right)$; **(b)** $-\frac{3}{4} \div \frac{3}{10}$.

SOLUTION   We divide by multiplying by the reciprocal of the divisor.

a) $-\dfrac{2}{3} \div \left(-\dfrac{5}{4}\right) = -\dfrac{2}{3} \cdot \left(-\dfrac{4}{5}\right) = \dfrac{8}{15}$ Multiplying by the reciprocal

Be careful not to change the sign when taking a reciprocal!

b) $-\dfrac{3}{4} \div \dfrac{3}{10} = -\dfrac{3}{4} \cdot \left(\dfrac{10}{3}\right) = -\dfrac{30}{12} = -\dfrac{5}{2} \cdot \dfrac{6}{6} = -\dfrac{5}{2}$ Removing a factor equal to 1: $\frac{6}{6} = 1$

To divide with decimal notation, it is usually easiest to carry out the division.

EXAMPLE 10   Divide: $27.9 \div (-3)$.

SOLUTION

$$27.9 \div (-3) = \frac{27.9}{-3} = -9.3$$ Dividing: $3)\overline{27.9}$ = 9.3. The answer is negative.

In Example 5(d), we explained why we cannot divide -17 by 0. To see why *no* nonzero number b can be divided by 0, remember that $b \div 0$ would have to be the number that when multiplied by 0 gives b. But since the product of 0 and any number is 0, not b, we say that $b \div 0$ is **undefined** for $b \neq 0$. In the special case of $0 \div 0$, we look for a number r such that $0 \div 0 = r$ and $r \cdot 0 = 0$. But, $r \cdot 0 = 0$ for *any* number r. For this reason, we say that $b \div 0$ is undefined for any choice of b.*

Finally, note that $0 \div 7 = 0$ since $0 \cdot 7 = 0$. This can be written $0/7 = 0$. It is important not to confuse division *by* 0 with division *into* 0.

EXAMPLE 11   Divide, if possible: **(a)** $\frac{0}{-2}$; **(b)** $\frac{5}{0}$.

SOLUTION

a) $\dfrac{0}{-2} = 0$ *Check:* $0(-2) = 0$.

b) $\dfrac{5}{0}$ is undefined.

---

*Sometimes $0 \div 0$ is said to be *indeterminate*.

Division Involving Zero   For any real number a, $\dfrac{a}{0}$ is undefined, and for $a \neq 0$, $\dfrac{0}{a} = 0$.

CAUTION!   It is important *not* to confuse *opposite* with *reciprocal*. Keep in mind that the opposite, or *additive inverse*, of a number is what we add to the number to get 0. The reciprocal, or *multiplicative inverse*, is what we multiply the number by to get 1.
Compare the following.

Number	Opposite (Change the sign.)	Reciprocal (Invert but do not change the sign.)	
$-\dfrac{3}{8}$	$\dfrac{3}{8}$	$-\dfrac{8}{3}$	$\left(-\dfrac{3}{8}\right)\left(-\dfrac{8}{3}\right) = 1$
19	-19	$\dfrac{1}{19}$	$-\dfrac{3}{8} + \dfrac{3}{8} = 0$
$\dfrac{18}{7}$	$-\dfrac{18}{7}$	$\dfrac{7}{18}$	
-7.9	7.9	$-\dfrac{1}{7.9}$, or $-\dfrac{10}{79}$	
0	0	Not defined	

1.7 EXERCISE SET

FOR EXTRA HELP

Math XP   MathXL   MyMathLab   InterAct Math   Tutor Center AW Math Tutor Center   Video Lectures on CD: Disc 1   Student's Solutions Manual

Concept Reinforcement   In each of Exercises 1–10, replace the blank with either 0 or 1 to match the description given.

1. The product of two reciprocals ____

2. The sum of a pair of opposites ____

3. The sum of a pair of additive inverses ____

4. The product of two multiplicative inverses ____

5. This number has no reciprocal. ____

6. This number is its own reciprocal. ____

7. This number is the multiplicative identity. ____

8. This number is the additive identity. ____

9. A nonzero number divided by itself ____

10. Division by this number is undefined. ____

Multiply.

11. $-3 \cdot 8$

12. $-3 \cdot 7$

13. $-8 \cdot 7$

14. $-9 \cdot 2$

15. $8 \cdot (-3)$

16. $9 \cdot (-5)$

17. $-9 \cdot 8$

18. $-10 \cdot 3$

19. $-6 \cdot (-7)$

20. $-2 \cdot (-5)$

21. $-5 \cdot (-9)$

22. $-9 \cdot (-2)$

23. $-19 \cdot (-10)$

24. $-12 \cdot (-10)$

25. $-12 \cdot 12$

26. $-13 \cdot (-15)$

27. $-25 \cdot (-48)$

28. $39 \cdot (-43)$

29. $-3.5 \cdot (-28)$

30. $97 \cdot (-2.1)$

31. $6 \cdot (-13)$

32. $7 \cdot (-9)$

33. $-7 \cdot (-3.1)$

34. $-4 \cdot (-3.2)$

35. $\frac{2}{3} \cdot \left(-\frac{3}{5}\right)$

36. $\frac{5}{7} \cdot \left(-\frac{2}{3}\right)$

37. $-\frac{3}{8} \cdot \left(-\frac{2}{9}\right)$

38. $-\frac{5}{8} \cdot \left(-\frac{2}{5}\right)$

39. $(-5.3)(2.1)$

40. $(-4.3)(9.5)$

41. $-\frac{5}{9} \cdot \frac{3}{4}$

42. $-\frac{8}{3} \cdot \frac{9}{4}$

43. $3 \cdot (-7) \cdot (-2) \cdot 6$

44. $9 \cdot (-2) \cdot (-6) \cdot 7$

Aha! **45.** $-27 \cdot (-34) \cdot 0$

46. $-43 \cdot (-74) \cdot 0$

47. $-\frac{1}{3} \cdot \frac{1}{4} \cdot \left(-\frac{3}{7}\right)$

48. $-\frac{1}{2} \cdot \frac{3}{5} \cdot \left(-\frac{2}{7}\right)$

49. $-2 \cdot (-5) \cdot (-3) \cdot (-5)$

50. $-3 \cdot (-5) \cdot (-2) \cdot (-1)$

51. $(-31) \cdot (-27) \cdot 0$

52. $7 \cdot (-6) \cdot 5 \cdot (-4) \cdot 3 \cdot (-2) \cdot 1 \cdot 0$

53. $(-8)(-9)(-10)$

54. $(-7)(-8)(-9)(-10)$

55. $(-6)(-7)(-8)(-9)(-10)$

56. $(-5)(-6)(-7)(-8)(-9)(-10)$

Divide, if possible, and check. If a quotient is undefined, state this.

57. $14 \div (-2)$

58. $\frac{24}{-3}$

59. $\frac{36}{-9}$

60. $26 \div (-13)$

61. $\frac{-56}{8}$

62. $-32 \div (-4)$

63. $\frac{-48}{-12}$

64. $-63 \div (-9)$

65. $\frac{-72}{9}$

66. $\frac{-50}{25}$

67. $-100 \div (-50)$

68. $\frac{-200}{8}$

69. $-108 \div 9$

70. $\frac{-64}{-7}$

71. $\frac{400}{-50}$

72. $-300 \div (-13)$

73. $\frac{28}{0}$

74. $\frac{0}{-5}$

75. $-4.8 \div 1.2$

76. $-3.9 \div 1.3$

77. $\frac{0}{-9}$

78. $0 \div (-47)$

Aha! **79.** $\frac{9.7(-2.8)0}{4.3}$

80. $\frac{(-4.9)(7.2)}{0}$

Write each number in two equivalent forms, as in Example 6.

81. $\frac{-8}{3}$

82. $\frac{-12}{7}$

83. $\frac{29}{-35}$

84. $\frac{9}{-14}$

85. $-\frac{7}{3}$

86. $-\frac{4}{15}$

87. $\frac{-x}{2}$

88. $\frac{9}{-a}$

Find the reciprocal of each number, if it exists.

89. $\frac{4}{-5}$

90. $\frac{2}{-9}$

91. $-\frac{47}{13}$

92. $-\frac{31}{12}$

93. -10

94. 34

95. 4.3

96. -1.7

97. $\dfrac{-9}{4}$

98. $\dfrac{-6}{11}$

99. 0

100. -1

Perform the indicated operation and, if possible, simplify. If a quotient is undefined, state this.

101. $\left(\dfrac{-7}{4}\right)\left(-\dfrac{3}{5}\right)$

102. $\left(-\dfrac{5}{6}\right)\left(\dfrac{-1}{3}\right)$

103. $\left(\dfrac{-6}{5}\right)\left(\dfrac{2}{-11}\right)$

104. $\left(\dfrac{7}{-2}\right)\left(\dfrac{-5}{6}\right)$

105. $\dfrac{-3}{8}+\dfrac{-5}{8}$

106. $\dfrac{-4}{5}+\dfrac{7}{5}$

Aha! **107.** $\left(\dfrac{-9}{5}\right)\left(\dfrac{5}{-9}\right)$

108. $\left(-\dfrac{2}{7}\right)\left(\dfrac{5}{-8}\right)$

109. $\left(-\dfrac{3}{11}\right)+\left(-\dfrac{6}{11}\right)$

110. $\left(-\dfrac{4}{7}\right)+\left(-\dfrac{2}{7}\right)$

111. $\dfrac{7}{8}\div\left(-\dfrac{1}{2}\right)$

112. $\dfrac{3}{4}\div\left(-\dfrac{2}{3}\right)$

113. $\dfrac{9}{5}\cdot\dfrac{-20}{3}$

114. $\dfrac{-5}{12}\cdot\dfrac{7}{15}$

115. $\left(-\dfrac{18}{7}\right)+\left(-\dfrac{3}{7}\right)$

116. $\left(-\dfrac{12}{5}\right)+\left(-\dfrac{3}{5}\right)$

Aha! **117.** $-\dfrac{5}{9}\div\left(-\dfrac{5}{9}\right)$

118. $-\dfrac{5}{4}\div\left(-\dfrac{3}{4}\right)$

119. $-44.1\div(-6.3)$

120. $-6.6\div3.3$

121. $\dfrac{5}{9}-\dfrac{7}{9}$

122. $\dfrac{2}{7}-\dfrac{6}{7}$

123. $\dfrac{-3}{10}+\dfrac{2}{5}$

124. $\dfrac{-5}{9}+\dfrac{2}{3}$

125. $\dfrac{7}{10}\div\left(\dfrac{-3}{5}\right)$

126. $\left(\dfrac{-3}{5}\right)\div\dfrac{6}{15}$

127. $\dfrac{5}{7}-\dfrac{1}{-7}$

128. $\dfrac{4}{9}-\dfrac{1}{-9}$

129. $\dfrac{-4}{15}+\dfrac{2}{-3}$

130. $\dfrac{3}{-10}+\dfrac{-1}{5}$

TW **131.** Most calculators have a key, often appearing as ⟨x⁻¹⟩, for finding reciprocals. To use this key, enter a number and then press ⟨x⁻¹⟩ ⟨ENTER⟩ to find its reciprocal. What should happen if you enter a number and then find the reciprocal twice? Why?

TW **132.** Multiplication can be regarded as repeated addition. Using this idea and a number line, explain why $3\cdot(-5)=-15$.

Skill Maintenance

133. Simplify: $\dfrac{264}{468}$. [1.3]

134. Determine whether 12 is a solution of $35-a=13$. [1.1]

Synthesis

TW **135.** If two nonzero numbers are opposites of each other, are their reciprocals opposites of each other? Why or why not?

TW **136.** If two numbers are reciprocals of each other, are their opposites reciprocals of each other? Why or why not?

137. Show that the reciprocal of a sum is *not* the sum of the two reciprocals.

138. Which real numbers are their own reciprocals?

Tell whether each expression represents a positive number or a negative number when m and n are negative.

139. $\dfrac{m}{-n}$

140. $\dfrac{-n}{-m}$

141. $-m\cdot\left(\dfrac{-n}{m}\right)$

142. $-\left(\dfrac{n}{-m}\right)$

143. $(m+n)\cdot\dfrac{m}{n}$

144. $(-n-m)\dfrac{n}{m}$

145. What must be true of m and n if $-mn$ is to be **(a)** positive? **(b)** zero? **(c)** negative?

146. The following is a proof that a positive number times a negative number is negative. Provide a reason for each step. Assume that $a>0$ and $b>0$.

$$a(-b)+ab=a[-b+b]$$
$$=a(0)$$
$$=0$$

Therefore, $a(-b)$ is the opposite of ab.

TW **147.** Is it true that for any numbers a and b, if a is larger than b, then the reciprocal of a is smaller than the reciprocal of b? Why or why not?

1.8 Exponential Notation and Order of Operations

Exponential Notation ◾ Order of Operations ◾
Simplifying and the Distributive Law ◾ The Opposite of a Sum

Algebraic expressions often contain *exponential notation*. In this section, we learn how to use exponential notation as well as rules for the *order of operations* in performing certain algebraic manipulations.

Exponential Notation

A product like $3 \cdot 3 \cdot 3 \cdot 3$, in which the factors are the same, is called a **power.** Powers occur often enough that a simpler notation called **exponential notation** is used. For

$$\underbrace{3 \cdot 3 \cdot 3 \cdot 3,}_{\text{4 factors}} \quad \text{we write} \quad 3^4.$$

Because $3^4 = 81$, we sometimes say that 81 "is a power of 3."

This is read "three to the fourth power," or simply, "three to the fourth." The number 4 is called an **exponent** and the number 3 a **base.**

Expressions like s^2 and s^3 are usually read "*s* squared" and "*s* cubed," respectively. This comes from the fact that a square with sides of length *s* has an area *A* given by $A = s^2$ and a cube with sides of length *s* has a volume *V* given by $V = s^3$.

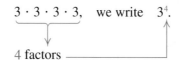

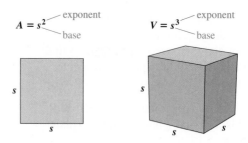

EXAMPLE 1 Write exponential notation for $10 \cdot 10 \cdot 10 \cdot 10 \cdot 10$.

SOLUTION

Exponential notation is 10^5.

5 is the exponent.
10 is the base.

EXAMPLE 2 Evaluate: **(a)** 5^2; **(b)** $(-5)^3$; **(c)** $(2n)^3$.

SOLUTION

a) $5^2 = 5 \cdot 5 = 25$ The exponent 2 indicates two factors of 5.

b) $(-5)^3 = (-5)(-5)(-5)$ The exponent 3 indicates three factors of -5.

$\qquad = 25(-5)$ Using the associative law of multiplication

$\qquad = -125$

c) $(2n)^3 = (2n)(2n)(2n)$ The exponent 3 indicates three factors of $2n$.

$\qquad = 2 \cdot 2 \cdot 2 \cdot n \cdot n \cdot n$ Using the associative and commutative laws of multiplication

$\qquad = 8n^3$

To determine what the exponent 1 will mean, look for a pattern in the following:

$$7 \cdot 7 \cdot 7 \cdot 7 = 7^4$$
$$7 \cdot 7 \cdot 7 = 7^3$$
$$7 \cdot 7 = 7^2$$
$$7 = 7^?$$

We divide by 7 each time.

The exponents decrease by 1 each time. To continue the pattern, we say that

$$7 = 7^1.$$

Exponential Notation   For any natural number n,

$$b^n \quad \text{means} \quad \overbrace{b \cdot b \cdot b \cdot b \cdots b}^{n \text{ factors}}.$$

Order of Operations

How should $4 + 2 \times 5$ be computed? If we multiply 2 by 5 and then add 4, the result is 14. If we add 2 and 4 first and then multiply by 5, the result is 30. Since these results differ, the order in which we perform operations matters. If grouping symbols such as parentheses ( ), brackets [ ], braces { }, absolute-value symbols | |, or fraction bars are used, they tell us what to do first. For example,

$$(4 + 2) \times 5 \quad \text{indicates} \quad 6 \times 5, \quad \text{resulting in 30},$$

and

$$4 + (2 \times 5) \quad \text{indicates} \quad 4 + 10, \quad \text{resulting in 14}.$$

In this text, we direct exploration of mathematical concepts using a graphing calculator in Interactive Discovery features such as the one that follows. Such explorations are part of the development of the material presented, and should be performed as you read the text.

Interactive Discovery

Use a calculator to compute $4 + 2 \times 5$.

1. What operation does the calculator perform first?
2. Insert parentheses in the expression $4 + 2 \times 5$ to indicate the order in which the calculator performs the addition and multiplication.

In addition to grouping symbols, the following conventions exist for determining the order in which operations should be performed. Most scientific and graphing calculators follow these rules when evaluating expressions.

Rules for Order of Operations

1. Calculate within the innermost grouping symbols, ( ), [ ], { }, | |, and above or below fraction bars.
2. Simplify all exponential expressions.
3. Perform all multiplication and division, working from left to right.
4. Perform all addition and subtraction, working from left to right.

Thus the correct way to compute $4 + 2 \times 5$ is to first multiply 2 by 5 and then add 4. The result is 14.

EXAMPLE 3   Simplify: $15 - 2 \times 5 + 3$.

SOLUTION   When no groupings or exponents appear, we *always* multiply or divide before adding or subtracting:

$$15 - 2 \times 5 + 3 = 15 - 10 + 3 \qquad \text{Multiplying}$$
$$= 5 + 3$$
$$= 8. \qquad \left. \begin{array}{c} \\ \\ \end{array} \right\} \quad \begin{array}{l} \text{Subtracting and adding} \\ \text{from left to right} \end{array}$$

Always calculate within parentheses first. When there are exponents and no parentheses, simplify powers before multiplying or dividing.

EXAMPLE 4   Simplify: **(a)** $(3 \cdot 4)^2$; **(b)** $3 \cdot 4^2$.

SOLUTION

a) $(3 \cdot 4)^2 = (12)^2$ **Working within parentheses first**
$$= 144$$

b) $3 \cdot 4^2 = 3 \cdot 16$ **Simplifying the power**
$$= 48 \qquad \text{Multiplying}$$

Note that $(3 \cdot 4)^2 \neq 3 \cdot 4^2$.

CAUTION! Example 4 illustrates that, in general, $(ab)^2 \neq ab^2$.

> **CAUTION!** Example 5 illustrates that, in general, $(-x)^2 \neq -x^2$.

Student Notes

The symbols ( ), [ ], and { } are all used in the same way. Used inside or next to each other, they make it easier to locate the left and right sides of a grouping. Try doing this in your own work to minimize mistakes.

EXAMPLE 5 Evaluate for $x = 5$: **(a)** $(-x)^2$; **(b)** $-x^2$.

SOLUTION

a) $(-x)^2 = (-5)^2 = (-5)(-5) = 25$ We square the opposite of 5.

b) $-x^2 = -5^2 = -25$ We square 5 and then find the opposite.

EXAMPLE 6 Evaluate $-15 \div 3(6 - a)^3$ for $a = 4$.

SOLUTION

$$
\begin{aligned}
-15 \div 3(6 - a)^3 &= -15 \div 3(6 - 4)^3 && \text{Substituting 4 for } a \\
&= -15 \div 3(2)^3 && \text{Working within parentheses first} \\
&= -15 \div 3 \cdot 8 && \text{Simplifying the exponential expression} \\
&= -5 \cdot 8 \Big\} && \text{Dividing and multiplying} \\
&= -40 \qquad && \text{from left to right}
\end{aligned}
$$

When combinations of grouping symbols are used, we begin with the innermost grouping symbols and work to the outside.

EXAMPLE 7 Simplify: $8 \div 4 + 3[9 + 2(3 - 5)^3]$.

SOLUTION

$$
\begin{aligned}
8 \div 4 + 3[9 + 2(3 - 5)^3] & \\
= 8 \div 4 + 3[9 + 2(-2)^3] && \text{Doing the calculations in the innermost parentheses first} \\
= 8 \div 4 + 3[9 + 2(-8)] && (-2)^3 = (-2)(-2)(-2) \\
&&= -8 \\
= 8 \div 4 + 3[9 + (-16)] & \\
= 8 \div 4 + 3[-7] && \text{Completing the calculations within the brackets} \\
= 2 + (-21) && \text{Multiplying and dividing from left to right} \\
= -19
\end{aligned}
$$

Exponents and Grouping Symbols

Exponents   To enter an exponential expression on most graphing calculators, enter the base, then press ⌃ and enter the exponent. x^2 is often used to enter an exponent of 2. Include parentheses around a negative base.

Grouping Symbols   Graphing calculators follow the rules for order of operations, so expressions can be entered as they are written. Grouping symbols such as brackets or braces are entered as parentheses. For fractions, write parentheses around the numerator and around the denominator.

▶ **EXAMPLE 8**   Calculate: $\dfrac{12(9-7)+4\cdot 5}{2^4+3^2}$.

SOLUTION   An equivalent expression with brackets is

$$[12(9-7)+4\cdot 5]\div[2^4+3^2].$$

In effect, we need to simplify the numerator, simplify the denominator, and then divide the results:

$$\frac{12(9-7)+4\cdot 5}{2^4+3^2}=\frac{12(2)+4\cdot 5}{16+9}=\frac{24+20}{25}=\frac{44}{25}.$$

To use a calculator, we enter the expression, replacing the brackets with parentheses, choose the Frac option of the MATH MATH submenu, and press **ENTER**, as shown in the figure at left.   ◢

```
(12(9-7)+4*5)/(2
^4+3²)▶Frac
                44/25
```

Example 8 demonstrates that graphing calculators do not make it any less important to understand the mathematics being studied. A solid understanding of the rules for order of operations is necessary in order to locate parentheses properly in an expression.

Simplifying and the Distributive Law

Sometimes we cannot simplify within grouping symbols. When a sum or difference is within parentheses, the distributive law provides a method for removing the grouping symbols.

▶ **EXAMPLE 9**   Simplify: $5x-9+2(4x+5)$.

SOLUTION

$$5x-9+2(4x+5)=5x-9+8x+10 \qquad \text{Using the distributive law}$$

$$=13x+1 \qquad \text{Combining like terms}$$
◢

Now that exponents have been introduced, we can make our definition of *like* or *similar terms* more precise. **Like,** or **similar, terms** are either constant terms or terms containing the same variable(s) raised to the same exponent(s). Thus, 5 and -7, $19xy$ and $2yx$, and $4a^3b$ and a^3b are all pairs of like terms.

▶ **EXAMPLE 10**   Simplify: $7x^2+3(x^2+2x)-5x$.

SOLUTION

$$7x^2+3(x^2+2x)-5x=7x^2+3x^2+6x-5x \qquad \text{Using the distributive law}$$

$$=10x^2+x \qquad \text{Combining like terms}$$
◢

The Opposite of a Sum

When a number is multiplied by -1, the result is the opposite of that number. For example, $-1(7) = -7$ and $-1(-5) = 5$.

> **The Property of -1**   For any real number a,
>
> $$-1 \cdot a = -a.$$
>
> (Negative one times a is the opposite of a.)

When grouping symbols are preceded by a "2" symbol, we can multiply the grouping by -1 and use the distributive law. In this manner, we can find the *opposite*, or *additive inverse*, of a sum.

EXAMPLE 11   Write an expression equivalent to $-(3x + 2y + 4)$ without using parentheses.

SOLUTION

$$-(3x + 2y + 4) = -1(3x + 2y + 4) \quad \text{Using the property of } -1$$
$$= -1(3x) + (-1)(2y) + (-1)4$$
$$\text{Using the distributive law}$$
$$= -3x - 2y - 4 \quad \text{Using the associative law and the property of } -1$$

Example 11 illustrates an important property of real numbers.

> **The Opposite of a Sum**   For any real numbers a and b,
>
> $$-(a + b) = -a + (-b).$$
>
> (The opposite of a sum is the sum of the opposites.)

To remove parentheses from an expression like $-(x - 7y + 5)$, we can first rewrite the subtraction as addition:

$$-(x - 7y + 5) = -(x + (-7y) + 5) \quad \text{Rewriting as addition}$$
$$= -x + 7y - 5. \quad \text{Taking the opposite of a sum}$$

This procedure is normally streamlined to one step in which we find the opposite by "removing parentheses and changing the sign of every term":

$$-(x - 7y + 5) = -x + 7y - 5.$$

EXAMPLE 12   Simplify: $3x - (4x + 2)$.

SOLUTION

$$\begin{aligned}
3x - (4x + 2) &= 3x + [-(4x + 2)] && \text{Adding the opposite of } 4x + 2 \\
&= 3x + [-4x - 2] && \text{Taking the opposite of } 4x + 2 \\
&= 3x + (-4x) + (-2) && \\
&= 3x - 4x - 2 && \text{Try to go directly to this step.} \\
&= -x - 2 && \text{Combining like terms}
\end{aligned}$$

In practice, the first three steps of Example 12 are usually skipped.

EXAMPLE 13   Simplify: $5t^2 - 2t - (4t^2 - 9t)$.

SOLUTION

$$\begin{aligned}
5t^2 - 2t - (4t^2 - 9t) &= 5t^2 - 2t - 4t^2 + 9t && \text{Removing parentheses and} \\
& && \text{changing the sign of each} \\
& && \text{term inside} \\
&= t^2 + 7t && \text{Combining like terms}
\end{aligned}$$

Expressions such as $7 - 3(x + 2)$ can be simplified as follows:

$$\begin{aligned}
7 - 3(x + 2) &= 7 + [-3(x + 2)] && \text{Adding the opposite of } 3\,(x + 2) \\
&= 7 + [-3x - 6] && \text{Multiplying } x + 2 \text{ by } -3 \\
&= 7 - 3x - 6 && \text{Try to go directly to this step.} \\
&= 1 - 3x. && \text{Combining like terms}
\end{aligned}$$

EXAMPLE 14   Simplify.

a) $3n - 2(4n - 5)$ **b)** $7x^3 + 2 - [5(x^3 - 1) + 8]$

SOLUTION

a)
$$\begin{aligned}
3n - 2(4n - 5) &= 3n - 8n + 10 && \text{Multiplying each term inside the} \\
& && \text{parentheses by } -2 \\
&= -5n + 10 && \text{Combining like terms}
\end{aligned}$$

b)
$$\begin{aligned}
7x^3 + 2 - [5(x^3 - 1) + 8] &= 7x^3 + 2 - [5x^3 - 5 + 8] && \text{Removing} \\
& && \text{parentheses} \\
&= 7x^3 + 2 - [5x^3 + 3] && \\
&= 7x^3 + 2 - 5x^3 - 3 && \text{Removing} \\
& && \text{brackets} \\
&= 2x^3 - 1 && \text{Combining like} \\
& && \text{terms}
\end{aligned}$$

We have seen in this chapter that certain problems can be translated to algebraic expressions that can in turn be simplified. In Chapter 2, we will solve problems that require us to solve an equation.

It is important that we be able to distinguish between the two tasks of **simplifying an expression** and **solving an equation.** In Chapter 1, we did not solve equations, but we did simplify expressions. This enabled us to write *equivalent expressions* that were simpler than the given expression. In Chapter 2, we will continue to simplify expressions, but we will also begin to solve equations.

1.8   EXERCISE SET

FOR EXTRA HELP

MathXL

MyMathLab

InterAct Math

AW Math Tutor Center

Video Lectures on CD: Disc 1

Student's Solutions Manual

   Concept Reinforcement   In each part of Exercises 1 and 2, name the operation that should be performed first. Do not perform the calculations.

1. a) $4 + 8 \div 2 \cdot 2$
   b) $7 - 9 + 15$
   c) $5 - 2(3 + 4)$
   d) $6 + 7 \cdot 3$
   e) $18 - 2[4 + (3 - 2)]$
   f) $\dfrac{5 - 6 \cdot 7}{2}$

2. a) $9 - 3 \cdot 4 \div 2$
   b) $8 + 7(6 - 5)$
   c) $5 \cdot [2 - 3(4 + 1)]$
   d) $8 - 7 + 2$
   e) $4 + 6 \div 2 \cdot 3$
   f) $\dfrac{37}{8 - 2 \cdot 2}$

Write exponential notation.

3. $2 \cdot 2 \cdot 2$ **4.** $6 \cdot 6 \cdot 6 \cdot 6$

5. $x \cdot x \cdot x \cdot x \cdot x \cdot x \cdot x$ **6.** $y \cdot y \cdot y \cdot y \cdot y \cdot y$

7. $3t \cdot 3t \cdot 3t \cdot 3t \cdot 3t$

8. $5m \cdot 5m \cdot 5m \cdot 5m \cdot 5m$

Simplify.

9. 3^2 **10.** 5^3

11. $(-4)^2$ **12.** $(-7)^2$

13. -4^2 **14.** -7^2

15. 4^3 **16.** 9^1

17. $(-5)^4$ **18.** 5^4

19. 7^1 **20.** $(-1)^7$

21. $(3t)^4$ **22.** $(5t)^2$

23. $(-7x)^3$ **24.** $(-5x)^4$

25. $5 + 3 \cdot 7$ **26.** $3 - 4 \cdot 2$

27. $8 \cdot 7 + 6 \cdot 5$ **28.** $10 \cdot 5 + 1 \cdot 1$

29. $19 - 5 \cdot 3 + 3$ **30.** $14 - 2 \cdot 6 + 7$

31. $9 \div 3 + 16 \div 8$ **32.** $32 - 8 \div 4 - 2$

Aha! **33.** $14 \cdot 19 \div (19 \cdot 14)$ **34.** $18 - 6 \div 3 \cdot 2 + 7$

35. $3(-10)^2 - 8 \div 2^2$ **36.** $9 - 3^2 \div 9(-1)$

37. $8 - (2 \cdot 3 - 9)$ **38.** $(8 - 2 \cdot 3) - 9$

39. $(8 - 2)(3 - 9)$ **40.** $32 \div (-2)^2 \cdot 4$

41. $13(-10)^2 + 45 \div (-5)$

42. $5 \cdot 3^2 - 4^2 \cdot 2$

43. $2^4 + 2^3 - 10 \div (-1)^4$

44. $112 \div 28 - 112 \div 28$

45. $5 + 3(2 - 9)^2$ **46.** $9 - (3 - 5)^3 - 4$

47. $[2 \cdot (5 - 8)]^2$ **48.** $3(5 - 7)^4 \div 4$

49. $\dfrac{7 + 2}{5^2 - 4^2}$ **50.** $\dfrac{5^2 - 3^2}{2 \cdot 6 - 4}$

51. $8(-7) + |6(-5)|$ **52.** $|10(-5)| + 1(-1)$

53. $\dfrac{(-2)^3 + 4^2}{3 - 5^2 + 3 \cdot 6}$

54. $\dfrac{7^2 - (-1)^5}{3 - 2 \cdot 3^2 + 5}$

55. $\dfrac{27 - 2 \cdot 3^2}{8 \div 2^2 - (-2)^2}$

56. $\dfrac{(-5)^2 - 4 \cdot 5}{3^2 + 4 \cdot 2(-1)^5}$

In Exercises 57–60, match the algebraic expression with the equivalent rewritten expression below. Check your answer by calculating the expression by hand and by using a calculator.

a) $(5(3 - 7) + 4 \wedge 3)/(-2 - 3)^2$
b) $(5(3 - 7) + 4 \wedge 3)/(-2 - 3^2)$
c) $(5(3 - 7) + 4) \wedge 3/-2 - 3^2$
d) $5(3 - 7) + 4 \wedge 3/(-2 - 3)^2$

57. $\dfrac{5(3 - 7) + 4^3}{(-2 - 3)^2}$

58. $(5(3 - 7) + 4)^3 \div (-2) - 3^2$

59. $5(3 - 7) + 4^3 \div (-2 - 3)^2$

60. $\dfrac{5(3 - 7) + 4^3}{(-2) - 3^2}$

■ *Simplify using a calculator. Round your answer to the nearest thousandth.*

61. $\dfrac{13.4 - 5|1.2 + 4.6|}{(9.3 - 5.4)^2}$

62. $|13.5 + 8(-4.7)|^3$

63. $134 - \sqrt{16 \cdot 35 - 48 \div 160}$

64. $-35 \div 700 \cdot 3 - \sqrt{-50 - (12 - 185)}$

Aha! **65.** $-12.86 - 5.2(-1.7 - 3.8)^2 \cdot 0$

66. $\dfrac{0 \cdot [(-1.2)^2 + (9.2 - 1.78)^3]}{-2.5 - (1.7 + 0.9)}$

Evaluate.

67. $9 - 4x$, for $x = 5$

68. $1 + x^3$, for $x = -2$

69. $24 \div t^3$, for $t = -2$

70. $20 \div a \cdot 4$, for $a = 5$

71. $45 \div 3 \cdot a$, for $a = -1$

72. $50 \div 2 \cdot t$, for $t = -5$

73. $5x \div 15x^2$, for $x = 3$

74. $6a \div 12a^3$, for $a = 2$

75. $45 \div 3^2 x(x - 1)$, for $x = 3$

76. $-30 \div t(t + 4)^2$, for $t = -6$

77. $-x^2 - 5x$, for $x = -3$

78. $(-x)^2 - 5x$, for $x = -3$

79. $\dfrac{3a - 4a^2}{a^2 - 20}$, for $a = 5$

80. $\dfrac{a^3 - 4a}{a(a - 3)}$, for $a = -2$

■ *Evaluate using a calculator.*

81. $13 - (y - 4)^3 + 10$, for $y = 6$

82. $(t + 4)^2 - 12 \div (19 - 17) + 68$, for $t = 5$

83. $3(m + 2n) \div m$, for $m = 1.6$ and $n = 5.9$

84. $1.5 + (2x - y)^2$, for $x = 9.25$ and $y = 1.7$

85. $\frac{1}{2}(x + 5/z)^2$, for $x = 141$ and $z = 0.2$

86. $a - \frac{3}{4}(2a - b)$, for $a = 213$ and $b = 165$

Write an equivalent expression without using grouping symbols.

87. $-(9x + 1)$

88. $-(3x + 5)$

89. $-[5 - 6x]$

90. $-(6x - 7)$

91. $-(4a - 3b + 7c)$

92. $-[5x - 2y - 3z]$

93. $-(3x^2 + 5x - 1)$

94. $-(8x^3 - 6x + 5)$

Simplify.

95. $8x - (6x + 7)$

96. $7y - (2y + 9)$

97. $2a - (5a - 9)$

98. $11n - (3n - 7)$

99. $2x + 7x - (4x + 6)$

100. $2a + 5a - (6a + 8)$

101. $9t - 5r - 2(3r + 6t)$

102. $4m - 9n - 3(2m - n)$

103. $15x - y - 5(3x - 2y + 5z)$

104. $4a - b - 4(5a - 7b + 8c)$

105. $3x^2 + 7 - (2x^2 + 5)$

106. $5x^4 + 3x - (5x^4 + 3x)$

107. $5t^3 + t - 3(t + 2t^3)$

108. $8n^2 + n - 2(n + 3n^2)$

109. $12a^2 - 3ab + 5b^2 - 5(-5a^2 + 4ab - 6b^2)$

110. $-8a^2 + 5ab - 12b^2 - 6(2a^2 - 4ab - 10b^2)$

111. $-7t^3 - t^2 - 3(5t^3 - 3t)$

112. $9t^4 + 7t - 5(9t^3 - 2t)$

113. $5(2x - 7) - [4(2x - 3) + 2]$

114. $3(6x - 5) - [3(1 - 8x) + 5]$

TW **115.** Some students use the mnemonic device PEMDAS to help remember the rules for the order of operations. Explain how this can be done and whether it is a valid approach.

TW **116.** Jake keys $18/2 \cdot 3$ into his calculator and expects the result to be 3. What mistake is he probably making?

Skill Maintenance

Translate to an algebraic expression. [1.1]

117. Nine more than twice a number

118. Half of the sum of two numbers

Synthesis

TW **119.** Write the sentence $(-x)^2 \neq -x^2$ in words. Explain why $(-x)^2$ and $-x^2$ are not equivalent.

TW **120.** Write the sentence $-|x| \neq -x$ in words. Explain why $-|x|$ and $-x$ are not equivalent.

Simplify.

121. $5t - \{7t - [4r - 3(t - 7)] + 6r\} - 4r$

122. $z - \{2z - [3z - (4z - 5z) - 6z] - 7z\} - 8z$

123. $\{x - [f - (f - x)] + [x - f]\} - 3x$

TW **124.** Is it true that for all real numbers a and b,
$$ab = (-a)(-b)?$$
Why or why not?

TW **125.** Is it true that for all real numbers a, b, and c,
$$a|b - c| = ab - ac?$$
Why or why not?

If $n > 0$, $m > 0$, and $n \neq m$, classify each of the following as either true or false.

126. $-n + m = -(n + m)$

127. $m - n = -(n - m)$

128. $n(-n - m) = -n^2 + nm$

129. $-m(n - m) = -(mn + m^2)$

130. $-n(-n - m) = n(n + m)$

Evaluate.

Aha! **131.** $[x + 3(2 - 5x) \div 7 + x](x - 3)$, for $x = 3$

Aha! **132.** $[x + 2 \div 3x] \div [x + 2 \div 3x]$, for $x = -7$

133. In Mexico, between 500 B.C. and 600 A.D., the Mayans represented numbers using powers of 20 and certain symbols (*Source*: National Council of Teachers of Mathematics, 1906 Association Drive, Reston, VA 22091). For example, the symbols

represent $4 \cdot 20^3 + 17 \cdot 20^2 + 10 \cdot 20^1 + 0 \cdot 20^0$. Evaluate this number.

134. Examine the Mayan symbols and the numbers in Exercise 133. What numbers do

•,   ⚊,   and   ⬭

each represent?

135. Calculate the volume of the tower shown below.

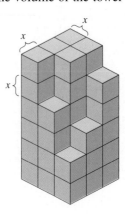

Collaborative Corner

Select the Symbols

Focus: Order of operations
Time: 15 minutes
Group Size: 2

One way to master the rules for the order of operations is to insert symbols within a display of numbers in order to obtain a predetermined result. For example, the display

$$1 \quad 2 \quad 3 \quad 4 \quad 5$$

can be used to obtain the result 21 as follows:

$$(1 + 2) \div 3 + 4 \cdot 5.$$

Note that without an understanding of the rules for the order of operations, solving a problem of this sort is impossible.

ACTIVITY

1. Each group should prepare an exercise similar to the example shown above. (Exponents are not allowed.) To do so, first select five single-digit numbers for display. Then insert operations and grouping symbols and calculate the result.

2. Pair with another group. Each group should give the other its result along with its five-number display, and challenge the other group to insert symbols that will make the display equal the result given.

3. Share with the entire class the various mathematical statements developed by each group.

1 Chapter Summary and Review

KEY TERMS AND DEFINITIONS

EXPRESSIONS

Algebraic expression, p. 2   A collection of **variables** and **constants** on which operations are performed.

Term, p. 18   A number, a variable, or a product or quotient of numbers and variables. Terms are separated in an expression by + signs.

Factor, p. 19   As a noun, multipliers in a product. As a verb, to write as a product.

Like terms, pp. 44, 68   Terms with exactly the same variable factors. These can be **combined** or **collected** in an expression.

Equivalent expressions, p. 15   Expressions that have the same value for all allowable replacements.

EQUATIONS AND INEQUALITIES

Equation, p. 6   A sentence formed by placing an equals sign between two expressions.

Inequality, p. 37   A sentence formed by placing an inequality sign between two expressions.

Solution, p. 7   A replacement that makes an equation true.

NUMBERS

Prime number, p. 22   A natural number having only itself and 1 as factors. **Composite numbers** have more than 2 factors.

Natural numbers, p. 22 $\{1, 2, 3, \ldots\}$

Whole numbers, p. 22 $\{0, 1, 2, 3, \ldots\}$

Integers, p. 32 $\{\ldots, -3, -2, -1, 0, 1, 2, 3, \ldots\}$

Rational numbers, p. 33   Numbers that can be expressed as an integer divided by a nonzero integer or as **repeating** decimals or **terminating** decimals.

Irrational numbers, p. 35   Real numbers that cannot be expressed as a ratio of integers.

Real numbers, p. 35   The set of all rational and irrational numbers.

INVERSES

Opposites, additive inverses, p. 48   Two numbers whose sum is 0; 2 and -2 are opposites, or additive inverses.

Reciprocals, multiplicative inverses, p. 24   Two numbers whose product is 1; 2 and $\frac{1}{2}$ are reciprocals, or multiplicative inverses.

GEOMETRY FORMULAS

Area of a rectangle: $A = lw$

Area of a triangle: $A = \frac{1}{2}bh$

Area of a parallelogram: $A = bh$

IMPORTANT CONCEPTS

[Section references appear in brackets.]

Concept	Example
An algebraic expression is **evaluated** by **substituting** values for the variables and carrying out the operations.	Evaluate $2 + 3x \div 5y$ for $x = -10$ and $y = 2$. $2 + 3x \div 5y = 2 + 3(-10) \div 5(2)$ $= 2 + (-30) \div 5(2)$ $= 2 + (-6)(2)$ $= 2 + (-12)$ $= -10$ [1.1], [1.8]
Commutative laws: $\quad a + b = b + a,$ $\qquad\qquad\qquad\qquad ab = ba$	$3 + (-5) = -5 + 3$ $8(10) = 10(8)$ [1.2]
Associative laws: $a + (b + c) = (a + b) + c,$ $a(bc) = (ab)c$	$-5 + (5 + 6) = (-5 + 5) + 6$ $2 \cdot (5 \cdot 9) = (2 \cdot 5) \cdot 9$ [1.2]
Distributive law: $\quad a(b + c) = ab + ac$	$4(x + 2) = 4 \cdot x + 4 \cdot 2 = 4x + 8$ [1.2]
The **prime factorization** of a composite number expresses that number as a product of prime numbers.	$600 = 2 \cdot 2 \cdot 2 \cdot 3 \cdot 5 \cdot 5$ [1.3]
The **absolute value** of a number is the number of units that number is from zero on the number line.	$\lvert 3 \rvert = 3$ since 3 is 3 units from 0. $\lvert -3 \rvert = 3$ since -3 is 3 units from 0. [1.4]
To **add** two real numbers, use the rules on p. 42.	$-8 + (-3) = -11$ $-8 + 3 = -5$ $8 + (-3) = 5$ $-8 + 8 = 0$ [1.5]
To **subtract** two real numbers, add the opposite of the number being subtracted.	$-10 - 12 = -10 + (-12) = -22$ $-10 - (-12) = -10 + 12 = 2$ [1.6]
To **multiply** or **divide** two real numbers, use the rules on p. 58. Division by 0 is undefined.	$(-5)(-2) = 10$ $30 \div (-6) = -5$ $-3 \div 0$ is undefined [1.7]

(continued)

Exponential notation: Exponent $\quad$ *n* factors $b^n = \underbrace{b \cdot b \cdot b \cdots b}$ Base	$6^2 = 6 \cdot 6 = 36$ $(-6)^2 = (-6) \cdot (-6) = 36$ $-6^2 = -6 \cdot 6 = -36$ $(6x)^2 = (6x) \cdot (6x) = 36x^2 \qquad [1.8]$
To perform multiple operations, use the **rules for order of operations** on p. 66.	$\begin{aligned} -3 + (3-5)^3 \div 4(-1) &= -3 + (-2)^3 \div 4(-1) \\ &= -3 + (-8) \div 4(-1) \\ &= -3 + (-2)(-1) \\ &= -3 + 2 \\ &= -1 \qquad [1.8] \end{aligned}$

Fraction notation for 1:	$\dfrac{a}{a} = 1$	$\dfrac{-2x}{-2x} = 1 \qquad [1.3]$
Identity property of 1:	$1 \cdot a = a \cdot 1 = a$	$1 \cdot (-10) = -10 \qquad [1.3]$
Identity property of 0:	$a + 0 = 0 + a = a$	$0 + (-5) = -5 \qquad [1.5]$
Opposite of an opposite:	$-(-a) = a$	$-(-10) = 10 \qquad [1.6]$
Law of opposites:	$a + (-a) = 0$	$2.8 + (-2.8) = 0 \qquad [1.5]$
Multiplicative property of 0:	$0 \cdot a = a \cdot 0 = 0$	$0 \cdot (-5) = 0 \qquad [1.7]$
Property of -1:	$-1 \cdot a = -a$	$-1 \cdot (-3x) = 3x \qquad [1.8]$
Opposite of a sum:	$-(a+b) = -a + (-b)$	$-(x - 2y + 6) = -x + (-(-2y)) + (-6) \quad [1.8]$
Division involving 0:	$\dfrac{0}{a} = 0; \ \dfrac{a}{0}$ is undefined.	$\dfrac{0}{7} = 0; \ \dfrac{7}{0}$ is undefined. $\qquad [1.7]$
$\dfrac{-a}{b} = \dfrac{a}{-b} = -\dfrac{a}{b}$		$\dfrac{-3}{4} = \dfrac{3}{-4} = -\dfrac{3}{4} \qquad [1.7]$
$\dfrac{-a}{-b} = \dfrac{a}{b}$		$\dfrac{-3}{-4} = \dfrac{3}{4} \qquad [1.7]$

Review Exercises

↪ *Concept Reinforcement* In each of Exercises 1–10, classify the statement as either true or false.

1. $4x - 5y$ and $12 - 7a$ are both algebraic expressions containing two terms. [1.2]

2. $3t + 1 = 7$ and $8 - 2 = 9$ are both equations. [1.1]

3. The fact that $2 + x$ is equivalent to $x + 2$ is an illustration of the associative law for addition. [1.2]

4. The statement $4(a + 3) = 4 \cdot a + 4 \cdot 3$ illustrates the distributive law. [1.2]

5. The number 2 is neither prime nor composite. [1.3]

6. Every irrational number can be written as a repeating or terminating decimal. [1.4]

7. Every natural number is a whole number and every whole number is an integer. [1.4]

8. The expressions $9r^2s$ and $5rs^2$ are like terms. [1.8]

9. The opposite of x, written $-x$, never represents a positive number. [1.6]

10. The number 0 has no reciprocal. [1.3]

Evaluate.

11. $5t$, for $t = 3$ [1.1]

12. $9 - y^2$, for $y = 4$ [1.8]

13. $-10 + a^2 \div (b + 1)$, for $a = 5$ and $b = 4$ [1.8]

Translate to an algebraic expression. [1.1]

14. 7 less than z

15. The product of x and z

16. One more than the product of two numbers

17. Determine whether 35 is a solution of $x/5 = 8$. [1.1]

18. Translate to an equation. Do not solve. [1.1]

According to Photo Marketing Association International, in 2006, 14.1 billion prints were made from film. This number is 3.2 billion more than the number of digital prints made. How many digital prints were made in 2006?

19. The following table lists the number of calories burned doing calisthenics for various lengths of time. Find an equation for the number c of calories burned by a person who does t hours of calisthenics. [1.1]

Number, t, of Hours Spent Doing Calisthenics	Number, c, of Calories Burned
1	300
2	600
4	1200
5	1500

20. Use the commutative law of multiplication to write an expression equivalent to $3t + 5$. [1.2]

21. Use the associative law of addition to write an expression equivalent to $(2x + y) + z$. [1.2]

22. Use the commutative and associative laws to write three expressions equivalent to $4(xy)$. [1.2]

Multiply. [1.2]

23. $6(3x + 5y)$

24. $8(5x + 3y + 2)$

Factor. [1.2]

25. $21x + 15y$

26. $35x + 14 + 7y$

27. Find the prime factorization of 52. [1.3]

Simplify. [1.3]

28. $\dfrac{20}{48}$

29. $\dfrac{18}{8}$

Perform the indicated operation and, if possible, simplify. [1.3]

30. $\dfrac{5}{12} + \dfrac{4}{9}$

31. $\dfrac{9}{16} \div 3$

32. $\dfrac{2}{3} - \dfrac{1}{15}$

33. $\dfrac{9}{10} \cdot \dfrac{16}{5}$

34. Tell which integers correspond to this situation: Renir has a debt of \$45 and Raoul has \$72 in his savings account. [1.4]

35. Graph on a number line: $\frac{-1}{3}$. [1.4]

36. Write an inequality with the same meaning as $-3 < x$. [1.4]

37. Classify as true or false: $2 \geq -8$. [1.4]

38. Classify as true or false: $0 \leq -1$. [1.4]

39. Find decimal notation: $-\frac{7}{8}$. [1.4]

40. Find the absolute value: $|-1|$. [1.4]

41. Find $-(-x)$ when x is -9. [1.6]

Simplify.

42. $4 + (-7)$ [1.5]

43. $-\frac{2}{3} + \frac{1}{12}$ [1.5]

44. $10 + (-9) + (-8) + 7$ [1.5]

45. $-3.8 + 5.1 + (-12) + (-4.3) + 10$ [1.5]

46. $-2 - (-7)$ [1.6]   **47.** $-\frac{9}{10} - \frac{1}{2}$ [1.6]

48. $-3.8 - 4.1$ [1.6]   **49.** $-9 \cdot (-6)$ [1.7]

50. $-2.7(3.4)$ [1.7]   **51.** $\frac{2}{3} \cdot \left(-\frac{3}{7}\right)$ [1.7]

52. $2 \cdot (-7) \cdot (-2) \cdot (-5)$ [1.7]

53. $35 \div (-5)$ [1.7]

54. $-5.1 \div 1.7$ [1.7]

55. $-\frac{3}{5} \div \left(-\frac{4}{5}\right)$ [1.7]

56. $|-3 \cdot 4 - 12 \cdot 2| - 8(-7)$ [1.8]

57. $\left|-12(-3) - 2^3 - (-9)(-10)\right|$ [1.8]

58. $120 - 6^2 \div 4 \cdot 8$ [1.8]

59. $(120 - 6^2) \div 4 \cdot 8$ [1.8]

60. $(120 - 6^2) \div (4 \cdot 8)$ [1.8]

61. $\dfrac{4(18 - 8) + 7 \cdot 9}{9^2 - 8^2}$ [1.8]

Combine like terms.

62. $11a + 2b + (-4a) + (-5b)$ [1.5]

63. $7x - 3y - 9x + 8y$ [1.6]

64. Find the opposite of -7. [1.6]

65. Find the reciprocal of -7. [1.7]

66. Write exponential notation for $2x \cdot 2x \cdot 2x \cdot 2x$. [1.8]

67. Simplify: $(-5x)^3$. [1.8]

Remove parentheses and simplify. [1.8]

68. $2a - (5a - 9)$

69. $3(b + 7) - 5b$

70. $3[11x - 3(4x - 1)]$

71. $2[6(y - 4) + 7]$

72. $[8(x + 4) - 10] - [3(x - 2) + 4]$

TW 73. Explain the difference between a constant and a variable. [1.1]

TW 74. Explain the difference between a term and a factor. [1.2]

Synthesis

TW 75. Describe at least three ways in which the distributive law was used in this chapter. [1.2]

TW 76. Devise a rule for determining the sign of a negative number raised to an exponent. [1.8]

77. Evaluate $a^{50} - 20a^{25}b^4 + 100b^8$ for $a = 1$ and $b = 2$. [1.8]

78. If $0.090909\ldots = \frac{1}{11}$ and $0.181818\ldots = \frac{2}{11}$, what rational number is named by each of the following?
   a) $0.272727\ldots$ [1.4]   **b)** $0.909090\ldots$ [1.4]

Simplify. [1.8]

79. $-\left|\frac{7}{8} - \left(-\frac{1}{2}\right) - \frac{3}{4}\right|$

80. $\left(\left|2.7 - 3\right| + 3^2 - \left|-3\right|\right) \div (-3)$

Match the phrase in the left column with the most appropriate choice from the right column.

81. _____ A number is nonnegative. [1.4]

82. _____ The reciprocal of a sum [1.7]

83. _____ A number squared [1.8]

84. _____ The opposite of a sum [1.8]

85. _____ The opposite of an opposite is the original number. [1.6]

86. _____ The order in which numbers are added does not change the result. [1.2]

87. _____ A number is negative. [1.4]

88. _____ The absolute value of a product [1.4]

89. _____ A sum of a number and its reciprocal [1.7]

90. _____ The square of a sum [1.8]

91. _____ The absolute value of one number is less than the absolute value of another number. [1.4]

a) a^2

b) $a + b = b + a$

c) $a < 0$

d) $a + \dfrac{1}{a}$

e) $|ab|$

f) $(a + b)^2$

g) $|a| < |b|$

h) $-(a + b)$

i) $a \geq 0$

j) $\dfrac{1}{a + b}$

k) $-(-a) = a$

Chapter Test 1

1. Evaluate $\dfrac{2x}{y}$ for $x = 10$ and $y = 5$.

2. Write an algebraic expression: Nine less than some number.

3. Find the area of a triangle when the height h is 30 ft and the base b is 16 ft.

4. Use the commutative law of addition to write an expression equivalent to $3p + q$.

5. Use the associative law of multiplication to write an expression equivalent to $x \cdot (4 \cdot y)$.

6. Determine whether 7 is a solution of $65 - x = 69$.

7. Translate to an equation. Do not solve.

   On a hot summer day, Green River Electric met a demand of 2518 megawatts. This is only 282 megawatts less than its maximum production capability. What is the maximum capability of production?

Multiply.

8. $7(5 - x)$ 9. $-5(y - 2)$

Factor.

10. $11 - 44x$ 11. $7x + 21 + 14y$

12. Find the prime factorization of 300.

13. Simplify: $\frac{10}{35}$.

Write a true sentence using either $<$ or $>$.

14. $-4 \ \blacksquare \ 0$ 15. $-3 \ \blacksquare \ -8$

Find the absolute value.

16. $\left|\frac{9}{4}\right|$ 17. $\left|-2.7\right|$

18. Find the opposite of $\frac{2}{3}$.

19. Find the reciprocal of $-\frac{4}{7}$.

20. Find $-x$ when x is -8.

21. Write an inequality with the same meaning as $x \le -2$.

Perform the indicated operations and, if possible, simplify.

22. $3.1 - (-4.7)$ 23. $-8 + 4 + (-7) + 3$

24. $3.2 - 5.7$ 25. $\frac{1}{8} - \left(-\frac{3}{4}\right)$

26. $4 \cdot (-12)$ 27. $-\frac{1}{2} \cdot \left(-\frac{3}{8}\right)$

28. $-54 \div 9$ 29. $-\frac{3}{5} \div \left(-\frac{4}{5}\right)$

30. $4.864 \div (-0.5)$

31. $-2(16) - |2(-8) - 5^3|$

32. $9 + 7 - 4 - (-3)$

33. $256 \div (-16) \div 4$

34. $2^3 - 10[4 - (-2 + 18)3]$

35. Combine like terms: $18y + 30a - 9a + 4y$.

36. Simplify: $(-2x)^4$.

Remove parentheses and simplify.

37. $4x - (3x - 7)$

38. $4(2a - 3b) + a - 7$

39. $4\{3[5(y - 3) + 9] + 2(y + 8)\}$

Synthesis

40. Evaluate $\dfrac{5y - x}{2}$ when $x = 20$ and y is 4 less than half of x.

41. Insert one pair of parentheses to make the following a true statement:
$$9 - 3 - 4 + 5 = 15.$$

Simplify.

42. $|-27 - 3(4)| - |-36| + |-12|$

43. $a - \{3a - [4a - (2a - 4a)]\}$

44. Classify the following as either true or false:
$$a|b - c| = |ab| - |ac|.$$

2

Equations, Inequalities, and Problem Solving

S olving equations and inequalities is a recurring theme in much of mathematics. In this chapter, we will study some of the principles used to solve equations and inequalities. We will then use equations and inequalities to solve applied problems.

APPLICATION *Aquariums.*

The following table lists the maximum recommended stocking density for fish in newly conditioned aquariums of various sizes. The density is calculated by adding the lengths of all fish in the aquarium. What size aquarium would be needed for 30 in. of fish?

SIZE OF AQUARIUM (IN GALLONS)	RECOMMENDED STOCKING DENSITY (IN INCHES OF FISH)
100	20
120	24
200	40
250	50

Source: *Aquarium Fish Magazine*, June 2002

This problem appears as Example 7 in Section 2.5.

2.1 Solving Equations

Equations and Solutions ▨ The Addition Principle ▨
The Multiplication Principle ▨ Selecting the Correct Approach

Solving equations is essential for problem solving in algebra. In this section, we study two of the most important principles used for this task.

Equations and Solutions

We have already seen that an equation is a number sentence stating that the expressions on either side of the equals sign represent the same number. Some equations, like $3 + 2 = 5$ or $2x + 6 = 2(x + 3)$, are *always* true and some, like $3 + 2 = 6$ or $x + 2 = x + 3$, are *never* true. In this text, we will concentrate on equations like $x + 6 = 13$ or $7x = 141$ that are *sometimes* true, depending on the replacement value for the variable.

> **Solution of an Equation**   Any replacement for the variable that makes an equation true is called a *solution* of the equation. To *solve* an equation means to find all of its solutions.

To determine whether a number is a solution, we substitute that number for the variable throughout the equation. If the values on both sides of the equals sign are the same, then the number that was substituted is a solution.

EXAMPLE 1   Determine whether 7 is a solution of $x + 6 = 13$.

SOLUTION   We have

$$\begin{array}{c|c} x + 6 = 13 & \text{Writing the equation} \\ \hline 7 + 6 \ \vrule\ 13 & \text{Substituting 7 for } x \\ 13 \overset{?}{=} 13 \quad \text{TRUE} & 13 = 13 \text{ is a true statement.} \end{array}$$

Since the left-hand and the right-hand sides are the same, 7 is a solution.

> **CAUTION!**   Note that in Example 1, the solution is 7, not 13.

EXAMPLE 2   Determine whether 3 is a solution of $7x - 2 = 4x + 5$.

SOLUTION   We have

$$\begin{array}{c|c} 7x - 2 = 4x + 5 & \text{Writing the equation} \\ \hline 7(3) - 2 \ \vrule\ 4(3) + 5 & \text{Substituting 3 for } x \\ 21 - 2 \ \vrule\ 12 + 5 & \text{Carrying out calculations on both sides} \\ 19 \overset{?}{=} 17 \quad \text{FALSE} & \text{The statement } 19 = 17 \text{ is false.} \end{array}$$

Since the left-hand and the right-hand sides differ, 3 is not a solution.

Editing and Evaluating Expressions

It is possible to correct an error or change a value in an expression by using the *arrow*, *insert*, and *delete* keys.

As the expression is being typed, use the arrow keys to move the cursor to the character you want to change. Pressing (INS) (the 2nd option associated with the **DEL** key) changes the calculator between the INSERT and OVERWRITE modes. The OVERWRITE mode is often indicated by a rectangular cursor and the INSERT mode by an underscore cursor. The following table shows how to make changes.

Insert a character in front of the cursor.	In the INSERT mode, press the character you wish to insert.
Replace the character under the cursor.	In the OVERWRITE mode, press the character you want as the replacement.
Delete the character under the cursor.	Press **DEL**.

After you have evaluated an expression by pressing **ENTER**, the expression can be recalled to the screen by pressing (ENTRY) (the 2nd option associated with the **ENTER** key.) Then it can be edited as described above. Pressing **ENTER** will then evaluate the edited expression.

Expressions such as $2xy - x$ can be entered into a graphing calculator as written. The calculator will evaluate an expression using the values that it has stored for the variables. Variable names can be entered by pressing **ALPHA** and then the key associated with that letter. The variable x can also be entered by pressing the **X,T,θ,n** key. Thus, for example, to store the value 1 as y, press (1) **STO)** **ALPHA** (Y) **ENTER**. The value for y can then be seen by pressing **ALPHA** (Y) **ENTER**.

EXAMPLE 3   Use a calculator to determine whether each of the following is a solution of $2x - 5 = -7$: $3, -1$.

SOLUTION   We evaluate the expression on the left side of the equals sign, $2x - 5$, for each given value of x. A number is a solution of the equation if the value of the expression on the left is the same as the right side of the equation, -7. We begin by substituting 3 for x. Since $2(3) - 5 \neq -7$, 3 is not a solution of the equation.

```
2(3)-5
                    1
2(-1)-5
                   -7
```

To check -1, we recall the last entry, $2(3) - 5$. Note that -1 has one more character than 3. We can insert a negative sign and then overwrite the 3 with a 1. We see that -1 is a solution of the equation.

For a more complicated expression, it may be easier to store different values of the variables and use the expression written with the variables. To illustrate, we check Example 3 using this method. First, we store 3 to the variable x by pressing ⟨3⟩ ⟨STO▸⟩ ⟨X,T,θ,n⟩ ⟨ENTER⟩. Then we enter the equation by pressing ⟨2⟩ ⟨X,T,θ,n⟩ ⟨−⟩ ⟨5⟩ ⟨ENTER⟩. We see that for $x = 3$, the value of $2x − 5$ is 1.

$$
\begin{array}{lr}
3 \to \text{X} & \\
 & 3 \\
2\text{X} - 5 & \\
 & 1
\end{array}
$$

Now we store $−1$ to x. We then press ⟨ENTRY⟩ repeatedly until $2x − 5$ appears again on the screen. Then we press ⟨ENTER⟩. We see that for $x = −1$, the value of $2x − 5$ is $−7$. Thus, $−1$ is a solution of $2x − 5 = −7$, but 3 is not.

$$
\begin{array}{lr}
^{-}1 \to \text{X} & \\
 & ^{-}1 \\
2\text{X} - 5 & \\
 & ^{-}7
\end{array}
$$

The Addition Principle

Consider the equation

$$x = 7.$$

We can easily see that the solution of this equation is 7. Replacing x with 7, we get

$$7 = 7, \quad \text{which is true.}$$

Now consider the equation

$$x + 6 = 13.$$

In Example 1, we found that the solution of $x + 6 = 13$ is also 7. Although the solution of $x = 7$ may seem more obvious, because the equations $x + 6 = 13$ and $x = 7$ have identical solutions, the equations are said to be **equivalent.**

Student Notes

Be sure to remember that equivalent equations are *equations*, whereas equivalent expressions are *expressions*, like $5a − 10$ or $5(a − 2)$.

> **Equivalent Equations**   Equations with the same solutions are called *equivalent equations*.

There are principles that enable us to begin with one equation and end up with an equivalent equation, like $x = 7$, for which the solution is obvious. One such principle concerns addition. The equation $a = b$ says that a and b stand for the same number. Suppose this is true, and some number c is added to a. We get the same result if we add c to b, because a and b are the same number.

The Addition Principle   For any real numbers a, b, and c,

$a = b$ is equivalent to $a + c = b + c$.

To visualize the addition principle, consider a balance similar to one a jeweler might use. When the two sides of the balance hold equal weight, the balance is level. If weight is then added or removed, equally, on both sides, the balance will remain level.

$a = b$ $a + c = b + c$

When using the addition principle, we often say that we "add the same number to both sides of an equation." We can also "subtract the same number from both sides," since subtraction can be regarded as the addition of an opposite.

EXAMPLE 4   Solve: $x + 5 = -7$.

SOLUTION   We can add any number we like to both sides. Since -5 is the opposite, or additive inverse, of 5, we add -5 to each side:

$$x + 5 = -7$$
$$x + 5 - 5 = -7 - 5 \qquad \text{Using the addition principle: adding } -5 \text{ to both sides or subtracting 5 from both sides}$$
$$x + 0 = -12 \qquad \text{Simplifying;}$$
$$x + 5 - 5 = x + 5 + (-5) = x + 0$$
$$x = -12. \qquad \text{Using the identity property of 0}$$

It is obvious that the solution of $x = -12$ is the number -12. To check the answer in the original equation, we substitute.

Check:
$$\begin{array}{c|c} x + 5 = -7 \\ \hline -12 + 5 & -7 \\ -7 \overset{?}{=} -7 & \text{TRUE} \end{array}$$

The solution of the original equation is -12.

In Example 4, note that because we added the *opposite*, or *additive inverse*, of 5, the left side of the equation simplified to x plus the *additive identity*, 0, or simply x. These steps effectively replaced the 5 on the left with a 0. When solving $x + a = b$ for x, we simply add $-a$ to (or subtract a from) both sides.

Student Notes

We can also think of "undoing" operations to isolate a variable. In Example 5, we began with $y - 8.4$ on the right side. To undo the subtraction, we *add* 8.4.

EXAMPLE 5   Solve: $-6.5 = y - 8.4$.

SOLUTION   The variable is on the right side this time. We can isolate y by adding 8.4 to each side:

$$-6.5 = y - 8.4$$ $y - 8.4$ can be regarded as $y + (-8.4)$.

$$-6.5 + 8.4 = y - 8.4 + 8.4$$ **Using the addition principle: Adding 8.4 to both sides "eliminates" -8.4 on the right side.**

$$1.9 = y.$$ $y - 8.4 + 8.4 = y + (-8.4) + 8.4$
$$= y + 0 = y$$

Check:

$$\begin{array}{c|c} -6.5 = y - 8.4 \\ \hline -6.5 & 1.9 - 8.4 \\ -6.5 \stackrel{?}{=} -6.5 \end{array}$$ TRUE $-6.5 = -6.5$ is true.

The solution is 1.9.

Note that the equations $a = b$ and $b = a$ have the same meaning. Thus, $-6.5 = y - 8.4$ could have been rewritten as $y - 8.4 = -6.5$.

The Multiplication Principle

A second principle for solving equations concerns multiplying. Suppose a and b are equal. If a and b are multiplied by some number c, then ac and bc will also be equal.

> **The Multiplication Principle**   For any real numbers a, b, and c, with $c \neq 0$,
>
> $$a = b \quad \text{is equivalent to} \quad a \cdot c = b \cdot c.$$

EXAMPLE 6   Solve: $\frac{5}{4}x = 10$.

SOLUTION   We can multiply both sides by any nonzero number we like. Since $\frac{4}{5}$ is the reciprocal of $\frac{5}{4}$, we multiply each side by $\frac{4}{5}$:

$$\frac{5}{4}x = 10$$

$$\frac{4}{5} \cdot \frac{5}{4}x = \frac{4}{5} \cdot 10$$ **Using the multiplication principle: Multiplying both sides by $\frac{4}{5}$ "eliminates" the $\frac{5}{4}$ on the left.**

$$1 \cdot x = 8$$ **Simplifying**

$$x = 8.$$ **Using the identity property of 1**

Check: $\dfrac{5}{4}x = 10$

$$\dfrac{5}{4} \cdot 8 \mid 10$$

$$10 \overset{?}{=} 10 \quad \text{TRUE} \qquad 10 = 10 \text{ is true.}$$

The solution is 8.

In Example 6, to get x alone, we multiplied by the *reciprocal*, or *multiplicative inverse* of $\frac{5}{4}$. We then simplified the left-hand side to x times the *multiplicative identity*, 1, or simply x. These steps effectively replaced the $\frac{5}{4}$ on the left with 1.

Because division is the same as multiplying by a reciprocal, the multiplication principle also tells us that we can "divide both sides by the same nonzero number." That is,

$$\text{if } a = b, \text{ then } \quad \dfrac{1}{c} \cdot a = \dfrac{1}{c} \cdot b \quad \text{and} \quad \dfrac{a}{c} = \dfrac{b}{c} \qquad (\text{provided } c \neq 0).$$

In a product like $3x$, the multiplier 3 is called the **coefficient.** *When the coefficient of the variable is an integer or a decimal, it is usually easiest to solve an equation by dividing on both sides. When the coefficient is in fraction notation, it is usually easier to multiply by the reciprocal.*

Student Notes

In Example 7(a), we can think of undoing the multiplication $-4 \cdot x$ by *dividing* by -4.

EXAMPLE 7   Solve.

a) $-4x = 92$ 　　　**b)** $-x = 9$ 　　　**c)** $\dfrac{2y}{9} = \dfrac{8}{3}$

SOLUTION

a) $-4x = 92$

$$\dfrac{-4x}{-4} = \dfrac{92}{-4} \qquad \text{Using the multiplication principle: Dividing both sides by } -4 \text{ is the same as multiplying by } -\tfrac{1}{4}.$$

$$1 \cdot x = -23 \qquad \text{Simplifying}$$

$$x = -23 \qquad \text{Using the identity property of 1}$$

Check:
$$\dfrac{-4x = 92}{-4(-23) \mid 92}$$
$$92 \overset{?}{=} 92 \quad \text{TRUE} \qquad 92 = 92 \text{ is true.}$$

The solution is -23.

b) To solve an equation like $-x = 9$, remember that when an expression is multiplied or divided by -1, its sign is changed. Here we multiply both sides by -1 to change the sign of $-x$:

$$-x = 9$$

$$(-1)(-x) = (-1)9 \qquad \text{Multiplying both sides by } -1 \text{ (dividing by } -1 \text{ would also work). Note that the reciprocal of } -1 \text{ is } -1.$$

$$x = -9. \qquad \text{Note that } (-1)(-x) \text{ is the same as } (-1)(-1)x.$$

Check:
$$\frac{-x = 9}{-(-9) \mid 9}$$
$$9 \overset{?}{=} 9 \quad \text{TRUE} \quad 9 = 9 \text{ is true.}$$

The solution is -9.

c) To solve an equation like $\dfrac{2y}{9} = \dfrac{8}{3}$, we rewrite the left-hand side as $\dfrac{2}{9} \cdot y$ and then use the multiplication principle, multiplying by the reciprocal of $\dfrac{2}{9}$:

$$\frac{2y}{9} = \frac{8}{3}$$

$$\frac{2}{9} \cdot y = \frac{8}{3} \qquad \text{Rewriting } \frac{2y}{9} \text{ as } \frac{2}{9} \cdot y$$

$$\frac{9}{2} \cdot \frac{2}{9} \cdot y = \frac{9}{2} \cdot \frac{8}{3} \qquad \text{Multiplying both sides by } \frac{9}{2}$$

$$1y = \frac{3 \cdot \cancel{3} \cdot \cancel{2} \cdot 4}{\cancel{2} \cdot \cancel{3}} \qquad \text{Removing a factor equal to 1: } \frac{3 \cdot 2}{2 \cdot 3} = 1$$

$$y = 12.$$

Check:
$$\frac{\dfrac{2y}{9} = \dfrac{8}{3}}{\begin{array}{c|c} \dfrac{2 \cdot 12}{9} & \dfrac{8}{3} \\ \dfrac{24}{9} & \end{array}}$$
$$\frac{8}{3} \overset{?}{=} \frac{8}{3} \quad \text{TRUE} \quad \tfrac{8}{3} = \tfrac{8}{3} \text{ is true.}$$

The solution is 12.

Selecting the Correct Approach

It is important that you be able to determine which principle should be used to solve a particular equation.

Seeking Help?

A variety of resources are available to help make studying easier and more enjoyable.

- **Textbook supplements.** See the preface for a description of the supplements that exist for this textbook: the *Student's Solutions Manual*, a complete set of lessons on CD, tutorial exercises on CD, and complete online courses in MathXL and MyMathLab.

- **Your college or university.** Your own college or university probably has resources to enhance your math learning: a learning lab or tutoring center for drop-in tutoring, study skills workshops or group tutoring sessions tailored for the specific course you are taking, or a bulletin board or network where you can locate the names of experienced private tutors.

- **Your instructor.** Find out your instructor's office hours and make it a point to visit when you need additional help. Many instructors also welcome student e-mail.

EXAMPLE 8 Solve: **(a)** $-\dfrac{2}{3} + x = \dfrac{5}{2}$; **(b)** $12.6 = 3t$.

SOLUTION

a) To undo addition of $-\frac{2}{3}$, we subtract $-\frac{2}{3}$ $\left(\text{or add } \frac{2}{3}\right)$ to both sides. Note that the opposite of *negative* $\frac{2}{3}$ is *positive* $\frac{2}{3}$.

$$-\tfrac{2}{3} + x = \tfrac{5}{2}$$
$$-\tfrac{2}{3} + x + \tfrac{2}{3} = \tfrac{5}{2} + \tfrac{2}{3} \qquad \text{Using the addition principle}$$
$$x = \tfrac{5}{2} + \tfrac{2}{3}$$
$$= \tfrac{5}{2} \cdot \tfrac{3}{3} + \tfrac{2}{3} \cdot \tfrac{2}{2} \qquad \text{Finding a common denominator}$$
$$= \tfrac{15}{6} + \tfrac{4}{6}$$
$$= \tfrac{19}{6}.$$

Check:

$$\begin{array}{c|c} -\frac{2}{3} + x = \frac{5}{2} & \\ \hline -\frac{2}{3} + \frac{19}{6} & \frac{5}{2} \\ -\frac{4}{6} + \frac{19}{6} & \\ \frac{15}{6} & \\ \frac{5 \cdot 3}{2 \cdot 3} & \\ \frac{5}{2} \overset{?}{=} \frac{5}{2} \quad \text{TRUE} & \end{array}$$

$-\frac{2}{3} \cdot \frac{2}{2} = -\frac{4}{6}$

Removing a factor equal to 1: $\frac{3}{3} = 1$

$\frac{5}{2} = \frac{5}{2}$ is true.

The solution is $\frac{19}{6}$.

b) To undo multiplication by 3, we either divide both sides by 3 or multiply both sides by $\frac{1}{3}$. Note that the reciprocal of *positive* 3 is *positive* $\frac{1}{3}$.

$$12.6 = 3t$$
$$\frac{12.6}{3} = \frac{3t}{3} \qquad \text{Using the multiplication principle}$$
$$4.2 = 1t$$
$$4.2 = t \qquad \text{Simplifying}$$

Check:

$$\begin{array}{c|c} 12.6 = 3t & \\ \hline 12.6 & 3(4.2) \\ 12.6 \overset{?}{=} 12.6 & \text{TRUE} \end{array}$$

$12.6 = 12.6$ is true.

The solution is 4.2.

2.1 EXERCISE SET

FOR EXTRA HELP

 MathXL   MyMathLab   InterAct Math   AW Math Tutor Center   Video Lectures on CD: Disc 1   Student's Solutions Manual

🖐 *Concept Reinforcement   For each of Exercises 1–6, match the statement with the most appropriate choice from the column on the right.*

1. _____ A replacement that makes an equation true

2. _____ The equations $x + 3 = 7$ and $6x = 24$

3. _____ The role of 9 in $9ab$

4. _____ The expressions $3(x - 2)$ and $3x - 6$

5. _____ The principle used to solve $\frac{2}{3} \cdot x = -4$

6. _____ The principle used to solve $\frac{2}{3} + x = -4$

a) Coefficient

b) Equivalent expressions

c) Equivalent equations

d) The multiplication principle

e) The addition principle

f) Solution

Solve using the addition principle. Don't forget to check!

7. $x + 6 = 23$

8. $x + 5 = 8$

9. $y + 7 = -4$

10. $t + 6 = 43$

11. $t + 9 = -12$

12. $y + 7 = -3$

13. $-6 = y + 25$

14. $-5 = x + 8$

15. $x - 8 = 5$

16. $x - 9 = 6$

17. $12 = -7 + y$

18. $15 = -8 + z$

19. $-5 + t = -9$

20. $-6 + y = -21$

21. $r + \frac{1}{3} = \frac{8}{3}$

22. $t + \frac{3}{8} = \frac{5}{8}$

23. $x + \frac{3}{5} = -\frac{7}{10}$

24. $x + \frac{2}{3} = -\frac{5}{6}$

25. $x - \frac{5}{6} = \frac{7}{8}$

26. $y - \frac{3}{4} = \frac{5}{6}$

27. $-\frac{1}{5} + z = -\frac{1}{4}$

28. $-\frac{1}{8} + y = -\frac{3}{4}$

29. $m + 3.9 = 5.4$

30. $y + 5.3 = 8.7$

31. $-9.7 = -4.7 + y$

32. $-7.8 = 2.8 + x$

Solve using the multiplication principle. Don't forget to check!

33. $5x = 70$

34. $3x = 39$

35. $9t = 36$

36. $6x = 72$

37. $84 = 7x$

38. $56 = 7t$

39. $-x = 23$

40. $100 = -x$

Aha! 41. $-t = -8$

42. $-68 = -r$

43. $7x = -49$

44. $9x = -36$

45. $-1.3a = -10.4$

46. $-3.4t = -20.4$

47. $\frac{y}{-8} = 11$

48. $\frac{a}{4} = 13$

49. $\frac{4}{5}x = 16$

50. $\frac{3}{4}x = 27$

51. $\frac{-x}{6} = 9$

52. $\frac{-t}{5} = 9$

53. $\frac{1}{9} = \frac{z}{5}$

54. $\frac{2}{7} = \frac{x}{3}$

Aha! 55. $-\frac{3}{5}r = -\frac{3}{5}$

56. $-\frac{2}{5}y = -\frac{4}{15}$

57. $\frac{-3r}{2} = -\frac{27}{4}$

58. $\frac{5x}{7} = -\frac{10}{14}$

Solve. The symbol 🖩 indicates an exercise designed to give practice using a calculator.

59. $4.5 + t = -3.1$

60. $\frac{3}{4}x = 18$

61. $-8.2x = 20.5$

62. $t - 7.4 = -12.9$

63. $x - 4 = -19$

64. $y - 6 = -14$

65. $3 + t = 21$

66. $9 + t = 3$

67. $-12x = 72$

68. $-15x = 105$

69. $48 = -\dfrac{3}{8}y$

70. $14 = t + 27$

71. $a - \dfrac{1}{6} = -\dfrac{2}{3}$

72. $-\dfrac{x}{7} = \dfrac{2}{9}$

73. $-24 = \dfrac{8x}{5}$

74. $\dfrac{1}{5} + y = -\dfrac{3}{10}$

75. $-\dfrac{4}{3}t = -16$

76. $\dfrac{17}{35} = -x$

▦ **77.** $-483.297 = -794.053 + t$

▦ **78.** $-0.2344x = 2028.732$

™ **79.** When solving an equation, how do you determine what number to add, subtract, multiply, or divide by on both sides of that equation?

™ **80.** What is the difference between equivalent expressions and equivalent equations?

Skill Maintenance

Simplify. [1.8]

81. $9 - 2 \cdot 5^2 + 7$

82. $10 \div 2 \cdot 3^2 - 4$

83. $16 \div (2 - 3 \cdot 2) + 5$

84. $12 - 5 \cdot 2^3 + 4 \cdot 3$

Synthesis

™ **85.** To solve $-3.5 = 14t$, Anita adds 3.5 to both sides. Will this form an equivalent equation? Will it help solve the equation? Explain.

™ **86.** Explain why it is not necessary to state a subtraction principle: For any real numbers a, b, and c, $a = b$ is equivalent to $a - c = b - c$.

Solve for x. Assume a, c, m ≠ 0.

87. $mx = 9.4m$

88. $x - 4 + a = a$

89. $cx + 5c = 7c$

90. $c \cdot \dfrac{21}{a} = \dfrac{7cx}{2a}$

91. $7 + |x| = 20$

92. $ax - 3a = 5a$

93. If $t - 3590 = 1820$, find $t + 3590$.

94. If $n + 268 = 124$, find $n - 268$.

▦ **95.** Lydia makes a calculation and gets an answer of 22.5. On the last step, she multiplies by 0.3 when she should have divided by 0.3. What should the correct answer be?

™ **96.** Are the equations $x = 5$ and $x^2 = 25$ equivalent? Why or why not?

2.2

Using the Principles Together

Applying Both Principles   ▨   Combining Like Terms   ▨
Clearing Fractions and Decimals   ▨   Contradictions and Identities

An important strategy for solving new problems is to find a way to make a new problem look like a problem that we already know how to solve. This is precisely the approach taken in this section. You will find that the last steps of the examples in this section are nearly identical to the steps used for solving the examples of Section 2.1. In the early steps of each example, we solve equations by using both principles together and by using the commutative, associative, and distributive laws to write equivalent expressions.

Applying Both Principles

The addition and multiplication principles, along with the laws discussed in Chapter 1, are our tools for solving equations. In this section, we will find that the sequence and manner in which these tools are used is especially important.

EXAMPLE 1   Solve: $5 + 3x = 17$.

SOLUTION   Were we to evaluate $5 + 3x$, the rules for the order of operations direct us to *first* multiply by 3 and *then* add 5. Because of this, we can isolate $3x$ and then x by reversing these operations: We first subtract 5 from both sides and then divide both sides by 3. Our goal is an equivalent equation of the form $x = a$.

$$5 + 3x = 17$$

$$5 + 3x - 5 = 17 - 5$$ Using the addition principle: subtracting 5 from both sides (adding -5)

$$5 + (-5) + 3x = 12$$ Using a commutative law. Try to perform this step mentally.

First isolate the x-term.

$$3x = 12$$ Simplifying

$$\frac{3x}{3} = \frac{12}{3}$$ Using the multiplication principle: dividing both sides by 3 (multiplying by $\frac{1}{3}$)

Then isolate x.

$$x = 4.$$ Simplifying

Check:

$$\begin{array}{c|c} 5 + 3x = 17 \\ \hline 5 + 3 \cdot 4 & 17 \\ 5 + 12 & \end{array}$$ We use the rules for order of operations: Find the product, $3 \cdot 4$, and then add.

$$17 \stackrel{?}{=} 17 \quad \text{TRUE}$$

The solution is 4.

EXAMPLE 2   Solve: $\frac{4}{3}x - 7 = 1$.

SOLUTION   In $\frac{4}{3}x - 7$, we multiply first and then subtract. To reverse these steps, we first add 7 and then either divide by $\frac{4}{3}$ or multiply by $\frac{3}{4}$.

$$\frac{4}{3}x - 7 = 1$$

$$\frac{4}{3}x - 7 + 7 = 1 + 7$$ Adding 7 to both sides

$$\frac{4}{3}x = 8$$

$$\frac{3}{4} \cdot \frac{4}{3}x = \frac{3}{4} \cdot 8$$ Multiplying both sides by $\frac{3}{4}$

$$\left. \begin{array}{c} 1 \cdot x = \dfrac{3 \cdot \cancel{4} \cdot 2}{\cancel{4}} \\ \\ x = 6 \end{array} \right\}$$ Simplifying

4/3*6 − 7

1

This time we check using a calculator. We replace *x* with 6 and evaluate the expression on the right side of the equation, as shown at left. Since $\frac{4}{3} \cdot 6 - 7 = 1$, the answer checks. The solution is 6.

EXAMPLE 3   Solve: $45 - t = 13$.

SOLUTION   We have

$$45 - t = 13$$
$$45 - t - 45 = 13 - 45 \qquad \text{Subtracting 45 from both sides}$$
$$\left. \begin{array}{l} 45 + (-t) + (-45) = 13 - 45 \\ 45 + (-45) + (-t) = 13 - 45 \end{array} \right\} \qquad \text{Try to do these steps mentally.}$$
$$-t = -32 \qquad \text{Try to go directly to this step.}$$
$$(-1)(-t) = (-1)(-32) \qquad \begin{array}{l} \text{Multiplying both sides by } -1. \\ \text{(Dividing by } -1 \text{ would also work.)} \end{array}$$
$$t = 32.$$

Check:
$$\begin{array}{c|c} 45 - t = 13 \\ \hline 45 - 32 & 13 \\ 13 \overset{?}{=} 13 & \text{TRUE} \end{array}$$

The solution is 32.

As our skills improve, many of the steps can be streamlined.

EXAMPLE 4   Solve: $16.3 - 7.2y = -8.18$.

SOLUTION   We have

$$16.3 - 7.2y = -8.18$$
$$16.3 - 7.2y - 16.3 = -8.18 - 16.3 \qquad \begin{array}{l} \text{Subtracting 16.3 from} \\ \text{both sides} \end{array}$$
$$-7.2y = -24.48 \qquad \text{Simplifying}$$
$$\frac{-7.2y}{-7.2} = \frac{-24.48}{-7.2} \qquad \text{Dividing both sides by } -7.2$$
$$y = 3.4. \qquad \text{Simplifying}$$

Check:
$$\begin{array}{c|c} 16.3 - 7.2y = -8.18 \\ \hline 16.3 - 7.2\,(3.4) & -8.18 \\ 16.3 - 24.48 & \\ -8.18 \overset{?}{=} -8.18 & \text{TRUE} \end{array}$$

The solution is 3.4.

Combining Like Terms

If like terms appear on the same side of an equation, we combine them and then solve. Should like terms appear on both sides of an equation, we can use the addition principle to rewrite all like terms on one side.

EXAMPLE 5 Solve.

a) $3x + 4x = -14$ **b)** $2x - 4 = -3x + 1$
c) $6x + 5 - 7x = 10 - 4x + 7$ **d)** $2 - 5(x + 5) = 3(x - 2) - 1$

SOLUTION

a) $3x + 4x = -14$

$$7x = -14 \qquad \text{Combining like terms}$$

$$\frac{7x}{7} = \frac{-14}{7} \qquad \text{Dividing both sides by 7}$$

$$x = -2$$

The check is left to the student. The solution is -2.

b) To solve $2x - 4 = -3x + 1$, we must first write only variable terms on one side and only constant terms on the other. This can be done by adding 4 to both sides, to get all constant terms on the right, and $3x$ to both sides, to get all variable terms on the left. We can add 4 first, or $3x$ first, or add both in one step.

$$2x - 4 = -3x + 1$$

$$2x - 4 + 4 = -3x + 1 + 4 \qquad \text{Adding 4 to both sides}$$

> Isolate variable terms on one side and constant terms on the other side.

$$2x = -3x + 5 \qquad \text{Simplifying}$$

$$2x + 3x = -3x + 3x + 5 \qquad \text{Adding } 3x \text{ to both sides}$$

$$5x = 5 \qquad \text{Combining like terms and simplifying}$$

$$\frac{5x}{5} = \frac{5}{5} \qquad \text{Dividing both sides by 5}$$

$$x = 1 \qquad \text{Simplifying}$$

Check:

$$\begin{array}{c|c} \multicolumn{2}{c}{2x - 4 = -3x + 1} \\ \hline 2 \cdot 1 - 4 & -3 \cdot 1 + 1 \\ 2 - 4 & -3 + 1 \\ -2 \overset{?}{=} -2 & \text{TRUE} \end{array}$$

The solution is 1.

c) $6x + 5 - 7x = 10 - 4x + 7$

$$-x + 5 = 17 - 4x \qquad \text{Combining like terms on both sides}$$

$$-x + 5 + 4x = 17 - 4x + 4x \qquad \text{Adding } 4x \text{ to both sides}$$

$$5 + 3x = 17 \qquad \text{Simplifying. This is identical to Example 1.}$$

$$3x = 12 \qquad \text{Subtracting 5 from both sides and simplifying}$$

$$x = 4 \qquad \text{Dividing both sides by 3 and simplifying}$$

```
6(4)+5−7(4)
                    1
10−4(4)+7
                    1
```

To check with a calculator, we evaluate the expressions on both sides of the equation for $x = 4$. The answer checks if both expressions have the same value. We see from the figure at left that 4 checks. The solution is 4.

d) $2 - 5(x + 5) = 3(x - 2) - 1$

$2 - 5x - 25 = 3x - 6 - 1$	Using the distributive law. This is now similar to part (c) above.
$-5x - 23 = 3x - 7$	Combining like terms on both sides
$-5x - 23 + 7 = 3x - 7 + 7$	Adding 7 to both sides
$-5x - 16 = 3x$	Simplifying
$-5x - 16 + 5x = 3x + 5x$	Adding $5x$ to both sides
$-16 = 8x$	Simplifying
$\dfrac{-16}{8} = \dfrac{8x}{8}$	Dividing both sides by 8
$-2 = x$	

The student can confirm that -2 checks and is the solution.

Clearing Fractions and Decimals

Equations are generally easier to solve when they do not contain fractions or decimals. The multiplication principle can be used to "clear" fractions or decimals, as shown here.

Clearing Fractions	Clearing Decimals
$\frac{1}{2}x + 5 = \frac{3}{4}$	$2.3x + 7 = 5.4$
$4\left(\frac{1}{2}x + 5\right) = 4 \cdot \frac{3}{4}$	$10(2.3x + 7) = 10 \cdot 5.4$
$2x + 20 = 3$	$23x + 70 = 54$

In each case, the resulting equation is equivalent to the original equation, but easier to solve.

An Equation-Solving Procedure

1. Use the multiplication principle to clear any fractions or decimals. (This is optional, but can ease computations. See Examples 6 and 7.)
2. If necessary, use the distributive law to remove parentheses. Then combine like terms on each side. (See Example 5.)
3. Use the addition principle, as needed, to get all variable terms on one side and all constant terms on the other. Then combine like terms. (See Examples 1–7.)
4. Multiply or divide to solve for the variable, using the multiplication principle. (See Examples 1–7.)
5. Check all possible solutions in the original equation. (See Examples 1–5.)

The easiest way to clear an equation of fractions is to multiply *both sides* of the equation by the smallest, or *least*, common denominator.

EXAMPLE 6   Solve: **(a)** $\frac{2}{3}x - \frac{1}{6} = 2x$; **(b)** $\frac{2}{5}(3x + 2) = 8$.

SOLUTION

a) We multiply both sides by 6, the least common denominator.

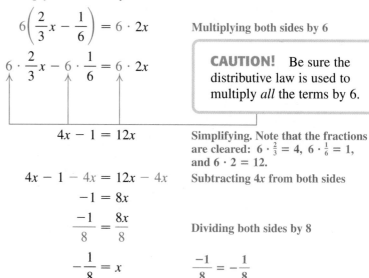

$$6\left(\frac{2}{3}x - \frac{1}{6}\right) = 6 \cdot 2x$$ **Multiplying both sides by 6**

$$6 \cdot \frac{2}{3}x - 6 \cdot \frac{1}{6} = 6 \cdot 2x$$

> **CAUTION!** Be sure the distributive law is used to multiply *all* the terms by 6.

$$4x - 1 = 12x$$ **Simplifying. Note that the fractions are cleared:** $6 \cdot \frac{2}{3} = 4$, $6 \cdot \frac{1}{6} = 1$, **and** $6 \cdot 2 = 12$.

$$4x - 1 - 4x = 12x - 4x$$ **Subtracting 4x from both sides**

$$-1 = 8x$$

$$\frac{-1}{8} = \frac{8x}{8}$$ **Dividing both sides by 8**

$$-\frac{1}{8} = x$$ $\dfrac{-1}{8} = -\dfrac{1}{8}$

The student can confirm that $-\frac{1}{8}$ checks and is the solution.

b) To solve $\frac{2}{5}(3x + 2) = 8$, we can multiply both sides by $\frac{5}{2}$ $\left(\text{or divide by } \frac{2}{5}\right)$ to "undo" the multiplication by $\frac{2}{5}$ on the left side.

$$\frac{5}{2} \cdot \frac{2}{5}(3x + 2) = \frac{5}{2} \cdot 8$$ **Multiplying both sides by** $\frac{5}{2}$

$$3x + 2 = 20$$ **Simplifying;** $\frac{5}{2} \cdot \frac{2}{5} = 1$ **and** $\frac{5}{2} \cdot \frac{8}{1} = 20$

$$3x + 2 - 2 = 20 - 2$$ **Subtracting 2 from both sides**

$$3x = 18$$

$$\frac{3x}{3} = \frac{18}{3}$$ **Dividing both sides by 3**

$$x = 6$$

The student can confirm that 6 checks and is the solution.

To clear an equation of decimals, we count the greatest number of decimal places in any one number. If the greatest number of decimal places is 1, we multiply both sides by 10; if it is 2, we multiply by 100; and so on.

Student Notes

Compare the steps of Examples 4 and 7. Note that although the two approaches differ, they yield the same solution. Whenever you can use two approaches to solve a problem, try to do so, both as a check and as a valuable learning experience.

EXAMPLE 7   Solve: $16.3 - 7.2y = -8.18$.

SOLUTION   The greatest number of decimal places in any one number is *two*. Multiplying by 100 will clear all decimals.

$$100(16.3 - 7.2y) = 100(-8.18)$$ Multiplying both sides by 100

$$100(16.3) - 100(7.2y) = 100(-8.18)$$ Using the distributive law

$$1630 - 720y = -818$$ Simplifying

$$1630 - 720y - 1630 = -818 - 1630$$ Subtracting 1630 from both sides

$$-720y = -2448$$ Combining like terms

$$\frac{-720y}{-720} = \frac{-2448}{-720}$$ Dividing both sides by -720

$$y = 3.4$$

In Example 4, the same solution was found without clearing decimals. Finding the same answer in two ways is a good check. The solution is 3.4.

Contradictions and Identities

All the equations we have examined so far had a solution. Equations that are true for some values (solutions), but not for others, are called **conditional equations.** Equations that have no solution, such as $x + 1 = x + 2$, are called **contradictions.** If, when solving an equation, we obtain an equation that is false for any value of x, the equation has no solution.

EXAMPLE 8   Solve: $3x - 5 = 3(x - 2) + 4$.

SOLUTION

$$3x - 5 = 3(x - 2) + 4$$

$$3x - 5 = 3x - 6 + 4$$ Using the distributive law

$$3x - 5 = 3x - 2$$ Combining like terms

$$-3x + 3x - 5 = -3x + 3x - 2$$ Using the addition principle

$$-5 = -2$$

Since the original equation is equivalent to $-5 = -2$, which is false for any choice of x, the original equation has no solution. There is no choice of x that will make $3x - 5 = 3(x - 2) + 4$ true. The equation is a contradiction. It is *never* true.

Some equations, like $x + 1 = x + 1$, are true for all replacements. Such an equation is called an **identity.**

EXAMPLE 9 Solve: $2x + 7 = 7(x + 1) - 5x$.

SOLUTION

$$2x + 7 = 7(x + 1) - 5x$$
$$2x + 7 = 7x + 7 - 5x \qquad \text{Using the distributive law}$$
$$2x + 7 = 2x + 7 \qquad \text{Combining like terms}$$

The equation $2x + 7 = 2x + 7$ is true regardless of the replacement for x, so all real numbers are solutions. Note that $2x + 7 = 2x + 7$ is equivalent to $2x = 2x$, $7 = 7$, or $0 = 0$. All real numbers are solutions and the equation is an identity.

In Sections 2.1 and 2.2, we have solved *linear equations.* A **linear equation** in one variable—say, x—is an equation equivalent to one of the form $ax = b$ with a and b constants and $a \neq 0$.

We will sometimes refer to the set of solutions, or **solution set,** of a particular equation. Thus the solution set for Example 7 is $\{3.4\}$. The solution set for Example 9 is simply $\mathbb{R}$, the set of all real numbers, and the solution set for Example 8 is the **empty set,** denoted $\varnothing$ or $\{\ \}$. As its name suggests, the empty set is the set containing no elements.

2.2 EXERCISE SET

FOR EXTRA HELP

MathXL | MyMathLab | InterAct Math | AW Math Tutor Center | Video Lectures on CD: Disc 1 | Student's Solutions Manual

↩ *Concept Reinforcement In each of Exercises 1–6, match the equation with an equivalent equation from the column on the right that could be the next step in finding a solution.*

1. _____ $3x - 1 = 7$

2. _____ $4x + 5x = 12$

3. _____ $6(x - 1) = 2$

4. _____ $7x = 9$

5. _____ $4x = 3 - 2x$

6. _____ $8x - 5 = 6 - 2x$

a) $6x - 6 = 2$

b) $4x + 2x = 3$

c) $3x = 7 + 1$

d) $8x + 2x = 6 + 5$

e) $9x = 12$

f) $x = \frac{9}{7}$

Solve and check. Label any contradictions or identities.

7. $2x + 9 = 25$

8. $3x + 6 = 30$

9. $6z + 4 = 46$

10. $6z + 3 = 57$

11. $7t - 8 = 27$

12. $6x - 3 = 15$

13. $3x - 9 = 33$

14. $5x - 9 = 41$

15. $8z + 2 = -54$

16. $4x + 3 = -21$

17. $-91 = 9t + 8$

18. $-39 = 1 + 8x$

19. $12 - 4x = 108$

20. $9 - 4x = 37$

21. $-6z - 18 = -132$

22. $-7x - 24 = -129$

23. $4x + 5x = 10$

24. $13 = 5x + 7x$

25. $32 - 7x = 11$

26. $27 - 6x = 99$

27. $\frac{3}{5}t - 1 = 8$

28. $\frac{2}{3}t - 1 = 5$

29. $4 + \frac{7}{2}x = -10$

30. $6 + \frac{5}{4}x = -4$

31. $-\dfrac{3a}{4} - 5 = 2$

32. $-\dfrac{7a}{8} - 2 = 1$

33. $2x = x + x$

34. $-3z + 8z = 45$

35. $4x - 6 = 6x$

36. $4x - x = 2x + x$

37. $5y - 2 = 28 - y$

38. $6x - 5 = 7 + 2x$

39. $7(2a - 1) = 21$

40. $5(2t - 2) = 30$

Aha! **41.** $8 = 8(x + 1)$

42. $9 = 3(5x - 2)$

43. $7r - (2r + 8) = 32$

44. $3(5 + 3m) - 8 = 88$

45. $6x + 3 = 2x + 3$

46. $6b - (3b + 8) = 16$

47. $5 - 2x = 3x - 7x + 25$

48. $10 - 3x = 2x - 8x + 40$

49. $7 + 3x - 6 = 3x + 5 - x$

50. $5 + 4x - 7 = 4x - 2 - x$

51. $4y - 4 + y + 24 = 6y + 20 - 4y$

52. $5y - 10 + y = 7y + 18 - 5y$

53. $13 - 3(2x - 1) = 4$

54. $5(d + 4) = 7(d - 2)$

55. $7(5x - 2) = 6(6x - 1)$

56. $5(t + 3) + 9 = 3(t - 2) + 6$

57. $19 - (2x + 3) = 2(x + 3) + x$

58. $13 - (2c + 2) = 2(c + 2) + 3c$

59. $3(x + 4) = 3(x - 1)$

60. $5(x - 7) = 3(x - 2) + 2x$

Clear fractions or decimals, solve, and check.

61. $\frac{5}{4}x + \frac{1}{4}x = 2x + \frac{1}{2} + \frac{3}{4}x$ **62.** $\frac{7}{8}x - \frac{1}{4} + \frac{3}{4}x = \frac{1}{16} + x$

63. $\frac{2}{3} + \frac{1}{4}t = 6$

64. $-\frac{1}{2} + x = -\frac{5}{6} - \frac{1}{3}$

65. $\frac{2}{3} + 4t = 6t - \frac{2}{15}$

66. $\frac{1}{2} + 4m = 3m - \frac{5}{2}$

67. $\frac{1}{3}x + \frac{2}{5} = \frac{4}{15} + \frac{3}{5}x - \frac{2}{3}$ **68.** $1 - \frac{2}{3}y = \frac{9}{5} - \frac{1}{5}y + \frac{3}{5}$

69. $2.1x + 45.2 = 3.2 - 8.4x$

70. $0.91 - 0.2z = 1.23 - 0.6z$

71. $0.76 + 0.21t = 0.96t - 0.49$

72. $1.7t + 8 - 1.62t = 0.4t - 0.32 + 8$

73. $\frac{2}{5}x - \frac{3}{2}x = \frac{3}{4}x + 2$ **74.** $\frac{5}{16}y + \frac{3}{8}y = 2 + \frac{1}{4}y$

75. $\frac{1}{3}(2x - 1) = 7$ **76.** $\frac{4}{3}(5x + 1) = 8$

77. $\frac{3}{4}(3t - 6) = 9$ **78.** $\frac{3}{2}(2x + 5) = -\frac{15}{2}$

79. $\frac{1}{6}\left(\frac{3}{4}x - 2\right) = -\frac{1}{5}$ **80.** $\frac{2}{3}\left(\frac{7}{8} - 4x\right) - \frac{5}{8} = \frac{3}{8}$

81. $0.7(3x + 6) = 1.1 - (x + 2)$

82. $0.9(2x + 8) = 20 - (x + 5)$

83. $a + (a - 3) = (a + 2) - (a + 1)$

84. $0.8 - 4(b - 1) = 0.2 + 3(4 - b)$

TW **85.** When an equation contains decimals, is it essential to clear the equation of decimals? Why or why not?

TW **86.** Why must the rules for the order of operations be understood before solving the equations in this section?

Skill Maintenance

Evaluate. [1.8]

87. $3 - 5a$, for $a = 2$ **88.** $12 \div 4 \cdot t$, for $t = 5$

89. $7x - 2x$, for $x = -3$ **90.** $t(8 - 3t)$, for $t = -2$

Synthesis

TW **91.** What procedure would you follow to solve an equation like $0.23x + \frac{17}{3} = -0.8 + \frac{3}{4}x$? Could your procedure be streamlined? If so, how?

TW **92.** Dave is determined to solve the equation $3x + 4 = -11$ by first using the multiplication principle to "eliminate" the 3. How should he proceed and why?

Solve. Label any contradictions or identities.

93. $8.43x - 2.5(3.2 - 0.7x) = -3.455x + 9.04$

94. $0.008 + 9.62x - 42.8 = 0.944x + 0.0083 - x$

95. $-2[3(x - 2) + 4] = 4(5 - x) - 2x$

96. $0 = y - (-14) - (-3y)$

97. $2|x| = -14$

98. $|3x| = 6$

99. $2x(x + 5) - 3(x^2 + 2x - 1) = 9 - 5x - x^2$

100. $x(x - 4) = 3x(x + 1) - 2(x^2 + x - 5)$

101. $9 - 3x = 2(5 - 2x) - (1 - 5x)$

102. $2(7 - x) - 20 = 7x - 3(2 + 3x)$

Aha! **103.** $[7 - 2(8 \div (-2))]x = 0$

104. $\dfrac{x}{14} - \dfrac{5x + 2}{49} = \dfrac{3x - 4}{7}$

105. $\dfrac{5x + 3}{4} + \dfrac{25}{12} = \dfrac{5 + 2x}{3}$

Collaborative Corner

Step-by-Step Solutions

Focus: Solving linear equations
Time: 20 minutes
Group size: 3

In general, there is more than one correct sequence of steps for solving an equation. This makes it important that you write your steps clearly and logically so that others can follow your approach.

ACTIVITY

1. Each group member should select a different one of the following equations and, on a fresh sheet of paper, perform the first step of the solution.

$$4 - 3(x - 3) = 7x + 6(2 - x)$$
$$5 - 7[x - 2(x - 6)] = 3x + 4(2x - 7) + 9$$
$$4x - 7[2 + 3(x - 5) + x] = 4 - 9(-3x - 19)$$

2. Pass the papers around so that the second and third steps of each solution are performed by the other two group members. Before writing, make sure that the previous step is correct. If a mistake is discovered, return the problem to the person who made the mistake for repairs. Continue passing the problems around until all equations have been solved.

3. Each group should reach a consensus on what the three solutions are and then compare their answers to those of other groups.

2.3 Formulas

Evaluating Formulas ■ Solving for a Letter

Many applications of mathematics involve relationships among two or more quantities. An equation that represents such a relationship will use two or more letters and is known as a **formula.** Although most of the letters in this book represent variables, some—like c in $E = mc^2$ or π in $C = \pi d$—represent constants.

Evaluating Formulas

EXAMPLE 1 Outdoor Concerts. The formula $d = 344t$ can be used to determine how far d, in meters, sound travels through room-temperature air in one second. In 2003, the Dave Matthews Band performed in New York City's Central Park. Fans near the back of the crowd experienced a 0.9-sec time lag between the time each word was pronounced on stage (as shown on large video monitors) and the time the sound reached their ears. How far were these fans from the stage?

SOLUTION We substitute 0.9 for t in $d = 344t$ and calculate d:

$$d = 344(0.9) = 309.6.$$

The fans were about 309.6 m from the stage.

Tables and Entering Equations

A formula can be evaluated for several values of a variable using the TABLE feature of a graphing calculator. In order to use the table, the formula must be entered using the Y= editor.

Y= Editor   To enter equations in a graphing calculator, first press **MODE** and make sure that the FUNC mode is selected. When the calculator is set in FUNCTION mode, pressing ⟨ Y= ⟩ accesses the Y= editor, as shown below. Formulas and other equations containing two variables can be entered into the calculator using this editor. The formula must be solved for a particular variable. We replace this variable with y and the other variable with x. To enter the formula, position the cursor after an = sign and enter the rest of the formula. Other formulas or equations can be entered using the other lines on the screen. They are distinguished by the subscripts 1, 2, and so on. The notation Y1 is read "y sub one."

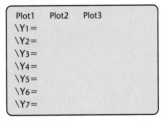

Table   After we enter an equation on the Y= editor screen, we can view a table of solutions. A table is set up by pressing ⟨ TBLSET ⟩ (the 2nd option associated with ⟨WINDOW⟩). Since the value of y *depends on* the choice of the value for x, we say that y is the **dependent** variable and x is the **independent** variable.

If we want to choose the values for the independent variable, we set Indpnt to Ask, as shown on the left below.

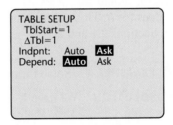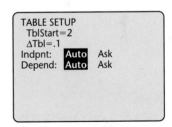

If Indpnt is set to Auto, the calculator will provide values for x, beginning with the value specified as TblStart. The symbol Δ (the Greek letter delta) often indicates a step or a change. Thus the value of ΔTbl is added to the preceding value of x. The figure on the right above shows a beginning value of 2 and an increment, or ΔTbl, of 0.1.

To view the values in a table, press ⟨ TABLE ⟩ (the 2nd option associated with ⟨ GRAPH ⟩). The up and down arrow keys allow us to scroll up and down the table.

EXAMPLE 2   Use the formula $d = 344t$, described in Example 1, to determine how far fans are from a stage when the time lag is 0.5 sec, 0.7 sec, and 1 sec.

SOLUTION   We replace d with y and t with x and enter the formula as $y = 344x$, as shown on the left below. Then we set up a table, with Indpnt set to Ask, and view the table. For the values of x, we enter 0.5, 0.7, and 1. As shown on the right below, the distances from the stage are 172 m, 240.8 m, and 344 m, respectively.

Plot1	Plot2	Plot3
\Y1█ 344X		
\Y2=		
\Y3=		
\Y4=		
\Y5=		
\Y6=		
\Y7=		

X	Y1	
0.5	172	
0.7	240.8	
1	344	
X =		

Solving for a Letter

In the Northeast, the formula $B = 30a$ is used to determine the minimum furnace output B, in British thermal units (Btu's), for a well-insulated home with a square feet of flooring. Suppose that a contractor has an extra furnace and wants to determine the size of the largest (well-insulated) house in which it can be used. The contractor can substitute the amount of the furnace's output in Btu's—say, 63,000—for B, and then solve for a:

$$63{,}000 = 30a \qquad \text{Replacing } B \text{ with 63,000}$$
$$2100 = a. \qquad \text{Dividing both sides by 30}$$

The home should have no more than 2100 ft^2 of flooring.

Were these calculations to be performed for a variety of furnaces, the contractor would find it easier to first solve $B = 30a$ for a, and *then* substitute values for B. This can be done in much the same way that we solved equations in Sections 2.1 and 2.2.

EXAMPLE 3   Solve for a: $B = 30a$.

SOLUTION   We have

$$B = 30a \qquad \text{We want this letter alone.}$$
$$\frac{B}{30} = a. \qquad \text{Dividing both sides by 30}$$

The equation $a = \dfrac{B}{30}$ gives a quick, easy way to determine the floor area of the largest (well-insulated) house that a furnace supplying B Btu's could heat.

To see how solving a formula is just like solving an equation, compare the following. In (A), we solve as usual; in (B), we show steps but do not simplify; and in (C), we *cannot* simplify since *a*, *b*, and *c* are unknown.

A. $5x + 2 = 12$

$5x = 12 - 2$

$5x = 10$

$x = \dfrac{10}{5} = 2$

B. $5x + 2 = 12$

$5x = 12 - 2$

$x = \dfrac{12 - 2}{5}$

C. $ax + b = c$

$ax = c - b$

$x = \dfrac{c - b}{a}$

▶ **EXAMPLE 4** Circumference of a Circle. The formula $C = 2\pi r$ gives the *circumference C* of a circle with radius *r*. Solve for *r*.

SOLUTION The **circumference** is the distance around a circle.

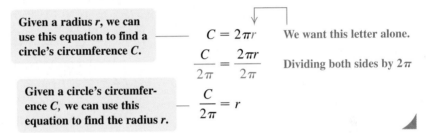

Given a radius r, we can use this equation to find a circle's circumference C.

$C = 2\pi r$ We want this letter alone.

$\dfrac{C}{2\pi} = \dfrac{2\pi r}{2\pi}$ Dividing both sides by 2π

Given a circle's circumference C, we can use this equation to find the radius r.

$\dfrac{C}{2\pi} = r$

For some applications, we may want to see a large table of values of a formula. A graphing calculator will generate such a table when the independent variable is set to Auto.

▶ **EXAMPLE 5** Use the formula $r = \dfrac{C}{2\pi}$, developed in Example 4, to create a table of values showing the radius of a circle given its circumference. Show values of the radius for circumferences of 1, 1.1, 1.2, and so on.

SOLUTION We first enter the formula, replacing *r* with *y* and *C* with *x*. The formula becomes $y = x/(2\pi)$.

Before creating the table, we set the independent variable to Auto. We will set the start of the *x*-values in the table to 1 (indicating a circumference of 1), and then set ΔTbl to 0.1. Once the setup is complete, we create the table of values by pressing (TABLE).

```
TABLE SETUP
  TblStart = 1
  ΔTbl = .01
Indpnt:  Auto  Ask
Depend:  Auto  Ask
```

X	Y1
1	.15915
1.1	.17507
1.2	.19099
1.3	.2069
1.4	.22282
1.5	.23873
1.6	.25465

Y1 = .159154943092

Note that values for Y_1 in the table are rounded to 5 decimal places. To see more digits for a value, we highlight that value in the table. We can now use the up and down arrow keys to scroll through the table to find the radius for other values of a circumference. ◢

EXAMPLE 6 Concert Promotions. Promoter Gene Shelton of DJGENO. com uses the formula $P = \dfrac{1.2X}{Z}$ to calculate a ticket price P for a concert with X dollars of expenses and Z predicted ticket sales (*Source: The Indianapolis Star*: 2/23/03). Solve for Z.

SOLUTION We use the multiplication principle to clear fractions and then solve for Z.

$$P = \frac{1.2X}{Z} \qquad \text{We want this letter alone.}$$

$$P \cdot Z = \frac{1.2X}{Z} \cdot Z \qquad \text{Multiplying both sides by } Z$$

$$PZ = \frac{1.2XZ}{Z} \qquad \frac{1.2X}{Z} \cdot Z = \frac{1.2X}{Z} \cdot \frac{Z}{1} = \frac{1.2XZ}{Z}$$

$$PZ = 1.2X \qquad \text{Removing a factor equal to 1: } \frac{Z}{Z} = 1.$$

$$\qquad\qquad\qquad \text{The equation is cleared of fractions.}$$

$$\frac{PZ}{P} = \frac{1.2X}{P} \qquad \text{Dividing both sides by } P$$

$$Z = \frac{1.2X}{P}$$

This formula can be used to estimate the predicted ticket sales when the expenses and the ticket price are known. ◢

EXAMPLE 7 Nutrition. The number of calories K needed each day by a moderately active woman who weighs w pounds, is h inches tall, and is a years old can be estimated using the formula

$$K = 917 + 6(w + h - a).*$$

Solve for w.

---

*Based on information from M. Parker (ed.), *She Does Math!* (Washington D.C.: Mathematical Association of America, 1995), p. 96.

SOLUTION   We reverse the order in which the operations occur on the right side:

We want this letter alone.

$$K = 917 + 6(w + h - a)$$

$$K - 917 = 6(w + h - a) \qquad \text{Subtracting 917 from both sides}$$

$$\frac{K - 917}{6} = w + h - a \qquad \text{Dividing both sides by 6}$$

$$\frac{K - 917}{6} + a - h = w. \qquad \text{Adding } a \text{ and subtracting } h \text{ on both sides}$$

This formula can be used to estimate a woman's weight, if we know her age, height, and caloric needs.

The above steps are similar to those used in Section 2.2 to solve equations. We use the addition and multiplication principles just as before. An important difference that we will see in the next example is that we will sometimes need to factor.

To Solve a Formula for a Given Letter

1. If the letter for which you are solving appears in a fraction, use the multiplication principle to clear fractions.
2. Isolate the term(s), with the letter you are solving for on one side of the equation.
3. If two or more terms contain the letter you are solving for, factor the letter out.
4. Multiply or divide to solve for the letter in question.

EXAMPLE 8   Solve for x: $y = ax + bx - 4$.

SOLUTION   We solve as follows:

$$y = ax + bx - 4 \qquad \text{We want this letter alone.}$$

$$y + 4 = ax + bx \qquad \text{Adding 4 to both sides}$$

$$y + 4 = x(a + b) \qquad \text{Using the distributive law to factor}$$

$$\frac{y + 4}{a + b} = x. \qquad \text{Dividing both sides by } a + b, \text{ or multiplying both sides by } 1/(a + b)$$

We can also write this as

$$x = \frac{y + 4}{a + b}.$$

Study Tip

How Did They Get That!?

The *Student's Solutions Manual* is an excellent resource if you need additional help with an exercise in the exercise sets. It contains step-by-step solutions to the odd-numbered exercises in each exercise set.

CAUTION! Had we performed the following steps in Example 8, we would *not* have solved for *x*:

$$y = ax + bx - 4$$

$$y - ax + 4 = bx \qquad \text{Subtracting } ax \text{ and adding 4 to both sides}$$

Two occurrences of *x*

$$\frac{y - ax + 4}{b} = x. \qquad \text{Dividing both sides by } b$$

The mathematics of each step is correct, but since *x* occurs on both sides of the formula, *we have not solved the formula for x*. Remember that the letter being solved for should be alone on one side of the equation, with no occurrence of that letter on the other side!

2.3 EXERCISE SET

FOR EXTRA HELP

MathXL · MyMathLab · InterAct Math · AW Math Tutor Center · Video Lectures on CD: Disc 1 · Student's Solutions Manual

1. *Distance from a Storm.* The formula $M = \frac{1}{5}t$ can be used to determine how far *M*, in miles, you are from lightning when its thunder takes *t* seconds to reach your ears. If it takes 10 sec for the sound of thunder to reach you after you have seen the lightning, how far away is the storm?

2. *Electrical Power.* The power rating *P*, in watts, of an electrical appliance is determined by

   $$P = I \cdot V,$$

   where *I* is the current, in amperes, and *V* is the voltage, measured in volts. If the appliances in a kitchen require 30 amps of current and the voltage in the house is 115 volts, what is the wattage of the kitchen?

3. *College Enrollment.* At many colleges, the number of "full-time-equivalent" students *f* is given by

   $$f = \frac{n}{15},$$

   where *n* is the total number of credits for which students have enrolled in a given semester.

Determine the number of full-time-equivalent students on a campus in which students registered for a total of 21,345 credits.

4. *Wavelength of a Musical Note.* The wavelength *w*, in meters per cycle, of a musical note is given by

   $$w = \frac{r}{f},$$

   where *r* is the speed of the sound, in meters per second, and *f* is the frequency, in cycles per second. The speed of sound in air is 344 m/sec. What is the wavelength of a note whose frequency in air is 24 cycles per second?

5. *Furnace Output.* Contractors in the Northeast use the formula $B = 30a$ to determine the minimum furnace output *B*, in British thermal units (Btu's), for a well-insulated house with *a* square feet of flooring (*Source*: U.S. Department of Energy). Determine the minimum furnace output for an 1800-ft² house that is well insulated.

6. *Calorie Density.* The calorie density D, in calories per ounce, of a food that contains c calories and weighs w ounces is given by

$$D = \frac{c}{w}.^*$$

Eight ounces of fat-free milk contains 84 calories. Find the calorie density of fat-free milk.

7. *Absorption of Ibuprofen.* When 400 mg of the painkiller ibuprofen is swallowed, the number of milligrams n in the bloodstream t hours later (for $0 \leq t \leq 6$) is estimated by

$$n = 0.5t^4 + 3.45t^3 - 96.65t^2 + 347.7t.$$

How many milligrams of ibuprofen remain in the blood 1 hr after 400 mg has been swallowed?

8. *Size of a League Schedule.* When all n teams in a league play every other team twice, a total of N games are played, where

$$N = n^2 - n.$$

If a soccer league has 7 teams and all teams play each other twice, how many games are played?

Solve each formula for the indicated letter.

9. $A = bh$, for b
(Area of parallelogram with base b and height h)

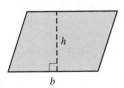

10. $A = bh$, for h

11. $d = rt$, for r
(Distance formula, where d is distance, r is speed, and t is time)

12. $d = rt$, for t

13. $I = Prt$, for P
(Simple-interest formula, where I is interest, P is principal, r is interest rate, and t is time)

14. $I = Prt$, for t

15. $H = 65 - m$, for m
(To determine the number of heating degree days H for a day with m degrees Fahrenheit as the average temperature)

16. $d = h - 64$, for h
(To determine how many inches d above average an h-inch-tall woman is)

17. $P = 2l + 2w$, for l
(Perimeter of a rectangle of length l and width w)

18. $P = 2l + 2w$, for w

19. $A = \pi r^2$, for π
(Area of a circle with radius r)

20. $A = \pi r^2$, for r^2

21. $A = \frac{1}{2}bh$, for h
(Area of a triangle with base b and height h)

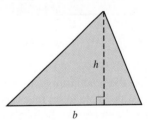

22. $A = \frac{1}{2}bh$, for b

23. $E = mc^2$, for m
(A relativity formula from physics)

24. $E = mc^2$, for c^2

25. $Q = \dfrac{c + d}{2}$, for d

26. $Q = \dfrac{p - q}{2}$, for p

27. $A = \dfrac{a + b + c}{3}$, for b

---

*Source: Nutrition Action Healthletter, March 2000, p. 9. Center for Science in the Public Interest, Suite 300; 1875 Connecticut Ave NW, Washington, D.C. 20008.

28. $A = \dfrac{a + b + c}{3}$, for c

29. $M = \dfrac{A}{s}$, for A

(To compute the Mach number M for speed A and speed of sound s)

30. $P = \dfrac{ab}{c}$, for b

31. $F = \dfrac{9}{5}C + 32$, for C

(To convert the Celsius temperature C to the Fahrenheit temperature F)

32. $M = \dfrac{3}{7}n + 29$, for n

33. $A = at + bt$, for t

34. $S = rx + sx$, for x

35. *Area of a Trapezoid.* The formula

$$A = \tfrac{1}{2}ah + \tfrac{1}{2}bh$$

can be used to find the area A of a trapezoid with bases a and b and height h. Solve for h. (*Hint:* First clear fractions.)

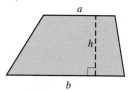

36. *Compounding Interest.* The formula

$$A = P + Prt$$

is used to find the amount A in an account when simple interest is added to an investment of P dollars (see Exercise 13). Solve for P.

37. *Chess Rating.* The formula

$$R = r + \dfrac{400(W - L)}{N}$$

is used to establish a chess player's rating R after that player has played N games, won W of them, and lost L of them. Here r is the average rating of the opponents (*Source:* The U.S. Chess Federation). Solve for L.

38. *Angle Measure.* The angle measure S, of a sector of a circle, is given by

$$S = \dfrac{360A}{\pi r^2},$$

where r is the radius, A is the area of the sector, and S is in degrees. Solve for r^2.

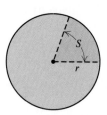

TW 39. Naomi has a formula that allows her to convert Celsius temperatures to Fahrenheit temperatures. She needs a formula for converting Fahrenheit temperatures to Celsius temperatures. What advice can you give her?

TW 40. Under what circumstances would it be useful to solve $d = rt$ for r? (See Exercise 11.)

Skill Maintenance

Multiply. [1.7]

Aha! 41. $0.79(38.4)0$

42. $(0.085)(108)$

Simplify. [1.8]

43. $20 \div (-4) \cdot 2 - 3$

44. $5|8 - (2 - 7)|$

Synthesis

TW 45. The equations

$$P = 2l + 2w \quad \text{and} \quad w = \dfrac{P}{2} - l$$

are equivalent formulas involving the perimeter P, length l, and width w of a rectangle. Devise a problem for which the second of the two formulas would be more useful.

TW 46. While solving $2A = ah + bh$ for h, Lea writes $\dfrac{2A - ah}{b} = h$. What is her mistake?

47. The number of calories K needed each day by a moderately active man who weighs w kilograms, is h centimeters tall, and is a years old, can be determined by

$$K = 19.18w + 7h - 9.52a + 92.4.*$$

If Janos is moderately active, weighs 82 kg, is 185 cm tall, and needs to consume 2627 calories a day, how old is he?

---

*Based on information from M. Parker (ed.), *She Does Math!* (Washington DC: Mathematical Association of America, 1995), p. 96.

48. *Altitude and Temperature.* Air temperature drops about 1° Celsius (C) for each 100-m rise above ground level, up to 12 km (*Source: A Sourcebook of School Mathematics*, Mathematical Association of America, 1980). If the ground level temperature is t°C, find a formula for the temperature T at an elevation of h meters.

49. *Surface Area of a Cube.* The surface area A of a cube with side s is given by

$$A = 6s^2.$$

If a cube's surface area is 54 in^2, find the volume of the cube.

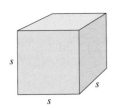

50. *Weight of a Fish.* An ancient fisherman's formula for estimating the weight of a fish is

$$w = \frac{lg^2}{800},$$

where w is the weight, in pounds, l is the length, in inches, and g is the girth (distance around the midsection), in inches. Estimate the girth of a 700-lb yellow tuna that is 8 ft long.

51. *Dosage Size.* Clark's rule for determining the size of a particular child's medicine dosage c is

$$c = \frac{w}{a} \cdot d,$$

where w is the child's weight, in pounds, and d is the usual adult dosage for an adult weighing a pounds (*Source:* Olsen, June Looby, et al., *Medical Dosage Calculations.* Redwood City, CA: Addison-Wesley, 1995). Solve for a.

Solve each formula for the given letter.

52. $\dfrac{y}{z} \div \dfrac{z}{t} = 1$, for y

53. $ac = bc + d$, for c

54. $qt = r(s + t)$, for t

55. $3a = c - a(b + d)$, for a

56. *Furnace Output.* The formula

$$B = 50a$$

is used in New England to estimate the minimum furnace output B, in Btu's, for an old, poorly insulated house with a square feet of flooring. Find an equation for determining the number of Btu's saved by insulating an old house. (*Hint:* See Exercise 5.)

57. Revise the formula in Example 7 so that a woman's weight in kilograms (2.2046 lb = 1 kg) and her height in centimeters (0.3937 in. = 1 cm) are used.

58. Revise the formula in Exercise 47 so that a man's weight in pounds (2.2046 lb = 1 kg) and his height in inches (0.3937 in. = 1 cm) are used.

2.4 Applications with Percent

Converting Between Percent Notation and Decimal Notation ■
Solving Percent Problems

Percent problems arise so frequently in everyday life that often we are not even aware of them. In this section, we will solve some real-world percent problems. Before doing so, however, we need to review a few basics.

Study Tip

Pace Yourself

Most instructors agree that it is better for a student to study for one hour four days in a week, than to study once a week for four hours. Of course, the total weekly study time will vary from student to student. It is common to expect an average of two hours of homework for each hour of class time.

Converting Between Percent Notation and Decimal Notation

Nutritionists recommend that no more than 30% of the calories in a person's diet come from fat. This means that of every 100 calories consumed, no more than 30 should come from fat. Thus, 30% is a ratio of 30 to 100.

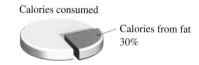

Calories consumed

Calories from fat 30%

The percent symbol % means "per hundred." We can regard the percent symbol as part of a name for a number. For example,

$$30\% \quad \text{is defined to mean} \quad \frac{30}{100}, \quad \text{or} \quad 30 \times \frac{1}{100}, \quad \text{or} \quad 30 \times 0.01.$$

Percent Notation

$$n\% \quad \text{means} \quad \frac{n}{100}, \quad \text{or} \quad n \times \frac{1}{100}, \quad \text{or} \quad n \times 0.01.$$

EXAMPLE 1 Convert to decimal notation: **(a)** 78%; **(b)** 1.3%.

SOLUTION

a) $78\% = 78 \times 0.01$ **Replacing % with ×0.01**

     $= 0.78$

b) $1.3\% = 1.3 \times 0.01$ **Replacing % with ×0.01**

     $= 0.013$

As shown above, multiplication by 0.01 simply moves the decimal point two places to the left.

To convert from percent notation to decimal notation, move the decimal point two places to the left and drop the percent symbol.

EXAMPLE 2 Convert the percent notation in the following sentence to decimal notation: 60% of the pollution from a typical car trip is emitted in the first few minutes (*Source*: Chittenden County Transportation Authority).

SOLUTION

$$60\% = 60.0\% \qquad 0.60.0 \qquad 60\% = 0.60, \text{ or simply } 0.6$$

Move the decimal point two places to the left.

The procedure used in Examples 1 and 2 can be reversed:

$$0.38 = 38 \times 0.01$$

$$= 38\%. \qquad \text{Replacing } \times 0.01 \text{ with } \%$$

To convert from decimal notation to percent notation, move the decimal point two places to the right and write a percent symbol.

EXAMPLE 3 Convert to percent notation: **(a)** 1.27; **(b)** $\frac{1}{4}$; **(c)** 0.3.

SOLUTION

a) We first move the decimal point
two places to the right: 1.27

and then write a % symbol: 127%   This is the same as multiplying 1.27 by 100 and writing %.

b) Note that $\frac{1}{4} = 0.25$. We move the
decimal point two places to the
right: 0.25

and then write a % symbol: 25%   Multiplying by 100 and writing %

c) We first move the decimal point
two places to the right (recall that
$0.3 = 0.30$):   0.30

and then write a % symbol: 30%   Multiplying by 100 and writing %

Solving Percent Problems

In solving percent problems, we first *translate* the problem to an equation. Then we *solve* the equation using the techniques discussed in Sections 2.1–2.3. The key words in the translation are as follows.

Key Words in Percent Translations

"**Of**" translates to " · " or " × ".   "**Is**" or "**Was**" translates to " = ".

"**What**" translates to a variable.   % translates to "$\times \frac{1}{100}$" or "$\times 0.01$".

Student Notes

A way of checking answers is by estimating as follows:

$$11\% \times 49 \approx 10\% \times 50$$

$$= 0.10 \times 50 = 5.$$

Since 5 is close to 5.39, our answer is reasonable.

EXAMPLE 4 What is 11% of 49?

SOLUTION

Translate: What is 11% of 49?

$$a \quad = \quad 0.11 \quad \cdot \quad 49 \qquad \text{"of" means multiply; } 11\% = 0.11$$

$$a = 5.39$$

Thus, 5.39 is 11% of 49. The answer is 5.39.

EXAMPLE 5   3 is 16 percent of what?

SOLUTION

Translate: 3   is   16 percent   of   what?
            ↓   ↓   ↓   ↓   ↓
            3   =   0.16   ·   y

$$\frac{3}{0.16} = y \qquad \textbf{Dividing both sides by 0.16}$$

$$18.75 = y$$

Thus, 3 is 16 percent of 18.75. The answer is 18.75.

EXAMPLE 6   What percent of $50 is $34?

SOLUTION

Translate: What percent   of $50   is $34?
                   ↓   ↓   ↓   ↓   ↓
                   n   ·   50   =   34

$$n = \frac{34}{50} \qquad \textbf{Dividing both sides by 50}$$

$$n = 0.68 = 68\% \qquad \textbf{Converting to percent notation}$$

Thus, 34 is 68% of 50. The answer is 68%.

Examples 4–6 represent the three basic types of percent problems.

EXAMPLE 7   *Coronary Heart Disease.*   In 2006, there were 299 million people in the United States. About 2.5% of them had heart disease (*Source*: American Heart Association). How many had heart disease?

SOLUTION   To solve the problem, we first reword and then translate. We let a = the number of people in the United States with heart disease.

   Rewording: What   is   2.5%   of   299?
                   ↓   ↓   ↓   ↓   ↓
   Translating: a   =   0.025 ×   299

The letter is by itself. To solve the equation, we need only multiply:

$$a = 0.025 \times 299 = 7.475.$$

Thus, 7.475 million is 2.5% of 299 million, so in 2006, about 7.475 million people in the United States had heart disease.

Student Notes

Always look for connections between examples. Here you should look for similarities between Examples 4 and 7 as well as between Examples 5 and 8 and between Examples 6 and 9.

EXAMPLE 8 Two-Year College Enrollments. In 2003, about 5.6 million students were enrolled in two-year institutions. This number was 38% of all enrollments in higher education. How many students in all were enrolled in higher education?

SOLUTION Before translating the problem to mathematics, we reword and let S represent the total number of students, in millions, enrolled in higher education.

Rewording: 5.6   is   38%   of S.

Translating: $5.6 = 0.38 \cdot S$

$$\frac{5.6}{0.38} = S$$

$$14.7 \approx S \qquad \text{The symbol} \approx \text{means } \textit{is approximately} \textit{ equal to.}$$

There were about 14.7 million students enrolled in higher education in 2003.

EXAMPLE 9 Automobile Prices. Recently, Harken Motors reduced the price of a 2007 Ford Focus from the manufacturer's suggested retail price (MSRP) of \$17,500 to \$15,925.

a) What percent of the MSRP does the sale price represent?

b) What is the percent of discount?

SOLUTION

a) We reword and translate, using n for the unknown percent.

Rewording: What percent   of   17,500   is   15,925?

Translating: $n \cdot 17{,}500 = 15{,}925$

$$n = \frac{15{,}925}{17{,}500} \qquad \text{Dividing both sides by 17,500}$$

$$n = 0.91 = 91\% \qquad \text{Converting to percent notation}$$

The sale price is 91% of the MSRP.

b) Since the original price of \$17,500 represents 100% of the MSRP, the sale price represents a discount of $(100 - 91)\%$, or 9%.

2.4 EXERCISE SET

FOR EXTRA HELP

   Tutor Center

MathXL   MyMathLab   InterAct Math   AW Math Tutor Center   Video Lectures on CD: Disc 1   Student's Solutions Manual

Concept Reinforcement   In each of Exercises 1–10, match the question with the most appropriate translation from the column on the right. Some choices are used more than once.

1. ____ What percent of 57 is 23?   **a)** $a = (0.57)23$

2. ____ What percent of 23 is 57?   **b)** $57 = 0.23y$

3. ____ 23 is 57% of what number?   **c)** $n \cdot 23 = 57$

4. ____ 57 is 23% of what number?   **d)** $n \cdot 57 = 23$

5. ____ 57 is what percent of 23?   **e)** $23 = 0.57y$

6. ____ 23 is what percent of 57?   **f)** $a = (0.23)57$

7. ____ What is 23% of 57?

8. ____ What is 57% of 23?

9. ____ 23% of what number is 57?

10. ____ 57% of what number is 23?

Convert the percent notation in each sentence to decimal notation.

11. *Energy Use.*   Heating accounts for 49% of household energy use (*Source*: Chevron).

12. *Energy Use.*   Water heating accounts for 15% of household energy use (*Source*: Chevron).

13. *Dehydration.*   A 2% drop in water content of the body can affect one's ability to study mathematics (*Source*: High Performance Nutrition).

14. *Left-Handed Golfers.*   Of those who golf, 7% are left-handed (*Source*: National Association of Left-Handed Golfers).

15. *Wealth in the Aged.*   Those 60 and older own 77% of the nation's financial assets.

16. *Wealth in the Aged.*   Those 60 and older make up 66% of the nation's stockholders.

17. *Women in the Workforce.*   Women comprise 20% of all database administrators (*Source*: U.S. Bureau of the Census).

18. *Women in the Workforce.*   Women comprise 60% of all accountants and auditors (*Source*: U.S. Bureau of the Census).

Convert to decimal notation.

19. 62.58%   20. 39.81%

21. 0.7%   22. 0.3%

23. 125%   24. 150%

Convert the decimal notation in each sentence to percent notation.

25. *NASCAR Fans.*   Of those who are fans of NASCAR racing, 0.64 of them have attended college or beyond (*Sources*: NASCAR; Goodyear).

26. *NASCAR Fans.*   Of those who are fans of NASCAR racing, 0.41 of them earn more than $50,000 per year (*Sources*: NASCAR; Goodyear).

27. *Foreign Student Enrollment.*   Of all the foreign students studying in the United States, 0.106 are from China (*Source*: Institute of International Education).

28. *Heart Attacks.*   Of those suffering heart attacks, 0.67 will survive (*Source*: American Heart Association).

29. *Women in the Workforce.*   Women comprise 0.42 of all college faculty.

30. *Women in the Workforce.*   Women comprise 0.19 of all architects.

31. *Water in Watermelon.*   Watermelon is 0.9 water.

32. *Tax Savings.*   The tax cut enacted in 2003 will save 0.88 of all U.S. taxpayers less than $100 (*Source*: Citizens for Tax Justice).

Convert to percent notation.

33. 0.0049

34. 0.0008

35. 1.08

36. 1.05

37. 2.3

38. 2.9

39. $\dfrac{4}{5}$

40. $\dfrac{3}{4}$

41. $\dfrac{8}{25}$

42. $\dfrac{3}{8}$

Solve.

43. What percent of 68 is 17?

44. What percent of 150 is 39?

45. What percent of 125 is 30?

46. What percent of 300 is 57?

47. 14 is 30% of what number?

48. 54 is 24% of what number?

49. 0.3 is 12% of what number?

50. 7 is 175% of what number?

51. What number is 35% of 240?

52. What number is 1% of one million?

53. What percent of 60 is 75?

Aha! **54.** What percent of 70 is 70?

55. What is 2% of 40?

56. What is 40% of 2?

Aha! **57.** 25 is what percent of 50?

58. 8 is 2% of what number?

Costs of Owning a Dog.   *The American Pet Products Manufacturers Association estimates that the total cost*

of owning a dog for its lifetime is $6600. *The following circle graph shows the relative costs of raising a dog from birth to death.*

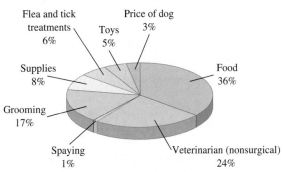

Costs of Owning a Dog

Flea and tick treatments 6%

Price of dog 3%

Toys 5%

Supplies 8%

Food 36%

Grooming 17%

Spaying 1%

Veterinarian (nonsurgical) 24%

Source: The American Pet Products Manufacturers Association

In each of Exercises 59–64, determine the cost of owning a dog for its lifetime.

59. Price of dog

60. Food

61. Veterinarian

62. Grooming

63. Supplies

64. Flea and tick treatments

65. *College Graduation.*   To obtain his bachelor's degree in nursing, Frank must complete 125 credit hours of instruction. If he has completed 60% of his requirement, how many credits did Frank complete?

66. *College Graduation.*   To obtain her bachelor's degree in journalism, Beth must complete 125 credit hours of instruction. If 20% of Beth's credit hours remain to be completed, how many credits does she still need to take?

67. *Batting Average.*   At one point in a recent season, Ichiro Suzuki of the Seattle Mariners had 194 hits. His batting average was 0.310, or 31% (*Source*: Major League Baseball). That is, of the total number of at-bats, 31% were hits. How many at-bats did he have?

68. *Pass Completions.*   At one point in a recent season, Peyton Manning of the Indianapolis Colts had completed 357 passes (*Source*: National Football League). This was 62.5% of his attempts. How many attempts did he make?

69. *Tipping.*   Leon left a $4 tip for a meal that cost $25.

a) What percent of the cost of the meal was the tip?

b) What was the total cost of the meal including the tip?

70. *Tipping.*   Selena left a $12.76 tip for a meal that cost $58.

a) What percent of the cost of the meal was the tip?

b) What was the total cost of the meal including the tip?

71. *Auto Sales.*   In June 2006, Dodge delivered 25,740 cars. Of these, 12,098 were Calibers. (*Source*: *The Wall Street Journal*, 7/13/06) What percent of Dodge deliveries were Calibers?

72. *Voting.*   Approximately 116.7 million Americans voted in the 2004 presidential election. In that election, John Kerry received 56 million votes. (*Source*: *The New York Times*, 11/9/04) What percent of the votes cast did Kerry receive?

73. *Student Loans.*   To finance her community college education, Sarah takes out a Stafford loan for $3500. After a year, Sarah decides to pay off the interest, which is 8% of $3500. How much will she pay?

74. *Student Loans.*   Paul takes out a subsidized federal Stafford loan for $2400. After a year, Paul decides to pay off the interest, which is 7% of $2400. How much will he pay?

75. *Infant Health.*   In a study of 300 pregnant women with "good-to-excellent" diets, 95% had babies in good or excellent health. How many women in this group had babies in good or excellent health?

76. *Infant Health.*   In a study of 300 pregnant women with "poor" diets, 8% had babies in good or excellent health. How many women in this group had babies in good or excellent health?

77. *Cost of Self-Employment.*   Because of additional taxes and fewer benefits, it has been estimated that a self-employed person must earn 20% more than a non–self-employed person performing the same task(s). If Joy earns $15 an hour working for Village Copy, how much would she need to earn on her own for a comparable income?

78. Refer to Exercise 77. Ray earns $12 an hour working for Round Edge stairbuilders. How much would Ray need to earn on his own for a comparable income?

79. *Triathletes.*   The number of USA Triathlon Members grew from 16,000 in 1993 to 40,000 in 2002 (*Source*: USA Triathlon). Calculate the percentage by which the number increased.

80. *Fastest Human.*   In 2002, Tim Montgomery of the United States set a world record by sprinting 100 m in 9.78 sec. This broke the record of 9.79 sec established in 1999 by Maurice Green, also of the United States. (*Source*: *Time Almanac 2006*) Calculate the percentage by which the record decreased.

81. A bill at Officeland totaled $37.80. How much did the merchandise cost if the sales tax is 5%?

82. Doreen's checkbook shows that she wrote a check for $987 for building materials, including the tax. What was the price of the materials if the sales tax is 5%?

83. *Deducting Sales Tax.*   A tax-exempt school group received a bill of $157.41 for educational software. The bill incorrectly included sales tax of 6%. How much should the school group pay?

84. *Deducting Sales Tax.*   A tax-exempt charity received a bill of $145.90 for a sump pump. The bill incorrectly included sales tax of 5%. How much does the charity owe?

85. *Body Fat.*   One author of this text exercises regularly at a local YMCA that recently offered a body-fat percentage test to its members. The device used measures the passage of a very low voltage of electricity through the body. The author's body-fat percentage was found to be 16.5% and he weighs 191 lb. What part, in pounds, of his body weight is fat?

86. *Areas of Alaska and Arizona.* The area of Arizona is 19% of the area of Alaska. The area of Alaska is 586,400 mi². What is the area of Arizona?

87. *Junk Mail.* The U.S. Postal Service reports that of the junk mail that is sent out, 78% is actually opened and read. A business sends out 9500 advertising brochures. How many of them can the business expect to be opened and read?

88. *Kissing and Colds.* In a medical study, it was determined that if 800 people kiss someone else who has a cold, only 56 will actually catch the cold. What percent is this?

89. *Calorie Content.* Pepperidge Farm Light Style 7 Grain Bread® has 140 calories in a 3-slice serving. This is 15% less than the number of calories in a serving of regular bread. How many calories are in a serving of regular bread?

90. *Fat Content.* Peek Freans Shortbread Reduced Fat Cookies® contain 35 calories of fat in each serving. This is 40% less than the fat content in the leading imported shortbread cookie. How many calories of fat are in a serving of the leading shortbread cookie?

TW **91.** Campus Bookbuyers pays $30 for a book and sells it for $60. Is this a 100% markup or a 50% markup? Explain.

TW **92.** If Julian leaves a $12 tip for a $90 dinner, is he being generous, stingy, or neither? Explain.

Skill Maintenance

Translate to an algebraic expression. [1.1]

93. 5 more than some number

94. 4 less than Tino's weight

95. The product of 8 and twice *a*

96. 1 more than the product of two numbers

Synthesis

TW **97.** Does the following advertisement provide a convincing argument that summertime is when most burglaries occur? Why or why not?

TW **98.** Erin is returning a tent that she bought during a 25%-off storewide sale that has ended. She is offered store credit for 125% of what she paid (not to be used on sale items). Is this fair to Erin? Why or why not?

99. The community of Bardville has 1332 left-handed females. If 48% of the community is female and 15% of all females are left-handed, how many people are in the community?

100. It has been determined that at the age of 15, a boy has reached 96.1% of his final adult height. Jaraan is 6 ft 4 in. at the age of 15. What will his final adult height be?

101. It has been determined that at the age of 10, a girl has reached 84.4% of her final adult height. Dana is 4 ft 8 in. at the age of 10. What will her final adult height be?

102. *Photography.*   A 6-in. by 8-in. photo is framed using a mat meant for a 5 in. by 7 in. photo. What percentage of the photo will be hidden by the mat?

103. *U.S. Birth Rate.*   Between 2002 and 2003, the U.S. birth rate rose from 13.9 to 14.1 babies per 1000 women (*Source*: National Center for Health Statistics). Calculate the percentage by which the birth rate rose and use that percentage to estimate birth rates for the United States in 2004 and in 2005.

TW Aha! 104. Would it be better to receive a 5% raise and then, a year later, an 8% raise or the other way around? Why?

TW 105. Herb is in the 30% tax bracket. This means that 30¢ of each dollar earned goes to taxes. Which would cost him the least: contributing $50 that is tax-deductible or contributing $40 that is not tax-deductible? Explain.

2.5   Problem Solving

Five Steps for Problem Solving   ■   Applying the Five Steps

Probably the most important use of algebra is as a tool for problem solving. In this section, we develop a problem-solving approach that is used throughout the remainder of the text.

Five Steps for Problem Solving

In Section 2.4, we solved several real-world problems. To solve them, we first *familiarized* ourselves with percent notation. We then *translated* each problem into an equation, *solved* the equation, *checked* the solution, and *stated* the answer.

Five Steps for Problem Solving in Algebra

1. *Familiarize* yourself with the problem.
2. *Translate* to mathematical language. (This often means writing an equation.)
3. *Carry out* some mathematical manipulation. (This often means *solving* an equation.)
4. *Check* your possible answer in the original problem.
5. *State* the answer clearly, using a complete English sentence.

Of the five steps, the most important is probably the first one: becoming familiar with the problem. Here are some hints for familiarization.

To Become Familiar with a Problem

1. Read the problem carefully. Try to visualize the problem.
2. Reread the problem, perhaps aloud. Make sure you understand all important words.
3. List the information given and the question(s) to be answered. Choose a variable (or variables) to represent the unknown and specify what the variable represents. For example, let L = length in centimeters, d = distance in miles, and so on.
4. Look for similarities between the problem and other problems you have already solved.
5. Find more information. Look up a formula in a book, at a library, or online. Consult a reference librarian or an expert in the field.
6. Make a table that uses all the information you have available. Look for patterns that may help in the translation.
7. Make a drawing and label it with known and unknown information, using specific units if given.
8. Think of a possible answer and check the guess. Note the manner in which the guess is checked.

Applying the Five Steps

EXAMPLE 1   Hiking.   In 1948, Earl Shaffer became the first person to hike all 2100 miles of the Appalachian trail—from Springer Mountain, Georgia, to Mt. Katahdin, Maine. Shaffer repeated the feat 50 years later, and at age 79 became the oldest person to hike the entire trail. When Shaffer stood atop Big Walker Mountain, Virginia, he was three times as far from the northern end of the trail as from the southern end. At that point, how far was he from each end of the trail?

SOLUTION

1. **Familiarize.** It may be helpful to make a drawing.

Earl Shaffer 1918–2002

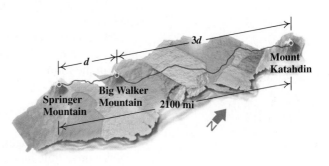

To gain some familiarity, let's suppose that Shaffer stood 600 mi from Springer Mountain. Three times 600 mi is 1800 mi. Since 600 mi + 1800 mi = 2400 mi and 2400 mi > 2100 mi, we see that our guess is too large. Rather than guess again, we let

$$d = \text{the distance, in miles, to the southern end}$$

and

$$3d = \text{the distance, in miles, to the northern end.}$$

(We could also let d = the distance to the northern end and $\frac{1}{3}d$ = the distance to the southern end.)

2. **Translate.** From the drawing, we see that the lengths of the two parts of the trail must add up to 2100 mi. This leads to our translation.

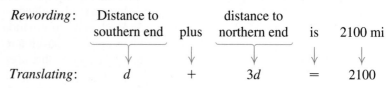

3. **Carry out.** We solve the equation:

$$d + 3d = 2100$$
$$4d = 2100 \qquad \text{Combining like terms}$$
$$d = 525. \qquad \text{Dividing both sides by 4}$$

4. **Check.** As predicted in the *Familiarize* step, d is less than 600 mi. If $d = 525$ mi, then $3d = 1575$ mi. Since 525 mi + 1575 mi = 2100 mi, we have a check.

5. **State.** Atop Big Walker Mountain, Shaffer stood 525 mi from Springer Mountain and 1575 mi from Mount Katahdin.

Before we solve the next problem, we need to learn some additional terminology regarding integers.

The following are examples of **consecutive integers:** 16, 17, 18, 19, 20; and −31, −30, −29, −28. Note that consecutive integers can be represented in the form $x, x + 1, x + 2$, and so on.

The following are examples of **consecutive even integers:** 16, 18, 20, 22, 24; and −52, −50, −48, −46. Note that consecutive even integers can be represented in the form $x, x + 2, x + 4$, and so on.

The following are examples of **consecutive odd integers:** 21, 23, 25, 27, 29; and −71, −69, −67, −65. Note that consecutive odd integers can be represented in the form $x, x + 2, x + 4$, and so on.

EXAMPLE 2   Interstate Mile Markers.   U.S. interstate highways post numbered markers at every mile to indicate location in case of an emergency (*Source:* Federal Highway Administration, Ed Rotalewski). The sum of two consecutive mile markers on I-70 in Kansas is 559. Find the numbers on the markers.

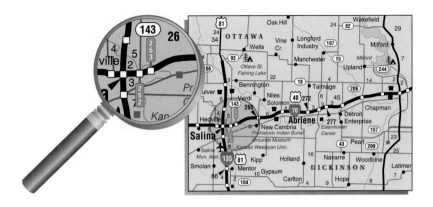

SOLUTION

1. **Familiarize.** The numbers on the mile markers are consecutive positive integers. Thus if we let $x =$ the smaller number, then $x + 1 =$ the larger number.

   The TABLE feature of a graphing calculator enables us to try many possible mile markers. We begin with a value for x. If we let $y_1 = x + 1$ represent the second mile marker, then $y_2 = x + (x + 1)$ is the sum of the markers. We enter both equations and set up a table. We will start the table at $x = 250$ and increase by 10's; that is, we set $\Delta\text{Tbl} = 10$.

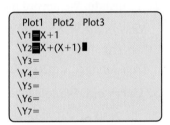

X	Y₁	Y₂
250	251	501
260	261	521
270	271	541
280	281	561
290	291	581
300	301	601
310	311	621

X = 250

   Since we are looking for a sum of 559, we see from the table that the value of x will be between 270 and 280. The problem could actually be solved by examining a table of values for x between 270 and 280, but let's work on developing our algebra skills.

2. **Translate.** We reword the problem and translate as follows.

   Rewording: First integer plus second integer is 559.

   Translating: x $+$ $(x + 1)$ $=$ 559

3. **Carry out.** We solve the equation:

$$x + (x + 1) = 559$$
$$2x + 1 = 559 \qquad \text{Using an associative law and combining like terms}$$
$$2x = 558 \qquad \text{Subtracting 1 from both sides}$$
$$x = 279. \qquad \text{Dividing both sides by 2}$$

If x is 279, then $x + 1$ is 280.

4. Check. Our possible answers are 279 and 280. These are consecutive positive integers and 279 + 280 = 559, so the answers check.

5. State. The mile markers are 279 and 280. ◢

EXAMPLE 3   Color Printers.   Egads Computer Corporation rents a Xerox Phaser 8400 Color Laser Printer for $300 a month (*Source*: egadscomputer. com). A new art gallery needs to lease a printer for a 2-month advertising campaign. The ink and paper for the brochures will cost 21.5¢ a copy. If the gallery allots a budget of $3000, how many brochures can they print?

SOLUTION

1. Familiarize. Suppose that the art gallery prints 20,000 brochures. Then the cost is the monthly charges plus ink and paper cost, or

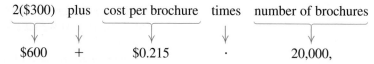

2($300)	plus	cost per brochure	times	number of brochures
$600	+	$0.215	·	20,000,

which is $4900. Our guess of 20,000 is too large, but we have familiarized ourselves with the way in which a calculation is made. Note that we convert 21.5¢ to $0.215 so that all information is in the same unit, dollars. We let c = the number of brochures that can be printed for $3000.

2. Translate. We reword the problem and translate as follows.

Rewording:	Monthly cost	plus	ink and paper cost	is	$3000
Translating:	2($300)	+	($0.215)$c$	=	$3000

3. Carry out. We solve the equation:

$$2(300) + 0.215c = 3000$$
$$600 + 0.215c = 3000$$
$$0.215c = 2400 \qquad \text{Subtracting 600 from both sides}$$
$$c = \frac{2400}{0.215} \qquad \text{Dividing both sides by 0.215}$$
$$c \approx 11{,}163. \qquad \text{Rounding to the nearest one}$$

4. Check. We check in the original problem. The cost for 11,163 brochures is 11,163($0.215) = $2400.045. The rental for 2 months is 2($300) = $600. The total cost is then $2400.045 + $600 ≈ $3000, which is the amount that was allotted. Our answer is less than 20,000, as we expected from the *Familiarize* step.

5. State. The art gallery can make 11,163 brochures with the rental allotment of $3000. ◢

EXAMPLE 4   Perimeter of NBA Court.   The perimeter of an NBA basketball court is 288 ft. The length is 44 ft longer than the width. (*Source*: National Basketball Association) Find the dimensions of the court.

Student Notes

For most students, the most challenging step is step (2), "Translate." The table on p. 5 (Section 1.1) can be helpful in this regard.

SOLUTION

1. **Familiarize.** Recall that the perimeter of a rectangle is twice the length plus twice the width. Suppose the court were 30 ft wide. The length would then be 30 + 44, or 74 ft, and the perimeter would be 2 · 30 ft + 2 · 74 ft, or 208 ft. This shows that in order for the perimeter to be 288 ft, the width must exceed 30 ft. Instead of guessing again, we let w = the width of the court, in feet. Since the court is "44 ft longer than it is wide," we let $w + 44 =$ the length of the court, in feet.

2. **Translate.** To translate, we use $w + 44$ as the length and 288 as the perimeter. To double the length, $w + 44$, parentheses are essential.

Rewording:	Twice the length	plus	twice the width	is	288 ft.
Translating:	$2(w + 44)$	$+$	$2w$	$=$	288

3. **Carry out.** We solve the equation:

$$2(w + 44) + 2w = 288$$
$$2w + 88 + 2w = 288 \qquad \text{Using the distributive law}$$
$$4w + 88 = 288 \qquad \text{Combining like terms}$$
$$4w = 200$$
$$w = 50.$$

The dimensions appear to be $w = 50$ ft, and $l = w + 44 = 94$ ft.

4. **Check.** If the width is 50 ft and the length is 94 ft, then the court is 44 ft longer than it is wide. The perimeter is $2(50\text{ ft}) + 2(94\text{ ft}) = 100$ ft + 188 ft, or 288 ft, as specified. We have a check.

5. **State.** An NBA court is 50 ft wide and 94 ft long.

CAUTION! Always be sure to answer the question in the original problem completely. For instance, in Example 1 we needed to find *two* numbers: the distances from *each* end of the trail to the hiker. Similarly, in Example 4 we needed to find two dimensions, not just the width. Be sure to label each answer with the proper unit.

EXAMPLE 5   Selling a Home.   The McCanns are planning to sell their home. If they want to be left with $117,500 after paying 6% of the selling price to a realtor as a commission, for how much must they sell the house?

SOLUTION

1. **Familiarize.** Suppose the McCanns sell the house for $120,000. A 6% commission can be determined by finding 6% of $120,000:

$$6\% \text{ of } \$120,000 = 0.06(\$120,000) = \$7200.$$

Subtracting this commission from $120,000 would leave the McCanns with

$$\$120,000 - \$7200 = \$112,800.$$

This shows that in order for the McCanns to clear $117,500, the house must sell for more than $120,000. To determine what the sale price must be, we let x = the selling price, in dollars. With a 6% commission, the realtor would receive $0.06x$.

2. **Translate.** We reword the problem and translate as follows.

Rewording:	Selling price	less	commission	is	amount remaining.
Translating:	x	$-$	$0.06x$	$=$	$117{,}500$

3. **Carry out.** We solve the equation:

$$x - 0.06x = 117{,}500$$
$$1x - 0.06x = 117{,}500$$
$$0.94x = 117{,}500 \qquad$$

> Combining like terms. Had we noted that after the commission has been paid, **94%** remains, we could have begun with this equation.

$$x = \frac{117{,}500}{0.94} \qquad$$

> Dividing both sides by **0.94**

$$x = 125{,}000.$$

4. **Check.** To check, we first find 6% of $125,000:

$$6\% \text{ of } \$125,000 = 0.06(\$125,000) = \$7500. \qquad$$

> This is the commission.

Next, we subtract the commission to find the remaining amount:

$$\$125,000 - \$7500 = \$117,500.$$

Since, after the commission, the McCanns are left with $117,500, our answer checks. Note that the $125,000 sale price is greater than $120,000, as predicted in the *Familiarize* step.

5. **State.** To be left with $117,500, the McCanns must sell the house for $125,000.

EXAMPLE 6   Cross Section of a Roof.   In a triangular gable end of a roof, the angle of the peak is twice as large as the angle on the back side of the house. The measure of the angle on the front side is 20° greater than the angle on the back side. How large are the angles?

SOLUTION

1. **Familiarize.** We make a drawing. In this case, the measure of the back angle is x, the measure of the front angle is $x + 20$, and the measure of the peak angle is $2x$.

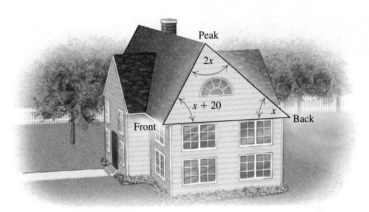

2. **Translate.** To translate, we need to recall that the sum of the measures of the angles in a triangle is 180°.

Rewording: Measure of   measure of   measure of
   back angle   +   front angle   +   peak angle   is 180°

Translating: x + $(x + 20)$ + $2x$ $= 180$

3. **Carry out.** We solve:

$$x + (x + 20) + 2x = 180$$
$$4x + 20 = 180$$
$$4x = 160$$
$$x = 40.$$

The measures for the angles appear to be:

Back angle: $x = 40°$,
Front angle: $x + 20 = 40 + 20 = 60°$,
Peak angle: $2x = 2(40) = 80°.$

4. **Check.** Consider 40°, 60°, and 80°. The measure of the front angle is 20° greater than the measure of the back angle, the measure of the peak angle is twice the measure of the back angle, and the sum is 180°. These numbers check.

5. **State.** The measures of the angles are 40°, 60°, and 80°.

EXAMPLE 7   Aquariums.   The following table lists the maximum recommended stocking density for fish in newly conditioned aquariums of various sizes. The density is calculated by adding the lengths of all fish in the aquarium. What size aquarium would be needed for 30 in. of fish?

Size of Aquarium (in gallons)	Recommended Stocking Density (in inches of fish)
100	20
120	24
200	40
250	50

Source: *Aquarium Fish Magazine*, June 2002

SOLUTION

1. **Familiarize.** We examine the values in the table to determine the relationship between the size of an aquarium and the recommended stocking density. Looking at the first and last rows of the table, we note that 100 is 5 times 20, and 250 is 5 times 50. This leads us to suspect that the recommended stocking density is $\frac{1}{5}$ the size of the aquarium. We check:

   $\frac{1}{5} \cdot 100 = 20,$

   $\frac{1}{5} \cdot 120 = 24,$ **Each recommended stocking density**

   $\frac{1}{5} \cdot 200 = 40,$ **is $\frac{1}{5}$ of the aquarium size.**

   $\frac{1}{5} \cdot 250 = 50.$

   We let x represent the size of the aquarium, in gallons. Then $\frac{1}{5} \cdot x$ gives the stocking density of that aquarium.

2. **Translate.** We reword the problem as follows.

   Rewording: Stocking density   is   30 in.

   Translating: $\frac{1}{5} \cdot x$ $=$ 30

3. **Carry out.** We solve the equation:

   $\frac{1}{5} \cdot x = 30$

   $5 \cdot \frac{1}{5} \cdot x = 5 \cdot 30$ **Multiplying both sides by 5**

   $x = 150.$ **Simplifying**

4. **Check.** We can check both our translation and solution by entering the equation in a graphing calculator and forming a table. We enter $y = \frac{1}{5}x$ and form a table, setting Indpnt to Ask. Then we enter the x-values 100, 120, 200, 250, and 150. We see that the equation gives the values for the stocking density given in the statement of the problem, and the x-value 150 gives a stocking density of 30. The answer checks.

5. **State.** A 150-gal aquarium would be needed for a stocking density of 30 in. of fish.

X	Y₁	
100	20	
120	24	
200	40	
250	50	
150	30	

X = 150

We close this section with some tips to aid you in problem solving.

Problem-Solving Tips

1. The more problems you solve, the more your skills will improve.

2. Look for patterns when solving problems. Each time you study an example in a text, you may observe a pattern for problems that you will encounter later in the exercise sets or in other practical situations.

3. When translating in mathematics, consider the dimensions of the variables and constants in the equation. The variables that represent length should all be in the same unit, those that represent money should all be in dollars or all in cents, and so on.

4. Make sure that units appear in the answer whenever appropriate and that you have completely answered the question in the original problem.

2.5 EXERCISE SET

FOR EXTRA HELP

MathXL   MyMathLab   InterAct Math   AW Math Tutor Center   Video Lectures on CD: Disc 1   Student's Solutions Manual

---

Solve. Even though you might find the answer quickly in some other way, practice using the five-step problem-solving process.

1. Two fewer than ten times a number is 78. What is the number?

2. Three less than twice a number is 19. What is the number?

3. Five times the sum of 3 and some number is 70. What is the number?

4. Twice the sum of 4 and some number is 34. What is the number?

5. *Price of Sneakers.*   Amy paid $72.25 for a pair of New Balance 765 running shoes during a 15%-off sale. What was the regular price?

6. *Price of an iPod.*   Vamar paid $67.20 for a 512-mB iPod shuffle during a 20% off sale. What was the original price of the iPod?

7. *Price of a Calculator.*   Evelyn paid $89.25, including 5% tax, for her graphing calculator. How much did the calculator itself cost?

8. *Price of a Printer.*   Jake paid $100.70, including 6% tax, for a color printer. How much did the printer itself cost?

9. *Running.*   In 1997, Yiannis Kouros of Australia set the record for the greatest distance run in 24 hr by running 188 mi. After 8 hr, he was approximately twice as far from the finish line as he was from the start (*Source*: *Guinness World Records 2004 Edition*). How far had he run?

10. *Sled-Dog Racing.* The Iditarod sled-dog race extends for 1049 mi from Anchorage to Nome. If a musher is twice as far from Anchorage as from Nome, how many miles has the musher traveled?

11. *Car Racing.* In May 2006, Sam Hornish, Jr., won the Indianapolis 500 in the second-closest finish of Indy 500 history, winning by 0.0635 sec. At one point, Hornish was 80 mi closer to the finish than the start. How far had Hornish traveled at that point?

12. *Track.* On May 31, 2004, Kenenisa Bekele of Ethiopia set a record for the men's 5000-m race with a time of 2:37.35. At one point, Bekele was 1200 m closer to the finish than to the start. How far had he run at that point?

13. *Apartment Numbers.* The apartments in Joan's apartment house are consecutively numbered on each floor. The sum of her number and her next-door neighbor's number is 2409. What are the two numbers?

14. *Apartment Numbers.* The apartments in Vincent's apartment house are numbered consecutively on each floor. The sum of his number and his next-door neighbor's number is 1419. What are the two numbers?

15. *Street Addresses.* The houses on the south side of Elm Street are consecutive even numbers. Wanda and Larry are next-door neighbors and the sum of their house numbers is 794. Find their house numbers.

16. *Street Addresses.* The houses on the west side of Lincoln Avenue are consecutive odd numbers. Sam and Colleen are next-door neighbors and the sum of their house numbers is 572. Find their house numbers.

17. The sum of three consecutive page numbers is 60. Find the numbers.

18. The sum of three consecutive page numbers is 99. Find the numbers.

19. *Oldest Bride.* The world's oldest bride was 19 yr older than her groom. Together, their ages totaled 185 yr (*Source*: *Guinness World Records 2004 Edition*). How old were the bride and the groom?

20. *Oldest Marrying Couple.* As half of the world's oldest marrying couple, the woman was 2 yr younger than the man. Together, their ages totaled 190 yr (*Source*: *Guinness World Records 2004 Edition*). How old were the man and the woman?

21. *Home Remodeling.* In a recent year, Americans spent a total of $35 billion to remodel bathrooms and kitchens. Twice as much was spent on kitchens as bathrooms. How much was spent on each?

22. *Women's Clothing.* Recently a total of $12.8 billion was spent on women's blouses and dresses. Of that total, $0.2 billion more was spent on blouses than on dresses. How much was spent on each type of clothing?

23. *Page Numbers.* The sum of the page numbers on the facing pages of a book is 281. What are the page numbers?

24. *Perimeter of a Triangle.* The perimeter of a triangle is 195 mm. If the lengths of the sides are consecutive odd integers, find the length of each side.

25. *Hancock Building Dimensions.* The top of the John Hancock Building in Chicago is a rectangle whose length is 60 ft more than the width. The perimeter is 520 ft. Find the width and the length of the rectangle. Find the area of the rectangle.

26. *Dimensions of a State.* The perimeter of the state of Wyoming is 1280 mi. The width is 90 mi less than the length. Find the width and the length.

27. *Perimeter of a High School Basketball Court.* The perimeter of a standard high school basketball court is 268 ft. The length is 34 ft longer than the width (*Source*: Indiana High School Athletic Association). Find the dimensions of the court.

28. A rectangular community garden is to be enclosed with 92 m of fencing. In order to allow for compost storage, the garden must be 4 m longer than it is wide. Determine the dimensions of the garden.

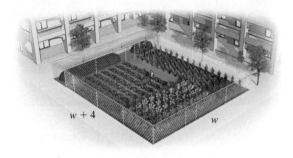

29. *Two-by-Four.* The perimeter of a cross section of a "two-by-four" piece of lumber is $10\frac{1}{2}$ in. The length is twice the width. Find the actual dimensions of the cross section of a two-by-four.

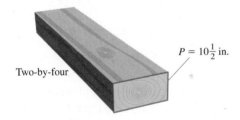

$P = 10\frac{1}{2}$ in.

Two-by-four

30. *Standard Billboard Sign.* A standard rectangular highway billboard sign has a perimeter of 124 ft. The length is 6 ft more than three times the width. Find the dimensions of the sign.

$3w + 6$

w

31. *Angles of a Triangle.* The second angle of an architect's triangle is three times as large as the first. The third angle is 30° more than the first. Find the measure of each angle.

32. *Angles of a Triangle.* The second angle of a triangular garden is four times as large as the first. The third angle is 45° less than the sum of the other two angles. Find the measure of each angle.

33. *Angles of a Triangle.* The second angle of a triangular building lot is three times as large as the first. The third angle is 10° more than the sum of the other two angles. Find the measure of the third angle.

34. *Angles of a Triangle.* The second angle of a triangular kite is four times as large as the first. The third angle is 5° more than the sum of the other two angles. Find the measure of the second angle.

35. *Rocket Sections.* A rocket is divided into three sections: the payload and navigation section in the top, the fuel section in the middle, and the rocket engine section in the bottom. The top section is one-sixth the length of the bottom section. The middle section is one-half the length of the bottom section. The total length is 240 ft. Find the length of each section.

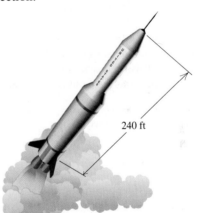

240 ft

36. *Gourmet Sandwiches.* Jenny, Demi, and Sarah buy an 18-in. long gourmet sandwich and take it back to their apartment. Since they have different appetites, Jenny cuts the sandwich so that Demi gets half of what Jenny gets and Sarah gets three-fourths of what Jenny gets. Find the length of each person's sandwich.

37. *Taxi Rates.* In Chicago, a taxi ride costs $2.25 plus $1.80 for each mile traveled. Debbie and Alex have budgeted $18 for a taxi ride (excluding tip). How far can they travel on their $18 budget?

38. *Taxi Fares.* In New York City, taxis charge $2.50 plus $2.00 per mile for off-peak fares (*Source: Burlington Free Press*, 3/31/04). How far can Ralph travel for $17.50 (assuming an off-peak fare)?

39. *Truck Rentals.* Truck-Rite Rentals rents trucks at a daily rate of $49.95 plus 39¢ per mile. Concert Productions has budgeted $100 for renting a truck to haul equipment to an upcoming concert. How far can they travel in one day and stay within their budget?

40. *Truck Rentals.* Fine Line Trucks rents an 18-ft truck for $42 plus 35¢ per mile. Judy needs a truck for one day to deliver a shipment of plants. How far can she drive and stay within a budget of $70?

41. *Complementary Angles.* The sum of the measures of two *complementary* angles is 90°. If one angle measures 15° more than twice the measure of its complement, find the measure of each angle.

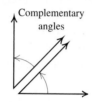

Complementary angles

42. *Supplementary Angles.* The sum of the measures of two *supplementary* angles is 180°. If one angle measures 45° less than twice the measure of its supplement, find the measure of each angle.

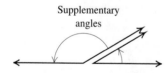

Supplementary angles

43. *Copier Paper.* The perimeter of standard-size copier paper is 99 cm. The width is 6.3 cm less than the length. Find the length and the width.

44. *Stock Prices.* Sarah's investment in Jet Blue stock grew 28% to $448. How much did she invest?

45. *Savings Interest.* Sharon invested money in a savings account at a rate of 6% simple interest. After 1 yr, she has $6996 in the account. How much did Sharon originally invest?

46. *Credit Cards.* The balance in Will's Mastercard® account grew 2%, to $870, in one month. What was his balance at the beginning of the month?

47. *Scrabble®.* The highest one-game score for a Scrabble player occurred in the same game in which the record margin of victory, 796 points, was established. The winning and losing scores totaled 1302 points (*Source: Guinness World Records 2004 Edition*). What was the winning score?

48. *Color Printers.* The art gallery in Example 3 decides to raise its budget to $5000 for the 2-month period. How many brochures can they make for $5000?

49. *Cricket Chirps and Temperature.* The equation $T = \frac{1}{4}N + 40$ can be used to determine the temperature T, in degrees Fahrenheit, given the number of times N a cricket chirps per minute. Determine the number of chirps per minute for a temperature of 80°F.

50. *Race Time.* The equation $R = -0.028t + 20.8$ can be used to predict the world record in the 200-m dash, where R is the record in seconds and t is the number of years since 1920. In what year will the record be 18.0 sec?

51. *Cell-Phone Usage.* Scot regularly checks sports scores using his cell phone. The following table lists his cost to check scores for various months. How many sports scores did he check if the cost was $1.92?

Month	Number of Sports Scores Checked	Cost
April	10	$0.60
May	30	1.80
June	15	0.90
July	28	1.68

52. *Internet Sales.*   The following table lists the cost, including shipping, for various orders from walgreens.com. If the total cost for an order was $25.68, what was the cost of the items before shipping?

Cost of Order Before Shipping	Cost Including Shipping
$15.50	$20.99
36.48	41.97
4.46	9.95
45.65	51.14

53. *Agriculture.*   The following table shows the weight of a prize-winning pumpkin for various days in August. On what day did the pumpkin weigh 920 lb?

Date	Weight of Pumpkin (in pounds)
August 1	380
August 2	410
August 3	440
August 4	470
August 11	680
August 25	1100

Source: USA Weekend, Oct. 19–21, 2001

54. *House Cleaning.*   The following table lists how much Susan and Leah charge for cleaning houses of various sizes. For what size house is the cleaning cost $145?

House Size (in square feet)	Cleaning Cost
1000	$ 60
1100	65
1200	70
2000	110
3000	160

TW 55. Sean claims he can solve most of the problems in this section by guessing. Is there anything wrong with this approach? Why or why not?

TW 56. When solving Exercise 19, Beth used a to represent the bride's age and Ben used a to represent the groom's age. Is one of these approaches preferable to the other? Why or why not?

Skill Maintenance

Write a true sentence using either $<$ or $>$. [1.4]

57. -9 ▢ 5

58. 1 ▢ 3

59. -4 ▢ 7

60. -9 ▢ -12

Synthesis

TW 61. Write a problem for a classmate to solve. Devise it so that the problem can be translated to the equation $x + (x + 2) + (x + 4) = 375$.

TW 62. Write a problem for a classmate to solve. Devise it so that the solution is "Audrey can drive the rental truck for 50 mi without exceeding her budget."

63. *Discounted Dinners.*   Kate's "Dining Card" entitles her to $10 off the price of a meal after a 15% tip has been added to the cost of the meal. If, after the discount, the bill is $32.55, how much did the meal originally cost?

64. *Test Scores.* Pam scored 78 on a test that had 4 fill-ins worth 7 points each and 24 multiple-choice questions worth 3 points each. She had one fill-in wrong. How many multiple-choice questions did Pam get right?

65. *Gettysburg Address.* Abraham Lincoln's 1863 Gettysburg Address refers to the year 1776 as "four *score* and seven years ago." Determine what a score is.

66. One number is 25% of another. The larger number is 12 more than the smaller. What are the numbers?

67. A storekeeper goes to the bank to get $10 worth of change. She requests twice as many quarters as half dollars, twice as many dimes as quarters, three times as many nickels as dimes, and no pennies or dollars. How many of each coin did the storekeeper get?

68. *Perimeter of a Rectangle.* The width of a rectangle is three-fourths of the length. The perimeter of the rectangle becomes 50 cm when the length and the width are each increased by 2 cm. Find the length and the width.

69. *Sharing Fruit.* Apples are collected in a basket for six people. One-third, one-fourth, one-eighth, and one-fifth of the apples are given to four people, respectively. The fifth person gets ten apples, and one apple remains for the sixth person. Find the original number of apples in the basket.

70. *Discounts.* In exchange for opening a new credit account, Kohl's Department Stores® subtracts 10% from all purchases made the day the account is established. Julio is opening an account and has a coupon for which he receives 10% off the first day's reduced price of a camera. If Julio's final price is $77.75, what was the price of the camera before the two discounts?

71. *Winning Percentage.* In a basketball league, the Falcons won 15 of their first 20 games. In order to win 60% of the total number of games, how many more games will they have to play, assuming they win only half of the remaining games?

72. *Cell-Phone Plan.* Tanya pays $14.99 a month for Cingular's MEdia Works Bundle. Under this plan, she gets 1000 messages, with a 5¢ charge for each additional message, and 5 MB (megabytes) of MEdia Net usage, with a 1¢ charge for each additional KB (kilobyte) of data. One month, Tanya's message and media bill was $27.03. Her MEdia Net usage was 6 MB. If there is 1024 KB in one MB, how many messages did she send or receive?

73. *Test Scores.* Ella has an average score of 82 on three tests. Her average score on the first two tests is 85. What was the score on the third test?

74. *Taxi Fares.* In New York City, a taxi ride costs $2.50 plus 40¢ per $\frac{1}{5}$ mile and 20¢ per minute stopped in traffic. Due to traffic, Glenda's taxi took 20 min to complete what is usually a 10-min drive. If she is charged $16.50 for the ride, how far did Glenda travel?

TW 75. A school purchases a piano and must choose between paying $2000 at the time of purchase or $2150 at the end of one year. Which option should the school select and why?

TW Aha! 76. Annette claims the following problem has no solution: "The sum of the page numbers on facing pages is 191. Find the page numbers." Is she correct? Why or why not?

77. The perimeter of a rectangle is 101.74 cm. If the length is 4.25 cm longer than the width, find the dimensions of the rectangle.

78. The second side of a triangle is 3.25 cm longer than the first side. The third side is 4.35 cm longer than the second side. If the perimeter of the triangle is 26.87 cm, find the length of each side.

2.6 Solving Inequalities

Solutions of Inequalities ◼ Graphs of Inequalities ◼
Interval and Set-Builder Notation ◼ Solving Inequalities Using
the Addition Principle ◼ Solving Inequalities Using the
Multiplication Principle ◼ Using the Principles Together

Many real-world situations translate to *inequalities*. For example, a student might need to register for *at least* 12 credits; an elevator might be designed to hold *at most* 2000 pounds; a tax credit might be allowable for families with incomes of *less than* $25,000; and so on. Before solving applications of this type, we must adapt our equation-solving principles to the solving of inequalities.

Solutions of Inequalities

Recall from Section 1.4 that an inequality is a number sentence containing $>$ (is greater than), $<$ (is less than), $\geq$ (is greater than or equal to), or $\leq$ (is less than or equal to). Inequalities like

$$-7 > x, \qquad t < 5, \qquad 5x - 2 \geq 9, \quad \text{and} \quad -3y + 8 \leq -7$$

are true for some replacements of the variable and false for others.

EXAMPLE 1 Determine whether the given number is a solution of $x < 2$: **(a)** -3; **(b)** 2.

SOLUTION

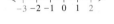

a) Since $-3 < 2$ is true, -3 is a solution.

b) Since $2 < 2$ is false, 2 is not a solution.

EXAMPLE 2 Determine whether the given number is a solution of $y \geq 6$: **(a)** 6; **(b)** -4.

SOLUTION

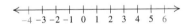

a) Since $6 \geq 6$ is true, 6 is a solution.

b) Since $-4 \geq 6$ is false, -4 is not a solution.

Graphs of Inequalities

Because the solutions of inequalities like $x < 2$ are too numerous to list, it is helpful to make a drawing that represents all the solutions. The **graph** of an inequality is such a drawing. Graphs of inequalities in one variable can be drawn on the number line by shading all points that are solutions. Parentheses are used to indicate endpoints that are *not* solutions and brackets indicate endpoints that *are* solutions.

EXAMPLE 3   Graph each inequality.

a) $x < 2$ **b)** $y \geq -3$ **c)** $-2 < x \leq 3$

SOLUTION

a) The solutions of $x < 2$ are those numbers less than 2. They are shown on the graph by shading all points to the left of 2. The right parenthesis at 2 and the shading to its left indicate that 2 is *not* part of the graph, but numbers like 1.2 and 1.99 are.

b) The solutions of $y \geq -3$ are shown on the number line by shading the point for -3 and all points to the right of -3. The left bracket at -3 indicates that -3 *is* part of the graph.

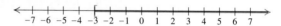

Student Notes

Note that $-2 < x < 3$ means $-2 < x$ and $x < 3$. Because of this, statements like $2 < x < 1$ make no sense—no number is both greater than 2 and less than 1.

c) The inequality $-2 < x \leq 3$ is read "-2 is less than x *and* x is less than or equal to 3," or "x is greater than -2 *and* less than or equal to 3." To be a solution of $-2 < x \leq 3$, a number must be a solution of both $-2 < x$ and $x \leq 3$. The number 1 is a solution, as are -0.5, 1.9, and 3. The parenthesis indicates that -2 is *not* a solution, whereas the bracket indicates that 3 *is* a solution. The other solutions are shaded.

Interval and Set-Builder Notation

The solutions of an inequality are numbers. In Example 3, $x < 2$, $y \geq -3$, and $-2 < x \leq 3$ are *inequalities*, not *solutions*. We will use two types of notation to write the **solution set** of an inequality: set-builder notation and interval notation.

One way to write the solution set of an inequality is **set-builder notation.** The solution set of Example 3(a), written in set-builder notation, is

$$\{x \mid x < 2\}.$$

This notation is read

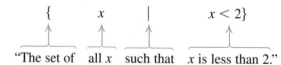

A number is in $\{x \mid x < 2\}$ if that number is less than 2. Thus, for example, 0 and -1 are in $\{x \mid x < 2\}$, but 2 and 3 are not.

Another way to write solutions of an inequality in one variable is to use **interval notation.** Interval notation uses parentheses, ( ), and brackets, [ ].

If a and b are real numbers such that $a < b$, we define the **open interval** (a, b) as the set of all numbers x for which $a < x < b$. This means that x can be any number between a and b, but it cannot be either a or b.

The **closed interval [a, b]** is defined as the set of all numbers x for which $a \leq x \leq b$. **Half-open intervals (a, b]** and **[a, b)** contain one endpoint and not the other.

We use the symbols ∞ and $-\infty$ to represent positive and negative infinity, respectively. Thus the notation (a, ∞) represents the set of all real numbers greater than a, and $(-\infty, a)$ represents the set of all real numbers less than a.

Interval notation for a set of numbers corresponds to its graph.

Interval Notation	Set-builder Notation	Graph*
(a, b) open interval	$\{x \mid a < x < b\}$	(a, b)
$[a, b]$ closed interval	$\{x \mid a \leq x \leq b\}$	$[a, b]$
$(a, b]$ half-open interval	$\{x \mid a < x \leq b\}$	$(a, b]$
$[a, b)$ half-open interval	$\{x \mid a \leq x < b\}$	$[a, b)$
(a, ∞)	$\{x \mid x > a\}$	
$[a, \infty)$	$\{x \mid x \geq a\}$	
$(-\infty, a)$	$\{x \mid x < a\}$	
$(-\infty, a]$	$\{x \mid x \leq a\}$	

CAUTION!   For interval notation, the order of the endpoints should mimic the number line, with the smaller number on the left and the larger on the right.

EXAMPLE 4   Graph $t \geq -2$ on the number line and write the solution set using both interval notation and set-builder notation.

*Some books use the representations ──○━━○── and ──●━━●── instead of, respectively, ──(━━)── and ──[━━]──.

SOLUTION Using interval notation, we write the solution set as $[-2, \infty)$. Using set-builder notation, we write the solution set as $\{t \mid t \geq -2\}$.

To graph the solution, we shade all numbers to the right of -2 and use a bracket to indicate that -2 is also a solution.

Solving Inequalities Using the Addition Principle

Consider a balance similar to one that appears in Section 2.1. When one side of the balance holds more weight than the other, the balance tips in that direction. If equal amounts of weight are then added to or subtracted from both sides of the balance, the balance remains tipped in the same direction.

The balance illustrates the idea that when a number, such as 2, is added to (or subtracted from) both sides of a true inequality, such as $3 < 7$, we get another true inequality:

$$3 + 2 < 7 + 2, \quad \text{or} \quad 5 < 9.$$

Similarly, if we add -4 to both sides of $x + 4 < 10$, we get an *equivalent* inequality:

$$x + 4 + (-4) < 10 + (-4), \quad \text{or} \quad x < 6.$$

We say that $x + 4 < 10$ and $x < 6$ are **equivalent,** which means that both inequalities have the same solution set.

The Addition Principle for Inequalities For any real numbers a, b, and c:

$a < b$ is equivalent to $a + c < b + c$;

$a \leq b$ is equivalent to $a + c \leq b + c$;

$a > b$ is equivalent to $a + c > b + c$;

$a \geq b$ is equivalent to $a + c \geq b + c$.

As with equations, our goal is to isolate the variable on one side.

EXAMPLE 5 Solve $x + 2 > 8$ and then graph the solution.

SOLUTION We use the addition principle, subtracting 2 from both sides:

$$x + 2 - 2 > 8 - 2 \qquad \text{Subtracting 2 from, or adding } -2 \text{ to, both sides}$$
$$x > 6.$$

From the inequality $x > 6$, we can determine the solutions easily. Any number greater than 6 makes $x > 6$ true and is a solution of that inequality as well as the inequality $x + 2 > 8$. The graph is as follows:

The solution set is $\{x \mid x > 6\}$, or, in interval notation, $(6, \infty)$.

Because most inequalities have an infinite number of solutions, we cannot possibly check them all. A partial check can be made using one of the possible solutions. For this example, we can substitute any number greater than 6—say, 6.1—into the original inequality:

$$\frac{x + 2 > 8}{6.1 + 2 \mid 8}$$
$$8.1 \overset{?}{>} 8 \quad \text{TRUE} \qquad 8.1 > 8 \text{ is a true statement.}$$

Since $8.1 > 8$ is true, 6.1 is a solution. Any number greater than 6 is a solution.

EXAMPLE 6 Solve $3x - 1 \le 2x - 5$ and then graph the solution.

SOLUTION We have

$$3x - 1 \le 2x - 5$$
$$3x - 1 + 1 \le 2x - 5 + 1 \qquad \text{Adding 1 to both sides}$$
$$3x \le 2x - 4 \qquad \text{Simplifying}$$
$$3x - 2x \le 2x - 4 - 2x \qquad \text{Subtracting } 2x \text{ from both sides}$$
$$x \le -4. \qquad \text{Simplifying}$$

The graph is as follows:

The graph is as follows:

Any number less than or equal to -4 is a solution, so the solution set is $\{x \mid x \le -4\}$, or, in interval notation, $(-\infty, -4]$.

As a partial check using the TABLE feature of a graphing calculator, we can let $y_1 = 3x - 1$ and $y_2 = 2x - 5$. By scrolling up or down, you can note that for $x \le -4$, we have $y_1 \le y_2$.

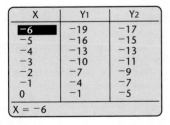

X	Y1	Y2
-6	-19	-17
-5	-16	-15
-4	-13	-13
-3	-10	-11
-2	-7	-9
-1	-4	-7
0	-1	-5

X = -6

Solving Inequalities Using the Multiplication Principle

There is a multiplication principle for inequalities similar to that for equations, but it must be modified when multiplying both sides by a negative number. Consider the true inequality

$$3 < 7.$$

If we multiply both sides by a *positive* number, say 2, we get another true inequality:

$$3 \cdot 2 < 7 \cdot 2, \quad \text{or} \quad 6 < 14. \qquad \text{TRUE}$$

If we multiply both sides by a negative number, say -2, we get a *false* inequality:

$$3 \cdot (-2) < 7 \cdot (-2), \quad \text{or} \quad -6 < -14. \qquad \text{FALSE}$$

The fact that $6 < 14$ is true, but $-6 < -14$ is false, stems from the fact that the negative numbers, in a sense, mirror the positive numbers.

Whereas 14 is to the *right* of 6, the number -14 is to the *left* of -6. Thus if we reverse the inequality symbol in $-6 < -14$, we get a true inequality:

$$-6 > -14. \qquad \text{TRUE}$$

The Multiplication Principle for Inequalities For any real numbers a and b, and for any *positive* number c:

$$a < b \quad \text{is equivalent to} \quad ac < bc, \quad \text{and}$$
$$a > b \quad \text{is equivalent to} \quad ac > bc.$$

For any real numbers a and b, and for any *negative* number c:

$$a < b \quad \text{is equivalent to} \quad ac > bc, \quad \text{and}$$
$$a > b \quad \text{is equivalent to} \quad ac < bc.$$

Similar statements hold for $\leq$ and $\geq$.

CAUTION! When multiplying or dividing both sides of an inequality by a negative number, don't forget to reverse the inequality symbol!

▶ **EXAMPLE 7**   Solve and graph each inequality: **(a)** $\frac{1}{4}x < 7$; **(b)** $-2y < 18$.

SOLUTION

a) $\frac{1}{4}x < 7$

$4 \cdot \frac{1}{4}x < 4 \cdot 7$ **Multiplying both sides by 4, the reciprocal of $\frac{1}{4}$**

The symbol stays the same, since 4 is positive.

$x < 28$ **Simplifying**

The solution set is $\{x \mid x < 28\}$, or, in interval notation, $(-\infty, 28)$. The graph is as follows:

```
    ←─────────────────┼──────────────┼──────→
                      0              28
```

b) $-2y < 18$

$\dfrac{-2y}{-2} > \dfrac{18}{-2}$ **Multiplying both sides by $-\frac{1}{2}$, or dividing both sides by -2**

At this step, we reverse the inequality, because $-\frac{1}{2}$ is negative.

$y > -9$ **Simplifying**

As a partial check, we substitute a number greater than -9, say -8, into the original inequality:

$$\frac{-2y < 18}{-2(-8) \mid 18}$$
$$16 \overset{?}{<} 18 \quad \text{TRUE} \qquad 16 < 18 \text{ is a true statement.}$$

The solution set is $\{y \mid y > -9\}$, or, in interval notation, $(-9, \infty)$. The graph is as follows:

```
    ←────────────┼──────────────┼──────────→
               -9              0
```

Using the Principles Together

We use the addition and multiplication principles together to solve inequalities much as we did when solving equations.

▶ **EXAMPLE 8**   Solve: **(a)** $6 - 5y > 7$; **(b)** $2x - 9 \leq 7x + 1$.

SOLUTION

a) $6 - 5y > 7$

$-6 + 6 - 5y > -6 + 7$ **Adding -6 to both sides**

$-5y > 1$ **Simplifying**

$-\frac{1}{5} \cdot (-5y) < -\frac{1}{5} \cdot 1$ **Multiplying both sides by $-\frac{1}{5}$, or dividing both sides by -5**

Remember to reverse the inequality symbol!

$y < -\frac{1}{5}$ **Simplifying**

As a partial check, we substitute a number smaller than $-\frac{1}{5}$, say -1, into the original inequality:

$$\frac{6 - 5y > 7}{\begin{array}{c|c} 6 - 5(-1) & 7 \\ 6 - (-5) & \end{array}}$$

$$11 \overset{?}{>} 7 \quad \text{TRUE} \qquad 11 > 7 \text{ is a true statement.}$$

The solution set is $\left\{ y \mid y < -\frac{1}{5} \right\}$, or, in interval notation, $\left(-\infty, -\frac{1}{5} \right)$.

b)
$$2x - 9 \leq 7x + 1$$

$$2x - 9 - 1 \leq 7x + 1 - 1 \qquad \text{Subtracting 1 from both sides}$$

$$2x - 10 \leq 7x \qquad \text{Simplifying}$$

$$2x - 10 - 2x \leq 7x - 2x \qquad \text{Subtracting } 2x \text{ from both sides}$$

$$-10 \leq 5x \qquad \text{Simplifying}$$

$$\frac{-10}{5} \leq \frac{5x}{5} \qquad \text{Dividing both sides by 5}$$

$$-2 \leq x \qquad \text{Simplifying}$$

The solution set is $\{x \mid -2 \leq x\}$, or $\{x \mid x \geq -2\}$, or, in interval notation, $[-2, \infty)$.

All of the equation-solving techniques used in Sections 2.1 and 2.2 can be used with inequalities provided we remember to reverse the inequality symbol when multiplying or dividing both sides by a negative number.

EXAMPLE 9   Solve.

a) $16.3 - 7.2p \leq -8.18$ **b)** $3(x - 9) - 1 < 2 - 5(x + 6)$

SOLUTION

a) The greatest number of decimal places in any one number is *two*. Multiplying both sides by 100 will clear decimals. Then we proceed as before.

$$16.3 - 7.2p \leq -8.18$$

$$100(16.3 - 7.2p) \leq 100(-8.18) \qquad \text{Multiplying both sides by 100}$$

$$100(16.3) - 100(7.2p) \leq 100(-8.18) \qquad \text{Using the distributive law}$$

$$1630 - 720p \leq -818 \qquad \text{Simplifying}$$

$$-720p \leq -818 - 1630 \qquad \text{Subtracting 1630 from both sides}$$

$$-720p \leq -2448 \qquad \text{Simplifying}$$

$$p \geq \frac{-2448}{-720} \qquad \text{Dividing both sides by } -720$$

Remember to reverse the symbol.

$$p \geq 3.4$$

The solution set is $\{p \mid p \geq 3.4\}$, or, in interval notation, $[3.4, \infty)$.

b) $3(x - 9) - 1 < 2 - 5(x + 6)$

$3x - 27 - 1 < 2 - 5x - 30$ Using the distributive law to remove parentheses

$3x - 28 < -5x - 28$ Simplifying

$3x - 28 + 28 < -5x - 28 + 28$ Adding 28 to both sides

$3x < -5x$

$3x + 5x < -5x + 5x$ Adding $5x$ to both sides

$8x < 0$

$x < 0$ Dividing both sides by 8

The solution set is $\{x \mid x < 0\}$, or, in interval notation, $(-\infty, 0)$.

2.6 EXERCISE SET

Concept Reinforcement *Insert the symbol* $<, >, \leq, or \geq$ *to make each pair of inequalities equivalent.*

1. $-5x \leq 30; x \ \blacksquare\ -6$

2. $-7t \geq 56; t \ \blacksquare\ -8$

3. $-2t > -14; t \ \blacksquare\ 7$

4. $-3x < -15; x \ \blacksquare\ 5$

Concept Reinforcement *Classify each pair of inequalities as "equivalent" or "not equivalent."*

5. $x < -2; -2 > x$

6. $t > -1; -1 < t$

7. $-4x - 1 \leq 15;$
    $-4x \leq 16$

8. $-2t + 3 \geq 11;$
    $-2t \geq 14$

Determine whether each number is a solution of the given inequality.

9. $x > -2$
    a) 5   **b)** 0   **c)** -1.9
    d) -7.3 **e)** 1.6

10. $y < 5$
    a) 0   **b)** 5   **c)** 4.99
    d) -13 **e)** $7\frac{1}{4}$

11. $x \geq 6$
    a) -6 **b)** 0   **c)** 6
    d) 6.01   **e)** $-3\frac{1}{2}$

12. $x \leq 10$
    a) 4   **b)** -10 **c)** 0
    d) 10.2   **e)** -4.7

Graph on the number line.

13. $x \leq 7$

14. $y < 2$

15. $t > -2$

16. $y > 4$

17. $1 \leq m$

18. $0 \leq t$

19. $-3 < x \leq 5$

20. $-5 \leq x < 2$

21. $0 < x < 3$

22. $-5 \leq x \leq 0$

Describe each graph using both set-builder notation and interval notation.

23.
$$-7\ -6\ -5\ -4\ -3\ -2\ -1\ 0\ 1\ 2\ 3\ 4\ 5\ 6\ 7$$

24.
$$-7\ -6\ -5\ -4\ -3\ -2\ -1\ 0\ 1\ 2\ 3\ 4\ 5\ 6\ 7$$

25.
$$-7\ -6\ -5\ -4\ -3\ -2\ -1\ 0\ 1\ 2\ 3\ 4\ 5\ 6\ 7$$

26.
$$-7\ -6\ -5\ -4\ -3\ -2\ -1\ 0\ 1\ 2\ 3\ 4\ 5\ 6\ 7$$

27.
$$-7\ -6\ -5\ -4\ -3\ -2\ -1\ 0\ 1\ 2\ 3\ 4\ 5\ 6\ 7$$

28.
$$-7\ -6\ -5\ -4\ -3\ -2\ -1\ 0\ 1\ 2\ 3\ 4\ 5\ 6\ 7$$

29.
$$-7\ -6\ -5\ -4\ -3\ -2\ -1\ 0\ 1\ 2\ 3\ 4\ 5\ 6\ 7$$

30.
$$-7\ -6\ -5\ -4\ -3\ -2\ -1\ 0\ 1\ 2\ 3\ 4\ 5\ 6\ 7$$

Solve using the addition principle. Graph and write both set-builder notation and interval notation for the answers.

31. $y + 2 > 9$ **32.** $y + 6 > 9$

33. $x + 8 \le -10$ **34.** $x + 9 \le -12$

35. $x - 3 < 7$ **36.** $x - 3 < 14$

37. $5 \le t + 8$ **38.** $4 \le t + 9$

39. $y - 7 > -12$ **40.** $y - 10 > -16$

41. $2x + 4 \le x + 9$ **42.** $2x + 4 \le x + 1$

Solve using the addition principle. Write the answers in both set-builder notation and interval notation.

43. $y + \frac{1}{3} \le \frac{5}{6}$

44. $x + \frac{1}{4} \le \frac{1}{2}$

45. $t - \frac{1}{8} > \frac{1}{2}$

46. $y - \frac{1}{3} > \frac{1}{4}$

47. $-9x + 17 > 17 - 8x$

48. $-8n + 12 > 12 - 7n$

Aha! **49.** $-23 < -t$ **50.** $19 < -x$

Solve using the multiplication principle. Graph and write both set-builder notation and interval notation for the answers.

51. $5x < 35$ **52.** $8x \ge 32$

53. $-7x < 13$ **54.** $8y < 17$

55. $-24 > 8t$ **56.** $-16x < -64$

Solve using the multiplication principle. Write the answers in both set-builder notation and interval notation.

57. $7y \ge -2$ **58.** $5x > -3$

59. $-2y \le \frac{1}{5}$ **60.** $-2x \ge \frac{1}{5}$

61. $-\frac{8}{5} > -2x$ **62.** $-\frac{5}{8} < -10y$

Solve using the addition and multiplication principles.

63. $7 + 3x < 34$ **64.** $5 + 4y < 37$

65. $6 + 5y \ge 26$ **66.** $7 + 8x \ge 71$

67. $4t - 5 \le 23$

68. $13x - 7 < -46$

69. $16 < 4 - 3y$

70. $22 < 6 - 8x$

71. $39 > 3 - 9x$

72. $5 > 5 - 7y$

73. $5 - 6y > 25$

74. $8 - 2y > 14$

75. $-3 < 8x + 7 - 7x$

76. $-5 < 9x + 8 - 8x$

77. $6 - 4y > 4 - 3y$

78. $7 - 8y > 5 - 7y$

79. $7 - 9y \le 4 - 8y$

80. $27 - 11x > 14x - 18$

81. $2.1x + 43.2 > 1.2 - 8.4x$

82. $0.96y - 0.79 \le 0.21y + 0.46$

83. $0.7n - 15 + n \ge 2n - 8 - 0.4n$

84. $1.7t + 8 - 1.62t < 0.4t - 0.32 + 8$

85. $\dfrac{x}{3} - 4 \le 1$

86. $\dfrac{2}{3} - \dfrac{x}{5} < \dfrac{4}{15}$

87. $3 < 5 - \dfrac{t}{7}$

88. $2 > 9 - \dfrac{x}{5}$

89. $4(2y - 3) < 36$

90. $3(2y - 3) > 21$

91. $3(t - 2) \geq 9(t + 2)$

92. $8(2t + 1) > 4(7t + 7)$

93. $3(r - 6) + 2 < 4(r + 2) - 21$

94. $5(t + 3) + 9 > 3(t - 2) + 6$

95. $\frac{2}{3}(2x - 1) \geq 10$

96. $\frac{4}{5}(3x + 4) \leq 20$

97. $\frac{3}{4}\left(3x - \frac{1}{2}\right) - \frac{2}{3} < \frac{1}{3}$

98. $\frac{2}{3}\left(\frac{7}{8} - 4x\right) - \frac{5}{8} < \frac{3}{8}$

TW **99.** Are the inequalities $x > -3$ and $3 > -x$ equivalent? Why or why not?

TW **100.** Are the inequalities $t > -7$ and $7 < -t$ equivalent? Why or why not?

Focused Review

Solve.

101. $4 - x = 8 - 5x$ [2.2]

102. $4 - x > 8 - 5x$ [2.6]

103. $2(5 - x) = \frac{1}{2}(x + 1)$ [2.2]

104. $2(5 - x) \leq \frac{1}{2}(x + 1)$ [2.6]

Synthesis

TW **105.** Explain how it is possible for the graph of an inequality to consist of just one number. (*Hint*: See Example 3c.)

TW **106.** Explain in your own words why it is necessary to reverse the inequality symbol when multiplying both sides of an inequality by a negative number.

Solve.

Aha! **107.** $x < x + 1$

108. $6[4 - 2(6 + 3t)] > 5[3(7 - t) - 4(8 + 2t)] - 20$

109. $27 - 4[2(4x - 3) + 7] \geq 2[4 - 2(3 - x)] - 3$

Solve for x.

110. $-(x + 5) \geq 4a - 5$

111. $\frac{1}{2}(2x + 2b) > \frac{1}{3}(21 + 3b)$

112. $y < ax + b$ (Assume $a > 0$.)

113. $y < ax + b$ (Assume $a < 0$.)

114. Graph the solutions of $|x| < 3$ on the number line.

Aha! **115.** Determine the solution set of $|x| > -3$.

116. Determine the solution set of $|x| < 0$.

2.7 Solving Applications with Inequalities

Translating to Inequalities   ■   Solving Problems

The five steps for problem solving can be used for problems involving inequalities.

Translating to Inequalities

Before solving problems that involve inequalities, we list some important phrases to look for. Sample translations are listed as well.

Important Words	Sample Sentence	Translation
is at least	Bill is at least 21 years old.	$b \geq 21$
is at most	At most 5 students dropped the course.	$n \leq 5$
cannot exceed	To qualify, earnings cannot exceed $12,000.	$r \leq 12{,}000$
must exceed	The speed must exceed 15 mph.	$s > 15$
is less than	Tucker's weight is less than 50 lb.	$w < 50$
is more than	Boston is more than 200 miles away.	$d > 200$
is between	The film was between 90 and 100 minutes long.	$90 < t < 100$
no more than	Bing weighs no more than 90 lb.	$w \leq 90$
no less than	Valerie scored no less than 8.3.	$s \geq 8.3$

The following phrases deserve special attention.

Translating "At Least" and "At Most"

The quantity x is at least some amount q: $x \geq q$.

  (If x is *at least* q, it cannot be less than q.)

The quantity x is at most some amount q: $x \leq q$.

  (If x is *at most* q, it cannot be more than q.)

Solving Problems

EXAMPLE 1   Catering Costs.   To cater a party, Curtis' Barbeque charges a $50 setup fee plus $15 per person. The cost of Hotel Pharmacy's end-of-season softball party cannot exceed $450. How many people can attend the party?

SOLUTION

1. **Familiarize.** Suppose that 20 people were to attend the party. The cost would then be $50 + $15 · 20, or $350. This shows that more than 20 people could attend without exceeding $450. Instead of making another guess, we let $n =$ the number of people in attendance.

2. **Translate.** The cost of the party will be $50 for the setup fee plus $15 times the number of people attending. We can reword as follows:

Rewording:	The setup fee	plus	the cost of the meals	cannot exceed	$450.
Translating:	50	+	$15 \cdot n$	$\leq$	450

3. **Carry out.** We solve for *n*:

$$50 + 15n \leq 450$$

$$15n \leq 400 \qquad \text{Subtracting 50 from both sides}$$

$$n \leq \frac{400}{15} \qquad \text{Dividing both sides by 15}$$

$$n \leq 26\frac{2}{3}. \qquad \text{Simplifying}$$

4. **Check.** The solution set of the inequality is all numbers less than or equal to $26\frac{2}{3}$. Since *n* represents the number of people in attendance, we round to a whole number. Since the nearest whole number, 27, is not part of the solution set, we round *down* to 26. If 26 people attend, the cost will be $50 + $15 \cdot 26$, or $440, and if 27 attend, the cost will exceed $450.

5. **State.** At most 26 people can attend the party.

> **CAUTION!** Solutions of problems should always be checked using the original wording of the problem. In some cases, answers might need to be whole numbers or integers or rounded off in a particular direction.

Some applications with inequalities involve *averages*, or *means*. You are already familiar with the concept of averages from grades in courses that you have taken.

> **Average, or Mean**
>
> To find the **average** or **mean** of a set of numbers, add the numbers and then divide by the number of addends.

EXAMPLE 2 Nutrition. The U.S. Department of Health and Human Services and the Department of Agriculture recommend that for a typical 2000-calorie daily diet, no more than 65 g of fat be consumed. In the first three days of a four-day vacation, Phil consumed 70 g, 62 g, and 80 g of fat.

Exercise

Grains Vegetables Fruit Oils Milk Meat and beans

Sources: U.S. Department of Health and Human Services and Department of Agriculture

Determine (in terms of an inequality) how many grams of fat Phil can consume on the fourth day if he is to average no more than 65 g of fat per day.

SOLUTION

1. **Familiarize.** Suppose Phil consumed 64 g of fat on the fourth day. His daily average for the vacation would then be

$$\frac{70 \text{ g} + 62 \text{ g} + 80 \text{ g} + 64 \text{ g}}{4} = 69 \text{ g.}$$ **There are 4 addends, so we divide by 4.**

This shows that Phil cannot consume 64 g of fat on the fourth day, if he is to average no more than 65 g of fat per day. Let's have x represent the number of grams of fat that Phil consumes on the fourth day.

We can use the TABLE feature of a graphing calculator to try a number of guesses. The total amount of fat consumed during the vacation is given by

$$y_1 = 70 + 62 + 80 + x.$$

The daily average for the weekend is then given by

$$y_2 = \frac{70 + 62 + 80 + x}{4}.$$

We enter y_1 and then y_2, using parentheses around the numerator.

Since we have already seen that Phil cannot consume 64 g of fat on the fourth day, we start the table at $x = 64$ and count down.

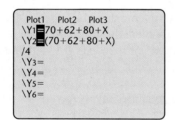

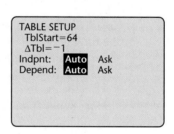

X	Y1	Y2
64	276	69
63	275	68.75
62	274	68.5
61	273	68.25
60	272	68
59	271	67.75
58	270	67.5
X = 64		

The entries in the Y2 column show that even if Phil consumes as few as 58 g on the fourth day, his average will still be greater than 65 g.

We could perhaps solve the problem by scrolling through the table, but instead let's translate to an inequality.

2. **Translate.** We reword the problem and translate as follows:

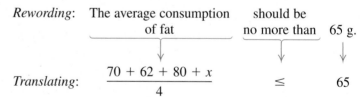

Rewording: The average consumption   should be
                           of fat   no more than   65 g.

Translating: $\dfrac{70 + 62 + 80 + x}{4}$ $\leq$ 65

3. **Carry out.** Because of the fraction, it is convenient to use the multiplication principle first:

$$\frac{70 + 62 + 80 + x}{4} \le 65$$

$$4\left(\frac{70 + 62 + 80 + x}{4}\right) \le 4 \cdot 65 \qquad \text{Multiplying both sides by 4}$$

$$70 + 62 + 80 + x \le 260$$

$$212 + x \le 260 \qquad \text{Simplifying}$$

$$x \le 48. \qquad \text{Subtracting 212 from both sides}$$

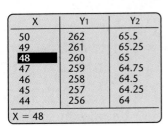

X	Y₁	Y₂
50	262	65.5
49	261	65.25
48	260	65
47	259	64.75
46	258	64.5
45	257	64.25
44	256	64

X = 48

4. **Check.** As a partial check, we scroll through the table of values that was created in the *Familiarize* step. We see from the table at left that when Phil consumes 48 g of fat on the fourth day, his average for the vacation is 65 g. When Phil consumes fewer than 48 g, his average for the vacation is fewer than 65 g. The answer is probably correct.

5. **State.** Phil's average fat intake for the vacation will not exceed 65 g per day if he consumes no more than 48 g of fat on the fourth day. ◢

EXAMPLE 3   Job Offers.   After graduation, Rose had two job offers in sales:

> Uptown Fashions: A salary of $600 per month, plus a commission of 4% of sales;
>
> Ergo Designs: A salary of $800 per month, plus a commission of 6% of sales in excess of $10,000.

If sales always exceed $10,000, for what amount of sales would Uptown Fashions provide higher pay?

SOLUTION

1. **Familiarize.** Listing the given information in a table will be helpful.

Uptown Fashions Monthly Income	Ergo Designs Monthly Income
$600 salary 4% of sales *Total*: $600 + 4% of sales	$800 salary 6% of sales over $10,000 *Total*: $800 + 6% of sales over $10,000

Next, suppose that Rose sold a certain amount—say, $12,000—in one month. Which plan would be better? Working for Uptown, she would earn $600 plus 4% of $12,000, or

$$600 + 0.04(12{,}000) = \$1080.$$

Since with Ergo Designs commissions are paid only on sales in excess of $10,000, Rose would earn $800 plus 6% of ($12,000 − $10,000), or

$$800 + 0.06(2000) = \$920.$$

This shows that for monthly sales of $12,000, Uptown pays better. Similar calculations will show that for sales of $30,000 a month, Ergo pays better. To determine *all* values for which Uptown pays more money, we must solve an inequality that is based on the calculations above.

2. **Translate.** We let S = the amount of monthly sales, in dollars, and will assume $S > 10,000$ so that both plans will pay a commission. Examining the calculations in the *Familiarize* step, we see that monthly income from Uptown is $600 + 0.04S$ and from Ergo is $800 + 0.06(S - 10,000)$. We want to find all values of S for which

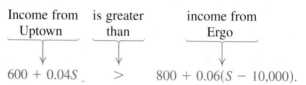

$$600 + 0.04S \quad > \quad 800 + 0.06(S - 10,000).$$

3. **Carry out.** We solve the inequality:

$$600 + 0.04S > 800 + 0.06(S - 10,000)$$

$$600 + 0.04S > 800 + 0.06S - 600 \qquad \text{Using the distributive law}$$

$$600 + 0.04S > 200 + 0.06S \qquad \text{Combining like terms}$$

$$400 > 0.02S \qquad \text{Subtracting 200 and } 0.04S \text{ from both sides}$$

$$20,000 > S, \text{ or } S < 20,000. \qquad \text{Dividing both sides by 0.02}$$

4. **Check.** The above steps indicate that income from Uptown Fashions is higher than income from Ergo Designs for sales less than $20,000. In the *Familiarize* step, we saw that for sales of $12,000, Uptown pays more. Since $12,000 < 20,000$, this is a partial check.

5. **State.** When monthly sales are less than $20,000, Uptown Fashions provides the higher pay.

2.7 EXERCISE SET

Concept Reinforcement   *In each of Exercises 1–8, match the sentence with one of the following:* $a < b$; $a \le b$; $b < a$; $b \le a$.

1. a is at least b.

2. a exceeds b.

3. a is at most b.

4. a is exceeded by b.

5. b is no more than a.

6. b is no less than a.

7. *b* is less than *a*.

8. *b* is more than *a*.

Translate to an inequality.

9. A number is at least 8.

10. A number is greater than or equal to 4.

11. The temperature is at most $-3°C$.

12. The average credit-card holder is more than $4000 in debt.

13. The price of Pat's PT Cruiser exceeded $21,900.

14. The time of the test was between 45 and 55 min.

15. Normandale Community College is no more than 15 mi away.

16. Tania's weight is less than 110 lb.

17. That number is greater than -2.

18. The costs of production of that CD-ROM cannot exceed $12,500.

19. Between 400,000 and 1,200,000 people attended the Million Man March.

20. The cost of gasoline was at most $2 per gallon.

Use an inequality and the five-step process to solve each problem.

21. *Furnace Repairs.* RJ's Plumbing and Heating charges $25 plus $30 per hour for emergency service. Gary remembers being billed over $100 for an emergency call. How long was RJ's there?

22. *College Tuition.* Karen's financial aid stipulates that her tuition not exceed $1000. If her local community college charges a $35 registration fee plus $375 per course, what is the greatest number of courses for which Karen can register?

23. *Mass Transit.* Local light rail service in Denver, Colorado, costs $1.15 per trip (one way). A student monthly pass costs $21. (*Source*: rtd-denver.com) Gail is a student at Community College of Denver.

Express as an inequality the number of trips per month that Gail should make if the pass is to save her money.

24. *Grade Average.* Nadia is taking a literature course in which four tests are given. To get a B, a student must average at least 80 on the four tests. Nadia scored 82, 76, and 78 on the first three tests. What scores on the last test will earn her at least a B?

25. *Quiz Average.* Rod's quiz grades are 73, 75, 89, and 91. What scores on a fifth quiz will make his average quiz grade at least 85?

26. *Nutrition.* Following the guidelines of the Food and Drug Administration, Dale tries to eat at least 5 servings of fruits or vegetables each day. For the first six days of one week, she had 4, 6, 7, 4, 6, and 4 servings. How many servings of fruits or vegetables should Dale eat on Saturday, in order to average at least 5 servings per day for the week?

27. *College Course Load.* To remain on financial aid, Millie needs to complete an average of at least 7 credits per quarter each year. In the first three quarters of 2006, Millie completed 5, 7, and 8 credits. How many credits of course work must Millie complete in the fourth quarter if she is to remain on financial aid?

28. *Music Lessons.* Band members at Colchester Middle School are expected to average at least 20 min of practice time per day. One week Monroe practiced 15 min, 28 min, 30 min, 0 min,

15 min, and 25 min. How long must he practice on the seventh day if he is to meet expectations?

29. *Electrician Visits.* Dot's Electric made 17 customer calls last week and 22 calls this week. How many calls must be made next week in order to maintain an average of at least 20 for the three-week period?

30. *Perimeter of a Rectangle.* The width of a rectangle is fixed at 8 ft. What lengths will make the perimeter at least 200 ft? at most 200 ft?

31. *Perimeter of a Triangle.* One side of a triangle is 2 cm shorter than the base. The other side is 3 cm longer than the base. What lengths of the base will allow the perimeter to be greater than 19 cm?

32. *Perimeter of a Pool.* The perimeter of a rectangular swimming pool is not to exceed 70 ft. The length is to be twice the width. What widths will meet these conditions?

33. *Well Drilling.* All Seasons Well Drilling offers two plans. Under the "pay-as-you-go" plan, they charge $500 plus $8 a foot for a well of any depth. Under their "guaranteed-water" plan, they charge a flat fee

of $4000 for a well that is guaranteed to provide adequate water for a household. For what depths would it save a customer money to use the pay-as-you-go plan?

34. *Cost of Road Service.* Rick's Automotive charges $50 plus $15 for each (15-min) unit of time when making a road call. Twin City Repair charges $70 plus $10 for each unit of time. Under what circumstances would it be more economical for a motorist to call Rick's?

35. *Insurance-Covered Repairs.* Most insurance companies will replace a vehicle if an estimated repair exceeds 80% of the "blue-book" value of the vehicle. Michelle's insurance company paid $8500 for repairs to her Subaru after an accident. What can be concluded about the blue-book value of the car?

36. *Insurance-Covered Repairs.* Following an accident, Jeff's Ford pickup was replaced by his insurance company because the damage was so extensive. Before the damage, the blue-book value of the truck was $21,000. How much would it have cost to repair the truck? (See Exercise 35.)

37. *Sizes of Envelopes.* Rhetoric Advertising is a direct-mail company. It determines that for a particular campaign, it can use any envelope with a fixed width of $3\frac{1}{2}$ in. and an area of at least $17\frac{1}{2}$ in². Determine (in terms of an inequality) those lengths that will satisfy the company constraints.

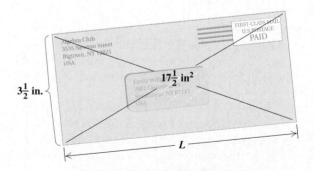

38. *Sizes of Packages.* An overnight delivery service accepts packages of up to 165 in. in length and girth combined. (Girth is the distance around the package.) A package has a fixed girth of 53 in.

Determine (in terms of an inequality) those lengths for which a package is acceptable.

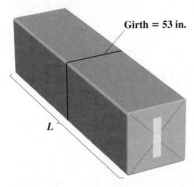

Girth = 53 in.

L

39. *Body Temperature.* A person is considered to be feverish when his or her temperature is higher than 98.6°F. The formula $F = \frac{9}{5}C + 32$ can be used to convert Celsius temperatures C to Fahrenheit temperatures F. For which Celsius temperatures is a person considered feverish?

40. *Gold Temperatures.* Gold stays solid at Fahrenheit temperatures below 1945.4°. Determine (in terms of an inequality) those Celsius temperatures for which gold stays solid. Use the formula given in Exercise 39.

41. *Area of a Triangular Flag.* As part of an outdoor education course, Wanda needs to make a bright-colored triangular flag with an area of at least 3 ft². What heights can the triangle be if the base is $1\frac{1}{2}$ ft?

$1\frac{1}{2}$ ft   ?

42. *Area of a Triangular Sign.* Zoning laws in Harrington prohibit displaying signs with areas exceeding 12 ft². If Flo's Marina is ordering a triangular sign with an 8-ft base, how tall can the sign be?

8 ft

Flo's Marina

?

43. *Fat Content in Foods.* Reduced Fat Skippy® peanut butter contains 12 g of fat per serving (*Source*: Best Foods). In order for a food to be labeled "reduced fat," it must have at least 25% less fat than the regular item. What can you conclude about the number of grams of fat in a serving of the regular Skippy peanut butter?

44. *Fat Content in Foods.* Reduced Fat Chips Ahoy!® cookies contain 5 g of fat per serving (*Source*: Nabisco Brands, Inc.). What can you conclude about the number of grams of fat in regular Chips Ahoy! cookies (see Exercise 43)?

45. *Weight Gain.* In the last weeks before the yearly Topsfield Weigh In, heavyweight pumpkins gain about 26 lb per day. Charlotte's heaviest pumpkin weighs 532 lb on September 5. For what dates will its weight exceed 818 lb?

46. *Pond Depth.* On July 1, Garrett's Pond was 25 ft deep. Since that date, the water level has dropped $\frac{2}{3}$ ft per week. For what dates will the water level not exceed 21 ft?

47. *Blueprints.* To make copies of blueprints, Vantage Reprographics charges a $6 setup fee plus $4 per copy. Myra can spend no more than $65 for the copying. What numbers of copies will allow her to stay within budget?

48. *Banquet Costs.* The women's volleyball team can spend at most $450 for its awards banquet at a local restaurant. If the restaurant charges a $40 setup fee plus $16 per person, at most how many can attend?

49. *World Records in the Mile Run.* The formula

$$R = -0.0065t + 4.3222$$

can be used to predict the world record, in minutes, for the 1-mi run t years after 1900 (*Source*: Based on information from Norris McWhirter, *Book of*

Historical Records. New York: Sterling, 2000). Determine (in terms of an inequality) those years for which the world record will be less than 3.7 min.

50. *World Records in the Women's 1500-m Run.*   The formula

$$R = -0.0223t + 4.4078$$

can be used to predict the world record, in minutes, for the 1500-m run t years after 1930. Determine (in terms of an inequality) those years for which the world record will be less than 3.5 min.

51. *Toll Charges.*   The equation $y = 0.03x + 0.21$ can be used to determine the approximate cost y, in dollars, of driving x miles on the Indiana toll road. For what mileages x will the cost be at most $6?

52. *Price of a Movie Ticket.*   The average price of a movie ticket can be estimated by the equation $P = 0.227Y - 448.71$, where Y is the year and P is the average price, in dollars. The price is lower than what might be expected due to senior-citizen discounts, children's prices, and special volume discounts. For what years will the average price of a movie ticket be at least $7? (Include the year in which the $7 ticket first occurs.)

Phone Rates.   In Vermont, Verizon recently charged customers $26.35 for monthly service. This included a call allowance of $8.00. Local calls made between 9 A.M. and 9 P.M. weekdays cost 2.2¢ per minute. The charge for off-peak local calls was 0.5¢.

53. Assume that only peak local calls were made. For how long can a customer speak on the phone and not exceed the $8.00 call allowance?

54. Assume that only off-peak calls were made. For how long can a customer speak on the phone and not exceed the $8.00 call allowance?

55. *Checking-Account Rates.*   The Hudson Bank offers two checking-account plans. Their Anywhere plan charges 20¢ per check whereas their Acu-checking plan costs $2 per month plus 12¢ per check. For what numbers of checks per month will the Acu-checking plan cost less?

56. *Moving Costs.*   Musclebound Movers charges $85 plus $40 an hour to move households across town. Champion Moving charges $60 an hour for cross-town moves. For what lengths of time is Champion more expensive?

57. *Wages.*   Toni can be paid in one of two ways:

> *Plan A:* A salary of $400 per month, plus a commission of 8% of gross sales;
>
> *Plan B:* A salary of $610 per month, plus a commission of 5% of gross sales.

For what amount of gross sales should Toni select plan A?

58. *Wages.*   Branford can be paid for his masonry work in one of two ways:

> *Plan A:* $300 plus $9.00 per hour;
>
> *Plan B:* Straight $12.50 per hour.

Suppose that the job takes n hours. For what values of n is plan B better for Branford?

59. Abriana rented a compact car for a business trip. At the time of rental, she was given the option of pre-paying for an entire tank of gasoline at $3.099 per gallon, or waiting until her return and paying $6.34 per gallon for enough gasoline to fill the tank. If the tank holds 14 gal, how many gallons can she use and still save money by choosing the second option?

60. Refer to Exercise 59. If Abriana's rental car gets 30 mpg, how many miles must she drive in order to make the first option more economical?

61. If f represents Fran's age and t represents Todd's age, write a sentence that would translate to $t + 3 < f$.

62. Explain how the meanings of "Five more than a number" and "Five is more than a number" differ.

Skill Maintenance

Simplify. [1.8]

63. $\dfrac{9 - 5}{6 - 4}$

64. $\dfrac{8 - 5}{12 - 6}$

65. $\dfrac{8 - (-2)}{1 - 4}$

66. $\dfrac{7 - 9}{4 - (-6)}$

Synthesis

TW 67. Write a problem for a classmate to solve. Devise the problem so the answer is "The Rothmans can drive 90 mi without exceeding their truck rental budget."

TW 68. Write a problem for a classmate to solve. Devise the problem so the answer is "At most 18 passengers can go on the boat." Design the problem so that at least one number in the solution must be rounded down.

69. *Wedding Costs.* The Arnold Inn offers two plans for wedding parties. Under plan A, the inn charges $30 for each person in attendance. Under plan B, the inn charges $1300 plus $20 for each person in excess of the first 25 who attend. For what size parties will plan B cost less? (Assume that more than 25 guests will attend.)

70. *Insurance Benefits.* Bayside Insurance offers two plans. Under plan A, Giselle would pay the first $50 of her medical bills and 20% of all bills after that. Under plan B, Giselle would pay the first $250 of bills, but only 10% of the rest. For what amount of medical bills will plan B save Giselle money? (Assume that her bills will exceed $250.)

71. *Parking Fees.* Mack's Parking Garage charges $4.00 for the first hour and $2.50 for each additional hour. For how long has a car been parked when the charge exceeds $16.50?

72. *Ski Wax.* Green ski wax works best between 5° and 15° Fahrenheit. Determine those Celsius temperatures for which green ski wax works best. (See Exercise 39.)

Aha! 73. The area of a square can be no more than 64 cm². What lengths of a side will allow this?

Aha! 74. The sum of two consecutive odd integers is less than 100. What is the largest pair of such integers?

75. *Nutritional Standards.* In order for a food to be labeled "lowfat," it must have fewer than 3 g of fat per serving. Reduced fat Tortilla Pops® contain 60% less fat than regular nacho cheese tortilla chips, but still cannot be labeled lowfat. What can you conclude about the fat content of a serving of nacho cheese tortilla chips?

76. *Parking Fees.* When asked how much the parking charge is for a certain car (see Exercise 71), Mack replies "between 14 and 24 dollars." For how long has the car been parked?

77. *Frequent Buyer Bonus.* Alice's Books allows customers to select one free book for every 10 books purchased. The price of that book cannot exceed the average cost of the 10 books. Neoma has bought 9 books that average $12 per book. How much should her tenth book cost if she wants to select a $15 book for free?

TW 78. *Grading.* After 9 quizzes, Blythe's average is 84. Is it possible for Blythe to improve her average by two points with the next quiz? Why or why not?

TW 79. *Discount Card.* Arnold and Diaz Booksellers offers a preferred-customer card for $25. The card entitles a customer to a 10% discount on all purchases for a period of one year. Under what circumstances would an individual save money by purchasing a card?

2 Chapter Summary and Review

KEY TERMS AND DEFINITIONS

EQUATIONS AND INEQUALITIES

Coefficient, p. 87   The numerical multiplier of the variable or variables in a term.

Equivalent equations, p. 84   Equations that have the same solution set. **Equivalent inequalities** are inequalities that have the same solution set.

Clear fractions or decimals, p. 95   To multiply both sides of an equation by a number that will eliminate fractions or decimals in the equation.

Dependent variable, p. 101   In an equation with two variables, the variable that depends on the value of the **independent variable.**

Solution set, p. 134   The set containing the solutions of an equation or an inequality. Sets that are graphed as an interval on the number line can be written in **set-builder notation** or **interval notation.**

APPLICATIONS

Percent notation, p. 110   A ratio of a quantity to 100.

$$n\% \quad \text{means} \quad \frac{n}{100} \quad \text{or} \quad n \times \frac{1}{100} \quad \text{or} \quad n \times 0.01.$$

Consecutive numbers, p. 120   Integers that are one unit apart. **Consecutive even integers** and **consecutive odd integers** are two units apart.

FIVE STEPS FOR PROBLEM SOLVING IN ALGEBRA

1. *Familiarize* yourself with the problem.
2. *Translate* to mathematical language. (This often means writing an equation.)
3. *Carry out* some mathematical manipulation. (This often means *solving* an equation.)
4. *Check* your possible answer in the original problem.
5. *State* the answer clearly.

IMPORTANT CONCEPTS

[Section references appear in brackets.]

Concept	Example
The Addition Principle for Equations $\quad a = b$ is equivalent to $\quad a + c = b + c.$	$x + 5 = -2$ is equivalent to $x + 5 + (-5) = -2 + (-5)$ [2.1]

(continued)

The Multiplication Principle for Equations $a = b$ is equivalent to $ac = bc$, for $c \neq 0$.	$-\frac{1}{3}x = 7$ is equivalent to $(-3)\left(-\frac{1}{3}x\right) = (-3)(7)$ [2.1]
Every equation is an **identity,** a **contradiction,** or a **conditional equation.**	*Identity*: $x + 1 = x + 1$ True for all replacements *Contradiction*: $x + 1 = x + 2$ Never true *Conditional equation*: $x + 1 = 2$ True for $x = 1$ [2.2]
The Addition Principle for Inequalities For any real numbers a, b, and c: $a < b$ is equivalent to $a + c < b + c$; $a > b$ is equivalent to $a + c > b + c$. Similar statements hold for $\leq$ and $\geq$.	$x + 3 \leq 5$ is equivalent to $x + 3 - 3 \leq 5 - 3$ $x \leq 2.$ [2.6]
The Multiplication Principle for Inequalities For any real numbers a and b, and for any positive number c, $a < b$ is equivalent to $ac < bc$; $a > b$ is equivalent to $ac > bc$. For any real numbers a and b, and for any *negative* number c, $a < b$ is equivalent to $ac > bc$; $a > b$ is equivalent to $ac < bc$. Similar statements hold for $\leq$ and $\geq$.	$3x > 9$ is equivalent to $\frac{1}{3} \cdot 3x > \frac{1}{3} \cdot 9$ The inequality symbol does not change, because $\frac{1}{3}$ is positive. $x > 3.$ $-3x > 9$ is equivalent to $-\frac{1}{3} \cdot -3x < -\frac{1}{3} \cdot 9$ The inequality symbol is reversed, because $-\frac{1}{3}$ is negative. $x < -3.$ [2.6]

Solution sets of inequalities can be **graphed** and written in **interval notation.**

Interval Notation	Set-builder Notation	Graph
(a, b)	$\{x \mid a < x < b\}$	
$[a, b]$	$\{x \mid a \leq x \leq b\}$	
$[a, b)$	$\{x \mid a \leq x < b\}$	
$(a, b]$	$\{x \mid a < x \leq b\}$	
(a, ∞)	$\{x \mid a < x\}$	
$(-\infty, a)$	$\{x \mid x < a\}$	

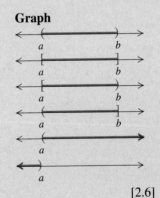

[2.6]

Review Exercises

Concept Reinforcement   *Classify each of the following as either true or false.*

1. $5x - 4 = 2x$ and $3x = 4$ are equivalent equations. [2.1]

2. $5 - 2t < 9$ and $t > 6$ are equivalent inequalities. [2.6]

3. Some equations have no solution. [2.1]

4. Consecutive odd integers are 2 units apart. [2.5]

5. For any number a, $a \le a$. [2.6]

6. The addition principle is always used before the multiplication principle. [2.2]

7. A 10% discount results in a sale price that is 90% of the original price. [2.4]

8. Often it is impossible to list all solutions of an inequality number by number. [2.6]

Solve. Label any contradictions or identities.

9. $x + 9 = -16$ [2.1]   10. $-8x = -56$ [2.1]

11. $-\dfrac{x}{5} = 13$ [2.1]   12. $n - 7 = -6$ [2.1]

13. $12x = -60$ [2.1]   14. $x - 0.1 = 1.01$ [2.1]

15. $-\frac{2}{3} + x = -\frac{1}{6}$ [2.1]   16. $\frac{4}{5}y = -\frac{3}{16}$ [2.1]

17. $5z + 3 = 41$ [2.2]   18. $5 - x = 13$ [2.2]

19. $5t + 9 = 3t - 1$ [2.2]   20. $7x - 6 = 25x$ [2.2]

21. $\frac{1}{4}x - \frac{5}{8} = \frac{3}{8}$ [2.2]

22. $14y = 23y - 17 - 10$ [2.2]

23. $0.22y - 0.6 = 0.12y + 3 - 0.8y$ [2.2]

24. $\frac{1}{4}x - \frac{1}{8}x = 3 - \frac{1}{16}x$ [2.2]

25. $3(x + 5) = 36$ [2.2]

26. $4(5x - 7) = -56$ [2.2]

27. $8(x - 2) = 5(x + 4)$ [2.2]

28. $-5x + 3(x + 8) = 16 - 2x$ [2.2]

Solve each formula for the given letter. [2.3]

29. $C = \pi d$, for d

30. $V = \dfrac{1}{3}Bh$, for B

31. $A = \dfrac{a + b}{2}$, for b

32. Find decimal notation: 0.9%. [2.4]

33. Find percent notation: $\frac{11}{25}$. [2.4]

34. What percent of 60 is 42? [2.4]

35. 42 is 30% of what number? [2.4]

Determine whether each number is a solution of $x \le 4$. [2.6]

36. -3

37. 7

38. 4

Graph on the number line. [2.6]

39. $5x - 6 < 2x + 3$ 40. $-2 < x \le 5$

41. $t > 0$

Solve. Write the answers in both set-builder notation and interval notation. [2.6]

42. $t + \frac{2}{3} \ge \frac{1}{6}$

43. $9x \ge 63$

44. $2 + 6y > 20$

45. $7 - 3y \ge 27 + 2y$

46. $3x + 5 < 2x - 6$

47. $-4y < 28$

48. $3 - 4x < 27$

49. $4 - 8x < 13 + 3x$

50. $13 \le -\frac{2}{3}t + 5$

51. $7 \le 1 - \frac{3}{4}x$

Solve.

52. A can of powdered infant formula makes 120 oz of formula. How many 6-oz bottles of formula will the can make? [2.5]

53. A 32-ft beam is cut into two pieces. One piece is 2 ft longer than the other. How long are the pieces? [2.5]

54. About 36% of all charitable contributions are made to religious organizations. In 2003, $86.4 billion was given to religious organizations. (*Source: Statistical Abstract of the United States*, 2006) How much was given to charities in general? [2.4]

55. The sum of two consecutive odd integers is 116. Find the integers. [2.5]

56. The perimeter of a rectangle is 56 cm. The width is 6 cm less than the length. Find the width and the length. [2.5]

57. After a 25% reduction, a picnic table is on sale for $120. What was the regular price? [2.4]

58. The expansion of the Marion County Central Library cost $153 million. This was 50% more than was budgeted for the project. (*Source: The Indianapolis Star*, 7/23/06) What was the original estimate for the expansion? [2.4]

59. The measure of the second angle of a triangle is 50° more than that of the first. The measure of the third angle is 10° less than twice the first. Find the measures of the angles. [2.5]

60. *Magazine Subscriptions.* The following table lists the cost of gift subscriptions for *Taste of Home*, a cooking magazine, during a recent subscription campaign (*Source: Taste of Home*). Tonya spent $73 for gift subscriptions. How many subscriptions did she purchase? [2.5]

Number of Gift Subscriptions	Total Cost of Subscriptions
1	$ 13
2	23
4	43
10	103

61. Lisa has budgeted an average of $95 a month for entertainment. For the first five months of the year, she has spent $98, $89, $110, $85, and $83. How much can Lisa spend in the sixth month without exceeding her average budget? [2.7]

62. The length of a rectangular frame is 43 cm. For what widths would the perimeter be greater than 120 cm? [2.7]

Synthesis

TW 63. How does the multiplication principle for equations differ from the multiplication principle for inequalities? [2.1], [2.6]

TW 64. Explain how checking the solutions of an equation differs from checking the solutions of an inequality. [2.1], [2.6]

65. Children (aged 2–17) with access to computers, video games, and television spend an average of 33 hr 36 min per week viewing a screen of some type. This represents 31% more time than that spent by children who view only television. (*Source:* Stanger, J. D., and N. Gridina, *Media in the Home 1999; the fourth annual survey of parents and children.* Philadelphia: Annenberg Public Policy Center, University of Pennsylvania, 1999) How much time each week does the average child spend in front of a television if he or she lives in a home without a computer or video games? [2.4]

66. The combined length of the Nile and Amazon Rivers is 12,986 km. If the Amazon were 394 km longer, it would be as long as the Nile. (*Source: Time Almanac 2006*) Find the length of each river. [2.5]

67. Consumer experts advise us never to pay the sticker price for a car. A rule of thumb is to pay the sticker price minus 20% of the sticker price, plus $200. A car is purchased for $15,080 using the rule. What was the sticker price? [2.4], [2.5]

Solve.

68. $2|n| + 4 = 50$ [1.4], [2.2]

69. $|3n| = 60$ [1.4], [2.1]

70. $y = 2a - ab + 3$, for a [2.3]

Chapter Test 2

Solve. Label any contradictions or identities.

1. $t + 7 = 16$

2. $t - 3 = 12$

3. $6x = -18$

4. $-\frac{4}{7}x = -28$

5. $3t + 7 = 2t - 5$

6. $\frac{1}{2}x - \frac{3}{5} = \frac{2}{5}$

7. $8 - y = 16$

8. $-\frac{2}{5} + x = -\frac{3}{4}$

9. $4(x + 2) = 36$

10. $9 - 3x = 6(x + 4)$

11. $\frac{5}{6}(3x + 1) = 20$

12. $3(2x - 8) = 6(x - 4)$

Solve. Write the answers in both set-builder notation and interval notation.

13. $x + 6 > 1$

14. $14x + 9 > 13x - 4$

15. $\frac{1}{3}x < \frac{7}{8}$

16. $-2y \leq 26$

17. $4y \leq -32$

18. $-5x \geq \frac{1}{4}$

19. $4 - 6x > 40$

20. $5 - 9x \geq 19 + 5x$

Solve each formula for the given letter.

21. $A = 2\pi rh$, for r

22. $w = \dfrac{P + l}{2}$, for l

23. Find decimal notation: 230%.

24. Find percent notation: 0.054.

25. What number is 32% of 50?

26. What percent of 75 is 33?

Graph on the number line.

27. $y < 4$

28. $-2 \leq x \leq 2$

Solve.

29. The perimeter of a rectangular calculator is 36 cm. The length is 4 cm greater than the width. Find the width and the length.

30. Kari is taking a 240-mi bicycle trip through Vermont. She has three times as many miles to go as she has already ridden. How many miles has she biked so far?

31. The perimeter of a triangle is 249 mm. If the sides are consecutive odd integers, find the length of each side.

32. By lowering the temperature of their electric hot-water heater from 140°F to 120°F, the Kellys' average electric bill dropped by 7% to $60.45. What was their electric bill before they lowered the temperature of their hot water?

33. *Van Rentals.*   Atlas rents a cargo van at a daily rate of $44.95 plus $0.39 per mile. A business has budgeted $250 for a one-day van rental. What mileages will allow the business to stay within budget? (Round to the nearest tenth of a mile.)

Synthesis

Solve.

34. $c = \dfrac{2cd}{a - d}$, for d

35. $3|w| - 8 = 37$

36. A concert promoter had a certain number of tickets to give away. Five people got the tickets. The first got one third of the tickets, the second got one fourth of the tickets, and the third got one fifth of the tickets. The fourth person got eight tickets, and there were five tickets left for the fifth person. Find the total number of tickets given away.

3

Introduction to Graphing and Functions

W e now begin our study of graphing. First, we examine graphs as they commonly appear in newspapers or magazines and develop some terminology. Following that, we graph certain equations and study the connection between rate and slope. We will use the graphing calculator as a tool in graphing equations and we will learn how graphs can be used as a problem-solving tool in many applications. Finally, we develop the idea of a function and look at some relationships between graphs and functions.

APPLICATION   *Registered Nurses.*

Data from the National Sample Survey of Registered Nurses, conducted every four years, indicate that the number of registered nurses in the United States who are working full time has increased steadily. Use linear regression to fit a linear function to the data, and use the function to predict the number of registered nurses working full time in 2008.

YEAR	NUMBER OF REGISTERED NURSES WORKING FULL TIME
1980	850,000
1984	975,000
1988	1,100,000
1992	1,225,000
1996	1,500,000
2000	1,600,000
2004	1,700,000

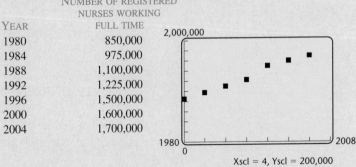

Xscl = 4, Yscl = 200,000

This problem appears as Example 10 in Section 3.7.

3.1 Reading Graphs, Plotting Points, and Scaling Graphs

Problem Solving with Bar, Circle, and Line Graphs ■ Points and Ordered Pairs ■ Axes and Windows

Today's print and electronic media make almost constant use of graphs. This can be attributed to the widespread availability of graphing software and the large quantity of information that a graph can display. In this section, we consider problem solving with bar graphs, line graphs, and circle graphs. Then we examine graphs that use a coordinate system.

Problem Solving with Bar, Circle, and Line Graphs

A *bar graph* is a convenient way of showing comparisons. In every bar graph, certain categories, such as body weight in the example below, are paired with certain numbers.

EXAMPLE 1 Driving under the Influence. A blood-alcohol level of 0.08% or higher makes driving illegal in the United States. This bar graph shows how many drinks a person of a certain weight would need to consume in 1 hr to achieve a blood-alcohol level of 0.08% (*Source*: Adapted from vsa.vassar.edu/~source/drugs/alcohol.html). Note that a 12-oz beer, a 5-oz glass of wine, or a cocktail containing $1\frac{1}{2}$ oz of distilled liquor all count as one drink.

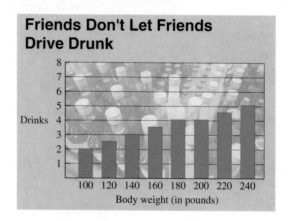

a) Approximately how many drinks would a 200-lb person have consumed if he or she had a blood-alcohol level of 0.08%?

b) What can be concluded about the weight of someone who can consume 3 drinks in an hour without reaching a blood-alcohol level of 0.08%?

SOLUTION

a) We go to the top of the bar that is above the body weight 200 lb. Then we move horizontally from the top of the bar to the vertical scale listing numbers of drinks. It appears that approximately 4 drinks will give a 200-lb person a blood-alcohol level of 0.08%.

b) By moving up the vertical scale to the number 3, and then moving horizontally, we see that the first bar to reach a height of 3 corresponds to a weight of 140 lb. Thus an individual should weigh over 140 lb if he or she wishes to consume 3 drinks in an hour without exceeding a blood-alcohol level of 0.08%.

Circle graphs, or *pie charts*, are often used to show what percent of the whole each particular item in a group represents.

EXAMPLE 2 Color Preference. The circle graph below shows the favorite colors of Americans and the percentage that prefers each color (*Source*: Vitaly Komar and Alex Melamid: The Most Wanted Paintings on the Web). Because of rounding, the total is slightly less than 100%. There are approximately 300 million Americans, and three quarters of them live in urban areas. Assuming that color preference is unaffected by geographical location, how many urban Americans choose red as their favorite color?

What is Your Favorite Color?

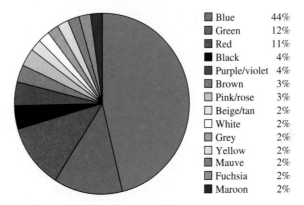

Blue	44%
Green	12%
Red	11%
Black	4%
Purple/violet	4%
Brown	3%
Pink/rose	3%
Beige/tan	2%
White	2%
Grey	2%
Yellow	2%
Mauve	2%
Fuchsia	2%
Maroon	2%

SOLUTION

1. Familiarize. The problem involves percents, so if we were unsure of how to solve percent problems, we might review Section 2.4. We are told that three quarters of all Americans live in urban areas. Since there are about 300 million Americans, this amounts to

$$\frac{3}{4} \cdot 300, \quad \text{or } 225 \text{ million urban Americans.}$$

The chart indicates that 11% of the U.S. population prefers red. We let r = the number of urban Americans who select red as their favorite color. We assume that the color preferences of urban Americans are typical of all Americans.

2. **Translate.** We reword and translate the problem as follows:

    Rewording: What   is   11%   of   225 million?

                      ↓   ↓   ↓   ↓   ↓

    Translating: r =   11%   ·   225,000,000

3. **Carry out.** We solve the equation:

    $r = 0.11 \cdot 225{,}000{,}000 = 24{,}750{,}000.$

4. **Check.** The check is left to the student.

5. **State.** About 24,750,000 urban Americans choose red as their favorite color.

◢

▼ **EXAMPLE 3**   Exercise and Pulse Rate.   The following line graph shows the relationship between a person's resting pulse rate and months of regular exercise.* Note that the symbol ⌇ is used to indicate that counting on the vertical axis begins at 50.

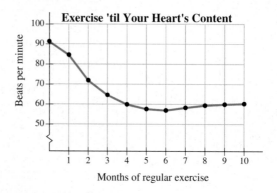

a) How many months of regular exercise are required to lower the pulse rate as much as possible?

b) How many months of regular exercise are needed to achieve a pulse rate of 65 beats per minute?

SOLUTION

a) The lowest point on the graph occurs above the number 6. Thus, after 6 months of regular exercise, the pulse rate is lowered as much as possible.

b) To determine how many months of exercise are needed to lower a person's resting pulse rate to 65, we locate 65 midway between 60 and 70 on the vertical axis. From that location, we move right until the line is

---

*Data from *Body Clock* by Dr. Martin Hughes (New York: Facts on File, Inc.), p. 60.

reached. At that point, we move down to the horizontal scale and read the number of months required, as shown.

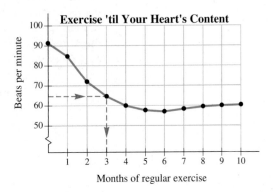

The pulse rate is 65 beats per minute after 3 months of regular exercise.

Points and Ordered Pairs

The line graph in Example 3 contains a collection of points. Each point pairs up a number of months of exercise with a pulse rate. To create such a graph, we **graph,** or **plot,** pairs of numbers on a plane. This is done using two perpendicular number lines called **axes** (pronounced "ak-sēz"; singular, **axis**). The point at which the axes cross is called the **origin.** Arrows on the axes indicate the positive directions.

Consider the pair (3, 4). The numbers in such a pair are called **coordinates.** The **first coordinate** in this case is 3 and the **second coordinate** is 4. To plot (3, 4), we start at the origin, move horizontally to the 3, move up vertically 4 units, and then make a "dot." Thus, (3, 4) is located above 3 on the first axis and to the right of 4 on the second axis.

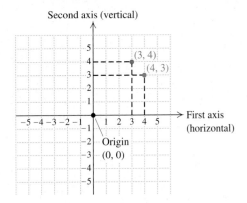

The point (4, 3) is also plotted in the figure above. Note that (3, 4) and (4, 3) are different points. For this reason, coordinate pairs are called **ordered pairs**—the order in which the numbers appear is important.

▶ **EXAMPLE 4**   Plot the point $(-3, 4)$.

SOLUTION   The first number, -3, is negative. Starting at the origin, we move 3 units in the negative horizontal direction (3 units to the left). The second number, 4, is positive, so we move 4 units in the positive vertical direction (up). The point $(-3, 4)$ is above -3 on the first axis and to the left of 4 on the second axis.

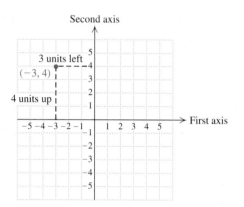

CAUTION!   Do not confuse the *ordered pair* (a, b) with the *interval* (a, b). The context in which the notation appears usually makes the meaning clear.

To find the coordinates of a point, we see how far to the right or left of the origin the point is and how far above or below the origin it is. Note that the coordinates of the origin itself are $(0, 0)$.

Student Notes

It is important that you remember that the first coordinate of an ordered pair is always represented on the horizontal axis. It is better to go slowly and plot points correctly than to go quickly and require more than one attempt.

▶ **EXAMPLE 5**   Find the coordinates of points A, B, C, D, E, F, and G.

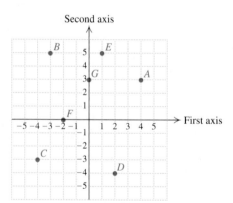

SOLUTION   Point A is 4 units to the right of the origin and 3 units above the origin. Its coordinates are (4, 3). The coordinates of the other points are as follows:

B: $(-3, 5)$; C: $(-4, -3)$; D: $(2, -4)$;

E: $(1, 5)$; F: $(-2, 0)$; G: $(0, 3)$.

Axes and Windows

We draw only part of the plane when we create a graph. Although it is standard to show the origin and parts of all quadrants, as in the graphs in Examples 4 and 5, for some applications it may be more practical to show a different portion of the plane. Note, too, that on the graphs in Examples 4 and 5, the marks on the axes are one unit apart. In this case, we say that the **scale** of each axis is 1.

Because the variable x is generally represented on the horizontal axis, we often call the first axis the **x-axis.** Similarly, because the variable y is generally represented on the vertical axis, the second axis is often called the **y-axis.**

Often it is necessary to graph a range of x-values and/or y-values that is too large to be displayed if each square of the grid is one unit wide and one unit high. For example, if we wish to plot the points $(-34, 450)$, $(48, 95)$, and $(10, -200)$ using a grid that is 10 squares wide and 10 squares high, we must make sure that x-values range, at least, from -34 to 48 and y-values extend, at least, from -200 to 450.

▶**EXAMPLE 6**   Use a grid 10 squares wide and 10 squares high to plot $(-34, 450)$, $(48, 95)$, and $(10, -200)$.

SOLUTION   Since x-values vary from a low of -34 to a high of 48, the 10 horizontal squares must span $48 - (-34)$, or 82 units. Because 82 is not a multiple of 10, we round *up* to the next multiple of 10, which is 90. Dividing 90 by 10, we note that if each square is 9 units wide (has a *scale* of 9), we could count backward from 0 four squares and forward six squares to include x-values from -36 to 54. However, since it is more convenient to count by 10's, we will instead count backward to -40 and forward to 60.

This is how we will arrange the *x*-axis.

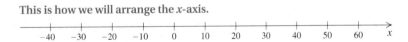

This is how we will arrange the *y*-axis.

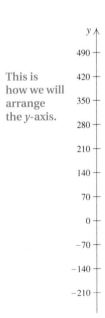

There is more than one correct way to cover the values from -34 to 48 using 10 increments. For instance, we could have counted from -60 to 90, using a scale of 15. In general, we try to use the smallest span and scale that will cover the given coordinates. Scales that are multiples of 2, 5, or 10 are especially convenient. It is essential that the numbering always starts from the origin.

Since we must be able to show y-values from -200 to 450, the 10 vertical squares must span $450 - (-200)$, or 650 units. For convenience, we round 650 *up* to 700 and then divide by 10: $700 \div 10 = 70$. Using 70 as the scale, we count *down* from 0 until we pass -200 and *up* from 0 until we pass 450, as shown at left.

Next, we combine our work with the *x*-values and the *y*-values to draw a graph in which the *x*-axis extends from −40 to 60 with a scale of 10 and the *y*-axis extends from −210 to 490 with a scale of 70. To correctly locate the axes on the grid, the two 0's must coincide where the axes cross. Finally, once the graph has been numbered, we plot the points as shown below.

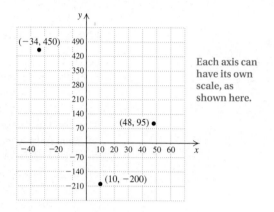

Each axis can have its own scale, as shown here.

EXAMPLE 7   Sketch the part of the plane showing the first, or horizontal, axis from −18 to 6 and the second, or vertical, axis from 2000 to 2500.

SOLUTION   We draw a pair of axes, making the horizontal axis longer in the negative direction than in the positive direction. Since we do not need to show the vertical axis from 0 to 2000, we draw a jagged break in the second axis. The horizontal axis is to show −18 to 6, so we choose a scale of 2 units per mark on the axis. The choice of scale may vary. We choose a scale of 100 for the vertical axis, labeling the marks on the axes carefully so that it is clear what scales are being used.

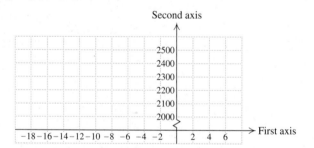

The horizontal and vertical axes divide the plane into four regions, or **quadrants,** as indicated by Roman numerals in the following figure. In region I (the *first quadrant*), both coordinates of any point are positive. In region II (the *second quadrant*), the first coordinate is negative and the second is positive. In region III (the *third quadrant*), both coordinates are negative. In region IV (the *fourth quadrant*), the first coordinate is positive and the second is negative.

Note that the point $(-4, 5)$ is in the second quadrant and the point $(5, -5)$ is in the fourth quadrant. The points $(3, 0)$ and $(0, 1)$ are on the axes and are not considered to be in any quadrant.

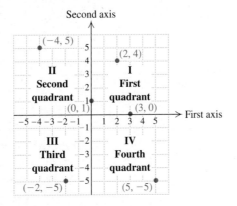

Windows

On a graphing calculator, the rectangular portion of the screen in which a graph appears is called the **viewing window.** Windows are described by four numbers of the form [L, R, B, T], representing the **L**eft and **R**ight endpoints of the *x*-axis and the **B**ottom and **T**op endpoints of the *y*-axis. The **standard viewing window** is the window described by $[-10, 10, -10, 10]$. On most graphing calculators, a standard viewing window can be set up quickly using the ZStandard option in the zoom menu.

Pressing ⟨WINDOW⟩, we set the window dimensions.

```
WINDOW
 Xmin = ⁻10
 Xmax = 10
 Xscl = 1
 Ymin = ⁻10
 Ymax = 10
 Yscl = 1
 Xres = 1
```

Xmin is the smallest *x*-value that will be displayed on the screen and Xmax is the largest. Similarly, the values of Ymin and Ymax determine the bottom and top endpoints of the vertical axis shown. The scales for the axes are set using Xscl and Yscl. Xres indicates the pixel resolution, which we generally set as Xres = 1.

(continued)

To see the viewing window, press (GRAPH).

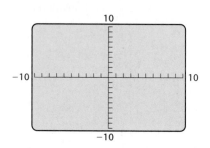

In the graph shown above, each mark on the axes represents 1 unit. In this text, the window dimensions are written outside the graphs. A scale other than 1 is indicated below the graph.

Inappropriate window dimensions may result in an error. For example, the message ERR:WINDOW RANGE occurs when Xmin is not less than Xmax.

EXAMPLE 8   Set up a $[-100, 100, -5, 5]$ viewing window on a graphing calculator, choosing appropriate scales for the axes.

SOLUTION   We are to show the x-axis from -100 to 100 and the y-axis from -5 to 5. The window dimensions are set as shown in the screen on the left below. Again, the choice of a scale for an axis may vary. If the scale chosen is too small, the marks on the axes will not be distinct. The resulting window is shown on the right below.

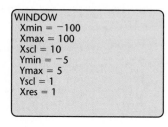

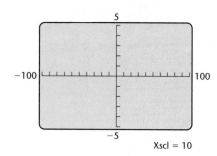

Xscl = 10

Note that the viewing window is not square. Also note that each mark on the x-axis represents 10 units, and each mark on the y-axis represents 1 unit.

Connecting the Concepts

INTERPRETING GRAPHS: READING GRAPHS

When a graph is used to model a problem, information can be read directly from the graph. Listed here are a few kinds of information that a graph contains.

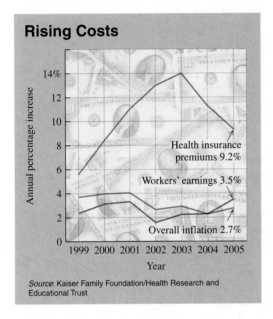

Rising Costs

Source: Kaiser Family Foundation/Health Research and Educational Trust

1. *Labels and units.* When reading a graph, look first at any labels. The title of the graph and the labels on the axes tell what information the graph represents. The scales on each axis indicate the numbers represented by a point. If the graph represents an equation, that equation will appear as a label.

   The title of the figure shown here indicates that it represents rising costs. There are three graphs shown in the figure. One represents health insurance premiums, one workers' earnings, and one overall inflation. The horizontal axis is labeled "Year," and the vertical axis is labeled "Annual percentage increase." Thus a second coordinate of 14 represents a 14% increase in cost.

2. *Relationships.* The coordinates of a point on a graph indicate a relationship between the quantities represented on the horizontal and vertical axes. For the graph labeled "Health insurance premiums," the point (2003, 14) indicates that health insurance premiums rose 14% in 2003. Since all three graphs are drawn on the same set of axes, the figure also shows relationships between increases in health insurance premiums, workers' earnings, and overall inflation. We can see that while workers' earnings have approximately kept pace with inflation, health insurance premiums have risen much more quickly.

3. *Trends.* As the graph continues from left to right, the quantity represented on the horizontal axis increases. The quantity represented on the vertical axis may increase, decrease, or remain constant. If the graph rises from left to right, the quantity on the vertical axis is increasing; if the graph falls from left to right, the quantity is decreasing; if the graph is level, the quantity is not changing, so it is constant.

   The health insurance premiums graph indicates that premiums increased at a higher rate each year between 1999 and 2003. For these years, the percentage increase itself increased. Between 2003 and 2005, the percentage increase decreased. Note that this does not mean that the premiums themselves decreased; it means instead that they increased at a slower rate.

4. *Values.* If we know one coordinate of a point on a graph, we can find the other coordinate. To find the percentage increase in overall inflation in 2001, we locate 2001 on the horizontal axis and move up to the graph labeled "Overall inflation." We then move horizontally to the corresponding value on the vertical axis. We see that in 2001, the overall inflation rate was 4%.

5. *Maximums and minimums.* A "high" point or a "low" point on a graph indicates a maximum or minimum value for the quantity on the vertical axis. The highest point on the health insurance premiums graph is approximately (2003, 14). Thus the maximum percentage increase for health insurance premiums was 14% in 2003.

3.1 EXERCISE SET

FOR EXTRA HELP

MathXL

MyMathLab

InterAct Math

Tutor Center

AW Math Tutor Center

Video Lectures on CD: Disc 2

Student's Solutions Manual

Concept Reinforcement In each of Exercises 1–4, match the set of coordinates with the graph from graphs (a)–(d) below that would be the best for plotting the points.

1. ___ $(-9, 3), (-2, -1), (1, 5)$

2. ___ $(-2, -1), (1, 5), (7, 3)$

3. ___ $(-2, -9), (2, 1), (4, -6)$

4. ___ $(-2, -1), (-9, 3), (-4, -6)$

a)
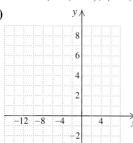

b)
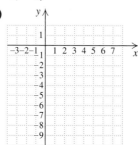

c)
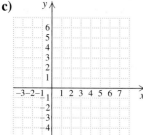

d)
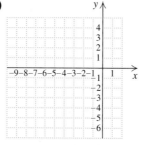

Blood Alcohol Level. Use the bar graph in Example 1 to answer Exercises 5–8.

5. Approximately how many drinks would a 100-lb person have consumed in 1 hr to reach a blood-alcohol level of 0.08%?

6. Approximately how many drinks would a 160-lb person have consumed in 1 hr to reach a blood-alcohol level of 0.08%?

7. What can you conclude about the weight of someone who has consumed 4 drinks in 1 hr without reaching a blood-alcohol level of 0.08%?

8. What can you conclude about the weight of someone who has consumed 5 drinks in 1 hr without reaching a blood-alcohol level of 0.08%?

Favorite Color. Use the information in Example 2 to answer Exercises 9–12. Assume that color preference is unaffected by geographical location or age.

9. About one-third of all Americans live in the South. How many Southerners choose brown as their favorite color?

10. About 50% of all Americans are at least 35 years old. How many Americans who are 35 or older choose purple/violet as their favorite color?

11. About one-eighth of all Americans are senior citizens. How many senior citizens choose black as their favorite color?

12. About one-fourth of all Americans are minors (under 18 years old). How many minors choose yellow as their favorite color?

Sorting Solid Waste. Use the following pie chart to answer Exercises 13–16.

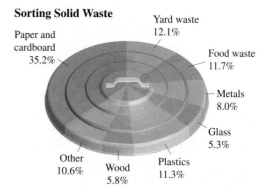

Sorting Solid Waste

Paper and cardboard 35.2%
Yard waste 12.1%
Food waste 11.7%
Metals 8.0%
Glass 5.3%
Plastics 11.3%
Wood 5.8%
Other 10.6%

Source: U.S. Environmental Protection Agency; based on 2003 data

13. In 2003, Americans generated 236 million tons of waste. How much of the waste was plastic?

14. In 2003, the average American generated 4.5 lb of waste per day. How much of that was paper and cardboard?

15. Americans are recycling about 22% of all glass that is in the waste stream. How much glass did Americans recycle in 2003? (See Exercise 13.)

16. Americans are recycling about 56.3% of all yard waste. How much yard waste did the average American recycle per day in 2003? (Use the information in Exercise 14.)

Cell-Phone Internet Access. *The graph below shows estimates of the number of cell phones with access to the Internet (Source: Forrester Research).*

Cell Phones with Web Access

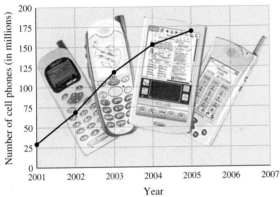

17. Approximately how many cell phones had Internet access in 2003?

18. Approximately how many cell phones had Internet access in 2002?

19. In what year did approximately 150 million cell phones have Internet access?

20. In what year did approximately 29 million cell phones have Internet access?

Plot each group of points.

21. $(1, 2), (-2, 3), (4, -1), (-5, -3), (4, 0), (0, -2)$

22. $(-2, -4), (4, -3), (5, 4), (-1, 0), (-4, 4), (0, 5)$

23. $(4, 4), (-2, 4), (5, -3), (-5, -5), (0, 4), (0, -4),$ $(3, 0), (-4, 0)$

24. $(2, 5), (-1, 3), (3, -2), (-2, -4), (0, 4), (0, -5),$ $(5, 0), (-5, 0)$

25. *Cell Phones.* Listed below are estimates of the number of cell phones with or without access to the Internet. Make a line graph of the data.

Year	Number of Cell Phones (in millions)
2001	119.8
2002	135.3
2003	150.2
2004	163.8
2005	176.9

Source: Forrester Research

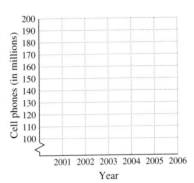

26. *Ozone Layer.* Make a line graph of the data in the following table, listing years on the horizontal scale.

Year	Ozone Level (in Dobson Units)
1996	290
1997	293
1998	294
1999	293
2000	292.1
2001	290.4
2002	292.6
2003	287.9
2004	284.3

Source: johnstonsarchive.net

Ozone Layer

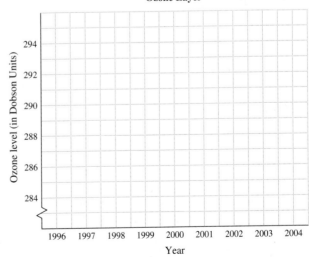

In Exercises 27–30, find the coordinates of points A, B, C, D, and E.

27.

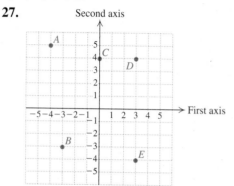

28.

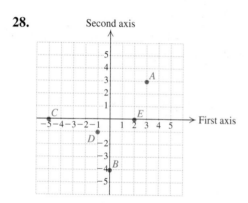

29.

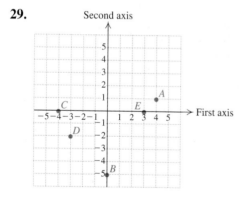

30.

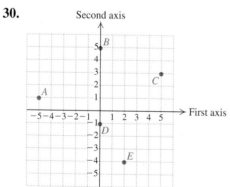

In Exercises 31–40, use a grid 10 squares wide and 10 squares high to plot the given coordinates. Choose your scale carefully. Scales may vary.

31. $(-75, 5)$, $(-18, -2)$, $(9, -4)$

32. $(-13, 3)$, $(48, -1)$, $(62, -4)$

33. $(-1, 83)$, $(-5, -14)$, $(5, 37)$

34. $(2, -79)$, $(4, -25)$, $(-4, 12)$

35. $(-10, -4)$, $(-16, 7)$, $(3, 15)$

36. $(5, -16)$, $(-7, -4)$, $(12, 3)$

37. $(-100, -5)$, $(350, 20)$, $(800, 37)$

38. $(750, -8)$, $(-150, 17)$, $(400, 32)$

39. $(-83, 491)$, $(-124, -95)$, $(54, -238)$

40. $(738, -89)$, $(-49, -6)$, $(-165, 53)$

In which quadrant is each point located?

41. $(7, -2)$ **42.** $(-1, -4)$ **43.** $(-4, -3)$

44. $(1, -5)$ **45.** $(2, 1)$ **46.** $(-4, 6)$

47. $(-4.9, 8.3)$ **48.** $(7.5, 2.9)$

49. In which quadrants are the first coordinates positive?

50. In which quadrants are the second coordinates negative?

51. In which quadrants do both coordinates have the same sign?

52. In which quadrants do the first and second coordinates have opposite signs?

TW **53.** The following graph was included in a mailing sent by Agway® to their oil customers in 2000. What information is missing from the graph and why is the graph misleading?

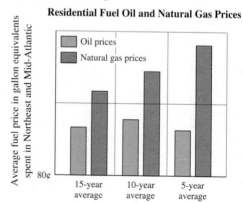

Residential Fuel Oil and Natural Gas Prices

Source: Energy Research Center, Inc. *3/1/99–2/29/00

TW **54.** What do all of the points on the vertical axis of a graph have in common?

Skill Maintenance

Simplify. [1.8]

55. $4 \cdot 3 - 6 \cdot 5$

56. $5(-2) + 3(-7)$

57. $-\frac{1}{2}(-6) + 3$

58. $-\frac{2}{3}(-12) - 7$

Solve for y. [2.3]

59. $3x - 2y = 6$

60. $7x - 4y = 14$

Synthesis

TW **61.** Describe what the result would be if the first and second coordinates of every point in the following graph of an arrow were interchanged.

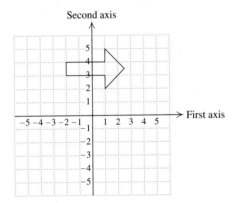

TW **62.** What advantage(s) does the use of a line graph have over that of a bar graph?

63. In which quadrant(s) could a point be located if its coordinates are opposites of each other?

64. In which quadrant(s) could a point be located if its coordinates are reciprocals of each other?

65. The points $(-1, 1)$, $(4, 1)$, and $(4, -5)$ are three vertices of a rectangle. Find the coordinates of the fourth vertex.

66. The pairs $(-2, -3)$, $(-1, 2)$, and $(4, -3)$ can serve as three (of four) vertices for three different parallelograms. Find the fourth vertex of each parallelogram.

67. Graph eight points such that the sum of the coordinates in each pair is 7. Answers may vary.

68. Find the perimeter of a rectangle if three of its vertices are $(5, -2)$, $(-3, -2)$, and $(-3, 3)$.

69. Find the area of a triangle whose vertices have coordinates $(0, 9)$, $(0, -4)$, and $(5, -4)$.

Coordinates on the Globe. *Coordinates can also be used to describe the location on a sphere: 0° latitude is the equator and 0° longitude is a line from the North Pole to the South Pole through France and Algeria. In the figure shown here, hurricane Clara is at a point about 260 mi northwest of Bermuda near latitude 36.0° North, longitude 69.0° West.*

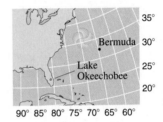

70. Approximate the latitude and the longitude of Bermuda.

71. Approximate the latitude and the longitude of Lake Okeechobee.

TW **72.** In the *Star Trek* science-fiction series, a three-dimensional coordinate system is used to locate objects in space. If the center of a planet is used as the origin, how many "quadrants" will exist? Why? If possible, sketch a three-dimensional coordinate system and label each "quadrant."

TW **73.** The graph accompanying Example 3 flattens out. Why do you think this occurs?

Collaborative Corner

You Sank My Battleship!

Focus: Graphing points
Time: 15–25 minutes
Group Size: 3–5
Materials: Graph paper

In the game Battleship®, a player places a miniature ship on a grid that only that player can see. An opponent guesses at coordinates that might "hit" the "hidden" ship. The following activity is similar to this game.

ACTIVITY

1. Using only integers from −10 to 10 (inclusive), one group member should secretly record the coordinates of a point on a slip of paper. (This point is the hidden "battleship.")

2. The other group members can then ask up to 10 "yes/no" questions in an effort to determine the coordinates of the secret point. Be sure to phrase each question mathematically (for example, "Is the *x*-coordinate negative?")

3. The group member who selected the point should answer each question. On the basis of the answer given, another group member should cross out the points no longer under consideration. All group members should check that this is done correctly.

4. If the hidden point has not been determined after 10 questions have been answered, the secret coordinates should be revealed to all group members.

5. Repeat parts (1)–(4) until each group member has had the opportunity to select the hidden point and answer questions.

3.2 Graphing Equations

Solutions of Equations ■ Graphing Linear Equations ■
Applications ■ Graphing Nonlinear Equations

We have seen how bar, line, and circle graphs can represent information. Now we begin to learn how graphs can be used to represent solutions of equations.

Solutions of Equations

When an equation contains two variables, solutions must be ordered pairs in which each number in the pair replaces a variable in the equation. Unless stated otherwise, the first number in each pair replaces the variable that occurs first alphabetically.

Study Tip

Talk Through Your Troubles

If you are finding it difficult to master a particular topic or concept, talk about it with a classmate. Verbalizing your questions about the material might help clarify it for you. If your classmate is also finding the material difficult, it is possible that the majority of the students in your class are confused and you can ask your instructor to explain the concept again.

EXAMPLE 1   Determine whether each of the following pairs is a solution of $4b - 3a = 22$: **(a)** (2, 7); **(b)** (1, 6).

SOLUTION

a) We substitute 2 for a and 7 for b (alphabetical order of variables):

$$
\begin{array}{c|c}
4b - 3a & = 22 \\
\hline
4 \cdot 7 - 3 \cdot 2 & 22 \\
28 - 6 & \\
22 \overset{?}{=} 22 & \text{TRUE}
\end{array}
$$

Since $22 = 22$ is *true*, the pair (2, 7) *is* a solution.

b) In this case, we replace a with 1 and b with 6:

$$
\begin{array}{c|c}
4b - 3a & = 22 \\
\hline
4 \cdot 6 - 3 \cdot 1 & 22 \\
24 - 3 & \\
21 \overset{?}{=} 22 & \text{FALSE}
\end{array}
$$

Since $21 = 22$ is *false*, the pair (1, 6) is *not* a solution.

EXAMPLE 2   Show that the pairs (3, 7), (0, 1), and (−3, −5) are solutions of $y = 2x + 1$. Then graph the three points to determine another pair that is a solution.

SOLUTION   To show that a pair is a solution, we substitute, replacing x with the first coordinate and y with the second coordinate of each pair.

$$
\begin{array}{c|c}
y & = 2x + 1 \\
\hline
7 & 2 \cdot 3 + 1 \\
& 6 + 1 \\
7 \overset{?}{=} 7 & \text{TRUE}
\end{array}
\qquad
\begin{array}{c|c}
y & = 2x + 1 \\
\hline
1 & 2 \cdot 0 + 1 \\
& 0 + 1 \\
1 \overset{?}{=} 1 & \text{TRUE}
\end{array}
\qquad
\begin{array}{c|c}
y & = 2x + 1 \\
\hline
-5 & 2(-3) + 1 \\
& -6 + 1 \\
-5 \overset{?}{=} -5 & \text{TRUE}
\end{array}
$$

In each of the three cases, the substitution results in a true equation. Thus the pairs (3, 7), (0, 1), and (−3, −5) are all solutions. We graph them at left, labeling the "first" axis x and the "second" axis y. Note that the three points appear to "line up." Will other points that line up with these points also represent solutions of $y = 2x + 1$? To find out, we use a ruler and draw a line passing through (−3, −5), (0, 1), and (3, 7).

The line appears to pass through (2, 5). Let's check if this pair is a solution of $y = 2x + 1$:

$$
\begin{array}{c|c}
y & = 2x + 1 \\
\hline
5 & 2 \cdot 2 + 1 \\
& 4 + 1 \\
5 \overset{?}{=} 5 & \text{TRUE}
\end{array}
$$

We see that (2, 5) *is* a solution. You should perform a similar check for at least one other point that appears to be on the line.

Example 2 leads us to suspect that *any* point on the line passing through $(3, 7)$, $(0, 1)$, and $(-3, -5)$ represents a solution of $y = 2x + 1$. In fact, every solution of $y = 2x + 1$ is represented by a point on this line and every point on this line represents a solution. The line is called the **graph** of the equation.

Connecting the Concepts

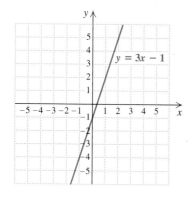

INTERPRETING GRAPHS: SOLUTIONS

A solution of an equation like $5 = 3x - 1$ is a number. For this particular equation, the solution is 2. Were we to graph the solution, it would be a point on the number line.

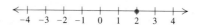

A solution of an equation like $y = 3x - 1$ is an ordered pair. One solution of this equation is $(2, 5)$. The graph of the solutions of this equation is a line in a coordinate plane, as shown in the figure at left.

Graphing Linear Equations

Equations like $y = 2x + 1$ or $4b - 3a = 22$ are said to be **linear** because the graph of each equation is a line. In general, any equation that can be written in the form $y = mx + b$ or $Ax + By = C$ (where m, b, A, B, and C are constants and A and B are not both 0) is linear. An equation of the form $Ax + By = C$ is said to be written in **standard form.**

To *graph* an equation is to make a drawing that represents its solutions. Linear equations can be graphed as follows.

To Graph a Linear Equation

1. Select a value for one coordinate and calculate the corresponding value of the other coordinate. Form an ordered pair. This pair is one solution of the equation.
2. Repeat step (1) to find a second ordered pair. A third ordered pair can be used as a check.
3. Plot the ordered pairs and draw a straight line passing through the points. The line represents all solutions of the equation.

EXAMPLE 3   Graph: $y = -3x + 1$.

SOLUTION   We select a convenient value for x, compute y, and form an ordered pair. Then we repeat the process for other choices of x.

If $x = 2$,   then $y = -3 \cdot 2 + 1 = -5$,   and $(2, -5)$ is a solution.

If $x = 0$,   then $y = -3 \cdot 0 + 1 = 1$,   and $(0, 1)$ is a solution.

If $x = -1$,   then $y = -3(-1) + 1 = 4$,   and $(-1, 4)$ is a solution.

Results are often listed in a table, as shown below. The points corresponding to each pair are then plotted.

$$y = -3x + 1$$

x	y	(x, y)
2	-5	$(2, -5)$
0	1	$(0, 1)$
-1	4	$(-1, 4)$

(1) Choose x.
(2) Compute y.
(3) Form the pair (x, y).
(4) Plot the points.

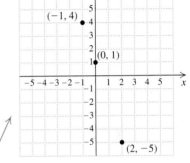

Note that all three points line up. If they didn't, we would know that we had made a mistake, because the equation is linear. When only two points are plotted, an error is more difficult to detect.

Finally, we use a ruler or other straightedge to draw a line. Every point on the line represents a solution of $y = -3x + 1$.

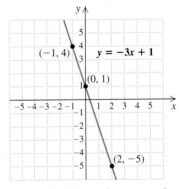

Note that the coordinates of *any* point on the line—for example, $(1, -2)$—satisfy the equation $y = -3x + 1$. Note too that the line continues indefinitely in both directions—only part of it is shown.

A graphing calculator also graphs equations by calculating solutions, plotting points, and connecting them. After entering an equation and setting a viewing window, we can view the graph of the equation.

Graphing Equations

Equations are entered using the equation-editor screen, often accessed by pressing (Y=). The first part of each equation, "Y=," is supplied by the calculator, including a subscript that identifies the equation.

(continued)

The variables used by the calculator are x and y. If an equation contains different variables, such as a and b, replace them with x and y before entering the equation. An equation can be cleared by positioning the cursor on the equation and pressing **CLEAR**. On many calculators, a symbol before the Y indicates the graph style, and a highlighted $=$ indicates that the equation selected is to be graphed. An equation can be selected or deselected by positioning the cursor on the $=$ and pressing **ENTER**. Selected equations are then graphed by pressing **GRAPH**. *Note*: If the window is not set appropriately, the graph may not appear on the screen at all.

A standard viewing window is not always the best window to use. Choosing an appropriate viewing window for a graph can be challenging. There is generally no one "correct" window; the choice can vary according to personal preference and can also be dictated by the portion of the graph that you need to see. It often involves trial and error, but as you learn more about graphs of equations, you will be able to determine an appropriate viewing window more directly. For now, start with a standard window, and change the dimensions if needed. The ZOOM menu can help in setting window dimensions; for example, choosing ZStandard from the menu will graph the selected equations using the standard viewing window.

```
ZOOM  MEMORY
1: ZBox
2: Zoom In
3: Zoom Out
4: ZDecimal
5: ZSquare
6: ZStandard
7↓ ZTrig
```

EXAMPLE 4 Graph $y = 2x$ using a graphing calculator.

SOLUTION After pressing **Y=** and clearing any other equations present, we enter $y = 2x$. We use **X,T,θ,n** to enter x. The standard $[-10, 10, -10, 10]$ window is a good choice for this graph. We check the window dimensions and then graph the equation by pressing **GRAPH**.

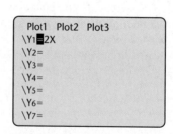

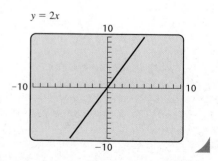

$y = 2x$

The equation in the next example is graphed both by hand and by using a graphing calculator. Let's compare the processes and the results.

▼ **EXAMPLE 5**   Graph the equation $y = -\frac{1}{2}x$.

SOLUTION

BY HAND

We find some ordered pairs that are solutions. This time we list the pairs in a table. To find an ordered pair, we can choose *any* number for x and then determine y. By choosing even numbers for x, we can avoid fractions when calculating y. For example, if we choose 4 for x, we get $y = \left(-\frac{1}{2}\right)(4)$, or -2. If x is -6, we get $y = \left(-\frac{1}{2}\right)(-6)$, or 3. We find several ordered pairs, plot them, and draw the line.

$$y = -\tfrac{1}{2}x$$

x	y	(x, y)
4	-2	$(4, -2)$
-6	3	$(-6, 3)$
0	0	$(0, 0)$
2	-1	$(2, -1)$

↑ Choose any x.

↑ Compute y.

↑ Form the pair.

Plot the points and
draw the line.

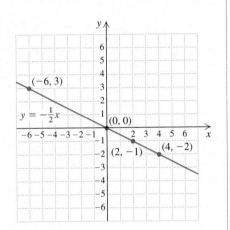

WITH A GRAPHING CALCULATOR

To enter the equation, we access the equation editor and clear any equations present. We then enter the equation as $y = -(1/2)x$. Remember to use the ⊖ key for the negative sign. Also note that some calculators require parentheses around a fraction coefficient. The standard viewing window is a good choice for this graph.

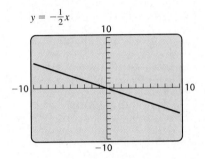

▼ **EXAMPLE 6**   Graph: $4x + 2y = 12$.

SOLUTION   To form ordered pairs, we can replace either variable with a number and then calculate the other coordinate:

If $y = 0$, we have $4x + 2 \cdot 0 = 12$
$$4x = 12$$
$$x = 3,$$

so $(3, 0)$ is a solution.

If $x = 0$, we have $4 \cdot 0 + 2y = 12$
$$2y = 12$$
$$y = 6,$$

so $(0, 6)$ is a solution.

If $y = 2$, we have $4x + 2 \cdot 2 = 12$

$$4x + 4 = 12$$
$$4x = 8$$
$$x = 2,$$

so $(2, 2)$ is a solution.

$4x + 2y = 12$

x	y	(x, y)
3	0	$(3, 0)$
0	6	$(0, 6)$
2	2	$(2, 2)$

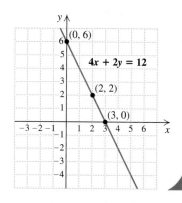

Note that in Examples 3–5 the variable y is isolated on one side of the equation. This generally simplifies calculations, so it is important to be able to solve for y before graphing. Also, equations must generally be solved for y before they can be entered on a graphing calculator.

EXAMPLE 7   Graph $3y = 2x$ by first solving for y.

SOLUTION   To isolate y, we divide both sides by 3, or multiply both sides by $\frac{1}{3}$:

$$3y = 2x$$

$\frac{1}{3} \cdot 3y = \frac{1}{3} \cdot 2x$ **Using the multiplication principle to multiply both sides by $\frac{1}{3}$**

$\left. \begin{array}{l} 1y = \frac{2}{3} \cdot x \\ y = \frac{2}{3}x. \end{array} \right\}$ **Simplifying**

Because all the equations above are equivalent, we can use $y = \frac{2}{3}x$ to draw the graph of $3y = 2x$.

To graph $y = \frac{2}{3}x$, we can select x-values that are multiples of 3. This will allow us to avoid fractions when the corresponding y-values are computed.

$\left. \begin{array}{ll} \text{If } x = 3, & \text{then } y = \frac{2}{3} \cdot 3 = 2. \\ \text{If } x = -3, & \text{then } y = \frac{2}{3}(-3) = -2. \\ \text{If } x = 6, & \text{then } y = \frac{2}{3} \cdot 6 = 4. \end{array} \right\}$ **Note that when multiples of 3 are substituted for x, the y-coordinates are not fractions.**

The following table lists these solutions. Next, we plot the points and see that they form a line. Finally, we draw and label the line.

$3y = 2x,$ or $y = \frac{2}{3}x$

x	y	(x, y)
3	2	$(3, 2)$
-3	-2	$(-3, -2)$
6	4	$(6, 4)$

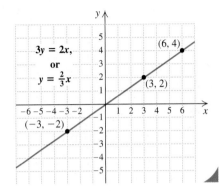

EXAMPLE 8 Graph $x + 5y = -10$ by first solving for y.

SOLUTION We have

$$x + 5y = -10$$
$$-x + x + 5y = -x - 10 \qquad \text{Adding } -x \text{ to both sides}$$
$$5y = -x - 10$$
$$\tfrac{1}{5} \cdot 5y = \tfrac{1}{5}(-x - 10) \qquad \text{Multiplying both sides by } \tfrac{1}{5}$$
$$y = -\tfrac{1}{5}x - 2. \qquad \text{Using the distributive law}$$

> **CAUTION!** You must multiply both $-x$ and -10 by $\frac{1}{5}$.

Thus, $x + 5y = -10$ is equivalent to $y = -\frac{1}{5}x - 2$. If we choose x-values that are multiples of 5, we can avoid fractions when calculating the corresponding y-values.

If $x = 5,$ then $y = -\frac{1}{5} \cdot 5 - 2 = -1 - 2 = -3.$
If $x = 0,$ then $y = -\frac{1}{5} \cdot 0 - 2 = 0 - 2 = -2.$
If $x = -5,$ then $y = -\frac{1}{5}(-5) - 2 = 1 - 2 = -1.$

$x + 5y = -10,$ or $y = -\frac{1}{5}x - 2$

x	y	(x, y)
5	-3	$(5, -3)$
0	-2	$(0, -2)$
-5	-1	$(-5, -1)$

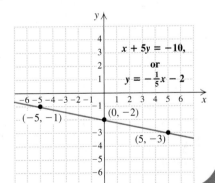

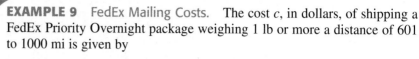

Applications

Linear equations appear in many real-life situations.

EXAMPLE 9 FedEx Mailing Costs. The cost c, in dollars, of shipping a FedEx Priority Overnight package weighing 1 lb or more a distance of 601 to 1000 mi is given by

$$c = 3.7w + 30.70,$$

where w is the package's weight in pounds (*Source*: Federal Express Corporation). Graph the equation and then use the graph to estimate the cost of shipping a $6\frac{1}{2}$-lb package.

SOLUTION We graph $c = 3.7w + 30.70$ by selecting values for w and then calculating the associated value c. In selecting values for w, note that, according to the problem, we must have $w \geq 1$.

If $w = 1$, then $c = 3.7(1) + 30.70 = 34.40$.

If $w = 4$, then $c = 3.7(4) + 30.70 = 45.50$.

If $w = 7$, then $c = 3.7(7) + 30.70 = 56.60$.

w	c
1	34.40
4	45.50
7	56.60

Because we are *selecting* values for w and *calculating* values for c, we represent w on the horizontal axis and c on the vertical axis. Counting by 2's horizontally and by 5's vertically will allow us to plot all three pairs, as shown on the left below.

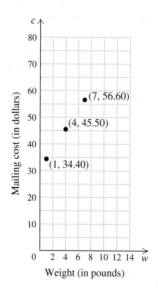

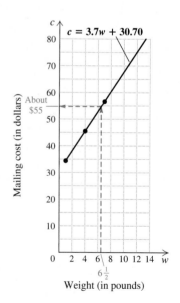

CAUTION! When the coordinates of a point are read from a graph, as in Example 9, values should not be considered exact.

Since the three points line up, our calculations are probably correct. We draw a line, beginning at (1, 34.40). To estimate the cost of shipping a $6\frac{1}{2}$-lb package, we locate the point on the line that is above $6\frac{1}{2}$ and then find the value on the c-axis that corresponds to that point, as shown on the right above. The cost of shipping a $6\frac{1}{2}$-lb package is about \$55.

To graph an equation like the one in Example 9 using a graphing calculator, we replace the variables in the equation with x and y. To graph $c = 3.7w + 30.70$, we enter the equation $y = 3.7x + 30.70$. The graph of the equation is shown below. We show only the first quadrant because neither x nor y can be negative. Note that the equation does not apply to weights less than 1 lb, but the graph shown does extend below $(1, 34.40)$.

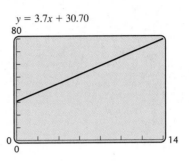

$y = 3.7x + 30.70$

Graphing Nonlinear Equations

We refer to any equation whose graph is a straight line as a **linear equation.** Many equations, however, are not linear. When ordered pairs that are solutions of such an equation are plotted, the pattern formed is not a straight line. Let's look at the graph of a **nonlinear equation.**

▶ **EXAMPLE 10**   Graph: $y = x^2 - 5$.

SOLUTION   We select numbers for x and find the corresponding values for y. For example, if we choose -2 for x, we get $y = (-2)^2 - 5 = 4 - 5 = -1$. The table lists several ordered pairs.

$y = x^2 - 5$

x	y
0	-5
-1	-4
1	-4
-2	-1
2	-1
-3	4
3	4

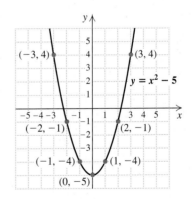

Next, we plot the points. The more points plotted, the clearer the shape of the graph becomes. Since the value of $x^2 - 5$ grows rapidly as x moves away from the origin, the graph rises steeply on either side of the y-axis.

Curves similar to the one in Example 10 are studied in detail in Chapter 11. Graphing calculators are especially helpful when drawing such nonlinear graphs.

3.2 EXERCISE SET

Determine whether each equation has the given ordered pair as a solution.

1. $y = 5x + 1$; $(0, 2)$

2. $y = 2x + 3$; $(0, 3)$

3. $3y + 2x = 12$; $(4, 2)$

4. $5x - 3y = 15$; $(0, 5)$

5. $4a - 3b = 11$; $(2, -1)$

6. $3q - 2p = -8$; $(1, -2)$

In Exercises 7–14, an equation and two ordered pairs are given. Show that each pair is a solution of the equation. Then graph the two pairs to determine another solution. See Example 2. Answers may vary.

7. $y = x + 3$; $(-1, 2), (4, 7)$

8. $y = x - 2$; $(3, 1), (-2, -4)$

9. $y = \frac{1}{2}x + 3$; $(4, 5), (-2, 2)$

10. $y = \frac{1}{2}x - 1$; $(6, 2), (0, -1)$

11. $y + 3x = 7$; $(2, 1), (4, -5)$

12. $2y + x = 5$; $(-1, 3), (7, -1)$

13. $4x - 2y = 10$; $(0, -5), (4, 3)$

14. $6x - 3y = 3$; $(1, 1), (-1, -3)$

Graph each equation by hand.

15. $y = x + 1$

16. $y = x - 1$

17. $y = -x$

18. $y = x$

19. $y = \frac{1}{3}x$

20. $y = \frac{1}{2}x$

21. $y = x + 3$

22. $y = x + 2$

23. $y = 2x + 2$

24. $y = 3x - 2$

25. $y = \frac{1}{3}x - 4$

26. $y = \frac{1}{2}x + 1$

27. $x + y = 4$

28. $x + y = -5$

29. $y = \frac{5}{2}x + 3$

30. $y = \frac{5}{3}x - 2$

31. $x + 2y = -6$

32. $x + 2y = 8$

33. $y = -\frac{2}{3}x + 4$

34. $y = \frac{3}{2}x + 1$

35. $8x - 4y = 12$

36. $6x - 3y = 9$

37. $6y + 2x = 8$

38. $8y + 2x = -4$

39. $y = x^2 + 1$

40. $y = x^2 - 3$

41. $y = 2 - x^2$

42. $y = 3 + x^2$

Graph the solutions of each equation.

43. **(a)** $x + 1 = 4$; **(b)** $x + 1 = y$

44. **(a)** $2x = 7$; **(b)** $2x = y$

45. **(a)** $y = 2x - 3$; **(b)** $-7 = 2x - 3$

46. **(a)** $y = x - 3$; **(b)** $-2 = x - 3$

Graph each equation using a graphing calculator. Remember to solve for y first if necessary.

47. $y = -\frac{3}{2}x + 1$

48. $y = -\frac{2}{3}x - 2$

49. $4y - 3x = 1$

50. $4y - 3x = 2$

51. $y = -2$

52. $y = 5$

53. $y = -x^2$

54. $y = x^2$

55. $x^2 - y = 3$

56. $y - 2 = x^2$

57. $y = x^3$

58. $y = x^3 - 2$

Graph each equation using both viewing windows indicated. Determine which window best shows both the shape of the graph and where the graph crosses the x- and y-axes.

59. $y = x - 15$
   a) $[-10, 10, -10, 10]$, Xscl = 1, Yscl = 1
   b) $[-20, 20, -20, 20]$, Xscl = 5, Yscl = 5

60. $y = -3x + 30$
   a) $[-10, 10, -10, 10]$, Xscl = 1, Yscl = 1
   b) $[-20, 20, -20, 40]$, Xscl = 5, Yscl = 5

61. $y = 5x^2 - 8$
   a) $[-10, 10, -10, 10]$, Xscl = 1, Yscl = 1
   b) $[-3, 3, -3, 3]$, Xscl = 1, Yscl = 1

62. $y = \frac{1}{10}x^2 + \frac{1}{3}$

 a) $[-10, 10, -10, 10]$, Xscl = 1, Yscl = 1
 b) $[-0.5, 0.5, -0.5, 0.5]$, Xscl = 0.1, Yscl = 0.1

63. $y = 4x^3 - 12$

 a) $[-10, 10, -10, 10]$, Xscl = 1, Yscl = 1
 b) $[-5, 5, -20, 10]$, Xscl = 1, Yscl = 5

64. $y = x^2 + 10$

 a) $[-10, 10, -10, 10]$, Xscl = 1, Yscl = 1
 b) $[-3, 3, 0, 20]$, Xscl = 1, Yscl = 5

Solve by graphing. Label all axes, and show where each solution is located on the graph.

65. *Health Insurance.* The number of full-time workers in the United States without health insurance continues to grow. This number, *n*, can be approximated by

$$n = 0.9t + 19,$$

where *n* is in millions and *t* is the number of years since 2001. (*Source: The New York Times*, 9/30/03) Graph the equation and use the graph to estimate the number of uninsured full-time workers in 2010.

66. *Value of a Color Copier.* The value of Dupliographic's color copier is given by

$$v = -0.68t + 3.4,$$

where *v* is the value, in thousands of dollars, *t* years from the date of purchase. Graph the equation and use the graph to estimate the value of the copier after $2\frac{1}{2}$ yr.

67. *Increasing Life Expectancy.* A smoker is 15 times more likely to die from lung cancer than a nonsmoker. An ex-smoker who stopped smoking *t* years ago is *w* times more likely to die from lung cancer than a nonsmoker, where

$$w = 15 - t.$$

(*Source:* Data from *Body Clock* by Dr. Martin Hughes, p. 60. New York: Facts on File, Inc.) Graph the equation and use the graph to estimate how much more likely it is for Sandy to die from lung cancer than Polly, if Polly never smoked and Sandy quit $2\frac{1}{2}$ years ago.

68. *Price of Printing.* The price *p*, in cents, of a photocopied and bound lab manual is given by $p = \frac{7}{2}n + 20$, where *n* is the number of pages in the manual. Graph the equation and use the graph to estimate the cost of a 25-page manual.

69. *Value of Computer Software.* The value *v* of a shopkeeper's inventory software program, in

hundreds of dollars, is given by $v = -\frac{3}{4}t + 6$, where *t* is the number of years since the shopkeeper first bought the program. Graph the equation and use the graph to estimate what the program is worth 4 yr after it was first purchased.

70. *Bottled Water.* The number of gallons of bottled water *w* consumed by an average American in one year is given by $w = 1.1t + 7.3$, where *t* is the number of years since 1990 (*Source:* Beverage Marketing Corporation). Graph the equation and use the graph to predict the number of gallons consumed per person in 2008.

71. *Cost of College.* The cost *T*, in hundreds of dollars, of tuition and fees at many community colleges can be approximated by $T = \frac{3}{4}c + 1$, where *c* is the number of credits for which a student registers. Graph the equation and use the graph to estimate the cost of tuition and fees when a student registers for 4 three-credit courses.

72. *Cost of College.* The cost *C*, in thousands of dollars, of a year at a private four-year college (all expenses) can be approximated by $C = \frac{6}{5}t + 21$, where *t* is the number of years since 1995 (*Source:* Based on information from U.S. National Center for Education Statistics). Graph the equation and use the graph to estimate the cost of a year at a private four-year college in 2009.

73. *Record Temperature Drop.* On January 22, 1943, the temperature *T*, in degrees Fahrenheit, in Spearfish, South Dakota, could be approximated by $T = -2m + 54$, where *m* is the number of minutes since 9:00 A.M. that morning (*Source:* Based on information from the National Oceanic and Atmospheric Administration). Graph the equation and use the graph to estimate the temperature at 9:15 A.M.

74. *Coffee Consumption.*   The number of gallons of coffee n consumed each year by the average U.S. consumer can be approximated by $n = \frac{6}{5}d + 20$, where d is the number of years since 1995 (*Source*: Based on information from the U.S. Department of Agriculture). Graph the equation and use the graph to estimate what the average coffee consumption was in 2004.

™ **75.** The equations $3x + 4y = 8$ and $y = -\frac{3}{4}x + 2$ are equivalent. Which equation would be easier to graph and why?

™ **76.** Suppose that a linear equation is graphed by plotting three points and that the three points line up with each other. Does this *guarantee* that the equation is being correctly graphed? Why or why not?

Skill Maintenance

Solve and check. [2.2]

77. $5x + 3 \cdot 0 = 12$ **78.** $2x - 5 \cdot 0 = 9$

79. $7 \cdot 0 - 4y = 10$

Solve. [2.3]

80. $pq + p = w$, for p

81. $Ax + By = C$, for y

82. $A = \dfrac{T + Q}{2}$, for Q

Synthesis

™ **83.** Janice consistently makes the mistake of plotting the x-coordinate of an ordered pair using the y-axis, and the y-coordinate using the x-axis. How will Janice's incorrect graph compare with the appropriate graph?

™ **84.** Explain how the graph in Example 9 can be used to determine when the cost of a shipment will reach $50.

85. *Bicycling.*   Long Beach Island in New Jersey is a long, narrow, flat island. For exercise, Lauren routinely bikes to the northern tip of the island and back. Because of the steady wind, she uses one gear going north and another for her return. Lauren's bike has 14 gears and the sum of the two gears used

on her ride is always 18. Write and graph an equation that represents the different pairings of gears that Lauren uses. Note that there are no fractional gears on a bicycle.

In Exercises 86–89, try to find an equation for the graph shown.

86.
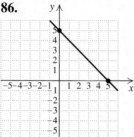

87.
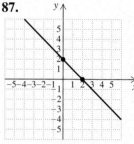

88.
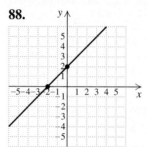

89.
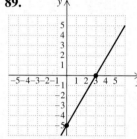

90. Translate to an equation:

d dimes and n nickels total $1.75.

Then graph the equation and use the graph to determine three different combinations of dimes and nickels that total $1.75 (see also Exercise 97).

91. Translate to an equation:

d $25 dinners and l $5 lunches total $225.

Then graph the equation and use the graph to determine three different combinations of lunches and dinners that total $225 (see also Exercise 97).

Use the suggested x-values $-3, -2, -1, 0, 1, 2$, and 3 to graph each equation. Check using a graphing calculator.

92. $y = |x|$ ᴬʰᵃ! **93.** $y = -|x|$

ᴬʰᵃ! **94.** $y = |x| - 2$ **95.** $y = -|x| + 2$

96. $y = |x| + 3$

™ **97.** Study the graph of Exercise 90 or 91. Does *every* point on the graph represent a solution of the associated problem? Why or why not?

3.3 Linear Equations and Intercepts

Recognizing Linear Equations ▨ Intercepts ▨ Using Intercepts to Graph ▨ Graphing Horizontal or Vertical Lines

As we saw in Section 3.2, a *linear equation* is an equation whose graph is a straight line. We can determine whether an equation is linear without graphing.

Recognizing Linear Equations

A linear equation may appear in different forms, but all linear equations can be written in the *standard form* $Ax + By = C$.

> **The Standard Form of a Linear Equation**   Any equation of the form $Ax + By = C$, where A, B, and C are real numbers and A and B are not both 0, is linear.
>
> Any equation of the form $Ax + By = C$ is said to be a linear equation in *standard form*.

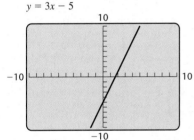

$y = 3x - 5$

EXAMPLE 1   Determine whether each of the following equations is linear:
(a) $y = 3x - 5$; **(b)** $y = x^2 - 5$; **(c)** $3y = 7$.

SOLUTION   We attempt to write each equation in the form $Ax + By = C$.

a) We have

$$y = 3x - 5$$
$$-3x + y = -5. \quad \text{Adding } -3x \text{ to both sides}$$

The equation $-3x + y = -5$ is in the form $Ax + By = C$, with $A = -3$, $B = 1$, and $C = -5$. The equation $y = 3x - 5$ is linear, as confirmed by the first graph at left.

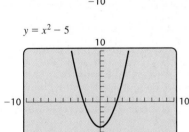

$y = x^2 - 5$

b) We attempt to put the equation in standard form:

$$y = x^2 - 5$$
$$-x^2 + y = -5. \quad \text{Adding } -x^2 \text{ to both sides}$$

This last equation is not linear because it has an x^2-term.
    We can see this as well from the graph of the equation shown at left.

c) Although there is no x-term in $3y = 7$, the definition of the standard form allows for either A or B to be 0. Thus if we write the equation as

$$0 \cdot x + 3y = 7,$$

the equation is in the standard form of a linear equation, with $A = 0$, $B = 3$, and $C = 7$. Thus, $3y = 7$ is linear, as we can see in the third graph at left.

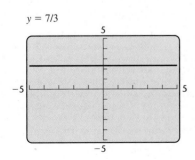

$y = 7/3$

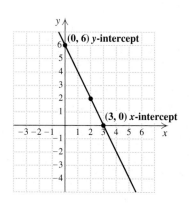

Once we have determined that an equation is linear, we can use any two points to graph the line. This is possible because, if we are given two points, only one line can be drawn through them. Unless a line is horizontal or vertical, it will cross both axes at points known as *intercepts*. Often, finding the intercepts of a graph is a convenient way to graph a linear equation.

Intercepts

In Example 6 of Section 3.2, we graphed $4x + 2y = 12$ by plotting the points $(3, 0)$, $(0, 6)$, and $(2, 2)$ and then drawing the line.

The point at which a graph crosses the y-axis is called the **y-intercept.** In the figure shown at left, the y-intercept is $(0, 6)$. The x-coordinate of a y-intercept is always 0.

The point at which a graph crosses the x-axis is called the **x-intercept.** In the figure shown at left, the x-intercept is $(3, 0)$. The y-coordinate of an x-intercept is always 0.

It is possible for the graph of a nonlinear curve to have more than one y-intercept or more than one x-intercept.

> **EXAMPLE 2** For the graph shown below, **(a)** give the coordinates of any x-intercepts and **(b)** give the coordinates of any y-intercepts.

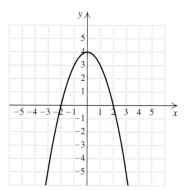

SOLUTION

a) The x-intercepts are the points at which the graph crosses the x-axis. For the graph shown, the x-intercepts are $(-2, 0)$ and $(2, 0)$.

b) The y-intercept is the point at which the graph crosses the y-axis. For the graph shown, the y-intercept is $(0, 4)$.

Using Intercepts to Graph

It is important to know how to locate a graph's intercepts from the equation being graphed.

> ### To Find Intercepts
>
> To find the y-intercept(s) of an equation's graph, replace x with 0 and solve for y.
>
> To find the x-intercept(s) of an equation's graph, replace y with 0 and solve for x.

EXAMPLE 3   Find the *y*-intercept and the *x*-intercept of the graph of $2x + 4y = 20$.

SOLUTION   To find the *y*-intercept, we let $x = 0$ and solve for *y*:

$$2 \cdot 0 + 4y = 20 \qquad \text{Replacing } x \text{ with } 0$$
$$4y = 20$$
$$y = 5.$$

Thus the *y*-intercept is $(0, 5)$.
  To find the *x*-intercept, we let $y = 0$ and solve for *x*:

$$2x + 4 \cdot 0 = 20 \qquad \text{Replacing } y \text{ with } 0$$
$$2x = 20$$
$$x = 10.$$

Thus the *x*-intercept is $(10, 0)$.

Intercepts can be used to graph a linear equation.

EXAMPLE 4   Graph $2x + 4y = 20$ using intercepts.

SOLUTION   In Example 3, we showed that the *y*-intercept is $(0, 5)$ and the *x*-intercept is $(10, 0)$. We plot both intercepts. Before drawing a line, we plot a third point as a check. We substitute any convenient value for *x* and solve for *y*.
  If we let $x = 5$, then

$$2 \cdot 5 + 4y = 20 \qquad \textbf{Substituting 5 for } x$$
$$10 + 4y = 20$$
$$4y = 10 \qquad \textbf{Subtracting 10 from both sides}$$
$$y = \tfrac{10}{4}, \text{ or } 2\tfrac{1}{2}. \qquad \textbf{Solving for } y$$

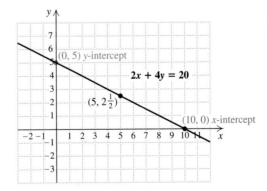

The point $\left(5, 2\tfrac{1}{2}\right)$ appears to line up with the intercepts, so our work is probably correct. To finish, we draw and label the line.

Note that when we solved for the *y*-intercept, we simplified $2x + 4y = 20$ to $4y = 20$. Thus, to find the *y*-intercept, we can momentarily ignore the *x*-term and solve the remaining equation.

In a similar manner, when we solved for the *x*-intercept, we simplified $2x + 4y = 20$ to $2x = 20$. Thus, to find the *x*-intercept, we can momentarily ignore the *y*-term and then solve this remaining equation.

EXAMPLE 5   Graph $3x - 2y = 60$ using intercepts.

SOLUTION   To find the *y*-intercept, we let $x = 0$. This amounts to temporarily ignoring the *x*-term and then solving:

$$-2y = 60 \qquad \text{For } x = 0, \text{ we have } 3 \cdot 0 - 2y, \text{ or simply } -2y.$$
$$y = -30.$$

The *y*-intercept is $(0, -30)$.

To find the *x*-intercept, we let $y = 0$. This amounts to temporarily disregarding the *y*-term and then solving:

$$3x = 60 \qquad \text{For } y = 0, \text{ we have } 3x - 2 \cdot 0, \text{ or simply } 3x.$$
$$x = 20.$$

The *x*-intercept is $(20, 0)$.

To find a third point, we replace *x* with 4 and solve for *y*:

$$3 \cdot 4 - 2y = 60 \qquad \text{Numbers other than 4 can be used for } x.$$
$$12 - 2y = 60$$
$$-2y = 48$$
$$y = -24. \qquad \text{This means that } (4, -24) \text{ is on the graph.}$$

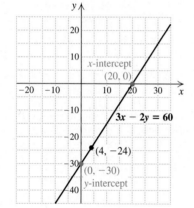

In order for us to graph all three points, the *y*-axis of our graph must go down to at least -30 and the *x*-axis must go up to at least 20. Using a scale of 5 units per square allows us to display both intercepts and $(4, -24)$, as well as the origin.

The point $(4, -24)$ appears to line up with the intercepts, so we draw and label the line, as shown at left.

EXAMPLE 6   Determine a viewing window that shows the intercepts of the graph of the equation $y = 2x + 15$.

SOLUTION   To find the *y*-intercept, we let $x = 0$:

$$y = 2 \cdot 0 + 15$$
$$y = 15.$$

The *y*-intercept is $(0, 15)$.

To find the *x*-intercept, we let $y = 0$ and solve for *x*:

$$0 = 2x + 15$$
$$-15 = 2x \qquad \text{Subtracting 15 from both sides}$$
$$-\tfrac{15}{2} = x.$$

The *x*-intercept is $\left(-\tfrac{15}{2}, 0\right)$.

A standard viewing window will not show the *y*-intercept, $(0, 15)$. Thus we adjust the Ymax value and then we choose a viewing window of

$[-10, 10, -10, 20]$, with Yscl = 5. Other choices of window dimensions are also possible.

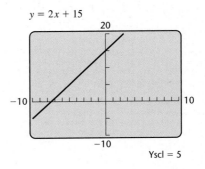

$y = 2x + 15$

Yscl = 5

Graphing Horizontal or Vertical Lines

The equations graphed in Examples 4 and 5 are both in the form $Ax + By = C$. We have already stated that any equation in the form $Ax + By = C$ is linear, provided A and B are not both zero. What if A or B (but not both) is zero? We will find that when A is zero, there is no x-term and the graph is a horizontal line. We will also find that when B is zero, there is no y-term and the graph is a vertical line.

Student Notes

Sometimes students draw horizontal lines when they should be drawing vertical lines and vice versa. To keep this straight, locate the correct number on the axis whose label is given and then draw a line perpendicular to that axis. Thus, to graph $x = 2$, we locate 2 on the x-axis and then draw a line perpendicular to that axis at that point. Note that the graph of $x = 2$ on a plane is a line, whereas the graph of $x = 2$ on a number line is a point.

EXAMPLE 7   Graph: $y = 3$.

SOLUTION   We can regard the equation $y = 3$ as $0 \cdot x + y = 3$. No matter what number we choose for x, we find that y must be 3 if the equation is to be solved. Consider the following table.

$y = 3$

Choose any number for x. →

x	y	(x, y)
-2	3	$(-2, 3)$
0	3	$(0, 3)$
4	3	$(4, 3)$

y must be 3.

All pairs will have 3 as the y-coordinate.

When we plot the ordered pairs $(-2, 3)$, $(0, 3)$, and $(4, 3)$ and connect the points, we obtain a horizontal line. Any ordered pair of the form $(x, 3)$ is a solution, so the line is parallel to the x-axis with y-intercept $(0, 3)$.

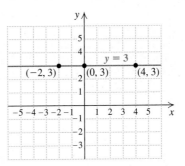

▸ **EXAMPLE 8**   Graph: $x = -4$.

SOLUTION   We can regard the equation $x = -4$ as $x + 0 \cdot y = -4$. We make up a table with all -4's in the x-column.

$x = -4$

	x	y	(x, y)	
x must be -4. →	-4	-5	$(-4, -5)$	All pairs will
	-4	1	$(-4, 1)$	have -4 as the
Choose any number for y.	-4	3	$(-4, 3)$	x-coordinate.

When we plot the ordered pairs $(-4, -5)$, $(-4, 1)$, and $(-4, 3)$ and connect them, we obtain a vertical line. Any ordered pair of the form $(-4, y)$ is a solution. The line is parallel to the y-axis with x-intercept $(-4, 0)$.

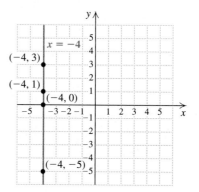

Linear Equations in One Variable

The graph of $y = b$ is a horizontal line, with y-intercept $(0, b)$.

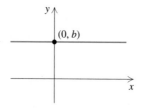

The graph of $x = a$ is a vertical line, with x-intercept $(a, 0)$.

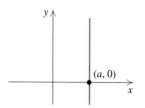

EXAMPLE 9   Write an equation for each graph.

a)

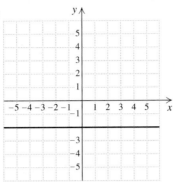

b)

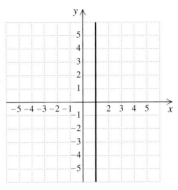

SOLUTION

a) Note that every point on the horizontal line passing through $(0, -2)$ has -2 as the y-coordinate. Thus the equation of the line is $y = -2$.

b) Note that every point on the vertical line passing through $(1, 0)$ has 1 as the x-coordinate. Thus the equation of the line is $x = 1$.

3.3 EXERCISE SET

FOR EXTRA HELP

MathXL   MyMathLab   InterAct Math   AW Math Tutor Center   Video Lectures on CD: Disc 2   Student's Solutions Manual

🌿 *Concept Reinforcement*   *In each of Exercises 1–6, match the phrase with the most appropriate choice from the column on the right.*

1. ____ A vertical line

2. ____ A horizontal line

3. ____ A y-intercept

4. ____ An x-intercept

5. ____ A third point as a check

6. ____ Use a scale of 10 units per square.

a) $2x + 5y = 100$

b) $(3, -2)$

c) $(1, 0)$

d) $(0, 2)$

e) $y = 3$

f) $x = -4$

Determine whether each equation is linear.

7. $5x - 3y = 15$

8. $2y + 10x = 5$

9. $7y = x - 5$

10. $3y = 4x$

11. $xy = 7$

12. $\dfrac{3y}{4x} = 2x$

13. $16 + 4y = 0$

14. $3x - 12 = 0$

15. $2y - \dfrac{3}{x} = 5$

16. $x + 6xy = 10$

For Exercises 17–24, list **(a)** *the coordinates of the y-intercept and* **(b)** *the coordinates of all x-intercepts.*

17.

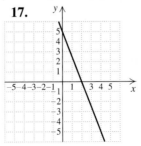

18.

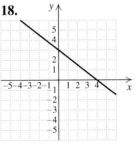

19.
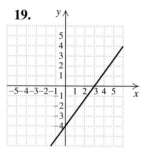

20.
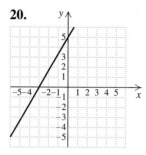

21.
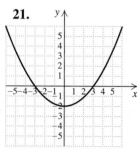

22.
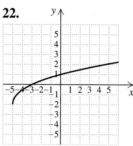

23.
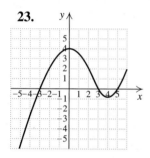

24.
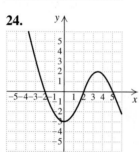

For Exercises 25–34, list **(a)** *the coordinates of any y-intercept and* **(b)** *the coordinates of any x-intercept. Do not graph.*

25. $5x + 3y = 15$ **26.** $5x + 2y = 20$

27. $7x - 2y = 28$ **28.** $4x - 3y = 24$

29. $-4x + 3y = 150$ **30.** $-2x + 3y = 80$

Aha! **31.** $y = 9$ **32.** $x = 8$

33. $x = -7$ **34.** $y = -1$

Find the intercepts. Then graph.

35. $3x + 2y = 12$ **36.** $x + 2y = 6$

37. $x + 3y = 6$ **38.** $6x + 9y = 36$

39. $-x + 2y = 8$ **40.** $-x + 3y = 9$

41. $3x + y = 9$ **42.** $2x - y = 8$

43. $y = 2x - 6$ **44.** $y = -3x + 6$

45. $3x - 9 = 3y$ **46.** $5x - 10 = 5y$

47. $2x - 3y = 6$ **48.** $2x - 5y = 10$

49. $4x + 5y = 20$ **50.** $6x + 2y = 12$

51. $3x + 2y = 8$ **52.** $x - 1 = y$

53. $2x + 4y = 6$ **54.** $5x - 6y = 18$

55. $5x + 3y = 180$ **56.** $10x + 7y = 210$

57. $y = -30 + 3x$ **58.** $y = -40 + 5x$

59. $-4x = 20y + 80$ **60.** $60 = 20x - 3y$

61. $y - 3x = 0$ **62.** $x + 2y = 0$

Graph.

63. $y = 5$ **64.** $y = 2$

65. $x = 4$ **66.** $x = 6$

67. $y = -2$ **68.** $y = -4$

69. $x = -1$ **70.** $x = -6$

71. $x = 18$ **72.** $x = -20$

73. $y = 0$ **74.** $y = \frac{3}{2}$

75. $x = -\frac{5}{2}$ **76.** $x = 0$

77. $-5y = -300$ **78.** $12x = -480$

79. $35 + 7y = 0$ **80.** $-3x - 24 = 0$

Write an equation for each graph.

81.
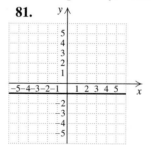

82.
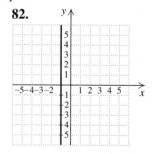

83.
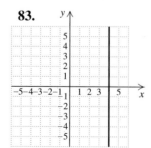

84.
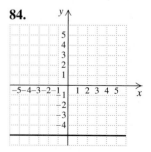

85.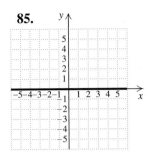

86.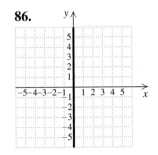

For each equation, find the x-intercept and the y-intercept. Then determine which of the given viewing windows will show both intercepts.

87. $y = 20 - 4x$
   a) $[-10, 10, -10, 10]$ b) $[-5, 10, -5, 10]$
   c) $[-10, 10, -10, 30]$ d) $[-10, 10, -30, 10]$

88. $y = 3x + 7$
   a) $[-10, 10, -10, 10]$ b) $[-1, 15, -1, 15]$
   c) $[-15, 5, -15, 5]$ d) $[-10, 10, -30, 0]$

89. $y = -35x + 7000$
   a) $[-10, 10, -10, 10]$
   b) $[-35, 0, 0, 7000]$
   c) $[-1000, 1000, -1000, 1000]$
   d) $[0, 500, 0, 10{,}000]$

90. $y = 0.2 - 0.01x$
   a) $[-10, 10, -10, 10]$ b) $[-5, 30, -1, 1]$
   c) $[-1, 1, -5, 30]$ d) $[0, 0.01, 0, 0.2]$

Using a graphing calculator, graph each equation so that both intercepts can be easily viewed. Adjust the window settings so that tick marks can be clearly seen on both axes.

91. $y = -0.72x - 15$ **92.** $y - 2.13x = 27$

93. $5x + 6y = 84$ **94.** $2x - 7y = 150$

95. $19x - 17y = 200$ **96.** $6x + 5y = 159$

TW 97. Explain in your own words why the graph of $y = 8$ is a horizontal line.

TW 98. Explain in your own words why the graph of $x = -4$ is a vertical line.

Skill Maintenance

Translate to an algebraic expression. [1.1]

99. 7 less than d

100. 5 more than w

101. The sum of 2 and a number

102. The product of 3 and a number

103. Twice the sum of two numbers

104. Half of the sum of two numbers

Synthesis

TW 105. Describe what the graph of $x + y = C$ will look like for any choice of C.

TW 106. If the graph of a linear equation has one point that is both the x- and the y-intercepts, what is that point? Why?

107. Write an equation for the x-axis.

108. Write an equation of the line parallel to the x-axis and passing through $(3, 5)$.

109. Write an equation of the line parallel to the y-axis and passing through $(-2, 7)$.

110. Find the coordinates of the point of intersection of the graphs of $y = x$ and $y = 6$.

111. Find the coordinates of the point of intersection of the graphs of the equations $x = -3$ and $y = x$.

112. Write an equation of the line shown in Exercise 17.

113. Write an equation of the line shown in Exercise 20.

114. Find the value of C such that the graph of $3x + C = 5y$ has an x-intercept of $(-4, 0)$.

115. Find the value of C such that the graph of $4x = C - 3y$ has a y-intercept of $(0, -8)$.

TW 116. For A and B nonzero, the graphs of $Ax + D = C$ and $By + D = C$ will be parallel to an axis. Explain why.

3.4   Rates

Rates of Change   ▪   Visualizing Rates

Rates of Change

Because graphs make use of two axes, they allow us to visualize how two quantities change with respect to each other. A number accompanied by units is used to represent this type of change and is referred to as a *rate*.

> **Rate**   A *rate* is a ratio that indicates how two quantities change with respect to each other.

Rates occur often in everyday life:

A town that grows by 3400 residents over a period of 2 yr has a *growth rate* of $\frac{3400}{2}$, or 1700, residents per year.

A person running 150 m in 20 sec is moving at a *rate* of $\frac{150}{20}$, or 7.5, m/sec (meters per second).

A class of 25 students pays a total of \$93.75 to visit a museum. The *rate* is $\frac{\$93.75}{25}$, or \$3.75, per student.

> **CAUTION!**   To calculate a rate, it is important to keep track of the units being used.

EXAMPLE 1   On January 3, Nell rented a Ford Focus with a full tank of gas and 9312 mi on the odometer. On January 7, she returned the car with 9630 mi on the odometer.* If the rental agency charged Nell \$108 for the rental and the necessary 12 gal of gas to fill up the gas tank, find the following rates:

a) The car's rate of gas consumption, in miles per gallon;
b) The average cost of the rental, in dollars per day;
c) The car's rate of travel, in miles per day.

---

*For all rental problems, assume that the pickup time was later in the day than the return time so that no late fees were applied.

SOLUTION

a) The rate of gas consumption, in miles per gallon, is found by dividing the number of miles traveled by the number of gallons used for that amount of driving:

$$\text{Rate, in miles per gallon} = \frac{9630 \text{ mi} - 9312 \text{ mi}}{12 \text{ gal}}$$

The word "per" indicates division.

$$= \frac{318 \text{ mi}}{12 \text{ gal}}$$

$$= 26.5 \text{ mi/gal} \qquad \text{Dividing}$$

$$= 26.5 \text{ miles per gallon.}$$

b) The average cost of the rental, in dollars per day, is found by dividing the cost of the rental by the number of days:

$$\text{Rate, in dollars per day} = \frac{108 \text{ dollars}}{4 \text{ days}}$$

From January 3 to January 7 is $7 - 3 = 4$ days.

$$= 27 \text{ dollars/day}$$

$$= \$27 \text{ per day.}$$

c) The car's rate of travel, in miles per day, is found by dividing the number of miles traveled by the number of days:

$$\text{Rate, in miles per day} = \frac{318 \text{ mi}}{4 \text{ days}}$$

$9630 \text{ mi} - 9312 \text{ mi} = 318 \text{ mi.}$
From January 3 to January 7 is $7 - 3 = 4$ days.

$$= 79.5 \text{ mi/day}$$

$$= 79.5 \text{ mi per day.}$$

Many problems involve a rate of travel, or *speed*. The **speed** of an object is found by dividing the distance traveled by the time required to travel that distance.

EXAMPLE 2 Transportation.   An Atlantic City Express bus makes regular trips between Paramus and Atlantic City, New Jersey. At 6:00 P.M., the bus is at mileage marker 40 on the Garden State Parkway, and at 8:00 P.M. it is at marker 170. Find the average speed of the bus.

SOLUTION Speed is the distance traveled divided by the time spent traveling:

$$\text{Bus speed} = \frac{\text{Distance traveled}}{\text{Time spent traveling}}$$

$$= \frac{\text{Change in mileage}}{\text{Change in time}}$$

$$= \frac{130 \text{ mi}}{2 \text{ hr}}$$

170 mi − 40 mi = 130 mi;
8:00 P.M. − 6:00 P.M. = 2 hr

$$= 65 \frac{\text{mi}}{\text{hr}}$$

$$= 65 \text{ miles per hour.}$$

This *average* speed does not indicate by how much the bus speed may vary along the route.

Visualizing Rates

Graphs allow us to visualize a rate of change. As a rule, the quantity listed in the numerator appears on the vertical axis and the quantity listed in the denominator appears on the horizontal axis.

EXAMPLE 3 Phone Lines. Between 1992 and 2002, the number of U.S. telephone lines (land and cellular combined) increased at a rate of approximately 17.5 million lines per year.* In 1992, there were about 155 million U.S. telephone lines. Draw a graph to represent this information.

SOLUTION To label the axes, note that the rate is given in millions of telephone lines per year. Thus we list *Number of telephone lines, in millions,* on the vertical axis and *Year* on the horizontal axis.

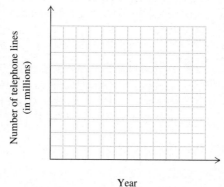

17.5 million per year is 17.5 million/yr; millions of customers is the vertical axis; year is the horizontal axis.

Next, we select a scale for each axis that allows us to plot the given information. If we count by increments of 40 million on the vertical axis, we can easily reach 155 million and beyond—allowing for growth. On the

*Based on information from the Federal Communications Commission and the Cellular Telecommunications and Internet Associations

horizontal axis, we list years, making certain that both 1992 and 2002 are included (see the figure on the left below).

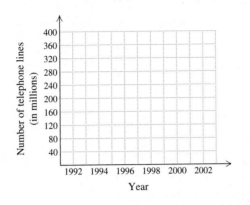

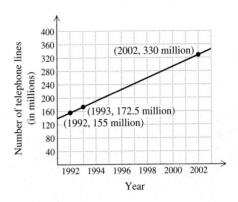

To display the given information, we plot the end point that corresponds to (1992, 155 million). Then, to display the rate of growth, we move from that point to a second point that represents 17.5 million more telephone lines one year later. The coordinates of this point are (1992 + 1, 155 + 17.5 million), or (1993, 172.5 million). To find the coordinates of the year 2002, note that 2002 is 10 years after 1992. Assuming a constant growth rate for those years, we calculate and plot (1992 + 10, 155 + 10 · 17.5 million), or (2002, 330 million). Finally, we draw a line through the three points.

EXAMPLE 4   Haircutting.   Sylvia's World of Hair has a graph displaying data from a recent day of work.

a) What rate can be determined from the graph?

b) What is that rate?

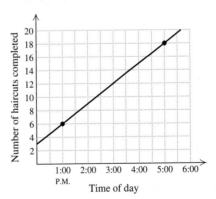

SOLUTION

a) Because the vertical axis shows the number of haircuts completed and the horizontal axis lists the time in hour-long increments, we can find the rate *Number of haircuts per hour.*

b) The points $(1:00, 6 \text{ haircuts})$ and $(5:00, 18 \text{ haircuts})$ are both on the graph. This tells us that in the 4 hr between 1:00 and 5:00, there were $18 - 6 = 12$ haircuts completed. Thus the rate is

$$\frac{18 \text{ haircuts} - 6 \text{ haircuts}}{5:00 - 1:00} = \frac{12 \text{ haircuts}}{4 \text{ hours}}$$

$$= 3 \text{ haircuts per hour}.$$

3.4 EXERCISE SET

FOR EXTRA HELP

 MathXL   MyMathLab   InterAct Math   Tutor Center   AW Math Tutor Center   Video Lectures on CD: Disc 2   Student's Solutions Manual

Solve. For Exercises 1–8, round answers to the nearest cent. For Exercises 1 and 2, assume that the pickup time was later in the day than the return time so that no late fees were applied.

1. *Van Rentals.* Late on June 5, Deb rented a Dodge Caravan with a full tank of gas and 13,741 mi on the odometer. On June 8, she returned the van with 14,014 mi on the odometer. The rental agency charged Deb $118 for the rental and the necessary 13 gal of gas to fill up the tank.
  a) Find the van's rate of gas consumption, in miles per gallon.
  b) Find the average cost of the rental, in dollars per day.
  c) Find the rate of travel, in miles per day.
  d) Find the rental rate, in cents per mile.

2. *Car Rentals.* On February 10, Oscar rented a Chevy Blazer with a full tank of gas and 13,091 mi on the odometer. On February 12, he returned the vehicle with 13,322 mi on the odometer. The rental agency charged $92 for the rental and the necessary 14 gal of gas to fill the tank.
  a) Find the Blazer's rate of gas consumption, in miles per gallon.
  b) Find the average cost of the rental, in dollars per day.
  c) Find the rate of travel, in miles per day.
  d) Find the rental rate, in cents per mile.

3. *Bicycle Rentals.* At 2:00, Perry rented a mountain bike from The Slick Rock Cyclery. He returned the bike at 5:00, after cycling 18 mi. Perry paid $12 for the rental.

  a) Find Perry's average speed, in miles per hour.
  b) Find the rental rate, in dollars per hour.
  c) Find the rental rate, in dollars per mile.

4. *Bicycle Rentals.* At 9:00, Jodi rented a mountain bike from The Bike Rack. She returned the bicycle at 11:00, after cycling 14 mi. Jodi paid $15 for the rental.
  a) Find Jodi's average speed, in miles per hour.
  b) Find the rental rate, in dollars per hour.
  c) Find the rental rate, in dollars per mile.

5. *Temporary Help.* A typist from Jobsite Services, Inc., reports to AKA International for work at 9:00 A.M. and leaves at 5:00 P.M. after having typed from the end of page 12 to the end of page 48 of a prospectus. AKA International pays $128 for the typist's services.
  a) Find the rate of pay, in dollars per hour.
  b) Find the average typing rate, in number of pages per hour.
  c) Find the rate of pay, in dollars per page.

6. *Temporary Help.* A typist for Kelly Services reports to 3E's Properties for work at 10:00 A.M. and leaves at 6:00 P.M. after having typed from the end of page 8 to the end of page 50 of a proposal. 3E's pays $120 for the typist's services.
  a) Find the rate of pay, in dollars per hour.
  b) Find the average typing rate, in number of pages per hour.
  c) Find the rate of pay, in dollars per page.

7. *Two-Year-College Tuition.* The average tuition at a public two-year college was $1380 in 2002 and approximately $1670 in 2004 (*Source*: U.S. National Center for Education Statistics). Find the rate at which tuition was increasing.

8. *Four-Year-College Tuition.*   The average tuition at a public four-year college was $4273 in 2002 and approximately $5368 in 2004 (*Source*: U.S. National Center for Education Statistics). Find the rate at which tuition was increasing.

9. *Elevators.*   At 2:38, Serge entered an elevator on the 34th floor of the Regency Hotel. At 2:40, he stepped off at the 5th floor.
 a) Find the elevator's average rate of travel, in number of floors per minute.
 b) Find the elevator's average rate of travel, in seconds per floor.

10. *Snow Removal.*   By 1:00 P.M., Erin had already shoveled 2 driveways, and by 6:00 P.M., the number was up to 7.
 a) Find Erin's shoveling rate, in number of driveways per hour.
 b) Find Erin's shoveling rate, in hours per driveway.

11. *Mountaineering.*   As part of an ill-fated expedition to climb Mt. Everest in 1996, author Jon Krakauer departed "The Balcony," elevation 27,600 ft, at 7:00 A.M. and reached the summit, elevation 29,028 ft, at 1:25 P.M. (*Source*: Krakauer, Jon, *Into Thin Air, the Illustrated Edition.* New York: Random House, 1998).

 a) Find Krakauer's average rate of ascent, in feet per minute.
 b) Find Krakauer's average rate of ascent, in minutes per foot.

12. *Mountaineering.*   The fastest ascent of Mt. Everest was accomplished by Lakpa Gelu Sherpa of Nepal in 2003. Lakpa Gelu Sherpa climbed from base camp, elevation 17,552 ft, to the summit, elevation 29,028 ft, in 10 hr 57 min (*Source*: *Guinness Book of World Records* 2004 Edition).
 a) Find Lakpa Gelu Sherpa's rate of ascent, in feet per minute.
 b) Find Lakpa Gelu Sherpa's rate of ascent, in minutes per foot.

In Exercises 13–22, draw a linear graph to represent the given information. Be sure to label and number the axes appropriately (see Example 3).

13. *Healthcare Costs.*   In 2003, the average copayment for a prescription drug on an insurance company's preferred list was $19, and the figure was rising at a rate of $2 per year (*Source*: *Daily Reporter*, Greenfield IN, 9/9/03).

14. *Healthcare Costs.*   In 2003, the average cost for health insurance for a family was about $9000, and the figure was rising at a rate of about $875 per year (*Source*: *Daily Reporter*, Greenfield IN, 9/9/03).

15. *Law Enforcement.*   In 2000, there were approximately 26 million crimes reported in the United States, and the figure was dropping at a rate of about 1.2 million per year (*Source*: Based on data from the Bureau of Justice Statistics).

16. *Fire Fighting.*   In 2001, there were approximately 1.7 million fires in the United States, and the figure was dropping at a rate of about 0.1 million per year (*Source*: Based on data from the *Statistical Abstract of the United States*, 2006).

17. *Train Travel.*   At 3:00 P.M., the Boston–Washington Metroliner had traveled 230 mi and was cruising at a rate of 90 miles per hour.

18. *Plane Travel.*   At 4:00 P.M., the Seattle–Los Angeles shuttle had traveled 400 mi and was cruising at a rate of 300 miles per hour.

19. *Wages.*   By 2:00 P.M., Diane had earned $50. She continued earning money at a rate of $15 per hour.

20. *Wages.*   By 3:00 P.M., Arnie had earned $70. He continued earning money at a rate of $12 per hour.

21. *Telephone Bills.*   Roberta's phone bill was already $7.50 when she made a call for which she was charged at a rate of $0.10 per minute.

22. *Telephone Bills.*   At 3:00 P.M., Larry's phone bill was $6.50 and increasing at a rate of 7¢ per minute.

In Exercises 23–32, use the graph provided to calculate a rate of change in which the units of the horizontal axis are used in the denominator.

23. *Hairdresser.*   Eve's Custom Cuts has a graph displaying data from a recent day of work. At what rate does Eve work?

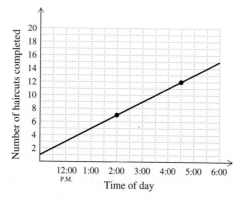

24. *Manicures.*   The following graph shows data from a recent day's work at the O'Hara School of Cosmetology. At what rate do they work?

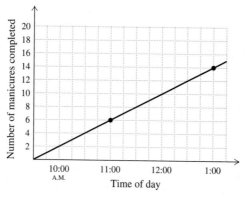

25. *Train Travel.*   The following graph shows data from a recent train ride from Chicago to St. Louis. At what rate did the train travel?

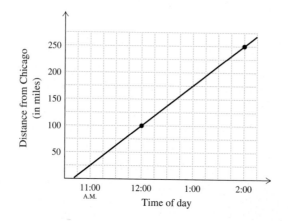

26. *Train Travel.*   The following graph shows data from a recent train ride from Denver to Kansas City. At what rate did the train travel?

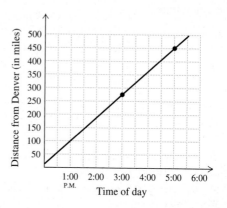

27. *Cost of a Telephone Call.*   The following graph shows data from a recent MCI phone call between San Francisco, CA, and Pittsburgh, PA. At what rate was the customer being billed?

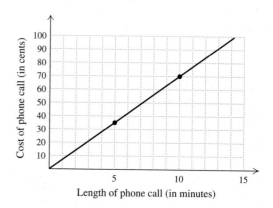

28. *Cost of a Telephone Call.*   The following graph shows data from a recent AT&T phone call between Burlington, VT, and Austin, TX. At what rate was the customer being billed?

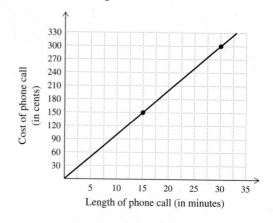

29. *Depreciation of an Office Machine.* Data regarding the value of a particular color copier is represented in the following graph. At what rate is the value changing?

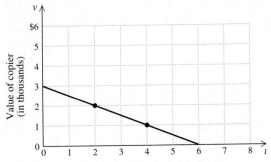

Time from date of purchase (in years)

30. *NASA Spending.* Spending on the National Aeronautics and Space Administration (NASA) during the late 1990s is represented in the following graph (*Source*: Based on information in the *Statistical Abstract of the United States*, 1999, 2002). At what rate was the amount spent on NASA changing?

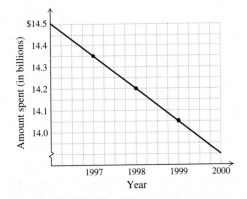

31. *Gas Mileage.* The following graph shows data for a Honda Odyssey driven on interstate highways. At what rate was the vehicle consuming gas?

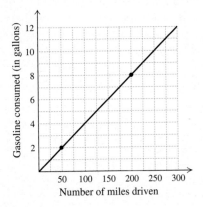

32. *Gas Mileage.* The following graph shows data for a Ford Explorer driven on city streets. At what rate was the vehicle consuming gas?

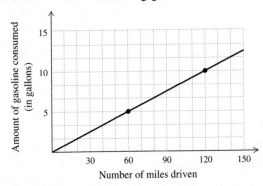

In each of Exercises 33–38, match the description with the most appropriate graph from the choices (a)–(f) below. Scales are intentionally omitted. Assume that of the three sports listed, swimming is the slowest and biking is the fastest.

33. _____ Robin trains for triathlons by running, biking, and then swimming every Saturday.

34. _____ Gene trains for triathlons by biking, running, and then swimming every Sunday.

35. _____ Shirley trains for triathlons by swimming, biking, and then running every Sunday.

36. _____ Evan trains for triathlons by swimming, running, and then biking every Saturday.

37. _____ Angie trains for triathlons by biking, swimming, and then running every Sunday.

38. _____ Mick trains for triathlons by running, swimming, and then biking every Saturday.

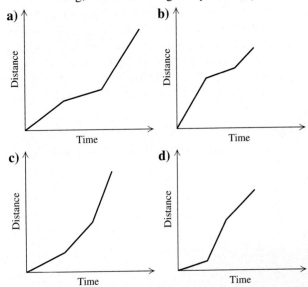

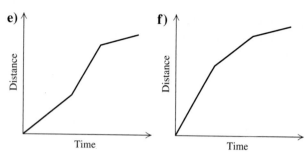

e) Distance / Time

f) Distance / Time

ᵀᵂ 39. What does a negative rate of travel indicate? Explain.

ᵀᵂ 40. Explain how to convert from kilometers per hour to meters per second.

Skill Maintenance

41. $-2 - (-7)$ [1.6]   **42.** $-9 - (-3)$ [1.6]

43. $\dfrac{5 - (-4)}{-2 - 7}$ [1.8]   **44.** $\dfrac{8 - (-4)}{2 - 11}$ [1.8]

45. $\dfrac{-4 - 8}{7 - (-2)}$ [1.8]   **46.** $\dfrac{-5 - 3}{6 - (-4)}$ [1.8]

Synthesis

ᵀᵂ 47. Write an exercise similar to Exercises 23–32 for a classmate to solve. Design the problem so that the solution is "The plane was traveling at a rate of 300 miles per hour."

ᵀᵂ 48. Write an exercise similar to Exercises 1–12 for a classmate to solve. Design the problem so that the solution is "The motorcycle's rate of gas consumption was 65 miles per gallon."

49. *Aviation.* A Boeing 737 climbs from sea level to a cruising altitude of 31,500 ft at a rate of 6300 ft/min. After cruising for 3 min, the jet is forced to land, descending at a rate of 3500 ft/min. Represent the flight with a graph in which altitude is measured on the vertical axis and time on the horizontal axis.

50. *Wages with Commissions.* Each salesperson at Mike's Bikes is paid $140 a week plus 13% of all sales up to $2000, and then 20% on any sales in excess of $2000. Draw a graph in which sales are measured on the horizontal axis and wages on the vertical axis. Then use the graph to estimate the wages paid when a salesperson sells $2700 in merchandise in one week.

51. *Taxi Fares.* The driver of a New York City Yellow Cab recently charged $2 plus 50¢ for each fifth of a mile traveled. Draw a graph that could be used to determine the cost of a fare.

52. *Gas Mileage.* Suppose that a Honda motorcycle goes twice as far as a Honda Odyssey on the same amount of gas (see Exercise 31). Draw a graph that reflects this information.

53. *Navigation.* In 3 sec, Penny walks 24 ft, to the bow (front) of a tugboat. The boat is cruising at a rate of 5 feet per second. What is Penny's rate of travel with respect to land?

54. *Aviation.* Tim's F-14 jet is moving forward at a deck speed of 95 mph aboard an aircraft carrier that is traveling 39 mph in the same direction. How fast is the jet traveling, in minutes per mile, with respect to the sea?

55. *Running.* Annette ran from the 4-km mark to the 7-km mark of a 10-km race in 15.5 min. At this rate, how long would it take Annette to run a 5-mi race? (*Hint:* 1 km ≈ 0.62 mi.)

56. *Running.* Jerod ran from the 2-mi marker to the finish line of a 5-mi race in 25 min. At this rate, how long would it take Jerod to run a 10-km race? (*Hint:* 1 mi ≈ 1.61 km.)

57. At 3:00 P.M., Catanya and Chad had already made 46 candles. By 5:00 P.M., the total reached 100 candles. Assuming a constant production rate, at what time did they make their 82nd candle?

58. Marcy picks apples twice as fast as Ryan. By 4:30, Ryan had already picked 4 bushels of apples. Fifty minutes later, his total reached $5\frac{1}{2}$ bushels. Find Marcy's picking rate. Give your answer in number of bushels per hour.

3.5   Slope

Rate and Slope   ■   Horizontal and Vertical Lines   ■
Applications

In Section 3.4, we introduced *rate* as a method of measuring how two quantities change with respect to each other. In this section, we will discuss how rate can be related to the slope, or steepness, of a line.

Rate and Slope

Suppose that a car manufacturer operates two plants: one in Michigan and one in Pennsylvania. Knowing that the Michigan plant produces 3 cars every 2 hours and the Pennsylvania plant produces 5 cars every 4 hours, we can set up tables listing the number of cars produced after various amounts of time.

Michigan Plant		Pennsylvania Plant	
Hours Elapsed	Cars Produced	Hours Elapsed	Cars Produced
0	0	0	0
2	3	4	5
4	6	8	10
6	9	12	15
8	12	16	20

By comparing the number of cars produced at each plant over a specified period of time, we can compare the two production rates. For example, the Michigan plant produces 3 cars every 2 hours, so its *rate* is $3 \div 2 = 1\frac{1}{2}$, or $\frac{3}{2}$ cars per hour. Since the Pennsylvania plant produces 5 cars every 4 hours, its rate is $5 \div 4 = 1\frac{1}{4}$, or $\frac{5}{4}$ cars per hour.

Let's now graph the pairs of numbers listed in the tables, using the horizontal axis for time and the vertical axis for the number of cars produced. Note that the rate in the Michigan plant is slightly greater so its graph is slightly steeper.

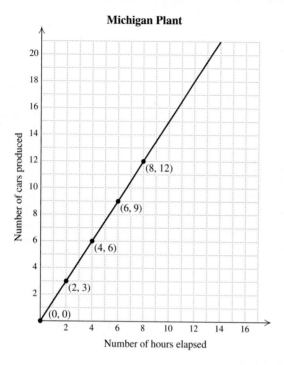

Michigan Plant

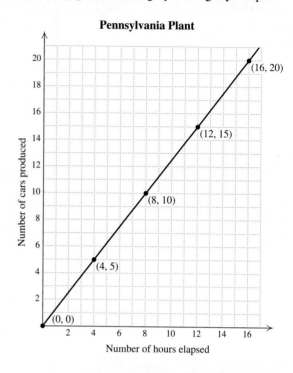

Pennsylvania Plant

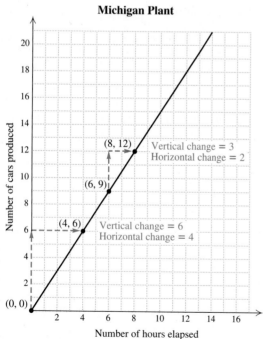

Michigan Plant

The rates $\frac{3}{2}$ and $\frac{5}{4}$ can also be found using the coordinates of any two points that are on the line. For example, we can use the points $(6, 9)$ and $(8, 12)$ to find the production rate for the Michigan plant. To do so, remember that these coordinates tell us that after 6 hr, 9 cars have been produced, and after 8 hr, 12 cars have been produced. In the 2 hr between the 6-hr and 8-hr points, $12 - 9$, or 3 cars were produced. Thus,

$$\text{Michigan production rate} = \frac{\text{change in number of cars produced}}{\text{corresponding change in time}}$$

$$= \frac{12 - 9 \text{ cars}}{8 - 6 \text{ hr}}$$

$$= \frac{3 \text{ cars}}{2 \text{ hr}} = \frac{3}{2} \text{ cars per hour.}$$

Because the line is straight, the same rate is found using *any* pair of points on the line. For example, using $(0, 0)$ and $(4, 6)$, we have

$$\text{Michigan production rate} = \frac{6 - 0 \text{ cars}}{4 - 0 \text{ hr}} = \frac{6 \text{ cars}}{4 \text{ hr}} = \frac{3}{2} \text{ cars per hour.}$$

Note that the rate is always the vertical change divided by the associated horizontal change.

EXAMPLE 1   Use the graph of car production at the Pennsylvania plant to find the rate of production.

SOLUTION   We can use any two points on the line, such as $(12, 15)$ and $(16, 20)$:

$$\begin{aligned}\text{Pennsylvania production rate} &= \frac{\text{change in number of cars produced}}{\text{corresponding change in time}} \\[1em] &= \frac{20 - 15 \text{ cars}}{16 - 12 \text{ hr}} \\[1em] &= \frac{5 \text{ cars}}{4 \text{ hr}} \\[1em] &= \frac{5}{4} \text{ cars per hour.}\end{aligned}$$

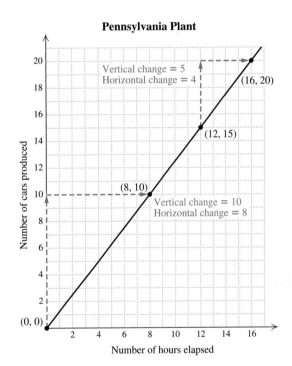

Pennsylvania Plant

As a check, we can use another pair of points, like $(0, 0)$ and $(8, 10)$:

$$\begin{aligned}\text{Pennsylvania production rate} &= \frac{10 - 0 \text{ cars}}{8 - 0 \text{ hr}} \\[1em] &= \frac{10 \text{ cars}}{8 \text{ hr}} \\[1em] &= \frac{5}{4} \text{ cars per hour.}\end{aligned}$$

When the axes of a graph are simply labeled x and y, it is useful to know the ratio of vertical change to horizontal change. This ratio is a measure of a line's slant, or **slope,** and is the rate at which y is changing with respect to x.

Consider a line passing through $(2, 3)$ and $(6, 5)$, as shown below. We find the ratio of vertical change, or *rise*, to horizontal change, or *run*, as follows:

$$\text{Ratio of vertical change to horizontal change} = \frac{\text{change in } y}{\text{change in } x} = \frac{\text{rise}}{\text{run}}$$

$$= \frac{5 - 3}{6 - 2}$$

$$= \frac{2}{4}, \text{ or } \frac{1}{2}.$$

Note that these calculations can be performed without viewing a graph.

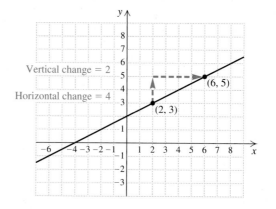

Thus the y-coordinates of points on this line increase at a rate of 2 units for every 4-unit increase in x, 1 unit for every 2-unit increase in x, or $\frac{1}{2}$ unit for every 1-unit increase in x. The slope of the line is $\frac{1}{2}$.

Slope   The *slope* of the line containing points (x_1, y_1) and (x_2, y_2) is given by

$$m = \frac{\text{change in } y}{\text{change in } x} = \frac{\text{rise}}{\text{run}} = \frac{y_2 - y_1}{x_2 - x_1}.$$

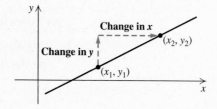

EXAMPLE 2   Graph the line containing the points $(-4, 3)$ and $(2, -6)$ and find the slope.

SOLUTION   The graph is shown below. From $(-4, 3)$ to $(2, -6)$, the change in y, or rise, is $-6 - 3$, or -9. The change in x, or run, is $2 - (-4)$, or 6. Thus,

$$\text{Slope} = \frac{\text{change in } y}{\text{change in } x}$$

$$= \frac{\text{rise}}{\text{run}}$$

$$= \frac{-6 - 3}{2 - (-4)}$$

$$= \frac{-9}{6}$$

$$= -\frac{9}{6}, \text{ or } -\frac{3}{2}.$$

Student Notes

You may wonder which point should be regarded as (x_1, y_1) and which should be (x_2, y_2). To see that the math works out the same either way, perform both calculations on your own.

CAUTION!   When we use the formula

$$m = \frac{y_2 - y_1}{x_2 - x_1},$$

it makes no difference which point is considered (x_1, y_1). What matters is that we subtract the y-coordinates in the same order that we subtract the x-coordinates.

To illustrate, we reverse *both* of the subtractions in Example 2. The slope is still $-\frac{3}{2}$:

$$\text{Slope} = \frac{\text{change in } y}{\text{change in } x} = \frac{3 - (-6)}{-4 - 2} = \frac{9}{-6} = -\frac{3}{2}.$$

As shown in the graphs below, a line with positive slope slants up from left to right, and a line with negative slope slants down from left to right. The larger the absolute value of the slope, the steeper the line.

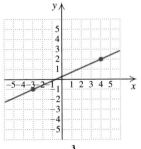

$m = \frac{3}{7}$

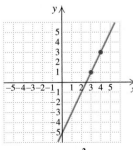

$m = 2$, or $\frac{2}{1}$

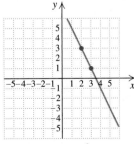

$m = -2$, or $\frac{-2}{1}$

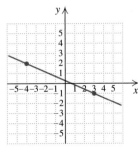

$m = -\frac{3}{7}$

Horizontal and Vertical Lines

What about the slope of a horizontal or a vertical line?

▶ **EXAMPLE 3**   Find the slope of the line $y = 4$.

SOLUTION   Consider the points $(2, 4)$ and $(-3, 4)$, which are on the line. The change in y, or the rise, is $4 - 4$, or 0. The change in x, or the run, is $-3 - 2$, or -5. Thus,

$$m = \frac{4 - 4}{-3 - 2}$$

$$= \frac{0}{-5}$$

$$= 0.$$

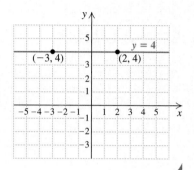

Any two points on a horizontal line have the same y-coordinate. Thus the change in y is 0, so the slope is 0.

A horizontal line has slope 0.

▶ **EXAMPLE 4**   Find the slope of the line $x = -3$.

SOLUTION   Consider the points $(-3, 4)$ and $(-3, -2)$, which are on the line. The change in y, or the rise, is $-2 - 4$, or -6. The change in x, or the run, is $-3 - (-3)$, or 0. Thus,

$$m = \frac{-2 - 4}{-3 - (-3)}$$

$$= \frac{-6}{0} \quad \text{(undefined)}.$$

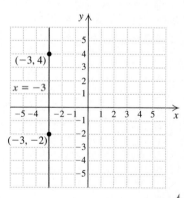

Since division by 0 is not defined, the slope of this line is not defined. The answer to a problem of this type is "The slope of this line is undefined."

The slope of a vertical line is undefined.

Applications

We have seen that slope has many real-world applications, ranging from car speed to production rate. Some applications use slope to measure steepness. For example, numbers like 2%, 3%, and 6% are often used to represent the

grade of a road, a measure of a road's steepness. That is, a 3% grade means that for every horizontal distance of 100 ft, the road rises or drops 3 ft. The concept of grade also occurs in skiing or snowboarding, where a 7% grade is considered very tame, but a 70% grade is considered steep.

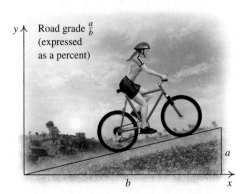

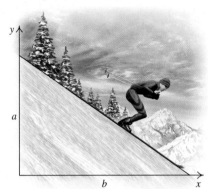

EXAMPLE 5 Skiing. Among the steepest skiable terrain in North America, the Headwall on Mount Washington, in New Hampshire, drops 720 ft over a horizontal distance of 900 ft. Find the grade of the Headwall.

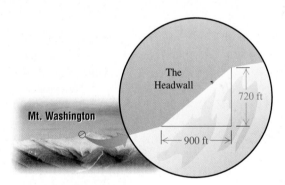

SOLUTION   The grade of the Headwall is its slope, expressed as a percent:

$$m = \frac{720}{900}$$
$$= \frac{8}{10}$$
$$= 80\%.$$

Grade is slope expressed as a percent.

Carpenters use slope when designing stairs, ramps, or roof pitches. Another application occurs in the engineering of a dam—the force or strength of a river depends on how much the river drops over a specified distance.

EXAMPLE 6 Running Speed. Stephanie runs 10 km during each workout. For the first 7 km, her pace is twice as fast as it is for the last 3 km. Which of the following graphs best describes Stephanie's workout?

A.
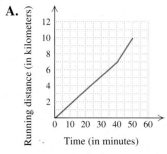

B.
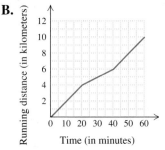

C.
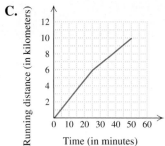

D.
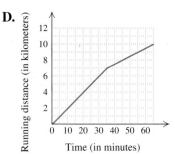

SOLUTION The slopes in graph A increase as we move to the right. This would indicate that Stephanie ran faster for the *last* part of her workout. Thus graph A is not the correct one.

The slopes in graph B indicate that Stephanie slowed down in the middle of her run and then resumed her original speed. Thus graph B does not correctly model the situation either.

According to graph C, Stephanie slowed down not at the 7-km mark, but at the 6-km mark. Thus graph C is also incorrect.

Graph D indicates that Stephanie ran the first 7 km in 35 min, a rate of 0.2 km/min. It also indicates that she ran the final 3 km in 30 min, a rate of 0.1 km/min. This means that Stephanie's rate was twice as fast for the first 7 km, so graph D provides a correct description of her workout.

3.5 EXERCISE SET

Concept Reinforcement   *State whether each of the following rates is positive, negative, or zero.*

1. The rate at which a teenager's height changes

2. The rate at which an elderly person's height changes

3. The rate at which a pond's water level changes during a drought

4. The rate at which a pond's water level changes during the rainy season

5. The rate at which the number of people in attendance at a basketball game changes in the moments before the opening tipoff

6. The rate at which the number of people in attendance at a basketball game changes in the moments after the final buzzer sounds

7. The rate at which a person's IQ changes during his or her sleep

8. The rate at which a gift shop's sales change as the holidays approach

9. The rate at which a bookstore's inventory changes during a liquidation sale

10. The rate at which a buoy's elevation changes while floating for an evening in a lake

11. Find the rate of change of the U.S. population (*Source*: Based on information in the *Statistical Abstract of the United States*, 2006).

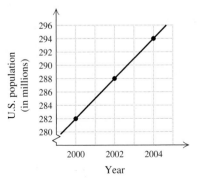

12. Find the rate at which a runner burns calories.

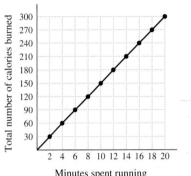

13. Find the rate of change in the percentage of high school students who admitted to cheating on an exam at least once in the past year (*Source*: Based on information from Josephson Institute of Ethics, as reported in *The New York Times*, 10/4/03).

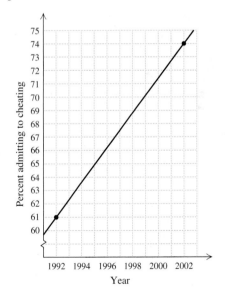

14. Find the rate of change in the percentage of college students who admitted to taking material from the Internet and claiming it as their own (*Source*: Based on information from Donald L. McCabe, as reported in *The New York Times*, 10/4/03).

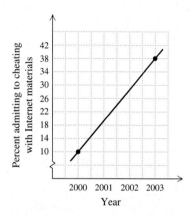

15. Find the rate of change in SAT math scores with respect to family income (*Source*: Based on data from 2004 college-bound seniors in Massachusetts).

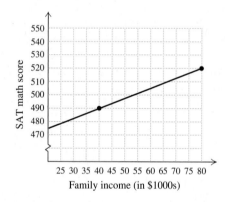

16. Find the rate of change in SAT verbal scores with respect to family income (*Source*: Based on data from 2004 college-bound seniors in Massachusetts).

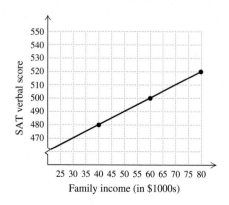

17. Find the rate of change in the temperature in Spearfish, South Dakota, on January 22, 1943, as shown below (*Source*: National Oceanic Atmospheric Administration).

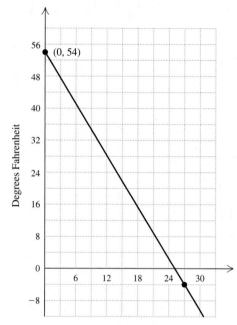

Number of minutes after 9 A.M.

18. Find the rate of change in the number of crimes reported in the United States (*Source*: Based on statistics from the U.S. Bureau of Justice).

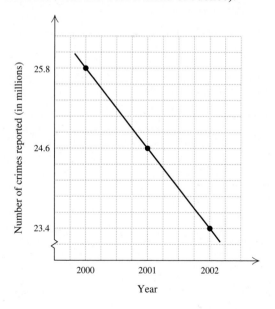

Find the slope, if it is defined, of each line. If the slope is undefined, state this.

19.

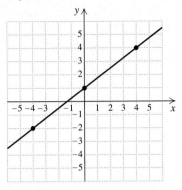

20.

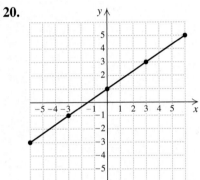

21.

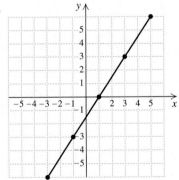

22.

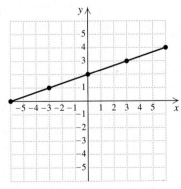

23.

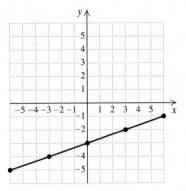

24.

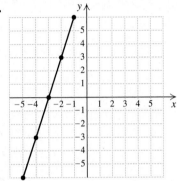

25.

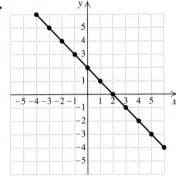

26.

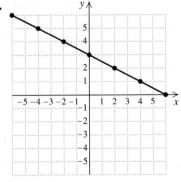

27.

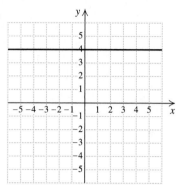

31.

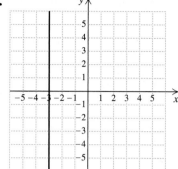

28.

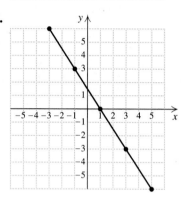

32.

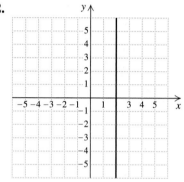

29.

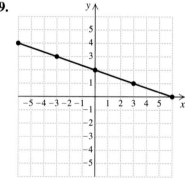

33.

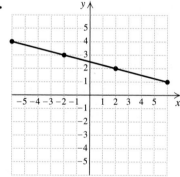

30.

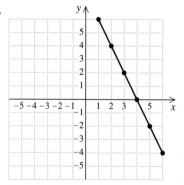

34.

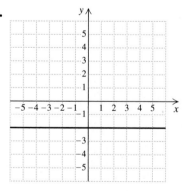

35.

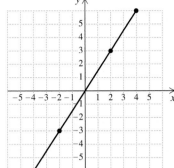

36.

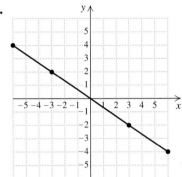

37.

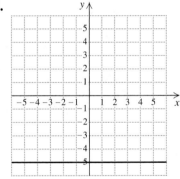

38.

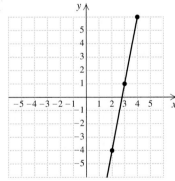

39.

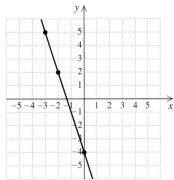

40.

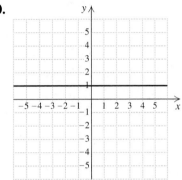

Find the slope of the line containing each given pair of points. If the slope is undefined, state this.

41. (1, 2) and (5, 8) **42.** (2, 1) and (6, 9)

43. (−2, 4) and (3, 0) **44.** (−4, 2) and (2, −3)

45. (−4, 0) and (5, 7) **46.** (3, 0) and (6, 2)

47. (0, 8) and (−3, 10) **48.** (0, 9) and (−5, 0)

49. (−2, 3) and (−6, 5) **50.** (−2, 4) and (6, −7)

Aha! **51.** $\left(-2, \frac{1}{2}\right)$ and $\left(-5, \frac{1}{2}\right)$ **52.** (−5, −1) and (2, 3)

53. (3, 4) and (9, −7)

54. (−10, 3) and (−10, 4)

55. (6, −4) and (6, 5)

56. (5, −2) and (−4, −2)

Find the slope of each line whose equation is given. If the slope is undefined, state this.

57. $x = -3$ **58.** $x = -4$

59. $y = 4$ **60.** $y = 17$

61. $x = 9$ **62.** $x = 6$

63. $y = -9$ **64.** $y = -4$

65. *Surveying.* Tucked between two ski areas, Vermont Route 108 rises 106 m over a horizontal distance of 1325 m. What is the grade of the road?

66. *Navigation.* Capital Rapids drops 54 ft vertically over a horizontal distance of 1080 ft. What is the slope of the rapids?

67. *Architecture.* To meet federal standards, a wheelchair ramp cannot rise more than 1 ft over a horizontal distance of 12 ft. Express this slope as a grade.

68. *Engineering.* At one point, Yellowstone's Beartooth Highway rises 315 ft over a horizontal distance of 4500 ft. Find the grade of the road.

69. *Carpentry.* Find the slope (or pitch) of the roof.

70. *Exercise.* Find the slope (or grade) of the treadmill.

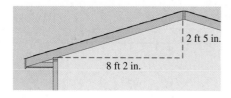

71. *Surveying.* From a base elevation of 9600 ft, Longs Peak, Colorado, rises to a summit elevation of 14,255 ft over a horizontal distance of 15,840 ft. Find the average grade of Longs Peak.

72. *Construction.* Public buildings regularly include steps with 7-in. risers and 11-in. treads. Find the grade of such a stairway.

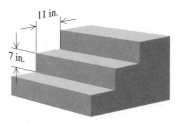

73. *Nursing.* Match each sentence with the most appropriate of the four graphs shown.

   a) The rate at which fluids were given intravenously was doubled after 3 hr.

   b) The rate at which fluids were given intravenously was gradually reduced to 0.

   c) The rate at which fluids were given intravenously remained constant for 5 hr.

   d) The rate at which fluids were given intravenously was gradually increased.

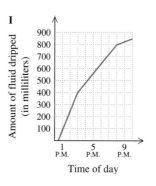

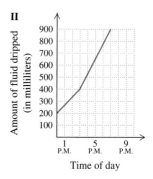

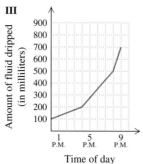

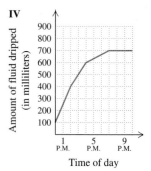

74. *Market Research.* Match each sentence with the most appropriate graph below.

   a) After January 1, daily sales continued to rise, but at a slower rate.

   b) After January 1, sales decreased faster than they ever grew.

c) The rate of growth in daily sales doubled after January 1.

d) After January 1, daily sales decreased at half the rate that they grew in December.

I

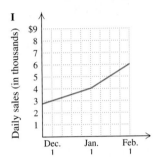

III

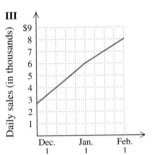

II

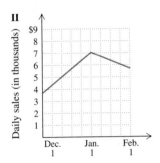

IV

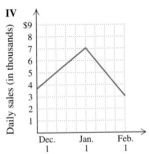

TW **75.** Explain why the order in which coordinates are subtracted to find slope does not matter so long as y-coordinates and x-coordinates are subtracted in the same order.

TW **76.** If one line has a slope of -3 and another has a slope of 2, which line is steeper? Why?

Skill Maintenance

Solve. [2.3]

77. $ax + by = c$, for y

78. $rx - mn = p$, for r

79. $ax - by = c$, for y

80. $rs + nt = q$, for t

Evaluate. [1.8]

81. $\dfrac{2}{3}x - 5$, for $x = 12$

82. $\dfrac{3}{5}x - 7$, for $x = 15$

Synthesis

TW **83.** The points $(-4, -3)$, $(1, 4)$, $(4, 2)$, and $(-1, -5)$ are vertices of a quadrilateral. Use slopes to explain why the quadrilateral is a parallelogram.

TW **84.** Can the points $(-4, 0)$, $(-1, 5)$, $(6, 2)$, and $(2, -3)$ be vertices of a parallelogram? Why or why not?

85. A line passes through $(4, -7)$ and never enters the first quadrant. What numbers could the line have for its slope?

86. A line passes through $(2, 5)$ and never enters the second quadrant. What numbers could the line have for its slope?

87. *Architecture.*   Architects often use the equation $x + y = 18$ to determine the height y, in inches, of the riser of a step when the tread is x inches wide. Express the slope of stairs designed with this equation without using the variable y.

In Exercises 88 and 89, the slope of each line is $-\dfrac{2}{3}$, but the numbering on one axis is missing. How many units should each tick mark on that unnumbered axis represent?

88.

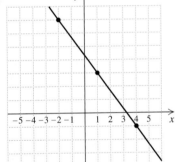

89.

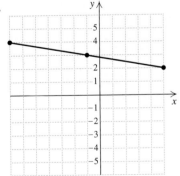

90. Match each sentence with the most appropriate graph below.

    a) Annie drove 2 mi to a lake, swam 1 mi, and then drove 3 mi to a store.

    b) During a preseason workout, Rico biked 2 mi, ran for 1 mi, and then walked 3 mi.

    c) James bicycled 2 mi to a park, hiked 1 mi over the notch, and then took a 3-mi bus ride back to the park.

    d) After hiking 2 mi, Marcy ran for 1 mi before catching a bus for the 3-mi ride into town.

91. The plans below are for a skateboard "Fun Box" (*Source*: www.heckler.com). For the ramps labeled A, find the slope or grade.

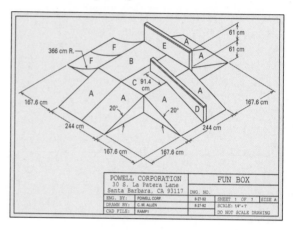

I

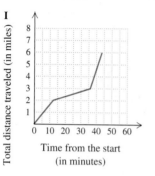

II

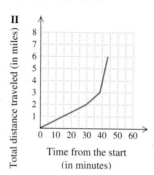

III

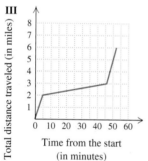

IV

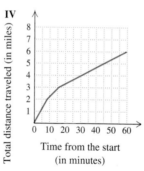

3.6 Slope–Intercept Form

Using the *y*-intercept and the Slope to Graph a Line ◼ Equations in Slope–Intercept Form ◼ Graphing and Slope–Intercept Form ◼ Parallel and Perpendicular Lines

Graphs are used in many important applications. Sections 3.4 and 3.5 have shown us that the slope of a line can be used to represent the rate at which the quantity measured on the vertical axis changes with respect to the quantity measured on the horizontal axis.

In this section, we return to the task of graphing equations. This time, however, our understanding of rates and slope will provide us with the tools necessary to develop shortcuts that will streamline our work.

If we know the slope and the *y*-intercept of a line, it is possible to graph the line. In this section, we will discover that a line's slope and *y*-intercept can be determined directly from the line's equation, provided the equation is written in a certain form.

Using the *y*-intercept and the Slope to Graph a Line

Let's modify the car production situation that first appeared in Section 3.5. Suppose that as a new workshift begins, 4 cars have already been produced. At the Michigan plant, 3 cars were being produced every 2 hours, a rate of $\frac{3}{2}$ cars per hour. If this rate remains the same regardless of how many cars have already been produced, the table and graph shown here can be made.

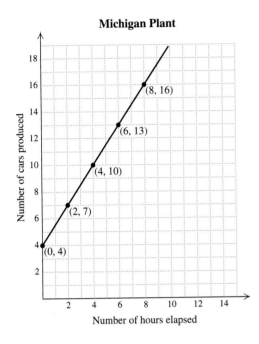

Michigan Plant

Michigan Plant	
Hours Elapsed	**Cars Produced**
0	4
2	7
4	10
6	13
8	16

To confirm that the production rate is still $\frac{3}{2}$, we calculate the slope. Recall that

$$\text{Slope} = \frac{\text{change in } y}{\text{change in } x} = \frac{\text{rise}}{\text{run}} = \frac{y_2 - y_1}{x_2 - x_1},$$

where (x_1, y_1) and (x_2, y_2) are any two points on the graphed line. Here we select $(0, 4)$ and $(2, 7)$:

$$\text{Slope} = \frac{\text{change in } y}{\text{change in } x} = \frac{7 - 4}{2 - 0} = \frac{3}{2}.$$

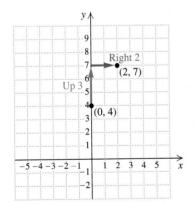

Knowing that the slope is $\frac{3}{2}$, we could have drawn the graph by plotting $(0, 4)$ and from there moving *up* 3 units and *to the right* 2 units. This would have located the point $(2, 7)$. Using $(0, 4)$ and $(2, 7)$, we can then draw the line. This is the method used in the next example.

EXAMPLE 1   Draw a line that has slope $\frac{1}{4}$ and *y*-intercept (0, 2).

SOLUTION   We plot (0, 2) and from there move *up* 1 unit and *to the right* 4 units. This locates the point (4, 3). We plot (4, 3) and draw a line passing through (0, 2) and (4, 3), as shown on the right below.

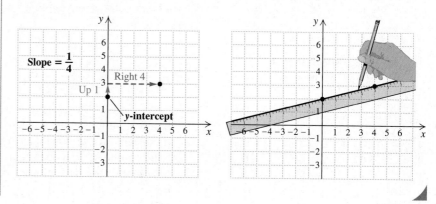

Equations in Slope–Intercept Form

Recall from Section 3.3 that an equation of the form $Ax + By = C$ is linear. A linear equation can also be written in a form from which the slope and the *y*-intercept are read directly.

Interactive Discovery

Graph each of the following equations using a standard viewing window:

$$y_1 = 2x, \qquad y_2 = -2x, \qquad y_3 = 0.3x.$$

1. Do the graphs appear to be linear?
2. At what point do they cross the *x*-axis?
3. At what point do they cross the *y*-axis?

The pattern you may have observed is true in general.

An equation of the form $y = mx$ is linear. Its graph is a straight line passing through the origin.

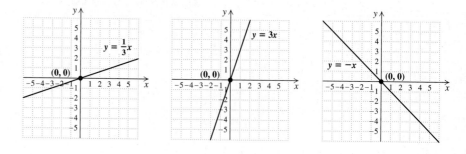

What happens to the graph of $y = mx$ if we add a number b to get an equation of the form $y = mx + b$?

Interactive Discovery

Graph the following equations on the same set of axes using a standard viewing window:

$$y_1 = 0.5x, \qquad y_2 = 0.5x + 5, \qquad y_3 = 0.5x - 7.$$

1. By examining the graphs and creating a table of values, explain how the values of y_2 and y_3 differ from those of y_1.

2. Predict what the graph of $y_4 = 0.5x + (-3.2)$ will look like. Test your description by graphing y_4 on the same set of axes as $y_1 = 0.5x$.

3. Describe what happens to the graph of $y_1 = 0.5x$ when a number b is added to $0.5x$.

The pattern you may have observed is also true for other values of m.

The graph of $y = mx + b, b \neq 0$, is a line parallel to $y = mx$, passing through the point $(0, b)$.

The point $(0, b)$ is called the **y-intercept.**

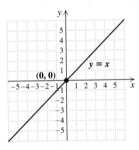   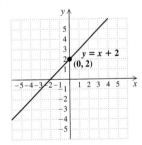   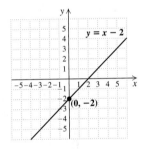

EXAMPLE 2   For each equation, find the y-intercept.

a) $y = -5x + 4$

b) $y = 5.3x - 12$

SOLUTION

a) The y-intercept is $(0, 4)$.

b) The y-intercept is $(0, -12)$.

Interactive Discovery

Graph the following equations on the same set of axes using a standard viewing window:

$$y_1 = 3x + 1, \quad y_2 = -2x + 1, \quad y_3 = 0.6x + 1.$$

1. Do the lines have the same y-intercept?
2. Do the lines have the same slope?

Deselect y_2 and y_3 and graph the following equations on the same set of axes using a standard viewing window:

$$y_1 = 3x + 1, \quad y_4 = 3x - 3, \quad y_5 = 3x + 2.5.$$

3. Do the lines have the same y-intercept?
4. Do the lines have the same slope?
5. On the basis of the graphs of the five equations given here, does it appear that the slope of a line $y = mx + b$ depends on the number m or the number b?

The number m, in the equation $y = mx + b$, is responsible for the slope of a line.

The slope of any line written in the form $y = mx + b$ is m.

A proof of this result is outlined in Exercise 91 on p. 233.

The letter m is traditionally used for slope. This usage has its roots in the French verb *monter*, to climb.

The Slope–Intercept Equation The equation $y = mx + b$ is called the *slope–intercept equation*. The equation represents a line of slope m with y-intercept $(0, b)$.

The equation of any nonvertical line can be written in this form.

EXAMPLE 3 Find the slope and the y-intercept of each line whose equation is given.

a) $y = \frac{4}{5}x - 8$ b) $2x + y = 5$
c) $3x - 4y = 7$

SOLUTION

a) We rewrite $y = \frac{4}{5}x - 8$ as $y = \frac{4}{5}x + (-8)$. Now we simply read the slope and the y-intercept from the equation:

$$y = \frac{4}{5}x + (-8).$$

The slope is $\frac{4}{5}$. The y-intercept is $(0, -8)$.

b) We first solve for y to find an equivalent equation in the form $y = mx + b$:

$$2x + y = 5$$
$$y = -2x + 5. \qquad \text{Adding } -2x \text{ to both sides}$$

The slope is -2. The y-intercept is $(0, 5)$.

c) We rewrite the equation in the form $y = mx + b$:

$$3x - 4y = 7$$
$$-4y = -3x + 7 \qquad \text{Adding } -3x \text{ to both sides}$$
$$y = -\tfrac{1}{4}(-3x + 7) \qquad \text{Multiplying both sides by } -\tfrac{1}{4}$$
$$y = \tfrac{3}{4}x - \tfrac{7}{4}. \qquad \text{Using the distributive law}$$

The slope is $\tfrac{3}{4}$. The y-intercept is $\left(0, -\tfrac{7}{4}\right)$.

EXAMPLE 4 A line has slope $-\tfrac{12}{5}$ and y-intercept $(0, 11)$. Find an equation of the line.

SOLUTION We use the slope–intercept equation, substituting $-\tfrac{12}{5}$ for m and 11 for b:

$$y = mx + b = -\tfrac{12}{5}x + 11.$$

The desired equation is $y = -\tfrac{12}{5}x + 11$.

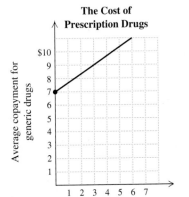

The Cost of Prescription Drugs

Average copayment for generic drugs

Number of years since 2000

EXAMPLE 5 Determine an equation for the graph shown at left (*Source*: Based on information from The Kaiser Family Foundation; Health Research and Education Trust).

SOLUTION To write an equation for a line, we can use slope–intercept form, provided the slope and the y-intercept are known. From the graph, we see that $(0, 7)$ is the y-intercept. Looking closely, we see that the line passes through $(3, 9)$. We can either count squares on the graph or use the formula to calculate the slope:

$$m = \frac{\text{change in } y}{\text{change in } x} = \frac{9 - 7}{3 - 0} = \frac{2}{3}.$$

The desired equation is

$$y = \frac{2}{3}x + 7, \qquad \text{Using } \tfrac{2}{3} \text{ for } m \text{ and 7 for } b$$

where y is the average cost, in dollars, of a copayment for generic drugs x years after 2000.

Graphing and Slope–Intercept Form

In Example 1, we drew a graph, knowing only the slope and the y-intercept. In Example 3, we determined the slope and the y-intercept of a line by examining its equation. We now combine the two procedures to develop a quick way to graph a linear equation.

EXAMPLE 6 Graph: **(a)** $y = \frac{3}{4}x + 5$; **(b)** $2x + 3y = 3$.

SOLUTION

a) From the equation $y = \frac{3}{4}x + 5$, we see that the slope of the graph is $\frac{3}{4}$ and the y-intercept is $(0, 5)$. We plot $(0, 5)$ and then consider the slope, $\frac{3}{4}$. Starting at $(0, 5)$, we plot a second point by moving *up* 3 units (since the numerator is *positive* and corresponds to the change in y) and *to the right* 4 units (since the denominator is *positive* and corresponds to the change in x). We reach a new point, $(4, 8)$.

We can also rewrite the slope as $\frac{-3}{-4}$. We again start at the y-intercept, $(0, 5)$, but move *down* 3 units (since the numerator is *negative* and corresponds to the change in y) and *to the left* 4 units (since the denominator is *negative* and corresponds to the change in x). We reach another point, $(-4, 2)$. Once two or three points have been plotted, the line representing all solutions of $y = \frac{3}{4}x + 5$ can be drawn.

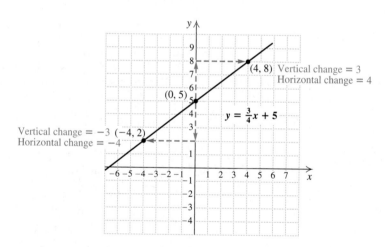

Student Notes

Graphing an equation in slope–intercept form can be done rather easily. First, identify and plot the y-intercept. Then identify the slope and use the slope to plot a second point. Finally, use a straightedge to draw a line passing through the two points.

b) To graph $2x + 3y = 3$, we first rewrite it in slope–intercept form:

$$2x + 3y = 3$$
$$3y = -2x + 3 \qquad \text{Adding } -2x \text{ to both sides}$$
$$y = \tfrac{1}{3}(-2x + 3) \qquad \text{Multiplying both sides by } \tfrac{1}{3}$$
$$y = -\tfrac{2}{3}x + 1. \qquad \text{Using the distributive law}$$

To graph $y = -\frac{2}{3}x + 1$, we first plot the y-intercept, $(0, 1)$. We can think of the slope as $\frac{-2}{3}$. Starting at $(0, 1)$ and using the slope, we find a second point by moving *down* 2 units (since the numerator is *negative*) and *to the right* 3 units (since the denominator is *positive*). We plot the new point, $(3, -1)$. In a similar manner, we can move from the point $(3, -1)$ to locate a third point, $(6, -3)$. The line can then be drawn.

Since $-\frac{2}{3} = \frac{2}{-3}$, an alternative approach is to again plot $(0, 1)$, but this time move *up* 2 units (since the numerator is *positive*) and *to the left* 3 units (since the denominator is *negative*). This leads to another point on the graph, $(-3, 3)$.

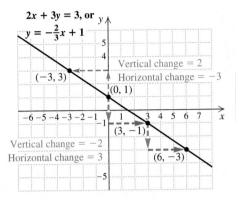

It is important to be able to use both $\frac{2}{-3}$ and $\frac{-2}{3}$ to draw the graph.

Slope–intercept form allows us to quickly determine the slope of a line by simply inspecting its equation. This can be especially helpful when attempting to decide whether two lines are parallel or perpendicular.

Parallel and Perpendicular Lines

Two lines are parallel if they lie in the same plane and do not intersect no matter how far they are extended. If two lines are vertical, they are parallel. How can we tell if nonvertical lines are parallel? The answer is simple: We look at their slopes.

Slope and Parallel Lines

Two lines with different y-intercepts are parallel if they have the same slope.

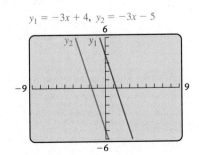

EXAMPLE 7   Determine whether the graphs of $y = -3x + 4$ and $6x + 2y = -10$ are parallel.

SOLUTION   When two lines have the same slope but different y-intercepts, they are parallel.

The slope of the line $y = -3x + 4$ is -3, and the y-intercept is $(0, 4)$. To find the slope of the other line, we first solve for y:

$$6x + 2y = -10$$
$$2y = -6x - 10 \qquad \text{Adding } -6x \text{ to both sides}$$
$$y = -3x - 5.$$

The slope is -3 and the y-intercept is $(0, -5)$. Since both lines have slope -3 but different y-intercepts, the graphs are parallel. The graphs of both equations are shown at left, and do appear parallel.

$y_1 = -\frac{7}{10}x + 5, \; y_2 = -\frac{4}{5}x - 5$

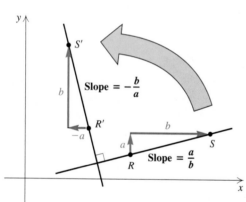

EXAMPLE 8   Determine whether the graphs of $7x + 10y = 50$ and $4x + 5y = -25$ are parallel.

SOLUTION   We find the slope of each line:

$$7x + 10y = 50$$
$$10y = -7x + 50 \qquad \text{Adding } -7x \text{ to both sides}$$
$$y = -\tfrac{7}{10}x + 5; \qquad \text{Multiplying both sides by } \tfrac{1}{10}$$
$$\text{The slope is } -\tfrac{7}{10}.$$

$$4x + 5y = -25$$
$$5y = -4x - 25 \qquad \text{Adding } -4x \text{ to both sides}$$
$$y = -\tfrac{4}{5}x - 5. \qquad \text{Multiplying both sides by } \tfrac{1}{5}$$
$$\text{The slope is } -\tfrac{4}{5}.$$

Since the slopes are not the same, the lines are not parallel. In decimal form, the slopes are -0.7 and -0.8. Since these slopes are close in value, in some viewing windows the graphs may appear parallel, as shown at left, when in reality they are not.

Two lines are perpendicular if they intersect at a right angle. If one line is vertical and another is horizontal, they are perpendicular. There are other instances in which two lines are perpendicular.

Consider a line $\overleftrightarrow{RS}$, as shown at left, with slope a/b. Then think of rotating the figure $90°$ to get a line $\overleftrightarrow{R'S'}$ perpendicular to $\overleftrightarrow{RS}$. For the new line, the rise and the run are interchanged, but the run is now negative. Thus the slope of the new line is $-b/a$. Let's multiply the slopes:

$$\frac{a}{b}\left(-\frac{b}{a}\right) = -1.$$

This can help us determine which lines are perpendicular.

> **Slope and Perpendicular Lines**
>
> Two lines are perpendicular if the product of their slopes is -1 or if one line is vertical and the other is horizontal.

Thus, if one line has slope m ($m \neq 0$), the slope of a line perpendicular to it is $-1/m$. That is, we take the reciprocal of m ($m \neq 0$) and change the sign.

EXAMPLE 9   Determine whether the graphs of $2x + y = 8$ and $y = \tfrac{1}{2}x + 7$ are perpendicular.

SOLUTION   First, we find the slope of each line. The second equation,

$$y = \tfrac{1}{2}x + 7,$$

is in slope–intercept form. It represents a line with slope $\tfrac{1}{2}$.

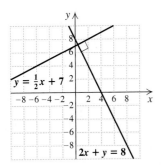

To find the slope of the other line, we solve for y:

$$2x + y = 8$$
$$y = -2x + 8. \quad \text{Adding } -2x \text{ to both sides}$$

The slope of the line is -2.

The lines are perpendicular if the product of their slopes is -1. Since

$$\tfrac{1}{2}(-2) = -1,$$

the graphs are perpendicular. The graphs of both equations are shown at left, and do appear perpendicular.

EXAMPLE 10   Write a slope–intercept equation for the line whose graph is described.

a) Parallel to the graph of $2x - 3y = 7$, with y-intercept $(0, -1)$

b) Perpendicular to the graph of $2x - 3y = 7$, with y-intercept $(0, -1)$

SOLUTION   We begin by determining the slope of the line represented by $2x - 3y = 7$:

$$2x - 3y = 7$$
$$-3y = -2x + 7 \quad \text{Adding } -2x \text{ to both sides}$$
$$y = \tfrac{2}{3}x - \tfrac{7}{3}. \quad \text{Dividing both sides by } -3$$

The slope is $\tfrac{2}{3}$.

a) A line parallel to the graph of $2x - 3y = 7$ has a slope of $\tfrac{2}{3}$. Since the y-intercept is $(0, -1)$, the slope–intercept equation is

$$y = \tfrac{2}{3}x - 1.$$

b) A line perpendicular to the graph of $2x - 3y = 7$ has a slope that is the negative reciprocal of $\tfrac{2}{3}$, or $-\tfrac{3}{2}$. Since the y-intercept is $(0, -1)$, the slope–intercept equation is

$$y = -\tfrac{3}{2}x - 1.$$

Student Notes

Although it is helpful to visualize parallel lines and perpendicular lines graphically, a graph cannot be used to *determine* whether two lines are parallel or perpendicular. Slopes must be used to determine this.

Squaring a Viewing Window

If the units on the x-axis are a different length than those on the y-axis, two lines that are perpendicular may not appear to be so when graphed. Finding a viewing window with units the same length on both axes is called *squaring* the viewing window.

Windows can be squared by choosing the ZSquare option in the ZOOM menu. They can be squared manually by choosing the portions of the axes shown in the correct proportion. For example, if the ratio of the length of a viewing window to its height is $3{:}2$, a window like $[-9, 9, -6, 6]$ will be squared.

Consider the graphs of $y = 3x - 7$ and $y = -\tfrac{1}{3}x + \tfrac{1}{3}$. The lines are perpendicular, since $3 \cdot \left(-\tfrac{1}{3}\right) = -1$. The graphs are shown in a standard viewing window on the left below, but the lines do not appear to be perpendicular.

(continued)

If we press ⬭ ZOOM ⬭ ⬭ 5 ⬭, the lines are graphed in a squared viewing window as shown in the graph on the right below, and they do appear perpendicular.

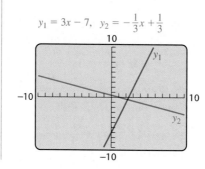

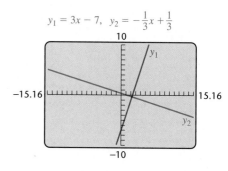

3.6 EXERCISE SET

⤹ *Concept Reinforcement*   *In each of Exercises 1–6, match the phrase with the most appropriate choice from the column on the right.*

1. ____ The y-intercept of the graph of $y = 2x - 3$

2. ____ The y-intercept of the graph of $y = 3x - 2$

3. ____ The slope of the graph of $y = 3x - 2$

4. ____ The slope of the graph of $y = 2x - 3$

5. ____ The slope of the graph of $y = \frac{2}{3}x + 3$

6. ____ The y-intercept of the graph of $y = \frac{2}{3}x + \frac{3}{4}$

a) $\left(0, \frac{3}{4}\right)$

b) 2

c) $(0, -3)$

d) $\frac{2}{3}$

e) $(0, -2)$

f) 3

Draw a line that has the given slope and y-intercept.

7. Slope $\frac{2}{5}$; y-intercept $(0, 1)$

8. Slope $\frac{3}{5}$; y-intercept $(0, -1)$

9. Slope $\frac{5}{3}$; y-intercept $(0, -2)$

10. Slope $\frac{5}{2}$; y-intercept $(0, 1)$

11. Slope $-\frac{3}{4}$; y-intercept $(0, 5)$

12. Slope $-\frac{4}{5}$; y-intercept $(0, 6)$

13. Slope 2; y-intercept $(0, -4)$

14. Slope -2; y-intercept $(0, -3)$

15. Slope -3; y-intercept $(0, 2)$

16. Slope 3; y-intercept $(0, 4)$

Find the slope and the y-intercept of each line whose equation is given.

17. $y = -\frac{2}{7}x + 5$

18. $y = -\frac{3}{8}x + 4$

19. $y = \frac{5}{8}x + 3$

20. $y = \frac{4}{5}x + 1$

21. $y = \frac{9}{5}x - 4$

22. $y = \frac{7}{4}x - 5$

23. $-3x + y = 7$

24. $-4x + y = 7$

25. $5x + 2y = 8$

26. $3x + 4y = 12$

Aha! **27.** $y = 4$

28. $y - 3 = 5$

29. $2x - 5y = -8$

30. $5x - 6y = 9$

31. Use the slope and the *y*-intercept of each line to match each equation with the correct graph.

a) $y = 3x - 5$
b) $y = 0.7x + 1$
c) $y = -0.25x - 3$
d) $y = -4x + 2$

I

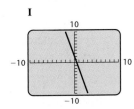

II

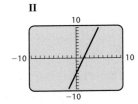

III

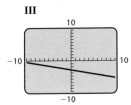

IV

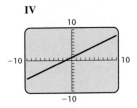

32. Use the slope and the *y*-intercept of each line to match each equation with the correct graph.

a) $y = \frac{1}{2}x - 5$
b) $y = 2x + 3$
c) $y = -3x + 1$
d) $y = -\frac{3}{4}x - 2$

I

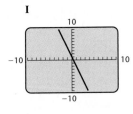

II

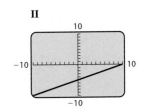

III

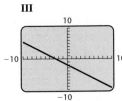

IV

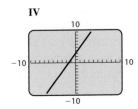

Find the slope–intercept equation for the line with the indicated slope and y-intercept.

33. Slope 3; *y*-intercept $(0, 7)$

34. Slope -4; *y*-intercept $(0, -2)$

35. Slope $\frac{7}{8}$; *y*-intercept $(0, -1)$

36. Slope $\frac{5}{7}$; *y*-intercept $(0, 4)$

37. Slope $-\frac{5}{3}$; *y*-intercept $(0, -8)$

38. Slope $\frac{3}{4}$; *y*-intercept $(0, 23)$

Aha! **39.** Slope 0; *y*-intercept $(0, 3)$

40. Slope 7; *y*-intercept $(0, 0)$

Determine an equation for each graph shown.

41.

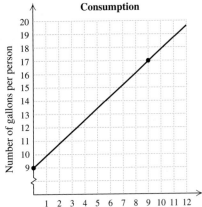

U.S. Bottled Water Consumption

Based on information from the International Bottled Water Association 2002

42.

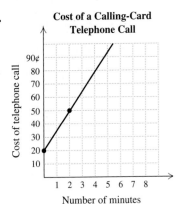

Cost of a Calling-Card Telephone Call

43.

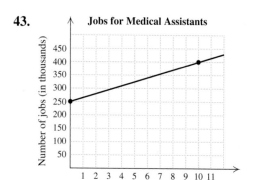

Jobs for Medical Assistants

Number of years since 1998

Estimates based on information from *Handbook of U.S. Labor*

44.

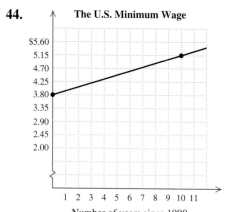

The U.S. Minimum Wage

Number of years since 1990

Based on information from United for a Fair Economy

Graph by hand.

45. $y = \frac{3}{5}x + 2$

46. $y = -\frac{3}{5}x - 1$

47. $y = -\frac{3}{5}x + 1$

48. $y = \frac{3}{5}x - 2$

49. $y = \frac{5}{3}x + 3$

50. $y = \frac{5}{3}x - 2$

51. $y = -\frac{3}{2}x - 2$

52. $y = -\frac{4}{3}x + 3$

53. $2x + y = 1$

54. $3x + y = 2$

55. $3x + y = 0$

56. $2x + y = 0$

57. $2x + 3y = 9$

58. $4x + 5y = 15$

59. $x - 4y = 12$

60. $x + 5y = 20$

Determine whether each pair of equations represents parallel lines.

61. $y = \frac{2}{3}x + 7$,
$y = \frac{2}{3}x - 5$

62. $y = -\frac{5}{4}x + 1$,
$y = \frac{5}{4}x + 3$

63. $y = 2x - 5$,
$4x + 2y = 9$

64. $y = -3x + 1$,
$6x + 2y = 8$

65. $3x + 4y = 8$,
$7 - 12y = 9x$

66. $3x = 5y - 2$,
$10y = 4 - 6x$

Determine whether each pair of equations represents perpendicular lines.

67. $y = 4x - 5$,
$4y = 8 - x$

68. $2x - 5y = -3$,
$2x + 5y = 4$

69. $x - 2y = 5$,
$2x + 4y = 8$

70. $y = -x + 7$,
$y - x = 3$

71. $2x + 3y = 1$,
$3x - 2y = 1$

72. $y = 5 - 3x$,
$3x - y = 8$

Write a slope–intercept equation of the line whose graph is described.

73. Parallel to the graph of $y = 5x - 7$; y-intercept $(0, 11)$

74. Parallel to the graph of $2x - y = 1$; y-intercept $(0, -3)$

75. Perpendicular to the graph of $2x + y = 0$; y-intercept $(0, 0)$

76. Perpendicular to the graph of $y = \frac{1}{3}x + 7$; y-intercept $(0, 5)$

Aha! **77.** Parallel to the graph of $y = x$; y-intercept $(0, 3)$

Aha! **78.** Perpendicular to the graph of $y = x$; y-intercept $(0, 0)$

79. Perpendicular to the graph of $x + y = 3$; y-intercept $(0, -4)$

80. Parallel to the graph of $3x + 2y = 5$; y-intercept $(0, -1)$

TW **81.** A student makes a mistake when using a graphing calculator to draw $4x + 5y = 12$ and the following screen appears. Use algebra to show that a mistake has been made. What do you think the mistake was?

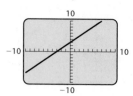

TW **82.** A student makes a mistake when using a graphing calculator to draw $5x - 2y = 3$ and the following screen appears. Use algebra to show that a mistake has been made. What do you think the mistake was?

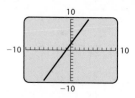

Focused Review

Graph.

83. $y = 3x - 2$ [3.6]

84. $x = -2$ [3.3]

85. $y = 4$ [3.3]

86. $2x - 4y = 8$ [3.3]

87. $y = -\frac{1}{2}x + 1$ [3.6]

88. $x + 5y = 5$ [3.3]

Synthesis

TW 89. Explain how it is possible for an incorrect graph to be drawn, even after plotting three points that line up.

TW 90. Which would you prefer, and why: graphing an equation of the form $y = mx + b$ or graphing an equation of the form $Ax + By = C$?

91. Show that the slope of the line given by $y = mx + b$ is m. (*Hint:* Substitute both 0 and 1 for x to find two pairs of coordinates. Then use the formula, Slope = change in y/change in x.)

92. Find k such that the line containing $(-3, k)$ and $(4, 8)$ is parallel to the line containing $(5, 3)$ and $(1, -6)$.

Solve.

93. *Refrigerator Size.* Kitchen designers recommend that a refrigerator be selected on the basis of the number of people in the household. For 1–2 people, a 16 ft³ model is suggested. For each additional person, an additional 1.5 ft³ is recommended. If x is the number of residents in excess of 2, find the slope–intercept equation for the recommended size of a refrigerator.

94. *Telephone Service.* In a recent promotion, AT&T charged a monthly fee of $4.95 plus 7¢ for each minute of long-distance phone calls. If x is the number of minutes of long-distance calls, find the slope–intercept equation for the monthly bill.

95. *Cost of a Speeding Ticket.* The penalty schedule shown below is used to determine the cost of a speeding ticket in certain states. Use this schedule to graph the cost of a speeding ticket as a function of the number of miles per hour over the limit that a driver is going.

STATE POLICE SPEEDING VIOLATION FINES

1–10 mph over limit: $5.00/mph plus $17.50 surcharge
11–20 mph over limit: $6.00/mph plus $17.50 surcharge
21–30 mph over limit: $7.00/mph plus $17.50 surcharge
31+ mph over limit: $8.00/mph plus $17.50 surcharge

Officer will enter mph over limit in line 5a on the front of this document.

In Exercises 96 and 97, assume that r, p, and s are constants and that x and y are variables. Determine the slope and the y-intercept.

96. $rx + py = s$

97. $rx + py = s - ry$

98. Graph the equations

$$y_1 = 1.4x + 2, \qquad y_2 = 0.6x + 2,$$
$$y_3 = 1.4x + 5, \quad \text{and} \quad y_4 = 0.6x + 5$$

using a graphing calculator. If possible, use the SIMULTANEOUS mode so that you cannot tell which equation is being graphed first. Then decide which line corresponds to each equation.

TW 99. *Aerobic Exercise.* The formula $T = -\frac{3}{4}a + 165$ can be used to determine the *target heart rate*, in beats per minute, for a person, a years old, participating in aerobic exercise. Graph the equation and interpret the significance of its slope.

3.7 Point–Slope Form; Introduction to Curve Fitting

Writing Equations in Point–Slope Form ◾ Graphing and Point–Slope Form ◾ Estimations and Predictions Using Two Points ◾ Curve Fitting

Specifying a line's slope and one point through which the line passes enables us to draw the line. In this section, we study how this same information can be used to produce an *equation* of the line.

Writing Equations in Point–Slope Form

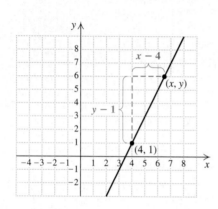

Consider a line with slope 2 passing through the point (4, 1), as shown in the figure. In order for a point (x, y) to be on the line, the coordinates x and y must be solutions of the slope equation

$$\frac{y - 1}{x - 4} = 2.$$

Take a moment to examine this equation. Pairs like (5, 3) and (3, −1) are solutions, since

$$\frac{3 - 1}{5 - 4} = 2 \quad \text{and} \quad \frac{-1 - 1}{3 - 4} = 2.$$

Note, however, that (4, 1) is not itself a solution of the equation:

$$\frac{1 - 1}{4 - 4} \neq 2.$$

To avoid this difficulty, we can use the multiplication principle:

$$(x - 4) \cdot \frac{y - 1}{x - 4} = 2(x - 4) \qquad \textbf{Multiplying both sides by } x - 4$$

$$y - 1 = 2(x - 4). \qquad \textbf{Removing a factor equal to 1: } \frac{x - 4}{x - 4} = 1$$

This is considered **point–slope form** for the line shown above. A point–slope equation can be written any time a line's slope and a point on the line are known.

> **The Point–Slope Equation** The equation $y - y_1 = m(x - x_1)$ is called the *point–slope equation* for the line with slope m that contains the point (x_1, y_1).

▶ **EXAMPLE 1** Write a point–slope equation for the line with slope $-\frac{4}{3}$ that contains the point $(1, -6)$.

SOLUTION We substitute $-\frac{4}{3}$ for m, 1 for x_1, and -6 for y_1:

$$y - y_1 = m(x - x_1) \quad \text{Using the point–slope equation}$$
$$y - (-6) = -\tfrac{4}{3}(x - 1). \quad \text{Substituting}$$

EXAMPLE 2 Write the slope–intercept equation for the line with slope 2 that contains the point (3, 1).

SOLUTION There are two parts to this solution. First, we write an equation in point–slope form:

$$y - y_1 = m(x - x_1)$$
$$y - 1 = 2(x - 3). \quad \text{Substituting}$$

Next, we find an equivalent equation of the form $y = mx + b$:

$$y - 1 = 2(x - 3)$$
$$y - 1 = 2x - 6 \quad \text{Using the distributive law}$$
$$y = 2x - 5. \quad \text{Adding 1 to both sides to get slope–intercept form}$$

Student Notes

There are several forms in which a line's equation can be written. For instance, as shown in Example 2, $y - 1 = 2(x - 3)$, $y - 1 = 2x - 6$, and $y = 2x - 5$ all are equations for the same line.

EXAMPLE 3 Consider the line given by the equation $8y = 7x - 24$.

a) Write the slope–intercept equation for a parallel line passing through $(-1, 2)$.

b) Write the slope–intercept equation for a perpendicular line passing through $(-1, 2)$.

SOLUTION Both parts (a) and (b) require us to find the slope of the line given by $8y = 7x - 24$. To do so, we solve for y to find slope–intercept form:

$$8y = 7x - 24.$$
$$y = \tfrac{7}{8}x - 3. \quad \text{Multiplying both sides by } \tfrac{1}{8}$$

The slope is $\frac{7}{8}$.

a) The slope of any parallel line will be $\frac{7}{8}$. The point–slope equation yields

$$y - 2 = \tfrac{7}{8}[x - (-1)] \quad \text{Substituting } \tfrac{7}{8} \text{ for the slope and } (-1, 2) \text{ for the point}$$
$$y - 2 = \tfrac{7}{8}[x + 1]$$
$$y = \tfrac{7}{8}x + \tfrac{7}{8} + 2 \quad \text{Using the distributive law and adding 2 to both sides}$$
$$y = \tfrac{7}{8}x + \tfrac{23}{8}.$$

b) The slope of a perpendicular line is given by the opposite of the reciprocal of $\frac{7}{8}$, or $-\frac{8}{7}$. The point–slope equation yields

$$y - 2 = -\tfrac{8}{7}[x - (-1)] \quad \text{Substituting } -\tfrac{8}{7} \text{ for the slope and } (-1, 2) \text{ for the point}$$
$$y - 2 = -\tfrac{8}{7}[x + 1]$$
$$y = -\tfrac{8}{7}x - \tfrac{8}{7} + 2 \quad \text{Using the distributive law and adding 2 to both sides}$$
$$y = -\tfrac{8}{7}x + \tfrac{6}{7}.$$

EXAMPLE 4   Write the slope–intercept equation for the line that contains the points $(-1, -5)$ and $(3, -2)$.

SOLUTION   We first determine the slope of the line and then use the point–slope equation. Note that

$$m = \frac{-5 - (-2)}{-1 - 3} = \frac{-3}{-4} = \frac{3}{4}.$$

Since the line passes through $(3, -2)$, we have

$$y - (-2) = \tfrac{3}{4}(x - 3) \qquad \text{Substituting into the point–slope equation}$$
$$y + 2 = \tfrac{3}{4}x - \tfrac{9}{4}. \qquad \text{Using the distributive law}$$

Finally, we solve for y to write the equation in slope–intercept form:

$$y = \tfrac{3}{4}x - \tfrac{9}{4} - 2 \qquad \text{Subtracting 2 from both sides}$$
$$y = \tfrac{3}{4}x - \tfrac{17}{4}. \qquad -\tfrac{9}{4} - \tfrac{8}{4} = -\tfrac{17}{4}$$

You can check that substituting $(-1, -5)$ instead of $(3, -2)$ in the point–slope equation will yield the same slope–intercept equation.

Graphing and Point–Slope Form

When we know a line's slope and a point that is on the line, we can draw the graph, much as we did in Section 3.6. For example, the information given in the statement of Example 2 is sufficient for drawing a graph.

EXAMPLE 5   Graph the line with slope 2 that passes through $(3, 1)$.

SOLUTION   We plot $(3, 1)$, move *up* 2 and *to the right* 1 $\left(\text{since } 2 = \tfrac{2}{1}\right)$, and draw the line.

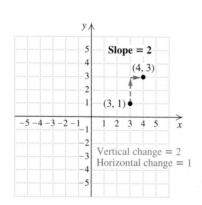

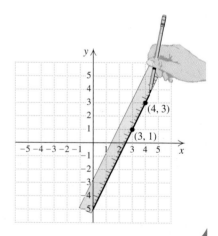

EXAMPLE 6   Graph: $y - 2 = 3(x - 4)$.

SOLUTION   Since $y - 2 = 3(x - 4)$ is in point–slope form, we know that the line has slope 3, or $\frac{3}{1}$, and passes through the point $(4, 2)$. We plot $(4, 2)$ and then find a second point by moving *up* 3 units and *to the right* 1 unit. The line can then be drawn, as shown below.

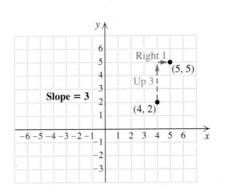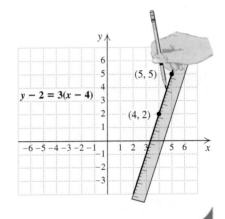

EXAMPLE 7   Graph: $y + 4 = -\frac{5}{2}(x + 3)$.

SOLUTION   Once we have written the equation in point–slope form, $y - y_1 = m(x - x_1)$, we can proceed much as we did in Example 6. To find an equivalent equation in point–slope form, we subtract opposites instead of adding:

$$y + 4 = -\tfrac{5}{2}(x + 3)$$
$$y - (-4) = -\tfrac{5}{2}(x - (-3)).$$ **Subtracting a negative instead of adding a positive. This is now in point–slope form.**

From this last equation, $y - (-4) = -\frac{5}{2}(x - (-3))$, we see that the line passes through $(-3, -4)$ and has slope $-\frac{5}{2}$, or $\frac{5}{-2}$.

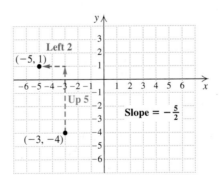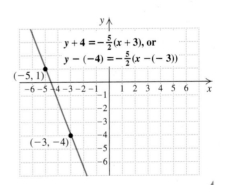

Estimations and Predictions Using Two Points

It is possible to use line graphs to estimate real-life quantities that are not already known. To do so, we calculate the coordinates of an unknown point by using two points with known coordinates. When the unknown point is located *between* the two points, this process is called **interpolation.*** Sometimes a graph passing through the known points is *extended* to predict future values. Making predictions in this manner is called **extrapolation.*** In statistics, methods exist for using a set of several points to interpolate or extrapolate values using curves other than lines.

▶ **EXAMPLE 8** Aerobic Exercise.   A person's target heart rate is the number of beats per minute that bring the most aerobic benefit to his or her heart. The target heart rate for a 20-year-old is 150 beats per minute and for a 60-year-old, 120 beats per minute.

a) Graph the given data and calculate the target heart rate for a 36-year-old.

b) Calculate the target heart rate for a 75-year-old.

c) For what age is the target heart rate 130 beats per minute?

SOLUTION

a) We first draw a horizontal axis for "Age" and a vertical axis for "Target heart rate." Next, we number the axes, using a scale that will permit us to view both the given and the desired data. The given information allows us to then plot (20, 150) and (60, 120).

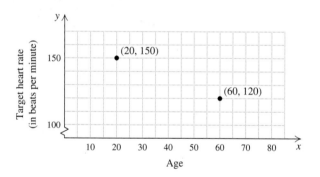

Next, we draw a line passing through both points. The slope of the line is given by

$$m = \frac{\text{change in } y}{\text{change in } x} = \frac{150 - 120 \text{ beats per minute}}{20 - 60 \text{ years}}$$

$$= \frac{30 \text{ beats per minute}}{-40 \text{ years}} = -\frac{3}{4} \text{ beat per minute per year.}$$

---

*Both interpolation and extrapolation can be performed using more than two known points and using curves other than lines.

The target heart rate drops at a rate of $\frac{3}{4}$ beat per minute for each year that we age. We can use either of the given points to write a point–slope equation for the line. Let's use (20, 150) and then use algebra to write an equivalent equation in slope–intercept form:

$$y - 150 = -\frac{3}{4}(x - 20) \qquad \text{This is a point–slope equation.}$$

$$y - 150 = -\frac{3}{4}x + 15 \qquad \text{Using the distributive law}$$

$$y = -\frac{3}{4}x + 165. \qquad \begin{array}{l}\text{Adding 150 to both sides. This is}\\ \text{slope–intercept form.}\end{array}$$

To calculate the target heart rate for a 36-year-old, we substitute 36 for x in the slope–intercept equation:

$$y = -\frac{3}{4} \cdot 36 + 165 = -27 + 165 = 138.$$

As the graph confirms, the target heart rate for a 36-year-old is about 138 beats per minute.

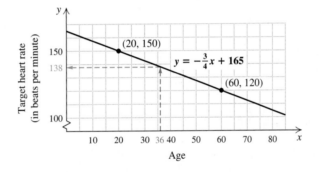

Because 36 is *between* the given values of 20 and 60, we are *interpolating* here.

b) To calculate the target heart rate for a 75-year-old, we again substitute for x in the slope–intercept equation:

$$y = -\frac{3}{4} \cdot 75 + 165 = -56.25 + 165 = 108.75 \approx 109.$$

As the graph confirms, the target heart rate for a 75-year-old is about 109 beats per minute.

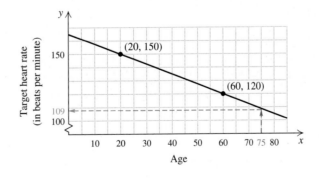

Because 75 is *beyond* the given values, we are *extrapolating* here.

c) Since x represents age and y represents the target heart rate, we substitute 130 for y and solve for x:

$$y = -\frac{3}{4}x + 165$$

$$130 = -\frac{3}{4}x + 165 \qquad \text{Substituting 130 for } y$$

$$-35 = -\frac{3}{4}x \qquad \text{Subtracting 165 from both sides}$$

$$47 \approx x. \qquad \text{Multiplying both sides by } -\tfrac{4}{3} \text{ and rounding}$$

As the graph confirms, the target heart rate is 130 beats per minute for a 47-year-old person.

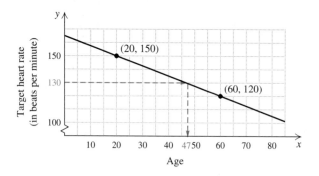

Connecting the Concepts

We have now studied the slope–intercept, point–slope, and standard forms of a linear equation. These are the most common ways in which linear equations are written. Depending on what information we are given and what information we are seeking, one form may be more useful than the others. A referenced summary is given below.

Slope–intercept form, $y = mx + b$	• Useful when an equation is needed and the slope and y-intercept are given. See Example 4 on p. 225. • Useful when a line's slope and y-intercept are needed. See Example 3 on pp. 224–225.
Standard form, $Ax + By = C$	• Allows for easy calculation of intercepts. See Example 3 on p. 189. • Will prove useful in future work. See Section 4.3.
Point–slope form, $y - y_1 = m(x - x_1)$	• Useful when an equation is needed and the slope and a point on the line are given. See Example 1 on pp. 234–235. • Useful when an equation is needed and two points on the line are given. See Example 4 on p. 236. • Will prove useful in future work with curves and tangents in calculus.

Curve Fitting

The process of understanding and interpreting *data*, or lists of information, is called *data analysis*. One helpful tool in data analysis is **curve fitting,** or finding an algebraic equation that describes the data. We fit a linear equation to two data points in Example 8.

One of the first steps in curve fitting is to decide what kind of equation will best describe the data. In Example 8, we assumed a linear relationship. Unless other information is known, two points are not sufficient to determine the pattern of the data. We gather all the data known and plot the points. If the data appear linear, we can fit a linear equation to the data.

EXAMPLE 9 Following are three graphs of sets of data. Determine whether each appears to be linear.

a) **Fiber for Children**

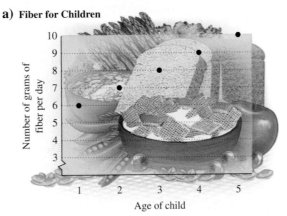

Source: Kellogg's

b) **Winter Heating Costs**

Source: Energy Information Administration, U.S. Department of Energy

c) **Registered Nurses**

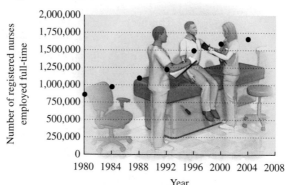

Source: National Sample Survey of Registered Nurses

SOLUTION In order for data to be linear, the points must lie, at least approximately, on a straight line. The rate of change is constant for a linear equation, so the change in the quantity on the vertical axis should be about the same for each unit on the horizontal axis.

a) The points lie on a straight line, so the data are linear. The rate of change is constant—1 gram of fiber per year of age.

b) Note that the average heating cost increased, then decreased, then began to increase, but not at a constant rate. The points do not lie on a straight line. The data are not linear.

c) The points lie approximately on a straight line. The data appear to be linear.

If data are linear, we can fit a linear equation to the data by choosing two points. However, if the data do not lie exactly on a straight line, the choice of those points will affect the equation found. The line that *best* describes the data may not actually go through any of the given points. Equation-fitting methods generally consider all the points, not just two, when fitting an equation to data. The most commonly used method is *linear regression*.

The development of the method of linear regression belongs to a later mathematics course, but most graphing calculators offer regression as a way of fitting a line or curve to a set of data. To use a REGRESSION feature, we must enter the data into the calculator.

Entering and Plotting Data

Coordinates of ordered pairs are entered as *data* in lists, using the STAT menu. To enter or change data, press **STAT** and then choose the EDIT option. The lists of numbers will appear as three columns on the screen. If there are already numbers in the lists, clear them by moving the cursor to the title of the list (L1, L2, and so on) and pressing **CLEAR** **ENTER**.

(continued)

To enter a number in a list, move the cursor to the correct position, type in the number, and press **ENTER**. Enter the first coordinates of the ordered pairs as one list and the second coordinates as another list. The coordinates of each point should be at the same position on both lists. Note that a DIM MISMATCH error will occur if there is not the same number of items in each list.

To plot the points, first turn on the STAT PLOT feature. Press (STAT PLOT). (STAT PLOT is the 2nd option associated with the (Y=) key.) The calculator allows several different sets of points to be displayed at once. See the screen on the left below. Choose the Plot you wish to define; if you are simply plotting one set of points, choose Plot1 by pressing (1). Then turn Plot1 on by positioning the cursor over On and pressing **ENTER**, as shown on the right below.

The remaining items on the Plot1 screen define the plot. Use the down arrow key to move to the next item.

The points entered can be used to make a line graph or a bar graph as well as to graph points. The screen on the right below shows six available types of graphs. To plot points, choose the first type of graph shown, a scatter diagram or scatterplot. The second option in the list is a line graph, in which the points are connected. The third type is a bar graph. The last three types will not be discussed in this course.

The next item on the screen, Xlist, should be the list in which the first coordinates were entered, probably L_1, and Ylist the list in which the second coordinates were entered, probably L_2. List names can be selected by pressing (LIST). (LIST is the 2nd option associated with the **STAT** key.)

The last choice on the screen is the type of mark used to plot the points. Different marks can be used to distinguish among several sets of data.

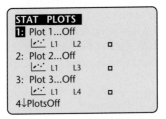
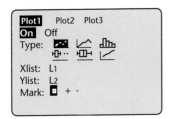

When the STAT PLOT feature has been set correctly, choose window dimensions that will allow all the points to be seen and press (GRAPH). The ZoomStat option of the ZOOM menu will choose an appropriate window automatically. When you no longer wish to plot data, turn off the plot. One way to do this is to move the cursor to the highlighted PLOT name at the top of the equation-editor screen and press **ENTER**. To turn off all the plots, choose option 4: PlotsOff on the STAT PLOTS screen and press **ENTER**.

Linear Regression

Fitting a curve to a set of data is done using the STAT menu. Choose EDIT from the STAT EDIT menu, and enter the values of the independent variable in one list and the corresponding values of the dependent variable in another list. Then press **STAT** again, and choose the CALC menu. To fit a line to the data, choose the LINREG option. After copying LinReg(ax + b) to the home screen, enter the list names containing the data separated by commas, with the independent values first. (The list names are 2nd options associated with the number keys ① through ⑥.)

To copy the equation found to the equation-editor screen, also list the Y-variable name, chosen in the VARS Y-VARS FUNCTION submenu, after the linear regression command. To execute the command, press **ENTER**.

The command at left indicates that L1 contains the values for the independent variable, L2 contains the values for the dependent variable, and the equation is to be copied to Y1. If no list names are entered, the calculator assumes that the first list is L1 and the second is L2.

Using CATALOG you can turn DiagnosticOn or DiagnosticOff. If diagnostics are turned on, values of r^2 and r will appear on the screen along with the regression equation. These give an indication of how well the regression line fits the data. When r^2 is close to 1, the line is a good fit. We call r the *coefficient of correlation.*

> Lin Reg (ax+b) L1, L2, Y1

EXAMPLE 10   *Registered Nurses.*   Data from the National Sample Survey of Registered Nurses, conducted every four years, indicate that the number of registered nurses in the United States who are working full time has increased steadily. The information from the survey is shown in the following table and graph. Use linear regression to fit a linear equation to the data. Graph the line with the data and use it to predict the number of registered nurses working full time in 2008.

Year	Number of Registered Nurses Working Full Time
1980	850,000
1984	975,000
1988	1,100,000
1992	1,225,000
1996	1,500,000
2000	1,600,000
2004	1,700,000

Registered Nurses

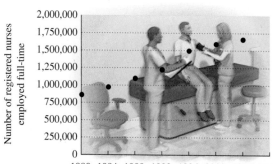

Source: National Sample Survey of Registered Nurses

SOLUTION   To make the numbers easier to work with, we will redefine the year t as the number of years since 1980. Then 1980 corresponds to $t = 0$, 1984 corresponds to $t = 4$, and so on. We also let N represent the number of registered nurses working full time, in thousands. Thus the point $(1980, 850{,}000)$ becomes $(0, 850)$. We are looking for an equation of the form $N = mt + b$.

We enter the data, with the number of years since 1980 as L1 and the number of full-time registered nurses, in thousands, as L2.

L1	L2	L3	1
0	850	------	
4	975		
8	1100		
12	1225		
16	1500		
20	1600		
24	1700		

L1(1) = 0

Next, we make sure that Plot1 is turned on, and clear any equations listed in the Y= screen. Since years vary from 0 to 24 and the number of nurses (in thousands) varies from 850 to 1700, we set a viewing window of $[0, 30, 0, 2000]$, with $\text{Xscl} = 4$ and $\text{Yscl} = 200$.

To calculate the equation, we choose the LinReg option in the STAT CALC menu. Since our lists are the default lists, L1 and L2, we need not give their names in the command. We select Y1 from the VARS Y-VARS FUNCTION menu and press **ENTER**.

The screen on the left below indicates that the equation is

$$y = 37.5x + 828.5714286, \quad \text{or}$$

$$N = 37.5t + 828.6. \quad \text{Substituting } N \text{ and } t \text{ and rounding}$$

The screen in the middle shows the equation copied as Y1. Note that there are even more decimal places in the value for b than are shown in the screen on the left. Pressing (GRAPH) gives the screen on the right below, showing the points plotted and the line graphed.

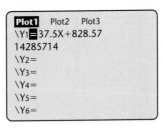

```
LinReg
y=ax+b
a=37.5
b=828.5714286
```

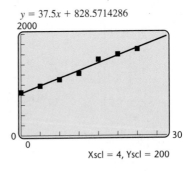

Xscl = 4, Yscl = 200

Since 2008 is 28 yr after 1980, we substitute 28 for t in order to predict the number of full-time registered nurses in 2008. We can do this using the VALUE option of the CALC menu for $x = 28$ or a table for $x = 28$. Both methods give a value of approximately 1879. Thus we predict that the number of full-time registered nurses in 2008 will be 1,879,000.

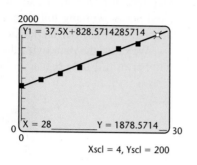

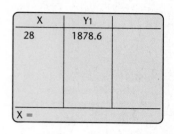

Xscl = 4, Yscl = 200

Connecting the Concepts

DATA INTERPRETATION: PREDICTIONS

In business and in many other fields, decisions are often made on the basis of a prediction of the future. Although extrapolation from past data is an important tool in making predictions, other factors should be considered as well.

Look, for example, at the following graph, which shows the number of pieces of priority mail handled by the U.S. Postal Service for 2000–2004.

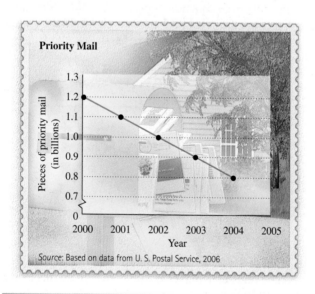

Source: Based on data from U. S. Postal Service, 2006

On the basis of past data, we might predict that the volume of priority mail will drop to 0.7 billion in 2005. However, it turns out that more pieces of priority mail were handled in 2005 than in 2004.

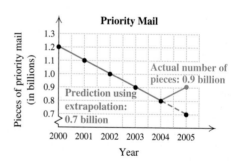

In this case, factors such as the economy, services offered, advertising, and competition contributed to the actual numbers. Although the past can be an excellent predictor of the future, any prediction is to some extent a guess.

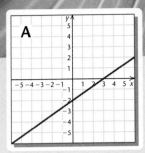

A

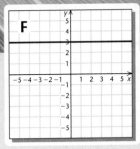

F

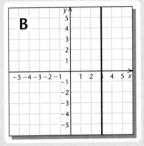

B

Visualizing the Graph

Match each equation or function with its graph.

1. $y = x + 4$

2. $y = 2x$

3. $y = 3$

4. $x = 3$

5. $y = -\frac{1}{2}x$

6. $2x - 3y = 6$

7. $y = -3x - 2$

8. $3x + 2y = 6$

9. $y - 3 = 2(x - 1)$

10. $y + 2 = \frac{1}{2}(x + 1)$

Answers on page A-13

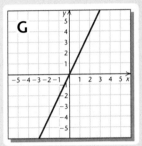

G

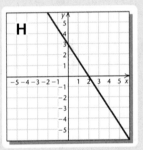

H

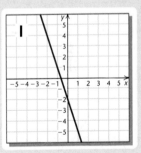

I

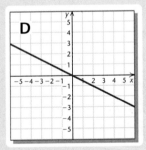

C

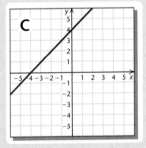

D

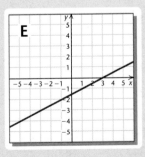

E

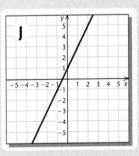

J

3.7 EXERCISE SET

 Concept Reinforcement *In each of Exercises 1–8, match the given information about a line with the appropriate equation from the column on the right.*

1. ____ Slope 5; includes (2, 3)

a) $y + 3 = 5(x + 2)$

b) $y - 2 = 5(x - 3)$

2. ____ Slope −5; includes (2, 3)

c) $y + 2 = 5(x + 3)$

d) $y - 3 = -5(x - 2)$

3. ____ Slope −5; includes (−2, −3)

e) $y + 3 = -5(x + 2)$

f) $y + 2 = -5(x + 3)$

4. ____ Slope 5; includes (−2, −3)

g) $y - 3 = 5(x - 2)$

h) $y - 2 = -5(x - 3)$

5. ____ Slope 5; includes (3, 2)

6. ____ Slope −5; includes (3, 2)

7. ____ Slope −5; includes (−3, −2)

8. ____ Slope 5; includes (−3, −2)

In each of Exercises 9–12, match the graph with the appropriate equation from the column on the right.

9.

a) $y - 4 = -\frac{3}{2}(x + 1)$

b) $y - 4 = \frac{3}{2}(x + 1)$

c) $y + 4 = -\frac{3}{2}(x - 1)$

d) $y + 4 = \frac{3}{2}(x - 1)$

10.

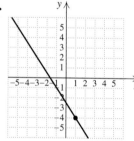

11.

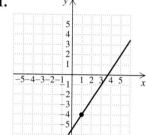

12.

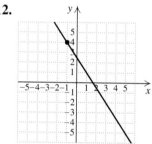

Write a point–slope equation for the line with the given slope that contains the given point.

13. $m = 5$; (6, 2)

14. $m = 4$; (3, 5)

15. $m = -4$; (3, 1)

16. $m = -5$; (6, 2)

17. $m = \frac{3}{2}$; (5, −4)

18. $m = \frac{4}{3}$; (7, −1)

19. $m = \frac{5}{4}$; (−2, 6)

20. $m = \frac{7}{2}$; (−3, 4)

21. $m = -2$; (−4, −1)

22. $m = -3$; (−2, −5)

23. $m = 1$; (−2, 8)

24. $m = -1$; (−3, 6)

For each point–slope equation listed, state the slope and a point on the graph.

25. $y - 9 = \frac{2}{7}(x - 8)$

26. $y - 3 = 9(x - 2)$

27. $y + 2 = -5(x - 7)$

28. $y - 4 = -\frac{2}{9}(x + 5)$

29. $y - 4 = -\frac{5}{3}(x + 2)$

30. $y + 7 = -4(x - 9)$

Aha! **31.** $y = \frac{4}{7}x$

32. $y = 3x$

Write the slope–intercept equation for the line with the given slope that contains the given point.

33. $m = 2$; (5, 7)

34. $m = 3$; (6, 2)

35. $m = \frac{7}{4}$; (4, −2)

36. $m = \frac{8}{3}$; (3, −4)

37. $m = -3;\ (-1, 6)$ **38.** $m = -2;\ (-1, 3)$

39. $m = -4;\ (-2, -1)$ **40.** $m = -5;\ (-1, -4)$

Aha! **41.** $m = -\frac{5}{6};\ (0, 4)$ **42.** $m = -\frac{3}{4};\ (0, 5)$

Write an equation of the line that contains the specified point and is parallel to the indicated line.

43. $(4, 7),\ x + 2y = 6$

44. $(1, 3),\ 3x - y = 7$

Aha! **45.** $(0, -7),\ y = 2x + 1$

46. $(-7, 0),\ 5x + 2y = 6$

47. $(2, -6),\ 5x - 3y = 8$

48. $(-4, -5),\ 2x + y = -3$

Write an equation of the line that contains the specified point and is perpendicular to the indicated line.

49. $(3, -2),\ 3x + 6y = 5$

50. $(-3, -5),\ 4x - 2y = 4$

51. $(-4, -7),\ 3x - 5y = 6$

52. $(-4, 5),\ 7x - 2y = 1$

Aha! **53.** $(0, 6),\ 2x - 5 = y$

54. $(4, 0),\ x - 3y = 0$

Write the slope–intercept equation for the line containing the given pair of points.

55. $(1, 5)$ and $(4, 2)$ **56.** $(3, 7)$ and $(4, 8)$

57. $(-3, 1)$ and $(3, 5)$ **58.** $(-2, 3)$ and $(2, 5)$

59. $(5, 0)$ and $(0, -2)$ **60.** $(-2, 0)$ and $(0, 3)$

61. $(-2, -4)$ and $(2, -1)$ **62.** $(-3, 5)$ and $(-1, -3)$

63. Graph the line with slope $\frac{4}{3}$ that passes through the point $(1, 2)$.

64. Graph the line with slope $\frac{2}{5}$ that passes through the point $(3, 4)$.

65. Graph the line with slope $-\frac{3}{4}$ that passes through the point $(2, 5)$.

66. Graph the line with slope $-\frac{3}{2}$ that passes through the point $(1, 4)$.

Graph.

67. $y - 2 = \frac{1}{2}(x - 1)$ **68.** $y - 5 = \frac{1}{3}(x - 2)$

69. $y - 1 = -\frac{1}{2}(x - 3)$ **70.** $y - 1 = -\frac{1}{4}(x - 3)$

71. $y + 4 = 3(x + 1)$ **72.** $y + 3 = 2(x + 1)$

73. $y + 3 = -(x + 2)$ **74.** $y + 3 = -1(x - 4)$

75. $y + 1 = -\frac{3}{5}(x + 2)$ **76.** $y + 2 = -\frac{2}{3}(x + 1)$

In Exercises 77–86, assume the relationship is linear, as in Example 8.

77. *Birth Rate among Teenagers.*   The birth rate among teenagers, measured in births per 1000 females age 15–19, fell steadily from 62.1 in 1991 to 45.9 in 2001 (*Source*: National Center for Health Statistics).

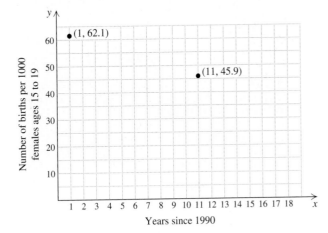

a) Find a linear equation that fits the data.
b) Calculate the birth rate among teenagers in 1999.
c) Predict the birth rate among teenagers in 2008.

78. *Food-Stamp Program Participation.*   Participation in the U.S. food-stamp program grew from approximately 17.1 million people in 2000 to approximately 19.1 million in 2002 (*Source*: U.S. Department of Agriculture).

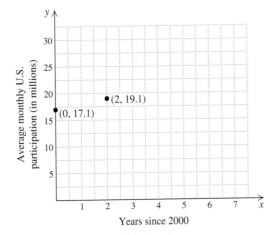

a) Find a linear equation that fits the data.
b) Calculate the number of participants in 2001.
c) Predict the number of participants in 2007.

79. *National Park Land.*   The number of acres in the National Park system has grown from 78.2 million acres in 2000 to 79.0 million acres in 2004 (*Source: Statistical Abstract of the United States*, 2006). Let *A* represent the number of acres, in millions, in the system and *t* the number of years after 2000.

a) Find a linear equation that fits the data.
b) Calculate the number of acres in the National Park system in 2003.
c) Predict the number of acres in the National Park system in 2010.

80. *Aging Population.*   The number of U.S. residents over the age of 65 was approximately 31 million in 1990 and 36.3 million in 2004 (*Source*: U.S. Bureau of the Census). Let *R* represent the number of U.S. residents over the age of 65, in millions, and *t* the number of years after 1990.

a) Find a linear equation that fits the data.
b) Calculate the number of U.S. residents over the age of 65 in 1996.
c) Predict the number of U.S. residents over the age of 65 in 2010.

81. *Dietary Trends.*   In 1971, the average American man consumed 2450 calories per day. By 2000, the figure had risen to 2618 calories per day. (*Source*: Centers for Disease Control and Prevention) Let *C* represent the average number of calories consumed per day by an American man and *t* the number of years after 1971.

a) Find a linear equation that fits the data.
b) Predict the number of calories consumed per day by an American man in 2008.
c) When will the average number of calories consumed per day reach 2750?

82. *Online Travel Plans.*   In 1999, about 48 million Americans used the Internet to find travel information. By 2006, that number had grown to about 79 million. (*Source*: Travel Industry Association of America) Let *N* represent the number of Americans using the Internet for travel information, in millions, and *t* the number of years after 1999.

a) Find a linear equation that fits the data.
b) Predict the number of Americans who will find travel information on the Internet in 2009.
c) In what year will 110 million Americans find travel information on the Internet?

83. *Life Expectancy of Females in the United States.* In 1993, the life expectancy at birth of females was 78.8 yr. In 2003, it was 80.1 yr. (*Source: Statistical Abstract of the United States*, 2006) Let *E* represent life expectancy and *t* the number of years since 1990.

a) Find a linear equation that fits the data.
b) Predict the life expectancy of females in 2010.

84. *Life Expectancy of Males in the United States.* In 1993, the life expectancy at birth of males was 72.2 yr. In 2003, it was 74.8 yr. (*Source: Statistical Abstract of the United States*, 2006) Let *E* represent life expectancy and *t* the number of years since 1990.

a) Find a linear equation that fits the data.
b) Predict the life expectancy of males in 2009.

85. *PAC Contributions.*   In 1994, Political Action Committees (PACs) contributed $189.6 million to congressional candidates. In 2004, the figure rose to $310.5 million. (*Source*: Congressional Research Service and Federal Election Commission) Let *N* represent the amount of political contributions, in millions of dollars, and *t* the number of years since 1994.

a) Find a linear equation that fits the data.
b) Predict the amount of PAC contributions in 2008.

86. *Recycling.* In 1996, Americans recycled 57.3 million tons of solid waste. In 2003, the figure grew to 72.3 million tons. (*Source: Statistical Abstract of the United States*, 2006) Let *N* represent the number of tons recycled, in millions, and *t* the number of years since 1996.

a) Find a linear equation that fits the data.

b) Predict the amount recycled in 2010.

Determine whether the data in each graph appear to be linear.

87.

FDIC-Insured Banking Offices

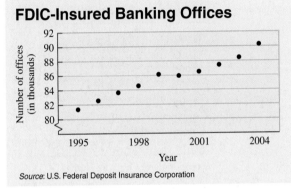

Source: U.S. Federal Deposit Insurance Corporation

88. U.S. Shopping Centers

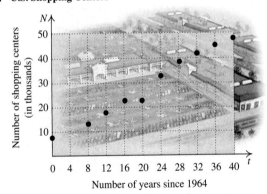

Source: International Council of Shopping Centers

89. Amtrak Riders

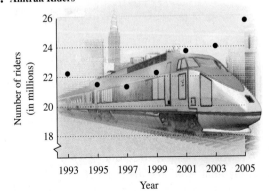

Source: National Association of Railroad Passengers

90.

U.S. Farming

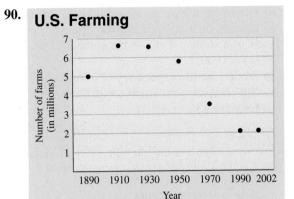

Source: Statistical Abstract of the United States, 2006

91.

Near-Term Births

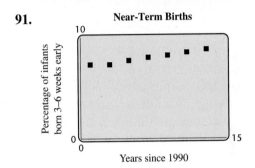

92.

Caesarean Births

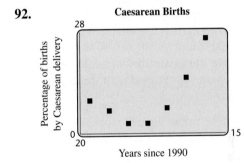

93. *Life Expectancy of Females in the United States.* The following table lists the life expectancy of women who were born in the United States in selected years.

Life expectancy of women

Year	Life Expectancy (in years)
1960	73.1
1970	74.4
1980	77.5
1990	78.8
2000	79.7
2003	80.1

Source: National Center for Health Statistics

a) Use linear regression to find a linear function that can be used to predict the life expectancy *W* of a woman. Let *x* = the number of years since 1900.

b) Predict the life expectancy of a woman in 2010 and compare your answer with the answer to Exercise 83.

94. *Life Expectancy of Males in the United States.* The following table lists the life expectancy of males born in the United States in selected years.

Life expectancy of men

Year	Life Expectancy (in years)
1960	66.6
1970	67.1
1980	70.0
1990	71.8
2000	74.4
2003	74.8

Source: National Center for Health Statistics

a) Use linear regression to find a linear function that can be used to predict the life expectancy *M* of a man. Let *x* = the number of years since 1900.

b) Predict the life expectancy of a man in 2009 and compare your answer with the answer to Exercise 84.

95. *Banking.* The following table lists the number of FDIC-insured institutions for several years.

Year	Number of FDIC-Insured Financial Institutions
1995	81,350
1996	82,578
1997	83,614
1998	84,587
1999	86,040
2000	85,952
2001	86,506
2002	87,429
2003	88,447
2004	90,267

Source: U.S. Federal Deposit Insurance Corporation

a) Use linear regression to find a linear equation that can be used to predict the number *B* of FDIC-insured financial institutions *x* years after 1995.

b) Estimate the number of FDIC-insured financial institutions in 2008.

96. *Near-Term Births.* The percent of infants who are born 3–6 weeks early has increased in recent years, as shown in the following table.

Year	Percent of Near-Term Births
1991	7.3
1993	7.3
1995	7.7
1997	8.0
1999	8.2
2001	8.5
2003	8.8

Source: National Center for Health Statistics

a) Use linear regression to find a linear equation that can be used to predict the percent *P* of near-term births *x* years after 1990.

b) Estimate the percent of near-term births in 2006.

97. Can equations for horizontal or vertical lines be written in point–slope form? Why or why not?

98. On the basis of your answers to Exercises 83 and 84, would you predict that at some point in the future the life expectancy of males will exceed that of females? Why or why not?

Focused Review

Find the slope of each line, if it exists.

99. $y = 4x - 7$ [3.6]

100. $x = 5$ [3.3]

101. $y = \frac{1}{3}$ [3.3]

102. $y - 2 = -6(x + 10)$ [3.7]

Write the slope–intercept equation for the line with the given characteristics.

103. Slope: 5; *y*-intercept: $(0, -1)$ [3.6]

104. Horizontal; contains the point $(6, -2)$ [3.3]

105. Contains the points $(1, 1)$ and $(2, -1)$ [3.7]

106. Parallel to the line $y = \frac{1}{2}x$; y-intercept: $(0, 3)$ [3.6]

107. Slope: -2; contains the point $(-1, 1)$ [3.7]

108. Perpendicular to the line $y = \frac{2}{3}x + 1$; contains the point $(3, 4)$ [3.7]

Synthesis

109. Why is slope–intercept form more useful than point–slope form when using a graphing calculator? How can point–slope form be modified so that it is more easily used with graphing calculators?

110. Any nonvertical line has many equations in point–slope form, but only one in slope–intercept form. Why is this?

Graph.

111. $y - 3 = 0(x - 52)$

112. $y + 4 = 0(x + 93)$

Write the slope–intercept equation for each line shown.

113.
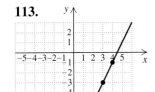

114.
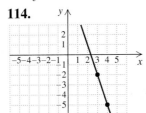

115.
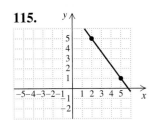

116.
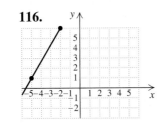

117. Write the slope–intercept equation of the line that has the same y-intercept as the line $x - 3y = 6$ and contains the point $(5, -1)$.

118. Write the slope–intercept equation of the line that contains the point $(-1, 5)$ and is parallel to the line passing through $(2, 7)$ and $(-1, -3)$.

119. Write the slope–intercept equation of the line that has x-intercept $(-2, 0)$ and is parallel to $4x - 8y = 12$.

120. *Depreciation of a Computer.*   After 6 months of use, the value of Pearl's computer had dropped to $900. After 8 months, the value had gone down to $750. How much did the computer cost originally?

121. *Temperature Conversion.*   Water freezes at 32° Fahrenheit and at 0° Celsius. Water boils at 212°F and at 100°C. What Celsius temperature corresponds to a room temperature of 70°F?

122. *Cell-Phone Charges.*   The total cost of Mel's cell phone was $230 after 5 months of service and $390 after 9 months. What costs had Mel already incurred when his service just began?

123. *Operating Expenses.*   The total cost for operating Ming's Wings was $7500 after 4 months and $9250 after 7 months. Predict the total cost after 10 months.

3.8 Functions

Functions and Graphs   ◼   Function Notation and Equations   ◼
Functions Defined Piecewise   ◼   Linear Functions and Applications

We now develop the idea of a *function*—one of the most important concepts in mathematics. A function is a special kind of correspondence between two sets. For example,

To each person in a class	there corresponds	a date of birth.
To each bar code in a store	there corresponds	a price.
To each real number	there corresponds	the cube of that number.

In each example, the first set is called the **domain.** The second set is called the **range.** For any member of the domain, there is *exactly one* member of the range to which it corresponds. This kind of correspondence is called a **function.**

Note in the list above that although two members of a class may have the same date of birth, and two bar codes may correspond to the same price, each correspondence is still a function since every member of the domain is paired with exactly one member of the range.

EXAMPLE 1   Determine whether each correspondence is a function.

a)

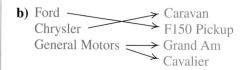

b) Ford ——————→ Caravan
Chrysler ——————→ F150 Pickup
General Motors ——→ Grand Am
——→ Cavalier

SOLUTION

a) The correspondence *is* a function because each member of the domain corresponds to *exactly one* member of the range.

b) The correspondence *is not* a function because a member of the domain (General Motors) corresponds to more than one member of the range.

Function

A *function* is a correspondence between a first set, called the *domain*, and a second set, called the *range*, such that each member of the domain corresponds to *exactly one* member of the range.

EXAMPLE 2   Determine whether each correspondence is a function.

	Domain	**Correspondence**	Range
a)	An elevator full of people	Each person's weight	A set of positive numbers
b)	$\{-2, 0, 1, 2\}$	Each number's square	$\{0, 1, 4\}$
c)	Authors of best-selling books	The titles of books written by each author	A set of book titles

SOLUTION

a) The correspondence *is* a function, because each person has *only one* weight.

b) The correspondence *is* a function, because each number has *only one* square.

c) The correspondence *is not* a function, because some authors have written *more than one* book.

Functions and Graphs

The functions in Examples 1(a) and 2(b) can be expressed as sets of ordered pairs. Example 1(a) can be written $\{(-3, 5), (1, 2), (4, 2)\}$ and Example 2(b) can be written $\{(-2, 4), (0, 0), (1, 1), (2, 4)\}$. We can graph these functions as follows.

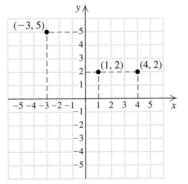

The function $\{(-3, 5), (1, 2), (4, 2)\}$
Domain is $\{-3, 1, 4\}$
Range is $\{5, 2\}$

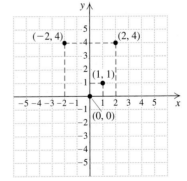

The function $\{(-2, 4), (0, 0), (1, 1), (2, 4)\}$
Domain is $\{-2, 0, 1, 2\}$
Range is $\{4, 0, 1\}$

When a function is given as a set of ordered pairs, the domain is the set of all first coordinates and the range is the set of all second coordinates.

Functions are generally represented by lower- or upper-case letters.

▶ **EXAMPLE 3** Find the domain and the range of the function *f* shown here.

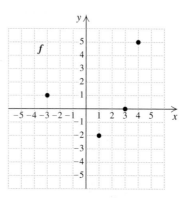

SOLUTION Here *f* can be written $\{(-3, 1), (1, -2), (3, 0), (4, 5)\}$. The domain is the set of all first coordinates, $\{-3, 1, 3, 4\}$, and the range is the set of all second coordinates, $\{1, -2, 0, 5\}$. ◀

We can also find the domain and the range directly from the graph, without first listing all pairs.

▶ **EXAMPLE 4** For the function *f* shown here, determine each of the following.

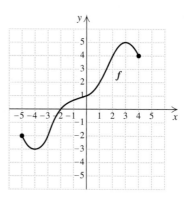

a) What member of the range is paired with 2

b) The domain of *f*

c) What member of the domain is paired with -3

d) The range of *f*

SOLUTION

a) To determine what member of the range is paired with 2, we locate 2 on the horizontal axis (this is where the domain is located). Next, we find the point directly above 2 on the graph of *f*. From that point, we can look to the vertical axis to find the corresponding *y*-coordinate, 4. The "input" 2 has the "output" 4.

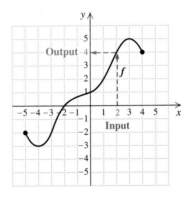

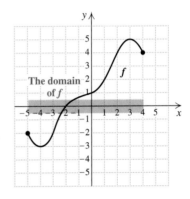

b) The domain of the function is the set of all *x*-values that are used in the points of the curve (see the figure on the right above). Because there are no breaks in the graph of *f*, these extend continuously from −5 to 4 and can be viewed as the curve's shadow, or *projection*, on the *x*-axis. Thus the domain is $\{x \mid -5 \leq x \leq 4\}$, or, using interval notation, $[-5, 4]$.

c) To determine what member of the domain is paired with −3, we locate −3 on the vertical axis. (This is where the range is located; see below.) From there we look left and right to the graph of *f* to find any points for which −3 is the second coordinate. One such point exists, $(-4, -3)$. We note that −4 is the only element of the domain paired with −3.

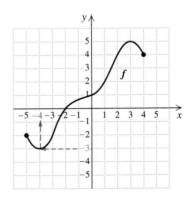

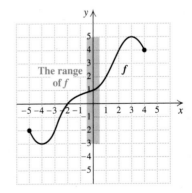

d) The range of the function is the set of all *y*-values that are in the graph. (See the graph on the right above.) These extend continuously from −3 to 5, and can be viewed as the curve's projection on the *y*-axis. Thus the range is $\{y \mid -3 \leq y \leq 5\}$, or, using interval notation, $[-3, 5]$.

The graphs of some functions have no endpoints.

> **EXAMPLE 5** Find the domain and the range of the function f shown here.

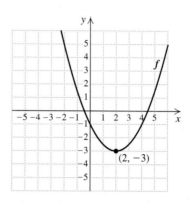

SOLUTION The domain is the set of all x-values that are in the graph. There are no endpoints, which indicates that the graph extends indefinitely. Only a portion of the graph is shown; in fact, it is impossible to show the entire graph. For any x-value, there is a point on the graph. Thus,

> Domain of $f = \{x \mid x \text{ is a real number}\}$, or, in interval notation, $(-\infty, \infty)$.

Again, this can be viewed as the curve's projection on the x-axis, as shown on the left below.

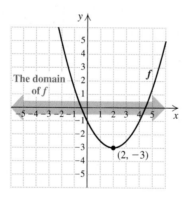 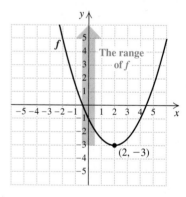

The range of the function is the set of all y-values that are in the graph. This can be viewed as the projection of the curve on the y-axis. The function has no y-values less than -3, and every y-value greater than or equal to -3 corresponds to at least one member of the domain. Thus,

> Range of $f = \{y \mid y \text{ is a real number } and \ y \geq -3\}$, or, in interval notation, $[-3, \infty)$.

The set of all real numbers is often abbreviated $\mathbb{R}$. Thus, in Example 5, we could write Domain of $f = \mathbb{R}$.

A closed dot, as in Example 4, emphasizes that a particular point is on a graph. An open dot, $\circ$, indicates that a particular point is *not* on a graph.

EXAMPLE 6   Find the domain and the range of the function *f* shown here.

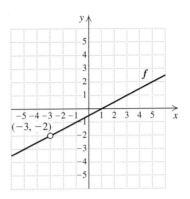

SOLUTION   The domain of *f* is the set of all *x*-values that are in the graph. The open dot in the graph at $(-3, -2)$ indicates that there is no *y*-value that corresponds to $x = -3$; that is, the function is not defined for $x = -3$. Thus, -3 is not in the domain of the function, and

Domain of $f = \{x \mid x$ is a real number *and* $x \neq -3\}$.

There is no function value at $(-3, -2)$, so -2 is not in the range of the function. Thus we have

Range of $f = \{y \mid y$ is a real number *and* $y \neq -2\}$.

Note that if a graph contains two or more points with the same first coordinate, that graph cannot represent a function (otherwise one member of the domain would correspond to more than one member of the range). This observation is the basis of the *vertical-line test*.

The Vertical-Line Test   If it is possible for a vertical line to cross a graph more than once, then the graph is not the graph of a function.

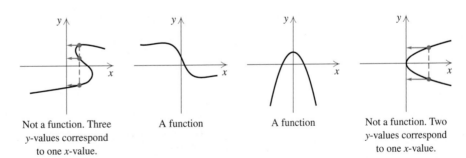

Not a function. Three *y*-values correspond to one *x*-value.

A function

A function

Not a function. Two *y*-values correspond to one *x*-value.

Graphs that do not represent functions still do represent **relations.**

> ### Relation
>
> A *relation* is a correspondence between a first set, called the *domain*, and a second set, called the *range*, such that each member of the domain corresponds to *at least one* member of the range.

Thus, although the correspondences and graphs above are not all functions, they *are* all relations.

Function Notation and Equations

We often think of an element of the domain of a function as an **input** and its corresponding element of the range as an **output.** In Example 3, the function f is written

$$\{(-3, 1), (1, -2), (3, 0), (4, 5)\}.$$

Thus, for an input of -3, the corresponding output is 1, and for an input of 3, the corresponding output is 0.

We use *function notation* to indicate what output corresponds to a given input. For the function f defined above, we write

$$f(-3) = 1, \qquad f(1) = -2, \qquad f(3) = 0, \quad \text{and} \quad f(4) = 5.$$

The notation $f(x)$ is read "f of x," "f at x," or "the value of f at x." If x is an input, then $f(x)$ is the corresponding output.

> **CAUTION!** $f(x)$ *does not mean* f *times* x.

Most functions are described by equations. For example, $f(x) = 2x + 3$ describes the function that takes an input x, multiplies it by 2, and then adds 3.

$$
\begin{array}{c}
\text{Input} \\
\downarrow \\
f(x) \;=\; 2x \;+\; 3 \\
\nearrow \qquad \underbrace{} \\
\text{Double} \quad \text{Add 3}
\end{array}
$$

To calculate the output $f(4)$, we take the input 4, double it, and add 3 to get 11. That is, we substitute 4 into the formula for $f(x)$:

$$f(4) = 2 \cdot 4 + 3$$
$$= 11. \longleftarrow \text{Output}$$

To understand function notation, it can help to imagine a "function machine." Think of putting an input into the machine. For the function

$f(x) = 2x + 3$, the machine will double the input and then add 3. The result will be the output.

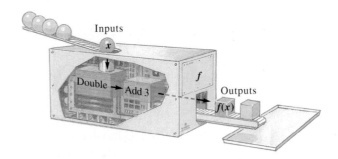

Sometimes, in place of $f(x) = 2x + 3$, we write $y = 2x + 3$, where it is understood that the value of y, the *dependent variable*, depends on our choice of x, the *independent variable*. To understand why $f(x)$ notation is so useful, consider two equivalent statements:

a) If $f(x) = 2x + 3$, then $f(4) = 11$.

b) If $y = 2x + 3$, then the value of y is 11 when x is 4.

Function notation also would have been more concise to use in Example 4. Example 4(a) can be restated as:

Find $f(2)$.

Example 4(c) can be restated as:

Find x such that $f(x) = -3$.

Function notation is not only more concise; it also emphasizes that x is the independent variable.

EXAMPLE 7   Find each indicated function value.

a) $f(5)$, for $f(x) = 3x + 2$
b) $g(-2)$, for $g(r) = 5r^2 + 3r$
c) $h(4)$, for $h(x) = 7$
d) $F(a + 1)$, for $F(x) = 3x + 2$
e) $F(a) + 1$, for $F(x) = 3x + 2$

SOLUTION   Finding function values is much like evaluating an algebraic expression.

a) $f(x) = 3x + 2$
         $f(5) = 3 \cdot 5 + 2 = 17$

b) $g(r) = 5r^2 + 3r$
      $g(-2) = 5(-2)^2 + 3(-2)$
            $= 5 \cdot 4 - 6 = 14$

c) For the function given by $h(x) = 7$, all inputs share the same output, 7. Therefore, $h(4) = 7$. The function h is an example of a *constant function*.

d) $F(x) = 3x + 2$

$F(a + 1) = 3(a + 1) + 2$ **The input is $a + 1$.**

$= 3a + 3 + 2 = 3a + 5$

e) $F(x) = 3x + 2$

$F(a) + 1 = [3(a) + 2] + 1$ **The input is a.**

$= [3a + 2] + 1 = 3a + 3$

Note that whether we write $f(x) = 3x + 2$, or $f(t) = 3t + 2$, or $f(\blacksquare) = 3\,\blacksquare + 2$, we still have $f(5) = 17$. The variable in parentheses (the independent variable) is the same as the variable in the algebraic expression. The letter chosen for the independent variable is not as important as the algebraic manipulations to which it is subjected.

Function notation is often used in formulas. For example, to emphasize that the area A of a circle is a function of its radius r, instead of

$$A = \pi r^2,$$

we can write

$$A(r) = \pi r^2.$$

Function Notation

The values of a function that is described by an equation can be found directly using a graphing calculator. For example, if

$$f(x) = x^2 - 6x + 7$$

is entered as

$$y_1 = x^2 - 6x + 7,$$

then $f(2)$ can be calculated by evaluating Y1(2). After entering the function, move to the home screen by pressing (QUIT). The notation "Y1" can be found by pressing **VARS**, choosing the Y-VARS submenu, selecting the FUNCTION option, and then selecting Y1. After Y1 appears on the home screen, press **(2)** **ENTER** to evaluate Y1(2).

Function values can also be found using a table with Indpnt set to Ask. They can also be read from the graph of a function using the VALUE option of the CALC menu.

EXAMPLE 8   For $f(a) = 2a^2 - 3a + 1$, find $f(3)$ and $f(-5.1)$.

SOLUTION   We first enter the function into the graphing calculator, replacing a with x and the notation $f(a)$ with Y1. The equation Y1 $= 2x^2 - 3x + 1$ represents the same function, with the understanding that the Y1-values are the outputs and the x-values are the inputs; Y1(x) is equivalent to $f(a)$.

To find $f(3)$, we enter Y1(3) and find that $f(3) = 10$.

Plot1 Plot2 Plot3
\Y1■2X²−3X+1
\Y2=
\Y3=

Y1(3)	
Y1(−5.1)	10
	68.32

X	Y1	
3	10	
−5.1	68.32	

To find $f(-5.1)$, we can press ⟨ENTRY⟩ and edit the previous entry. Or, using a table, we can find both $f(3)$ and $f(-5.1)$, as shown on the right above. We see that $f(-5.1) = 68.32$.   ◢

We can also find function values from the graph of the function.

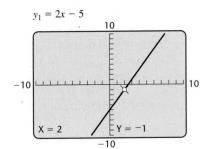

$y_1 = 2x - 5$

EXAMPLE 9   Find $g(2)$ for $g(x) = 2x - 5$.

SOLUTION   We enter and graph the function, using the standard viewing window. Most graphing calculators have a VALUE option in the CALC menu that will calculate a function's value given a value of x. For most calculators, that x-value must be shown in the viewing window; that is, Xmin $\le x \le$ Xmax. In this case, when $x = 2$, $y = -1$, so $g(2) = -1$, as shown at left.   ◢

When we find a function value, we are determining an output that corresponds to a given input. To do this, we evaluate the expression for the given value of the input.

Sometimes we want to determine an input that corresponds to a given output. To do this, we solve an equation.

EXAMPLE 10   Let $f(x) = 3x - 7$.

a) What output corresponds to an input of 5?

b) What input corresponds to an output of 5?

SOLUTION

a) We find $f(5)$:

$$f(x) = 3x - 7$$
$$f(5) = 3(5) - 7 \qquad \text{The input is 5. We substitute 5 for } x.$$
$$= 15 - 7$$
$$= 8. \qquad \text{Carrying out the calculations}$$

The output 8 corresponds to the input 5; that is, $f(5) = 8$.

b) We find the value of x for which $f(x) = 5$:

$$f(x) = 3x - 7$$
$$5 = 3x - 7 \qquad \text{The output is 5. We substitute 5 for } f(x).$$
$$12 = 3x$$
$$4 = x. \qquad \text{Solving for } x$$

The input 4 corresponds to the output 5; that is, $f(4) = 5$.   ◢

When a function is described by an equation, we assume that the domain is the set of all real numbers for which function values can be calculated. If an x-value is not in the domain of a function, no point on the graph will have that x-value as its first coordinate.

EXAMPLE 11 For each equation, determine the domain of f.

a) $f(x) = |x|$ **b)** $f(x) = \dfrac{7}{2x - 6}$

SOLUTION

a) We ask ourselves, "Is there any number x for which we cannot compute $|x|$?" Since we can find the absolute value of *any* number, the answer is no. Thus the domain of f is $\mathbb{R}$, the set of all real numbers.

b) Is there any number x for which $\dfrac{7}{2x - 6}$ cannot be computed? Since $\dfrac{7}{2x - 6}$ cannot be computed when the denominator $2x - 6$ is 0, the answer is yes. To determine what x-value causes the denominator to be 0, we set up and solve an equation:

$$2x - 6 = 0 \qquad \text{Setting the denominator equal to 0}$$
$$2x = 6 \qquad \text{Adding 6 to both sides}$$
$$x = 3. \qquad \text{Dividing both sides by 2}$$

$y_1 = 7/(2x - 6)$

X	Y₁	
3	**ERROR**	

Y₁ = ERROR

Thus, 3 is *not* in the domain of f, whereas all other real numbers are. The table at left indicates that $f(3)$ is not defined. The domain of f is $\{x \mid x \text{ is a real number } and\ x \neq 3\}$.

Functions Defined Piecewise

Some functions are defined by different equations for various parts of their domains. Such functions are said to be **piecewise** defined. For example, the function given by $f(x) = |x|$ is described by

$$f(x) = \begin{cases} x, & \text{if } x \geq 0, \\ -x, & \text{if } x < 0. \end{cases}$$

To evaluate a piecewise-defined function for an input a, we first determine what part of the domain a belongs to. Then we use the appropriate formula for that part of the domain.

EXAMPLE 12   Find each function value for the function f given by

$$f(x) = |x| = \begin{cases} x, & \text{if } x \geq 0, \\ -x, & \text{if } x < 0. \end{cases}$$

a) $f(4)$ **b)** $f(-10)$

SOLUTION

a) Since $4 \geq 0$, we use the equation $f(x) = x$. Thus, $f(4) = 4$.

b) Since $-10 < 0$, we use the equation $f(x) = -x$. Thus,
    $f(-10) = -(-10) = 10$.

EXAMPLE 13   Find each function value for the function g given by

$$g(x) = \begin{cases} x + 2, & \text{if } x \leq -2, \\ x^2, & \text{if } -2 < x \leq 5, \\ 3x, & \text{if } x > 5. \end{cases}$$

a) $g(-2)$ **b)** $g(3)$ **c)** $g(10)$

SOLUTION

a) Since $-2 \leq -2$, we use the first equation, $g(x) = x + 2$:

$$g(-2) = -2 + 2 = 0.$$

b) Since $-2 < 3 \leq 5$, we use the second equation, $g(x) = x^2$:

$$g(3) = 3^2 = 9.$$

c) Since $10 > 5$, we use the last equation, $g(x) = 3x$:

$$g(10) = 3 \cdot 10 = 30.$$

Linear Functions and Applications

Any nonvertical line passes the vertical-line test and is the graph of a function. Such a function is called a **linear function** and can be written in the form $f(x) = mx + b$.

Linear functions arise continually in today's world. As with rate problems, it is critical to use proper units in all answers.

EXAMPLE 14   Salvage Value.   Tyline Electric uses the function

$$S(t) = -700t + 3500$$

to determine the *salvage value* $S(t)$, in dollars, of a color photocopier t years after its purchase.

a) What do the numbers -700 and 3500 signify?

b) How long will it take the copier to *depreciate* completely?

c) What is the domain of S?

SOLUTION   Drawing, or at least visualizing, a graph can be useful here.

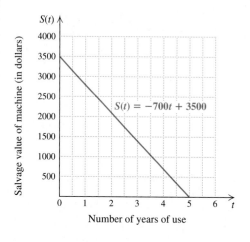

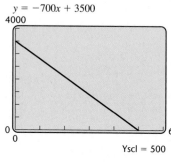

$y = -700x + 3500$

Yscl = 500

a) At time $t = 0$, we have $S(0) = -700 \cdot 0 + 3500 = 3500$. Thus the number 3500 signifies the original cost of the copier, in dollars.

This function is written in slope–intercept form. Since the output is measured in dollars and the input in years, the number -700 signifies that the value of the copier is declining at a rate of $700 per year.

b) The copier will have depreciated completely when its value drops to 0. To learn when this occurs, we determine when $S(t) = 0$:

$$S(t) = 0 \qquad \text{A graph is not always available.}$$
$$-700t + 3500 = 0 \qquad \text{Substituting } -700t + 3500 \text{ for } S(t)$$
$$-700t = -3500 \qquad \text{Subtracting 3500 from both sides}$$
$$t = 5. \qquad \text{Dividing both sides by } -700$$

The copier will have depreciated completely in 5 yr.

c) Neither the number of years of service nor the salvage value can be negative. In part (b) we found that after 5 yr the salvage value will have dropped to 0. Thus the domain of S is $\{t \,|\, 0 \le t \le 5\}$, or $[0, 5]$. Either graph above serves as a visual check of this result.

3.8 EXERCISE SET

FOR EXTRA HELP

 MathXL   MyMathLab   InterAct Math   AW Math Tutor Center   Video Lectures on CD: Disc 3   Student's Solutions Manual

🪝 *Concept Reinforcement* *Complete each of the following sentences.*

1. For any function, the set of all inputs, or first values, is called the _____.

2. For any function, the set of all outputs, or second values, is called the _____.

3. In any function, each member of the domain is paired with _____ one member of the range.

4. In a relation, a member of the domain may be paired with _____ than one member of the range.

5. When a function is graphed, members of the domain are located on the _____ axis.

6. When a function is graphed, members of the range are located on the _____ axis.

7. The notation $f(3)$ is read _____

8. The vertical-line test can be used to determine whether or not a graph represents a(n) _____.

Determine whether each correspondence is a function.

9.
2 ⟶ a
4 ⟶ b
6 ⟶ c
8 ⟶ d
10

10.
3 ⟶ 9
6 ⟶ 8
9 ⟶ 7
12 ⟶ 6

11. *Girl's age (in months)*   *Average daily weight gain (in grams)*

2 ⟶ 21.8
9 ⟶ 11.7
16 ⟶ 8.5
23 ⟶ 7.0

Source: American Family Physician, December 1993, p. 1435

12. *Boy's age (in months)*   *Average daily weight gain (in grams)*

2 ⟶ 24.8
9 ⟶ 11.7
16 ⟶ 8.2
23 ⟶ 7.0

Source: American Family Physician, December 1993, p. 1435

13. *Celebrity*   *Birthday*

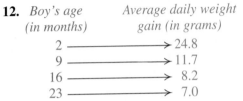

Sigourney Weaver
Jesse Jackson ⟶ October 8
Chevy Chase
Muhammad Ali
Jim Carrey ⟶ January 17

Source: www.kidsparties.com

14. *Predator*   *Prey*

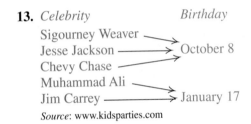

cat
fish
dog
tiger
bat ⟶ mosquito

dog
worm
cat
fish

15. *Birthday*   *Celebrity*

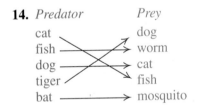

July 24
July 18

Jennifer Lopez
Barry Bonds
Pam Tillis
Vin Diesel
John Glenn
Nelson Mandela

Source: www.leannesbirthdays.com and www.kidsparties.com

16. *State*   *Neighboring state*

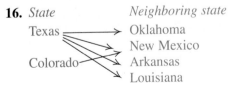

Texas
Colorado

Oklahoma
New Mexico
Arkansas
Louisiana

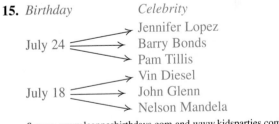

Determine whether each of the following is a function. Identify any relations that are not functions.

	Domain	Correspondence	Range
17.	A yard full of pumpkins	The price of each pumpkin	A set of prices
18.	The members of a rock band	An instrument the person can play	A set of instruments
19.	The players on a team	The uniform number of each player	A set of numbers
20.	A set of triangles	The area of each triangle	A set of numbers

For each graph of a function, determine **(a)** $f(1)$; **(b)** *the domain;* **(c)** *any x-values for which* $f(x) = 2$; *and* **(d)** *the range.*

21.

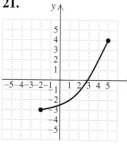

22.

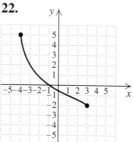

23.

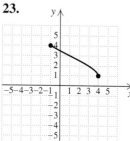

24.

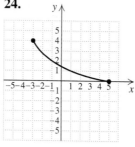

25.

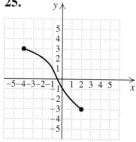

26.

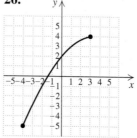

27.

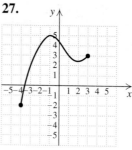

28.

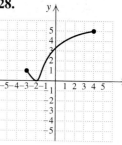

29.

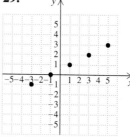

30.

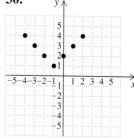

31.

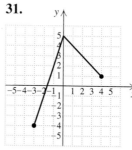

32.

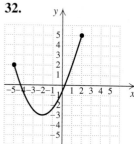

33.

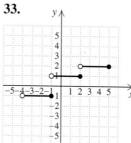

34.

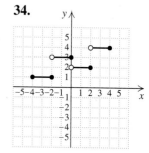

Determine the domain and the range of each function.

35.

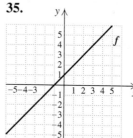

36.

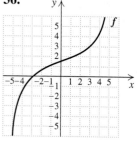

37.

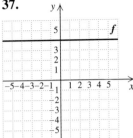

38.

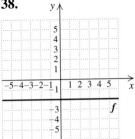

39.

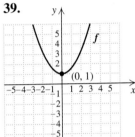

40.

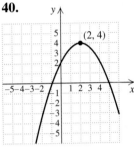

41.

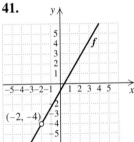

42.

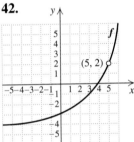

43.

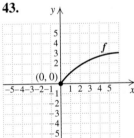

44.

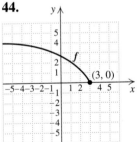

Determine whether each of the following is the graph of a function.

45.

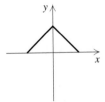

46.

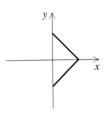

47.

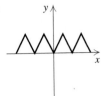

48.

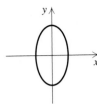

49.

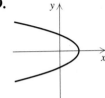

50.

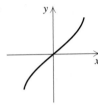

51.

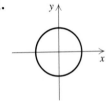

52.

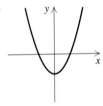

Find the function values.

53. $g(x) = 2x + 3$

  a) $g(0)$ **b)** $g(-4)$ **c)** $g(-7)$
  d) $g(8)$ **e)** $g(a + 2)$ **f)** $g(a) + 2$

54. $h(x) = 3x - 2$

  a) $h(4)$ **b)** $h(8)$ **c)** $h(-3)$
  d) $h(-4)$ **e)** $h(a - 1)$ **f)** $h(a) - 1$

55. $f(n) = 5n^2 + 4n$

  a) $f(0)$ **b)** $f(-1)$ **c)** $f(3)$
Aha!   **d)** $f(t)$ **e)** $f(2a)$ **f)** $2 \cdot f(a)$

56. $g(n) = 3n^2 - 2n$

  a) $g(0)$ **b)** $g(-1)$ **c)** $g(3)$
  d) $g(t)$ **e)** $g(2a)$ **f)** $2 \cdot g(a)$

57. $f(x) = \dfrac{x - 3}{2x - 5}$

  a) $f(0)$ **b)** $f(4)$ **c)** $f(-1)$
  d) $f(3)$ **e)** $f(x + 2)$

58. $s(x) = \dfrac{3x - 4}{2x + 5}$

  a) $s(10)$ **b)** $s(2)$ **c)** $s\left(\tfrac{1}{2}\right)$
  d) $s(-1)$ **e)** $s(x + 3)$

The function A described by $A(s) = s^2 \dfrac{\sqrt{3}}{4}$ gives the area of an equilateral triangle with side s.

59. Find the area when a side measures 4 cm.

60. Find the area when a side measures 6 in.

The function V described by $V(r) = 4\pi r^2$ gives the surface area of a sphere with radius r.

61. Find the surface area when the radius is 3 in.

62. Find the surface area when the radius is 5 cm.

63. *Pressure at Sea Depth.* The function $P(d) = 1 + (d/33)$ gives the pressure, in *atmospheres* (atm), at a depth of d feet in the sea. Note that $P(0) = 1$ atm, $P(33) = 2$ atm, and so on. Find the pressure at 20 ft, at 30 ft, and at 100 ft.

64. *Melting Snow.* The function $W(d) = 0.112d$ approximates the amount, in centimeters, of water that results from d centimeters of snow melting. Find the amount of water that results from snow melting from depths of 16 cm, 25 cm, and 100 cm.

Fill in the missing values in each table.

$f(x) = 2x - 5$	
x	**f(x)**
65. 8	
66.	13
67.	−5
68. −4	

$f(x) = \frac{1}{3}x + 4$	
x	**f(x)**
69.	$\frac{1}{2}$
70.	$-\frac{1}{3}$
71. $\frac{1}{2}$	
72. $-\frac{1}{3}$	

73. If $f(x) = 4 - x$, for what input is the output 7?

74. If $f(x) = 5x + 1$, for what input is the output $\frac{1}{2}$?

75. If $f(x) = 0.1x - 0.5$, for what input is the output -3?

76. If $f(x) = 2.3 - 1.5x$, for what input is the output 10?

Find the domain of f.

77. $f(x) = \dfrac{5}{x - 3}$

78. $f(x) = \dfrac{7}{6 - x}$

79. $f(x) = \dfrac{3}{2x - 1}$

80. $f(x) = \dfrac{9}{4x + 3}$

81. $f(x) = 2x + 1$

82. $f(x) = x^2 + 3$

83. $f(x) = |5 - x|$

84. $f(x) = |3x - 4|$

85. $f(x) = \dfrac{5x}{x - 9}$

86. $f(x) = \dfrac{x}{x + 1}$

87. $f(x) = x^2 - 9$

88. $f(x) = x^2 - 2x + 1$

89. $f(x) = \dfrac{2x - 7}{5}$

90. $f(x) = \dfrac{x + 5}{8}$

Find the indicated function values for each function.

91. $f(x) = \begin{cases} x, & \text{if } x < 0, \\ 2x + 1, & \text{if } x \geq 0 \end{cases}$

    a) $f(-5)$ **b)** $f(0)$ **c)** $f(10)$

92. $g(x) = \begin{cases} x - 5, & \text{if } x \leq 5, \\ 3x, & \text{if } x > 5 \end{cases}$

    a) $g(0)$ **b)** $g(5)$ **c)** $g(6)$

93. $G(x) = \begin{cases} x - 5, & \text{if } x < -1, \\ x, & \text{if } -1 \le x \le 2, \\ x + 2, & \text{if } x > 2 \end{cases}$

   a) $G(0)$ **b)** $G(2)$ **c)** $G(5)$

94. $F(x) = \begin{cases} 2x, & \text{if } x \le 0, \\ x, & \text{if } 0 < x \le 3, \\ -5x, & \text{if } x > 3 \end{cases}$

   a) $F(-1)$ **b)** $F(3)$ **c)** $F(10)$

95. $f(x) = \begin{cases} x^2 - 10, & \text{if } x < -10, \\ x^2, & \text{if } -10 \le x \le 10, \\ x^2 + 10, & \text{if } x > 10 \end{cases}$

   a) $f(-10)$ **b)** $f(10)$ **c)** $f(11)$

96. $f(x) = \begin{cases} 2x^2 - 3, & \text{if } x \le 2, \\ x^2, & \text{if } 2 < x < 4, \\ 5x - 7, & \text{if } x \ge 4 \end{cases}$

   a) $f(0)$ **b)** $f(3)$ **c)** $f(6)$

In Exercises 97–110, each model is of the form
$f(x) = mx + b$. *In each case, determine what m and b signify.*

97. *Catering.* When catering a party for x people, Jennette's Catering uses the formula $C(x) = 25x + 75$, where $C(x)$ is the cost of the party, in dollars.

98. *Weekly Pay.* Each salesperson at Knobby's Furniture is paid $P(x)$ dollars per week, where $P(x) = 0.05x + 200$ and x is the value of the salesperson's sales for the week.

99. *Hair Growth.* After Ty gets a "buzz cut," the length $L(t)$ of his hair, in inches, is given by $L(t) = \frac{1}{2}t + 1$, where t is the number of months after he gets the haircut.

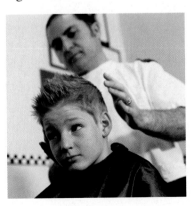

100. *Natural Gas Demand.* The demand, in quadrillions of Btu's, for natural gas is approximated by $D(t) = \frac{9}{25}t + 22$, where t is the number of years after 2000 (*Source*: Based on data from the U.S. Energy Information Administration).

101. *Life Expectancy of American Women.* The life expectancy of American women t years after 1970 is given by $A(t) = \frac{1}{7}t + 75.5$.

102. *Landscaping.* After being cut, the length $G(t)$ of the lawn, in inches, at Great Harrington Community College is given by $G(t) = \frac{1}{8}t + 2$, where t is the number of days since the lawn was cut.

103. *Cost of a Movie Ticket.* The average price $P(t)$, in dollars, of a movie ticket is given by $P(t) = 0.227t + 4.29$, where t is the number of years since 1995.

104. *Sales of Cotton Goods.* The function given by $f(t) = -0.5t + 7$ can be used to estimate the cash receipts, in billions of dollars, from farm marketing of cotton t years after 1995.

105. *Cost of a Taxi Ride.* The cost, in dollars, of a taxi ride during off-peak hours in New York City is given by $C(d) = 2d + 2.5$, where d is the number of miles traveled.

106. *Cost of Renting a Truck.* The cost, in dollars, of a one-day truck rental is given by $C(d) = 0.3d + 20$, where d is the number of miles driven.

107. *Salvage Value.* Green Glass Recycling uses the function given by $F(t) = -5000t + 90,000$ to determine the salvage value $F(t)$, in dollars, of a waste removal truck t years after it has been put into use.

   a) What do the numbers -5000 and $90,000$ signify?

   b) How long will it take the truck to depreciate completely?

   c) What is the domain of F?

108. *Salvage Value.* Consolidated Shirt Works uses the function given by $V(t) = -2000t + 15,000$ to determine the salvage value $V(t)$, in dollars, of a color separator t years after it has been put into use.

   a) What do the numbers -2000 and $15,000$ signify?

   b) How long will it take the machine to depreciate completely?

   c) What is the domain of V?

109. *Trade-In Value.* The trade-in value of a Homelite snowblower can be determined using the function given by $v(n) = -150n + 900$. Here $v(n)$ is the trade-in value, in dollars, after n winters of use.

  a) What do the numbers -150 and 900 signify?
  b) When will the trade-in value of the snowblower be $300?
  c) What is the domain of v?

110. *Trade-In Value.* The trade-in value of a John Deere riding lawnmower can be determined using the function given by $T(x) = -300x + 2400$. Here $T(x)$ is the trade-in value, in dollars, after x summers of use.

  a) What do the numbers -300 and 2400 signify?
  b) When will the value of the mower be $1200?
  c) What is the domain of T?

TW 111. Explain why the domain of the function given by $f(x) = \dfrac{x + 3}{2}$ is $\mathbb{R}$, but the domain of the function given by $g(x) = \dfrac{2}{x + 3}$ is not $\mathbb{R}$.

TW 112. Alayna asserts that for a function described by a set of ordered pairs, the range of the function will always have the same number of elements as there are ordered pairs. Is she correct? Why or why not?

Focused Review

Graph by hand.

113. $y = 2x + 3$ [3.6]   **114.** $y - 3 = -(x + 4)$ [3.7]

115. $y = -1$ [3.3]   **116.** $f(x) = \frac{1}{2}x - 2$ [3.6], [3.8]

Graph using a graphing calculator. [3.2]

117. $y = x^2$ **118.** $y = 0.1x - 5.7$

119. $y = |x|$ **120.** $y = \frac{1}{3}x - 15$

Synthesis

TW 121. For the function given by $n(z) = ab + wz$, what is the independent variable? How can you tell?

TW 122. Explain in your own words why every function is a relation, but not every relation is a function.

For Exercises 123 and 124, let $f(x) = 3x^2 - 1$ and $g(x) = 2x + 5$.

123. Find $f(g(-4))$ and $g(f(-4))$.

124. Find $f(g(-1))$ and $g(f(-1))$.

125. If f represents the function in Exercise 14, find $f(f(f(f(tiger))))$.

Pregnancy. *For Exercises 126–129, use the following graph of a woman's "stress test." This graph shows the size of a pregnant woman's contractions as a function of time.*

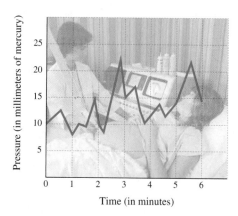

126. How large is the largest contraction that occurred during the test?

127. At what time during the test did the largest contraction occur?

TW 128. On the basis of the information provided, how large a contraction would you expect 60 sec after the end of the test? Why?

129. What is the frequency of the largest contraction?

130. The *greatest integer function* $f(x) = [\![x]\!]$ is defined as follows: $[\![x]\!]$ is the greatest integer that is less than or equal to x. For example, if $x = 3.74$, then $[\![x]\!] = 3$; and if $x = -0.98$, then $[\![x]\!] = -1$. Graph the greatest integer function for $-5 \le x \le 5$. (The notation $f(x) = \text{int}(x)$, used in many graphing calculators, is often found in the MATH NUM submenu.)

131. Suppose that a function g is such that $g(-1) = -7$ and $g(3) = 8$. Find a formula for g if $g(x)$ is of the form $g(x) = mx + b$, where m and b are constants.

Given that $f(x) = mx + b$, classify each of the following as true or false.

132. $f(c + d) = f(c) + f(d)$

133. $f(cd) = f(c)f(d)$

134. $f(kx) = kf(x)$

135. $f(c - d) = f(c) - f(d)$

3 Chapter Summary and Review

KEY TERMS AND DEFINITIONS

GRAPHS

Graph, p. 160   A picture or diagram of the data in a table, such as a **bar graph, circle graph,** or **line graph.** A line, curve, or collection of points that represents all the solutions of an equation.

LINEAR EQUATIONS

Linear equation, p. 183   An equation whose graph is a straight line. **Nonlinear equations** have graphs that are not straight lines.

***x*-intercept,** p. 188   A point at which a graph crosses the *x*-axis.

***y*-intercept,** p. 188   A point at which a graph crosses the *y*-axis.

Interpolation, p. 238   Estimating a value between two given values.

Extrapolation, p. 238   Predicting a future value based on given data.

Linear regression, p. 242   A method of finding a "best" line to fit a set of data.

FUNCTIONS

Function, p. 254   A correspondence between a set called the **domain** and a set called the **range,** such that each member of the domain corresponds to exactly one member of the range. A function is a special kind of **relation.**

$f(x) = 5$ is a **constant function.**

$g(x) = 2x + 3$ is a **linear function.**

Domain, p. 254   The set of all **inputs** of a function.

Range, p. 254   The set of all **outputs** of a function.

For $f(x) = x - 5$:

x is the **independent variable;**

values of *x* are the **inputs;**

values of $f(x)$ are the **outputs;**

domain of $f = \mathbb{R}$;

range of $f = \mathbb{R}$.

The vertical-line test, p. 259   A graph represents a function if it is not possible to draw a vertical line that intersects the graph more than once.

IMPORTANT CONCEPTS

[Section references appear in brackets.]

Concept	Example
To **graph** an equation means to make a drawing that represents all of its solutions. Solutions of equations are represented with **ordered pairs** and graphed on a **coordinate system.**	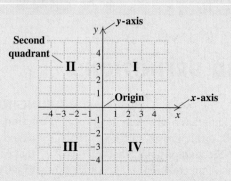 [3.1]
Slope $= m = \dfrac{\text{rise}}{\text{run}} = \dfrac{\text{difference in } y}{\text{difference in } x}$ $= \dfrac{y_2 - y_1}{x_2 - x_1}$	The slope of the line containing the points $(-1, -4)$ and $(2, -6)$ is $\dfrac{-4 - (-6)}{-1 - 2} = \dfrac{2}{-3} = -\dfrac{2}{3}.$ [3.5]
Horizontal line: Slope 0; equation $y = b$ Vertical line: Slope is undefined; equation $x = a$	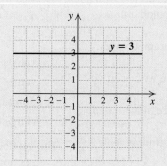 Horizontal line; slope is 0. Vertical line; slope is undefined. [3.3]
Parallel lines: The slopes are equal. Perpendicular lines: The product of the slopes is -1.	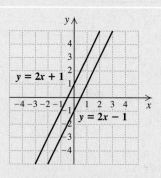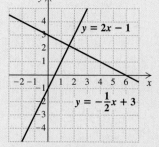 Both lines have slope 2. The product of the slopes The lines are parallel. is $2\left(-\frac{1}{2}\right) = -1$. The lines are perpendicular. [3.6]

(continued)

Slope–intercept form: $y = mx + b$

Point–slope form: $y - y_1 = m(x - x_1)$

Standard form: $Ax + By = C$

The following are equations for the same line, in different forms:

Slope–intercept form: $y = -\frac{2}{3}x - 5$ [3.6]
   The slope is $-\frac{2}{3}$ and the y-intercept is $(0, -5)$.

Point–slope form: $y - (-7) = -\frac{2}{3}(x - 3)$ [3.7]
   The slope is $-\frac{2}{3}$, and $(3, -7)$ is a point on the line.

Standard form: $2x + 3y = -15$ [3.3]
   The x-intercept is $\left(-\frac{15}{2}, 0\right)$ and the y-intercept is $(0, -5)$.

Graphing lines using a point and a slope

Graph: $f(x) = \frac{2}{3}x - 1$.
y-intercept is $(0, -1)$.
Slope is $\frac{2}{3}$.
Plot $(0, -1)$, and count off a slope: up 2 units and to the right 3 units. Draw the line.

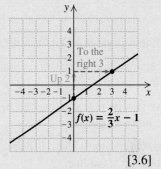

[3.6]

Graphing lines using two points

Graph: $3x + y = 3$.
x-intercept is $(1, 0)$.
y-intercept is $(0, 3)$.
Plot the points and draw the line.

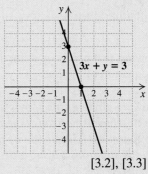

[3.2], [3.3]

Review Exercises

Classify each statement as either true or false.

1. Not every ordered pair lies in one of the four quadrants. [3.1]

2. The equation of a vertical line cannot be written in slope–intercept form. [3.6]

3. Equations for lines written in slope–intercept form appear in the form $Ax + By = C$. [3.6]

4. Every horizontal line has an x-intercept. [3.3]

5. A line's slope is a measure of rate. [3.5]

6. A positive rate of inflation means prices are rising. [3.4]

7. Any two points on a line can be used to determine the line's slope. [3.5]

8. Knowing a line's slope is enough to write the equation of the line. [3.6]

9. Parallel lines that are not vertical have the same slope. [3.6]

10. Knowing two points on a line is enough to write the equation of the line. [3.7]

The following circle graph shows a breakdown of the charities to which Americans donated in a recent year (Source: American Association of Fund-Raising Counsel Trust for Philanthropy Giving USA 2003). Use the graph for Exercises 11 and 12.

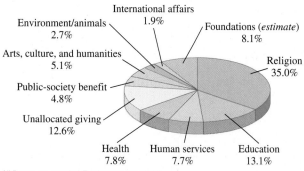

International affairs 1.9%
Environment/animals 2.7%
Arts, culture, and humanities 5.1%
Public-society benefit 4.8%
Unallocated giving 12.6%
Foundations (*estimate*) 8.1%
Religion 35.0%
Health 7.8%
Human services 7.7%
Education 13.1%

All figures are rounded. Total may not be 100%.

11. The citizens of Pittsford are typical of Americans in general with regard to charitable contributions. If the citizens donated a total of $2 million to charities, how much was given to the environment/animals? [3.1]

12. About 5% of the Velez's income of $70,000 goes to charity. If their giving habits are typical, approximate the amount that they contribute to human services. [3.1]

Plot each point. [3.1]

13. $(5, -1)$ 14. $(2, 3)$ 15. $(-4, 0)$

In which quadrant is each point located? [3.1]

16. $(2, -6)$ 17. $(-16.5, -20.3)$

18. $(-3, 12)$

Find the coordinates of each point in the figure. [3.1]

19. A 20. B 21. C

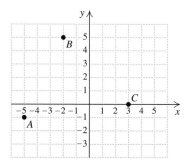

22. Use a grid 10 squares wide and 10 squares high to plot $(-65, -2)$, $(-10, 6)$, and $(25, 7)$. Choose the scale carefully. Scales may vary. [3.1]

23. Determine whether the equation $y = 2x - 5$ has *each* ordered pair as a solution: **(a)** $(3, 1)$; **(b)** $(-3, 1)$. [3.2]

24. Show that the ordered pairs $(0, -3)$ and $(2, 1)$ are solutions of the equation $2x - y = 3$. Then use the graph of the two points to determine another solution. Answers may vary. [3.2]

Graph by hand.

25. $y = x - 5$ [3.2]   26. $y = -\frac{1}{4}x$ [3.2]

27. $y = -x + 4$ [3.2]   28. $4x + y = 3$ [3.2]

29. $4x + 5 = 3$ [3.3]   30. $5x - 2y = 10$ [3.3]

Graph using a graphing calculator. [3.2]

31. $y = x^2 + 1$ 32. $2y - x = 8$

33. The percentage of households engaged in flower planting can be approximated by $F = -2t + 48$, where t is the number of years since 2000 (*Source*:

Based on data from The National Gardening Association). Graph the equation and use the graph to estimate the percentage of households engaged in flower planting in 2010. [3.2]

Determine whether each equation is linear. [3.3]

34. $y = x^2 - 2$ **35.** $y = x - 2$

36. *Meal Service.* At 8:30 A.M., the Colchester Boy Scouts had served 47 people at their annual pancake breakfast. By 9:15, the total served had reached 67.
   a) Find the Boy Scouts' serving rate, in number of meals per minute. [3.4]
   b) Find the Boy Scouts' serving rate, in minutes per meal. [3.4]

37. *Gas Mileage.* The following graph shows data for a Ford Explorer driven on city streets. At what rate was the vehicle consuming gas? [3.4]

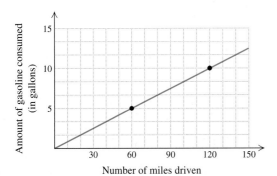

For each line in Exercises 38–40, list **(a)** *the coordinates of the y-intercept,* **(b)** *the coordinates of any x-intercepts, and* **(c)** *the slope.* [3.3], [3.5]

38.

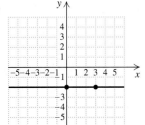

39.

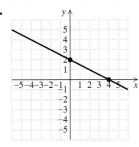

40.

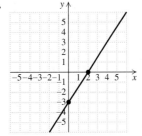

Find the slope of the line containing the given pair of points. If the slope is undefined, state this. [3.5]

41. $(6, 8)$ and $(-2, -4)$ **42.** $(5, 1)$ and $(-1, 1)$

43. $(-3, 0)$ and $(-3, 5)$

44. $(-8.3, 4.6)$ and $(-9.9, 1.4)$

45. Lombard Street in San Francisco, sometimes called the world's most crooked street, at one point drops 35 ft over a 125-ft horizontal distance. Find the grade. [3.5]

Find the slope of each line. If the slope is undefined, state this.

46. $y = -\frac{7}{10}x - 3$ [3.6]   **47.** $y = 5$ [3.5]

48. $x = -\frac{1}{3}$ [3.5]   **49.** $3x - 2y = 6$ [3.6]

50. Find the *x*-intercept and the *y*-intercept of the line given by $3x + 2y = 18$. [3.3]

51. Find the slope and the *y*-intercept of the line given by $2x + 4y = 20$. [3.6]

Determine whether each pair of lines is parallel, perpendicular, or neither. [3.6]

52. $y + 5 = -x,$ **53.** $3x - 5 = 7y,$
     $x - y = 2$ $7y - 3x = 7$

54. Write the slope–intercept equation of the line with slope $-\frac{3}{4}$ and *y*-intercept $(0, 6)$. [3.6]

55. Write a point–slope equation for the line with slope $-\frac{1}{2}$ that contains the point $(3, 6)$. [3.7]

56. Write the slope–intercept equation for the line that contains the points $(1, -2)$ and $(-3, -7)$. [3.7]

57. Write the slope–intercept equation for the line that is perpendicular to the line $3x - 5y = 9$ and that contains the point $(2, -5)$. [3.7]

58. Total revenue for the motion picture and recording industry rose from \$56 billion in 1997 to \$77 billion in 2002 (*Source*: U.S. Bureau of the Census).
   a) Assuming that the growth was linear, find a linear equation that fits the data. Let r represent the revenue, in billions of dollars, and t the number of years since 1997. [3.7]
   b) Calculate the revenue in 2000. [3.7]
   c) Predict the revenue in 2005. [3.7]

Graph.

59. $y = \frac{2}{3}x - 5$ [3.6]

60. $2x + y = 4$ [3.3]

61. $y = 6$ [3.3]

62. $x = -2$ [3.3]

63. $y + 2 = -\frac{1}{2}(x - 3)$ [3.7]

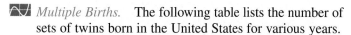 *Multiple Births.* The following table lists the number of sets of twins born in the United States for various years.

Year	Number of Twin Births
1994	97,064
1996	100,750
1998	110,670
2000	118,916
2003	128,665

Source: Centers for Disease Control and Prevention

64. Graph the data and determine whether the relationship appears to be linear. [3.7]

65. Use linear regression to find a linear equation that fits the data. Let $B =$ the number of twin births and t the number of years after 1990. [3.7]

66. Use the equation from Exercise 65 to estimate the number of twin births in 2009. [3.7]

67. For the graph of f shown here, determine **(a)** $f(2)$; **(b)** the domain of f; **(c)** any x-values for which $f(x) = 2$; and **(d)** the range of f. [3.8]

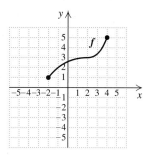

68. Find $g(-3)$ for $g(x) = \dfrac{x}{x + 1}$. [3.8]

69. Find $f(2a)$ for $f(x) = x^2 + 2x - 3$. [3.8]

70. The function $A(t) = 0.233t + 5.87$ can be used to estimate the median age of cars in the United States t years after 1990 (*Source*: The Polk Co.). (In this context, a median age of 3 yr means that half the cars are more than 3 yr old and half are less.)
   a) Predict the median age of cars in 2010; that is, find $A(20)$. [3.8]
   b) Determine what 0.233 and 5.87 signify. [3.8]

For each of the graphs in Exercises 71–74, **(a)** *determine whether the graph represents a function and* **(b)** *if so, determine the domain and the range of the function.* [3.8]

71.

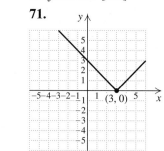

72.

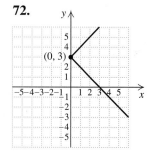

73.

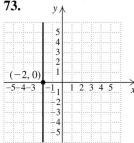

74.

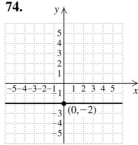

Find the domain of each function. [3.8]

75. $f(x) = 3x^2 - 7$ **76.** $g(x) = \dfrac{x^2}{x-1}$

77. For the function given by

$$f(x) = \begin{cases} 2 - x, & \text{for } x \le -2, \\ x^2, & \text{for } -2 < x \le 5, \\ x + 10, & \text{for } x > 5, \end{cases}$$

find **(a)** $f(-3)$; **(b)** $f(-2)$; **(c)** $f(4)$; and **(d)** $f(25)$. [3.8]

Synthesis

TW 78. Can two perpendicular lines share the same y-intercept? Why or why not? [3.3], [3.6]

TW 79. If two functions have the same domain and range, are the functions identical? Why or why not? [3.5]

80. Find the value of m in $y = mx + 3$ such that $(-2, 5)$ is on the graph. [3.2]

81. Find the value of b in $y = -5x + b$ such that $(3, 4)$ is on the graph. [3.2]

82. Find the area and the perimeter of a rectangle for which $(-2, 2)$, $(7, 2)$, and $(7, -3)$ are three of the vertices. [3.1]

83. Find three solutions of $y = 4 - |x|$. [3.2]

84. Determine the domain and the range of the function graphed below. [3.8]

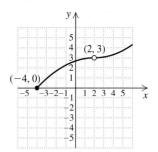

Chapter Test 3

Use of Tax Dollars. *The following pie chart shows how federal income tax dollars were spent in 2003 (Source: Based on information from www.taxfoundation.com).*

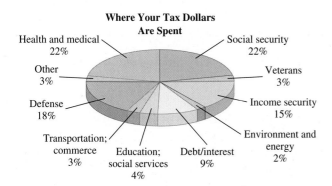

1. Debbie pays 15% of her taxable income of \$34,200 in taxes. How much of her income will go to education and social services?

2. Larry pays 10% of his taxable income of \$29,000 in taxes. How much of his income will go to social security?

In which quadrant is each point located?

3. $\left(-\frac{1}{2}, 7\right)$ **4.** $(-5, -6)$

Find the coordinates of each point in the figure.

5. A

6. B

7. C

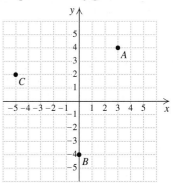

Graph by hand.

8. $y = 2x - 1$

9. $2x - 4y = -8$

10. $y + 4 = -\frac{1}{2}(x - 3)$

11. $y = \frac{3}{4}x$

12. $y = 7$

13. $2x - y = 3$

14. $x = -1$

 15. Graph using a graphing calculator: $1.2x - y = 5$.

16. Find the x- and y-intercepts of $5x - 3y = 30$. Do not graph.

17. Find the slope of the line containing the points $(4, -1)$ and $(6, 8)$.

18. *Running.* Ted reached the 3-km mark of a race at 2:15 P.M. and the 6-km mark at 2:24 P.M. What is his running rate in kilometers/minute?

19. At one point Filbert Street, the steepest street in San Francisco, drops 63 ft over a horizontal distance of 200 ft. Find the road grade.

20. Find the slope and the y-intercept of the line given by $y - 3x = 7$.

Determine without graphing whether each pair of lines is parallel, perpendicular, or neither.

21. $4y + 2 = 3x,$
$-3x + 4y = -12$

22. $y = -2x + 5,$
$2y - x = 6$

23. Write a point–slope equation for the line of slope -3 that contains the point $(6, 8)$.

24. Write the slope–intercept equation of the line that is perpendicular to the line $x - y = 3$ and that contains the point $(3, 7)$.

25. The average urban commuter was sitting in traffic for 16 hr in 1982 and for 47 hr in 2003 (*Source:* Urban Mobility Report).

a) Assuming a linear relationship, find an equation that fits the data. Let $c =$ the number of hours the average urban commuter is sitting in traffic and t the number of years after 1982.

b) Calculate the number of hours the average urban commuter was sitting in traffic in 1998.

c) Predict the number of hours the average urban commuter will be sitting in traffic in 2010.

The following table lists the number of DVDs sold in the United States for various years.

Year	DVD Sales (in millions)
2001	235
2002	510
2003	875
2004	1297
2005	1620

Source: "Diagnosing the DVD Disappointment: A Life Cycle View," Coplan, Judson, April 3, 2006.

 26. Use linear regression to find a linear equation that fits the data. Let $d =$ DVD sales, in millions, and t the number of years after 2000.

 27. Use the equation from Exercise 26 to estimate DVD sales in 2007.

For each of the graphs in Exercises 28 and 29, **(a)** *determine whether the graph represents a function and* **(b)** *if so, determine the domain and the range of the function.*

28.

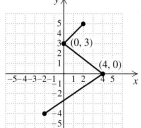

29.

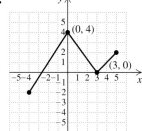

30. Use the graph in Exercise 29 to find each of the following.

a) $f(3)$

b) Any inputs x such that $f(x) = -2$

31. If $g(x) = \dfrac{x - 6}{2x + 1}$, find each of the following.

   a) $g(1)$
   b) The domain of g

32. For the function given by
$$f(x) = \begin{cases} x, & \text{for } x \geq 0, \\ -x, & \text{for } x < 0, \end{cases}$$
find **(a)** $f(3)$; **(b)** $f(-10)$; **(c)** $f(0)$.

Synthesis

33. Write an equation of the line that is parallel to the graph of $2x - 5y = 6$ and has the same y-intercept as the graph of $3x + y = 9$.

34. A diagonal of a square connects the points $(-3, -1)$ and $(2, 4)$. Find the area and the perimeter of the square.

35. List the coordinates of three other points that are on the same line as $(-2, 14)$ and $(17, -5)$. Answers may vary.

1–3	**Cumulative Review**

Solve.

1. U.S. bicycle sales rose from 15 million in 1995 to 20 million in 2005 (*Sources*: National Bicycle Dealers Association; U.S. Department of Transportation). Find the rate of change of bicycle sales. [3.4]

2. A 150-lb person will burn 240 calories an hour when riding a bicycle at 6 mph. The same person will burn 410 calories an hour when cycling at 12 mph. (*Source*: American Heart Association) [3.7]

   a) Find a linear equation that fits the data. Let c represent the number of calories burned and r the rate at which the person is cycling.

b) Use the equation of part (a) to estimate the number of calories burned by a 150-lb person cycling at 10 mph.

3. The following table lists the number of calories burned per hour while riding a bicycle at 12 mph for persons of various weights.

Weight	Calories Burned Per Hour at 12 mph
100	270
150	410
200	534

Source: American Heart Association

a) Use linear regression to find an equation that fits the data. Let c represent the number of calories burned and w represent weight, in pounds. [3.7]

b) Use the equation of part (a) to estimate the number of calories burned per hour by a 135-lb person cycling at 12 mph. [3.7]

4. Americans spent an estimated $238 billion on home remodeling in 2006. This was $\frac{17}{15}$ of the amount spent on remodeling in 2005. (*Source*: National Association of Home Builders' Remodelers Council) How much was spent on remodeling in 2005? [2.5]

5. In 2003, the mean earnings of individuals with some college education but no degree was $29,533. This was about 58% of the mean earnings of those with a bachelor's degree. (*Source*: *Statistical Abstract of the United States*) What were the mean earnings of individuals with a bachelor's degree in 2003? [2.4]

6. *Blood Types.*   There were recently 117 million Americans with either O-positive or O-negative blood. Those with O-positive blood outnumbered those with O-negative blood by 85.8 million. How many Americans had O-negative blood? [2.5]

7. Tina paid $126 for a cordless drill, including a 5% sales tax. How much did the drill itself cost? [2.4]

8. A 143-m wire is cut into three pieces. The second is 3 m longer than the first. The third is four-fifths as long as the first. How long is each piece? [2.5]

9. Tori's contract stipulates that she cannot work more than 40 hr per week. For the first 4 days of one week, she worked 7, 10, 9, and 6 hr. How many hours can she work the fifth day without violating her contract? [2.5]

10. *Fund Raising.*   The following graph shows data from a recent road race held to raise money for Amnesty International. At what rate was money raised? [3.4]

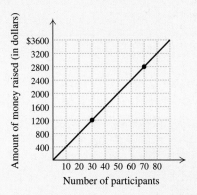

11. Evaluate $\dfrac{x}{2y}$ for $x = 70$ and $y = 5$. [1.1]

12. Multiply: $3(4x - 5y + 7)$. [1.2]

13. Factor: $15x - 9y + 3$. [1.2]

14. Find the prime factorization of 42. [1.3]

15. Find decimal notation: $\frac{9}{20}$. [1.4]

16. Find the absolute value: $|-4|$. [1.4]

17. Find the opposite of $-\frac{1}{4}$. [1.6]

18. Find the reciprocal of $-\frac{1}{4}$. [1.7]

19. Combine like terms: $2x - 5y + (-3x) + 4y$. [1.6]

20. Find decimal notation: 78.5%. [2.4]

Simplify.

21. $\frac{3}{5} - \frac{5}{12}$ [1.3]   **22.** $3.4 + (-0.8)$ [1.5]

23. $(-2)(-1.4)(2.6)$ [1.7]   **24.** $\frac{3}{8} \div \left(-\frac{9}{10}\right)$ [1.7]

25. $1 - [32 \div (4 + 2^2)]$ [1.8]

26. $-5 + 16 \div 2 \cdot 4$ [1.8]

27. $y - (3y + 7)$ [1.8]

28. $3(x - 1) - 2[x - (2x + 7)]$ [1.8]

Solve.

29. $2.6 = 3.8 + x$ [2.1]

30. $\frac{2}{7}x = -6$ [2.1]

31. $5x - 8 = 37$ [2.2]

32. $\frac{2}{3} = \frac{-m}{10}$ [2.1]

33. $5.4 - 1.9x = 0.8x$ [2.2]

34. $x - \frac{7}{8} = \frac{3}{4}$ [2.1]

35. $2(2 - 3x) = 3(5x + 7)$ [2.2]

36. $\frac{1}{4}x - \frac{2}{3} = \frac{3}{4} + \frac{1}{3}x$ [2.2]

37. $y + 5 - 3y = 5y - 9$ [2.2]

38. $x - 28 < 20 - 2x$ [2.6]

39. $2(x + 2) \geq 5(2x + 3)$ [2.6]

40. Solve $A = 2\pi rh + \pi r^2$ for h. [2.3]

41. In which quadrant is the point $(3, -1)$ located? [3.1]

42. Graph on a number line: $-1 < x \leq 2$. [2.6]

43. Use a grid 10 squares wide and 10 squares high to plot $(-150, -40)$, $(40, -7)$, and $(0, 6)$. Choose the scale carefully. [3.1]

Graph by hand.

44. $x = 3$ [3.3]   **45.** $2x - 5y = 10$ [3.3]

46. $y = -2x + 1$ [3.2]   **47.** $y = \frac{2}{3}x$ [3.2]

48. $y = -\frac{3}{4}x + 2$ [3.6]   **49.** $y - 1 = -(x + 2)$ [3.7]

Find the coordinates of the x- and y-intercepts. Do not graph. [3.3]

50. $2x - 7y = 21$ **51.** $y = 4x + 5$

52. Find the slope of the line containing the points $(-4, 1)$ and $(2, -1)$. [3.5]

53. Write an equation of the line with slope $\frac{2}{7}$ and y-intercept $(0, -4)$. [3.6]

54. Find the slope and the y-intercept of the line given by $2x + 6y = 18$. [3.6]

55. Write the slope–intercept equation of the line parallel to the line $x + 2y = 3$ and containing the point $(2, 4)$. [3.7]

56. For the graph of f shown, determine the domain, the range, $f(-3)$, and any value of x for which $f(x) = 5$. [3.8]

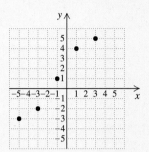

57. For the function given by $f(x) = \dfrac{7}{2x - 1}$, determine each of the following.
    a) $f(0)$ [3.8]
    b) The domain of f [3.8]

Synthesis

58. *Yearly Earnings.*   Paula's salary at the end of a year is $26,780. This reflects a 4% salary increase in February and then a 3% cost-of-living adjustment in June. What was her salary at the beginning of the year? [2.4]

Solve. Label any contradictions or identities.

59. $4|x| - 13 = 3$ [1.4], [2.2]

60. $4(x + 2) = 9(x - 2) + 16$ [2.2]

61. $2(x + 3) + 4 = 0$ [2.2]

62. $\dfrac{2 + 5x}{4} = \dfrac{11}{28} + \dfrac{8x + 3}{7}$ [2.2]

63. $5(7 + x) = (x + 6)5$ [2.2]

64. Solve $p = \dfrac{2}{m + Q}$ for Q. [2.3]

65. The points $(-3, 0)$, $(0, 7)$, $(3, 0)$, and $(0, -7)$ are vertices of a parallelogram. Find four equations of lines that intersect to form the parallelogram. [3.6]

66. Given that $f(x) = mx + b$ and that $f(5) = -3$ when $f(-4) = 2$, find m and b. [3.7], [3.8]

Systems of Equations in Two Variables

n fields ranging from business to zoology, problems arise that are most easily solved using a *system of equations*. In this chapter, we solve systems and applications using graphing and two algebraic methods. The graphing method is then applied to the solution of equations.

APPLICATION *Bottled-Water Consumption.*

Americans are buying more bottled water and less regular (not diet) soft drink. The following table and graph show per-capita consumption of both beverages for several years. When will Americans consume as much bottled water per year as regular soft drink?

YEAR	BOTTLED WATER (IN GALLONS)	SOFT DRINK (IN GALLONS)
2000	16.7	39.4
2001	18.2	39.0
2002	20.1	38.4
2003	21.6	37.5
2004	23.2	36.5

Source: USDA/Economic Research Service

$$y_1 = 1.64x + 16.68,$$
$$y_2 = -0.73x + 39.62$$

This problem appears as Example 6 in Section 4.1.

4.1 Systems of Equations and Graphing

Solutions of Systems   ■   Solving Systems of Equations by Graphing   ■   Models

Solutions of Systems

A **system of equations** is a set of two or more equations that are to be solved simultaneously. We will find that it is often easier to translate real-world situations to a system of two equations that use two variables, or unknowns, than it is to represent the situation with one equation using one variable. To see this, let's see how the following problem from Section 2.5 can be represented by two equations using two unknowns:

> The perimeter of an NBA basketball court is 288 ft. The length is 44 ft longer than the width. (*Source*: National Basketball Association) Find the dimensions of the court.

If we let w = the width of the court, in feet, and l = the length of the court, in feet, the problem translates to the following system of equations:

$$2l + 2w = 288, \qquad \text{The perimeter is 288.}$$
$$l = w + 44. \qquad \text{The length is 44 more than the width.}$$

To solve a system of equations is to find all ordered pairs (if any exist) for which *both* equations are true. Thus a solution of the system listed above is an ordered pair of the form (l, w). Note that variables are listed in alphabetical order within an ordered pair unless stated otherwise.

> **EXAMPLE 1**   Consider the system from above:
>
> $$2l + 2w = 288,$$
> $$l = w + 44.$$
>
> Determine whether each pair is a solution of the system: **(a)** $(94, 50)$; **(b)** $(90, 46)$.

SOLUTION

a) Unless stated otherwise, we use alphabetical order of the variables. Thus we replace l with 94 and w with 50.

$$\begin{array}{c|c} 2l + 2w = 288 & \\ \hline 2 \cdot 94 + 2 \cdot 50 & 288 \\ 188 + 100 & \\ 288 \overset{?}{=} 288 & \text{TRUE} \end{array} \qquad \begin{array}{c|c} l = w + 44 & \\ \hline 94 & 50 + 44 \\ 94 \overset{?}{=} 94 & \text{TRUE} \end{array}$$

Since (94, 50) checks in *both* equations, it is a solution of the system.

b) We substitute 90 for l and 46 for w:

$$\frac{2l + 2w = 288}{\begin{array}{c|c} 2 \cdot 90 + 2 \cdot 46 & 288 \\ 180 + 92 & \\ & 272 \stackrel{?}{=} 288 \quad \text{FALSE} \end{array}} \qquad \frac{l = w + 44}{\begin{array}{c|c} 90 & 46 + 44 \\ 90 \stackrel{?}{=} 90 \quad \text{TRUE} \end{array}}$$

Since (90, 46) is not a solution of *both* equations, it is not a solution of the system.

A solution of a system of two equations in two variables is an ordered pair of numbers that makes both equations true. The numbers in the ordered pair correspond to the variables in alphabetical order.

In Example 1, we demonstrated that (94, 50) is a solution of the system, but we did not show how the pair (94, 50) was found. One way to find such a solution uses graphs.

Solving Systems of Equations by Graphing

Recall that a graph of an equation is a set of points representing its solution set. Each point on the graph corresponds to an ordered pair that is a solution of the equation. By graphing two equations using one set of axes, we can identify a solution of both equations by looking for a point of intersection.

▶ **EXAMPLE 2**   Solve this system of equations by graphing:

$$x + y = 7,$$
$$y = 3x - 1.$$

SOLUTION   We graph the equations using any method studied earlier. The equation $x + y = 7$ can be graphed easily using the intercepts, (0, 7) and (7, 0). The equation $y = 3x - 1$ is in slope–intercept form, so it can be graphed by plotting its y-intercept, $(0, -1)$, and "counting off" a slope of 3.

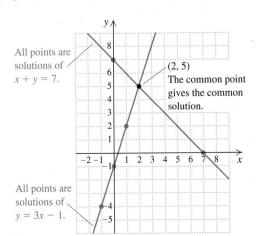

All points are solutions of $x + y = 7$.

(2, 5) The common point gives the common solution.

All points are solutions of $y = 3x - 1$.

The "apparent" solution of the system, (2, 5), should be checked in both equations. Note that the solution is an ordered pair in the form (x, y).

Check:

$$\frac{x + y = 7}{2 + 5 \mid 7}$$
$$7 \overset{?}{=} 7 \quad \text{TRUE}$$

$$\frac{y = 3x - 1}{5 \mid 3 \cdot 2 - 1}$$
$$5 \overset{?}{=} 5 \quad \text{TRUE}$$

Since it checks in both equations, (2, 5) is a solution of the system.

Point of Intersection

A graphing calculator can be used to determine the point of intersection of graphs. Often it provides a more accurate solution than graphs drawn by hand. We can trace along the graph of one of the functions to find the coordinates of the point of intersection, enlarging the graph using a ZOOM feature if necessary. Most graphing calculators can find the point of intersection directly using an INTERSECT feature, found in the CALC menu.

To find the point of intersection of two graphs, press ⬚CALC⬚ and choose the INTERSECT option. Because more than two equations may be graphed, the questions FIRST CURVE? and SECOND CURVE? are used to identify the graphs with which we are concerned. Position the cursor on each graph, in turn, and press **ENTER**, using the up and down arrow keys if necessary.

The calculator then asks a third question, GUESS?. Since graphs may intersect at more than one point, we must visually identify the point of intersection in which we are interested and indicate the general location of that point. Enter a guess either by moving the cursor near the point of intersection and pressing **ENTER** or by typing a guess and pressing **ENTER**. The calculator then returns the coordinates of the point of intersection. At this point, the calculator variables X and Y contain the coordinates of the point of intersection. These values can be used to check an answer, and often they can be converted to fraction notation from the home screen using the FRAC option of the MATH menu. We can check that these values satisfy the original system.

Most pairs of lines have exactly one point in common. We will soon see, however, that this is not always the case.

EXAMPLE 3   Solve graphically:

$$y - x = 1,$$
$$y + x = 3.$$

SOLUTION

BY HAND

We graph each equation using any method studied in Chapter 3. All ordered pairs from line L_1 in the graph below are solutions of the first equation. All ordered pairs from line L_2 are solutions of the second equation. The point of intersection has coordinates that make *both* equations true. Apparently, $(1, 2)$ is the solution. Graphing is not always accurate, so solving by graphing may yield approximate answers. Our check below shows that $(1, 2)$ is indeed the solution.

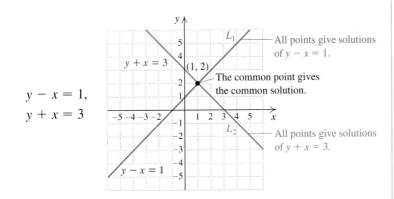

$$y - x = 1,$$
$$y + x = 3$$

Check:

$y - x = 1$
$2 - 1 \mid 1$
$1 \overset{?}{=} 1$ TRUE

$y + x = 3$
$2 + 1 \mid 3$
$3 \overset{?}{=} 3$ TRUE

WITH A GRAPHING CALCULATOR

We first solve each equation for y:

$$y - x = 1 \qquad\qquad y + x = 3$$
$$y = x + 1; \qquad\qquad y = 3 - x.$$

Then we enter and graph $y_1 = x + 1$ and $y_2 = 3 - x$ using the same viewing window. After choosing the INTERSECT option, we indicate which two graphs, or *curves*, we are considering. Then we enter a guess. The coordinates of the point of intersection then appear at the bottom of the screen.

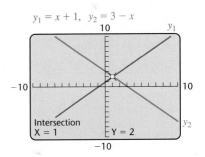

The point $(1, 2)$ must make both equations true in order to be a solution of the system. Since, in the calculator, X now has the value 1 and Y has the value 2, we can check from the home screen. The solution is $(1, 2)$.

EXAMPLE 4 Solve each system graphically.

a) $y = -3x + 5,$
$\quad y = -3x - 2$

b) $\quad 3y - 2x = 6,$
$\quad -12y + 8x = -24$

SOLUTION

a) We graph the equations. The lines have the same slope, -3, and different y-intercepts, so they are parallel. There is no point at which they cross, so the system has no solution.

$$y = -3x + 5,$$
$$y = -3x - 2$$

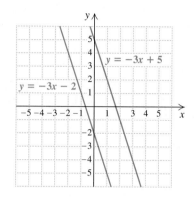

$y_1 = -3x + 5, \quad y_2 = -3x - 2$

What happens when we try to solve this system using a graphing calculator? We enter and graph both equations as shown at left, noting that the graphs appear to be parallel. (This must be verified algebraically.) When we attempt to find the intersection using the INTERSECT feature, we get an error message.

b) We graph the equations and find that the same line is drawn twice. Thus any solution of one equation is a solution of the other. Each equation has an infinite number of solutions, so the system itself has an infinite number of solutions. We check one solution, $(0, 2)$, which is the y-intercept of each equation.

Student Notes

Although the system in Example 4(b) is true for an infinite number of ordered pairs, those pairs must be of a certain form. Only pairs that are solutions of $3y - 2x = 6$ or $-12y + 8x = -24$ are solutions of the system. It is incorrect to think that *all* ordered pairs are solutions.

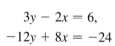

$$3y - 2x = 6,$$
$$-12y + 8x = -24$$

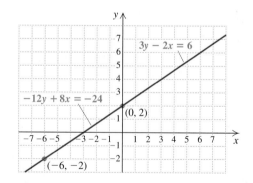

Check:

$$\begin{array}{c|c} 3y - 2x = 6 \\ \hline 3(2) - 2(0) & 6 \\ 6 - 0 & \\ & 6 \overset{?}{=} 6 \quad \text{TRUE} \end{array}$$

$$\begin{array}{c|c} -12y + 8x = -24 \\ \hline -12(2) + 8(0) & -24 \\ -24 + 0 & \\ & -24 \overset{?}{=} -24 \quad \text{TRUE} \end{array}$$

We can check that $(-6, -2)$ is another solution of both equations. In fact, any pair that is a solution of one equation is a solution of the other equation as well. Thus the solution set is

$$\{(x, y) \mid 3y - 2x = 6\}$$

or, in words, "the set of all pairs (x, y) for which $3y - 2x = 6$." Since the two equations are equivalent, we could have written instead $\{(x, y) \mid -12y + 8x = -24\}$.

If we attempt to find the intersection of the graphs of the equations in this system using INTERSECT, the graphing calculator will return as the intersection whatever point we choose as the guess. This is a point of intersection, as is any other point on the graph of the lines. ◢

When we graph a system of two linear equations in two variables, one of the following three outcomes will occur.

1. The lines have one point in common, and that point is the only solution of the system (see Example 3). Any system that has at least one solution is said to be **consistent.**

2. The lines are parallel, with no point in common, and the system has no solution (see Example 4a). This type of system is called **inconsistent.**

3. The lines coincide, sharing the same graph. Because every solution of one equation is a solution of the other, the system has an infinite number of solutions (see Example 4b). Since it has a solution, this type of system is also consistent.

When one equation in a system can be obtained by multiplying both sides of another equation by a constant, the two equations are said to be **dependent.** Thus the equations in Example 4(b) are dependent, but those in Examples 3 and 4(a) are **independent.** For systems of three or more equations, the definitions of dependent and independent must be slightly modified.

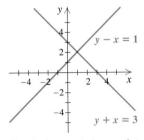

Graphs intersect at one point.
The system is *consistent* and has one solution. Since neither equation is a multiple of the other, they are *independent.*

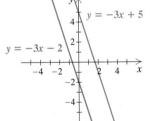

Graphs are parallel.
The system is *inconsistent* because there is no solution. Since the equations are not equivalent, they are *independent.*

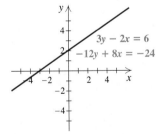

Equations have the same graph.
The system is *consistent* and has an infinite number of solutions. The equations are *dependent* since they are equivalent.

Graphing calculators are especially useful when equations contain fractions or decimals or when the coordinates of the intersection are not integers.

EXAMPLE 5   Solve graphically:

$$3.45x + 4.21y = 8.39,$$
$$7.12x - 5.43y = 6.18.$$

SOLUTION   First, we solve for y in each equation:

$$3.45x + 4.21y = 8.39$$
$$4.21y = 8.39 - 3.45x \qquad \text{Subtracting } 3.45x \text{ from both sides}$$
$$y = (8.39 - 3.45x)/4.21; \qquad \text{Dividing both sides by } 4.21$$

$$7.12x - 5.43y = 6.18$$
$$-5.43y = 6.18 - 7.12x \qquad \text{Subtracting } 7.12x \text{ from both sides}$$
$$y = (6.18 - 7.12x)/(-5.43). \qquad \text{Dividing both sides by } -5.43$$

It is not necessary to simplify further. We have the system

$$y = (8.39 - 3.45x)/4.21,$$
$$y = (6.18 - 7.12x)/(-5.43).$$

Next, we enter both equations and graph using the same viewing window. By using the INTERSECT feature in the CALC menu, we see that, to the nearest hundredth, the solution is $(1.47, 0.79)$. Note that the coordinates found by the calculator are approximations.

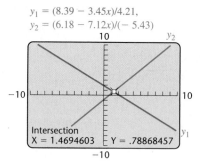

$y_1 = (8.39 - 3.45x)/4.21,$
$y_2 = (6.18 - 7.12x)/(-5.43)$

Intersection
X = 1.4694603   Y = .78868457

Models

Sometimes two or more sets of data can be modeled by a system of linear equations.

EXAMPLE 6   Bottled-Water Consumption.   Americans are buying more bottled water and less regular (not diet) soft drink. The following table lists per-capita consumption of both beverages for several years. When will Americans consume as much bottled water per year as regular soft drink?

Year	Bottled Water (in gallons)	Soft Drink (in gallons)
2000	16.7	39.4
2001	18.2	39.0
2002	20.1	38.4
2003	21.6	37.5
2004	23.2	36.5

Source: USDA/Economic Research Service

SOLUTION   We first enter and graph the data using a graphing calculator. To make the numbers smaller, we will let x represent the number of years after 2000. Thus we enter the number of years after 2000 as L1, the number of gallons of bottled water as L2, and the number of gallons of soft drink as L3. A good viewing window, based on the data, is $[0, 8, 0, 40]$, with Yscl $= 5$. The left and middle screens below show the setup of the plots, with Plot1 graphing the bottled water data with boxes and Plot2 graphing the soft drink data with plus signs.

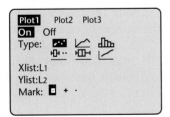

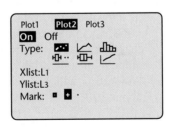

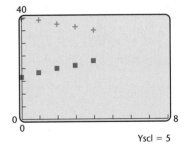

Both sets of data are approximately linear, and we can see that bottled water consumption is increasing and soft drink consumption is decreasing. It appears likely that the numbers will be equal in the future, if the trend continues. We use linear regression to fit a line to each set of data.

Since bottled water is entered as L2, we first find the linear regression line when L1 is the independent variable and L2 is the dependent variable and store this equation as Y1. Since L1 and L2 are the default lists, we did not need to specify them, although they are shown in the figure on the left below.

To find the second regression line, we must specify that L1 is the independent variable and L3 is the dependent variable. We store this equation to Y2, as shown in the figure on the right below.

LinReg(ax+b) L1,
L2, Y1

LinReg(ax+b) L1,
L3, Y2

The screen on the left below gives us the two equations

$$y_1 = 1.64x + 16.68 \quad \text{and} \quad y_2 = -0.73x + 39.62.$$

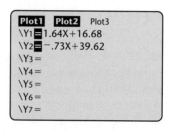

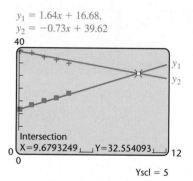

We graph the equations along with the data and determine the coordinates of the point of intersection. The *x*-coordinate of the point of intersection must be within the window dimensions, so we extend the window, as shown on the right above. The point of intersection is approximately (9.68, 32.55). Since *x* represents the number of years after 2000, this tells us that Americans will drink about 32.55 gal of bottled water and the same amount of soft drink about 9.7 yr after 2000. Rounding, we state that Americans will consume as much bottled water as regular soft drink in about 2010.

Graphing is helpful when solving systems because it allows us to "see" the solution. It can also be used on systems of nonlinear equations, and in many applications, it provides a satisfactory answer. However, graphing often lacks precision, especially when fraction or decimal solutions are involved. In Sections 4.2 and 4.3, we will develop two algebraic methods of solving systems. Both methods produce exact answers.

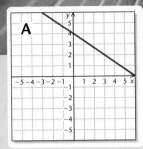

A

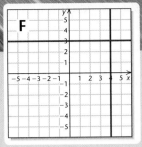

F

Visualizing the Graph

Match each equation or system of equations with its graph.

1. $x + y = 2,$
   $x - y = 2$

2. $y = \frac{1}{3}x - 5$

3. $4x - 2y = -8$

4. $2x + y = 1,$
   $x + 2y = 1$

5. $8y + 32 = 0$

6. $y = -x + 4$

7. $\frac{2}{3}x + y = 4$

8. $x = 4,$
   $y = 3$

9. $y = \frac{1}{2}x + 3,$
   $2y - x = 6$

10. $y = -x + 5,$
    $y = 3 - x$

Answers on page A-17

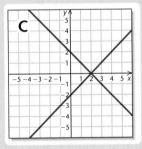

B

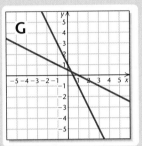

G

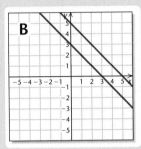

C

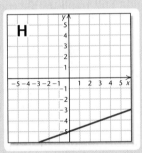

H

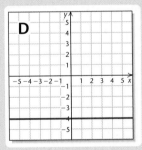

D

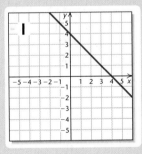

I

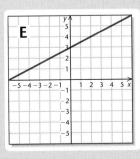

E

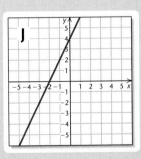

J

4.1 EXERCISE SET

Concept Reinforcement *Classify each statement as either true or false.*

1. Solutions of systems of equations in two variables are ordered pairs.

2. Every system of equations has at least one solution.

3. It is possible for a system of equations to have an infinite number of solutions.

4. The graphs of the equations in a system of two equations may coincide.

5. The graphs of the equations in a system of two equations could be parallel lines.

6. Any system of equations that has at most one solution is said to be consistent.

7. Any system of equations that has more than one solution is said to be inconsistent.

8. If one equation in a system can be obtained by multiplying both sides of another equation in that system by a constant, the two equations are said to be dependent.

Determine whether the ordered pair is a solution of the given system of equations. Remember to use alphabetical order of variables.

9. $(1, 2)$; $4x - y = 2$,
   $10x - 3y = 4$

10. $(-1, -2)$; $2x + y = -4$,
    $x - y = 1$

11. $(2, 5)$; $y = 3x - 1$,
    $2x + y = 4$

12. $(-1, -2)$; $x + 3y = -7$,
    $3x - 2y = 12$

13. $(1, 5)$; $x + y = 6$,
    $y = 2x + 3$

14. $(5, 2)$; $a + b = 7$,
    $2a - 8 = b$

Aha! 15. $(3, 1)$; $3x + 4y = 13$,
    $6x + 8y = 26$

16. $(4, -2)$; $-3x - 2y = -8$,
    $8 = 3x + 2y$

Solve each system graphically. Be sure to check your solution. If a system has an infinite number of solutions, use set-builder notation to write the solution set. If a system has no solution, state this. Where appropriate, round to the nearest hundredth.

17. $x - y = 3$,
    $x + y = 5$

18. $x + y = 4$,
    $x - y = 2$

19. $3x + y = 5$,
    $x - 2y = 4$

20. $2x - y = 4$,
    $5x - y = 13$

21. $4y = x + 8$,
    $3x - 2y = 6$

22. $4x - y = 9$,
    $x - 3y = 16$

23. $x = y - 1$,
    $2x - 3y$

24. $a = 1 + b$,
    $b = 5 - 2a$

25. $x = -3$,
    $y = 2$

26. $x = 4$,
    $y = -5$

27. $t + 2s = -1$,
    $s = t + 10$

28. $b + 2a = 2$,
    $a = -3 - b$

29. $2b + a = 11$,
    $a - b = 5$

30. $y = -\frac{1}{3}x - 1$,
    $4x - 3y = 18$

31. $y = -\frac{1}{4}x + 1$,
    $2y = x - 4$

32. $6x - 2y = 2$,
    $9x - 3y = 1$

33. $y - x = 5$,
    $2x - 2y = 10$

34. $y = -x - 1$,
    $4x - 3y = 24$

35. $y = 3 - x$,
    $2x + 2y = 6$

36. $2x - 3y = 6$,
    $3y - 2x = -6$

37. $y = -5.43x + 10.89$,
    $y = 6.29x - 7.04$

38. $y = 123.52x + 89.32,$
$\quad y = -89.22x + 33.76$

39. $2.6x - 1.1y = 4,$
$\quad 1.32y = 3.12x - 5.04$

40. $2.18x + 7.81y = 13.78,$
$\quad 5.79x - 3.45y = 8.94$

41. $0.2x - y = 17.5,$
$\quad 2y - 10.6x = 30$

42. $1.9x = 4.8y + 1.7,$
$\quad 12.92x + 23.8 = 32.64y$

43. For the systems in the odd-numbered exercises 17–41, which are consistent?

44. For the systems in the even-numbered exercises 18–42, which are consistent?

45. For the systems in the odd-numbered exercises 17–41, which contain dependent equations?

46. For the systems in the even-numbered exercises 18–42, which contain dependent equations?

47. *College Faculty.*   The number of part-time faculty in institutions of higher learning is growing rapidly. The following table lists the number of full-time faculty and the number of part-time faculty for various years. Use linear regression to fit a line to each set of data, and use those equations to predict the year in which the number of part-time faculty and the number of full-time faculty will be the same.

Year	Number of Full-time Faculty (in thousands)	Number of Part-time Faculty (in thousands)
1980	450	236
1985	459	256
1991	536	291
1995	551	381
1999	591	437
2003	632	543

Source: U.S. National Center for Education Statistics

48. *Milk Cows.*   The number of milk cows in Minnesota has decreased since 2000, while the number in New Mexico has increased, as shown in the following table. Use linear regression to fit a line to each set of data, and use those equations to predict the year in which the number of milk cows in the two states will be the same.

Year	Number of Milk Cows in Minnesota (in thousands)	Number of Milk Cows in New Mexico (in thousands)
2000	534	250
2001	510	268
2002	487	301
2003	473	317
2004	463	326

Source: U.S. Department of Agriculture

49. *Recycling.*   In the United States, the amount of paper being recycled is slowly catching up to the amount of paper waste being generated, as shown in the following table. Use linear regression to fit a line to each set of data, and use those equations to predict the year in which the amount recycled will equal the amount generated.

Year	Amount of Paper Waste Generated (in millions of tons)	Amount of Paper Recycled (in millions of tons)
1990	72.7	27.8
1994	80.0	36.5
1998	84.1	41.6
2001	82.7	45.6
2003	83.1	48.1

Source: U.S. Bureau of Labor Statistics

50. *Grocery Stores.*   Conventional grocery stores in the United States are being replaced by other types of food stores. One such type of store is the combination food and drug store, which contains a pharmacy and a greater variety of health and beauty aids than a conventional grocery store. Use linear regression to fit a line to each set of data in the following table, and use those equations to predict the year in which there will be the same number of combination food and drug stores as conventional grocery stores.

Year	Number of Conventional Grocery Stores (in thousands)	Number of Combination Food and Drug Stores (in thousands)
1990	13.2	1.6
1995	12.3	2.7
2000	9.9	3.7
2002	8.3	4.5
2003	8.9	5.0

Source: U.S. Department of Agriculture

TW **51.** Suppose that the equations in a system of two linear equations are dependent. Does it follow that the system is consistent? Why or why not?

TW **52.** Why is slope–intercept form especially useful when solving systems of equations by graphing?

Skill Maintenance

Solve. [2.2]

53. $2(4x - 3) - 7x = 9$

54. $6y - 3(5 - 2y) = 4$

55. $4x - 5x = 8x - 9 + 11x$

56. $8x - 2(5 - x) = 7x + 3$

Solve. [2.3]

57. $3x + 4y = 7$, for y

58. $2x - 5y = 9$, for y

Synthesis

Presidential Primaries. *For Exercises 59 and 60, consider the graph at the top of the next column, which shows the results of a poll in which Iowans were asked which Democrat candidate for president they favored.*

TW **59.** At what point in time could it have been said that no one was in fourth place? Explain.

TW **60.** At what point in time was there no clear leader? Explain how you reach this conclusion.

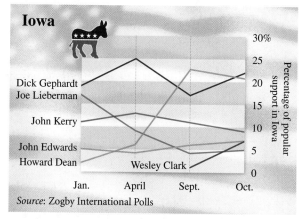

Iowa

Source: Zogby International Polls

61. For each of the following conditions, write a system of equations.

a) (5, 1) is a solution.

b) There is no solution.

c) There is an infinite number of solutions.

62. A system of linear equations has $(1, -1)$ and $(-2, 3)$ as solutions. Determine:

a) a third point that is a solution, and

b) how many solutions there are.

63. The solution of the following system is $(4, -5)$. Find A and B.

$$Ax - 6y = 13,$$
$$x - By = -8.$$

Solve graphically.

64.
$$y = |x|,$$
$$x + 4y = 15$$

65. $x - y = 0,$
$$y = x^2$$

In Exercises 66–69, match each system with the appropriate graph from the selections given.

a)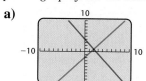

b)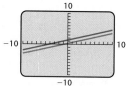

c)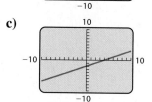

d)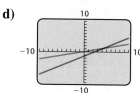

66. $x = 4y,$
$$3x - 5y = 7$$

67. $2x - 8 = 4y,$
$$x - 2y = 4$$

68. $8x + 5y = 20,$
$$4x - 3y = 6$$

69. $x = 3y - 4,$
$$2x + 1 = 6y$$

70. *Copying Costs.* Shelby occasionally goes to The UPS Store® with small copying jobs. He can purchase a "copy card" for $20 that will entitle him to 500 copies, or he can simply pay 6¢ per page.

a) Create cost equations for each method of paying for a number (up to 500) of copies.

b) Graph both cost equations on the same set of axes.

c) Use the graph to determine how many copies Shelby must make if the card is to be more economical.

71. *Photography.* In 2000, about 33.1 billion photos were printed from film and that number was declining at a rate of 1.1 billion per year. There were 2.8 billion digital photos printed or stored in 2000 and that number was growing at a rate of 2.9 billion per year. (*Source*: *Deseret Morning News*, Nov. 16, 2003)

a) Write two equations that can be used to predict n, the number of filmed or digital photos, in billions, t years after 2000.

b) Use a graphing calculator to determine the year in which the numbers of filmed photos and digital photos are the same.

4.2 Systems of Equations and Substitution

The Substitution Method ■ Solving for the Variable First ■
Problem Solving

Near the end of Section 4.1, we mentioned that graphing can be an imprecise method for solving systems. In this section and the next, we develop methods of finding exact solutions using algebra.

The Substitution Method

One method for solving systems is known as the **substitution method.** It uses algebra instead of graphing and is thus considered an *algebraic* method.

EXAMPLE 1 Solve the system

$$x + y = 7, \quad (1)$$
$$y = 3x - 1. \quad (2)$$

We have numbered the equations (1) and (2) for easy reference.

Study Tip

Learn from Your Mistakes

Immediately after each quiz or test, write out a step-by-step solution to any questions you missed. Visit your professor during office hours or consult with a tutor for help with problems that are still giving you trouble. Misconceptions tend to resurface if they are not corrected as soon as possible.

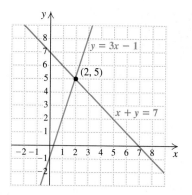

SOLUTION   The second equation says that y and $3x - 1$ represent the same value. Thus, in the first equation, we can substitute $3x - 1$ for y:

$$x + y = 7, \qquad \text{Equation (1)}$$
$$x + 3x - 1 = 7. \qquad \begin{array}{l}\text{Substituting } 3x - 1 \text{ (from equation 2) for } y \\ \text{(in equation 1)}\end{array}$$

The equation $x + 3x - 1 = 7$ has only one variable, for which we now solve:

$$4x - 1 = 7 \qquad \text{Combining like terms}$$
$$4x = 8 \qquad \text{Adding 1 to both sides}$$
$$x = 2. \qquad \text{Dividing both sides by 4}$$

We have found the x-value of the solution. To find the y-value, we return to the original pair of equations. Substituting into either equation will give us the y-value. We choose equation (1):

$$x + y = 7 \qquad \text{Equation (1)}$$
$$2 + y = 7 \qquad \text{Substituting 2 for } x$$
$$y = 5. \qquad \text{Subtracting 2 from both sides}$$

The ordered pair $(2, 5)$ appears to be a solution. We check:

$$
\begin{array}{c|c}
x + y = 7 & y = 3x - 1 \\
\hline
2 + 5 \mid 7 & 5 \mid 3 \cdot 2 - 1 \\
7 \overset{?}{=} 7 \quad \text{TRUE} & 5 \overset{?}{=} 5 \qquad \text{TRUE}
\end{array}
$$

Since $(2, 5)$ checks, it is the solution. For this particular system, we can also check by examining the graph from Example 2 in Section 4.1, as shown at left.

> **CAUTION!**   A solution of a system of equations in two variables is an ordered *pair* of numbers. Once you have solved for one variable, don't forget the other. A common mistake is to solve for only one variable.

EXAMPLE 2   Solve:

$$x = 3 - 2y, \qquad (1)$$
$$y - 3x = 5. \qquad (2)$$

SOLUTION   We substitute $3 - 2y$ for x in the second equation:

$$y - 3x = 5 \qquad \text{Equation (2)}$$
$$y - 3(3 - 2y) = 5. \qquad \begin{array}{l}\text{Substituting } 3 - 2y \text{ for } x. \text{ The parentheses are} \\ \text{very important.}\end{array}$$

Now we solve for y:

$$y - 9 + 6y = 5 \qquad \text{Using the distributive law}$$
$$\left.\begin{array}{r} 7y - 9 = 5 \\ 7y = 14 \\ y = 2. \end{array}\right\} \quad \text{Solving for } y$$

Next, we substitute 2 for y in equation (1) of the original system:

$$x = 3 - 2y \qquad \text{Equation (1)}$$
$$x = 3 - 2 \cdot 2 \qquad \text{Substituting 2 for } y$$
$$x = -1. \qquad \text{Simplifying}$$

We check the ordered pair $(-1, 2)$.

Check:

$$
\begin{array}{c|c}
\multicolumn{2}{c}{x = 3 - 2y} \\
\hline
-1 & 3 - 2 \cdot 2 \\
 & 3 - 4 \\
\multicolumn{2}{c}{-1 \overset{?}{=} -1 \qquad \text{TRUE}}
\end{array}
\qquad
\begin{array}{c|c}
\multicolumn{2}{c}{y - 3x = 5} \\
\hline
2 - 3(-1) & 5 \\
2 + 3 & \\
\multicolumn{2}{c}{5 \overset{?}{=} 5 \qquad \text{TRUE}}
\end{array}
$$

The pair $(-1, 2)$ is the solution. A graph is shown at left as another check.

Solving for the Variable First

Sometimes neither equation has a variable alone on one side. In that case, we solve one equation for one of the variables and then proceed as before.

EXAMPLE 3 Solve:

$$x - 2y = 6, \qquad (1)$$
$$3x + 2y = 4. \qquad (2)$$

SOLUTION We can solve either equation for either variable. Since the coefficient of x is 1 in equation (1), it is easier to solve that equation for x:

$$x - 2y = 6 \qquad\qquad \text{Equation (1)}$$
$$x = 6 + 2y. \qquad (3) \qquad \text{Adding 2y to both sides}$$

We substitute $6 + 2y$ for x in equation (2) of the original pair and solve for y:

$$3x + 2y = 4 \qquad\qquad \text{Equation (2)}$$
$$3(6 + 2y) + 2y = 4 \qquad\qquad \text{Substituting } 6 + 2y \text{ for } x$$

Remember to use parentheses when you substitute.

$$18 + 6y + 2y = 4 \qquad\qquad \text{Using the distributive law}$$
$$18 + 8y = 4 \qquad\qquad \text{Combining like terms}$$
$$8y = -14 \qquad\qquad \text{Subtracting 18 from both sides}$$
$$y = \frac{-14}{8} = -\frac{7}{4}. \qquad \text{Dividing both sides by 8}$$

To find x, we can substitute $-\frac{7}{4}$ for y in equation (1), (2), or (3). Because it is generally easier to use an equation that has already been solved for a specific variable, we decide to use equation (3):

$$x = 6 + 2y = 6 + 2\left(-\frac{7}{4}\right) = 6 - \frac{7}{2} = \frac{12}{2} - \frac{7}{2} = \frac{5}{2}.$$

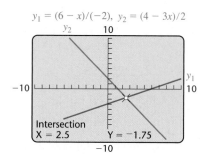

$y_1 = (6 - x)/(-2), \ y_2 = (4 - 3x)/2$

We check the ordered pair $\left(\frac{5}{2}, -\frac{7}{4}\right)$.

Check:

$$
\begin{array}{c|c}
x - 2y = 6 & \\
\hline
\frac{5}{2} - 2\left(-\frac{7}{4}\right) & 6 \\
\frac{5}{2} + \frac{7}{2} & \\
\frac{12}{2} & \\
6 \overset{?}{=} 6 & \text{TRUE}
\end{array}
\qquad
\begin{array}{c|c}
3x + 2y = 4 & \\
\hline
3 \cdot \frac{5}{2} + 2\left(-\frac{7}{4}\right) & 4 \\
\frac{15}{2} - \frac{7}{2} & \\
\frac{8}{2} & \\
4 \overset{?}{=} 4 & \text{TRUE}
\end{array}
$$

To check using a graphing calculator, we must first solve each equation for y. When we do so, equation (1) becomes $y = (6 - x)/(-2)$ and equation (2) becomes $y = (4 - 3x)/2$. The graph at left confirms the solution. Since $\left(\frac{5}{2}, -\frac{7}{4}\right)$ checks, it is the solution.

Some systems have no solution and some have an infinite number of solutions.

EXAMPLE 4 Solve each system.

a) $y = \frac{5}{2}x + 4,$ (1)   **b)** $2y = 6x + 4,$ (1)

     $y = \frac{5}{2}x - 3$ (2) $y = 3x + 2$ (2)

SOLUTION

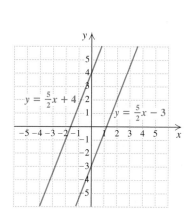

a) Since the lines have the same slope, $\frac{5}{2}$, and different y-intercepts, they are parallel. Let's see what happens if we try to solve this system by substituting $\frac{5}{2}x - 3$ for y in the first equation:

$$
\begin{aligned}
y &= \tfrac{5}{2}x + 4 && \text{Equation (1)} \\
\tfrac{5}{2}x - 3 &= \tfrac{5}{2}x + 4 && \text{Substituting } \tfrac{5}{2}x - 3 \text{ for } y \\
-3 &= 4. && \text{Subtracting } \tfrac{5}{2}x \text{ from both sides}
\end{aligned}
$$

When we subtract $\frac{5}{2}x$ from both sides, we obtain a *false* equation, or contradiction. In such a case, when solving algebraically leads to a false equation, we state that the system has no solution and thus is inconsistent.

b) If we use substitution to solve the system, we can substitute $3x + 2$ for y in equation (1):

$$
\begin{aligned}
2y &= 6x + 4 && \text{Equation (1)} \\
2(3x + 2) &= 6x + 4 && \text{Substituting } 3x + 2 \text{ for } y \\
6x + 4 &= 6x + 4.
\end{aligned}
$$

This last equation is true for *any* choice of x, indicating that for any choice of x, a solution can be found. When the algebraic solution of a system of two equations leads to an identity—that is, an equation that is true for all real numbers—any pair that is a solution of equation (1) is also a solution of equation (2). The equations are dependent and the solution set is infinite:

$$\{(x, y) \,|\, y = 3x + 2\}, \quad \text{or equivalently,} \quad \{(x, y) \,|\, 2y = 6x + 4\}.$$

> **To Solve a System Using Substitution**
>
> 1. Solve for a variable in either one of the equations if neither equation already has a variable isolated.
> 2. Using the result of step (1), substitute in the *other* equation for the variable isolated in step (1).
> 3. Solve the equation from step (2).
> 4. Substitute the solution from step (3) into one of the other equations to solve for the other variable.
> 5. Check that the ordered pair resulting from steps (3) and (4) checks in both of the original equations.

Problem Solving

Now let's use the substitution method in problem solving.

EXAMPLE 5 Supplementary Angles. Two angles are supplementary. One angle measures 30° more than twice the other. Find the measures of the two angles.

SOLUTION

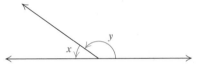

Supplementary angles

1. **Familiarize.** Recall that two angles are supplementary if the sum of their measures is 180°. We could try to guess a solution, but instead we make a drawing and translate. Let x and y represent the measures of the two angles.

2. **Translate.** Since we are told that the angles are supplementary, one equation is

$$x + y = 180. \qquad (1)$$

The second sentence can be rephrased and translated as follows:

Rewording: One angle is 30° more than two times the other.

Translating: y $=$ $2x + 30$ (2)

We now have a system of two equations in two unknowns:

$$x + y = 180, \qquad (1)$$
$$y = 2x + 30. \qquad (2)$$

3. **Carry out.** We substitute $2x + 30$ for y in equation (1):

$$x + y = 180 \qquad \text{Equation (1)}$$
$$x + (2x + 30) = 180 \qquad \text{Substituting}$$
$$3x + 30 = 180$$
$$3x = 150 \qquad \text{Subtracting 30 from both sides}$$
$$x = 50. \qquad \text{Dividing both sides by 3}$$

Substituting 50 for x in equation (1) then gives us

$$x + y = 180 \qquad \textbf{Equation (1)}$$
$$50 + y = 180 \qquad \textbf{Substituting 50 for } x$$
$$y = 130.$$

4. **Check.** If one angle is 50° and the other is 130°, then the sum of the measures is 180°. Thus the angles are supplementary. If 30° is added to twice the measure of the smaller angle, we have $2 \cdot 50° + 30°$, or 130°, which is the measure of the other angle. The numbers check.

5. **State.** One angle measures 50° and the other 130°.

4.2 EXERCISE SET

↪ *Concept Reinforcement* *Classify each statement as either true or false.*

1. The substitution method requires that we solve for the variables in the order in which they occur alphabetically.

2. The substitution method often requires us to first solve for a variable, much as we did when solving for a letter in a formula.

3. When solving a system of equations algebraically leads to a false equation, the system has no solution.

4. When solving a system of two equations algebraically leads to an equation that is always true, the system has an infinite number of solutions.

Solve each system using the substitution method. If a system has an infinite number of solutions, use set-builder notation to write the solution set. If a system has no solution, state this.

5. $x + y = 7,$
   $y = x + 3$

6. $x + y = 9,$
   $x = y + 1$

7. $x = y + 1,$
   $x + 2y = 4$

8. $y = x - 3,$
   $3x + y = 5$

9. $y = 2x - 5,$
   $3y - x = 5$

10. $y = 2x + 1,$
    $x + y = 4$

11. $3x + y = 2,$
    $y = -2x$

12. $r = -3s,$
    $r + 4s = 10$

13. $2x + 3y = 8,$
    $x = y - 6$

14. $x = y - 8,$
    $3x + 2y = 1$

15. $x = 2y + 1,$
    $3x - 6y = 2$

16. $y = 3x - 1,$
    $6x - 2y = 2$

17. $s + t = -4,$
    $s - t = 2$

18. $x - y = 6,$
    $x + y = -2$

19. $x - y = 5,$
    $x + 2y = 7$

20. $y - 2x = -6,$
    $2y - x = 5$

21. $x - 2y = 7,$
    $3x - 21 = 6y$

22. $x - 4y = 3,$
    $2x - 6 = 8y$

23. $y = 2x + 5,$
    $-2y = -4x - 10$

24. $y = -2x + 3,$
    $3y = -6x + 9$

25. $2x + 3y = -2,$
    $2x - y = 9$

26. $x + 2y = 10,$
    $3x + 4y = 8$

27. $x - y = -3,$
    $2x + 3y = -6$

28. $3a + 2b = 2,$
    $-2a + b = 8$

29. $r - 2s = 0,$
    $4r - 3s = 15$

30. $y - 2x = 0,$
    $3x + 7y = 17$

31. $x - 3y = 7,$
    $-4x + 12y = 28$

32. $8x + 2y = 6,$
    $y = 3 - 4x$

33. $x - 2y = 5,$
$2y - 3x = 1$

34. $x - 3y = -1,$
$5y - 2x = 4$

Aha! **35.** $2x - y = 0,$
$2x - y = -2$

36. $5x = y - 3,$
$5x = y + 5$

Solve using a system of equations.

37. The sum of two numbers is 83. One number is 5 more than the other. Find the numbers.

38. The sum of two numbers is 76. One number is 2 more than the other. Find the numbers.

39. Find two numbers for which the sum is 93 and the difference is 9.

40. Find two numbers for which the sum is 76 and the difference is 12.

41. The difference between two numbers is 16. Three times the larger number is seven times the smaller. What are the numbers?

42. The difference between two numbers is 18. Twice the smaller number plus three times the larger is 74. What are the numbers?

43. *Supplementary Angles.*   Two angles are supplementary. One angle is 15° more than twice the other. Find the measure of each angle.

44. *Supplementary Angles.*   Two angles are supplementary. One angle is 8° less than three times the other. Find the measure of each angle.

45. *Complementary Angles.*   Two angles are complementary. Their difference is 18°. Find the measure of each angle. (*Complementary angles* are angles for which the sum is 90°.)

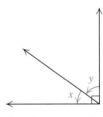
Complementary angles

46. *Complementary Angles.*   Two angles are complementary. One angle is 42° more than one-half the other. Find the measure of each angle.

47. *Two-by-Four.*   The perimeter of a cross section of a "two-by-four" piece of lumber is $10\frac{1}{2}$ in. The length is twice the width. Find the actual dimensions of the cross section of a two-by-four.

Two-by-four

$P = 10\frac{1}{2}$ in.

48. *Standard Billboard.*   A standard rectangular highway billboard has a perimeter of 124 ft. The length is 34 ft more than the width. (*Source*: Eller Sign Company) Find the length and the width.

49. *Dimensions of Colorado.*   The state of Colorado is roughly in the shape of a rectangle whose perimeter is 1300 mi. The width is 110 mi less than the length. Find the length and the width.

50. *Dimensions of Wyoming.*   The state of Wyoming is a rectangle with a perimeter of 1280 mi. The width is 90 mi less than the length. Find the length and the width.

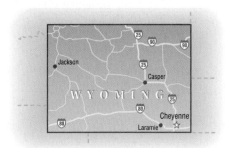

51. *Soccer.* The perimeter of a soccer field is 280 yd. The width is 5 more than half the length. Find the length and the width.

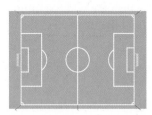

52. *Racquetball.* A regulation racquetball court has a perimeter of 120 ft, with a length that is twice the width. Find the length and the width of a court.

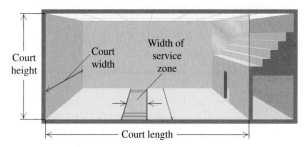

53. *Racquetball.* The height of the front wall of a standard racquetball court is four times the width of the service zone (see the figure). Together, these measurements total 25 ft. Find the height and the width.

54. *Lacrosse.* The perimeter of a lacrosse field is 340 yd. The length is 10 yd less than twice the width. Find the length and the width.

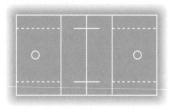

TW 55. Joel solves every system of two equations (in x and y) by first solving for y in the first equation and then substituting into the second equation. Is he using the best approach? Why or why not?

TW 56. Describe two advantages of the substitution method over the graphing method for solving systems of equations.

Skill Maintenance

Simplify. [1.8]

57. $2(5x + 3y) - 3(5x + 3y)$

58. $5(2x + 3y) - 3(7x + 5y)$

59. $4(5x + 6y) - 5(4x + 7y)$

60. $2(7x + 5 - 3y) - 7(2x + 5)$

61. $2(5x - 3y) - 5(2x + y)$

62. $4(2x + 3y) + 3(5x - 4y)$

Synthesis

TW 63. Janine can tell by inspection that the system

$$x = 2y - 1,$$
$$x = 2y + 3$$

has no solution. How can she tell?

TW 64. Under what circumstances can a system of equations be solved more easily by graphing than by substitution?

Solve by the substitution method.

65. $\dfrac{1}{6}(a + b) = 1,$
$\dfrac{1}{4}(a - b) = 2$

66. $\dfrac{x}{5} - \dfrac{y}{2} = 3,$
$\dfrac{x}{4} + \dfrac{3y}{4} = 1$

67. $y + 5.97 = 2.35x,$
$2.14y - x = 4.88$

68. $a + 4.2b = 25.1,$
$9a - 1.8b = 39.78$

69. *Age at Marriage.* Trudy is 20 yr younger than Dennis. She feels that she needs to be 7 more than half of Dennis's age before they can marry. What is the youngest age at which Trudy can marry Dennis and honor this requirement?

Exercises 70 and 71 contain systems of three equations in three variables. A solution is an ordered triple of the form (x, y, z). Use the substitution method to solve.

70. $x + y + z = 4,$
$x - 2y - z = 1,$
$y = -1$

71. $x + y + z = 180,$
$x = z - 70,$
$2y - z = 0$

72. *Softball.*   The perimeter of a softball diamond is two-thirds of the perimeter of a baseball diamond. Together, the two perimeters measure 200 yd. Find the distance between the bases in each sport.

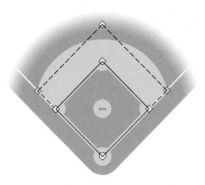

TW 73. Solve Example 3 by first solving for $2y$ in equation (1) and then substituting for $2y$ in equation (2). Is this method easier than the procedure used in Example 3? Why or why not?

74. Write a system of two linear equations that can be solved more quickly—but still precisely—by a graphing calculator than by substitution. Time yourself using both methods to solve the system.

4.3   Systems of Equations and Elimination

Solving by the Elimination Method   ■   Problem Solving

We have seen that graphing is not always a precise method of solving a system of equations, especially when fraction solutions are involved. The substitution method, considered in Section 4.2, is precise but sometimes difficult to use. For example, to solve the system

$$2x + 3y = 13, \quad (1)$$
$$4x - 3y = 17 \quad (2)$$

by substitution, we would need to first solve for a variable in one of the equations. Were we to solve equation (1) for y, we would find (after several steps) that $y = \frac{13}{3} - \frac{2}{3}x$. We could then use the expression $\frac{13}{3} - \frac{2}{3}x$ in equation (2) as a replacement for y:

$$4x - 3\left(\tfrac{13}{3} - \tfrac{2}{3}x\right) = 17.$$

As you will see, although substitution *could* be used to solve this system, another method, *elimination*, is simpler to use on problems like this.

Solving by the Elimination Method

The **elimination method** for solving systems of equations makes use of the addition principle. To see how it works, we use it to solve the system above.

EXAMPLE 1 Solve the system

$$2x + 3y = 13, \quad (1)$$
$$4x - 3y = 17. \quad (2)$$

SOLUTION According to equation (2), $4x - 3y$ and 17 are the same number. Thus we can add $4x - 3y$ to the left side of equation (1) and 17 to the right side:

$$2x + 3y = 13 \quad (1)$$
$$\underline{4x - 3y = 17} \quad (2)$$
$$6x + 0y = 30. \quad \text{Adding. Note that } y \text{ has been "eliminated."}$$

The resulting equation has just one variable:

$$6x = 30.$$

Dividing both sides of this equation by 6, we find that $x = 5$.

Next, we substitute 5 for x in either of the original equations:

$$2x + 3y = 13 \quad \text{Equation (1)}$$
$$2 \cdot 5 + 3y = 13 \quad \text{Substituting 5 for } x$$
$$10 + 3y = 13$$
$$3y = 3$$
$$y = 1. \quad \text{Solving for } y$$

We check the ordered pair $(5, 1)$. The graph shown at left also serves as a check.

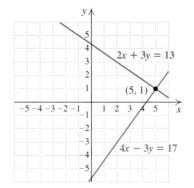

Check:

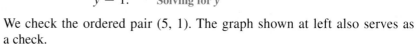

$$\begin{array}{c|c} 2x + 3y = 13 \\ \hline 2(5) + 3(1) & 13 \\ 10 + 3 & \\ & 13 \overset{?}{=} 13 \quad \text{TRUE} \end{array} \qquad \begin{array}{c|c} 4x - 3y = 17 \\ \hline 4(5) - 3(1) & 17 \\ 20 - 3 & \\ & 17 \overset{?}{=} 17 \quad \text{TRUE} \end{array}$$

Since $(5, 1)$ checks in both equations, it is the solution.

The system in Example 1 is easier to solve by elimination than by substitution because two terms, $-3y$ in equation (2) and $3y$ in equation (1), are opposites. Most systems have no pair of terms that are opposites. When this occurs, we can multiply one or both of the equations by appropriate numbers to create a pair of terms that are opposites.

EXAMPLE 2 Solve:

$$2x + 3y = 8, \quad (1)$$
$$x + 3y = 7. \quad (2)$$

SOLUTION Adding these equations as they now appear will not eliminate a variable. However, if the $3y$ were $-3y$ in one equation, we could eliminate y. We multiply both sides of equation (2) by -1 to find an equivalent equation that contains $-3y$, and then add:

$$
\begin{array}{ll}
2x + 3y = 8 & \text{Equation (1)} \\
\underline{-x - 3y = -7} & \text{Multiplying both sides of equation (2) by } -1 \\
x = 1. & \text{Adding}
\end{array}
$$

Next, we substitute 1 for x in either of the original equations:

$$
\begin{array}{ll}
x + 3y = 7 & \text{Equation (2)} \\
1 + 3y = 7 & \text{Substituting 1 for } x \\
\left.\begin{array}{l} 3y = 6 \\ y = 2. \end{array}\right\} & \text{Solving for } y
\end{array}
$$

We can check the ordered pair $(1, 2)$. The graph shown at left is also a check.

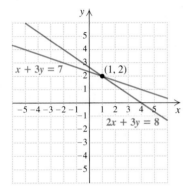

Check:

$$
\begin{array}{c|c}
\multicolumn{2}{c}{2x + 3y = 8} \\
\hline
2 \cdot 1 + 3 \cdot 2 & 8 \\
2 + 6 & \\
& 8 \stackrel{?}{=} 8 \quad \text{TRUE}
\end{array}
\qquad
\begin{array}{c|c}
\multicolumn{2}{c}{x + 3y = 7} \\
\hline
1 + 3 \cdot 2 & 7 \\
1 + 6 & \\
& 7 \stackrel{?}{=} 7 \quad \text{TRUE}
\end{array}
$$

Since $(1, 2)$ checks in both equations, it is the solution. ◢

When deciding which variable to eliminate, we inspect the coefficients in both equations. If one coefficient is a multiple of the coefficient of the same variable in the other equation, that is the easiest variable to eliminate.

EXAMPLE 3 Solve:

$$
\begin{array}{ll}
3x + 6y = -6, & (1) \\
5x - 2y = 14. & (2)
\end{array}
$$

SOLUTION No terms are opposites, but if both sides of equation (2) are multiplied by 3 (or if both sides of equation (1) are multiplied by $\frac{1}{3}$), the coefficients of y will be opposites. Note that 6 is the LCM of 2 and 6:

$$
\begin{array}{ll}
3x + 6y = -6 & \text{Equation (1)} \\
\underline{15x - 6y = 42} & \text{Multiplying both sides of equation (2) by 3} \\
18x = 36 & \text{Adding} \\
x = 2. & \text{Solving for } x
\end{array}
$$

We then substitute 2 for x in either equation (1) or equation (2):

$$
\begin{array}{ll}
3 \cdot 2 + 6y = -6 & \text{Substituting 2 for } x \text{ in equation (1)} \\
\left.\begin{array}{r} 6 + 6y = -6 \\ 6y = -12 \\ y = -2. \end{array}\right\} & \text{Solving for } y
\end{array}
$$

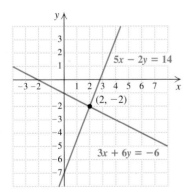

We leave it to the student to confirm that $(2, -2)$ checks and is the solution. The graph at left also serves as a check. ◢

Sometimes both equations must be multiplied to find the least common multiple of two coefficients.

▸ **EXAMPLE 4** Solve:

$$3y + 1 + 2x = 0, \qquad (1)$$
$$5x = 7 - 4y. \qquad (2)$$

SOLUTION It is often helpful to write both equations in the form $Ax + By = C$ before attempting to eliminate a variable:

$$2x + 3y = -1, \qquad (3) \qquad$$ Subtracting 1 from both sides and rearranging the terms of the first equation

$$5x + 4y = 7. \qquad (4) \qquad$$ Adding $4y$ to both sides of equation (2)

Since neither coefficient of x is a multiple of the other and neither coefficient of y is a multiple of the other, we use the multiplication principle with *both* equations. Note that we can eliminate the x-term by multiplying both sides of equation (3) by 5 and both sides of equation (4) by -2:

$$\begin{array}{ll} 10x + 15y = -5 & \text{Multiplying both sides of equation (3)} \\ & \text{by 5} \\ \underline{-10x - 8y = -14} & \text{Multiplying both sides of equation (4)} \\ & \text{by } -2 \\ 7y = -19 & \text{Adding} \\ y = \frac{-19}{7} = -\frac{19}{7}. & \text{Dividing by 7} \end{array}$$

We substitute $-\frac{19}{7}$ for y in equation (3):

$$\begin{array}{ll} 2x + 3y = -1 & \text{Equation (3)} \\ 2x + 3\left(-\frac{19}{7}\right) = -1 & \text{Substituting } -\frac{19}{7} \text{ for } y \\ 2x - \frac{57}{7} = -1 & \\ 2x = -1 + \frac{57}{7} & \text{Adding } \frac{57}{7} \text{ to both sides} \\ 2x = -\frac{7}{7} + \frac{57}{7} = \frac{50}{7} & \\ x = \frac{50}{7} \cdot \frac{1}{2} = \frac{25}{7}. & \text{Solving for } x \end{array}$$

We check the ordered pair $\left(\frac{25}{7}, -\frac{19}{7}\right)$.

Student Notes

Before diving into a problem, spend some time deciding which variable to eliminate. Always try to select the approach that will minimize your calculations.

Check:

$$\begin{array}{c|c} 3y + 1 + 2x = 0 \\ \hline 3\left(-\frac{19}{7}\right) + 1 + 2 \cdot \frac{25}{7} & 0 \\ -\frac{57}{7} + \frac{7}{7} + \frac{50}{7} & \\ & 0 \overset{?}{=} 0 \quad \text{TRUE} \end{array} \qquad \begin{array}{c|c} 5x = 7 - 4y \\ \hline 5 \cdot \frac{25}{7} & 7 - 4\left(-\frac{19}{7}\right) \\ \frac{125}{7} & \frac{49}{7} + \frac{76}{7} \\ \frac{125}{7} \overset{?}{=} \frac{125}{7} & \text{TRUE} \end{array}$$

The solution is $\left(\frac{25}{7}, -\frac{19}{7}\right)$. ◢

Next, we consider a system with no solution and see what happens when the elimination method is used.

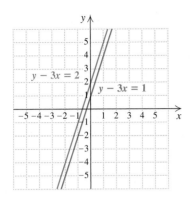

EXAMPLE 5   Solve:

$$y - 3x = 2, \qquad (1)$$
$$y - 3x = 1. \qquad (2)$$

SOLUTION   To eliminate y, we multiply both sides of equation (2) by -1. Then we add:

$$y - 3x = 2$$
$$\underline{-y + 3x = -1} \qquad \text{Multiplying both sides of equation (2) by } -1$$
$$ 0 = 1. \qquad \text{Adding}$$

Note that in eliminating y, we eliminated x as well. The resulting equation, $0 = 1$, is false for any pair (x, y), so there is *no solution*.

Sometimes there is an infinite number of solutions.

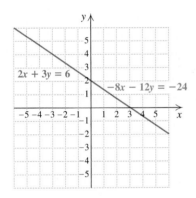

EXAMPLE 6   Solve:

$$2x + 3y = 6, \qquad (1)$$
$$-8x - 12y = -24. \qquad (2)$$

SOLUTION   To eliminate x, we multiply both sides of equation (1) by 4 and then add the two equations:

$$8x + 12y = 24 \qquad \text{Multiplying both sides of equation (1) by 4}$$
$$\underline{-8x - 12y = -24}$$
$$ 0 = 0. \qquad \text{Adding}$$

Again, we have eliminated *both* variables. The resulting equation, $0 = 0$, is always true, indicating that the equations are dependent. Such a system has an infinite solution set:

$$\{(x, y) \mid 2x + 3y = 6\}, \quad \text{or equivalently,} \quad \{(x, y) \mid -8x - 12y = -24\}.$$

When decimals or fractions appear, we can first multiply to clear them. Then we proceed as before.

EXAMPLE 7   Solve:

$$\tfrac{1}{2}x + \tfrac{3}{4}y = 2, \qquad (1)$$
$$x + 3y = 7. \qquad (2)$$

SOLUTION   The number 4 is the LCD for equation (1). Thus we multiply both sides of equation (1) by 4 to clear fractions:

$$4\left(\tfrac{1}{2}x + \tfrac{3}{4}y\right) = 4 \cdot 2 \qquad \text{Multiplying both sides of equation (1) by 4}$$
$$4 \cdot \tfrac{1}{2}x + 4 \cdot \tfrac{3}{4}y = 8 \qquad \text{Using the distributive law}$$
$$2x + 3y = 8.$$

The resulting system is

$$2x + 3y = 8, \qquad \text{This equation is equivalent to equation (1).}$$
$$x + 3y = 7.$$

As we saw in Example 2, the solution of this system is $(1, 2)$.

Problem Solving

We now use the elimination method to solve a problem.

EXAMPLE 8   Phone Cards.   A calling-card company offers two prepaid cards for domestic calls, the Liberty Prepaid Card and the USA Calling Card. The Liberty Card has a 50¢ connection fee per call and a 1¢ per minute rate. The USA Card has a 13.6¢ connection fee per call and a 2.2¢ per minute rate. (*Source*: www.PrepaidPhoneCardsUSA.com) For what length of call will the costs be the same?

SOLUTION

1. **Familiarize.** To become familiar with the problem, we make and check a guess of 20 min. A 20-min Liberty call would cost $50 + 1 \cdot 20$, or 70¢. Using the USA card, a 20-min call would cost $13.6 + 2.2 (20)$, or 57.6¢. Since $70 \neq 57.6$, our guess is incorrect. However, from the check, we can see how equations can be written to model the situation. We let $m =$ the length of a phone call, in minutes, and $c =$ the cost of the call, in cents.

2. **Translate.** We reword the problem and translate as follows:

Rewording:	Liberty's cost	is	50¢	plus	1¢	times	the length of the call, in minutes.
Translating:	c	$=$	50	$+$	1	$\cdot$	m

   Note that cents are used as the monetary unit. We write a second equation representing the cost of a call on the USA card:

Rewording:	USA's cost	is	13.6¢	plus	2.2¢	times	the length of the call, in minutes.
Translating:	c	$=$	13.6	$+$	2.2	$\cdot$	m

   We now have the system of equations

   $$c = 50 + m,$$
   $$c = 13.6 + 2.2m.$$

3. **Carry out.** To solve the system, we multiply the second equation by -1 and add to eliminate c:

   $$\begin{array}{rcl} c = & 50 & + & m \\ -c = & -13.6 & - & 2.2m \\ \hline 0 = & 36.4 & - & 1.2m. \end{array}$$

We can now solve for m:

$$1.2m = 36.4$$
$$m = 30.\overline{3}$$
$$m \approx 30.$$

4. **Check.** For 30 min, the cost of a Liberty Card call would be

$$50 + 1 \cdot 30, \quad \text{or} \quad 50 + 30¢, \quad \text{or} \quad 80¢,$$

and the cost of a USA Card call would be

$$13.6 + 2.2 \cdot 30, \quad \text{or} \quad 13.6 + 66¢, \quad \text{or} \quad 80¢. \qquad \text{Rounding to the nearest cent}$$

Thus the costs are the same for a 30-min call.

5. **State.** For a 30-min call, the costs are the same.

4.3 EXERCISE SET

FOR EXTRA HELP

 MathXL

MyMathLab

 InterAct Math

 Tutor Center AW Math Tutor Center

 Video Lectures on CD: Disc 2

Student's Solutions Manual

Concept Reinforcement   Classify each statement as either true or false.

1. The elimination method is never easier to use than the substitution method.

2. The elimination method works especially well when the coefficients of one variable are opposites of each other.

3. When the elimination method yields an equation, like $0 = 0$, that is always true, there is an infinite number of solutions.

4. When the elimination method yields an equation, like $0 = 3$, that is never true, the system has no solution.

Solve using the elimination method. If a system has an infinite number of solutions, use set-builder notation to write the solution set. If a system has no solution, state this.

5. $x - y = 6,$
$x + y = 12$

6. $x + y = 3,$
$x - y = 7$

7. $x + y = 6,$
$-x + 3y = -2$

8. $x + y = 6,$
$-x + 2y = 15$

9. $4x - y = 1,$
$3x + y = 13$

10. $3x - y = 9,$
$2x + y = 6$

11. $5a + 4b = 7,$
$-5a + b = 8$

12. $7c + 4d = 16,$
$c - 4d = -4$

13. $8x - 5y = -9,$
$3x + 5y = -2$

14. $3a - 3b = -15,$
$-3a - 3b = -3$

15. $3a - 6b = 8,$
$-3a + 6b = -8$

16. $8x + 3y = 4,$
$-8x - 3y = -4$

17. $-x - y = 8,$
$2x - y = -1$

18. $x + y = -7,$
$3x + y = -9$

19. $x + 3y = 19,$
$x - y = -1$

20. $3x - y = 8,$
$x + 2y = 5$

21. $x + y = 5,$
$4x - 3y = 13$

22. $x - y = 7,$
$3x - 5y = 15$

23. $2w - 3z = -1,$
$-4w + 6z = 5$

24. $7p + 5q = 2,$
$8p - 9q = 17$

25. $2a + 3b = -1,$
$3a + 5b = -2$

26. $3x - 4y = 16,$
$5x + 6y = 14$

27. $3y = x,$
$5x + 14 = y$

28. $5a = 2b,$
$2a + 11 = 3b$

29. $4x - 10y = 13,$
$-2x + 5y = 8$

30. $2p + 5q = 9,$
$3p - 2q = 4$

31. $8n + 6 - 3m = 0,$
$32 = m - n$

32. $6x - 8 + y = 0,$
$11 = 3y - 8x$

33. $3x + 5y = 4,$
$-2x + 3y = 10$

34. $2x + y = 13,$
$4x + 2y = 23$

35. $0.06x + 0.05y = 0.07,$
$0.4x - 0.3y = 1.1$

36. $x - \frac{3}{2}y = 13,$
$\frac{3}{2}x - y = 17$

37. $x + \frac{9}{2}y = \frac{15}{4},$
$\frac{9}{10}x - y = \frac{9}{20}$

38. $1.8x - 2y = 0.9,$
$0.04x + 0.18y = 0.15$

Solve.

39. *Local Truck Rentals.*   Budget rents a 15-ft truck for $39.95 plus 79¢ per mile. Penske rents a 15-ft van for $49.95 plus 59¢ per mile. (*Sources*: Budget Truck Rental; Penske Truck Leasing) For what mileage is the cost the same?

40. *Local Truck Rentals.*   U-Haul rents a 14-ft truck for $29.95 plus 79¢ per mile. Penske rents a 15-ft van for $49.95 plus 59¢ per mile. (*Sources*: Penske Truck Leasing; U-Haul International) For what mileage is the cost the same?

41. *Complementary Angles.*   Two angles are complementary. One angle is 12° more than twice the other. Find the measure of each angle.

42. *Complementary Angles.*   Two angles are complementary. Their difference is 26°. Find the measure of each angle.

43. *Phone Rates.*   Recently, AT&T offered two long-distance calling plans. The One-Rate® 7¢ Plus Plan costs $4.95 per month plus 7¢ a minute. Another plan has no monthly fee, but costs 10¢ a minute. For what number of minutes will the two plans cost the same?

44. *Phone Rates.*   Recently, MCI offered one calling plan that charges 45¢ a minute for calling-card calls. Sprint offered a plan charging 25¢ a minute for calling-card calls, but costs an additional $1 per month. For what number of minutes will the two plans cost the same?

45. *Supplementary Angles.*   Two angles are supplementary. One angle measures 5° less than four times the measure of the other. Find the measure of each angle.

46. *Supplementary Angles.*   Two angles are supplementary. One angle measures 45° more than twice the measure of the other. Find the measure of each angle.

47. *Baking.*   Maple Branch Bakers sells 175 loaves of bread each day—some white and the rest whole-wheat. Because of a regular order from a local sandwich shop, Maple Branch consistently bakes 9 more loaves of white bread than whole-wheat. How many loaves of each type of bread do they bake?

48. *Planting Grapes.*   South Wind Vineyards uses 820 acres to plant Chardonnay and Riesling grapes. The vintner knows the profits will be greatest by planting 140 more acres of Chardonnay than Riesling. How many acres of each type of grape should be planted?

49. *Framing.*   Angel has 18 ft of molding from which he needs to make a rectangular frame. Because of the dimensions of the mirror being framed, the frame must be twice as long as it is wide. What should the dimensions of the frame be?

50. *Gardening.*   Patrice has 108 ft of fencing for a rectangular garden. If the garden's length is to be $1\frac{1}{2}$ times its width, what should the garden's dimensions be?

TW 51. Describe a method that could be used for writing a system that contains dependent equations.

TW 52. Describe a method that could be used for writing an inconsistent system of equations.

Focused Review

Solve using the indicated method. If a system has an infinite number of solutions, use set-builder notation to write the solution set. If a system has no solution, state this.

Solve by graphing. [4.1]

53. $y = \frac{1}{2}x + 1,$
$\quad x - y = 1$

54. $2x - 3y = 6,$
$\quad y = x - 3$

Solve using substitution. [4.2]

55. $2x - 3y = 7,$
$\quad y = x + 4$

56. $2x - y = 1,$
$\quad x - 2y = 3$

Solve using elimination. [4.3]

57. $3x + 2y = 5,$
$\quad 5x - 2y = 3$

58. $x + 3y = 7,$
$\quad 4x - y = 2$

Solve using any method. [4.1], [4.2], [4.3]

59. $x = y + 1,$
$\quad y = x + 1$

60. $\frac{1}{2}x - \frac{1}{4}y = 1,$
$\quad y + 4 = 2x$

61. $4x - 5y = 10,$
$\quad 3x + 5y = 4$

62. $y = x - 7,$
$\quad 2x + y = 3$

Synthesis

TW 63. If a system has an infinite number of solutions, does it follow that *any* ordered pair is a solution? Why or why not?

TW 64. Explain how the multiplication and addition principles are used in this section. Then count the number of times that these principles are used in Example 4.

Solve using substitution, elimination, or graphing.

65. $x + y = 7,$
$\quad 3(y - x) = 9$

66. $y = 3x + 4,$
$\quad 3 + y = 2(y - x)$

67. $2(5a - 5b) = 10,$
$\quad -5(2a + 6b) = 10$

68. $0.05x + y = 4,$
$\quad \dfrac{x}{2} + \dfrac{y}{3} = 1\dfrac{1}{3}$

Aha! 69. $y = -\frac{2}{7}x + 3,$
$\quad y = \frac{4}{5}x + 3$

70. $y = \frac{2}{5}x - 7,$
$\quad y = \frac{2}{5}x + 4$

Solve for x and y.

71. $y = ax + b,$
$\quad y = x + c$

72. $ax + by + c = 0,$
$\quad ax + cy + b = 0$

73. *Caged Rabbits and Pheasants.* Several ancient Chinese books included problems that can be solved by translating to systems of equations. *Arithmetical Rules in Nine Sections* is a book of 246 problems compiled by a Chinese mathematician, Chang Tsang, who died in 152 B.C. One of the problems is: Suppose there are a number of rabbits and pheasants confined in a cage. In all, there are 35 heads and 94 feet. How many rabbits and how many pheasants are there? Solve the problem.

74. *Age.* Patrick's age is 20% of his mother's age. Twenty years from now, Patrick's age will be 52% of his mother's age. How old are Patrick and his mother now?

75. *Age.* If 5 is added to a man's age and the total is divided by 5, the result will be his daughter's age. Five years ago, the man's age was eight times his daughter's age. Find their present ages.

76. *Dimensions of a Triangle.* When the base of a triangle is increased by 1 ft and the height is increased by 2 ft, the height changes from being two-thirds of the base to being four-fifths of the base. Find the original dimensions of the triangle.

4.4 More Applications Using Systems

Total-Value and Mixture Problems ▪ Motion Problems

Connecting the Concepts

We now have three distinctly different ways to solve a system. Each method has certain strengths and weaknesses.

Method	Strengths	Weaknesses
Graphical	Solutions are displayed visually. Works with any system that can be graphed.	Inexact when solutions involve numbers that are not integers or are very large and off the graph.
Substitution	Always yields exact solutions. Easy to use when a variable is alone on one side of an equation.	Introduces extensive computations with fractions when solving more complicated systems. Solutions are not graphically displayed.
Elimination	Always yields exact solutions. Easy to use when fractions or decimals appear in the system. The preferred method for systems of 3 or more equations in 3 or more variables (see Section 9.1)	Solutions are not graphically displayed.

When selecting the best method to use for a particular system, consider the strengths and weaknesses listed above. As you gain experience with these methods, it will become easier to choose the best method for any given system.

The five steps for problem solving and the methods for solving systems of equations can be used in a variety of applications.

EXAMPLE 1 Endangered Species. The number of species listed as threatened or endangered has more than tripled in the past 20 yr. In 2005, there were 749 species of plants considered threatened or endangered. The number considered threatened, or likely to become endangered in the foreseeable future, was 1 less than one-fourth of the number considered endangered, or in danger of becoming extinct. (*Source*: U.S. Fish and Wildlife Service) How many plant species were considered endangered and how many were considered threatened in 2005?

SOLUTION

1. **Familiarize.** Often statements of problems contain information that has no bearing on the question asked. In this case, the fact that the number of threatened or endangered species has tripled in the past 20 yr does not help us solve the problem. Instead, we need to focus on the number of endangered species and the number of threatened species in 2005. We let d represent the number of endangered plant species and t represent the number of threatened plant species in 2005.

Endangered Piper's Bellflowers. The endangered species Piper's bellflowers growing against a rock wall. Olympic National Park, Washington USA. © Darrell Gulin/CORBIS

2. **Translate.** There are two statements to translate. First, we look at the total number of endangered or threatened species of plants:

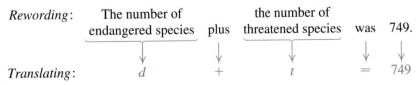

The second statement compares the two amounts, d and t:

Rewording: $\underbrace{\text{The number of threatened species}}$ was $\underbrace{\text{one less than one-fourth of the number of endangered species.}}$

Translating: t $=$ $\dfrac{1}{4}d - 1$

We have now translated the problem to the system of equations

$$d + t = 749,$$

$$t = \frac{1}{4}d - 1.$$

3. Carry out. We solve the system of equations both algebraically and graphically.

ALGEBRAIC APPROACH

Since one equation already has a variable isolated, let's use the substitution method:

$$d + t = 749$$

$$d + \tfrac{1}{4}d - 1 = 749 \qquad \text{Substituting } \tfrac{1}{4}d - 1 \text{ for } t$$

$$\tfrac{5}{4}d - 1 = 749 \qquad \text{Combining like terms}$$

$$\tfrac{5}{4}d = 750 \qquad \text{Adding 1 to both sides}$$

$$d = \tfrac{4}{5} \cdot 750 \qquad \text{Multiplying both sides by } \tfrac{4}{5}$$

$$d = 600. \qquad \text{Simplifying}$$

Next, using either of the original equations, we substitute and solve for t:

$$t = \tfrac{1}{4} \cdot 600 - 1 = 150 - 1 = 149.$$

We have $d = 600$, $t = 149$.

GRAPHICAL APPROACH

We let

$$d = x \quad \text{and} \quad t = y$$

and substitute. Then we graph $y_1 = 749 - x$ and $y_2 = \tfrac{1}{4}x - 1$ and find the point of intersection of the graphs. Since $x + y = 749$, and only positive values of x and y make sense in this problem, we choose a viewing window of $[0, 750, 0, 750]$, with Xscl $= 50$ and Yscl $= 50$.

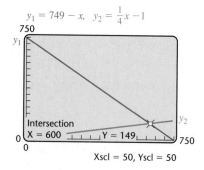

Xscl = 50, Yscl = 50

We have a solution of $(600, 149)$.

4. Check. The sum of 600 and 149 is 749, so the total number of species is correct. Since 1 less than one-fourth of 600 is $150 - 1$, or 149, the numbers check.

5. State. In 2005, there were 600 plant species considered endangered and 149 considered threatened.

Total-Value and Mixture Problems

EXAMPLE 2 Purchasing. Recently the Woods County Art Center purchased 120 stamps for $35.55. If the stamps were a combination of 24¢ postcard stamps and 39¢ first-class stamps, how many of each type were bought?

SOLUTION

1. **Familiarize.** To familiarize ourselves with this problem, let's guess that the art center bought 60 stamps at 24¢ each and 60 stamps at 39¢ each. The total cost would then be

$$60 \cdot \$0.24 + 60 \cdot \$0.39 = \$14.40 + \$23.40, \text{ or } \$37.80.$$

Since $\$37.80 \neq \$35.55$, our guess is incorrect. Rather than guess again, let's see how algebra can be used to translate the problem.

2. **Translate.** We let $p =$ the number of postcard stamps and $f =$ the number of first-class stamps. The information can be organized in a table, which will help with the translating.

Type of Stamp	Postcard	First-class	Total	
Number Sold	p	f	120	→ $p + f = 120$
Price	$0.24	$0.39		
Amount	$0.24p	$0.39f	$35.55	→ $0.24p + 0.39f = 35.55$

The first row of the table and the first sentence of the problem indicate that a total of 120 stamps were bought:

$$p + f = 120.$$

Since each postcard stamp cost $0.24 and p stamps were bought, $0.24p$ represents the amount paid, in dollars, for the postcard stamps. Similarly, $0.39f$ represents the amount paid, in dollars, for the first-class stamps. This leads to a second equation:

$$0.24p + 0.39f = 35.55.$$

Multiplying both sides by 100, we can clear the decimals. This gives the following system of equations as the translation:

$$p + f = 120,$$
$$24p + 39f = 3555.$$

3. **Carry out.** We are to solve the system of equations

$$p + f = 120, \qquad (1)$$
$$24p + 39f = 3555, \qquad (2) \qquad \text{**Working in cents rather than dollars**}$$

where p is the number of postcard stamps bought and f is the number of first-class stamps bought.

ALGEBRAIC APPROACH

Because both equations are in the form $Ax + By = C$, let's use the elimination method to solve the system. We can eliminate p by multiplying both sides of equation (1) by -24 and adding them to the corresponding sides of equation (2):

$$-24p - 24f = -2880 \qquad \text{Multiplying both sides of equation (1) by } -24$$

$$\underline{24p + 39f = 3555}$$

$$15f = 675 \qquad \text{Adding}$$

$$f = 45. \qquad \text{Solving for } f$$

To find p, we substitute 45 for f in equation (1) and then solve for p:

$$p + f = 120 \qquad \text{Equation (1)}$$

$$p + 45 = 120 \qquad \text{Substituting 45 for } f$$

$$p = 75. \qquad \text{Solving for } p$$

We obtain $(45, 75)$, or $f = 45, p = 75$.

GRAPHICAL APPROACH

We replace f with x and p with y and solve for y:

$$y + x = 120 \qquad \text{Solving for } y \text{ in equation (1)}$$

$$y = 120 - x$$

and

$$24y + 39x = 3555 \qquad \text{Solving for } y \text{ in equation (2)}$$

$$24y = 3555 - 39x$$

$$y = (3555 - 39x)/24.$$

Since the number of each kind of stamp is between 0 and 120, an appropriate viewing window is $[0, 120, 0, 120]$.

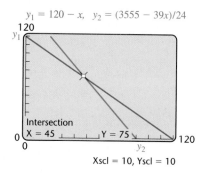

$$y_1 = 120 - x, \quad y_2 = (3555 - 39x)/24$$

Intersection
X = 45 Y = 75

Xscl = 10, Yscl = 10

The point of intersection is $(45, 75)$. Since f was replaced with x and p with y, we have $f = 45, p = 75$.

Student Notes

It is very important that you clearly label precisely what each variable represents. Not only will this assist you in writing equations, but it will help you to identify and state solutions.

4. Check. We check in the original problem. Recall that f is the number of first-class stamps and p the number of postcard stamps.

Number of stamps: $f + p = 45 + 75 = 120$

Cost of first-class stamps: $\$0.39f = 0.39 \times 45 = \$17.55$

Cost of postcard stamps: $\$0.24p = 0.24 \times 75 = \underline{\$18.00}$

Total $= \$35.55$

The numbers check.

5. State. The art center bought 45 first-class stamps and 75 postcard stamps.

Example 2 involved two types of items (first-class stamps and postcard stamps), the quantity of each type bought, and the total value of the items. We refer to this type of problem as a *total-value problem*.

EXAMPLE 3   Blending Teas.   Sonya's House of Tea sells loose Lapsang Souchong tea for 95¢ an ounce and Assam Gingia for $1.43 an ounce. Sonya wants to make a 20-oz mixture of the two types, called Dragon Blend, that sells for $1.10 an ounce. How much tea of each type should Sonya use?

• Assam Gingia $1.43 oz
• Lapsang Souchong .95 oz
• Dragon Blend   1.10 oz

SOLUTION

1. **Familiarize.** This problem is similar to Example 2. Rather than postcard stamps and first-class stamps, we have ounces of Assam Gingia and ounces of Lapsang Souchong. Instead of a different price for each type of stamp, we have a different price per ounce for each type of tea. Finally, rather than knowing the total cost of the stamps, we know the weight and the price per ounce of the mixture. Thus we can find the total value of the blend by multiplying 20 ounces times $1.10, or 110¢ per ounce. Although we could make and check a guess, we proceed to let l = the number of ounces of Lapsang Souchong and a = the number of ounces of Assam Gingia.

2. **Translate.** Since a 20-oz batch is being made, we must have

$$l + a = 20.$$

To find a second equation, note that the total value of the 20-oz blend must match the combined value of the separate ingredients:

Rewording:	The value of the Lapsang Souchong	plus	the value of the Assam Gingia	is	the value of the Dragon Blend.
Translating:	$l \cdot 95$	$+$	$a \cdot 143$	$=$	$20 \cdot 110$

These equations can also be obtained from a table.

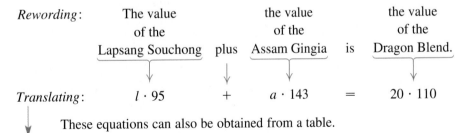

	Lapsang Souchong	**Assam Gingia**	**Dragon Blend**	
Number of Ounces	l	a	20	→ $l + a = 20$
Price per Ounce	95¢	143¢	110¢	
Value of Tea	$95l$	$143a$	$20 \cdot 110$, or 2200¢	→ $95l + 143a = 2200$

We have translated to a system of equations:

$$l + a = 20, \qquad (1)$$
$$95l + 143a = 2200. \qquad (2)$$

3. Carry out. We solve the system both algebraically and graphically.

ALGEBRAIC APPROACH

We can solve using substitution. When equation (1) is solved for l, we have $l = 20 - a$. Substituting $20 - a$ for l in equation (2), we find a:

$95(20 - a) + 143a = 2200$	Substituting
$1900 - 95a + 143a = 2200$	Using the distributive law
$48a = 300$	Combining like terms; subtracting 1900 from both sides
$a = 6.25.$	Dividing both sides by 48

We have $a = 6.25$ and, from equation (1) above, $l + a = 20$. Thus, $l = 13.75$.

GRAPHICAL APPROACH

We replace a with x and l with y and solve for y, which gives us the system of equations

$$y = 20 - x,$$
$$y = (2200 - 143x)/95.$$

We know from the problem that the number of ounces of each kind of tea is between 0 and 20, so we choose the viewing window [0, 20, 0, 20], graph the equations, and find the point of intersection.

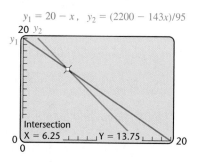

$y_1 = 20 - x, \quad y_2 = (2200 - 143x)/95$

Intersection
X = 6.25   Y = 13.75

The point of intersection is (6.25, 13.75). Since a was replaced with x and l with y, we have $a = 6.25$, $l = 13.75$.

4. Check. If 13.75 oz of Lapsang Souchong and 6.25 oz of Assam Gingia are combined, a 20-oz blend will result. The value of 13.75 oz of Lapsang Souchong is 13.75(95¢) or 1306.25¢. The value of 6.25 oz of Assam Gingia is 6.25(143¢), or 893.75¢. Thus the combined value of the blend is 1306.25¢ + 893.75¢, or 2200¢, which is $22. A 20-oz blend priced at $1.10 an ounce would also be worth $22, so our answer checks.

5. State. The Dragon Blend should be made by combining 13.75 oz of Lapsang Souchong with 6.25 oz of Assam Gingia.

EXAMPLE 4   Student Loans.   Ranjay's student loans totaled $9600. Part was a Perkins loan made at 5% interest and the rest was a Stafford Loan made at 8% interest. After one year, Ranjay's loans accumulated $633 in interest. What was the original amount of each loan?

SOLUTION

1. Familiarize. We begin with a guess. If $7000 was borrowed at 5% and $2600 was borrowed at 8%, the two loans would total $9600. The interest would then be 0.05($7000), or $350, and 0.08($2600), or $208, for a total of only $558 in interest. Our guess was wrong, but checking the guess familiarized us with the problem. More than $2600 was borrowed at the higher rate.

2. **Translate.** We let p = the amount of the Perkins loan and f = the amount of the Stafford loan. Next, we organize a table in which the entries in each column come from the formula for simple interest:

$Principal \cdot Rate \cdot Time = Interest.$

	Perkins Loan	Stafford Loan	Total	
Principal	p	f	$9600	$\longrightarrow p + f = 9600$
Rate of Interest	5%	8%		
Time	1 yr	1 yr		
Interest	0.05p	0.08f	$633	$\longrightarrow 0.05p + 0.08f = 633$

The total amount borrowed is found in the first row of the table:

$p + f = 9600.$

A second equation, representing the accumulated interest, can be found in the last row:

$0.05p + 0.08f = 633,$ or $5p + 8f = 63{,}300.$ **Clearing decimals**

3. **Carry out.** The system can be solved by elimination:

$$p + f = 9600.$$
$$5p + 8f = 63{,}300.$$

$\xrightarrow{\text{Multiplying both sides by } -5}$

$$-5p - 5f = -48{,}000$$
$$\underline{5p + 8f = 63{,}300}$$
$$3f = 15{,}300$$

$$p + f = 9600 \longleftarrow f = 5100$$
$$p + 5100 = 9600$$
$$p = 4500.$$

We find that $p = 4500$ and $f = 5100$.

4. **Check.** The total amount borrowed is $4500 + $5100, or $9600. The interest on $4500 at 5% for 1 yr is 0.05($4500), or $225. The interest on $5100 at 8% for 1 yr is 0.08($5100), or $408. The total amount of interest is $225 + $408, or $633, so the numbers check.

5. **State.** The Perkins loan was for $4500 and the Stafford loan was for $5100. ◢

Before proceeding to Example 5, briefly scan Examples 2–4 for similarities. Note that in each case, one of the equations in the system is a simple sum while the other equation represents a sum of products. Example 5 continues this pattern with what is commonly called a *mixture problem*.

Problem-Solving Tip

When solving a problem, see if it is patterned or modeled after a problem that you have already solved.

EXAMPLE 5 Mixing Fertilizers. Sky Meadow Gardening, Inc., carries two brands of fertilizer containing nitrogen and water. "Gently Green" is 5% nitrogen and "Sun Saver" is 15% nitrogen. Sky Meadow Gardening needs to combine the two types of solutions in order to make 90 L of a solution that is 12% nitrogen. How much of each brand should be used?

SOLUTION

1. **Familiarize.** We make a drawing and then make a guess to gain familiarity with the problem.

Suppose that 40 L of Gently Green and 50 L of Sun Saver are mixed. The resulting mixture will be the right size, 90 L, but will it be the right strength? To find out, note that 40 L of Gently Green would contribute $0.05(40) = 2$ L of nitrogen to the mixture while 50 L of Sun Saver would contribute $0.15(50) = 7.5$ L of nitrogen to the mixture. The total amount of nitrogen in the mixture would then be $2 + 7.5$, or 9.5 L. But we want 12% of 90, or 10.8 L, to be nitrogen. Thus our guess of 40 L and 50 L is incorrect. Still, checking our guess has familiarized us with the problem.

2. **Translate.** Let g = the number of liters of Gently Green and s = the number of liters of Sun Saver. The information can be organized in a table.

	Gently Green	Sun Saver	Mixture	
Number of Liters	g	s	90	→ $g + s = 90$
Percent of Nitrogen	5%	15%	12%	
Amount of Nitrogen	$0.05g$	$0.15s$	0.12×90, or 10.8 liters	→ $0.05g + 0.15s = 10.8$

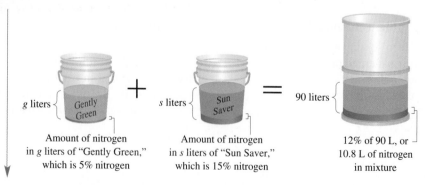

If we add g and s in the first row, we get one equation. It represents the total amount of mixture: $g + s = 90$.

If we add the amounts of nitrogen listed in the third row, we get a second equation. This equation represents the amount of nitrogen in the mixture: $0.05g + 0.15s = 10.8$.

After clearing decimals, we have translated the problem to the system

$$g + s = 90, \qquad (1)$$
$$5g + 15s = 1080. \qquad (2)$$

3. **Carry out.** We use the elimination method to solve the system:

$-5g - 5s = -450$	**Multiplying both sides of**
$\underline{5g + 15s = 1080}$	**equation (1) by -5**
$10s = 630$	**Adding**
$s = 63;$	**Solving for s**
$g + 63 = 90$	**Substituting into equation (1)**
$g = 27.$	**Solving for g**

4. **Check.** Remember, g is the number of liters of Gently Green and s is the number of liters of Sun Saver.

Total amount of mixture: $\qquad\qquad\qquad g + s = 27 + 63 = 90$

Total amount of nitrogen: $\quad$ 5% of 27 + 15% of 63 = 1.35 + 9.45 = 10.8

Percentage of nitrogen in mixture: $\quad \dfrac{\text{Total amount of nitrogen}}{\text{Total amount of mixture}} = \dfrac{10.8}{90} = 12\%$

The numbers check in the original problem. We can also check graphically, as shown at left, where x replaces g and y replaces s.

5. **State.** Sky Meadow Gardening should mix 27 L of Gently Green with 63 L of Sun Saver.

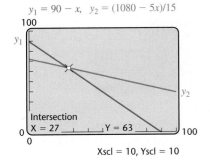

$y_1 = 90 - x, \; y_2 = (1080 - 5x)/15$

Intersection
$X = 27 \qquad Y = 63$
$Xscl = 10, \; Yscl = 10$

Motion Problems

When a problem deals with distance, speed (rate), and time, recall the following.

Distance, Rate, and Time Equations   If r represents rate, t represents time, and d represents distance, then

$$d = rt, \qquad r = \frac{d}{t}, \quad \text{and} \quad t = \frac{d}{r}.$$

Be sure to remember at least one of these equations. The others can be obtained by multiplying or dividing on both sides as needed.

EXAMPLE 6 Train Travel. A Vermont Railways freight train, loaded with logs, leaves Boston, heading to Washington D.C., at a speed of 60 km/h. Two hours later, an Amtrak® Metroliner leaves Boston, bound for Washington D.C., on a parallel track at 90 km/h. At what point will the Metroliner catch up to the freight train?

SOLUTION

1. **Familiarize.** Let's make a guess—say, 180 km—and check to see if it is correct. The freight train, traveling 60 km/h, would travel 180 km in $\frac{180}{60} = 3$ hr. The Metroliner, traveling 90 km/h, would travel 180 km in $\frac{180}{90} = 2$ hr. Since 3 hr is *not* two hours more than 2 hr, our guess of 180 km is incorrect. Although our guess is wrong, we see that the time that the trains are running and the point at which they meet are both unknown. We let t = the number of hours that the freight train is running before they meet and d = the distance at which the trains meet. Since the freight train has a 2-hr head start, the Metroliner runs for $t - 2$ hours before catching up to the freight train, at which point both trains have traveled the same distance.

2. **Translate.** We can organize the information in a chart. The formula *Distance = Rate · Time* guides our choice of rows and columns.

	Distance	**Rate**	**Time**	
Freight Train	d	60	t	→ $d = 60t$
Metroliner	d	90	$t - 2$	→ $d = 90(t - 2)$

Using *Distance = Rate · Time* twice, we get two equations:

$$d = 60t, \qquad (1)$$
$$d = 90(t - 2). \qquad (2)$$

3. **Carry out.** We solve the system both algebraically and graphically.

ALGEBRAIC APPROACH

We solve the system using the substitution method:

$$60t = 90(t - 2)$$ **Substituting $60t$ for d in equation (2)**

$$60t = 90t - 180$$ **Using the distributive law**

$$-30t = -180$$

$$t = 6.$$

The time for the freight train is 6 hr, which means that the time for the Metroliner is $6 - 2$, or 4 hr. Remember that it is distance, not time, that the problem asked for. Thus for $t = 6$, we have $d = 60 \cdot 6 = 360$ km.

GRAPHICAL APPROACH

The variables used in the equations are d and t. Note that in both equations, d is given in terms of t. If we replace d with y and t with x, we can enter the equations directly. Since $y = $ distance and $x = $ time, we use a viewing window of $[0, 10, 0, 500]$. We graph $y_1 = 60x$ and $y_2 = 90(x - 2)$ and find the point of intersection.

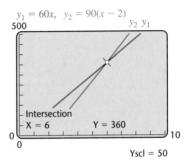

The point of intersection is $(6, 360)$. The problem asks for the distance that the trains travel. Recalling that $y = $ distance, we see that the distance that both trains travel is 360 km.

4. **Check.** At 60 km/h, the freight train will travel $60 \cdot 6$, or 360 km, in 6 hr. At 90 km/h, the Metroliner will travel $90 \cdot (6 - 2) = 360$ km in 4 hr. The numbers check.

5. **State.** The freight train will catch up to the Metroliner at a point 360 km from Boston.

EXAMPLE 7   Jet Travel.   A Boeing 747-400 jet flies 4 hr west with a 60-mph tailwind. Returning *against* the wind takes 5 hr. Find the speed of the plane with no wind.

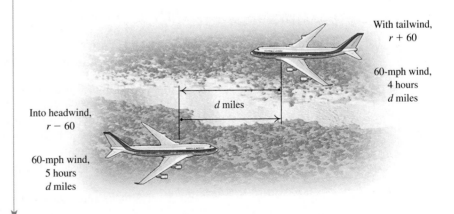

With tailwind,
$r + 60$

60-mph wind,
4 hours
d miles

Into headwind,
$r - 60$

60-mph wind,
5 hours
d miles

d miles

SOLUTION

1. **Familiarize.** We imagine the situation and make a drawing. Note that the wind *speeds up* the jet on the outbound flight, but *slows down* the jet on the return flight. Since the distances traveled each way must be the same, we can check a guess of the jet's speed with no wind. Suppose the speed of the jet with no wind is 400 mph. The jet would then fly $400 + 60 = 460$ mph with the wind and $400 - 60 = 340$ mph into the wind. In 4 hr, the jet would travel $460 \cdot 4 = 1840$ mi with the wind and $340 \cdot 5 = 1700$ mi against the wind. Since $1840 \neq 1700$, our guess of 400 mph is incorrect. Rather than guess again, let's have $r =$ the speed, in miles per hour, of the jet in still air. Then $r + 60 =$ the jet's speed with the wind and $r - 60 =$ the jet's speed against the wind. We also let $d =$ the distance traveled, in miles.

2. **Translate.** The information can be organized in a chart. The distances traveled are the same, so we use *Distance = Rate* (or *Speed*) · *Time*. Each row of the chart gives an equation.

	Distance	Rate	Time	
With Wind	d	$r + 60$	4	$\longrightarrow\ d = (r + 60)4$
Against Wind	d	$r - 60$	5	$\longrightarrow\ d = (r - 60)5$

The two equations constitute a system:

$$d = (r + 60)4, \qquad (1)$$
$$d = (r - 60)5. \qquad (2)$$

3. **Carry out.** We solve the system using substitution:

$$(r - 60)5 = (r + 60)4 \qquad \text{Substituting } (r - 60)5 \text{ for } d \text{ in equation (1)}$$
$$5r - 300 = 4r + 240 \qquad \text{Using the distributive law}$$
$$r = 540. \qquad \text{Solving for } r$$

4. **Check.** When $r = 540$, the speed with the wind is $540 + 60 = 600$ mph, and the speed against the wind is $540 - 60 = 480$ mph. The distance with the wind, $600 \cdot 4 = 2400$ mi, matches the distance into the wind, $480 \cdot 5 = 2400$ mi, so we have a check.

5. **State.** The speed of the jet with no wind is 540 mph.

Tips for Solving Motion Problems

1. Draw a diagram using an arrow or arrows to represent distance and the direction of each object in motion.
2. Organize the information in a chart.
3. Look for times, distances, or rates that are the same. These often can lead to an equation.
4. Translating to a system of equations allows for the use of two variables.
5. Always make sure that you have answered the question asked.

4.4 EXERCISE SET

1. *Scholastic Aptitude Test.* Many high-school students take the Scholastic Aptitude Test. Each student receives two scores, a *verbal* score and a *math* score. In 2004–2005, the average total score of students was 1028, with the average math score exceeding the average verbal score by 12 points. (*Source*: College Entrance Examination Board). What was the average verbal score and what was the average math score?

2. *Nontoxic Furniture Polish.* A nontoxic wood furniture polish can be made by mixing mineral (or olive) oil with vinegar (*Sources*: Based on information from Chittenden Solid Waste District and *Clean House, Clean Planet* by Karen Logan). To make a 16-oz batch for a squirt bottle, Mabel uses an amount of mineral oil that is 4 oz more than twice the amount of vinegar. How much of each ingredient is required?

3. *Real Estate.* In 1996, the Simon Property Group and the DeBartolo Realty Corporation merged to form the largest real estate company in the United States, owning 183 shopping centers in 32 states. Prior to merging, Simon owned twice as many properties as DeBartolo. How many properties did each company own before the merger?

The Indianapolis Star
REAL ESTATE GIANTS MERGE

DeBartolo shareholders would receive 0.68 share of Simon common stock for each share of DeBartolo common stock. Simon also would agree to repay $1.5 billion in DeBartolo debt. At Tuesday's closing price of $23.625 a share for common stock, the transaction is valued at roughly $3 billion.

Executives say the proposed company, Simon DeBartolo Group, would be the largest real estate company in the United States, worth $7.5 billion. **Not included in the deal:** DeBartolo's ownership stake in the San Francisco 49ers, or the Indiana Pacers, owned separately by the Simon

4. *Hockey Rankings.* Hockey teams receive 2 points for a win and 1 point for a tie. The Wildcats once won a championship with 60 points. They won 9 more games than they tied. How many wins and how many ties did the Wildcats have?

5. *College Credits.* Each course at Pease County Community College is worth either 3 or 4 credits. The members of the women's swim team are taking a total of 27 courses that are worth a total of 89 credits. How many 3-credit courses and how many 4-credit courses are being taken?

6. *College Credits.* Each course at Mt. Regis College is worth either 3 or 4 credits. The members of the men's soccer team are taking a total of 48 courses that are worth a total of 155 credits. How many 3-credit courses and how many 4-credit courses are being taken?

7. *Returnable Bottles.* As part of a fundraiser, the Cobble Hill Daycare collected 430 returnable bottles and cans, some worth 5 cents each and the rest worth 10 cents each. If the total value of the cans and bottles was $26.20, how many 5-cent bottles or cans and how many 10-cent bottles or cans were collected?

8. *Ice Cream Cones.* A busload of campers stopped at a dairy stand for ice cream. They ordered 40 cones, some soft-serve at $1.75 and the rest hard-pack at $2.00. If the total bill was $74, how many of each type of cone were ordered?

9. *Yellowstone Park Admissions.* Entering Yellowstone National Park costs $20 for a car and $15 for a motorcycle (*Source*: National Park Service, U.S. Department of the Interior). On a typical day, 5950 cars or motorcycles enter and pay a total of $107,875. How many motorcycles enter on a typical day?

10. *Zoo Admissions.*   During the summer months, the Bronx Zoo charges $11 for adults and $8 for children and seniors (*Source*: Bronx Zoo). One July day, a total of $8700 was collected from 960 admissions. How many adult admissions were there?

11. *Printing.*   King Street Printing recently charged 1.9¢ per sheet of paper, but 2.4¢ per sheet for paper made of recycled fibers. Darren's bill for 150 sheets of paper was $3.41. How many sheets of each type were used?

12. *Photocopying.*   Quick Copy recently charged 6¢ a page for copying pages that can be machine-fed and 18¢ a page for copying pages that must be hand-placed on the copier. If Lea's bill for 90 copies was $9.24, how many copies of each type were made?

13. *Lighting.*   Booth Bros. Hardware charges $7.50 for a General Electric Biax Energy Saver light bulb and $5 for an SLi Lighting Cool White Energy Saver bulb. If Paul County Hospital purchased 200 such bulbs for $1150, how many of each type did they purchase?

14. *Office Supplies.*   Barlow's Office Supply charges $16.75 for a box of Erase-A-Gel™ pens and $14.25 for a box of Icy™ automatic pencils. If Letsonville Community College purchased 120 such boxes for $1790, how many boxes of each type did they purchase?

15. *Sales.*   Staples® recently sold a black Epson Stylus R340 ink cartridge for $17.35 and a black HP Officejet Pro K550 cartridge for $19.99. At the start of a recent fall semester, a total of 50 of these cartridges was sold for a total of $907.10. How many of each type were purchased?

16. *Sales.*   Staples® recently sold a wirebound graph-paper notebook for $2.50 and a college-ruled note-book made of recycled paper for $2.30. At the start

of a recent spring semester, a combination of 50 of these notebooks was sold for a total of $118.60. How many of each type were sold?

17. *Blending Coffees.*   The Bean Counter charges $9.00 per pound for Kenyan French Roast coffee and $8.00 per pound for Sumatran coffee. How much of each type should be used to make a 20-lb blend that sells for $8.40 per pound?

18. *Mixed Nuts.*   Oh Nuts! sells cashews for $6.75 per pound and Brazil nuts for $5.00 per pound. How much of each type should be used to make a 50-lb mixture that sells for $5.70 per pound?

19. *Seed Mix.*   Sunflower seed is worth $1.00 per pound and rolled oats are worth $1.35 per pound. How much of each would you use to make 50 lb of a mixture worth $1.14 per pound?

20. *Coffee Blends.*   Cafe Europa mixes Brazilian coffee worth $19 per kilogram with Turkish coffee worth $22 per kilogram. The mixture should be worth $20 per kilogram. How much of each type of coffee should be used to make a 300-kg mixture?

21. *Acid Mixtures.*   Jerome's experiment requires him to mix a 50%-acid solution with an 80%-acid solution to create 200 mL of a 68%-acid solution. How much 50%-acid solution and how much 80%-acid solution should he use? Complete the following table as part of the *Translate* step.

Type of Solution	50%-Acid	80%-Acid	68%-Acid Mix
Amount of Solution	x	y	
Percent Acid	50%		68%
Amount of Acid in Solution		$0.8y$	

22. *Ink Remover.*   Etch Clean Graphics uses one cleanser that is 25% acid and a second that is 50% acid. How many liters of each should be mixed to get 30 L of a solution that is 40% acid?

23. *Blending Granola.*   Deep Thought Granola is 25% nuts and dried fruit. Oat Dream Granola is 10%

nuts and dried fruit. How much of Deep Thought and how much of Oat Dream should be mixed to form a 20-lb batch of granola that is 19% nuts and dried fruit?

24. *Livestock Feed.* Soybean meal is 16% protein and corn meal is 9% protein. How many pounds of each should be mixed to get a 350-lb mixture that is 12% protein?

25. *Student Loans.* Lomasi's two student loans totaled $12,000. One of her loans was at 6% simple interest and the other at 9%. After one year, Lomasi owed $855 in interest. What was the amount of each loan?

26. *Investments.* An executive nearing retirement made two investments totaling $15,000. In one year, these investments yielded $1432 in simple interest. Part of the money was invested at 9% and the rest at 10%. How much was invested at each rate?

27. *Automotive Maintenance.* "Arctic Antifreeze" is 18% alcohol and "Frost No-More" is 10% alcohol. How many liters of each should be mixed to get 20 L of a mixture that is 15% alcohol?

28. *Catering.* Casella's Catering is planning a wedding reception. The bride and groom would like to serve a nut mixture containing 25% peanuts. Casella has available mixtures that are either 40% or 10% peanuts. How much of each type should be mixed to get a 20-lb mixture that is 25% peanuts?

29. *Octane Ratings.* The octane rating of a gasoline is a measure of the amount of isooctane in the gas. When a tanker delivers gas to a gas station, it brings only two grades of gasoline, the highest and the lowest, filling two large underground tanks. If you purchase a middle grade, the pump's computer mixes the other two grades appropriately. The 2002

Dodge Neon RT requires 91-octane gasoline (*Sources*: Champlain Electric and Petroleum Equipment; Goss Dodge). How much 87-octane gas and 93-octane gas should Kasey mix in order to make 12 gal of 91-octane gas for her Neon RT?

30. *Octane Ratings.* The octane rating of a gasoline is a measure of the amount of isooctane in the gas. The 2005 Chrysler Crossfire requires 93-octane gasoline (*Sources*: Champlain Electric and Petroleum Equipment; Freedom Chrysler Plymouth). How much 87-octane gas and 95-octane gas should Ken mix in order to make 10 gal of 93-octane gas for his Crossfire?

31. *Food Science.* The following bar graph shows the milk fat percentages in three dairy products. How many pounds each of whole milk and cream should be mixed to form 200 lb of milk for cream cheese?

32. *Food Science.* How much lowfat (1% fat) milk and how much whole milk (4% fat) should be mixed to make 5 gal of reduced fat (2% fat) milk?

33. *Basketball Scoring.* Wilt Chamberlain once scored a record 100 points on a combination of 64 foul shots (each worth one point) and two-pointers. How many shots of each type did he make?

34. *Basketball Scoring.* Shaquille O'Neal recently scored 34 points on a combination of 21 foul shots and two-pointers. How many shots of each type did he make?

35. *Train Travel.* A train leaves Danville Junction and travels north at a speed of 75 km/h. Two hours later, an express train leaves on a parallel track and travels north at 125 km/h. How far from the station will they meet?

36. *Car Travel.* Two cars leave Salt Lake City, traveling in opposite directions. One car travels at a speed of 80 km/h and the other at 96 km/h. In how many hours will they be 528 km apart?

37. *Boating.* Mia's motorboat took 3 hr to make a trip downstream with a 6-mph current. The return trip against the same current took 5 hr. Find the speed of the boat in still water.

38. *Canoeing.* Alvin paddled for 4 hr with a 6-km/h current to reach a campsite. The return trip against the same current took 10 hr. Find the speed of Alvin's canoe in still water.

39. *Point of No Return.* A plane flying the 3458-mi trip from New York City to London has a 50-mph tailwind. The flight's *point of no return* is the point at which the flight time required to return to New York is the same as the time required to continue to London. If the speed of the plane in still air is 360 mph, how far is New York from the point of no return?

40. *Point of No Return.* A plane is flying the 2553-mi trip from Los Angeles to Honolulu into a 60-mph headwind. If the speed of the plane in still air is 310 mph, how far from Los Angeles is the plane's point of no return? (See Exercise 39.)

41. *Radio Airplay.* Roscoe must play 12 commercials during his 1-hr radio show. Each commercial is either 30 sec or 60 sec long. If the total commercial time during that hour is 10 min, how many commercials of each type does Roscoe play?

42. *DVD Rentals.* J. P.'s Video rents general-interest DVDs for $3.00 each and children's DVDs for $1.50 each. In one day, a total of $213 was taken in from the rental of 77 DVDs. How many of each type of DVD was rented?

43. *Making Change.* Cecilia makes a $9.25 purchase at the bookstore with a $20 bill. The store has no bills and gives her the change in quarters and fifty-cent pieces. There are 30 coins in all. How many of each kind are there?

44. *Teller Work.* Ashford goes to a bank and gets change for a $50 bill consisting of all $5 bills and $1 bills. There are 22 bills in all. How many of each kind are there?

TW 45. In what ways are Examples 3 and 4 similar? In what sense are their systems of equations similar?

TW 46. Write at least three study tips of your own for someone beginning this exercise set.

Focused Review

Solve.

47. *Diabetes.* In 2000, there were 11 million diagnosed cases of diabetes in the United States. This is expected to rise 165% by 2050. (*Source:* Centers for Disease Control and Prevention) How many diagnosed cases of diabetes are expected in 2050? [2.4]

48. *Fuel Costs.* The total fuel bill for the trucking industry in 2005 was $85 billion. This was a 37% increase from the cost in 2004. (*Source: The Indianapolis Star,* 5/2/06) What was the total fuel bill in 2004? [2.4]

49. *Gifts.* In 2005, $9 billion was spent on Father's Day gifts. This was $\frac{2}{3}$ of what was spent on Mother's Day gifts. (*Source:* National Retail Federation) How much was spent on Mother's Day gifts in 2005? [2.5]

50. *Hurricanes.* There were 26 named storms in the Atlantic Ocean in 2005. This was 4 more than twice the number forecast. (*Sources:* NOAA and the Tropical Meteorology Project) How many named storms were forecast for 2005? [2.5]

51. *Real Estate.* The perimeter of a rectangular ocean-front lot is 190 m. The width is one-fourth of the length. Find the dimensions. [4.2]

52. *Retail Sales.* Paint Town sold 45 paintbrushes, one kind at $8.50 each and another at $9.75 each. In all, $398.75 was taken in for the brushes. How many of each kind were sold? [4.4]

Synthesis

53. Suppose that in Example 3 you are asked only for the amount of Assam Gingia needed for the Dragon Blend. Would the method of solving the problem change? Why or why not?

54. Write a problem similar to Example 2 for a classmate to solve. Design the problem so that the solution is "The florist sold 14 hanging plants and 9 flats of petunias."

55. *Recycled Paper.*   Unable to purchase 60 reams of paper that contains 20% post-consumer fiber, the Naylor School bought paper that was either 0% post-consumer fiber or 30% post-consumer fiber. How many reams of each should be purchased in order to use the same amount of post-consumer fiber as if the 20% post-consumer fiber paper were available?

56. *Retail.*   Some of the world's best and most expensive coffee is Hawaii's Kona coffee. In order for coffee to be labeled "Kona Blend," it must contain at least 30% Kona beans. Bean Town Roasters has 40 lb of Mexican coffee. How much Kona coffee must they add if they wish to market it as Kona Blend?

57. *Automotive Maintenance.*   The radiator in Michelle's car contains 6.3 L of antifreeze and water. This mixture is 30% antifreeze. How much of this mixture should she drain and replace with pure antifreeze so that there will be a mixture of 50% antifreeze?

58. *Exercise.*   Natalie jogs and walks to school each day. She averages 4 km/h walking and 8 km/h jogging. From home to school is 6 km and Natalie makes the trip in 1 hr. How far does she jog in a trip?

59. *Book Sales.*   A limited edition of a book published by a historical society was offered for sale to members. The cost was one book for $12 or two books for $20 (maximum of two per member). The society sold 880 books, for a total of $9840. How many members ordered two books?

60. The tens digit of a two-digit positive integer is 2 more than three times the units digit. If the digits are interchanged, the new number is 13 less than half the given number. Find the given integer. (*Hint*: Let x = the tens-place digit and y = the units-place digit; then $10x + y$ is the number.)

61. *Wood Stains.*   Williams' Custom Flooring has 0.5 gal of stain that is 20% brown and 80% neutral. A customer orders 1.5 gal of a stain that is 60% brown and 40% neutral. How much pure brown stain and how much neutral stain should be added to the original 0.5 gal in order to make up the order?*

62. *Train Travel.*   A train leaves Union Station for Central Station, 216 km away, at 9 A.M. One hour later, a train leaves Central Station for Union Station. They meet at noon. If the second train had started at 9 A.M. and the first train at 10:30 A.M., they would still have met at noon. Find the speed of each train.

63. *Fuel Economy.*   Grady's station wagon gets 18 miles per gallon (mpg) in city driving and 24 mpg in highway driving. The car is driven 465 mi on 23 gal of gasoline. How many miles were driven in the city and how many were driven on the highway?

64. *Biochemistry.*   Industrial biochemists routinely use a machine to mix a buffer of 10% acetone by adding 100% acetone to water. One day, instead of adding 5 L of acetone to create a vat of buffer, a machine added 10 L. How much additional water was needed to bring the concentration down to 10%?

*This problem was suggested by Professor Chris Burditt of Yountville, California.

Collaborative Corner

How Many Two's? How Many Three's?

Focus: Systems of linear equations
Time: 20 minutes
Group Size: 3

The box score at right, from a 2006 NBA playoff game between the Miami Heat and the Detroit Pistons, contains information on how many field goals and free throws each player attempted and made. For example, the line "Wade 11-20 8-8 32" means that Miami's Dwayne Wade made 11 field goals out of 20 attempts and 8 free throws out of 8 attempts, for a total of 32 points. (Each free throw is worth 1 point and each field goal is worth either 2 or 3 points, depending on how far from the basket it was shot.)

ACTIVITY

1. Work as a group to develop a system of two equations in two unknowns that can be used to determine how many 2-pointers and how many 3-pointers were made by Detroit.

2. Each group member should solve the system from part (1) in a different way: one person algebraically, one person by making a table and

methodically checking all combinations of 2- and 3-pointers, and one person by guesswork. Compare answers when this has been completed.

3. Determine, as a group, how many 2- and 3-pointers Miami made.

Miami (88)
Walker 3-12 4-4 11, Haslem 1-5 0-0 2, S. O'Neal 9-16 3-5 21, Wade 11-20 8-8 32, Williams 3-8 0-0 7, Posey 2-4 0-0 6, Payton 1-6 5-6 7, Mourning 1-2 0-0 2, S. Anderson 0-0 0-0 0
Totals 31-73 20-23 88
Detroit (92)
Prince 10-20 3-5 24, R. Wallace 6-12 2-2 16, B. Wallace 4-4 1-2 9, Hamilton 7-18 6-7 22, Billups, 5-13 7-7 18, Delfino 0-1 0-0 0, McDyess 1-4 1-4 3, D. Davis 0-0 0-0 0, Hunter 0-3 0-0 0, Delk 0-0 0-0 0
Totals 33-75 20-27 92

Miami	12	25	19	32	88
Detroit	25	23	22	22	92

4.5 Solving Equations by Graphing

Solving Equations Graphically: The Intersect Method ◼ Solving Equations Graphically: The Zero Method ◼ Applications

Recall that to *solve* an equation or inequality means to find all the replacements for the variable that make the equation or inequality true. We have seen how to solve algebraically; we now use a graphical method to solve.

Solving Equations Graphically: The Intersect Method

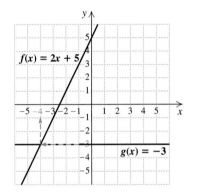

To see how solutions of equations are related to graphs, consider the graphs of the functions given by $f(x) = 2x + 5$ and $g(x) = -3$, shown at left.

At the point at which the graphs intersect, $f(x) = g(x)$. Thus, for that particular x-value, we have $2x + 5 = -3$. In other words, the solution of $2x + 5 = -3$ is the x-coordinate of the point of intersection of the graphs of $f(x) = 2x + 5$ and $g(x) = -3$. Careful inspection suggests that -4 is that x-value. To check, note that $f(-4) = 2(-4) + 5 = -3$.

EXAMPLE 1 Solve graphically: $\frac{1}{2}x + 3 = 2$.

SOLUTION To find the x-value for which $\frac{1}{2}x + 3$ will equal 2, we graph $f(x) = \frac{1}{2}x + 3$ and $g(x) = 2$ on the same set of axes. Since the intersection appears to be $(-2, 2)$, the solution is apparently -2.

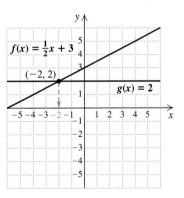

Check:

$$\frac{1}{2}x + 3 = 2$$

$$\frac{\frac{1}{2}(-2) + 3 \;\Big|\; 2}{\phantom{\frac{1}{2}}-1 + 3 \;\Big|\;}$$

$$2 \stackrel{?}{=} 2 \quad \text{TRUE}$$

The solution is -2.

Connecting the Concepts

INTERPRETING GRAPHS: POINT OF INTERSECTION

In Section 4.1, we solved systems of linear equations by graphing two equations and determining their point of intersection. For example, to solve the system

$$y = x - 3,$$
$$y = 3x - 5,$$

we graph $y = x - 3$ and $y = 3x - 5$. The point of intersection gives the solution, the ordered pair $(1, -2)$. Since there are two variables in the system, the solution will contain a value for each variable.

In this section, we are solving linear equations in one variable by graphing two functions and looking for points of intersection. For example, to solve

$$x - 3 = 3x - 5,$$

we graph $f(x) = x - 3$ and $g(x) = 3x - 5$. The x-coordinate of the point of intersection, 1, gives the solution of the equation. Since there is one variable in the equation, the solution is a single number.

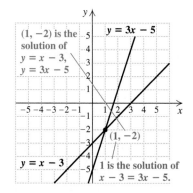

EXAMPLE 2   Solve: $-\frac{3}{4}x + 6 = 2x - 1$.

ALGEBRAIC APPROACH

We have

$$-\frac{3}{4}x + 6 = 2x - 1$$

$$-\frac{3}{4}x + 6 - 6 = 2x - 1 - 6 \qquad \text{Subtracting 6 from both sides}$$

$$-\frac{3}{4}x = 2x - 7 \qquad \text{Simplifying}$$

$$-\frac{3}{4}x - 2x = 2x - 7 - 2x \qquad \text{Subtracting } 2x \text{ from both sides}$$

$$-\frac{3}{4}x - \frac{8}{4}x = -7 \qquad \text{Simplifying}$$

$$-\frac{11}{4}x = -7 \qquad \text{Combining like terms}$$

$$-\frac{4}{11}\left(-\frac{11}{4}x\right) = -\frac{4}{11}(-7) \qquad \text{Multiplying both sides by } -\frac{4}{11}$$

$$x = \frac{28}{11}. \qquad \text{Simplifying}$$

Check:

$$\begin{array}{c|c} \multicolumn{2}{c}{-\frac{3}{4}x + 6 = 2x - 1} \\ \hline -\frac{3}{4}\left(\frac{28}{11}\right) + 6 & 2\left(\frac{28}{11}\right) - 1 \\ -\frac{21}{11} + \frac{66}{11} & \frac{56}{11} - \frac{11}{11} \\ \frac{45}{11} \overset{?}{=} \frac{45}{11} \end{array} \qquad \text{TRUE}$$

The solution is $\frac{28}{11}$.

GRAPHICAL APPROACH

We graph $f(x) = -\frac{3}{4}x + 6$ and $g(x) = 2x - 1$. It appears that the lines intersect at $(2.5, 4)$.

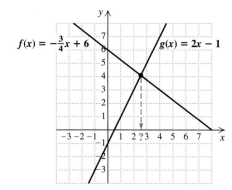

Check:

$$\begin{array}{c|c} \multicolumn{2}{c}{-\frac{3}{4}x + 6 = 2x - 1} \\ \hline -\frac{3}{4}(2.5) + 6 & 2(2.5) - 1 \\ -1.875 + 6 & 5 - 1 \\ 4.125 \overset{?}{=} 4 \end{array} \qquad \text{FALSE}$$

Our check shows that 2.5 is *not* the solution. To find the exact solution graphically, we need a method that will determine coordinates more precisely.

CAUTION!   When using a hand-drawn graph to solve an equation, it is important to use graph paper and to work as neatly as possible. Use a straightedge when drawing lines and be sure to erase any mistakes completely.

We can use the INTERSECT option of the CALC menu on a graphing calculator to find the point of intersection.

EXAMPLE 3   Solve using a graphing calculator: $-\frac{3}{4}x + 6 = 2x - 1$.

SOLUTION   In Example 2, we saw that the graphs of $f(x) = -\frac{3}{4}x + 6$ and $g(x) = 2x - 1$ intersect near the point $(2.5, 4)$. To determine more precisely the point of intersection, we graph $y_1 = -\frac{3}{4}x + 6$ and $y_2 = 2x - 1$ using the same viewing window. After choosing the INTERSECT feature, we indicate which two graphs, or *curves*, we are considering. Then we enter a guess. The coordinates of the point of intersection then appear at the bottom of the screen.

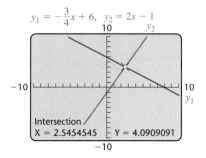

It appears from the screen above that the solution is 2.5454545. To check, we evaluate both sides of the equation $-\frac{3}{4}x + 6 = 2x - 1$ for this value of x. The first coordinate of the point of intersection is stored as X, so we evaluate Y1(X) and Y2(X) from the home screen, as shown on the left below.

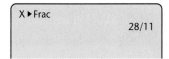

Although the check shows that 2.5454545 is the solution, it is actually an approximation of the solution. In some cases, the calculator can give us the exact answer. Since the x-coordinate of the point of intersection is stored as X, converting X to fraction notation will give an exact solution. From the screen on the right above, we see that $\frac{28}{11}$ is the solution of the equation.   ◢

Solving Equations Graphically: The Zero Method

When we are solving an equation graphically, it can be challenging to determine a portion of the x, y-coordinate plane that contains the point of intersection. The Zero method makes that determination easier because we are interested only in the point at which a graph crosses the x-axis.

To solve an equation using the Zero method, we use the addition principle to get zero on one side of the equation. Then we look for the intersection of the line and the x-axis, or the *x-intercept* of the graph. If using function notation, we look for the *zero* of the function, or the input x that makes the output zero.

Zero of a Function   A zero of a function is an input whose corresponding output is 0. If a is a zero of the function f, then $f(a) = 0$.

The y-coordinate of a point is 0 when the point is on the x-axis. Thus a zero of a function is the x-coordinate of any point at which its graph crosses or touches the x-axis.

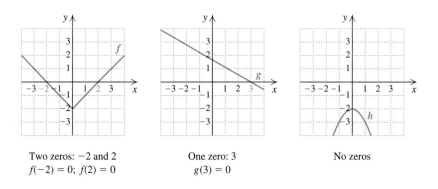

Two zeros: -2 and 2 One zero: 3 No zeros
$f(-2) = 0$; $f(2) = 0$ $g(3) = 0$

Zeros of a Function

We can determine any zeros of a function using the ZERO option in the CALC menu of a graphing calculator. After graphing the function and choosing the ZERO option, we will be prompted for a Left Bound, a Right Bound, and a Guess. Since there may be more than one zero of a function, the left and right bounds indicate which zero we are currently finding. We examine the graph to find any places where it appears that the graph touches or crosses the x-axis, and then find those x-values one at a time. By using the arrow keys or entering a value on the keyboard, we choose an x-value less than the zero for the left bound, an x-value more than the zero for the right bound, and a value close to the zero for the Guess.

> **EXAMPLE 4** Find the zeros of the function given by $f(x) = \frac{2}{3}x + 5$.

ALGEBRAIC APPROACH	GRAPHICAL APPROACH

ALGEBRAIC APPROACH

We want to find any x-values for which $f(x) = 0$, so we substitute 0 for $f(x)$ and solve:

$$f(x) = \frac{2}{3}x + 5$$
$$0 = \frac{2}{3}x + 5 \qquad \text{Substituting 0 for } f(x)$$
$$-\frac{2}{3}x = 5 \qquad \text{Subtracting } -\frac{2}{3}x \text{ from both sides}$$
$$\left(-\frac{3}{2}\right)\left(-\frac{2}{3}x\right) = \left(-\frac{3}{2}\right)(5) \qquad \text{Multiplying by the reciprocal of } -\frac{2}{3}$$
$$x = -\frac{15}{2}. \qquad \text{Simplifying}$$

The zero of the function is $-\frac{15}{2}$, or -7.5.

GRAPHICAL APPROACH

We graph the function and look for any points at which the graph crosses the x-axis. From the graph, there appears to be one zero of the function, approximately -7.

We press ⬭CALC ⬭2 to choose the ZERO option of the CALC menu. When prompted, we use the number keys to choose -10 for a left bound, -5 for a right bound, and -7 for a guess. We can then read from the screen that -7.5 is the zero of the function.

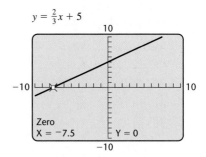

Note that the answer to Example 4 is a value for x. A zero of a function is an x-value, not a function value or an ordered pair.

> **EXAMPLE 5** Solve graphically, using the Zero method: $2x - 5 = 4x - 11$.

SOLUTION We first get 0 on one side of the equation:

$$2x - 5 = 4x - 11$$
$$-2x - 5 = -11 \qquad \text{Subtracting } 4x \text{ from both sides}$$
$$-2x + 6 = 0. \qquad \text{Adding 11 to both sides}$$

We then graph $f(x) = -2x + 6$, and find the x-intercept of the graph.

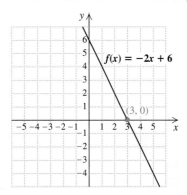

The x-intercept of the graph appears to be $(3, 0)$. We check 3 in the original equation.

Check:

$$\begin{array}{c|c} 2x - 5 = 4x - 11 \\ \hline 2 \cdot 3 - 5 & 4 \cdot 3 - 11 \\ 6 - 5 & 12 - 11 \\ 1 \stackrel{?}{=} 1 & \text{TRUE} \end{array}$$

The solution is 3.

EXAMPLE 6 Solve $3 - 8x = 5 - 7x$ using both the Intersect and Zero methods.

GRAPHICAL SOLUTION: INTERSECT METHOD

We graph $y_1 = 3 - 8x$ and $y_2 = 5 - 7x$ and determine the coordinates of any point of intersection.

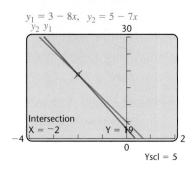

$y_1 = 3 - 8x, \quad y_2 = 5 - 7x$

The point of intersection is $(-2, 19)$.
The solution is -2.

GRAPHICAL SOLUTION: ZERO METHOD

We first get zero on one side of the equation:

$$3 - 8x = 5 - 7x$$
$$-2 - 8x = -7x \qquad \text{Subtracting 5 from both sides}$$
$$-2 - x = 0. \qquad \text{Adding } 7x \text{ to both sides}$$

Then we graph $y = -2 - x$ and determine the point for which $y = 0$, or the x-intercept of the graph. We do this using the ZERO option in the CALC menu.

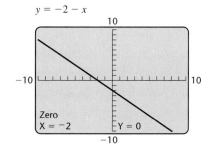

$y = -2 - x$

The solution is the first coordinate of the x-intercept, or -2.

Applications

EXAMPLE 7 Cost Projections. Cleartone Communications charges $50 for a cell phone and $40 per month for calls made under its Call Anywhere plan. Formulate and graph a mathematical model for the cost. Then use the model to estimate the time required for the total cost to reach $250.

SOLUTION

1. **Familiarize.** The problem describes a situation in which a monthly fee is charged after an initial purchase has been made. After 1 month of service, the total cost will be $50 + $40 = $90. After 2 months, the total cost will be $50 + $40 \cdot 2 = $130. This can be generalized in a model if we let $C(t)$ represent the total cost, in dollars, for t months of service.

2. **Translate.** We reword and translate as follows:

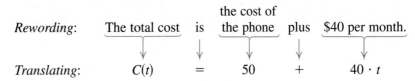

Rewording: The total cost   is   the cost of the phone   plus $40 per month.

Translating: $C(t)$ $=$ 50 $+$ $40 \cdot t$

where $t \geq 0$ (since there cannot be a negative number of months).

3. **Carry out.** To estimate the time required for the total cost to reach $250, we are estimating the solution of

$$40t + 50 = 250. \quad \textbf{Replacing } C(t) \textbf{ with 250}$$

We do this by graphing $C(t) = 40t + 50$ and $y = 250$ and looking for the point of intersection. On a graphing calculator, we let $y_1 = 40x + 50$ and $y_2 = 250$, and adjust the window dimensions to include the point of intersection. Using either a hand-drawn graph or one generated by a graphing calculator, we find a point of intersection at $(5, 250)$.

Thus we estimate that it takes 5 months for the total cost to reach $250.

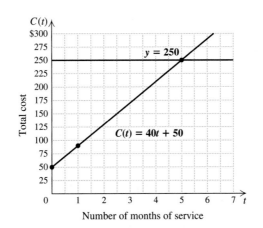

Number of months of service

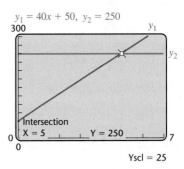

4. **Check.** We evaluate:

$$C(5) = 40 \cdot 5 + 50$$
$$= 200 + 50$$
$$= 250.$$

Our estimate turns out to be precise.

5. **State.** It takes 5 months for the total cost to reach $250.

4.5 EXERCISE SET

FOR EXTRA HELP

MathXL

MyMathLab

InterAct
Math

Tutor
Center
AW Math
Tutor Center

Video Lectures
on CD: Disc 2

Student's
Solutions Manual

Concept Reinforcement   *Exercises 1–6 refer to the following graph. Match each description with the appropriate answer from the column on the right.*

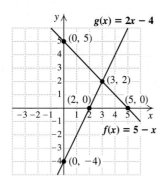

1. The solution of $5 - x = 2x - 4$

2. The zero of g

3. The solution of $5 - x = 0$

4. The x-intercept of the graph of f

5. The y-intercept of the graph of g

6. The solution of $y = 5 - x$,
$\qquad y = 2x - 4$

a) $(0, -4)$

b) $(5, 0)$

c) $(3, 2)$

d) 2

e) 3

f) 5

Estimate the solution of each equation from the associated graph.

7. $2x - 1 = -5$

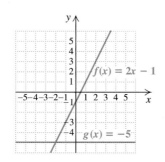

8. $\frac{1}{2}x + 1 = 3$

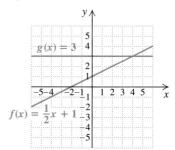

9. $2x + 3 = x - 1$

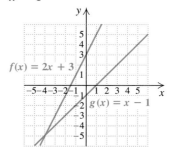

10. $2 - x = 3x - 2$

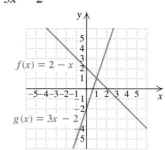

11. $\frac{1}{2}x + 3 = x - 1$

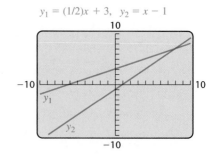

12. $2x - \frac{1}{2} = x + \frac{5}{2}$

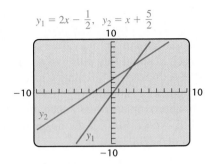

13. $f(x) = g(x)$

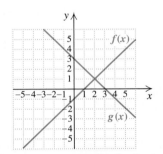

14. $f(x) = g(x)$

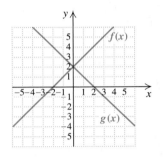

15. Estimate the value of x for which $y_1 = y_2$.

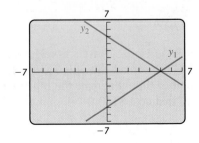

16. Estimate the value of x for which $y_1 = y_2$.

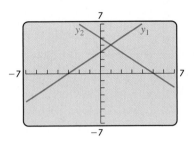

In Exercises 17–30, determine the zeros, if any, of each function.

17.

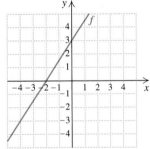

18.

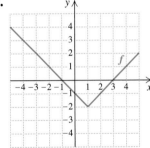

19.

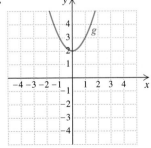

20.

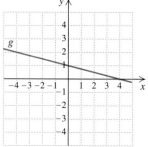

21.

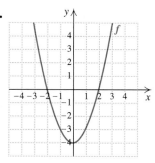

22.

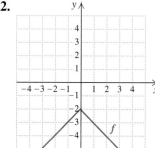

23. $f(x) = x - 5$

24. $f(x) = x + 3$

25. $f(x) = \frac{1}{2}x + 10$

26. $f(x) = \frac{2}{3}x - 6$

27. $f(x) = 2.7 - x$

28. $f(x) = 0.5 - x$

29. $f(x) = 3x + 7$

30. $f(x) = 5x - 8$

Solve graphically.

31. $x - 3 = 4$

32. $x + 4 = 6$

33. $2x + 1 = 7$

34. $3x - 5 = 1$

35. $\frac{1}{3}x - 2 = 1$

36. $\frac{1}{2}x + 3 = -1$

37. $x + 3 = 5 - x$

38. $x - 7 = 3x - 3$

39. $5 - \frac{1}{2}x = x - 4$

40. $3 - x = \frac{1}{2}x - 3$

41. $2x - 1 = -x + 3$

42. $-3x + 4 = 3x - 4$

Use a graph to estimate the solution in each of the following. Be sure to use graph paper and a straightedge if graphing by hand.

43. *Healthcare Costs.* Under one particular CIGNA health-insurance plan, an individual pays the first $100 of each hospital stay plus $\frac{1}{5}$ of all charges in excess of $100 (*Source*: VT State Chamber of Commerce membership information from VACE, 2004). By approximately how much did Gerry's hospital bill exceed $100, if the hospital stay cost him a total of $2200?

44. *Indiana Toll Road.* To drive an automobile on the Indiana Toll Road costs approximately 21¢ plus 3¢ per mile (*Source*: www.in.gov/dot/motoristinfo). Estimate how far Gina has driven on the Toll Road if her toll is $2.50.

45. *Telephone Charges.* Skytone Calling charges $20 for a telephone and $25 per month under its economy plan. Estimate the time required for the total cost to reach $145.

46. *Cell-Phone Charges.* The Cellular Connection charges $30 for a cell phone and $40 per month under its economy plan. Estimate the time required for the total cost to reach $190.

47. *Parking Fees.* Karla's Parking charges $3.00 to park plus 50¢ for each 15-min unit of time. Estimate how long someone can park for $7.50.*

48. *Cost of a Road Call.* Dave's Foreign Auto Village charges $50 for a road call plus $15 for each 15-min unit of time. Estimate the time required for a road call that cost $140.*

_____

*More precise, nonlinear models of Exercises 47 and 48 appear in Exercises 69 and 70, respectively.

49. *Cost of a FedEx Delivery.*   In 2006, for Standard delivery to the closest zone of packages weighing from 100 to 499 lb, FedEx charged $114 plus $1.14 for each pound over 100 (*Source*: www.fedex.com). Estimate the weight of a package that cost $171 to ship.

50. *Copying Costs.*   For each copy of a town report, a FedEx Kinko's Office and Print Center charged $3.45 for a coil binding and 15¢ for each page that was a double-sided copy (*Source*: FedEx Kinko's price list brochure). Estimate the number of pages in a coil-bound town report that cost $6.90 per copy. Assume that every page in the report was a double-sided copy.

TW 51. Explain the difference between finding $f(0)$ and finding the the zeros of f.

TW 52. Darnell used a graphical method to solve $2x - 3 = x + 4$. He stated that the solution was $(7, 11)$. Can this answer be correct? What mistake do you think Darnell is making?

Skill Maintenance

Simplify. [1.8]

53. $0.5x - 2.34 + 2.4x - (7.8x - 9)$

54. $3(1 - 2^3 \div 4 \cdot 2) - 7 + 3 \cdot 4^2$

Solve.

55. $0.5x - 2.34 + 2.4x = 7.8x - 9$ [2.2]

56. $5x + 7x \le -144$ [2.6]

57. $x - y = 10,$
    $y = 3 - x$ [4.2]

58. $2x - y = 3,$
    $4x + y = 5$ [4.3]

Synthesis

TW 59. Explain why, when we are solving an equation graphically, the x-coordinate of the point of intersection gives the solution of the equation.

TW 60. Explain the difference between "solving by graphing" and "graphing the solution set."

Estimate the solution(s) from the associated graph.

61. $f(x) = g(x)$

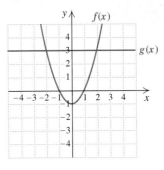

62. $f(x) = g(x)$

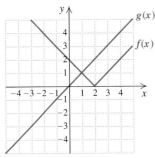

Solve graphically. Be sure to check.

63. $2x = |x + 1|$

64. $x - 1 = |4 - 2x|$

65. $\frac{1}{2}x = 3 - |x|$

66. $2 - |x| = 1 - 3x$

67. $x^2 = x + 2$

68. $x^2 = x$

69. (Refer to Exercise 47.) It costs as much to park at Karla's for 16 min as it does for 29 min. Thus the linear graph drawn in the solution of Exercise 47 is not a precise representation of the situation. Draw a graph with a series of "steps" that more accurately reflects the situation.

70. (Refer to Exercise 48.) A 32-min road call with Dave's costs the same as a 44-min road call. Thus the linear graph drawn in the solution of Exercise 48 is not a precise representation of the situation. Draw a graph with a series of "steps" that more accurately reflects the situation.

4   Chapter Summary and Review

KEY TERMS AND DEFINITIONS

SYSTEMS

System of equations, p. 286   Two or more equations that are to be solved simultaneously. A system is **consistent** if it has at least one solution. Otherwise it is **inconsistent.** The equations in a system are **dependent** if one of them can be removed without changing the solution set. Otherwise, they are **independent.**

APPLICATIONS

Supplementary angles, p. 303   Two angles are supplementary if the sum of their measures is 180°. If the sum of the measures of two angles is 90°, the angles are **complementary.**

Distance d**, rate** r**, and time** t (p. 325) are related by

$$d = rt, \quad r = \frac{d}{t}, \quad \text{and} \quad t = \frac{d}{r}.$$

IMPORTANT CONCEPTS

[Section references appear in brackets.]

Concept	Example

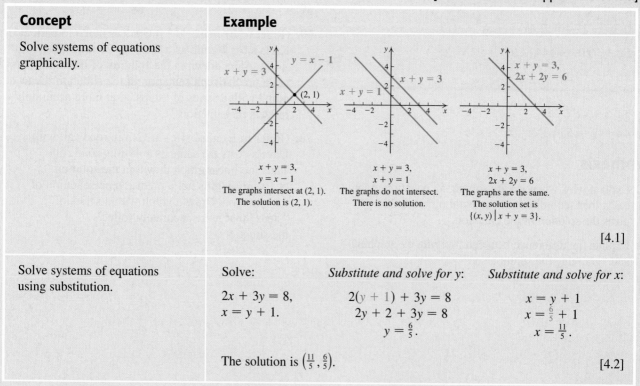

Solve systems of equations graphically.

$x + y = 3,$
$y = x - 1$
The graphs intersect at (2, 1). The solution is (2, 1).

$x + y = 3,$
$x + y = 1$
The graphs do not intersect. There is no solution.

$x + y = 3,$
$2x + 2y = 6$
The graphs are the same. The solution set is $\{(x, y) \mid x + y = 3\}$.

[4.1]

Solve systems of equations using substitution.

Solve:

$2x + 3y = 8,$
$x = y + 1.$

Substitute and solve for y:

$2(y + 1) + 3y = 8$
$2y + 2 + 3y = 8$
$y = \frac{6}{5}.$

Substitute and solve for x:

$x = y + 1$
$x = \frac{6}{5} + 1$
$x = \frac{11}{5}.$

The solution is $\left(\frac{11}{5}, \frac{6}{5}\right)$.

[4.2]

(*continued*)

Solve systems of equations using elimination.	Solve: $4x - 2y = 6,$ $3x + y = 7.$	*Eliminate y and solve for x:* $4x - 2y = 6$ $\underline{6x + 2y = 14}$ $10x = 20$ $x = 2.$

Substitute and solve for y:

$3x + y = 7$
$3 \cdot 2 + y = 7$
$y = 1.$

The solution is $(2, 1)$. [4.3]

A **zero** of a function is an input x such that $f(x) = 0$. Graphically, it is the x-coordinate of an x-intercept of the graph of f.

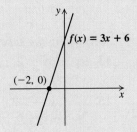

The x-intercept is $(-2, 0)$.
The zero of f is -2 because $f(-2) = 0$. [4.5]

Equations can be solved graphically using either the **Intersect** or **Zero** method.

Solve graphically: $x + 1 = 2x - 3.$

Intersect method

Graph $y = x + 1$
and $y = 2x - 3$.

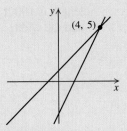

The solution is 4.

Zero method

$x + 1 = 2x - 3$
$0 = x - 4$
Graph $y = x - 4$.

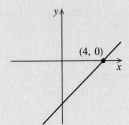

The solution is 4. [4.5]

Review Exercises

↪ *Concept Reinforcement*   *Complete each of the following sentences.*

1. The system
$$5x + 3y = 7,$$
$$y = 2x + 1$$
is most easily solved using the _____ method. [4.2]

2. The system
$$-2x + 3y = 8,$$
$$2x + 2y = 7$$
is most easily solved using the _____ method. [4.3]

3. A weakness in using graphs to solve a system is that when solutions involve fractions or decimals, the graph may yield only a(n) _____ solution. [4.4]

4. When one equation in a system is a multiple of another equation in that system, the equations are said to be _____. [4.1]

5. A system for which there is no solution is said to be _____. [4.1]

6. When using elimination to solve a system of two equations, if an identity is obtained, we know that there is a(n) _____ number of solutions. [4.3]

7. When we are graphing to solve a system of two equations, if there is no solution, the lines will be _____. [4.1]

8. If $f(10) = 0$, 10 is a _____ of f. [4.5]

9. The numbers in an ordered pair that is a solution of a system correspond to the variables in _____ order. [4.1]

10. When we are solving an equation graphically, the solution is the _____ of the point of intersection of the graphs. [4.5]

For Exercises 11–19, if a system has an infinite number of solutions, use set-builder notation to write the solution set. If a system has no solution, state this.

Solve graphically. [4.1]

11. $3x + 2y = -4,$
$y = 3x + 7$

12. $16x - 7y = 25,$
$8x + 3y = 19$

Solve using the substitution method. [4.2]

13. $9x - 6y = 2,$
$x = 4y + 5$

14. $y = x + 2,$
$y - x = 8$

15. $x - 3y = -2,$
$7y - 4x = 6$

Solve using the elimination method. [4.3]

16. $8x - 2y = 10,$
$-4y - 3x = -17$

17. $4x - 7y = 18,$
$9x + 14y = 40$

18. $3x - 5y = -4,$
$5x - 3y = 4$

19. $1.5x - 3 = -2y,$
$3x + 4y = 6$

Solve.

20. *Architecture.*   The rectangular ground floor of the John Hancock building has a perimeter of 860 ft. The length is 100 ft more than the width. Find the length and the width. [4.2]

$x + 100$

x

21. Luther bought two equally priced DVDs and one CD for $48. If he had purchased one DVD and two CDs, he would have spent $3 less. What is the price of a DVD? What is the price of a CD? [4.3]

22. *Music Lessons.* Alice charges $25 for a private guitar lesson and $18 for a group guitar lesson. One day in August, Alice earned $265 from 12 students. How many students of each type did Alice teach? [4.4]

23. *Geometry.* Two angles are supplementary. One angle is 3° less than twice the other. Find the measures of the angles. [4.2]

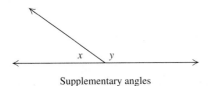

Supplementary angles

24. A freight train leaves Houston at midnight traveling north at a speed of 44 mph. One hour later, a passenger train, going 55 mph, travels north from Houston on a parallel track. How many hours will the passenger train travel before it overtakes the freight train? [4.4]

25. Yolanda wants 14 L of fruit punch that is 10% juice. At the store, she finds punch that is 15% juice and punch that is 8% juice. How much of each should she purchase? [4.4]

26. *Printing.* Using some pages that hold 1300 words per page and others that hold 1850 words per page, a typesetter is able to completely fill 12 pages with an 18,350-word document. How many pages of each kind were used? [4.4]

27. Use the following graph to solve $x - 1 = 2x + 1$. [4.5]

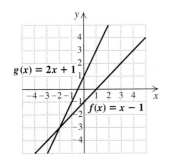

28. Use the graph in Exercise 27 to determine any zeros of f. [4.5]

29. Determine any zeros of $f(x) = 4 - 7x$. [4.5]

30. Solve graphically: $3x - 2 = x + 4$. [4.5]

Synthesis

TW 31. Explain why any solution of a system of equations is a point of intersection of the graphs of each equation in the system. [4.1]

TW 32. Explain the difference between solving a system of equations graphically and solving an equation graphically. [4.1], [4.5]

33. Solve graphically:

$$y = x + 2,$$
$$y = x^2 + 2. \ [4.1]$$

34. The solution of the following system is $(6, 2)$. Find C and D.

$$2x - Dy = 6,$$
$$Cx + 4y = 14 \ [4.1]$$

35. For a two-digit number, the sum of the ones digit and the tens digit is 6. When the digits are reversed, the new number is 18 more than the original number. Find the original number. [4.3]

36. A farmhand agrees to work for one year in exchange for $24,000 and a horse. After 7 months, the farm is sold and the farmhand receives the horse and $10,000. What was the value of her yearly salary? [4.4]

Chapter Test 4

For Exercises 1–9, if a system has an infinite number of solutions, use set-builder notation to write the solution set. If a system has no solution, state this.

Solve by graphing.

1. $2x + y = 8,$
$\quad y - x = 2$

2. $2y - x = 7,$
$\quad 2x - 4y = 4$

Solve using the substitution method.

3. $x + 3y = -8,$
$\quad 4x - 3y = 23$

4. $2x + 4y = -6,$
$\quad y = 3x - 9$

5. $x = 5y - 10,$
$\quad 15y = 3x + 30$

Solve using the elimination method.

6. $4x - 6y = 3,$
$\quad 6x - 4y = -3$

7. $4y + 2x = 18,$
$\quad 3x + 6y = 26$

8. $4x + 5y = 5,$
$\quad 6x + 7y = 7$

9. $\frac{3}{2}x - y = 24,$
$\quad 2x + \frac{3}{2}y = 15$

Solve.

10. *Court Dimensions.*   The perimeter of a standard basketball court is 288 ft. The length is 44 ft longer than the width. Find the dimensions.

$P = 288$ ft

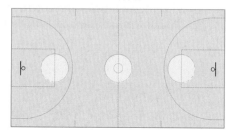

11. *Geometry.*   Two angles are complementary. The sum of the measures of the first angle and half the second angle is 64°. Find the measures of the angles.

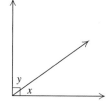

y

x

Complementary angles

12. *Books.*   The yearly book sale for the Northfield Public Library charges $1.25 for used hardbacks and 50¢ for used paperbacks. Keith purchased a total of 23 books for $19.75. How many books of each type did he buy?

13. Pepperidge Farm® Goldfish is a snack food for which 40% of its calories come from fat. Rold Gold® Pretzels receive 9% of their calories from fat. How many grams of each would be needed to make 620 g of a snack mix for which 15% of the calories are from fat?

14. *Boating.*   Kylie's motorboat took 3 hr to make a trip downstream on a river flowing at 5 mph. The return trip against the same current took 5 hr. Find the speed of the boat in still water.

15. *Coin Value.*   A collection of quarters and nickels is worth $1.25. There are 13 coins in all. How many of each are there?

16. Solve graphically: $2x - 5 = 3x - 3.$

17. Determine any zeros of the function given by $f(x) = \frac{1}{2}x - 5.$

Synthesis

18. Solve:

$$3(x - y) = 4 + x,$$
$$x = 5y + 2.$$

19. Find the numbers C and D such that $(-2, 3)$ is a solution of the system

$$Cx - 4y = 7,$$
$$3x + Dy = 8.$$

20. You are in line at a ticket window. There are 2 more people ahead of you in line than there are behind you. In the entire line, there are three times as many people as there are behind you. How many are in the line?

21. The graph of the function $f(x) = mx + b$ contains the points $(-1, 3)$ and $(-2, -4)$. Find m and b.

5

Polynomials

Our work in Chapters 3 and 4 concentrated on using graphs to represent solutions of equations in two variables. We looked primarily at linear equations.

Here in Chapter 5 we will focus on finding equivalent expressions, not on solving equations. We will learn how to manipulate algebraic expressions called polynomials, and we will begin the study of polynomial functions and their graphs.

APPLICATION   *Blind-spot Radar.*

A vehicle's blind spot is the area around the vehicle that cannot be seen by the driver. Some cars are equipped with radar to warn the driver of an object in that car's blind spot. The following table lists the projected need for cars fitted with blind-spot radar, in millions. The data can be modeled by the polynomial function $f(x) = 0.07x^2 + 0.17x + 0.07$, where x is the number of years after 2006. Estimate the number of cars that will be fitted with blind-spot radar in 2014.

YEARS AFTER 2006	CARS WORLDWIDE FITTED WITH BLIND-SPOT RADAR (IN MILLIONS)
0	0.1
2	0.6
4	2.0
6	3.65

Source: "Strategy Analytics," *New Scientist*, January 14, 2006

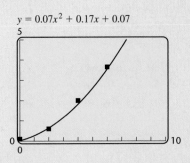

$$y = 0.07x^2 + 0.17x + 0.07$$

This problem appears as Example 7 in Section 5.3.

5.1 Exponents and Their Properties

Multiplying Powers with Like Bases ■ Dividing Powers with Like Bases ■ Zero as an Exponent ■ Raising a Power to a Power ■ Raising a Product or a Quotient to a Power

In Section 5.3, we begin our study of polynomials. Before doing so, however, we must develop some rules for manipulating exponents.

Multiplying Powers with Like Bases

Recall from Section 1.8 that an expression like a^3 means $a \cdot a \cdot a$. We can use this fact to find the product of two expressions that have the same base:

$$a^3 \cdot a^2 = (a \cdot a \cdot a)(a \cdot a) \qquad \text{There are three factors of } a \text{ in } a^3;$$
$$\text{two factors of } a \text{ in } a^2.$$
$$= a \cdot a \cdot a \cdot a \cdot a \qquad \text{Using an associative law}$$
$$= a^5.$$

Note that the exponent in a^5 is the sum of the exponents in $a^3 \cdot a^2$. That is, $3 + 2 = 5$. Similarly,

$$b^4 \cdot b^3 = (b \cdot b \cdot b \cdot b)(b \cdot b \cdot b)$$
$$= b^7, \quad \text{where } 4 + 3 = 7.$$

Adding the exponents gives the correct result.

> **The Product Rule** For any number a and any positive integers m and n,
>
> $$a^m \cdot a^n = a^{m+n}.$$
>
> (To multiply powers with the same base, keep the base and add the exponents.)

EXAMPLE 1 Multiply and simplify each of the following. (Here "simplify" means express the product as one base to a power whenever possible.)

a) $2^3 \cdot 2^8$ **b)** $5 \cdot 5^8 \cdot 5^3$

c) $(r + s)^7 (r + s)^6$ **d)** $(a^3 b^2)(a^3 b^5)$

SOLUTION

a) $2^3 \cdot 2^8 = 2^{3+8}$ **Adding exponents:** $a^m \cdot a^n = a^{m+n}$
$$= 2^{11}$$

> **CAUTION!** The base is unchanged:
> $$2^3 \cdot 2^8 \neq 4^{11}.$$

b) $5 \cdot 5^8 \cdot 5^3 = 5^1 \cdot 5^8 \cdot 5^3$ Recall that $x^1 = x$ for any number x.

$\qquad\qquad\quad = 5^{1+8+3}$ Adding exponents

$\qquad\qquad\quad = 5^{12}$

> **CAUTION!** $5^{12} \neq 5 \cdot 12.$

c) $(r + s)^7 (r + s)^6 = (r + s)^{7+6}$ The base here is $r + s$.

$\qquad\qquad\qquad\quad = (r + s)^{13}$

> **CAUTION!** $(r + s)^{13} \neq r^{13} + s^{13}.$

d) $(a^3 b^2)(a^3 b^5) = a^3 b^2 a^3 b^5$ Using an associative law

$\qquad\qquad\qquad = a^3 a^3 b^2 b^5$ Using a commutative law

$\qquad\qquad\qquad = a^6 b^7$ Adding exponents

Dividing Powers with Like Bases

Recall that any expression that is divided or multiplied by 1 is unchanged. This, together with the fact that anything (besides 0) divided by itself is 1, can lead to a rule for division:

$$\frac{a^5}{a^2} = \frac{a \cdot a \cdot a \cdot a \cdot a}{a \cdot a} = \frac{a \cdot a \cdot a}{1} \cdot \frac{a \cdot a}{a \cdot a}$$

$$= \frac{a \cdot a \cdot a}{1} \cdot 1$$

$$= a \cdot a \cdot a = a^3.$$

Note that the exponent in a^3 is the difference of the exponents in $\dfrac{a^5}{a^2}$.

Similarly,

$$\frac{x^4}{x^3} = \frac{x \cdot x \cdot x \cdot x}{x \cdot x \cdot x} = \frac{x}{1} \cdot \frac{x \cdot x \cdot x}{x \cdot x \cdot x} = \frac{x}{1} \cdot 1 = x^1, \text{ or } x.$$

Subtracting the exponents gives the correct result.

> **The Quotient Rule**   For any nonzero number a and any positive integers m and n for which $m > n$,
>
> $$\frac{a^m}{a^n} = a^{m-n}.$$
>
> (To divide powers with the same base, subtract the exponent of the denominator from the exponent of the numerator.)

EXAMPLE 2   Divide and simplify. (Here "simplify" means express the quotient as one base to an exponent whenever possible.)

a) $\dfrac{x^8}{x^2}$

b) $\dfrac{7^9}{7^4}$

c) $\dfrac{(5a)^{12}}{(5a)^4}$

d) $\dfrac{4p^5q^7}{6p^2q}$

SOLUTION

a) $\dfrac{x^8}{x^2} = x^{8-2}$ Subtracting exponents: $\dfrac{a^m}{a^n} = a^{m-n}$

$\quad = x^6$

b) $\dfrac{7^9}{7^4} = 7^{9-4}$

$\quad = 7^5$

> **CAUTION!**   The base is unchanged:
> $$\dfrac{7^9}{7^4} \neq 1^5.$$

c) $\dfrac{(5a)^{12}}{(5a)^4} = (5a)^{12-4} = (5a)^8$ The base here is $5a$.

d) $\dfrac{4p^5q^7}{6p^2q} = \dfrac{4}{6} \cdot \dfrac{p^5}{p^2} \cdot \dfrac{q^7}{q^1}$ Note that the 4 and the 6 are factors, not exponents!

$\quad = \dfrac{2}{3} \cdot p^{5-2} \cdot q^{7-1} = \dfrac{2}{3}p^3q^6$ Simplifying; using the quotient rule twice

Zero as an Exponent

The quotient rule can be used to help determine what 0 should mean when it appears as an exponent. Consider a^4/a^4, where a is nonzero. Since the numerator and the denominator are the same,

$$\dfrac{a^4}{a^4} = 1.$$

On the other hand, using the quotient rule would give us

$$\dfrac{a^4}{a^4} = a^{4-4} = a^0.$$ Subtracting exponents

Since $a^0 = a^4/a^4 = 1$, this suggests that $a^0 = 1$ for any nonzero value of a.

> **The Exponent Zero**   For any real number a, $a \neq 0$,
> $$a^0 = 1.$$
> (Any nonzero number raised to the 0 exponent is 1.)

Note that in the above box, 0^0 is not defined. For this text, we will assume that expressions like a^m do not represent 0^0.

EXAMPLE 3 Simplify: **(a)** 1948^0; **(b)** $(-9)^0$; **(c)** $(3x)^0$; **(d)** $(-1)9^0$; **(e)** -9^0.

SOLUTION

a) $1948^0 = 1$ Any nonzero number raised to the 0 exponent is 1.

b) $(-9)^0 = 1$ Any nonzero number raised to the 0 exponent is 1. The base here is -9.

c) $(3x)^0 = 1$, for any $x \neq 0$.   The parentheses indicate that the base is $3x$.

d) Recall that, unless there are calculations within parentheses, exponents are calculated before multiplication:

$$(-1)9^0 = (-1)1 = -1.$$ The base here is 9.

e) -9^0 is read "the opposite of 9^0" and is equivalent to $(-1)9^0$:

$$-9^0 = (-1)9^0 = (-1)1 = -1.$$

Note from parts (b), (d), and (e) that $-9^0 = (-1)9^0$ and $-9^0 \neq (-9)^0$.

CAUTION! $-9^0 \neq (-9)^0$; $-a^n$ and $(-a)^n$ are not equivalent expressions.

Raising a Power to a Power

Consider an expression like $(7^2)^4$:

$$
\begin{aligned}
(7^2)^4 &= (7^2)(7^2)(7^2)(7^2) && \text{There are four factors of } 7^2. \\
&= (7 \cdot 7)(7 \cdot 7)(7 \cdot 7)(7 \cdot 7) && \text{We could also use the} \\
& && \text{product rule.} \\
&= 7 \cdot 7 \cdot 7 \cdot 7 \cdot 7 \cdot 7 \cdot 7 \cdot 7 && \text{Using an associative law} \\
&= 7^8.
\end{aligned}
$$

Note that the exponent in 7^8 is the product of the exponents in $(7^2)^4$. Similarly,

$$
\begin{aligned}
(y^5)^3 &= y^5 \cdot y^5 \cdot y^5 && \text{There are three factors of } y^5. \\
&= (y \cdot y \cdot y \cdot y \cdot y)(y \cdot y \cdot y \cdot y \cdot y)(y \cdot y \cdot y \cdot y \cdot y) \\
&= y^{15}.
\end{aligned}
$$

Once again, we get the same result if we multiply exponents:

$$(y^5)^3 = y^{5 \cdot 3} = y^{15}.$$

The Power Rule   For any number a and any whole numbers m and n,

$$(a^m)^n = a^{mn}.$$

(To raise a power to a power, multiply the exponents and leave the base unchanged.)

Remember that for this text we assume that 0^0 is not considered.

Student Notes

There are several rules for manipulating exponents in this section. One way to remember them all is to replace variables with small numbers and see what the results suggest. For example, multiplying $2^2 \cdot 2^3$ and examining the result is a fine way of reminding yourself that $a^m \cdot a^n = a^{m+n}$.

EXAMPLE 4 Simplify: **(a)** $(m^2)^5$; **(b)** $(3^5)^4$.

SOLUTION

a) $(m^2)^5 = m^{2 \cdot 5}$ **Multiplying exponents:** $(a^m)^n = a^{mn}$

           $= m^{10}$

b) $(3^5)^4 = 3^{5 \cdot 4}$

           $= 3^{20}$

Raising a Product or a Quotient to a Power

When an expression inside parentheses is raised to a power, the inside expression is the base. Let's compare $2a^3$ and $(2a)^3$:

$$2a^3 = 2 \cdot a \cdot a \cdot a; \qquad \text{The base is } a.$$

$$(2a)^3 = (2a)(2a)(2a) \qquad\qquad \text{The base is } 2a.$$
$$= (2 \cdot 2 \cdot 2)(a \cdot a \cdot a) \qquad \text{Using an associative and a}$$
$$\qquad\qquad\qquad\qquad\qquad\qquad \text{commutative law}$$
$$= 2^3 a^3$$
$$= 8a^3.$$

We see that $2a^3$ and $(2a)^3$ are *not* equivalent. Note too that $(2a)^3$ can be simplified by cubing each factor. This leads to the following rule for raising a product to a power.

> **Raising a Product to a Power**   For any numbers a and b and any whole number n,
>
> $$(ab)^n = a^n b^n.$$
>
> (To raise a product to a power, raise each factor to that power.)

EXAMPLE 5 Simplify: **(a)** $(4a)^3$; **(b)** $(-5x^4)^2$; **(c)** $(a^7b)^2(a^3b^4)$.

SOLUTION

a) $(4a)^3 = 4^3 a^3 = 64a^3$ **Raising each factor to the third power and simplifying**

b) $(-5x^4)^2 = (-5)^2(x^4)^2$ **Raising each factor to the second power. Parentheses are important here.**

           $= 25x^8$ **Simplifying $(-5)^2$ and using the power rule**

c) $(a^7b)^2(a^3b^4) = (a^7)^2 b^2 a^3 b^4$ **Raising a product to a power**

                $= a^{14} b^2 a^3 b^4$ **Multiplying exponents**

                $= a^{17} b^6$ **Adding exponents**

> **CAUTION!**   The rule $(ab)^n = a^n b^n$ applies only to *products* raised to a power, not to sums or differences. For example, $(3 + 4)^2 \neq 3^2 + 4^2$ since $7^2 \neq 9 + 16$.

There is a similar rule for raising a quotient to a power.

Raising a Quotient to a Power   For any numbers a and b, $b \neq 0$, and any whole number n,

$$\left(\frac{a}{b}\right)^n = \frac{a^n}{b^n}.$$

(To raise a quotient to a power, raise the numerator to the power and divide by the denominator to the power.)

EXAMPLE 6   Simplify: **(a)** $\left(\frac{x}{5}\right)^2$; **(b)** $\left(\frac{5}{a^4}\right)^3$; **(c)** $\left(\frac{3a^4}{b^3}\right)^2$.

SOLUTION

a) $\left(\frac{x}{5}\right)^2 = \frac{x^2}{5^2} = \frac{x^2}{25}$ Squaring the numerator and the denominator

b) $\left(\frac{5}{a^4}\right)^3 = \frac{5^3}{(a^4)^3}$ Raising a quotient to a power

$= \frac{125}{a^{4 \cdot 3}} = \frac{125}{a^{12}}$ Using the power rule and simplifying

c) $\left(\frac{3a^4}{b^3}\right)^2 = \frac{(3a^4)^2}{(b^3)^2}$ Raising a quotient to a power

$= \frac{3^2(a^4)^2}{b^{3 \cdot 2}} = \frac{9a^8}{b^6}$ Raising a product to a power and using the power rule

In the following summary of definitions and rules, we assume that no denominators are 0 and 0^0 is not considered.

Definitions and Properties of Exponents

For any whole numbers m and n,

1 as an exponent:	$a^1 = a$
0 as an exponent:	$a^0 = 1$
The Product Rule:	$a^m \cdot a^n = a^{m+n}$
The Quotient Rule:	$\frac{a^m}{a^n} = a^{m-n}$
The Power Rule:	$(a^m)^n = a^{mn}$
Raising a product to a power:	$(ab)^n = a^n b^n$
Raising a quotient to a power:	$\left(\frac{a}{b}\right)^n = \frac{a^n}{b^n}$

5.1 EXERCISE SET

FOR EXTRA HELP

MathXL   MyMathLab   InterAct Math   AW Math Tutor Center   Video Lectures on CD: Disc 3   Student's Solutions Manual

Concept Reinforcement *In each of Exercises 1–8, complete the sentence using the most appropriate phrase from the column on the right.*

1. To raise a power to a power, ____

2. To raise a quotient to a power, ____

3. To raise a product to a power, ____

4. To divide powers with the same base, ____

5. Any nonzero number raised to the 0 exponent ____

6. To multiply powers with the same base, ____

7. To square a fraction, ____

8. To square a product, ____

a) keep the base and add the exponents.

b) multiply the exponents and leave the base unchanged.

c) square the numerator and square the denominator.

d) square each factor.

e) raise each factor to that power.

f) raise the numerator to the power and divide by the denominator to the power.

g) is one.

h) subtract the exponent of the denominator from the exponent of the numerator.

Simplify. Assume that no denominator is zero and 0^0 is not considered.

9. $r^4 \cdot r^6$

10. $8^4 \cdot 8^3$

11. $9^5 \cdot 9^3$

12. $n^3 \cdot n^{20}$

13. $a^6 \cdot a$

14. $y^7 \cdot y^9$

15. $8^4 \cdot 8^7$

16. $t^0 \cdot t^{16}$

17. $(3y)^4(3y)^8$

18. $(2t)^8(2t)^{17}$

19. $(5t)(5t)^6$

20. $(8x)^0(8x)^1$

21. $(a^2b^7)(a^3b^2)$

22. $(m-3)^4(m-3)^5$

23. $(x+1)^5(x+1)^7$

24. $(a^8b^3)(a^4b)$

25. $r^3 \cdot r^7 \cdot r^0$

26. $s^4 \cdot s^5 \cdot s^2$

27. $(xy^4)(xy)^3$

28. $(a^3b)(ab)^4$

29. $\dfrac{7^5}{7^2}$

30. $\dfrac{4^7}{4^3}$

31. $\dfrac{x^{15}}{x^3}$

32. $\dfrac{a^{10}}{a^2}$

33. $\dfrac{t^5}{t}$

34. $\dfrac{x^7}{x}$

35. $\dfrac{(5a)^7}{(5a)^6}$

36. $\dfrac{(3m)^9}{(3m)^8}$

Aha! **37.** $\dfrac{(x+y)^8}{(x+y)^8}$

38. $\dfrac{(a-b)^4}{(a-b)^3}$

39. $\dfrac{6m^5}{8m^2}$

40. $\dfrac{10n^7}{15n^3}$

41. $\dfrac{8a^9b^7}{2a^2b}$

42. $\dfrac{12r^{10}s^7}{4r^2s}$

43. $\dfrac{m^9n^8}{m^0n^4}$

44. $\dfrac{a^{10}b^{12}}{a^2b^0}$

Simplify.

45. x^0 when $x = 13$

46. y^0 when $y = 38$

47. $5x^0$ when $x = -4$

48. $7m^0$ when $m = 1.7$

49. $7^0 + 4^0$

50. $(8 + 5)^0$

51. $(-3)^1 - (-3)^0$

52. $(-4)^0 - (-4)^1$

Simplify. Assume that no denominator is zero and 0^0 is not considered.

53. $(x^4)^7$

54. $(a^3)^8$

55. $(5^8)^2$

56. $(2^5)^3$

57. $(m^7)^5$

58. $(n^9)^2$

59. $(t^{20})^4$

60. $(t^3)^9$

61. $(7x)^2$

62. $(5a)^2$

63. $(-2a)^3$

64. $(-3x)^3$

65. $(4m^3)^2$

66. $(5n^4)^2$

67. $(a^2b)^7$

68. $(xy^4)^9$

69. $(x^3y)^2(x^2y^5)$

70. $(a^4b^6)(a^2b)^5$

71. $(2x^5)^3(3x^4)$

72. $(5x^3)^2(2x^7)$

73. $\left(\dfrac{a}{4}\right)^3$

74. $\left(\dfrac{3}{x}\right)^4$

75. $\left(\dfrac{7}{5a}\right)^2$

76. $\left(\dfrac{5x}{2}\right)^3$

77. $\left(\dfrac{a^4}{b^3}\right)^5$

78. $\left(\dfrac{x^5}{y^2}\right)^7$

79. $\left(\dfrac{y^3}{2}\right)^2$

80. $\left(\dfrac{a^5}{2}\right)^3$

81. $\left(\dfrac{x^2y}{z^3}\right)^4$

82. $\left(\dfrac{x^3}{y^2z}\right)^5$

83. $\left(\dfrac{a^3}{-2b^5}\right)^4$

84. $\left(\dfrac{x^5}{-3y^3}\right)^4$

85. $\left(\dfrac{5x^7y}{2z^4}\right)^3$

86. $\left(\dfrac{4a^2b}{3c^7}\right)^3$

Aha! **87.** $\left(\dfrac{4x^3y^5}{3z^7}\right)^0$

88. $\left(\dfrac{5a^7}{2b^5c}\right)^0$

TW 89. Explain in your own words why $-5^2 \neq (-5)^2$.

TW 90. Under what circumstances should exponents be added?

Skill Maintenance

Factor. [1.2]

91. $3s - 3r + 3t$

92. $-7x + 7y - 7z$

Combine like terms. [1.6]

93. $9x + 2y - x - 2y$

94. $5a - 7b - 8a + b$

Graph.

95. $y = x - 5$ [3.6]

96. $2x + y = 8$ [3.3]

Synthesis

TW 97. Under what conditions does a^n represent a negative number? Why?

TW 98. Using the quotient rule, explain why 9^0 is 1.

TW 99. Suppose that the width of a square is three times the width of a second square (see the figure below). How do the areas of the squares compare? Why?

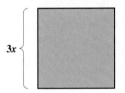

TW 100. Suppose that the width of a cube is twice the width of a second cube. How do the volumes of the cubes compare? Why?

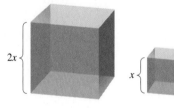

Find a value of the variable that shows that the two expressions are not equivalent. Answers may vary.

101. $(a + 5)^2$; $a^2 + 5^2$

102. $3x^2$; $(3x)^2$

103. $\dfrac{a + 7}{7}$; a

104. $\dfrac{t^6}{t^2}$; t^3

Simplify.

105. $a^{10k} \div a^{2k}$

106. $y^{4x} \cdot y^{2x}$

107. $\dfrac{\left(\frac{1}{2}\right)^3\left(\frac{2}{3}\right)^4}{\left(\frac{5}{6}\right)^3}$

108. $\dfrac{x^{5t}(x^t)^2}{(x^{3t})^2}$

109. Solve for x: $\dfrac{t^{26}}{t^x} = t^x$.

Replace ▨ with >, <, or = to write a true sentence.

110. 3^5 ▨ 3^4

111. 4^2 ▨ 4^3

112. 4^3 ▨ 5^3

113. 4^3 ▨ 3^4

114. 9^7 ▨ 3^{13}

115. 25^8 ▨ 125^5

▦ *Use the fact that $10^3 \approx 2^{10}$ to estimate each of the following powers of 2. Then compute the power of 2 with a calculator and find the difference between the exact value and the approximation.*

116. 2^{14}

117. 2^{22}

118. 2^{26}

119. 2^{31}

▦ *In computer science, 1 KB of memory refers to 1 kilobyte, or 1×10^3 bytes, of memory. This is really an approximation of 1×2^{10} bytes (since computer memory uses powers of 2).*

120. The TI-84 Plus graphing calculator has 480 KB of "FLASH ROM." How many bytes is this?

121. The TI-84 Plus Silver Edition graphing calculator has 1.5 MB (megabytes) of FLASH ROM, where 1 MB is 1000 KB. How many bytes of FLASH ROM does this calculator have?

5.2 Negative Exponents and Scientific Notation

Negative Integers as Exponents ▪ Scientific Notation ▪
Multiplying, Dividing, and Significant Digits ▪ Problem Solving
Using Scientific Notation

We now attach a meaning to negative exponents. Once we understand both positive and negative exponents, we can study a method of writing numbers known as *scientific notation*.

Negative Integers as Exponents

Let's define negative exponents so that the rules that apply to whole-number exponents will hold for all integer exponents. To do so, consider a^{-5} and the rule for adding exponents:

$$a^{-5} = a^{-5} \cdot 1 \qquad \text{Using the identity property of 1}$$

$$= \frac{a^{-5}}{1} \cdot \frac{a^5}{a^5} \qquad \text{Writing 1 as } \frac{a^5}{a^5} \text{ and } a^{-5} \text{ as } \frac{a^{-5}}{1}$$

$$= \frac{a^{-5+5}}{a^5} \qquad \text{Adding exponents}$$

$$= \frac{1}{a^5}. \qquad -5 + 5 = 0 \text{ and } a^0 = 1$$

This leads to our definition of negative exponents.

Negative Exponents   For any real number a that is nonzero and any integer n,

$$a^{-n} = \frac{1}{a^n}.$$

(The numbers a^{-n} and a^n are reciprocals of each other.)

EXAMPLE 1   Express using positive exponents and, if possible, simplify.

a) m^{-3} b) 4^{-2}

c) $(-3)^{-2}$ d) ab^{-1}

SOLUTION

a) $m^{-3} = \dfrac{1}{m^3}$ m^{-3} is the reciprocal of m^3.

b) $4^{-2} = \dfrac{1}{4^2} = \dfrac{1}{16}$ 4^{-2} is the reciprocal of 4^2.
   Note that $4^{-2} \neq 4(-2)$.

c) $(-3)^{-2} = \dfrac{1}{(-3)^2} = \dfrac{1}{(-3)(-3)} = \dfrac{1}{9}$ $\begin{cases} (-3)^{-2} \text{ is the reciprocal of } (-3)^2. \\ \text{Note that } (-3)^{-2} \neq -\dfrac{1}{3^2}. \end{cases}$

d) $ab^{-1} = a\left(\dfrac{1}{b^1}\right) = a\left(\dfrac{1}{b}\right) = \dfrac{a}{b}$ b^{-1} is the reciprocal of b^1.

Study Tip

Connect the Dots

Whenever possible, look for connections between concepts covered in different sections or chapters. For example, both Sections 5.1 and 5.2 discuss exponents, and both Chapters 5 and 6 cover polynomials.

CAUTION!   A negative exponent does not, in itself, indicate that an expression is negative. As shown in Example 1,

$$4^{-2} \neq 4(-2) \quad \text{and} \quad (-3)^{-2} \neq -\frac{1}{3^2}.$$

The following is another way to illustrate why negative exponents are defined as they are.

On this side, we divide by 5 at each step.

$$
\begin{aligned}
125 &= 5^3 \\
25 &= 5^2 \\
5 &= 5^1 \\
1 &= 5^0 \\
\frac{1}{5} &= 5^? \\
\frac{1}{25} &= 5^?
\end{aligned}
$$

On this side, the exponents decrease by 1.

To continue the pattern, it follows that

$$\frac{1}{5} = \frac{1}{5^1} = 5^{-1}, \qquad \frac{1}{25} = \frac{1}{5^2} = 5^{-2}, \quad \text{and, in general,} \quad \frac{1}{a^n} = a^{-n}.$$

EXAMPLE 2 Express $\dfrac{1}{x^7}$ using negative exponents.

SOLUTION We know that $\dfrac{1}{a^n} = a^{-n}$. Thus, $\dfrac{1}{x^7} = x^{-7}$.

The rules for exponents still hold when exponents are negative.

EXAMPLE 3 Simplify. Do not use negative exponents in the answer.

a) $t^5 \cdot t^{-2}$ **b)** $(5x^{-2}y^3)^{-4}$ **c)** $\dfrac{x^{-4}}{x^{-5}}$

d) $\dfrac{1}{t^{-5}}$ **e)** $\dfrac{s^{-3}}{t^{-5}}$ **f)** $\dfrac{-10x^{-3}y}{5x^2y^5}$

SOLUTION

a) $t^5 \cdot t^{-2} = t^{5+(-2)} = t^3$ Adding exponents

b) $(5x^{-2}y^3)^{-4} = 5^{-4}(x^{-2})^{-4}(y^3)^{-4}$ **Raising each factor to the exponent of -4**

$\qquad = \dfrac{1}{5^4}x^8y^{-12} = \dfrac{x^8}{625y^{12}}$ **Multiplying exponents; writing with positive exponents**

c) $\dfrac{x^{-4}}{x^{-5}} = x^{-4-(-5)} = x^1 = x$ **We subtract exponents even if the exponent in the denominator is negative.**

d) Since $\dfrac{1}{a^n} = a^{-n}$, we have $\dfrac{1}{t^{-5}} = t^{-(-5)} = t^5$.

e) $\dfrac{s^{-3}}{t^{-5}} = s^{-3} \cdot \dfrac{1}{t^{-5}} = \dfrac{1}{s^3} \cdot t^5 = \dfrac{t^5}{s^3}$ Using the result from part (d) above

f) $\dfrac{-10x^{-3}y}{5x^2y^5} = \dfrac{-10}{5} \cdot \dfrac{x^{-3}}{x^2} \cdot \dfrac{y^1}{y^5}$ Note that the -10 and 5 are factors.

$\qquad = -2 \cdot x^{-3-2} \cdot y^{1-5}$ Using the quotient rule twice

$\qquad = -2x^{-5}y^{-4} = \dfrac{-2}{x^5y^4}$ Simplifying

The result from Example 3(e) can be generalized.

Factors and Negative Exponents   For any nonzero real numbers a and b and any integers m and n,

$$\frac{a^{-n}}{b^{-m}} = \frac{b^m}{a^n}.$$

(A factor can be moved to the other side of the fraction bar if the sign of the exponent is changed.)

EXAMPLE 4 Simplify: $\dfrac{-15x^{-7}}{5y^2z^{-4}}$.

SOLUTION The factors in the numerator are -15 and x^{-7}. Note that the base of the exponent -7 is x. Similarly, the factors in the denominator are 5, y^2, and z^{-4}. We can move the factors x^{-7} and z^{-4} to the other side of the fraction bar if we change the sign of each exponent:

$$\frac{-15x^{-7}}{5y^2z^{-4}} = \frac{-3z^4}{y^2x^7}, \quad \text{or} \quad \frac{-3z^4}{x^7y^2}.$$

When possible, be sure to simplify the constant factors: $\frac{-15}{5} = -3$.

Another way to change the sign of the exponent is to take the reciprocal of the base. To understand why this is true, note that

$$\left(\frac{s}{t}\right)^{-5} = \frac{s^{-5}}{t^{-5}} = \frac{t^5}{s^5} = \left(\frac{t}{s}\right)^5.$$

This often provides the easiest way to simplify an expression containing a negative exponent.

Reciprocals and Negative Exponents For any nonzero real numbers a and b and any integer n,

$$\left(\frac{a}{b}\right)^{-n} = \left(\frac{b}{a}\right)^n.$$

(Any base to an exponent is equal to the reciprocal of the base raised to the opposite exponent.)

EXAMPLE 5 Simplify: $\left(\dfrac{x^4}{2y}\right)^{-3}$.

SOLUTION

$$\left(\frac{x^4}{2y}\right)^{-3} = \left(\frac{2y}{x^4}\right)^3$$

Taking the reciprocal of the base and changing the sign of the exponent

$$= \frac{(2y)^3}{(x^4)^3}$$

Raising a quotient to an exponent by raising both the numerator and denominator to the exponent

$$= \frac{2^3y^3}{x^{12}}$$

Raising a product to an exponent; using the power rule in the denominator

$$= \frac{8y^3}{x^{12}}$$

Cubing 2

Scientific Notation

When we are working with the very large or very small numbers that frequently occur in science, **scientific notation** provides a useful way of writing numbers. The following are examples of scientific notation.

The mass of the earth:

6.0×10^{24} kilograms (kg) $= 6{,}000{,}000{,}000{,}000{,}000{,}000{,}000{,}000$ kg

The mass of a hydrogen atom:

1.7×10^{-24} g $= 0.0000000000000000000000017$ g

Scientific Notation *Scientific notation* for a number is an expression of the type

$$N \times 10^m,$$

where N is at least 1 but less than 10 ($1 \leq N < 10$), N is expressed in decimal notation, and m is an integer.

Converting from scientific notation to decimal notation involves multiplying by a power of 10. Consider the following.

Scientific Notation $N \times 10^m$	Multiplication	Decimal Notation
4.52×10^2	4.52×100	452.
4.52×10^1	4.52×10	45.2
4.52×10^0	4.52×1	4.52
4.52×10^{-1}	4.52×0.1	0.452
4.52×10^{-2}	4.52×0.01	0.0452

Note that when m, the power of 10, is positive, the decimal point moves right m places in decimal notation. When m is negative, the decimal point moves left $|m|$ places.

EXAMPLE 6 Convert to decimal notation.

a) 7.893×10^5 **b)** 4.7×10^{-8}

SOLUTION

a) Since the exponent is positive, the decimal point moves to the right:

7.89300. $7.893 \times 10^5 = 789{,}300$ **The decimal point moves 5 places to the right.**

5 places

b) Since the exponent is negative, the decimal point moves to the left:

0.00000004.7 $4.7 \times 10^{-8} = 0.000000047$ **The decimal point moves 8 places to the left.**

8 places

To convert from decimal notation to scientific notation, this procedure is reversed.

EXAMPLE 7 Write in scientific notation: **(a)** 83,000; **(b)** 0.0327.

SOLUTION

a) We need to find m such that $83{,}000 = 8.3 \times 10^m$. To change 8.3 to 83,000 requires moving the decimal point 4 places to the right. This can be accomplished by multiplying by 10^4. Thus,

$$83{,}000 = 8.3 \times 10^4. \quad \text{This is scientific notation.}$$

b) We need to find m such that $0.0327 = 3.27 \times 10^m$. To change 3.27 to 0.0327 requires moving the decimal point 2 places to the left. This can be accomplished by multiplying by 10^{-2}. Thus,

$$0.0327 = 3.27 \times 10^{-2}. \qquad \text{This is scientific notation.} \qquad \blacktriangle$$

Conversions to and from scientific notation are often made mentally. Remember that positive exponents are used when representing large numbers and negative exponents are used when representing numbers between 0 and 1.

Multiplying, Dividing, and Significant Digits

In the world of science, it is important to know just how accurate a measurement is. For example, the measurement 5.12×10^3 km is more precise than the measurement 5.1×10^3 km. We say that 5.12×10^3 has three **significant digits** whereas 5.1×10^3 has only two significant digits. If 5.1×10^3, or 5100, includes no rounding in the tens column, we would indicate that by writing 5.10×10^3.

> When two or more measurements written in scientific notation are multiplied or divided, the result should be rounded so that it has the same number of significant digits as the measurement with the fewest significant digits. Rounding should be performed at the *end* of the calculation.

Thus,

$$\underset{\substack{\uparrow \\ 2 \text{ digits}}}{(3.1 \times 10^{-3} \text{ mm})} \, \underset{\substack{\uparrow \\ 3 \text{ digits}}}{(2.45 \times 10^{-4} \text{ mm})} = 7.595 \times 10^{-7} \text{ mm}^2$$

should be rounded to

$$\overset{2 \text{ digits}}{\overbrace{7.6}} \times 10^{-7} \text{ mm}^2.$$

> When two or more measurements written in scientific notation are added or subtracted, the result should be rounded so that it has as many decimal places as the measurement with the fewest decimal places.

For example,

$$\underset{\substack{\uparrow \\ 4 \text{ decimal} \\ \text{places}}}{1.6354 \times 10^4 \text{ km}} + \underset{\substack{\uparrow \\ 3 \text{ decimal} \\ \text{places}}}{2.078 \times 10^4 \text{ km}} = 3.7134 \times 10^4 \text{ km}$$

should be rounded to

$$\overset{\substack{3 \text{ decimal} \\ \text{places}}}{\overbrace{3.713}} \times 10^4 \text{ km}.$$

EXAMPLE 8 Multiply and write scientific notation for the answer:

$$(7.2 \times 10^5)(4.3 \times 10^9).$$

SOLUTION We have

$$(7.2 \times 10^5)(4.3 \times 10^9) = (7.2 \times 4.3)(10^5 \times 10^9) \qquad \text{Using the commutative and associative laws}$$

$$= 30.96 \times 10^{14}. \qquad \text{Adding exponents}$$

To find scientific notation for this result, we convert 30.96 to scientific notation and simplify:

$$30.96 \times 10^{14} = (3.096 \times 10^1) \times 10^{14}$$
$$= 3.096 \times 10^{15}$$
$$\approx 3.1 \times 10^{15}. \qquad \text{Rounding to 2 significant digits}$$

EXAMPLE 9 Divide and write scientific notation for the answer:

$$\frac{3.48 \times 10^{-7}}{4.64 \times 10^6}.$$

SOLUTION

$$\frac{3.48 \times 10^{-7}}{4.64 \times 10^6} = \frac{3.48}{4.64} \times \frac{10^{-7}}{10^6} \qquad \text{Separating factors. Our answer must have 3 significant digits.}$$

$$= 0.75 \times 10^{-13} \qquad \text{Subtracting exponents; simplifying}$$

$$= (7.5 \times 10^{-1}) \times 10^{-13} \qquad \text{Converting 0.75 to scientific notation}$$

$$= 7.50 \times 10^{-14} \qquad \text{Adding exponents. We write 7.50 to indicate 3 significant digits.}$$

Exponents and Scientific Notation

To simplify an exponential expression like 3^5, we can use a calculator's exponentiation key, usually labeled **^**. If the exponent is 2, we can also use the **x²** key. If it is -1, we can use the **x⁻¹** key. If the exponent is a single number, not an expression, we do not need to enclose it in parentheses.

Graphing calculators will accept entries using scientific notation, and will normally write very large or very small numbers using scientific notation. The **EE** key, the 2nd option associated with the **,** key, is used to enter scientific notation. On the calculator screen, a notation like E22 represents $\times 10^{22}$.

> **EXAMPLE 10**   Use a graphing calculator to calculate 3^5, $(-4.7)^2$, and $(-8)^{-1}$.
>
> **SOLUTION**   To calculate 3^5, we press ③ ⌃ ⑤ **ENTER**.
> To calculate $(-4.7)^2$, we press ((−) ④ . ⑦) ⌃
> ② **ENTER** or ((−) ④ . ⑦) **x²** **ENTER**.
> To calculate $(-8)^{-1}$, we press ((−) ⑧) **x⁻¹** **ENTER** or
> ((−) ⑧) ⌃ (−) ① **ENTER**.

```
3^5
                  243
(-4.7)^2
                22.09
(-4.7)2
                22.09
```

```
(-8)^-1
                 -.125
(-8)^-1
                 -.125
Ans▶Frac
                  -1/8
```

> We see that $3^5 = 243$, $(-4.7)^2 = 22.09$, and $(-8)^{-1} = -0.125$, or $-\frac{1}{8}$.

> **EXAMPLE 11**   Use a graphing calculator to calculate
>
> $$(7.5 \times 10^8)(1.2 \times 10^{-14}).$$
>
> **SOLUTION**   We press ⑦ . ⑤ **EE** ⑧ × ① . ② **EE**
> (−) ① ④ **ENTER**.

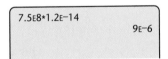
```
7.5E8*1.2E-14
                  9E-6
```

> The result shown is then read as 9×10^{-6}.

Problem Solving Using Scientific Notation

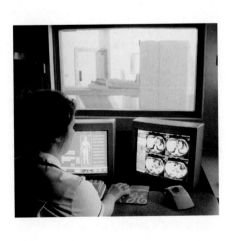

> **EXAMPLE 12**   *Information Technology.* In 2003, the University of California, Berkeley, estimated that in the previous year, approximately 5 exabytes of new information were generated by the worldwide population of 6.3 billion people. If 1 exabyte is 10^{12} megabytes, find the average number of megabytes of information generated per person in 2002.
>
> **SOLUTION**
>
> 1. **Familiarize.** If necessary, we can consult an outside reference to confirm that one billion is 1,000,000,000, or 10^9. Thus, 6.3 billion is 6.3×10^9. Note also that to find an average we need to divide. We let $a =$ the average number of megabytes of information generated, per person, in 2002.

2. **Translate.** To find the average amount of information generated per person, we divide the total amount of information generated by the number representing the world population:

$$a = \frac{5.0 \times 10^{12} \text{ megabytes}}{6.3 \times 10^{9} \text{ people}}.$$

3. **Carry out.** We calculate and write scientific notation for the result:

$$a = \frac{5.0 \times 10^{12} \text{ megabytes}}{6.3 \times 10^{9} \text{ people}}$$

$$= \frac{5.0}{6.3} \times \frac{10^{12} \text{ megabytes}}{10^{9} \text{ people}}$$

$$\approx 0.79 \times 10^{3} \text{ megabytes/person} \qquad \text{Rounding to 2 significant digits}$$

$$\left.\begin{array}{l} \approx 7.9 \times 10^{-1} \times 10^{3} \text{ megabytes/person} \\ \approx 7.9 \times 10^{2} \text{ megabytes/person.} \end{array}\right\} \begin{array}{l} \text{Writing scientific notation} \end{array}$$

4. **Check.** To check, we multiply our answer, the average number of megabytes per person, by the worldwide population:

$$\underbrace{(7.9 \times 10^{2} \text{ megabytes per person})}_{\substack{\text{Average amount of} \\ \text{information generated}}}\underbrace{(6.3 \times 10^{9} \text{ people})}_{\text{Worldwide population}}$$

$$= (7.9 \times 6.3)(10^{2} \times 10^{9})\frac{\text{megabytes}}{\text{person}} \cdot \text{people}$$

$$= 49.77 \times 10^{11} \text{ megabytes}$$

$$\approx 5.0 \times 10^{1} \times 10^{11} \text{ megabytes} \qquad \text{Rounding to 2 significant digits}$$

$$\approx 5.0 \times 10^{12} \text{ megabytes.}$$

Our answer checks.

5. **State.** An average of 7.9×10^{2} megabytes of information was generated by each person in the world in 2002.

5.2 EXERCISE SET

FOR EXTRA HELP

MathXL · MyMathLab · InterAct Math · Tutor Center · AW Math Tutor Center · Video Lectures on CD: Disc 3 · Student's Solutions Manual

 Concept Reinforcement *State whether scientific notation for each of the following numbers would include a positive or a negative power of 10.*

1. The length of an Olympic marathon, in centimeters

2. The thickness of a cat's whisker, in meters

3. The mass of a hydrogen atom, in grams

4. The mass of a pickup truck, in grams

5. The time between leap years, in seconds

6. The time between a bird's heartbeats, in hours

Write an equivalent expression without negative exponents and, if possible, simplify.

7. 7^{-2} **8.** 10^{-4} **9.** $(-2)^{-6}$

10. $(-3)^{-4}$ **11.** a^{-3} **12.** n^{-6}

13. $\dfrac{1}{5^{-3}}$ **14.** $\dfrac{1}{2^{-6}}$ **15.** 7^{-1}

16. 3^{-1} **17.** $8x^{-3}$ **18.** $7x^{-3}$

19. $3a^8b^{-6}$ **20.** $5a^{-7}b^4$ **21.** $\dfrac{z^{-4}}{3x^5}$

22. $\dfrac{y^{-5}}{x^{-3}}$ **23.** $\dfrac{x^{-2}y^7}{z^{-4}}$ **24.** $\dfrac{y^4z^{-3}}{x^{-2}}$

25. $\left(\dfrac{a}{2}\right)^{-3}$ **26.** $\left(\dfrac{x}{3}\right)^{-4}$

Write an equivalent expression with negative exponents.

27. $\dfrac{1}{8^4}$ **28.** $\dfrac{1}{t^5}$ **29.** $\dfrac{1}{(-5)^6}$

30. $\dfrac{1}{(2x)^3}$ **31.** $\dfrac{1}{x}$ **32.** $\dfrac{1}{2}$

33. x^5 **34.** 4^7 **35.** $4x^3$

36. $-4y^5$ **37.** $\dfrac{1}{5y^3}$ **38.** $\dfrac{1}{3y^4}$

Simplify. If negative exponents appear in the answer, write a second answer using only positive exponents.

39. $8^{-2} \cdot 8^{-4}$ **40.** $9^{-1} \cdot 9^{-6}$

41. $b^2 \cdot b^{-5}$ **42.** $a^4 \cdot a^{-3}$

43. $a^{-3} \cdot a^4 \cdot a$ **44.** $x^{-6} \cdot x^5 \cdot x$

45. $(5a^{-2}b^{-3})(2a^{-4}b)$ **46.** $(3a^{-5}b^{-7})(2ab^{-2})$

47. $\dfrac{10^{-3}}{10^6}$ **48.** $\dfrac{12^{-4}}{12^8}$

49. $\dfrac{2^{-7}}{2^{-5}}$ **50.** $\dfrac{9^{-4}}{9^{-6}}$

51. $\dfrac{y^4}{y^{-5}}$ **52.** $\dfrac{a^3}{a^{-2}}$

53. $\dfrac{24a^5b}{-8a^4b^2}$ **54.** $\dfrac{9a^2}{3ab^3}$

55. $\dfrac{-6x^{-2}y^4z^8}{24x^{-5}y^6z^{-3}}$ **56.** $\dfrac{8a^6b^{-4}c^8}{32a^{-4}b^5c^9}$

57. $(9^3)^{-4}$ **58.** $(8^4)^{-3}$

59. $(t^{-8})^{-5}$ **60.** $(x^{-4})^{-3}$

61. $(x^6y^{-2})^{-3}$ **62.** $(x^{-2}y^7)^{-5}$

63. $\dfrac{(x^5)^2(x^{-3})^4}{(x^2)^3}$ **64.** $\dfrac{(a^{-2})^3(a^4)^2}{(a^3)^{-3}}$

65. $\dfrac{(2a^3)^3 4a^{-3}}{(a^2)^5}$ **66.** $\dfrac{(3x^2)^3 2x^{-4}}{(x^4)^2}$

Aha! **67.** $(8x^{-3}y^2)^{-4}(8x^{-3}y^2)^4$

68. $(2a^{-1}b^3)^{-2}(2a^{-1}b^3)^{-2}$

69. $\left(\dfrac{a^4}{3}\right)^{-2}$ **70.** $\left(\dfrac{7}{x^{-3}}\right)^{-4}$

71. $\left(\dfrac{m^{-1}}{n^{-4}}\right)^3$ **72.** $\left(\dfrac{x^2y}{z^{-5}}\right)^3$

73. $\left(\dfrac{-4x^4y^{-2}}{5x^{-1}y^4}\right)^{-4}$ **74.** $\left(\dfrac{2x^3y^{-2}}{3y^{-3}}\right)^3$

Aha! **75.** $\left(\dfrac{4a^3b^{-9}}{6a^{-2}b^5}\right)^0$ **76.** $\left(\dfrac{5x^0y^{-7}}{2x^{-2}y^4}\right)^1$

📱 *Evaluate using a calculator.*

77. -8^4 **78.** $(-8)^4$ **79.** $(-2)^{-4}$

80. -2^{-4} **81.** $3^4 5^{-3}$ **82.** $\left(\dfrac{2}{3}\right)^{-5}$

Convert to decimal notation.

83. 4×10^{-4} **84.** 5×10^{-5}

85. 6.73×10^8 **86.** 9.24×10^7

87. 8.923×10^{-10} **88.** 7.034×10^{-2}

89. 9.03×10^{10} **90.** 9.001×10^{10}

Convert to scientific notation.

91. 47,000,000,000   **92.** 2,600,000,000,000

93. 0.000000016   **94.** 0.000000263

95. 407,000,000,000   **96.** 3,090,000,000,000

97. 0.000000603   **98.** 0.00000000802

📱 *Write scientific notation for the number represented on each calculator screen.*

99.

5.02E18

100.
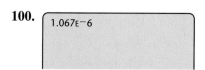

1.067E−6

101.

−3.05E−10

102.

−5.968E27

Simplify and write scientific notation for the answer. Use the correct number of significant digits.

103. $(2.3 \times 10^6)(4.2 \times 10^{-11})$

104. $(6.5 \times 10^3)(5.2 \times 10^{-8})$

105. $(2.34 \times 10^{-8})(5.7 \times 10^{-4})$

106. $(4.26 \times 10^{-6})(8.2 \times 10^{-6})$

Aha! **107.** $(2.0 \times 10^6)(3.02 \times 10^{-6})$

108. $(7.04 \times 10^{-9})(9.01 \times 10^{-7})$

109. $\dfrac{5.1 \times 10^6}{3.4 \times 10^3}$ **110.** $\dfrac{8.5 \times 10^8}{3.4 \times 10^5}$

111. $\dfrac{7.5 \times 10^{-9}}{2.5 \times 10^{-4}}$ **112.** $\dfrac{12.6 \times 10^8}{4.2 \times 10^{-3}}$

113. $\dfrac{1.23 \times 10^8}{6.87 \times 10^{-13}}$

114. $\dfrac{4.95 \times 10^{-3}}{1.64 \times 10^{10}}$

115. $5.9 \times 10^{23} + 6.3 \times 10^{23}$

116. $7.8 \times 10^{-34} + 5.4 \times 10^{-34}$

Solve. Write the answers using scientific notation.

117. *Radioactivity.* The lightest known particle in the universe, a neutrino has a maximum mass of 1.8×10^{-36} kg (*Source: Guinness Book of World Records* 2004). What is the smallest number of neutrinos that could have the same mass as an alpha particle of mass 3.62×10^{-27} kg that results from the decay of radon?

118. *Printing and Engraving.* A ton of five-dollar bills is worth $4,540,000. How many pounds does a five-dollar bill weigh?

119. *Astronomy.* The average distance of Earth from the sun is about 9.3×10^7 mi. About how far does Earth travel in a yearly orbit about the sun? (Assume a circular orbit.)

120. *Astronomy.* The brightest star in the night sky, Sirius, is about 4.704×10^{13} mi from Earth. The distance light travels in one year, or one light year, is 5.88×10^{12} mi. How many light years is it from Earth to Sirius?

121. *Astronomy.* The diameter of the Milky Way galaxy is approximately 5.88×10^{17} mi. The distance light travels in one year, or one light year, is 5.88×10^{12} mi. How many light years is it from one end of the galaxy to the other?

122. *Astronomy.* The diameter of Jupiter is about 1.43×10^5 km. A day on Jupiter lasts about 10 hr. At what speed is Jupiter's equator spinning?

123. *Telecommunications.* A wire will be used for 375 km of transmission line. The wire has a diameter of 1.2 cm. What is the volume of wire needed for the line? (*Hint:* The volume of a cylinder is given by $V = \pi r^2 h$.)

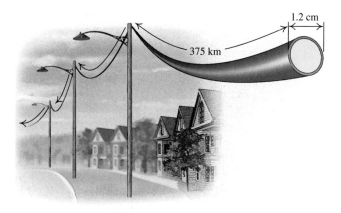

124. *High-Tech Fibers.* A carbon nanotube is a thin cylinder of carbon atoms that, pound for pound, is stronger than steel and may one day be used in clothing (*Source: The Indianapolis Star,* 6/15/03). With a diameter of about 4.0×10^{-10} in., a fiber can be made 100 yd long. Find the volume of such a fiber.

125. *Biology.* An average of 4.55×10^{11} bacteria live in each pound of U.S. mud. There are 60.0 drops in one teaspoon and 6.0 teaspoons in an ounce. (*Source: Harper's Magazine,* April 1996, p. 13) How many bacteria live in a drop of U.S. mud?

126. *Astronomy.*   If a star 5.9×10^{14} mi from the earth were to explode today, its light would not reach us for 100 yr. How far does light travel in 13 weeks?

127. *Home Maintenance.*   The thickness of a sheet of plastic is measured in *mils*, where $1 \text{ mil} = \frac{1}{1000}$ in. To help conserve heat, the foundation of a 24-ft by 32-ft rectangular home is covered with a 4-ft high sheet of 8-mil plastic. Find the volume of plastic used.

128. *Office Supplies.*   A ream of copier paper weighs 2.25 kg. How much does a sheet of copier paper weigh?

TW 129. Without performing actual computations, explain why 3^{-29} is smaller than 2^{-29}.

TW 130. List two advantages of using scientific notation. Answers may vary.

Skill Maintenance

Simplify. [1.8]

131. $2x - 7 - (3x + 1)$

132. $3x - 2[x - (1 - x)]$

Solve.

133. $2x - 7 = 3x + 1$ [2.2]

134. $3x = 2[x - (1 - x)]$ [2.2]

135. $2x - 7 \leq 3x + 1$ [2.6]

136. $3x > 2[x - (1 - x)]$ [2.6]

Synthesis

TW 137. Explain why $(-17)^{-8}$ is positive.

TW 138. Is the following true or false, and why?
$$5^{-6} > 4^{-9}$$

TW 139. Some numbers exceed the limits of the calculator. Enter 1.3×10^{-1000} and 1.3×10^{1000} and explain the results.

Simplify.

140. $(7^{-12})^2 \cdot 7^{25}$

141. $\dfrac{125^{-4}(25^2)^4}{125}$

142. $\dfrac{27^{-2}(81^2)^3}{9^8}$

143. $\dfrac{4.2 \times 10^8[(2.5 \times 10^{-5}) \div (5.0 \times 10^{-9})]}{3.0 \times 10^{-12}}$

144. Write $8^{-3} \cdot 32 \div 16^2$ as a power of 2.

145. Write $81^3 \cdot 27 \div 9^2$ as a power of 3.

146. Compare 8×10^{-90} and 9×10^{-91}. Which is the larger value? How much larger? Write scientific notation for the difference.

147. Write the reciprocal of 8.00×10^{-23} in scientific notation.

148. Write $\frac{4}{32}$ in decimal notation, simplified fraction notation, and scientific notation.

149. A grain of sand is placed on the first square of a chessboard, two grains on the second square, four grains on the third, eight on the fourth, and so on. Without a calculator, use scientific notation to approximate the number of grains of sand required for the 64th square. (*Hint*: Use the fact that $2^{10} \approx 10^3$.)

5.3   Polynomials and Polynomial Functions

Terms ■ Types of Polynomials ■ Degree and Coefficients ■ Combining Like Terms ■ Polynomial Functions ■ Graphs of Polynomial Functions

We now examine an important algebraic expression known as a *polynomial.* Certain polynomials have appeared earlier in this text so you already have some experience working with them.

Terms

At this point, we have seen a variety of algebraic expressions like

$$3a^2b^4, \qquad 2l + 2w, \quad \text{and} \quad 5x^2 + x - 2.$$

Of these, $3a^2b^4$, $2l$, $2w$, $5x^2$, x, and -2 are examples of *terms*. A **term** can be a number (like -2), a variable (like x), a product of numbers and/or variables (like $3a^2b^4$, $2l$, or $5x^2$), or a quotient of numbers and/or variables $\left(\text{like } \dfrac{7}{t} \text{ or } \dfrac{a^2b^3}{4c} \right)$.

Types of Polynomials

A term that is a product of constants and/or variables is called a **monomial.*** All the terms listed above, except for the quotients, are monomials. Note that variables in monomials are raised to only whole-number exponents. Other examples of monomials are

$$7, \qquad t, \qquad 23x^2y, \quad \text{and} \quad \tfrac{3}{7}a^5.$$

A **polynomial** is a monomial or a sum of monomials. The following are examples of polynomials:

$$4x + 7, \quad \tfrac{2}{3}t^2, \quad 6a + 7, \quad -5n^2 + n - 1, \quad 42r^5, \quad x, \quad \text{and} \quad 0.$$

The following algebraic expressions are *not* polynomials:

$$\textbf{(1)} \ \frac{x+3}{x-4}, \qquad \textbf{(2)} \ 5x^3 - 2x^2 + \frac{1}{x}, \qquad \textbf{(3)} \ \frac{1}{x^3 - 2}.$$

Expressions (1) and (3) are not polynomials because they represent quotients, not sums. Expression (2) is not a polynomial because $1/x$ is not a monomial. ($1/x = x^{-1}$, and -1 is not a whole-number exponent.)

When a polynomial is written as a sum of monomials, each monomial is called a *term of the polynomial.*

---

*Note that a term, but not a monomial, can include division by a variable.

EXAMPLE 1   Identify the terms of the polynomial $3t^4 - 5t^6 - 4t + 2$.

SOLUTION   The terms are $3t^4$, $-5t^6$, $-4t$, and 2. We can see this by rewriting all subtractions as additions of opposites:

$$3t^4 - 5t^6 - 4t + 2 = 3t^4 + (-5t^6) + (-4t) + 2.$$

These are the terms of the polynomial.

A polynomial that is composed of two terms is called a **binomial,** whereas those composed of three terms are called **trinomials.** Polynomials with four or more terms have no special name.

Monomials	Binomials	Trinomials	No Special Name
$4x^2$	$2x + 4$	$3t^3 + 4t + 7$	$4x^3 - 5x^2 + xy - 8$
9	$3a^5 + 6bc$	$6x^7 - 8z^2 + 4$	$z^5 + 2z^4 - z^3 + 7z + 3$
$-7a^{19}b^5$	$-9x^7 - 6$	$4x^2 - 6x - \frac{1}{2}$	$4x^6 - 3x^5 + x^4 - x^3 + 2x - 1$

Degree and Coefficients

The **degree of a term** of a polynomial is the number of variable factors in that term. Thus the degree of $7t^2$ is 2 because $7t^2$ has two variable factors: $7t^2 = 7 \cdot t \cdot t$. We will revisit the meaning of degree in Section 5.7 when polynomials in several variables are examined.

EXAMPLE 2   Determine the degree of each term: **(a)** $8x^4$; **(b)** $3x$; **(c)** 7.

SOLUTION

a) The degree of $8x^4$ is 4. x^4 represents 4 variable factors: $x \cdot x \cdot x \cdot x$.

b) The degree of $3x$ is 1.   There is 1 variable factor.

c) The degree of 7 is 0.   There is no variable factor.

The part of a term that is a constant factor is the **coefficient** of that term. Thus the coefficient of $3x$ is 3, and the coefficient for the term 7 is simply 7.

EXAMPLE 3   Identify the coefficient of each term in the polynomial

$$4x^3 - 7x^2y + x - 8.$$

SOLUTION

The coefficient of $4x^3$ is 4.

The coefficient of $-7x^2y$ is -7.

The coefficient of the third term is 1, since $x = 1x$.

The coefficient of -8 is simply -8.

The **leading term** of a polynomial is the term of highest degree. Its coefficient is called the **leading coefficient** and its degree is referred to as the **degree of the polynomial.** To see how this terminology is used, consider the polynomial

$$3x^2 - 8x^3 + 5x^4 + 7x - 6.$$

The *terms* are	$3x^2,$	$-8x^3,$	$5x^4,$	$7x,$	and $-6.$
The *coefficients* are	3,	$-8,$	5,	7,	and $-6.$
The *degree of each term* is	2,	3,	4,	1,	and 0.

The *leading term* is $5x^4$ and the *leading coefficient* is 5.

The *degree of the polynomial* is 4.

Combining Like Terms

Recall from Section 1.8 that *like*, or *similar*, *terms* are either constant terms or terms containing the same variable(s) raised to the same power(s). To simplify certain polynomials, we can often *combine*, or *collect*, like terms.

EXAMPLE 4   Identify the like terms in $4x^3 + 5x - 7x^2 + 2x^3 + x^2.$

SOLUTION

Like terms:	$4x^3$	and $2x^3$	Same variable and exponent
Like terms:	$-7x^2$	and x^2	Same variable and exponent

EXAMPLE 5   Combine like terms.

a) $2x^3 - 6x^3$ 　　　　　　　　　　　　**b)** $5x^2 + 7 + 2x^4 + 4x^2 - 11 - 2x^4$

c) $7a^3 - 5a^2 + 9a^3 + a^2$ 　　　　　**d)** $\frac{2}{3}x^4 - x^3 - \frac{1}{6}x^4 + \frac{2}{5}x^3 - \frac{3}{10}x^3$

SOLUTION

a) $2x^3 - 6x^3 = (2 - 6)x^3$ 　　**Using the distributive law**

$$= -4x^3$$

b) $5x^2 + 7 + 2x^4 + 4x^2 - 11 - 2x^4$

$$= 5x^2 + 4x^2 + 2x^4 - 2x^4 + 7 - 11$$
$$= (5 + 4)x^2 + (2 - 2)x^4 + (7 - 11)$$
$$= 9x^2 - 0x^4 + (-4)$$
$$= 9x^2 - 4$$

These steps are often done mentally.

c) $7a^3 - 5a^2 + 9a^3 + a^2 = 7a^3 - 5a^2 + 9a^3 + 1a^2$

When a term appears without a coefficient, we can write in 1.

$$= 16a^3 - 4a^2$$

d) $\frac{2}{3}x^4 - x^3 - \frac{1}{6}x^4 + \frac{2}{5}x^3 - \frac{3}{10}x^3 = \left(\frac{2}{3} - \frac{1}{6}\right)x^4 + \left(-1 + \frac{2}{5} - \frac{3}{10}\right)x^3$

$$= \left(\frac{4}{6} - \frac{1}{6}\right)x^4 + \left(-\frac{10}{10} + \frac{4}{10} - \frac{3}{10}\right)x^3$$
$$= \frac{3}{6}x^4 - \frac{9}{10}x^3$$
$$= \frac{1}{2}x^4 - \frac{9}{10}x^3$$

Student Notes

Remember that when we combine like terms, we are forming equivalent expressions. In these examples, there are no equations to solve.

Note in Example 5 that the solutions are written so that the term of highest degree appears first, followed by the term of next highest degree, and so on. This is known as **descending order** and is the form in which answers will normally appear.

Polynomial Functions

A *polynomial function* is a function in which outputs are determined by evaluating a polynomial. For example, the function P given by

$$P(x) = 5x^7 + 3x^5 - 4x^2 - 5$$

is an example of a polynomial function.

If the polynomial describing a function has degree 1, we say that the function is *linear*. A **quadratic function** is described by a polynomial of degree 2, a **cubic function** by a polynomial of degree 3, and a **quartic function** by a polynomial of degree 4. We studied linear functions in Chapter 3, and we will study quadratic functions in Chapter 10.

To evaluate a polynomial function, we substitute a number for the variable just as we did in Chapter 3. The result will be a number.

EXAMPLE 6   Find $P(-5)$ for the polynomial function given by $P(x) = -x^2 + 4x - 1$.

SOLUTION   We evaluate the function using several methods discussed in Chapter 3.

Using algebraic substitution. We substitute -5 for x and carry out the operations using the rules for order of operations:

$$P(-5) = -(-5)^2 + 4(-5) - 1$$
$$= -25 - 20 - 1$$
$$= -46.$$

> **CAUTION!**   Note that $-(-5)^2 = -25$. We square the input first and then take its opposite.

Using function notation on a graphing calculator. We let $y_1 = -x^2 + 4x - 1$. We enter the function into the graphing calculator and evaluate $Y_1(-5)$ using the Y-VARS menu.

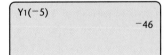

Using a table. If $y_1 = -x^2 + 4x - 1$, we can find the value of y_1 for $x = -5$ by setting Indpnt to Ask in the TABLE SETUP. If Depend is set to Auto, the y-value will appear when the x-value is entered. If Depend is set to Ask, we position the cursor in the y-column and press **ENTER** to see the corresponding y-value.

```
TABLE SETUP
  TblStart=1
  ΔTbl=1
  Indpnt:  Auto  Ask
  Depend:  Auto  Ask
```

X	Y1	
-5	-46	

X =

EXAMPLE 7 Blind-spot Radar. A vehicle's blind spot is the area around the vehicle that cannot be seen by the driver. Some cars are equipped with radar to warn the driver of an object in that car's blind spot. The following table lists the projected need for cars fitted with blind-spot radar, in millions. The data can be modeled by the polynomial function

$$f(x) = 0.07x^2 + 0.17x + 0.07,$$

where x is the number of years after 2006.

a) Use the given function to estimate the number of cars that will be fitted with blind-spot radar in 2014.

b) Use the graph to estimate $f(5)$ and tell what that number represents.

Year	Years After 2006	Cars Worldwide Fitted with Blind-spot Radar (in millions)
2006	0	0.1
2008	2	0.6
2010	4	2.0
2012	6	3.65

Source: "Strategy Analytics," *New Scientist*, January 14, 2006

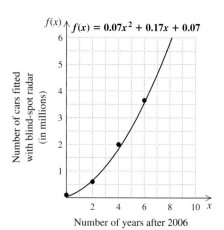

SOLUTION

a) We evaluate the function for $x = 8$, since 2014 is 8 years after 2006:

$$f(8) = 0.07(8)^2 + 0.17(8) + 0.07$$
$$= 0.07(64) + 0.17(8) + 0.07$$
$$= 4.48 + 1.36 + 0.07$$
$$= 5.91.$$

According to this model, there will be about 5,910,000 cars worldwide fitted with blind-spot radar in 2014.

b) To estimate $f(5)$ using the graph, we locate 5 on the horizontal axis. From there we move vertically to the graph of the function and then horizontally to the $f(x)$-axis, as shown below. This locates a value of about 2.7.

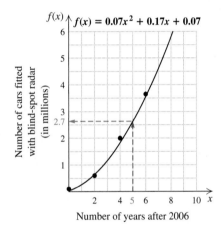

Number of years after 2006

Since $f(x)$ represents the number of cars, in millions, fitted with blind-spot radar x years after 2006, $f(5)$, or 2.7, indicates a prediction of 2.7 million cars fitted with blind-spot radar 5 years after 2006, or in 2011.

Note in Example 7 that 2.7 is a good estimate of the value of f at $x = 5$ found by evaluating the function:

$$f(5) = 0.07(5)^2 + 0.17(5) + 0.07 = 2.67.$$

Graphs of Polynomial Functions

Connecting the Concepts

FAMILIES OF GRAPHS OF FUNCTIONS

Often, the shape of the graph of a function can be predicted by examining the equation describing the function.

We examined *linear functions* and their graphs in Chapter 3. Functions f that can be expressed in the form $f(x) = mx + b$ have graphs that are straight lines. Their slope, or rate of change, is constant. We can think of linear functions as a "family" of functions, with the function $f(x) = x$ being the simplest such function.

We have also studied some non-linear functions. An *absolute-value function* of the form $f(x) = |ax + b|$, where $a \neq 0$, has a graph similar to that of $f(x) = |x|$.

Another nonlinear function that we will consider in Chapter 10 is a square-root function. A *square-root function* of the form $f(x) = \sqrt{ax + b}$, where $a \neq 0$, has a graph similar to the graph of $f(x) = \sqrt{x}$.

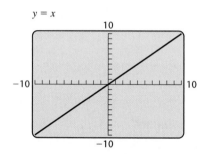

$y = x$

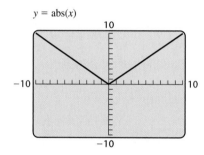

$y = \text{abs}(x)$

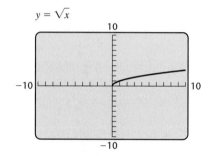

$y = \sqrt{x}$

These basic functions, along with others that we will develop throughout the text, form a **library of functions.** Knowing the general shape of the graphs of different types of functions will help in analyzing both data and graphs.

The graphs of polynomial functions have certain common characteristics.

Interactive Discovery

The following table lists some polynomial functions and some nonpolynomial functions. Graph each function and compare the graphs.

Polynomial Functions	Nonpolynomial Functions
$f(x) = x^2 + 3x + 5$ $f(x) = 4$	$f(x) = |x - 4|$ $f(x) = 1 + \sqrt{2x - 5}$
$f(x) = -0.5x^4 + 5x - 2.3$	$f(x) = \dfrac{x - 7}{2x}$

What are some characteristics of graphs of polynomial functions?

You may have noticed the following:

The graph of a polynomial function is "smooth," that is, there are no sharp corners.

The graph of a polynomial function is continuous, that is, there are no holes or breaks.

The domain of a polynomial function, unless otherwise specified, is all real numbers.

Recall from Section 3.8 that the domain of a function, when not specified, is the set of all real numbers for which the function is defined. If no restrictions are given, a polynomial function is defined for all real numbers. In other words,

The domain of a polynomial function is $(-\infty, \infty)$.

Note in Example 7 that the domain of the function f is restricted by the context of the problem. Since the function is defined for the number of years after 2006, negative values of x are not in the domain of the function. Thus the domain of f is $[0, \infty)$.

The range of a function is the set of all possible outputs (y-values) that correspond to inputs from the domain (x-values). For a polynomial function with an unrestricted domain, the range may be $(-\infty, \infty)$, or there may be a maximum value or a minimum value of the function. We can estimate the range of a function from its graph.

EXAMPLE 8   Estimate the range of each of the following functions from its graph.

a)

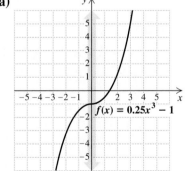

$f(x) = 0.25x^3 - 1$

b)

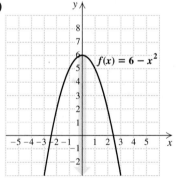

$f(x) = 6 - x^2$

c)

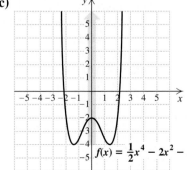

$f(x) = \frac{1}{2}x^4 - 2x^2 - 2$

SOLUTION The domain of each function represented in the graphs is $(-\infty, \infty)$.

a) Since there is no maximum or minimum y-value indicated on the graph of the first function, we estimate its range to be $(-\infty, \infty)$. The range is indicated by the shading on the y-axis.

b) The second function has a maximum value of about 6, and no minimum is indicated, so its range is about $(-\infty, 6]$.

c) The third function has a minimum value of about -4 and no maximum value, so its range is $[-4, \infty)$.

When we are estimating the range of a polynomial function from its graph, care must be taken to show enough of the graph to see its behavior. As you learn more about graphs of polynomial functions, you will be able to more readily choose appropriate scales or viewing windows. In this chapter, we will supply the graph or an appropriate viewing window for the graph of a polynomial function.

EXAMPLE 9 Graph each of the following functions in the standard viewing window and estimate the range of the function.

a) $f(x) = x^3 - 4x^2 + 5$ **b)** $g(x) = x^4 - 4x^2 + 5$

SOLUTION

a) The graph of $y = x^3 - 4x^2 + 5$ is shown on the left below. It appears that the graph extends downward and upward indefinitely, so the range of the function is $(-\infty, \infty)$.

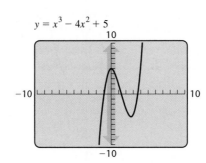 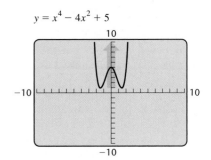

b) The graph of $y = x^4 - 4x^2 + 5$ is shown on the right above. The graph extends upward indefinitely but does not go below a y-value of 1. (This can be confirmed by tracing along the graph.) Since there is a minimum function value of 1 and no maximum function value, the range of the function is $[1, \infty)$.

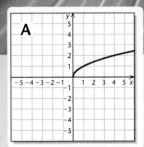

A

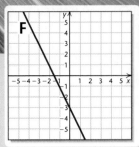

F

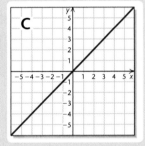

B

Visualizing the Graph

Match each equation with its graph.

1. $y = x$

2. $y = |x|$

3. $y = \sqrt{x}$

4. $y = x^2$

5. $y = 2x$

6. $y = -2x$

7. $y = 2x + 3$

8. $y = -2x - 3$

9. $y = x + 3$

10. $y = x - 3$

Answers on page A-19

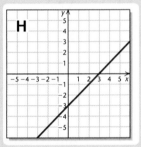

G

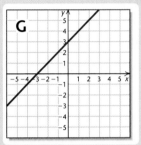

H

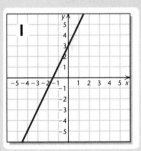

I

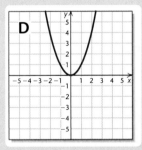

C

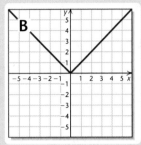

D

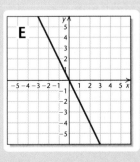

E

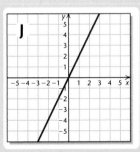

J

5.3 EXERCISE SET

🖐 *Concept Reinforcement In each of Exercises 1–8, match the description with the most appropriate algebraic expression from the column on the right.*

1. ____ A polynomial with four terms

2. ____ A trinomial written in descending order

3. ____ A polynomial with degree 5

4. ____ A polynomial with 7 as its leading coefficient

5. ____ A binomial with degree 7

6. ____ An expression with two terms that is not a binomial

7. ____ An expression with three terms that is not a trinomial

8. ____ A monomial of degree 0

a) $8x^3 + \dfrac{2}{x^2}$

b) $5x^4 + 3x^3 - 4x + 7$

c) $\dfrac{3}{x} - 6x^2 + 9$

d) $8t - 4t^5$

e) 5

f) $6x^2 + 7x^4 - 2x^3$

g) $4t - 2t^7$

h) $3t^2 + 4t + 7$

Determine whether each expression is a polynomial.

9. $3x - 7$

10. $-2x^5 + 9 - 7x^2$

11. $\dfrac{x^2 + x + 1}{x^3 - 7}$

12. -10

13. $\frac{1}{4}x^{10} - 8.6$

14. $\dfrac{3}{x^4} - \dfrac{1}{x} + 13$

Identify the terms of each polynomial.

15. $7x^4 + x^3 - 5x + 8$

16. $5a^3 + 4a^2 - a - 7$

17. $-t^4 + 7t^3 - 3t^2 + 6$

18. $n^5 - 4n^3 + 2n - 8$

Determine the coefficient and the degree of each term in each polynomial.

19. $4x^5 + 7x$

20. $9a^3 - 4a^2$

21. $9t^2 - 3t + 4$

22. $7x^4 + 5x - 3$

23. $x^4 - x^3 + 4x - 3$

24. $3a^4 - a^3 + a - 9$

*For each of the following polynomials, **(a)** list the degree of each term; **(b)** determine the leading term and the leading coefficient; and **(c)** determine the degree of the polynomial.*

25. $2a^3 + 7a^5 + a^2$

26. $5x - 9x^2 + 3x^6$

27. $9x^4 + x^2 + x^7 + 4$

28. $8 + 6x^2 - 3x - x^5$

29. $9a - a^4 + 3 + 2a^3$

30. $-x + 2x^5 - 5x^2 + x^6$

31. Complete the following table for the polynomial
$$7x^2 + 8x^5 - 4x^3 + 6 - \tfrac{1}{2}x^4.$$

Term	Coefficient	Degree of the Term	Degree of the Polynomial
		5	
$-\frac{1}{2}x^4$			
	-4		
		2	
	6		

32. Complete the following table for the polynomial $-3x^4 + 6x^3 - 2x^2 + 8x + 7$.

Term	Coefficient	Degree of the Term	Degree of the Polynomial
	-3		
$6x^3$			
		2	
		1	
	7		

Classify each polynomial as a monomial, binomial, trinomial, or none of these.

33. $x^2 - 23x + 17$ **34.** $-9x^2$

35. $x^3 - 7x^2 + 2x - 4$ **36.** $t^3 + 4$

37. $8t^2 + 5t$ **38.** $4x^2 + 12x + 9$

39. 17

40. $2x^4 - 7x^3 + x^2 + x - 6$

Combine like terms. Write all answers in descending order.

41. $7x^2 + 3x + 4x^2$

42. $5a + 7a^2 + 3a$

43. $3a^4 - 2a + 2a + a^4$

44. $9b^5 + 3b^2 - 2b^5 - 3b^2$

45. $2x^2 - 6x + 3x + 4x^2$

46. $3x^4 - 7x + x^4 - 2x$

47. $9x^3 + 2x - 4x^3 + 5 - 3x$

48. $6x^2 + 2x^4 - 2x^2 - x^4 - 4x^2$

49. $10x^2 + 2x^3 - 3x^3 - 4x^2 - 6x^2 - x^4$

50. $8x^5 - x^4 + 2x^5 + 5x^4 - 4x^4 - x^6$

51. $\frac{1}{5}x^4 + 7 - 2x^2 + 3 - \frac{2}{15}x^4 + 2x^2$

52. $\frac{1}{6}x^3 + 3x^2 - \frac{1}{3}x^3 + 7 + x^2 - 10$

53. $5.9x^2 - 2.1x + 6 + 3.4x - 2.5x^2 - 0.5$

54. $7.4x^3 - 4.9x + 2.9 - 3.5x - 4.3 + 1.9x^3$

55. $6t - 9t^3 + 8t^4 + 4t + 2t^4 + 7t - 3t^3$

56. $5b^2 - 3b + 7b^2 - 4b^3 + 4b - 9b^2 + 10b^3$

Evaluate each polynomial for $x = 3$ and for $x = -3$.

57. $-7x + 4$

58. $-5x + 7$

59. $2x^2 - 3x + 7$

60. $4x^2 - 6x + 9$

61. $-2x^3 - 3x^2 + 4x + 2$

62. $-3x^3 + 7x^2 - 4x - 8$

63. $\frac{1}{3}x^4 - 2x^3$

64. $2x^4 - \frac{1}{9}x^3$

65. $-x^4 - x^3 - x^2$

66. $-x^2 - 3x^3 - x^4$

Find the specified function values.

67. Find $P(4)$ and $P(0)$: $P(x) = 3x^2 - 2x + 7$.

68. Find $Q(3)$ and $Q(-1)$: $Q(x) = -4x^3 + 7x^2 - 6$.

69. Find $P(-2)$ and $P\left(\frac{1}{3}\right)$: $P(y) = 8y^3 - 12y - 5$.

70. Find $Q(-3)$ and $Q(0)$:
$Q(y) = -8y^3 + 7y^2 - 4y - 9$.

Evaluate each polynomial function for $x = -1$.

71. $f(x) = -5x^3 + 3x^2 - 4x - 3$

72. $g(x) = -4x^3 + 2x^2 + 5x - 7$

Consumer Debt.   The amount of debt, in trillions of dollars, held by U.S. consumers can be estimated by the polynomial

$$0.15t + 1.42,$$

where t is the number of years since 2000 (Source: Based on information from the Federal Reserve).

73. Estimate U.S. consumer debt in 2005.

74. Estimate U.S. consumer debt in 2008.

75. *Skydiving.*   During the first 13 sec of a jump, the number of feet that a skydiver falls in t seconds is approximated by the polynomial

$11.12t^2$.

Approximately how far has a skydiver fallen 10 sec after jumping from a plane?

76. *Skydiving.*   For jumps that exceed 13 sec, the polynomial $173t - 369$ can be used to approximate the distance, in feet, that a skydiver has fallen in t seconds. Approximately how far has a skydiver fallen 20 sec after jumping from a plane?

Circumference.   *The circumference of a circle of radius r is given by the polynomial $2\pi r$, where π is an irrational number. For an approximation of π, use 3.14.*

77. Find the circumference of a circle with radius 10 cm.

78. Find the circumference of a circle with radius 5 ft.

Area of a Circle.   *The area of a circle of radius r is given by the polynomial πr^2. Use 3.14 for π.*

79. Find the area of a circle with radius 7 m.

80. Find the area of a circle with radius 6 ft.

81. *Freestyle Skiing.*   During the 2004 Winter X-Games, Simon Dumont established a world record by launching himself 22 ft above the edge of a halfpipe. His speed $v(t)$, in miles per hour, upon re-entering the pipe can be approximated by $v(t) = 10.9t$, where t is the number of seconds for which he was airborne. Dumont was airborne for approximately 2.34 sec. (*Source*: Based on data from EXPN.com) How fast was he going when he re-entered the halfpipe?

82. *Cliff Diving.*   The speed $v(t)$, in miles per hour, at which a swimmer enters the water can be approximated by $v(t) = 21.82t$, where t is the number of seconds the diver is falling. Cliff divers in Acapulco, Mexico, often are in the air for a total of 5 sec. (*Source*: Based on information from www.guinnessworldrecords.com) Estimate the speed of such a cliff diver when he or she enters the water.

83. *Skateboarding.*   The distance $s(t)$, in feet, traveled by a body falling freely from rest in t seconds is approximated by

$s(t) = 16t^2$.

In 2002, Adil Dyani set a record for the highest skateboard drop into a quarterpipe. Dyani was airborne for about 1.36 sec. (*Source*: Based on data from *Guinness Book of World Records* 2004) How far did he drop?

84. A paintbrush falls from a scaffold and takes 3 sec to hit the ground (see Exercise 83). How high is the scaffold?

NASCAR Attendance.   *Attendance at NASCAR auto races has grown rapidly over a 10-yr span. Attendance A, in millions, can be approximated by the polynomial function given by*

$A(x) = 0.0024x^3 - 0.005x^2 + 0.31x + 3,$

where x is the number of years since 1989. Use the following graph for Exercises 85–88.

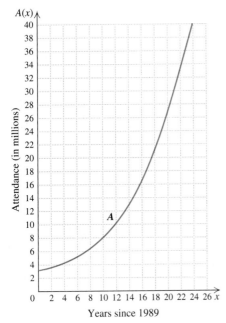

Years since 1989

Source: NASCAR

85. Estimate the attendance at NASCAR races in 2009.

86. Estimate the attendance at NASCAR races in 2006.

87. Approximate $A(8)$.

88. Approximate $A(12)$.

89. *Stacking Spheres.* In 2004, the journal *Annals of Mathematics* accepted a proof of the so-called Kepler Conjecture: that the most efficient way to pack spheres is in the shape of a square pyramid (*Source*: *The New York Times* 4/6/04).

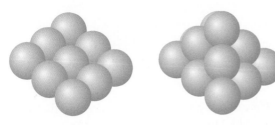

Bottom layer   Second layer   Top layer

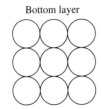      

The number N of balls in the stack is given by the polynomial function

$$N(x) = \tfrac{1}{3}x^3 + \tfrac{1}{2}x^2 + \tfrac{1}{6}x,$$

where x is the number of layers. Use both the function and the figure to find $N(3)$. Then calculate the number of oranges in a pyramid with 5 layers.

90. *Stacking Cannonballs.* The function in Exercise 89 was discovered by Thomas Harriot, assistant to Sir Walter Raleigh, when preparing for an expedition at sea (*Source*: *The New York Times* 4/7/04). How many cannonballs did they pack if there were 10 layers to their pyramid?

Veterinary Science. *Gentamicin is an antibiotic frequently used by veterinarians. The concentration, in micrograms per milliliter (mcg/mL), of Gentamicin in a horse's bloodstream t hours after injection can be approximated by the polynomial function*

$$C(t) = -0.005t^4 + 0.003t^3 + 0.35t^2 + 0.5t$$

(*Source*: Michele Tulis, DVM, telephone interview). *Use the following graph for Exercises 91–94.*

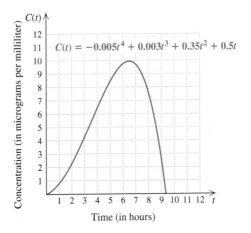

Time (in hours)

91. Estimate the concentration, in mcg/mL, of Gentamicin in the bloodstream 2 hr after injection.

92. Estimate the concentration, in mcg/mL, of Gentamicin in the bloodstream 4 hr after injection.

93. Approximate the range of C.

94. Approximate the domain of C.

Estimate the range of each function from its graph.

95.
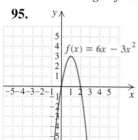

96.
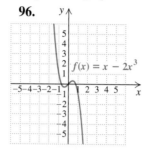

97.
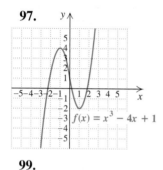

98.
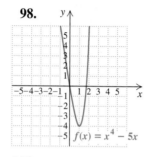

99.

$y = x^2 - 4$
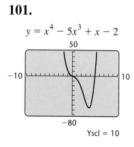

100.

$y = 0.3x^3$
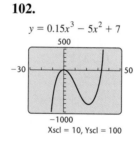

101.

$y = x^4 - 5x^3 + x - 2$

Yscl = 10

102.

$y = 0.15x^3 - 5x^2 + 7$

Xscl = 10, Yscl = 100

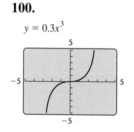

Use a graphing calculator to graph each polynomial function in the indicated viewing window, and estimate its range.

103. $f(x) = x^2 + 2x + 1, [-10, 10, -10, 10]$

104. $p(x) = x^2 + x - 6, [-10, 10, -10, 10]$

105. $q(x) = -2x^2 + 5, [-10, 10, -10, 10]$

106. $g(x) = 1 - x^2, [-10, 10, -10, 10]$

107. $p(x) = -2x^3 + x + 5, [-10, 10, -10, 10]$

108. $f(x) = -x^4 + 2x^3 - 10, [-5, 5, -30, 10]$, Yscl = 5

109. $g(x) = x^4 + 2x^3 - 5, [-5, 5, -10, 10]$

110. $q(x) = x^5 - 2x^4, [-5, 5, -5, 5]$

TW 111. Explain how it is possible for a term to not be a monomial.

TW 112. Is it possible to evaluate polynomials without understanding the rules for order of operations? Why or why not?

Skill Maintenance

Simplify.

113. $-19 + 24$ [1.5]

114. $5 - 14$ [1.6]

Factor.

115. $5x + 15$ [1.2]

116. $7a - 21$ [1.2]

117. A family spent $2011 to drive a car one year, during which the car was driven 14,800 mi. The family spent $972 for insurance and $114 for registration and oil. The only other cost was for gasoline. How much did gasoline cost per mile? [3.4]

118. The sum of the page numbers on the facing pages of a book is 549. What are the page numbers? [2.5]

Synthesis

TW 119. Is it easier to evaluate a polynomial before or after like terms have been combined? Why?

TW 120. A student who is trying to graph

$$p(x) = 0.05x^4 - x^2 + 5$$

gets the following screen. How can the student tell at a glance that a mistake has been made?

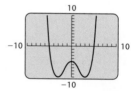

121. Construct a polynomial in x (meaning that x is the variable) of degree 5 with four terms and coefficients that are consecutive even integers.

122. Construct a trinomial in y of degree 4 with coefficients that are rational numbers.

123. What is the degree of $(5m^5)^2$?

124. Construct three like terms of degree 4.

Simplify.

125. $\frac{9}{2}x^8 + \frac{1}{9}x^2 + \frac{1}{2}x^9 + \frac{9}{2}x + \frac{9}{2}x^9 + \frac{8}{9}x^2 + \frac{1}{2}x - \frac{1}{2}x^8$

126. $(3x^2)^3 + 4x^2 \cdot 4x^4 - x^4(2x)^2 + ((2x)^2)^3 - 100x^2(x^2)^2$

127. A polynomial in x has degree 3. The coefficient of x^2 is 3 less than the coefficient of x^3. The coefficient of x is three times the coefficient of x^2. The remaining constant is 2 more than the coefficient of x^3. The sum of the coefficients is -4. Find the polynomial.

128. *Daily Accidents.* The average number of accidents per day involving drivers of age r can be approximated by the polynomial

$$0.4r^2 - 40r + 1039.$$

For what age is the number of daily accidents smallest?

Semester Averages. Professor Kopecki calculates a student's average for her course using

$$A = 0.3q + 0.4t + 0.2f + 0.1h,$$

with q, t, f, and h representing a student's quiz average, test average, final exam score, and homework average, respectively. In Exercises 129 and 130, find the given student's course average rounded to the nearest tenth.

129. Mary Lou: quizzes: 60, 85, 72, 91; final exam: 84; tests: 89, 93, 90; homework: 88

130. Nigel: quizzes: 95, 99, 72, 79; final exam: 91; tests: 68, 76, 92; homework: 86

In Exercises 131 and 132, complete the table for the given choices of t. Then plot the points and connect them with a smooth curve representing the graph of the polynomial.

131.

t	$-t^2 + 10t - 18$
3	
4	
5	
6	
7	

132.

t	$-t^2 + 6t - 4$
1	
2	
3	
4	
5	

133. *Path of the Olympic Arrow.* The Olympic flame at the 1992 Summer Olympics was lit by a flaming arrow. As the arrow moved d meters horizontally from the archer, its height h, in meters, was approximated by the polynomial

$$-0.0064d^2 + 0.8d + 2.$$

Complete the table for the choices of d given. Then plot the points and draw a graph representing the path of the arrow.

d	$-0.0064d^2 + 0.8d + 2$
0	
30	
60	
90	
120	

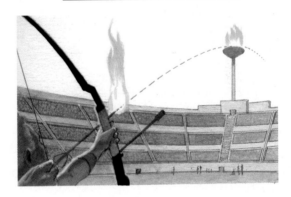

5.4 Addition and Subtraction of Polynomials

Addition of Polynomials ▣ Opposites of Polynomials ▣
Subtraction of Polynomials ▣ Problem Solving

In many ways, polynomials are to algebra as numbers are to arithmetic. Like numbers, polynomials can be added, subtracted, multiplied, and divided. In Sections 5.4–5.8, we examine how these operations are performed. Thus we are again learning how to write equivalent expressions, not how to solve equations.

Addition of Polynomials

To add two polynomials, we write a plus sign between them and combine like terms.

EXAMPLE 1   Add.

a) $(-5x^3 + 6x - 1) + (4x^3 + 3x^2 + 2)$
b) $\left(\frac{2}{3}x^4 + 3x^2 - 7x + \frac{1}{2}\right) + \left(-\frac{1}{3}x^4 + 5x^3 - 3x^2 + 3x - \frac{1}{2}\right)$

SOLUTION
a) $(-5x^3 + 6x - 1) + (4x^3 + 3x^2 + 2)$
$= (-5 + 4)x^3 + 3x^2 + 6x + (-1 + 2)$ Combining like terms; using the distributive law
$= -x^3 + 3x^2 + 6x + 1$ Note that $-1x^3 = -x^3$.

b) $\left(\frac{2}{3}x^4 + 3x^2 - 7x + \frac{1}{2}\right) + \left(-\frac{1}{3}x^4 + 5x^3 - 3x^2 + 3x - \frac{1}{2}\right)$
$= \left(\frac{2}{3} - \frac{1}{3}\right)x^4 + 5x^3 + (3 - 3)x^2 + (-7 + 3)x + \left(\frac{1}{2} - \frac{1}{2}\right)$
Combining like terms
$= \frac{1}{3}x^4 + 5x^3 - 4x$

After some practice, polynomial addition is often performed mentally.

EXAMPLE 2   Add: $(2 - 3x + x^2) + (-5 + 7x - 3x^2 + x^3)$.

SOLUTION   We have

$(2 - 3x + x^2) + (-5 + 7x - 3x^2 + x^3)$ Note that $x^2 = 1x^2$.
$= (2 - 5) + (-3 + 7)x + (1 - 3)x^2 + x^3$ You might do this step mentally.
$= -3 + 4x - 2x^2 + x^3$.   Then you would write only this.

The polynomials in the last example are written with the terms arranged according to degree, from least to greatest. Such an arrangement is called *ascending order*. As a rule, answers are written in ascending order when the polynomials being added are given in ascending order. When the polynomials being added are given in descending order, the answer is written in descending order.

Checking

A graphing calculator can be used to check whether two algebraic expressions given by y_1 and y_2 are equivalent. Some of the methods listed below have been described before; here they are listed together as a summary.

1. *Comparing graphs.* Graph both y_1 and y_2 on the same set of axes using the SEQUENTIAL mode. If the expressions are equivalent, the graphs will be identical. To distinguish the graphs, we can change the GRAPHSTYLE by locating the cursor on the icon before Y1= or Y2= and pressing **ENTER**. Using the PATH graphstyle, indicated by a small circle, for y_2 allows us to see if the graph of y_2 is being traced over the graph of y_1. Because two different graphs may appear identical in certain windows, this provides only a partial check.

2. *Subtracting expressions.* Let $y_3 = y_1 - y_2$. If the expressions are equivalent, the graph of y_3 will be $y = 0$, or the x-axis. Using the PATH graphstyle or TRACE allows us to see if the graph of y_3 is the x-axis.

3. *Comparing values.* Viewing a table allows us to compare values of y_1 and y_2. If the expressions are equivalent, the values will be the same for any given x-value. Again, this is only a partial check, but it becomes more certain as we use more x-values.

4. *Comparing both graphs and values.* Many graphing calculators will split the viewing screen and show both a graph and a table. This choice is usually made using the MODE key. In the bottom line of the screen on the left below, the FULL mode indicates a full-screen table or graph. In the HORIZ mode, the screen is split in half horizontally, and in the G-T mode, the screen is split vertically. In the screen in the middle below, a graph and a table of values are shown for $y = x^2 - 1$ using the HORIZ mode. The screen on the right shows the same graph and table in the G-T mode.

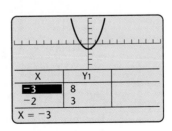

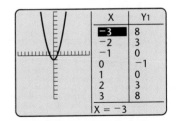

Interactive Discovery

1. Use the method of comparing graphs to determine whether the following addition is correct, and explain your answer.
$$(x^2 - 4x) + (x^2 - 3x) = 2x^2 - x$$

2. Use the method of subtracting expressions to determine whether the following addition is correct, and explain your answer.
$$(2x^2 + 5) + (x^2 - 8) = 3x^2 - 3$$

3. Use the HORIZ mode to compare both graphs and values to determine whether the following addition is correct, and explain your answer.
$$(x^2 - 5x) + (3x^2 - 6) = 4x^2 - 11x$$

Graphs and tables of values provide good checks of the results of algebraic manipulation. When checking by graphing, remember that two different graphs might not appear different in some viewing windows. It is also possible for two different functions to have the same value for several entries in a table. However, when enough values of two polynomial functions are the same, a table does provide a foolproof check.

> If the values of two nth-degree polynomial functions are the same for $n + 1$ or more values, then the polynomials are equivalent. For example, if the values of two cubic functions ($n = 3$) are the same for 4 or more values, the cubic polynomials are equivalent.

To add using columns, we write the polynomials one under the other, listing like terms under one another and leaving spaces for missing terms.

EXAMPLE 3   Add: $4x^3 + 4x - 5$ and $-x^3 + 7x^2 - 2$. Check using a table of values.

SOLUTION   We have

$$\begin{array}{r} 4x^3 + 4x - 5 \\ -x^3 + 7x^2 - 2 \\ \hline 3x^3 + 7x^2 + 4x - 7 \end{array}$$

To check, we let
$$y_1 = (4x^3 + 4x - 5) + (-x^3 + 7x^2 - 2)$$
and $y_2 = 3x^3 + 7x^2 + 4x - 7.$

Because the polynomials are of degree 3, we need to show that $y_1 = y_2$ for four different x-values. This can be accomplished quickly by creating a table of values, as shown at left. The answer checks.

X	Y1	Y2
0	−7	−7
1	7	7
2	53	53
3	149	149
4	313	313
5	563	563
6	917	917
X = 0		

Note that if a graphing calculator were not available, we would need to show, manually, that $y_1 = y_2$ for four different x-values.

> **CAUTION!** Note that equations like those in Examples 1–3 are written to show how one expression can be rewritten in an equivalent form. This is very different from solving an equation.

Opposites of Polynomials

In Section 1.8, we used the property of -1 to show that the opposite of a sum is the sum of the opposites. This idea can be extended.

> **The Opposite of a Polynomial** To find an equivalent polynomial for the *opposite*, or *additive inverse*, of a polynomial, change the sign of every term. This is the same as multiplying the polynomial by -1.

EXAMPLE 4 Write the opposite of $4x^5 - 7x^3 - 8x + \frac{5}{6}$ in two different forms.

SOLUTION

i) $-\left(4x^5 - 7x^3 - 8x + \frac{5}{6}\right)$

ii) $-4x^5 + 7x^3 + 8x - \frac{5}{6}$ **Changing the sign of every term**

Thus, $-\left(4x^5 - 7x^3 - 8x + \frac{5}{6}\right)$ and $-4x^5 + 7x^3 + 8x - \frac{5}{6}$ are equivalent. Both expressions represent the opposite of $4x^5 - 7x^3 - 8x + \frac{5}{6}$. ◢

EXAMPLE 5 Simplify: $-\left(-7x^4 - \frac{5}{9}x^3 + 8x^2 - x + 67\right)$.

SOLUTION

$$-\left(-7x^4 - \tfrac{5}{9}x^3 + 8x^2 - x + 67\right) = 7x^4 + \tfrac{5}{9}x^3 - 8x^2 + x - 67 \quad ◢$$

Subtraction of Polynomials

We can now subtract one polynomial from another by adding the opposite of the polynomial being subtracted.

EXAMPLE 6 Subtract: $(9x^5 + x^3 - 2x^2 + 4) - (-2x^5 + x^4 - 4x^3 - 3x^2)$.

SOLUTION We have

$$(9x^5 + x^3 - 2x^2 + 4) - (-2x^5 + x^4 - 4x^3 - 3x^2)$$
$$= 9x^5 + x^3 - 2x^2 + 4 + 2x^5 - x^4 + 4x^3 + 3x^2 \quad \text{Adding the opposite}$$
$$= 11x^5 - x^4 + 5x^3 + x^2 + 4. \quad \text{Combining like terms}$$

We can check the result of a subtraction using graphs or a table of values. ◢

EXAMPLE 7 Subtract: $(7x^5 + x^3 - 9x) - (3x^5 - 4x^3 + 5)$.

SOLUTION

$$(7x^5 + x^3 - 9x) - (3x^5 - 4x^3 + 5)$$
$$= 7x^5 + x^3 - 9x + (-3x^5) + 4x^3 - 5 \qquad \text{Adding the opposite}$$
$$= 7x^5 + x^3 - 9x - 3x^5 + 4x^3 - 5 \qquad \text{Try to go directly to this step.}$$
$$= 4x^5 + 5x^3 - 9x - 5 \qquad \text{Combining like terms}$$

To subtract using columns, we first replace the coefficients in the polynomial being subtracted with their opposites. We then add as before.

EXAMPLE 8 Write in columns and subtract:

$$(5x^2 - 3x + 6) - (9x^2 - 5x - 3).$$

SOLUTION

i) $\quad 5x^2 - 3x + 6$
$\quad \underline{-(9x^2 - 5x - 3)} \qquad$ Writing similar terms in columns

ii) $\quad 5x^2 - 3x + 6$
$\quad \underline{-9x^2 + 5x + 3} \qquad$ Changing signs and removing parentheses

iii) $\quad 5x^2 - 3x + 6$
$\quad \underline{-9x^2 + 5x + 3}$
$\quad -4x^2 + 2x + 9 \qquad$ Adding

If you can do so without error, you can arrange the polynomials in columns, mentally find the opposite of each term being subtracted, and write the answer. Lining up like terms is important and may require leaving some blanks.

EXAMPLE 9 Write in columns and subtract:

$$(x^3 + x^2 + 2x - 12) - (-2x^3 + x^2 - 3x).$$

SOLUTION We have

$$\quad x^3 + x^2 + 2x - 12$$
$$\quad \underline{-(-2x^3 + x^2 - 3x \qquad)} \qquad \text{Leaving space for the missing term}$$
$$\quad 3x^3 \qquad\quad + 5x - 12.$$

Problem Solving

EXAMPLE 10 Find a polynomial for the sum of the areas of rectangles A, B, C, and D.

SOLUTION

1. **Familiarize.** Recall that the area of a rectangle is the product of its length and width.

2. **Translate.** We translate the problem to mathematical language. The sum of the areas is a sum of products. We find each product and then add:

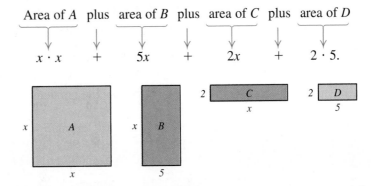

Area of A plus area of B plus area of C plus area of D

$x \cdot x$ $+$ $5x$ $+$ $2x$ $+$ $2 \cdot 5.$

3. **Carry out.** We combine like terms:

$$x^2 + 5x + 2x + 10 = x^2 + 7x + 10.$$

4. **Check.** A partial check is to replace x with a number, say, 3. Then we evaluate $x^2 + 7x + 10$ and compare that result with an alternative calculation:

$$3^2 + 7 \cdot 3 + 10 = 9 + 21 + 10 = 40.$$

When we substitute 3 for x and calculate the total area by regarding the figure as one large rectangle, we should also get 40:

$$\text{Total area} = (x + 5)(x + 2) = (3 + 5)(3 + 2) = 8 \cdot 5 = 40.$$

Our check is only partial, since it is possible for an incorrect answer to equal 40 when evaluated for $x = 3$. This would be unlikely, especially if a second choice of x, say $x = 5$, also checks. We leave that check to the student.

5. **State.** A polynomial for the sum of the areas is $x^2 + 7x + 10$.

EXAMPLE 11   An 8-ft by 8-ft shed is placed on a square lawn x ft on a side. Find a polynomial for the remaining area.

SOLUTION

1. **Familiarize.** We make a drawing of the situation as follows.

2. **Translate.** We reword the problem and translate as follows.

Rewording: Area of lawn   −   area of shed   =   area left over

Translating: x ft · x ft   −   8 ft · 8 ft   =   Area left over

3. **Carry out.** We carry out the multiplication:

$$x^2 \text{ ft}^2 - 64 \text{ ft}^2 = \text{Area left over.}$$

4. **Check.** As a partial check, note that the units in the answer are square feet (ft^2), a measure of area, as expected.

5. **State.** The remaining area in the yard is $(x^2 - 64)$ ft^2.

5.4 EXERCISE SET

FOR EXTRA HELP

MathXL   MyMathLab   InterAct Math   AW Math Tutor Center   Video Lectures on CD: Disc 3   Student's Solutions Manual

Add.

1. $(3x + 2) + (-5x + 4)$

2. $(5x + 1) + (-9x + 4)$

3. $(-6x + 2) + (x^2 + x - 3)$

4. $(x^2 - 5x + 4) + (8x - 9)$

5. $(7t^2 - 3t + 6) + (2t^2 + 8t - 9)$

6. $(8a^2 + 4a - 5) + (6a^2 - 3a - 1)$

7. $(2m^3 - 4m^2 + m - 7) + (4m^3 + 7m^2 - 4m - 2)$

8. $(5n^3 - n^2 + 4n - 3) + (2n^3 - 4n^2 + 3n - 4)$

9. $(3 + 6a + 7a^2 + 8a^3) + (4 + 7a - a^2 + 6a^3)$

10. $(7 + 4t - 5t^2 + 6t^3) + (2 + t + 6t^2 - 4t^3)$

11. $(9x^8 - 7x^4 + 2x^2 + 5) + (8x^7 + 4x^4 - 2x)$

12. $(4x^5 - 6x^3 - 9x + 1) + (6x^3 + 9x^2 + 9x)$

13. $\left(\frac{1}{4}x^4 + \frac{2}{3}x^3 + \frac{5}{8}x^2 + 7\right) + \left(-\frac{3}{4}x^4 + \frac{3}{8}x^2 - 7\right)$

14. $\left(\frac{1}{3}x^9 + \frac{1}{5}x^5 - \frac{1}{2}x^2 + 7\right) + \left(-\frac{1}{5}x^9 + \frac{1}{4}x^4 - \frac{3}{5}x^5\right)$

15. $(5.3t^2 - 6.4t - 9.1) + (4.2t^3 - 1.8t^2 + 7.3)$

16. $(4.9a^3 + 3.2a^2 - 5.1a) + (2.1a^2 - 3.7a + 4.6)$

17. $\begin{array}{r} -3x^4 + 6x^2 + 2x - 1 \\ -\ 3x^2 + 2x + 1 \\ \hline \end{array}$

18. $\begin{array}{r} -4x^3 + 8x^2 + 3x - 2 \\ -\ 4x^2 + 3x + 2 \\ \hline \end{array}$

19. $\begin{array}{l} 0.15x^4 + 0.10x^3 - 0.9x^2 \\ \quad\quad\ -0.01x^3 + 0.01x^2 + x \\ 1.25x^4 \quad\quad\quad + 0.11x^2 \quad\quad + 0.01 \\ \quad\quad\ 0.27x^3 \quad\quad\quad\quad\quad + 0.99 \\ \underline{-0.35x^4 \quad\quad\quad\quad + 15x^2 \quad - 0.03} \end{array}$

20. $\begin{array}{l} 0.05x^4 + 0.12x^3 - 0.5x^2 \\ \quad\quad\ -0.02x^3 + 0.02x^2 + 2x \\ 1.5x^4 \quad\quad\quad + 0.01x^2 \quad\quad + 0.15 \\ \quad\quad\ 0.25x^3 \quad\quad\quad\quad\quad + 0.85 \\ \underline{-0.25x^4 \quad\quad\quad\quad + 10x^2 \quad - 0.04} \end{array}$

Write the opposite of each polynomial in two different forms, as in Example 4.

21. $-t^3 + 4t^2 - 9$

22. $-4x^3 - 5x^2 + 2x$

23. $12x^4 - 3x^3 + 3$

24. $5a^3 + 2a - 17$

Simplify.

25. $-(8x - 9)$

26. $-(-6x + 5)$

27. $-(3a^4 - 5a^2 + 9)$

28. $-(-6a^3 + 2a^2 - 7)$

29. $-\left(-4x^4 + 6x^2 + \frac{3}{4}x - 8\right)$

30. $-(-5x^4 + 4x^3 - x^2 + 0.9)$

Subtract.

31. $(7x + 4) - (-2x + 1)$

32. $(5x + 6) - (-2x + 4)$

33. $(-5t + 6) - (t^2 + 3t - 1)$

34. $(a^2 - 5a + 2) - (3a^2 + 2a - 4)$

35. $(6x^4 + 3x^3 - 1) - (4x^2 - 3x + 3)$

36. $(-4x^2 + 2x) - (3x^3 - 5x^2 + 3)$

37. $(1.2x^3 + 4.5x^2 - 3.8x) - (-3.4x^3 - 4.7x^2 + 23)$

38. $(0.5x^4 - 0.6x^2 + 0.7) - (2.3x^4 + 1.8x - 3.9)$

Aha! **39.** $(7x^3 - 2x^2 + 6) - (7x^3 - 2x^2 + 6)$

40. $(8x^5 + 3x^4 + x - 1) - (8x^5 + 3x^4 - 1)$

41. $(3 + 5a + 3a^2 - a^3) - (2 + 3a - 4a^2 + 2a^3)$

42. $(7 + t - 5t^2 + 2t^3) - (1 + 2t - 4t^2 + 5t^3)$

43. $\left(\frac{5}{8}x^3 - \frac{1}{4}x - \frac{1}{3}\right) - \left(-\frac{1}{8}x^3 + \frac{1}{4}x - \frac{1}{3}\right)$

44. $\left(\frac{1}{5}x^3 + 2x^2 - \frac{3}{10}\right) - \left(-\frac{2}{5}x^3 + 2x^2 + \frac{7}{1000}\right)$

45. $(0.07t^3 - 0.03t^2 + 0.01t) - (0.02t^3 + 0.04t^2 - 1)$

46. $(0.9a^3 + 0.2a - 5) - (0.7a^4 - 0.3a - 0.1)$

47. $\begin{array}{l} x^2 + 5x + 6 \\ -(x^2 + 2x + 1) \end{array}$ **48.** $\begin{array}{l} x^3 + 3x^2 + 1 \\ -(x^3 + x^2 - 5) \end{array}$

49. $\begin{array}{l} 5x^4 + 6x^3 - 9x^2 \\ -(-6x^4 - 6x^3 + x^2) \end{array}$ **50.** $\begin{array}{l} 5x^4 - 2x^3 + 6x^2 \\ -(7x^4 + 6x^3 + 7x^2) \end{array}$

51. Solve.
  a) Find a polynomial for the sum of the areas of the rectangles shown in the figure.
  b) Find the sum of the areas when $x = 5$ and $x = 7$.

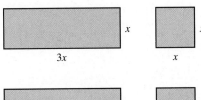
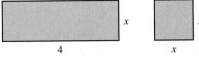

52. Solve.
  a) Find a polynomial for the sum of the areas of the circles shown in the figure.
  b) Find the sum of the areas when $r = 5$ and $r = 11.3$.

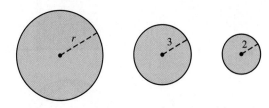

Find a polynomial for the perimeter of each figure in Exercises 53 and 54.

53.

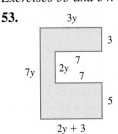

54.

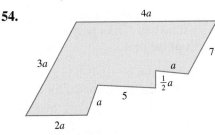

Find two algebraic expressions for the area of each figure. First, regard the figure as one large rectangle, and then regard the figure as a sum of four smaller rectangles.

55. **56.**

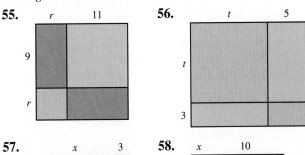

57. **58.**

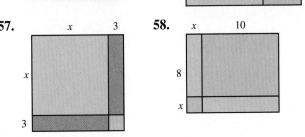

Find a polynomial for the shaded area of each figure.

59.
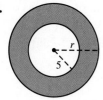

60.
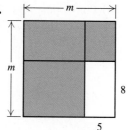

61.
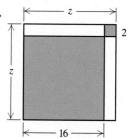

62.
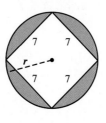

63. A 2-ft by 6-ft bath enclosure is installed in a new bathroom measuring x ft by x ft. Find a polynomial for the remaining floor area.

64. A 5-ft by 7-ft Jacuzzi™ is installed on an outdoor deck measuring y ft by y ft. Find a polynomial for the remaining area of the deck.

65. An 8-ft–wide round hot tub is installed in a patio measuring z ft by z ft. Find a polynomial for the remaining area of the patio.

66. A 10-ft–wide round water trampoline is floating in a pool measuring x ft by x ft. Find a polynomial for the remaining surface area of the pool.

67. A 12-m by 12-m mat includes a circle of diameter d meters for wrestling. Find a polynomial for the area of the mat outside the wrestling circle.

68. A 2-m by 3-m rug is spread inside a tepee that has a diameter of x meters. Find a polynomial for the area of the tepee's floor that is not covered.

Use a graphing calculator to determine whether each addition or subtraction is correct.

69. $(3x^2 - x^3 + 5x) + (4x^3 - x^2 + 7) = 7x^3 - 2x^2 + 5x + 7$

70. $(-5x + 3) + (4x - 7) + (10 - x) = -2x + 6$

71. $\left(x^2 - x - 3x^3 + \frac{1}{2}\right) - \left(x^3 - x^2 - x - \frac{1}{2}\right) = -4x^3 + 2x^2 + 1$

72. $(8x^2 - 6x - 5) - (4x^3 - 6x + 2) = 4x^2 - 7$

73. $(4x^2 + 3) + (x^2 - 5x) - (8 - 3x) = 5x^2 - 8x - 5$

74. $(4.1x^2 - 1.3x - 2.7) - (3.7x - 4.8 - 6.5x^2) = 0.4x^2 + 3.5x + 3.8$

TW 75. Is the sum of two trinomials always a trinomial? Why or why not?

TW 76. What advice would you offer to a student who is successful at adding, but not subtracting, polynomials?

Skill Maintenance

Simplify. [1.8]

77. $5(4 + 3) - 5 \cdot 4 - 5 \cdot 3$

78. $7(2 + 6) - 7 \cdot 2 - 7 \cdot 6$

79. $2(5t + 7) + 3t$

80. $3(4t - 5) + 2t$

Solve. [2.6]

81. $2(x + 3) > 5(x - 3) + 7$

82. $7(x - 8) \le 4(x - 5)$

Synthesis

TW 83. What can be concluded about two polynomials whose sum is zero?

TW 84. Which, if any, of the commutative, associative, and distributive laws are needed for adding polynomials? Why?

Simplify.

85. $(6t^2 - 7t) + (3t^2 - 4t + 5) - (9t - 6)$

86. $(3x^2 - 4x + 6) - (-2x^2 + 4) + (-5x - 3)$

87. $(-8y^2 - 4) - (3y + 6) - (2y^2 - y)$

88. $(5x^3 - 4x^2 + 6) - (2x^3 + x^2 - x) + (x^3 - x)$

89. $(-y^4 - 7y^3 + y^2) + (-2y^4 + 5y - 2) - (-6y^3 + y^2)$

90. $(-4 + x^2 + 2x^3) - (-6 - x + 3x^3) - (-x^2 - 5x^3)$

91. $(345.099x^3 - 6.178x) - (94.508x^3 - 8.99x)$

Find a polynomial for the surface area of each right rectangular solid.

92.

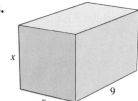

93.

94.

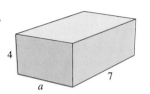

95.

96. Find a polynomial for the total length of all edges in the figure appearing in Exercise 95.

97. Find a polynomial for the total length of all edges in the figure appearing in Exercise 92.

98. *Total Profit.*   Hadley Electronics is marketing a new kind of stereo. Total revenue is the total amount of money taken in. The firm determines that when it sells x stereos, its total revenue is given by

$$R(x) = 280x - 0.4x^2.$$

Total cost is the total cost of producing x stereos. Hadley Electronics determines that the total cost of producing x stereos is given by

$$C(x) = 5000 + 0.6x^2.$$

The total profit $P(x)$ is

$$(\text{Total Revenue}) - (\text{Total Cost}) = R(x) - C(x).$$

a) Find a polynomial function for total profit.
b) What is the total profit on the production and sale of 75 stereos?
c) What is the total profit on the production and sale of 100 stereos?

TW 99. Does replacing each occurrence of the variable x in $4x^7 - 6x^3 + 2x$ with its opposite result in the opposite of the polynomial? Why or why not?

5.5 Multiplication of Polynomials

Multiplying Monomials ■ Multiplying a Monomial and a Polynomial ■ Multiplying Any Two Polynomials

We now multiply polynomials using techniques based on the distributive, associative, and commutative laws and the rules for exponents.

Multiplying Monomials

Consider $(3x)(4x)$. We multiply as follows:

$$
\begin{aligned}
(3x)(4x) &= 3 \cdot x \cdot 4 \cdot x && \text{Using an associative law}\\
&= 3 \cdot 4 \cdot x \cdot x && \text{Using a commutative law}\\
&= (3 \cdot 4) \cdot x \cdot x && \text{Using an associative law}\\
&= 12x^2.
\end{aligned}
$$

To Multiply Monomials   To find an equivalent expression for the product of two monomials, multiply the coefficients and then multiply the variables using the product rule for exponents.

Student Notes

Remember that when we compute $(3 \cdot 5)(2 \cdot 4)$, each factor is used only once, even if we change the order:

$$(3 \cdot 5)(2 \cdot 4) = (5 \cdot 2)(3 \cdot 4)$$
$$= 10 \cdot 12 = 120.$$

Some students mistakenly "reuse" a factor, as if the distributive law applied.

EXAMPLE 1   Multiply: **(a)** $(5x)(6x)$; **(b)** $(3a)(-a)$; **(c)** $(-7x^5)(4x^3)$.

SOLUTION

a) $(5x)(6x) = (5 \cdot 6)(x \cdot x)$ **Multiplying the coefficients; multiplying the variables**

$\qquad\quad = 30x^2$ **Simplifying**

b) $(3a)(-a) = (3a)(-1a)$ **Writing $-a$ as $-1a$ can ease calculations.**

$\qquad\qquad = (3)(-1)(a \cdot a)$ **Using an associative and a commutative law**

$\qquad\qquad = -3a^2$

c) $(-7x^5)(4x^3) = (-7 \cdot 4)(x^5 \cdot x^3)$

$\qquad\qquad = -28x^{5+3}$

$\qquad\qquad = -28x^8$ **Using the product rule for exponents**

After some practice, you can try writing only the answer.

Multiplying a Monomial and a Polynomial

To find an equivalent expression for the product of a monomial, such as $5x$, and a polynomial, such as $2x^2 - 3x + 4$, we use the distributive law.

EXAMPLE 2   Multiply: **(a)** x and $x + 3$; **(b)** $5x(2x^2 - 3x + 4)$.

SOLUTION

a) $x(x + 3) = x \cdot x + x \cdot 3$ **Using the distributive law**

$\qquad\quad = x^2 + 3x$

b) $5x(2x^2 - 3x + 4) = (5x)(2x^2) - (5x)(3x) + (5x)(4)$ **Using the distributive law**

$\qquad\qquad = 10x^3 - 15x^2 + 20x$ **Performing the three multiplications**

We can use graphs or tables to check the multiplication of polynomials. For example, we can check the product found in Example 2(a) by letting

$$y_1 = x(x + 3)$$
and $$y_2 = x^2 + 3x.$$

X	Y₁	Y₂
0	0	0
1	4	4
2	10	10
3	18	18
4	28	28
5	40	40
6	54	54

X = 0

The table of values for y_1 and y_2 shown at left indicates that the answer found is correct.

The product in Example 2(a) can be visualized as the area of a rectangle with width x and length $x + 3$.

Note that the total area can be expressed as $x(x + 3)$ or, by adding the two smaller areas, $x^2 + 3x$.

The Product of a Monomial and a Polynomial   To multiply a monomial and a polynomial, multiply each term of the polynomial by the monomial.

Try to do this mentally, when possible.

EXAMPLE 3   Multiply: $2x^2(x^3 - 7x^2 + 10x - 4)$.

SOLUTION

Think: $\underbrace{2x^2 \cdot x^3} - \underbrace{2x^2 \cdot 7x^2} + \underbrace{2x^2 \cdot 10x} - \underbrace{2x^2 \cdot 4}$

$2x^2(x^3 - 7x^2 + 10x - 4) = \quad 2x^5 \quad - \quad 14x^4 \quad + \quad 20x^3 \quad - \quad 8x^2$

Multiplying Any Two Polynomials

Before considering the product of *any* two polynomials, let's look at the product of two binomials.

To find an equivalent expression for the product of two binomials, we again begin by using the distributive law. This time, however, it is a *binomial* rather than a monomial that is being distributed.

EXAMPLE 4   Multiply each of the following.

a) $x + 5$ and $x + 4$ **b)** $4x - 3$ and $x - 2$

SOLUTION

a) $(x + 5)\ (x + 4) = (x + 5)\ x + (x + 5)\ 4$ Using the distributive law

$\qquad\qquad\qquad = x(x + 5) + 4(x + 5)$ Using the commutative law for multiplication

$\qquad\qquad\qquad = x \cdot x + x \cdot 5 + 4 \cdot x + 4 \cdot 5$ Using the distributive law (twice)

$\qquad\qquad\qquad = x^2 + 5x + 4x + 20$ Multiplying the monomials

$\qquad\qquad\qquad = x^2 + 9x + 20$ Combining like terms

b) $(4x - 3)(x - 2) = (4x - 3)x - (4x - 3)2$ Using the distributive law

$= x(4x - 3) - 2(4x - 3)$ Using the commutative law for multiplication. This step is often omitted.

$= x \cdot 4x - x \cdot 3 - 2 \cdot 4x - 2(-3)$ Using the distributive law (twice)

$= 4x^2 - 3x - 8x + 6$ Multiplying the monomials

$= 4x^2 - 11x + 6$ Combining like terms

To visualize the product in Example 4(a), consider a rectangle of length $x + 5$ and width $x + 4$.

The total area can be expressed as $(x + 5)(x + 4)$ or, by adding the four smaller areas, $x^2 + 5x + 4x + 20$.

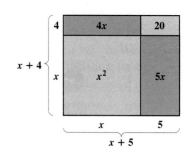

Let's consider the product of a binomial and a trinomial. Again we make repeated use of the distributive law.

EXAMPLE 5   Multiply: $(x^2 + 2x - 3)(x + 4)$.

SOLUTION

$(x^2 + 2x - 3)(x + 4)$

$= (x^2 + 2x - 3)x + (x^2 + 2x - 3)4$ Using the distributive law

$= x(x^2 + 2x - 3) + 4(x^2 + 2x - 3)$ Using the commutative law

$= x \cdot x^2 + x \cdot 2x - x \cdot 3 + 4 \cdot x^2 + 4 \cdot 2x - 4 \cdot 3$ Using the distributive law (twice)

$= x^3 + 2x^2 - 3x + 4x^2 + 8x - 12$ Multiplying the monomials

$= x^3 + 6x^2 + 5x - 12$ Combining like terms

Perhaps you have discovered the following in the preceding examples.

The Product of Two Polynomials   To multiply two polynomials P and Q, select one of the polynomials, say P. Then multiply each term of P by every term of Q and combine like terms.

To use columns for long multiplication, multiply each term in the top row by every term in the bottom row. We write like terms in columns, and then add the results. Such multiplication is like multiplying with whole numbers:

$$
\begin{array}{r}
3\ 2\ 1 \\
\times\ 1\ 2 \\
\hline
6\ 4\ 2 \\
3\ 2\ 1 \\
\hline
3\ 8\ 5\ 2
\end{array}
$$

$$
\begin{array}{rl}
300 + 20 + 1 & \\
\times\ \qquad\ 10 + 2 & \\
\hline
600 + 40 + 2 & \text{Multiplying the top row by 2} \\
3000 + 200 + 10 & \text{Multiplying the top row by 10} \\
\hline
3000 + 800 + 50 + 2 & \text{Adding}
\end{array}
$$

EXAMPLE 6   Multiply: $(4x^3 - 2x^2 + 3x)(x^2 + 2x)$.

SOLUTION

$$
\begin{array}{rl}
\left.\begin{array}{r} 4x^3 - 2x^2 + 3x \\ x^2 + 2x \end{array}\right\} & \text{Note that each polynomial is} \\
& \text{written in descending order.} \\
\hline
8x^4 - 4x^3 + 6x^2 & \text{Multiplying the top row by } 2x \\
4x^5 - 2x^4 + 3x^3 & \text{Multiplying the top row by } x^2 \\
\hline
4x^5 + 6x^4 - \ x^3 + 6x^2 & \text{Combining like terms}
\end{array}
$$

Line up like terms in columns.

If a term is missing, it helps to leave space for it so that like terms can be easily aligned.

EXAMPLE 7   Multiply: $(-2x^2 - 3)(5x^3 - 3x + 4)$.

SOLUTION

$$
\begin{array}{rl}
\left.\begin{array}{r} 5x^3 \qquad -3x + \ 4 \\ -\ 2x^2 \qquad\quad -\ 3 \end{array}\right\} & \text{Leaving space for missing terms} \\
\hline
-15x^3 \qquad + 9x - 12 & \text{Multiplying by } -3 \\
-10x^5 + \ 6x^3 - 8x^2 & \text{Multiplying by } -2x^2 \\
\hline
-10x^5 - \ 9x^3 - 8x^2 + 9x - 12 & \text{Combining like terms}
\end{array}
$$

Sometimes we multiply horizontally, while still aligning like terms.

EXAMPLE 8   Multiply: $(2x^3 + 3x^2 - 4x + 6)(3x + 5)$.

SOLUTION

Multiplying by $3x$

$$
(2x^3 + 3x^2 - 4x + 6)(3x + 5) = 6x^4 + \ 9x^3 - 12x^2 + 18x \\
+\ 10x^3 + 15x^2 - 20x + 30
$$

Multiplying by 5

$$
= 6x^4 + 19x^3 + 3x^2 - 2x + 30
$$

X	Y1	Y2
−3	36	36
−2	−10	−10
−1	22	22
0	30	30
1	56	56
2	286	286
3	1050	1050

X = −3

To check the multiplication in Example 8 by evaluating or using a table of values, we must compare values for at least 5 x-values, because the degree is 4. The table at left shows values of

$$y_1 = (2x^3 + 3x^2 - 4x + 6)(3x + 5) \quad \text{and}$$
$$y_2 = 6x^4 + 19x^3 + 3x^2 - 2x + 30$$

for 7 x-values. Thus the multiplication checks.

5.5 EXERCISE SET

Concept Reinforcement In each of Exercises 1–4, match the product with the correct result from the column on the right.

1. ___ $4x^4 \cdot 2x^2$

2. ___ $3x^2 \cdot 2x^4$

3. ___ $4x^3 \cdot 2x^5$

4. ___ $3x^5 \cdot 2x^3$

a) $6x^8$

b) $8x^6$

c) $6x^6$

d) $8x^8$

Multiply.

5. $(4x^3)9$

6. $(5x^4)6$

7. $(-x^2)(-x)$

8. $(-x^3)(x^4)$

9. $(-x^6)(x^2)$

10. $(-x^3)(-x^5)$

11. $(7t^5)(4t^3)$

12. $(10a^2)(3a^2)$

13. $(-0.1x^6)(0.2x^4)$

14. $(0.3x^3)(-0.4x^6)$

15. $\left(-\frac{1}{5}x^3\right)\left(-\frac{1}{3}x\right)$

16. $\left(-\frac{1}{4}x^4\right)\left(\frac{1}{5}x^8\right)$

17. $19t^2 \cdot 0$

18. $(-5n^3)(-1)$

19. $(-4y^5)(6y^2)(-3y^3)$

20. $7x^2(-2x^3)(2x^6)$

21. $3x(-x + 5)$

22. $2x(4x - 6)$

23. $4x(x + 1)$

24. $3x(x + 2)$

25. $(a - 7)4a$

26. $(a + 9)3a$

27. $x^2(x^3 + 1)$

28. $-2x^3(x^2 - 1)$

29. $3x(2x^2 - 6x + 1)$

30. $-4x(2x^3 - 6x^2 - 5x + 1)$

31. $5t^2(3t + 6)$

32. $7t^2(2t + 1)$

33. $-6x^2(x^2 + x)$

34. $-4x^2(x^2 - x)$

35. $\frac{2}{3}a^4\left(6a^5 - 12a^3 - \frac{5}{8}\right)$

36. $\frac{3}{4}t^5\left(8t^6 - 12t^4 + \frac{12}{7}\right)$

37. $(x + 6)(x + 1)$

38. $(x + 5)(x + 2)$

39. $(x + 5)(x - 2)$

40. $(x + 6)(x - 2)$

41. $(a - 6)(a - 7)$

42. $(a - 4)(a - 8)$

43. $(x + 3)(x - 3)$

44. $(x + 6)(x - 6)$

45. $(5 - x)(5 - 2x)$

46. $(3 + x)(6 + 2x)$

47. $\left(t + \frac{3}{2}\right)\left(t + \frac{4}{3}\right)$

48. $\left(a - \frac{2}{5}\right)\left(a + \frac{5}{2}\right)$

49. $\left(\frac{1}{4}a + 2\right)\left(\frac{3}{4}a - 1\right)$

50. $\left(\frac{2}{5}t - 1\right)\left(\frac{3}{5}t + 1\right)$

Draw and label rectangles similar to those following Examples 2 and 4 to illustrate each product.

51. $x(x + 5)$

52. $x(x + 2)$

53. $(x + 1)(x + 2)$

54. $(x + 3)(x + 1)$

55. $(x + 5)(x + 3)$

56. $(x + 4)(x + 6)$

Multiply and check.

57. $(x^2 - x + 5)(x + 1)$

58. $(x^2 + x - 7)(x + 2)$

59. $(2a + 5)(a^2 - 3a + 2)$

60. $(3t + 4)(t^2 - 5t + 1)$

61. $(y^2 - 7)(2y^3 + y + 1)$

62. $(a^2 + 4)(5a^3 - 3a - 1)$

aha! **63.** $(3x + 2)(5x + 4x + 7)$

64. $(4x - 5x - 3)(1 + 2x^2)$

65. $(x^2 - 3x + 2)(x^2 + x + 1)$

66. $(x^2 + 5x - 1)(x^2 - x + 3)$

67. $(2t^2 - 5t - 4)(3t^2 - t + 1)$

68. $(5t^2 - t + 1)(2t^2 + t - 3)$

69. $(x + 1)(x^3 + 7x^2 + 5x + 4)$

70. $(x + 2)(x^3 + 5x^2 + 9x + 3)$

TW **71.** Is it possible to understand polynomial multiplication without understanding the distributive law? Why or why not?

TW **72.** The polynomials

$$(a + b + c + d) \quad \text{and} \quad (r + s + m + p)$$

are multiplied. Without performing the multiplication, determine how many terms the product will contain. Provide a justification for your answer.

Focused Review

Perform the indicated operations.

73. $(3x + 2) + (4x^2 - 6x - 7)$ [5.4]

74. $(3x + 2) - (4x^2 - 6x - 7)$ [5.4]

75. $(3x + 2)(4x^2 - 6x - 7)$ [5.5]

76. $(-3x - 2)(4x^2 - 6x - 7)$ [5.5]

Synthesis

TW **77.** Under what conditions will the product of two binomials be a trinomial?

TW **78.** How can the following figure be used to show that $(x + 3)^2 \neq x^2 + 9$?

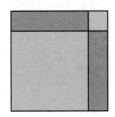

Find a polynomial for the shaded area of each figure.

79.

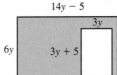

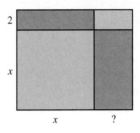

80.

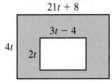

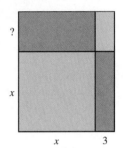

For each figure, determine what the missing number must be in order for the figure to have the given area.

81. Area is $x^2 + 7x + 10$ **82.** Area is $x^2 + 8x + 15$

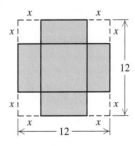

83. A box with a square bottom is to be made from a 12-in.-square piece of cardboard. Squares with side x are cut out of the corners and the sides are folded up. Find the polynomials for the volume and the outside surface area of the box.

84. A rectangular garden is twice as long as it is wide and is surrounded by a sidewalk that is 4 ft wide (see the figure below). The area of the sidewalk is 256 ft^2. Find the dimensions of the garden.

85. An open wooden box is a cube with side x cm. The box, including its bottom, is made of wood that is 1 cm thick. Find a polynomial for the interior volume of the cube.

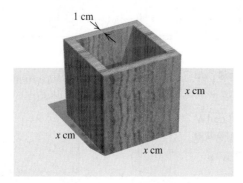

86. A side of a cube is $(x + 2)$ cm long. Find a polynomial for the volume of the cube.

87. Find a polynomial for the volume of the solid shown below.

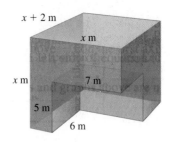

Compute and simplify.

88. $(x + 3)(x + 6) + (x + 3)(x + 6)$

Aha! **89.** $(x - 2)(x - 7) - (x - 7)(x - 2)$

90. $(x + 5)^2 - (x - 3)^2$

Aha! **91.** Extend the pattern and simplify
$$(x - a)(x - b)(x - c)(x - d) \cdots (x - z).$$

5.6   Special Products

Products of Two Binomials ◾ Multiplying Sums and Differences of Two Terms ◾ Squaring Binomials ◾ Multiplications of Various Types

Certain products of two binomials occur so often that it is helpful to be able to compute them quickly. In this section, we develop methods for doing so.

Products of Two Binomials

In Section 5.5, we found the product $(x + 5)(x + 4)$ by using the distributive law a total of three times (see p. 399). Note that each term in $x + 5$ is multiplied by each term in $x + 4$. To shorten our work, we can go right to this step:

$$(x + 5)(x + 4) = x \cdot x + x \cdot 4 + 5 \cdot x + 5 \cdot 4$$
$$= x^2 + 4x + 5x + 20$$
$$= x^2 + 9x + 20.$$

Note that $x \cdot x$ is found by multiplying the *First* terms of each binomial, $x \cdot 4$ is found by multiplying the *Outer* terms of the two binomials, $5 \cdot x$ is the product of the *Inner* terms of the two binomials, and $5 \cdot 4$ is the product of the *Last* terms of each binomial:

$$\underbrace{\text{First}}_{\text{terms}} \quad \underbrace{\text{Outer}}_{\text{terms}} \quad \underbrace{\text{Inner}}_{\text{terms}} \quad \underbrace{\text{Last}}_{\text{terms}}$$

$$(x + 5)(x + 4) = x \cdot x + 4 \cdot x + 5 \cdot x + 5 \cdot 4.$$

Student Notes

The special products discussed in this section are developed by recognizing patterns.

Looking for patterns often aids in understanding, remembering, and applying the material you are studying.

To remember this shortcut for multiplying, we use the initials **FOIL.**

The FOIL Method   To multiply two binomials, $A + B$ and $C + D$, multiply the First terms AC, the Outer terms AD, the Inner terms BC, and then the Last terms BD. Then combine like terms, if possible.

$$(A + B)(C + D) = AC + AD + BC + BD$$

1. Multiply First terms: AC.
2. Multiply Outer terms: AD.
3. Multiply Inner terms: BC.
4. Multiply Last terms: BD.

FOIL

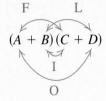

Because addition is commutative, the individual multiplications can be performed in any order. Both FLOI and FIOL yield the same result as FOIL, but FOIL is most easily remembered and most widely used.

EXAMPLE 1   Multiply: $(x + 8)(x^2 + 5)$.

SOLUTION

$$(x + 8)(x^2 + 5) = x^3 + 5x + 8x^2 + 40 \qquad \text{There are no like terms.}$$
$$= x^3 + 8x^2 + 5x + 40 \qquad \textbf{Writing in descending order}$$

After multiplying, remember to combine any like terms.

EXAMPLE 2   Multiply.

a) $(x + 7)(x + 4)$ **b)** $(y + 3)(y - 2)$
c) $(4t^3 + 5t)(3t^2 - 2)$ **d)** $(3 - 4x)(7 - 5x^3)$

SOLUTION

a) $(x + 7)(x + 4) = x^2 + 4x + 7x + 28$ **Using FOIL**
$$= x^2 + 11x + 28 \qquad \textbf{Combining like terms}$$

b) $(y + 3)(y - 2) = y^2 - 2y + 3y - 6$
$$= y^2 + y - 6$$

c) $(4t^3 + 5t)(3t^2 - 2) = 12t^5 - 8t^3 + 15t^3 - 10t$ **Remember to add exponents when multiplying terms with the same base.**

$$= 12t^5 + 7t^3 - 10t$$

$y_1 = (x + 7)(x + 4),$
$y_2 = x^2 + 11x + 28,$
$y_3 = y_1 - y_2$

X	Y3
0	0
1	0
2	0
3	0
4	0
5	0
6	0

X = 0

d) $(3 - 4x)(7 - 5x^3) = 21 - 15x^3 - 28x + 20x^4$
$$= 21 - 28x - 15x^3 + 20x^4$$

In general, if the original binomials are in *ascending* order, we also write the answer that way. ◢

We can check these multiplications by using a table of values or by graphing. To check Example 2(a), we let $y_1 = (x + 7)(x + 4)$, $y_2 = x^2 + 11x + 28$, and $y_3 = y_1 - y_2$. The graph and table at left show that the difference of the original expression and the resulting product is 0.

Multiplying Sums and Differences of Two Terms

Consider the product of the sum and the difference of the same two terms, such as

$$(x + 5)(x - 5).$$

Since this is the product of two binomials, we can use FOIL. In doing so, we find that the "outer" and "inner" products are opposites:

$$(x + 5)(x - 5) = x^2 - 5x + 5x - 25$$
$$= x^2 - 25.$$

The "outer" and "inner" terms "drop out." Their sum is zero.

Interactive Discovery

Determine whether each of the following is an identity.

1. $(x + 3)(x - 3) = x^2 - 9$
2. $(x + 3)(x - 3) = x^2 + 9$
3. $(x + 3)(x - 3) = x^2 - 6x + 9$
4. $(x + 3)(x - 3) = x^2 + 6x + 9$

The pattern you may have observed is true in general. Note the following:

$$\overset{\text{F}}{\downarrow}\quad\overset{\text{O}}{\downarrow}\quad\overset{\text{I}}{\downarrow}\quad\overset{\text{L}}{\downarrow}$$
$$(A + B)(A - B) = A^2 - AB + AB - B^2$$
$$= A^2 - B^2. \qquad -AB + AB = 0$$

The Product of a Sum and a Difference The product of the sum and the difference of the same two terms is the square of the first term minus the square of the second term:

$$(A + B)(A - B) = \underline{A^2 - B^2}.$$

This is called a *difference of squares*.

EXAMPLE 3   Multiply.

a) $(x + 4)(x - 4)$
b) $(5 + 2w)(5 - 2w)$
c) $(3a^4 - 5)(3a^4 + 5)$

SOLUTION

$$(A + B)(A - B) = A^2 - B^2$$

a) $(x + 4)(x - 4) = x^2 - 4^2$ Saying the words can help: "The square of the first term, x^2, minus the square of the second, 4^2"

$$= x^2 - 16$$ Simplifying

b) $(5 + 2w)(5 - 2w) = 5^2 - (2w)^2$

$$= 25 - 4w^2$$ Squaring both 5 and $2w$

c) $(3a^4 - 5)(3a^4 + 5) = (3a^4)^2 - 5^2$

$$= 9a^8 - 25$$ Using the rules for exponents. Remember to multiply exponents when raising a power to a power.

Squaring Binomials

Consider the square of a binomial, such as $(x + 3)^2$. This can be expressed as $(x + 3)(x + 3)$. Since this is the product of two binomials, we can use FOIL. But again, this product occurs so often that a faster method has been developed. Look for a pattern in the following.

Interactive Discovery

Determine whether each of the following is an identity.

1. $(x + 3)^2 = x^2 + 9$
2. $(x + 3)^2 = x^2 + 6x + 9$
3. $(x - 3)^2 = x^2 - 6x - 9$
4. $(x - 3)^2 = x^2 - 9$
5. $(x - 3)^2 = x^2 - 6x + 9$

A fast method for squaring binomials can be developed using FOIL:

$$(A + B)^2 = (A + B)(A + B)$$
$$= A^2 + AB + AB + B^2$$ Note that AB occurs twice.
$$= A^2 + 2AB + B^2;$$

$$(A - B)^2 = (A - B)(A - B)$$
$$= A^2 - AB - AB + B^2$$ Note that $-AB$ occurs twice.
$$= A^2 - 2AB + B^2.$$

The Square of a Binomial   The square of a binomial is the square of the first term, plus twice the product of the two terms, plus the square of the last term:

$$(A + B)^2 = A^2 + 2AB + B^2;$$
$$(A - B)^2 = A^2 - 2AB + B^2.$$

These are called *perfect-square trinomials.**

EXAMPLE 4   Multiply.

a) $(x + 7)^2$ **b)** $(t - 5)^2$

c) $(3a + 0.4)^2$ **d)** $(5x - 3x^4)^2$

SOLUTION

$$(A + B)^2 = A^2 + 2 \cdot A \cdot B + B^2$$

a) $(x + 7)^2 = x^2 + 2 \cdot x \cdot 7 + 7^2$

Saying the words can help: "The square of the first term, x^2, plus twice the product of the terms, $2 \cdot 7x$, plus the square of the second term, 7^2"

$$= x^2 + 14x + 49$$

b) $(t - 5)^2 = t^2 - 2 \cdot t \cdot 5 + 5^2$
$$= t^2 - 10t + 25$$

c) $(3a + 0.4)^2 = (3a)^2 + 2 \cdot 3a \cdot 0.4 + 0.4^2$
$$= 9a^2 + 2.4a + 0.16$$

d) $(5x - 3x^4)^2 = (5x)^2 - 2 \cdot 5x \cdot 3x^4 + (3x^4)^2$
$$= 25x^2 - 30x^5 + 9x^8 \quad \text{Using the rules for exponents}$$

CAUTION!   Although the square of a product is the product of the squares, the square of a sum is *not* the sum of the squares. That is, $(AB)^2 = A^2B^2$, but

The term $2AB$ is missing.

$$(A + B)^2 \neq A^2 + B^2.$$

To confirm this inequality, note that

$$(7 + 5)^2 = 12^2 = 144,$$

whereas

$$7^2 + 5^2 = 49 + 25 = 74, \quad \text{and} \quad 74 \neq 144.$$

*In some books, these are called *trinomial squares.*

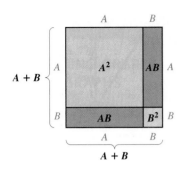

Geometrically, $(A + B)^2$ can be viewed as the area of a square with sides of length $A + B$:

$$(A + B)(A + B) = (A + B)^2.$$

This is equal to the sum of the areas of the four smaller regions:

$$A^2 + AB + AB + B^2 = A^2 + 2AB + B^2.$$

Thus,

$$(A + B)^2 = A^2 + 2AB + B^2.$$

Multiplications of Various Types

Recognizing patterns often helps when new problems are encountered. To simplify a new multiplication problem, always examine what type of product it is so that the best method for finding that product can be used. To do this, ask yourself questions similar to the following.

Multiplying Two Polynomials

1. Is the multiplication the product of a monomial and a polynomial? If so, multiply each term of the polynomial by the monomial.

2. Is the multiplication the product of two binomials? If so:

a) Is it the product of the sum and the difference of the *same* two terms? If so, use the pattern

$$(A + B)(A - B) = A^2 - B^2.$$

b) Is the product the square of a binomial? If so, use the pattern

$$(A + B)(A + B) = (A + B)^2 = A^2 + 2AB + B^2,$$

or

$$(A - B)(A - B) = (A - B)^2 = A^2 - 2AB + B^2.$$

c) If neither (a) nor (b) applies, use FOIL.

3. Is the multiplication the product of two polynomials other than those above? If so, multiply each term of one by every term of the other. Use columns if you wish.

EXAMPLE 5   Multiply.

a) $(x + 3)(x - 3)$ **b)** $(t + 7)(t - 5)$

c) $(x + 7)(x + 7)$ **d)** $2x^3(9x^2 + x - 7)$

e) $(p + 3)(p^2 + 2p - 1)$ **f)** $\left(3x + \frac{1}{4}\right)^2$

SOLUTION

a) $(x + 3)(x - 3) = x^2 - 9$ This is the product of the sum and the difference of the same two terms.

b) $(t + 7)(t - 5) = t^2 - 5t + 7t - 35$ Using FOIL

$\qquad\qquad\qquad = t^2 + 2t - 35$

c) $(x + 7)(x + 7) = x^2 + 14x + 49$ This is the square of a binomial, $(x + 7)^2$.

d) $2x^3(9x^2 + x - 7) = 18x^5 + 2x^4 - 14x^3$ Multiplying each term of the trinomial by the monomial

e)

$$
\begin{array}{r}
p^2 + 2p - 1 \\
p + 3 \\
\hline
3p^2 + 6p - 3 \\
p^3 + 2p^2 - p \\
\hline
p^3 + 5p^2 + 5p - 3
\end{array}
$$

Using columns to multiply a binomial and a trinomial

Multiplying by 3

Multiplying by p

f) $\left(3x + \frac{1}{4}\right)^2 = 9x^2 + 2(3x)\left(\frac{1}{4}\right) + \frac{1}{16}$ Squaring a binomial
$= 9x^2 + \frac{3}{2}x + \frac{1}{16}$

5.6 EXERCISE SET

FOR EXTRA HELP

MathXL   MyMathLab   InterAct Math   Tutor Center AW Math Tutor Center   Video Lectures on CD: Disc 3   Student's Solutions Manual

 Concept Reinforcement   *Identify each statement as either true or false.*

1. FOIL is simply a memory device for finding the product of two binomials.

2. Once FOIL is used, it is always possible to combine like terms.

3. The square of a binomial cannot be found using FOIL.

4. The square of $A + B$ is not the sum of the squares of A and B.

Multiply.

5. $(x + 3)(x^2 + 5)$ **6.** $(x^2 - 3)(x - 1)$

7. $(x^3 + 6)(x + 2)$ **8.** $(x^4 + 2)(x + 12)$

9. $(y + 2)(y - 3)$ **10.** $(a + 2)(a + 2)$

11. $(3x + 2)(3x + 5)$ **12.** $(4x + 1)(2x + 7)$

13. $(5x - 6)(x + 2)$ **14.** $(t - 9)(t + 9)$

15. $(1 + 3t)(2 - 3t)$ **16.** $(7 - a)(2 + 3a)$

17. $(2x - 7)(2x + 7)$ **18.** $(2x - 1)(3x + 1)$

19. $\left(p - \frac{1}{4}\right)\left(p + \frac{1}{4}\right)$ **20.** $\left(q + \frac{3}{4}\right)\left(q + \frac{3}{4}\right)$

21. $(x + 0.1)(x + 0.1)$ **22.** $(x + 0.3)(x - 0.4)$

23. $(-2x + 1)(x + 6)$ **24.** $(-x + 4)(2x - 5)$

25. $(a + 9)(a + 9)$ **26.** $(2y + 7)(2y + 7)$

27. $(1 + 3t)(1 - 5t)$ **28.** $(1 + 2t)(1 - 3t^2)$

29. $(x^2 + 3)(x^3 - 1)$ **30.** $(x^4 - 3)(2x + 1)$

31. $(3x^2 - 2)(x^4 - 2)$ **32.** $(x^{10} + 3)(x^{10} - 3)$

33. $(2t^3 + 5)(2t^3 + 3)$ **34.** $(5t^2 + 1)(2t^2 + 3)$

35. $(8x^3 + 5)(x^2 + 2)$ **36.** $(4 - 2x)(5 - 2x^2)$

37. $(4x^2 + 3)(x - 3)$ **38.** $(7x - 2)(2x - 7)$

Multiply. Try to recognize what type of product each multiplication is before multiplying.

39. $(x + 7)(x - 7)$ **40.** $(x + 1)(x - 1)$

41. $(2x + 1)(2x - 1)$ **42.** $(x^2 + 1)(x^2 - 1)$

43. $(5m - 2)(5m + 2)$ **44.** $(3x^4 + 2)(3x^4 - 2)$

45. $(3x^4 - 1)(3x^4 + 1)$ **46.** $(t^2 - 0.2)(t^2 + 0.2)$

47. $(x^4 + 7)(x^4 - 7)$ **48.** $(t^3 + 4)(t^3 - 4)$

49. $\left(t - \frac{3}{4}\right)\left(t + \frac{3}{4}\right)$ **50.** $\left(m - \frac{2}{3}\right)\left(m + \frac{2}{3}\right)$

51. $(x + 2)^2$ **52.** $(2x - 1)^2$

53. $(3x^5 + 1)^2$ **54.** $(4x^3 + 1)^2$

55. $\left(a - \frac{2}{5}\right)^2$ **56.** $\left(t - \frac{1}{5}\right)^2$

57. $(x^2 + 1)(x^2 - x + 2)$

58. $(a + 3)(a^2 + 6a + 1)$

59. $(2 - 3x^4)^2$ **60.** $(5 - 2t^3)^2$

61. $(5 + 6t^2)^2$ **62.** $(3p^2 - p)^2$

63. $(7x - 0.3)^2$ **64.** $(4a - 0.6)^2$

65. $5a^3(2a^2 - 1)$ **66.** $9x^3(2x^2 - 5)$

67. $(a - 3)(a^2 + 2a - 4)$

68. $(x^2 - 5)(x^2 + x - 1)$

69. $(3 - 2x^3)^2$ **70.** $(x - 4x^3)^2$

71. $4x(x^2 + 6x - 3)$ **72.** $8x(-x^5 + 6x^2 + 9)$

73. $(-t^3 + 1)^2$ **74.** $(-x^2 + 1)^2$

75. $3t^2(5t^3 - t^2 + t)$ **76.** $-5x^3(x^2 + 8x - 9)$

77. $(6x^4 - 3)^2$ **78.** $(8a^3 + 5)^2$

79. $(3x + 2)(4x^2 + 5)$ **80.** $(2x^2 - 7)(3x^2 + 9)$

81. $(5 - 6x^4)^2$ **82.** $(3 - 4t^5)^2$

83. $(a + 1)(a^2 - a + 1)$

84. $(x - 5)(x^2 + 5x + 25)$

Find the total area of all shaded rectangles.

85.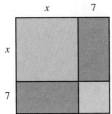

86.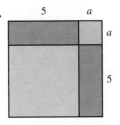

87.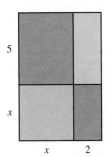

88.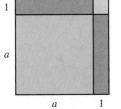

89.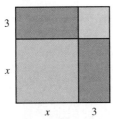

90.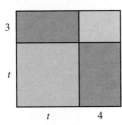

91.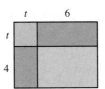

92.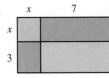

93.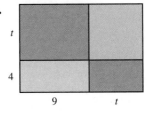

94.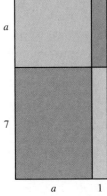

95.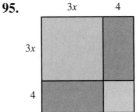

96.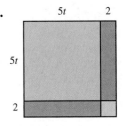

Draw and label rectangles similar to those in Exercises 85–96 to illustrate each of the following.

97. $(x + 5)^2$ **98.** $(x + 8)^2$

99. $(t + 9)^2$ **100.** $(a + 12)^2$

101. $(3 + x)^2$ **102.** $(7 + t)^2$

ᴛᴡ **103.** Bryce feels that since he can find the product of any two binomials using FOIL, he needn't study the other special products. What advice would you give him?

ᴛᴡ **104.** Under what conditions is the product of two binomials a binomial?

Skill Maintenance

105. *Energy Use.* In an apartment, lamps, an air conditioner, and a television set are all operating at the same time. The lamps take 10 times as many watts as the television set, and the air conditioner takes 40 times as many watts as the television set. The total wattage used in the apartment is 2550 watts. How many watts are used by each appliance? [2.5]

106. In what quadrant is the point $(-3, 4)$ located? [3.1]

Solve. [2.3]

107. $5xy = 8$, for y **108.** $3ab = c$, for a

109. $ax - b = c$, for x **110.** $st + r = u$, for t

Synthesis

ᴛᴡ 111. Tyra claims that by writing $19 \cdot 21$ as $(20 - 1)(20 + 1)$, she can find the product mentally. How is this possible?

ᴛᴡ 112. The product $(A + B)^2$ can be regarded as the sum of the areas of four regions (as shown following Example 4). How might one visually represent $(A + B)^3$? Why?

Multiply.

Aha! 113. $(4x^2 + 9)(2x + 3)(2x - 3)$

114. $(9a^2 + 1)(3a - 1)(3a + 1)$

Aha! 115. $(3t - 2)^2(3t + 2)^2$

116. $(5a + 1)^2(5a - 1)^2$

117. $(t^3 - 1)^4(t^3 + 1)^4$

118. $(32.41x + 5.37)^2$

Calculate as the difference of squares.

119. 18×22 [*Hint:* $(20 - 2)(20 + 2)$.]

120. 93×107

Solve.

121. $(x + 2)(x - 5) = (x + 1)(x - 3)$

122. $(2x + 5)(x - 4) = (x + 5)(2x - 4)$

123. Find $(y - 2)^2$ by subtracting the white areas from y^2.

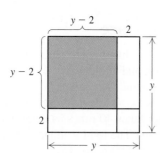

124. Find $(10 - 2x)^2$ by subtracting the white areas from 10^2.

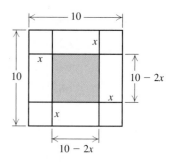

The height of a box is 1 more than its length l, and the length is 1 more than its width w. Find a polynomial for the volume V in terms of the following.

125. The length l **126.** The width w

Find a polynomial for the total shaded area in each figure.

127.

128.

129.

130. Find three consecutive integers for which the sum of the squares is 65 more than three times the square of the smallest integer.

5.7 Polynomials in Several Variables

Evaluating Polynomials ◾ Like Terms and Degree ◾
Addition and Subtraction ◾ Multiplication ◾
Function Notation

Thus far, the polynomials that we have studied have had only one variable. Polynomials such as

$$5x + x^2y - 3y + 7, \qquad 9ab^2c - 2a^3b^2 + 8a^2b^3, \quad \text{and} \quad 4m^2 - 9n^2$$

contain two or more variables. In this section, we will add, subtract, multiply, and evaluate such **polynomials in several variables.**

Evaluating Polynomials

To evaluate a polynomial in two or more variables, we substitute numbers for the variables. Then we compute, using the rules for order of operations.

EXAMPLE 1 Evaluate the polynomial $4 + 3x + xy^2 + 8x^3y^3$ for $x = -2$ and $y = 5$.

SOLUTION We substitute -2 for x and 5 for y:

$$4 + 3x + xy^2 + 8x^3y^3 = 4 + 3(-2) + (-2) \cdot 5^2 + 8(-2)^3 \cdot 5^3$$
$$= 4 - 6 - 50 - 8000 = -8052.$$

We can evaluate polynomials in several variables using a graphing calculator. To evaluate the polynomial in Example 1 for $x = -2$ and $y = 5$, we store -2 to X and 5 to Y using the **STO▸** key and then enter the polynomial. The result is shown in the screen below.

```
-2 → X
                          -2
5 → Y
                           5
4+3X+XY²+8X^3Y^3
                        -8052
■
```

EXAMPLE 2 Surface Area of a Right Circular Cylinder. The surface area of a right circular cylinder is given by the polynomial

$$2\pi rh + 2\pi r^2,$$

where h is the height and r is the radius of the base. A 12-oz can has a height of 4.7 in. and a radius of 1.2 in. Approximate its surface area.

SOLUTION   We evaluate the polynomial for $h = 4.7$ in. and $r = 1.2$ in. If 3.14 is used to approximate π, we have

$$2\pi rh + 2\pi r^2 \approx 2(3.14)(1.2 \text{ in.})(4.7 \text{ in.}) + 2(3.14)(1.2 \text{ in.})^2$$
$$\approx 2(3.14)(1.2 \text{ in.})(4.7 \text{ in.}) + 2(3.14)(1.44 \text{ in}^2)$$
$$\approx 35.4192 \text{ in}^2 + 9.0432 \text{ in}^2 \approx 44.4624 \text{ in}^2.$$

If the π key of a calculator is used, we have

$$2\pi rh + 2\pi r^2 \approx 2(3.141592654)(1.2 \text{ in.})(4.7 \text{ in.}) + 2(3.141592654)(1.2 \text{ in.})^2$$
$$\approx 44.48495197 \text{ in}^2.$$

Note that the unit in the answer (square inches) is a unit of area. The surface area is about 44.5 in^2 (square inches).

Like Terms and Degree

Recall that the degree of a term is the number of variable factors in the term. For example, the degree of $5x^2$ is 2 because there are two variable factors in $5 \cdot x \cdot x$. Similarly, the degree of $5a^2b^4$ is 6 because there are 6 variable factors in $5 \cdot a \cdot a \cdot b \cdot b \cdot b \cdot b$. Note that 6 can be found by adding the exponents 2 and 4.

As we learned in Section 5.3, the degree of a polynomial is the degree of the term of highest degree.

EXAMPLE 3   Identify the coefficient and the degree of each term and the degree of the polynomial

$$9x^2y^3 - 14xy^2z^3 + xy + 4y + 5x^2 + 7.$$

SOLUTION

Term	Coefficient	Degree	Degree of the Polynomial
$9x^2y^3$	9	5	
$-14xy^2z^3$	-14	6	
xy	1	2	6
$4y$	4	1	
$5x^2$	5	2	
7	7	0	

Note in Example 3 that although both xy and $5x^2$ have degree 2, they are *not* like terms. *Like*, or *similar*, *terms* either have exactly the same variables with exactly the same exponents or are constants. For example,

$$8a^4b^7 \text{ and } 5b^7a^4 \text{ are like terms}$$

and

$$-17 \text{ and } 3 \text{ are like terms,}$$

but

$$-2x^2y \text{ and } 9xy^2 \text{ are } not \text{ like terms.}$$

As always, combining like terms is based on the distributive law:

$$-5x^2y^3 + 3x^2y^3 = (-5 + 3)x^2y^3 = -2x^2y^3.$$

EXAMPLE 4 Combine like terms.

a) $9x^2y + 3xy^2 - 5x^2y - xy^2$
b) $7ab - 5ab^2 + 3ab^2 + 6a^3 + 9ab - 11a^3 + b - 1$

SOLUTION

a) $9x^2y + 3xy^2 - 5x^2y - xy^2 = (9 - 5)x^2y + (3 - 1)xy^2$
$$= 4x^2y + 2xy^2 \quad \text{Try to go directly to this step.}$$

b) $7ab - 5ab^2 + 3ab^2 + 6a^3 + 9ab - 11a^3 + b - 1$
$$= -5a^3 - 2ab^2 + 16ab + b - 1 \quad \text{We choose to write descending powers of } a. \text{ Other, equivalent, forms can also be used.}$$

Addition and Subtraction

The procedure used for adding polynomials in one variable is used to add polynomials in several variables.

EXAMPLE 5 Add.

a) $(-5x^3 + 3y - 5y^2) + (8x^3 + 4x^2 + 7y^2)$
b) $(5ab^2 - 4a^2b + 5a^3 + 2) + (3ab^2 - 2a^2b + 3a^3b - 5)$

SOLUTION

a) $(-5x^3 + 3y - 5y^2) + (8x^3 + 4x^2 + 7y^2)$
$$= (-5 + 8)x^3 + 4x^2 + 3y + (-5 + 7)y^2 \quad \text{Try to do this step mentally.}$$
$$= 3x^3 + 4x^2 + 3y + 2y^2$$

b) $(5ab^2 - 4a^2b + 5a^3 + 2) + (3ab^2 - 2a^2b + 3a^3b - 5)$
$$= 8ab^2 - 6a^2b + 5a^3 + 3a^3b - 3$$

When subtracting a polynomial, remember to find the opposite of each term in that polynomial and then add.

EXAMPLE 6 Subtract:
$$(4x^2y + x^3y^2 + 3x^2y^3 + 6y) - (4x^2y - 6x^3y^2 + x^2y^2 - 5y).$$

SOLUTION

$$(4x^2y + x^3y^2 + 3x^2y^3 + 6y) - (4x^2y - 6x^3y^2 + x^2y^2 - 5y)$$
$$= 4x^2y + x^3y^2 + 3x^2y^3 + 6y - 4x^2y + 6x^3y^2 - x^2y^2 + 5y$$
$$= 7x^3y^2 + 3x^2y^3 - x^2y^2 + 11y \quad \text{Combining like terms}$$

Student Notes

Always read the problem carefully. The difference between

$(-5x^3 + 3y - 5y^2) +$
$\qquad (8x^3 + 4x^2 + 7y^2)$

and

$(-5x^3 + 3y - 5y^2)(8x^3 + 4x^2 + 7y^2)$

is enormous. To avoid wasting time working on an incorrectly copied exercise, be sure to double-check that you have written the correct problem in your notebook.

Multiplication

To multiply polynomials in several variables, multiply each term of one polynomial by every term of the other, just as we did in Sections 5.5 and 5.6.

EXAMPLE 7 Multiply: $(3x^2y - 2xy + 3y)(xy + 2y)$.

SOLUTION

$$
\begin{array}{r}
3x^2y - 2xy + 3y \\
xy + 2y \\
\hline
6x^2y^2 - 4xy^2 + 6y^2 \\
3x^3y^2 - 2x^2y^2 + 3xy^2 \\
\hline
3x^3y^2 + 4x^2y^2 - xy^2 + 6y^2
\end{array}
$$

Multiplying by $2y$
Multiplying by xy
Adding

The special products discussed in Section 5.6 can speed up our work.

EXAMPLE 8 Multiply.

a) $(p + 5q)(2p - 3q)$ **b)** $(3x + 2y)^2$
c) $(a^3 - 7a^2b)^2$ **d)** $(3x^2y + 2y)(3x^2y - 2y)$
e) $(-2x^3y^2 + 5t)(2x^3y^2 + 5t)$ **f)** $(2x + 3 - 2y)(2x + 3 + 2y)$

SOLUTION

$$
\begin{array}{cccc}
& \text{F} & \text{O} \quad \text{I} & \text{L}
\end{array}
$$

a) $(p + 5q)(2p - 3q) = 2p^2 - 3pq + 10pq - 15q^2$

$\qquad\qquad\qquad\quad = 2p^2 + 7pq - 15q^2$ **Combining like terms**

$$(A + B)^2 = A^2 + 2 \cdot A \cdot B + B^2$$

b) $(3x + 2y)^2 = (3x)^2 + 2(3x)(2y) + (2y)^2$ **Using the pattern for squaring a binomial**

$\qquad\qquad = 9x^2 + 12xy + 4y^2$

$$(A - B)^2 = A^2 - 2 \cdot A \cdot B + B^2$$

c) $(a^3 - 7a^2b)^2 = (a^3)^2 - 2(a^3)(7a^2b) + (7a^2b)^2$ **Squaring a binomial**

$\qquad\qquad = a^6 - 14a^5b + 49a^4b^2$ **Using the rules for exponents**

$$(A + B)(A - B) = A^2 - B^2$$

d) $(3x^2y + 2y)(3x^2y - 2y) = (3x^2y)^2 - (2y)^2$ **Using the pattern for multiplying the sum and the difference of two terms**

$\qquad\qquad\qquad = 9x^4y^2 - 4y^2$ **Using the rules for exponents**

e) $(-2x^3y^2 + 5t)(2x^3y^2 + 5t) = (5t - 2x^3y^2)(5t + 2x^3y^2)$ Using the commutative law for addition twice

$$= (5t)^2 - (2x^3y^2)^2$$ Multiplying the sum and the difference of the same two terms

$$= 25t^2 - 4x^6y^4$$

$$(\quad A \quad - \quad B)(\quad A \quad + \quad B) = \quad A^2 \quad - \quad B^2$$

f) $(\ \boxed{2x + 3}\ - 2y)(\ \boxed{2x + 3}\ + 2y) = (\ \boxed{2x + 3}\ )^2 - (2y)^2$ Multiplying a sum and a difference

$$= 4x^2 + 12x + 9 - 4y^2$$ Squaring a binomial

In Example 8, we recognized patterns that might not be obvious, particularly in parts (e) and (f). In part (e), we can use FOIL, and in part (f), we can use long multiplication, but doing so would be slower. By carefully inspecting a problem before "jumping in," we can often save ourselves considerable work. At least one instructor refers to this as "working smart" instead of "working hard."*

Function Notation

Our work with multiplying can be used when evaluating functions.

EXAMPLE 9   Given $f(x) = x^2 - 4x + 5$, find and simplify each of the following.

a) $f(a + 3)$

b) $f(a + h) - f(a)$

SOLUTION

a) To find $f(a + 3)$, we replace x with $a + 3$. Then we simplify:

$$f(a + 3) = (a + 3)^2 - 4(a + 3) + 5$$
$$= a^2 + 6a + 9 - 4a - 12 + 5$$
$$= a^2 + 2a + 2.$$

b) To find $f(a + h)$ and $f(a)$, we replace x with $a + h$ and a, respectively.

$$f(a + h) - f(a) = [(a + h)^2 - 4(a + h) + 5] - [a^2 - 4a + 5]$$
$$= a^2 + 2ah + h^2 - 4a - 4h + 5 - a^2 + 4a - 5$$
$$= 2ah + h^2 - 4h$$

---

*Thanks to Pauline Kirkpatrick of Wharton County Junior College for this language.

5.7 EXERCISE SET

FOR EXTRA HELP

MathXL
MathXL

MyMathLab
MyMathLab

InterAct Math

Tutor Center
AW Math Tutor Center

Video Lectures on CD: Disc 3

Student's Solutions Manual

Concept Reinforcement *Each of the expressions in Exercises 1–8 can be regarded as either* **(a)** *the square of a binomial,* **(b)** *the product of the sum and difference of the same two terms, or* **(c)** *neither* (a) *nor* (b). *Select the appropriate choice for each expression.*

1. $(2x - 7y)^2$

2. $(4x - 9y)(4x + 9y)$

3. $(5a + 6b)(-6b + 5a)$

4. $(7a - 2b)(7a - 2b)$

5. $(r - 3s)(5r + 3s)$

6. $(2x - 7y)(7y - 2x)$

7. $(4x - 9y)(4x - 9y)$

8. $(2r - 3t)^2$

Evaluate each polynomial for $x = 5$ *and* $y = -2$.

9. $x^2 - 3y^2 + 2xy$

10. $x^2 + 5y^2 - 4xy$

Evaluate each polynomial for $x = 2$, $y = -3$, *and* $z = -4$.

11. $xyz^2 - z$

12. $xy - xz + yz$

Lung Capacity. *The polynomial*
$$0.041h - 0.018A - 2.69$$
can be used to estimate the lung capacity, in liters, of a female with height h, in centimeters, and age A, in years.

13. Find the lung capacity of a 50-year-old woman who is 160 cm tall.

14. Find the lung capacity of a 20-year-old woman who is 165 cm tall.

15. *Male Caloric Needs.* The number of calories needed each day by a moderately active man who weighs *w* kilograms, is *h* centimeters tall, and is *a* years old can be estimated by the polynomial
$$19.18w + 7h - 9.52a + 92.4$$
(*Source*: Parker, M., *She Does Math*. Mathematical Association of America). One of the authors of this text is moderately active, weighs 87 kg, is 185 cm tall, and is 59 years old. What are his daily caloric needs?

16. *Female Caloric Needs.* The number of calories needed each day by a moderately active woman who weighs *w* pounds, is *h* inches tall, and is *a* years old can be estimated by the polynomial
$$917 + 6w + 6h - 6a$$
(*Source*: Parker, M., *She Does Math*. Mathematical Association of America). Christine is moderately active, weighs 125 lb, is 64 in. tall, and is 27 years old. What are her daily caloric needs?

Surface Area of a Silo. *A silo is a structure that is shaped like a right circular cylinder with a half sphere on top. The surface area of a silo of height h and radius r (including the area of the base) is given by the polynomial* $2\pi rh + \pi r^2$. (*Note that h is the height of the entire silo.*)

17. A $1\frac{1}{2}$-oz bottle of roll-on deodorant has a height of 4 in. and a radius of $\frac{3}{4}$ in. Find the surface area of the bottle if the bottle is shaped like a silo. Use 3.14 for π.

18. A container of tennis balls is silo-shaped, with a height of $7\frac{1}{2}$ in. and a radius of $1\frac{1}{4}$ in. Find the surface area of the container. Use 3.14 for π.

Altitude of a Launched Object. *The altitude of an object, in meters, is given by the polynomial*
$$h + vt - 4.9t^2,$$
where h is the height, in meters, at which the launch occurs, v is the initial upward speed (or velocity), in meters per second, and t is the number of seconds for which the object is airborne.

19. A bocce ball is thrown upward with an initial speed of 18 m/sec by a person atop the Leaning

Tower of Pisa, which is 50 m above the ground. How high will the ball be 2 sec after it is thrown?

20. A golf ball is launched upward with an initial speed of 30 m/sec by a golfer atop the Washington Monument, which is 160 m above the ground. How high above the ground will the ball be after 3 sec?

Identify the coefficient and the degree of each term of each polynomial. Then find the degree of each polynomial.

21. $x^3y - 2xy + 3x^2 - 5$

22. $xy^2 - y^2 + 9x^2y + 7$

23. $17x^2y^3 - 3x^3yz - 7$

24. $6 - xy + 8x^2y^2 - y^5$

Combine like terms.

25. $7a + b - 4a - 3b$

26. $8r + s - 5r - 4s$

27. $3x^2y - 2xy^2 + x^2 + 5x$

28. $m^3 + 2m^2n - 3m^2 + 3mn^2$

29. $2u^2v - 3uv^2 + 6u^2v - 2uv^2 + 7u^2$

30. $3x^2 + 6xy + 3y^2 - 5x^2 - 10xy$

31. $5a^2c - 2ab^2 + a^2b - 3ab^2 + a^2c - 2ab^2$

32. $3s^2t + r^2t - 4st^2 - s^2t + 3st^2 - 7r^2t$

Add or subtract, as indicated.

33. $(4x^2 - xy + y^2) + (-x^2 - 3xy + 2y^2)$

34. $(2r^3 + 3rs - 5s^2) - (5r^3 + rs + 4s^2)$

35. $(3a^4 - 5ab + 6ab^2) - (9a^4 + 3ab - ab^2)$

36. $(2r^2t - 5rt + rt^2) - (7r^2t + rt - 5rt^2)$

Aha! **37.** $(5r^2 - 4rt + t^2) + (-6r^2 - 5rt - t^2) + (-5r^2 + 4rt - t^2)$

38. $(2x^2 - 3xy + y^2) + (-4x^2 - 6xy - y^2) + (4x^2 + 6xy + y^2)$

39. $(x^3 - y^3) - (-2x^3 + x^2y - xy^2 + 2y^3)$

40. $(a^3 + b^3) - (-5a^3 + 2a^2b - ab^2 + 3b^3)$

41. $(2y^4x^2 - 5y^3x) + (5y^4x^2 - y^3x) + (3y^4x^2 - 2y^3x)$

42. $(5a^2b + 7ab) + (9a^2b - 5ab) + (a^2b - 6ab)$

43. Subtract $7x + 3y$ from the sum of $4x + 5y$ and $-5x + 6y$.

44. Subtract $5a + 2b$ from the sum of $2a + b$ and $3a - 4b$.

Multiply.

45. $(3z - u)(2z + 3u)$ **46.** $(5x + y)(2x - 3y)$

47. $(xy + 7)(xy - 4)$ **48.** $(ab + 3)(ab - 5)$

49. $(2a - b)(2a + b)$ **50.** $(a - 3b)(a + 3b)$

51. $(5rt - 2)(3rt + 1)$ **52.** $(3xy - 1)(4xy + 2)$

53. $(m^3n + 8)(m^3n - 6)$ **54.** $(3 - c^2d^2)(4 + c^2d^2)$

55. $(6x - 2y)(5x - 3y)$ **56.** $(7a - 6b)(5a + 4b)$

57. $(pq + 0.1)(-pq + 0.1)$

58. $(rt + 0.2)(-rt + 0.2)$

59. $(x + h)^2$ **60.** $(r + t)^2$

61. $(4a + 5b)^2$ **62.** $(3x + 2y)^2$

63. $(c^2 - d)(c^2 + d)$ **64.** $(p^3 - 5q)(p^3 + 5q)$

65. $(ab + cd^2)(ab - cd^2)$ **66.** $(xy + pq)(xy - pq)$

Aha! **67.** $(a + b - c)(a + b + c)$

68. $(x + y + z)(x + y - z)$

69. $[a + b + c][a - (b + c)]$

70. $(a + b + c)(a - b - c)$

Find the total area of each shaded region.

71.
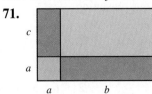

72.
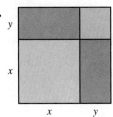

73.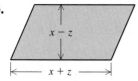

74.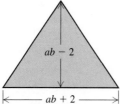

75.

76.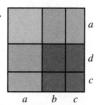

77.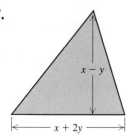

78.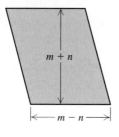

Draw and label rectangles similar to those in Exercises 71, 72, 75, and 76 to illustrate each product.

79. $(r + s)(u + v)$ **80.** $(m + r)(n + v)$

81. $(a + b + c)(a + d + f)$

82. $(r + s + t)^2$

83. Given $f(x) = x^2 + 5$, find and simplify.
   a) $f(t - 1)$
   b) $f(a + h) - f(a)$
   c) $f(a) - f(a - h)$

84. Given $f(x) = x^2 + 7$, find and simplify.
   a) $f(p + 1)$
   b) $f(a + h) - f(a)$
   c) $f(a) - f(a - h)$

TW **85.** Is it possible for a polynomial in 4 variables to have a degree less than 4? Why or why not?

TW **86.** A fourth-degree polynomial is multiplied by a third-degree polynomial. What is the degree of the product? Explain your reasoning.

Skill Maintenance

Simplify. [1.8]

87. $5 + \dfrac{7 + 4 + 2 \cdot 5}{3}$ **88.** $9 - \dfrac{2 + 6 \cdot 3 + 4}{6}$

89. $(4 + 3 \cdot 5 + 8) \div 3 \cdot 3$

90. $(5 + 2 \cdot 7 + 5) \div 2 \cdot 3$

91. $[3 \cdot 5 - 4 \cdot 2 + 7(-3)] \div (-2)$

92. $(7 - 3 \cdot 9 - 2 \cdot 5) \div (-6)$

Synthesis

TW **93.** Explain how it is possible for the sum of two trinomials in several variables to be a binomial in one variable.

TW **94.** Explain how it is possible for the sum of two trinomials in several variables to be a trinomial in one variable.

Find a polynomial for the shaded area. (Leave results in terms of π where appropriate.)

95.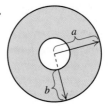

96.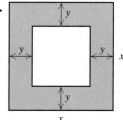

97.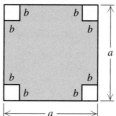

98.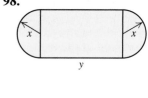

99. Find a polynomial for the total volume of the figure shown.

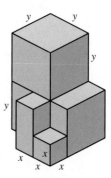

100. Find the shaded area in this figure using each of the approaches given below. Then check that both answers match.

a) Find the shaded area by subtracting the area of the unshaded square from the total area of the figure.

b) Find the shaded area by adding the areas of the three shaded rectangles.

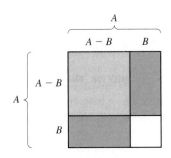

Find a polynomial for the surface area of each solid object shown. (Leave results in terms of π.)

101.

102.

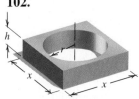

 103. The observatory at Danville University is shaped like a silo that is 40 ft high and 30 ft wide (see Exercise 17). The Heavenly Bodies Astronomy Club is to paint the exterior of the observatory using paint that covers 250 ft² per gallon. How many gallons should they purchase? Explain your reasoning.

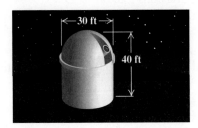

104. Multiply: $(x + a)(x - b)(x - a)(x + b)$.

105. *Interest Compounded Annually.*   An amount of money P that is invested at the yearly interest rate r grows to the amount

$$P(1 + r)^t$$

after t years. Find a polynomial that can be used to determine the amount to which P will grow after 2 yr.

106. *Yearly Depreciation.*   An investment P that drops in value at the yearly rate r drops in value to

$$P(1 - r)^t$$

after t years. Find a polynomial that can be used to determine the value to which P has dropped after 2 yr.

107. Suppose that $10,400 is invested at 8.5% compounded annually. How much is in the account at the end of 5 yr? (See Exercise 105.)

108. A $90,000 investment in computer hardware is depreciating at a yearly rate of 12.5%. How much is the investment worth after 4 yr? (See Exercise 106.)

5.8 Division of Polynomials

Dividing by a Monomial ◼ Dividing by a Binomial ◼
Synthetic Division

In this section, we study division of polynomials. We will find that polynomial division is similar to division in arithmetic.

Dividing by a Monomial

We first consider division by a monomial. When dividing a monomial by a monomial, we use the quotient rule of Section 5.1 to subtract exponents when bases are the same. For example,

$$\frac{15x^{10}}{3x^4} = 5x^{10-4}$$

$$= 5x^6$$

> **CAUTION!** The coefficients are divided but the exponents are subtracted.

and

$$\frac{42a^2b^5}{-3ab^2} = \frac{42}{-3}a^{2-1}b^{5-2} \qquad \text{Recall that } a^m/a^n = a^{m-n}.$$

$$= -14ab^3.$$

To divide a polynomial by a monomial, we note that since

$$\frac{A}{C} + \frac{B}{C} = \frac{A+B}{C},$$

it follows that

$$\frac{A+B}{C} = \frac{A}{C} + \frac{B}{C}. \qquad \text{Switching the left and right sides of the equation}$$

This is actually how we perform divisions like $86 \div 2$:

$$\frac{86}{2} = \frac{80+6}{2} = \frac{80}{2} + \frac{6}{2} = 40 + 3.$$

Similarly, to divide a polynomial by a monomial, we divide each term by the monomial:

$$\frac{80x^5 + 6x^3}{2x^2} = \frac{80x^5}{2x^2} + \frac{6x^3}{2x^2}$$

$$= \frac{80}{2}x^{5-2} + \frac{6}{2}x^{3-2} \qquad \text{Dividing coefficients and subtracting exponents}$$

$$= 40x^3 + 3x.$$

EXAMPLE 1   Divide $x^4 + 15x^3 - 6x^2$ by $3x$.

SOLUTION   We have

$$\frac{x^4 + 15x^3 - 6x^2}{3x} = \frac{x^4}{3x} + \frac{15x^3}{3x} - \frac{6x^2}{3x}$$

$$= \frac{1}{3}x^{4-1} + \frac{15}{3}x^{3-1} - \frac{6}{3}x^{2-1} \qquad \text{Dividing coeffi-cients and sub-tracting exponents}$$

$$= \frac{1}{3}x^3 + 5x^2 - 2x. \qquad \text{This is the quotient.}$$

To check, we multiply our answer by $3x$, using the distributive law:

$$3x\left(\frac{1}{3}x^3 + 5x^2 - 2x\right) = 3x \cdot \frac{1}{3}x^3 + 3x \cdot 5x^2 - 3x \cdot 2x$$

$$= x^4 + 15x^3 - 6x^2.$$

This is the polynomial that was being divided, so our answer, $\frac{1}{3}x^3 + 5x^2 - 2x$, checks.

EXAMPLE 2   Divide and check: $(10a^5b^4 - 2a^3b^2 + 6a^2b) \div (-2a^2b)$.

SOLUTION   We have

$$\frac{10a^5b^4 - 2a^3b^2 + 6a^2b}{-2a^2b} = \frac{10a^5b^4}{-2a^2b} - \frac{2a^3b^2}{-2a^2b} + \frac{6a^2b}{-2a^2b}$$

$$= -\frac{10}{2}a^{5-2}b^{4-1} - \left(-\frac{2}{2}\right)a^{3-2}b^{2-1} + \left(-\frac{6}{2}\right)$$

$$\text{Dividing coefficients and subtracting exponents}$$

$$= -5a^3b^3 + ab - 3.$$

Check: $-2a^2b(-5a^3b^3 + ab - 3) = -2a^2b(-5a^3b^3) + (-2a^2b)(ab) + (-2a^2b)(-3)$

$$= 10a^5b^4 - 2a^3b^2 + 6a^2b$$

Our answer, $-5a^3b^3 + ab - 3$, checks.

Dividing by a Binomial

For divisors with more than one term, we use long division, much as we do in arithmetic. Polynomials are written in descending order and any missing terms are written in, using 0 for the coefficients.

Student Notes

Since long division of polynomials requires many steps, we recommend that you double-check each step of your work as you move forward.

EXAMPLE 3 Divide $x^2 + 5x + 6$ by $x + 3$.

SOLUTION We have

$$x + 3 \overline{)x^2 + 5x + 6}$$

Divide the first term, x^2, by the first term in the divisor: $x^2/x = x$. Ignore the term 3 for the moment.

$$x$$

$$-(x^2 + 3x)$$ ← Multiply $x + 3$ by x, using the distributive law.

$$2x$$ ← Subtract by changing signs and adding: $x^2 + 5x - (x^2 + 3x) = 2x$.

Now we "bring down" the next term of the dividend—in this case, 6. To divide the remainder, $2x + 6$, by $x + 3$, we focus on $2x$ and x.

$$\begin{array}{r} x + 2 \\ x + 3 \overline{)x^2 + 5x + 6} \\ -(x^2 + 3x) \end{array}$$

$$2x + 6$$ Divide $2x$ by x: $2x/x = 2$.

$$-(2x + 6)$$ ← Multiply 2 by the divisor, $x + 3$, using the distributive law.

$$0$$ ← Subtract: $(2x + 6) - (2x + 6) = 0$.

The quotient is $x + 2$. The notation R 0 indicates a remainder of 0, although a remainder of 0 is generally not listed in an answer.

Check: To check, we multiply the quotient by the divisor and add any remainder to see if we get the dividend:

Divisor	Quotient		Remainder		Dividend
$(x + 3)$	$(x + 2)$	$+$	0	$=$	$x^2 + 5x + 6.$

Our answer, $x + 2$, checks.

EXAMPLE 4 Divide: $(2x^2 + 5x - 1) \div (2x - 1)$.

SOLUTION We have

$$2x - 1 \overline{)2x^2 + 5x - 1}$$

Divide the first term by the first term: $2x^2/(2x) = x$.

$$x$$

$$-(2x^2 - x)$$ ← Multiply $2x - 1$ by x.

$$6x$$ ← Subtract by changing signs and adding: $2x^2 + 5x - (2x^2 - x) = 6x$.

Now, we bring down the next term of the dividend, -1.

$$\begin{array}{r} x + 3 \\ 2x - 1 \overline{)2x^2 + 5x - 1} \\ -(2x^2 - x) \end{array}$$

$$6x - 1$$ Divide $6x$ by $2x$: $6x/(2x) = 3$.

$$-(6x - 3)$$ ← Multiply 3 by the divisor, $2x - 1$.

$$2$$ ← Note that $-1 - (-3) = -1 + 3 = 2$.

The answer is $x + 3$ with R 2.

Another way to write $x + 3$ R 2 is as

$$\underbrace{x + 3}_{\text{Quotient}} + \underbrace{\frac{2}{2x - 1}}_{\text{Divisor}}.$$

Quotient — $x + 3$ — Remainder, $2x - 1$ — Divisor

(This is the way answers will be given at the back of the book.)

Check: To check, we multiply the divisor by the quotient and add the remainder:

$$(2x - 1)(x + 3) + 2 = 2x^2 + 5x - 3 + 2$$
$$= 2x^2 + 5x - 1. \quad \text{Our answer checks.}$$

   Our division procedure ends when the degree of the remainder is less than that of the divisor. Check that this was indeed the case in Example 4.

EXAMPLE 5   Divide each of the following.

a) $(x^3 + 1) \div (x + 1)$ **b)** $(9a^2 + a^3 - 5) \div (a^2 - 1)$

SOLUTION

a)
$$
\begin{array}{r}
x^2 - x + 1 \\
x + 1 \overline{\smash{)}\, x^3 + 0x^2 + 0x + 1} \\
\end{array}
$$
←── Writing in the missing terms

$-(x^3 + x^2)$

$-x^2 + 0x$ ←──── Subtracting $x^3 + x^2$ from $x^3 + 0x^2$ and bringing down the $0x$

$-(-x^2 - x)$

$x + 1$ ←── Subtracting $-x^2 - x$ from $-x^2 + 0x$ and bringing down the 1

$-(x + 1)$

0

The answer is $x^2 - x + 1$.

Check: $(x + 1)(x^2 - x + 1) = x^3 - x^2 + x + x^2 - x + 1$
$$= x^3 + 1.$$

b)   We rewrite the problem in descending order:

$$(a^3 + 9a^2 - 5) \div (a^2 - 1).$$

Thus,

$$
\begin{array}{r}
a + 9 \\
a^2 - 1 \overline{\smash{)}\, a^3 + 9a^2 + 0a - 5} \\
\end{array}
$$
←── Writing in the missing term

$a^3 \qquad - a$

$9a^2 + a - 5$ ←── Subtracting $a^3 - a$ from $a^3 + 9a^2 + 0a$ and bringing down the -5

$9a^2 \qquad - 9$

$a + 4$ The degree of the remainder is less than the degree of the divisor, so we are finished.

The answer is $a + 9 + \dfrac{a + 4}{a^2 - 1}$.

Check: $(a^2 - 1)(a + 9) + a + 4 = a^3 + 9a^2 - a - 9 + a + 4$
$$= a^3 + 9a^2 - 5.$$

Synthetic Division

To divide a polynomial by a binomial of the type $x - a$, we can streamline the usual procedure to develop a process called *synthetic division*.

Compare the following. When a polynomial is written in descending order, the coefficients provide the essential information:

$$
\begin{array}{r}
4x^2 + 5x\ + 11 \\
x - 2\overline{)4x^3 - 3x^2 +\ \ x\ +\ 7} \\
\underline{4x^3 - 8x^2} \\
5x^2 +\ \ \ x\ +\ 7 \\
\underline{5x^2 - 10x} \\
11x\ +\ 7 \\
\underline{11x - 22} \\
29
\end{array}
\qquad
\begin{array}{r}
4 + 5 + 11 \\
1 - 2\overline{)4 - 3 +\ 1\ +\ 7} \\
\underline{4 - 8} \\
5 +\ 1\ +\ 7 \\
\underline{5 - 10} \\
11\ +\ 7 \\
\underline{11 - 22} \\
29
\end{array}
$$

Because the coefficient of x is 1 in the divisor, each time we multiply the divisor by a term in the answer, the leading coefficient of that product duplicates a coefficient in the answer. The process of synthetic division eliminates the duplication. To simplify the process further, we reverse the sign of the constant in the divisor and add rather than subtract.

EXAMPLE 6 Use synthetic division to divide:

$$(x^3 + 6x^2 - x - 30) \div (x - 2).$$

SOLUTION

$$
\begin{array}{r|rrrr}
2 & 1 & 6 & -1 & -30 \\
\hline
 & 1 & & &
\end{array}
$$
Write the 2 of $x - 2$ and the coefficients of the dividend.

Bring down the first coefficient.

$$
\begin{array}{r|rrrr}
2 & 1 & 6 & -1 & -30 \\
 & & 2 & & \\
\hline
 & 1 & 8 & &
\end{array}
$$
Multiply 1 by 2 to get 2.

Add 6 and 2.

$$
\begin{array}{r|rrrr}
2 & 1 & 6 & -1 & -30 \\
 & & 2 & 16 & \\
\hline
 & 1 & 8 & 15 &
\end{array}
$$
Multiply 8 by 2.

Add -1 and 16.

$$
\begin{array}{r|rrrr}
2 & 1 & 6 & -1 & -30 \\
 & & 2 & 16 & 30 \\
\hline
 & 1 & 8 & 15 & 0
\end{array}
$$
Multiply 15 by 2 and add.

$\underbrace{}_{\text{Quotient}}\quad \underbrace{}_{\text{Remainder}}$

The last number, 0, is the remainder. The other numbers are the coefficients of the quotient. The degree of the quotient is 1 less than the degree of the dividend.

$$
\begin{array}{ccc|c}
1 & 8 & 15 & 0
\end{array}
\;\longleftarrow\text{This is the remainder.}
$$

This is the zero-degree coefficient.
This is the first-degree coefficient.
This is the second-degree coefficient.

Since the remainder is 0, we have

$$(x^3 + 6x^2 - x - 30) \div (x - 2) = x^2 + 8x + 15.$$

We can check this with a table, letting

$$y_1 = (x^3 + 6x^2 - x - 30) \div (x - 2) \quad \text{and} \quad y_2 = x^2 + 8x + 15.$$

The values of both expressions are the same except for $x = 2$, when y_1 is not defined.

The answer is $x^2 + 8x + 15$ with R0, or just $x^2 + 8x + 15$.

X	Y₁	Y₂
-2	3	3
-1	8	8
0	15	15
1	24	24
2	ERROR	35
3	48	48
4	63	63

X = -2

Remember that in order for this method to work, the divisor must be of the form $x - a$, that is, a variable minus a constant. The coefficient of the variable must be 1.

EXAMPLE 7   Use synthetic division to divide.

a) $(2x^3 + 7x^2 - 5) \div (x + 3)$

b) $(10x^2 - 13x + 3x^3 - 20) \div (4 + x)$

SOLUTION

a) $(2x^3 + 7x^2 - 5) \div (x + 3)$

The dividend has no x-term, so we need to write 0 for its coefficient of x. Note that $x + 3 = x - (-3)$, so we write -3 inside the $\rfloor$.

$$
\begin{array}{r|rrrr}
-3 & 2 & 7 & 0 & -5 \\
   &   & -6 & -3 & 9 \\
\hline
   & 2 & 1 & -3 & 4
\end{array}
$$

The answer is $2x^2 + x - 3$, with R4, or $2x^2 + x - 3 + \dfrac{4}{x + 3}$.

b) We first rewrite $(10x^2 - 13x + 3x^3 - 20) \div (4 + x)$ in descending order:

$$(3x^3 + 10x^2 - 13x - 20) \div (x + 4).$$

Next, we use synthetic division. Note that $x + 4 = x - (-4)$.

$$
\begin{array}{r|rrrr}
-4 & 3 & 10 & -13 & -20 \\
   &   & -12 & 8 & 20 \\
\hline
   & 3 & -2 & -5 & 0
\end{array}
$$

The answer is $3x^2 - 2x - 5$.

5.8 EXERCISE SET

Divide and check.

1. $\dfrac{32x^5 - 24x}{8}$

2. $\dfrac{12a^4 - 3a^2}{6}$

3. $\dfrac{u - 2u^2 + u^7}{u}$

4. $\dfrac{50x^5 - 7x^4 + x^2}{x}$

5. $(15t^3 - 24t^2 + 6t) \div (3t)$

6. $(20t^3 - 15t^2 + 30t) \div (5t)$

7. $(24t^5 - 40t^4 + 6t^3) \div (4t^3)$

8. $(18t^6 - 27t^5 - 3t^3) \div (9t^3)$

9. $(15x^7 - 21x^4 - 3x^2) \div (-3x^2)$

10. $(16x^6 + 32x^5 - 8x^2) \div (-8x^2)$

11. $\dfrac{4x^7 + 6x^5 + 4x^2}{4x^3}$

12. $\dfrac{10x^6 + 15x^3 + 5x^2}{5x^3}$

13. $\dfrac{9r^2s^2 + 3r^2s - 6rs^2}{-3rs}$

14. $\dfrac{4x^4y - 8x^6y^2 + 12x^8y^6}{4x^4y}$

15. $(a^2b - a^3b^3 - a^5b^5) \div (a^2b)$

16. $(x^3y^2 - x^3y^3 - x^4y^2) \div (x^2y^2)$

17. $(x^2 + 10x + 21) \div (x + 7)$

18. $(y^2 - 8y + 16) \div (y - 4)$

19. $(a^2 - 8a - 16) \div (a + 4)$

20. $(y^2 - 10y - 25) \div (y - 5)$

21. $(x^2 - 9x + 21) \div (x - 5)$

22. $(x^2 - 11x + 23) \div (x - 7)$

Aha! 23. $(y^2 - 25) \div (y + 5)$

24. $(a^2 - 81) \div (a - 9)$

25. $\dfrac{a^3 + 8}{a + 2}$

26. $\dfrac{t^3 + 27}{t + 3}$

27. $\dfrac{t^2 - 13}{t - 4}$

28. $\dfrac{a^2 - 21}{a - 5}$

29. $\dfrac{2t^3 - 9t^2 + 11t - 3}{2t - 3}$

30. $\dfrac{8t^3 - 22t^2 - 5t + 12}{4t + 3}$

31. $(6a^2 + 17a + 8) \div (2a + 5)$

32. $(10a^2 + 19a + 9) \div (2a + 3)$

33. $(t^3 + t - t^2 - 1) \div (t + 1)$

34. $(x^3 + x - x^2 - 1) \div (x - 1)$

35. $(t^4 + 4t^2 + 3t - 6) \div (5 + t^2)$

36. $(t^4 - 2t^2 + 4t - 5) \div (t^2 - 3)$

37. $(4x^4 - 3 - x - 4x^2) \div (2x^2 - 3)$

38. $(x + 6x^4 - 4 - 3x^2) \div (1 + 2x^2)$

Use synthetic division to divide.

39. $(x^3 - 2x^2 + 2x - 7) \div (x + 1)$

40. $(x^3 - 2x^2 + 2x - 7) \div (x - 1)$

41. $(a^2 + 8a + 11) \div (a + 3)$

42. $(a^2 + 8a + 11) \div (a + 5)$

43. $(x^3 - 7x^2 - 13x + 3) \div (x - 2)$

44. $(x^3 - 7x^2 - 13x + 3) \div (x + 2)$

45. $(3x^3 + 7x^2 - 4x + 3) \div (x + 3)$

46. $(3x^3 + 7x^2 - 4x + 3) \div (x - 3)$

47. $(y^3 - 3y + 10) \div (y - 2)$

48. $(x^3 - 2x^2 + 8) \div (x + 2)$

49. $(x^5 - 32) \div (x - 2)$

50. $(y^5 - 1) \div (y - 1)$

51. $(3x^3 + 1 - x + 7x^2) \div \left(x + \frac{1}{3}\right)$

52. $(8x^3 - 1 + 7x - 6x^2) \div \left(x - \frac{1}{2}\right)$

TW **53.** How is the distributive law used when dividing a polynomial by a binomial?

TW **54.** On an assignment, Emmy Lou *incorrectly* writes

$$\frac{12x^3 - 6x}{3x} = 4x^2 - 6x.$$

What mistake do you think she is making and how might you convince her that a mistake has been made?

Skill Maintenance

Simplify.

55. $-4 + (-13)$ [1.5]

56. $-8 + (-15)$ [1.5]

57. $-9 - (-7)$ [1.6]

58. $-2 - (-7)$ [1.6]

59. The perimeter of a rectangle is 640 ft. The length is 15 ft greater than the width. Find the length of the rectangle. [2.5]

60. Solve: $3(2x - 1) = 7x - 5$. [2.2]

61. Graph: $3x - 2y = 12$. [3.3]

62. Plot the points $(4, -1)$, $(0, 5)$, $(-2, 3)$, and $(-3, 0)$. [3.1]

Synthesis

TW **63.** Explain how to construct a trinomial that has a remainder of 2 when divided by $x - 5$.

TW **64.** Explain how the quotient of two binomials can have more than two terms.

Divide.

65. $(10x^{9k} - 32x^{6k} + 28x^{3k}) \div (2x^{3k})$

66. $(45a^{8k} + 30a^{6k} - 60a^{4k}) \div (3a^{2k})$

67. $(6t^{3h} + 13t^{2h} - 4t^h - 15) \div (2t^h + 3)$

68. $(x^4 + a^2) \div (x + a)$

69. $(5a^3 + 8a^2 - 23a - 1) \div (5a^2 - 7a - 2)$

70. $(15y^3 - 30y + 7 - 19y^2) \div (3y^2 - 2 - 5y)$

71. Divide the sum of $4x^5 - 14x^3 - x^2 + 3$ and $2x^5 + 3x^4 + x^3 - 3x^2 + 5x$ by $3x^3 - 2x - 1$.

72. Divide $5x^7 - 3x^4 + 2x^2 - 10x + 2$ by the sum of $(x - 3)^2$ and $5x - 8$.

If the remainder is 0 when one polynomial is divided by another, the divisor is a factor of the dividend. Find the value(s) of c for which $x - 1$ is a factor of each polynomial.

73. $x^2 - 4x + c$

74. $2x^2 - 3cx - 8$

75. $c^2x^2 + 2cx + 1$

76. Let

$$f(x) = \frac{3x + 7}{x + 2}.$$

a) Use division to find an expression equivalent to $f(x)$. Then graph f.

b) On the same set of axes, sketch both $g(x) = 1/(x + 2)$ and $h(x) = 1/x$.

c) How do the graphs of f, g, and h compare?

TW **77.** Jamaladeen incorrectly states that

$$(x^3 + 9x^2 - 6) \div (x^2 - 1) = x + 9 + \frac{x + 4}{x^2 - 1}.$$

Without performing any long division, how could you show Jamaladeen that his division cannot possibly be correct?

5.9 The Algebra of Functions

The Sum, Difference, Product, or Quotient of Two Functions ■
Domains and Graphs

We now examine four ways in which functions can be combined.

The Sum, Difference, Product, or Quotient of Two Functions

Suppose that a is in the domain of two functions, f and g. The input a is paired with $f(a)$ by f and with $g(a)$ by g. The outputs can then be added to get $f(a) + g(a)$.

For example, if $f(x) = x + 4$ and $g(x) = x^2 + 1$, then we can find $f(2) + g(2)$:

$$f(2) = 2 + 4 = 6$$

and

$$g(2) = 2^2 + 1 = 5,$$

so

$$f(2) + g(2) = 6 + 5 = 11.$$

We can also add the expressions defining f and g:

$$f(x) + g(x) = (x + 4) + (x^2 + 1) = x^2 + x + 5.$$

This can then be regarded as a "new" function, written $(f + g)(x)$.

The Algebra of Functions If f and g are functions and x is in the domain of both functions, then:

1. $(f + g)(x) = f(x) + g(x)$;
2. $(f - g)(x) = f(x) - g(x)$;
3. $(f \cdot g)(x) = f(x) \cdot g(x)$;
4. $(f/g)(x) = f(x)/g(x)$, provided $g(x) \neq 0$.

EXAMPLE 1 For $f(x) = x^2 - x$ and $g(x) = x + 2$, find the following.

a) $(f + g)(3)$
b) $(f - g)(x)$ and $(f - g)(-1)$
c) $(f/g)(x)$ and $(f/g)(-4)$
d) $(f \cdot g)(3)$

SOLUTION

a) Since $f(3) = 3^2 - 3 = 6$ and $g(3) = 3 + 2 = 5$, we have

$$(f + g)(3) = f(3) + g(3)$$
$$= 6 + 5 \qquad \text{Substituting}$$
$$= 11.$$

Alternatively, we could first find $(f + g)(x)$:

$$(f + g)(x) = f(x) + g(x)$$
$$= x^2 - x + x + 2$$
$$= x^2 + 2. \qquad \text{Combining like terms}$$

Thus,

$$(f + g)(3) = 3^2 + 2 = 11. \qquad \text{Our results match.}$$

b) We have

$$(f - g)(x) = f(x) - g(x)$$
$$= x^2 - x - (x + 2) \qquad \text{Substituting}$$
$$= x^2 - 2x - 2. \qquad \begin{array}{l}\text{Removing parentheses and}\\ \text{combining like terms}\end{array}$$

Thus,

$$(f - g)(-1) = (-1)^2 - 2(-1) - 2 \qquad \begin{array}{l}\text{Using } (f - g)(x) \text{ is faster}\\ \text{than using } f(x) - g(x).\end{array}$$
$$= 1. \qquad \text{Simplifying}$$

c) We have

$$(f/g)(x) = f(x)/g(x)$$
$$= \frac{x^2 - x}{x + 2}. \qquad \text{We assume that } x \neq -2.$$

Thus,

$$(f/g)(-4) = \frac{(-4)^2 - (-4)}{-4 + 2} \qquad \text{Substituting}$$
$$= \frac{20}{-2} = -10.$$

d) Using our work in part (a), we have

$$(f \cdot g)(3) = f(3) \cdot g(3)$$
$$= 6 \cdot 5$$
$$= 30.$$

Alternatively, we could first find $(f \cdot g)(x)$:

$$(f \cdot g)(x) = f(x) \cdot g(x)$$
$$= (x^2 - x)(x + 2)$$
$$= x^3 + x^2 - 2x. \qquad \begin{array}{l}\text{Multiplying and combining}\\ \text{like terms}\end{array}$$

Then

$$(f \cdot g)(3) = 3^3 + 3^2 - 2 \cdot 3$$
$$= 27 + 9 - 6$$
$$= 30.$$

Domains and Graphs

Although applications involving products and quotients of functions rarely appear in newspapers, situations involving sums or differences of functions often do appear in print. For example, the following graphs are similar to those published by the California Department of Education to promote breakfast programs in which students eat a balanced meal of fruit or juice, toast or cereal, and 2% or whole milk. The combination of carbohydrate, protein, and fat gives a sustained release of energy, delaying the onset of hunger for several hours.

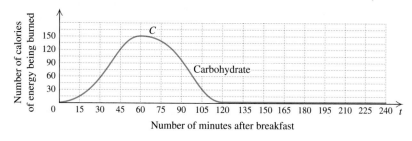

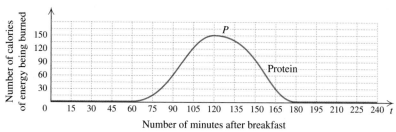

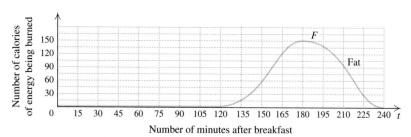

When the three graphs are superimposed, and the calorie expenditures added, it becomes clear that a balanced meal results in a steady, sustained supply of energy.

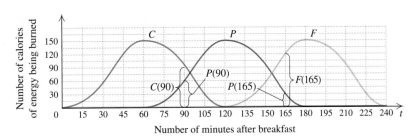

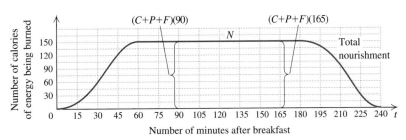

Note that for $t > 120$, we have $C(t) = 0$; for $t < 60$ or $t > 180$, we have $P(t) = 0$; and for $t < 120$, we have $F(t) = 0$. For any point on this last graph, we have

$$N(t) = (C + P + F)(t) = C(t) + P(t) + F(t).$$

To find $(f + g)(a)$, $(f - g)(a)$, $(f \cdot g)(a)$, or $(f/g)(a)$, we must first be able to find $f(a)$ and $g(a)$. Thus we need to ensure that a is in the domain of both f and g.

Interactive Discovery

Let $f(x) = \dfrac{5}{x}$ and $g(x) = \dfrac{2x - 6}{x + 1}$. Enter $y_1 = f(x)$, $y_2 = g(x)$, and $y_3 = y_1 + y_2$. Create a table of values for the functions with TblStart $= -3$, ΔTbl $= 1$, and Indpnt set to Auto.

1. What number is not in the domain of f?
2. What number is not in the domain of g?
3. What numbers are not in the domain of $f + g$?

Now enter $y_4 = y_1 - y_2$, $y_5 = y_1 \cdot y_2$, and $y_6 = y_1/y_2$. Create a table of values for the functions.

4. Does the domain of $f - g$ appear to be the same as the domain of $f + g$?
5. Does the domain of $f \cdot g$ appear to be the same as the domain of $f + g$?
6. Does the domain of f/g appear to be the same as the domain of $f + g$?

In the Interactive Discovery above, because division by 0 is not defined, we have

the domain of $f = \{x \mid x$ is a real number *and* $x \neq 0\}$

and

the domain of $g = \{x \mid x$ is a real number *and* $x \neq -1\}$.

In order to find $f(a) + g(a), f(a) - g(a),$ or $f(a) \cdot g(a)$, we must know that a is in *both* of the above domains. Thus,

the domain of $f + g =$ the domain of $f - g =$ the domain of $f \cdot g$
$= \{x \mid x$ is a real number *and* $x \neq 0$ *and* $x \neq -1\}$.

The domain of f/g also excludes the number 3, because $g(3) = 0$:

the domain of $f/g = \{x \mid x$ is a real number *and*
$x \neq 0$ *and* $x \neq -1$ *and* $x \neq 3\}$.

Determining the Domain

The domain of $f + g, f - g,$ or $f \cdot g$ is the set of all values common to the domains of f and g.

The domain of f/g is the set of all values common to the domains of f and g, excluding any values for which $g(x)$ is 0.

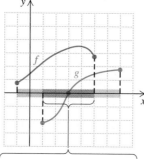

Domain of $f + g, f - g,$ and $f \cdot g$

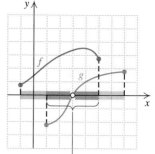

Domain of f/g

EXAMPLE 2   Given $f(x) = \dfrac{1}{x}$ and $g(x) = 2x - 7$, find the domains of $f + g, f - g, f \cdot g,$ and f/g.

SOLUTION   The domain of f is $\{x \mid x$ is a real number *and* $x \neq 0\}$. The domain of g is $\mathbb{R}$. The domains of $f + g, f - g,$ and $f \cdot g$ are the set of all elements common to both the domain of f and the domain of g. We have

$$\text{the domain of } f + g = \text{the domain of } f - g = \text{the domain of } f \cdot g$$
$$= \{x \mid x \text{ is a real number } and \ x \neq 0\}.$$

The domain of f/g is $\{x \mid x$ is a real number *and* $x \neq 0\}$, *with the additional restriction* that $g(x) \neq 0$. To determine what x-values would make $g(x) = 0$, we solve:

$$2x - 7 = 0 \qquad \text{\small Replacing } g(x) \text{ with } 2x - 7$$
$$2x = 7$$
$$x = \tfrac{7}{2}.$$

Since $g(x) = 0$ for $x = \tfrac{7}{2}$,

$$\text{the domain of } f/g = \left\{x \mid x \text{ is a real number } and \ x \neq 0 \ and \ x \neq \tfrac{7}{2}\right\}.$$

Student Notes

The concern over a denominator being 0 arises throughout this course. Try to develop the habit of checking for any possible input-values that would create a denominator of 0 whenever you work with functions.

EXAMPLE 3   Let $f(x) = 125x^3 - 8$ and $g(x) = 5x - 2$. If $F(x) = (f/g)(x)$, find a simplified expression for $F(x)$.

SOLUTION   Since $(f/g)(x) = f(x)/g(x)$,

$$F(x) = \frac{125x^3 - 8}{5x - 2}.$$

We have

$$
\begin{array}{r}
25x^2 + 10x + 4 \\
5x - 2 \overline{)\, 125x^3 + 0x^2 + 0x - 8} \\
\underline{125x^3 - 50x^2} \\
50x^2 + 0x \\
\underline{50x^2 - 20x} \\
20x - 8 \\
\underline{20x - 8} \\
0.
\end{array}
$$

Writing in the missing terms

Subtracting $125x^3 - 50x^2$ from $125x^3 + 0x^2$ and bringing down the $0x$

Subtracting $50x^2 - 20x$ from $50x^2 + 0x$ and bringing down the -8

Note that, because $F(x) = f(x)/g(x)$, $g(x)$ cannot be 0. Since $g(x)$ is 0 for $x = \tfrac{2}{5}$ (check this), we have

$$F(x) = 25x^2 + 10x + 4, \quad \text{provided } x \neq \tfrac{2}{5}.$$

Division by 0 is not the only condition that can force restrictions on the domain of a function. In Chapter 10, we will examine functions similar to that given by $f(x) = \sqrt{x}$, for which the concern is taking the square root of a negative number.

5.9 EXERCISE SET

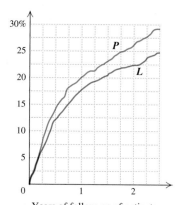

Concept Reinforcement *Make each of the following sentences true by selecting the correct word for each blank.*

1. If f and g are functions and x is in the
_____ of both functions, then
   range/domain
$(f + g)(x) = f(x) + g(x)$.

2. One way to compute $(f - g)(2)$ is to
_____ $g(2)$ from $f(2)$.
   erase/subtract

3. One way to compute $(f - g)(2)$ is to simplify
$f(x) - g(x)$ and then _____ the
      evaluate/substitute
result for $x = 2$.

4. The domain of $f + g, f - g,$ and $f \cdot g$ is the set of
all values common to the _____
      domains/ranges
of f and g.

5. The domain of f/g is the set of all values common
to the domains of f and g, _____
      including/excluding
any values for which $g(x)$ is 0.

6. The height of $(f + g)(a)$ on a graph is the
_____ of the heights of $f(a)$ and $g(a)$.
   product/sum

Let $f(x) = -3x + 1$ and $g(x) = x^2 + 2$. Find the following.

7. $f(2) + g(2)$ **8.** $f(-1) + g(-1)$

9. $f(5) - g(5)$ **10.** $f(4) - g(4)$

11. $f(-1) \cdot g(-1)$ **12.** $f(-2) \cdot g(-2)$

13. $f(-4)/g(-4)$ **14.** $f(3)/g(3)$

15. $g(1) - f(1)$ **16.** $g(2)/f(2)$

17. $(f + g)(x)$ **18.** $(g - f)(x)$

19. $(f \cdot g)(x)$ **20.** $(g/f)(x)$

Let $F(x) = x^2 - 2$ and $G(x) = 5 - x$. Find the following.

21. $(F + G)(x)$ **22.** $(F + G)(a)$

23. $(F + G)(-4)$ **24.** $(F + G)(-5)$

25. $(F - G)(3)$ **26.** $(F - G)(2)$

27. $(F \cdot G)(-3)$ **28.** $(F \cdot G)(-4)$

29. $(F/G)(x)$ **30.** $(F \cdot G)(x)$

31. $(F - G)(x)$ **32.** $(G - F)(x)$

33. $(F/G)(-2)$ **34.** $(F/G)(-1)$

In 2004, a study comparing high doses of the cholesterol-lowering drugs Lipitor and Pravachol indicated that patients taking Lipitor were significantly less likely to have heart attacks or require angioplasty or surgery.

In the graph below, L(t) is the percentage of patients on Lipitor (80 mg) and P(t) is the percentage of patients on Pravachol (40 mg) who suffered heart problems or death t years after beginning to take the medication (Source: New York Times, March 9, 2004).

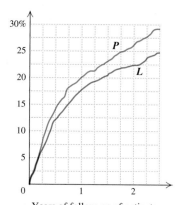

Years of follow-up of patients

Source: New England Journal of Medicine

35. Use estimates of $P(2)$ and $L(2)$ to estimate
$(P - L)(2)$.

36. Use estimates of $P(1)$ and $L(1)$ to estimate
$(P - L)(1)$.

The following graph shows the number of women, in millions, who had a child the previous year. Here $W(t)$ represents the number of women under 30 who gave birth in year t, $R(t)$ the number of women 30 and older who gave birth in year t, and N(t) the total number of women who gave birth in year t.

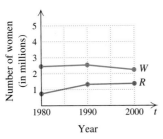

Source: U.S Bureau of the Census

37. Use estimates of $R(2000)$ and $W(2000)$ to estimate $N(2000)$.

38. Use estimates of $R(1990)$ and $W(1990)$ to estimate $N(1990)$.

Often function addition is represented by stacking the individual functions directly on top of each other. The graph below indicates how the three major airports servicing New York City have been utilized. The braces indicate the values of the individual functions.

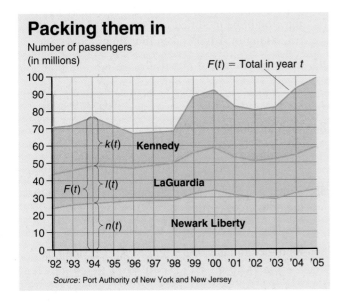

Source: Port Authority of New York and New Jersey

39. Estimate $(n + l)('98)$. What does it represent?

40. Estimate $(k + l)('98)$. What does it represent?

41. Estimate $F('02)$. What does it represent?

42. Estimate $F('01)$. What does it represent?

43. Estimate $(F - k)('02)$. What does it represent?

44. Estimate $(F - k)('01)$. What does it represent?

For each pair of functions f and g, determine the domain of the sum, difference, and product of the two functions.

45. $f(x) = x^2,$
    $g(x) = 7x - 4$

46. $f(x) = 5x - 1,$
    $g(x) = 2x^2$

47. $f(x) = \dfrac{1}{x - 3},$
    $g(x) = 4x^3$

48. $f(x) = 3x^2,$
    $g(x) = \dfrac{1}{x - 9}$

49. $f(x) = \dfrac{2}{x},$
    $g(x) = x^2 - 4$

50. $f(x) = x^3 + 1,$
    $g(x) = \dfrac{5}{x}$

51. $f(x) = x + \dfrac{2}{x - 1},$
    $g(x) = 3x^3$

52. $f(x) = 9 - x^2,$
    $g(x) = \dfrac{3}{x - 6} + 2x$

53. $f(x) = \dfrac{3}{x - 2},$
    $g(x) = \dfrac{5}{4 - x}$

54. $f(x) = \dfrac{5}{x - 3},$
    $g(x) = \dfrac{1}{x - 2}$

For each pair of functions f and g, determine the domain of f/g.

55. $f(x) = x^4,$
    $g(x) = x - 3$

56. $f(x) = 2x^3,$
    $g(x) = 5 - x$

57. $f(x) = 3x - 2,$
    $g(x) = 2x - 8$

58. $f(x) = 5 + x,$
    $g(x) = 6 - 2x$

59. $f(x) = \dfrac{3}{x - 4},$
    $g(x) = 5 - x$

60. $f(x) = \dfrac{1}{2 - x},$
    $g(x) = 7 - x$

61. $f(x) = \dfrac{2x}{x + 1},$
    $g(x) = 2x + 5$

62. $f(x) = \dfrac{7x}{x - 2},$
    $g(x) = 3x + 7$

For Exercises 63–66, f(x) and g(x) are as given. Find a simplified expression for F(x) if F(x) = (f/g)(x). (See Example 3.)

63. $f(x) = 8x^3 + 27,\ g(x) = 2x + 3$

64. $f(x) = 64x^3 - 8,\ g(x) = 4x - 2$

65. $f(x) = 6x^2 - 11x - 10,\ g(x) = 3x + 2$

66. $f(x) = 8x^2 - 22x - 21,\ g(x) = 2x - 7$

For Exercises 67–74, consider the functions F and G as shown.

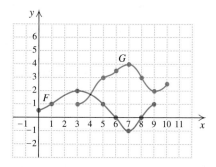

67. Determine $(F + G)(5)$ and $(F + G)(7)$.

68. Determine $(F \cdot G)(6)$ and $(F \cdot G)(9)$.

69. Determine $(G - F)(7)$ and $(G - F)(3)$.

70. Determine $(F/G)(3)$ and $(F/G)(7)$.

71. Find the domains of F, G, $F + G$, and F/G.

72. Find the domains of $F - G$, $F \cdot G$, and G/F.

73. Graph $F + G$.

74. Graph $G - F$.

*In the following graph, W(t) represents the number of gallons of whole milk, L(t) the number of gallons of lowfat milk, and S(t) the number of gallons of skim milk consumed by the average American in year t.**

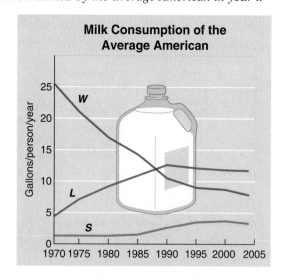

Milk Consumption of the Average American

Sources: Copyright 1990, CSPI. Adapted from *Nutrition Action Healthletter* (1875 Connecticut Avenue, N.W., Suite 300, Washington, DC 20009-5728. $24.00 for 10 issues); USDA Agricultural Fact Book 2000, USDA Economic Research Service; ers.usda.gov

TW 75. From 1970 to 2004, did American milk consumption increase or decrease? Explain how you determined this.

TW 76. Examine the graphs in Exercises 39–44. To what do you attribute the decline in $F(t)$ from 2000 to 2002?

Skill Maintenance

Solve. [2.3]

77. $4x - 7y = 8$, for x

78. $3x - 8y = 5$, for y

79. $5x + 2y = -3$, for y

80. $6x + 5y = -2$, for x

Synthesis

TW 81. If $f(x) = c$, where c is some positive constant, describe how the graphs of $y = g(x)$ and $y = (f + g)(x)$ will differ.

TW 82. Examine the graphs showing number of calories expended following Example 1 and explain how they might be modified to represent the absorption of 200 mg of Advil® taken four times a day.

83. Find the domain of f/g, if

$$f(x) = \frac{3x}{2x + 5} \quad \text{and} \quad g(x) = \frac{x^4 - 1}{3x + 9}.$$

84. Find the domain of F/G, if

$$F(x) = \frac{1}{x - 4} \quad \text{and} \quad G(x) = \frac{x^2 - 4}{x - 3}.$$

85. Sketch the graph of two functions f and g such that the domain of f/g is

$$\{x \mid -2 \le x \le 3 \text{ and } x \ne 1\}.$$

86. Find the domains of $f + g$, $f - g$, $f \cdot g$, and f/g, if

$$f = \{(-2, 1), (-1, 2), (0, 3), (1, 4), (2, 5)\}$$

and

$$g = \{(-4, 4), (-3, 3), (-2, 4), (-1, 0), (0, 5), (1, 6)\}.$$

87. Find the domain of m/n, if

$$m(x) = 3x \text{ for } -1 < x < 5$$

and

$$n(x) = 2x - 3.$$

88. For f and g as defined in Exercise 86, find $(f + g)(-2)$, $(f \cdot g)(0)$, and $(f/g)(1)$.

89. Write equations for two functions f and g such that the domain of $f + g$ is

$$\{x \mid x \text{ is a real number } and \ x \neq -2 \ and \ x \neq 5\}.$$

90. Using the window $[-5, 5, -1, 9]$, graph $y_1 = 5$, $y_2 = x + 2$, and $y_3 = \sqrt{x}$. Then predict what shape the graphs of $y_1 + y_2, y_1 + y_3$, and $y_2 + y_3$ will take. Use a graph to check each prediction.

91. Let $y_1 = 2.5x + 1.5$, $y_2 = x - 3$, and $y_3 = y_1/y_2$. For many calculators, depending on whether the CONNECTED or DOT mode is used, the graph of y_3 appears as follows.

CONNECTED MODE
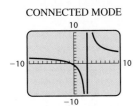

DOT MODE
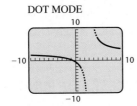

Use algebra to determine which graph more accurately represents y_3.

92. Use the graphs of f and g, shown at the top of the next column, to match each of $(f + g)(x)$, $(f - g)(x)$, $(f \cdot g)(x)$, and $(f/g)(x)$ with its graph.

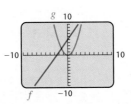

a) $(f + g)(x)$
b) $(f - g)(x)$
c) $(f \cdot g)(x)$
d) $(f/g)(x)$

I
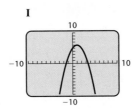

II
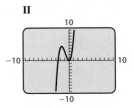

III
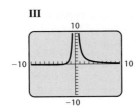

IV
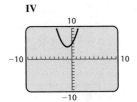

Collaborative Corner

Time On Your Hands

Focus: The algebra of functions
Time: 10–15 minutes
Group Size: 2–3

The graph and data at right chart the average retirement age $R(x)$ and life expectancy $E(x)$ of U.S. citizens in year x (*Source*: Bureau of Labor Statistics, courtesy of the Insurance Advisory Board).

ACTIVITY

1. Working as a team, perform the appropriate calculations and then graph $E - R$.
2. What does $(E - R)(x)$ represent? In what fields of study or business might the function $E - R$ prove useful?

3. Should E and R really be calculated separately for men and women? Why or why not?
4. What advice would you give to someone considering early retirement?

Year:	1955	1965	1975	1985	1995	2005e
Average Retirement Age:	67.3	64.9	63.2	62.8	62.7	61.5
Average Life Expectancy:	73.9	75.5	77.2	78.5	79.1	80.1

e = estimated

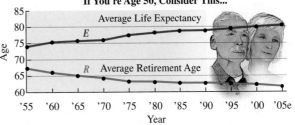

5 Chapter Summary and Review

KEY TERMS AND DEFINITIONS

POLYNOMIALS

Polynomial, p. 372   A monomial or a sum of monomials. All exponents of variables in a polynomial are whole numbers. The domain of a **polynomial function** is $(-\infty, \infty)$.

Term, p. 372   A number, a variable, or a product or a quotient of numbers and/or variables. The number in a term is its **coefficient.** The **degree** of a term is the number of variable factors.

Term	Coefficient	Degree
$-8t^2$	-8	2
$\frac{1}{3}xy^4$	$\frac{1}{3}$	5
7	7	0

Leading term, p. 374   The term in a polynomial with the highest degree. The coefficient of the leading term is the **leading coefficient,** and the degree of the leading term is the **degree of the polynomial.**

Descending order, p. 375   An arrangement of a polynomial with terms of higher degree written first. If the terms with lower degree are written first, the polynomial is in **ascending order.**

THE ALGEBRA OF FUNCTIONS

1. $(f + g)(x) = f(x) + g(x)$
2. $(f - g)(x) = f(x) - g(x)$
3. $(f \cdot g)(x) = f(x) \cdot g(x)$
4. $(f/g)(x) = f(x)/g(x), g(x) \neq 0$

IMPORTANT CONCEPTS

[Section references appear in brackets.]

Concept	Example	
Definitions of Exponents		
(Assume that no denominators are 0 and that 0^0 is not considered.)		
For any integers m and n:		
1 as an exponent: $a^1 = a$	$3^1 = 3$	[5.1]
0 as an exponent: $a^0 = 1$	$3^0 = 1$	[5.1]
Negative exponents: $a^{-n} = \dfrac{1}{a^n}$,	$3^{-2} = \dfrac{1}{3^2} = \dfrac{1}{9}$	[5.2]
$\dfrac{a^{-n}}{b^{-m}} = \dfrac{b^m}{a^n}$,	$\dfrac{3^{-7}}{x^{-5}} = \dfrac{x^5}{3^7}$	[5.2]
$\left(\dfrac{a}{b}\right)^{-n} = \left(\dfrac{b}{a}\right)^n$	$\left(\dfrac{3}{x}\right)^{-2} = \left(\dfrac{x}{3}\right)^2$	[5.2]
Properties of Exponents		
(Assume that no denominators are 0 and that 0^0 is not considered.)		
For any integers m and n:		
The Product Rule: $a^m \cdot a^n = a^{m+n}$	$3^{-5} \cdot 3^9 = 3^{-5+9} = 3^4$	
The Quotient Rule: $\dfrac{a^m}{a^n} = a^{m-n}$	$\dfrac{3^4}{3^{-1}} = 3^{4-(-1)} = 3^5$	
The Power Rule: $(a^m)^n = a^{mn}$	$(3^{-4})^2 = 3^{(-4)(2)} = 3^{-8} = \dfrac{1}{3^8}$	
Raising a product to a power: $(ab)^n = a^n b^n$	$(3x^5)^4 = 3^4(x^5)^4 = 81x^{20}$	
Raising a quotient to a power: $\left(\dfrac{a}{b}\right)^n = \dfrac{a^n}{b^n}$	$\left(\dfrac{3}{x}\right)^6 = \dfrac{3^6}{x^6}$	[5.1], [5.2]
Scientific notation: $N \times 10^m, 1 \leq N < 10$	$4100 = 4.1 \times 10^3$; $0.005 = 5 \times 10^{-3}$	[5.2]
Polynomials are classified by the number of terms.	**Monomial** (one term): $4x^3$	
	Binomial (two terms): $x^2 - 5$	
	Trinomial (three terms): $3t^3 + 2t - 10$	[5.3]

(continued)

Polynomials are classified by degree.	**Constant** (degree 0): -9 **Linear** (degree 1): $\frac{1}{3}x + 8$ **Quadratic** (degree 2): $x^2 + 2x + 3$ **Cubic** (degree 3): $-x^3 - 5x$ **Quartic** (degree 4): $15x^4 + x^2 + 1$ [5.3]
Add polynomials by combining like terms.	$(2x^2 - 3x + 7) + (5x^3 + 3x - 9)$ $= 2x^2 + (-3x) + 7 + 5x^3 + 3x + (-9)$ $= 5x^3 + 2x^2 - 2$ [5.4]
Subtract polynomials by adding the opposite of the polynomial being subtracted.	$(2x^2 - 3x + 7) - (5x^3 + 3x - 9)$ $= 2x^2 - 3x + 7 - 5x^3 - 3x + 9$ $= -5x^3 + 2x^2 - 6x + 16$ [5.4]
Multiply polynomials by multiplying each term of one polynomial by each term of the other.	$(x + 2)(x^2 - x - 1)$ $= x \cdot x^2 - x \cdot x - x \cdot 1 + 2 \cdot x^2 - 2 \cdot x - 2 \cdot 1$ $= x^3 - x^2 - x + 2x^2 - 2x - 2$ $= x^3 + x^2 - 3x - 2$ [5.5]
Divide polynomials using long division or synthetic division.	$\begin{array}{r} x + 8 \\ x - 3 \overline{)\, x^2 + 5x - 2} \\ \underline{x^2 - 3x} \\ 8x - 2 \\ \underline{8x - 24} \\ 22 \end{array}$ $\quad \begin{array}{r} 3 \,\rfloor\; 1 \quad 5 \quad -2 \\ 3 \quad 24 \\ \hline 1 \quad 8 \;\rfloor\; 22 \end{array}$ $(x^2 + 5x - 2) \div (x - 3) = x + 8 + \dfrac{22}{x - 3}$ [5.8]

Special Products

FOIL:

$(A + B)(C + D) = AC + AD + BC + BD$ $\quad (x + 3)(x - 2) = x^2 - 2x + 3x - 6 = x^2 + x - 6$

The product of a sum and a difference:

$(A + B)(A - B) = A^2 - B^2$ $\quad (t^3 + 5)(t^3 - 5) = (t^3)^2 - 5^2 = t^6 - 25$

$A^2 - B^2$ is called a **difference of squares.**

The square of a binomial:

$(A + B)^2 = A^2 + 2AB + B^2$ $\quad (5x + 3)^2 = (5x)^2 + 2(5x)(3) + 3^2 = 25x^2 + 30x + 9$

$(A - B)^2 = A^2 - 2AB + B^2$ $\quad (5x - 3)^2 = (5x)^2 - 2(5x)(3) + 3^2 = 25x^2 - 30x + 9$

$A^2 + 2AB + B^2$ and $A^2 - 2AB + B^2$ are called **perfect-square trinomials.**

[5.6]

Review Exercises

🖎 *Concept Reinforcement*   *Classify each statement as either true or false.*

1. When two polynomials that are written in descending order are added, the result is generally written in ascending order. [5.4]

2. The product of the sum and the difference of the same two terms is a difference of squares. [5.6]

3. When a binomial is squared, the result is a perfect-square trinomial. [5.6]

4. FOIL can be used whenever two polynomials are being multiplied. [5.6]

5. The degree of a polynomial can exceed the value of the polynomial's leading coefficient. [5.3]

6. Scientific notation is used only for extremely large numbers. [5.2]

7. FOIL can be used with polynomials in several variables. [5.7]

8. A positive number raised to a negative exponent can never represent a negative number. [5.2]

Simplify. [5.1]

9. $y^7 \cdot y^3 \cdot y$

10. $(3x)^5 \cdot (3x)^9$

11. $t^8 \cdot t^0$

12. $\dfrac{4^5}{4^2}$

13. $\dfrac{(a + b)^4}{(a + b)^4}$

14. $\left(\dfrac{3t^4}{2s^3}\right)^2$

15. $(-2xy^2)^3$

16. $(2x^3)(-3x)^2$

17. $(a^2b)(ab)^5$

18. Express using a positive exponent: m^{-7}. [5.2]

19. Express using a negative exponent: $\dfrac{1}{t^8}$. [5.2]

Simplify. [5.2]

20. $7^2 \cdot 7^{-4}$

21. $\dfrac{a^{-5}b}{a^8b^8}$

22. $(x^3)^{-4}$

23. $(2x^{-3}y)^{-2}$

24. $\left(\dfrac{2x}{y}\right)^{-3}$

25. Convert to decimal notation: 8.3×10^6. [5.2]

26. Convert to scientific notation: 0.0000328. [5.2]

Multiply or divide and write scientific notation for the result. Use the correct number of significant digits. [5.2]

27. $(3.8 \times 10^4)(5.5 \times 10^{-1})$

28. $\dfrac{1.28 \times 10^{-8}}{2.5 \times 10^{-4}}$

29. *Blood Donors.* Every 4–6 weeks, one of the authors of this text donates 1.14×10^6 cubic millimeters (two pints) of blood platelets to the American Red Cross. In one cubic millimeter of blood, there are about 2×10^5 platelets. Approximate the number of platelets in a typical donation by this author. [5.2]

Identify the terms of each polynomial. [5.3]

30. $3x^2 + 6x + \frac{1}{2}$

31. $-4y^5 + 7y^2 - 3y - 2$

List the coefficients of the terms in each polynomial. [5.3]

32. $7x^2 - x + 7$

33. $4x^3 + 6x^2 - 5x + \frac{5}{3}$

For each polynomial, **(a)** *list the degree of each term;* **(b)** *determine the leading term and the leading coefficient; and* **(c)** *determine the degree of the polynomial.* [5.3]

34. $4t^2 + 6 + 15t^5$

35. $-2x^5 + x^4 - 3x^2 + x$

Classify each polynomial as a monomial, a binomial, a trinomial, or none of these. [5.3]

36. $4x^3 - 1$

37. $4 - 9t^3 - 7t^4 + 10t^2$

38. $7y^2$

Combine like terms and write in descending order. [5.3]

39. $5x - x^2 + 4x$

40. $\frac{3}{4}x^3 + 4x^2 - x^3 + 7$

41. $-2x^4 + 16 + 2x^4 + 9 - 3x^5$

42. $3x^2 - 2x + 3 - 5x^2 - 1 - x$

43. $-x + \frac{1}{2} + 14x^4 - 7x^2 - 1 - 4x^4$

44. Find $P(-1)$ for $P(x) = x^2 - 3x + 6$. [5.3]

45. Evaluate $3 - 5x$ for $x = -5$. [5.3]

46. Graph $p(x) = 2 - x^2$ in a standard viewing window and estimate the range of p. [5.3]

Medicine.   Ibuprofen is a medication used to relieve pain. The polynomial function

$$M(t) = 0.5t^4 + 3.45t^3 - 96.65t^2 + 347.7t,$$
$$0 \le t \le 6$$

can be used to estimate the number of milligrams of ibuprofen in the bloodstream t hours after 400 mg of the medication has been swallowed (Source: Based on data from Dr. P. Carey, Burlington, VT). Use the following graph for Exercises 47–50. [5.3]

Time (in hours)	Milligrams of Ibuprofen in Bloodstream
0.5	150
1	255
2.5	340
4.5	125
5.5	20

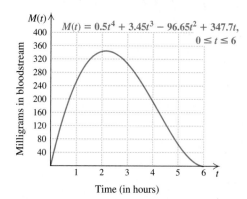

Time (in hours)

47. Use the graph above to estimate the number of milligrams of ibuprofen in the bloodstream 2 hr after 400 mg has been swallowed.

48. Use the graph above to estimate the number of milligrams of ibuprofen in the bloodstream 4 hr after 400 mg has been swallowed.

49. Approximate the range of M.

50. Give the domain of M.

Add or subtract. [5.4]

51. $(3x^4 - x^3 + x - 4) + (x^5 + 7x^3 - 3x - 5)$

52. $(5x^2 - 4x + 1) - (3x^2 + 7)$

53. $(3x^5 - 4x^4 + 2x^2 + 3) - (2x^5 - 4x^4 + 3x^3 + 4x^2 - 5)$

54.
$$\begin{array}{l} -\frac{3}{4}x^4 + \frac{1}{2}x^3 \qquad\qquad\quad + \frac{7}{8} \\ \qquad\quad -\frac{1}{4}x^3 - x^2 - \frac{7}{4}x \\ +\frac{3}{2}x^4 \qquad\quad + \frac{2}{3}x^2 \qquad\quad - \frac{1}{2} \end{array}$$

55.
$$\begin{array}{l} 2x^5 \qquad\quad - x^3 \qquad\quad + x + 3 \\ -(3x^5 - x^4 + 4x^3 + 2x^2 - x + 3) \end{array}$$

56. The length of a rectangle is 3 m greater than its width.

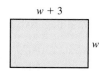

$w + 3$

w

a) Find a polynomial for the perimeter. [5.4]
b) Find a polynomial for the area. [5.5]

Multiply.

57. $3x(-4x^2)$ [5.5]

58. $(7x + 1)^2$ [5.6]

59. $(a - 7)(a + 4)$ [5.6]

60. $(m + 5)(m - 5)$ [5.6]

61. $(4x^2 - 5x + 1)(3x - 2)$ [5.5]

62. $(x - 9)^2$ [5.6]

63. $3t^2(5t^3 - 2t^2 + 4t)$ [5.5]

64. $(a - 7)(a + 7)$ [5.6]

65. $(x - 0.3)(x - 0.75)$ [5.6]

66. $(x^4 - 2x + 3)(x^3 + x - 1)$ [5.5]

67. $(3x - 5)^2$ [5.6]

68. $(2t^2 + 3)(t^2 - 7)$ [5.6]

69. $\left(a - \frac{1}{2}\right)\left(a + \frac{2}{3}\right)$ [5.6]

70. $(3x^2 + 4)(3x^2 - 4)$ [5.6]

71. $(2 - x)(2 + x)$ [5.6]

72. $(2x + 3y)(x - 5y)$ [5.7]

73. Evaluate $2 - 5xy + y^2 - 4xy^3 + x^6$ for $x = -1$ and $y = 2$. [5.7]

Identify the coefficient and the degree of each term of each polynomial. Then find the degree of each polynomial. [5.7]

74. $x^5y - 7xy + 9x^2 - 8$

75. $x^2y^5z^9 - y^{40} + x^{13}z^{10}$

Combine like terms. [5.7]

76. $y + w - 2y + 8w - 5$

77. $6m^3 + 3m^2n + 4mn^2 + m^2n - 5mn^2$

Add or subtract. [5.7]

78. $(5x^2 - 7xy + y^2) + (-6x^2 - 3xy - y^2)$

79. $(6x^3y^2 - 4x^2y - 6x) - (-5x^3y^2 + 4x^2y + 6x^2 - 6)$

Multiply. [5.7]

80. $(p - q)(p^2 + pq + q^2)$ **81.** $\left(3a^4 - \frac{1}{3}b^3\right)^2$

82. Find a polynomial for the shaded area. [5.7]

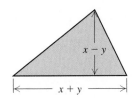

Divide. [5.8]

83. $(10x^3 - x^2 + 6x) \div (2x)$

84. $(6x^3 - 5x^2 - 13x + 13) \div (2x + 3)$

85. $\dfrac{t^4 + t^3 + 2t^2 - t - 3}{t + 1}$

Let $g(x) = 3x - 6$ and $h(x) = x^2 + 1$. Find the following. [5.9]

86. $(g \cdot h)(4)$

87. $(g - h)(-2)$

88. $(g/h)(-1)$

89. $(g + h)(x)$

90. $(g \cdot h)(x)$

91. $(h/g)(x)$

92. The domains of $g + h$ and $g \cdot h$

93. The domain of h/g

Synthesis

TW 94. Explain why $5x^3$ and $(5x)^3$ are not equivalent expressions.

TW 95. If two polynomials of degree n are added, is the sum also of degree n? Why or why not?

96. How many terms are there in each of the following? [5.3], [5.6]
   a) $(x - a)(x - b) + (x - a)(x - b)$
   b) $(x + a)(x - b) + (x - a)(x + b)$

97. Combine like terms: [5.1], [5.3]
   $-3x^5 \cdot 3x^3 - x^6(2x)^2 + (3x^4)^2 + (2x^2)^4 - 40x^2(x^3)^2$.

98. A polynomial has degree 4. The x^2-term is missing. The coefficient of x^4 is two times the coefficient of x^3. The coefficient of x is 3 less than the coefficient of x^4. The remaining coefficient is 7 less than the coefficient of x. The sum of the coefficients is 15. Find the polynomial. [5.3]

Aha! 99. Multiply: $[(x - 5) - 4x^3][(x - 5) + 4x^3]$. [5.6]

100. Solve: $(x - 7)(x + 10) = (x - 4)(x - 6)$. [2.2], [5.6]

Chapter Test 5

Simplify.

1. $t^2 \cdot t^5 \cdot t$

2. $(x + 3)^5(x + 3)^6$

3. $\dfrac{3^5}{3^2}$

4. $\dfrac{(2x)^5}{(2x)^5}$

5. $(x^3)^2$

6. $(-3y^2)^3$

7. $(3x^2)(-2x^5)^3$

8. $(a^3b^2)(ab)^3$

9. Express using a positive exponent: 5^{-3}.

10. Express using a negative exponent: $\dfrac{1}{y^8}$.

Simplify.

11. $t^{-4} \cdot t^{-2}$

12. $\dfrac{x^3y^2}{x^8y^{-3}}$

13. $(2a^3b^{-1})^{-4}$

14. $\left(\dfrac{ab}{c}\right)^{-3}$

15. Convert to scientific notation: 3,900,000,000.

16. Convert to decimal notation: 5×10^{-8}.

Multiply or divide and write scientific notation for the result. Use the correct number of significant digits.

17. $\dfrac{5.6 \times 10^6}{3.2 \times 10^{-11}}$

18. $(2.4 \times 10^5)(5.4 \times 10^{16})$

19. A CD-ROM can contain about 600 million pieces of information. How many sound files, each needing 40,000 pieces of information, can a CD-ROM hold? Write scientific notation for the answer.

20. Classify the polynomial as a monomial, a binomial, a trinomial, or none of these:
$$6t^2 - 9t.$$

21. Identify the coefficient of each term of the polynomial:
$$\tfrac{1}{3}x^5 - x + 7.$$

22. Determine the degree of each term, the leading term and the leading coefficient, and the degree of the polynomial:
$$2t^3 - t + 7t^5 + 4.$$

23. Find $p(-2)$ for $p(x) = x^2 + 5x - 1$.

Combine like terms and write in descending order.

24. $4a^2 - 6 + a^2$

25. $y^2 - 3y - y + \tfrac{3}{4}y^2$

26. $3 - x^2 + 2x^3 + 5x^2 - 6x - 2x + x^5$

27. Graph $f(x) = x^3 - x + 1$ in the standard viewing window and estimate the range of f.

Add or subtract.

28. $\left(x^4 + \tfrac{2}{3}x + 5\right) + \left(4x^4 + 5x^2 + \tfrac{1}{3}x\right)$

29. $(2x^4 + x^3 - 8x^2 - 6x - 3) - (6x^4 - 8x^2 + 2x)$

30. $(x^3 - 0.4x^2 - 12) - (x^5 - 0.3x^3 + 0.4x^2 - 9)$

Multiply.

31. $-3x^2(4x^2 - 3x - 5)$

32. $\left(x - \tfrac{1}{3}\right)^2$

33. $(5t - 7)(5t + 7)$

34. $(3b + 5)(b - 3)$

35. $(x^6 - 4)(x^8 + 4)$

36. $(8 - y)(6 + 5y)$

37. $(2x + 1)(3x^2 - 5x - 3)$

38. $(8a + 3)^2$

39. Combine like terms:
$$x^3y - y^3 + xy^3 + 8 - 6x^3y - x^2y^2 + 11.$$

40. Subtract:
$$(8a^2b^2 - ab + b^3) - (-6ab^2 - 7ab - ab^3 + 5b^3).$$

41. Multiply: $(3x^5 - 4y^5)(3x^5 + 4y^5)$.

Divide.

42. $(12x^4 + 9x^3 - 15x^2) \div (3x^2)$

43. $(6x^3 - 8x^2 - 14x + 13) \div (x + 2)$

Find each of the following, given that $g(x) = -3x - 4$ and $h(x) = x^2 + 1$.

44. $(g \cdot h)(3)$

45. $(g \cdot h)(x)$

46. The domain of h/g

Synthesis

47. The height of a box is 1 less than its length, and the length is 2 more than its width. Express the volume in terms of the length.

48. Solve: $x^2 + (x - 7)(x + 4) = 2(x - 6)^2$.

6

Polynomial Factorizations and Equations

*I*n Chapter 1, we learned that factoring is writing an expression as a product. In this chapter, we factor polynomials to find equivalent expressions and use factoring to solve equations, many of which arise from real-world problems. Factoring polynomials requires a solid command of the multiplication methods studied in Chapter 5.

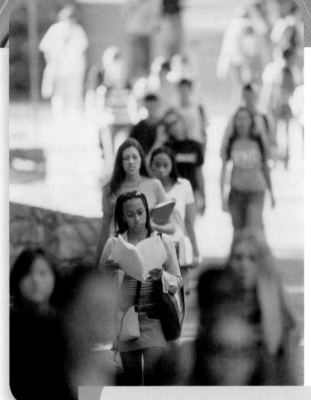

APPLICATION *College Tuition.*

The average in-state tuition for a public four-year college for various years is shown in the following table and graph. Fit a cubic polynomial function to the data, and use the function to predict public-college tuition in 2008.

YEAR	IN-STATE PUBLIC COLLEGE TUITION
1985	$1386
1988	1726
1991	2159
1994	2820
1997	3323
2000	3768
2003	4686
2005*	5948

*Preliminary data

Source: National Center for Education Statistics, *Digest of Education Statistics*

Xscl = 5, Yscl = 1000

This problem appears as Example 4 in Section 6.7.

6.1 Introduction to Polynomial Factorizations and Equations

Graphical Solutions   ◾   The Principle of Zero Products   ◾
Terms with Common Factors   ◾   Factoring by Grouping   ◾
Factoring and Equations

Whenever two polynomials are set equal to each other, the result is a **polynomial equation.** In this section, we learn how to solve such equations both graphically and algebraically by *factoring*.

Graphical Solutions

EXAMPLE 1   Solve: $x^2 = 6x$.

SOLUTION   We can find real-number solutions of a polynomial equation by finding the points of intersection of two graphs.

Alternatively, we can rewrite the equation so that one side is 0 and then find the *x*-intercepts of one graph, or the zeros of a function. Both methods were discussed in Section 4.5.

INTERSECT METHOD

To solve $x^2 = 6x$ using the first method, we graph $y_1 = x^2$ and $y_2 = 6x$ and find the coordinates of any points of intersection.

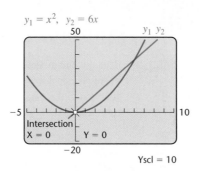

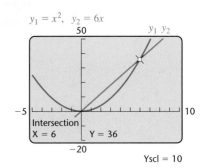

There are two points of intersection of the graphs, so there are two solutions of the equation. The *x*-coordinates of the points of intersection are 0 and 6, so the solutions of the equation are 0 and 6.

ZERO METHOD

Using the second method, we rewrite the equation so that one side is 0:

$$x^2 = 6x$$

$$x^2 - 6x = 0. \quad \text{Adding } -6x \text{ to both sides}$$

To solve $x^2 - 6x = 0$, we graph the function $f(x) = x^2 - 6x$ and look for values of x for which $f(x) = 0$, or the zeros of f. These correspond to the x-intercepts of the graph.

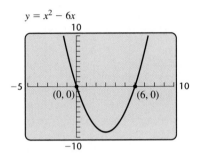

Student Notes

Recall that a zero of a function is a number, not an ordered pair. In Example 1, the zeros are the first coordinates of the x-intercepts.

The solutions of the equation are the x-coordinates of the x-intercepts of the graph, 0 and 6. Both 0 and 6 check in the original equation.

In Example 1, the values 0 and 6 are called **zeros** of the function $f(x) = x^2 - 6x$. They are also referred to as **roots** of the equation $f(x) = 0$.

Zeros and Roots   The x-values for which a function $f(x)$ is 0 are called the *zeros* of the function.

The x-values for which an equation such as $f(x) = 0$ is true are called the *roots* of the equation.

In this chapter, we consider only the real-number zeros of functions. We can solve, or find the roots of, the equation $f(x) = 0$ by finding the zeros of the function f.

EXAMPLE 2   Find the zeros of the function given by

$$f(x) = x^3 - 3x^2 - 4x + 12.$$

SOLUTION First, we graph the equation $y = x^3 - 3x^2 - 4x + 12$, choosing a viewing window that shows the x-intercepts of the graph. It may require trial and error to choose an appropriate viewing window. In the standard viewing window, shown on the left below, it appears that there are three zeros. However, we cannot see the shape of the graph for x-values between -2 and 1, so Ymax should be increased. A viewing window of $[-10, 10, -100, 100]$, with Yscl $= 10$, as shown in the middle below, gives a better idea of the overall shape of the graph, but we cannot see all three x-intercepts clearly.

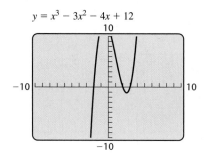

$y = x^3 - 3x^2 - 4x + 12$

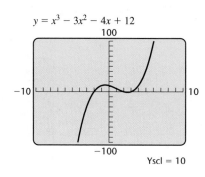

$y = x^3 - 3x^2 - 4x + 12$

Yscl $= 10$

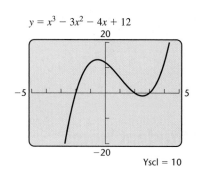

$y = x^3 - 3x^2 - 4x + 12$

Yscl $= 10$

We magnify the portion of the graph close to the origin by making the window dimensions smaller. The window shown on the right above, $[-5, 5, -20, 20]$, is a good choice for viewing the zeros of this function. There are other good choices as well. The zeros of the function seem to be about -2, 2, and 3.

To find the zero that appears to be about -2, we first choose the ZERO option from the CALC menu. We then choose a Left Bound to the left of -2 on the x-axis. Next, we choose a Right Bound to the right of -2 on the x-axis. For a Guess, we choose an x-value close to -2. We see that -2 is indeed a zero of the function f.

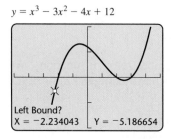

$y = x^3 - 3x^2 - 4x + 12$

Left Bound?
X = -2.234043 | Y = -5.186654

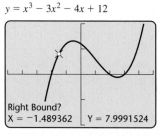

$y = x^3 - 3x^2 - 4x + 12$

Right Bound?
X = -1.489362 | Y = 7.9991524

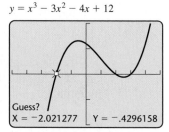

$y = x^3 - 3x^2 - 4x + 12$

Guess?
X = -2.021277 | Y = $-.4296158$

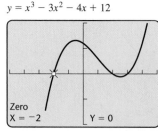

$y = x^3 - 3x^2 - 4x + 12$

Zero
X = -2 | Y = 0

Using the same procedure for each of the other two zeros, we find that the zeros of the function $f(x) = x^3 - 3x^2 - 4x + 12$ are -2, 2, and 3.

We saw in the discussion after Example 1 that $f(x) = x^2 - 6x$ has 2 zeros. We also saw in Example 2 that $f(x) = x^3 - 3x^2 - 4x + 12$ has 3 zeros. We can tell something about the number of zeros of a polynomial function from its degree.

[**Interactive
Discovery**

Each of the following is a second-degree polynomial (quadratic) function. Use a graph to determine the *number* of real-number zeros of each function.

1. $f(x) = x^2 - 2x + 1$
2. $g(x) = x^2 + 9$
3. $h(x) = 2x^2 + 7x - 4$

Each of the following is a third-degree polynomial (cubic) function. Use a graph to determine the *number* of real-number zeros of each function.

4. $f(x) = 2x^3 - 5x^2 - 3x$
5. $g(x) = x^3 - 7x^2 + 16x - 12$
6. $h(x) = x^3 - 4x^2 + 5x - 20$
7. Compare the degree of each function with the number of real-number zeros of that function. What conclusion can you draw?

A second-degree polynomial function will have 0, 1, or 2 real-number zeros. A third-degree polynomial function will have 1, 2, or 3 real-number zeros. This result can be generalized, although we will not prove it here.

> An *n*th-degree polynomial function will have at most *n* zeros.

"Seeing" all the zeros of a polynomial function when solving an equation graphically depends on a good choice of viewing window. Many polynomial equations can be solved algebraically. One principle used in solving polynomial equations is the *principle of zero products*.

The Principle of Zero Products

When we multiply two or more numbers, the product is 0 if any one of those numbers (factors) is 0. Conversely, if a product is 0, then at least one of the factors must be 0. This property of 0 gives us a new principle for solving equations.

> **The Principle of Zero Products**   For any real numbers a and b:
> If $ab = 0$, then $a = 0$ or $b = 0$. If $a = 0$ or $b = 0$, then $ab = 0$.

When a polynomial is written as a product, we say that it is *factored*. The product is called a *factorization* of the polynomial. For example,

$3x(x + 4)$ is a factorization of $3x^2 + 12x$

since $3x(x + 4)$ is a product and

$3x(x + 4) = 3x^2 + 12x$.

The polynomials $3x$ and $x + 4$ are *factors* of $3x^2 + 12x$. The zeros of a polynomial function are related to the factorization of the polynomial.

Consider the function f given by $f(x) = 3x^2 + 12x$. If $g(x) = 3x$ and $h(x) = x + 4$, then $f(x) = g(x) \cdot h(x)$. The graph on the left below shows that the zero of g is 0. The graph in the middle shows that the zero of h is -4. On the right, we see that the zeros of f are 0 and -4.

$y = 3x$

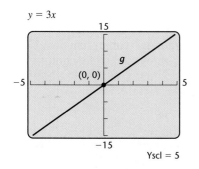

$y = x + 4$

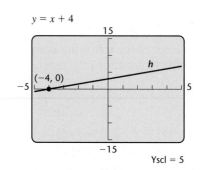

$y = 3x(x + 4)$

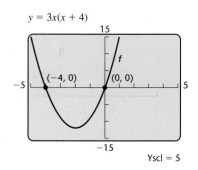

The zeros of a polynomial function are the zeros of the functions described by the factors of the polynomial.

Thus we can use the principle of zero products to solve a polynomial equation. In order to do so, we must have 0 on one side of the equation and the other side must be factored.

EXAMPLE 3   Solve: $(x - 3)(x + 2) = 0$.

SOLUTION The principle of zero products says that in order for $(x - 3)(x + 2)$ to be 0, at least one factor must be 0. Thus,

$$x - 3 = 0 \quad or \quad x + 2 = 0. \quad \text{Using the principle of zero products}$$

Each of these linear equations is then solved separately:

$$x = 3 \quad or \quad x = -2.$$

We check both algebraically and graphically, as follows.

ALGEBRAIC CHECK

For 3:
$$\frac{(x - 3)(x + 2) = 0}{(3 - 3)(3 + 2) \;\big|\; 0}$$
$$0(5) \;\big|$$
$$0 \overset{?}{=} 0 \quad \text{TRUE}$$

For -2:
$$\frac{(x - 3)(x + 2) = 0}{(-2 - 3)(-2 + 2) \;\big|\; 0}$$
$$(-5)(0) \;\big|$$
$$0 \overset{?}{=} 0 \quad \text{TRUE}$$

GRAPHICAL CHECK

$y = (x - 3)(x + 2)$

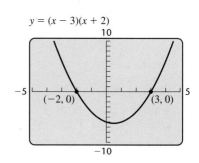

The solutions are -2 and 3.

EXAMPLE 4   Given that $f(x) = x(3x + 2)$, find all the values of a for which $f(a) = 0$.

SOLUTION   We are looking for all numbers a for which $f(a) = 0$. Since $f(a) = a(3a + 2)$, we must have

$$a(3a + 2) = 0 \qquad \text{Setting } f(a) \text{ equal to 0}$$
$$a = 0 \quad or \quad 3a + 2 = 0 \qquad \text{Using the principle of zero products}$$
$$a = 0 \quad or \qquad a = -\tfrac{2}{3}.$$

We check by evaluating $f(0)$ and $f\left(-\tfrac{2}{3}\right)$. One way to check with a graphing calculator is to enter $y_1 = x(3x + 2)$ and calculate Y1(0) and Y1($-2/3$).

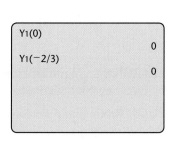

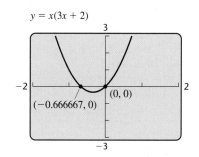

We can also graph $f(x) = x(3x + 2)$ and observe that its zeros are $-\tfrac{2}{3}$ and 0. Thus, to have $f(a) = 0$, we must have $a = 0$ or $a = -\tfrac{2}{3}$.

If the polynomial in an equation of the form $p(x) = 0$ is not in factored form, we must factor it before we can use the principle of zero products to solve the equation. To **factor** an expression means to write an equivalent expression that is a product.

Terms with Common Factors

When factoring a polynomial, we look for factors common to every term and then use the distributive law.

EXAMPLE 5   Factor out a common factor: $6y^2 - 18$.

SOLUTION   We have

$$6y^2 - 18 = 6 \cdot y^2 - 6 \cdot 3 \qquad \text{Noting that 6 is a common factor}$$
$$= 6(y^2 - 3). \qquad \text{Using the distributive law}$$

Check: $6(y^2 - 3) = 6y^2 - 18$.

Suppose in Example 5 that the common factor 2 were used:

$$6y^2 - 18 = 2 \cdot 3y^2 - 2 \cdot 9 \qquad \text{2 is a common factor.}$$
$$= 2(3y^2 - 9). \qquad \text{Using the distributive law}$$

Note that $3y^2 - 9$ itself has a common factor, 3. It is standard practice to factor out the *largest*, or *greatest*, *common factor*, so that the polynomial factor cannot be factored any further. Thus, by now factoring out 3, we can complete the factorization:

$$6y^2 - 18 = 2(3y^2 - 9)$$
$$= 2 \cdot 3(y^2 - 3) = 6(y^2 - 3).$$

Remember to multiply the two common factors: $2 \cdot 3 = 6$.

To find the greatest common factor of a polynomial, we multiply the greatest common factor of the coefficients by the greatest common factor of the variable factors appearing in every term. Thus, to find the greatest common factor of $30x^4 + 20x^5$, we multiply the greatest common factor of 30 and 20, which is 10, by the greatest common factor of x^4 and x^5, which is x^4:

$$30x^4 + 20x^5 = 10 \cdot 3 \cdot x^4 + 10 \cdot 2 \cdot x^4 \cdot x$$
$$= 10x^4(3 + 2x).$$

The greatest common factor is $10x^4$.

EXAMPLE 6   Write an expression equivalent to $8p^6q^2 - 4p^5q^3 + 10p^4q^4$ by factoring out the greatest common factor.

SOLUTION   First, we look for the greatest positive common factor in the coefficients:

$$8, -4, 10 \longrightarrow \text{Greatest common factor} = 2.$$

Second, we look for the greatest common factor in the powers of p:

$$p^6, p^5, p^4 \longrightarrow \text{Greatest common factor} = p^4.$$

Third, we look for the greatest common factor in the powers of q:

$$q^2, q^3, q^4 \longrightarrow \text{Greatest common factor} = q^2.$$

Thus, $2p^4q^2$ is the greatest common factor of the given polynomial. Then

$$8p^6q^2 - 4p^5q^3 + 10p^4q^4 = 2p^4q^2 \cdot 4p^2 - 2p^4q^2 \cdot 2pq + 2p^4q^2 \cdot 5q^2$$
$$= 2p^4q^2(4p^2 - 2pq + 5q^2).$$

As a final check, note that

$$2p^4q^2(4p^2 - 2pq + 5q^2) = 2p^4q^2 \cdot 4p^2 - 2p^4q^2 \cdot 2pq + 2p^4q^2 \cdot 5q^2$$
$$= 8p^6q^2 - 4p^5q^3 + 10p^4q^4,$$

which is the original polynomial. Since $4p^2 - 2pq + 5q^2$ has no common factor, we know that $2p^4q^2$ is the greatest common factor.

The polynomials in Examples 5 and 6 have been **factored completely.** They cannot be factored further. The factors in the resulting factorizations are said to be **prime polynomials.**

EXAMPLE 7   Factor: $15x^5 - 12x^4 + 27x^3 - 3x^2$.

SOLUTION   We have

$$15x^5 - 12x^4 + 27x^3 - 3x^2$$

$$= 3x^2 \cdot 5x^3 - 3x^2 \cdot 4x^2 + 3x^2 \cdot 9x - 3x^2 \cdot 1$$ **Try to do this mentally.**

$$= 3x^2(5x^3 - 4x^2 + 9x - 1).$$ **Factoring out $3x^2$**

CAUTION!   Don't forget the term -1. The check below shows why it is essential.

Since $5x^3 - 4x^2 + 9x - 1$ has no common factor, we are finished, except for a check:

$$3x^2(5x^3 - 4x^2 + 9x - 1) = 15x^5 - 12x^4 + 27x^3 - 3x^2.$$ **Our factorization checks.**

The factorization is $3x^2(5x^3 - 4x^2 + 9x - 1)$.

When the leading coefficient is a negative number, we generally factor out a common factor with a negative coefficient.

EXAMPLE 8   Write an equivalent expression by factoring out a common factor with a negative coefficient.

a) $-4x - 24$ **b)** $-2x^3 + 6x^2 - 2x$

SOLUTION

a) $-4x - 24 = -4(x + 6)$

b) $-2x^3 + 6x^2 - 2x = -2x(x^2 - 3x + 1)$ **The 1 is essential.**

EXAMPLE 9   Height of a Thrown Object.   Suppose that a baseball is thrown upward with an initial velocity of 64 ft/sec and an initial height of 0 ft. Its height in feet, $h(t)$, after t seconds is given by

$$h(t) = -16t^2 + 64t.$$

Find an equivalent expression for $h(t)$ by factoring out a common factor with a negative coefficient.

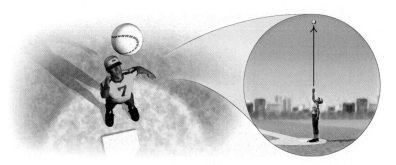

SOLUTION We factor out $-16t$ as follows:

$$h(t) = -16t^2 + 64t = -16t(t - 4). \qquad Check: \ -16t \cdot t = -16t^2 \text{ and} \\ -16t(-4) = 64t.$$

Note that we can obtain function values using either expression for $h(t)$, since factoring forms equivalent expressions. For example,

$$h(1) = -16 \cdot 1^2 + 64 \cdot 1 = 48$$

and $\qquad h(1) = -16 \cdot 1(1 - 4) = 48.$ **Using the factorization** ◢

$y_1 = -16x^2 + 64x,$
$y_2 = (-16x)(x - 4)$

X	Y1	Y2
0	0	0
1	48	48
2	64	64
3	48	48
4	0	0
5	−80	−80
6	−192	−192

X = 0

We can evaluate the expressions $-16t^2 + 64t$ and $-16t(t - 4)$ in Example 9 using any value for t. The results should always match. Thus a quick partial check of any factorization is to evaluate the factorization and the original polynomial for one or two convenient replacements. The check in Example 9 becomes foolproof if three replacements are used. Recall that, in general, an nth-degree factorization is correct if it checks for $n + 1$ different replacements. The table shown at left confirms that the factorization is correct.

Factoring by Grouping

The largest common factor is sometimes a binomial.

▶ **EXAMPLE 10** Factor: $(a - b)(x + 5) + (a - b)(x - y^2)$.

SOLUTION Here the largest common factor is the binomial $a - b$:

$$(a - b)(x + 5) + (a - b)(x - y^2) = (a - b)[(x + 5) + (x - y^2)]$$
$$= (a - b)[2x + 5 - y^2]. \qquad ◢$$

Often, in order to identify a common binomial factor in a polynomial with four terms, we must regroup into two groups of two terms each.

▶ **EXAMPLE 11** Write an equivalent expression by factoring.

a) $y^3 + 3y^2 + 4y + 12$ **b)** $4x^3 - 15 + 20x^2 - 3x$

SOLUTION

a) $y^3 + 3y^2 + 4y + 12 = (y^3 + 3y^2) + (4y + 12)$ Each grouping has a common factor.

$\qquad\qquad\qquad\qquad = y^2(y + 3) + 4(y + 3)$ Factoring out a common factor from each binomial

$\qquad\qquad\qquad\qquad = (y + 3)(y^2 + 4)$ Factoring out $y + 3$

b) When we try grouping $4x^3 - 15 + 20x^2 - 3x$ as

$$(4x^3 - 15) + (20x^2 - 3x),$$

we are unable to factor $4x^3 - 15$. When this happens, we can rearrange the polynomial and try a different grouping:

$$4x^3 - 15 + 20x^2 - 3x = 4x^3 + 20x^2 - 3x - 15 \qquad \text{Using the commutative law to rearrange the terms}$$

$$= 4x^2(x + 5) - 3(x + 5) \qquad \text{By factoring out } -3, \text{ we see that } x + 5 \text{ is a common factor.}$$

$$= (x + 5)(4x^2 - 3).$$

We can use the distributive law to show that the expressions

$$b - a, \qquad -(a - b), \quad \text{and} \quad -1(a - b)$$

are equivalent. Remembering this can help anytime we wish to reverse subtraction (see the third step below).

EXAMPLE 12   Factor: $ax - bx + by - ay$.

SOLUTION   We have

$$ax - bx + by - ay = (ax - bx) + (by - ay) \qquad \text{Grouping}$$

$$= x(a - b) + y(b - a) \qquad \text{Factoring each binomial}$$

$$= x(a - b) + y(-1)(a - b) \qquad \text{Factoring out } -1 \text{ to reverse } b - a$$

$$= x(a - b) - y(a - b) \qquad \text{Simplifying}$$

$$= (a - b)(x - y). \qquad \text{Factoring out } a - b$$

Check: To check, note that $a - b$ and $x - y$ are both prime and that

$$(a - b)(x - y) = ax - ay - bx + by = ax - bx + by - ay.$$

Some polynomials with four terms, like $x^3 + x^2 + 3x - 3$, are prime. Not only is there no common monomial factor, but no matter how we group terms, there is no common binomial factor:

$$x^3 + x^2 + 3x - 3 = x^2(x + 1) + 3(x - 1); \qquad \text{No common factor}$$

$$x^3 + 3x + x^2 - 3 = x(x^2 + 3) + (x^2 - 3); \qquad \text{No common factor}$$

$$x^3 - 3 + x^2 + 3x = (x^3 - 3) + x(x + 3). \qquad \text{No common factor}$$

Student Notes

In Example 12, make certain that you understand why -1 or $-y$ must be factored from $by - ay$.

Factoring and Equations

Factoring can help us solve polynomial equations.

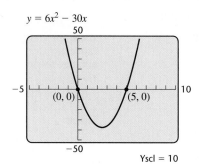

$y = 6x^2 - 30x$

(0, 0)   (5, 0)

Yscl = 10

▸ **EXAMPLE 13**   Solve: $6x^2 = 30x$.

SOLUTION   We can use the principle of zero products if there is a 0 on one side of the equation and the other side is in factored form:

$$6x^2 = 30x$$
$$6x^2 - 30x = 0 \qquad \text{Subtracting } 30x. \text{ One side is now 0.}$$
$$6x(x - 5) = 0 \qquad \text{Factoring}$$
$$6x = 0 \quad or \quad x - 5 = 0 \qquad \text{Using the principle of zero products}$$
$$x = 0 \quad or \qquad x = 5.$$

We check by substitution or graphically, as shown in the figure at left. The solutions are 0 and 5. ◢

CAUTION!   Remember that we can *factor expressions* and *solve equations*. In Example 13, we factored the expression $6x^2 - 30x$ in order to solve the equation $6x^2 - 30x = 0$.

The principle of zero products can be used to show that an *n*th-degree polynomial function can have at most *n* zeros. In Example 13, we wrote a quadratic polynomial as a product of two linear factors. Each linear factor corresponded to one zero of the polynomial function. In general, a polynomial function of degree *n* can have at most *n* linear factors, so a polynomial function of degree *n* can have at most *n* zeros. Thus, when solving an *n*th-degree polynomial equation, we need not look for more than *n* zeros.

To Use the Principle of Zero Products

1. Write an equivalent equation with 0 on one side, using the addition principle.
2. Factor the nonzero side of the equation.
3. Set each factor that is not a constant equal to 0.
4. Solve the resulting equations.

6.1 EXERCISE SET

FOR EXTRA HELP

  MathXL MyMathLab InterAct Math AW Math Tutor Center Video Lectures on CD: Disc 3 Student's Solutions Manual

Concept Reinforcement Classify each statement as either true or false.

1. The largest common factor of $10x^4 + 15x^2$ is $5x$.

2. The largest common factor of a polynomial always has the same degree as the polynomial itself.

3. It is possible for a polynomial to contain several different common factors.

4. When the leading coefficient of a polynomial is negative, we generally factor out a common factor with a negative coefficient.

5. A polynomial is not prime if it contains a common factor other than 1 or -1.

6. It is possible for a polynomial with four terms to factor into a product of two binomials.

7. The expressions $b - a$, $-(a - b)$, and $-1(a - b)$ are all equivalent.

8. The complete factorization of $12x^3 - 20x^2$ is $4x(3x^2 - 5x)$.

In Exercises 9 and 10, use the graph to solve $f(x) = 0$.

9.

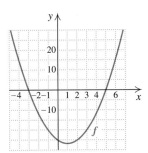

10.

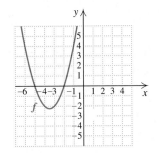

In Exercises 11 and 12, use the graph to find the zeros of the function f.

11.

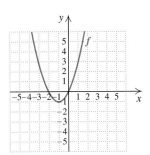

12.

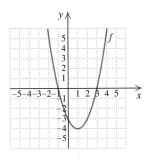

13. Use the following graph to solve $x^2 + 2x = 3$.

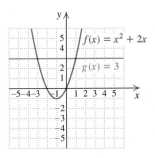

14. Use the following graph to solve $x^2 = 4$.

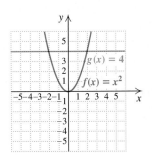

15. Use the following graph to solve $x^2 + 2x - 8 = 0$.

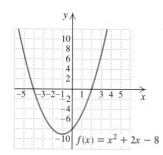

$f(x) = x^2 + 2x - 8$

16. Use the following graph to find the zeros of the function given by $f(x) = x^2 - 2x + 1$.

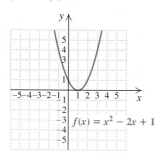

$f(x) = x^2 - 2x + 1$

Solve using a graphing calculator.

17. $x^2 = 5x$

18. $2x^2 = 20x$

19. $4x = x^2 + 3$

20. $x^2 = 1$

21. $x^2 + 150 = 25x$

22. $2x^2 + 25 = 51x$

23. $x^3 - 3x^2 + 2x = 0$

24. $x^3 + 2x^2 = x + 2$

25. $x^3 - 3x^2 - 198x + 1080 = 0$

26. $2x^3 + 25x^2 - 282x + 360 = 0$

27. $21x^2 + 2x - 3 = 0$

28. $66x^2 - 49x - 5 = 0$

Find the zeros of each function.

29. $f(x) = x^2 - 4x - 45$

30. $g(x) = x^2 + x - 20$

31. $p(x) = 2x^2 - 13x - 7$

32. $f(x) = 6x^2 + 17x + 6$

33. $f(x) = x^3 - 2x^2 - 3x$

34. $r(x) = 3x^3 - 12x$

Match each graph to the corresponding function in Exercises 35–38.

I

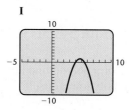

II

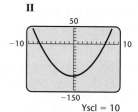

Yscl = 10

III

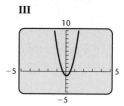

IV

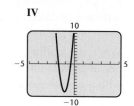

35. $f(x) = (2x - 1)(3x + 1)$

36. $f(x) = (2x + 15)(x - 7)$

37. $f(x) = (4 - x)(2x - 11)$

38. $f(x) = (5x + 2)(4x + 7)$

Tell whether each of the following is an expression or an equation.

39. $x^2 + 6x + 9$

40. $x^3 = x^2 - x + 3$

41. $3x^2 = 3x$

42. $x^4 + 3x^3 + x^2$

43. $2x^3 + x^2 = 0$

44. $5x^4 + 5x$

Write an equivalent expression by factoring out the greatest common factor.

45. $2t^2 + 8t$

46. $3y^2 + 6y$

47. $y^3 + 9y^2$

48. $x^3 + 8x^2$

49. $15x^2 - 5x^4$

50. $8y^2 + 4y^4$

51. $4x^2y - 12xy^2$

52. $5x^2y^3 + 15x^3y^2$

53. $3y^2 - 3y - 9$

54. $5x^2 - 5x + 15$

55. $6ab - 4ad + 12ac$

56. $8xy + 10xz - 14xw$

57. $9x^3y^6z^2 - 12x^4y^4z^4 + 15x^2y^5z^3$

58. $14a^4b^3c^5 + 21a^3b^5c^4 - 35a^4b^4c^3$

Write an equivalent expression by factoring out a factor with a negative coefficient.

59. $-5x + 35$

60. $-5x - 40$

61. $-6y - 72$

62. $-8t + 72$

63. $-2x^2 + 4x - 12$

64. $-2x^2 + 12x + 40$

65. $3y - 24x$

66. $7x - 56y$

67. $7s - 14t$

68. $5r - 10s$

69. $-x^2 + 5x - 9$

70. $-p^3 - 4p^2 + 11$

71. $-a^4 + 2a^3 - 13a$

72. $-m^3 - m^2 + m - 2$

Write an equivalent expression by factoring.

73. $a(b - 5) + c(b - 5)$

74. $r(t - 3) - s(t - 3)$

75. $(x + 7)(x - 1) + (x + 7)(x - 2)$

76. $(a + 5)(a - 2) + (a + 5)(a + 1)$

77. $a^2(x - y) + 5(y - x)$

78. $5x^2(x - 6) + 2(6 - x)$

79. $ac + ad + bc + bd$

80. $xy + xz + wy + wz$

81. $b^3 - b^2 + 2b - 2$

82. $y^3 - y^2 + 3y - 3$

83. $a^3 - 3a^2 + 6 - 2a$

84. $t^3 + 6t^2 - 2t - 12$

85. $72x^3 - 36x^2 + 24x$

86. $12a^4 - 21a^3 - 9a^2$

87. $x^6 - x^5 - x^3 + x^4$

88. $y^4 - y^3 - y + y^2$

89. $2y^4 + 6y^2 + 5y^2 + 15$

90. $2xy - x^2y - 6 + 3x$

91. *Height of a Baseball.*   A baseball is popped up with an upward velocity of 72 ft/sec. Its height in feet, $h(t)$, after t seconds is given by
$$h(t) = -16t^2 + 72t.$$
a) Find an equivalent expression for $h(t)$ by factoring out a common factor with a negative coefficient.
b) Perform a partial check of part (a) by evaluating both expressions for $h(t)$ at $t = 1$.

92. *Height of a Rocket.*   A model rocket is launched upward with an initial velocity of 96 ft/sec. Its height in feet, $h(t)$, after t seconds is given by
$$h(t) = -16t^2 + 96t.$$
a) Find an equivalent expression for $h(t)$ by factoring out a common factor with a negative coefficient.
b) Check your factoring by evaluating both expressions for $h(t)$ at $t = 1$.

93. *Airline Routes.*   When an airline links n cities so that from any one city it is possible to fly directly to each of the other cities, the total number of direct routes is given by
$$R(n) = n^2 - n.$$
Find an equivalent expression for $R(n)$ by factoring out a common factor.

94. *Surface Area of a Silo.* A silo is a structure that is shaped like a right circular cylinder with a half sphere on top. The surface area of a silo of height h and radius r (including the area of the base) is given by the polynomial $2\pi rh + \pi r^2$. (Note that h is the height of the entire silo.) Find an equivalent expression by factoring out a common factor.

95. *Total Profit.* When x hundred CD players are sold, Rolics Electronics collects a profit of $P(x)$, where

$$P(x) = x^2 - 3x,$$

and $P(x)$ is in thousands of dollars. Find an equivalent expression by factoring out a common factor.

96. *Total Profit.* After t weeks of production, Claw Foot, Inc., is making a profit of $P(t) = t^2 - 5t$ from sales of their surfboards. Find an equivalent expression by factoring out a common factor.

97. *Total Revenue.* Urban Sounds is marketing a new MP3 player. The firm determines that when it sells x units, the total revenue R is given by the polynomial function

$$R(x) = 280x - 0.4x^2 \text{ dollars.}$$

Find an equivalent expression for $R(x)$ by factoring out $0.4x$.

98. *Total Cost.* Urban Sounds determines that the total cost C of producing x MP3 players is given by the polynomial function

$$C(x) = 0.18x + 0.6x^2.$$

Find an equivalent expression for $C(x)$ by factoring out $0.6x$.

99. *Counting Spheres in a Pile.* The number N of spheres in a triangular pile like the one shown here is a polynomial function given by

$$N(x) = \tfrac{1}{6}x^3 + \tfrac{1}{2}x^2 + \tfrac{1}{3}x,$$

where x is the number of layers and $N(x)$ is the number of spheres. Find an equivalent expression for $N(x)$ by factoring out $\tfrac{1}{6}$.

100. *Number of Games in a League.* If there are n teams in a league and each team plays every other team once, we can find the total number of games played by using the polynomial function $f(n) = \tfrac{1}{2}n^2 - \tfrac{1}{2}n$. Find an equivalent expression by factoring out $\tfrac{1}{2}$.

101. *High-fives.* When a team of n players all give each other high-fives, a total of $H(n)$ hand slaps occurs, where

$$H(n) = \tfrac{1}{2}n^2 - \tfrac{1}{2}n.$$

Find an equivalent expression by factoring out $\tfrac{1}{2}n$.

102. *Number of Diagonals.* The number of diagonals of a polygon having n sides is given by the polynomial function

$$P(n) = \tfrac{1}{2}n^2 - \tfrac{3}{2}n.$$

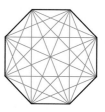

Find an equivalent expression for $P(n)$ by factoring out $\tfrac{1}{2}$.

Solve using the principle of zero products.

103. $x(x + 1) = 0$

104. $5x(x - 2) = 0$

105. $x^2 - 3x = 0$

106. $2x^2 + 8x = 0$

107. $-5x^2 = 15x$

108. $2x - 4x^2 = 0$

109. $12x^4 + 4x^3 = 0$

110. $21x^3 = 7x^2$

TW 111. Under what conditions would it be easier to evaluate a polynomial *after* it has been factored?

TW 112. Gabrielle claims that the zeros of the function given by $f(x) = x^4 - 3x^2 + 7x + 20$ are -1, 1, 2, 4, and 5. How can you tell, without performing any calculations, that she cannot be correct?

Skill Maintenance

Simplify. [1.8]

113. $2(-3) + 4(-5)$

114. $-7(-2) + 5(-3)$

115. $4(-6) - 3(2)$

116. $5(-3) + 2(-2)$

117. *Geometry.*   The perimeter of a triangle is 174. The lengths of the three sides are consecutive even numbers. What are the lengths of the sides of the triangle? [2.5]

118. *Shopping Centers.*   In 2005, there were 48,695 shopping centers in the United States. Of these, there were 12,025 more small centers (under 100,000 square feet) than there were large centers (over 100,000 square feet). (*Source*: U.S. Bureau of the Census, *2007 Statistical Abstract*) How many shopping centers over 100,000 square feet were there in 2005? [4.4]

Synthesis

TW 119. Is it true that if the coefficients and exponents of a polynomial are all prime numbers, then the polynomial itself is prime? Why or why not?

TW 120. Following Example 9, we stated that checking the factorization of a second-degree polynomial by making a single replacement is only a *partial* check. Write an *incorrect* factorization and explain how evaluating both the polynomial and the factorization might not catch the mistake.

121. Use the results of Exercise 15 to factor $x^2 + 2x - 8$.

122. Use the results of Exercise 16 to factor $x^2 - 2x + 1$.

Complete each of the following.

123. $x^5y^4 + \underline{\quad} = x^3y(\underline{\quad} + xy^5)$

124. $a^3b^7 - \underline{\quad} = \underline{\quad}(ab^4 - c^2)$

Write an equivalent expression by factoring.

125. $rx^2 - rx + 5r + sx^2 - sx + 5s$

126. $3a^2 + 6a + 30 + 7a^2b + 14ab + 70b$

127. $a^4x^4 + a^4x^2 + 5a^4 + a^2x^4 + a^2x^2 + 5a^2 + 5x^4 + 5x^2 + 25$

(*Hint*: Use three groups of three.)

Write an equivalent expression by factoring out the smallest power of x in each of the following.

128. $x^{-8} + x^{-4} + x^{-6}$

129. $x^{-6} + x^{-9} + x^{-3}$

130. $x^{3/4} + x^{1/2} - x^{1/4}$

131. $x^{1/3} - 5x^{1/2} + 3x^{3/4}$

132. $x^{-3/2} + x^{-1/2}$

133. $x^{-5/2} + x^{-3/2}$

134. $x^{-3/4} - x^{-5/4} + x^{-1/2}$

135. $x^{-4/5} - x^{-7/5} + x^{-1/3}$

Write an equivalent expression by factoring. Assume that all exponents are natural numbers.

136. $2x^{3a} + 8x^a + 4x^{2a}$

137. $3a^{n+1} + 6a^n - 15a^{n+2}$

138. $4x^{a+b} + 7x^{a-b}$

139. $7y^{2a+b} - 5y^{a+b} + 3y^{a+2b}$

Factoring Trinomials of the Type $x^2 + bx + c$ ■ Equations Containing Trinomials ■ Zeros and Factoring

In this section, we expand our list of the types of polynomials that we can factor so that we can solve a wider variety of polynomial equations. We begin by factoring trinomials of the type $x^2 + bx + c$ and then solve equations containing such trinomials. We then see how to use zeros of a function to factor a polynomial.

Factoring Trinomials of the Type $x^2 + bx + c$

When trying to factor trinomials of the type $x^2 + bx + c$, we can use a trial-and-error procedure.

Constant Term Positive

Recall the FOIL method of multiplying two binomials:

$$(x + 3)(x + 5) = x^2 + \underbrace{5x + 3x}_{} + 15$$
$$= x^2 + 8x + 15.$$

Because the leading coefficient in each binomial is 1, the leading coefficient in the product is also 1. To factor $x^2 + 8x + 15$, we think of FOIL: The first term, x^2, is the product of the First terms of two binomial factors, so the first term in each binomial must be x. The challenge is to find two numbers p and q such that

$$x^2 + 8x + 15 = (x + p)(x + q)$$
$$= x^2 + qx + px + pq.$$

Note that the Outer and Inner products, qx and px, can be written as $(p + q)x$. The Last product, pq, will be a constant. Thus the numbers p and q must be selected so that their product is 15 and their sum is 8. In this case, we know from above that these numbers are 3 and 5. The factorization is

$$(x + 3)(x + 5), \quad \text{or} \quad (x + 5)(x + 3). \qquad \text{Using a commutative law}$$

In general, to factor $x^2 + (p + q)x + pq$, we use FOIL in reverse:

$$x^2 + (p + q)x + pq = (x + p)(x + q).$$

EXAMPLE 1 Write an equivalent expression by factoring: $x^2 + 9x + 8$.

SOLUTION We think of FOIL in reverse. The first term of each factor is x. We are looking for numbers p and q such that:

$$x^2 + 9x + 8 = (x + p)(x + q) = x^2 + (p + q)x + pq.$$

Thus we search for factors of 8 whose sum is 9.

Pair of Factors	Sum of Factors
2, 4	6
1, 8	9

The numbers we need are 1 and 8.

The factorization is thus $(x + 1)(x + 8)$. The student should check by multiplying to confirm that the product is the original trinomial.

When factoring trinomials with a leading coefficient of 1, it suffices to consider all pairs of factors along with their sums, as we did above. At times, however, you may be tempted to form factors without calculating any sums. It is essential that you check any attempt made in this manner! For example, if we attempt the factorization

$$x^2 + 9x + 8 \overset{?}{=} (x + 2)(x + 4),$$

a check reveals that

$$(x + 2)(x + 4) = x^2 + 6x + 8 \neq x^2 + 9x + 8.$$

This type of trial-and-error procedure becomes easier to use with time. As you gain experience, you will find that many trials can be performed mentally.

When the constant term of a trinomial is positive, the constant terms in the binomial factors must either both be positive or both be negative. This ensures a positive product. The sign used is that of the trinomial's middle term.

EXAMPLE 2 Factor: $t^2 - 9t + 20$.

SOLUTION Since the constant term is positive and the coefficient of the middle term is negative, we look for a factorization of 20 in which both factors are negative. Their sum must be -9.

$y_1 = x^2 - 9x + 20,$
$y_2 = (x - 4)(x - 5)$

X	Y1	Y2
0	20	20
1	12	12
2	6	6
3	2	2
4	0	0
5	0	0
6	2	2

X = 0

Pair of Factors	Sum of Factors
$-1, -20$	-21
$-2, -10$	-12
$-4, -5$	-9

The numbers we need are -4 and -5.

The factorization is $(t - 4)(t - 5)$. We check by comparing values using a table, as shown at left.

Constant Term Negative

When the constant term of a trinomial is negative, we look for one negative factor and one positive factor. The sum of the factors must still be the coefficient of the middle term.

EXAMPLE 3   Factor: $x^3 - x^2 - 30x$.

SOLUTION   *Always* look first for a common factor! This time there is one, x. We factor it out:

$$x^3 - x^2 - 30x = x(x^2 - x - 30).$$

Now we consider $x^2 - x - 30$. We need a factorization of -30 in which one factor is positive, the other factor is negative, and the sum of the factors is -1. Since the sum is to be negative, the negative factor must have the greater absolute value. Thus we need consider only the following pairs of factors.

Pair of Factors	Sum of Factors
1, −30	−29
3, −10	−7
5, −6	−1 ⟵ ——— The numbers we need are 5 and −6.

The factorization of $x^2 - x - 30$ is $(x + 5)(x - 6)$. *Don't forget to include the factor that was factored out earlier!* In this case, the factorization of the original trinomial is $x(x + 5)(x - 6)$. We check by graphing $y_1 = x^3 - x^2 - 30x$ and $y_2 = x(x + 5)(x - 6)$, as shown at left.

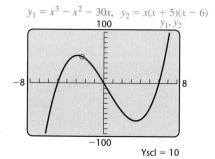

$y_1 = x^3 - x^2 - 30x, \quad y_2 = x(x + 5)(x - 6)$

Yscl = 10

EXAMPLE 4   Factor: $2x^2 + 34x - 220$.

SOLUTION   *Always* look first for a common factor! This time we can factor out 2:

$$2x^2 + 34x - 220 = 2(x^2 + 17x - 110).$$

We next look for a factorization of -110 in which one factor is positive, the other factor is negative, and the sum of the factors is 17. Since the sum is to be positive, we examine only pairs of factors in which the positive term has the larger absolute value.

Pair of Factors	Sum of Factors
−1, 110	109
−2, 55	53
−5, 22	17 ⟵ ——— The numbers we need are −5 and 22.

The factorization of $x^2 + 17x - 110$ is $(x - 5)(x + 22)$. The factorization of the original trinomial, $2x^2 + 34x - 220$, is $2(x - 5)(x + 22)$.

Check: $2(x - 5)(x + 22) = 2[x^2 + 17x - 110]$
$$= 2x^2 + 34x - 220.$$

Some polynomials are not factorable using integers.

EXAMPLE 5   Factor: $x^2 - x - 7$.

SOLUTION   There are no factors of -7 whose sum is -1. This trinomial is *not* factorable into binomials with integer coefficients. Although $x^2 - x - 7$ can be factored using more advanced techniques, for our purposes the polynomial is **prime.**

Tips for Factoring $x^2 + bx + c$

1. If necessary, rewrite the trinomial in descending order.
2. Find a pair of factors that have c as their product and b as their sum. Remember the following:

   - If c is positive, its factors will have the same sign as b.
   - If c is negative, one factor will be positive and the other will be negative. Select the factors such that the factor with the larger absolute value is the factor with the same sign as b.
   - If the sum of the two factors is the opposite of b, changing the signs of both factors will give the desired factors whose sum is b.

3. Check the result by multiplying the binomials.

These tips also apply when a trinomial has more than one variable.

EXAMPLE 6   Factor: $x^2 - 2xy - 48y^2$.

SOLUTION   We look for numbers p and q such that

$$x^2 - 2xy - 48y^2 = (x + py)(x + qy).$$ The x's and y's can be written in the binomials in advance.

Our thinking is much the same as if we were factoring $x^2 - 2x - 48$. We look for factors of -48 whose sum is -2. Those factors are 6 and -8. Thus,

$$x^2 - 2xy - 48y^2 = (x + 6y)(x - 8y).$$

The check is left to the student.

Equations Containing Trinomials

We can now use our new factoring skills to solve some polynomial equations.

EXAMPLE 7   Solve: $x^2 + 9x + 8 = 0$.

CAUTION!   In Example 7, we are solving an equation. Do not try to solve an expression!

ALGEBRAIC APPROACH

We use the principle of zero products:

$$x^2 + 9x + 8 = 0$$

$(x + 1)(x + 8) = 0$ **Using the factorization from Example 1**

$x + 1 = 0$ *or* $x + 8 = 0$ **Using the principle of zero products**

$x = -1$ *or* $x = -8.$ **Solving each equation for x**

The solutions are -1 and -8.

GRAPHICAL APPROACH

The real-number solutions of $x^2 + 9x + 8 = 0$ are the first coordinates of the x-intercepts of the graph of $f(x) = x^2 + 9x + 8$.

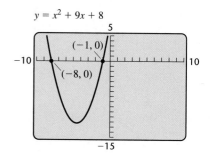

$y = x^2 + 9x + 8$

We find that -1 and -8 are solutions.

Check:

For -1:

$$
\begin{array}{c|c}
x^2 + 9x + 8 = 0 & \\
\hline
(-1)^2 + 9(-1) + 8 & 0 \\
1 - 9 + 8 & \\
& 0 \overset{?}{=} 0 \quad \text{TRUE}
\end{array}
$$

For -8:

$$
\begin{array}{c|c}
x^2 + 9x + 8 = 0 & \\
\hline
(-8)^2 + 9(-8) + 8 & 0 \\
64 - 72 + 8 & \\
& 0 \overset{?}{=} 0 \quad \text{TRUE}
\end{array}
$$

The solutions are -1 and -8.

EXAMPLE 8   Solve: $(t - 10)(t + 1) = -24$.

ALGEBRAIC APPROACH

Note that the left side of the equation is factored. It may be tempting to set both factors equal to -24 and solve, but this is *not* correct. The principle of zero products requires 0 on one side of the equation. We begin by multiplying the left side.

$$(t - 10)(t + 1) = -24$$

$t^2 - 9t - 10 = -24$ **Multiplying**

$t^2 - 9t + 14 = 0$ **Adding 24 to both sides to get 0 on one side**

$(t - 2)(t - 7) = 0$ **Factoring**

$t - 2 = 0$ *or* $t - 7 = 0$ **Using the principle of zero products**

$t = 2$ *or* $t = 7$

The solutions are 2 and 7.

GRAPHICAL APPROACH

There is no need to multiply. We let $y_1 = (x - 10)(x + 1)$ and $y_2 = -24$ and look for any points of intersection of the graphs. The x-coordinates of these points are the solutions of the equation.

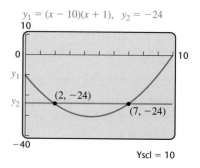

$y_1 = (x - 10)(x + 1), \ y_2 = -24$

Yscl = 10

We find that 2 and 7 are solutions.

EXAMPLE 9   Solve: $x^3 = x^2 + 30x$.

ALGEBRAIC APPROACH

In order to solve $x^3 = x^2 + 30x$ using the principle of zero products, we must rewrite the equation with 0 on one side.
    We have

$$x^3 = x^2 + 30x$$

$x^3 - x^2 - 30x = 0$ **Getting 0 on one side**

$x(x - 6)(x + 5) = 0$ **Using the factorization from Example 3**

$x = 0 \quad or \quad x - 6 = 0 \quad or \quad x + 5 = 0$ **Using the principle of zero products**

$x = 0 \quad or \qquad x = 6 \quad or \qquad x = -5.$

The solutions are 0, 6, and -5.

GRAPHICAL APPROACH

We let $y_1 = x^3$ and $y_2 = x^2 + 30x$ and look for points of intersection of the graphs.

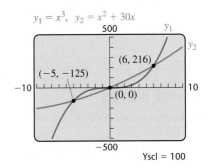

The viewing window $[-10, 10, -500, 500]$ shows three points of intersection. Since the polynomial equation is of degree 3, we know there will be no more than 3 solutions. The x-coordinates of the points of intersection are -5, 0, and 6.

We check by substituting $-5, 0,$ and 6 into the original equation. The solutions are $-5, 0,$ and 6.

Zeros and Factoring

Note in Example 7 the relationship between the factors of $x^2 + 9x + 8$ and the solutions of the equation $x^2 + 9x + 8 = 0$. We can use the principle of zero products "in reverse" to factor a polynomial and to write a function with given zeros.

EXAMPLE 10   Factor: $x^2 - 13x - 608$.

SOLUTION   The factorization of the trinomial $x^2 - 13x - 608$ will be in the form

$$x^2 - 13x - 608 = (x + p)(x + q).$$

We could use trial and error to find p and q, but there are many factors of 608. Thus we factor the trinomial using the principle of zero products.
    Note that the roots of the equation

$$x^2 - 13x - 608 = 0$$

are the zeros of the function $f(x) = x^2 - 13x - 608$. We graph the function and find the zeros.

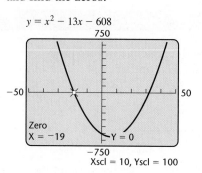

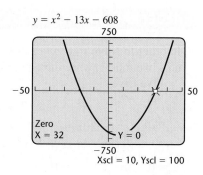

The zeros are -19 and 32. We also know that -19 is a zero of the linear function $g(x) = x + 19$ and that 32 is a zero of the linear function $h(x) = x - 32$. This suggests that $x + 19$ and $x - 32$ are factors of $x^2 - 13x - 608$. Using the principle of zero products in reverse, we now have

$$x^2 - 13x - 608 = (x + 19)(x - 32).$$

Multiplication indicates that the factorization is correct.

Not every trinomial is factorable. The roots of $x^2 - 2x - 2 = 0$ are irrational; thus $x^2 - 2x - 2$ cannot be factored using integers. The equation $x^2 - x + 1 = 0$ has no real roots, and is also prime. We will study equations like these in Chapter 11.

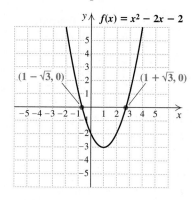

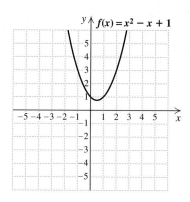

We can also use the principle of zero products to write a function whose zeros are given.

EXAMPLE 11 Write a polynomial function $f(x)$ whose zeros are -1, 0, and 3.

SOLUTION Each zero of the polynomial function f is a zero of a linear factor of the polynomial. We write a linear function for each given zero:

$$-1 \text{ is a zero of } g(x) = x + 1;$$
$$0 \text{ is a zero of } h(x) = x;$$
and $$3 \text{ is a zero of } k(x) = x - 3.$$

Student Notes

A factorization of $x^2 - 13x - 608$ will be of the form $(x - \underline{\quad})(x - \underline{\quad})$. The zeros of $f(x) = x^2 - 13x - 608$ will fill in the blanks:

$$x^2 - 13x - 608$$
$$= (x - (-19))(x - 32).$$

Thus a polynomial function with zeros -1, 0, and 3 is

$$f(x) = (x + 1) \cdot x \cdot (x - 3);$$

multiplying gives us

$$f(x) = x^3 - 2x^2 - 3x.$$

6.2 EXERCISE SET

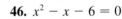 *Concept Reinforcement*   *Classify each of the following as either true or false.*

1. When factoring any polynomial, it is always best to look first for a common factor.

2. Whenever the sum of a negative number and a positive number is negative, the negative number has the greater absolute value.

3. Whenever the product of a pair of factors is negative, the factors have the same sign.

4. If $p + q = -17$, then $-p + (-q) = 17$.

5. If b and c are positive and $x^2 + bx + c$ can be factored as $(x + p)(x + q)$, then it follows that p and q are both negative.

6. If b is negative and c is positive and $x^2 + bx + c$ can be factored as $(x + p)(x + q)$, then it follows that p and q are both positive.

7. If 1 is a zero of a polynomial function, then $x - 1$ is a factor of the polynomial.

8. If $x - 2$ is a factor of $p(x)$, then $p(2) = 0$.

Factor. If a polynomial is prime, state this.

9. $x^2 + 8x + 12$
10. $x^2 + 6x + 5$
11. $t^2 + 8t + 15$
12. $y^2 + 12y + 27$
13. $x^2 - 27 - 6x$
14. $t^2 - 15 - 2t$
15. $2n^2 - 20n + 50$
16. $2a^2 - 16a + 32$
17. $a^3 - a^2 - 72a$
18. $x^3 + 3x^2 - 54x$
19. $14x + x^2 + 45$
20. $12y + y^2 + 32$
21. $3x + x^2 - 10$
22. $x + x^2 - 6$
23. $3x^2 + 15x + 18$
24. $5y^2 + 40y + 35$

25. $56 + x - x^2$
26. $32 + 4y - y^2$
27. $32y + 4y^2 - y^3$
28. $56x + x^2 - x^3$
29. $x^4 + 11x^3 - 80x^2$
30. $y^4 + 5y^3 - 84y^2$
31. $x^2 + 12x + 13$
32. $x^2 - 3x + 7$
33. $p^2 - 5pq - 24q^2$
34. $x^2 + 12xy + 27y^2$
35. $y^2 + 8yz + 16z^2$
36. $x^2 - 14xy + 49y^2$
37. $p^4 + 80p^3 + 79p^2$
38. $x^4 + 50x^3 + 49x^2$

39. Use the results of Exercise 9 to solve $x^2 + 8x + 12 = 0$.

40. Use the results of Exercise 10 to solve $x^2 + 6x + 5 = 0$.

41. Use the results of Exercise 15 to solve $2n^2 + 50 = 20n$.

42. Use the results of Exercise 26 to solve $32 + 4y = y^2$.

43. Use the results of Exercise 17 to solve $a^3 - a^2 = 72a$.

44. Use the results of Exercise 28 to solve $56x + x^2 = x^3$.

In Exercises 45–48, use the graph to solve the given equation. Check by substituting into the equation.

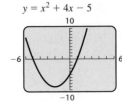 **45.** $x^2 + 4x - 5 = 0$

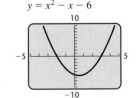 **46.** $x^2 - x - 6 = 0$

47. $x^2 + x - 6 = 0$

$y = x^2 + x - 6$

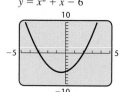

48. $x^2 + 8x + 15 = 0$

$y = x^2 + 8x + 15$

Find the zeros of each function.

49. $f(x) = x^2 - 4x - 45$

50. $f(x) = x^2 + x - 20$

51. $r(x) = x^3 - 2x^2 - 3x$

52. $g(x) = 3x^2 + 21x + 30$

Solve.

53. $x^2 + 4x = 45$

54. $t^2 - 3t = 28$

55. $x^2 - 9x = 0$

56. $a^2 + 18a = 0$

57. $a^3 - 3a^2 = 40a$

58. $x^3 - 2x^2 = 63x$

59. $(x - 3)(x + 2) = 14$

60. $(z + 4)(z - 2) = -5$

61. $35 - x^2 = 2x$

62. $40 - x^2 + 3x = 0$

In Exercises 63 and 64, use the graph to factor the given polynomial.

63. $x^2 + 10x - 264$

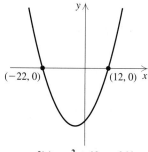

$(-22, 0)$ $(12, 0)$ x

$f(x) = x^2 + 10x - 264$

64. $x^2 + 16x - 336$

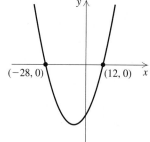

$(-28, 0)$ $(12, 0)$ x

$f(x) = x^2 + 16x - 336$

In Exercises 65–68, use a graph to help factor each polynomial.

65. $x^2 + 40x + 384$

66. $x^2 - 13x - 300$

67. $x^2 + 26x - 2432$

68. $x^2 - 46x + 504$

Write a polynomial function that has the given zeros. Answers may vary.

69. $-1, 2$

70. $2, 5$

71. $-7, -10$

72. $8, -3$

73. $0, 1, 2$

74. $-3, 0, 5$

TW 75. Sandra says that she will never miss a point of intersection when solving graphically because she always uses a $[-100, 100, -100, 100]$ window. Is she correct? Why or why not?

TW 76. How can one conclude that $x^2 - 59x + 6$ is a prime polynomial without performing any trials?

Focused Review

Factor completely.

77. $9x^5 - 24x^3 + 3x^2$ [6.1]

78. $x^4 + 3x^3 + 2x^2$ [6.2]

79. $xy + 3x + 2y + 6$ [6.1]

80. $(x + 2)(x - 7) + (x - 5)(x - 7)$ [6.1]

81. $45x^3y^4 + 30x^2y^5 + 60x^4y^3$ [6.1]

82. $-2t^3 - 4t + 8$ [6.1]

83. $xw - xz - wy + yz$ [6.1]

84. $x^2 + 15x + 50$ [6.2]

Synthesis

TW 85. Explain how the following graph of

$$y = x^2 + 3x - 2 - (x - 2)(x + 1)$$

can be used to show that

$$x^2 + 3x - 2 \neq (x - 2)(x + 1).$$

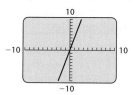

TW 86. Explain why Example 11 reads "Write *a* polynomial function $f(x)$" and not "Write *the* polynomial function $f(x)$."

87. Use the following graph of $f(x) = x^2 - 2x - 3$ to solve $x^2 - 2x - 3 = 0$ and to solve $x^2 - 2x - 3 < 5$.

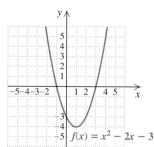

$f(x) = x^2 - 2x - 3$

88. Use the following graph of $g(x) = -x^2 - 2x + 3$ to solve $-x^2 - 2x + 3 = 0$ and to solve $-x^2 - 2x + 3 \geq -5$.

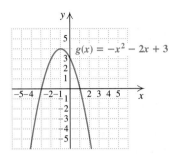

89. Find a polynomial function f for which $f(2) = 0$, $f(-1) = 0$, $f(3) = 0$, and $f(0) = 30$.

90. Find a polynomial function g for which $g(-3) = 0$, $g(1) = 0$, $g(5) = 0$, and $g(0) = 45$.

In Exercises 91–94, use a graphing calculator to find any solutions that exist accurate to two decimal places.

91. $-x^2 + 13.80x = 47.61$

92. $-x^2 + 3.63x + 34.34 = x^2$

93. $x^3 - 3.48x^2 + x = 3.48$

94. $x^2 + 4.68 = 1.2x$

Factor. Assume that variables in exponents represent positive integers.

95. $2a^4b^6 - 3a^2b^3 - 20ab^2$

96. $5x^8y^6 + 35x^4y^3 + 60$

97. $x^2 - \frac{4}{25} + \frac{3}{5}x$

98. $y^2 - \frac{8}{49} + \frac{2}{7}y$

99. $y^2 + 0.4y - 0.05$

100. $x^{2a} + 5x^a - 24$

101. $x^2 + ax + bx + ab$

102. $bdx^2 + adx + bcx + ac$

103. $a^2p^{2a} + a^2p^a - 2a^2$

Aha! **104.** $(x + 3)^2 - 2(x + 3) - 35$

105. Find all integers m for which $x^2 + mx + 75$ can be factored.

106. Find all integers q for which $x^2 + qx - 32$ can be factored.

107. One factor of $x^2 - 345x - 7300$ is $x + 20$. Find the other factor.

6.3   Equations Containing Trinomials of the Type $ax^2 + bx + c$

Factoring Trinomials of the Type $ax^2 + bx + c$ ■
Equations and Functions

Factoring Trinomials of the Type $ax^2 + bx + c$

Now we look at trinomials in which the leading coefficient is not 1. We consider two factoring methods. Use what works best for you or what your instructor chooses for you.

Method 1: Reversing FOIL

We first consider the **FOIL method** for factoring trinomials of the type

$$ax^2 + bx + c, \text{ where } a \neq 1.$$

Consider the following multiplication.

$$\underset{}{(3x + 2)(4x + 5)} = 12x^2 + \underset{}{\underbrace{15x + 8x}} + 10$$

$$= 12x^2 + 23x + 10$$

To factor $12x^2 + 23x + 10$, we must reverse what we just did. We look for two binomials whose product is this trinomial. The product of the First terms must be $12x^2$. The product of the Outer terms plus the product of the Inner terms must be $23x$. The product of the Last terms must be 10. We know from the preceding discussion that the factorization is

$$(3x + 2)(4x + 5).$$

In general, however, finding such an answer involves trial and error. We use the following method.

To Factor $ax^2 + bx + c$ by Reversing FOIL

1. Factor out the largest common factor, if one exists. Here we assume none does.
2. Find two **First** terms whose product is ax^2:

$$(\blacksquare x +)(\blacksquare x +) = ax^2 + bx + c.$$
$$\text{FOIL}$$

3. Find two **Last** terms whose product is c:

$$(x + \blacksquare)(x + \blacksquare) = ax^2 + bx + c.$$
$$\text{FOIL}$$

4. Repeat steps (2) and (3) until a combination is found for which the sum of the **Outer** and **Inner** products is bx:

$$(\blacksquare x + \blacksquare)(\blacksquare x + \blacksquare) = ax^2 + bx + c.$$
$$\text{I} \qquad \text{FOIL}$$
$$\text{O}$$

EXAMPLE 1 Factor: $3x^2 + 10x - 8$.

SOLUTION

1. First, note that there is no common factor (other than 1 or -1).
2. Next, factor the first term, $3x^2$. The only possibility for factors is $3x \cdot x$. Thus, if a factorization exists, it must be of the form

$$(3x + \blacksquare)(x + \blacksquare).$$

We need to find the right numbers for the blanks.

3. Note that the constant term, -8, can be factored as $(-8)(1)$, $8(-1)$, $(-2)4$, and $2(-4)$, as well as $(1)(-8)$, $(-1)8$, $4(-2)$, and $(-4)2$.

4. Find a pair of factors for which the sum of the Outer and Inner products is the middle term, $10x$. Each possibility should be checked by multiplying:

$$(3x - 8)(x + 1) = 3x^2 - 5x - 8.   O + I = 3x + (-8x) = -5x$$

This gives a middle term with a negative coefficient. Since a positive coefficient is needed, a second possibility must be tried:

$$(3x + 8)(x - 1) = 3x^2 + 5x - 8.   O + I = -3x + 8x = 5x$$

Note that changing the signs of the two constant terms changes only the sign of the middle term. We try again:

$$(3x - 2)(x + 4) = 3x^2 + 10x - 8.   \text{This is what we wanted.}$$

Thus the desired factorization is $(3x - 2)(x + 4)$.

Student Notes

Keep your work organized so that you can see what you have already considered. When factoring $6x^2 - 19x + 10$, we can list all possibilities and cross out those in which a common factor appears:

$(3x - 10)(2x - 1),$
$\cancel{(3x - 1)(2x - 10)},$
$\cancel{(3x - 5)(2x - 2)},$
$(3x - 2)(2x - 5),$
$\cancel{(6x - 10)(x - 1)},$
$(6x - 1)(x - 10),$
$(6x - 5)(x - 2),$
$\cancel{(6x - 2)(x - 5)}.$

By being organized and not erasing, we can see that there are only four possible factorizations.

$y_1 = 6x^6 - 19x^5 + 10x^4,$
$y_2 = x^4(3x - 2)(2x - 5)$

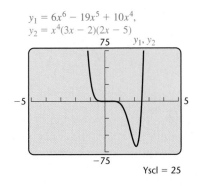

Yscl = 25

EXAMPLE 2   Factor: $6x^6 - 19x^5 + 10x^4$.

SOLUTION

1. First, factor out the common factor x^4:

    $$x^4(6x^2 - 19x + 10).$$

2. Note that $6x^2 = 6x \cdot x$ and $6x^2 = 3x \cdot 2x$. Thus, $6x^2 - 19x + 10$ may factor into

    $$(3x + \blacksquare)(2x + \blacksquare)   \text{or}   (6x + \blacksquare)(x + \blacksquare).$$

3. We factor the last term, 10. The possibilities are $10 \cdot 1, (-10)(-1), 5 \cdot 2$, and $(-5)(-2)$, as well as $1 \cdot 10, (-1)(-10), 2 \cdot 5$, and $(-2)(-5)$.

4. There are 8 possibilities for *each* factorization in step (2). We need factors for which the sum of the Outer and Inner products is the middle term, $-19x$. Since the x-coefficient is negative, we consider pairs of negative factors. Each possible factorization must be checked by multiplying:

    $$(3x - 10)(2x - 1) = 6x^2 - 23x + 10.$$

We try again:

$$(3x - 5)(2x - 2) = 6x^2 - 16x + 10.$$

Actually this last attempt could have been rejected by simply noting that $2x - 2$ has a common factor, 2. Since the *largest* common factor was removed in step (1), no other common factors can exist. We try again, reversing the -5 and -2:

$$(3x - 2)(2x - 5) = 6x^2 - 19x + 10.   \text{This is what we wanted.}$$

The factorization of $6x^2 - 19x + 10$ is $(3x - 2)(2x - 5)$. *But do not forget the common factor!* We must include it to get the complete factorization of the original trinomial:

$$6x^6 - 19x^5 + 10x^4 = x^4(3x - 2)(2x - 5).$$

The graphs of the original polynomial and the factorization coincide, as shown in the figure at left.

> **Tips for Factoring $ax^2 + bx + c$ with FOIL**
>
> 1. If the largest common factor has been factored out of the original trinomial, then no binomial factor can have a common factor (other than 1 or -1).
> 2. If a and c are both positive, then the signs in the factors will be the same as the sign of b.
> 3. When a possible factoring produces the opposite of the desired middle term, reverse the signs of the constants in the factors.
> 4. Be systematic about your trials. Keep track of those possibilities that you have tried and those that you have not.

Keep in mind that this method of factoring involves trial and error. With practice, you will find yourself making fewer and better guesses.

Method 2: The *ac*-Method

The second method for factoring trinomials of the type $ax^2 + bx + c$, $a \neq 1$, is known as the **ac-method.** It involves not only trial and error and FOIL but also factoring by grouping. We know that

$$x^2 + 7x + 10 = x^2 + 2x + 5x + 10$$
$$= x(x + 2) + 5(x + 2)$$
$$= (x + 2)(x + 5),$$

but what if the leading coefficient is not 1? Consider $6x^2 + 23x + 20$. The method is similar to what we just did with $x^2 + 7x + 10$, but we need two more steps.* First, multiply the leading coefficient, 6, and the constant, 20, to get 120. Then find a factorization of 120 in which the sum of the factors is the coefficient of the middle term: 23. The middle term is then split into a sum or difference using these factors.

$6x^2 + 23x + 20$

(1) Multiply 6 and 20: $6 \cdot 20 = 120$.
(2) Factor 120: $120 = 8 \cdot 15$, and $8 + 15 = 23$.
(3) Split the middle term: $23x = 8x + 15x$.
(4) Factor by grouping.

We factor by grouping as follows:

$$6x^2 + 23x + 20 = 6x^2 + 8x + 15x + 20$$
$$= 2x(3x + 4) + 5(3x + 4)$$
$$= (3x + 4)(2x + 5).$$

Factoring by grouping

---
*The rationale behind these steps is outlined in Exercise 115.

> **To Factor $ax^2 + bx + c$ Using the ac-Method**
> 1. Make sure that any common factors have been factored out.
> 2. Multiply the leading coefficient a and the constant c.
> 3. Find a pair of factors, p and q, such that $pq = ac$ and $p + q = b$.
> 4. Rewrite the middle term of the trinomial, bx, as $px + qx$.
> 5. Factor by grouping.

Student Notes

A table can be used in step (3) of the ac-method.

Factors of -24	Sum of Factors
$-1, 24$	23
$-2, 12$	10
$-3, 8$	5
$-4, 6$	2
$-6, 4$	-2
$-8, 3$	-5
$-12, 2$	-10
$-24, 1$	-23

EXAMPLE 3   Factor: $3x^2 + 10x - 8$.

SOLUTION

1. First, look for a common factor. There is none (other than 1 or -1).
2. Multiply the leading coefficient and the constant, 3 and -8.

$$3(-8) = -24.$$

3. Try to factor -24 so that the sum of the factors is 10:

$$-24 = 12(-2) \quad \text{and} \quad 12 + (-2) = 10.$$

4. Split $10x$ using the results of step (3):

$$10x = 12x - 2x.$$

5. Finally, factor by grouping:

$$3x^2 + 10x - 8 = 3x^2 + 12x - 2x - 8 \quad \text{Substituting } 12x - 2x \text{ for } 10x$$

$$= 3x(x + 4) - 2(x + 4)$$
$$= (x + 4)(3x - 2). \qquad \text{Factoring by grouping}$$

We check the solution by multiplying and using a table.

ALGEBRAIC CHECK

$$(x + 4)(3x - 2) = 3x^2 - 2x + 12x - 8$$
$$= 3x^2 + 10x - 8$$

CHECK BY EVALUATING

$y_1 = 3x^2 + 10x - 8$,
$y_2 = (x + 4)(3x - 2)$

X	Y₁	Y₂
0	-8	-8
1	5	5
2	24	24
3	49	49
4	80	80
5	117	117
6	160	160

X = 0

Pairs of Factors	Sum of Factors
1, −120	−119
2, −60	−58
3, −40	−37
4, −30	−26
5, −24	−19
6, −20	−14
8, −15	−7
10, −12	−2

EXAMPLE 4 Factor: $6x^4 - 116x^3 - 80x^2$.

SOLUTION

1. First, factor out the greatest common factor, if any. The expression $2x^2$ is common to all three terms: $2x^2(3x^2 - 58x - 40)$.
2. To factor $3x^2 - 58x - 40$, we first multiply the leading coefficient, 3, and the constant, −40: $3(-40) = -120$.
3. Next, try to factor −120 so that the sum of the factors is −58. Since −58 is negative, the negative factor of −120 must have the larger absolute value. We see from the table at left that the factors we need are 2 and −60.
4. Split the middle term, $-58x$, using the results of step (3): $-58x = 2x - 60x$.
5. Factor by grouping:

$$3x^2 - 58x - 40 = 3x^2 + 2x - 60x - 40 \quad \text{Substituting } 2x - 60x \text{ for } -58x$$
$$= x(3x + 2) - 20(3x + 2) \quad \bigg\} \text{ Factoring by}$$
$$= (3x + 2)(x - 20). \quad \text{grouping}$$

The factorization of $3x^2 - 58x - 40$ is $(3x + 2)(x - 20)$. *But don't forget the common factor!* We must include it to factor the original trinomial:

$$6x^4 - 116x^3 - 80x^2 = 2x^2(3x + 2)(x - 20).$$

Check: $2x^2(3x + 2)(x - 20) = 2x^2(3x^2 - 58x - 40)$
$$= 6x^4 - 116x^3 - 80x^2.$$

The complete factorization is $2x^2(3x + 2)(x - 20)$.

Equations and Functions

We now use our new factoring skill to solve a polynomial equation. We factor a polynomial *expression* and use the principle of zero products to solve the *equation*.

EXAMPLE 5 Solve: $6x^6 - 19x^5 + 10x^4 = 0$.

SOLUTION We note at the outset that the polynomial is of degree 6, so there will be at most 6 solutions of the equation.

ALGEBRAIC APPROACH

We solve as follows:

$$6x^6 - 19x^5 + 10x^4 = 0$$
$$x^4(3x - 2)(2x - 5) = 0 \quad \text{Factoring as in Example 2}$$
$$x^4 = 0 \quad or \quad 3x - 2 = 0 \quad or \quad 2x - 5 = 0 \quad \text{Using the principle of zero products}$$
$$x = 0 \quad or \quad x = \tfrac{2}{3} \quad or \quad x = \tfrac{5}{2}. \quad \text{Solving for } x; \; x^4 = x \cdot x \cdot x \cdot x = 0 \text{ when } x = 0$$

The solutions are $0, \tfrac{2}{3},$ and $\tfrac{5}{2}$.

GRAPHICAL APPROACH

We find the *x*-intercepts of the function

$$f(x) = 6x^6 - 19x^5 + 10x^4$$

using the ZERO option of the CALC menu.

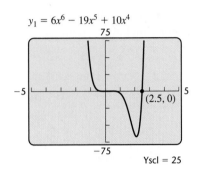

$y_1 = 6x^6 - 19x^5 + 10x^4$

Yscl = 25

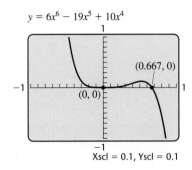

$y = 6x^6 - 19x^5 + 10x^4$

Xscl = 0.1, Yscl = 0.1

Using the viewing window $[-5, 5, -75, 75]$ shown on the left above, we see that 2.5 is a zero of the function. From this window, we cannot tell how many times the graph of the function intersects the *x*-axis between -1 and 1. We magnify that portion of the *x*-axis, as shown on the right above. The *x*-coordinates of the *x*-intercepts are 0, 0.667, and 2.5. Since $\frac{5}{2} = 2.5$ and $\frac{2}{3} \approx 0.667$, the solutions are 0, $\frac{2}{3}$, and $\frac{5}{2}$.

Example 5 illustrates two disadvantages of solving equations graphically: (1) It can be easy to "miss" solutions in some viewing windows and (2) solutions may be given as approximations.

Our work with factoring can help us when we are working with functions.

EXAMPLE 6   Given that $f(x) = 3x^2 - 4x$, find all values of *a* for which $f(a) = 4$.

SOLUTION   We want all numbers *a* for which $f(a) = 4$. Since $f(a) = 3a^2 - 4a$, we must have

$$\begin{aligned}
3a^2 - 4a &= 4 && \text{Setting } f(a) \text{ equal to 4} \\
3a^2 - 4a - 4 &= 0 && \text{Getting 0 on one side} \\
(3a + 2)(a - 2) &= 0 && \text{Factoring} \\
3a + 2 = 0 \quad &or \quad a - 2 = 0 \\
a = -\tfrac{2}{3} \quad &or \qquad a = 2.
\end{aligned}$$

To check using a graphing calculator, we enter $y_1 = 3x^2 - 4x$ and calculate $Y_1(-2/3)$ and $Y_1(2)$. To have $f(a) = 4$, we must have $a = -\frac{2}{3}$ or $a = 2$.

Y₁(−2/3)

4

Y₁(2)

4

▼**EXAMPLE 7** Find the domain of F if $F(x) = \dfrac{x - 2}{x^2 + 2x - 15}$.

SOLUTION The domain of F is the set of all values for which

$$\frac{x - 2}{x^2 + 2x - 15}$$

is a real number. Since division by 0 is undefined, we *exclude* any x-value for which the denominator, $x^2 + 2x - 15$, is 0. We solve:

$$x^2 + 2x - 15 = 0 \qquad \text{Setting the denominator equal to 0}$$
$$(x - 3)(x + 5) = 0 \qquad \text{Factoring}$$
$$x - 3 = 0 \quad \text{or} \quad x + 5 = 0$$
$$x = 3 \quad \text{or} \qquad x = -5. \qquad \text{These are the values to } \textit{exclude.}$$

We can check to see whether either 3 or -5 is in the domain of F using a graphing calculator, as shown at left. We enter $y = (x - 2)/(x^2 + 2x - 15)$ and set up a table with Indpnt set to Ask. When the x-values of 3 and -5 are supplied, the graphing calculator should return an error message, indicating that those numbers are not in the domain of the function. There will be a function value given for any other value of x.

The domain of F is $\{x \,|\, x$ is a real number *and* $x \neq -5$ *and* $x \neq 3\}$. ◢

X	Y₁	
3	ERROR	
−5	ERROR	
0	.13333	
.5	.10909	
8	.09231	
127	.00764	
−3	.41667	
X = 3		

6.3 EXERCISE SET

↪ *Concept Reinforcement* *In each of Exercises 1–8, match the polynomial with one of its factors from the column on the right.*

1. $2x^2 - 7x - 15$

2. $3x^2 + 4x - 7$

3. $6x^2 + 7x + 2$

4. $10x^2 - x - 2$

5. $3x^2 + 4x - 15$

6. $2x^2 + 9x + 7$

7. $10x^2 + 9x - 7$

8. $15x^2 + 14x + 3$

a) $5x + 2$

b) $5x + 3$

c) $3x + 7$

d) $2x + 7$

e) $3x + 2$

f) $2x + 3$

g) $3x - 5$

h) $5x + 7$

Factor.

9. $6x^2 - 5x - 25$

10. $3x^2 - 16x - 12$

11. $10y^3 - 12y - 7y^2$

12. $6x^3 - 15x - x^2$

13. $24a^2 - 14a + 2$

14. $3a^2 - 10a + 8$

15. $35y^2 + 34y + 8$

16. $9a^2 + 18a + 8$

17. $4t + 10t^2 - 6$

18. $8x + 30x^2 - 6$

19. $8x^2 - 16 - 28x$

20. $18x^2 - 24 - 6x$

21. $14x^4 - 19x^3 - 3x^2$

22. $70x^4 - 68x^3 + 16x^2$

23. $12a^2 - 4a - 16$

24. $12a^2 - 14a - 20$

25. $9x^2 + 15x + 4$

26. $6y^2 - y - 2$

Aha! **27.** $4x^2 + 15x + 9$

28. $2y^2 + 7y + 6$

29. $-8t^2 - 8t + 30$

30. $-36a^2 + 21a - 3$

31. $8 - 6z - 9z^2$

32. $3 + 35a - 12a^2$

33. $18xy^3 + 3xy^2 - 10xy$

34. $3x^3y^2 - 5x^2y^2 - 2xy^2$

35. $24x^2 - 2 - 47x$

36. $15z^2 - 10 - 47z$

37. $63x^3 + 111x^2 + 36x$

38. $50t^3 + 115t^2 + 60t$

39. $48x^4 + 4x^3 - 30x^2$

40. $40y^4 + 4y^2 - 12$

41. $12a^2 - 17ab + 6b^2$

42. $20p^2 - 23pq + 6q^2$

43. $2x^2 + xy - 6y^2$

44. $8m^2 - 6mn - 9n^2$

45. $6x^2 - 29xy + 28y^2$

46. $10p^2 + 7pq - 12q^2$

47. $9x^2 - 30xy + 25y^2$

48. $4p^2 + 12pq + 9q^2$

49. $9x^2y^2 + 5xy - 4$

50. $7a^2b^2 + 13ab + 6$

51. Use the results of Exercise 9 to solve $6x^2 - 5x - 25 = 0$.

52. Use the results of Exercise 10 to solve $3x^2 - 16x - 12 = 0$.

53. Use the results of Exercise 31 to solve $9z^2 + 6z = 8$.

54. Use the results of Exercise 32 to solve $3 + 35a = 12a^2$.

55. Use the results of Exercise 37 to solve $63x^3 + 111x^2 + 36x = 0$.

56. Use the results of Exercise 38 to solve $50t^3 + 115t^2 + 60t = 0$.

Solve.

57. $3x^2 - 8x + 4 = 0$

58. $9x^2 - 15x + 4 = 0$

59. $4t^3 + 11t^2 + 6t = 0$

60. $8n^3 + 10n^2 + 3n = 0$

61. $6x^2 = 13x + 5$

62. $40x^2 + 43x = 6$

63. $x(5 + 12x) = 28$

64. $a(1 + 21a) = 10$

65. Find the zeros of the function given by $f(x) = 2x^2 - 13x - 7$.

66. Find the zeros of the function given by $g(x) = 6x^2 + 13x + 6$.

67. Let $f(x) = x^2 + 12x + 40$. Find a such that $f(a) = 8$.

68. Let $f(x) = x^2 + 14x + 50$. Find a such that $f(a) = 5$.

69. Let $g(x) = 2x^2 + 5x$. Find a such that $g(a) = 12$.

70. Let $g(x) = 2x^2 - 15x$. Find a such that $g(a) = -7$.

71. Let $h(x) = 12x + x^2$. Find a such that $h(a) = -27$.

72. Let $h(x) = 4x - x^2$. Find a such that $h(a) = -32$.

Find the domain of the function f given by each of the following.

73. $f(x) = \dfrac{3}{x^2 - 4x - 5}$

74. $f(x) = \dfrac{2}{x^2 - 7x + 6}$

75. $f(x) = \dfrac{x - 5}{9x - 18x^2}$

76. $f(x) = \dfrac{1 + x}{3x - 15x^2}$

77. $f(x) = \dfrac{7}{5x^3 - 35x^2 + 50x}$

78. $f(x) = \dfrac{3}{2x^3 - 2x^2 - 12x}$

TW 79. Emily has factored a polynomial as $(a - b)(x - y)$, while Jorge has factored the same polynomial as $(b - a)(y - x)$. Who is correct, and why?

TW 80. Austin says that the domain of the function
$$F(x) = \frac{x + 3}{3x^2 - x - 2}$$
is $\left\{-\frac{2}{3}, 1\right\}$. Is he correct? Why or why not?

Focused Review

Factor.

81. $x^2 + 4x + 3$ [6.2]

82. $2x^2 - 3x - 5$ [6.3]

83. $x^2y + 2xy + xy^2 + 2y^2$ [6.1]

84. $15x^3 + 4x^2 - 3x$ [6.3]

85. $12x^2 - 5x - 3$ [6.3]

86. $2x^2 - 10x - 28$ [6.2]

87. $ac - ad - bc + bd$ [6.1]

88. $12x^3y^4 - 24x^2y^5 + 4x^2y^3$ [6.1]

89. $(a - 2)(a + b) + (a - 2)(2a - b)$ [6.1]

90. $-8x^2 + 8x + 8$ [6.1]

Synthesis

TW 91. Suppose that $(rx + p)(sx - q) = ax^2 - bx + c$ is true. Explain how this can be used to factor $ax^2 + bx + c$.

TW 92. Describe in your own words an approach that can be used to factor any "nonprime" trinomial of the form $ax^2 + bx + c$.

Use a graph to help factor each polynomial.

93. $4x^2 + 120x + 675$ **94.** $4x^2 + 164x + 1197$

95. $3x^3 + 150x^2 - 3672x$ **96.** $5x^4 + 20x^3 - 1600x^2$

Solve.

97. $(8x + 11)(12x^2 - 5x - 2) = 0$

98. $(x + 1)^3 = (x - 1)^3 + 26$

99. $(x - 2)^3 = x^3 - 2$

Factor. Assume that variables in exponents represent positive integers. If a polynomial is prime, state this.

100. $9x^2y^2 - 12xy - 2$

101. $18a^2b^2 - 3ab - 10$

102. $16x^2y^3 + 20xy^2 + 4y$

103. $16a^2b^3 + 25ab^2 + 9$

104. $9t^8 + 12t^4 + 4$

105. $25t^{10} - 10t^5 + 1$

106. $-15x^{2m} + 26x^m - 8$

107. $20x^{2n} + 16x^n + 3$

108. $3(a + 1)^{n+1}(a + 3)^2 - 5(a + 1)^n(a + 3)^3$

109. $7(t - 3)^{2n} + 5(t - 3)^n - 2$

110. $6(x - 7)^2 + 13(x - 7) - 5$

111. $2a^4b^6 - 3a^2b^3 - 20$

112. $5x^8y^6 + 35x^4y^3 + 60$

113. $4x^{2a} - 4x^a - 3$

114. $2ar^2 + 4asr + as^2 - asr$

115. To better understand factoring $ax^2 + bx + c$ by the ac-method, suppose that
$$ax^2 + bx + c = (mx + r)(nx + s).$$
Show that if $P = ms$ and $Q = rn$, then $P + Q = b$ and $PQ = ac$.

6.4 Equations Containing Perfect-Square Trinomials and Differences of Squares

Perfect-Square Trinomials   ▪   Differences of Squares   ▪
More Factoring by Grouping   ▪   Solving Equations

We now introduce a faster way to factor trinomials that are squares of binomials. A method for factoring differences of squares is also developed. These factoring methods enable us to solve more types of polynomial equations using the principle of zero products.

Perfect-Square Trinomials

Consider the trinomial

$$x^2 + 6x + 9.$$

To factor it, we can proceed as in Section 6.2 and look for factors of 9 that add to 6. These factors are 3 and 3 and the factorization is

$$x^2 + 6x + 9 = (x + 3)(x + 3) = (x + 3)^2.$$

Note that the result is the square of a binomial. Because of this, we call $x^2 + 6x + 9$ a **perfect-square trinomial.** Although trial and error can be used to factor a perfect-square trinomial, once recognized, a perfect-square trinomial can be quickly factored.

Student Notes

If you're not already quick to recognize that $1^2 = 1$, $2^2 = 4$, $3^2 = 9$, $4^2 = 16$, $5^2 = 25$, $6^2 = 36$, $7^2 = 49$, $8^2 = 64$, $9^2 = 81$, $10^2 = 100$, $11^2 = 121$, and $12^2 = 144$, this is a good time to familiarize yourself with these numbers.

To Recognize a Perfect-Square Trinomial

- Two terms must be squares, such as A^2 and B^2.
- There must be no minus sign before A^2 or B^2.
- The remaining term must be $2AB$ or its opposite, $-2AB$.

EXAMPLE 1   Determine whether each polynomial is a perfect-square trinomial.

a) $x^2 + 10x + 25$ **b)** $4x + 16 + 3x^2$ **c)** $100y^2 + 81 - 180y$

SOLUTION

a) • Two of the terms in $x^2 + 10x + 25$ are squares: x^2 and 25.
- There is no minus sign before either x^2 or 25.
- The remaining term, $10x$, is twice the product of the square roots, x and 5.

Thus, $x^2 + 10x + 25$ *is* a perfect square.

b) In $4x + 16 + 3x^2$, only one term, 16, is a square ($3x^2$ is not a square because 3 is not a perfect square; $4x$ is not a square because x is not a square).

Thus, $4x + 16 + 3x^2$ *is not* a perfect square.

c) It can help to first write the polynomial in descending order:

$$100y^2 - 180y + 81.$$

- Two of the terms, $100y^2$ and 81, are squares.
- There is no minus sign before either $100y^2$ or 81.
- If the product of the square roots, $10y$ and 9, is doubled, we get the opposite of the remaining term: $2(10y)(9) = 180y$ (the opposite of $-180y$).

Thus, $100y^2 + 81 - 180y$ *is* a perfect-square trinomial.

To factor a perfect-square trinomial, we reuse the patterns that we learned in Section 5.6.

Study Tip

Fill In Your Blanks

Don't hesitate to write out any missing steps that you'd like to see included. For instance, in Example 1(c), we state that $100y^2$ is a square. To solidify your understanding, you may want to write in $10y \cdot 10y = 100y^2$.

> **Factoring a Perfect-Square Trinomial**
> $$A^2 + 2AB + B^2 = (A + B)^2;$$
> $$A^2 - 2AB + B^2 = (A - B)^2$$

> **CAUTION!** In Example 2, we are factoring expressions. These are *not* equations and thus we do not try to solve them.

EXAMPLE 2 Factor.

a) $x^2 - 10x + 25$ **b)** $16y^2 + 49 + 56y$ **c)** $-20xy + 4y^2 + 25x^2$

SOLUTION

a) $x^2 - 10x + 25 = (x - 5)^2$ We find the square terms and write the square roots with a minus sign between them.

  Note the sign!

As always, any factorization can be checked by multiplying:

$$(x - 5)^2 = (x - 5)(x - 5) = x^2 - 5x - 5x + 25 = x^2 - 10x + 25.$$

b) $16y^2 + 49 + 56y = 16y^2 + 56y + 49$ Using a commutative law

$$= (4y + 7)^2$$ We find the square terms and write the square roots with a plus sign between them.

The check is left to the student.

c) $-20xy + 4y^2 + 25x^2 = 4y^2 - 20xy + 25x^2$ Writing descending order with respect to y

$$= (2y - 5x)^2$$

This square can also be expressed as

$$25x^2 - 20xy + 4y^2 = (5x - 2y)^2.$$

The student should confirm that both factorizations check.

When factoring, always look first for a factor common to all the terms.

EXAMPLE 3 Factor: **(a)** $2x^2 - 12xy + 18y^2$; **(b)** $-4y^2 - 144y^8 + 48y^5$.

SOLUTION

a) We first look for a common factor. This time, there is a common factor, 2.

$$2x^2 - 12xy + 18y^2 = 2(x^2 - 6xy + 9y^2)$$ Factoring out the 2

$$= 2(x - 3y)^2$$ Factoring the perfect-square trinomial

The check is left to the student.

b) $-4y^2 - 144y^8 + 48y^5 = -4y^2(1 + 36y^6 - 12y^3)$ Factoring out the common factor

$$= -4y^2(36y^6 - 12y^3 + 1)$$ Changing order. Note that $(y^3)^2 = y^6$.

$$= -4y^2(6y^3 - 1)^2$$ Factoring the perfect-square trinomial

Check: $-4y^2(6y^3 - 1)^2 = -4y^2(6y^3 - 1)(6y^3 - 1)$
$$= -4y^2(36y^6 - 12y^3 + 1)$$
$$= -144y^8 + 48y^5 - 4y^2$$
$$= -4y^2 - 144y^8 + 48y^5$$

The factorization $-4y^2(6y^3 - 1)^2$ checks.

Differences of Squares

When an expression like $x^2 - 9$ is recognized as a difference of two squares, we can reverse another pattern first seen in Section 5.6.

Factoring a Difference of Two Squares

$$A^2 - B^2 = (A + B)(A - B)$$

To factor a difference of two squares, write the product of the sum and the difference of the quantities being squared.

EXAMPLE 4   Factor: **(a)** $x^2 - 9$; **(b)** $25y^6 - 49x^2$.

SOLUTION

a) $x^2 - 9 = x^2 - 3^2 = (x + 3)(x - 3)$

$$\begin{array}{ccccccccc} A^2 & - & B^2 & = & (A & + & B)(A & - & B) \\ \downarrow & & \downarrow & & \downarrow & & \downarrow & & \downarrow \end{array}$$

b) $25y^6 - 49x^2 = (5y^3)^2 - (7x)^2 = (5y^3 + 7x)(5y^3 - 7x)$

As always, the first step in factoring is to look for common factors.

EXAMPLE 5   Factor: **(a)** $5 - 5x^2y^6$; **(b)** $16x^4y - 81y$.

SOLUTION

a) $5 - 5x^2y^6 = 5(1 - x^2y^6)$ Factoring out the common factor
$$= 5[1^2 - (xy^3)^2]$$ Rewriting x^2y^6 as a quantity squared
$$= 5(1 + xy^3)(1 - xy^3)$$ Factoring the difference of squares

Check: $5(1 + xy^3)(1 - xy^3) = 5(1 - xy^3 + xy^3 - x^2y^6)$
$$= 5(1 - x^2y^6) = 5 - 5x^2y^6$$

The factorization $5(1 + xy^3)(1 - xy^3)$ checks.

b) $16x^4y - 81y = y(16x^4 - 81)$ Factoring out the common factor
$$= y[(4x^2)^2 - 9^2]$$
$$= y(4x^2 + 9)(4x^2 - 9)$$ Factoring the difference of squares
$$= y(4x^2 + 9)(2x + 3)(2x - 3)$$ Factoring $4x^2 - 9$, which is itself a difference of squares

The check is left to the student.

Note in Example 5(b) that $4x^2 - 9$ *could* be factored further. Whenever a factor itself can be factored, do so. We say that we have factored completely when none of the factors can be factored further.

More Factoring by Grouping

Sometimes, when factoring a polynomial with four terms, we may be able to factor further.

EXAMPLE 6 Factor: $x^3 + 3x^2 - 4x - 12$.

SOLUTION

$$
\begin{aligned}
x^3 + 3x^2 - 4x - 12 &= x^2(x + 3) - 4(x + 3) && \text{Factoring by grouping} \\
&= (x + 3)(x^2 - 4) && \text{Factoring out } x + 3 \\
&= (x + 3)(x + 2)(x - 2) && \text{Factoring } x^2 - 4
\end{aligned}
$$

A difference of squares can have four or more terms. For example, one of the squares may be a trinomial. In this case, a type of grouping can be used.

EXAMPLE 7 Factor: **(a)** $x^2 + 6x + 9 - y^2$; **(b)** $a^2 - b^2 + 8b - 16$.

SOLUTION

a)
$$
\begin{aligned}
x^2 + 6x + 9 - y^2 &= (x^2 + 6x + 9) - y^2 && \begin{array}{l}\text{Grouping as a perfect-} \\ \text{square trinomial minus } y^2 \text{ to} \\ \text{show a difference of squares}\end{array} \\
&= (x + 3)^2 - y^2 \\
&= (x + 3 + y)(x + 3 - y)
\end{aligned}
$$

b) Grouping $a^2 - b^2 + 8b - 16$ into two groups of two terms does not yield a common binomial factor, so we look for a perfect-square trinomial. In this case, the perfect-square trinomial is being subtracted from a^2:

$$
\begin{aligned}
a^2 - b^2 + 8b - 16 &= a^2 - (b^2 - 8b + 16) && \begin{array}{l}\text{Factoring out } -1 \\ \text{and rewriting as} \\ \text{subtraction}\end{array} \\
&= a^2 - (b - 4)^2 && \begin{array}{l}\text{Factoring the perfect-square} \\ \text{trinomial}\end{array} \\
&= (a + (b - 4))(a - (b - 4)) && \begin{array}{l}\text{Factoring a} \\ \text{difference} \\ \text{of squares}\end{array} \\
&= (a + b - 4)(a - b + 4) && \begin{array}{l}\text{Removing} \\ \text{parentheses}\end{array}
\end{aligned}
$$

Solving Equations

We can now solve polynomial equations involving differences of squares and perfect-square trinomials.

EXAMPLE 8 Solve: $x^3 + 3x^2 = 4x + 12$.

SOLUTION We first note that the equation is a third-degree polynomial equation. Thus it will have 3 or fewer solutions.

ALGEBRAIC APPROACH

We have

$$x^3 + 3x^2 = 4x + 12$$

$$x^3 + 3x^2 - 4x - 12 = 0 \quad \text{Getting 0 on one side}$$

$$(x + 3)(x + 2)(x - 2) = 0 \quad \text{Factoring; using the results of Example 6}$$

$$x + 3 = 0 \quad or \quad x + 2 = 0 \quad or \quad x - 2 = 0$$

Using the principle of zero products

$$x = -3 \quad or \quad x = -2 \quad or \quad x = 2.$$

The solutions are -3, -2, and 2.

GRAPHICAL APPROACH

We let $f(x) = x^3 + 3x^2$ and $g(x) = 4x + 12$ and look for any points of intersection of the graphs of f and g.

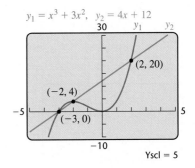

The x-coordinates of the points of intersection are -3, -2, and 2. These are the solutions of the equation.

EXAMPLE 9   Find the zeros of the function given by $f(x) = x^3 + x^2 - x - 1$.

SOLUTION   This function is a third-degree polynomial function, so there will be at most 3 zeros. To find the zeros of the function, we find the roots of the equation $x^3 + x^2 - x - 1 = 0$.

ALGEBRAIC APPROACH

We first factor the polynomial:

$$x^3 + x^2 - x - 1 = 0$$

$$x^2(x + 1) - 1(x + 1) = 0 \quad \text{Factoring by grouping}$$

$$(x + 1)(x^2 - 1) = 0$$

$$(x + 1)(x + 1)(x - 1) = 0. \quad \text{Factoring a difference of squares}$$

Then we use the principle of zero products:

$$x + 1 = 0 \quad or \quad x + 1 = 0 \quad or \quad x - 1 = 0$$

$$x = -1 \quad or \quad x = -1 \quad or \quad x = 1.$$

The solutions are -1 and 1. The numbers -1 and 1 check. The zeros are -1 and 1.

GRAPHICAL APPROACH

We graph $f(x) = x^3 + x^2 - x - 1$ and find the zeros of the function.

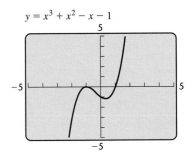

The zeros are -1 and 1.

In Example 9, the factor $(x + 1)$ appeared twice in the factorization. When a factor appears two or more times, we say that we have a **repeated root.** If the factor occurs twice, we say that we have a **double root,** or a **root of multiplicity two.**

In this chapter, we have emphasized the relationship between factoring and equation solving. There are many polynomial equations, however, that cannot be solved by factoring. Nonetheless, we can find approximations of any real-number solutions using a graphing calculator.

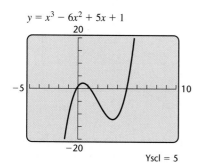

$y = x^3 - 6x^2 + 5x + 1$

X	Y1	
−.166	9E−5	
1.217	9.5E−4	
4.949	.00328	

X = −.166

▶ **EXAMPLE 10**   Solve: $x^3 - 6x^2 + 5x + 1 = 0$.

SOLUTION   The polynomial $x^3 - 6x^2 + 5x + 1$ is prime; it cannot be factored using rational coefficients. We solve the equation by graphing $f(x) = x^3 - 6x^2 + 5x + 1$ and finding the zeros of the function. Since the polynomial is of degree 3, we need not look for more than 3 zeros.

There are three x-intercepts of the graph. We use the ZERO feature to find each root, or zero. Since the solutions are irrational, the calculator will give approximations. Rounding each to the nearest thousandth, we have the solutions -0.166, 1.217, and 4.949. We check the solutions using a table (see the figure at left). Note that the function values are close to 0 for each approximate solution.   ◢

We have considered both algebraic and graphical methods of solving polynomial equations. It is important to understand and be able to use both methods. Some of the advantages and disadvantages of each method are given in the following table.

	Advantages	**Disadvantages**
Algebraic Method	• Can find exact answers • Works well when the polynomial is in factored form or can be readily factored • Can be used to find solutions that are not real numbers (see Chapter 11)	• Cannot be used if the polynomial is not factorable • Can be difficult to use if factorization is not readily apparent
Graphical Method	• Does not require the polynomial to be factored • Can visualize solutions	• Easy to miss solutions if an appropriate viewing window is not chosen • Gives approximations of solutions • For most graphing calculators, only real-number solutions can be found.

6.4 EXERCISE SET

⤷ *Concept Reinforcement   Classify each of the following as a perfect-square trinomial, a difference of two squares, a polynomial having a common factor, or none of these.*

1. $9t^2 - 49$

2. $25x^2 - 20x + 4$

3. $36x^2 - 12x + 1$

4. $36a^2 - 25$

5. $4r^2 + 8r + 9$

6. $9x^2 - 12$

7. $4x^2 + 8x + 10$

8. $t^2 - 6t + 8$

9. $4t^2 + 9s^2 + 12st$

10. $9rt^2 - 5rt + 6r$

Factor completely.

11. $t^2 + 6t + 9$

12. $x^2 - 8x + 16$

13. $a^2 - 14a + 49$

14. $a^2 + 16a + 64$

15. $4a^2 - 16a + 16$

16. $2a^2 + 8a + 8$

17. $y^2 + 36 + 12y$

18. $y^2 + 36 - 12y$

19. $-18y^2 + y^3 + 81y$

20. $24a^2 + a^3 + 144a$

21. $2x^2 - 40x + 200$

22. $32x^2 + 48x + 18$

23. $1 - 8d + 16d^2$

24. $64 + 25y^2 - 80y$

25. $y^3 + 8y^2 + 16y$

26. $a^3 - 10a^2 + 25a$

27. $0.25x^2 + 0.30x + 0.09$

28. $0.04x^2 - 0.28x + 0.49$

29. $p^4 + 2p^2q^2 + q^4$

30. $m^{10} + 2m^5n^5 + n^{10}$

31. $25a^2 + 30ab + 9b^2$

32. $49p^2 - 84pq + 36q^2$

33. $5a^2 - 10ab + 5b^2$

34. $4t^2 - 8tr + 4r^2$

35. $y^2 - 100$

36. $x^2 - 16$

37. $m^2 - 64$

38. $p^2 - 49$

39. $p^2q^2 - 25$

40. $a^2b^2 - 81$

41. $8x^2 - 8y^2$

42. $6x^2 - 6y^2$

43. $7xy^4 - 7xz^4$

44. $25ab^4 - 25az^4$

45. $4a^3 - 49a$

46. $9x^4 - 25x^2$

47. $3x^8 - 3y^8$

48. $9a^4 - a^2b^2$

49. $9a^4 - 25a^2b^4$

50. $16x^6 - 121x^2y^4$

51. $\frac{1}{49} - x^2$

52. $\frac{1}{16} - y^2$

53. $(a + b)^2 - 9$

54. $(p + q)^2 - 25$

55. $x^2 - 6x + 9 - y^2$

56. $a^2 - 8a + 16 - b^2$

57. $t^3 + 8t^2 - t - 8$

58. $x^3 - 7x^2 - 4x + 28$

59. $r^3 - 3r^2 - 9r + 27$

60. $t^3 + 2t^2 - 4t - 8$

61. $m^2 - 2mn + n^2 - 25$

62. $x^2 + 2xy + y^2 - 9$

63. $36 - (x + y)^2$

64. $49 - (a + b)^2$

65. $r^2 - 2r + 1 - 4s^2$

66. $c^2 + 4cd + 4d^2 - 9p^2$

Aha! **67.** $16 - a^2 - 2ab - b^2$

68. $9 - x^2 - 2xy - y^2$

69. $x^3 + 5x^2 - 4x - 20$

70. $t^3 + 6t^2 - 9t - 54$

71. $a^3 - ab^2 - 2a^2 + 2b^2$

72. $p^2q - 25q + 3p^2 - 75$

Solve. Round any irrational solutions to the nearest thousandth.

73. $a^2 + 1 = 2a$

74. $r^2 + 16 = 8r$

75. $2x^2 - 24x + 72 = 0$

76. $-t^2 - 16t - 64 = 0$

77. $x^2 - 9 = 0$

78. $r^2 - 64 = 0$

79. $a^2 = \frac{1}{25}$

80. $x^2 = \frac{1}{100}$

81. $8x^3 + 1 = 4x^2 + 2x$

82. $27x^3 + 18x^2 = 12x + 8$

83. $x^3 + 3 = 3x^2 + x$

84. $x^3 + x^2 = 16x + 16$

85. $x^2 - 3x - 7 = 0$ **86.** $x^2 - 5x + 1 = 0$

87. $2x^2 + 8x + 1 = 0$ **88.** $3x^2 + x - 1 = 0$

89. $x^3 + 3x^2 + x - 1 = 0$

90. $x^3 + x^2 + x - 1 = 0$

91. Let $f(x) = x^2 - 12x$. Find a such that $f(a) = -36$.

92. Let $g(x) = x^2$. Find a such that $g(a) = 144$.

Find the zeros of each function.

93. $f(x) = x^2 - 16$

94. $f(x) = x^2 - 8x + 16$

95. $f(x) = 2x^2 + 4x + 2$

96. $f(x) = 3x^2 - 27$

97. $f(x) = x^3 - 2x^2 - x + 2$

98. $f(x) = x^3 + x^2 - 4x - 4$

TW 99. Are the product and power rules for exponents (see Section 5.1) important when factoring differences of squares? Why or why not?

TW 100. Describe a procedure that could be used to find a polynomial with four terms that can be factored as a difference of two squares.

Focused Review

Factor completely.

101. $6x^2 - 17x + 5$ [6.3]

102. $16x^2 - 1$ [6.4]

103. $16x^2 + 8x + 1$ [6.4]

104. $3x^3 - 3x^2 - 6x$ [6.2]

105. $x^3 + 3x^2 - 2x - 6$ [6.1]

106. $(2x + 1)^2 - 16$ [6.4]

107. $\frac{1}{64} - x^2$ [6.4]

108. $100 - t^2 - 2st - s^2$ [6.4]

109. $12x^3y - 24x^2y^2 + 12xy^3$ [6.4]

110. $-3x^4 + 18x^3 + 3x$ [6.1]

Synthesis

TW 111. Without finding the entire factorization, determine the number of factors of $x^{256} - 1$. Explain how you arrived at your answer.

112. Under what conditions can a sum of two squares be factored?

Factor completely. Assume that variables in exponents represent positive integers.

TW 113. $-\frac{8}{27}r^2 - \frac{10}{9}rs - \frac{1}{6}s^2 + \frac{2}{3}rs$

114. $\frac{1}{36}x^8 + \frac{2}{9}x^4 + \frac{4}{9}$

115. $0.09x^8 + 0.48x^4 + 0.64$

116. $a^2 + 2ab + b^2 - c^2 + 6c - 9$

117. $r^2 - 8r - 25 - s^2 - 10s + 16$

118. $x^{2a} - y^2$

119. $x^{4a} - y^{2b}$

Aha! 120. $4y^{4a} + 20y^{2a} + 20y^{2a} + 100$

121. $25y^{2a} - (x^{2b} - 2x^b + 1)$

122. $8(a - 3)^2 - 64(a - 3) + 128$

123. $3(x + 1)^2 + 12(x + 1) + 12$

124. $5c^{100} - 80d^{100}$

125. $9x^{2n} - 6x^n + 1$

126. $c^{2w+1} + 2c^{w+1} + c$

127. $(t + 1)^2 - 4(t + 1) + 3$

128. $(a - 3)^2 - 8(a - 3) + 16$

129. $m^2 + 4mn + 4n^2 + 5m + 10n$

130. $s^2 - 4st + 4t^2 + 4s - 8t + 4$

131. If $P(x) = x^2$, use factoring to simplify
$$P(a + h) - P(a).$$

132. If $P(x) = x^4$, use factoring to simplify
$$P(a + h) - P(a).$$

133. *Volume of Carpeting.* The volume of a carpet that is rolled up can be estimated by the polynomial $\pi R^2 h - \pi r^2 h$.

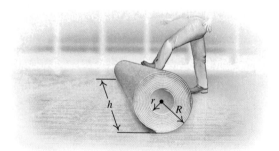

a) Factor the polynomial.

b) Use both the original and the factored forms to find the volume of a roll for which $R = 50$ cm, $r = 10$ cm, and $h = 4$ m. Use 3.14 for π.

6.5   Equations Containing Sums or Differences of Cubes

Factoring Sums or Differences of Cubes   ■   Solving Equations

Factoring Sums or Differences of Cubes

We have seen that a difference of two squares can be factored but (unless a common factor exists) a *sum* of two squares is usually prime. The situation is different with cubes: The difference *or* sum of two cubes can always be factored. To see this, consider the following products:

$$(A + B)(A^2 - AB + B^2) = A(A^2 - AB + B^2) + B(A^2 - AB + B^2)$$
$$= A^3 - A^2B + AB^2 + A^2B - AB^2 + B^3$$
$$= A^3 + B^3 \qquad \text{Combining like terms}$$

and

$$(A - B)(A^2 + AB + B^2) = A(A^2 + AB + B^2) - B(A^2 + AB + B^2)$$
$$= A^3 + A^2B + AB^2 - A^2B - AB^2 - B^3$$
$$= A^3 - B^3. \qquad \text{Combining like terms}$$

These products allow us to factor a sum or a difference of two cubes. Note how the location of the $+$ and $-$ signs changes.

> **Factoring a Sum or a Difference of Two Cubes**
> $$A^3 + B^3 = (A + B)(A^2 - AB + B^2);$$
> $$A^3 - B^3 = (A - B)(A^2 + AB + B^2)$$

When factoring a sum or a difference of cubes, it can be helpful to remember that $1^3 = 1, 2^3 = 8, 3^3 = 27, 4^3 = 64, 5^3 = 125, 6^3 = 216$, and so on. We say that 2 is the *cube root* of 8, that 3 is the cube root of 27, and so on.

Study Tip

A Good Night's Sleep

Don't try studying when you are tired. A good night's sleep or a 10-minute "power nap" can often make a problem suddenly seem much easier to solve.

EXAMPLE 1   Write an equivalent expression by factoring: $x^3 - 27$.

SOLUTION   We first observe that

$$x^3 - 27 = x^3 - 3^3.$$

Next, in one set of parentheses, we write the first cube root, x, minus the second cube root, 3:

$$(x - 3)(\qquad).$$

To get the other factor, we think of $x - 3$ and do the following:

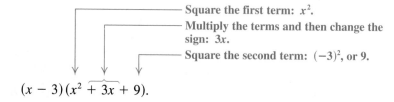

$$(x - 3)(x^2 + 3x + 9).$$

Check: $(x - 3)(x^2 + 3x + 9) = x^3 + 3x^2 + 9x - 3x^2 - 9x - 27$

$$= x^3 - 27 \qquad \text{Combining like terms}$$

Thus, $x^3 - 27 = (x - 3)(x^2 + 3x + 9)$.

   In Example 1, note that $x^2 + 3x + 9$ cannot be factored further. As a general rule, unless A and B share a common factor, the trinomials $A^2 + AB + B^2$ and $A^2 - AB + B^2$ are prime.

EXAMPLE 2   Factor.

a) $125x^3 + y^3$ **b)** $m^6 + 64$

c) $128y^7 - 250x^6y$ **d)** $r^6 - s^6$

SOLUTION

a) We have

$$125x^3 + y^3 = (5x)^3 + y^3.$$

In one set of parentheses, we write the cube root of the first term, $5x$, plus the cube root of the second term, y:

$$(5x + y)(\qquad).$$

To get the other factor, we think of $5x + y$ and do the following:

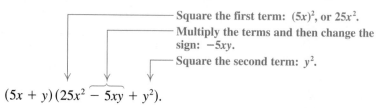

$$(5x + y)(25x^2 - 5xy + y^2).$$

Student Notes

If you think of $A^3 - B^3$ as $A^3 + (-B)^3$, it is then sufficient to remember only the pattern for factoring a sum of two cubes. Be sure to simplify your result if you do this.

Check: $(5x + y)(25x^2 - 5xy + y^2) = 125x^3 - 25x^2y + 5xy^2$
$$+ 25x^2y - 5xy^2 + y^3$$
$$= 125x^3 + y^3 \qquad \textbf{Combining}$$
$$\textbf{like terms}$$

Thus, $125x^3 + y^3 = (5x + y)(25x^2 - 5xy + y^2)$.

b) We have

$$m^6 + 64 = (m^2)^3 + 4^3. \qquad \textbf{Rewriting as quantities cubed}$$

Next, we reuse the pattern used in part (a) above:

$$A^3 + B^3 = (A + B)(A^2 - A \cdot B + B^2)$$

$$(m^2)^3 + 4^3 = (m^2 + 4)((m^2)^2 - m^2 \cdot 4 + 4^2)$$
$$= (m^2 + 4)(m^4 - 4m^2 + 16).$$

The check is left to the student.

c) We have

$$128y^7 - 250x^6y = 2y(64y^6 - 125x^6) \qquad \textbf{Remember:} \textit{Always} \textbf{ look for a common factor.}$$
$$= 2y[(4y^2)^3 - (5x^2)^3]. \qquad \textbf{Rewriting as quantities cubed}$$

> **CAUTION!**   Use parentheses when evaluating A^2 and B^2:
>
> $$(4y^2)^2 = 16y^4 \quad \text{and}$$
> $$(5x^2)^2 = 25x^4.$$

To factor $(4y^2)^3 - (5x^2)^3$, we again use the pattern in Example 1:

$$A^3 - B^3 = (A - B)(A^2 + A \cdot B + B^2)$$

$$(4y^2)^3 - (5x^2)^3 = (4y^2 - 5x^2)((4y^2)^2 + 4y^2 \cdot 5x^2 + (5x^2)^2)$$
$$= (4y^2 - 5x^2)(16y^4 + 20x^2y^2 + 25x^4).$$

The check is left to the student. We have

$$128y^7 - 250x^6y = 2y(4y^2 - 5x^2)(16y^4 + 20x^2y^2 + 25x^4).$$

d) We have

$$r^6 - s^6 = (r^3)^2 - (s^3)^2$$
$$= (r^3 + s^3)(r^3 - s^3) \qquad \textbf{Factoring a difference of two} \textit{squares}$$
$$= (r + s)(r^2 - rs + s^2)(r - s)(r^2 + rs + s^2).$$
$$\textbf{Factoring the sum and difference of two cubes}$$

To check, read the steps in reverse order and inspect the multiplication.

In Example 2(d), suppose we first factored $r^6 - s^6$ as a difference of two cubes:

$$(r^2)^3 - (s^2)^3 = (r^2 - s^2)(r^4 + r^2s^2 + s^4)$$
$$= (r + s)(r - s)(r^4 + r^2s^2 + s^4).$$

In this case, we might have missed some factors; $r^4 + r^2s^2 + s^4$ can be factored as $(r^2 - rs + s^2)(r^2 + rs + s^2)$, but we probably would never have suspected that such a factorization exists. Given a choice, it is generally better to factor as a difference of squares before factoring as a sum or a difference of cubes.

Try to remember the following.

Useful Factoring Facts

Sum of cubes: $A^3 + B^3 = (A + B)(A^2 - AB + B^2)$

Difference of cubes: $A^3 - B^3 = (A - B)(A^2 + AB + B^2)$

Difference of squares: $A^2 - B^2 = (A + B)(A - B)$

There is no formula for factoring a sum of squares.

Solving Equations

We can now solve equations involving sums or differences of cubes.

EXAMPLE 3   Solve: $x^3 = 27$.

SOLUTION   We rewrite the equation to get 0 on one side and factor:

$$x^3 = 27$$
$$x^3 - 27 = 0 \qquad \text{Subtracting 27 from both sides}$$
$$(x - 3)(x^2 + 3x + 9) = 0. \qquad \text{Using the factorization from Example 1}$$

The second-degree factor, $x^2 + 3x + 9$, does not factor using real coefficients. When we apply the principle of zero products, we have

$$x - 3 = 0 \quad or \quad x^2 + 3x + 9 = 0. \qquad \text{Using the principle of zero products}$$
$$x = 3 \qquad\qquad\qquad\qquad \text{We cannot factor } x^2 + 3x + 9.$$

We have one solution, 3. In Chapter 11, we will learn how to solve equations involving prime quadratic polynomials like $x^2 + 3x + 9 = 0$. For now, we can find the real-number solutions of the equation $x^3 = 27$ graphically, by looking for any points of intersection of the graphs of $y_1 = x^3$ and $y_2 = 27$.

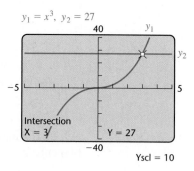

The real-number solution of $x^3 = 27$ is 3.

6.5 EXERCISE SET

↪ *Concept Reinforcement* *Classify each binomial as either a sum of two cubes, a difference of two cubes, a difference of two squares, both a difference of two cubes and a difference of two squares, a prime polynomial, or none of these.*

1. $8x^3 - 27$

2. $64t^3 + 27$

3. $9x^4 - 25$

4. $9x^2 + 25$

5. $1000t^3 + 1$

6. $m^3 - 8n^3$

7. $25x^2 + 8x$

8. $64t^6 - 1$

9. $s^{12} - t^6$

10. $14x^3 - 2x$

Factor completely.

11. $x^3 + 64$

12. $t^3 + 27$

13. $z^3 - 1$

14. $x^3 - 8$

15. $x^3 - 27$

16. $m^3 - 64$

17. $27x^3 + 1$

18. $8a^3 + 1$

19. $64 - 125x^3$

20. $27 - 8t^3$

21. $27y^3 + 64$

22. $8x^3 + 27$

23. $x^3 - y^3$

24. $y^3 - z^3$

25. $a^3 + \frac{1}{8}$

26. $x^3 + \frac{1}{27}$

27. $8t^3 - 8$

28. $2y^3 - 128$

29. $54x^3 + 2$

30. $8a^3 + 1000$

31. $ab^3 + 125a$

32. $rs^3 + 64r$

33. $5x^3 - 40z^3$

34. $2y^3 - 54z^3$

35. $x^3 + 0.001$

36. $y^3 + 0.125$

37. $64x^6 - 8t^6$

38. $125c^6 - 8d^6$

39. $2y^4 - 128y$

40. $3z^5 - 3z^2$

41. $z^6 - 1$

42. $t^6 + 1$

43. $t^6 + 64y^6$

44. $p^6 - q^6$

45. $x^{12} - y^3z^{12}$

46. $a^9 + b^{12}c^{15}$

Solve.

47. $x^3 + 1 = 0$

48. $t^3 - 8 = 0$

49. $8x^3 = 27$

50. $64x^3 + 27 = 0$

51. $2t^3 - 2000 = 0$

52. $375 = 24x^3$

TW **53.** How could you use factoring to convince someone that $x^3 + y^3 \neq (x + y)^3$?

TW **54.** Is the following statement true or false and why? If A^3 and B^3 have a common factor, then A and B have a common factor.

Focused Review

Factor completely.

55. $5x^3 - 5z^3$ [6.5]

56. $x^2 + 4x - 21$ [6.2]

57. $-8x^3z + 24x^2z^2 - 16xz^3 - 8xz$ [6.1]

58. $1 + 1000x^3$ [6.5]

59. $3x^4 - 3z^4$ [6.4]

60. $4x^2 - 20x + 25$ [6.4]

61. $x^6 - 1$ [6.5]

62. $x^3 + x^2 + x + 1$ [6.1]

63. $3x^3 + 60x^2 + 300x$ [6.4]

64. $x^4y^3 + 8x^3y^3 + 12x^2y^3$ [6.2]

65. $12x^2 + 25x + 12$ [6.3]

66. $x^2 + 2x + 1 - y^2$ [6.4]

Synthesis

TW **67.** Explain how the geometric model below can be used to verify the formula for factoring $a^3 - b^3$.

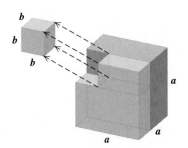

TW **68.** Explain how someone could construct a binomial that is both a difference of two cubes and a difference of two squares.

Factor.

69. $x^{6a} - y^{3b}$

Aha! **71.** $(x + 5)^3 + (x - 5)^3$

73. $5x^3y^6 - \frac{5}{8}$

75. $x^{6a} - (x^{2a} + 1)^3$

77. $t^4 - 8t^3 - t + 8$

70. $2x^{3a} + 16y^{3b}$

72. $\frac{1}{16}x^{3a} + \frac{1}{2}y^{6a}z^{9b}$

74. $x^3 - (x + y)^3$

76. $(x^{2a} - 1)^3 - x^{6a}$

78. If $P(x) = x^3$, use factoring to simplify
$$P(a + h) - P(a).$$

79. If $Q(x) = x^6$, use factoring to simplify
$$Q(a + h) - Q(a).$$

80. Using one viewing window, graph the following.
  a) $f(x) = x^3$
  b) $g(x) = x^3 - 8$
  c) $h(x) = (x - 2)^3$

6.6 Factoring: A General Strategy

Choosing the Right Method

Thus far, each section in this chapter has examined one or two different methods for factoring polynomials. In practice, when the need for factoring a polynomial arises, we must decide on our own which method to use. As preparation for such a situation, we now encounter polynomials of various types, in random order. We concentrate in this section on factoring expressions and return to solving equations in Section 6.7.

We can use the following guidelines to factor polynomials.

To Factor a Polynomial

A. Always look for a common factor first. If there is one, factor out the largest common factor. Be sure to include it in your final answer.

B. Then look at the number of terms.

Two terms: Try factoring as a difference of squares first: $A^2 - B^2 = (A - B)(A + B)$. Next, try factoring as a sum or a difference of cubes: $A^3 + B^3 = (A + B)(A^2 - AB + B^2)$ and $A^3 - B^3 = (A - B)(A^2 + AB + B^2)$.

Three terms: If the trinomial is a perfect-square trinomial, factor accordingly: $A^2 + 2AB + B^2 = (A + B)^2$ or $A^2 - 2AB + B^2 = (A - B)^2$. If it is not a perfect-square trinomial, try using FOIL or grouping.

Four terms: Try factoring by grouping.

C. Always *factor completely*. When a factor can itself be factored, be sure to factor it. Remember that some polynomials, like $A^2 + B^2$, are prime.

D. Check by multiplying.

Choosing the Right Method

EXAMPLE 1   Factor: $5t^4 - 80$.

SOLUTION

A. We look for a common factor:

$$5t^4 - 80 = 5(t^4 - 16). \qquad \text{5 is the largest common factor.}$$

B. The factor $t^4 - 16$ is a difference of squares: $(t^2)^2 - 4^2$. We factor it, being careful to rewrite the 5 from step (A):

$$5t^4 - 80 = 5(t^2 + 4)(t^2 - 4). \qquad t^4 - 16 = (t^2 + 4)(t^2 - 4)$$

C. Since $t^2 - 4$ is not prime, we continue factoring:

$$5t^4 - 80 = 5(t^2 + 4)(t^2 - 4) = 5(t^2 + 4)(t - 2)(t + 2).$$

This is a sum of squares with no common factor. It cannot be factored!

D. *Check:* $\quad 5(t^2 + 4)(t - 2)(t + 2) = 5(t^2 + 4)(t^2 - 4)$
$$= 5(t^4 - 16) = 5t^4 - 80.$$

The factorization is $5(t^2 + 4)(t - 2)(t + 2)$.

EXAMPLE 2   Factor: $2x^3 + 10x^2 + x + 5$.

SOLUTION

A. We look for a common factor. There is none (other than 1 or -1).

B. Because there are four terms, we try factoring by grouping:

$$2x^3 + 10x^2 + x + 5$$
$$= (2x^3 + 10x^2) + (x + 5) \qquad \text{Separating into two binomials}$$
$$= 2x^2(x + 5) + 1(x + 5) \qquad \text{Factoring out the largest common factor from each binomial. The 1 serves as an aid.}$$
$$= (x + 5)(2x^2 + 1). \qquad \text{Factoring out the common factor, } x + 5$$

C. Nothing can be factored further, so we have factored completely.

D. *Check:* $\quad (x + 5)(2x^2 + 1) = 2x^3 + x + 10x^2 + 5$
$$= 2x^3 + 10x^2 + x + 5.$$

The factorization is $(x + 5)(2x^2 + 1)$.

EXAMPLE 3   Factor: $-x^5 + 2x^4 + 35x^3$.

SOLUTION

A. We note that there is a common factor, $-x^3$:

$$-x^5 + 2x^4 + 35x^3 = -x^3(x^2 - 2x - 35).$$

B. The factor $x^2 - 2x - 35$ is not a perfect-square trinomial. We factor it using trial and error:

$$-x^5 + 2x^4 + 35x^3 = -x^3(x^2 - 2x - 35)$$
$$= -x^3(x - 7)(x + 5).$$

C. Nothing can be factored further, so we have factored completely.

D. *Check:* $-x^3(x - 7)(x + 5) = -x^3(x^2 - 2x - 35)$
$$= -x^5 + 2x^4 + 35x^3.$$

The factorization is $-x^3(x - 7)(x + 5)$.

Student Notes

Quickly checking the leading and constant terms to see if they are squares can save you time. If they aren't both squares, the trinomial can't possibly be a perfect-square trinomial.

EXAMPLE 4   Factor: $x^2 - 20x + 100$.

SOLUTION

A. We look first for a common factor. There is none.

B. This polynomial is a perfect-square trinomial. We factor it accordingly:

$$x^2 - 20x + 100 = x^2 - 2 \cdot x \cdot 10 + 10^2 \qquad \text{Try to do this step mentally.}$$
$$= (x - 10)^2.$$

C. Nothing can be factored further, so we have factored completely.

D. *Check:* $(x - 10)(x - 10) = x^2 - 20x + 100.$

The factorization is $(x - 10)(x - 10)$, or $(x - 10)^2$.

EXAMPLE 5   Factor: $6x^2y^4 - 21x^3y^5 + 3x^2y^6$.

SOLUTION

A. We first factor out the largest common factor, $3x^2y^4$:

$$6x^2y^4 - 21x^3y^5 + 3x^2y^6 = 3x^2y^4(2 - 7xy + y^2).$$

B. There are three terms in $2 - 7xy + y^2$. Since only y^2 is a square, we do not have a perfect-square trinomial. Note that x appears only in $-7xy$. If $2 - 7xy + y^2$ could be factored into a form like $(1 - y)(2 - y)$, there would be no x in the middle term. Thus, $2 - 7xy + y^2$ cannot be factored.

C. Nothing can be factored further, so we have factored completely.

D. *Check:* $3x^2y^4(2 - 7xy + y^2) = 6x^2y^4 - 21x^3y^5 + 3x^2y^6.$

The factorization is $3x^2y^4(2 - 7xy + y^2)$.

EXAMPLE 6 Factor: $x^6 - 64$.

SOLUTION

A. Look for a common factor. There is none (other than 1 or -1).

B. There are two terms, a difference of squares: $(x^3)^2 - (8)^2$. We factor it:

$$x^6 - 64 = (x^3 + 8)(x^3 - 8). \qquad \text{Note that } x^6 = (x^3)^2.$$

C. One factor is a sum of two cubes, and the other factor is a difference of two cubes. We factor both:

$$x^6 - 64 = (x + 2)(x^2 - 2x + 4)(x - 2)(x^2 + 2x + 4).$$

The factorization is complete because no factor can be factored further.

D. *Check:*

$$(x + 2)(x^2 - 2x + 4)(x - 2)(x^2 + 2x + 4) = (x^3 + 8)(x^3 - 8)$$
$$= x^6 - 64.$$

The factorization is $(x + 2)(x^2 - 2x + 4)(x - 2)(x^2 + 2x + 4)$.

EXAMPLE 7 Factor: $25x^2 + 20xy + 4y^2$.

SOLUTION

A. We look first for a common factor. There is none.

B. There are three terms. Note that the first term and the last term are squares:

$$25x^2 = (5x)^2 \quad \text{and} \quad 4y^2 = (2y)^2.$$

We see that twice the product of $5x$ and $2y$ is the middle term,

$$2 \cdot 5x \cdot 2y = 20xy,$$

so the trinomial is a perfect square.

To factor, we write a binomial squared. The binomial is the sum of the terms being squared:

$$25x^2 + 20xy + 4y^2 = (5x + 2y)^2.$$

C. Nothing can be factored further, so we have factored completely.

D. *Check:* $(5x + 2y)(5x + 2y) = 25x^2 + 20xy + 4y^2$.

The factorization is $(5x + 2y)(5x + 2y)$, or $(5x + 2y)^2$.

EXAMPLE 8   Factor: $p^2q^2 + 7pq + 12$.

SOLUTION

A. We look first for a common factor. There is none.

B. Since only one term is a square, we do not have a perfect-square trinomial. We use trial and error, thinking of the product pq as a single variable:

$$(pq + \quad)(pq + \quad).$$

We factor the last term, 12. All the signs are positive, so we consider only positive factors. Possibilities are 1, 12 and 2, 6 and 3, 4. The pair 3, 4 gives a sum of 7 for the coefficient of the middle term. Thus,

$$p^2q^2 + 7pq + 12 = (pq + 3)(pq + 4).$$

C. Nothing can be factored further, so we have factored completely.

D. *Check:* $(pq + 3)(pq + 4) = p^2q^2 + 7pq + 12$.

The factorization is $(pq + 3)(pq + 4)$.

Compare the variables appearing in Example 7 with those appearing in Example 8. Note that when one variable appears in the leading term and another variable appears in the last term, as in Example 7, each binomial contains two variable terms. When two variables appear in the leading term, as in Example 8, each binomial contains just one variable term.

EXAMPLE 9   Factor: $2a^3 + 12a^2 + 18a - 8ab^2$.

SOLUTION

A. We note that there is a common factor, $2a$:

$$2a^3 + 12a^2 + 18a - 8ab^2 = 2a(a^2 + 6a + 9 - 4b^2).$$

B. There are four terms. Grouping into two groups of two terms does not yield a common binomial factor, but we can group into a difference of squares:

$$\begin{aligned}
2a^3 + 12a^2 + 18a - 8ab^2 &= 2a[(a^2 + 6a + 9) - 4b^2] \\
&= 2a[(a + 3)^2 - (2b)^2] \\
&= 2a[(a + 3 + 2b)(a + 3 - 2b)].
\end{aligned}$$

C. Nothing can be factored further, so we have factored completely.

D. *Check:* $\begin{aligned}[t] 2a(a + 3 + 2b)(a + 3 - 2b) &= 2a[(a + 3)^2 - (2b)^2] \\
&= 2a[a^2 + 6a + 9 - 4b^2] \\
&= 2a^3 + 12a^2 + 18a - 8ab^2. \end{aligned}$

The factorization is $2a(a + 3 + 2b)(a + 3 - 2b)$.

6.6 EXERCISE SET

Concept Reinforcement   In each of Exercises 1–4, complete the sentence.

1. As a first step when factoring polynomials, always check for a _____.

2. When factoring a trinomial, if two terms are not squares, it cannot be a _____.

3. If a polynomial has four terms and no common factor, it may be possible to factor by _____.

4. It is always possible to check a factorization by _____.

Factor completely. If a polynomial is prime, state this.

5. $10a^2 - 640$

6. $5x^2 - 45$

7. $y^2 + 49 - 14y$

8. $a^2 + 25 + 10a$

9. $2t^2 + 11t + 12$

10. $8t^2 - 18t - 5$

11. $x^3 - 18x^2 + 81x$

12. $x^3 - 24x^2 + 144x$

13. $x^3 - 5x^2 - 25x + 125$

14. $x^3 + 3x^2 - 4x - 12$

15. $27t^3 - 3t$

16. $98t^2 - 18$

17. $9x^3 + 12x^2 - 45x$

18. $20x^3 - 4x^2 - 72x$

19. $t^2 + 25$

20. $x^2 + 4$

21. $6x^2 + 3x - 45$

22. $2t^2 + 6t - 20$

23. $-2a^6 + 8a^5 - 8a^4$

24. $-x^5 - 14x^4 - 49x^3$

25. $5x^5 - 80x$

26. $4x^4 - 64$

27. $t^4 - 9$

28. $9 + t^8$

29. $-x^6 + 2x^5 - 7x^4$

30. $-x^5 + 4x^4 - 3x^3$

31. $x^3 - y^3$

32. $8t^3 + 1$

33. $ax^2 + ay^2$

34. $12n^2 + 24n^3$

35. $36mn - 9m^2n^2$

36. $ab^2 - a^2b$

37. $2\pi rh + 2\pi r^2$

38. $4\pi r^2 + 2\pi r$

Aha! 39. $(a + b)5a + (a + b)3b$

40. $5c(a^3 + b) - (a^3 + b)$

41. $x^2 + x + xy + y$

42. $n^2 + 2n + np + 2p$

43. $n^2 - 10n + 25 - 9m^2$

44. $t^2 + 2t + 1 - 81r^2$

45. $3x^2 + 13xy - 10y^2$

46. $-x^2 - y^2 - 2xy$

47. $4b^2 + a^2 - 4ab$

48. $a^2 - 7a - 6$

49. $16x^2 + 24xy + 9y^2$

50. $7p^4 - 7q^4$

51. $t^2 - 8t + 10$

52. $25z^2 + 10zy + y^2$

53. $64t^6 - 1$

54. $m^6 - 1$

55. $4p^2q - pq^2 + 4p^3$

56. $a^2 + ab + 2b^2$

57. $3b^2 + 17ab - 6a^2$

58. $2mn - 360n^2 + m^2$

59. $-12 - x^2y^2 - 8xy$

60. $m^2n^2 - 4mn - 32$

61. $t^8 - s^{10} - 12s^5 - 36$

62. $p^4 - 1 + 2q - q^2$

63. $54a^4 + 16ab^3$

64. $54x^3y - 250y^4$

65. $x^6 + x^5y - 2x^4y^2$

66. $2s^6t^2 + 10s^3t^3 + 12t^4$

67. $36a^2 - 15a + \frac{25}{16}$

68. $a^2 + 2a^2bc + a^2b^2c^2$

69. $\frac{1}{81}x^2 - \frac{8}{27}x + \frac{16}{9}$

70. $\frac{1}{4}a^2 + \frac{1}{3}ab + \frac{1}{9}b^2$

71. $1 - 16x^{12}y^{12}$

72. $b^4a - 81a^5$

73. $4a^2b^2 + 12ab + 9$

74. $9c^2 + 6cd + d^2$

75. $a^4 + 8a^2 + 8a^3 + 64a$

76. $t^4 + 7t^2 - 3t^3 - 21t$

TW 77. Kelly factored $16 - 8x + x^2$ as $(x - 4)^2$, while Tony factored it as $(4 - x)^2$. Are they both correct? Why or why not?

TW 78. Describe in your own words a strategy for factoring polynomials.

Skill Maintenance

79. Show that the ordered pairs $(-1, 11)$, $(0, 7)$, and $(3, -5)$ are solutions of $y = -4x + 7$. [3.2]

Solve. [2.2]

80. $5x - 4 = 0$

81. $3x + 7 = 0$

82. $2x + 9 = 0$

83. $4x - 9 = 0$

84. Graph: $y = -\frac{1}{2}x + 4$. [3.2]

Synthesis

85. There are third-degree polynomials in x that we are not yet able to factor, despite the fact that they are not prime. Explain how such a polynomial could be created.

86. Describe a method that could be used to find a binomial of degree 16 that can be expressed as the product of prime binomial factors.

Factor.

87. $-(x^5 + 7x^3 - 18x)$ **88.** $18 + a^3 - 9a - 2a^2$

89. $-3a^4 + 15a^2 - 12$ **90.** $-x^4 + 7x^2 + 18$

Aha! **91.** $y^2(y + 1) - 4y(y + 1) - 21(y + 1)$

92. $y^2(y - 1) - 2y(y - 1) + (y - 1)$

93. $6(x - 1)^2 + 7y(x - 1) - 3y^2$

94. $(y + 4)^2 + 2x(y + 4) + x^2$

95. $2(a + 3)^4 - (a + 3)^3(b - 2) - (a + 3)^2(b - 2)^2$

96. $5(t - 1)^5 - 6(t - 1)^4(s - 1) + (t - 1)^3(s - 1)^2$

Collaborative Corner

Matching Factorizations*

Focus: Factoring

Time: 20 minutes

Group Size: Begin with the entire class. The end result is pairs of students. If there is an odd number of students, the instructor should participate.

Materials: Prepared sheets of paper, pins or tape. On half of the sheets, the instructor writes a polynomial. On the remaining sheets, the instructor writes the factorization of those polynomials. The activity is more interesting if the polynomials and factorizations are similar; for example,

$$x^2 - 2x - 8, \quad (x - 2)(x - 4),$$
$$x^2 - 6x + 8, \quad (x - 1)(x - 8),$$
$$x^2 - 9x + 8, \quad (x + 2)(x - 4).$$

ACTIVITY

1. As class members enter the room, the instructor pins or tapes either a polynomial or a factorization to the back of each student. Class members are told only whether their sheet of paper contains a polynomial or a factorization. All students should remain quiet and not tell others what is on their sheet of paper.

2. After all students are wearing a sheet of paper, they should mingle with one another, attempting to match up their factorization with the appropriate polynomial or vice versa. They may ask "yes/no" questions of one another that relate to factoring and polynomials. Answers to the questions should be yes or no. For example, a legitimate question might be "Is my last term negative?", "Do my factors have opposite signs?", or "Does my factorization include the factors of 6?"

3. The game is over when all factorization/polynomial pairs have "found" one another.

---

*Thanks to Jann MacInnes of Florida Community College at Jacksonville–Kent Campus for suggesting this activity.

6.7   Applications of Polynomial Equations

Problem Solving   ■   Fitting Polynomial Functions to Data

Polynomial equations and functions occur frequently in problem solving and in data analysis.

Problem Solving

Some problems can be translated to polynomial equations that we can now solve. The problem-solving process is the same as that used for other kinds of problems.

▎**EXAMPLE 1**   Prize Tee Shirts.   During intermission at sporting events, it has become common for team mascots to launch tightly rolled tee shirts into the stands. The height $h(t)$, in feet, of an airborne tee shirt t seconds after being launched can be approximated by

$$h(t) = -15t^2 + 75t + 10.$$

After peaking, a rolled-up tee shirt is caught by a fan 70 ft above ground level. How long was the tee shirt in the air?

SOLUTION

1. **Familiarize.** We make a drawing and label it, using the information provided. If we wanted to, we could evaluate $h(t)$ for a few values of t. Note that t cannot be negative, since it represents time from launch.

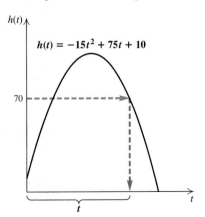

2. **Translate.** The relevant function has been provided. Since we are asked to determine how long it will take for the shirt to reach someone 70 ft above ground level, we are interested in the value of t for which $h(t) = 70$:

$$-15t^2 + 75t + 10 = 70.$$

3. **Carry out.** We solve both algebraically and graphically, as follows.

ALGEBRAIC APPROACH

We solve by factoring:

$$-15t^2 + 75t + 10 = 70$$
$$-15t^2 + 75t - 60 = 0 \qquad \text{Subtracting 70 from both sides}$$
$$\left.\begin{array}{l} -15(t^2 - 5t + 4) = 0 \\ -15(t - 4)(t - 1) = 0 \end{array}\right\} \quad \text{Factoring}$$
$$t - 4 = 0 \quad or \quad t - 1 = 0$$
$$t = 4 \quad or \quad t = 1.$$

The solutions of the equation are 4 and 1.

GRAPHICAL APPROACH

We graph $y_1 = -15x^2 + 75x + 10$ and $y_2 = 70$ and look for any points of intersection of the graphs.

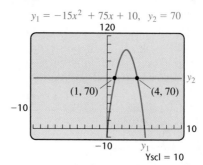

Using the INTERSECT option of the CALC menu, we see that the graphs intersect at $(1, 70)$ and $(4, 70)$. Thus the solutions of the equation are 1 and 4.

4. **Check.** We have

$$h(4) = -15 \cdot 4^2 + 75 \cdot 4 + 10 = -240 + 300 + 10 = 70 \text{ ft};$$
$$h(1) = -15 \cdot 1^2 + 75 \cdot 1 + 10 = -15 + 75 + 10 = 70 \text{ ft}.$$

Both 4 and 1 check. However, the problem states that the tee shirt is caught after peaking. Thus we reject 1 since that would indicate when the height of the tee shirt was 70 ft on the way *up*.

5. **State.** The tee shirt was in the air for 4 sec before being caught 70 ft above ground level.

The following problem involves the **Pythagorean theorem,** which relates the lengths of the sides of a right triangle. A **right triangle** has a 90°, or right, angle, which is denoted in the triangle by the symbol ⌐ or ⌐. The longest side, opposite the 90° angle, is called the **hypotenuse.** The other sides, called **legs,** form the two sides of the right angle.

The Pythagorean Theorem

In any right triangle, if a and b are the lengths of the legs and c is the length of the hypotenuse, then

$$a^2 + b^2 = c^2.$$

The symbol ⌐ denotes a 90° angle.

EXAMPLE 2   Carpentry.   In order to build a deck at a right angle to their house, Lucinda and Felipe decide to place a stake in the ground a precise distance from the back wall of their house. This stake will combine with two marks on the house to form a right triangle. From a course in geometry, Lucinda remembers that there are three consecutive integers that can work as sides of a right triangle. Find the measurements of that triangle.

SOLUTION

1. **Familiarize.** Recall that x, $x + 1$, and $x + 2$ can be used to represent three unknown consecutive integers. Since $x + 2$ is the largest number, it must represent the hypotenuse. The legs serve as the sides of the right angle, so one leg must be formed by the marks on the house. We make a drawing in which

   $x =$ the distance between the marks on the house,

   $x + 1 =$ the length of the other leg,

   and

   $x + 2 =$ the length of the hypotenuse.

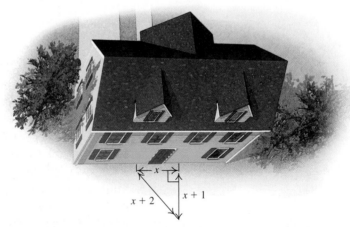

2. **Translate.** Applying the Pythagorean theorem, we translate as follows:

   $$a^2 + b^2 = c^2$$
   $$x^2 + (x + 1)^2 = (x + 2)^2.$$

3. **Carry out.** We solve both algebraically and graphically.

ALGEBRAIC APPROACH	GRAPHICAL APPROACH

ALGEBRAIC APPROACH

We solve the equation as follows:

$$x^2 + (x + 1)^2 = (x + 2)^2$$

$x^2 + (x^2 + 2x + 1) = x^2 + 4x + 4$ Squaring the binomials

$2x^2 + 2x + 1 = x^2 + 4x + 4$ Combining like terms

$x^2 - 2x - 3 = 0$ Subtracting $x^2 + 4x + 4$ from both sides

$(x - 3)(x + 1) = 0$ Factoring

$x - 3 = 0$ *or* $x + 1 = 0$

    Using the principle of zero products

$x = 3$ *or* $x = -1$.

GRAPHICAL APPROACH

We graph $y_1 = x^2 + (x + 1)^2$ and $y_2 = (x + 2)^2$ and look for the points of intersection of the graphs. A standard $[-10, 10, -10, 10]$ viewing window shows one point of intersection.

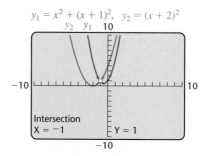

To see if the graphs intersect at another point, we use a larger viewing window. The window $[-10, 10, -10, 50]$ shows that there is indeed another point of intersection.

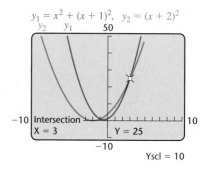

The points of intersection are $(-1, 1)$ and $(3, 25)$, so the solutions are -1 and 3.

4. **Check.** The integer -1 cannot be a length of a side because it is negative. For $x = 3$, we have $x + 1 = 4$, and $x + 2 = 5$. Since $3^2 + 4^2 = 5^2$, the lengths 3, 4, and 5 determine a right triangle. Thus, 3, 4, and 5 check.

5. **State.** Lucinda and Felipe should use a triangle with sides having a ratio of $3:4:5$. Thus, if the marks on the house are 3 yd apart, they should locate the stake at the point in the yard that is precisely 4 yd from one mark and 5 yd from the other mark.

EXAMPLE 3   Display of a Sports Card.   A valuable sports card is 4 cm wide and 5 cm long. The card is to be sandwiched by two pieces of Lucite, each of which is $5\frac{1}{2}$ times the area of the card. Determine the dimensions of the Lucite that will ensure a uniform border around the card.

SOLUTION

1. **Familiarize.** We make a drawing and label it, using x to represent the width of the border, in centimeters. Since the border extends uniformly around the entire card, the length of the Lucite must be $5 + 2x$ and the width must be $4 + 2x$.

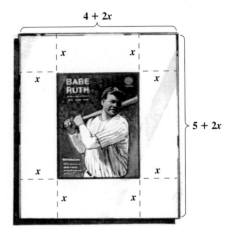

2. **Translate.** We rephrase the information given and translate as follows:

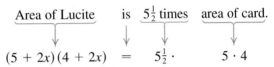

$$(5 + 2x)(4 + 2x) = 5\frac{1}{2} \cdot \quad 5 \cdot 4$$

3. **Carry out.** We solve both algebraically and graphically, as follows.

ALGEBRAIC APPROACH

We solve the equation:

$$(5 + 2x)(4 + 2x) = 5\frac{1}{2} \cdot 5 \cdot 4$$

$20 + 10x + 8x + 4x^2 = 110$	Multiplying
$4x^2 + 18x - 90 = 0$	Finding standard form
$2x^2 + 9x - 45 = 0$	Multiplying by $\frac{1}{2}$ on both sides
$(2x + 15)(x - 3) = 0$	Factoring
$2x + 15 = 0 \quad or \quad x - 3 = 0$	
	Principle of zero products
$x = -7\frac{1}{2} \quad or \quad x = 3.$	

GRAPHICAL APPROACH

We graph $y_1 = (5 + 2x)(4 + 2x)$ and $y_2 = 5.5(5)(4)$ and look for the points of intersection of the graphs. Since we have $y_2 = 110$, we use a viewing window with Ymax > 110.

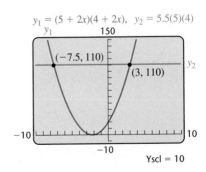

Using INTERSECT, we find the points of intersection $(-7.5, 110)$ and $(3, 110)$. The solutions are -7.5 and 3.

4. **Check.** We check 3 in the original problem. (Note that $-7\frac{1}{2}$ is not a solution because measurements cannot be negative.) If the border is 3 cm wide, the Lucite will have a length of $5 + 2 \cdot 3$, or 11 cm, and a width of $4 + 2 \cdot 3$, or 10 cm. The area of the Lucite is thus $11 \cdot 10$, or 110 cm². Since the area of the card is 20 cm² and 110 cm² is $5\frac{1}{2}$ times 20 cm², the number 3 checks.

5. **State.** Each piece of Lucite should be 11 cm long and 10 cm wide. ◢

Fitting Polynomial Functions to Data

In Chapter 3, we modeled data using linear functions, or polynomial functions of degree 1. Some data that are not linear can be modeled using polynomial functions with higher degrees.

When data are not linear, it can be difficult to choose what type of function would best model the data. Graphs of quadratic functions are *parabolas*. We will study these graphs in detail in Chapter 11. Data that, when graphed, follow one of these patterns may best be modeled with a quadratic function.

Graphs of quadratic equations

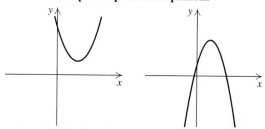

Graphs of cubic functions can follow any of the general shapes below. Graphs of polynomials of higher degree have even more possible shapes. Choosing a model may involve forming the model and noting how well it matches the data. Another consideration is whether estimates and predictions made using the model are reasonable. In this section, we will indicate what type of model to use for each set of data.

Graphs of cubic equations

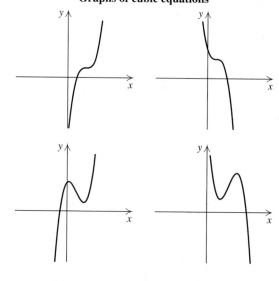

EXAMPLE 4   College Tuition.   The average in-state tuition for a public four-year college for various years is shown in the following table and graph.

a) Fit a quartic (degree 4) polynomial function to the data.

b) Use the function found in part (a) to estimate public-college tuition in 2004 and to predict public-college tuition in 2008.

c) In what year will public-college tuition be $12,000?

Year	In-State Public-College Tuition
1985	$1386
1988	1726
1991	2159
1994	2820
1997	3323
2000	3768
2003	4686
2005*	5948

*Preliminary data
Source: National Center for Education Statistics, *Digest of Education Statistics*

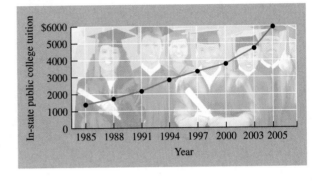

SOLUTION

a) We enter the data in a graphing calculator. To avoid large numbers, we let $x =$ the number of years since 1985. The years are entered as L1 and the tuition as L2.

L1	L2	L3	1
0	1386	------	
3	1726		
6	2159		
9	2820		
12	3323		
15	3768		
18	4686		

L1(7) = 18

QuarticReg
y = ax4 + bx3 + ... + e
a = .1326718311
b = −4.604061642
c = 52.1190821
d = −35.92556258
e = 1405.265136

From the graph, we see that the data are not linear. We fit a quartic function to the data, choosing the QuartReg option from the STAT CALC menu. Pressing **STAT** ⊳ **7** **VARS** ⊳ **1** **1** **ENTER** will calculate the regression equation and copy it to Y1. Rounding the coefficients, shown in the figure on the right above, to the nearest thousandth, we find that the quartic function that fits the data is

$$f(x) = 0.133x^4 - 4.604x^3 + 52.119x^2 - 35.926x + 1405.265.$$

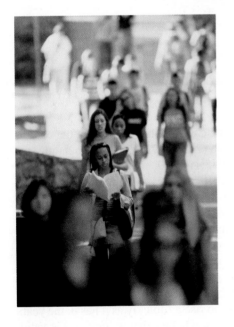

Finishing a Chapter

Try to be aware of when you near the end of a chapter. Sometimes the end of a chapter signals the arrival of a quiz or test. Almost always, the end of a chapter indicates the end of a particular area of study. Because future work will likely rely on your mastery of the chapter completed, make use of the chapter review and test to solidify your understanding before moving onward.

b) To estimate and predict tuition for years not given, we will use a table with Indpnt set to Ask. Since 2004 is 19 yr after 1985, and 2008 is 23 yr after 1985, we enter 19 and 23 for X in the table. On the basis of this model, we estimate that public-college tuition was $5248 in 2004 and will be $9259 in 2008.

X	Y₁	
19	5248.3	
23	9259.4	

c) To determine the year in which public-college tuition will be $12,000, we solve the equation $f(x) = 12,000$. Since $f(x)$ has been copied as Y₁, we let Y₂ = 12,000 and graph both equations. To see the scatterplot of the data as well, we turn on Plot1 and check that Xlist is L₁ and Ylist is L₂. We choose window settings that will show the given data as well as the intersection of the graphs of Y₁ and Y₂. For this situation, we use a setting of [0, 30, 0, 15000], with Xscl = 5 and Yscl = 1000.

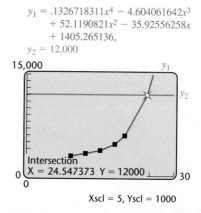

$$y_1 = .1326718311x^4 - 4.604061642x^3 + 52.1190821x^2 - 35.92556258x + 1405.265136,$$
$$y_2 = 12,000$$

Using the INTERSECT option of the CALC menu, we find that the coordinates of the point of intersection are (24.547373, 12,000). Rounding, we have $x = 25$. Since 25 yr after 1985 is 2010, we estimate that public-college tuition will be $12,000 in 2010.

When we use data to predict values not given, we assume that the trends shown by the data are consistent for values not shown. In general, estimating values between given data points (interpolation) is more reliable than predicting values beyond the last data point (extrapolation). However, the results of both interpolation and extrapolation depend on the assumptions we made.

In Example 4, we predicted that public-college tuition will be $12,000 in 2010. This assumes that tuition continues to rise as quickly as it has in the years before 2005. Before 2010, we cannot know whether that assumption is valid.

Many personal, business, and government decisions are made on the basis of predictions of the future. Mathematics and curve fitting can be helpful in making those predictions, but we must keep in mind the underlying assumptions and the knowledge that a prediction is not a fact.

A

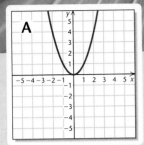

B

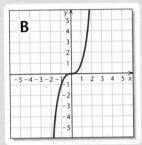

C

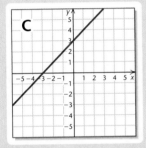

D

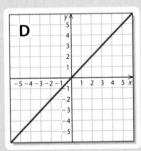

E

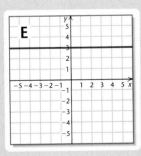

F

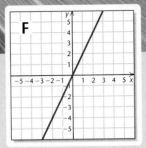

G

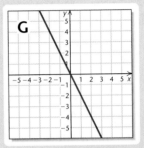

H

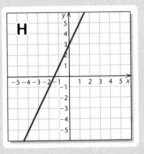

I

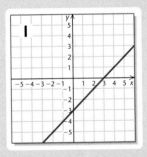

J

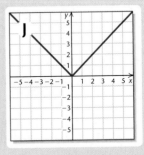

Visualizing the Graph

Match each equation with its graph.

1. $y = x$

2. $y = |x|$

3. $y = x^2$

4. $y = x^3$

5. $y = 3$

6. $y = x + 3$

7. $y = x - 3$

8. $y = 2x$

9. $y = -2x$

10. $y = 2x + 3$

Answers on page A-26

6.7 EXERCISE SET

Solve.

1. The square of a number plus the number is 132. What is the number?

2. The square of a number plus the number is 156. What is the number?

3. A photo is 5 cm longer than it is wide. Find the length and the width if the area is 84 cm².

4. An envelope is 4 cm longer than it is wide. The area is 96 cm². Find the length and the width.

5. *Geometry.* If each of the sides of a square is lengthened by 4 m, the area becomes 49 m². Find the length of a side of the original square.

6. *Geometry.* If each of the sides of a square is lengthened by 6 cm, the area becomes 144 cm². Find the length of a side of the original square.

7. *Framing a Picture.* A picture frame measures 12 cm by 20 cm, and 84 cm² of picture shows. Find the width of the frame.

8. *Framing a Picture.* A picture frame measures 14 cm by 20 cm, and 160 cm² of picture shows. Find the width of the frame.

9. *Landscaping.* A rectangular lawn measures 60 ft by 80 ft. Part of the lawn is torn up to install a sidewalk of uniform width around it. The area of the new lawn is 2400 ft². How wide is the sidewalk?

10. *Landscaping.* A rectangular garden is 30 ft by 40 ft. Part of the garden is removed in order to install a walkway of uniform width around it. The area of the new garden is one-half the area of the old garden. How wide is the walkway?

11. Three consecutive even integers are such that the square of the third is 76 more than the square of the second. Find the three integers.

12. Three consecutive even integers are such that the square of the first plus the square of the third is 136. Find the three integers.

13. *Tent Design.* The triangular entrance to a tent is 2 ft taller than it is wide. The area of the entrance is 12 ft². Find the height and the base.

Area = 12 ft²

14. *Antenna Wires.* A wire is stretched from the ground to the top of an antenna tower, as shown. The wire is 20 ft long. The height of the tower is 4 ft greater than the distance *d* from the tower's base to the bottom of the wire. Find the distance *d* and the height of the tower.

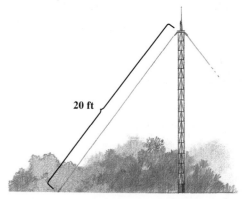

20 ft

15. *Sailing.* A triangular sail is 9 m taller than it is wide. The area is 56 m². Find the height and the base of the sail.

Area = 56 m²

16. *Parking Lot Design.* A rectangular parking lot is 50 ft longer than it is wide. Determine the dimensions of the parking lot if it measures 250 ft diagonally.

17. *Ladder Location.* The foot of an extension ladder is 9 ft from a wall. The height that the ladder reaches on the wall and the length of the ladder are consecutive integers. How long is the ladder?

18. *Ladder Location.* The foot of an extension ladder is 10 ft from a wall. The ladder is 2 ft longer than the height that it reaches on the wall. How far up the wall does the ladder reach?

19. *Garden Design.* Ignacio is planning a garden that is 25 m longer than it is wide. The garden will have an area of 7500 m². What will its dimensions be?

20. *Garden Design.* A flower bed is to be 3 m longer than it is wide. The flower bed will have an area of 108 m². What will its dimensions be?

21. *Cabinet Making.* Dovetail Woodworking determines that the revenue R, in thousands of dollars, from the sale of x sets of cabinets is given by $R(x) = 2x^2 + x$. If the cost C, in thousands of dollars, of producing x sets of cabinets is given by $C(x) = x^2 - 2x + 10$, how many sets must be produced and sold in order for the company to break even?

22. *Camcorder Production.* Suppose that the cost of making x video cameras is $C(x) = \frac{1}{9}x^2 + 2x + 1$, where $C(x)$ is in thousands of dollars. If the revenue from the sale of x video cameras is given by $R(x) = \frac{5}{36}x^2 + 2x$, where $R(x)$ is in thousands of dollars, how many cameras must be sold in order for the firm to break even?

23. *Prize Tee Shirts.* Using the model in Example 1, determine how long a tee shirt has been airborne if it is caught on the way *up* by a fan 100 ft above ground level.

24. *Prize Tee Shirts.* Using the model in Example 1, determine how long a tee shirt has been airborne if it is caught on the way *down* by a fan 10 ft above ground level.

25. *Fireworks Displays.* Fireworks are typically launched from a mortar with an upward velocity (initial speed) of about 64 ft/sec. The height $h(t)$, in feet, of a "weeping willow" display, t seconds after having been launched from an 80-ft-high rooftop, is given by

$$h(t) = -16t^2 + 64t + 80.$$

After how long will the cardboard shell from the fireworks reach the ground?

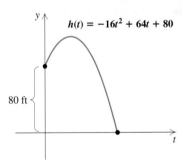

26. *Safety Flares.* Suppose that a flare is launched upward with an initial velocity of 80 ft/sec from a height of 224 ft. Its height in feet, $h(t)$, after t seconds is given by

$$h(t) = -16t^2 + 80t + 224.$$

After how long will the flare reach the ground?

27. *Home-healthcare.* The number of home-healthcare patients in the United States per 10,000 people, t years after 1992, can be approximated by $N(t) = -2t^2 + 15t + 51$ (*Source*: Based on data from Centers for Disease Control, 2003). How many years after 1992 were there about 69 home-healthcare patients per 10,000 people?

28. *Publishing.* The number of books B, in thousands, published in the United States t years after 2000 can be approximated by $B(t) = -8t^2 + 62t + 60$ (*Source*: Based on data from R. R. Bowker). How many years after 2000 were there about 170,000 books published in the United States?

29. *Baseball.* The following table lists the average value, in millions of dollars, of multiyear contracts for free-agent pitchers.

Year	Average Value of Multiyear Contract of Free-agent MLB Pitchers (in millions)
2001	$21.4
2002	17.8
2003	10.3
2004	12.1
2005	17.2
2006	22.6

Source: USA TODAY 12/7/05

a) Use regression to find a quadratic polynomial function C that can be used to estimate the average value of a multiyear contract x years after 2000.
b) Estimate the average value of a multiyear contract in 2008.
c) In what year after 2000 will the average value of a multiyear contract for a free-agent pitcher be $30 million?

30. *Olympics.* The following table lists the number of female athletes participating in the winter Olympic games for various years.

Year	Number of Female Athletes in the Winter Olympic Games
1924	11
1936	80
1952	109
1964	199
1976	231
1984	274
1994	522
2006	960

Source: Olympics.org

a) Use regression to find a cubic polynomial function F that can be used to estimate the number of female athletes competing in the winter Olympic games x years after 1900.
b) Estimate the number of female athletes competing in 2010.
c) In what year will 1200 female athletes compete in the winter Olympic games?

31. *Life Insurance Premiums.* The following table lists monthly life insurance premiums for a $500,000 policy for females of various ages.

Age	Monthly Premium for $500,000 Life Insurance Policy for Females
35	$14
40	18
45	24
50	33
55	48
60	70
65	110
70	188

Source: Insure.com

a) Use regression to find a cubic polynomial function L that can be used to estimate the monthly life insurance premium for a female of age x.
b) Estimate the monthly premium for a female of age 58.
c) For what age is the monthly premium $100?

32. *Health Insurance Premiums.*   The following table lists the percent of increase from the previous year in health insurance premiums in the United States.

Year	Percent of Increase in Health Insurance Premiums
1988	12.0%
1989	18.0
1990	14.0
1993	8.5
1996	0.8
1999	5.3
2000	8.2
2001	10.9
2003	13.9
2004	11.2

Source: Kaiser Family Foundation

a) Use regression to find a quartic polynomial function H that can be used to estimate the percent of increase in health insurance premiums x years after 1988.
b) Estimate the percent of increase in premiums in 2002.
c) In what year or years after 1988 is the percent of increase 5%?

33. *Biological Sciences.*   The following table lists the number of bachelor's degrees earned in the biological and biomedical sciences for various years.

Year	Number of Bachelor's Degrees Earned in Biological and Biomedical Sciences
1990	37,204
1992	42,781
1994	51,157
1996	60,750
1998	65,583
2000	63,005
2002	59,415
2004	61,509

Source: U.S. National Council for Education Statistics

a) Use regression to find a quartic polynomial function B that can be used to estimate the number of bachelor's degrees earned in the biological and biomedical sciences x years after 1990.
b) Estimate the number of degrees earned in 2006.
c) In what year or years were approximately 62,000 degrees earned?

34. *Engineering.*   The following table lists the number of bachelor's degrees earned in engineering and engineering technologies for various years.

Year	Number of Bachelor's Degrees Earned in Engineering and Engineering Technologies
1986	97,122
1990	82,480
1994	78,662
1998	74,649
2002	74,679
2004	78,227

Source: U.S. National Council for Education Statistics

a) Use regression to find a quadratic polynomial function E that can be used to estimate the number of bachelor's degrees earned in engineering and engineering technologies x years after 1980.
b) Estimate the number of degrees earned in 2010.
c) In what year after 1986 will aproximately 100,000 degrees be earned?

35. Tyler disregards any negative solutions that he finds when solving applied problems. Is this approach correct? Why or why not?

36. Write a chart of the population of two imaginary cities. Devise the numbers in such a way that one city has linear growth and the other has nonlinear growth.

Focused Review

Solve.

37. $3(x - 7) + 5x = 4x - (7 - x)$ [2.2]

38. $2x - 14 + 9x > -8x + 16 + 10x$ [2.6]

39. $3x^2 = 5x + 2$ [6.3]

40. $2x - y = 5,$
$y = x - 7$ [4.2]

41. $x^3 = 8$ [6.5]

42. $5x^2 + 10x + 5 = 0$ [6.4]

43. $2x + y = 5,$
$3x - 2y = 8$ [4.3]

44. $2x - 3y = 8,$
$x + 3y = 10$ [4.3]

45. $x^2 = \frac{1}{9}$ [6.4]

46. $x^2 - 6 = 5x$ [6.2]

Synthesis

TW 47. Explain how you might decide what kind of function would fit a particular set of data.

TW 48. Compare the data in Exercise 33 with that in Exercise 34. Describe the trends occurring between 1990 and 2004. How might this affect decisions made by college administrations?

49. *Box Construction.* A rectangular piece of tin is twice as long as it is wide. Squares 2 cm on a side are cut out of each corner, and the ends are turned up to make a box whose volume is 480 cm³. What are the dimensions of the piece of tin?

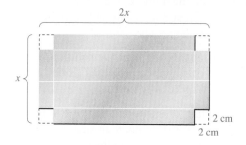

50. *Navigation.* A tugboat and a freighter leave the same port at the same time at right angles. The freighter travels 7 km/h slower than the tugboat. After 4 hr, they are 68 km apart. Find the speed of each boat.

51. *Skydiving.* During the first 13 sec of a jump, a skydiver falls approximately $11.12t^2$ feet in t seconds. A small heavy object (with less wind resistance) falls about $15.4t^2$ feet in t seconds. Suppose that a skydiver jumps from 30,000 ft, and 1 sec later a camera falls out of the airplane. How long will it take the camera to catch up to the skydiver?

52. *Olympics.* The following table lists the number of male athletes participating in the winter Olympic games for various years.

Year	Number of Male Athletes in the Winter Olympic Games
1924	247
1936	566
1952	585
1964	892
1976	892
1984	998
1994	1215
2006	1548

Source: Olympics.org

a) Use regression to find a linear function M that can be used to estimate the number of male athletes competing in the winter Olympic games x years after 1900.

b) Using the model found in part (a) and that found in Exercise 30, predict the first year in which there will be more female athletes than male athletes in the winter Olympic games.

6 Chapter Summary and Review

KEY TERMS AND DEFINITIONS

FACTORING

Factor completely, p. 454   To write a polynomial as a product in which no factor can itself be factored.

Prime polynomial, p. 454   A polynomial that cannot be factored.

SOLVING

Root, p. 449   A solution of an equation that is set equal to 0.

Zero, p. 449   An input for which the function output is 0.

Repeated root, p. 488   A root of an equation that appears in more than one factor. If the root is repeated twice, it is a **root of multiplicity two.**

IMPORTANT CONCEPTS

[Section references appear in brackets.]

Concept	Example
Some polynomials can be factored by **grouping.**	$\begin{aligned} x^3 - x^2 - 2x + 2 &= x^2(x - 1) - 2(x - 1) \\ &= (x - 1)(x^2 - 2) \end{aligned}$ [6.1]
Factoring by Reversing FOIL Form products of two binomials. The product of the First terms is ax^2. The product of the Last terms is c. Binomials with common factors can be eliminated. Check remaining products.	$\begin{aligned} 6x^2 + x - 2 & \\ = (3x + 2)(2x - 1) & \end{aligned}$ $\begin{array}{ll} \cancel{(6x + 2)(x - 1)} & \text{Common factor} \\ \cancel{(6x - 2)(x + 1)} & \text{Common factor} \\ (6x + 1)(x - 2) & 6x^2 - 11x - 2 \\ (6x - 1)(x + 2) & 6x^2 + 11x - 2 \end{array}$ Correct product $\longrightarrow$ $\begin{array}{ll} (3x + 2)(2x - 1) & 6x^2 + x - 2 \\ (3x - 2)(2x + 1) & 6x^2 - x - 2 \\ \cancel{(3x + 1)(2x - 2)} & \text{Common factor} \\ \cancel{(3x - 1)(2x + 2)} & \text{Common factor [6.3]} \end{array}$
Factoring Using the *ac*-Method Multiply a and c. Factor ac so that the sum of the factors is b. Split the middle term. Factor by grouping.	$\begin{aligned} 15x^2 - 2x - 8 & \\ = 15x^2 - 12x + 10x - 8 & \\ = 3x(5x - 4) + 2(5x - 4) & \\ = (5x - 4)(3x + 2) & \end{aligned}$ $\begin{aligned} 15 \cdot (-8) &= -120 \\ -120 &= 10 \cdot (-12) \\ 10 + (-12) &= -2 \\ & \quad [6.3] \end{aligned}$

(continued)

Factoring Formulas

Trinomial Square:

$$A^2 + 2AB + B^2 = (A + B)^2$$
$$A^2 - 2AB + B^2 = (A - B)^2$$

$$x^2 + 20x + 100 = (x + 10)^2$$
$$4y^2z^2 - 4yz + 1 = (2yz - 1)^2$$

Difference of Two Squares:

$$A^2 - B^2 = (A + B)(A - B)$$

$$x^2 - 81 = (x + 9)(x - 9)$$

Sum of Two Cubes:

$$A^3 + B^3 = (A + B)(A^2 - AB + B^2)$$

$$y^3 + 27 = (y + 3)(y^2 - 3y + 9)$$

Difference of Two Cubes:

$$A^3 - B^3 = (A - B)(A^2 + AB + B^2)$$

$$1000 - t^3 = (10 - t)(100 + 10t + t^2)$$ [6.4], [6.5]

The Principle of Zero Products

For any real numbers a and b:

If $ab = 0$, then $a = 0$ or $b = 0$.
If $a = 0$ or $b = 0$, then $ab = 0$.

$$6x^2 + x = 2$$
$$6x^2 + x - 2 = 0 \qquad \text{Writing with 0 on one side}$$
$$(3x + 2)(2x - 1) = 0 \qquad \text{Factoring}$$
$$3x + 2 = 0 \quad or \quad 2x - 1 = 0 \qquad \text{Principle of zero products}$$
$$3x = -2 \quad or \quad 2x = 1$$
$$x = -\tfrac{2}{3} \quad or \quad x = \tfrac{1}{2}$$

The solutions are $-\tfrac{2}{3}$ and $\tfrac{1}{2}$. [6.3]

Pythagorean Theorem

In any right triangle, if a and b are the lengths of the legs and c is the length of the hypotenuse, then $a^2 + b^2 = c^2$.

For the right triangle shown,

$$x^2 + (x + 1)^2 = 5^2.$$

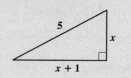

[6.7]

The **roots** of an equation, the **zeros** of a function, and the **x-intercepts** of the graph of the function are related.

The roots of $x^2 - 4 = 0$ are -2 and 2.
The zeros of $f(x) = x^2 - 4$ are -2 and 2.
The x-intercepts of the graph of $f(x) = x^2 - 4$ are $(-2, 0)$ and $(2, 0)$.

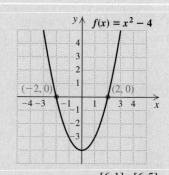

[6.1]–[6.5]

Review Exercises

Concept Reinforcement *Classify each statement as either true or false.*

1. Every polynomial has a common factor other than 1 or −1. [6.1]

2. A prime polynomial has no common factor other than 1 or −1. [6.1]

3. Every perfect-square trinomial can be expressed as a binomial squared. [6.4]

4. Every binomial can be regarded as a difference of squares. [6.4]

5. In a right triangle, the hypotenuse is always longer than either leg. [6.7]

6. Every quadratic equation has two different solutions. [6.1]

7. The principle of zero products can be applied whenever a product equals 0. [6.1]

8. The Pythagorean theorem can be applied to any triangle that has an angle measuring at least 90°. [6.7]

Factor.

9. $7x^2 + 6x$ [6.1]

10. $9y^4 - 3y^2$ [6.1]

11. $15x^4 - 18x^3 + 21x^2 - 9x$ [6.1]

12. $a^2 - 12a + 27$ [6.2]

13. $3m^2 + 14m + 8$ [6.3]

14. $25x^2 + 20x + 4$ [6.4]

15. $4y^2 - 16$ [6.4]

16. $5x^2 + x^3 - 14x$ [6.2]

17. $ax + 2bx - ay - 2by$ [6.1]

18. $3y^3 + 6y^2 - 5y - 10$ [6.1]

19. $a^4 - 81$ [6.4]

20. $4x^4 + 4x^2 + 20$ [6.1]

21. $27x^3 - 8$ [6.5]

22. $0.064b^3 - 0.125c^3$ [6.5]

23. $y^5 + y$ [6.1]

24. $2z^8 - 16z^6$ [6.1]

25. $54x^6y - 2y$ [6.5]

26. $36x^2 - 120x + 100$ [6.4]

27. $6t^2 + 17pt + 5p^2$ [6.3]

28. $x^3 + 2x^2 - 9x - 18$ [6.4]

29. $a^2 - 2ab + b^2 - 4t^2$ [6.4]

30. The following graph is that of a polynomial function $p(x)$. Use the graph to find the zeros of p. [6.1]

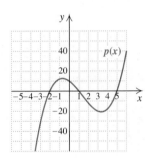

31. Find the zeros of the function given by
$$f(x) = x^2 - 11x + 28.\ [6.2]$$

Solve. Where appropriate, round answers to the nearest thousandth.

32. $x^2 - 20x = -100$ [6.4]

33. $6b^2 - 13b + 6 = 0$ [6.3]

34. $8t^2 = 14t$ [6.1]

35. $r^2 = 16$ [6.4]

36. $a^3 = 4a^2 + 21a$ [6.2]

37. $(x - 1)(x - 4) = 10$ [6.2]

38. $x^3 - 5x^2 - 16x + 80 = 0$ [6.4]

39. $x^2 + 180 = 27x$ [6.2]

40. $x^2 - 2x = 6$ [6.2]

41. Let $f(x) = x^2 - 7x - 40$. Find a such that $f(a) = 4$. [6.3]

42. Find the domain of the function f given by

$$f(x) = \frac{x - 3}{3x^2 + 19x - 14}.\ [6.3]$$

43. The area of a square is 5 more than four times the length of a side. What is the length of a side of the square? [6.7]

44. The sum of the squares of three consecutive odd numbers is 83. Find the numbers. [6.7]

45. A photograph is 3 in. longer than it is wide. When a 2-in. border is placed around the photograph, the total area of the photograph and the border is 108 in². Find the dimensions of the photograph. [6.7]

46. Tim is designing a rectangular garden with a width of 8 ft. The path that leads diagonally across the garden is 2 ft longer than the length of the garden. How long is the path? [6.7]

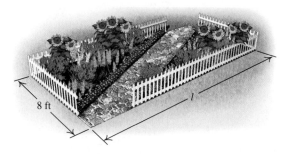

8 ft

l

47. *Labor Force.* The following table lists the percent of the U.S. population age 65 and older participating in the labor force. [6.7]

Year	Labor Force Participation Rate, 65 and Older
1970	25.2
1980	16.8
1990	15.7
2000	17.3
2004	20.3

Source: U.S. Bureau of Labor Statistics

a) Use regression to find a quadratic polynomial function P that can be used to estimate the labor participation rate for those 65 and older x years after 1970.

b) Estimate the participation rate in 2010.

c) In what year or years is the participation rate 24%?

Synthesis

TW 48. Explain how to find the roots of a polynomial function from its graph. [6.1]

TW 49. Explain in your own words why there must be a 0 on one side of an equation before you can use the principle of zero products. [6.1]

Factor. [6.5]

50. $128x^6 - 2y^6$

51. $(x - 1)^3 - (x + 1)^3$

Multiply.

52. $[a - (b - 1)][(b - 1)^2 + a(b - 1) + a^2]$
[5.5], [6.5]

53. Solve: $(x + 1)^3 = x^2(x + 1)$. [6.4]

Chapter Test 6

Factor.

1. $15x^2 - 5x^4$

2. $y^3 + 5y^2 - 4y - 20$

3. $p^2 - 12p - 28$

4. $12m^2 + 20m + 3$

5. $9y^2 - 25$

6. $3r^3 - 3$

7. $9x^2 + 25 - 30x$

8. $x^8 - y^8$

9. $y^2 + 8y + 16 - 100t^2$

10. $20a^2 - 5b^2$

11. $24x^2 - 46x + 10$

12. $16a^7b + 54ab^7$

13. $4y^4x + 36yx^2 + 8y^2x^3 - 16xy$

14. The graph shown below is that of the polynomial function $p(x)$. Use the graph to determine the zeros of p.

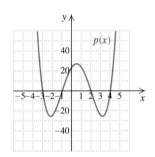

15. Find the zeros of the function given by
$f(x) = 2x^2 - 11x - 40$.

Solve. Where appropriate, round solutions to the nearest thousandth.

16. $x^2 - 18 = 3x$

17. $5t^2 = 125$

18. $2x^2 + 21 = -17x$

19. $9x^2 + 3x = 0$

20. $x^2 + 81 = 18x$

21. $x^2(x + 1) = 8x$

22. Let $f(x) = 3x^2 - 15x + 11$. Find a such that $f(a) = 11$.

23. Find the domain of the function f given by
$$f(x) = \frac{3 - x}{x^2 + 2x + 1}.$$

24. A photograph is 3 cm longer than it is wide. Its area is 40 cm². Find its length and its width.

25. To celebrate a town's centennial, fireworks are launched over a lake off a dam 36 ft above the water. The height of a display, t seconds after it has been launched, is given by
$$h(t) = -16t^2 + 64t + 36.$$

After how long will the shell from the fireworks reach the water?

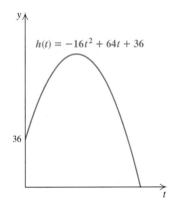

26. *IRS Revenue.* The following table lists the amount of IRS enforcement revenue, in billions of dollars, for various years.

Year	IRS Enforcement Revenue (in billions)
2001	$33.8
2002	34.1
2003	37.6
2004	43.1
2005	47.3

Source: INC. Magazine, March 2006

a) Use regression to find a quadratic polynomial function E that can be used to estimate the IRS enforcement revenue x years after 2000.
b) Estimate the enforcement revenue in 2009.
c) In what year will the enforcement revenue be $100 billion?

Synthesis

27. a) Multiply: $(x^2 + x + 1)(x^3 - x^2 + 1)$.
   b) Factor: $x^5 + x + 1$.

28. Factor: $6x^{2n} - 7x^n - 20$.

1-6 Cumulative Review

1. Employer surveys indicate that salary raises are based on employee performance. A high-performing employee who started at $50,000 in 2001 was earning $72,078 by 2005. This was approximately 25% more than an average employee starting at the same salary was earning. (*Source*: Hewitt Associates) How much was an average-performing employee who started at $50,000 in 2001 earning by 2005? [2.4]

2. The Greene County Art Museum has budgeted $250 for engraved invitations to a banquet. The invitations cost $85 for the first 20, and $1.08 for each additional invitation. How many invitations can the museum order and stay within its budget? [2.7]

3. Wilt Chamberlain once scored 100 points, setting a record for points scored in an NBA game. Chamberlain took only two-point shots and (one-point) foul shots and made a total of 64 shots. How many shots of each type did he make? [4.4]

4. In 2006, the Diabetic Express charged $80.86 for a vial of Humalog insulin and $83.70 for a vial of Novolog insulin. If a total of $4125.36 was collected for 50 vials of insulin, how many vials of each type were sold? [4.4]

5. If the sides of a square are increased by 2 ft, the area of the original square plus the area of the enlarged square is 452 ft². Find the length of a side of the original square. [6.7]

6. The sum of two consecutive even integers is -554. Find the integers. [2.5]

7. Emily and Andrew co-wrote a book of advice for new business owners. They agreed that Emily should get twice as much in royalty income as Andrew. If royalties for the first year are $1260, how much should each person receive? [2.5]

Determine whether a linear function could be used to model each set of data. [3.7]

8.

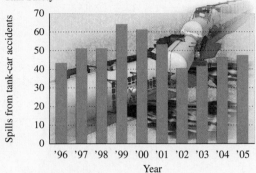

Rail Safety

Source: *The Wall Street Journal*, 9/19/06

9.

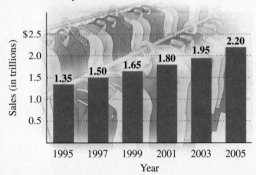

Retail Industry

Source: National Retail Federation

10. A large truck traveling 55 mph gets 7.6 miles per gallon of fuel. The same truck traveling 70 mph gets 6.1 miles per gallon. (*Source*: Kenworth Truck Co.)

a) Graph the data and determine an equation that describes the miles per gallon y in terms of the speed x. [3.7]

b) Use the equation to estimate the miles per gallon for a truck traveling 60 mph. [3.7]

11. One of the sets of data in Exercises 8 and 9 is linear. Use linear regression to find a function that can be used to model the data. [3.7]

Evaluate.

12. $3m - 4n + 6/n$, for $m = 4$ and $n = -2$ [1.8]

13. $f(-2)$, for $f(x) = -0.1x^3 - 3x^2 + 10x + 0.5$ [3.8]

Simplify.

14. $-4[2(x - 5) - 3]$ [1.8]

15. $2\left[4\left(x - \frac{1}{2}\right) - 5\left(x - \frac{1}{5}\right) - 3(2 - x)\right] - 5x + 7$ [1.8]

16. $-\frac{1}{2} + \frac{3}{8} + (-6) + \frac{3}{4}$ [1.5]

17. $-\frac{72}{108} \div \left(-\frac{2}{3}\right)$ [1.7]

Solve.

18. $5(x - 2) = 40$ [2.2]

19. $49 = x^2$ [6.4]

20. $-4x < -18$ [2.6]

21. $5x + 7 = -3x - 9$ [2.2]

22. $x^2 + 11x + 10 = 0$ [6.2]

23. $\frac{1}{3}x - \frac{2}{9} = \frac{2}{3} + \frac{4}{9}x$ [2.2]

24. $3x + 2y = 5,$
$x - 3y = 1$ [4.2]

25. $2x^2 + 7x = 4$ [6.3]

26. $4(x + 7) < 5(x - 3)$ [2.6]

27. $2y = 4x - 3,$
$4x = 1 - 5y$ [4.3]

28. $(2x + 7)(x - 5) = 0$ [6.1]

29. Solve $3a - b + 9 = c$, for b. [2.3]

30. Solve $\frac{3}{4}(x + 2y) = z$, for y. [2.3]

Combine like terms.

31. $x + 2y - 2z + \frac{1}{2}x - z$ [1.6]

32. $2x^3 - 7 + \frac{3}{7}x^2 - 6x^3 - \frac{4}{7}x^2 + 5$ [1.6]

Graph by hand on a plane.

33. $y = 1 - \frac{1}{2}x$ [3.6]

34. $x = -3$ [3.3]

35. $x - 6y = 6$ [3.3]

36. $y = 6$ [3.3]

37. Graph using a graphing calculator: $y = x^2 - 4$. [3.2]

38. Find the slope of the line containing the points $(1, 5)$ and $(2, 3)$. [3.5]

39. Determine the slope and the *y*-intercept of the graph of $3x + 4y = 8$. [3.6]

40. Find a linear equation whose graph passes through $(2, 1)$ and $(4, -2)$. [3.7]

41. Find an equation for the line with slope $\frac{1}{2}$ that contains $(0, -7)$. [3.6]

42. Find an equation for the line that is perpendicular to $y = x + 3$ and has the same *y*-intercept. [3.6]

Simplify.

43. $\dfrac{x^{-5}}{x^{-3}}$ [5.2]

44. $y^2 \cdot y^{-10}$ [5.2]

45. $-(2a^2b^7)^2$ [5.1]

46. Subtract:
$$(-8y^2 - y + 2) - (y^3 - 6y^2 + y - 5). \ [5.4]$$

Multiply.

47. $5(3a - 2b + c)$ [5.7]

48. $(2x^2 - 1)(x^3 + x - 3)$ [5.5]

49. $(6x - 5y)^2$ [5.7]

50. $(x + 3)(2x - 7)$ [5.6]

51. $(2x^3 + 1)(2x^3 - 1)$ [5.6]

Divide. [5.8]

52. $\dfrac{15x^4 - 12x^3 + 6x^2 + 2x + 18}{3x^2}$

53. $(x^4 + 2x^3 + 6x^2 + 2x + 18) \div (x + 3)$

Factor.

54. $6x - 2x^2 - 24x^4$ [6.1]

55. $16x^2 - 81$ [6.4]

56. $t^2 - 10t + 24$ [6.2]

57. $8x^2 + 10x + 3$ [6.3]

58. $6x^2 - 28x + 16$ [6.3]

59. $2x^3 + 250$ [6.5]

60. $16x^2 + 40x + 25$ [6.4]

61. $3x^2 + 10x - 8$ [6.3]

62. $x^4 + 2x^3 - 3x - 6$ [6.1]

Estimate the domain and the range of each function from its graph. [3.8]

63.

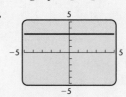

64.

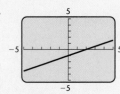

65.

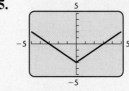

66.

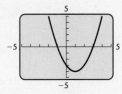

Synthesis

67. Simplify: $(x + 7)(x - 4) - (x + 8)(x - 5)$. [5.4], [5.5]

68. Multiply: $[4y^3 - (y^2 - 3)][4y^3 + (y^2 - 3)]$. [5.6]

69. Factor: $2a^{32} - 13{,}122b^{40}$. [6.4]

70. Solve: $x(x^2 + 3x - 28) - 12(x^2 + 3x - 28) = 0$. [6.1], [6.2]

71. Simplify: $-\left|0.875 - \left(-\frac{1}{8}\right) - 8\right|$. [1.4], [1.6]

72. Find all roots for $f(x) = x^4 - 34x^2 + 225$. [6.2], [6.4]

7

Rational Expressions, Equations, and Functions

Like fractions in arithmetic, a rational expression is an expression that indicates division. In this chapter, we add, subtract, multiply, and divide rational expressions and use them in equations and functions. We then use rational expressions to solve problems that we could not have solved before.

APPLICATION *Commuter Travel.*

One factor influencing urban planning is VMT, or vehicle miles traveled. Areas with a high population density have fewer VMT per household than those with a lower density. This occurs in part because employment and shopping are available closer to higher density areas. The following table lists annual VMT per household for various densities for a typical urban area. Find an equation of variation that describes the data.

POPULATION DENSITY	ANNUAL VMT
25	12,000
50	6,000
100	3,000
200	1,500

Source: Based on information from http://www.sflcv.org/density

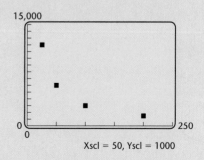

15,000

0

0   250

Xscl = 50, Yscl = 1000

This problem appears as Example 10 in Section 7.8.

7.1 Rational Expressions and Functions

Rational Functions ∎ Simplifying Rational Expressions
and Functions ∎ Factors That Are Opposites ∎
Vertical Asymptotes

An expression that consists of a polynomial divided by a nonzero polynomial is called a **rational expression.** The following are examples of rational expressions:

$$\frac{3}{4}, \quad \frac{x}{y}, \quad \frac{9}{a+b}, \quad \frac{x^2 + 7xy - 4}{x^3 - y^3}, \quad \frac{1 + z^3}{1 - z^6}.$$

Rational Functions

Like polynomials, certain rational expressions are used to describe functions. Such functions are called **rational functions.**

Following is the graph of the rational function

$$f(x) = \frac{1}{x}.$$

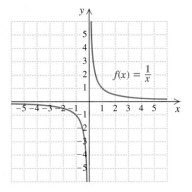

Graphs of rational functions vary widely in shape, but some general statements can be drawn from the graph shown above.

1. Graphs of rational functions may not be continuous—that is, there may be a break in the graph. The graph of $f(x) = 1/x$ consists of two unconnected parts, with a break at $x = 0$.
2. The domain of a rational function may not include all real numbers. The domain of $f(x) = 1/x$ does not include 0.

Although we will not consider graphs of rational functions in detail in this course, you should add the graph of $f(x) = 1/x$ to your library of functions.

EXAMPLE 1   The function given by

$$H(t) = \frac{t^2 + 5t}{2t + 5}$$

gives the time, in hours, for two machines, working together, to complete a job that the first machine could do alone in t hours and the other machine could do in $t + 5$ hours. How long will the two machines, working together, require for the job if the first machine alone would take **(a)** 1 hour? **(b)** 7 hours?

SOLUTION

a) $H(1) = \dfrac{1^2 + 5 \cdot 1}{2 \cdot 1 + 5} = \dfrac{1 + 5}{2 + 5} = \dfrac{6}{7}$ hr

b) $H(7) = \dfrac{7^2 + 5 \cdot 7}{2 \cdot 7 + 5} = \dfrac{49 + 35}{14 + 5} = \dfrac{84}{19}$, or $4\frac{8}{19}$ hr

$y_1 = (x^2 + 5x)/(2x + 5)$

X	Y₁	
−2.5	ERR:	

In Section 6.3, we found that the domain of a rational function must exclude any numbers for which the denominator is 0. For a function like H above, the denominator is 0 when t is $-\frac{5}{2}$, so the domain of H is $\{t \mid t$ is a real number *and* $t \neq -\frac{5}{2}\}$. Note from the table at left that H is not defined when $t = -\frac{5}{2}$.

EXAMPLE 2   Find all numbers for which the rational expression

$$\frac{x + 4}{x^2 - 3x - 10}$$

is undefined.

SOLUTION   The value of the numerator has no bearing on whether or not a rational expression is defined. To determine which numbers make the rational expression undefined, we set the *denominator* equal to 0 and solve:

$y_1 = (x + 4)/(x^2 - 3x - 10)$

X	Y₁	
5	ERR:	
−2	ERR:	

$$x^2 - 3x - 10 = 0$$

$(x - 5)(x + 2) = 0$ **Factoring**

$x - 5 = 0$ *or* $x + 2 = 0$ **Using the principle of zero products**

$x = 5$ *or* $x = -2.$ **Solving each equation**

To check, we let $y_1 = (x + 4)/(x^2 - 3x - 10)$ and evaluate y_1 for $x = 5$ and for $x = -2$. The table at left confirms that the expression is undefined when $x = 5$ and when $x = -2$.

In Example 2, we found that the expression

$$\frac{x + 4}{x^2 - 3x - 10}$$

is undefined when $x = 5$ and when $x = -2$. Thus the expression *is* defined for all other values of x. This tells us that the domain of the function given by

$$f(x) = \frac{x + 4}{x^2 - 3x - 10}$$

is $\{x \mid x$ is a real number *and* $x \neq -2$ *and* $x \neq 5\}$.

Simplifying Rational Expressions and Functions

Simplifying rational expressions is similar to simplifying the fraction expressions studied in Section 1.3. We saw, for example, that an expression like $\frac{15}{40}$ can be simplified as follows:

$$\frac{15}{40} = \frac{3 \cdot 5}{8 \cdot 5} \quad \text{Factoring the numerator and the denominator.}$$
$$\text{Note the common factor, 5.}$$

$$= \frac{3}{8} \cdot \frac{5}{5} \quad \text{Rewriting as a product of two fractions}$$

$$= \frac{3}{8} \cdot 1 \quad \frac{5}{5} = 1$$

$$= \frac{3}{8}. \quad \text{Using the identity property of 1 to remove the factor 1}$$

Similar steps are followed when simplifying rational expressions: We factor and remove a factor equal to 1, using the fact that

$$\frac{ab}{cb} = \frac{a}{c} \cdot \frac{b}{b}.$$

Student Notes

When using a graphing calculator or tutorial software, use parentheses around the numerator and around the denominator of rational expressions. Note that

$$5/3x \quad \text{means} \quad \frac{5}{3}x$$

and

$$5/(3x) \quad \text{means} \quad \frac{5}{3x}.$$

EXAMPLE 3 Simplify: $\dfrac{8x^2}{24x}$.

SOLUTION

$$\frac{8x^2}{24x} = \frac{8 \cdot x \cdot x}{3 \cdot 8 \cdot x} \quad \text{Factoring the numerator and the denominator.}$$
$$\text{Note the common factor of } 8 \cdot x.$$

$$= \frac{x}{3} \cdot \frac{8x}{8x} \quad \text{Rewriting as a product of two rational expressions}$$

$$= \frac{x}{3} \cdot 1 \quad \frac{8x}{8x} = 1$$

$$= \frac{x}{3} \quad \text{Removing the factor 1}$$

EXAMPLE 4   Simplify: $\dfrac{7a^2 + 21a}{14a}$.

SOLUTION   We first factor the numerator and the denominator, looking for the largest factor common to both. Once the greatest common factor is found, we use it to write 1 and simplify:

$$\frac{7a^2 + 21a}{14a} = \frac{7a(a + 3)}{7 \cdot 2 \cdot a} \qquad \text{Factoring. The greatest common factor is } 7a.$$

$$= \frac{7a}{7a} \cdot \frac{a + 3}{2} \qquad \text{Rewriting as a product of two rational expressions}$$

$$= 1 \cdot \frac{a + 3}{2} \qquad \frac{7a}{7a} = 1; \text{ try to do this step mentally.}$$

$$= \frac{a + 3}{2}. \qquad \text{Removing the factor 1}$$

As a partial check, we let $y_1 = (7x^2 + 21x)/(14x)$ and $y_2 = (x + 3)/2$ and compare values in a table. Note that the y-values are the same for any given x-value except 0. We exclude 0 because although $(x + 3)/2$ is defined when $x = 0$, $(7x^2 + 21x)/(14x)$ is not. Both expressions represent the same value when x is replaced with a number that can be used in *either* expression, so they are equivalent.

X	Y₁	Y₂
−3	0	0
−2	.5	.5
−1	1	1
0	ERROR	1.5
1	2	2
2	2.5	2.5
3	3	3

X = −3

A rational expression is said to be **simplified** when no factors equal to 1 can be removed. If the largest common factor is not found, simplifying may require two or more steps. For example, suppose we remove 7/7 instead of $(7a)/(7a)$ in Example 4. We would then have

$$\frac{7a^2 + 21a}{14a} = \frac{7(a^2 + 3a)}{7 \cdot 2a}$$

$$= \frac{a^2 + 3a}{2a}. \qquad \left.\right\} \quad \text{Removing a factor equal to 1: } \frac{7}{7} = 1$$

Here, since 7 is not the *greatest* common factor, we need to simplify further:

$$\frac{a^2 + 3a}{2a} = \frac{a(a + 3)}{a \cdot 2}$$

$$= \frac{a + 3}{2}. \qquad \left.\right\} \quad \text{Removing another factor equal to 1: } a/a = 1. \\ \text{The rational expression is now simplified.}$$

Sometimes the common factor has two or more terms.

EXAMPLE 5   Simplify.

a) $\dfrac{6x + 12}{7x + 14}$

b) $\dfrac{x^2 - 1}{x^2 + 3x + 2}$

SOLUTION

a) $\dfrac{6x + 12}{7x + 14} = \dfrac{6(x + 2)}{7(x + 2)}$ Factoring the numerator and the denominator

$= \dfrac{6}{7} \cdot \dfrac{x + 2}{x + 2}$ Rewriting as a product of two rational expressions

$= \dfrac{6}{7} \cdot 1$ $\dfrac{x + 2}{x + 2} = 1$

$= \dfrac{6}{7}$ Removing the factor 1

b) $\dfrac{x^2 - 1}{x^2 + 3x + 2} = \dfrac{(x + 1)(x - 1)}{(x + 1)(x + 2)}$ Factoring

$= \dfrac{x + 1}{x + 1} \cdot \dfrac{x - 1}{x + 2}$ Rewriting as a product of two rational expressions

$= 1 \cdot \dfrac{x - 1}{x + 2}$ $\dfrac{x + 1}{x + 1} = 1$

$= \dfrac{x - 1}{x + 2}$ Removing the factor 1

Canceling

Canceling is a shortcut that can be used—and easily *misused*—when we are working with rational expressions. As we stated in Section 1.3, canceling must be done with care and understanding. Essentially, canceling streamlines the steps in which we remove a factor equal to 1. Example 5(b) could have been streamlined as follows:

$\dfrac{x^2 - 1}{x^2 + 3x + 2} = \dfrac{(x + 1)(x - 1)}{(x + 1)(x + 2)}$ When a factor that equals 1 is noted, it is "canceled": $\dfrac{x + 1}{x + 1} = 1.$

$= \dfrac{x - 1}{x + 2}.$ Simplifying

CAUTION!   Canceling is often performed incorrectly:

$\dfrac{\cancel{x} + 3}{\cancel{x}} = 3, \quad \dfrac{\cancel{4}x + 3}{\cancel{2}} = 2x + 3, \quad \dfrac{\cancel{5}}{\cancel{5} + x} = \dfrac{1}{x}$ To check that these are not equivalent, substitute a number for x or compare graphs.

Incorrect!   Incorrect!   Incorrect!

In each situation, the expressions canceled are *not* factors that equal 1. Factors are parts of products. For example, in $x \cdot 3$, x and 3 are factors, but in $x + 3$, x and 3 are *not* factors, but terms. If you can't factor, you can't cancel! If in doubt, don't cancel!

EXAMPLE 6   Simplify: $\dfrac{3x^2 - 2x - 1}{x^2 - 3x + 2}$.

SOLUTION   We factor the numerator and the denominator and look for common factors:

$$\frac{3x^2 - 2x - 1}{x^2 - 3x + 2} = \frac{(3x + 1)(x - 1)}{(x - 2)(x - 1)}$$

$$= \frac{3x + 1}{x - 2}.$$

> **Try to visualize this as**
> $$\frac{3x + 1}{x - 2} \cdot \frac{x - 1}{x - 1}.$$
> **Removing a factor equal to 1:**
> $$\frac{x - 1}{x - 1} = 1$$

In Example 4, we found that

$$\frac{7x^2 + 21x}{14x} = \frac{x + 3}{2}.$$

These expressions are equivalent; their values are equal for every value of x except $x = 0$, for which the expression on the left is not defined. Special care must be taken, however, when simplifying a rational expression within a function. If

$$f(x) = \frac{7x^2 + 21x}{14x} \quad \text{and} \quad g(x) = \frac{x + 3}{2},$$

we cannot say that $f = g$, because the domains of the functions are not the same. If it is necessary to simplify a rational expression that defines a function, the domain must be carefully specified.

EXAMPLE 7   Given the function $f(x) = \dfrac{x^2 - 4}{2x^2 - 3x - 2}$.

a) Find the domain of f.

b) Simplify the rational expression.

c) Write the simplified form of the function, specifying any restrictions on the domain.

SOLUTION

a) To find the domain of f, we set the denominator equal to 0 and solve:

$$2x^2 - 3x - 2 = 0$$

$$(2x + 1)(x - 2) = 0 \qquad \text{Factoring the denominator}$$

$$2x + 1 = 0 \quad or \quad x - 2 = 0 \qquad \text{Using the principle of zero products}$$

$$x = -\tfrac{1}{2} \quad or \qquad x = 2.$$

The numbers $-\tfrac{1}{2}$ and 2 are not in the domain of the function. Thus the domain of f is $\left\{ x \mid x \text{ is a real number } and\ x \neq -\tfrac{1}{2} \text{ and } x \neq 2 \right\}$.

b) We have

$$\frac{x^2 - 4}{2x^2 - 3x - 2} = \frac{(x - 2)(x + 2)}{(2x + 1)(x - 2)}$$ Factoring the numerator and the denominator

$$= \frac{x - 2}{x - 2} \cdot \frac{x + 2}{2x + 1}$$ Factoring the rational expression; $\frac{x - 2}{x - 2} = 1$

$$= \frac{x + 2}{2x + 1}.$$ Removing a factor equal to 1

The simplified form of the expression is $\dfrac{x + 2}{2x + 1}$.

c) To write the function in simplified form, we list the numbers that are not in the domain of f:

$$f(x) = \frac{x + 2}{2x + 1}, \quad x \neq -\frac{1}{2}, \quad x \neq 2.$$

Factors That Are Opposites

Consider

$$\frac{x - 4}{8 - 2x}, \quad \text{or, equivalently,} \quad \frac{x - 4}{2(4 - x)}.$$

At first glance, the numerator and the denominator do not appear to have any common factors. But $x - 4$ and $4 - x$ are opposites, or additive inverses, of each other. Thus we can find a common factor by factoring out -1 in one expression.

EXAMPLE 8   Simplify $\dfrac{x - 4}{8 - 2x}$ and check by evaluating.

SOLUTION   We have

$$\frac{x - 4}{8 - 2x} = \frac{x - 4}{2(4 - x)}$$ Factoring

$$= \frac{x - 4}{2(-1)(x - 4)}$$ Note that $4 - x = -(x - 4)$.

$$= \frac{x - 4}{-2(x - 4)}$$ Had we originally factored out -2, we could have gone directly to this step.

$$= \frac{1}{-2} \cdot \frac{x - 4}{x - 4} = -\frac{1}{2}.$$ Removing a factor equal to 1: $(x - 4)/(x - 4) = 1$

As a partial check, note that for any choice of x other than 4, the value of the rational expression is $-\frac{1}{2}$. For example, if $x = 6$, then

$$\frac{x - 4}{8 - 2x} = \frac{6 - 4}{8 - 2 \cdot 6} = \frac{2}{-4} = -\frac{1}{2}.$$

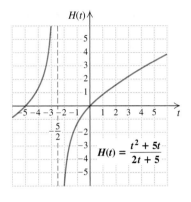

Vertical Asymptotes

Consider the graph of the function in Example 1,

$$H(t) = \frac{t^2 + 5t}{2t + 5}.$$

Note that the graph consists of two unconnected "branches." Since $-\frac{5}{2}$ is not in the domain of H, a vertical line drawn at $-\frac{5}{2}$ does not touch the graph of H. The line $t = -\frac{5}{2}$ is called a **vertical asymptote.** As t gets closer to, or *approaches*, $-\frac{5}{2}$, $H(t)$ approaches the asymptote.

Now consider the graph of

$$H(t) = \frac{t^2 + 5t}{2t + 5}$$

as drawn using some graphing calculators. The vertical line that appears on the screen on the left below is not part of the graph, nor should it be considered a vertical asymptote. Since a graphing calculator graphs an equation by plotting points and connecting them, the vertical line is actually connecting a point just to the left of the line $x = -\frac{5}{2}$ with a point just to the right of the line $x = -\frac{5}{2}$. Such a vertical line may or may not appear for values not in the domain of a function.

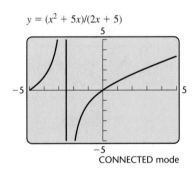

CONNECTED mode

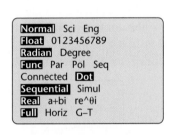

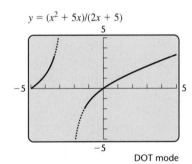

DOT mode

If we tell the calculator not to connect the points that it plots, the vertical line will not appear. Pressing **MODE** and changing from CONNECTED to DOT gives us the graph on the right above.

Not every number excluded from the domain of a function corresponds to a vertical asymptote.

Interactive Discovery

Let

$$f(x) = \frac{x^2 - 4}{2x^2 - 3x - 2} \quad \text{and} \quad g(x) = \frac{x + 2}{2x + 1}.$$

As we saw in Example 7, the rational expressions describing these functions are equivalent but the domains of the functions are different.

1. Graph $f(x)$. How many vertical asymptotes does the graph appear to have? What is the vertical asymptote?

2. Graph $g(x)$ and determine the vertical asymptote.

If a function $f(x)$ is described by a *simplified* rational expression, and a is a number that makes the denominator 0, then $x = a$ is a vertical asymptote of the graph of $f(x)$.

EXAMPLE 9   Determine the vertical asymptotes of the graph of

$$f(x) = \frac{9x^2 + 6x - 3}{12x^2 - 12}.$$

SOLUTION   We first simplify the rational expression describing the function:

$$\frac{9x^2 + 6x - 3}{12x^2 - 12} = \frac{3(x + 1)(3x - 1)}{3 \cdot 4(x + 1)(x - 1)} \qquad \text{Factoring the numerator and the denominator}$$

$$= \frac{3(x + 1)}{3(x + 1)} \cdot \frac{3x - 1}{4(x - 1)} \qquad \text{Factoring the rational expression}$$

$$= \frac{3x - 1}{4(x - 1)}. \qquad \text{Removing a factor equal to 1}$$

The denominator of the simplified expression, $4(x - 1)$, is 0 when $x = 1$. Thus, $x = 1$ is a vertical asymptote of the graph. Although the domain of the function also excludes -1, there is no asymptote at $x = -1$, only a "hole." This hole should be indicated on a hand-drawn graph but it may not be obvious on a graphing-calculator screen. A table will show that -1 is not in the domain of this function.

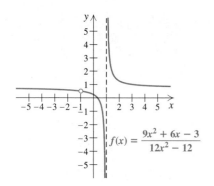

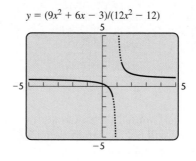

$y = (9x^2 + 6x - 3)/(12x^2 - 12)$

X	Y₁
−3	.625
−2	.58333
−1	ERROR
0	.25
1	ERROR
2	1.25
3	1
X = −3	

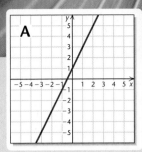

A

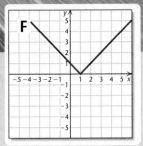

F

Visualizing the Graph

Match each function with its graph.

1. $f(x) = x^2 - 3$

2. $f(x) = 3$

3. $f(x) = \dfrac{1}{x}$

4. $f(x) = 2x + 1$

5. $f(x) = 1 - 2x$

6. $f(x) = \dfrac{3}{x + 2}$

7. $f(x) = \sqrt{x - 1}$

8. $f(x) = \dfrac{2}{x - 1}$

9. $f(x) = x - 1$

10. $f(x) = |x - 1|$

Answers on page A-27

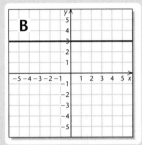

B

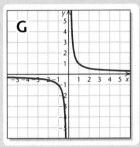

G

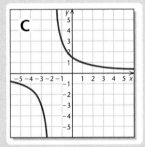

C

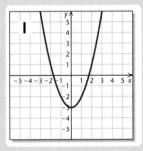

H

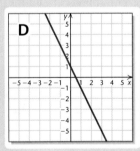

D

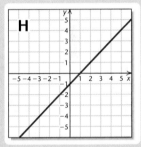

I

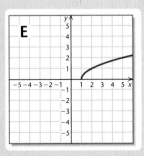

E

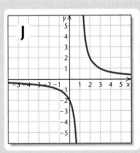

J

7.1 EXERCISE SET

FOR EXTRA HELP

MathXL   MyMathLab   InterAct Math   Tutor Center   Video Lectures on CD: Disc 4   Student's Solutions Manual

Concept Reinforcement   *In each of Exercises 1–10, match the function described with the appropriate domain from the column on the right.*

1. ___ $f(x) = \dfrac{2 - x}{x - 5}$

2. ___ $g(x) = \dfrac{x + 2}{x + 5}$

3. ___ $h(x) = \dfrac{x - 5}{x - 2}$

4. ___ $f(x) = \dfrac{x + 5}{x + 2}$

5. ___ $g(x) = \dfrac{x - 3}{(x - 2)(x - 5)}$

6. ___ $h(x) = \dfrac{x - 3}{(x + 2)(x + 5)}$

7. ___ $f(x) = \dfrac{x + 3}{(x - 2)(x + 5)}$

8. ___ $g(x) = \dfrac{x + 3}{(x + 2)(x - 5)}$

9. ___ $h(x) = \dfrac{(x - 2)(x - 3)}{x + 3}$

10. ___ $f(x) = \dfrac{(x + 2)(x + 3)}{x - 3}$

a) $\{x \mid x \neq -5, x \neq 2\}$

b) $\{x \mid x \neq 3\}$

c) $\{x \mid x \neq -2\}$

d) $\{x \mid x \neq -3\}$

e) $\{x \mid x \neq 5\}$

f) $\{x \mid x \neq -2, x \neq 5\}$

g) $\{x \mid x \neq 2\}$

h) $\{x \mid x \neq -2, x \neq -5\}$

i) $\{x \mid x \neq 2, x \neq 5\}$

j) $\{x \mid x \neq -5\}$

Photo Developing.   *Rik usually takes 3 hr more than Pearl does to process a day's orders at Liberty Place Photo. If Pearl takes t hr to process a day's orders, the function given by*

$$H(t) = \frac{t^2 + 3t}{2t + 3}$$

can be used to determine how long it would take if they worked together.

11. How long will it take them, working together, to complete a day's orders if Pearl can process the orders alone in 5 hr?

12. How long will it take them, working together, to complete a day's orders if Pearl can process the orders alone in 7 hr?

For each rational function, find the function values indicated, provided the value exists.

13. $v(t) = \dfrac{4t^2 - 5t + 2}{t + 3}$; $v(0), v(-2), v(7)$

14. $f(x) = \dfrac{5x^2 + 4x - 12}{6 - x}$; $f(0), f(-1), f(3)$

15. $g(x) = \dfrac{2x^3 - 9}{x^2 - 4x + 4}$; $g(0), g(2), g(-1)$

16. $r(t) = \dfrac{t^2 - 5t + 4}{t^2 - 9}$; $r(1), r(2), r(-3)$

List all numbers for which each rational expression is undefined.

17. $\dfrac{25}{-7x}$

18. $\dfrac{14}{-5y}$

19. $\dfrac{t - 3}{t + 8}$

20. $\dfrac{a - 8}{a + 7}$

21. $\dfrac{a - 4}{3a - 12}$

22. $\dfrac{x^2 - 9}{4x - 12}$

23. $\dfrac{x^2 - 16}{x^2 - 3x - 28}$

24. $\dfrac{p^2 - 9}{p^2 - 7p + 10}$

25. $\dfrac{m^3 - 2m}{m^2 - 25}$

26. $\dfrac{7 - 3x + x^2}{49 - x^2}$

Simplify by removing a factor equal to 1.

27. $\dfrac{15x}{5x^2}$

28. $\dfrac{7a^3}{21a}$

29. $\dfrac{18t^3}{27t^7}$

30. $\dfrac{8y^5}{4y^9}$

31. $\dfrac{2a - 10}{2}$

32. $\dfrac{3a + 12}{3}$

33. $\dfrac{3x - 12}{3x + 15}$

34. $\dfrac{4y - 20}{4y + 12}$

35. $\dfrac{6a^2 - 3a}{7a^2 - 7a}$

36. $\dfrac{3m^2 + 3m}{6m^2 + 9m}$

37. $\dfrac{3a - 1}{2 - 6a}$

38. $\dfrac{6 - 5a}{10a - 12}$

39. $\dfrac{3a^2 + 9a - 12}{6a^2 - 30a + 24}$

40. $\dfrac{2t^2 - 6t + 4}{4t^2 + 12t - 16}$

41. $\dfrac{x^2 + 8x + 16}{x^2 - 16}$

42. $\dfrac{x^2 - 25}{x^2 - 10x + 25}$

43. $\dfrac{t^2 - 1}{t + 1}$

44. $\dfrac{a^2 - 1}{a - 1}$

45. $\dfrac{y^2 + 4}{y + 2}$

46. $\dfrac{x^2 + 1}{x + 1}$

47. $\dfrac{5x^2 - 20}{10x^2 - 40}$

48. $\dfrac{6x^2 - 54}{4x^2 - 36}$

49. $\dfrac{x - 8}{8 - x}$

50. $\dfrac{6 - x}{x - 6}$

51. $\dfrac{2t - 1}{1 - 4t^2}$

52. $\dfrac{3a - 2}{4 - 9a^2}$

53. $\dfrac{a^2 - 25}{a^2 + 10a + 25}$

54. $\dfrac{a^2 - 16}{a^2 - 8a + 16}$

Aha! **55.** $\dfrac{7s^2 - 28t^2}{28t^2 - 7s^2}$

56. $\dfrac{9m^2 - 4n^2}{4n^2 - 9m^2}$

57. $\dfrac{x^3 - 1}{x^2 - 1}$

58. $\dfrac{a^3 + 8}{a^2 - 4}$

59. $\dfrac{3y^3 + 24}{y^2 - 2y + 4}$

60. $\dfrac{x^3 - 27}{5x^2 + 15x + 45}$

Write simplified form for each of the following. Be sure to list all restrictions on the domain, as in Example 7.

61. $f(x) = \dfrac{3x + 21}{x^2 + 7x}$

62. $f(x) = \dfrac{5x + 20}{x^2 + 4x}$

63. $g(x) = \dfrac{x^2 - 9}{5x + 15}$

64. $g(x) = \dfrac{8x - 16}{x^2 - 4}$

65. $h(x) = \dfrac{4 - x}{5x - 20}$

66. $h(x) = \dfrac{7 - x}{3x - 21}$

67. $f(t) = \dfrac{t^2 - 16}{t^2 - 8t + 16}$

68. $f(t) = \dfrac{t^2 - 25}{t^2 + 10t + 25}$

69. $g(t) = \dfrac{21 - 7t}{3t - 9}$

70. $g(t) = \dfrac{12 - 6t}{5t - 10}$

71. $h(t) = \dfrac{t^2 + 5t + 4}{t^2 - 8t - 9}$

72. $h(t) = \dfrac{t^2 - 3t - 4}{t^2 + 9t + 8}$

73. $f(x) = \dfrac{9x^2 - 4}{3x - 2}$

74. $f(x) = \dfrac{4x^2 - 1}{2x - 1}$

75. $g(t) = \dfrac{16 - t^2}{t^2 - 8t + 16}$

76. $g(p) = \dfrac{25 - p^2}{p^2 + 10p + 25}$

Determine the vertical asymptotes of the graph of each function.

77. $f(x) = \dfrac{3x - 12}{3x + 15}$

78. $f(x) = \dfrac{4x - 20}{4x + 12}$

79. $g(x) = \dfrac{12 - 6x}{5x - 10}$

80. $r(x) = \dfrac{21 - 7x}{3x - 9}$

81. $t(x) = \dfrac{x^3 + 3x^2}{x^2 + 6x + 9}$

82. $g(x) = \dfrac{x^2 - 4}{2x^2 - 5x + 2}$

83. $f(x) = \dfrac{x^2 - x - 6}{x^2 - 6x + 8}$

84. $f(x) = \dfrac{x^2 + 2x + 1}{x^2 - 2x + 1}$

In Exercises 85–90, match each function with one of the following graphs.

a)
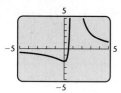

b)
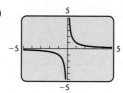

c)
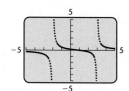

d)
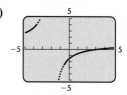

e)
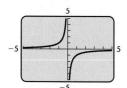

f)
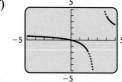

85. $h(x) = \dfrac{1}{x}$

86. $q(x) = -\dfrac{1}{x}$

87. $f(x) = \dfrac{x}{x - 3}$

88. $g(x) = \dfrac{x - 3}{x + 2}$

89. $r(x) = \dfrac{4x - 2}{x^2 - 2x + 1}$

90. $t(x) = \dfrac{x - 1}{x^2 - x - 6}$

TW 91. Explain why the graphs of $f(x) = 5x$ and
$g(x) = \dfrac{5x^2}{x}$ differ.

TW 92. Nancy *incorrectly* simplifies $\dfrac{x + 2}{x}$ as
$$\frac{x + 2}{x} = \frac{\not{x} + 2}{\not{x}} = 1 + 2 = 3.$$

She insists this is correct because it checks when x is replaced with 1. Explain her misconception.

Skill Maintenance

Simplify.

93. $\dfrac{3}{10} - \dfrac{8}{15}$ [1.6]

94. $\dfrac{3}{8} - \dfrac{7}{10}$ [1.6]

95. $\dfrac{2}{3} \cdot \dfrac{5}{7} - \dfrac{5}{7} \cdot \dfrac{1}{6}$ [1.8]

96. $\dfrac{4}{7} \cdot \dfrac{1}{5} - \dfrac{3}{10} \cdot \dfrac{2}{7}$ [1.8]

97. $(8x^3 - 5x^2 + 6x + 2) - (4x^3 + 2x^2 - 3x + 7)$ [5.3]

98. $(6t^4 + 9t^3 - t^2 + 4t) - (8t^4 - 2t^3 - 6t + 3)$ [5.3]

Synthesis

TW 99. To check Example 4, Kara lets
$$y_1 = \frac{7x^2 + 21x}{14x} \quad \text{and} \quad y_2 = \frac{x + 3}{2}.$$

Since the graphs of y_1 and y_2 appear to be identical, Kara believes that the domains of the functions described by y_1 and y_2 are the same, $\mathbb{R}$. How could you convince Kara otherwise?

TW 100. Tony *incorrectly* argues that since

$$\frac{a^2 - 4}{a - 2} = \frac{a^2}{a} + \frac{-4}{-2} = a + 2$$

is correct, it follows that

$$\frac{x^2 + 9}{x + 1} = \frac{x^2}{x} + \frac{9}{1} = x + 9.$$

Explain his misconception.

101. Calculate the slope of the line passing through $(a, f(a))$ and $(a + h, f(a + h))$ for the function f given by $f(x) = x^2 + 5$. Be sure your answer is simplified.

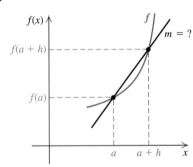

102. Calculate the slope of the line passing through the points $(a, f(a))$ and $(a + h, f(a + h))$ for the function f given by $f(x) = 3x^2$. Be sure your answer is simplified.

Simplify.

103. $\dfrac{x^4 - y^4}{(y - x)^4}$

104. $\dfrac{16y^4 - x^4}{(x^2 + 4y^2)(x - 2y)}$

105. $\dfrac{(x - 1)(x^4 - 1)(x^2 - 1)}{(x^2 + 1)(x - 1)^2(x^4 - 2x^2 + 1)}$

106. $\dfrac{m^2 - t^2}{m^2 + t^2 + m + t + 2mt}$

107. $\dfrac{a^3 - 2a^2 + 2a - 4}{a^3 - 2a^2 - 3a + 6}$

108. $\dfrac{x^3 + x^2 - y^3 - y^2}{x^2 - 2xy + y^2}$

109. $\dfrac{(t + 2)^3(t^2 + 2t + 1)(t + 1)}{(t + 1)^3(t^2 + 4t + 4)(t + 2)}$

110. $\dfrac{x^5 - x^3 + x^2 - 1 - (x^3 - 1)(x + 1)^2}{(x^2 - 1)^2}$

Determine the domain and the range of each function from its graph.

111.
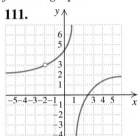

112.
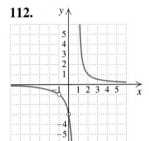

113.
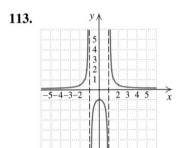

114. Graph the function given by

$$f(x) = \frac{x^2 - 9}{x - 3}.$$

(*Hint*: Determine the domain of f and simplify.)

TW 115. Select any number x, multiply by 2, add 5, multiply by 5, subtract 25, and divide by 10. What do you get? Explain how this procedure can be used for a number trick.

7.2 Multiplication and Division

Multiplication ■ Division

Multiplication and division of rational expressions is similar to multiplication and division with fractions.

Multiplication

Recall that to multiply fractions, we simply multiply their numerators and multiply their denominators. Rational expressions are multiplied in a similar way.

> **Products of Rational Expressions** To multiply two rational expressions, multiply numerators and multiply denominators:
>
> $$\frac{A}{B} \cdot \frac{C}{D} = \frac{AC}{BD}, \quad \text{where } B \neq 0, D \neq 0.$$

For example,

$$\frac{3}{5} \cdot \frac{8}{11} = \frac{24}{55} \quad \text{and} \quad \frac{x}{3} \cdot \frac{x+2}{y} = \frac{x(x+2)}{3y}.$$

Fraction bars are grouping symbols, so parentheses are needed when writing some products. Because we generally simplify, we often leave parentheses in the product. There is no need to multiply further.

> ▸ **EXAMPLE 1** Multiply. (Write the product as a single rational expression.) Then simplify by removing a factor equal to 1.
>
> **a)** $\dfrac{5a^3}{4} \cdot \dfrac{2}{5a}$ **b)** $\dfrac{x^2 + 6x + 9}{x^2 - 4} \cdot \dfrac{x - 2}{x + 3}$
>
> **c)** $\dfrac{1 - a^3}{a^2} \cdot \dfrac{a^5}{a^2 - 1}$
>
> **SOLUTION**
>
> **a)** $\dfrac{5a^3}{4} \cdot \dfrac{2}{5a} = \dfrac{5a^3(2)}{4(5a)}$ Forming the product of the numerators and the product of the denominators
>
> $\qquad\qquad = \dfrac{2 \cdot 5 \cdot a \cdot a \cdot a}{2 \cdot 2 \cdot 5 \cdot a}$ Factoring the numerator and the denominator
>
> $\qquad\qquad = \dfrac{2 \cdot 5 \cdot a \cdot a \cdot a}{2 \cdot 2 \cdot 5 \cdot a}$ ⎫
>
> $\qquad\qquad = \dfrac{a^2}{2}$ ⎬   Removing a factor equal to 1: $\dfrac{2 \cdot 5 \cdot a}{2 \cdot 5 \cdot a} = 1$

b) $\dfrac{x^2 + 6x + 9}{x^2 - 4} \cdot \dfrac{x - 2}{x + 3} = \dfrac{(x^2 + 6x + 9)(x - 2)}{(x^2 - 4)(x + 3)}$ Multiplying the numerators and the denominators

$= \dfrac{(x + 3)(x + 3)(x - 2)}{(x + 2)(x - 2)(x + 3)}$ Factoring the numerator and the denominator

$= \dfrac{\cancel{(x + 3)}(x + 3)\cancel{(x - 2)}}{(x + 2)\cancel{(x - 2)}\cancel{(x + 3)}}$ ⎫

$= \dfrac{x + 3}{x + 2}$ ⎬   Removing a factor equal to 1: $\dfrac{(x + 3)(x - 2)}{(x + 3)(x - 2)} = 1$

c) $\dfrac{1 - a^3}{a^2} \cdot \dfrac{a^5}{a^2 - 1} = \dfrac{(1 - a^3)a^5}{a^2(a^2 - 1)}$

$= \dfrac{(1 - a)(1 + a + a^2)a^5}{a^2(a - 1)(a + 1)}$ Factoring a difference of cubes and a difference of squares

$= \dfrac{-1(a - 1)(1 + a + a^2)a^5}{a^2(a - 1)(a + 1)}$ *Important!* $a - b = -1(b - a)$

$= \dfrac{\cancel{(a - 1)}\cancel{a^2} \cdot a^3(-1)(1 + a + a^2)}{\cancel{(a - 1)}\cancel{a^2}(a + 1)}$ Rewriting a^5 as $a^2 \cdot a^3$; removing a factor equal to 1: $\dfrac{(a - 1)a^2}{(a - 1)a^2} = 1$

$= \dfrac{-a^3(1 + a + a^2)}{a + 1}$ Simplifying ◢

Division

Two expressions are reciprocals of each other if their product is 1. As in arithmetic, to find the reciprocal of a rational expression, we interchange numerator and denominator.

The reciprocal of $\dfrac{x}{x^2 + 3}$ is $\dfrac{x^2 + 3}{x}$.

The reciprocal of $y - 8$ is $\dfrac{1}{y - 8}$.

Student Notes

The procedures covered in this chapter are by their nature rather long. As is often the case in mathematics, it usually helps to write out each step as you do the problems. Use plenty of paper and, if you are using lined paper, consider using two spaces at a time, writing the fraction bar on a line of the paper.

Quotients of Rational Expressions   For any rational expressions A/B and C/D, with $B, C, D \neq 0$,

$$\frac{A}{B} \div \frac{C}{D} = \frac{A}{B} \cdot \frac{D}{C}.$$

(To divide two rational expressions, multiply by the reciprocal of the divisor. We often say that we "*invert* and multiply.")

▼**EXAMPLE 2**   Divide: **(a)** $\dfrac{x}{5} \div \dfrac{7}{y}$; **(b)** $(x + 2) \div \dfrac{x - 1}{x + 3}$.

SOLUTION

a) $\dfrac{x}{5} \div \dfrac{7}{y} = \dfrac{x}{5} \cdot \dfrac{y}{7}$ **Multiplying by the reciprocal of the divisor**

$\phantom{\dfrac{x}{5} \div \dfrac{7}{y}} = \dfrac{xy}{35}$ **Multiplying rational expressions**

b) $(x + 2) \div \dfrac{x - 1}{x + 3} = \dfrac{x + 2}{1} \cdot \dfrac{x + 3}{x - 1}$ **Multiplying by the reciprocal of the divisor. Writing $x + 2$ as $\dfrac{x + 2}{1}$ can be helpful.**

$\phantom{(x + 2) \div \dfrac{x - 1}{x + 3}} = \dfrac{(x + 2)(x + 3)}{x - 1}$

As usual, we should simplify when possible. Often that will require that we factor one or more polynomials. Our hope is to discover a common factor that appears in both the numerator and the denominator.

EXAMPLE 3   Divide and simplify: $\dfrac{x + 1}{x^2 - 1} \div \dfrac{x + 1}{x^2 - 2x + 1}$.

SOLUTION

$\dfrac{x + 1}{x^2 - 1} \div \dfrac{x + 1}{x^2 - 2x + 1} = \dfrac{x + 1}{x^2 - 1} \cdot \dfrac{x^2 - 2x + 1}{x + 1}$ **Multiplying by the reciprocal of the divisor**

$ = \dfrac{(x + 1)(x - 1)(x - 1)}{(x + 1)(x - 1)(x + 1)}$ **Multiplying rational expressions and factoring numerators and denominators**

$ = \dfrac{(\cancel{x + 1})(\cancel{x - 1})(x - 1)}{(\cancel{x + 1})(\cancel{x - 1})(x + 1)}$ **Removing a factor equal to 1: $\dfrac{(x + 1)(x - 1)}{(x + 1)(x - 1)} = 1$**

$ = \dfrac{x - 1}{x + 1}$

EXAMPLE 4   Divide and, if possible, simplify.

a) $\dfrac{a^2 + 3a + 2}{a^2 + 4} \div (5a^2 + 10a)$ **b)** $\dfrac{x^2 - 2x - 3}{x^2 - 4} \div \dfrac{x + 1}{x + 5}$

SOLUTION

a) $\dfrac{a^2 + 3a + 2}{a^2 + 4} \div (5a^2 + 10a)$

$ = \dfrac{a^2 + 3a + 2}{a^2 + 4} \cdot \dfrac{1}{5a^2 + 10a}$ **Multiplying by the reciprocal of the divisor**

$ = \dfrac{(a + 2)(a + 1)}{(a^2 + 4)5a(a + 2)}$ **Multiplying rational expressions and factoring**

$ = \dfrac{(\cancel{a + 2})(a + 1)}{(a^2 + 4)5a(\cancel{a + 2})}$ **Removing a factor equal to 1: $\dfrac{a + 2}{a + 2} = 1$**

$ = \dfrac{a + 1}{(a^2 + 4)5a}$

b)
$$\frac{x^2 - 2x - 3}{x^2 - 4} \div \frac{x + 1}{x + 5} = \frac{x^2 - 2x - 3}{x^2 - 4} \cdot \frac{x + 5}{x + 1}$$
Multiplying by the reciprocal of the divisor

$$= \frac{(x - 3)(x + 1)(x + 5)}{(x - 2)(x + 2)(x + 1)}$$
Multiplying rational expressions and factoring

$$= \frac{(x - 3)\cancel{(x + 1)}(x + 5)}{(x - 2)(x + 2)\cancel{(x + 1)}}$$
Removing a factor equal to 1: $\frac{x + 1}{x + 1} = 1$

$$= \frac{(x - 3)(x + 5)}{(x - 2)(x + 2)}$$

We can perform a partial check of the result using a graphing calculator, taking care when placing parentheses. We enter the original expression as

$$y_1 = ((x^2 - 2x - 3)/(x^2 - 4))/((x + 1)/(x + 5))$$

and the simplified expression as

$$y_2 = ((x - 3)(x + 5))/((x - 2)(x + 2)).$$

Note that parentheses are placed around the entire numerator and around the entire denominator of each expression. Comparing values of y_1 and y_2 in the table shown at left, we see that the simplification is probably correct.

X	Y₁	Y₂
−5	ERROR	0
−4	−.5833	−.5833
−3	−2.4	−2.4
−2	ERROR	ERROR
−1	ERROR	5.3333
0	3.75	3.75
1	4	4

X = −5

7.2   EXERCISE SET

Multiply. Leave each answer in factored form.

1. $\dfrac{7x}{5} \cdot \dfrac{x - 5}{2x + 1}$

2. $\dfrac{3x}{4} \cdot \dfrac{5x + 2}{x - 1}$

3. $\dfrac{a - 4}{a + 6} \cdot \dfrac{a + 2}{a + 6}$

4. $\dfrac{a + 3}{a + 6} \cdot \dfrac{a + 3}{a - 1}$

5. $\dfrac{2x + 3}{4} \cdot \dfrac{x + 1}{x - 5}$

6. $\dfrac{x + 2}{3x - 4} \cdot \dfrac{4}{5x + 6}$

Multiply and, if possible, simplify.

7. $\dfrac{5a^4}{6a} \cdot \dfrac{2}{a}$

8. $\dfrac{10}{t^7} \cdot \dfrac{3t^2}{25t}$

9. $\dfrac{3c}{d^2} \cdot \dfrac{8d}{6c^3}$

10. $\dfrac{3x^2y}{2} \cdot \dfrac{4}{xy^3}$

11. $\dfrac{y^2 - 16}{4y + 12} \cdot \dfrac{y + 3}{y - 4}$

12. $\dfrac{m^2 - n^2}{4m + 4n} \cdot \dfrac{m + n}{m - n}$

13. $\dfrac{x^2 - 3x - 10}{(x - 2)^2} \cdot \dfrac{x - 2}{x - 5}$

14. $\dfrac{t + 2}{t - 2} \cdot \dfrac{t^2 - 5t + 6}{(t + 2)^2}$

15. $\dfrac{a^2 + 25}{a^2 - 4a + 3} \cdot \dfrac{a - 5}{a + 5}$

16. $\dfrac{x + 3}{x^2 + 9} \cdot \dfrac{x^2 + 5x + 4}{x + 9}$

17. $\dfrac{a^2 - 9}{a^2} \cdot \dfrac{5a}{a^2 + a - 12}$

18. $\dfrac{x^2 + 10x - 11}{5x} \cdot \dfrac{x^3}{x + 11}$

19. $\dfrac{4a^2}{3a^2 - 12a + 12} \cdot \dfrac{3a - 6}{2a}$

20. $\dfrac{5v + 5}{v - 2} \cdot \dfrac{2v^2 - 8v + 8}{v^2 - 1}$

21. $\dfrac{t^2 + 2t - 3}{t^2 + 4t - 5} \cdot \dfrac{t^2 - 3t - 10}{t^2 + 5t + 6}$

22. $\dfrac{x^2 + 5x + 4}{x^2 - 6x + 8} \cdot \dfrac{x^2 + 5x - 14}{x^2 + 8x + 7}$

23. $\dfrac{5a^2 - 180}{10a^2 - 10} \cdot \dfrac{20a + 20}{2a - 12}$

24. $\dfrac{2t^2 - 98}{4t^2 - 4} \cdot \dfrac{8t + 8}{16t - 112}$

Aha! **25.** $\dfrac{x^2 + 4x + 4}{(x - 1)^2} \cdot \dfrac{x^2 - 2x + 1}{(x + 2)^2}$

26. $\dfrac{x + 5}{(x + 2)^2} \cdot \dfrac{x^2 + 7x + 10}{(x + 5)^2}$

27. $\dfrac{t^2 + 8t + 16}{(t + 4)^3} \cdot \dfrac{(t + 2)^3}{t^2 + 4t + 4}$

28. $\dfrac{(y - 1)^3}{y^2 - 2y + 1} \cdot \dfrac{y^2 - 4y + 4}{(y - 2)^3}$

29. $\dfrac{7a - 14}{4 - a^2} \cdot \dfrac{5a^2 + 6a + 1}{35a + 7}$

30. $\dfrac{a^2 - 1}{2 - 5a} \cdot \dfrac{15a - 6}{a^2 + 5a - 6}$

31. $\dfrac{t^3 - 4t}{t - t^4} \cdot \dfrac{t^4 - t}{4t - t^3}$

32. $\dfrac{x^2 - 6x + 9}{12 - 4x} \cdot \dfrac{x^6 - 9x^4}{x^3 - 3x^2}$

33. $\dfrac{c^3 + 8}{c^5 - 4c^3} \cdot \dfrac{c^6 - 4c^5 + 4c^4}{c^2 - 2c + 4}$

34. $\dfrac{x^3 - 27}{x^4 - 9x^2} \cdot \dfrac{x^5 - 6x^4 + 9x^3}{x^2 + 3x + 9}$

35. $\dfrac{a^3 - b^3}{3a^2 + 9ab + 6b^2} \cdot \dfrac{a^2 + 2ab + b^2}{a^2 - b^2}$

36. $\dfrac{x^3 + y^3}{x^2 + 2xy - 3y^2} \cdot \dfrac{x^2 - y^2}{3x^2 + 6xy + 3y^2}$

Find the reciprocal of each expression.

37. $\dfrac{3x}{7}$

38. $\dfrac{3 - x}{x^2 + 4}$

39. $a^3 - 8a$

40. $\dfrac{7}{a^2 - b^2}$

Divide and, if possible, simplify.

41. $\dfrac{5}{8} \div \dfrac{3}{7}$

42. $\dfrac{4}{9} \div \dfrac{5}{7}$

43. $\dfrac{x}{4} \div \dfrac{5}{x}$

44. $\dfrac{5}{x} \div \dfrac{x}{12}$

45. $\dfrac{9x^5}{8y^2} \div \dfrac{3x}{16y^9}$

46. $\dfrac{16a^7}{3b^5} \div \dfrac{8a^3}{6b}$

47. $\dfrac{y + 5}{4} \div \dfrac{y}{2}$

48. $\dfrac{a + 2}{a - 3} \div \dfrac{a - 1}{a + 3}$

49. $\dfrac{5x + 10}{x^8} \div \dfrac{x + 2}{x^3}$

50. $\dfrac{3y + 15}{y^7} \div \dfrac{y + 5}{y^2}$

51. $\dfrac{4y - 8}{y + 2} \div \dfrac{y - 2}{y^2 - 4}$

52. $\dfrac{x^2 - 1}{x} \div \dfrac{x + 1}{x - 1}$

53. $\dfrac{a}{a - b} \div \dfrac{b}{b - a}$

54. $\dfrac{x - y}{6} \div \dfrac{y - x}{3}$

55. $(y^2 - 9) \div \dfrac{y^2 - 2y - 3}{y^2 + 1}$

56. $(x^2 - 5x - 6) \div \dfrac{x^2 - 1}{x + 6}$

57. $\dfrac{-3 + 3x}{16} \div \dfrac{x - 1}{5}$

58. $\dfrac{-12 + 4x}{12} \div \dfrac{-6 + 2x}{6}$

59. $\dfrac{a + 2}{a - 1} \div \dfrac{3a + 6}{a - 5}$

60. $\dfrac{t - 3}{t + 2} \div \dfrac{4t - 12}{t + 1}$

61. $\dfrac{25x^2 - 4}{x^2 - 9} \div \dfrac{2 - 5x}{x + 3}$

62. $\dfrac{4a^2 - 1}{a^2 - 4} \div \dfrac{2a - 1}{2 - a}$

63. $(2x - 1) \div \dfrac{2x^2 - 11x + 5}{4x^2 - 1}$

64. $(a + 7) \div \dfrac{3a^2 + 14a - 49}{a^2 + 8a + 7}$

65. $\dfrac{a^2 - 10a + 25}{a^2 + 7a + 12} \div \dfrac{a^2 - a - 20}{a^2 + 6a + 9}$

66. $\dfrac{3a^2 + 7a - 20}{a^2 - 2a + 1} \div \dfrac{3a^2 - 8a + 5}{a^2 - 5a - 6}$

67. $\dfrac{c^2 + 10c + 21}{c^2 - 2c - 15} \div (5c^2 + 32c - 21)$

68. $\dfrac{z^2 - 2z + 1}{z^2 - 1} \div (4z^2 - z - 3)$

69. $\dfrac{x^3 - 64}{x^3 + 64} \div \dfrac{x^2 - 16}{x^2 - 4x + 16}$

70. $\dfrac{8y^3 - 27}{64y^3 - 1} \div \dfrac{4y^2 - 9}{16y^2 + 4y + 1}$

71. $\dfrac{8a^3 + b^3}{2a^2 + 3ab + b^2} \div \dfrac{8a^2 - 4ab + 2b^2}{4a^2 + 4ab + b^2}$

72. $\dfrac{x^3 + 8y^3}{2x^2 + 5xy + 2y^2} \div \dfrac{x^3 - 2x^2y + 4xy^2}{8x^2 - 2y^2}$

TW 73. Why is it important to insert parentheses when multiplying rational expressions in which the numerators and the denominators contain more than one term?

TW 74. A student claims to be able to divide, but not multiply, rational expressions. Why is this claim difficult to believe?

Skill Maintenance

Simplify.

75. $\dfrac{3}{4} + \dfrac{5}{6}$ [1.3]

76. $\dfrac{7}{8} + \dfrac{5}{6}$ [1.3]

77. $\dfrac{2}{9} - \dfrac{1}{6}$ [1.3]

78. $\dfrac{3}{10} - \dfrac{7}{15}$ [1.5]

79. $\dfrac{2}{5} - \left(\dfrac{3}{2}\right)^2$ [1.8]

80. $\dfrac{5}{9} + \dfrac{2}{3} \cdot \dfrac{4}{5}$ [1.8]

Synthesis

TW 81. Is the reciprocal of a product the product of the two reciprocals? Why or why not?

TW 82. Explain why, in the check for Example 4(b), there were three error messages in the Y1-column but only one error in the Y2-column.

Simplify.

Aha! 83. $\dfrac{3x - y}{2x + y} \div \dfrac{3x - y}{2x + y}$

84. $\dfrac{2a^2 - 5ab}{c - 3d} \div (4a^2 - 25b^2)$

85. $(x - 2a) \div \dfrac{a^2x^2 - 4a^4}{a^2x + 2a^3}$

86. $\dfrac{3a^2 - 5ab - 12b^2}{3ab + 4b^2} \div (3b^2 - ab)^2$

Aha! 87. $\dfrac{a^2 - 3b}{a^2 + 2b} \cdot \dfrac{a^2 - 2b}{a^2 + 3b} \cdot \dfrac{a^2 + 2b}{a^2 - 3b}$

88. $\dfrac{z^2 - 8z + 16}{z^2 + 8z + 16} \div \dfrac{(z - 4)^5}{(z + 4)^5} \div \dfrac{3z + 12}{z^2 - 16}$

Perform the indicated operations and simplify.

89. $\left[\dfrac{r^2 - 4s^2}{r + 2s} \div (r + 2s)\right] \cdot \dfrac{2s}{r - 2s}$

90. $\left[\dfrac{d^2 - d}{d^2 - 6d + 8} \cdot \dfrac{d - 2}{d^2 + 5d}\right] \div \dfrac{5d}{d^2 - 9d + 20}$

Aha! 91. $\left[\dfrac{6t^2 - 26t + 30}{8t^2 - 15t - 21} \cdot \dfrac{5t^2 - 9t - 15}{6t^2 - 14t - 20}\right] \div \dfrac{5t^2 - 9t - 15}{6t^2 - 14t - 20}$

92. Let
$$g(x) = \dfrac{2x + 3}{4x - 1}.$$
Determine each of the following.
a) $g(x + h)$
b) $g(2x - 2) \cdot g(x)$
c) $g\left(\tfrac{1}{2}x + 1\right) \cdot g(x)$

93. Let
$$f(x) = \dfrac{4}{x^2 - 1} \quad \text{and} \quad g(x) = \dfrac{4x^2 + 8x + 4}{x^3 - 1}.$$
Find each of the following.
a) $(f \cdot g)(x)$
b) $(f/g)(x)$
c) $(g/f)(x)$

7.3 Addition, Subtraction, and Least Common Denominators

- Addition When Denominators Are the Same ■
- Subtraction When Denominators Are the Same ■
- Least Common Multiples and Denominators

X	Y₁	Y₂
-2	-2.5	-2.5
-1	-6	-6
0	ERROR	ERROR
1	8	8
2	4.5	4.5
3	3.3333	3.3333
4	2.75	2.75
X = -2		

Addition When Denominators Are the Same

Recall that to add fractions having the same denominator, like $\frac{2}{7}$ and $\frac{3}{7}$, we add the numerators and keep the common denominator: $\frac{2}{7} + \frac{3}{7} = \frac{5}{7}$. The same procedure is used when rational expressions share a common denominator.

The Sum of Two Rational Expressions To add when the denominators are the same, add the numerators and keep the common denominator:

$$\frac{A}{B} + \frac{C}{B} = \frac{A + C}{B}, \quad \text{where } B \neq 0.$$

EXAMPLE 1 Add: $\dfrac{3 + x}{x} + \dfrac{4}{x}$.

SOLUTION We have

$$\frac{3 + x}{x} + \frac{4}{x} = \frac{3 + x + 4}{x} = \frac{x + 7}{x}.$$

Because x is not a factor of both the numerator and the denominator, the result cannot be simplified.

To check, we let $y_1 = (3 + x)/x + 4/x$ and $y_2 = (x + 7)/x$. The table at left shows that $y_1 = y_2$ for all x not equal to 0.

CAUTION! Using a table can indicate that two expressions are equivalent, but it does not verify that one of them is completely simplified. A rational expression is simplified when its numerator and its denominator do not contain any common factors.

EXAMPLE 2   Add. Simplify if possible.

a) $\dfrac{2x^2 + 3x - 7}{2x + 1} + \dfrac{x^2 + x - 8}{2x + 1}$ b) $\dfrac{x - 5}{x^2 - 9} + \dfrac{2}{x^2 - 9}$

SOLUTION

a) $\dfrac{2x^2 + 3x - 7}{2x + 1} + \dfrac{x^2 + x - 8}{2x + 1} = \dfrac{(2x^2 + 3x - 7) + (x^2 + x - 8)}{2x + 1}$

$= \dfrac{3x^2 + 4x - 15}{2x + 1}$ **Combining like terms in the numerator**

b) $\dfrac{x - 5}{x^2 - 9} + \dfrac{2}{x^2 - 9} = \dfrac{x - 3}{x^2 - 9}$ **Combining like terms in the numerator:**
$x - 5 + 2 = x - 3$

$= \dfrac{x - 3}{(x - 3)(x + 3)}$ **Factoring**

$= \dfrac{1 \cdot \cancel{(x - 3)}}{\cancel{(x - 3)}(x + 3)}$ **Removing a factor equal to 1:** $\dfrac{x - 3}{x - 3} = 1$

$= \dfrac{1}{x + 3}$

Subtraction When Denominators Are the Same

When two fractions have the same denominator, we subtract one numerator from the other and keep the common denominator: $\frac{5}{7} - \frac{2}{7} = \frac{3}{7}$. The same procedure is used with rational expressions.

The Difference of Two Rational Expressions   To subtract when the denominators are the same, subtract the second numerator from the first and keep the common denominator:

$$\dfrac{A}{B} - \dfrac{C}{B} = \dfrac{A - C}{B}, \quad \text{where } B \neq 0.$$

CAUTION!   A fraction bar under a numerator is a grouping symbol, just like parentheses. When a numerator is subtracted, it is important to subtract *every* term in that numerator.

EXAMPLE 3   If

$$f(x) = \dfrac{4x + 5}{x + 3} - \dfrac{x - 2}{x + 3},$$

find a simplified form of $f(x)$ and list all restrictions on the domain.

SOLUTION

$$f(x) = \frac{4x + 5}{x + 3} - \frac{x - 2}{x + 3} \qquad \text{Note that } x \neq -3.$$

$$= \frac{4x + 5 - (x - 2)}{x + 3} \qquad \begin{array}{l}\text{The parentheses remind us to}\\\text{subtract } both \text{ terms.}\end{array}$$

$$= \frac{4x + 5 - x + 2}{x + 3}$$

$$= \frac{3x + 7}{x + 3}, \quad x \neq -3$$

EXAMPLE 4   Subtract: $\dfrac{x^2}{x - 4} - \dfrac{x + 12}{x - 4}$.

SOLUTION

$$\frac{x^2}{x - 4} - \frac{x + 12}{x - 4} = \frac{x^2 - (x + 12)}{x - 4} \qquad \text{Remember the parentheses!}$$

$$= \frac{x^2 - x - 12}{x - 4} \qquad \begin{array}{l}\text{Removing parentheses}\\\text{(using the distributive law)}\end{array}$$

$$= \frac{(x - 4)(x + 3)}{x - 4} \qquad \begin{array}{l}\text{Factoring, in hopes of}\\\text{simplifying}\end{array}$$

$$= \frac{(x - 4)(x + 3)}{(x - 4)} \qquad \begin{array}{l}\text{Removing a factor equal to 1:}\\\frac{x - 4}{x - 4} = 1\end{array}$$

$$= x + 3$$

Least Common Multiples and Denominators

Thus far, every pair of rational expressions that we have added or subtracted shared a common denominator. To add or subtract rational expressions that have different denominators, we must first find equivalent rational expressions that *do* have a common denominator.

In algebra, we find a common denominator much as we do in arithmetic. Recall that to add $\frac{1}{12}$ and $\frac{7}{30}$, we first identify the smallest number that contains both 12 and 30 as factors. Such a number, the **least common multiple (LCM)** of the denominators, is then used as the **least common denominator (LCD).**

Let's find the LCM of 12 and 30 using a method that can also be used with polynomials. We begin by writing the prime factorization of 12:

$$12 = 2 \cdot 2 \cdot 3.$$

Next, we write the prime factorization of 30:

$$30 = 2 \cdot 3 \cdot 5.$$

The LCM must include the factors of each number, so it must include each prime factor the greatest number of times that it appears in either of the factorizations. To find the LCM for 12 and 30, we select one factorization, say

$$2 \cdot 2 \cdot 3,$$

and note that because it lacks a factor of 5, it does not contain the entire factorization of 30. If we multiply $2 \cdot 2 \cdot 3$ by 5, every prime factor occurs just often enough to contain both 12 and 30 as factors.

$$\text{LCM} = 2 \cdot 2 \cdot 3 \cdot 5$$

12 is a factor of the LCM.

30 is a factor of the LCM.

Note that each prime factor—2, 3, and 5—is used the greatest number of times that it appears in either of the individual factorizations. The factor 2 occurs twice and the factors 3 and 5 once each.

> **To Find the Least Common Denominator (LCD)**
>
> **1.** Write the prime factorization of each denominator.
> **2.** Select one of the factorizations and inspect it to see if it contains the other.
>
> > **a)** If it does, it represents the LCM of the denominators.
> > **b)** If it does not, multiply that factorization by any factors of the other denominator that it lacks. The final product is the LCM of the denominators.
>
> The LCD is the LCM of the denominators. It should contain each factor the greatest number of times that it occurs in any of the individual factorizations.

Let's finish adding $\dfrac{1}{12}$ and $\dfrac{7}{30}$:

$$\frac{1}{12} + \frac{7}{30} = \frac{1}{2 \cdot 2 \cdot 3} + \frac{7}{2 \cdot 3 \cdot 5}. \qquad \text{The least common denominator (LCD) is } 2 \cdot 2 \cdot 3 \cdot 5.$$

We found above that the LCD is $2 \cdot 2 \cdot 3 \cdot 5$. To get the LCD, we see that the first denominator needs a factor of 5, and the second denominator needs another factor of 2. Thus we multiply $\frac{1}{12}$ by 1, using $\frac{5}{5}$, and we multiply $\frac{7}{30}$ by 1, using $\frac{2}{2}$. We can do this because $a \cdot 1 = a$, for any number a:

$$\frac{1}{12} + \frac{7}{30} = \frac{1}{2 \cdot 2 \cdot 3} \cdot \frac{5}{5} + \frac{7}{2 \cdot 3 \cdot 5} \cdot \frac{2}{2} \qquad \frac{5}{5} = 1 \text{ and } \frac{2}{2} = 1$$

$$= \frac{5}{60} + \frac{14}{60} \qquad \qquad \text{Both denominators are now the LCD.}$$

$$= \frac{19}{60}. \qquad \qquad \text{Adding the numerators and keeping the LCD}$$

Expressions like $\dfrac{5}{36x^2}$ and $\dfrac{7}{24x}$ are added in much the same manner.

EXAMPLE 5 Find the LCD of $\dfrac{5}{36x^2}$ and $\dfrac{7}{24x}$.

SOLUTION

1. We begin by writing the prime factorizations of $36x^2$ and $24x$:

$$36x^2 = 2 \cdot 2 \cdot 3 \cdot 3 \cdot x \cdot x;$$
$$24x = 2 \cdot 2 \cdot 2 \cdot 3 \cdot x.$$

2. Except for a third factor of 2, the factorization of $36x^2$ contains the entire factorization of $24x$. To find the smallest product that contains both $36x^2$ and $24x$ as factors, we multiply $36x^2$ by a third factor of 2.

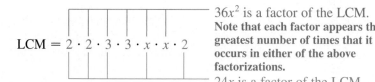

$$\text{LCM} = 2 \cdot 2 \cdot 3 \cdot 3 \cdot x \cdot x \cdot 2$$

$36x^2$ is a factor of the LCM. **Note that each factor appears the greatest number of times that it occurs in either of the above factorizations.** $24x$ is a factor of the LCM.

The LCM of the denominators is thus $2^3 \cdot 3^2 \cdot x^2$, or $72x^2$, so the LCD is $72x^2$.

We can now add $\dfrac{5}{36x^2}$ and $\dfrac{7}{24x}$:

$$\frac{5}{36x^2} + \frac{7}{24x} = \frac{5}{2 \cdot 2 \cdot 3 \cdot 3 \cdot x \cdot x} + \frac{7}{2 \cdot 2 \cdot 2 \cdot 3 \cdot x}.$$

In Example 5, we found that the LCD is $2 \cdot 2 \cdot 2 \cdot 3 \cdot 3 \cdot x \cdot x$. To obtain equivalent expressions with this LCD, we multiply each expression by 1, using the missing factors of the LCD to write 1:

$$\frac{5}{36x^2} + \frac{7}{24x} = \frac{5}{2 \cdot 2 \cdot 3 \cdot 3 \cdot x \cdot x} \cdot \frac{2}{2} + \frac{7}{2 \cdot 2 \cdot 2 \cdot 3 \cdot x} \cdot \frac{3 \cdot x}{3 \cdot x}$$

The LCD requires another factor of 2.   The LCD requires additional factors of 3 and x.

$$= \frac{10}{72x^2} + \frac{21x}{72x^2} \qquad \text{Both denominators are now the LCD.}$$

$$= \frac{21x + 10}{72x^2}.$$

You now have the "big" picture of why LCMs are needed when adding rational expressions. For the remainder of this section, we will practice finding LCMs and rewriting rational expressions so that they have the LCD as the denominator. In Section 7.4, we will return to the addition and subtraction of rational expressions.

EXAMPLE 6   For each pair of polynomials, find the least common multiple.

a) $15a$ and $35b$

b) $21x^3y^6$ and $7x^5y^2$

c) $x^2 + 5x - 6$ and $x^2 - 1$

SOLUTION

a) We write the prime factorizations and then construct the LCM:

$$15a = 3 \cdot 5 \cdot a$$
$$35b = 5 \cdot 7 \cdot b$$

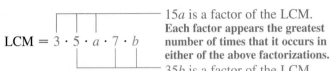

$15a$ is a factor of the LCM.
Each factor appears the greatest number of times that it occurs in either of the above factorizations.
$35b$ is a factor of the LCM.

The LCM is $3 \cdot 5 \cdot a \cdot 7 \cdot b$, or $105ab$.

b) $21x^3y^6 = 3 \cdot 7 \cdot x \cdot x \cdot x \cdot y \cdot y \cdot y \cdot y \cdot y \cdot y$ **Try to visualize the factors of x and y mentally.**
   $7x^5y^2 = 7 \cdot x \cdot x \cdot x \cdot x \cdot x \cdot y \cdot y$

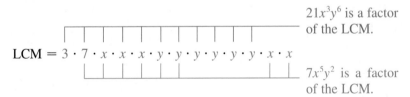

$21x^3y^6$ is a factor of the LCM.

$7x^5y^2$ is a factor of the LCM.

Note that we used the highest power of each factor in $21x^3y^6$ and $7x^5y^2$. The LCM is $21x^5y^6$.

c) $x^2 + 5x - 6 = (x - 1)(x + 6)$
     $x^2 - 1 = (x - 1)(x + 1)$

$$LCM = (x - 1)(x + 6)(x + 1)$$

$x^2 + 5x - 6$ is a factor of the LCM.

$x^2 - 1$ is a factor of the LCM.

The LCM is $(x - 1)(x + 6)(x + 1)$. There is no need to multiply this out.

The above procedure can be used to find the LCM of three polynomials as well. We factor each polynomial and then construct the LCM using each factor the greatest number of times that it appears in any one factorization.

Student Notes

If you prefer, the LCM for a group of three polynomials can be found by finding the LCM of two of them and then finding the LCM of that result and the remaining polynomial.

EXAMPLE 7 For each group of polynomials, find the LCM.

a) $12x$, $16y$, and $8xyz$ **b)** $x^2 + 4$, $x + 1$, and 5

SOLUTION

a) $12x = 2 \cdot 2 \cdot 3 \cdot x$
$16y = 2 \cdot 2 \cdot 2 \cdot 2 \cdot y$
$8xyz = 2 \cdot 2 \cdot 2 \cdot x \cdot y \cdot z$

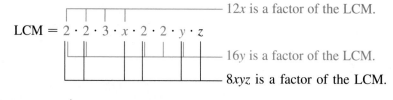

$$\text{LCM} = 2 \cdot 2 \cdot 3 \cdot x \cdot 2 \cdot 2 \cdot y \cdot z$$

12x is a factor of the LCM.

16y is a factor of the LCM.

8xyz is a factor of the LCM.

The LCM is $2^4 \cdot 3 \cdot xyz$, or $48xyz$.

b) Since $x^2 + 4$, $x + 1$, and 5 are not factorable, the LCM is their product: $5(x^2 + 4)(x + 1)$.

To add or subtract rational expressions with different denominators, we must be able to write equivalent expressions that have the LCD. Doing so usually requires multiplying each rational expression.

EXAMPLE 8 Find equivalent expressions that have the LCD:

$$\frac{x + 3}{x^2 + 5x - 6}, \qquad \frac{x + 7}{x^2 - 1}.$$

SOLUTION From Example 6(c), we know that the LCD is

$$(x + 6)(x - 1)(x + 1).$$

Since

$$x^2 + 5x - 6 = (x + 6)(x - 1),$$

the factor of the LCD that is missing from the first denominator is $x + 1$. We multiply by 1 using $(x + 1)/(x + 1)$:

$$\left.\begin{array}{l} \dfrac{x + 3}{x^2 + 5x - 6} = \dfrac{x + 3}{(x + 6)(x - 1)} \cdot \dfrac{x + 1}{x + 1} \\[2mm] \qquad\qquad = \dfrac{(x + 3)(x + 1)}{(x + 6)(x - 1)(x + 1)}. \end{array}\right\}$$ Finding an equivalent expression that has the least common denominator

For the second expression, we have $x^2 - 1 = (x + 1)(x - 1)$. The factor of the LCD that is missing is $x + 6$. We multiply by 1 using $(x + 6)/(x + 6)$:

$$\left.\begin{array}{l} \dfrac{x + 7}{x^2 - 1} = \dfrac{x + 7}{(x + 1)(x - 1)} \cdot \dfrac{x + 6}{x + 6} \\[2mm] \qquad\qquad = \dfrac{(x + 7)(x + 6)}{(x + 1)(x - 1)(x + 6)}. \end{array}\right\}$$ Finding an equivalent expression that has the least common denominator

We leave the results in factored form. In Section 7.4, we will carry out the actual addition and subtraction of such rational expressions.

7.3 EXERCISE SET

FOR EXTRA HELP

MathXL

MyMathLab

InterAct
Math

Tutor
Center
AW Math
Tutor Center

Video Lectures
on CD: Disc 4

Student's
Solutions Manual

🖐 *Concept Reinforcement   Use one or more words to complete each of the following sentences.*

1. To add two rational expressions when the denominators are the same, add _____ and keep the common _____.

2. When a numerator is being subtracted, be sure to treat its fraction bar as a _____ symbol and subtract every _____ in that numerator.

3. The least common multiple of two denominators is usually referred to as the _____ and is abbreviated _____.

4. The least common denominator of two fractions must contain the prime _____ of both _____.

Perform the indicated operation. Simplify, if possible.

5. $\dfrac{6}{x} + \dfrac{4}{x}$

6. $\dfrac{4}{a^2} + \dfrac{9}{a^2}$

7. $\dfrac{x}{12} + \dfrac{2x+5}{12}$

8. $\dfrac{a}{7} + \dfrac{3a-4}{7}$

9. $\dfrac{4}{a+3} + \dfrac{5}{a+3}$

10. $\dfrac{5}{x+2} + \dfrac{8}{x+2}$

11. $\dfrac{8}{a+2} - \dfrac{2}{a+2}$

12. $\dfrac{8}{x+7} - \dfrac{2}{x+7}$

13. $\dfrac{3y+8}{2y} - \dfrac{y+1}{2y}$

14. $\dfrac{5+3t}{4t} - \dfrac{2t+1}{4t}$

15. $\dfrac{7x+8}{x+1} + \dfrac{4x+3}{x+1}$

16. $\dfrac{3a+13}{a+4} + \dfrac{2a+7}{a+4}$

17. $\dfrac{7x+8}{x+1} - \dfrac{4x+3}{x+1}$

18. $\dfrac{3a+13}{a+4} - \dfrac{2a+7}{a+4}$

19. $\dfrac{a^2}{a-4} + \dfrac{a-20}{a-4}$

20. $\dfrac{x^2}{x+5} + \dfrac{7x+10}{x+5}$

21. $\dfrac{x^2}{x-2} - \dfrac{6x-8}{x-2}$

22. $\dfrac{a^2}{a+3} - \dfrac{2a+15}{a+3}$

Aha! 23. $\dfrac{t^2-5t}{t-1} + \dfrac{5t-t^2}{t-1}$

24. $\dfrac{y^2+6y}{y+2} + \dfrac{2y+12}{y+2}$

25. $\dfrac{x-6}{x^2+5x+6} + \dfrac{9}{x^2+5x+6}$

26. $\dfrac{x-5}{x^2-4x+3} + \dfrac{2}{x^2-4x+3}$

27. $\dfrac{3a^2+14}{a^2+5a-6} - \dfrac{13a}{a^2+5a-6}$

28. $\dfrac{2a^2+15}{a^2-7a+12} - \dfrac{11a}{a^2-7a+12}$

29. $\dfrac{t^2-3t}{t^2+6t+9} + \dfrac{2t-12}{t^2+6t+9}$

30. $\dfrac{y^2-7y}{y^2+8y+16} + \dfrac{6y-20}{y^2+8y+16}$

31. $\dfrac{2x^2+x}{x^2-8x+12} - \dfrac{x^2-2x+10}{x^2-8x+12}$

32. $\dfrac{2x^2+3}{x^2-6x+5} - \dfrac{3+2x^2}{x^2-6x+5}$

33. $\dfrac{3-2x}{x^2-6x+8} + \dfrac{7-3x}{x^2-6x+8}$

34. $\dfrac{1-2t}{t^2-5t+4} + \dfrac{4-3t}{t^2-5t+4}$

35. $\dfrac{x-9}{x^2+3x-4} - \dfrac{2x-5}{x^2+3x-4}$

36. $\dfrac{5-3x}{x^2-2x+1} - \dfrac{x+1}{x^2-2x+1}$

Find the LCM.

37. 15, 27

38. 10, 15

39. 8, 9

40. 12, 15

41. 6, 9, 21

42. 8, 36, 40

Find the LCM.

43. $12x^2$, $6x^3$

44. $10t^3$, $5t^4$

45. $15a^4b^7$, $10a^2b^8$

46. $6a^2b^7$, $9a^5b^2$

47. $2(y-3)$, $6(y-3)$

48. $4(x-1)$, $8(x-1)$

49. $x^2 - 4$, $x^2 + 5x + 6$

50. $x^2 + 3x + 2$, $x^2 - 4$

51. $t^3 + 4t^2 + 4t$, $t^2 - 4t$

52. $y^3 - y^2$, $y^4 - y^2$

53. $10x^2y$, $6y^2z$, $5xz^3$

54. $8x^3z$, $12xy^2$, $4y^5z^2$

55. $a + 1$, $(a-1)^2$, $a^2 - 1$

56. $x^2 - 9$, $x + 3$, $(x-3)^2$

57. $m^2 - 5m + 6$, $m^2 - 4m + 4$

58. $2x^2 + 5x + 2$, $2x^2 - x - 1$

Aha! **59.** $t - 3$, $t + 3$, $(t^2 - 9)^2$

60. $a - 5$, $(a^2 - 10a + 25)^2$

61. $6x^3 - 24x^2 + 18x$, $4x^5 - 24x^4 + 20x^3$

62. $9x^3 - 9x^2 - 18x$, $6x^5 - 24x^4 + 24x^3$

63. $t^3 - 1$, $t^2 - 1$

64. $5n + 5$, $n^3 + 1$

Find equivalent expressions that have the LCD.

65. $\dfrac{5}{6x^5}$, $\dfrac{y}{12x^3}$

66. $\dfrac{3}{10a^3}$, $\dfrac{b}{5a^6}$

67. $\dfrac{3}{2a^2b}$, $\dfrac{7}{8ab^2}$

68. $\dfrac{7}{3x^4y^2}$, $\dfrac{4}{9xy^3}$

69. $\dfrac{2x}{x^2 - 4}$, $\dfrac{4x}{x^2 + 5x + 6}$

70. $\dfrac{5x}{x^2 - 9}$, $\dfrac{2x}{x^2 + 11x + 24}$

TW 71. If the LCM of two numbers is their product, what can you conclude about the two numbers?

TW 72. Explain why the product of two numbers is not always their least common multiple.

Skill Maintenance

Write each number in two equivalent forms. [1.7]

73. $\dfrac{7}{-9}$

74. $-\dfrac{3}{2}$

Simplify. [1.6]

75. $\dfrac{5}{18} - \dfrac{7}{12}$

76. $\dfrac{8}{15} - \dfrac{13}{20}$

Find a polynomial that can represent the shaded area of each figure. [5.6]

77. 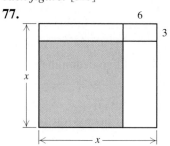   **78.**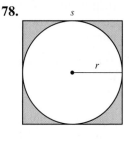

Synthesis

TW 79. If the LCM of a binomial and a trinomial is the trinomial, what relationship exists between the two expressions?

TW 80. If the LCM of two third-degree polynomials is a sixth-degree polynomial, what can be concluded about the two polynomials?

Perform the indicated operations. Simplify, if possible.

81. $\dfrac{6x - 1}{x - 1} + \dfrac{3(2x + 5)}{x - 1} + \dfrac{3(2x - 3)}{x - 1}$

82. $\dfrac{2x + 11}{x - 3} \cdot \dfrac{3}{x + 4} + \dfrac{-1}{4 + x} \cdot \dfrac{6x + 3}{x - 3}$

83. $\dfrac{x^2}{3x^2 - 5x - 2} - \dfrac{2x}{3x + 1} \cdot \dfrac{1}{x - 2}$

84. $\dfrac{x + y}{x^2 - y^2} + \dfrac{x - y}{x^2 - y^2} - \dfrac{2x}{x^2 - y^2}$

South African Artistry. **In South Africa, the design of** *every woven handbag, or* gipatsi (*plural,* sipatsi) *is created by repeating two or more geometric patterns. Each pattern encircles the bag, sharing the strands of fabric with any pattern above or below. The length, or period, of each pattern is the number of strands required to construct the pattern. For a gipatsi to be considered beautiful, each individual pattern must fit a whole number of times around the bag.* (*Source: Gerdes, Paulus,* Women, Art and Geometry in Southern Africa. *Asmara, Eritrea: Africa World Press, Inc., p. 5*)

85. A weaver is using two patterns to create a gipatsi. Pattern A is 10 strands long, and pattern B is 3 strands long. What is the smallest number of strands that can be used to complete the gipatsi?

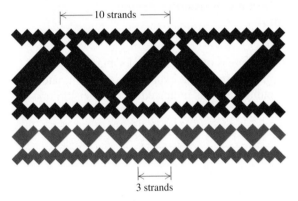

← 10 strands →

← 3 strands →

86. A weaver is using a four-strand pattern, a six-strand pattern, and an eight-strand pattern. What is the smallest number of strands that can be used to complete the gipatsi?

87. For technical reasons, the number of strands is generally a multiple of 4. Answer Exercise 85 with this additional requirement in mind.

Find the LCM.

88. 72, 90, 96

89. $8x^2 - 8$, $6x^2 - 12x + 6$, $10 - 10x$

90. $9x^2 - 16$, $6x^2 - x - 12$, $16 - 24x + 9x^2$

91. *Copiers.* The Brother® DCP-1000 copier can print 10 pages per minute. The Sharp® AL-1540CS copier can print 14 pages per minute. If both machines begin printing at the same instant, how long will it be until they again begin copying a page at exactly the same time?

92. *Running.* Kim and Jed leave the starting point of a fitness loop at the same time. Kim jogs a lap in 6 min and Jed jogs one in 8 min. Assuming they continue to run at the same pace, when will they next meet at the starting place?

93. *Bus Schedules.* Beginning at 5:00 A.M., a hotel shuttle bus leaves Salton Airport every 25 min, and the downtown shuttle bus leaves the airport every 35 min. What time will it be when both shuttles again leave at the same time?

94. *Appliances.* Dishwashers last an average of 10 yr, clothes washers an average of 14 yr, and refrigerators an average of 20 yr (*Source: Energy Savers: Tips on Saving Energy and Money at Home,* produced for the U.S. Department of Energy by the National Renewable Energy Laboratory, 1998). If an apartment house is equipped with new dishwashers, clothes washers, and refrigerators in 2002, in what year will all three appliances need to be replaced at once?

TW 95. Explain how evaluating can be used to perform a partial check on the result of Example 2(b):

$$\frac{x - 5}{x^2 - 9} + \frac{2}{x^2 - 9} = \frac{1}{x + 3}.$$

TW 96. On p. 549, the second step in finding an LCD is to select one of the factorizations of the denominators. Does it matter which one is selected? Why or why not?

7.4 Addition and Subtraction with Unlike Denominators

Adding and Subtracting with LCDs ▪ When Factors Are Opposites

Adding and Subtracting with LCDs

We now know how to rewrite two rational expressions in equivalent forms that use the LCD. Once rational expressions share a common denominator, they can be added or subtracted just as in Section 7.3.

> **To Add or Subtract Rational Expressions Having Different Denominators**
>
> 1. Find the LCD.
> 2. Multiply each rational expression by a form of 1 made up of the factors of the LCD missing from that expression's denominator.
> 3. Add or subtract the numerators, as indicated. Write the sum or difference over the LCD.
> 4. Simplify, if possible.

EXAMPLE 1 Add: $\dfrac{5x^2}{8} + \dfrac{7x}{12}$.

SOLUTION

1. First, we find the LCD:

$$\left. \begin{array}{l} 8 = 2 \cdot 2 \cdot 2 \\ 12 = 2 \cdot 2 \cdot 3 \end{array} \right\} \quad \text{LCD} = 2 \cdot 2 \cdot 2 \cdot 3, \text{ or } 24.$$

2. The denominator 8 needs to be multiplied by 3 in order to obtain the LCD. The denominator 12 needs to be multiplied by 2 in order to obtain the LCD. Thus we multiply the first expression by $\frac{3}{3}$ and the second expression by $\frac{2}{2}$ to get the LCD:

$$\frac{5x^2}{8} + \frac{7x}{12} = \frac{5x^2}{2 \cdot 2 \cdot 2} + \frac{7x}{2 \cdot 2 \cdot 3}$$

$$= \frac{5x^2}{2 \cdot 2 \cdot 2} \cdot \frac{3}{3} + \frac{7x}{2 \cdot 2 \cdot 3} \cdot \frac{2}{2} \qquad \text{Multiplying each expression by a form of 1 to get the LCD}$$

$$= \frac{15x^2}{24} + \frac{14x}{24}.$$

3. Next, we add the numerators:

$$\frac{15x^2}{24} + \frac{14x}{24} = \frac{15x^2 + 14x}{24}.$$

4. Since $15x^2 + 14x$ and 24 have no common factor,

$$\frac{15x^2 + 14x}{24}$$

cannot be simplified any further. ◢

Subtraction is performed in much the same way.

EXAMPLE 2 Subtract: $\dfrac{7}{8x} - \dfrac{5}{12x^2}$.

SOLUTION We follow the four steps shown above. First, we find the LCD:

$$\left.\begin{array}{l} 8x = 2 \cdot 2 \cdot 2 \cdot x \\ 12x^2 = 2 \cdot 2 \cdot 3 \cdot x \cdot x \end{array}\right\} \quad \text{LCD} = 2 \cdot 2 \cdot 3 \cdot x \cdot x \cdot 2, \text{ or } 24x^2.$$

The denominator $8x$ must be multiplied by $3x$ in order to obtain the LCD. The denominator $12x^2$ must be multiplied by 2 in order to obtain the LCD. Thus we multiply by $\dfrac{3x}{3x}$ and $\dfrac{2}{2}$ to get the LCD. Then we subtract and, if possible, simplify.

$$\begin{aligned} \frac{7}{8x} - \frac{5}{12x^2} &= \frac{7}{8x} \cdot \frac{3x}{3x} - \frac{5}{12x^2} \cdot \frac{2}{2} \\[2mm] &= \frac{21x}{24x^2} - \frac{10}{24x^2} \quad \longleftarrow \end{aligned}$$

CAUTION! Do not simplify *these* rational expressions or you will lose the LCD.

$$= \frac{21x - 10}{24x^2} \qquad \text{This cannot be simplified, so we are done.} \quad ◢$$

When denominators contain polynomials with two or more terms, the same steps are used.

EXAMPLE 3 Add: $\dfrac{2a}{a^2 - 1} + \dfrac{1}{a^2 + a}$.

SOLUTION First, we find the LCD:

$$\left.\begin{array}{l} a^2 - 1 = (a - 1)(a + 1) \\ a^2 + a = a(a + 1) \end{array}\right\} \quad \text{LCD} = (a - 1)(a + 1)a.$$

We multiply by a form of 1 to get the LCD in each expression:

$$\frac{2a}{a^2 - 1} + \frac{1}{a^2 + a} = \frac{2a}{(a-1)(a+1)} \cdot \frac{a}{a} + \frac{1}{a(a+1)} \cdot \frac{a-1}{a-1}$$

Multiplying by $\dfrac{a}{a}$ and $\dfrac{a-1}{a-1}$ to get the LCD

$$= \frac{2a^2}{(a-1)(a+1)a} + \frac{a-1}{a(a+1)(a-1)}$$

$$= \frac{2a^2 + a - 1}{a(a-1)(a+1)}$$

Adding numerators

$$= \frac{(2a-1)\cancel{(a+1)}}{a(a-1)\cancel{(a+1)}}$$

$$= \frac{2a-1}{a(a-1)}.$$

Simplifying by factoring and removing a factor equal to 1: $\dfrac{a+1}{a+1} = 1$

EXAMPLE 4   Subtract: $\dfrac{x+4}{x-2} - \dfrac{x-7}{x+5}$.

SOLUTION   First, we find the LCD. It is just the product of the denominators:

$$\text{LCD} = (x-2)(x+5).$$

We multiply by a form of 1 to get the LCD in each expression. Then we subtract and try to simplify.

$$\frac{x+4}{x-2} - \frac{x-7}{x+5} = \frac{x+4}{x-2} \cdot \frac{x+5}{x+5} - \frac{x-7}{x+5} \cdot \frac{x-2}{x-2}$$

$$= \frac{x^2 + 9x + 20}{(x-2)(x+5)} - \frac{x^2 - 9x + 14}{(x-2)(x+5)}$$

Multiplying out numerators (but not denominators)

$$= \frac{x^2 + 9x + 20 - (x^2 - 9x + 14)}{(x-2)(x+5)}$$

When subtracting a numerator with more than one term, parentheses are important.

$$= \frac{x^2 + 9x + 20 - x^2 + 9x - 14}{(x-2)(x+5)}$$

Removing parentheses and subtracting every term

$$= \frac{18x + 6}{(x-2)(x+5)}$$

Although $18x + 6$ can be factored as $6(3x + 1)$, doing so will not enable us to simplify our result.

When possible, we simplify rational expressions before finding the LCD.

EXAMPLE 5   Add: $\dfrac{x^2}{x^2 + 2xy + y^2} + \dfrac{2x - 2y}{x^2 - y^2}$.

SOLUTION   To find the LCD, we first factor the denominators. We also factor numerators to see if we can simplify either expression. In this case, the rightmost rational expression can be simplified:

$$\frac{x^2}{x^2 + 2xy + y^2} + \frac{2x - 2y}{x^2 - y^2} = \frac{x^2}{(x+y)(x+y)} + \frac{2(x-y)}{(x+y)(x-y)} \quad \text{Factoring}$$

$$= \frac{x^2}{(x+y)(x+y)} + \frac{2}{x+y}. \quad \begin{array}{l}\text{Removing a factor} \\ \text{equal to 1:} \\ \frac{x-y}{x-y} = 1\end{array}$$

Note that the LCM of $(x+y)(x+y)$ and $(x+y)$ is $(x+y)(x+y)$. The factor missing in the denominator of the second expression is $x + y$. We multiply that expression by 1, using $(x+y)/(x+y)$. Then we add and, if possible, simplify.

$$\frac{x^2}{(x+y)(x+y)} + \frac{2}{x+y} = \frac{x^2}{(x+y)(x+y)} + \frac{2}{x+y}\cdot\frac{x+y}{x+y}$$

$$= \frac{x^2}{(x+y)(x+y)} + \frac{2x+2y}{(x+y)(x+y)} \quad \begin{array}{l}\text{We now} \\ \text{have the} \\ \text{LCD.}\end{array}$$

$$= \frac{x^2 + 2x + 2y}{(x+y)(x+y)} \quad \begin{array}{l}\text{Since the numerator} \\ \text{cannot be factored, we} \\ \text{cannot simplify further.}\end{array}$$

EXAMPLE 6   Subtract: $\dfrac{x}{x^2 + 5x + 6} - \dfrac{2}{x^2 + 3x + 2}.$

SOLUTION

$$\frac{x}{x^2 + 5x + 6} - \frac{2}{x^2 + 3x + 2}$$

$$= \frac{x}{(x+2)(x+3)} - \frac{2}{(x+2)(x+1)} \quad \begin{array}{l}\text{Factoring denominators. The} \\ \text{LCD is }(x+2)(x+3)(x+1).\end{array}$$

$$= \frac{x}{(x+2)(x+3)}\cdot\frac{x+1}{x+1} - \frac{2}{(x+2)(x+1)}\cdot\frac{x+3}{x+3}$$

$$= \frac{x^2 + x}{(x+2)(x+3)(x+1)} - \frac{2x+6}{(x+2)(x+3)(x+1)}$$

$$= \frac{x^2 + x - (2x+6)}{(x+2)(x+3)(x+1)} \quad \text{Don't forget the parentheses!}$$

$$= \frac{x^2 + x - 2x - 6}{(x+2)(x+3)(x+1)} \quad \begin{array}{l}\text{Remember to subtract} \\ \text{each term in }2x+6.\end{array}$$

$$= \frac{x^2 - x - 6}{(x+2)(x+3)(x+1)} \quad \begin{array}{l}\text{Combining like terms in} \\ \text{the numerator}\end{array}$$

$$= \frac{(x+2)(x-3)}{(x+2)(x+3)(x+1)} \quad \begin{array}{l}\text{Factoring and simplifying;} \\ \frac{x+2}{x+2} = 1\end{array}$$

$$= \frac{x-3}{(x+3)(x+1)}$$

Student Notes

As you can see, adding or subtracting rational expressions can involve many steps. Therefore, it is important to double-check each step of your work as you work through each problem. Waiting to check your work at the end of the problems is usually a less efficient use of your time.

When Factors Are Opposites

Expressions of the form $a - b$ and $b - a$ are opposites of each other. When either of these binomials is multiplied by -1, the result is the other binomial:

$$\left.\begin{array}{l} -1(a - b) = -a + b = b + (-a) = b - a; \\ -1(b - a) = -b + a = a + (-b) = a - b. \end{array}\right\}$$ Multiplication by -1 reverses the order in which subtraction occurs.

When one denominator is the opposite of the other, we can multiply either expression by 1 using $\dfrac{-1}{-1}$.

EXAMPLE 7 Add: $\dfrac{3}{8a} + \dfrac{1}{-8a}$.

SOLUTION

$$\frac{3}{8a} + \frac{1}{-8a} = \frac{3}{8a} + \frac{-1}{-1} \cdot \frac{1}{-8a}$$

$$= \frac{3}{8a} + \frac{-1}{8a} = \frac{2}{8a}$$

$$= \frac{\cancel{2} \cdot 1}{\cancel{2} \cdot 4a} = \frac{1}{4a}$$

> When denominators are opposites, we multiply one rational expression by $-1/-1$ to get the LCD.

Simplifying by removing a factor equal to 1: $\dfrac{2}{2} = 1$

EXAMPLE 8 Subtract: $\dfrac{5x}{x - 2y} - \dfrac{3y - 7}{2y - x}$.

SOLUTION

$$\frac{5x}{x - 2y} - \frac{3y - 7}{2y - x} = \frac{5x}{x - 2y} - \frac{-1}{-1} \cdot \frac{3y - 7}{2y - x}$$ Note that $x - 2y$ and $2y - x$ are opposites.

$$= \frac{5x}{x - 2y} - \frac{7 - 3y}{x - 2y}$$ Performing the multiplication. *Note*: $-1(2y - x) = -2y + x$ $= x - 2y$.

$$\left.\begin{array}{l} = \dfrac{5x - (7 - 3y)}{x - 2y} \\[2mm] = \dfrac{5x - 7 + 3y}{x - 2y} \end{array}\right\}$$ Subtracting. The parentheses are important.

EXAMPLE 9 Find simplified form for the function given by

$$f(x) = \frac{2x}{x^2 - 4} + \frac{5}{2 - x} - \frac{1}{2 + x}$$

and list all restrictions on the domain.

SOLUTION We have

$$\frac{2x}{x^2 - 4} + \frac{5}{2 - x} - \frac{1}{2 + x}$$

$$= \frac{2x}{(x-2)(x+2)} + \frac{5}{2-x} - \frac{1}{2+x} \qquad \text{Factoring. Note that } x \neq -2, 2.$$

$$= \frac{2x}{(x-2)(x+2)} + \frac{-1}{-1} \cdot \frac{5}{(2-x)} - \frac{1}{x+2} \qquad \begin{array}{l}\text{Multiplying by}\\ -1/-1 \text{ since } 2 - x \text{ is}\\ \text{the opposite of } x - 2\end{array}$$

$$= \frac{2x}{(x-2)(x+2)} + \frac{-5}{x-2} - \frac{1}{x+2} \qquad \text{The LCD is } (x-2)(x+2).$$

$$= \frac{2x}{(x-2)(x+2)} + \frac{-5}{x-2} \cdot \frac{x+2}{x+2} - \frac{1}{x+2} \cdot \frac{x-2}{x-2} \qquad \begin{array}{l}\text{Multiplying by 1}\\ \text{to get the LCD}\end{array}$$

$$= \frac{2x - 5(x+2) - (x-2)}{(x-2)(x+2)}$$

$$= \frac{2x - 5x - 10 - x + 2}{(x-2)(x+2)}$$

$$= \frac{-4x - 8}{(x-2)(x+2)}$$

$$= \frac{-4(x+2)}{(x-2)(x+2)}$$

$$= \frac{-4\cancel{(x+2)}}{(x-2)\cancel{(x+2)}} \qquad \begin{array}{l}\text{Removing a factor equal to 1:}\\ \frac{x+2}{x+2} = 1, x \neq -2\end{array}$$

$$= \frac{-4}{x-2}, \text{ or } -\frac{4}{x-2}, x \neq \pm 2.$$

An equivalent answer is $\dfrac{4}{2-x}$. It is found by writing $-\dfrac{4}{x-2}$ as $\dfrac{4}{-(x-2)}$ and then using the distributive law to remove parentheses. ◢

Our work in Example 9 indicates that if

$$f(x) = \frac{2x}{x^2 - 4} + \frac{5}{2 - x} - \frac{1}{2 + x} \quad \text{and} \quad g(x) = \frac{-4}{x - 2},$$

then, for $x \neq -2$ and $x \neq 2$, we have $f = g$. Note that whereas the domain of f includes all real numbers except -2 or 2, the domain of g excludes only 2. This is illustrated in the following graphs. Methods for drawing such graphs by hand are discussed in more advanced courses. The graphs are for visualization only.

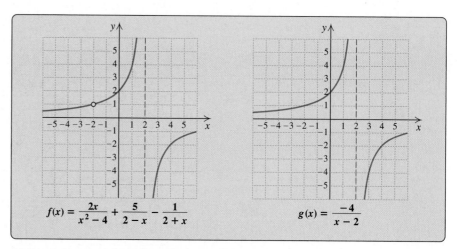

A computer-generated visualization of Example 9

A quick, partial check of any simplification is to evaluate both the original and the simplified expressions for a convenient choice of x. For instance, to check Example 9, if $x = 1$, we have

$$f(1) = \frac{2 \cdot 1}{1^2 - 4} + \frac{5}{2 - 1} - \frac{1}{2 + 1}$$

$$= \frac{2}{-3} + \frac{5}{1} - \frac{1}{3} = 5 - \frac{3}{3} = 4$$

and

$$g(1) = \frac{-4}{1 - 2} = \frac{-4}{-1} = 4.$$

Since both functions include the pair $(1, 4)$, our algebra was *probably* correct. Although this is only a partial check (on rare occasions, an incorrect answer might "check"), because it is so easy to perform, it is nonetheless very useful. Further evaluation provides a more definitive check.

7.4 EXERCISE SET

Concept Reinforcement   *In Exercises 1–4, the four steps for adding rational expressions with different denominators are listed. Fill in the missing word or words for each step.*

1. To add or subtract when the denominators are different, first find the ——————.

2. Multiply each rational expression by a form of 1 made up of the factors of the LCD that are —————— from that expression's ——————.

3. Add or subtract the ——————, as indicated. Write the sum or difference over the ——————.

4. ——————, if possible.

Perform the indicated operation. Simplify, if possible.

5. $\dfrac{4}{x} + \dfrac{9}{x^2}$

6. $\dfrac{5}{x} + \dfrac{6}{x^2}$

7. $\dfrac{1}{6r} - \dfrac{3}{8r}$

8. $\dfrac{4}{9t} - \dfrac{7}{6t}$

9. $\dfrac{2}{c^2d} + \dfrac{7}{cd^3}$

10. $\dfrac{4}{xy^2} + \dfrac{2}{x^2y}$

11. $\dfrac{-2}{3xy^2} - \dfrac{6}{x^2y^3}$

12. $\dfrac{8}{9t^3} - \dfrac{5}{6t^2}$

13. $\dfrac{x-4}{9} + \dfrac{x+5}{6}$

14. $\dfrac{x+5}{8} + \dfrac{x-3}{12}$

15. $\dfrac{a+2}{2} - \dfrac{a-4}{4}$

16. $\dfrac{x-2}{6} - \dfrac{x+1}{3}$

17. $\dfrac{2a-1}{3a^2} + \dfrac{5a+1}{9a}$

18. $\dfrac{a+4}{16a} + \dfrac{3a+4}{4a^2}$

19. $\dfrac{x-1}{4x} - \dfrac{2x+3}{x}$

20. $\dfrac{4z-9}{3z} - \dfrac{3z-8}{4z}$

21. $\dfrac{2c-d}{c^2d} + \dfrac{c+d}{cd^2}$

22. $\dfrac{x+y}{xy^2} + \dfrac{3x+y}{x^2y}$

23. $\dfrac{5x+3y}{2x^2y} - \dfrac{3x+4y}{xy^2}$

24. $\dfrac{4x+2t}{3xt^2} - \dfrac{5x-3t}{x^2t}$

25. $\dfrac{5}{x-1} + \dfrac{5}{x+1}$

26. $\dfrac{3}{x-2} + \dfrac{3}{x+2}$

27. $\dfrac{4}{z-1} - \dfrac{2}{z+1}$

28. $\dfrac{5}{x+5} - \dfrac{3}{x-5}$

29. $\dfrac{2}{x+5} + \dfrac{3}{4x}$

30. $\dfrac{3}{x+1} + \dfrac{2}{3x}$

31. $\dfrac{8}{3t^2-15t} - \dfrac{3}{2t-10}$

32. $\dfrac{3}{2t^2-2t} - \dfrac{5}{2t-2}$

33. $\dfrac{4x}{x^2-25} + \dfrac{x}{x+5}$

34. $\dfrac{2x}{x^2-16} + \dfrac{x}{x-4}$

35. $\dfrac{t}{t-3} - \dfrac{5}{4t-12}$

36. $\dfrac{6}{z+4} - \dfrac{2}{3z+12}$

37. $\dfrac{2}{x+3} + \dfrac{4}{(x+3)^2}$

38. $\dfrac{3}{x-1} + \dfrac{2}{(x-1)^2}$

39. $\dfrac{3}{x+2} - \dfrac{8}{x^2-4}$

40. $\dfrac{2t}{t^2-9} - \dfrac{3}{t-3}$

41. $\dfrac{3a}{4a-20} + \dfrac{9a}{6a-30}$

42. $\dfrac{4a}{5a-10} + \dfrac{3a}{10a-20}$

Aha! 43. $\dfrac{x}{x-5} + \dfrac{x}{5-x}$

44. $\dfrac{x+4}{x} + \dfrac{x}{x+4}$

45. $\dfrac{7}{a^2+a-2} + \dfrac{5}{a^2-4a+3}$

46. $\dfrac{x}{x^2+2x+1} + \dfrac{1}{x^2+5x+4}$

47. $\dfrac{x}{x^2+9x+20} - \dfrac{4}{x^2+7x+12}$

48. $\dfrac{x}{x^2+5x+6} - \dfrac{2}{x^2+3x+2}$

49. $\dfrac{3z}{z^2-4z+4} + \dfrac{10}{z^2+z-6}$

50. $\dfrac{3}{x^2 - 9} + \dfrac{2}{x^2 - x - 6}$

Aha! 51. $\dfrac{-5}{x^2 + 17x + 16} - \dfrac{0}{x^2 + 9x + 8}$

52. $\dfrac{x}{x^2 + 15x + 56} - \dfrac{1}{x^2 + 13x + 42}$

53. $\dfrac{4x}{5} - \dfrac{x - 3}{-5}$

54. $\dfrac{x}{4} - \dfrac{3x - 5}{-4}$

55. $\dfrac{y^2}{y - 3} + \dfrac{9}{3 - y}$

56. $\dfrac{t^2}{t - 2} + \dfrac{4}{2 - t}$

57. $\dfrac{t^2 + 3}{t^4 - 16} + \dfrac{7}{16 - t^4}$

58. $\dfrac{y^2 - 5}{y^4 - 81} + \dfrac{4}{81 - y^4}$

59. $\dfrac{m - 3n}{m^3 - n^3} - \dfrac{2n}{n^3 - m^3}$

60. $\dfrac{r - 6s}{r^3 - s^3} - \dfrac{5s}{s^3 - r^3}$

61. $\dfrac{y + 2}{y - 7} + \dfrac{3 - y}{49 - y^2}$

62. $\dfrac{4 - p}{25 - p^2} + \dfrac{p + 1}{p - 5}$

63. $\dfrac{3x + 2}{3x + 6} + \dfrac{x}{4 - x^2}$

64. $\dfrac{a}{a^2 - 1} + \dfrac{2a}{a - a^2}$

65. $\dfrac{4 - a^2}{a^2 - 9} - \dfrac{a - 2}{3 - a}$

66. $\dfrac{4x}{x^2 - y^2} - \dfrac{6}{y - x}$

Perform the indicated operations. Simplify, if possible.

67. $\dfrac{x - 3}{2 - x} - \dfrac{x + 3}{x + 2} + \dfrac{x + 6}{4 - x^2}$

68. $\dfrac{t - 5}{1 - t} - \dfrac{t + 4}{t + 1} + \dfrac{t + 2}{t^2 - 1}$

69. $\dfrac{x + 5}{x + 3} + \dfrac{x + 7}{x + 2} - \dfrac{7x + 19}{(x + 3)(x + 2)}$

70. $\dfrac{2x + 5}{x + 1} + \dfrac{x + 7}{x + 5} - \dfrac{5x + 17}{(x + 1)(x + 5)}$

71. $\dfrac{1}{x + y} + \dfrac{1}{x - y} - \dfrac{2x}{x^2 - y^2}$

72. $\dfrac{2r}{r^2 - s^2} + \dfrac{1}{r + s} - \dfrac{1}{r - s}$

Find simplified form for f(x) and list all restrictions on the domain.

73. $f(x) = 2 + \dfrac{x}{x - 3} - \dfrac{18}{x^2 - 9}$

74. $f(x) = 5 + \dfrac{x}{x + 2} - \dfrac{8}{x^2 - 4}$

75. $f(x) = \dfrac{3x - 1}{x^2 + 2x - 3} - \dfrac{x + 4}{x^2 - 16}$

76. $f(x) = \dfrac{3x - 2}{x^2 + 2x - 24} - \dfrac{x - 3}{x^2 - 9}$

Aha! 77. $f(x) = \dfrac{1}{x^2 + 5x + 6} - \dfrac{2}{x^2 + 3x + 2} - \dfrac{1}{x^2 + 5x + 6}$

78. $f(x) = \dfrac{2}{x^2 - 5x + 6} - \dfrac{4}{x^2 - 2x - 3} + \dfrac{2}{x^2 + 4x + 3}$

TW 79. Janine found that the sum of two rational expressions was $(3 - x)/(x - 5)$. The answer given at the back of the book is $(x - 3)/(5 - x)$. Is Janine's answer incorrect? Why or why not?

TW 80. When two rational expressions are added or subtracted, should the numerator of the result be factored? Why or why not?

Skill Maintenance

Simplify. Use only positive exponents in your answer. [5.2]

81. $\dfrac{15x^{-7}y^{12}z^4}{35x^{-2}y^6z^{-3}}$

82. $\dfrac{21a^{-4}b^6c^8}{27a^{-2}b^{-5}c}$

Simplify.

83. $\dfrac{\frac{2}{3}}{\frac{6}{7}}$ [1.3]

84. $\dfrac{-\frac{5}{8}}{\frac{3}{4}}$ [1.7]

Graph. [3.2]

85. $y = -\dfrac{1}{2}x - 5$

86. $y = \dfrac{1}{2}x - 5$

Synthesis

TW 87. How could you convince someone that

$$\dfrac{1}{3 - x} \quad \text{and} \quad \dfrac{1}{x - 3}$$

are opposites of each other?

88. Are parentheses as important for adding rational expressions as they are for subtracting rational expressions? Why or why not?

Write expressions for the perimeter and the area of each rectangle.

89.
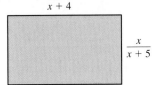

$\dfrac{3}{x+4}$

$\dfrac{2}{x-5}$

90.

$\dfrac{x}{x+4}$

$\dfrac{x}{x+5}$

Perform the indicated operations and simplify.

91. $\dfrac{2x+11}{x-3} \cdot \dfrac{3}{x+4} + \dfrac{2x+1}{4+x} \cdot \dfrac{3}{3-x}$

92. $\dfrac{x^2}{3x^2-5x-2} - \dfrac{2x}{3x+1} \cdot \dfrac{1}{x-2}$

Aha! 93. $\left(\dfrac{x}{x+7} - \dfrac{3}{x+2}\right)\left(\dfrac{x}{x+7} + \dfrac{3}{x+2}\right)$

94. $\dfrac{1}{ay-3a+2xy-6x} - \dfrac{xy+ay}{a^2-4x^2}\left(\dfrac{1}{y-3}\right)^2$

95. $\dfrac{2x^2+5x-3}{2x^2-9x+9} + \dfrac{x+1}{3-2x} + \dfrac{4x^2+8x+3}{x-3} \cdot \dfrac{x+3}{9-4x^2}$

96. $\left(\dfrac{a}{a-b} + \dfrac{b}{a+b}\right)\left(\dfrac{1}{3a+b} + \dfrac{2a+6b}{9a^2-b^2}\right)$

97. $5(x-3)^{-1} + 4(x+3)^{-1} - 2(x+3)^{-2}$

98. $4(y-1)(2y-5)^{-1} + 5(2y+3)(5-2y)^{-1} + (y-4)(2y-5)^{-1}$

99. Express

$$\dfrac{a-3b}{a-b}$$

as a sum of two rational expressions with denominators that are opposites of each other. Answers may vary.

100. Determine the domain and the range of the function graphed below.

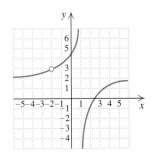

If

$$f(x) = \dfrac{x^3}{x^2-4} \quad and \quad g(x) = \dfrac{x^2}{x^2+3x-10},$$

find each of the following.

101. $(f+g)(x)$ **102.** $(f-g)(x)$

103. $(f \cdot g)(x)$ **104.** $(f/g)(x)$

105. The domain of $f+g$

106. The domain of f/g

Determine the domain and estimate the range of each function.

107. $f(x) = 2 + \dfrac{x-3}{x+1}$

108. $g(x) = \dfrac{2}{(x+1)^2} + 5$

109. $r(x) = \dfrac{1}{x^2} + \dfrac{1}{(x-1)^2}$

7.5 Complex Rational Expressions

Multiplying by 1 ■ Dividing Two Rational Expressions

A **complex rational expression** is a rational expression that contains rational expressions within its numerator and/or its denominator. Here are some examples:

$$\dfrac{x + \dfrac{5}{x}}{4x}, \quad \dfrac{\dfrac{x-y}{x+y}}{\dfrac{2x-y}{3x+y}}, \quad \dfrac{\dfrac{7x}{3} - \dfrac{4}{x}}{\dfrac{5x}{6} + \dfrac{8}{3}}, \quad \dfrac{\dfrac{r}{6} + \dfrac{r^2}{144}}{\dfrac{r}{12}}.$$

The rational expressions within each complex rational expression are red.

Complex rational expressions arise in a variety of real-world applications. For example, the last complex rational expression in the list above is used when calculating the size of certain loan payments.

Two methods are used to simplify complex rational expressions. Determining restrictions on variables may now require the solution of equations not yet studied. *Thus for this section we will not state restrictions on variables.*

Method 1: Multiplying by 1

One method of simplifying a complex rational expression is to multiply the entire expression by 1. To write 1, we use the LCD of the rational expressions within the complex rational expression.

Student Notes

In Example 1, we found the LCD for four rational expressions. To find this LCD, you may prefer to (1) find one LCD for the two expressions in the numerator, (2) find a second LCD for the expressions in the denominator, and (3) find the LCM of those two LCDs.

EXAMPLE 1 Simplify:

$$\dfrac{\dfrac{1}{a^3b} + \dfrac{1}{b}}{\dfrac{1}{a^2b^2} - \dfrac{1}{b^2}}.$$

SOLUTION The denominators within the complex rational expression are a^3b, b, a^2b^2, and b^2. Thus the LCD is a^3b^2. We multiply by 1, using $(a^3b^2)/(a^3b^2)$:

$$\dfrac{\dfrac{1}{a^3b} + \dfrac{1}{b}}{\dfrac{1}{a^2b^2} - \dfrac{1}{b^2}} = \dfrac{\dfrac{1}{a^3b} + \dfrac{1}{b}}{\dfrac{1}{a^2b^2} - \dfrac{1}{b^2}} \cdot \dfrac{a^3b^2}{a^3b^2}$$

Multiplying by 1, using the LCD

$$= \dfrac{\left(\dfrac{1}{a^3b} + \dfrac{1}{b}\right)a^3b^2}{\left(\dfrac{1}{a^2b^2} - \dfrac{1}{b^2}\right)a^3b^2}$$

Multiplying the numerator and the denominator. Remember to use parentheses.

$$= \frac{\dfrac{1}{a^3b} \cdot a^3b^2 + \dfrac{1}{b} \cdot a^3b^2}{\dfrac{1}{a^2b^2} \cdot a^3b^2 - \dfrac{1}{b^2} \cdot a^3b^2}$$ **Using the distributive law to carry out the multiplications**

$$= \frac{\dfrac{\cancel{a^3b}}{\cancel{a^3b}} \cdot b + \dfrac{\cancel{b}}{\cancel{b}} \cdot a^3b}{\dfrac{\cancel{a^2b^2}}{\cancel{a^2b^2}} \cdot a - \dfrac{\cancel{b^2}}{\cancel{b^2}} \cdot a^3}$$ **Removing factors that equal 1. Study this carefully.**

$$= \frac{b + a^3b}{a - a^3}$$ **Simplifying**

$$= \frac{b(1 + a^3)}{a(1 - a^2)}$$ **Factoring**

$$= \frac{b\cancel{(1 + a)}(1 - a + a^2)}{a\cancel{(1 + a)}(1 - a)}$$ **Factoring further and identifying a factor that equals 1**

$$= \frac{b(1 - a + a^2)}{a(1 - a)}.$$ **Simplifying**

Using Multiplication by 1 to Simplify a Complex Rational Expression

1. Find the LCD of all rational expressions *within* the complex rational expression.
2. Multiply the complex rational expression by 1, writing 1 as the LCD divided by itself.
3. Distribute and simplify so that the numerator and the denominator of the complex rational expression are polynomials.
4. Factor and, if possible, simplify.

Note that use of the LCD when multiplying by a form of 1 clears the numerator and the denominator of the complex rational expression of all rational expressions.

EXAMPLE 2   Simplify:

$$\frac{\dfrac{3}{2x - 2} - \dfrac{1}{x + 1}}{\dfrac{1}{x - 1} + \dfrac{x}{x^2 - 1}}.$$

SOLUTION   In this case, to find the LCD, we have to factor first:

$$\frac{\dfrac{3}{2x - 2} - \dfrac{1}{x + 1}}{\dfrac{1}{x - 1} + \dfrac{x}{x^2 - 1}} = \frac{\dfrac{3}{2(x - 1)} - \dfrac{1}{x + 1}}{\dfrac{1}{x - 1} + \dfrac{x}{(x - 1)(x + 1)}}$$ **The LCD is** $2(x - 1)(x + 1)$.

Student Notes

Writing $2(x - 1)(x + 1)$ as

$$\frac{2(x - 1)(x + 1)}{1}$$

may help with lining up numerators and denominators for multiplying.

$$= \frac{\dfrac{3}{2(x - 1)} - \dfrac{1}{x + 1}}{\dfrac{1}{x - 1} + \dfrac{x}{(x - 1)(x + 1)}} \cdot \frac{2(x - 1)(x + 1)}{2(x - 1)(x + 1)} \quad \begin{array}{l}\textbf{Multiplying by 1,}\\ \textbf{using the LCD}\end{array}$$

$$= \frac{\dfrac{3}{2(x - 1)} \cdot 2(x - 1)(x + 1) - \dfrac{1}{x + 1} \cdot 2(x - 1)(x + 1)}{\dfrac{1}{x - 1} \cdot 2(x - 1)(x + 1) + \dfrac{x}{(x - 1)(x + 1)} \cdot 2(x - 1)(x + 1)}$$

Using the distributive law

$$= \frac{\dfrac{2(x - 1)}{2(x - 1)} \cdot 3(x + 1) - \dfrac{x + 1}{x + 1} \cdot 2(x - 1)}{\dfrac{x - 1}{x - 1} \cdot 2(x + 1) + \dfrac{(x - 1)(x + 1)}{(x - 1)(x + 1)} \cdot 2x} \quad \begin{array}{l}\textbf{Removing factors}\\ \textbf{that equal 1}\end{array}$$

$$= \frac{3(x + 1) - 2(x - 1)}{2(x + 1) + 2x} \quad \textbf{Simplifying}$$

$$= \frac{3x + 3 - 2x + 2}{2x + 2 + 2x} \quad \textbf{Using the distributive law}$$

$$= \frac{x + 5}{4x + 2}. \quad \textbf{Combining like terms}$$

Entering Rational Expressions

We can check the simplification of rational expressions in one variable using a graphing calculator. To enter a complex rational expression, we use parentheses around the entire numerator and parentheses around the entire denominator. If necessary, we also use parentheses around numerators and denominators of rational expressions within the complex rational expression. Following are some examples of complex rational expressions rewritten with parentheses.

Complex Rational Expression	Expression Rewritten with Parentheses
$\dfrac{x + \dfrac{5}{x}}{4x}$	$(x + 5/x)/(4x)$
$\dfrac{\dfrac{3}{2x - 2} - \dfrac{1}{x + 1}}{\dfrac{1}{x - 1} + \dfrac{x}{x^2 - 1}}$	$(3/(2x - 2) - 1/(x + 1))/(1/(x - 1) + x/(x^2 - 1))$

The following may help in placing parentheses properly.

1. Close each set of parentheses: For every left parenthesis, there should be a right parenthesis.
2. Remember the rules for order of operations when deciding how to place parentheses.
3. When in doubt, place parentheses around every numerator and around every denominator. These extra parentheses will make the expression look more complicated, but will not change its value.

Method 2: Dividing Two Rational Expressions

Another method for simplifying complex rational expressions involves first adding or subtracting, as necessary, to obtain one rational expression in the numerator and one rational expression in the denominator. The problem is then a division of two rational expressions.

EXAMPLE 3   Simplify:

$$\frac{\dfrac{3}{x} - \dfrac{2}{x^2}}{\dfrac{3}{x-2} + \dfrac{1}{x^2}}.$$

SOLUTION

$$\frac{\dfrac{3}{x} - \dfrac{2}{x^2}}{\dfrac{3}{x-2} + \dfrac{1}{x^2}} = \frac{\dfrac{3}{x} \cdot \dfrac{x}{x} - \dfrac{2}{x^2}}{\dfrac{3}{x-2} \cdot \dfrac{x^2}{x^2} + \dfrac{1}{x^2} \cdot \dfrac{x-2}{x-2}}$$

Multiplying $3/x$ by 1 to obtain x^2 as a common denominator

Multiplying by 1, twice, to obtain $x^2(x-2)$ as a common denominator

$$= \frac{\dfrac{3x}{x^2} - \dfrac{2}{x^2}}{\dfrac{3x^2}{(x-2)x^2} + \dfrac{x-2}{x^2(x-2)}}$$

There is now a common denominator in the numerator and a common denominator in the denominator of the complex rational expression.

$$= \frac{\dfrac{3x-2}{x^2}}{\dfrac{3x^2 + x - 2}{(x-2)x^2}}$$

Subtracting in the numerator and adding in the denominator. We now have one rational expression divided by another rational expression.

$$= \frac{3x-2}{x^2} \div \frac{3x^2 + x - 2}{(x-2)x^2}$$

Rewriting with a division symbol

$$= \frac{3x-2}{x^2} \cdot \frac{(x-2)x^2}{3x^2 + x - 2}$$

To divide, multiply by the reciprocal of the divisor.

$$= \frac{(3x-2)(x-2)x^2}{x^2(3x-2)(x+1)}$$

Factoring and removing a factor equal to 1: $\dfrac{x^2(3x-2)}{x^2(3x-2)} = 1$

$$= \frac{x-2}{x+1}$$

$y_1 = (3/x - 2/x^2)/(3/(x - 2) + 1/x^2),$
$y_2 = (x - 2)/(x + 1)$

X	Y₁	Y₂
−3	2.5	2.5
−2	4	4
−1	ERROR	ERROR
0	ERROR	−2
1	−.5	−.5
2	ERROR	0
3	.25	.25

X = −3

As a partial check, we let $y_1 = (3/x - 2/x^2)/(3/(x - 2) + 1/x^2)$ and $y_2 = (x - 2)/(x + 1)$. A table of values, like that shown at left, indicates that the values of y_1 and y_2 are the same for all x-values for which both expressions are defined.

Using Division to Simplify a Complex Rational Expression

1. Add or subtract, as necessary, to get one rational expression in the numerator.
2. Add or subtract, as necessary, to get one rational expression in the denominator.
3. Perform the indicated division (invert the divisor and multiply).
4. Simplify, if possible, by removing any factors that equal 1.

EXAMPLE 4   Simplify:

$$\frac{1 + \dfrac{2}{x}}{1 - \dfrac{4}{x^2}}.$$

SOLUTION   We have

$$\frac{1 + \dfrac{2}{x}}{1 - \dfrac{4}{x^2}} = \frac{\dfrac{x}{x} + \dfrac{2}{x}}{\dfrac{x^2}{x^2} - \dfrac{4}{x^2}}$$ Finding a common denominator

Finding a common denominator

$$= \frac{\dfrac{x + 2}{x}}{\dfrac{x^2 - 4}{x^2}}$$ Adding in the numerator

Subtracting in the denominator

$$= \frac{x + 2}{x} \cdot \frac{x^2}{x^2 - 4}$$ Multiplying by the reciprocal of the divisor

$$= \frac{(x + 2) \cdot x^2}{x(x + 2)(x - 2)}$$ Factoring. Remember to simplify when possible.

$$= \frac{(x + 2)x \cdot x}{x(x + 2)(x - 2)}$$ Removing a factor equal to 1: $\dfrac{(x + 2)x}{(x + 2)x} = 1$

$$= \frac{x}{x - 2}.$$ Simplifying

As a quick, partial check, we select a convenient value for x—say, 1:

$$\frac{1 + \dfrac{2}{1}}{1 - \dfrac{4}{1^2}} = \frac{1 + 2}{1 - 4} = \frac{3}{-3} = -1$$ We evaluated the original expression for $x = 1$.

and

$$\frac{1}{1-2} = \frac{1}{-1} = -1.$$

> We evaluated the simplified expression for $x = 1$.

Since both expressions yield the same result, our simplification is probably correct. More evaluation would provide a more definitive check and can be done using the TABLE feature of a graphing calculator.

If negative exponents occur, we first find an equivalent expression using positive exponents and then proceed as in the preceding examples.

EXAMPLE 5   Simplify:

$$\frac{a^{-1} + b^{-1}}{a^{-3} + b^{-3}}.$$

SOLUTION

$$\frac{a^{-1} + b^{-1}}{a^{-3} + b^{-3}} = \frac{\dfrac{1}{a} + \dfrac{1}{b}}{\dfrac{1}{a^3} + \dfrac{1}{b^3}}$$

> **Rewriting with positive exponents. We continue, using method 2.**

$$= \frac{\dfrac{1}{a} \cdot \dfrac{b}{b} + \dfrac{1}{b} \cdot \dfrac{a}{a}}{\dfrac{1}{a^3} \cdot \dfrac{b^3}{b^3} + \dfrac{1}{b^3} \cdot \dfrac{a^3}{a^3}}$$

> **Finding a common denominator**
> **Finding a common denominator**

$$= \frac{\dfrac{b}{ab} + \dfrac{a}{ab}}{\dfrac{b^3}{a^3b^3} + \dfrac{a^3}{a^3b^3}}$$

$$= \frac{\dfrac{b+a}{ab}}{\dfrac{b^3 + a^3}{a^3b^3}}$$

> **Adding in the numerator**
> **Adding in the denominator**

$$= \frac{b+a}{ab} \cdot \frac{a^3b^3}{b^3 + a^3}$$

> **Multiplying by the reciprocal of the divisor**

$$= \frac{(b+a) \cdot ab \cdot a^2b^2}{ab(b+a)(b^2 - ab + a^2)}$$

> **Factoring and looking for common factors**

$$= \frac{\cancel{(b+a)} \cdot \cancel{ab} \cdot a^2b^2}{\cancel{ab}\cancel{(b+a)}(b^2 - ab + a^2)}$$

> **Removing a factor equal to 1:** $\dfrac{(b+a)ab}{(b+a)ab} = 1$

$$= \frac{a^2b^2}{b^2 - ab + a^2}$$

There is no one method that is best to use. When it is little or no work to write an expression as a quotient of two rational expressions, the second method is probably easier to use. On the other hand, some expressions require fewer steps if we use the first method. Either method can be used with any complex rational expression.

7.5 EXERCISE SET

Concept Reinforcement In each of Exercises 1–6, match the complex rational expression with an equivalent expression from the column on the right.

1. ___
$$\dfrac{\dfrac{5}{x} + \dfrac{3}{y}}{\dfrac{3}{x} - \dfrac{2}{y}}$$

2. ___
$$\dfrac{\dfrac{5}{x+y}}{\dfrac{3}{x-y}}$$

3. ___
$$\dfrac{\dfrac{3}{x} + \dfrac{2}{y}}{\dfrac{5}{x} - \dfrac{3}{y}}$$

4. ___
$$\dfrac{\dfrac{3}{x} - \dfrac{2}{y}}{\dfrac{5}{x} + \dfrac{3}{y}}$$

5. ___
$$\dfrac{\dfrac{5}{x} - \dfrac{3}{y}}{\dfrac{3}{x} + \dfrac{2}{y}}$$

6. ___
$$\dfrac{\dfrac{3}{x-y}}{\dfrac{5}{x+y}}$$

a) $\dfrac{3}{x-y} \div \dfrac{5}{x+y}$

b) $\dfrac{\dfrac{5}{x}\cdot\dfrac{y}{y} - \dfrac{3}{y}\cdot\dfrac{x}{x}}{\dfrac{3}{x}\cdot\dfrac{y}{y} + \dfrac{2}{y}\cdot\dfrac{x}{x}}$

c) $\dfrac{\dfrac{3}{x} - \dfrac{2}{y}}{\dfrac{5}{x} + \dfrac{3}{y}} \cdot \dfrac{xy}{xy}$

d) $\dfrac{5}{x+y} \div \dfrac{3}{x-y}$

e) $\dfrac{\dfrac{5}{x}\cdot\dfrac{y}{y} + \dfrac{3}{y}\cdot\dfrac{x}{x}}{\dfrac{3}{x}\cdot\dfrac{y}{y} - \dfrac{2}{y}\cdot\dfrac{x}{x}}$

f) $\dfrac{\dfrac{3}{x} + \dfrac{2}{y}}{\dfrac{5}{x} - \dfrac{3}{y}} \cdot \dfrac{xy}{xy}$

Simplify. If possible, use a second method, evaluation, or a graphing calculator as a check.

7. $\dfrac{\dfrac{x+2}{x-1}}{\dfrac{x+4}{x-3}}$

8. $\dfrac{\dfrac{x-1}{x+3}}{\dfrac{x-6}{x+2}}$

9. $\dfrac{\dfrac{5}{a} - \dfrac{4}{b}}{\dfrac{2}{a} + \dfrac{3}{b}}$

10. $\dfrac{\dfrac{2}{r} - \dfrac{3}{t}}{\dfrac{4}{r} + \dfrac{5}{t}}$

11. $\dfrac{\dfrac{3}{z} + \dfrac{2}{y}}{\dfrac{4}{z} - \dfrac{1}{y}}$

12. $\dfrac{\dfrac{6}{x} + \dfrac{7}{y}}{\dfrac{7}{x} - \dfrac{6}{y}}$

13. $\dfrac{\dfrac{a^2 - b^2}{ab}}{\dfrac{a-b}{b}}$

14. $\dfrac{\dfrac{x^2 - y^2}{xy}}{\dfrac{x-y}{y}}$

15. $\dfrac{1 - \dfrac{2}{3x}}{x - \dfrac{4}{9x}}$

16. $\dfrac{\dfrac{3x}{y} - x}{2y - \dfrac{y}{x}}$

17. $\dfrac{x^{-1} + y^{-1}}{\dfrac{x^2 - y^2}{xy}}$

18. $\dfrac{a^{-1} + b^{-1}}{\dfrac{a^2 - b^2}{ab}}$

19. $\dfrac{\dfrac{1}{a-h}-\dfrac{1}{a}}{h}$

20. $\dfrac{\dfrac{1}{x+h}-\dfrac{1}{x}}{h}$

21. $\dfrac{\dfrac{a^2-4}{a^2+3a+2}}{\dfrac{a^2-5a-6}{a^2-6a-7}}$

22. $\dfrac{\dfrac{x^2-x-12}{x^2-2x-15}}{\dfrac{x^2+8x+12}{x^2-5x-14}}$

23. $\dfrac{\dfrac{x}{x^2+3x-4}-\dfrac{1}{x^2+3x-4}}{\dfrac{x}{x^2+6x+8}+\dfrac{3}{x^2+6x+8}}$

24. $\dfrac{\dfrac{x}{x^2+5x-6}+\dfrac{6}{x^2+5x-6}}{\dfrac{x}{x^2-5x+4}-\dfrac{2}{x^2-5x+4}}$

25. $\dfrac{\dfrac{1}{y}+2}{\dfrac{1}{y}-3}$

26. $\dfrac{7+\dfrac{1}{a}}{\dfrac{1}{a}-3}$

27. $\dfrac{y+y^{-1}}{y-y^{-1}}$

28. $\dfrac{x-x^{-1}}{x+x^{-1}}$

29. $\dfrac{\dfrac{1}{x-2}+\dfrac{3}{x-1}}{\dfrac{2}{x-1}+\dfrac{5}{x-2}}$

30. $\dfrac{\dfrac{2}{y-3}+\dfrac{1}{y+1}}{\dfrac{3}{y+1}+\dfrac{4}{y-3}}$

31. $\dfrac{a(a+3)^{-1}-2(a-1)^{-1}}{a(a+3)^{-1}-(a-1)^{-1}}$

32. $\dfrac{a(a+2)^{-1}-3(a-3)^{-1}}{a(a+2)^{-1}-(a-3)^{-1}}$

33. $\dfrac{\dfrac{2}{a^2-1}+\dfrac{1}{a+1}}{\dfrac{3}{a^2-1}+\dfrac{2}{a-1}}$

34. $\dfrac{\dfrac{3}{a^2-9}+\dfrac{2}{a+3}}{\dfrac{4}{a^2-9}+\dfrac{1}{a+3}}$

35. $\dfrac{\dfrac{5}{x^2-4}-\dfrac{3}{x-2}}{\dfrac{4}{x^2-4}-\dfrac{2}{x+2}}$

36. $\dfrac{\dfrac{4}{x^2-1}-\dfrac{3}{x+1}}{\dfrac{5}{x^2-1}-\dfrac{2}{x-1}}$

37. $\dfrac{\dfrac{y}{y^2-4}+\dfrac{5}{4-y^2}}{\dfrac{y^2}{y^2-4}+\dfrac{25}{4-y^2}}$

38. $\dfrac{\dfrac{y}{y^2-1}+\dfrac{3}{1-y^2}}{\dfrac{y^2}{y^2-1}+\dfrac{9}{1-y^2}}$

39. $\dfrac{\dfrac{y^2}{y^2-25}-\dfrac{y}{y-5}}{\dfrac{y}{y^2-25}-\dfrac{1}{y+5}}$

40. $\dfrac{\dfrac{y^2}{y^2-9}-\dfrac{y}{y+3}}{\dfrac{y}{y^2-9}-\dfrac{1}{y-3}}$

41. $\dfrac{\dfrac{a}{a+2}+\dfrac{5}{a}}{\dfrac{a}{2a+4}+\dfrac{1}{3a}}$

42. $\dfrac{\dfrac{a}{a+3}+\dfrac{4}{5a}}{\dfrac{a}{2a+6}+\dfrac{3}{a}}$

43. $\dfrac{\dfrac{1}{x^2-3x+2}+\dfrac{1}{x^2-4}}{\dfrac{1}{x^2+4x+4}+\dfrac{1}{x^2-4}}$

44. $\dfrac{\dfrac{1}{x^2+3x+2}+\dfrac{1}{x^2-1}}{\dfrac{1}{x^2-1}+\dfrac{1}{x^2-4x+3}}$

45. $\dfrac{\dfrac{3}{a^2-4a+3}+\dfrac{3}{a^2-5a+6}}{\dfrac{3}{a^2-3a+2}+\dfrac{3}{a^2+3a-10}}$

46. $\dfrac{\dfrac{1}{a^2+7a+10}-\dfrac{2}{a^2-7a+12}}{\dfrac{2}{a^2-a-6}-\dfrac{1}{a^2+a-20}}$

Aha! **47.** $\dfrac{\dfrac{y}{y^2 - 4} - \dfrac{2y}{y^2 + y - 6}}{\dfrac{2y}{y^2 + y - 6} - \dfrac{y}{y^2 - 4}}$

48. $\dfrac{\dfrac{y}{y^2 - 1} - \dfrac{3y}{y^2 + 5y + 4}}{\dfrac{3y}{y^2 - 1} - \dfrac{y}{y^2 - 4y + 3}}$

49. $\dfrac{\dfrac{3}{x^2 + 2x - 3} - \dfrac{1}{x^2 - 3x - 10}}{\dfrac{3}{x^2 - 6x + 5} - \dfrac{1}{x^2 + 5x + 6}}$

50. $\dfrac{\dfrac{1}{a^2 + 7a + 12} + \dfrac{1}{a^2 + a - 6}}{\dfrac{1}{a^2 + 2a - 8} + \dfrac{1}{a^2 + 5a + 4}}$

TW 51. Michael *incorrectly* simplifies

$$\frac{a + b^{-1}}{a + c^{-1}} \quad \text{as} \quad \frac{a + c}{a + b}.$$

What mistake is he making and how could you convince him that this is incorrect?

TW 52. To simplify a complex rational expression in which the sum of two fractions is divided by the difference of the same two fractions, which method is easier? Why?

Focused Review

Simplify.

53. $\dfrac{-6 + 3x}{5} \div \dfrac{4x - 8}{25}$ [7.2]

54. $\dfrac{t}{t - 3} - \dfrac{3}{4t - 12}$ [7.4]

55. $\dfrac{7}{a^2 + a - 2} + \dfrac{a}{a^2 - 4a + 3}$ [7.4]

56. $\dfrac{t^2 + 2t - 3}{t^2 + 4t - 5} \cdot \dfrac{t^2 - 3t - 10}{t^2 + 5t + 6}$ [7.2]

57. $\dfrac{\dfrac{2}{x^2 y} + \dfrac{3}{xy^2}}{\dfrac{3}{xy^2} + \dfrac{2}{x^2 y}}$ [7.5]

58. $\dfrac{x - 2 + \dfrac{1}{x}}{x - 5 + \dfrac{4}{x}}$ [7.5]

Synthesis

TW 59. In arithmetic, we are taught that

$$\frac{a}{b} \div \frac{c}{d} = \frac{a}{b} \cdot \frac{d}{c}$$

(to divide by a fraction, we invert and multiply). Use method 1 to explain *why* we do this.

TW 60. Use algebra to determine the domain of the function given by

$$f(x) = \frac{\dfrac{1}{x - 2}}{\dfrac{x}{x - 2} - \dfrac{5}{x - 2}}.$$

Then explain how a graphing calculator could be used to check your answer.

Simplify.

61. $\dfrac{5x^{-2} + 10x^{-1}y^{-1} + 5y^{-2}}{3x^{-2} - 3y^{-2}}$

62. $(a^2 - ab + b^2)^{-1}(a^2 b^{-1} + b^2 a^{-1}) \times (a^{-2} - b^{-2})(a^{-2} + 2a^{-1}b^{-1} + b^{-2})^{-1}$

63. *Astronomy.* When two galaxies are moving in opposite directions at velocities v_1 and v_2, an observer in one of the galaxies would see the other galaxy receding at speed

$$\frac{v_1 + v_2}{1 + \dfrac{v_1 v_2}{c^2}},$$

where c is the speed of light. Determine the observed speed if v_1 and v_2 are both one-fourth the speed of light.

Find and simplify

$$\frac{f(x + h) - f(x)}{h}$$

for each rational function f in Exercises 64–67.

64. $f(x) = \dfrac{2}{x^2}$

65. $f(x) = \dfrac{3}{x}$

66. $f(x) = \dfrac{x}{1-x}$

67. $f(x) = \dfrac{2x}{1+x}$

68. If

$$F(x) = \dfrac{3 + \dfrac{1}{x}}{2 - \dfrac{8}{x^2}},$$

find the domain of F.

69. If

$$G(x) = \dfrac{x - \dfrac{1}{x^2-1}}{\dfrac{1}{9} - \dfrac{1}{x^2-16}},$$

find the domain of G.

70. Find the reciprocal of y if

$$y = x^2 + x + 1 + \dfrac{1}{x} + \dfrac{1}{x^2}.$$

71. For $f(x) = \dfrac{2}{2+x}$, find $f(f(a))$.

72. For $g(x) = \dfrac{x+3}{x-1}$, find $g(g(a))$.

73. Let

$$f(x) = \left[\dfrac{\dfrac{x+3}{x-3} + 1}{\dfrac{x+3}{x-3} - 1} \right]^4.$$

Find a simplified form of $f(x)$ and specify the domain of f.

74. *Financial Planning.* Alexis wishes to invest a portion of each month's pay in an account that pays 7.5% interest. If he wants to have \$30,000 in the account after 10 yr, the amount invested each month is given by

$$\dfrac{30,000 \cdot \dfrac{0.075}{12}}{\left(1 + \dfrac{0.075}{12}\right)^{120} - 1}.$$

Find the amount of Alexis' monthly investment.

Collaborative Corner

Which Method Is Better?

Focus: Complex rational expressions
Time: 10–15 minutes
Group Size: 3–4

ACTIVITY

Consider the steps in Examples 2 and 3 for simplifying a complex rational expression by each of the two methods. Then, work as a group to simplify

$$\dfrac{\dfrac{5}{x+1} - \dfrac{1}{x}}{\dfrac{2}{x^2} + \dfrac{4}{x}}$$

subject to the following conditions.

  1. The group should predict which method will more easily simplify this expression.

  2. Using the method selected in part (1), one group member should perform the first step in the simplification and then pass the problem on to another member of the group. That person then checks the work, performs the next step, and passes the problem on to another group member. If a mistake is found, the problem should be passed to the person who made the mistake for repair. This process continues until, eventually, the simplification is complete.

  3. At the same time that part (2) is being performed, another group member should perform the first step of the solution using the method not selected in part (1). He or she should then pass the problem to another group member and so on, just as in part (2).

  4. What method *was* easier? Why? Compare your responses with those of other groups.

7.6 Rational Equations

Solving Rational Equations

Connecting the Concepts

It is not unusual to learn a skill that enables us to write equivalent expressions and then to put that skill to work solving a new type of equation. That is precisely what we are now doing: Sections 7.1–7.5 have been devoted to writing equivalent expressions in which rational expressions appeared. There we used least common denominators to add or subtract. Here in Section 7.6, we return to the task of solving equations, but this time the equations contain rational expressions and the LCD is used as part of the multiplication principle.

Study Tip

Does More Than One Solution Exist?

Keep in mind that many problems—in math and elsewhere—have more than one solution. When asked to solve an equation, we are expected to find any and all solutions of the equation.

Solving Rational Equations

In Sections 7.1–7.5, we learned how to *simplify expressions*. We now learn to *solve* a new type of *equation*. A **rational equation** is an equation that contains one or more rational expressions. Here are some examples:

$$\frac{2}{3} - \frac{5}{6} = \frac{1}{t}, \qquad \frac{a-1}{a-5} = \frac{4}{a^2 - 25}, \qquad x^3 + \frac{6}{x} = 5.$$

As you will see in Section 7.7, equations of this type occur frequently in applications. To solve rational equations, recall that one way to *clear fractions* from an equation is to multiply both sides of the equation by the LCD.

> ### To Solve a Rational Equation
> Multiply both sides of the equation by the LCD. This is called *clearing fractions* and produces an equation similar to those we have already solved.

Recall that division by 0 is undefined. Note, too, that variables usually appear in at least one denominator of a rational equation. Thus certain numbers can often be ruled out as possible solutions before we even attempt to solve a given rational equation.

EXAMPLE 1   Solve: $\dfrac{x+4}{3x} + \dfrac{x+8}{5x} = 2$.

ALGEBRAIC APPROACH

Because the left side of this equation is undefined when x is 0, we state at the outset that $x \neq 0$. Next, we multiply both sides of the equation by the LCD, $3 \cdot 5 \cdot x$, or $15x$:

$$15x\left(\frac{x+4}{3x} + \frac{x+8}{5x}\right) = 15x \cdot 2 \quad \begin{array}{l}\text{Multiplying by the}\\ \text{LCD to clear}\\ \text{fractions}\end{array}$$

$$15x \cdot \frac{x+4}{3x} + 15x \cdot \frac{x+8}{5x} = 15x \cdot 2 \quad \begin{array}{l}\text{Using the}\\ \text{distributive law}\end{array}$$

$$\frac{5 \cdot 3x \cdot (x+4)}{3x} + \frac{3 \cdot 5x \cdot (x+8)}{5x} = 30x \quad \begin{array}{l}\text{Locating factors}\\ \text{equal to 1}\end{array}$$

$$\rightarrow 5(x+4) + 3(x+8) = 30x \quad \begin{array}{l}\text{Removing factors}\\ \text{equal to 1:}\\ \dfrac{3x}{3x} = 1; \dfrac{5x}{5x} = 1\end{array}$$

We have solved equations like this before.

$$5x + 20 + 3x + 24 = 30x \quad \begin{array}{l}\text{Using the}\\ \text{distributive law}\end{array}$$

$$8x + 44 = 30x$$

$$44 = 22x$$

$$2 = x. \quad \begin{array}{l}\text{This should check}\\ \text{since } x \neq 0.\end{array}$$

GRAPHICAL APPROACH

We let $y_1 = (x + 4)/(3x) + (x + 8)/(5x)$ and $y_2 = 2$. The solutions of the equation will be the x-coordinates of any points of intersection.

$y_1 = (x + 4)/(3x) + (x + 8)/(5x), \ y_2 = 2$

It appears from the graph that there is one point of intersection, $(2, 2)$. The solution is 2.

Student Notes

When we are solving rational equations, as in Examples 1–5, the LCD is never actually used as a denominator. Rather, it is used as a multiplier to eliminate the denominators in the equation. This differs from how the LCD is used in Section 7.4.

Check:
$$\frac{x+4}{3x} + \frac{x+8}{5x} = 2$$

$$\begin{array}{c|c}\dfrac{2+4}{3 \cdot 2} + \dfrac{2+8}{5 \cdot 2} & 2 \\[2mm] \dfrac{6}{6} + \dfrac{10}{10} & \\[2mm] 2 \overset{?}{=} 2 & \text{TRUE}\end{array}$$

The number 2 is the solution.

It is important that when we clear fractions, all denominators are eliminated. This produces an equation without rational expressions, which is easier to solve.

EXAMPLE 2   Solve: $\dfrac{x-1}{x-5} = \dfrac{4}{x-5}$.

ALGEBRAIC APPROACH

To ensure that neither denominator is 0, we state at the outset the restriction that $x \neq 5$. Then we proceed as before, multiplying both sides by the LCD, $x - 5$:

$$(x - 5) \cdot \frac{x - 1}{x - 5} = (x - 5) \cdot \frac{4}{x - 5}$$

$$x - 1 = 4$$

$$x = 5. \qquad \textbf{But recall that } x \neq 5.$$

In this case, it is important to remember that, because of the restriction above, 5 cannot be a solution. A check confirms the necessity of that restriction.

Check:
$$\frac{x - 1}{x - 5} = \frac{4}{x - 5}$$

$$\frac{5 - 1}{5 - 5} \,\bigg|\, \frac{4}{5 - 5}$$

$$\frac{4}{0} \stackrel{?}{=} \frac{4}{0} \qquad \textbf{Division by 0 is undefined.}$$

This equation has no solution.

GRAPHICAL APPROACH

We graph $y_1 = (x - 1)/(x - 5)$ and $y_2 = 4/(x - 5)$ and look for any points of intersection of the graphs.

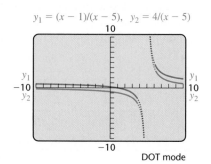

$y_1 = (x - 1)/(x - 5), \quad y_2 = 4/(x - 5)$

DOT mode

It appears that the graphs may intersect around $x = 4$ or $x = 6$. If we use INTERSECT, the calculator returns an error message. This in itself does not tell us that there is no point of intersection, because we do not see all the graph in the viewing window. **In order to state positively that there is no solution, we must solve the equation with algebra.**

To see why 5 is not a solution of Example 2, note that the multiplication principle for equations requires that we multiply both sides by a *nonzero* number. When both sides of an equation are multiplied by an expression containing variables, it is possible that certain replacements will make that expression equal to 0. Thus it is safe to say that *if* a solution of

$$\frac{x - 1}{x - 5} = \frac{4}{x - 5}$$

exists, then it is also a solution of $x - 1 = 4$. We *cannot* conclude that every solution of $x - 1 = 4$ is a solution of the original equation.

> **CAUTION!** When solving rational equations, do not forget to list any restrictions as part of the first step. Refer to the restriction(s) as you proceed.

Example 2 illustrates a disadvantage of graphical methods of solving equations. It can be difficult to tell whether two graphs actually intersect. Also, it is easy to "miss" seeing a solution. This is especially true for rational equations, where there are usually two or more branches of the graph. Solving by graphing does make a good check, however, of an algebraic solution.

EXAMPLE 3 Solve: $\dfrac{x^2}{x-3} = \dfrac{9}{x-3}$.

SOLUTION Note that $x \neq 3$. We multiply both sides by the LCD, $x - 3$:

$$(x-3) \cdot \frac{x^2}{x-3} = (x-3) \cdot \frac{9}{x-3}$$

$$x^2 = 9 \qquad \text{Simplifying}$$

$$x^2 - 9 = 0 \qquad \text{Getting 0 on one side}$$

$$(x-3)(x+3) = 0 \qquad \text{Factoring}$$

$$x = 3 \quad or \quad x = -3. \qquad \text{Using the principle of zero products}$$

Although 3 is a solution of $x^2 = 9$, it must be rejected as a solution of the rational equation because of the restriction stated in red above. You should check that -3 *is* a solution despite the fact that 3 is not. This check can also be done graphically, as shown at left.

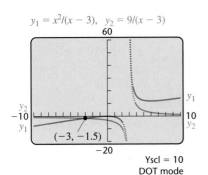

$y_1 = x^2/(x-3), \ y_2 = 9/(x-3)$

$(-3, -1.5)$

Yscl = 10
DOT mode

EXAMPLE 4 Solve: $\dfrac{2}{x+5} + \dfrac{1}{x-5} = \dfrac{16}{x^2-25}$.

SOLUTION To find all restrictions and to assist in finding the LCD, we factor:

$$\frac{2}{x+5} + \frac{1}{x-5} = \frac{16}{(x+5)(x-5)}. \qquad \text{Factoring } x^2 - 25$$

Note that $x \neq -5$ and $x \neq 5$. We multiply by the LCD, $(x+5)(x-5)$, and then use the distributive law:

$$(x+5)(x-5)\left(\frac{2}{x+5} + \frac{1}{x-5}\right) = (x+5)(x-5) \cdot \frac{16}{(x+5)(x-5)}$$

$$(x+5)(x-5)\frac{2}{x+5} + (x+5)(x-5)\frac{1}{x-5} = \frac{(x+5)(x-5)16}{(x+5)(x-5)}$$

$$2(x-5) + (x+5) = 16$$

$$2x - 10 + x + 5 = 16$$

$$3x - 5 = 16$$

$$3x = 21$$

$$x = 7.$$

A check will confirm that the solution is 7.

EXAMPLE 5 Let $f(x) = x + \dfrac{6}{x}$. Find all values of a for which $f(a) = 5$.

SOLUTION Since $f(a) = a + \dfrac{6}{a}$, the problem asks that we find all values of a for which

$$a + \frac{6}{a} = 5.$$

First note that $a \neq 0$. To solve for a, we multiply both sides of the equation by the LCD, a:

$$a\left(a + \frac{6}{a}\right) = 5 \cdot a \qquad \text{Multiplying both sides by } a.$$
Parentheses are important.

$$a \cdot a + a \cdot \frac{6}{a} = 5a \qquad \text{Using the distributive law}$$

$$a^2 + 6 = 5a \qquad \text{Simplifying}$$

$$a^2 - 5a + 6 = 0 \qquad \text{Getting 0 on one side}$$

$$(a - 3)(a - 2) = 0 \qquad \text{Factoring}$$

$$a = 3 \quad or \quad a = 2. \qquad \text{Using the principle of zero products}$$

Check: $\quad f(3) = 3 + \dfrac{6}{3} = 3 + 2 = 5;$

$$f(2) = 2 + \frac{6}{2} = 2 + 3 = 5.$$

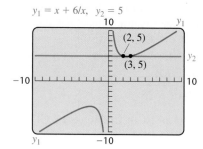

$y_1 = x + 6/x, \quad y_2 = 5$

The solutions are 2 and 3.

The graph at left shows the points of intersection of the graph of $f(x)$ and the line $y = 5$. It also shows that we have two solutions of $f(x) = 5$, 2 and 3. For $a = 2$ or $a = 3$, we have $f(a) = 5$.

7.6 EXERCISE SET

FOR EXTRA HELP

 MathXL   MyMathLab   InterAct Math   Tutor Center   Video Lectures on CD: Disc 4   Student's Solutions Manual

Tutor Center
AW Math
Tutor Center

🦢 *Concept Reinforcement* Classify each of the following as either an expression or an equation.

1. $\dfrac{5x}{x + 2} - \dfrac{3}{x} = 7$

2. $\dfrac{3}{x - 4} + \dfrac{2}{x + 4}$

3. $\dfrac{4}{t^2 - 1} + \dfrac{3}{t + 1}$

4. $\dfrac{2}{t^2 - 1} + \dfrac{3}{t + 1} = 5$

5. $\dfrac{2}{x + 7} + \dfrac{6}{5x} = 4$

6. $\dfrac{5}{2x} - \dfrac{3}{x^2} = 7$

7. $\dfrac{t + 3}{t - 4} = \dfrac{t - 5}{t - 7}$

8. $\dfrac{7t}{2t - 3} \div \dfrac{3t}{2t + 3}$

9. $\dfrac{5x}{x^2 - 4} \cdot \dfrac{7}{x^2 - 5x + 4}$

10. $\dfrac{7x}{2 - x} = \dfrac{3}{4 - x}$

Solve. If no solution exists, state this.

11. $\dfrac{t}{2} + \dfrac{t}{3} = 7$

12. $\dfrac{y}{4} + \dfrac{y}{3} = 8$

13. $\dfrac{5}{8} - \dfrac{1}{x} = \dfrac{3}{4}$

14. $\dfrac{1}{3} - \dfrac{1}{t} = \dfrac{5}{6}$

15. $\dfrac{2x + 1}{3} + \dfrac{x + 2}{5} = \dfrac{4x}{15}$

16. $\dfrac{3x + 2}{2} + \dfrac{x + 4}{8} = \dfrac{2x}{5}$

17. $\dfrac{2}{3} - \dfrac{1}{t} = \dfrac{7}{3t}$

18. $\dfrac{1}{2} - \dfrac{2}{t} = \dfrac{3}{2t}$

19. $\dfrac{4}{x - 1} + \dfrac{5}{6} = \dfrac{2}{3x - 3}$

20. $\dfrac{3}{2x + 10} + \dfrac{5}{4} = \dfrac{7}{x + 5}$

Aha! **21.** $\dfrac{2}{6} + \dfrac{1}{2x} = \dfrac{1}{3}$

22. $\dfrac{12}{15} - \dfrac{1}{3x} = \dfrac{4}{5}$

23. $y + \dfrac{4}{y} = -5$

24. $t + \dfrac{6}{t} = -5$

25. $x - \dfrac{12}{x} = 4$

26. $y - \dfrac{14}{y} = 5$

27. $\dfrac{t + 10}{7 - t} = \dfrac{3}{t - 7}$

28. $\dfrac{4}{8 - a} = \dfrac{4 - a}{a - 8}$

29. $\dfrac{x}{x - 5} = \dfrac{25}{x^2 - 5x}$

30. $\dfrac{t}{t - 6} = \dfrac{36}{t^2 - 6t}$

31. $\dfrac{5}{4t} = \dfrac{7}{5t - 2}$

32. $\dfrac{3}{x - 2} = \dfrac{5}{x + 4}$

Aha! **33.** $\dfrac{x^2 + 4}{x - 1} = \dfrac{5}{x - 1}$

34. $\dfrac{x^2 - 1}{x + 2} = \dfrac{3}{x + 2}$

35. $\dfrac{6}{a + 1} = \dfrac{a}{a - 1}$

36. $\dfrac{4}{a - 7} = \dfrac{-2a}{a + 3}$

37. $\dfrac{60}{t - 5} - \dfrac{18}{t} = \dfrac{40}{t}$

38. $\dfrac{50}{t - 2} - \dfrac{16}{t} = \dfrac{30}{t}$

39. $\dfrac{3}{x - 3} + \dfrac{5}{x + 2} = \dfrac{5x}{x^2 - x - 6}$

40. $\dfrac{2}{x - 2} + \dfrac{1}{x + 4} = \dfrac{x}{x^2 + 2x - 8}$

41. $\dfrac{3}{x} + \dfrac{x}{x + 2} = \dfrac{4}{x^2 + 2x}$

42. $\dfrac{x}{x + 1} + \dfrac{5}{x} = \dfrac{1}{x^2 + x}$

43. $\dfrac{5}{x + 2} - \dfrac{3}{x - 2} = \dfrac{2x}{4 - x^2}$

44. $\dfrac{y + 3}{y + 2} - \dfrac{y}{y^2 - 4} = \dfrac{y}{y - 2}$

45. $\dfrac{3}{x^2 - 6x + 9} + \dfrac{x - 2}{3x - 9} = \dfrac{x}{2x - 6}$

46. $\dfrac{3 - 2y}{y + 1} - \dfrac{10}{y^2 - 1} = \dfrac{2y + 3}{1 - y}$

In Exercises 47–52, a rational function f is given. Find all values of a for which f(a) is the indicated value.

47. $f(x) = 2x - \dfrac{15}{x}$; $f(a) = 7$

48. $f(x) = 2x - \dfrac{6}{x}$; $f(a) = 1$

49. $f(x) = \dfrac{x - 5}{x + 1}$; $f(a) = \dfrac{3}{5}$

50. $f(x) = \dfrac{x - 3}{x + 2}$; $f(a) = \dfrac{1}{5}$

51. $f(x) = \dfrac{12}{x} - \dfrac{12}{2x}$; $f(a) = 8$

52. $f(x) = \dfrac{6}{x} - \dfrac{6}{2x}$; $f(a) = 5$

For each pair of functions f and g, find all values of a for which f(a) = g(a).

53. $f(x) = \dfrac{3x - 1}{x^2 - 7x + 10}$,

$g(x) = \dfrac{x - 1}{x^2 - 4} + \dfrac{2x + 1}{x^2 - 3x - 10}$

54. $f(x) = \dfrac{2x + 5}{x^2 + 4x + 3}$,

$g(x) = \dfrac{x + 2}{x^2 - 9} + \dfrac{x - 1}{x^2 - 2x - 3}$

55. $f(x) = \dfrac{2}{x^2 - 8x + 7}$,

$g(x) = \dfrac{3}{x^2 - 2x - 3} - \dfrac{1}{x^2 - 1}$

56. $f(x) = \dfrac{4}{x^2 + 3x - 10}$,

$g(x) = \dfrac{3}{x^2 - x - 12} + \dfrac{1}{x^2 + x - 6}$

TW **57.** Below are unlabeled graphs of $f(x) = x + 2$ and $g(x) = (x^2 - 4)/(x - 2)$. How could you determine which graph represents f and which graph represents g?

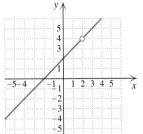
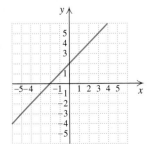

TW **58.** Explain how one can easily produce rational equations for which no solution exists. (*Hint*: Examine Example 2.)

Focused Review

Solve.

59. $3x - 2(5 - x) = 4(x + 7)$ [2.2]

60. $5 - 3x < 7x - 2$ [2.6]

61. $x^2 - 1 = 0$ [5.5]

62. $(2x + 1)(x - 8) = 0$ [5.1]

63. $2x - y = 8,$
$\quad y = 3x - 5$ [4.2]

64. $2x - y = 3,$
$\quad 5x + y = 4$ [4.3]

65. $\dfrac{4}{x} = 10$ [7.5]

66. $\dfrac{6}{x} + 1 = x$ [7.5]

67. $(x + 3)(x - 4) = 8$ [5.2]

68. $2x^3 + 9x^2 = 5x$ [5.4]

Synthesis

TW **69.** Karin and Kyle are working together to solve the equation

$$\frac{x}{x^2 - x - 6} = 2x - 1.$$

Karin obtains the first graph below using the DOT mode of a graphing calculator, while Kyle obtains the second graph at the top of the next column using the CONNECTED mode. Which approach is the better method of showing the solutions? Why?

$y_1 = x/(x^2 - x - 6), \quad y_2 = 2x - 1$

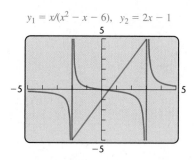

$y_1 = x/(x^2 - x - 6), \quad y_2 = 2x - 1$

TW **70.** Is the following statement true or false: "For any real numbers a, b, and c, if $ac = bc$, then $a = b$"? Explain why you answered as you did.

For each pair of functions f and g, find all values of a for which $f(a) = g(a)$.

71. $f(x) = \dfrac{x - \dfrac{2}{3}}{x + \dfrac{1}{2}}, \quad g(x) = \dfrac{x + \dfrac{2}{3}}{x - \dfrac{3}{2}}$

72. $f(x) = \dfrac{2 - \dfrac{x}{4}}{2}, \quad g(x) = \dfrac{\dfrac{x}{4} - 2}{\dfrac{x}{2} + 2}$

73. $f(x) = \dfrac{x + 3}{x + 2} - \dfrac{x + 4}{x + 3}, \quad g(x) = \dfrac{x + 5}{x + 4} - \dfrac{x + 6}{x + 5}$

74. $f(x) = \dfrac{1}{1 + x} + \dfrac{x}{1 - x}, \quad g(x) = \dfrac{1}{1 - x} - \dfrac{x}{1 + x}$

75. $f(x) = \dfrac{0.793}{x} + 18.15, \quad g(x) = \dfrac{6.034}{x} - 43.17$

76. $f(x) = \dfrac{2.315}{x} - \dfrac{12.6}{17.4}, \quad g(x) = \dfrac{6.71}{x} + 0.763$

Recall that identities are true for any possible replacement of the variable(s). Determine whether each of the following equations is an identity.

77. $\dfrac{x^2 + 6x - 16}{x - 2} = x + 8, \quad x \neq 2$

78. $\dfrac{x^3 + 8}{x^2 - 4} = \dfrac{x^2 - 2x + 4}{x - 2}, \quad x \neq -2, x \neq 2$

7.7 Applications Using Rational Equations and Proportions

Problems Involving Work ▪ Problems Involving Motion ▪
Problems Involving Proportions

Now that we are able to solve rational equations, it is possible to solve problems that we could not have handled before. The five problem-solving steps remain the same.

Problems Involving Work

EXAMPLE 1 The roof of Grady and Cleo's townhouse needs to be reshingled. Grady can do the job alone in 8 hr and Cleo can do the job alone in 10 hr. How long will it take the two of them, working together, to reshingle the roof?

SOLUTION

1. **Familiarize.** We familiarize ourselves with the problem by considering two *incorrect* ways of translating the problem to mathematical language.

   a) One *incorrect* way to translate the problem is to add the two times:

      8 hr + 10 hr = 18 hr.

      Think about this. Grady can do the job *alone* in 8 hr. If Grady and Cleo work together, whatever time it takes them must be *less* than 8 hr.

   b) Another *incorrect* approach is to assume that each person reshingles half of the roof. Were this the case,

      Grady would reshingle $\frac{1}{2}$ the roof in $\frac{1}{2}(8$ hr), or 4 hr,

      and

      Cleo would reshingle $\frac{1}{2}$ the roof in $\frac{1}{2}(10$ hr), or 5 hr.

      But time would be wasted since Grady would finish 1 hr before Cleo. Were Grady to help Cleo after completing his half, the entire job would take between 4 and 5 hr. This information provides a partial check on any answer we get—the answer should be between 4 and 5 hr.

      Let's consider how much of the job each person completes in 1 hr, 2 hr, 3 hr, and so on. Since Grady takes 8 hr to reshingle the entire roof, in 1 hr he reshingles $\frac{1}{8}$ of the roof. Since Cleo takes 10 hr to reshingle the entire roof, in 1 hr she reshingles $\frac{1}{10}$ of the roof. Thus Grady works at a rate of $\frac{1}{8}$ roof per hour, and Cleo works at a rate of $\frac{1}{10}$ roof per hour.

      Working together, Grady and Cleo reshingle $\frac{1}{8} + \frac{1}{10}$ roof in 1 hr, so their rate, as a team, is $\frac{1}{8} + \frac{1}{10} = \frac{5}{40} + \frac{4}{40} = \frac{9}{40}$ roof per hour.

In 2 hr, Grady reshingles $\frac{1}{8} \cdot 2$ of the roof and Cleo reshingles $\frac{1}{10} \cdot 2$ of the roof. Working together, they reshingle

$\frac{1}{8} \cdot 2 + \frac{1}{10} \cdot 2$, or $\frac{9}{20}$ of the roof in 2 hr.   Note that $\frac{9}{40} \cdot 2 = \frac{9}{20}$.

We are looking for the time required to reshingle 1 entire roof, not just a part of it. To find this time, we form a table and extend the pattern developed above.

Time	Fraction of the Roof Reshingled		
	By Grady	**By Cleo**	**Together**
1 hr	$\frac{1}{8}$	$\frac{1}{10}$	$\frac{1}{8} + \frac{1}{10}$, or $\frac{9}{40}$
2 hr	$\frac{1}{8} \cdot 2$	$\frac{1}{10} \cdot 2$	$\left(\frac{1}{8} + \frac{1}{10}\right)2$, or $\frac{9}{40} \cdot 2$, or $\frac{9}{20}$
3 hr	$\frac{1}{8} \cdot 3$	$\frac{1}{10} \cdot 3$	$\left(\frac{1}{8} + \frac{1}{10}\right)3$, or $\frac{9}{40} \cdot 3$, or $\frac{27}{40}$
t hr	$\frac{1}{8} \cdot t$	$\frac{1}{10} \cdot t$	$\left(\frac{1}{8} + \frac{1}{10}\right)t$, or $\frac{9}{40} \cdot t$

2. **Translate.** From the table, we see that t must be some number for which

Fraction of roof done by Grady in t hr $\frac{1}{8} \cdot t + \frac{1}{10} \cdot t = 1$,   Fraction of roof done by Cleo in t hr

or $\frac{t}{8} + \frac{t}{10} = 1$.

3. **Carry out.** We solve the equation both algebraically and graphically.

ALGEBRAIC APPROACH

We solve the equation:

$$\frac{t}{8} + \frac{t}{10} = 1$$

$$40\left(\frac{t}{8} + \frac{t}{10}\right) = 40 \cdot 1 \qquad \text{Multiplying by the LCD}$$

$$\frac{40t}{8} + \frac{40t}{10} = 40 \qquad \text{Distributing the 40}$$

$$5t + 4t = 40 \qquad \text{Simplifying}$$

$$9t = 40$$

$$t = \frac{40}{9}, \text{ or } 4\frac{4}{9}.$$

GRAPHICAL APPROACH

We graph $y_1 = x/8 + x/10$ and $y_2 = 1$.

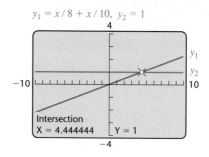

$y_1 = x/8 + x/10, \ y_2 = 1$

Intersection
X = 4.444444   Y = 1

We can convert X to fraction notation and write the point of intersection as $\left(\frac{40}{9}, 1\right)$. The solution is $\frac{40}{9}$, or about 4.4.

4. **Check.** In $\frac{40}{9}$ hr, Grady reshingles $\frac{1}{8} \cdot \frac{40}{9}$, or $\frac{5}{9}$, of the roof and Cleo reshingles $\frac{1}{10} \cdot \frac{40}{9}$, or $\frac{4}{9}$, of the roof. Together, they reshingle $\frac{5}{9} + \frac{4}{9}$, or 1 roof. The fact that our solution is between 4 and 5 hr (see step 1 above) is also a check.

5. **State.** It will take $4\frac{4}{9}$ hr for Grady and Cleo, working together, to re-shingle the roof.

EXAMPLE 2   It takes Pepe 9 hr longer than Wendy to rebuild an engine. Working together, they can do the job in 20 hr. How long would it take each, working alone, to rebuild an engine?

SOLUTION

1. **Familiarize.** Unlike Example 1, this problem does not provide us with the times required by the individuals to do the job alone. Let's have w = the number of hours it would take Wendy working alone and $w + 9$ = the number of hours it would take Pepe working alone.

2. **Translate.** Using the same reasoning as in Example 1, we see that Wendy completes $\dfrac{1}{w}$ of the job in 1 hr and Pepe completes $\dfrac{1}{w + 9}$ of the job in 1 hr. In 2 hr, Wendy completes $\dfrac{1}{w} \cdot 2$ of the job and Pepe completes $\dfrac{1}{w + 9} \cdot 2$ of the job. We are told that, working together, Wendy and Pepe can complete the entire job in 20 hr. This gives the following:

Fraction of job done by Wendy in 20 hr $\dfrac{1}{w} \cdot 20 + \dfrac{1}{w + 9} \cdot 20 = 1,$ Fraction of job done by Pepe in 20 hr

or $\dfrac{20}{w} + \dfrac{20}{w + 9} = 1.$

3. **Carry out.** We solve the equation both algebraically and graphically.

ALGEBRAIC APPROACH

We have

$$\frac{20}{w} + \frac{20}{w + 9} = 1$$

$$w(w + 9)\left(\frac{20}{w} + \frac{20}{w + 9}\right) = w(w + 9)1 \qquad \text{Multiplying by the LCD}$$

$$(w + 9)20 + w \cdot 20 = w(w + 9) \qquad \text{Distributing and simplifying}$$

$$40w + 180 = w^2 + 9w$$

$$0 = w^2 - 31w - 180 \qquad \text{Getting 0 on one side}$$

$$0 = (w - 36)(w + 5) \qquad \text{Factoring}$$

$$w - 36 = 0 \quad \textit{or} \quad w + 5 = 0 \qquad \text{Principle of zero products}$$

$$w = 36 \quad \textit{or} \qquad w = -5.$$

GRAPHICAL APPROACH

We graph $y_1 = 20/x + 20/(x + 9)$ and $y_2 = 1$ and find the points of intersection. If we use the window $[-8, 8, -10, 10]$, we see that there is a point of intersection near the x-value -5. It also appears that the graphs may intersect at a point off the screen to the right. Thus we adjust the window to $[-8, 50, 0, 5]$ and see that there is indeed another point of intersection.

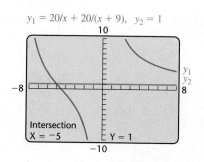

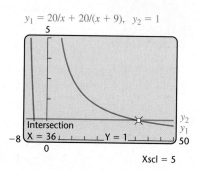

We have $x = 36$ or $x = -5$.

4. **Check.** Since negative time has no meaning in the problem, -5 is not a solution to the original problem. The number 36 checks since, if Wendy takes 36 hr alone and Pepe takes $36 + 9 = 45$ hr alone, in 20 hr they would have rebuilt

$$\frac{20}{36} + \frac{20}{45} = \frac{5}{9} + \frac{4}{9} = 1 \text{ complete engine.}$$

5. **State.** It would take Wendy 36 hr to rebuild an engine alone, and Pepe 45 hr.

The equations used in Examples 1 and 2 can be generalized as follows.

Modeling Work Problems

If

a = the time needed for A to complete the work alone,

b = the time needed for B to complete the work alone, and

t = the time needed for A and B to complete the work together,

then

$$\frac{t}{a} + \frac{t}{b} = 1.$$

The following are equivalent equations that can also be used:

$$\frac{1}{a} \cdot t + \frac{1}{b} \cdot t = 1 \quad \text{and} \quad \frac{1}{a} + \frac{1}{b} = \frac{1}{t}.$$

Problems Involving Motion

Problems dealing with distance, rate (or speed), and time are called **motion problems.** To translate them, we use either the basic motion formula, $d = rt$, or the formulas $r = d/t$ or $t = d/r$, which can be derived from $d = rt$.

EXAMPLE 3   On her road bike, Ignacia bikes 15 km/h faster than Franklin does on his mountain bike. In the time it takes Ignacia to travel 80 km, Franklin travels 50 km. Find the speed of each bicyclist.

SOLUTION

1. **Familiarize.** Let's guess that Franklin is going 10 km/h. Ignacia would then be traveling $10 + 15$, or 25 km/h. At 25 km/h, she would travel 80 km in $\frac{80}{25} = 3.2$ hr. Going 10 km/h, Franklin would cover 50 km in $\frac{50}{10} = 5$ hr. Since $3.2 \neq 5$, our guess was wrong, but we can see that if r = the rate, in kilometers per hour, of Franklin's bike, then the rate of Ignacia's bike $= r + 15$.

   Making a drawing and constructing a table can be helpful.

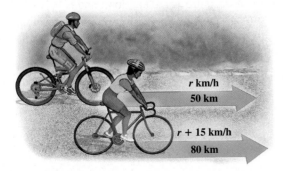

	Distance	Speed	Time
Franklin's Mountain Bike	50	r	t
Ignacia's Road Bike	80	$r + 15$	t

2. **Translate.** By looking at how we checked our guess, we see that in the **Time** column of the table, the *t*'s can be replaced, using the formula *Time = Distance/Rate*, as follows.

	Distance	Speed	Time
Franklin's Mountain Bike	50	r	$50/r$
Ignacia's Road Bike	80	$r + 15$	$80/(r + 15)$

Since we are told that the times must be the same, we can write an equation:

$$\frac{50}{r} = \frac{80}{r + 15}.$$

3. **Carry out.** We solve the equation both algebraically and graphically.

ALGEBRAIC APPROACH

We have

$$\frac{50}{r} = \frac{80}{r + 15}$$

$$r(r + 15)\,\frac{50}{r} = r(r + 15)\,\frac{80}{r + 15} \qquad \text{Multiplying by the LCD}$$

$$50r + 750 = 80r \qquad \text{Simplifying}$$

$$750 = 30r$$

$$25 = r.$$

GRAPHICAL APPROACH

We graph the equations $y_1 = 50/x$ and $y_2 = 80/(x + 15)$, and look for a point of intersection. By adjusting the window dimensions and using the INTERSECT feature, we see that the graphs intersect at the point $(25, 2)$.

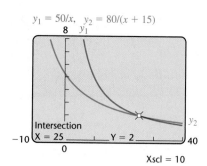

The solution is 25.

4. **Check.** If our answer checks, Franklin's mountain bike is going 25 km/h and Ignacia's road bike is going $25 + 15 = 40$ km/h.

Traveling 80 km at 40 km/h, Ignacia is riding for $\frac{80}{40} = 2$ hr. Traveling 50 km at 25 km/h, Franklin is riding for $\frac{50}{25} = 2$ hr. Our answer checks since the two times are the same.

5. **State.** Ignacia's speed is 40 km/h, and Franklin's speed is 25 km/h.

In the following example, although the distance is the same in both directions, the key to the translation lies in an additional piece of given information.

> **EXAMPLE 4** A Hudson River tugboat goes 10 mph in still water. It travels 24 mi upstream and 24 mi back in a total time of 5 hr. What is the speed of the current? (*Sources*: Based on information from the Department of the Interior, U.S. Geological Survey, and *The Tugboat Captain*, Montgomery County Community College)

SOLUTION

1. **Familiarize.** Let's guess that the speed of the current is 4 mph. The tugboat would then be moving $10 - 4 = 6$ mph upstream and $10 + 4 = 14$ mph downstream. The tugboat would require $\frac{24}{6} = 4$ hr to travel 24 mi upstream and $\frac{24}{14} = 1\frac{5}{7}$ hr to travel 24 mi downstream. Since the total time, $4 + 1\frac{5}{7} = 5\frac{5}{7}$ hr, is not the 5 hr mentioned in the problem, we know that our guess is wrong.

   Suppose that the current's speed $= c$ mph. The tugboat's speed would then be $10 - c$ mph going upstream and $10 + c$ mph going downstream.

   A drawing and table can help display the information.

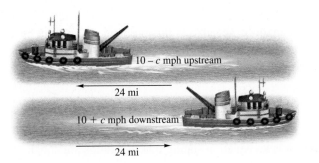

	Distance	Speed	Time
Upstream	24	$10 - c$	t_1
Downstream	24	$10 + c$	t_2

2. **Translate.** From examining our guess, we see that the time traveled can be represented using the formula *Time = Distance/Rate*:

	Distance	Speed	Time
Upstream	24	$10 - c$	$24/(10 - c)$
Downstream	24	$10 + c$	$24/(10 + c)$

   Since the total time upstream and back is 5 hr, we use the last column of the table to form an equation:

$$\frac{24}{10 - c} + \frac{24}{10 + c} = 5.$$

3. **Carry out.** We solve the equation both algebraically and graphically.

ALGEBRAIC APPROACH	GRAPHICAL APPROACH

We have

$$\frac{24}{10-c} + \frac{24}{10+c} = 5$$

$$(10-c)(10+c)\left[\frac{24}{10-c} + \frac{24}{10+c}\right] = (10-c)(10+c)5$$

Multiplying by the LCD

$$24(10+c) + 24(10-c) = (100-c^2)5$$

$$480 = 500 - 5c^2 \qquad \text{Simplifying}$$

$$5c^2 - 20 = 0$$

$$5(c^2 - 4) = 0$$

$$5(c-2)(c+2) = 0$$

$$c = 2 \quad or \quad c = -2.$$

We graph

$$y_1 = 24/(10-x) + 24/(10+x) \quad \text{and}$$
$$y_2 = 5$$

and see that there are two points of intersection.

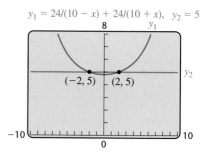

The solutions are -2 and 2.

4. Check. Since speed cannot be negative in this problem, -2 cannot be a solution. You should confirm that 2 checks in the original problem.

5. State. The speed of the current is 2 mph.

Problems Involving Proportions

A **ratio** of two quantities is their quotient. For example, 37% is the ratio of 37 to 100, or $\frac{37}{100}$. A **proportion** is an equation stating that two ratios are equal.

> **Proportion**   An equality of ratios, $A/B = C/D$, is called a *proportion*. The numbers within a proportion are said to be *proportional* to each other.

Proportions arise in geometry when we are studying *similar triangles*. If two triangles are **similar,** then their corresponding angles have the same measure and their corresponding sides are proportional. To illustrate, consider triangles ABC and RST at the top of the following page. If triangle ABC is similar to triangle RST, then angles A and R have the same measure, angles B and S have the same measure, angles C and T have the same measure, and

$$\frac{a}{r} = \frac{b}{s} = \frac{c}{t}.$$

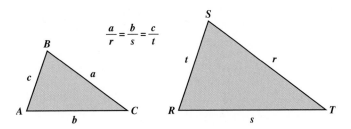

$$\frac{a}{r} = \frac{b}{s} = \frac{c}{t}$$

EXAMPLE 5 Similar Triangles. Triangles *ABC* and *XYZ* are similar. Solve for *z* if *x* = 10, *a* = 8, and *c* = 5.

SOLUTION We make a drawing, write a proportion, and then solve. Note that side *a* is always opposite angle *A*, side *x* is always opposite angle *X*, and so on.

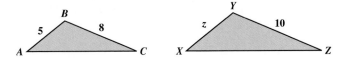

We have

$$\frac{z}{5} = \frac{10}{8} \qquad \text{The proportions } \frac{5}{z} = \frac{8}{10}, \frac{5}{8} = \frac{z}{10}, \text{ or}$$

$$\frac{8}{5} = \frac{10}{z} \text{ could also be used.}$$

$$z = \frac{10}{8} \cdot 5 \qquad \text{Multiplying both sides by 5}$$

$$z = \frac{50}{8}, \text{ or } 6.25.$$

EXAMPLE 6 Architecture. A *blueprint* is a scale drawing of a building representing an architect's plans. Emily is adding 12 ft to the length of an apartment and needs to indicate the addition on an existing blueprint. If a 10-ft long bedroom is represented by $2\frac{1}{2}$ in. on the blueprint, how much longer should Emily make the drawing in order to represent the addition?

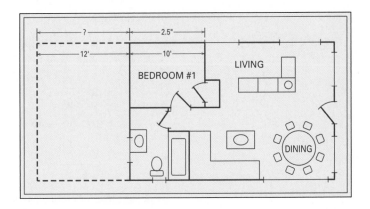

SOLUTION We let w represent the width, in inches, of the addition that Emily is drawing. Because the drawing must be to scale, we have

$$\text{Inches on drawing} \longrightarrow \frac{w}{12} = \frac{2.5}{10}. \longleftarrow \text{Inches on drawing}$$
$$\text{Feet in real life} \longrightarrow \qquad \qquad \longleftarrow \text{Feet in real life}$$

To solve for w, we multiply both sides by the LCM of the denominators, 60:

$$60 \cdot \frac{w}{12} = 60 \cdot \frac{2.5}{10}$$

$$5w = 6 \cdot 2.5 \qquad \text{Simplifying}$$

$$w = \frac{15}{5}, \text{ or } 3.$$

Emily should make the blueprint 3 in. longer.

Proportions can be used to solve a variety of applied problems.

EXAMPLE 7 Environmental Science. To determine the number of humpback whales in a pod, a marine biologist, using tail markings, identifies 27 members of the pod. Several weeks later, 40 whales from the pod are randomly sighted. Of the 40 sighted, 12 are from the 27 originally identified. Estimate the number of whales in the pod.

SOLUTION

1. **Familiarize.** If we knew that the 27 whales that were first identified constituted, say, 10% of the pod, we could easily calculate the pod's population from the proportion

$$\frac{27}{W} = \frac{10}{100},$$

where W is the size of the pod's population. Unfortunately, we are *not* told the percentage of the pod identified. We must reread the problem, looking for numbers that could be used to approximate this percentage.

2. **Translate.** Since 12 of the 40 whales that were later sighted were among those originally identified, the ratio 12/40 estimates the percentage of the pod originally identified. We can then translate to a proportion:

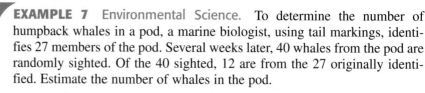

$$\text{Whales originally identified} \longrightarrow \frac{27}{W} = \frac{12}{40}. \longleftarrow \text{Original whales sighted later}$$
$$\text{Entire pod} \longrightarrow \qquad \qquad \longleftarrow \text{Whales sighted later}$$

3. **Carry out.** To solve the proportion, we multiply by the LCD, $40W$:

$$40W \cdot \frac{27}{W} = 40W \cdot \frac{12}{40} \qquad \text{Multiplying both sides by } 40W$$

$$40 \cdot 27 = W \cdot 12 \qquad \text{Removing factors equal to 1:} \\ W/W = 1 \text{ and } 40/40 = 1$$

$$\frac{40 \cdot 27}{12} = W \quad \text{or} \quad W = 90. \qquad \text{Dividing both sides by 12}$$

4. **Check.** The check is left to the student.

5. **State.** There are about 90 whales in the pod.

7.7 EXERCISE SET

FOR EXTRA HELP

| MathXL | MyMathLab | InterAct Math | AW Math Tutor Center | Video Lectures on CD: Disc 4 | Student's Solutions Manual |

Solve.

1. The reciprocal of 3, plus the reciprocal of 6, is the reciprocal of what number?

2. The reciprocal of 5, plus the reciprocal of 7, is the reciprocal of what number?

3. The sum of a number and 6 times its reciprocal is -5. Find the number.

4. The sum of a number and 21 times its reciprocal is -10. Find the number.

5. The reciprocal of the product of two consecutive integers is $\frac{1}{42}$. Find the two integers.

6. The reciprocal of the product of two consecutive integers is $\frac{1}{72}$. Find the two integers.

7. *Photo Processing.*   Nadine can develop a day's accumulation of film orders in 5 hr. Willy, a new employee, needs 9 hr to complete the same job. Working together, how long will it take them to do the job?

8. *Home Restoration.*   Cedric can refinish the floor of an apartment in 8 hr. Carolyn can refinish the floor in 6 hr. How long will it take them, working together, to refinish the floor?

9. *Filling a Pool.*   The San Paulo community swimming pool can be filled in 12 hr if water enters through a pipe alone or in 30 hr if water enters through a hose alone. If water is entering through both the pipe and the hose, how long will it take to fill the pool?

10. *Filling a Tank.*   A community water tank can be filled in 18 hr by the town office well alone and in 22 hr by the high school well alone. How long will it take to fill the tank if both wells are working?

11. *Pumping Water.*   A $\frac{1}{3}$ HP Craftsman Thermoplastic sump pump can remove water from Helen's flooded basement in 44 min. The $\frac{1}{2}$ HP Simer Thermoplastic sump pump can complete the same job in 36 min. (*Sources:* Based on data from

Sears, Northern Tool & Electric Co., and Home Depot) How long would it take the two pumps together to pump out the basement?

12. *Hotel Management.*   The Honeywell HQ17 air cleaner can clean the air in a 12-ft by 14-ft conference room in 10 min. The HQ174 can clean the air in a room of the same size in 6 min. How long would it take the two machines together to clean the air in such a room?

13. *Photocopiers.*   The HP Officejet 6110 takes twice the time required by the Canon Imageclass D680 to photocopy brochures for the New Bretton Arts Council year-end concert (*Source:* Manufacturers' marketing brochures). If working together the two machines can complete the job in 24 min, how long would it take each machine, working alone, to copy the brochures?

14. *Computer Printers.*   The HP Laser Jet 9000 works twice as fast as the Laser Jet 2300 (*Source:* www.hewlettpackard.com). If the machines work together, a university can produce all its staff manuals in 15 hr. Find the time it would take each machine, working alone, to complete the same job.

15. *Hotel Management.*   The Blueair 402 can purify the air in a conference hall in 10 fewer minutes than it takes the Panasonic F-P20HU1 to do the same job (*Source:* Based on information from manufacturers' websites). Together the two machines can purify the air in the conference hall in $\frac{120}{7}$, or $17\frac{1}{7}$ min. How long would it take each machine, working alone, to purify the air in the room?

16. *Cutting Firewood.*   Jake can cut and split a cord of firewood in 6 fewer hr than Skyler can. When they work together, it takes them 4 hr. How long would it take each of them to do the job alone?

17. *Forest Fires.*   The Erickson Air-Crane helicopter can scoop water and douse a certain forest fire four times as fast as an S-58T helicopter (*Sources*: Based on information from www.emergency.com and www.arishelicopters.com). Working together, the two helicopters can douse the fire in 8 hr. How long would it take each helicopter, working alone, to douse the fire?

18. *Waxing a Car.*   Rosita can wax her car in 2 hr. When she works together with Helga, they can wax the car in 45 min. How long would it take Helga, working by herself, to wax the car?

19. *Newspaper Delivery.*   Zsuzanna can deliver papers three times as fast as Stan can. If they work together, it takes them 1 hr. How long would it take each to deliver the papers alone?

20. *Sorting Recyclables.*   Together, it takes John and Deb 2 hr 55 min to sort recyclables. Alone, John would require 2 more hr than Deb. How long would it take Deb to do the job alone? (*Hint*: Convert minutes to hours or hours to minutes.)

21. *Paving.*   Together, Larry and Mo require 4 hr 48 min to pave a driveway. Alone, Larry would require 4 hr more than Mo. How long would it take Mo to do the job alone? (*Hint*: Convert minutes to hours.)

22. *Painting.*   Sara takes 3 hr longer to paint a floor than it takes Kate. When they work together, it takes them 2 hr. How long would each take to do the job alone?

23. *Kayaking.*   The speed of the current in Catamount Creek is 3 mph. Zeno can kayak 4 mi upstream in the same time it takes him to kayak 10 mi downstream. What is the speed of Zeno's kayak in still water?

24. *Boating.*   The current in the Lazy River moves at a rate of 4 mph. Monica's dinghy motors 6 mi upstream in the same time it takes to motor 12 mi downstream. What is the speed of the dinghy in still water?

25. *Moving Sidewalks.*   Newark Airport's moving sidewalk moves at a speed of 1.7 ft/sec. Walking on the moving sidewalk, Benny can travel 120 ft forward in the same time it takes to travel 52 ft in the opposite direction. How fast would Benny be walking on a nonmoving sidewalk?

26. *Moving Sidewalks.*   The moving sidewalk at O'Hare Airport in Chicago moves 1.8 ft/sec. Walking on the moving sidewalk, Camille travels 105 ft forward in the time it takes to travel 51 ft in the opposite direction. How fast would Camille be walking on a nonmoving sidewalk?

27. *Train Speeds.*   A B&M freight train is traveling 14 km/h slower than an AMTRAK passenger train. The B&M train travels 330 km in the same time that it takes the AMTRAK train to travel 400 km. Find their speeds. Complete the following table as part of the familiarization.

Distance =	*Rate*	·	*Time*
	Distance (in km)	Speed (in km/h)	Time (in hours)
B&M	330		
AMTRAK	400	r	$\dfrac{400}{r}$

28. *Speed of Travel.*   A loaded Roadway truck is moving 40 mph faster than a New York Railways freight train. In the time that it takes the train to travel 150 mi, the truck travels 350 mi. Find their speeds. Complete the following table as part of the familiarization.

Distance =	*Rate*	·	*Time*
	Distance (in miles)	Speed (in miles per hour)	Time (in hours)
Truck	350	r	$\dfrac{350}{r}$
Train	150		

Aha! **29.** *Bus Travel.*   A local bus travels 7 mph slower than the express. The express travels 45 mi in the time it takes the local to travel 38 mi. Find the speed of each bus.

30. *Walking.*   Rosanna walks 2 mph slower than Simone. In the time it takes Simone to walk 8 mi, Rosanna walks 5 mi. Find the speed of each person.

31. *Boating.*   Laverne's Mercruiser travels 15 km/h in still water. She motors 140 km downstream in the same time it takes to travel 35 km upstream. What is the speed of the river?

32. *Boating.*   Audrey's paddleboat travels 2 km/h in still water. The boat is paddled 4 km downstream in the same time it takes to go 1 km upstream. What is the speed of the river?

33. *Shipping.*   A barge moves 7 km/h in still water. It travels 45 km upriver and 45 km downriver in a total time of 14 hr. What is the speed of the current?

34. *Moped Speed.*   Jaime's moped travels 8 km/h faster than Mara's. Jaime travels 69 km in the same time that Mara travels 45 km. Find the speed of each person's moped.

35. *Aviation.*   A Citation II Jet travels 350 mph in still air and flies 487.5 mi into the wind and 487.5 mi with the wind in a total of 2.8 hr (*Source*: Eastern Air Charter). Find the wind speed.

36. *Canoeing.*   Al paddles 55 m per minute in still water. He paddles 150 m upstream and 150 m downstream in a total time of 5.5 min. What is the speed of the current?

37. *Train Travel.*   A freight train covered 120 mi at a certain speed. Had the train been able to travel 10 mph faster, the trip would have been 2 hr shorter. How fast did the train go?

38. *Boating.*   Julia's Boston Whaler cruised 45 mi upstream and 45 mi back in a total of 8 hr. The speed of the river is 3 mph. Find the speed of the boat in still water.

Geometry.   For each pair of similar triangles, find the value of the indicated letter.

39. *b*

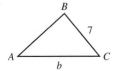

40. *a*

41. *f*

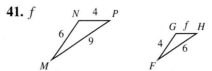

42. *r*

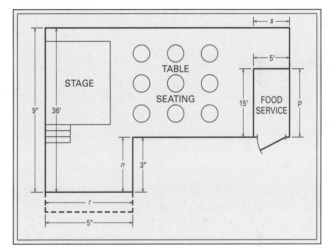

Architecture.   Use the blueprint below to find the indicated length.

43. *p*, in inches on blueprint

44. *s*, in inches on blueprint

45. *r*, in feet on actual building

46. *n*, in feet on actual building

Construction.   Find the indicated length.

47. *h*

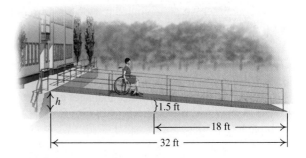

48. *l*

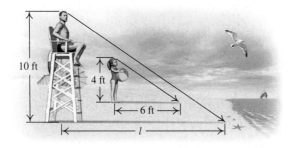

Graphing. **Find the indicated length.**

49. *r*

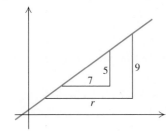

50. *s*

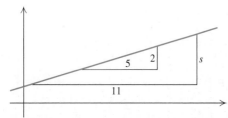

51. *Walking Speed.* For adults with an average stride that is $2\frac{1}{2}$ ft long, walking at a rate of 140 steps per minute corresponds to a speed of 4 mph (*Source:* Runnersworld.com). How many steps per minute would correspond to a speed of 3 mph?

52. *Burning Calories.* The average 150-lb adult burns about 360 calories walking 4 mi (*Source:* Runnersworld.com). How far should the average 150-lb adult walk in order to burn 20 calories?

Aha! **53.** *Photography.* Wanda snapped 234 photos over a period of 14 days. At this rate, how many would she take in 42 days?

54. *Hemoglobin.* A normal 10-cc specimen of human blood contains 1.2 g of hemoglobin. How much hemoglobin would 16 cc of the same blood contain?

55. *Wages.* For comparable work, U.S. women earn 77 cents for each dollar earned by a man (*Source:* U.S. Bureau of the Census, Public Information Office, 9/26/03). If a male sales manager earns $42,000, how much would a female earn for comparable work?

Aha! **56.** *Money.* The ratio of the weight of copper to the weight of zinc in a U.S. penny is $\frac{1}{39}$ (*Source:* United States Mint). If 50 kg of zinc is being turned into pennies, how much copper is needed?

57. *Firecrackers.* A sample of 144 firecrackers contained 9 "duds." How many duds would you expect in a sample of 320 firecrackers?

58. *Light Bulbs.* A sample of 184 light bulbs contained 6 defective bulbs. How many defective bulbs would you expect in a sample of 1288 bulbs?

59. *Moose Population.* To determine the size of Pine County's moose population, naturalists catch 69 moose, tag them, and then set them free. Months later, 40 moose are caught, of which 15 have tags. Estimate the size of the moose population.

60. *Deer Population.* To determine the number of deer in the Great Gulf Wilderness, a game warden catches 318 deer, tags them, and lets them loose. Later, 168 deer are caught; 56 of them have tags. Estimate the number of deer in the preserve.

61. *Weight on the Moon.* The ratio of the weight of an object on the moon to the weight of that object on Earth is 0.16 to 1.
   a) How much would a 12-ton rocket weigh on the moon?
   b) How much would a 180-lb astronaut weigh on the moon?

62. *Weight on Mars.* The ratio of the weight of an object on Mars to the weight of that object on Earth is 0.4 to 1.
   a) How much would a 12-ton rocket weigh on Mars?
   b) How much would a 120-lb astronaut weigh on Mars?

TW **63.** Two steamrollers are paving a parking lot. Working together, will the two steamrollers take less than half as long as the slower steamroller would working alone? Why or why not?

TW **64.** Two fuel lines are filling a freighter with oil. Will the faster fuel line take more or less than twice as long to fill the freighter by itself? Why?

Skill Maintenance

Simplify.

65. $\dfrac{35a^6b^8}{7a^2b^2}$ [5.1]

66. $\dfrac{20x^9y^6}{4x^3y^2}$ [5.1]

67. $\dfrac{36s^{15}t^{10}}{9s^5t^2}$ [5.1]

68. $6x^4 - 3x^2 + 9x - (8x^4 + 4x^2 - 2x)$ [5.3]

69. $2(x^3 + 4x^2 - 5x + 7) - 5(2x^3 - 4x^2 + 3x - 1)$ [5.3]

70. $9x^4 + 7x^3 + x^2 - 8 - (-2x^4 + 3x^2 + 4x + 2)$ [5.3]

Synthesis

TW **71.** Write a work problem for a classmate to solve. Devise the problem so that the solution is "Liane and Michele will take 4 hr to complete the job, working together."

TW **72.** Write a work problem for a classmate to solve. Devise the problem so that the solution is "Jen takes 5 hr and Pablo takes 6 hr to complete the job alone."

73. *Filling a Bog.* The Norwich cranberry bog can be filled in 9 hr and drained in 11 hr. How long will it take to fill the bog if the drainage gate is left open?

74. *Filling a Tub.* Justine's hot tub can be filled in 10 min and drained in 8 min. How long will it take to empty a full tub if the water is left on?

75. Refer to Exercise 24. How long will it take Monica to motor 3 mi downstream?

76. Refer to Exercise 23. How long will it take Zeno to kayak 5 mi downstream?

77. *Escalators.* Together, a 100-cm wide escalator and a 60-cm wide escalator can empty a 1575-person auditorium in 14 min. The wider escalator moves twice as many people as the narrower one. (*Source: McGraw-Hill Encyclopedia of Science and Technology*). How many people per hour does the 60-cm wide escalator move?

78. *Aviation.* A Coast Guard plane has enough fuel to fly for 6 hr, and its speed in still air is 240 mph. The plane departs with a 40-mph tailwind and returns to the same airport flying into the same wind. How far from the airport can the plane travel under these conditions? (Assume that the plane can use all its fuel.)

79. *Boating.* Shoreline Travel operates a 3-hr paddle-boat cruise on the Missouri River. If the speed of the boat in still water is 12 mph, how far upriver can the pilot travel against a 5-mph current before it is time to turn around?

80. *Travel by Car.* Melissa drives to work at 50 mph and arrives 1 min late. She drives to work at 60 mph and arrives 5 min early. How far does Melissa live from work?

81. *Photocopying.* The printer in an admissions office can print a 500-page document in 50 min, while the printer in the business office can print the same document in 40 min. If the two printers work together to print the document, with the faster machine starting on page 1 and the slower machine working backwards from page 500, at what page will the two machines meet to complete the job?

82. At what time after 4:00 will the minute hand and the hour hand of a clock first be in the same position?

83. At what time after 10:30 will the hands of a clock first be perpendicular?

Average speed is defined as total distance divided by total time.

84. Lenore drove 200 km. For the first 100 km of the trip, she drove at a speed of 40 km/h. For the second half of the trip, she traveled at a speed of 60 km/h. What was the average speed of the entire trip? (It was *not* 50 km/h.)

85. For the first 50 mi of a 100-mi trip, Chip drove 40 mph. What speed would he have to travel for the last half of the trip so that the average speed for the entire trip would be 45 mph?

7.8 Formulas, Applications, and Variation

Formulas ■ Direct Variation ■ Inverse Variation ■ Joint and Combined Variation ■ Models

Formulas

Formulas occur frequently as mathematical models. Many formulas contain rational expressions, and to solve such formulas for a specified letter, we proceed as when solving rational equations.

EXAMPLE 1 Electronics. The formula

$$\frac{1}{R} = \frac{1}{r_1} + \frac{1}{r_2}$$

is used by electricians to determine the resistance R of two resistors r_1 and r_2 connected in parallel.* Solve for r_1.

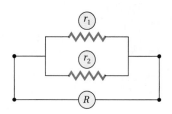

SOLUTION We use the same approach as in Section 7.6:

$$Rr_1r_2 \cdot \frac{1}{R} = Rr_1r_2 \cdot \left(\frac{1}{r_1} + \frac{1}{r_2}\right) \quad \text{**Multiplying both sides by the LCD**}$$

$$Rr_1r_2 \cdot \frac{1}{R} = Rr_1r_2 \cdot \frac{1}{r_1} + Rr_1r_2 \cdot \frac{1}{r_2} \quad \text{**Multiplying to remove parentheses**}$$

$$r_1r_2 = Rr_2 + Rr_1. \quad \text{**Simplifying by removing factors equal to 1:** } \frac{R}{R} = 1; \frac{r_1}{r_1} = 1; \frac{r_2}{r_2} = 1$$

At this point it is tempting to multiply by $1/r_2$ to get r_1 alone on the left, *but* note that there is an r_1 on the right. We must get all the terms involving r_1 on the *same side* of the equation.

$$r_1r_2 - Rr_1 = Rr_2 \quad \text{**Subtracting Rr_1 from both sides**}$$

$$r_1(r_2 - R) = Rr_2 \quad \text{**Factoring out r_1 in order to combine like terms**}$$

$$r_1 = \frac{Rr_2}{r_2 - R} \quad \text{**Dividing both sides by $r_2 - R$ to get r_1 alone**}$$

This formula can be used to calculate r_1 whenever R and r_2 are known.

---

*Recall that the subscripts 1 and 2 merely indicate that r_1 and r_2 are different variables representing similar quantities.

EXAMPLE 2   Astronomy.   The formula

$$\frac{V^2}{R^2} = \frac{2g}{R + h}$$

is used to find a satellite's *escape velocity* V, where R is a planet's radius, h is the satellite's height above the planet, and g is the planet's gravitational constant. Solve for h.

SOLUTION   We first clear fractions by multiplying by the LCD, which is $R^2(R + h)$:

$$\frac{V^2}{R^2} = \frac{2g}{R + h}$$

$$R^2(R + h)\frac{V^2}{R^2} = R^2(R + h)\frac{2g}{R + h}$$

$$\frac{R^2(R + h)V^2}{R^2} = \frac{R^2(R + h)2g}{R + h}$$

$$(R + h)V^2 = R^2 \cdot 2g. \qquad \text{Removing factors equal to 1:}$$
$$\frac{R^2}{R^2} = 1 \text{ and } \frac{R + h}{R + h} = 1$$

Remember: We are solving for h. Although we *could* distribute V^2, since h appears only within the factor $R + h$, it is easier to divide both sides by V^2:

$$\frac{(R + h)V^2}{V^2} = \frac{2R^2g}{V^2} \qquad \text{Dividing both sides by } V^2$$

$$R + h = \frac{2R^2g}{V^2} \qquad \text{Removing a factor equal to 1: } \frac{V^2}{V^2} = 1$$

$$h = \frac{2R^2g}{V^2} - R. \qquad \text{Subtracting } R \text{ from both sides}$$

The last equation can be used to determine the height of a satellite above a planet when the planet's radius and gravitational constant, along with the satellite's escape velocity, are known.

EXAMPLE 3   Acoustics (the Doppler Effect).   The formula

$$f = \frac{sg}{s + v}$$

is used to determine the frequency *f* of a sound that is moving at velocity *v* toward a listener who hears the sound as frequency *g*. Here *s* is the speed of sound in a particular medium. Solve for *s*.

SOLUTION   We first clear fractions by multiplying by the LCD, $s + v$:

$$f \cdot (s + v) = \frac{sg}{s + v}(s + v)$$

$$fs + fv = sg.$$ **The variable for which we are solving, *s*, appears on both sides, forcing us to distribute on the left side.**

Next, we must get all terms containing *s* on one side:

$$fv = sg - fs$$ **Subtracting *fs* from both sides**

$$fv = s(g - f)$$ **Factoring out *s***

$$\frac{fv}{g - f} = s.$$ **Dividing both sides by $g - f$**

Since *s* is isolated on one side, we have solved for *s*. This last equation can be used to determine the speed of sound whenever *f*, *v*, and *g* are known.

Student Notes

To Solve a Rational Equation for a Specified Variable

1. If necessary, multiply both sides by the LCD to clear fractions.
2. Multiply, as needed, to remove parentheses.
3. Get all terms with the specified variable alone on one side.
4. Factor out the specified variable if it is in more than one term.
5. Multiply or divide on both sides to isolate the specified variable.

Variation

To extend our study of formulas and functions, we now examine three real-world situations: direct variation, inverse variation, and combined variation.

Direct Variation

A registered nurse earns $22 per hour. In 1 hr, $22 is earned. In 2 hr, $44 is earned. In 3 hr, $66 is earned, and so on. This gives rise to a set of ordered pairs:

$$(1, 22), (2, 44), (3, 66), (4, 88), \text{ and so on.}$$

Note that the ratio of earnings *E* to time *t* is $\frac{22}{1}$ in every case.

   If a situation gives rise to pairs of numbers in which the ratio is constant, we say that there is **direct variation.** Here earnings *vary directly* as the time:

We have $\dfrac{E}{t} = 22$, so $E = 22t$ or, using function notation, $E(t) = 22t$.

Direct Variation   When a situation gives rise to a linear function of the form $f(x) = kx$, or $y = kx$, where k is a nonzero constant, we say that there is *direct variation*, that *y varies directly as x*, or that *y is proportional to x*. The number k is called the *variation constant*, or *constant of proportionality*.

Note that for $k > 0$, any equation of the form $y = kx$ indicates that as x increases, y increases as well.

EXAMPLE 4   Find the variation constant and an equation of variation if y varies directly as x, and $y = 32$ when $x = 2$.

SOLUTION   We know that $(2, 32)$ is a solution of $y = kx$. Therefore,

$32 = k \cdot 2$ **Substituting**

$\dfrac{32}{2} = k$,   or $k = 16$.   **Solving for k**

The variation constant is 16. The equation of variation is $y = 16x$. The notation $y(x) = 16x$ or $f(x) = 16x$ is also used.

EXAMPLE 5   Water from Melting Snow.   The number of centimeters W of water produced from melting snow varies directly as the number of centimeters S of snow. Meteorologists know that under certain conditions, 150 cm of snow will melt to 16.8 cm of water. The average annual snowfall in Alta, Utah, is 500 in. Assuming the above conditions, how much water will replace the 500 in. of snow?

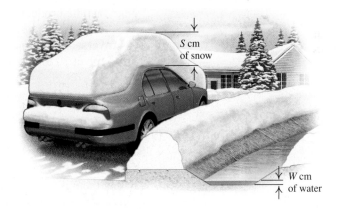

SOLUTION

1. **Familiarize.** Because of the phrase "W ... varies directly as ... S," we express the amount of water as a function of the amount of snow. Thus, $W(S) = kS$, where k is the variation constant. Knowing that 150 cm of snow becomes 16.8 cm of water, we have $W(150) = 16.8$. Because we are using ratios, it does not matter whether we work in inches or centimeters, provided the same units are used for W and S.

Alta, Utah

2. **Translate.** We find the variation constant using the data and then use it to write the equation of variation:

$$W(S) = kS$$
$$W(150) = k \cdot 150 \qquad \text{Replacing } S \text{ with 150}$$
$$16.8 = k \cdot 150 \qquad \text{Replacing } W(150) \text{ with 16.8}$$
$$\frac{16.8}{150} = k \qquad \text{Solving for } k$$
$$0.112 = k. \qquad \text{This is the variation constant.}$$

The equation of variation is $W(S) = 0.112S$. This is the translation.

3. **Carry out.** To find how much water 500 in. of snow will become, we compute $W(500)$:

$$W(S) = 0.112S$$
$$W(500) = 0.112(500) \qquad \text{Substituting 500 for } S$$
$$W = 56.$$

4. **Check.** To check, we could reexamine all our calculations. Note that our answer seems reasonable since $500/56$ and $150/16.8$ are equal.

5. **State.** Alta's 500 in. of snow will become 56 in. of water.

Inverse Variation

To see what we mean by inverse variation, suppose a bus is traveling 20 mi. At 20 mph, the trip will take 1 hr. At 40 mph, it will take $\frac{1}{2}$ hr. At 60 mph, it will take $\frac{1}{3}$ hr, and so on. This gives rise to pairs of numbers, all having the same product:

$$(20, 1), \left(40, \tfrac{1}{2}\right), \left(60, \tfrac{1}{3}\right), \left(80, \tfrac{1}{4}\right), \quad \text{and so on.}$$

Note that the product of each pair of numbers is 20. Whenever a situation gives rise to pairs of numbers for which the product is constant, we say that there is **inverse variation.** Since $r \cdot t = 20$, the time t, in hours, required for the bus to travel 20 mi at r mph is given by

$$t = \frac{20}{r} \quad \text{or, using function notation,} \quad t(r) = \frac{20}{r}.$$

Inverse Variation   When a situation gives rise to a rational function of the form $f(x) = k/x$, or $y = k/x$, where k is a nonzero constant, we say that there is *inverse variation*, that *y varies inversely as x*, or that *y is inversely proportional to x*. The number k is called the *variation constant*, or *constant of proportionality*.

Note that for $k > 0$, any equation of the form $y = k/x$ indicates that as x increases, y decreases.

EXAMPLE 6   Find the variation constant and an equation of variation if y varies inversely as x, and $y = 32$ when $x = 0.2$.

SOLUTION   We know that $(0.2, 32)$ is a solution of

$$y = \frac{k}{x}.$$

Therefore,

$$32 = \frac{k}{0.2} \quad \textbf{Substituting}$$

$$(0.2)32 = k$$

$$6.4 = k. \quad \textbf{Solving for } k$$

The variation constant is 6.4. The equation of variation is

$$y = \frac{6.4}{x}.$$

Many real-life problems translate to an equation of inverse variation.

EXAMPLE 7   Ultraviolet Index.   The ultraviolet, or UV, index is a measure that indicates the strength of the sun's rays in a particular locale. For those people whose skin is quite sensitive, a UV rating of 6 will cause sunburn after 10 min (*Source*: *The Electronic Textbook of Dermatology* found at www.telemedicine.org, January 2004). Given that the number of minutes it takes to burn, t, varies inversely as the UV rating u, how long will it take a highly sensitive person to burn on a day with a UV rating of 4?

SOLUTION

1. **Familiarize.** Because of the phrase "… varies inversely as the UV rating," we express the amount of time needed to burn as a function of the UV rating: $t(u) = k/u$.

2. **Translate.** We use the given information to solve for k. Then we use that result to write the equation of variation.

$$t(u) = \frac{k}{u} \quad \textbf{Using function notation}$$

$$t(6) = \frac{k}{6} \quad \textbf{Replacing } u \textbf{ with 6}$$

$$10 = \frac{k}{6} \quad \textbf{Replacing } t(6) \textbf{ with 10}$$

$$60 = k \quad \textbf{Solving for } k, \textbf{ the variation constant}$$

The equation of variation is $t(u) = 60/u$. This is the translation.

3. **Carry out.** To find how long it would take a highly sensitive person to burn on a day with a UV index of 4, we calculate $t(4)$:

$$t(4) = \frac{60}{4} = 15. \quad t = 15 \text{ when } u = 4$$

4. **Check.** We could now recheck each step. Note that, as expected, as the UV rating goes *down*, the time it takes to burn goes *up*.

5. **State.** On a day with a UV rating of 4, a highly sensitive person will begin to burn after 15 min of exposure. ◢

Joint and Combined Variation

When a variable varies directly with more than one other variable, we say that there is *joint variation*. For example, in the formula for the volume of a right circular cylinder, $V = \pi r^2 h$, we say that V varies *jointly* as h and the square of r.

> **Joint Variation** y varies *jointly* as x and z if, for some nonzero constant k, $y = kxz$.

EXAMPLE 8   Find an equation of variation if y varies jointly as x and z, and $y = 30$ when $x = 2$ and $z = 3$.

SOLUTION   We have

$$y = kxz,$$

so

$$30 = k \cdot 2 \cdot 3$$
$$k = 5. \qquad \text{The variation constant is 5.}$$

The equation of variation is $y = 5xz$. ◢

Joint variation is one form of *combined variation*. In general, when a variable varies directly and/or inversely, at the same time, with more than one other variable, there is **combined variation.** Examples 8 and 9 are both examples of combined variation.

EXAMPLE 9   Find an equation of variation if y varies jointly as x and z and inversely as the square of w, and $y = 105$ when $x = 3$, $z = 20$, and $w = 2$.

SOLUTION   The equation of variation is of the form

$$y = k \cdot \frac{xz}{w^2},$$

so, substituting, we have

$$105 = k \cdot \frac{3 \cdot 20}{2^2}$$
$$105 = k \cdot 15$$
$$k = 7.$$

Thus,

$$y = 7 \cdot \frac{xz}{w^2}.$$

Models

We may be able to recognize from a set of data whether two quantities vary directly or inversely. A graph like the one on the left below indicates that the quantities represented vary directly. The graph on the right below represents quantities that vary inversely. If we know the type of variation involved, we can choose one data point and calculate an equation of variation.

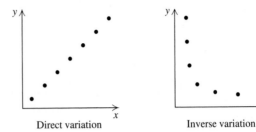

Direct variation   Inverse variation

EXAMPLE 10 Commuter Travel.   One factor influencing urban planning is VMT, or vehicle miles traveled. Areas with a high population density have fewer VMT per household than those with a lower density. This occurs in part because employment and shopping are available closer to higher density areas. The following table lists annual VMT per household for various densities for a typical urban area.

Population Density (in number of households per residential acre)	Annual VMT per Household
25	12,000
50	6,000
100	3,000
200	1,500

Source: Based on information from http://www.sflcv.org/density

a) Determine whether the data indicate direct variation or inverse variation.

b) Find an equation of variation that describes the data.

c) Use the equation to estimate the annual VMT per household for areas with 10 households per residential acre.

SOLUTION

a) We graph the data, letting x represent the population density, in number of households per residential acre, and y the annual VMT per household.

The points approximate the graph of a function of the type $f(x) = k/x$. Annual VMT per household varies inversely as the population density.

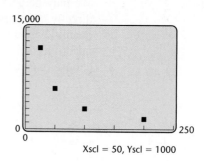

Xscl = 50, Yscl = 1000

b) To find an equation of variation, we choose one point. We will use the point $(50, 6000)$.

$$y = \frac{k}{x} \qquad \text{An equation of inverse variation}$$

$$6000 = \frac{k}{50} \qquad \text{Substituting 50 for } x \text{ and 6000 for } y$$

$$300{,}000 = k \qquad \text{Multiplying both sides by 50}$$

We now have an equation of inverse variation, $y = 300{,}000/x$. This equation can also be written $y = 300{,}000x^{-1}$. The PwrReg option of the STAT CALC menu will return equations of this type, as shown in the figure on the right below. If we graph the equation along with the data, we can see that it does match the data.

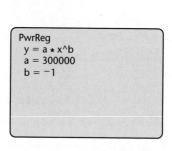

$y = 300{,}000/x$

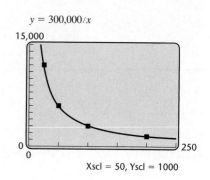

Xscl = 50, Yscl = 1000

c) To find the annual VMT per household when the population density is 10 households per residential acre, we substitute 10 for x in the equation of variation and calculate y:

$$y = \frac{300{,}000}{x} \qquad \text{The equation of variation}$$

$$y = \frac{300{,}000}{10} \qquad \text{Substituting}$$

$$y = 30{,}000.$$

When the population density is 10 households per residential acre, each household will drive 30,000 miles per year.

7.8 EXERCISE SET

🖐 *Concept Reinforcement*   *Complete each of the following statements.*

1. To solve a formula for a variable appearing in a denominator, we multiply both sides of the equation by the _____.

2. All terms containing the variable being solved for must ultimately be on the _____ side of the equation.

3. If the variable being solved for appears in more than one term on the same side of the equation, it is usually necessary to _____.

4. If y varies directly as x, then $y = kx$ and k is called the _____ of _____.

🖐 *Concept Reinforcement*   *Determine whether each situation represents direct or inverse variation.*

5. Two painters can scrape a house in 9 hr, whereas three painters can scrape the house in 6 hr.

6. Fanny planted 5 bulbs in 20 min and 7 bulbs in 28 min.

7. Glen swam 2 laps in 7 min and 6 laps in 21 min.

8. It took 2 band members 80 min to set up for a show; with 4 members working, it took 40 min.

9. It took 3 hr for 4 volunteers to wrap the campus' collection of Toys for Tots, but only 1.5 hr with 8 volunteers working.

10. Fatuma's air conditioner cooled off 1000 ft³ in 10 min and 3000 ft³ in 30 min.

Solve each formula for the specified variable.

11. $f = \dfrac{L}{d}$; d

12. $\dfrac{W_1}{W_2} = \dfrac{d_1}{d_2}$; W_1

13. $s = \dfrac{(v_1 + v_2)t}{2}$; v_1

14. $s = \dfrac{(v_1 + v_2)t}{2}$; t

15. $\dfrac{t}{a} + \dfrac{t}{b} = 1$; b

16. $\dfrac{1}{R} = \dfrac{1}{r_1} + \dfrac{1}{r_2}$; R

17. $I = \dfrac{2V}{R + 2r}$; R

18. $I = \dfrac{2V}{R + 2r}$; r

19. $R = \dfrac{gs}{g + s}$; g

20. $K = \dfrac{rt}{r - t}$; t

21. $I = \dfrac{nE}{R + nr}$; n

22. $I = \dfrac{nE}{R + nr}$; r

23. $\dfrac{1}{p} + \dfrac{1}{q} = \dfrac{1}{f}$; q

24. $\dfrac{1}{p} + \dfrac{1}{q} = \dfrac{1}{f}$; p

25. $S = \dfrac{H}{m(t_1 - t_2)}$; t_1

26. $S = \dfrac{H}{m(t_1 - t_2)}$; H

27. $\dfrac{E}{e} = \dfrac{R + r}{r}$; r

28. $\dfrac{E}{e} = \dfrac{R + r}{R}$; R

29. $S = \dfrac{a}{1 - r}$; r

30. $S = \dfrac{a - ar^n}{1 - r}$; a

31. $c = \dfrac{f}{(a + b)c}$; $a + b$ 32. $d = \dfrac{g}{d(c + f)}$; $c + f$

33. *Interest.*   The formula

$$P = \frac{A}{1 + r}$$

is used to determine what principal P should be invested for one year at $(100 \cdot r)\%$ simple interest in order to have A dollars after a year. Solve for r.

34. *Taxable Interest.*   The formula

$$I_t = \frac{I_f}{1 - T}$$

gives the *taxable interest rate* I_t equivalent to the *tax-free interest rate* I_f for a person in the $(100 \cdot T)\%$ tax bracket. Solve for T.

35. *Average Speed.*   The formula

$$v = \frac{d_2 - d_1}{t_2 - t_1}$$

gives an object's average speed v when that object has traveled d_1 miles in t_1 hours and d_2 miles in t_2 hours. Solve for t_2.

36. *Average Acceleration.*   The formula

$$a = \frac{v_2 - v_1}{t_2 - t_1}$$

gives a vehicle's *average acceleration* when its velocity changes from v_1 at time t_1 to v_2 at time t_2. Solve for t_1.

37. *Planetary Orbits.*   The formula

$$\frac{x^2}{a^2} + \frac{y^2}{b^2} = 1$$

can be used to plot a planet's elliptical orbit of width $2a$ and length $2b$. Solve for b^2.

38. *Work Rate.*   The formula

$$\frac{1}{t} = \frac{1}{a} + \frac{1}{b}$$

gives the total time t required for two workers to complete a job, if the workers' individual times are a and b. Solve for t.

39. *Semester Average.*   The formula

$$A = \frac{2Tt + Qq}{2T + Q}$$

gives a student's average A after T tests and Q quizzes, where each test counts as 2 quizzes, t is the test average, and q is the quiz average. Solve for Q.

40. *Astronomy.*   The formula

$$L = \frac{dR}{D - d},$$

where D is the diameter of the sun, d is the diameter of the earth, R is the earth's distance from the sun, and L is some fixed distance, is used in calculating when lunar eclipses occur. Solve for D.

Find the variation constant and an equation of variation if y varies directly as x and the following conditions apply.

41. $y = 28$ when $x = 4$

42. $y = 5$ when $x = 12$

43. $y = 3.4$ when $x = 2$

44. $y = 2$ when $x = 5$

45. $y = 2$ when $x = \frac{1}{3}$

46. $y = 0.9$ when $x = 0.5$

Find the variation constant and an equation of variation in which y varies inversely as x, and the following conditions exist.

47. $y = 3$ when $x = 20$

48. $y = 16$ when $x = 4$

49. $y = 28$ when $x = 4$

50. $y = 9$ when $x = 5$

51. $y = 27$ when $x = \frac{1}{3}$

52. $y = 81$ when $x = \frac{1}{9}$

53. *Use of Aluminum Cans.*   The number N of aluminum cans used each year varies directly as the number of people using the cans. If 250 people use 60,000 cans in one year, how many cans are used each year in Dallas, which has a population of 1,008,000?

54. *Hooke's Law.*   Hooke's law states that the distance d that a spring is stretched by a hanging object varies directly as the mass m of the object. If the distance is 20 cm when the mass is 3 kg, what is the distance when the mass is 5 kg?

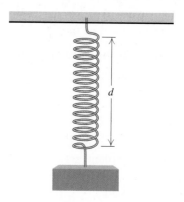

55. *Ohm's Law.*   The electric current I, in amperes, in a circuit varies directly as the voltage V. When 15 volts are applied, the current is 5 amperes. What is the current when 18 volts are applied?

56. *Pumping Rate.*   The time t required to empty a tank varies inversely as the rate r of pumping. If a Briggs and Stratton pump can empty a tank in 45 min at the rate of 600 kL/min, how long will it take the pump to empty the tank at 1000 kL/min?

57. *Work Rate.*   The time T required to do a job varies inversely as the number of people P working. It takes 5 hr for 7 volunteers to pick up rubbish from 1 mi of roadway. How long would it take 10 volunteers to complete the job?

58. *Weekly Allowance.*   According to Fidelity Investments *Investment Vision Magazine*, the average weekly allowance A of children varies directly as their grade level, G. In a recent year, the average allowance of a 9th-grade student was $13.50 per week (*Source*: Based on data from www.kidsmoney.org). What was the average weekly allowance of a 4th-grade student?

Aha! **59.** *Mass of Water in a Human.*   The number of kilograms W of water in a human body varies directly as the mass of the body. A 96-kg person contains 64 kg of water. How many kilograms of water are in a 48-kg person?

60. *Weight on Mars.*   The weight M of an object on Mars varies directly as its weight E on Earth. A person who weighs 95 lb on Earth weighs 38 lb on Mars. How much would a 100-lb person weigh on Mars?

61. *Bicycling.*   The number of calories burned by a bicyclist is directly proportional to the time spent bicycling. At 10 mph, it takes 30 min to burn 150 calories (*Source*: Physical Activity and Health: A Report of the Surgeon General). How long would it take to burn 250 calories when biking 10 mph?

62. *Wavelength and Frequency.*   The wavelength W of a radio wave varies inversely as its frequency F. A wave with a frequency of 1200 kilohertz has a length of 300 meters. What is the length of a wave with a frequency of 800 kilohertz?

63. *Ultraviolet Index.*   At an ultraviolet, or UV, rating of 4, those people who are moderately sensitive to the sun will burn in 70 min (*Source*: The Electronic Textbook of Dermatology at www.telemedicine.org). Given that the number of minutes it takes to burn, t, varies inversely with the UV rating, u, how long will it take moderately sensitive people to burn when the UV rating is 14?

64. *Current and Resistance.*   The current I in an electrical conductor varies inversely as the resistance R of the conductor. If the current is $\frac{1}{2}$ ampere when the resistance is 240 ohms, what is the current when the resistance is 540 ohms?

65. *Air Pollution.*   The average U.S. household of 2.6 people released 1.1 tons of carbon monoxide into the environment in a recent year (*Sources*: Based on data from the U.S. Environmental Protection Agency and the U.S. Bureau of the Census). How many tons were released nationally? Use 289,000,000 as the U.S. population.

66. *Relative Aperture.*   The relative aperture, or f-stop, of a 23.5-mm lens is directly proportional to the focal length F of the lens. If a lens with a 150-mm focal length has an f-stop of 6.3, find the f-stop of a 23.5-mm lens with a focal length of 80 mm.

Find an equation of variation in which:

67. y varies directly as the square of x, and $y = 6$ when $x = 3$.

68. y varies directly as the square of x, and $y = 0.15$ when $x = 0.1$.

69. y varies inversely as the square of x, and $y = 6$ when $x = 3$.

70. y varies inversely as the square of x, and $y = 0.15$ when $x = 0.1$.

71. y varies jointly as x and the square of z, and $y = 105$ when $x = 14$ and $z = 5$.

72. y varies jointly as x and z and inversely as w, and $y = \frac{3}{2}$ when $x = 2$, $z = 3$, and $w = 4$.

73. y varies jointly as w and the square of x and inversely as z, and $y = 49$ when $w = 3$, $x = 7$, and $z = 12$.

74. y varies directly as x and inversely as w and the square of z, and $y = 4.5$ when $x = 15$, $w = 5$, and $z = 2$.

75. *Electrical Safety.*   The amount of time t needed for an electrical shock to stop a 150-lb person's heart from beating varies inversely as the square of the current flowing through the body. It is known that a 0.089-amp current is deadly to a 150-lb person after 3.4 sec. (*Source*: Safety Consulting Services) How long would it take a 0.096-amp current to be deadly?

76. *Stopping Distance of a Car.* The stopping distance *d* of a car after the brakes have been applied varies directly as the square of the speed *r* (*Source*: Based on data from Edmunds.com). Once the brakes are applied, a car traveling 60 mph can stop in 138 ft. What stopping distance corresponds to a speed of 40 mph?

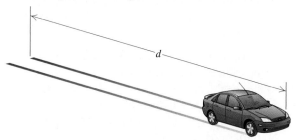

77. *Volume of a Gas.* The volume *V* of a given mass of a gas varies directly as the temperature *T* and inversely as the pressure *P*. If $V = 231 \text{ cm}^3$ when $T = 300°\text{K}$ (Kelvin) and $P = 20 \text{ lb/cm}^2$, what is the volume when $T = 320°\text{K}$ and $P = 16 \text{ lb/cm}^2$?

78. *Intensity of a Signal.* The intensity *I* of a television signal varies inversely as the square of the distance *d* from the transmitter. If the intensity is 25 W/m^2 at a distance of 2 km, what is the intensity 6.25 km from the transmitter?

79. *Atmospheric Drag.* Wind resistance, or atmospheric drag, tends to slow down moving objects. Atmospheric drag *W* varies jointly as an object's surface area *A* and velocity *v*. If a car traveling at a speed of 40 mph with a surface area of 37.8 ft^2 experiences a drag of 222 N (Newtons), how fast must a car with 51 ft^2 of surface area travel in order to experience a drag force of 430 N?

80. *Drag Force.* The drag force *F* on a boat varies jointly as the wetted surface area *A* and the square of the velocity of the boat. If a boat traveling 6.5 mph experiences a drag force of 86 N when the wetted surface area is 41.2 ft^2, find the wetted surface area of a boat traveling 8.2 mph with a drag force of 94 N.

81. *Ultraviolet Index.* The following table shows the safe exposure time for people with less sensitive skin (see Example 7).

UV Index	Safe Exposure Time (in minutes)
2	120
4	75
6	50
8	35
10	25

a) Determine whether the data indicate direct variation or inverse variation.
b) Find an equation of variation that approximates the data. Use the data point (6, 50).
c) Use the equation to predict the safe exposure time for people with less sensitive skin when the UV rating is 3.

82. *Perceived Height.* The following table shows the perceived height *y* of a pole when the observer is *x* feet from the pole.

Distance from Pole (in feet)	Perceived Height of Pole (in feet)
5	4
10	2
15	$1\frac{1}{3}$
20	1
40	0.5

a) Determine whether the data indicate direct variation or inverse variation.
b) Find an equation of variation that approximates the data. Use the data point (10, 2).
c) Use the equation to predict the perceived height of the pole when the observer is 2 ft from the pole.

83. *Mail Order.* The following table lists the number of persons *y* ordering merchandise by telephone and the number of persons *x* ordering merchandise by mail for various age groups.

Age	Number Ordering by Mail (in millions)	Number Ordering by Telephone (in millions)
18–24	6.146	5.255
25–34	11.813	12.775
35–44	14.294	16.650
45–54	11.294	13.703
55–64	6.948	8.275
65+	10.541	9.358

Source: Simmons Market Research Bureau, *Study of Media and Markets*, New York, NY

a) Determine whether the number of people ordering by mail varies directly or inversely as the number ordering by telephone.
b) Find an equation of variation that approximates the data. Use the data point (11.813, 12.775).
c) Use the equation to predict the number of persons ordering by telephone if 8 million order by mail.

84. *Motor Vehicle Registrations.* The following table lists the number of automobile registrations y and the number of drivers licensed x for various states.

State	Number of Driver's Licenses (in millions)	Number of Automobiles Registered (in millions)
Alabama	2.063	3.434
Colorado	1.843	2.946
Georgia	4.033	5.316
Idaho	0.502	0.863
Maryland	2.622	3.178
Texas	7.456	13.323
Virginia	3.774	4.787

a) Determine whether the number of automobiles registered varies directly or inversely as the number of driver's licenses.
b) Find an equation of variation that approximates the data. Use the data point (2.063, 3.434).
c) Use the equation to estimate the number of automobiles registered in Hawaii, where there are 0.45 million licensed drivers.

85. If two quantities vary directly, as in Exercise 83, does this mean that one is "caused" by the other? Why or why not?

86. If y varies directly as x, does doubling x cause y to be doubled as well? Why or why not?

Skill Maintenance

Find the domain of f. [3.8]

87. $f(x) = \dfrac{2x - 1}{x^2 + 1}$

88. $f(x) = |2x - 1|$

89. $f(x) = \dfrac{x}{x - 3}$

90. If $f(x) = x^3 - x$, find $f(2a)$. [3.8]

91. Factor: $t^3 + 8b^3$. [6.5]

92. Solve: $6x^2 = 11x + 35$. [6.3]

Synthesis

93. Suppose that the number of customer complaints is inversely proportional to the number of employees hired. Will a firm reduce the number of complaints more by expanding from 5 to 10 employees, or from 20 to 25? Explain. Consider using a graph to help justify your answer.

94. Why do you think subscripts are used in Exercises 13 and 25 but not in Exercises 27 and 28?

95. *Escape Velocity.* A satellite's escape velocity is 6.5 mi/sec, the radius of the earth is 3960 mi, and the earth's gravitational constant is 32.2 ft/sec². How far is the satellite from the surface of the earth? (See Example 2.)

96. The *harmonic mean* of two numbers a and b is a number M such that the reciprocal of M is the average of the reciprocals of a and b. Find a formula for the harmonic mean.

97. *Health-care.* Young's rule for determining the size of a particular child's medicine dosage c is

$$c = \frac{a}{a + 12} \cdot d,$$

where a is the child's age and d is the typical adult dosage (*Source*: Olsen, June Looby, Leon J. Ablon, and Anthony Patrick Giangrasso, *Medical Dosage Calculations*, 6th ed.). If a child's age is doubled, the dosage increases. Find the ratio of the larger dosage to the smaller dosage. By what percent does the dosage increase?

98. Solve for x:
$$x^2\left(1 - \frac{2pq}{x}\right) = \frac{2p^2q^3 - pq^2x}{-q}.$$

99. *Average Acceleration.*   The formula
$$a = \frac{\dfrac{d_4 - d_3}{t_4 - t_3} - \dfrac{d_2 - d_1}{t_2 - t_1}}{t_4 - t_2}$$

can be used to approximate average acceleration, where the d's are distances and the t's are the corresponding times. Solve for t_1.

100. If y varies inversely as the cube of x and x is multiplied by 0.5, what is the effect on y?

101. *Intensity of Light.*   The intensity I of light from a bulb varies directly as the wattage of the bulb and inversely as the square of the distance d from the bulb. If the wattage of a light source and its distance from reading matter are both doubled, how does the intensity change?

102. Describe in words the variation represented by $W = \dfrac{km_1M_1}{d^2}$. Assume k is a constant.

103. *Tension of a Musical String.*   The tension T on a string in a musical instrument varies jointly as the string's mass per unit length m, the square of its length l, and the square of its fundamental frequency f. A 2-m long string of mass 5 gm/m with a fundamental frequency of 80 has a tension of 100 N. How long should the same string be if its tension is going to be changed to 72 N?

104. *Volume and Cost.*   A peanut butter jar in the shape of a right circular cylinder is 4 in. high and 3 in. in diameter and sells for \$1.20. If we assume that cost is proportional to volume, how much should a jar 6 in. high and 6 in. in diameter cost?

105. *Golf Distance Finder.*   A device used in golf to estimate the distance d to a hole measures the size s that the 7-ft pin *appears* to be in a viewfinder. The viewfinder uses the principle, diagrammed here, that s gets bigger when d gets smaller. If $s = 0.56$ in. when $d = 50$ yd, find an equation of variation that expresses d as a function of s. What is d when $s = 0.40$ in.?

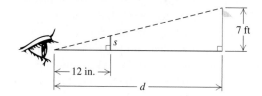

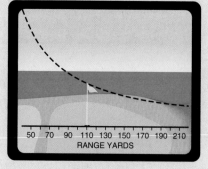

HOW IT WORKS:

Just sight the flagstick through the viewfinder...
fit flag between top dashed line and the solid line below...
...read the distance, 50 – 220 yards.

Nothing to focus.
•
Gives you exact distance that your ball lies from the flagstick.
•
Choose proper club on every approach shot.
•
Figure new pin placement instantly.
•
Train your naked eye for formal and tournament play.
•
Eliminate the need to remember every stake, tree, and bush on the course.

7 Chapter Summary and Review

KEY TERMS AND DEFINITIONS

RATIONAL EXPRESSIONS AND EQUATIONS

Rational expression, p. 526   A quotient of two polynomials. The domain of a **rational function** does not include values that make a denominator zero.

Least common denominator, LCD, p. 548   The least common multiple of the denominators.

APPLICATIONS

Variation

y varies directly as x if there is some nonzero constant k such that $y = kx$.

y varies inversely as x if there is some nonzero constant k such that $y = k/x$.

y varies jointly as x and z if there is some nonzero constant k such that $y = kxz$.

Modeling Work Problems

If

a = the time needed for A to complete the work alone,

b = the time needed for B to complete the work alone, and

t = the time needed for A and B to complete the work together,

then

$$\frac{t}{a} + \frac{t}{b} = 1 \quad \text{and} \quad \frac{1}{a} \cdot t + \frac{1}{b} \cdot t = 1$$

and $\dfrac{1}{a} + \dfrac{1}{b} = \dfrac{1}{t}$.

Motion Formula

$$d = rt \quad \text{or} \quad r = d/t \quad \text{or} \quad t = d/r$$

IMPORTANT CONCEPTS

[Section references appear in brackets.]

Concept	Example
The graph of a rational function may approach one or more **vertical asymptotes.**	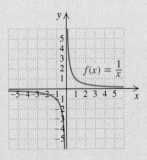 Domain: $\{x \mid x$ is a real number *and* $x \neq 0\}$ Range: $\{y \mid y$ is a real number *and* $y \neq 0\}$ Vertical asymptote: $x = 0$ [7.1]

(continued)

Addition: $\dfrac{A}{C} + \dfrac{B}{C} = \dfrac{A+B}{C}$

$\dfrac{3}{x} + \dfrac{2}{x+1} = \dfrac{3x+3}{x(x+1)} + \dfrac{2x}{x(x+1)} = \dfrac{5x+3}{x(x+1)}$ [7.3], [7.4]

Subtraction: $\dfrac{A}{C} - \dfrac{B}{C} = \dfrac{A-B}{C}$

$\dfrac{2x+5}{x-3} - \dfrac{x-7}{x-3} = \dfrac{(2x+5)-(x-7)}{x-3}$

You must have a common denominator to add or subtract.

$= \dfrac{2x+5-x+7}{x-3} = \dfrac{x+12}{x-3}$ [7.3], [7.4]

Multiplication: $\dfrac{A}{B} \cdot \dfrac{C}{D} = \dfrac{AC}{BD}$

$\dfrac{2x}{x+3} \cdot \dfrac{x^2+3x}{4x^2} = \dfrac{2 \cdot \cancel{x} \cdot \cancel{x} \cdot \cancel{(x+3)}}{\cancel{(x+3)} \cdot 2 \cdot 2 \cdot \cancel{x} \cdot \cancel{x}} = \dfrac{1}{2}$ [7.2]

Division: $\dfrac{A}{B} \div \dfrac{C}{D} = \dfrac{A}{B} \cdot \dfrac{D}{C}$

$\dfrac{x}{x+1} \div \dfrac{3}{x+1} = \dfrac{x}{x+1} \cdot \dfrac{x+1}{3} = \dfrac{x\cancel{(x+1)}}{\cancel{(x+1)}3} = \dfrac{x}{3}$ [7.2]

Complex rational expressions contain one or more rational expressions within the numerator and/or denominator. They can be simplified either by multiplying by a form of 1 to clear the fractions or by dividing two rational expressions.

Multiplying by 1:

$\dfrac{\dfrac{4}{x} + \dfrac{4}{y}}{\dfrac{2}{x} + \dfrac{6}{y}} = \dfrac{\dfrac{4}{x} + \dfrac{4}{y}}{\dfrac{2}{x} + \dfrac{6}{y}} \cdot \dfrac{xy}{xy} = \dfrac{\dfrac{4\cancel{x}y}{\cancel{x}} + \dfrac{4x\cancel{y}}{\cancel{y}}}{\dfrac{2\cancel{x}y}{\cancel{x}} + \dfrac{6x\cancel{y}}{\cancel{y}}}$

$= \dfrac{4y+4x}{2y+6x} = \dfrac{\cancel{2} \cdot 2 \cdot (y+x)}{\cancel{2}(y+3x)}$

$= \dfrac{2(y+x)}{y+3x}$

Dividing two rational expressions:

$\dfrac{\dfrac{4}{x} + \dfrac{4}{y}}{\dfrac{2}{x} + \dfrac{6}{y}} = \dfrac{\dfrac{4y}{xy} + \dfrac{4x}{xy}}{\dfrac{2y}{xy} + \dfrac{6x}{xy}} = \dfrac{\dfrac{4y+4x}{xy}}{\dfrac{2y+6x}{xy}}$

$= \dfrac{4y+4x}{xy} \cdot \dfrac{xy}{2y+6x}$

$= \dfrac{\cancel{2} \cdot 2(y+x) \cdot \cancel{x} \cdot \cancel{y}}{\cancel{x} \cdot \cancel{y} \cdot \cancel{2}(y+3x)}$

$= \dfrac{2(y+x)}{y+3x}$ [7.5]

Multiplying a **rational equation** by the LCD will clear the equation of fractions. All potential solutions must be checked in the original equation.

$\dfrac{1}{x} = \dfrac{x}{4}$

$4x\left(\dfrac{1}{x}\right) = 4x\left(\dfrac{x}{4}\right)$

$4 = x^2$

$x = 2 \quad or \quad x = -2$ **Both 2 and −2 check. The solutions are 2 and −2.** [7.6]

Review Exercises

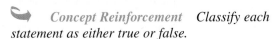 *Concept Reinforcement* *Classify each statement as either true or false.*

1. Every rational expression can be simplified. [7.1]

2. The expression $(t - 3)/(t^2 - 4)$ is undefined for $t = 2$. [7.1]

3. The expression $(t - 3)/(t^2 - 4)$ is undefined for $t = 3$. [7.1]

4. To multiply rational expressions, a common denominator is never required. [7.2]

5. To divide rational expressions, a common denominator is never required. [7.2]

6. To add rational expressions, a common denominator is never required. [7.3]

7. To subtract rational expressions, a common denominator is never required. [7.3]

8. Every rational equation has at least one solution. [7.6]

9. If Carlie swims 5 km/h in still water and heads into a current of 2 km/h, her speed will change to 3 km/h. [7.7]

10. If x varies inversely as y, then there exists some constant k for which $x = k/y$. [7.8]

11. If
$$f(t) = \frac{t^2 - 3t + 2}{t^2 - 9},$$
find the following function values. [7.1]
a) $f(0)$
b) $f(-1)$
c) $f(2)$

List all numbers for which each expression is undefined. [7.1]

12. $\dfrac{17}{-x^2}$

13. $\dfrac{9}{a - 4}$

14. $\dfrac{x - 5}{x^2 - 36}$

15. $\dfrac{x^2 + 3x + 2}{x^2 + x - 30}$

Simplify. [7.1]

16. $\dfrac{4x^2 - 8x}{4x^2 + 4x}$

17. $\dfrac{(y - 5)^2}{y^2 - 25}$

18. $\dfrac{5x^2 - 20y^2}{2y - x}$

Multiply or divide and, if possible, simplify. [7.2]

19. $\dfrac{a^2 - 36}{10a} \cdot \dfrac{2a}{a + 6}$

20. $\dfrac{8t + 8}{2t^2 + t - 1} \cdot \dfrac{t^2 - 1}{t^2 - 2t + 1}$

21. $\dfrac{16 - 8t}{3} \div \dfrac{t - 2}{12t}$

22. $\dfrac{4x^4}{x^2 - 1} \div \dfrac{2x^3}{x^2 - 2x + 1}$

23. $\dfrac{x^2 + 1}{x - 2} \cdot \dfrac{2x + 1}{x + 1}$

24. $(t^2 + 3t - 4) \div \dfrac{t^2 - 1}{t + 4}$

Find the LCD. [7.3]

25. $\dfrac{5}{6x^3}, \dfrac{15}{16x^2}$

26. $\dfrac{x - 2}{x^2 + x - 20}, \dfrac{x - 3}{x^2 + 3x - 10}$

Add or subtract and, if possible, simplify.

27. $\dfrac{x + 6}{x + 3} + \dfrac{9 - 4x}{x + 3}$ [7.3]

28. $\dfrac{3}{3x - 9} + \dfrac{x - 2}{3 - x}$ [7.4]

29. $\dfrac{6x - 3}{x^2 - x - 12} - \dfrac{2x - 15}{x^2 - x - 12}$ [7.3]

30. $\dfrac{3x - 1}{2x} - \dfrac{x - 3}{x}$ [7.4]

31. $\dfrac{x + 5}{x - 2} - \dfrac{x}{2 - x}$ [7.4]

32. $\dfrac{2a}{a + 1} - \dfrac{4a}{1 - a^2}$ [7.4]

33. $\dfrac{d^2}{d-c} + \dfrac{c^2}{c-d}$ [7.4]

34. $\dfrac{1}{x^2-25} - \dfrac{x-5}{x^2-4x-5}$ [7.4]

35. $\dfrac{2}{5x} + \dfrac{3}{2x+4}$ [7.4]

Find simplified form for f(x) and list all restrictions on the domain.

36. $f(x) = \dfrac{14x^2 - x - 3}{2x^2 - 7x + 3}$ [7.1]

37. $f(x) = \dfrac{3x}{x+2} - \dfrac{x}{x-2} + \dfrac{8}{x^2-4}$ [7.4]

Simplify. [7.5]

38. $\dfrac{\dfrac{4}{x} - 4}{\dfrac{9}{x} - 9}$

39. $\dfrac{\dfrac{3}{a} + \dfrac{3}{b}}{\dfrac{6}{a^3} + \dfrac{6}{b^3}}$

40. $\dfrac{\dfrac{y^2 + 4y - 77}{y^2 - 10y + 25}}{\dfrac{y^2 - 5y - 14}{y^2 - 25}}$

41. $\dfrac{\dfrac{5}{x^2-9} - \dfrac{3}{x+3}}{\dfrac{4}{x^2+6x+9} + \dfrac{2}{x-3}}$

Solve. [7.6]

42. $\dfrac{3}{x} + \dfrac{7}{x} = 5$

43. $\dfrac{5}{3x+2} = \dfrac{3}{2x}$

44. $\dfrac{4x}{x+1} + \dfrac{4}{x} + 9 = \dfrac{4}{x^2+x}$

45. $\dfrac{x+6}{x^2+x-6} + \dfrac{x}{x^2+4x+3} = \dfrac{x+2}{x^2-x-2}$

46. If
$$f(x) = \dfrac{2}{x-1} + \dfrac{2}{x+2},$$
find all a for which $f(a) = 1$. [7.6]

Solve. [7.7]

47. Kim can arrange the books for a book sale in 9 hr. Kelly can set up for the same book sale in 12 hr. How long would it take them, working together, to set up for the book sale?

48. A research company uses personal computers to process data while the owner is not using the computer. A Pentium 4 3.0-gigahertz processor can process a megabyte of data in 15 sec less time than a Celeron 2.53-gigahertz processor. Working together, the computers can process a megabyte of data in 18 sec. How long does it take each computer to process one megabyte of data?

49. The Black River's current is 6 mph. A boat travels 50 mi downstream in the same time that it takes to travel 30 mi upstream. What is the speed of the boat in still water?

50. A car and a motorcycle leave a rest area at the same time, with the car traveling 8 mph faster than the motorcycle. The car then travels 105 mi in the time it takes the motorcycle to travel 93 mi. Find the speed of each vehicle.

51. To estimate the frog population in a pond, a scientist captures, tags, and releases 24 frogs. Weeks later, 20 frogs are caught and 8 of them are wearing tags. Estimate the frog population of the pond.

52. Triangles *ABC* and *XYZ* are similar. Find the value of *x*.

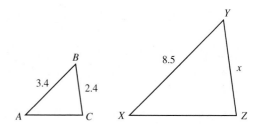

Solve. [7.8]

53. $R = \dfrac{gs}{g+s}$, for *s*

54. $S = \dfrac{H}{m(t_1 - t_2)}$, for *m*

55. $\dfrac{1}{ac} = \dfrac{2}{ab} - \dfrac{3}{bc}$, for *c*

56. $T = \dfrac{A}{v(t_2 - t_1)}$, for t_1

57. The amount of waste generated by a family varies directly as the number of people in the family. The average U.S. family has 3.14 people and generates 13.8 lb of waste daily (*Sources:* Based on data from the U.S. Bureau of the Census and the *U.S. Statistical Abstract* 2003). How many pounds of waste would be generated daily by a family of 5?

58. A warning dye is used by people in lifeboats to aid search planes. The volume V of the dye used varies directly as the square of the diameter d of the circular patch of water formed by the dye. If 4 L of dye is required for a 10-m wide circle, how much dye is needed for a 40-m wide circle?

59. Find an equation of variation in which y varies inversely as x, and $y = 3$ when $x = \frac{1}{4}$.

60. The following table lists the size y of one serving of breakfast cereal for the corresponding number of servings x obtained from a given box. [7.8]

Number of Servings	Size of Serving (in ounces)
1	24
4	6
6	4
12	2
16	1.5

a) Determine whether the data indicate direct variation or inverse variation.
b) Find an equation of variation that fits the data. Use the data point $(12, 2)$.

c) Use the equation to estimate the size of each serving when 8 servings are obtained from the box.

Synthesis

TW 61. Discuss at least three different uses of the LCD studied in this chapter. [7.4], [7.5], [7.6]

TW 62. Explain the difference between a rational expression and a rational equation. [7.1]

Solve.

63. $\dfrac{5}{x-13} - \dfrac{5}{x} = \dfrac{65}{x^2-13x}$ [7.6]

64. $\dfrac{\dfrac{x}{x^2-25} + \dfrac{2}{x-5}}{\dfrac{3}{x-5} - \dfrac{4}{x^2-10x+25}} = 1$ [7.5], [7.6]

65. A Xeon 3.6-gigahertz processor can process a megabyte of data in 20 sec. How long would it take a Xeon working together with the Pentium 4 and Celeron processors (see Exercise 48) to process a megabyte of data? [7.7]

Chapter Test 7

List all numbers for which each expression is undefined.

1. $\dfrac{2-x}{5x}$

2. $\dfrac{x^2+x-30}{x^2-3x+2}$

3. Find the LCM:
$$y^2-9, \quad y^2+10y+21, \quad y^2+4y-21.$$

Multiply or divide and, if possible, simplify.

4. $\dfrac{a^2-25}{9a} \cdot \dfrac{6a}{5-a}$

5. $\dfrac{25y^2-1}{9y^2-6y} \div \dfrac{5y^2+9y-2}{3y^2+y-2}$

6. $\dfrac{4x^2-1}{x^2-2x+1} \div \dfrac{x-2}{x^2+1}$

7. $(x^2+6x+9) \cdot \dfrac{(x-3)^2}{x^2-9}$

Add or subtract, and, if possible, simplify.

8. $\dfrac{2+x}{x^3} + \dfrac{7-4x}{x^3}$

9. $\dfrac{5-t}{t^2+1} - \dfrac{t-3}{t^2+1}$

10. $\dfrac{x-4}{x-3} + \dfrac{x-1}{3-x}$

11. $\dfrac{x-4}{x-3} - \dfrac{x-1}{3-x}$

12. $\dfrac{7}{t-2} + \dfrac{4}{t}$

13. $\dfrac{1}{x^2-16} - \dfrac{x+4}{x^2-3x-4}$

14. $\dfrac{6}{x^3-64} - \dfrac{4}{x^2-16}$

Find simplified form for $f(x)$ and list all restrictions on the domain.

15. $f(x) = \dfrac{6x^2+17x+7}{2x^2+7x+3}$

16. $f(x) = \dfrac{4}{x+3} - \dfrac{x}{x-2} + \dfrac{x^2+4}{x^2+x-6}$

Simplify.

17. $\dfrac{\dfrac{2}{a} + \dfrac{3}{b}}{\dfrac{5}{ab} + \dfrac{1}{a^2}}$

18. $\dfrac{\dfrac{x^2-5x-36}{x^2-36}}{\dfrac{x^2+x-12}{x^2-12x+36}}$

19. $\dfrac{\dfrac{2}{x+3} - \dfrac{1}{x^2-3x+2}}{\dfrac{3}{x-2} + \dfrac{4}{x^2+2x-3}}$

Solve.

20. $\dfrac{4}{2x-5} = \dfrac{6}{5x+3}$

21. $\dfrac{t+11}{t^2-t-12} + \dfrac{1}{t-4} = \dfrac{4}{t+3}$

For Exercises 22 and 23, let $f(x) = \dfrac{x+3}{x-1}$.

22. Find $f(2)$ and $f(-3)$.

23. Find all a for which $f(a) = 7$.

24. Solve $A = \dfrac{h(b_1+b_2)}{2}$ for b_1.

25. The product of the reciprocals of two consecutive integers is $\frac{1}{30}$. Find the integers.

26. Emma bicycles 12 mph with no wind. Against the wind, she bikes 8 mi in the same time that it takes to bike 14 mi with the wind. What is the speed of the wind?

27. Kyla can install a vinyl kitchen floor in 3.5 hr. Brock can perform the same job in 4.5 hr. How long will it take them, working together, to install the vinyl?

28. A recipe for pizza crust calls for $3\frac{1}{2}$ cups of whole wheat flour and $1\frac{1}{4}$ cups of warm water. If 6 cups of whole wheat flour are used, how much water should be used?

29. The number of workers n needed to clean a stadium after a game varies inversely as the amount of time t allowed for the cleanup. If it takes 25 workers to clean the stadium when there are 6 hr allowed for the job, how many workers are needed if the stadium must be cleaned in 5 hr?

30. The surface area of a balloon varies directly as the square of its radius. The area is 325 in² when the radius is 5 in. What is the surface area when the radius is 7 in.?

Synthesis

31. Let
$$f(x) = \dfrac{1}{x+3} + \dfrac{5}{x-2}.$$
Find all a for which $f(a) = f(a+5)$.

32. Solve: $\dfrac{6}{x-15} - \dfrac{6}{x} = \dfrac{90}{x^2-15x}$.

33. Find the x- and y-intercepts for the function given by
$$f(x) = \dfrac{\dfrac{5}{x+4} - \dfrac{3}{x-2}}{\dfrac{2}{x-3} + \dfrac{1}{x+4}}.$$

34. One summer, Hans mowed 4 lawns for every 3 lawns mowed by his brother Franz. Together, they mowed 98 lawns. How many lawns did each mow?

8

Inequalities

Inequalities are mathematical sentences containing symbols such as < (is less than).

In this chapter, we use the principles for solving inequalities developed in Chapter 2 to solve various kinds of inequalities. We also combine our knowledge of inequalities and systems of equations to solve systems of inequalities.

8.1 Graphical Solutions and Compound Inequalities

8.2 Absolute-Value Equations and Inequalities

8.3 Inequalities in Two Variables

8.4 Polynomial and Rational Inequalities

APPLICATION *Earnings Ratio.*

The ratio of median earnings of women to median earnings of men is called the female-to-male earnings ratio. A ratio of 0.77, or 77:100, means that women earn $77 for every $100 that men earn. This ratio is shown for various years in the following table. Use linear regression to find a linear function that can be used to estimate the female-to-male earnings ratio x years after 1980. Then use the function to predict the years in which the female-to-male earnings ratio will be 0.9 or higher.

YEAR	FEMALE-TO-MALE EARNINGS RATIO
1980	0.60
1985	0.64
1990	0.71
1995	0.72
2000	0.75
2004	0.77

Source: U.S. Bureau of the Census

Xscl = 5, Yscl = .1

This problem appears as Example 2 in Section 8.1.

<table>
<tr><td>

8.1

</td><td>

Graphical Solutions and Compound Inequalities

</td><td>

Solving Inequalities Graphically   ■   Intersections of Sets and
Conjunctions of Sentences   ■   Unions of Sets and Disjunctions
of Sentences   ■   Interval Notation and Domains

</td></tr>
</table>

Recall from Chapter 1 that an **inequality** is any sentence containing $<$, $>$, $\leq$, $\geq$, or $\neq$ (see Section 1.4)—for example,

$$-2 < a, \quad x > 4, \quad x + 3 \leq 6, \quad 7y \geq 10y - 4, \quad \text{and} \quad 5x \neq 10.$$

Any replacement for the variable that makes an inequality true is called a **solution.** The set of all solutions is called the **solution set.** When all solutions of an inequality are found, we say that we have **solved** the inequality.

We solved inequalities algebraically in Chapter 2, using the addition and multiplication principles for inequalities. These principles are similar to those used to solve equations with one important difference: The direction of the inequality symbol must be reversed when multiplying or dividing by a negative number.

We now look at a graphical method for solving inequalities.

Solving Inequalities Graphically

Consider the graphs of the functions $f(x) = 2x + 4$ and $g(x) = -x + 1$. The function values are *equal* at the point of intersection, $(-1, 2)$.

Graphs of $f(x)$ and $g(x)$

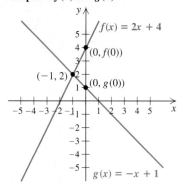

Graph of solution set of $f(x) = g(x)$

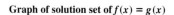

Graph of solution set of $f(x) < g(x)$

Graph of solution set of $f(x) > g(x)$

At all x-values except -1, either

$$f(x) > g(x) \quad \text{or} \quad f(x) < g(x).$$

Note from the graphs that $f(x) > g(x)$ when the graph of f lies above the graph of g. Also, $f(x) < g(x)$ when the graph of f lies below the graph of g.

Compare $f(0)$ and $g(0)$. Note from the graphs that $f(0)$ lies above $g(0)$. In fact, for all values of x greater than -1, $f(x) > g(x)$. For all values of x less than -1, $f(x) < g(x)$. In this way, the point of intersection of the graphs marks the endpoint of the solution set of an inequality.

EXAMPLE 1 Solve graphically: $16 - 7x \geq 10x - 4$.

SOLUTION We let $y_1 = 16 - 7x$ and $y_2 = 10x - 4$, and graph y_1 and y_2 in the window $[-5, 5, -5, 15]$.

From the graph, we see that $y_1 = y_2$ at the point of intersection. To the left of the point of intersection, $y_1 > y_2$. You can check this by calculating the values of both y_1 and y_2 for an x-value to the left of the point of intersection— say, for $x = 0.5$. Since $12.5 > 1$, we know that $y_1 > y_2$ when $x = 0.5$. Similarly, we see that $y_1 < y_2$ to the right of the point of intersection.

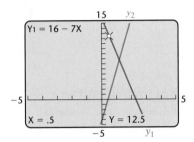 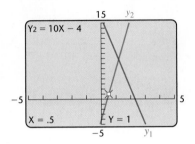

The solution set will be all x-values to the left of the point of intersection, as well as the x-coordinate of the point of intersection. Using INTERSECT, we find that the x-coordinate of the point of intersection is approximately 1.1764706. Thus the solution set is approximately $(-\infty, 1.1764706]$, or converting to fraction notation, $\left(-\infty, \frac{20}{17}\right]$.

On many graphing calculators, the interval that is the solution set can be indicated by using the VARS and TEST keys. To find the solution of Example 1, we also enter and graph $y_3 = y_1 \geq y_2$ (the "$\geq$" symbol can be found in the TEST menu). Where this is true, the value of y_3 will be 1, and where it is false, the value will be 0. The solution set is thus displayed as an interval, shown by a horizontal line 1 unit above the x-axis. The endpoint of the interval corresponds to the intersection of the graphs of the equations. For some calculators, using DOT mode for y_3 will result in a more accurate graph.

$y_1 = 16 - 7x, \quad y_2 = 10x - 4,$
$y_3 = y_1 \geq y_2$

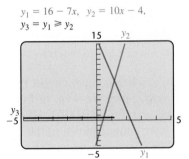

EXAMPLE 2 Earnings Ratio. The ratio of median earnings of women to median earnings of men is called the female-to-male earnings ratio. A ratio of 0.77, or 77:100, means that women earn $77 for every $100 that men earn. This ratio is shown for various years in the following table. Use linear regression to find a linear function that can be used to estimate the female-to-male

earnings ratio *x* years after 1980. Then use the function to predict the years in which the female-to-male earnings ratio will be 0.9 or higher.

Year	Female-to-Male Earnings Ratio
1980	0.60
1985	0.64
1990	0.71
1995	0.72
2000	0.75
2004	0.77

Source: U.S. Bureau of the Census

SOLUTION We let *x* represent the number of years after 1980 and *y* the female-to-male earnings ratio, and graph the data to determine whether the relationship appears to be linear.

The data are approximately linear. We find a linear equation that fits the data, as shown in the figure on the left below. Since we want to find the years in which the ratio is 0.9 or higher, we graph $y_1 = 0.0070241935x + 0.6117016129$ and $y_2 = 0.9$. The point of intersection, shown in the figure on the right below, tells us that the ratio will be 0.9 approximately 41 yr after 1980, or in about 2021. Since $y_1 > y_2$ to the right of the point of intersection, we predict that the female-to-male earnings ratio will be 0.9 or higher in 2021 and after, or for years after 2020.

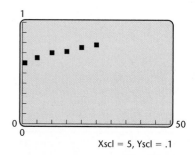

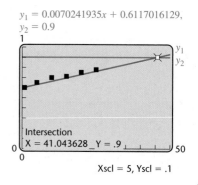

$y_1 = 0.0070241935x + 0.6117016129,$
$y_2 = 0.9$

Two inequalities joined by the word "and" or the word "or" are called **compound inequalities.** Thus, "$2x - 7 < 3$ *or* $x - 1 > 4$" and "$7x < 9$ *and* $x - 1 > -5$" are two examples of compound inequalities. In order to discuss how to solve compound inequalities, we must first study ways in which sets can be combined.

Intersections of Sets and Conjunctions of Sentences

The **intersection** of two sets A and B is the set of all elements that are common to both A and B. We denote the intersection of sets A and B as

$$A \cap B.$$

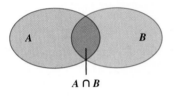

$A \cap B$

The intersection of two sets is represented by the purple region shown in the figure at left. For example, if A = {all students who are more than $5'4''$ tall} and B = {all students who weigh more than 120 lb}, then $A \cap B$ = {all students who are more than $5'4''$ tall and weigh more than 120 lb}.

> **EXAMPLE 3**   Find the intersection: $\{1, 2, 3, 4, 5\} \cap \{-2, -1, 0, 1, 2, 3\}$.
>
> **SOLUTION**   The numbers 1, 2, and 3 are common to both sets, so the intersection is $\{1, 2, 3\}$. ◢

When two or more sentences are joined by the word *and* to make a compound sentence, the new sentence is called a **conjunction** of the sentences. The following is a conjunction of inequalities:

$$-2 < x \quad and \quad x < 1.$$

The word *and* means that *both* sentences must be true if the conjunction is to be true.

A number is a solution of a conjunction if it is a solution of *both* of the separate parts. For example, -1 is a solution because it is a solution of $-2 < x$ as well as $x < 1$.

Below we show the graph of $-2 < x$, followed by the graph of $x < 1$, and finally the graph of the conjunction $-2 < x$ *and* $x < 1$. *Note that the solution set of a conjunction is the intersection of the solution sets of the individual sentences.*

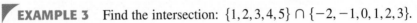

$\{x \mid -2 < x\}$ $(-2, \infty)$

$\{x \mid x < 1\}$ $(-\infty, 1)$

$\{x \mid -2 < x\} \cap \{x \mid x < 1\}$ $(-2, 1)$
$= \{x \mid -2 < x \text{ and } x < 1\}$

Because there are numbers that are both greater than -2 and less than 1, the conjunction $-2 < x$ *and* $x < 1$ can be abbreviated by $-2 < x < 1$. Thus the interval $(-2, 1)$ can be represented as $\{x \mid -2 < x < 1\}$, the set of all numbers that are *simultaneously* greater than -2 *and* less than 1. Note that for $a < b$,

$$a < x \quad \textit{and} \quad x < b \quad \textbf{can be abbreviated} \quad a < x < b;$$

and, equivalently,

$$b > x \quad \textit{and} \quad x > a \quad \textbf{can be abbreviated} \quad b > x > a.$$

EXAMPLE 4   Solve and graph: $-1 \le 2x + 5 < 13$.

SOLUTION   This inequality is an abbreviation for the conjunction

$$-1 \le 2x + 5 \quad \textit{and} \quad 2x + 5 < 13.$$

We solve both algebraically and graphically.

ALGEBRAIC APPROACH

The word *and* corresponds to set *intersection.* To solve the conjunction, we solve each of the two inequalities separately and then find the intersection of the solution sets:

$$-1 \le 2x + 5 \quad \textit{and} \quad 2x + 5 < 13$$

$-6 \le 2x \quad \textit{and} \quad 2x < 8$ **Subtracting 5 from both sides of each inequality**

$-3 \le x \quad \textit{and} \quad x < 4.$ **Dividing both sides of each inequality by 2**

These steps are sometimes combined as follows:

$$-1 \le 2x + 5 < 13$$
$$-1 - 5 \le 2x + 5 - 5 < 13 - 5 \quad \text{\textbf{Subtracting 5 from all three regions}}$$
$$-6 \le 2x < 8$$
$$-3 \le x < 4. \quad \text{\textbf{Dividing by 2 in all three regions}}$$

The solution set is $\{x \mid -3 \le x < 4\}$, or, in interval notation, $[-3, 4)$. The graph is the intersection of the two separate solution sets.

$\{x \mid -3 \le x\}$, or $[-3, \infty)$
$\begin{array}{c} \longleftarrow\!+\!+\!+\!+\!+\![\!+\!+\!+\!+\!+\!+\!+\!+\!+\!+\!\longrightarrow \\ {\scriptstyle -7\ -6\ -5\ -4\ -3\ -2\ -1\ \ 0\ \ 1\ \ 2\ \ 3\ \ 4\ \ 5\ \ 6\ \ 7}\end{array}$

$\{x \mid x < 4\}$, or $(-\infty, 4)$
$\begin{array}{c} \longleftarrow\!+\!+\!+\!+\!+\!+\!+\!+\!+\!+\!+\!)\!+\!+\!+\!\longrightarrow \\ {\scriptstyle -7\ -6\ -5\ -4\ -3\ -2\ -1\ \ 0\ \ 1\ \ 2\ \ 3\ \ 4\ \ 5\ \ 6\ \ 7}\end{array}$

$\{x \mid -3 \le x\} \cap \{x \mid x < 4\}$
$= \{x \mid -3 \le x < 4\}$, or
$[-3, 4)$
$\begin{array}{c} \longleftarrow\!+\!+\!+\!+\![\!+\!+\!+\!+\!+\!+\!+\!)\!+\!+\!+\!\longrightarrow \\ {\scriptstyle -7\ -6\ -5\ -4\ -3\ -2\ -1\ \ 0\ \ 1\ \ 2\ \ 3\ \ 4\ \ 5\ \ 6\ \ 7}\end{array}$

GRAPHICAL APPROACH

We graph the equations $y_1 = -1$, $y_2 = 2x + 5$, and $y_3 = 13$, and determine those x-values for which $y_1 \le y_2$ *and* $y_2 < y_3$.

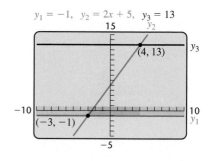

$y_1 = -1, \quad y_2 = 2x + 5, \quad y_3 = 13$

From the graph, we see that $y_1 < y_2$ for x-values greater than -3, as indicated by the purple and blue shading on the x-axis. (The shading does not appear on the calculator, but was added here to illustrate the intervals.) We also see that $y_2 < y_3$ for x-values less than 4, as shown by the purple and red shading on the x-axis. The solution set, indicated by the purple shading, is the intersection of these sets, as well as the number -3. It includes all x-values for which the line $y_2 = 2x + 5$ is both on or above the line $y_1 = -1$ *and* below the line $y_3 = 13$. This can be written $\{x \mid -3 \le x < 4\}$, or $[-3, 4)$.

CAUTION!   The abbreviated form of a conjunction, like $-3 \le x < 4$, can be written only if both inequality symbols point in the same direction.

◤**EXAMPLE 5**   Solve and graph: $2x - 5 \ge -3$ *and* $5x + 2 \ge 17$.

SOLUTION   We first solve each inequality separately, retaining the word *and*:

$$2x - 5 \ge -3 \quad and \quad 5x + 2 \ge 17$$
$$2x \ge 2 \quad and \quad 5x \ge 15$$
$$x \ge 1 \quad and \quad x \ge 3.$$

Next, we find the intersection of the two separate solution sets.

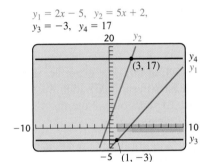

$\{x \,|\, x \ge 1\}$

$\{x \,|\, x \ge 3\}$

$\{x \,|\, x \ge 1\} \cap \{x \,|\, x \ge 3\}$
$= \{x \,|\, x \ge 3\}$

$y_1 = 2x - 5, \ y_2 = 5x + 2,$
$y_3 = -3, \ y_4 = 17$

The numbers common to both sets are those greater than or equal to 3. Thus the solution set is $\{x \,|\, x \ge 3\}$, or, in interval notation, $[3, \infty)$. You should check that any number in $[3, \infty)$ satisfies the conjunction whereas numbers outside $[3, \infty)$ do not. The graph at left serves as another check.   ◢

Mathematical Use of the Word "and"   The word "and" corresponds to "intersection" and to the symbol "∩". Any solution of a conjunction must make each part of the conjunction true.

Sometimes there is no way to solve both parts of a conjunction at once.

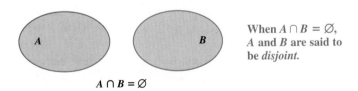

When $A \cap B = \varnothing$, A and B are said to be *disjoint*.

$A \cap B = \varnothing$

◤**EXAMPLE 6**   Solve and graph: $2x - 3 > 1$ *and* $3x - 1 < 2$.

SOLUTION   We solve each inequality separately:

$$2x - 3 > 1 \quad and \quad 3x - 1 < 2$$
$$2x > 4 \quad and \quad 3x < 3$$
$$x > 2 \quad and \quad x < 1.$$

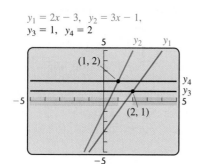

$y_1 = 2x - 3, \quad y_2 = 3x - 1,$
$y_3 = 1, \quad y_4 = 2$

The solution set is the intersection of the individual inequalities.

$\{x \mid x > 2\}$ (graph on number line)   **(2, ∞)**

$\{x \mid x < 1\}$ (graph on number line)   **(−∞, 1)**

$\{x \mid x > 2\} \cap \{x \mid x < 1\}$
$= \{x \mid x > 2 \ and \ x < 1\} = \varnothing$ (graph on number line) $\varnothing$

Since no number is both greater than 2 and less than 1, the solution set is the empty set, $\varnothing$. The graph at left confirms that the solution set is empty.

Unions of Sets and Disjunctions of Sentences

The **union** of two sets A and B is the collection of elements belonging to A and/or B. We denote the union of A and B by

$$A \cup B.$$

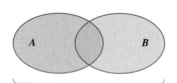

$A \cup B$

The union of two sets is often pictured as shown at left. For example, if $A =$ {all parents} and $B =$ {all people who are at least 30 years old}, then $A \cup B =$ {all people who are parents *or* who are at least 30 years old}. Note that this set includes people who are parents *and* at least 30 years old.

EXAMPLE 7   Find the union: $\{2, 3, 4\} \cup \{3, 5, 7\}$.

SOLUTION   The numbers in either or both sets are 2, 3, 4, 5, and 7, so the union is $\{2, 3, 4, 5, 7\}$.

Student Notes

Remember that the union or intersection of two sets is itself a set and should be written with set braces.

When two or more sentences are joined by the word *or* to make a compound sentence, the new sentence is called a **disjunction** of the sentences. Here is an example:

$$x < -3 \quad or \quad x > 3.$$

The word *or* is generally used to mean "one or the other but not both." You may have to decide whether to take an algebra class at 10:00 *or* at 2:00; you would not take it at both times. Mathematicians, however, use *or* to mean "one or the other or possibly both." Sometimes we do use *or* this way. Your major may require you to take a philosophy course *or* a psychology course. You must take at least one of these courses, but you would still fulfill the requirements if you took both.

A number is a solution of a disjunction if it is a solution of at least one of the separate parts. For example, -5 is a solution of the disjunction above since -5 is a solution of $x < -3$. Below we show the graph of $x < -3$, followed by the graph of $x > 3$, and finally the graph of the disjunction $x < -3 \ or \ x > 3$. *Note that the solution set of a disjunction is the union of the solution sets of the individual sentences.*

$\{x \mid x < -3\}$  $(-\infty, -3)$

$\{x \mid x > 3\}$ $(3, \infty)$

$\{x \mid x < -3\} \cup \{x \mid x > 3\}$  $(-\infty, -3) \cup (3, \infty)$
$= \{x \mid x < -3 \ or \ x > 3\}$

The solution set of $x < -3 \ or \ x > 3$ is $\{x \mid x < -3 \ or \ x > 3\}$, or, in interval notation, $(-\infty, -3) \cup (3, \infty)$. There is no simpler way to write the solution.

Mathematical Use of the Word "or"   The word "or" corresponds to "union" and to the symbol "∪". For a number to be a solution of a disjunction, it must be in *at least one* of the solution sets of the individual sentences.

EXAMPLE 8   Solve and graph: $7 + 2x < -1 \ or \ 13 - 5x \leq 3$.

SOLUTION   We solve each inequality separately, retaining the word *or*:

$$7 + 2x < -1 \quad or \quad 13 - 5x \leq 3$$
$$2x < -8 \quad or \quad -5x \leq -10$$

Dividing by a negative and reversing the symbol

$$x < -4 \quad or \quad x \geq 2.$$

To find the solution set of the disjunction, we consider the individual graphs. We graph $x < -4$ and then $x \geq 2$. Then we take the union of the graphs.

$y_1 = 7 + 2x, \ y_2 = 13 - 5x,$
$y_3 = -1, \ y_4 = 3$

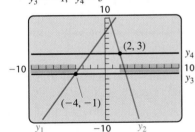

$\{x \mid x < -4\}$ $(-\infty, -4)$

$\{x \mid x \geq 2\}$ $[2, \infty)$

$\{x \mid x < -4\} \cup \{x \mid x \geq 2\}$ 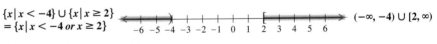 $(-\infty, -4) \cup [2, \infty)$
$= \{x \mid x < -4 \ or \ x \geq 2\}$

The solution set is $\{x \mid x < -4 \ or \ x \geq 2\}$, or $(-\infty, -4) \cup [2, \infty)$. This is confirmed by the graph at left.

CAUTION!   A compound inequality like

$$x < -4 \quad or \quad x \geq 2,$$

as in Example 8, *cannot* be expressed as $2 \leq x < -4$ because to do so would be to say that x is *simultaneously* less than -4 and greater than or equal to 2. No number is both less than -4 *and* greater than 2, but many are less than -4 *or* greater than 2.

EXAMPLE 9   Solve: $-2x - 5 < -2 \ or \ x - 3 < -10$.

SOLUTION   We solve the individual inequalities separately, retaining the word *or*:

$$-2x - 5 < -2 \quad or \quad x - 3 < -10$$
$$-2x < 3 \quad or \quad x < -7$$

Dividing by a negative and reversing the symbol ⟶   ⟶ Keep the word "or."

$$x > -\tfrac{3}{2} \quad or \quad x < -7.$$

The solution set is $\left\{x \mid x < -7 \ or \ x > -\tfrac{3}{2}\right\}$, or $(-\infty, -7) \cup \left(-\tfrac{3}{2}, \infty\right)$.

EXAMPLE 10   Solve: $3x - 11 < 4 \ or \ 4x + 9 \geq 1$.

SOLUTION   We solve the individual inequalities separately, retaining the word *or:*

$$3x - 11 < 4 \quad or \quad 4x + 9 \geq 1$$
$$3x < 15 \quad or \quad 4x \geq -8$$
$$x < 5 \quad or \quad x \geq -2.$$

⟵ Keep the word "or."

To find the solution set, we first look at the individual graphs.

$\{x \mid x < 5\}$ $(-\infty, 5)$

$\{x \mid x \geq -2\}$ $[-2, \infty)$

$\{x \mid x < 5\} \cup \{x \mid x \geq -2\}$
$= \{x \mid x < 5 \ or \ x \geq -2\}$ $(-\infty, \infty) = \mathbb{R}$

Since *all* numbers are less than 5 or greater than or equal to -2, the two sets fill the entire number line. Thus the solution set is $\mathbb{R}$, the set of all real numbers.

Interval Notation and Domains

In Section 6.3, we saw that if $g(x) = \dfrac{5x - 2}{3x - 7}$, then the domain of $g = \left\{x \mid x \text{ is} \right.$ a real number *and* $\left. x \neq \tfrac{7}{3}\right\}$. We can now represent such a set using interval notation:

$$\left\{x \mid x \text{ is a real number } and \ x \neq \tfrac{7}{3}\right\} = \left(-\infty, \tfrac{7}{3}\right) \cup \left(\tfrac{7}{3}, \infty\right).$$

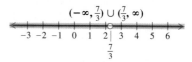

EXAMPLE 11 Use interval notation to write the domain of f if $f(x) = \sqrt{x + 2}$.

SOLUTION The expression $\sqrt{x + 2}$ is not a real number when $x + 2$ is negative. Thus the domain of f is the set of all x-values for which $x + 2 \geq 0$:

$$x + 2 \geq 0 \qquad \text{$x + 2$ cannot be negative.}$$
$$x \geq -2. \qquad \text{Adding -2 to both sides}$$

We have the domain of $f = \{x \mid x \geq -2\} = [-2, \infty)$.

We can partially check that $[-2, \infty)$ is the domain of f either by using a table of values or by viewing the graph of $f(x)$, as shown at left. By tracing the curve, we can confirm that no y-value is given for x-values less than -2.

$y = \sqrt{(x + 2)}$

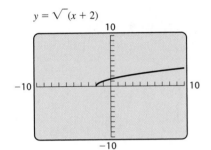

The check in Example 11 is a partial check, because we cannot check every number in the interval. Also, if a function is not defined at a point, that "hole" in the graph will not be seen on a graphing calculator. Keeping those limitations in mind, know that a graph is still an efficient and visual check of a domain.

Interactive Discovery

Consider the functions $f(x) = \sqrt{3 - x}$ and $g(x) = \sqrt{x + 1}$. Enter these in a graphing calculator as $y_1 = \sqrt{(3 - x)}$ and $y_2 = \sqrt{(x + 1)}$.

1. Determine algebraically the domain of f and the domain of g. Then graph y_1 and y_2, and trace each curve to verify the domains.
2. Graph $y_1 + y_2$, $y_1 - y_2$, and $y_1 \cdot y_2$. Use the graphs to determine the domains of $f + g$, $f - g$, and $f \cdot g$.
3. How can the domains of the sum, the difference, and the product of f and g be found algebraically?

Domain of the Sum, Difference, or Product of Functions

The domain of the sum, the difference, or the product of the functions f and g is the intersection of the domains of f and g.

EXAMPLE 12 Find the domain of $f + g$ if $f(x) = \sqrt{2x - 5}$ and $g(x) = \sqrt{x + 1}$.

SOLUTION We first find the domain of f and the domain of g. The domain of f is the set of all x-values for which $2x - 5 \geq 0$, or $\left\{x \mid x \geq \frac{5}{2}\right\}$, or $\left[\frac{5}{2}, \infty\right)$. Similarly, the domain of g is $\{x \mid x \geq -1\}$, or $[-1, \infty)$. The intersection of the domains is $\left\{x \mid x \geq \frac{5}{2}\right\}$, or $\left[\frac{5}{2}, \infty\right)$. We can confirm this at least approximately by tracing the graph of $f + g$.

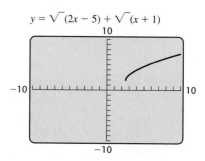

Thus the

$$\text{Domain of } f + g = \left\{x \mid x \geq \tfrac{5}{2}\right\}, \text{ or } \left[\tfrac{5}{2}, \infty\right).$$

8.1 EXERCISE SET

↪ *Concept Reinforcement* *In each of Exercises 1–10, match the set with the most appropriate choice from the column on the right.*

1. ____ $\{x \mid x < -2 \text{ or } x > 2\}$

2. ____ $\{x \mid x < -2 \text{ and } x > 2\}$

3. ____ $\{x \mid x > -2\} \cap \{x \mid x < 2\}$

4. ____ $\{x \mid x \leq -2\} \cup \{x \mid x \geq 2\}$

5. ____ $\{x \mid x \leq -2\} \cup \{x \mid x \leq 2\}$

6. ____ $\{x \mid x \leq -2\} \cap \{x \mid x \leq 2\}$

7. ____ $\{x \mid x \geq -2\} \cap \{x \mid x \geq 2\}$

8. ____ $\{x \mid x \geq -2\} \cup \{x \mid x \geq 2\}$

9. ____ $\{x \mid x \leq 2\} \text{ and } \{x \mid x \geq -2\}$

10. ____ $\{x \mid x \leq 2\} \text{ or } \{x \mid x \geq -2\}$

a) -2 2

b) 2

c) -2 2

d) -2

e) 2

f) -2 2

g) -2

h) -2 2

i) $\mathbb{R}$

j) $\varnothing$

Solve each inequality using the given graph.

11. $f(x) \geq g(x)$

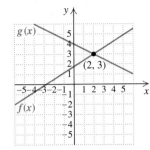

12. $f(x) < g(x)$

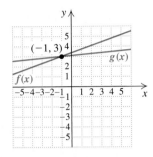

13. $y_1 < y_2$

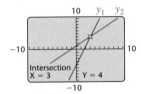

14. $y_1 \geq y_2$

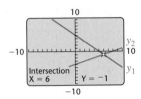

15. The graphs of $f(x) = 2x + 1$, $g(x) = -\frac{1}{2}x + 3$, and $h(x) = x - 1$ are as shown below. Solve each inequality, referring only to the figure.

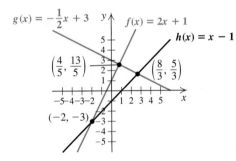

a) $2x + 1 \leq x - 1$
b) $x - 1 > -\frac{1}{2}x + 3$
c) $-\frac{1}{2}x + 3 < 2x + 1$

16. The graphs of $y_1 = -\frac{1}{2}x + 5$, $y_2 = x - 1$, and $y_3 = 2x - 3$ are as shown below. Solve each inequality, referring only to the figure.

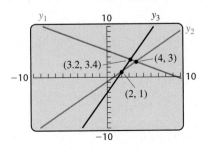

a) $-\frac{1}{2}x + 5 > x - 1$
b) $x - 1 \leq 2x - 3$
c) $2x - 3 \geq -\frac{1}{2}x + 5$

Solve graphically.

17. $x - 3 < 4$

18. $x + 4 \geq 6$

19. $2x - 3 \geq 1$

20. $3x + 1 < 1$

21. $x + 3 > 2x - 5$

22. $3x - 5 \leq 3 - x$

23. $\frac{1}{2}x - 2 \leq 1 - x$

24. $x + 5 > \frac{1}{3}x - 1$

25. $4x + 7 \leq 3 - 5x$

26. $5x + 6 < 8x - 11$

Use a graph to estimate the solution in each of the following. Be sure to use graph paper and a straightedge if graphing by hand.

27. *Show Business.*   A band receives $750 plus 15% of receipts over $750 for playing a club date. If a club charges a $6 cover charge, how many people must attend in order for the band to receive at least $1200?

28. *Temperature Conversion.*   The function
$$C(F) = \tfrac{5}{9}(F - 32)$$
can be used to find the Celsius temperature $C(F)$ that corresponds to $F°$ Fahrenheit.

a) Gold is solid at Celsius temperatures less than 1063°C. Find the Fahrenheit temperatures for which gold is solid.
b) Silver is solid at Celsius temperatures less than 960.8°C. Find the Fahrenheit temperatures for which silver is solid.

29. *Winter Olympic Games.*   The number of nations participating in the Winter Olympic Games has been increasing over the years, as shown in the following table. Use linear regression to find a linear function that can be used to predict the number of nations participating x years after 1988.

Then predict those years in which more than 100 nations will participate in the games.

Year	Number of Nations Participating in Winter Olympic Games
1988	57
1992	64
1994	67
1998	72
2002	77
2006	80

Source: Olympics.org

30. *Trampoline Injuries.* The following table lists the number of trampoline injuries in the United States for various years. Use linear regression to find a linear function that can be used to estimate the number t of trampoline injuries x years after 1990. Then predict those years in which the number of trampoline injuries will exceed 150,000.

Year	Number of Injuries
1990	29,600
1992	39,000
1995	58,400
1997	82,722
2002	89,393

Source: National Safety Council, Itasca, IL, *Injury Facts*

31. *Media Usage.* The following table lists the number of hours spent per person per year reading a newspaper and playing video games. Use linear

regression to find two linear functions that can be used to estimate the number of hours n spent reading a newspaper and the number of hours v spent playing video games x years after 2000. Then predict those years in which a person will spend more time playing video games than reading a newspaper.

Year	Daily Newspapers (in hours per person per year)	Video Games (in hours per person per year)
2000	180	59
2001	177	60
2002	175	64
2003	171	69
2004*	169	71
2005*	168	75
2006*	165	81
2007*	165	86
2008*	164	98

*Projected

Source: Veronis Suhler Stevenson, New York, NY

32. *Basketball.* The following table lists the attendance for NCAA and professional basketball for various years. Use linear regression to find two linear functions that can be used to estimate the attendance n for the NCAA and the attendance p for professional basketball x years after 1985. Then predict those years in which professional basketball attendance will exceed NCAA basketball attendance.

Year	NCAA Basketball Attendance (in thousands)	Professional Basketball Attendance (in thousands)
1985	26,584	11,534
1990	28,741	18,586
1995	28,548	19,883
2000	29,025	21,503
2002	29,395	21,571
2004	30,761	22,953

Sources: NCAA; NBA

33. *High Jump.* The following table lists the world record for the men's high jump for various years. Use linear regression to find a linear function that can be used to predict the high-jump world record *x* years after 1900. Then predict those years in which the world record will exceed 2.5 m.

Year	High Jump (in meters)	Athlete
1912	2.00	George L. Horine (USA)
1924	2.04	Harold M. Osborn (USA)
1934	2.06	Walter Marty (USA)
1941	2.11	Lester Steers (USA)
1957	2.162	Yuriy Styepanov (USSR)
1970	2.29	Ni Zhinquin (China)
1980	2.36	Gerd Wessig (Germany)
1993	2.45	Javier Sotomayor (Cuba)

Source: www.iaaf.org

34. *Mile Run.* The following table lists the world record for the mile run for various years. Use linear regression to find a linear function that can be used to predict the world record for the mile run *x* years after 1954. Then predict those years in which the world record will be less than 3.5 min.

Year	Time for Mile (in minutes)	Athlete
1954	3.99	Sir Roger Bannister (Great Britain)
1962	3.9017	Peter Snell (New Zealand)
1975	3.8233	John Walker (New Zealand)
1981	3.7888	Sebastian Coe (Great Britain)
1993	3.7398	Noureddine Morceli (Algeria)
1999	3.7188	Hican El Guerrouj (Morocco)

Source: www.iaaf.org

Find each indicated intersection or union.

35. $\{5, 9, 11\} \cap \{9, 11, 18\}$

36. $\{2, 4, 8\} \cup \{8, 9, 10\}$

37. $\{0, 5, 10, 15\} \cup \{5, 15, 20\}$

38. $\{2, 5, 9, 13\} \cap \{5, 8, 10\}$

39. $\{a, b, c, d, e, f\} \cap \{b, d, f\}$

40. $\{a, b, c\} \cup \{a, c\}$

41. $\{r, s, t\} \cup \{r, u, t, s, v\}$

42. $\{m, n, o, p\} \cap \{m, o, p\}$

43. $\{3, 6, 9, 12\} \cap \{5, 10, 15\}$

44. $\{1, 5, 9\} \cup \{4, 6, 8\}$

45. $\{3, 5, 7\} \cup \varnothing$

46. $\{3, 5, 7\} \cap \varnothing$

Graph and write interval notation for each compound inequality.

47. $3 < x < 7$

48. $0 \leq y \leq 4$

49. $-6 \leq y \leq -2$

50. $-9 \leq x < -5$

51. $x < -1 \text{ or } x > 4$

52. $x < -5 \text{ or } x > 1$

53. $x \leq -2 \text{ or } x > 1$

54. $x \leq -5 \text{ or } x > 2$

55. $-4 \leq -x < 2$

56. $x > -7 \text{ and } x < -2$

57. $x > -2 \text{ and } x < 4$

58. $3 > -x \geq -1$

59. $5 > a \text{ or } a > 7$

60. $t \geq 2 \text{ or } -3 > t$

61. $x \geq 5 \text{ or } -x \geq 4$

62. $-x < 3 \text{ or } x < -6$

63. $7 > y \text{ and } y \geq -3$

64. $6 > -x \geq 0$

65. $x < 7 \text{ and } x \geq 3$

66. $x \geq -3 \text{ and } x < 3$

Aha! **67.** $t < 2 \text{ or } t < 5$

68. $t > 4 \text{ or } t > -1$

Solve and graph each solution set.

69. $-2 < t + 1 < 8$

70. $-3 < t + 1 \le 5$

71. $2 < x + 3 \text{ and } x + 1 \le 5$

72. $-1 < x + 2 \text{ and } x - 4 < 3$

73. $-7 \le 2a - 3 \text{ and } 3a + 1 < 7$

74. $-4 \le 3n + 5 \text{ and } 2n - 3 \le 7$

Aha! **75.** $x + 7 \le -2 \text{ or } x + 7 \ge -3$

76. $x + 5 < -3 \text{ or } x + 5 \ge 4$

77. $5 > \dfrac{x - 3}{4} > 1$

78. $3 \ge \dfrac{x - 1}{2} \ge -4$

79. $-7 \le 4x + 5 \le 13$

80. $-4 \le 2x + 3 \le 15$

81. $2 \le f(x) \le 8$, where $f(x) = 3x - 1$

82. $7 \ge g(x) \ge -2$, where $g(x) = 3x - 5$

83. $-21 \le f(x) < 0$, where $f(x) = -2x - 7$

84. $4 > g(t) \ge 2$, where $g(t) = -3t - 8$

85. $f(x) \le 2 \text{ or } f(x) \ge 8$, where $f(x) = 3x - 1$

86. $g(x) \le -2 \text{ or } g(x) \ge 10$, where $g(x) = 3x - 5$

87. $f(x) < -3 \text{ or } f(x) > 5$, where $f(x) = 2x - 7$

88. $g(x) < -7 \text{ or } g(x) > 7$, where $g(x) = 3x + 5$

89. $6 > 2a - 1 \text{ or } -4 \le -3a + 2$

90. $3a - 7 > -10 \text{ or } 5a + 2 \le 22$

91. $a + 3 < -2 \text{ and } 3a - 4 < 8$

92. $1 - a < -2 \text{ and } 2a + 1 > 9$

93. $3x + 2 < 2 \text{ or } 4 - 2x < 14$

94. $2x - 1 > 5 \text{ or } 3 - 2x \ge 7$

95. $2t - 7 \le 5 \text{ or } 5 - 2t > 3$

96. $5 - 3a \le 8 \text{ or } 2a + 1 > 7$

97. Use the accompanying graph of $f(x) = 2x - 5$ to solve $-7 < 2x - 5 < 7$.

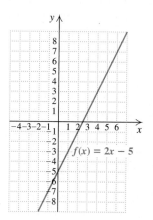

98. Use the accompanying graph of $g(x) = 4 - x$ to solve $4 - x < -2 \text{ or } 4 - x > 7$.

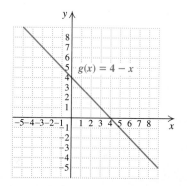

For $f(x)$ as given, use interval notation to write the domain of f.

99. $f(x) = \dfrac{9}{x + 8}$

100. $f(x) = \dfrac{2}{x + 3}$

101. $f(x) = \sqrt{x - 6}$

102. $f(x) = \sqrt{x - 2}$

103. $f(x) = \dfrac{x + 3}{(2x - 8)(x - 5)}$

104. $f(x) = \dfrac{x - 1}{(3x + 6)(x - 3)}$

105. $f(x) = \sqrt{2x + 7}$

106. $f(x) = \sqrt{8 - 5x}$

107. $f(x) = \sqrt{8 - 2x}$

108. $f(x) = \sqrt{10 - 2x}$

Use interval notation to write each domain.

109. The domain of $f + g$, if $f(x) = \sqrt{x - 5}$ and $g(x) = \sqrt{\frac{1}{2}x + 1}$

110. The domain of $f - g$, if $f(x) = \sqrt{x + 3}$ and $g(x) = \sqrt{2x - 1}$

111. The domain of $f \cdot g$, if $f(x) = \sqrt{3 - x}$ and $g(x) = \sqrt{3x - 2}$

112. The domain of $f + g$, if $f(x) = \sqrt{3 - 4x}$ and $g(x) = \sqrt{x + 2}$

TW 113. Why can the conjunction $2 < x$ *and* $x < 5$ be rewritten as $2 < x < 5$, but the disjunction $2 < x$ *or* $x < 5$ cannot be rewritten as $2 < x < 5$?

TW 114. Can the solution set of a disjunction be empty? Why or why not?

Skill Maintenance

Graph.

115. $y = 5$ [3.3]

116. $y = -2$ [3.3]

117. $f(x) = |x|$ [3.8]

118. $g(x) = x - 1$ [3.8]

Solve each system graphically. [4.1]

119. $y = x - 3,$
$y = 5$

120. $y = x + 2,$
$y = -3$

Synthesis

TW 121. What can you conclude about a, b, c, and d, if $[a, b] \cup [c, d] = [a, d]$? Why?

TW 122. What can you conclude about a, b, c, and d, if $[a, b] \cap [c, d] = [a, b]$? Why?

123. *Childless Women.*   On the basis of trends from the late 1900s, the function given by

$$P(t) = 0.44t + 10.2$$

can be used to estimate the percentage of U.S. women age 40–44, $P(t)$, t years after 1980, who have not given birth (*Sources*: Based on data from the U.S. Bureau of the Census; *Deseret Morning News*, 11/03). For what years will the percentage of childless 40–44-year-old women be between 19 and 30 percent?

124. *Pressure at Sea Depth.*   The function given by

$$P(d) = 1 + \frac{d}{33}$$

gives the pressure, in atmospheres (atm), at a depth of d feet in the sea. For what depths d is the pressure at least 1 atm and at most 7 atm?

125. *Converting Dress Sizes.*   The function given by

$$f(x) = 2(x + 10)$$

can be used to convert dress sizes x in the United States to dress sizes $f(x)$ in Italy. For what dress sizes in the United States will dress sizes in Italy be between 32 and 46?

126. *Solid-Waste Generation.*   The function given by

$$w(t) = 0.01t + 4.3$$

can be used to estimate the number of pounds of solid waste, $w(t)$, produced daily, on average, by each person in the United States, t years after 1991. For what years will waste production range from 4.5 to 4.75 lb per person per day?

127. *Records in the Women's 100-m Dash.* Florence Griffith Joyner set a world record of 10.49 sec in the women's 100-m dash in 1988. The function given by

$$R(t) = -0.0433t + 10.49$$

can be used to predict the world record in the women's 100-m dash t years after 1988. (*Sources: Guinness Book of World Records 2004*; www.Runnersworld.com) Predict (using an inequality) those years for which the world record was between 11.5 and 10.8 sec. (Measure from the middle of 1988.)

128. *Minimizing Tolls.* A $3.00 toll is charged to cross the bridge from Sanibel Island to mainland Florida. A six-month pass, costing $15.00, reduces the toll to $0.50. A one-year pass, costing $150, allows for free crossings. (*Source:* www.leewayinfo.com) How many crossings per year does it take, on average, for two consecutive six-month passes to be the most economical choice? Assume a constant number of trips per month.

Solve and graph.

129. $4m - 8 > 6m + 5$ *or* $5m - 8 < -2$

130. $4a - 2 \leq a + 1 \leq 3a + 4$

131. $3x < 4 - 5x < 5 + 3x$

132. $x - 10 < 5x + 6 \leq x + 10$

Determine whether each sentence is true or false for all real numbers a, b, and c.

133. If $-b < -a$, then $a < b$.

134. If $a \leq c$ and $c \leq b$, then $b > a$.

135. If $a < c$ and $b < c$, then $a < b$.

136. If $-a < c$ and $-c > b$, then $a > b$.

For f(x) as given, use interval notation to write the domain of f.

137. $f(x) = \dfrac{\sqrt{3 - 4x}}{x + 7}$

138. $f(x) = \dfrac{\sqrt{5 + 2x}}{x - 1}$

139. On many graphing calculators, the TEST key provides access to inequality symbols, while the LOGIC option of that same key accesses the conjunction *and* and the disjunction *or*. Thus, if $y_1 = x > -2$ and $y_2 = x < 4$, Exercise 57 can be checked by forming the expression $y_3 = y_1$ *and* y_2. The interval(s) in the solution set appears as a horizontal line 1 unit above the x-axis. (Be careful to "deselect" y_1 and y_2 so that only y_3 is drawn.)

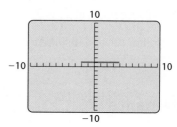

8.2 Absolute-Value Equations and Inequalities

Equations with Absolute Value   ◼   Inequalities with Absolute Value

Equations with Absolute Value

Recall from Section 1.4 the definition of absolute value.

> **Absolute Value**   The absolute value of x, denoted $|x|$, is defined as
> $$|x| = \begin{cases} x, & \text{if } x \geq 0, \\ -x, & \text{if } x < 0. \end{cases}$$
> (When x is nonnegative, the absolute value of x is x. When x is negative, the absolute value of x is the opposite of x.)

To better understand this definition, suppose x is -5. Then $|x| = |-5| = 5$, and 5 is the opposite of -5. This shows that when x represents a negative number, the absolute value of x is positive.

Since distance is always nonnegative, we can think of a number's absolute value as its distance from zero on the number line.

> **EXAMPLE 1**   Find the solution set: **(a)** $|x| = 4$; **(b)** $|x| = 0$; **(c)** $|x| = -7$.

SOLUTION

a) We interpret $|x| = 4$ to mean that the number x is 4 units from zero on the number line. There are two such numbers, 4 and -4. Thus the solution set is $\{-4, 4\}$.

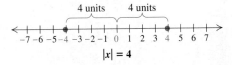

A second way to visualize this problem is to graph $f(x) = |x|$. We also graph $g(x) = 4$. The x-values of the points of intersection are the solutions of $|x| = 4$.

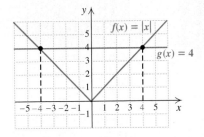

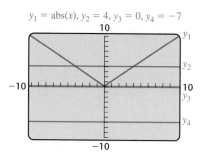

$y_1 = \text{abs}(x),\ y_2 = 4,\ y_3 = 0,\ y_4 = -7$

b) We interpret $|x| = 0$ to mean that x is 0 units from zero on the number line. The only number that satisfies this is 0 itself. Thus the solution set is $\{0\}$.

c) Since distance is always nonnegative, it doesn't make sense to talk about a number that is -7 units from zero. Remember: The absolute value of a number is never negative. Thus, $|x| = -7$ has no solution; the solution set is $\varnothing$.

The graph at left illustrates that, in Example 1, there are two solutions of $|x| = 4$, one solution of $|x| = 0$, and no solutions of $|x| = -7$.

Other equations involving absolute value can be solved graphically.

EXAMPLE 2   Solve: $|x - 3| = 5$.

SOLUTION   We let $y_1 = \text{abs}(x - 3)$ and $y_2 = 5$. The solution set of the equation consists of the x-coordinates of any points of intersection of the graphs of y_1 and y_2. We see that the graphs intersect at two points. We use INTERSECT twice to find the coordinates of the points of intersection. Each time, we enter a GUESS that is close to the point we are looking for.

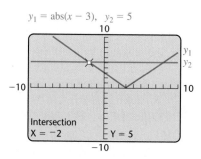

$y_1 = \text{abs}(x - 3),\ y_2 = 5$

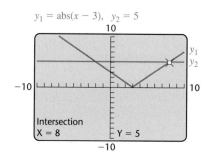

$y_1 = \text{abs}(x - 3),\ y_2 = 5$

The graphs intersect at $(-2, 5)$ and $(8, 5)$. Thus the solutions are -2 and 8. To check, we substitute each in the original equation.

Check: For -2:

$$|x - 3| = 5$$

$$\begin{array}{c|c} |-2 - 3| & 5 \\ |-5| & \\ 5 \overset{?}{=} 5 & \text{TRUE} \end{array}$$

For 8:

$$|x - 3| = 5$$

$$\begin{array}{c|c} |8 - 3| & 5 \\ |5| & \\ 5 \overset{?}{=} 5 & \text{TRUE} \end{array}$$

The solution set is $\{-2, 8\}$.

We can solve equations involving absolute value algebraically using the following principle.

Student Notes

The Absolute-Value Principle for Equations

For any positive number p and any algebraic expression X:

a) The solutions of $|X| = p$ are those numbers that satisfy

$$X = -p \quad or \quad X = p.$$

b) The equation $|X| = 0$ is equivalent to the equation $X = 0$.

c) The equation $|X| = -p$ has no solution.

EXAMPLE 3   Find the solution set: **(a)** $|2x + 5| = 13$; **(b)** $|4 - 7x| = -8$.

a)

ALGEBRAIC APPROACH

We use the absolute-value principle:

$$|X| = p$$
$$|2x + 5| = 13 \qquad \textbf{Substituting}$$
$$2x + 5 = -13 \quad or \quad 2x + 5 = 13$$
$$2x = -18 \quad or \quad \qquad 2x = 8$$
$$x = -9 \quad or \quad \qquad x = 4.$$

The solutions are -9 and 4.

Check:

For -9:

$$\begin{array}{c|c} |2x + 5| = 13 & \\ \hline |2(-9) + 5| & 13 \\ |-18 + 5| & \\ |-13| & \\ & 13 \overset{?}{=} 13 \quad \text{TRUE} \end{array}$$

For 4:

$$\begin{array}{c|c} |2x + 5| = 13 & \\ \hline |2 \cdot 4 + 5| & 13 \\ |8 + 5| & \\ |13| & \\ & 13 \overset{?}{=} 13 \quad \text{TRUE} \end{array}$$

The number $2x + 5$ is 13 units from zero if x is replaced with -9 or 4. The solution set is $\{-9, 4\}$.

GRAPHICAL APPROACH

We graph $y_1 = \text{abs}(2x + 5)$ and $y_2 = 13$ and use INTERSECT to find any points of intersection. The graphs intersect at $(-9, 13)$ and $(4, 13)$. The x-coordinates, -9 and 4, are the solutions.

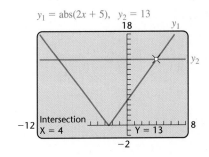

$y_1 = \text{abs}(2x + 5)$, $y_2 = 13$

The algebraic solution serves as a check. The solution set is $\{-9, 4\}$.

b)

ALGEBRAIC APPROACH

The absolute-value principle reminds us that absolute value is always nonnegative. Thus the equation $|4 - 7x| = -8$ has no solution. The solution set is $\varnothing$.

GRAPHICAL APPROACH

We graph $y_1 = \text{abs}(4 - 7x)$ and $y_2 = -8$. There are no points of intersection, so the solution set is $\varnothing$.

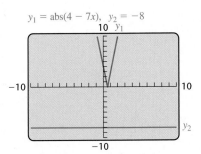

$y_1 = \text{abs}(4 - 7x)$, $y_2 = -8$

To use the absolute-value principle, we must be sure that the absolute-value expression is alone on one side of the equation.

> **EXAMPLE 4**   Given that $f(x) = 2|x + 3| + 1$, find all x for which $f(x) = 15$.

ALGEBRAIC APPROACH

Since we are looking for $f(x) = 15$, we substitute:

$$f(x) = 15$$
$$2|x + 3| + 1 = 15 \qquad \text{Replacing } f(x) \text{ with } 2|x+3| + 1$$
$$2|x + 3| = 14 \qquad \text{Subtracting 1 from both sides}$$
$$|x + 3| = 7 \qquad \text{Dividing both sides by 2}$$
$$x + 3 = -7 \quad or \quad x + 3 = 7 \qquad \begin{array}{l}\text{Replacing } X \text{ with } x + 3 \\ \text{and } p \text{ with 7 in the} \\ \text{absolute-value principle}\end{array}$$
$$x = -10 \quad or \qquad x = 4.$$

Check: $f(-10) = 2|-10 + 3| + 1 = 2|-7| + 1$
$$= 2 \cdot 7 + 1 = 15;$$
$$f(4) = 2|4 + 3| + 1 = 2|7| + 1$$
$$= 2 \cdot 7 + 1 = 15$$

The solution set is $\{-10, 4\}$.

GRAPHICAL APPROACH

We graph $y_1 = 2\, \text{abs}(x + 3) + 1$ and $y_2 = 15$ and determine the coordinates of any points of intersection. Choosing a viewing window that includes all points of intersection may involve some trial and error.

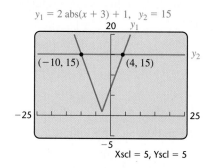

The solutions are the x-coordinates of the points of intersection, -10 and 4. The solution set is $\{-10, 4\}$.

> **EXAMPLE 5**   Solve: $|x - 2| = 3$.
>
> **SOLUTION**   Because this equation is of the form $|a - b| = c$, it can be solved two ways.
>
> *Method 1.*   We interpret $|x - 2| = 3$ as stating that the number $x - 2$ is 3 units from zero. Using the absolute-value principle, we replace X with $x - 2$ and p with 3:
>
> $$|X| = p$$
> $$|x - 2| = 3$$
> $$x - 2 = -3 \quad or \quad x - 2 = 3 \qquad \text{Using the absolute-value principle}$$
> $$x = -1 \quad or \qquad x = 5.$$
>
> *Method 2.*   This approach is helpful in calculus. The expressions $|a - b|$ and $|b - a|$ can be used to represent the *distance between a and b* on the number line. For example, the distance between 7 and 8 is given by $|8 - 7|$ or $|7 - 8|$. From this viewpoint, the equation $|x - 2| = 3$ states that the

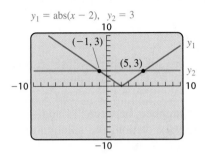

$y_1 = \text{abs}(x - 2), \ \ y_2 = 3$

distance between x and 2 is 3 units. We draw a number line and locate all numbers that are 3 units from 2.

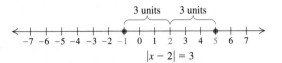

$|x - 2| = 3$

The solutions of $|x - 2| = 3$ are -1 and 5.

Check: The check consists of observing that both methods give the same solutions. The graphical solution shown at left serves as another check. The solution set is $\{-1, 5\}$.

Sometimes an equation has two absolute-value expressions. Consider $|a| = |b|$. This means that a and b are the same distance from zero.

If a and b are the same distance from zero, then either they are the same number or they are opposites.

For any algebraic expressions X and Y:

If $|X| = |Y|$,   then $X = Y$ or $X = -Y$.

EXAMPLE 6   Solve: $|2x - 3| = |x + 5|$.

ALGEBRAIC APPROACH

The given equation tells us that $2x - 3$ and $x + 5$ are the same distance from zero. This means that they are either the same number or opposites.

This assumes these numbers are the same.   This assumes these numbers are opposites.

$$2x - 3 = x + 5 \quad or \quad 2x - 3 = -(x + 5)$$
$$x - 3 = 5 \quad\ \ or \quad 2x - 3 = -x - 5$$
$$x = 8 \quad\ \ or \quad 3x - 3 = -5$$
$$3x = -2$$
$$x = -\tfrac{2}{3}$$

The solutions are 8 and $-\tfrac{2}{3}$. The solution set is $\left\{-\tfrac{2}{3}, 8\right\}$.

GRAPHICAL APPROACH

We graph $y_1 = \text{abs}(2x - 3)$ and $y_2 = \text{abs}(x + 5)$ and look for any points of intersection. The x-coordinates of the points of intersection are -0.6666667 and 8.

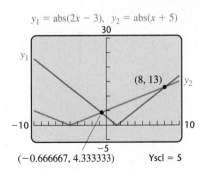

$y_1 = \text{abs}(2x - 3), \ \ y_2 = \text{abs}(x + 5)$

$(8, 13)$

$(-0.666667, 4.333333)$ Yscl = 5

The solution set is $\{-0.6666667, 8\}$. Converted to fraction notation, the solution set is $\left\{-\tfrac{2}{3}, 8\right\}$.

Inequalities with Absolute Value

Our methods for solving equations with absolute value can be adapted for solving inequalities. Inequalities of this sort arise regularly in more advanced courses.

EXAMPLE 7 Solve $|x| < 4$. Then graph.

SOLUTION The solutions of $|x| < 4$ are all numbers whose *distance from zero is less than* 4. By substituting or by looking at the number line, we can see that numbers like $-3, -2, -1, -\frac{1}{2}, -\frac{1}{4}, 0, \frac{1}{4}, \frac{1}{2}, 1, 2,$ and 3 are all solutions. In fact, the solutions are all the numbers between -4 and 4. The solution set is $\{x \mid -4 < x < 4\}$. In interval notation, the solution set is $(-4, 4)$. The graph is as follows:

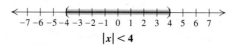

$|x| < 4$

We can also visualize Example 7 by graphing $f(x) = |x|$ and $g(x) = 4$, as in Example 1. The solution set consists of all x-values for which $(x, f(x))$ is below the horizontal line $g(x) = 4$. These x-values comprise the interval $(-4, 4)$.

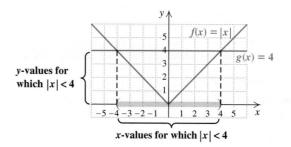

EXAMPLE 8 Solve $|x| \geq 4$. Then graph.

SOLUTION The solutions of $|x| \geq 4$ are all numbers that are at least 4 units from zero—in other words, those numbers x for which $x \leq -4$ *or* $4 \leq x$. The solution set is $\{x \mid x \leq -4 \ or \ x \geq 4\}$. In interval notation, the solution set is $(-\infty, -4] \cup [4, \infty)$. We can check mentally with numbers like $-4.1, -5, 4.1,$ and 5. The graph is as follows:

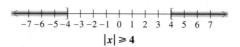

$|x| \geq 4$

As with Examples 1 and 7, Example 8 can be visualized by graphing $f(x) = |x|$ and $g(x) = 4$. The solution set of $|x| \geq 4$ consists of all x-values for which $(x, f(x))$ is on or above the horizontal line $g(x) = 4$. These x-values comprise $(-\infty, -4] \cup [4, \infty)$.

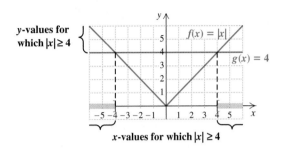

Examples 1, 7, and 8 illustrate three types of problems in which absolute-value symbols appear. The following is a general principle for solving such problems.

Principles for Solving Absolute-Value Problems   For any positive number p and any expression X:

a) The solutions of $|X| = p$ are those numbers that satisfy
$X = -p \ or \ X = p.$

b) The solutions of $|X| < p$ are those numbers that satisfy
$-p < X < p.$

c) The solutions of $|X| > p$ are those numbers that satisfy
$X < -p \ or \ p < X.$

Student Notes

Resist the temptation to simply ignore the absolute-value symbol and solve the resulting equation or inequality. Doing so will lead to only *part* of the solution.

Of course, if p is negative, any value of X will satisfy the inequality $|X| > p$ because absolute value is never negative. By the same reasoning, $|X| < p$ has no solution when p is not positive. Thus, $|2x - 7| > -3$ is true for any real number x, and $|2x - 7| < -3$ has no solution.

Note that an inequality of the form $|X| < p$ corresponds to a *conjunction*, whereas an inequality of the form $|X| > p$ corresponds to a *disjunction*.

▸ **EXAMPLE 9** Solve: $|3x - 2| < 4$.

ALGEBRAIC APPROACH	**GRAPHICAL APPROACH**

ALGEBRAIC APPROACH

The number $3x - 2$ must be less than 4 units from zero. This is of the form $|X| < p$, so part (b) of the priniciples listed above applies.

$$|X| < p$$
$$|3x - 2| < 4 \quad \text{Replacing } X \text{ with } 3x - 2 \text{ and } p \text{ with } 4$$
$$-4 < 3x - 2 < 4 \quad \text{The number } 3x - 2 \text{ must be within 4 units of zero.}$$
$$-2 < \quad 3x \quad < 6 \quad \text{Adding 2}$$
$$-\tfrac{2}{3} < \quad x \quad < 2 \quad \text{Multiplying by } \tfrac{1}{3}$$

The solution set is $\left\{x \mid -\tfrac{2}{3} < x < 2\right\}$, or, in interval notation, $\left(-\tfrac{2}{3}, 2\right)$. The graph is as follows:

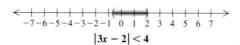

$$|3x - 2| < 4$$

GRAPHICAL APPROACH

We graph $y_1 = \text{abs}(3x - 2)$ and $y_2 = 4$.

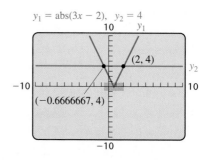

The solution set consists of the x-values for which $y_1 < y_2$, and y_1 is less than y_2 between the points $(-0.6666667, 4)$ and $(2, 4)$. The solution set of the inequality is thus the interval $(-0.6666667, 2)$, as indicated by the shading on the x-axis. Converted to fraction notation, the solution set is $\left(-\tfrac{2}{3}, 2\right)$. ◢

▸ **EXAMPLE 10** Given that $f(x) = |4x + 2|$, find all x for which $f(x) \geq 6$.

ALGEBRAIC APPROACH	**GRAPHICAL APPROACH**

ALGEBRAIC APPROACH

We have
$$f(x) \geq 6,$$
or $\quad |4x + 2| \geq 6$. **Substituting**

To solve, we use part (c) of the principles listed above. In this case, X is $4x + 2$ and p is 6.

$$|X| \geq p$$
$$|4x + 2| \geq 6 \quad \text{Replacing } X \text{ with } 4x + 2 \text{ and } p \text{ with } 6$$
$$4x + 2 \leq -6 \quad or \quad 6 \leq 4x + 2$$

The number $4x + 2$ must be at least 6 units from zero.

$$4x \leq -8 \quad or \quad 4 \leq 4x \quad \text{Adding } -2$$
$$x \leq -2 \quad or \quad 1 \leq x \quad \text{Multiplying by } \tfrac{1}{4}$$

The solution set is $\{x \mid x \leq -2 \text{ or } x \geq 1\}$, or, in interval notation, $(-\infty, -2] \cup [1, \infty)$. The graph is as follows:

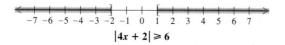

$$|4x + 2| \geq 6$$

GRAPHICAL APPROACH

To find all values of x for which $|4x + 2| \geq 6$, we graph $y_1 = \text{abs}(4x + 2)$ and $y_2 = 6$ and determine the points of intersection of the graphs.

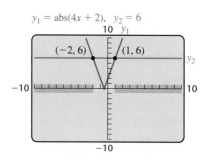

The solution set consists of all x-values for which the graph of $y_1 = |4x + 2|$ is *on or above* the graph of $y_2 = 6$. The graph of y_1 lies *on* the graph of y_2 when $x = -2$ or $x = 1$. The graph of y_1 is *above* the graph of y_2 when $x < -2$ or $x > 1$. Thus the solution set is $(-\infty, -2] \cup [1, \infty)$, as indicated by the shading on the x-axis. ◢

8.2 EXERCISE SET

⤵ *Concept Reinforcement*   *Classify each of the following as either true or false.*

1. If x is negative, then $|x| = -x$.

2. $|x|$ is never negative.

3. $|x|$ is always positive.

4. The distance between a and b can be expressed as $|a - b|$.

5. The number a is $|a|$ units from 0.

6. There are two solutions of $|3x - 8| = 17$.

7. There is no solution of $|4x + 9| > -5$.

8. All real numbers are solutions of $|2x - 7| < -3$.

Use the following graph to solve Exercises 9–14.

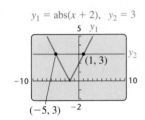

$y_1 = \text{abs}(x + 2), \quad y_2 = 3$

9. $|x + 2| = 3$

10. $|x + 2| \leq 3$

11. $|x + 2| < 3$

12. $|x + 2| > 3$

13. $|x + 2| \geq 3$

14. $|x + 2| = -1$

Solve.

15. $|x| = 7$

16. $|x| = 9$

Aha! 17. $|x| = -6$

18. $|x| = -3$

19. $|p| = 0$

20. $|y| = 7.3$

21. $|t| = 5.5$

22. $|m| = 0$

23. $|2x - 3| = 4$

24. $|5x + 2| = 7$

25. $|3x - 5| = -8$

26. $|7x - 2| = -9$

27. $|x - 2| = 6$

28. $|x - 3| = 8$

29. $|x - 5| = 3$

30. $|x - 6| = 1$

31. $|x - 7| = 9$

32. $|x - 4| = 5$

33. $|5x| - 3 = 37$

34. $|2y| - 5 = 13$

35. $7|q| - 2 = 9$

36. $7|z| + 2 = 16$

37. $\left|\dfrac{2x - 1}{3}\right| = 5$

38. $\left|\dfrac{4 - 5x}{6}\right| = 3$

39. $|m + 5| + 9 = 16$

40. $|t - 7| + 1 = 4$

41. $5 - 2|3x - 4| = -5$

42. $3|2x - 5| - 7 = -1$

43. Let $f(x) = |2x + 6|$. Find all x for which $f(x) = 8$.

44. Let $f(x) = |2x + 4|$. Find all x for which $f(x) = 10$.

45. Let $f(x) = |x| - 3$. Find all x for which $f(x) = 5.7$.

46. Let $f(x) = |x| + 7$. Find all x for which $f(x) = 18$.

47. Let $f(x) = \left|\dfrac{3x - 2}{5}\right|$. Find all x for which $f(x) = 2$.

48. Let $f(x) = \left|\dfrac{1 - 2x}{3}\right|$. Find all x for which $f(x) = 1$.

Solve.

49. $|x + 4| = |2x - 7|$

50. $|3x + 5| = |x - 6|$

51. $|x + 4| = |x - 3|$

52. $|x - 9| = |x + 6|$

53. $|3a - 1| = |2a + 4|$

54. $|5t + 7| = |4t + 3|$

Aha! 55. $|n - 3| = |3 - n|$

56. $|y - 2| = |2 - y|$

57. $|7 - a| = |a + 5|$

58. $|6 - t| = |t + 7|$

59. $\left|\tfrac{1}{2}x - 5\right| = \left|\tfrac{1}{4}x + 3\right|$

60. $\left|2 - \tfrac{2}{3}x\right| = \left|4 + \tfrac{7}{8}x\right|$

Solve and graph.

61. $|a| \leq 9$

62. $|x| < 2$

63. $|x| > 8$

64. $|a| \geq 3$

65. $|t| > 0$

66. $|t| \geq 1.7$

67. $|x - 1| < 4$

68. $|x - 1| < 3$

69. $|x + 2| \leq 6$

70. $|x + 4| \leq 1$

71. $|x - 3| + 2 > 7$

72. $|x - 4| + 5 > 2$

Aha! **73.** $|2y - 9| > -5$

74. $|3y - 4| > 8$

75. $|3a - 4| + 2 \geq 8$

76. $|2a - 5| + 1 \geq 9$

77. $|y - 3| < 12$

78. $|p - 2| < 3$

79. $9 - |x + 4| \leq 5$

80. $12 - |x - 5| \leq 9$

81. $|4 - 3y| > 8$

82. $|7 - 2y| < -6$

Aha! **83.** $|5 - 4x| < -6$

84. $7 + |4a - 5| \leq 26$

85. $\left| \dfrac{2 - 5x}{4} \right| \geq \dfrac{2}{3}$

86. $\left| \dfrac{1 + 3x}{5} \right| > \dfrac{7}{8}$

87. $|m + 3| + 8 \leq 14$

88. $|t - 7| + 3 \geq 4$

89. $25 - 2|a + 3| > 19$

90. $30 - 4|a + 2| > 12$

91. Let $f(x) = |2x - 3|$. Find all x for which $f(x) \leq 4$.

92. Let $f(x) = |5x + 2|$. Find all x for which $f(x) \leq 3$.

93. Let $f(x) = 5 + |3x - 4|$. Find all x for which $f(x) \geq 16$.

94. Let $f(x) = |2 - 9x|$. Find all x for which $f(x) \geq 25$.

95. Let $f(x) = 7 + |2x - 1|$. Find all x for which $f(x) < 16$.

96. Let $f(x) = 5 + |3x + 2|$. Find all x for which $f(x) < 19$.

TW **97.** Explain in your own words why -7 is not a solution of $|x| < 5$.

TW **98.** Explain in your own words why $[6, \infty)$ is only part of the solution of $|x| \geq 6$.

Focused Review

Solve.

99. $3(x + 1) = 5(x - 2) + 1$ [2.2]

100. $3(x + 1) \leq 5(x - 2) + 1$ [2.6]

101. $y = 3(x + 1),$
$y = 5(x - 2) + 1$ [4.2]

102. $3(x + 1) \leq 2$ *and* $5(x - 2) + 1 \leq -4$ [8.1]

103. $|x + 5| < 6$ [8.2]

104. $|x + 5| = 6$ [8.2]

Synthesis

TW **105.** Is it possible for an equation in x of the form $|ax + b| = c$ to have exactly one solution? Why or why not?

TW **106.** Isabel is using the following graph to solve $|x - 3| < 4$. How can you tell that a mistake has been made?

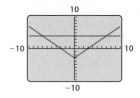

107. From the definition of absolute value, $|x| = x$ only when $x \geq 0$. Solve $|3t - 5| = 3t - 5$ using this same reasoning.

Solve.

108. $|3x - 5| = x$

109. $|x + 2| > x$

110. $2 \leq |x - 1| \leq 5$

111. $|5t - 3| = 2t + 4$

112. $t - 2 \leq |t - 3|$

Find an equivalent inequality with absolute value.

113. $-3 < x < 3$

114. $-5 \leq y \leq 5$

115. $x \leq -6 \ or \ 6 \leq x$

116. $x < -4 \ or \ 4 < x$

117. $x < -8 \ or \ 2 < x$

118. $-5 < x < 1$

119. x is less than 2 units from 7.

120. x is less than 1 unit from 5.

Write an absolute-value inequality for which the interval shown is the solution.

121. ◄─┼──┼──┼──┼──┼──┼──┼──┼━━━━━━━━━━━━━━━━━►
$\ \ \ -7 \ -6 \ -5 \ -4 \ -3 \ -2 \ -1 \ \ \ 0 \ \ \ 1 \ \ \ 2 \ \ \ 3 \ \ \ 4 \ \ \ 5 \ \ \ 6 \ \ \ 7$

122. ◄─┼━━━━━━━━━━━━━━━━━━━━━━━━━━━━━┼─►
$\ \ \ -5 \ -4 \ -3 \ -2 \ -1 \ \ \ 0 \ \ \ 1 \ \ \ 2 \ \ \ 3 \ \ \ 4 \ \ \ 5 \ \ \ 6 \ \ \ 7 \ \ \ 8 \ \ \ 9$

123. ◄━━━━━━━━━━━━━━━┼──┼──┼──┼──┼──┼──┼──┼──►
$\ \ \ -7 \ -6 \ -5 \ -4 \ -3 \ -2 \ -1 \ \ \ 0 \ \ \ 1 \ \ \ 2 \ \ \ 3 \ \ \ 4 \ \ \ 5 \ \ \ 6 \ \ \ 7$

124. ◄─┼──┼━━━━━━━━━━━━━━━━━━━━━━━━━━┼──┼─►
$\ \ \ 0 \ \ \ 1 \ \ \ 2 \ \ \ 3 \ \ \ 4 \ \ \ 5 \ \ \ 6 \ \ \ 7 \ \ \ 8 \ \ \ 9 \ \ 10 \ 11 \ 12 \ 13 \ 14$

125. *Bungee Jumping.*   A bungee jumper is bouncing up and down so that her distance d above a river satisfies the inequality $|d\text{-}60\text{ ft}| \leq 10\text{ ft}$ (see the figure below). If the bridge from which she jumped is 150 ft above the river, how far is the bungee jumper from the bridge at any given time?

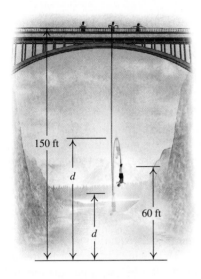

126. *Water Level.*   Depending on how dry or wet the weather has been, water in a well will rise and fall. The distance d that a well's water level is below the ground satisfies the inequality $|d - 15| \leq 2.5$ (see the figure below).

a) Solve for d.

b) How tall a column of water is in the well at any given time?

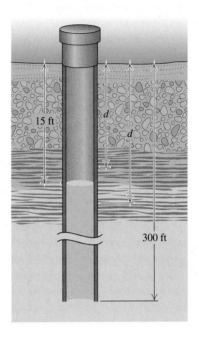

8.3   Inequalities in Two Variables

Graphs of Linear Inequalities   ■   Systems of Linear Inequalities

We have graphed inequalities in one variable on the number line. Now we graph inequalities in two variables on a plane.

Graphs of Linear Inequalities

When the equals sign in a linear equation is replaced with an inequality sign, a **linear inequality** is formed. Solutions of linear inequalities are ordered pairs.

Study Tip

Improve Your Study Skills

The time you spend learning to study better will be returned many times over. Study skills resources such as books and videos are available, your school may offer a class on study skills, or you can find Web sites that offer tips and instruction.

EXAMPLE 1 Determine whether $(-3, 2)$ and $(6, -7)$ are solutions of the inequality $5x - 4y > 13$.

SOLUTION Below, on the left, we replace x with -3 and y with 2. On the right, we replace x with 6 and y with -7.

$$
\begin{array}{c|c}
5x - 4y > 13 & \\
\hline
5(-3) - 4 \cdot 2 & 13 \\
-15 - 8 & \\
-23 \overset{?}{>} 13 & \text{FALSE}
\end{array}
$$

Since $-23 > 13$ is false, $(-3, 2)$ is not a solution.

$$
\begin{array}{c|c}
5x - 4y > 13 & \\
\hline
5(6) - 4(-7) & 13 \\
30 + 28 & \\
58 \overset{?}{>} 13 & \text{TRUE}
\end{array}
$$

Since $58 > 13$ is true, $(6, -7)$ is a solution.

The graph of a linear equation is a straight line. The graph of a linear inequality is a *half-plane*, bordered by the graph of the *related equation*. To find an inequality's related equation, we simply replace the inequality sign with an equals sign.

EXAMPLE 2 Graph: $y \le x$.

SOLUTION We first graph the related equation $y = x$. Every solution of $y = x$ is an ordered pair, like $(3, 3)$, in which both coordinates are the same. The graph of $y = x$ is shown on the left below. Since the inequality symbol is $\le$, the line is drawn solid and is part of the graph of $y \le x$.

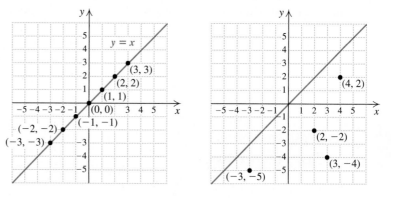

Note that in the graph on the right above each ordered pair on the half-plane below $y = x$ contains a y-coordinate that is less than the x-coordinate. All these pairs represent solutions of $y \le x$. We check one pair, $(4, 2)$, as follows:

$$
\begin{array}{c|c}
y \le x & \\
\hline
2 & 4 \quad \text{TRUE}
\end{array}
$$

It turns out that *any* point on the same side of $y = x$ as $(4, 2)$ is also a solution. Thus, if one point in a half-plane is a solution, then *all* points in that half-plane are solutions. We finish drawing the solution set by shading the half-plane below $y = x$. The complete solution set consists of the shaded half-plane as well as the boundary line itself.

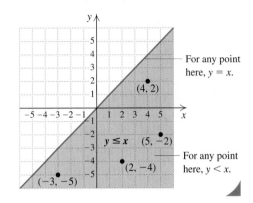

For any point here, $y = x$.

For any point here, $y < x$.

From Example 2, we see that for any inequality of the form $y \le f(x)$ or $y < f(x)$, we shade *below* the graph of $y = f(x)$.

EXAMPLE 3   Graph: $8x + 3y > 24$.

SOLUTION   First, we sketch the graph of $8x + 3y = 24$. Since the inequality sign is $>$, points on this line do not represent solutions of the inequality, so the line is drawn dashed. Points representing solutions of $8x + 3y > 24$ are in either the half-plane above the line or the half-plane below the line. To determine which, we select a point that is not on the line and determine whether it is a solution of $8x + 3y > 24$. Let's use $(-3, 4)$ as this *test point*:

$$
\begin{array}{c|c}
\multicolumn{2}{c}{8x + 3y > 24} \\
\hline
8(-3) + 3 \cdot 4 & 24 \\
-24 + 12 & \\
-12 \overset{?}{>} 24 & \text{FALSE}
\end{array}
$$

Since $-12 > 24$ is *false*, $(-3, 4)$ is not a solution. Thus no point in the half-plane containing $(-3, 4)$ is a solution. The points in the other half-plane *are* solutions, so we shade that half-plane and obtain the graph shown at right.

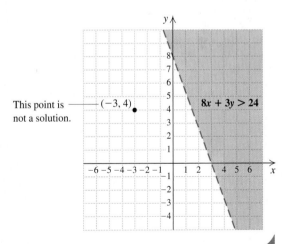

This point is not a solution.

$8x + 3y > 24$

Steps for Graphing Linear Inequalities

1. Replace the inequality sign with an equals sign and graph this line as the boundary. If the inequality symbol is $<$ or $>$, draw the line dashed. If the inequality symbol is $\leq$ or $\geq$, draw the line solid.
2. The graph consists of a half-plane on one side of the line and, if the line is solid, the line as well.

   a) If the inequality is of the form $y < mx + b$ or $y \leq mx + b$, shade *below* the line. If the inequality is of the form $y > mx + b$ or $y \geq mx + b$, shade *above* the line.
   b) If y is not isolated, either solve for y and proceed as in part (a) or select a test point not on the line. If the test point represents a solution of the inequality, shade the half-plane containing the point. If it does not, shade the other half-plane.

Linear Inequalities

On most graphing calculators, an inequality like $y < \frac{6}{5}x + 3.49$ can be drawn by entering $(6/5)x + 3.49$ as y_1, moving the cursor to the GraphStyle icon just to the left of y_1, pressing **ENTER** until ◤ appears, and then pressing ⬭ GRAPH ⬭.

Many newer calculators have an INEQUALZ program that is accessed using the **APPS** key. Running this program allows us to write inequalities at the ⬭ Y= ⬭ screen by pressing **ALPHA** and then one of the five keys just below the screen, as shown on the left below.

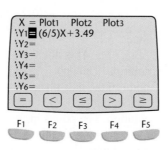

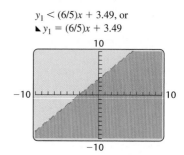

$y_1 < (6/5)x + 3.49$, or
◤ $y_1 = (6/5)x + 3.49$

Although the graphs should be identical regardless of the method used, on the newer calculators the boundary line appears dashed when $<$ or $>$ is selected. The graph of the inequality is shown on the right above. Most calculators will shade this area with black vertical lines. Inequalities containing $\leq$ or $\geq$ are handled in a similar manner.

When you are finished with the INEQUALZ application, select it again from the **APPS** menu and quit the program.

EXAMPLE 4   Graph: $6x - 2y < 12$.

SOLUTION   We graph both by hand and using a graphing calculator.

| BY HAND | USING A GRAPHING CALCULATOR |

BY HAND

We could graph $6x - 2y = 12$ and use a test point, as in Example 3. Instead, let's solve $6x - 2y < 12$ for y:

$6x - 2y < 12$
$\quad -2y < -6x + 12$ Adding $-6x$ to both sides
$\quad\quad y > 3x - 6$.   Dividing both sides by -2 and reversing the $<$ symbol

The graph consists of the half plane above the dashed boundary line $y = 3x - 6$ (see the graph below).

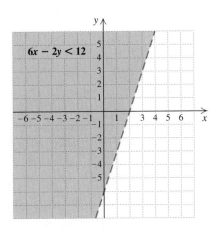

USING A GRAPHING CALCULATOR

We solve the inequality for y:

$6x - 2y < 12$
$\quad -2y < -6x + 12$
$\quad\quad y > 3x - 6$.   Reversing the $<$ symbol

We enter the boundary equation, $y = 3x - 6$, and use the INEQUALZ application to graph.

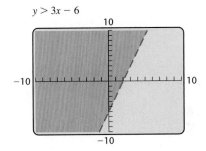

Many calculators do not draw a dashed line, so the graph of $y > 3x - 6$ appears to be the same as the graph of $y \geq 3x - 6$.

Quit the INEQUALZ application when you are finished.

Although a graphing calculator can be a very useful tool, some equations and inequalities are actually quicker to graph by hand. It is important to be able to quickly sketch basic linear equations and inequalities by hand. Also, when we are using a calculator to graph equations, knowing the basic shape of the graph can serve as a check that the equation was entered correctly.

EXAMPLE 5   Graph $x > -3$ on a plane.

SOLUTION   There is a missing variable in this inequality. If we graph the inequality on a line, its graph is as follows:

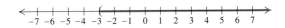

However, we can also write this inequality as $x + 0y > -3$ and graph it on a plane. We can use the same technique as in the examples above. First, we graph the related equation $x = -3$ on a plane. Then we test some point, say, $(2, 5)$:

$$\frac{x + 0y > -3}{2 + 0 \cdot 5 \mid -3}$$
$$2 \overset{?}{>} -3 \quad \text{TRUE}$$

Since $(2, 5)$ is a solution, all points in the half-plane containing $(2, 5)$ are solutions. We shade that half-plane. Another approach is to simply note that the solutions of $x > -3$ are all pairs with first coordinates greater than -3.

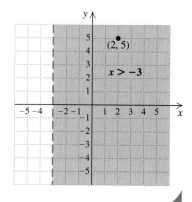

Although many graphing calculators can graph this inequality from the DRAW menu, it is simpler to graph it by hand.

EXAMPLE 6 Graph $y \le 4$ on a plane.

SOLUTION The inequality is of the form $y \le mx + b$ (with $m = 0$), so we shade below the solid horizontal line representing $y = 4$.

This inequality can also be graphed by drawing $y = 4$ and testing a point above or below the line. The half-plane below $y = 4$ should be shaded.

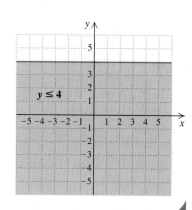

Systems of Linear Inequalities

To graph a system of equations, we graph the individual equations and then find the intersection of the individual graphs. We do the same thing for a system of inequalities—that is, we graph each inequality and find the intersection of the individual graphs.

EXAMPLE 7 Graph the system

$$x + y \le 4,$$
$$x - y < 4.$$

SOLUTION To graph $x + y \le 4$, we graph $x + y = 4$ using a solid line. Since the test point $(0, 0)$ *is* a solution and $(0, 0)$ is below the line, we shade the half-plane below the graph red. The arrows near the ends of the line are another way of indicating the half-plane containing solutions.

Next, we graph $x - y < 4$. We graph $x - y = 4$ using a dashed line and consider $(0, 0)$ as a test point. Again, $(0, 0)$ is a solution, so we shade that side of the line blue. The solution set of the system is the region that is shaded purple (both red and blue) and part of the line $x + y = 4$.

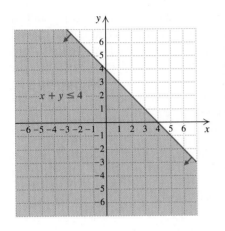

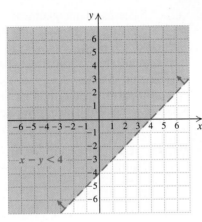

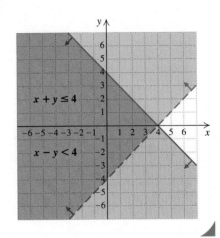

Student Notes

If you don't use differently colored pencils or pens to shade different regions, consider using a pencil to make slashes that tilt in different directions in each region, similar to the way shading is done on a graphing calculator. You may also find it useful to attach arrows to the lines, as in the examples shown.

EXAMPLE 8   Graph: $-2 < x \le 3$.

SOLUTION   This is a system of inequalities:

$$-2 < x,$$
$$x \le 3.$$

We graph the equation $-2 = x$, and see that the graph of the first inequality is the half-plane to the right of the boundary $-2 = x$. It is shaded red.

We graph the second inequality, starting with the line $x = 3$, and find that its graph is the line and also the half-plane to its left. It is shaded blue.

The solution set of the system is the region that is the intersection of the individual graphs. Since it is shaded both blue and red, it appears to be purple. All points in this region have x-coordinates that are greater than -2 but do not exceed 3.

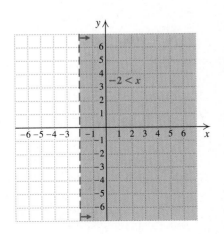

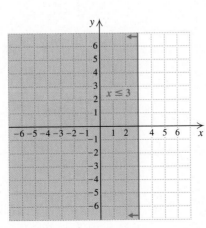

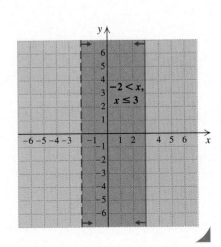

A system of inequalities may have a graph that consists of a polygon and its interior. The corners of such a graph are called *vertices* (singular, *vertex*).

Systems of Linear Inequalities

Systems of inequalities can be graphed by solving for y and then graphing each inequality. The graph of the system will be the intersection of the shaded regions. To graph systems directly using the INEQUALZ application, enter the correct inequalities, press GRAPH, and then press ALPHA and Shades (F1 or F2). At the SHADES menu, select Ineq Intersection to see the final graph. To find the vertices, or points of intersection, select PoI-Trace from the graph menu.

EXAMPLE 9 Graph the system of inequalities. Find the coordinates of any vertices formed.

$$6x - 2y \leq 12, \qquad (1)$$
$$y - 3 \leq 0, \qquad (2)$$
$$x + y \geq 0. \qquad (3)$$

SOLUTION We graph both by hand and using a graphing calculator.

BY HAND

We graph the boundaries

$$6x - 2y = 12,$$
$$y - 3 = 0,$$
and $\qquad x + y = 0$

using solid lines. The regions for each inequality are indicated by the arrows near the ends of the lines. We note where the regions overlap and shade the region of solutions purple.

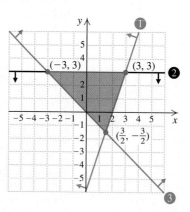

To find the vertices, we solve three different systems of two equations. The system of boundary equations from inequalities (1) and (2) is

$$6x - 2y = 12,$$
$$y - 3 = 0.$$

Solving, we obtain the vertex $(3, 3)$.

The system of boundary equations from inequalities (1) and (3) is

$$6x - 2y = 12,$$
$$x + y = 0.$$

Solving, we obtain the vertex $\left(\frac{3}{2}, -\frac{3}{2}\right)$.

The system of boundary equations from inequalities (2) and (3) is

$$y - 3 = 0,$$
$$x + y = 0.$$

Solving, we obtain the vertex $(-3, 3)$.

USING A GRAPHING CALCULATOR

First, we solve each inequality for y and obtain the equivalent system of inequalities

$$y_1 \geq (12 - 6x)/(-2),$$
$$y_2 \leq 3,$$
$$y_3 \geq -x.$$

We run the INEQUALZ application, enter each inequality, and press GRAPH. We then press ALPHA and F1 or F2 to choose Shades. Selecting the Ineq Intersection option results in the graph shown on the left below.

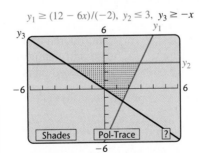
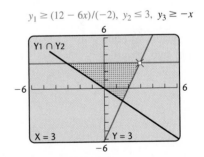

The vertices are the points of intersection of the graphs of the boundary equations y_1 and y_2, y_1 and y_3, and y_2 and y_3. We can find them by choosing PoI-Trace. Pressing the arrow keys moves the cursor to different vertices. The vertices are $(3, 3)$, $(-3, 3)$, and $(1.5, -1.5)$.

Graphs of systems of inequalities and the coordinates of vertices are used to solve a variety of problems using a branch of mathematics called *linear programming*.

Connecting the Concepts

We have now solved a variety of equations, inequalities, systems of equations, and systems of inequalities. In each case, there are different ways to represent the solution. Below is a list of the different types of problems we have solved, along with illustrations of each type.

Type	Example	Solution	Graph
Linear equations in one variable	$2x - 8 = 3(x + 5)$	A number	
Linear inequalities in one variable	$-3x + 5 > 2$	A set of numbers; an interval	
Linear equations in two variables	$2x + y = 7$	A set of ordered pairs; a line	
Linear inequalities in two variables	$x + y \geq 4$	A set of ordered pairs; a half-plane	
System of equations in two variables	$x + y = 3,$ $5x - y = -27$	An ordered pair or a (possibly empty) set of ordered pairs	
System of inequalities in two variables	$6x - 2y \leq 12,$ $y - 3 \leq 0,$ $x + y \geq 0$	A set of ordered pairs; a region of a plane	

Keeping in mind how these solutions vary and what their graphs look like will help you as you progress further in this book and in mathematics in general.

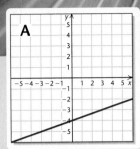

A

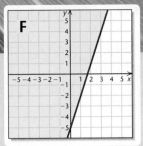

F

Visualizing the Graph

Match each equation, inequality, or system of equations or inequalities to its graph.

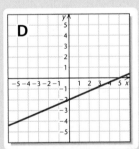

B

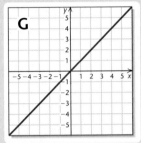

G

1. $x - y = 3,$
   $2x + y = 1$

2. $3x - y \leq 5$

3. $x > -3$

4. $y = \dfrac{1}{3}x - 4$

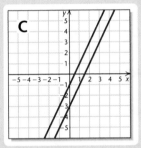

C

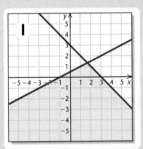

H

5. $y > \dfrac{1}{3}x - 4,$
   $y \leq x$

6. $x = y$

7. $y = 2x - 1,$
   $y = 2x - 3$

8. $2x - 5y = 10$

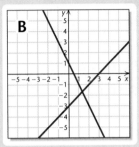

D

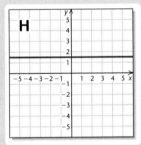

I

9. $x + y \leq 3,$
   $2y \leq x + 1$

10. $y = \dfrac{3}{2}$

Answers on page A-32

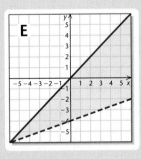

E

J

8.3 EXERCISE SET

Concept Reinforcement   *In each of Exercises 1–6, match the phrase with the most appropriate choice from the column on the right.*

1. ____ A solution of a linear inequality

2. ____ The graph of a linear inequality

3. ____ The graph of a system of linear inequalities

4. ____ Often a convenient test point

5. ____ The name for the corners of a graph of a system of linear inequalities

6. ____ A dashed line

a) $(0, 0)$

b) Vertices

c) A half-plane

d) The intersection of two or more half-planes

e) An ordered pair that satisfies the inequality

f) Indicates the line is not part of the solution

Determine whether each ordered pair is a solution of the given inequality.

7. $(-4, 2)$; $2x + 3y < -1$

8. $(3, -6)$; $4x + 2y \leq -2$

9. $(8, 14)$; $2y - 3x \geq 9$

10. $(7, 20)$; $3x - y > -1$

Graph on a plane.

11. $y > \frac{1}{2}x$

12. $y > 2x$

13. $y \geq x - 3$

14. $y < x + 3$

15. $y \leq x + 5$

16. $y > x - 2$

17. $x - y \leq 4$

18. $x + y < 4$

19. $2x + 3y < 6$

20. $3x + 4y \leq 12$

21. $2y - x \leq 4$

22. $2y - 3x > 6$

23. $2x - 2y \geq 8 + 2y$

24. $3x - 2 \leq 5x + y$

25. $y \geq 3$

26. $x < -5$

27. $x \leq 6$

28. $y > -3$

29. $-2 < y < 7$

30. $-4 < y < -1$

31. $-4 \leq x \leq 2$

32. $-3 \leq y \leq 4$

33. $0 \leq y \leq 3$

34. $0 \leq x \leq 6$

 Graph using a graphing calculator.

35. $y > x + 3.5$

36. $7y \leq 2x + 5$

37. $8x - 2y < 11$

38. $11x + 13y + 4 \geq 0$

Graph each system.

39. $y > -x,$
$y < x + 2$

40. $y < x,$
$y > -x + 1$

41. $y \geq x,$
$y \geq 2x - 4$

42. $y \geq x,$
$y \leq -x + 4$

43. $y \leq -3,$
$x \geq -1$

44. $y \geq -3,$
$x \geq 1$

45. $x > -4,$
$y < -2x + 3$

46. $x < 3,$
$y > -3x + 2$

47. $y \leq 5,$
$y \geq -x + 4$

48. $y \geq -2,$
$y \geq x + 3$

49. $x + y \leq 6,$
$x - y \leq 4$

50. $x + y < 1,$
$x - y < 2$

51. $y + 3x > 0,$
$y + 3x < 2$

52. $y - 2x \geq 1,$
$y - 2x \leq 3$

Graph each system of inequalities. Find the coordinates of any vertices formed.

53. $y \leq 2x - 3,$
$y \geq -2x + 1,$
$x \leq 5$

54. $2y - x \leq 2,$
$y - 3x \geq -4,$
$y \geq -1$

55. $x + 2y \leq 12,$
$2x + y \leq 12,$
$x \geq 0,$
$y \geq 0$

56. $x - y \leq 2,$
$x + 2y \geq 8,$
$y \leq 4$

57. $8x + 5y \leq 40,$
$x + 2y \leq 8,$
$x \geq 0,$
$y \geq 0$

58. $4y - 3x \geq -12,$
$4y + 3x \geq -36,$
$y \leq 0,$
$x \leq 0$

59. $y - x \geq 2,$
$y - x \leq 4,$
$2 \leq x \leq 5$

60. $3x + 4y \geq 12,$
$5x + 6y \leq 30,$
$1 \leq x \leq 3$

TW 61. In Example 7, is the point $(4, 0)$ part of the solution set? Why or why not?

TW 62. When graphing linear inequalities, Ron makes a habit of always shading above the line when the symbol $\geq$ is used. Is this wise? Why or why not?

Skill Maintenance

Solve.

63. *Catering.* Sandy's Catering needs to provide 10 lb of mixed nuts for a wedding reception. Peanuts cost \$2.50 per pound and fancy nuts cost \$7 per pound. If \$40 has been allocated for nuts, how many pounds of each type should be mixed? [4.4]

64. *Household Waste.* The Hendersons generate two and a half times as much trash as their neighbors, the Savickis. Together, the two households produce 14 bags of trash each month. How much trash does each household produce? [4.4]

65. *Paid Admissions.* There were 203 tickets sold for a volleyball game. For activity-card holders the price was \$2.50, and for noncard holders the price was \$4. The total amount of money collected was \$620. How many of each type of ticket were sold? [4.4]

66. *Paid Admissions.* There were 200 tickets sold for a women's basketball game. Tickets for students were \$4 each and for adults were \$6 each. The total amount collected was \$1060. How many of each type of ticket were sold? [4.4]

67. *Landscaping.* Grass seed is being spread on a triangular traffic island. If the grass seed can cover an area of 200 ft^2 and the island's base is 16 ft long, how tall a triangle can the seed fill? [2.5]

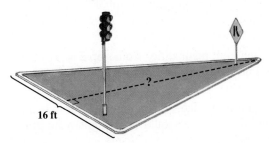

16 ft

68. *Interest Rate.* What rate of interest is required for a principal of \$1280 to earn \$17.60 in half a year? [2.4]

Synthesis

TW 69. Explain how a system of linear inequalities could have a solution set containing exactly one pair.

TW 70. Do all systems of linear inequalities have solutions? Why or why not?

Graph.

71. $x + y > 8$,
$x + y \leq -2$

72. $x + y \geq 1$,
$-x + y \geq 2$,
$x \geq -2$,
$y \geq 2$,
$y \leq 4$,
$x \leq 2$

73. $x - 2y \leq 0$,
$-2x + y \leq 2$,
$x \leq 2$,
$y \leq 2$,
$x + y \leq 4$

74. Write four systems of four inequalities that describe a 2-unit by 2-unit square that has $(0, 0)$ as one of the vertices.

75. *Luggage Size.* Unless an additional fee is paid, most major airlines will not check any luggage for which the sum of the item's length, width, and height exceeds 62 in. The U.S. Postal Service will ship a package only if the sum of the package's length and girth (distance around its midsection) does not exceed 130 in. (*Sources*: U.S. Postal Service; www.case2go.com) Video Promotions is ordering several 30-in. long cases that will be both mailed and checked as luggage. Using w and h for width and height (in inches), respectively, write and graph an inequality that represents all acceptable combinations of width and height.

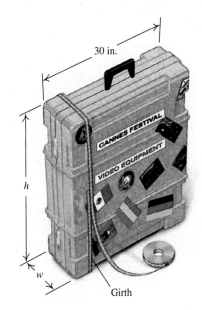

30 in.

h

w

Girth

76. *Hockey Wins and Losses.* The Skating Stars figure that they need at least 60 points for the season in order to make the playoffs. A win is worth 2 points and a tie is worth 1 point. Graph a system of inequalities that describes the situation. (*Hint*: Let w = the number of wins and t = the number of ties.)

77. *Elevators.* Many elevators have a capacity of 1 metric ton (1000 kg). Suppose that c children, each weighing 35 kg, and a adults, each 75 kg, are on an elevator. Graph a system of inequalities that indicates when the elevator is overloaded.

78. *Widths of a Basketball Floor.* Sizes of basketball floors vary due to building sizes and other constraints such as cost. The length L is to be at most 94 ft and the width W is to be at most 50 ft. Graph a system of inequalities that describes the possible dimensions of a basketball floor.

Write a system of inequalities for each region shown.

79.

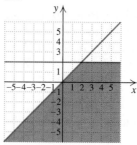

80.

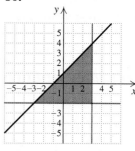

81.

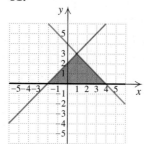

82.

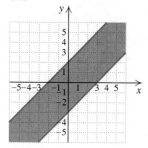

Collaborative Corner

How Old Is Old Enough?

Focus: Linear inequalities
Time: 15–25 minutes
Group Size: 2

It is not unusual for the ages of a bride and groom to differ significantly. Yet is it possible for the difference in age to be too great? In answer to this question, the following rule of thumb has emerged: *The younger spouse's age should be at least seven more than half the age of the older spouse.* (*Source*: http://home.earthlink.net/~mybrainhurts/2002_06_01_archive.html)

ACTIVITY

1. Let b = the age of the bride, in years, and g = the age of the groom, in years. One group member should write an equation for calculating the bride's minimum age if the groom's age is known. The other group member should write an equation for finding the groom's minimum age if the bride's age is known. The equations should look similar.

2. Convert each equation into an inequality by selecting the appropriate symbol from $<$, $>$, $\leq$, and $\geq$. Be sure to reflect the rule of thumb stated above.

3. Graph both inequalities from step 2 as a system of linear inequalities. What does the solution set represent?

4. If your group feels that a minimum or maximum age for marriage should exist, adjust your graph accordingly.

5. Compare your finished graph with those of other groups.

8.4 Polynomial and Rational Inequalities

Quadratic and Other Polynomial Inequalities   ◼
Rational Inequalities

Quadratic and Other Polynomial Inequalities

Inequalities like the following are called *polynomial inequalities*:

$$x^3 - 5x > x^2 + 7, \qquad 4x - 3 < 9, \qquad 5x^2 - 3x + 2 \geq 0.$$

Second-degree polynomial inequalities in one variable are called *quadratic inequalities*. To solve polynomial inequalities, we often focus attention on where the outputs of a polynomial function are positive and where they are negative.

Study Tip

Map Out Your Day

As the semester winds down and term papers are due, it becomes more critical than ever that you manage your time wisely. If you aren't already doing so, consider writing out an hour-by-hour schedule for each day and then abide by it as much as possible.

▸ **EXAMPLE 1**   Solve: $x^2 + 3x - 10 > 0$.

SOLUTION   Consider the "related" function $f(x) = x^2 + 3x - 10$ and its graph.

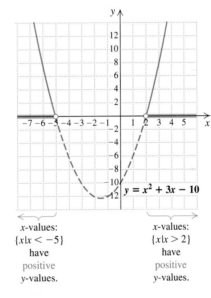

$y = x^2 + 3x - 10$

| *x*-values: $\{x \mid x < -5\}$ have positive *y*-values. | *x*-values: $\{x \mid x > 2\}$ have positive *y*-values. |

When $x^2 + 3x - 10 > 0$, $f(x)$, or y, will be positive.

Values of y will be positive to the left and right of the x-intercepts, as shown. To find the intercepts, we set the polynomial equal to 0 and solve:

$$x^2 + 3x - 10 = 0$$
$$(x + 5)(x - 2) = 0$$
$$x + 5 = 0 \quad or \quad x - 2 = 0$$
$$x = -5 \quad or \qquad x = 2.$$

Thus the solution set of the inequality is

$$\{x \mid x < -5 \ or \ x > 2\}, \quad \text{or} \quad (-\infty, -5) \cup (2, \infty).$$

Any inequality with 0 on one side can be solved by considering a graph of the related function and finding intercepts as in Example 1.

EXAMPLE 2   Solve: $5x^2 - 2x \leq 7$.

SOLUTION   We first find standard form with 0 on one side:

$$5x^2 - 2x - 7 \leq 0. \qquad \text{This is equivalent to the original inequality.}$$

If $f(x) = 5x^2 - 2x - 7$, then $5x^2 - 2x - 7 \leq 0$ when $f(x)$ is 0 or negative. The graph of $f(x) = 5x^2 - 2x - 7$ is a parabola opening upward, as shown in the graphs below. Values of $f(x)$ are negative for x-values between the x-intercepts. We find the x-intercepts by solving $f(x) = 0$:

$$5x^2 - 2x - 7 = 0$$
$$(5x - 7)(x + 1) = 0$$
$$5x - 7 = 0 \quad or \quad x + 1 = 0$$
$$5x = 7 \quad or \quad x = -1$$
$$x = \frac{7}{5} \quad or \quad x = -1.$$

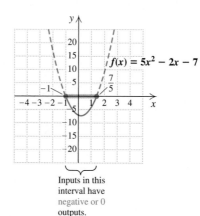

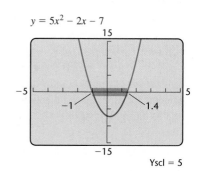

We can also find the x-intercepts using a graphing calculator. At the x-intercepts, the value of $f(x)$ is 0. Thus the solution set of the inequality is

$$\left\{ x \mid -1 \leq x \leq \frac{7}{5} \right\}, \quad or \quad \left[-1, \frac{7}{5} \right], \quad or \quad [-1, 1.4].$$

In Example 2, it was not essential to draw the graph. The important information came from finding the x-intercepts and the sign of $f(x)$ on each side of those intercepts. We now solve a third-degree polynomial inequality, without graphing, by locating the x-intercepts, or *zeros*, of f and then using *test points* to determine the sign of $f(x)$ over each interval of the x-axis.

EXAMPLE 3 For $f(x) = 5x^3 + 10x^2 - 15x$, find all x-values for which $f(x) > 0$.

SOLUTION We first solve the related equation:

$$5x^3 + 10x^2 - 15x = 0$$
$$5x(x^2 + 2x - 3) = 0$$
$$5x(x + 3)(x - 1) = 0$$
$$5x = 0 \quad or \quad x + 3 = 0 \quad or \quad x - 1 = 0$$
$$x = 0 \quad or \qquad\quad x = -3 \quad or \qquad\quad x = 1.$$

The zeros of f are -3, 0, and 1. These zeros divide the number line, or x-axis, into four intervals: A, B, C, and D.

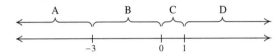

Next, selecting one convenient test value from each interval, we determine the sign of $f(x)$ for that interval. Within each interval, the sign of $f(x)$ cannot change. If it did, there would need to be another zero in that interval.

We choose -4 for a test value from interval A, -1 from interval B, 0.5 from interval C, and 2 from interval D. We enter $y_1 = 5x^3 + 10x^2 - 15x$ and use a table to evaluate the polynomial for each test value.

X	Y₁
−4	−100
−1	20
.5	−4.375
2	50
X = −4	

We are interested only in the signs of each function value. From the table, we see that $f(-4)$ and $f(0.5)$ are negative and that $f(-1)$ and $f(2)$ are positive. We indicate on the number line the sign of $f(x)$ in each interval.

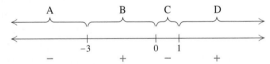

Recall that we are looking for all x for which $5x^3 + 10x^2 - 15x > 0$. The calculations above indicate that $f(x)$ is positive for any number in intervals B and D. The solution set of the original inequality is

$$(-3, 0) \cup (1, \infty), \quad or \quad \{x \mid -3 < x < 0 \text{ or } x > 1\}.$$

The method of Example 3 works because polynomial function values can change signs only when the graph of the function crosses the x-axis. The graph of $f(x) = 5x^3 + 10x^2 - 15x$ illustrates the solution of Example 3. We can see from the graph at left that the function values are positive in the intervals $(-3, 0)$ and $(1, \infty)$.

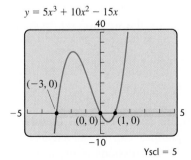

$y = 5x^3 + 10x^2 - 15x$

Yscl = 5

EXAMPLE 4   Solve: $x^4 + x^3 - 2x^2 < 0$.

SOLUTION   We first solve the related equation:

$$x^4 + x^3 - 2x^2 = 0$$
$$x^2(x^2 + x - 2) = 0$$
$$x^2(x + 2)(x - 1) = 0$$
$$x^2 = 0 \quad or \quad x + 2 = 0 \quad or \quad x - 1 = 0$$
$$x = 0 \quad or \quad x = -2 \quad or \quad x = 1.$$

The function $p(x) = x^4 + x^3 - 2x^2$ has zeros at -2, 0, and 1. We graph the function and determine the sign of $p(x)$ over each interval of the number line. The solution will include those intervals where $p(x)$ is negative.

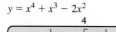
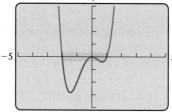
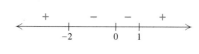

We see that $p(x)$ is negative in the intervals $(-2, 0)$ and $(0, 1)$. The solution set is $(-2, 0) \cup (0, 1)$.

In Example 4, if the inequality symbol were $\leq$, the endpoints of the intervals would be included in the solution set. The solution of $x^4 + x^3 - 2x^2 \leq 0$ is $[-2, 0] \cup [0, 1]$, or simply $[-2, 1]$.

To Solve a Polynomial Inequality

1. Add or subtract to get 0 on one side and solve the related polynomial equation $p(x) = 0$.
2. Use the numbers found in step (1) to divide the number line into intervals.
3. Using a test value from each interval or the graph of the related function, determine the sign of $p(x)$ over each interval.
4. Select the interval(s) for which the inequality is satisfied and write set-builder notation or interval notation for the solution set. Include the endpoints of the intervals when $\leq$ or $\geq$ is used.

Rational Inequalities

Inequalities involving rational expressions are called **rational inequalities.** Like polynomial inequalities, rational inequalities can be solved using test values. Unlike polynomials, however, rational expressions often have values for which the expression is undefined.

EXAMPLE 5   Solve: $\dfrac{x-3}{x+4} \geq 2$.

SOLUTION   We write the related equation by changing the $\geq$ symbol to $=$:

$$\frac{x-3}{x+4} = 2. \quad \text{Note that } x \neq -4.$$

We show both algebraic and graphical approaches.

ALGEBRAIC APPROACH

We first solve the related equation:

$$(x+4) \cdot \frac{x-3}{x+4} = (x+4) \cdot 2 \quad \begin{array}{l}\textbf{Multiplying both sides} \\ \textbf{by the LCD, } x+4\end{array}$$

$$x - 3 = 2x + 8$$

$$-11 = x. \qquad \textbf{Solving for } x$$

In the case of rational inequalities, we must always find any values that make the denominator 0. As noted at the beginning of the example, $x \neq -4$.

Now we use -11 and -4 to divide the number line into intervals:

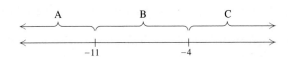

We test a number in each interval to see where the original inequality is satisfied:

$$\frac{x-3}{x+4} \geq 2.$$

If we enter $y_1 = (x-3)/(x+4)$ and $y_2 = 2$, the inequality is satisfied when $y_1 \geq y_2$.

X	Y1	Y2
-15	1.6364	2
-8	2.75	2
1	-.4	2
X =		

$y_1 < y_2$; -15 *is not* a solution.
$y_1 > y_2$; -8 *is* a solution.
$y_1 < y_2$; 1 *is not* a solution.

The solution set includes the interval B. The endpoint -11 is included because the inequality symbol is $\geq$ and -11 is a solution of the related equation. The number -4 is *not* included because $(x-3)/(x+4)$ is undefined for $x = -4$. Thus the solution set of the original inequality is

$$[-11, -4), \quad \text{or} \quad \{x \,|\, -11 \leq x < -4\}.$$

GRAPHICAL APPROACH

We graph $y_1 = (x-3)/(x+4)$ and $y_2 = 2$ using DOT mode. The inequality is true for those values of x for which the graph of y_1 is above or intersects the graph of y_2, or where $y_1 \geq y_2$.

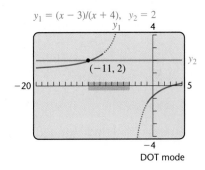

We see that

$$y_1 = y_2 \quad \text{when } x = -11$$

and

$$y_1 > y_2 \quad \text{when } -11 < x < -4.$$

Thus the solution set is $[-11, -4)$.

To Solve a Rational Inequality

1. Change the inequality symbol to an equals sign and solve the related equation.
2. Find any replacements for which the rational expression is undefined.
3. Use the numbers found in steps (1) and (2) to divide the number line into intervals.
4. Substitute a test value from each interval into the inequality. If the number is a solution, then the interval to which it belongs is part of the solution set.
5. Select the interval(s) and any endpoints for which the inequality is satisfied and write set-builder or interval notation for the solution set. If the inequality symbol is ≤ or ≥, then the solutions from step (1) are also included in the solution set. Those numbers found in step (2) should be excluded from the solution set, even if they are solutions from step (1).

To Solve a Rational Inequality: Graphical Approach

1. Let $y_1 =$ one side of the inequality and $y_2 =$ the other side of the inequality. Examine the graph to determine the intervals that satisfy the inequality.
2. Select the intervals for which the inequality is satisfied. If the inequality symbol is ≤ or ≥, include any endpoints that also satisfy the inequality. Write set-builder notation or interval notation for the solution set.

8.4 EXERCISE SET

Concept Reinforcement *Classify each of the following as either true or false.*

1. The solution of $(x-3)(x+2) \le 0$ is $[-2, 3]$.

2. The solution of $(x+5)(x-4) \ge 0$ is $[-5, 4]$.

3. The solution of $(x-1)(x-6) > 0$ is $\{x | x < 1 \text{ or } x > 6\}$.

4. The solution of $(x+4)(x+2) < 0$ is $(-4, -2)$.

5. To solve $\dfrac{x-5}{x+4} \ge 0$ using intervals, we divide the number line into the intervals $(-\infty, -4)$, $(-4, 5)$, and $(5, \infty)$.

6. To solve $\dfrac{x+2}{x-3} < 0$ using intervals, we divide the number line into the intervals $(-\infty, -2)$, $(-2, 3)$, and $(3, \infty)$.

7. The solution of $\dfrac{3}{x-5} \leq 0$ is $[5, \infty)$.

8. The solution of $\dfrac{-2}{x+3} \geq 0$ is $(-\infty, -3]$.

Determine the solution set of each inequality from the given graph.

9. $p(x) \leq 0$

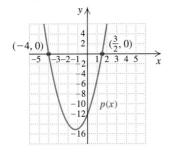

10. $p(x) < 0$

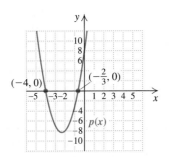

11. $x^4 + 12x > 3x^3 + 4x^2$

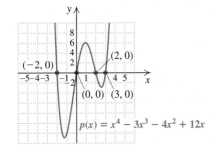

12. $x^4 + x^3 \geq 6x^2$

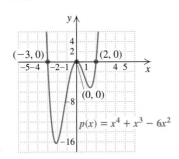

13. $\dfrac{x-1}{x+2} < 3$

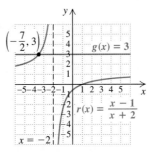

14. $\dfrac{2x-1}{x-5} \geq 1$

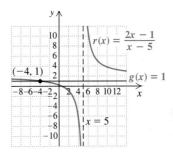

Solve.

15. $(x+4)(x-3) < 0$

16. $(x-5)(x+2) > 0$

17. $(x+7)(x-2) \geq 0$

18. $(x-1)(x+4) \leq 0$

19. $x^2 - x - 2 > 0$

20. $x^2 + x - 2 < 0$

Aha! **21.** $x^2 + 4x + 4 < 0$

22. $x^2 + 6x + 9 < 0$

23. $x^2 - 4x < 12$

24. $x^2 + 6x > -8$

25. $3x(x+2)(x-2) < 0$

26. $5x(x+1)(x-1) > 0$

27. $(x-1)(x+2)(x-4) \geq 0$

28. $(x+3)(x+2)(x-1) < 0$

29. $4.32x^2 - 3.54x - 5.34 \leq 0$

30. $7.34x^2 - 16.55x - 3.89 \geq 0$

31. $x^3 - 2x^2 - 5x + 6 < 0$

32. $\frac{1}{3}x^3 - x + \frac{2}{3} > 0$

33. For $f(x) = x^2 - 1$, find all x-values for which $f(x) \leq 3$.

34. For $f(x) = x^2 - 20$, find all x-values for which $f(x) > 5$.

35. For $g(x) = (x - 2)(x - 3)(x + 1)$, find all x-values for which $g(x) > 0$.

36. For $g(x) = (x + 3)(x - 2)(x + 1)$, find all x-values for which $g(x) < 0$.

37. For $F(x) = x^3 - 7x^2 + 10x$, find all x-values for which $F(x) \leq 0$.

38. For $G(x) = x^3 - 8x^2 + 12x$, find all x-values for which $G(x) \geq 0$.

Solve.

39. $\dfrac{1}{x + 5} < 0$

40. $\dfrac{1}{x + 4} > 0$

41. $\dfrac{x + 1}{x - 3} \geq 0$

42. $\dfrac{x - 2}{x + 4} \leq 0$

43. $\dfrac{x + 1}{x + 6} \geq 1$

44. $\dfrac{x - 1}{x - 2} \leq 1$

45. $\dfrac{(x - 2)(x + 1)}{x - 5} \leq 0$

46. $\dfrac{(x + 4)(x - 1)}{x + 3} \geq 0$

47. $\dfrac{x}{x + 3} \geq 0$

48. $\dfrac{x - 2}{x} \leq 0$

49. $\dfrac{x - 5}{x} < 1$

50. $\dfrac{x}{x - 1} > 2$

51. $\dfrac{x - 1}{(x - 3)(x + 4)} \leq 0$

52. $\dfrac{x + 2}{(x - 2)(x + 7)} \geq 0$

53. For $f(x) = \dfrac{5 - 2x}{4x + 3}$, find all x-values for which $f(x) \geq 0$.

54. For $g(x) = \dfrac{2 + 3x}{2x - 4}$, find all x-values for which $g(x) \geq 0$.

55. For $G(x) = \dfrac{1}{x - 2}$, find all x-values for which $G(x) \leq 1$.

56. For $F(x) = \dfrac{1}{x - 3}$, find all x-values for which $F(x) \leq 2$.

57. Explain how any quadratic inequality can be solved by examining a parabola.

58. Describe a method for creating a quadratic inequality for which there is no solution.

Focused Review

Solve.

59. $2(x - 3) \geq 4 - (x + 1)$ [2.6]

60. $|2x + 3| > 8$ [8.2]

61. $|2 - 5x| \leq 10$ [8.2]

62. $|1.8x - 5.6| < -3$ [8.2]

63. $x^2 - 6x < 7$ [8.4]   **64.** $\dfrac{1}{x} \leq 1$ [8.4]

Synthesis

65. Step (5) on p. 666 states that even when the inequality symbol is $\leq$ or $\geq$, the solutions from step (1) are not always part of the solution set. Why?

66. Describe a method that could be used to create quadratic inequalities that have $(-\infty, a] \cup [b, \infty)$ as the solution set.

Find each solution set.

67. $x^4 + x^2 < 0$ **68.** $x^4 + 2x^2 \geq 0$

69. $x^4 + 3x^2 \leq 0$ **70.** $\left|\dfrac{x + 2}{x - 1}\right| \leq 3$

71. *Total Profit.* Derex, Inc., determines that its total-profit function is given by

$$P(x) = -3x^2 + 630x - 6000.$$

  a) Find all values of x for which Derex makes a profit.
  b) Find all values of x for which Derex loses money.

72. *Height of a Thrown Object.* The function

$$S(t) = -16t^2 + 32t + 1920$$

gives the height S, in feet, of an object thrown from a cliff that is 1920 ft high. Here t is the time, in seconds, that the object is in the air.

  a) For what times does the height exceed 1920 ft?
  b) For what times is the height less than 640 ft?

73. *Number of Handshakes.*   There are n people in a room. The number N of possible handshakes by the people is given by the function

$$N(n) = \frac{n(n-1)}{2}.$$

For what number of people n is $66 \le N \le 300$?

74. *Number of Diagonals.*   A polygon with n sides has D diagonals, where D is given by the function

$$D(n) = \frac{n(n-3)}{2}.$$

Find the number of sides n if

$$27 \le D \le 230.$$

Use a graphing calculator to graph each function and find solutions of $f(x) = 0$. Then solve the inequalities $f(x) < 0$ and $f(x) > 0$.

75. $f(x) = x + \dfrac{1}{x}$

76. $f(x) = x - \sqrt{x},\ x \ge 0$

77. $f(x) = \dfrac{x^3 - x^2 - 2x}{x^2 + x - 6}$

78. $f(x) = x^4 - 4x^3 - x^2 + 16x - 12$

79. *HMOs.*   The percent of the population of the United States enrolled in an HMO (Health Maintenance Organization) increased in the 1990s but began to decrease in 2000. The following table lists the percent enrolled for several years.

Year	Percent Enrolled in an HMO
1994	17.3%
1996	22.3
1998	28.6
2000	30.0
2002	26.4
2004	23.4

Source: Centers for Disease Control and Prevention

a) Use regression to find a quadratic function that can be used to estimate the percent h of the population enrolled in an HMO x years after 1994.

b) Use the function to predict the years after 1994 for which less than 20% of the population will be enrolled in an HMO.

8 Chapter Summary and Review

KEY TERMS AND DEFINITIONS

INEQUALITIES

Inequality, p. 620   Any sentence containing $<$, $>$, $\leq$, $\geq$, or $\neq$.

Compound inequality, p. 623   Two or more inequalities that are combined using the word *and* or the word *or*.

Conjunction, p. 623   Two statements joined by the word *and*. The solution set of a conjunction is the **intersection,** indicated by $\cap$, of the solution sets of the individual statements.

Disjunction. p. 626   Two statements joined by the word *or*. The solution set of a disjunction is the **union,** indicated by $\cup$, of the solution sets of the individual statements.

GRAPHS OF INEQUALITIES

Linear inequality, p. 647   An inequality whose related equation is a linear equation.

Half-plane, p. 648   The region to one side of a boundary line.

Vertices (singular, vertex), p. 654   The points of intersection of boundary lines in a system of inequalities.

IMPORTANT CONCEPTS

[Section references appear in brackets.]

Concept	Example
The Absolute-Value Principles for Equations and Inequalities For any positive number p and any algebraic expression X: **a)** The solutions of $\lvert X \rvert = p$ are those numbers that satisfy $X = -p$ *or* $X = p$. **b)** The solutions of $\lvert X \rvert < p$ are those numbers that satisfy $-p < X < p$. **c)** The solutions of $\lvert X \rvert > p$ are those numbers that satisfy $X < -p$ *or* $p < X$. If $\lvert X \rvert = 0$, then $X = 0$. If p is negative, then $\lvert X \rvert = p$ and $\lvert X \rvert < p$ have no solution, and any value of X will satisfy $\lvert X \rvert > p$.	Using part (a): $$\lvert x + 3 \rvert = 4$$ $$x + 3 = 4 \quad or \quad x + 3 = -4$$ Using part (b): $$\lvert x + 3 \rvert < 4$$ $$-4 < x + 3 < 4$$ Using part (c): $$\lvert x + 3 \rvert \geq 4$$ $$x + 3 \leq -4 \quad or \quad 4 \leq x + 3 \qquad \text{[8.2]}$$

(continued)

Polynomial and **rational inequalities** are solved by dividing the x-axis into intervals and using test values from each interval to identify the solution set.

$(x + 2)(x - 3) < 0$

Let $f(x) = (x + 2)(x - 3)$.
Solve: $(x + 2)(x - 3) = 0$.
$$x = -2 \quad or \quad x = 3.$$

Test $-3, 0, 4$:
A: $f(-3)$ is positive.
B: $f(0)$ is negative.
C: $f(4)$ is positive.

Any number in interval B is a solution.
The endpoints are not solutions.
The solution set is $(-2, 3)$, or $\{x | -2 < x < 3\}$.   [8.4]

Review Exercises

Concept Reinforcement   *Classify each of the following as either true or false.*

1. The solution of $|3x - 5| \leq 8$ is a closed interval. [8.1]

2. The inequality $2 < 5x + 1 < 9$ is equivalent to $2 < 5x + 1 \ or \ 5x + 1 < 9$. [8.1]

3. The solution set of a disjunction is the union of two solution sets. [8.1]

4. In mathematics, the word "or" means "one or the other or both." [8.1]

5. The domain of the sum of two functions is the union of the domains of the individual functions. [8.1]

6. The equation $|x| = -p$ has no solution when p is positive. [8.2]

7. $|f(x)| > 3$ is equivalent to $f(x) < -3 \ or \ f(x) > 3$. [8.2]

8. A test point is used to determine whether the line in a linear inequality is drawn solid or dashed. [8.3]

9. The graph of a system of linear inequalities is always a half-plane. [8.3]

10. To solve a polynomial inequality, we often must solve a polynomial equation. [8.4]

Solve graphically. [8.1]

11. $x - 3 < 3x + 5$

12. $x + 1 \geq \frac{1}{2}x - 2$

13. Find the intersection:
$$\{1, 2, 5, 6, 9\} \cap \{1, 3, 5, 9\}. \ [8.1]$$

14. Find the union:
$$\{1, 2, 5, 6, 9\} \cup \{1, 3, 5, 9\}. \ [8.1]$$

Graph and write interval notation. [8.1]

15. $x \leq 3 \ and \ x > -5$

16. $x \leq 3 \ or \ x > -5$

Solve and graph each solution set. [8.1]

17. $-4 < x + 8 \leq 5$

18. $-15 < -4x - 5 < 0$

19. $3x < -9 \ or \ -5x < -5$

20. $2x + 5 < -17 \ or \ -4x + 10 \leq 34$

21. $2x + 7 \leq -5 \ or \ x + 7 \geq 15$

22. $f(x) < -5 \ or \ f(x) > 5$, where $f(x) = 3 - 5x$

For f(x) as given, use interval notation to write the domain of f. [8.1]

23. $f(x) = \dfrac{2x}{x - 8}$

24. $f(x) = \sqrt{x + 5}$

25. $f(x) = \sqrt{8 - 3x}$

Solve. [8.2]

26. $|x| = 5$

27. $|t| \geq 3.5$

28. $|x - 3| = 7$

29. $|2x + 5| < 12$

30. $|3x - 4| \geq 15$

31. $|2x + 5| = |x - 9|$

32. $|5n + 6| = -8$

33. $\left|\dfrac{x + 4}{6}\right| \leq 2$

34. $2|x - 5| - 7 > 3$

35. Let $f(x) = |3x - 5|$. Find all x for which $f(x) < 0$. [8.2]

36. Graph $x - 2y \geq 6$ on a plane. [8.3]

Graph each system of inequalities. Find the coordinates of any vertices formed. [8.3]

37. $x + 3y > -1$,
$x + 3y < 4$

38. $x - 3y \leq 3$,
$x + 3y \geq 9$,
$y \leq 6$

Solve. [8.4]

39. $x^3 - 3x > 2x^2$

40. $\dfrac{x - 5}{x + 3} \leq 0$

Synthesis

TW 41. Explain in your own words why $|X| = p$ has two solutions when p is positive and no solution when p is negative. [8.2]

TW 42. Explain why the graph of the solution of a system of linear inequalities is the intersection, not the union, of the individual graphs. [8.3]

43. Solve: $|2x + 5| \leq |x + 3|$. [8.2]

44. Just-For-Fun manufactures marbles with a 1.1-cm diameter and a ± 0.03-cm manufacturing tolerance, or allowable variation in diameter. Write the tolerance as an inequality with absolute value. [8.2]

45. The Twinrocker Paper Company makes paper by hand. Each sheet is between 18 thousandths and 25 thousandths of an inch thick. Write the thickness t of a sheet of handmade paper as an inequality with absolute value. [8.2]

Chapter Test 8

Solve graphically.

1. $2x + 1 \geq 3x - 2$

2. $x + 2 > \frac{1}{3}x$

3. Find the intersection:
$$\{1, 3, 5, 7, 9\} \cap \{3, 5, 11, 13\}.$$

4. Find the union:
$$\{1, 3, 5, 7, 9\} \cup \{3, 5, 11, 13\}.$$

Write the domain of f using interval notation.

5. $f(x) = \dfrac{x}{(x + 3)(x - 1)}$

6. $f(x) = \sqrt{8 - 2x}$

Solve and graph each solution set.

7. $-2 < x - 3 < 5$

8. $-11 \leq -5t - 2 < 0$

9. $3x - 2 < 7 \text{ or } x - 2 > 4$

10. $-3x > 12 \text{ or } 4x > -10$

11. $-\frac{1}{3} \leq \frac{1}{6}x - 1 < \frac{1}{4}$ 12. $|x| = 13$

13. $|a| > 7$ 14. $|3x - 1| < 7$

15. $|-5t - 3| \geq 10$ 16. $|2 - 5x| = -12$

17. $g(x) < -3 \text{ or } g(x) > 3$, where $g(x) = 4 - 2x$

18. Let $f(x) = |x + 10|$ and $g(x) = |x - 12|$. Find all values of x for which $f(x) = g(x)$.

Graph each system of inequalities. Find the coordinates of any vertices formed.

19. $x + y \geq 3,$
$x - y \geq 5$

20. $2y - x \geq -7,$
$2y + 3x \leq 15,$
$y \leq 0,$
$x \leq 0$

Solve.

21. $x^2 + 5x \leq 6$

22. $x - \dfrac{1}{x} > 0$

Synthesis

Solve. Write the solution set using interval notation.

23. $|2x - 5| \leq 7 \text{ and } |x - 2| \geq 2$

24. $7x < 8 - 3x < 6 + 7x$

25. Write an absolute-value inequality for which the interval shown is the solution.

9

More on Systems

In this chapter, we extend the study of systems of equations that we began in Chapter 4. We solve systems of equations in three variables, solve systems using matrices and determinants, and look at some applications from economics.

APPLICATION *Aircraft Production.*

In October 2006, due to increased cost, Airbus revised its estimated break-even point for the new A380. The break-even point rose from 270 aircraft to 420 aircraft. The following table lists the number of aircraft on order at various points in time. (*Sources*: *The Register; Newswire Today;* www.dw-word.de) Assuming linear growth, when will 420 aircraft be on order?

MONTHS AFTER JANUARY 1, 2004	AIRBUS A380 ORDERS
5	129
13	149
16	139
18	144
29	159
30	167

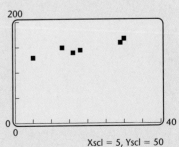

Xscl = 5, Yscl = 50

This problem appears as Exercise 46 in Exercise Set 9.5.

9.1 Systems of Equations in Three Variables

Identifying Solutions ▣ Solving Systems in Three Variables ▣ Dependency, Inconsistency, and Geometric Considerations

As we saw in Chapter 4, some problems translate directly to a system of two equations. Others call for a translation to three or more equations. Here we learn how to solve systems of three linear equations. Later, we will use such systems in problem-solving situations.

Identifying Solutions

A **linear equation in three variables** is an equation equivalent to one in the form $Ax + By + Cz = D$, where A, B, C, and D are real numbers. We refer to the form $Ax + By + Cz = D$ as *standard form* for a linear equation in three variables.

A solution of a system of three equations in three variables is an ordered triple (x, y, z) that makes *all three* equations true.

EXAMPLE 1   Determine whether $\left(\frac{3}{2}, -4, 3\right)$ is a solution of the system

$$4x - 2y - 3z = 5,$$
$$-8x - y + z = -5,$$
$$2x + y + 2z = 5.$$

SOLUTION   We substitute $\left(\frac{3}{2}, -4, 3\right)$ into the three equations, using alphabetical order.

| BY HAND | USING A GRAPHING CALCULATOR |

We have the following:

$$4x - 2y - 3z = 5$$

$$\frac{4 \cdot \frac{3}{2} - 2(-4) - 3 \cdot 3 \quad | \quad 5}{}$$
$$6 + 8 - 9 \quad |$$
$$5 \stackrel{?}{=} 5 \quad \text{TRUE}$$

$$-8x - y + z = -5$$

$$\frac{-8 \cdot \frac{3}{2} - (-4) + 3 \quad | \quad -5}{}$$
$$-12 + 4 + 3 \quad |$$
$$-5 \stackrel{?}{=} -5 \quad \text{TRUE}$$

$$2x + y + 2z = 5$$

$$\frac{2 \cdot \frac{3}{2} + (-4) + 2 \cdot 3 \quad | \quad 5}{}$$
$$3 - 4 + 6 \quad |$$
$$5 \stackrel{?}{=} 5 \quad \text{TRUE}$$

The triple makes all three equations true, so it is a solution.

We store $\frac{3}{2}$ as X, -4 as Y, and 3 as Z, using the **ALPHA** key to enter Y and Z.

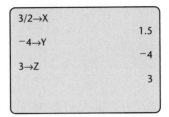

Now we enter the expression on the left side of each equation and press **ENTER**. If the value is the same as the right side of the equation, the ordered triple makes the equation true.

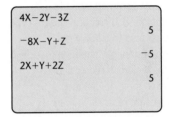

The triple makes all three equations true, so it is a solution.

Solving Systems in Three Variables

In Chapter 4, we solved systems of two equations using graphing, substitution, and elimination. Graphical methods for solving linear equations in three variables are problematic, because a three-dimensional coordinate system is required and the graph of a linear equation in three variables is a plane. The substitution method *can* be used but becomes very cumbersome unless one or more of the equations has only two variables. Fortunately, the elimination method allows us to manipulate a system of three equations in three variables so that a simpler system of two equations in two variables is formed. Once that simpler system has been solved, we can substitute into one of the three original equations and solve for the third variable.

EXAMPLE 2   Solve the following system of equations:

$$x + y + z = 4, \quad (1)$$
$$x - 2y - z = 1, \quad (2)$$
$$2x - y - 2z = -1. \quad (3)$$

SOLUTION   We select *any* two of the three equations and work to get one equation in two variables. Let's add equations (1) and (2):

$$
\begin{aligned}
x + y + z &= 4 \qquad (1)\\
\underline{x - 2y - z} &= \underline{1} \qquad (2)\\
2x - y \phantom{{}+z} &= 5. \qquad (4) \qquad \text{Adding to eliminate } z
\end{aligned}
$$

Next, we select a different pair of equations and eliminate the *same variable* we did above. Let's use equations (1) and (3) to again eliminate z. Be careful here! A common error is to eliminate a different variable in this step.

$$
\begin{array}{l}
x + y + z = 4, \\
2x - y - 2z = -1
\end{array}
\xrightarrow[\text{equation (1) by 2}]{\text{Multiplying both sides of}}
\begin{array}{l}
2x + 2y + 2z = 8 \\
\underline{2x - y - 2z = -1} \\
4x + y \phantom{{}+2z} = 7 \qquad (5)
\end{array}
$$

Now we solve the resulting system of equations (4) and (5). That solution will give us two of the numbers in the solution of the original system.

$$
\begin{aligned}
2x - y &= 5 \qquad (4)\\
\underline{4x + y} &= \underline{7} \qquad (5)\\
6x \phantom{{}+y} &= 12 \qquad \text{Adding}\\
x &= 2
\end{aligned}
$$

Note that we now have two equations in two variables. Had we not eliminated the same variable in both of the above steps, this would not be the case.

We can use either equation (4) or (5) to find y. We choose equation (5):

$$
\begin{aligned}
4x + y &= 7 \qquad (5)\\
4 \cdot 2 + y &= 7 \qquad \text{Substituting 2 for } x \text{ in equation (5)}\\
8 + y &= 7 \\
y &= -1.
\end{aligned}
$$

We now have $x = 2$ and $y = -1$. To find the value for z, we use any of the original three equations and substitute to find the third number, z. Let's use equation (1) and substitute our two numbers in it:

$$
\begin{aligned}
x + y + z &= 4 \qquad (1)\\
2 + (-1) + z &= 4 \qquad \text{Substituting 2 for } x \text{ and } -1 \text{ for } y\\
1 + z &= 4 \\
z &= 3.
\end{aligned}
$$

We have obtained the triple $(2, -1, 3)$. It should check in *all three* equations:

$$
\begin{array}{lll}
\dfrac{x + y + z = 4}{2 + (-1) + 3 \,\big|\, 4} & \dfrac{x - 2y - z = 1}{2 - 2(-1) - 3 \,\big|\, 1} & \dfrac{2x - y - 2z = -1}{2 \cdot 2 - (-1) - 2 \cdot 3 \,\big|\, -1}\\[2mm]
\qquad\quad 4 \overset{?}{=} 4 \;\; \text{TRUE} & \qquad\quad 1 \overset{?}{=} 1 \;\; \text{TRUE} & \qquad\qquad -1 \overset{?}{=} -1 \;\; \text{TRUE}
\end{array}
$$

The solution is $(2, -1, 3)$.

Solving Systems of Three Linear Equations

To use the elimination method to solve systems of three linear equations:

1. Write all equations in the standard form $Ax + By + Cz = D$.
2. Clear any decimals or fractions.
3. Choose a variable to eliminate. Then select two of the three equations and work to get one equation in which the selected variable is eliminated.
4. Next, use a different pair of equations and eliminate the same variable that you did in step (3).
5. Solve the system of equations that resulted from steps (3) and (4).
6. Substitute the solution from step (5) into one of the original three equations and solve for the third variable. Then check.

Student Notes

Because solving systems of three equations can be lengthy, it is important that you use plenty of paper, work in pencil, and double-check each step as you proceed.

EXAMPLE 3 Solve the system

$$4x - 2y - 3z = 5, \quad (1)$$
$$-8x - y + z = -5, \quad (2)$$
$$2x + y + 2z = 5. \quad (3)$$

SOLUTION

1., 2. The equations are already in standard form with no fractions or decimals.

3. Next, select a variable to eliminate. We decide on y because the y-terms are opposites of each other in equations (2) and (3). We add:

$$
\begin{array}{ll}
-8x - y + z = -5 & (2) \\
\underline{2x + y + 2z = 5} & (3) \\
-6x + 3z = 0. & (4) \quad \text{Adding}
\end{array}
$$

4. We use another pair of equations to create a second equation in x and z. That is, we eliminate the same variable, y, as in step (3). We use equations (1) and (3):

$$
\begin{array}{l}
4x - 2y - 3z = 5, \\
2x + y + 2z = 5
\end{array}
\xrightarrow[\text{of equation (3) by 2}]{\text{Multiplying both sides}}
\begin{array}{l}
4x - 2y - 3z = 5 \\
\underline{4x + 2y + 4z = 10} \\
8x + z = 15. \quad (5)
\end{array}
$$

5. Now we solve the resulting system of equations (4) and (5). That allows us to find two parts of the ordered triple.

$$
\begin{array}{l}
-6x + 3z = 0, \\
8x + z = 15
\end{array}
\xrightarrow[\text{of equation (5) by } -3]{\text{Multiplying both sides}}
\begin{array}{l}
-6x + 3z = 0 \\
\underline{-24x - 3z = -45} \\
-30x = -45 \\
x = \frac{-45}{-30} = \frac{3}{2}
\end{array}
$$

We use equation (5) to find z:

$$8x + z = 15 \qquad (5)$$
$$8 \cdot \tfrac{3}{2} + z = 15 \qquad \text{Substituting } \tfrac{3}{2} \text{ for } x$$
$$12 + z = 15$$
$$z = 3.$$

6. Finally, we use any of the original equations and substitute to find the third number, y. We choose equation (3):

$$2x + y + 2z = 5 \qquad (3)$$
$$2 \cdot \tfrac{3}{2} + y + 2 \cdot 3 = 5 \qquad \text{Substituting } \tfrac{3}{2} \text{ for } x \text{ and } 3 \text{ for } z$$
$$3 + y + 6 = 5$$
$$y + 9 = 5$$
$$y = -4.$$

The solution is $\left(\tfrac{3}{2}, -4, 3\right)$. The check was performed as Example 1.

Sometimes, certain variables are missing at the outset.

EXAMPLE 4   Solve the system

$$x + \ y + z = 180, \qquad (1)$$
$$x \qquad\ - z = -70, \qquad (2)$$
$$2y - z = \quad 0. \qquad (3)$$

SOLUTION

1., 2. The equations appear in standard form with no fractions or decimals.

3., 4. Note that there is no x-term in equation (3) and no y-term in equation (2). It will save some steps if we eliminate one of these variables; we choose to eliminate y. Since, at the outset, we already have one equation with no y, we need only one more equation with y eliminated. We use equations (1) and (3):

$$x + y + z = 180, \qquad \xrightarrow[\text{of equation (1) by } -2]{\text{Multiplying both sides}} \qquad -2x - 2y - 2z = -360$$
$$2y - z = \quad 0 \qquad\qquad\qquad\qquad\qquad \underline{ 2y - \quad z = \qquad 0}$$
$$-2x \qquad\ - 3z = -360. \qquad (4)$$

5., 6. Now we solve the resulting system of equations (2) and (4):

$$x - \quad z = \ -70, \qquad \xrightarrow[\text{of equation (2) by } 2]{\text{Multiplying both sides}} \qquad 2x - 2z = -140$$
$$-2x - 3z = -360 \qquad\qquad\qquad\qquad\qquad \underline{-2x - 3z = -360}$$
$$-5z = -500$$
$$z = \quad 100.$$

Continuing as in Examples 2 and 3, we get the solution $(30, 50, 100)$. The check is left to the student.

Dependency, Inconsistency, and Geometric Considerations

Each equation in Examples 2, 3, and 4 has a graph that is a plane in three dimensions. The solutions are points common to the planes of each system. Since three planes can have an infinite number of points in common or no points at all in common, we need to generalize the concept of *consistency*.

Planes intersect at one point. System is *consistent* and has one solution.

Planes intersect along a common line. System is *consistent* and has an infinite number of solutions.

Three parallel planes. System is *inconsistent;* it has no solution.

Planes intersect two at a time, with no point common to all three. System is *inconsistent;* it has no solution.

Consistency

A system of equations that has at least one solution is said to be **consistent.**

A system of equations that has no solution is said to be **inconsistent.**

EXAMPLE 5   Solve:

$$
\begin{aligned}
y + 3z &= 4, &&(1)\\
-x - y + 2z &= 0, &&(2)\\
x + 2y + z &= 1. &&(3)
\end{aligned}
$$

SOLUTION   The variable x is missing in equation (1). By adding equations (2) and (3), we can find a second equation in which x is missing:

$$
\begin{aligned}
-x - y + 2z &= 0 &&(2)\\
\underline{x + 2y + z} &= \underline{1} &&(3)\\
y + 3z &= 1. &&(4) \quad \text{Adding}
\end{aligned}
$$

Equations (1) and (4) form a system in y and z. We solve as before:

$$
\begin{aligned}
y + 3z &= 4, & \xrightarrow{\substack{\text{Multiplying both sides}\\\text{of equation (1) by } -1}} & & -y - 3z &= -4\\
y + 3z &= 1 & & & \underline{y + 3z} &= \underline{1}\\
& & \text{This is a contradiction.} \longrightarrow & & 0 &= -3. \quad \text{Adding}
\end{aligned}
$$

Since we end up with a *false* equation, or contradiction, we know that the system has no solution. It is *inconsistent*.

The notion of *dependency* from Section 4.1 can also be extended.

EXAMPLE 6   Solve:

$$2x + y + z = 3, \quad (1)$$
$$x - 2y - z = 1, \quad (2)$$
$$3x + 4y + 3z = 5. \quad (3)$$

SOLUTION   Our plan is to first use equations (1) and (2) to eliminate z. Then we will select another pair of equations and again eliminate z:

$$2x + y + z = 3$$
$$\underline{x - 2y - z = 1}$$
$$3x - y \quad\quad = 4. \quad (4)$$

Next, we use equations (2) and (3) to eliminate z again:

$$
\begin{array}{l}
x - 2y - z = 1, \\
3x + 4y + 3z = 5
\end{array}
\quad
\xrightarrow[\text{of equation (2) by 3}]{\text{Multiplying both sides}}
\quad
\begin{array}{l}
3x - 6y - 3z = 3 \\
\underline{3x + 4y + 3z = 5} \\
6x - 2y \quad\quad = 8. \quad (5)
\end{array}
$$

We now try to solve the resulting system of equations (4) and (5):

$$
\begin{array}{l}
3x - y = 4, \\
6x - 2y = 8
\end{array}
\quad
\xrightarrow[\text{of equation (4) by} -2]{\text{Multiplying both sides}}
\quad
\begin{array}{l}
-6x + 2y = -8 \\
\underline{6x - 2y = 8} \\
0 = 0. \quad (6)
\end{array}
$$

Equation (6), which is an identity, indicates that equations (1), (2), and (3) are *dependent*. This means that the original system of three equations is equivalent to a system of two equations. Removing equation (3) from the system does not affect the solution of the system.* In writing an answer to this problem, we simply state that "the equations are dependent."

Recall that when dependent equations appeared in Section 4.1, the solution sets were always infinite in size and were written in set-builder notation. There, all systems of dependent equations were *consistent*. This is not always the case for systems of three or more equations. The following figures illustrate some possibilities geometrically.

The planes intersect along a common line. The equations are dependent and the system is consistent. There is an infinite number of solutions.

The planes coincide. The equations are dependent and the system is consistent. There is an infinite number of solutions.

Two planes coincide. The third plane is parallel. The equations are dependent and the system is inconsistent. There is no solution.

---

*A set of equations is dependent if at least one equation can be expressed as a sum of multiples of other equations in that set.

9.1 EXERCISE SET

Concept Reinforcement   *Classify each statement as either true or false.*

1. $3x + 5y + 4z = 7$ is a linear equation in three variables.

2. It is not difficult to solve a system of three equations in three unknowns by graphing.

3. Every system of three equations in three unknowns has at least one solution.

4. If, when we are solving a system of three equations, a false equation results from adding a multiple of one equation to another, the system is inconsistent.

5. If, when we are solving a system of three equations, an identity results from adding a multiple of one equation to another, the equations are dependent.

6. Whenever a system of three equations contains dependent equations, there is an infinite number of solutions.

7. Determine whether $(2, -1, -2)$ is a solution of the system
$$x + y - 2z = 5,$$
$$2x - y - z = 7,$$
$$-x - 2y + 3z = 6.$$

8. Determine whether $(1, -2, 3)$ is a solution of the system
$$x + y + z = 2,$$
$$x - 2y - z = 2,$$
$$3x + 2y + z = 2.$$

Solve each system. If a system's equations are dependent or if there is no solution, state this.

9. $2x - y + z = 10,$
$4x + 2y - 3z = 10,$
$x - 3y + 2z = 8$

10. $x + y + z = 6,$
$2x - y + 3z = 9,$
$-x + 2y + 2z = 9$

11. $x - y + z = 6,$
$2x + 3y + 2z = 2,$
$3x + 5y + 4z = 4$

12. $2x - y - 3z = -1,$
$2x - y + z = -9,$
$x + 2y - 4z = 17$

13. $6x - 4y + 5z = 31,$
$5x + 2y + 2z = 13,$
$x + y + z = 2$

14. $2x - 3y + z = 5,$
$x + 3y + 8z = 22,$
$3x - y + 2z = 12$

15. $x + y + z = 0,$
$2x + 3y + 2z = -3,$
$-x - 2y - z = 1$

16. $3a - 2b + 7c = 13,$
$a + 8b - 6c = -47,$
$7a - 9b - 9c = -3$

17. $2x + y - 3z = -4,$
$4x - 2y + z = 9,$
$3x + 5y - 2z = 5$

18. $4x + y + z = 17,$
$x - 3y + 2z = -8,$
$5x - 2y + 3z = 5$

19. $2x + y + 2z = 11,$
$3x + 2y + 2z = 8,$
$x + 4y + 3z = 0$

20. $2x + y + z = -2,$
$2x - y + 3z = 6,$
$3x - 5y + 4z = 7$

21. $-2x + 8y + 2z = 4,$
$x + 6y + 3z = 4,$
$3x - 2y + z = 0$

22. $x - y + z = 4,$
$5x + 2y - 3z = 2,$
$4x + 3y - 4z = -2$

23. $4x - y - z = 4,$
$2x + y + z = -1,$
$6x - 3y - 2z = 3$

24. $a + 2b + c = 1,$
$7a + 3b - c = -2,$
$a + 5b + 3c = 2$

25. $r + \frac{3}{2}s + 6t = 2,$
$2r - 3s + 3t = 0.5,$
$r + s + t = 1$

26. $5x + 3y + \frac{1}{2}z = \frac{7}{2},$
$0.5x - 0.9y - 0.2z = 0.3,$
$3x - 2.4y + 0.4z = -1$

27. $4a + 9b = 8,$
$8a + 6c = -1,$
$6b + 6c = -1$

28. $3p + 2r = 11,$
$q - 7r = 4,$
$p - 6q = 1$

29. $x + y + z = 57,$
$-2x + y = 3,$
$x - z = 6$

30. $x + y + z = 105,$
$10y - z = 11,$
$2x - 3y = 7$

31. $a - 3c = 6,$
$b + 2c = 2,$
$7a - 3b - 5c = 14$

32. $2a - 3b = 2,$
$7a + 4c = \frac{3}{4},$
$2c - 3b = 1$

Aha! **33.** $x + y + z = 83,$
$y = 2x + 3,$
$z = 40 + x$

34. $l + m = 7,$
$3m + 2n = 9,$
$4l + n = 5$

35. $x \qquad + z = 0,$
$x + y + 2z = 3,$
$y + z = 2$

36. $x + y \qquad = 0,$
$x \qquad + z = 1,$
$2x + y + z = 2$

37. $x + y + z = 1,$
$-x + 2y + z = 2,$
$2x - y \qquad = -1$

38. $\qquad y + z = 1,$
$x + y + z = 1,$
$x + 2y + 2z = 2$

TW **39.** Describe a method for writing an inconsistent system of three equations in three variables.

TW **40.** Abbie recommends that a frustrated classmate double- and triple-check each step of work when attempting to solve a system of three equations. Is this good advice? Why or why not?

Focused Review

41. Solve by graphing by hand. [4.1]
$$x + y = 5,$$
$$2x + 3y = 6$$

42. Solve using a graphing calculator. [4.1]
$$2x - y = 5,$$
$$y = 3x + 7$$

43. Solve using substitution. [4.2]
$$y = \tfrac{1}{2}x + 3,$$
$$5x - 2y = 1$$

44. Solve using elimination. [4.3]
$$2x + y = 8,$$
$$3x - 2y = 1$$

Solve.

45. $x + y - z = 4,$
$2x - y - z = 7,$
$x + 3y = 4z$ [9.1]

46. $1.5x - 3.2y = -10,$
$20x + 25y = 205$ [4.3]

Synthesis

TW **47.** Is it possible for a system of three linear equations to have exactly two ordered triples in its solution set? Why or why not?

TW **48.** Describe a procedure that could be used to solve a system of four equations in four variables.

Solve.

49. $\dfrac{x + 2}{3} - \dfrac{y + 4}{2} + \dfrac{z + 1}{6} = 0,$

$\dfrac{x - 4}{3} + \dfrac{y + 1}{4} - \dfrac{z - 2}{2} = -1,$

$\dfrac{x + 1}{2} + \dfrac{y}{2} + \dfrac{z - 1}{4} = \dfrac{3}{4}$

50. $w + x + y + z = 2,$
$w + 2x + 2y + 4z = 1,$
$w - x + y + z = 6,$
$w - 3x - y + z = 2$

51. $w + x - y + z = 0,$
$w - 2x - 2y - z = -5,$
$w - 3x - y + z = 4,$
$2w - x - y + 3z = 7$

For Exercises 52 and 53, let u represent $1/x$, v represent $1/y$, and w represent $1/z$. Solve for u, v, and w, and then solve for x, y, and z.

52. $\dfrac{2}{x} - \dfrac{1}{y} - \dfrac{3}{z} = -1,$

$\dfrac{2}{x} - \dfrac{1}{y} + \dfrac{1}{z} = -9,$

$\dfrac{1}{x} + \dfrac{2}{y} - \dfrac{4}{z} = 17$

53. $\dfrac{2}{x} + \dfrac{2}{y} - \dfrac{3}{z} = 3,$

$\dfrac{1}{x} - \dfrac{2}{y} - \dfrac{3}{z} = 9,$

$\dfrac{7}{x} - \dfrac{2}{y} + \dfrac{9}{z} = -39$

Determine k so that each system is dependent.

54. $x - 3y + 2z = 1,$
$2x + y - z = 3,$
$9x - 6y + 3z = k$

55. $5x - 6y + kz = -5,$
$x + 3y - 2z = 2,$
$2x - y + 4z = -1$

In each case, three solutions of an equation in x, y, and z are given. Find the equation.

56. $Ax + By + Cz = 12;$
$\left(1, \tfrac{3}{4}, 3\right), \left(\tfrac{4}{3}, 1, 2\right),$ and $(2, 1, 1)$

57. $z = b - mx - ny;$
$(1, 1, 2), (3, 2, -6),$ and $\left(\tfrac{3}{2}, 1, 1\right)$

58. Write an inconsistent system of equations that contains dependent equations.

Collaborative Corner

Finding the Preferred Approach

Focus: Systems of three linear equations
Time: 10–15 minutes
Group Size: 3

Consider the six steps outlined on p. 679 along with the following system:

$$2x + 4y = 3 - 5z,$$
$$0.3x = 0.2y + 0.7z + 1.4,$$
$$0.04x + 0.03y = 0.07 + 0.04z.$$

ACTIVITY

1. Working independently, each group member should solve the system above. One person should begin by eliminating x, one should first eliminate y, and one should first eliminate z. Write neatly so that others can follow your steps.

2. Once all group members have solved the system, compare your answers. If the answers do not check, exchange notebooks and check each other's work. If a mistake is detected, allow the person who made the mistake to make the repair.

3. Decide as a group which of the three approaches above (if any) ranks as easiest and which (if any) ranks as most difficult. Then compare your rankings with the other groups in the class.

9.2 Solving Applications: Systems of Three Equations

Applications of Three Equations in Three Unknowns

Study Tip

You've Got Mail

If your instructor makes his or her e-mail address available, consider using it to get help. Often, just the act of writing out your question brings some clarity. If you do use e-mail, allow some time for your instructor to reply.

Solving systems of three or more equations is important in many applications. Such systems arise in the natural and social sciences, business, and engineering. In mathematics, purely numerical applications also arise.

EXAMPLE 1   The sum of three numbers is 4. The first number minus twice the second, minus the third is 1. Twice the first number minus the second, minus twice the third is -1. Find the numbers.

SOLUTION

1. **Familiarize.** There are three statements involving the same three numbers. Let's label these numbers x, y, and z.

2. **Translate.** We can translate directly as follows.

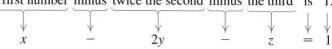

$$x + y + z = 4$$

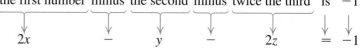

$$x - 2y - z = 1$$

The first number minus twice the second minus twice the third is -1.

$$2x - y - 2z = -1$$

We now have a system of three equations:

$$x + y + z = 4,$$
$$x - 2y - z = 1,$$
$$2x - y - 2z = -1.$$

3. **Carry out.** We need to solve the system of equations. Note that we found the solution, $(2, -1, 3)$, in Example 2 of Section 9.1.

4. **Check.** The first statement of the problem says that the sum of the three numbers is 4. That checks, because $2 + (-1) + 3 = 4$. The second statement says that the first number minus twice the second, minus the third is 1: $2 - 2(-1) - 3 = 1$. That checks. The check of the third statement is left to the student.

5. **State.** The three numbers are 2, -1, and 3.

EXAMPLE 2   Architecture.   In a triangular cross section of a roof, the largest angle is 70° greater than the smallest angle. The largest angle is twice as large as the remaining angle. Find the measure of each angle.

SOLUTION

1. **Familiarize.** The first thing we do is make a drawing, or a sketch.

Since we don't know the size of any angle, we use x, y, and z to represent the three measures, from smallest to largest. Recall that the measures of the angles in any triangle add up to 180°.

2. **Translate.** This geometric fact about triangles gives us one equation:

$$x + y + z = 180.$$

Two of the statements can be translated almost directly.

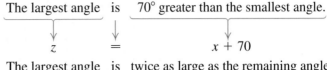

The largest angle   is $70°$ greater than the smallest angle.

$$z = x + 70$$

The largest angle   is   twice as large as the remaining angle.

$$z = 2y$$

We now have a system of three equations:

$$
\begin{array}{lll}
x + y + z = 180, & \quad x + y + z = 180, & \\
x + 70 = z, & \text{or} \quad x \qquad - z = -70, & \textbf{Rewriting in} \\
2y = z; & \quad\quad\quad 2y - z = 0. & \textbf{standard form}
\end{array}
$$

3. **Carry out.** The system was solved in Example 4 of Section 9.1. The solution is (30, 50, 100).

4. **Check.** The sum of the numbers is 180, so that checks. The measure of the largest angle, $100°$, is $70°$ greater than the measure of the smallest angle, $30°$, so that checks. The measure of the largest angle is also twice the measure of the remaining angle, $50°$. Thus we have a check.

5. **State.** The angles in the triangle measure $30°$, $50°$, and $100°$.

EXAMPLE 3   Cholesterol Levels.   Recent studies indicate that a child's intake of cholesterol should be no more than 300 mg per day. By eating 1 egg, 1 cupcake, and 1 slice of pizza, a child consumes 302 mg of cholesterol. A child who eats 2 cupcakes and 3 slices of pizza takes in 65 mg of cholesterol. By eating 2 eggs and 1 cupcake, a child consumes 567 mg of cholesterol. How much cholesterol is in each item?

SOLUTION

1. **Familiarize.** After reading the problem, it becomes clear that an egg contains considerably more cholesterol than the other foods. Let's guess that one egg contains 200 mg of cholesterol and one cupcake contains 50 mg. Because of the second sentence in the problem, it would follow that a slice of pizza contains 52 mg of cholesterol since $200 + 50 + 52 = 302$.

    To see if our guess satisfies the other statements in the problem, we find the amount of cholesterol that 2 cupcakes and 3 slices of pizza would contain: $2 \cdot 50 + 3 \cdot 52 = 256$. Since this does not match the 65 mg listed in the problem, our guess was incorrect. Rather than guess again, we examine how we checked our guess and let g, c, and $s = $ the number of milligrams of cholesterol in an egg, a cupcake, and a slice of pizza, respectively.

2. Translate. Rewording some of the sentences, we can translate as follows:

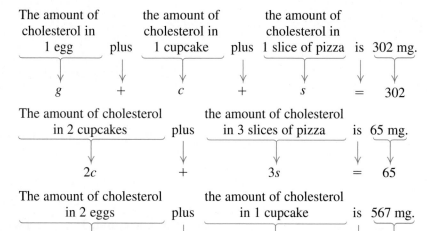

We now have a system of three equations:

$$g + c + s = 302,$$
$$2c + 3s = 65,$$
$$2g + c = 567.$$

3. Carry out. We solve and get $g = 274$, $c = 19$, and $s = 9$.

4. Check. The sum of 274, 19, and 9 is 302 so the total cholesterol in 1 egg, 1 cupcake, and 1 slice of pizza checks. Two cupcakes and three slices of pizza would contain $2 \cdot 19 + 3 \cdot 9 = 65$ mg, while two eggs and one cupcake would contain $2 \cdot 274 + 19 = 567$ mg of cholesterol. The answer checks.

5. State. An egg contains 274 mg of cholesterol, a cupcake contains 19 mg of cholesterol, and a slice of pizza contains 9 mg of cholesterol.

9.2 EXERCISE SET

Solve.

1. The sum of three numbers is 57. The second is 3 more than the first. The third is 6 more than the first. Find the numbers.

2. The sum of three numbers is 5. The first number minus the second plus the third is 1. The first minus the third is 3 more than the second. Find the numbers.

3. The sum of three numbers is 26. Twice the first minus the second is 2 less than the third. The third is the second minus three times the first. Find the numbers.

4. The sum of three numbers is 105. The third is 11 less than ten times the second. Twice the first is 7 more than three times the second. Find the numbers.

5. *Geometry.* In triangle *ABC*, the measure of angle *B* is three times that of angle *A*. The measure of angle *C* is 20° more than that of angle *A*. Find the angle measures.

6. *Geometry.* In triangle *ABC*, the measure of angle *B* is twice the measure of angle *A*. The measure of angle *C* is 80° more than that of angle *A*. Find the angle measures.

7. *Health Insurance.* In 2004, UNICARE® health insurance for adults under the age of 30 cost $121/month for a couple, $107/month for an adult with one child, and $164/month for a couple with one child (*Source*: UNICARE Life and Health Insurance Company® advertisement). On the basis of the information given, find the monthly rate for an individual adult, a spouse, and a child.

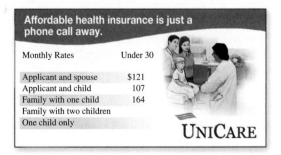

8. *Health Insurance.* In 2004, UNICARE® health insurance cost $160/month for a 35–39-year-old adult and spouse, $145/month for a 35–39-year-old adult with one child, and $245/month for a 35–39-year-old adult with a spouse and child (*Source*: UNICARE Life and Health Insurance Company® advertisement). On the basis of the information given, find the monthly rates for a 39-year-old adult, a 39-year-old's spouse, and a 39-year-old's child.

9. *Nutrition.* Most nutritionists now agree that a healthy adult diet should include 25–35 g of fiber each day (*Sources*: usda.gov; InteliHealth.com). A breakfast of 2 bran muffins, 1 banana, and a 1-cup serving of Wheaties® contains 9 g of fiber; a breakfast of 1 bran muffin, 2 bananas, and a 1-cup serving of Wheaties® contains 10.5 g of fiber; and a breakfast of 2 bran muffins and a 1-cup serving of Wheaties® contains 6 g of fiber. How much fiber is in each of these foods?

10. *Nutrition.* Refer to Exercise 9. A breakfast consisting of 2 pancakes and a 1-cup serving of strawberries contains 4.5 g of fiber, whereas a breakfast of 2 pancakes and a 1-cup serving of Cheerios® contains 4 g of fiber. When a meal consists of 1 pancake, a 1-cup serving of Cheerios®, and a 1-cup serving of strawberries, it contains 7 g of fiber. (*Source*: InteliHealth.com) How much fiber is in each of these foods?

Aha! **11.** *Automobile Pricing.*   The basic model of a 2006 Jeep Grand Cherokee Laredo (2WD) with a power sunroof cost $28,215. When equipped with four-wheel drive (4WD) and a sunroof, the vehicle's price rose to $30,185. The cost of the basic model with 4WD was $29,385. Find the basic price, the cost of 4WD, and the cost of a sunroof.

12. *Lens Production.*   When Sight-Rite's three polishing machines, A, B, and C, are all working, 5700 lenses can be polished in one week. When only A and B are working, 3400 lenses can be polished in one week. When only B and C are working, 4200 lenses can be polished in one week. How many lenses can be polished in a week by each machine?

13. *Welding Rates.*   Elrod, Dot, and Wendy can weld 74 linear feet per hour when working together. Elrod and Dot together can weld 44 linear feet per hour, while Elrod and Wendy can weld 50 linear feet per hour. How many linear feet per hour can each weld alone?

14. *Telemarketing.*   Sven, Tillie, and Isaiah can process 740 telephone orders per day. Sven and Tillie together can process 470 orders, while Tillie and Isaiah together can process 520 orders per day. How many orders can each person process alone?

15. *Coffee Prices.*   Roz worked at a Starbucks® coffee shop where a 12-oz cup of coffee cost $1.40, a 16-oz cup cost $1.60, and a 20-oz cup cost $1.70. During one busy period, Roz served 55 cups of coffee, emptying six 144-oz "brewers" while collecting a total of $85.90. How many cups of each size did Roz fill?

12 oz   16 oz   20 oz
$1.40 $1.60 $1.70

16. *Advertising.*   In a recent year, U.S. companies spent a total of $106.5 billion on newspaper, television, and radio ads. The total amount spent on television and radio ads was $18.7 billion more than the amount spent on newspaper ads alone. The amount spent on newspaper ads was $28.8 billion more than what was spent on radio ads. (*Sources*: NAA (newspapers); McCann–Erickson Inc. (television and radio)) How much was spent on each form of advertising?

17. *Restaurant Management.*   McDonald's® recently sold small soft drinks for $1, medium soft drinks for $1.15, and large soft drinks for $1.30. During a lunch-time rush, Chris sold 40 soft drinks for a total of $45.25. The number of small and large drinks, combined, was 10 fewer than the number of medium drinks. How many drinks of each size were sold?

18. *Investments.*   A business class divided an imaginary investment of $80,000 among three mutual funds. The first fund grew by 10%, the second by 6%, and the third by 15%. Total earnings were $8850. The earnings from the first fund were $750 more than the earnings from the third. How much was invested in each fund?

19. *Nutrition.*   A dietician in a hospital prepares meals under the guidance of a physician. Suppose that for a particular patient a physician prescribes a meal to have 800 calories, 55 g of protein, and 220 mg of vitamin C. The dietician prepares a meal of roast beef, baked potatoes, and broccoli according to the data in the following table.

Serving Size	Calories	Protein (in grams)	Vitamin C (in milligrams)
Roast Beef, 3 oz	300	20	0
Baked Potato, 1	100	5	20
Broccoli, 156 g	50	5	100

How many servings of each food are needed in order to satisfy the doctor's orders?

20. *Nutrition.*   Repeat Exercise 19 but replace the broccoli with asparagus, for which a 180-g serving contains 50 calories, 5 g of protein, and 44 mg of vitamin C. Which meal would you prefer eating?

21. *World Population Growth.* The world population is projected to be 9.1 billion in 2050. At that time, there are expected to be approximately 3 billion more people in Asia than in Africa. The population for the rest of the world will be approximately 0.1 billion more than half the population of Asia. (*Sources*: U.S. Bureau of the Census; *Burlington Free Press* 3/23/04) Find the projected populations of Asia, Africa, and the rest of the world in 2050.

22. *Crying Rate.* The sum of the average number of times a man, a woman, and a one-year-old child cry each month is 56.7. A woman cries 3.9 more times than a man. The average number of times a one-year-old cries per month is 43.3 more than the average number of times combined that a man and a woman cry. What is the average number of times per month that each cries?

23. *Basketball Scoring.* The New York Knicks recently scored a total of 92 points on a combination of 2-point field goals, 3-point field goals, and 1-point foul shots. Altogether, the Knicks made 50 baskets and 19 more 2-pointers than foul shots. How many shots of each kind were made?

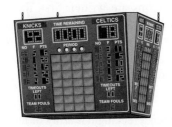

24. *History.* Find the year in which the first U.S. transcontinental railroad was completed. The following are some facts about the number. The sum of the digits in the year is 24. The ones digit is 1 more than the hundreds digit. Both the tens and the ones digits are multiples of 3.

TW 25. Problems like Exercises 15 and 17 could be classified as total-value problems. How do these problems differ from the total-value problems of Section 4.4?

TW 26. Write a problem for a classmate to solve. Design the problem so that it translates to a system of three equations in three variables.

Skill Maintenance

Simplify. [1.8]

27. $5(-3) + 7$

28. $-4(-6) + 9$

29. $-6(8) + (-7)$

30. $7(-9) + (-8)$

31. $-7(2x - 3y + 5z)$

32. $-6(4a + 7b - 9c)$

33. $-4(2a + 5b) + 3a + 20b$

34. $3(2x - 7y) + 5x + 21y$

Synthesis

TW 35. Consider Exercise 23. Suppose there were no foul shots made. Would there still be a solution? Why or why not?

TW 36. Consider Exercise 15. Suppose Roz collected $46. Could the problem still be solved? Why or why not?

37. *Health Insurance.* In 2004, UNICARE® health insurance for a 35–39-year-old and his or her spouse cost $160/month. That rate increased to $203/month if a child were included and $243/month if two children were included. The rate dropped to $145/month for just the applicant and one child. (*Source*: UNICARE Life and Health Insurance Company® advertisement) Find the separate costs for insuring the applicant, the spouse, the first child, and the second child.

38. Find a three-digit positive integer such that the sum of all three digits is 14, the tens digit is 2 more than the ones digit, and if the digits are reversed, the number is unchanged.

39. *Ages.* Tammy's age is the sum of the ages of Carmen and Dennis. Carmen's age is 2 more than the sum of the ages of Dennis and Mark. Dennis's age is four times Mark's age. The sum of all four ages is 42. How old is Tammy?

40. *Ticket Revenue.* A magic show's audience of 100 people consists of adults, students, and children. The ticket prices are $10 for adults, $3 for students, and 50¢ for children. The total amount of money taken in is $100. How many adults, students, and children are in attendance? Does there seem to be some information missing? Do some more careful reasoning.

41. *Sharing Raffle Tickets.* Hal gives Tom as many raffle tickets as Tom first had and Gary as many as Gary first had. In like manner, Tom then gives Hal and Gary as many tickets as each then has. Similarly, Gary gives Hal and Tom as many tickets as each then has. If each finally has 40 tickets, with how many tickets does Tom begin?

42. Find the sum of the angle measures at the tips of the star in this figure.

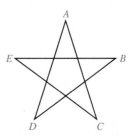

9.3 Elimination Using Matrices

Matrices and Systems ■ Row-Equivalent Operations

In solving systems of equations, we perform computations with the constants. The variables play no important role until the end. Thus we can simplify writing a system by omitting the variables. For example, the system

$$3x + 4y = 5, \qquad \text{simplifies to} \qquad \begin{array}{ccc} 3 & 4 & 5 \\ 1 & -2 & 1 \end{array}$$
$$x - 2y = 1$$

if we do not write the variables, the operation of addition, and the equals signs. Note that the coefficients of the *x*-terms are written first, the coefficients of the *y*-terms are written next, and the constant terms are written last. Equations must be written in standard form $Ax + By = C$ before they are written without variables.

Matrices and Systems

In the example above, we have written a rectangular array of numbers. Such an array is called a **matrix** (plural, **matrices**). We ordinarily write brackets around matrices. The following are matrices:

$$\begin{bmatrix} -3 & 1 \\ 0 & 5 \end{bmatrix}, \quad \begin{bmatrix} 2 & 0 & -1 & 3 \\ -5 & 2 & 7 & -1 \\ 4 & 5 & 3 & 0 \end{bmatrix}, \quad \begin{bmatrix} 2 & 3 \\ 7 & 15 \\ -2 & 23 \\ 4 & 1 \end{bmatrix}.$$

The individual numbers are called *elements* or *entries*.

The **rows** of a matrix are horizontal, and the **columns** are vertical.

$$A = \begin{bmatrix} 5 & -2 & -2 \\ 1 & 0 & 1 \\ 4 & -3 & 2 \end{bmatrix} \begin{matrix} \leftarrow \text{row 1} \\ \leftarrow \text{row 2} \\ \leftarrow \text{row 3} \end{matrix}$$

$$\qquad\quad \uparrow \qquad \uparrow \qquad \uparrow$$
$$\text{column 1} \;\; \text{column 2} \;\; \text{column 3}$$

Let's see how matrices can be used to solve a system.

▶ **EXAMPLE 1**   Solve the system

$$5x - 4y = -1,$$
$$-2x + 3y = 2.$$

SOLUTION   We write a matrix using only coefficients and constants, listing x-coefficients in the first column and y-coefficients in the second. Note that in each matrix a dashed line separates the coefficients from the constants:

$$\begin{bmatrix} 5 & -4 & \vdots & -1 \\ -2 & 3 & \vdots & 2 \end{bmatrix}.$$

Our goal is to transform

$$\begin{bmatrix} 5 & -4 & \vdots & -1 \\ -2 & 3 & \vdots & 2 \end{bmatrix} \quad \text{into the form} \quad \begin{bmatrix} a & b & \vdots & c \\ 0 & d & \vdots & e \end{bmatrix}.$$

The variables x and y can then be reinserted to form equations from which we can complete the solution.

We do calculations that are similar to those that we would do if we wrote the entire equations. The first step is to multiply and/or interchange the rows so that each number in the first column below the first number is a multiple of that number. Here that means multiplying Row 2 by 5. This corresponds to multiplying both sides of the second equation by 5.

$$\begin{bmatrix} 5 & -4 & \vdots & -1 \\ -10 & 15 & \vdots & 10 \end{bmatrix}$$ **New Row 2 = 5(Row 2 from above)**

Next, we multiply the first row by 2, add this to Row 2, and write that result as the "new" Row 2. This corresponds to multiplying the first equation by 2 and adding the result to the second equation in order to eliminate a variable. Write out these computations as necessary—we perform them mentally.

$$\begin{bmatrix} 5 & -4 & \vdots & -1 \\ 0 & 7 & \vdots & 8 \end{bmatrix}$$

$2(5 \quad -4 \;\vdots\; -1) = (10 \quad -8 \;\vdots\; -2)$ **and**
$(10 \quad -8 \;\vdots\; -2) + (-10 \quad 15 \;\vdots\; 10) = (0 \quad 7 \;\vdots\; 8)$
New Row 2 = 2(Row 1) + (Row 2)

If we now reinsert the variables, we have

$$5x - 4y = -1, \qquad (1)$$
$$7y = 8. \qquad (2)$$

We can now proceed as before, solving equation (2) for y:

$$7y = 8 \qquad (2)$$
$$y = \tfrac{8}{7}.$$

As an aid for understanding, we list the corresponding system in the margin.

$$5x - 4y = -1,$$
$$-2x + 3y = 2$$

$$5x - 4y = -1,$$
$$-10x + 15y = 10$$

$$5x - 4y = -1,$$
$$7y = 8$$

Next, we substitute $\frac{8}{7}$ for y in equation (1):

$$5x - 4y = -1 \qquad (1)$$
$$5x - 4 \cdot \tfrac{8}{7} = -1 \qquad \text{Substituting } \tfrac{8}{7} \text{ for } y \text{ in equation (1)}$$
$$x = \tfrac{5}{7}. \qquad \text{Solving for } x$$

The solution is $\left(\frac{5}{7}, \frac{8}{7}\right)$. The check is left to the student.

All the systems of equations shown in the margin by Example 1 are **equivalent systems of equations;** that is, they all have the same solution.

EXAMPLE 2   Solve the system

$$2x - y + 4z = -3,$$
$$x \qquad - 4z = 5,$$
$$6x - y + 2z = 10.$$

SOLUTION   We first write a matrix, using only the constants. Where there are missing terms, we must write 0's:

$$2x - y + 4z = -3,$$
$$x \qquad - 4z = 5,$$
$$6x - y + 2z = 10$$

$$\left[\begin{array}{ccc|c} 2 & -1 & 4 & -3 \\ 1 & 0 & -4 & 5 \\ 6 & -1 & 2 & 10 \end{array}\right].$$

Note that the x-coefficients are in column 1, the y-coefficients in column 2, the z-coefficients in column 3, and the constant terms in the last column.

Our goal is to transform the matrix to one of the form

$$ax + by + cz = d,$$
$$ey + fz = g,$$
$$hz = i$$

$$\left[\begin{array}{ccc|c} a & b & c & d \\ 0 & e & f & g \\ 0 & 0 & h & i \end{array}\right].$$

A matrix of this form can be rewritten as a system of equations that is equivalent to the original system, and from which a solution can be easily found.

The first step is to multiply and/or interchange the rows so that each number in the first column is a multiple of the first number in the first row. In this case, we do so by interchanging Rows 1 and 2:

$$x \qquad - 4z = 5,$$
$$2x - y + 4z = -3,$$
$$6x - y + 2z = 10$$

$$\left[\begin{array}{ccc|c} 1 & 0 & -4 & 5 \\ 2 & -1 & 4 & -3 \\ 6 & -1 & 2 & 10 \end{array}\right]$$

This corresponds to interchanging the first two equations.

Next, we multiply the first row by -2, add it to the second row, and replace Row 2 with the result:

$$x \qquad - 4z = 5,$$
$$-y + 12z = -13,$$
$$6x - y + 2z = 10$$

$$\left[\begin{array}{ccc|c} 1 & 0 & -4 & 5 \\ 0 & -1 & 12 & -13 \\ 6 & -1 & 2 & 10 \end{array}\right].$$

$-2(1 \;\; 0 \;\; -4 \mid 5) = (-2 \;\; 0 \;\; 8 \mid -10)$ and $(-2 \;\; 0 \;\; 8 \mid -10) + (2 \;\; -1 \;\; 4 \mid -3) = (0 \;\; -1 \;\; 12 \mid -13)$

Now we multiply the first row by -6, add it to the third row, and replace Row 3 with the result:

$$x \qquad - 4z = 5,$$
$$-y + 12z = -13,$$
$$-y + 26z = -20$$

$$\left[\begin{array}{ccc|c} 1 & 0 & -4 & 5 \\ 0 & -1 & 12 & -13 \\ 0 & -1 & 26 & -20 \end{array}\right].$$

$-6(1 \;\; 0 \;\; -4 \mid 5) = (-6 \;\; 0 \;\; 24 \mid -30)$ and $(-6 \;\; 0 \;\; 24 \mid -30) + (6 \;\; -1 \;\; 2 \mid 10) = (0 \;\; -1 \;\; 26 \mid -20)$

Next, we multiply Row 2 by -1, add it to the third row, and replace Row 3 with the result:

$$x \quad - 4z = 5,$$
$$-y + 12z = -13,$$
$$14z = -7$$

$$\begin{bmatrix} 1 & 0 & -4 & \vdots & 5 \\ 0 & -1 & 12 & \vdots & -13 \\ 0 & 0 & 14 & \vdots & -7 \end{bmatrix}. \qquad \begin{array}{l} -1(0 \quad -1 \quad 12 \; \vdots \; -13) = (0 \quad 1 \quad -12 \; \vdots \; 13) \\ \text{and } (0 \quad 1 \quad -12 \; \vdots \; 13) + (0 \quad -1 \quad 26 \; \vdots \; -20) = \\ (0 \quad 0 \quad 14 \; \vdots \; -7) \end{array}$$

Reinserting the variables gives us

$$x \quad - 4z = 5,$$
$$-y + 12z = -13,$$
$$14z = -7.$$

We now solve this last equation for z and get $z = -\frac{1}{2}$. Next, we substitute $-\frac{1}{2}$ for z in the preceding equation and solve for y: $-y + 12\left(-\frac{1}{2}\right) = -13$, so $y = 7$. Since there is no y-term in the first equation of this last system, we need only substitute $-\frac{1}{2}$ for z to solve for x: $x - 4\left(-\frac{1}{2}\right) = 5$, so $x = 3$. The solution is $\left(3, 7, -\frac{1}{2}\right)$. The check is left to the student.

The operations used in the preceding example correspond to those used to produce equivalent systems of equations. We call the matrices **row-equivalent** and the operations that produce them **row-equivalent operations.**

Row-Equivalent Operations

> **Row-Equivalent Operations**
>
> Each of the following row-equivalent operations produces a row-equivalent matrix:
>
> **a)** Interchanging any two rows.
> **b)** Multiplying all elements of a row by a nonzero constant.
> **c)** Replacing a row with the sum of that row and a multiple of another row.

The best overall method for solving systems of equations is by row-equivalent matrices; even computers are programmed to use them. Matrices are part of a branch of mathematics known as linear algebra. They are also studied in many courses in finite mathematics.

When we solved the systems in Examples 1 and 2, we used row-equivalent operations to write an equivalent system of equations that we could solve without using elimination. We can continue to use row-equivalent operations to write a row-equivalent matrix in **reduced row-echelon form,** from which the solution of the system can often be read directly.

Reduced Row-Echelon Form

A matrix is in reduced row-echelon form if:

1. All rows consisting entirely of zeros are at the bottom of the matrix.
2. The first nonzero number in any nonzero row is 1, called a leading 1.
3. The leading 1 in any row is farther to the left than the leading 1 in any lower row.
4. Each column that contains a leading 1 has zeros everywhere else.

EXAMPLE 3 Solve the system

$$5x - 4y = -1,$$
$$-2x + 3y = 2$$

by writing a reduced row-echelon matrix.

SOLUTION We began this solution in Example 1, and used row-equivalent operations to write the row-equivalent matrix:

$$\begin{bmatrix} 5 & -4 & | & -1 \\ 0 & 7 & | & 8 \end{bmatrix}.$$

$$5x - 4y = -1,$$
$$7y = 8$$

Before we can write reduced row-echelon form, the first nonzero number in any nonzero row must be 1. We multiply Row 1 by $\frac{1}{5}$ and Row 2 by $\frac{1}{7}$:

$$\begin{bmatrix} 1 & -\frac{4}{5} & | & -\frac{1}{5} \\ 0 & 1 & | & \frac{8}{7} \end{bmatrix}.$$

New Row 1 = $\frac{1}{5}$(Row 1 from above)
New Row 2 = $\frac{1}{7}$(Row 2 from above)

$$x - \frac{4}{5}y = -\frac{1}{5},$$
$$y = \frac{8}{7}$$

There are no zero rows, there is a leading 1 in each row, and the leading 1 in Row 1 is farther to the left than the leading 1 in Row 2. Thus conditions (1)–(3) for reduced row-echelon form have been met.

To satisfy condition (4), we must obtain a 0 in Row 1, Column 2, above the leading 1 in Row 2. We multiply Row 2 by $\frac{4}{5}$ and add it to Row 1:

$$x = \frac{5}{7},$$
$$y = \frac{8}{7}$$

$$\begin{bmatrix} 1 & 0 & | & \frac{5}{7} \\ 0 & 1 & | & \frac{8}{7} \end{bmatrix}.$$

$\frac{4}{5}\begin{pmatrix} 0 & 1 & | & \frac{8}{7} \end{pmatrix} = \begin{pmatrix} 0 & \frac{4}{5} & | & \frac{32}{35} \end{pmatrix}$ and
$\begin{pmatrix} 0 & \frac{4}{5} & | & \frac{32}{35} \end{pmatrix} + \begin{pmatrix} 1 & -\frac{4}{5} & | & -\frac{1}{5} \end{pmatrix} = \begin{pmatrix} 1 & 0 & | & \frac{5}{7} \end{pmatrix}$
New Row 1 = $\frac{4}{5}$(Row 2) + Row 1

This matrix is in reduced row-echelon form. If we reinsert the variables, we have

$$x = \frac{5}{7},$$
$$y = \frac{8}{7}.$$

The solution, $\left(\frac{5}{7}, \frac{8}{7}\right)$, can be read directly from the last column of the matrix.

Finding reduced row-echelon form often involves extensive calculations with fractions or decimals. Thus it is common to use a computer or a graphing calculator to store and manipulate matrices.

Matrix Operations

On many graphing calculators, matrix operations are accessed by pressing (MATRIX). (On some calculators, MATRIX is the 2nd feature associated with the x^{-1} key.) The MATRIX menu has three submenus: NAMES, MATH, and EDIT.

The EDIT menu allows matrices to be entered. For example, to enter

$$A = \begin{bmatrix} 1 & 2 & 5 \\ -3 & 0 & 4 \end{bmatrix},$$

choose option [A] from the MATRIX EDIT menu. Enter the **dimensions,** or number of rows and columns, of the matrix first, listing the number of rows before the number of columns. Thus the dimensions of A are 2×3, read "2 by 3." Then enter each element of A by pressing the number and ENTER. The notation $2, 3 = 4$ indicates that the 3rd entry of the 2nd row is 4. Matrices are generally entered row by row rather than column by column.

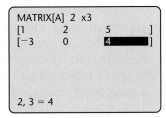

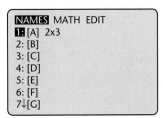

After entering the matrix, exit the matrix editor by pressing **2ND** (QUIT). To access a matrix once it has been entered, use the MATRIX NAMES submenu as shown above on the right. The brackets around A indicate that it is a matrix.

The submenu MATRIX MATH lists operations that can be performed on matrices.

EXAMPLE 4   Solve the following system using a graphing calculator:

$$2x + 5y - 8z = 7,$$
$$3x + 4y - 3z = 8,$$
$$5y - 2x = 9.$$

SOLUTION   Before writing a matrix to represent this system, we rewrite the third equation in the form $ax + by + cz = d$:

$$2x + 5y - 8z = 7,$$
$$3x + 4y - 3z = 8,$$
$$-2x + 5y + 0z = 9.$$

The matrix that represents this system is thus

$$\begin{bmatrix} 2 & 5 & -8 & \vdots & 7 \\ 3 & 4 & -3 & \vdots & 8 \\ -2 & 5 & 0 & \vdots & 9 \end{bmatrix}.$$

We enter the matrix as A using the MATRIX EDIT menu, noting that its dimensions are 3×4. Once we have entered each element of the matrix, we return to the home screen to perform operations. The contents of matrix A can be displayed on the screen, using the MATRIX NAMES menu.

```
[A]
            [[2  5  -8  7]
             [3  4  -3  8]
             [-2  5  0  9]]
```

Each of the row-equivalent operations can be performed using the MATRIX MATH menu. On many calculators, it is also possible to go directly to the reduced row-echelon form. To find this form, from the home screen, we choose the RREF option from the MATRIX MATH menu and then choose [A] from the MATRIX NAMES menu. Finally, press **MATH** **1** **ENTER** to write the entries in the reduced row-echelon form using fraction notation.

```
rref ( [A] ) ▶ Frac
   [[1 0 0 1/2]
    [0 1 0 2]
    [0 0 1 1/2]]
```

The reduced row-echelon form shown on the screen above is equivalent to the system of equations

$$x = \tfrac{1}{2},$$
$$y = 2,$$
$$z = \tfrac{1}{2}.$$

The solution of the system is thus $\left(\tfrac{1}{2}, 2, \tfrac{1}{2}\right)$, which can be read directly from the last column of the reduced row-echelon matrix.

Recall that some systems of equations are inconsistent, and some sets of equations are dependent. When these cases occur, the reduced row-echelon form of the corresponding matrix will have a row or a column of zeros. For example, the matrix

$$\begin{bmatrix} 1 & 0 & | & 4 \\ 0 & 0 & | & 0 \end{bmatrix}$$

is in reduced row-echelon form. The second row translates to the equation $0 = 0$, and indicates that the equations in the system are dependent. The second row of the matrix

$$\begin{bmatrix} 1 & 0 & | & 4 \\ 0 & 0 & | & 6 \end{bmatrix}$$

translates to the equation $0 = 6$. This system is inconsistent.

9.3 EXERCISE SET

FOR EXTRA HELP

MathXL

MathXL   MyMathLab   InterAct Math   AW Math Tutor Center   Video Lectures on CD: Disc 5   Student's Solutions Manual

🖐 *Concept Reinforcement*   *Complete each of the following statements.*

1. The rows of a matrix are _____ and the _____ are vertical.

2. Multiplying the numbers in a row of a matrix by a constant corresponds to multiplying both sides of a(n) _____ by a constant.

3. Each number in a matrix is called a(n) _____ or element.

4. The plural of the word matrix is _____.

5. To solve a system using matrices, we can replace any row by the sum of that row and a(n) _____ of another row.

6. In the final step of solving a system of equations, the leftmost column has zeros in all rows except the _____ one.

Solve using matrices.

7. $9x - 2y = 5,$
$3x - 3y = 11$

8. $4x + y = 7,$
$5x - 3y = 13$

9. $x + 4y = 8,$
$3x + 5y = 3$

10. $x + 4y = 5,$
$-3x + 2y = 13$

11. $6x - 2y = 4,$
$7x + y = 13$

12. $3x + 4y = 7,$
$-5x + 2y = 10$

13. $3x + 2y + 2z = 3,$
$x + 2y - z = 5,$
$2x - 4y + z = 0$

14. $4x - y - 3z = 19,$
$8x + y - z = 11,$
$2x + y + 2z = -7$

15. $p - 2q - 3r = 3,$
$2p - q - 2r = 4,$
$4p + 5q + 6r = 4$

16. $x + 2y - 3z = 9,$
$2x - y + 2z = -8,$
$3x - y - 4z = 3$

17. $3p + 2r = 11,$
$q - 7r = 4,$
$p - 6q = 1$

18. $4a + 9b = 8,$
$8a + 6c = -1,$
$6b + 6c = -1$

19. $2x + 2y - 2z - 2w = -10,$
$w + y + z + x = -5,$
$x - y + 4z + 3w = -2,$
$w - 2y + 2z + 3x = -6$

20. $-w - 3y + z + 2x = -8,$
$x + y - z - w = -4,$
$w + y + z + x = 22,$
$x - y - z - w = -14$

Solve using matrices.

21. *Coin Value.*   A collection of 42 coins consists of dimes and nickels. The total value is $3.00. How many dimes and how many nickels are there?

22. *Coin Value.*   A collection of 43 coins consists of dimes and quarters. The total value is $7.60. How many dimes and how many quarters are there?

23. *Mixed Granola.*   Grace sells two kinds of granola. One is worth $4.05 per pound and the other is worth $2.70 per pound. She wants to blend the two granolas to get a 15-lb mixture worth $3.15 per pound. How much of each kind of granola should be used?

24. *Trail Mix.*   Phil mixes nuts worth $1.60 per pound with oats worth $1.40 per pound to get 20 lb of trail mix worth $1.54 per pound. How many pounds of nuts and how many pounds of oats should be used?

25. *Investments.*   Elena receives $212 per year in simple interest from three investments totaling $2500. Part is invested at 7%, part at 8%, and part at 9%. There is $1100 more invested at 9% than at 8%. Find the amount invested at each rate.

26. *Investments.*   Miguel receives $306 per year in simple interest from three investments totaling $3200. Part is invested at 8%, part at 9%, and part at 10%. There is $1900 more invested at 10% than at 9%. Find the amount invested at each rate.

TW 27. Explain how you can recognize dependent equations when solving with matrices.

TW 28. Explain how you can recognize an inconsistent system when solving with matrices.

Skill Maintenance

Simplify. [1.8]

29. $5(-3) - (-7)4$

30. $8(-5) - (-2)9$

31. $-2(5 \cdot 3 - 4 \cdot 6) - 3(2 \cdot 7 - 15)$
$$+ 4(3 \cdot 8 - 5 \cdot 4)$$

32. $6(2 \cdot 7 - 3(-4)) - 4(3(-8) - 10)$
$$+ 5(4 \cdot 3 - (-2)7)$$

Synthesis

TW 33. If the matrices

$$\begin{bmatrix} a_1 & b_1 & c_1 \\ d_1 & e_1 & f_1 \end{bmatrix} \quad \text{and} \quad \begin{bmatrix} a_2 & b_2 & c_2 \\ d_2 & e_2 & f_2 \end{bmatrix}$$

share the same solution, does it follow that the corresponding entries are all equal to each other ($a_1 = a_2$, $b_1 = b_2$, etc.)? Why or why not?

TW 34. Explain how the row-equivalent operations make use of the addition, multiplication, and distributive properties.

35. The sum of the digits in a four-digit number is 10. Twice the sum of the thousands digit and the tens digit is 1 less than the sum of the other two digits. The tens digit is twice the thousands digit. The ones digit equals the sum of the thousands digit and the hundreds digit. Find the four-digit number.

36. Solve for x and y:
$$ax + by = c,$$
$$dx + ey = f.$$

9.4 Determinants and Cramer's Rule

Determinants of 2×2 Matrices ■ Cramer's Rule: 2×2
Systems ■ Cramer's Rule: 3×3 Systems

Determinants of 2×2 Matrices

When a matrix has m rows and n columns, it is called an "m by n" matrix. Thus its *dimensions* are denoted by $m \times n$. If a matrix has the same number of rows and columns, it is called a **square matrix**. Associated with every square matrix is a number called its **determinant,** defined as follows for 2×2 matrices.

2×2 Determinants   The determinant of a two-by-two matrix
$\begin{bmatrix} a & c \\ b & d \end{bmatrix}$ is denoted $\begin{vmatrix} a & c \\ b & d \end{vmatrix}$ and is defined as follows:

$$\begin{vmatrix} a & c \\ b & d \end{vmatrix} = ad - bc.$$

EXAMPLE 1   Evaluate: $\begin{vmatrix} 2 & -5 \\ 6 & 7 \end{vmatrix}$.

SOLUTION   We multiply and subtract as follows:

$$\begin{vmatrix} 2 & -5 \\ 6 & 7 \end{vmatrix} = 2 \cdot 7 - 6 \cdot (-5) = 14 + 30 = 44.$$

Cramer's Rule: 2×2 Systems

One of the many uses for determinants is in solving systems of linear equations in which the number of variables is the same as the number of equations and the constants are not all 0. Let's consider a system of two equations:

$$a_1 x + b_1 y = c_1,$$
$$a_2 x + b_2 y = c_2.$$

If we use the elimination method, a series of steps can show that

$$x = \frac{c_1 b_2 - c_2 b_1}{a_1 b_2 - a_2 b_1} \quad \text{and} \quad y = \frac{a_1 c_2 - a_2 c_1}{a_1 b_2 - a_2 b_1}.$$

These fractions can be rewritten using determinants.

Cramer's Rule: 2 × 2 Systems   The solution of the system

$$a_1x + b_1y = c_1,$$
$$a_2x + b_2y = c_2,$$

if it is unique, is given by

$$x = \frac{\begin{vmatrix} c_1 & b_1 \\ c_2 & b_2 \end{vmatrix}}{\begin{vmatrix} a_1 & b_1 \\ a_2 & b_2 \end{vmatrix}}, \qquad y = \frac{\begin{vmatrix} a_1 & c_1 \\ a_2 & c_2 \end{vmatrix}}{\begin{vmatrix} a_1 & b_1 \\ a_2 & b_2 \end{vmatrix}}.$$

These formulas apply only if the denominator is not 0. If the denominator *is* 0, then one of two things happens:

1. If the denominator is 0 and the numerators are also 0, then the equations in the system are dependent.
2. If the denominator is 0 and at least one numerator is not 0, then the system is inconsistent.

To use Cramer's rule, we find the determinants and compute x and y as shown above. Note that the denominators are identical and the coefficients of x and y appear in the same position as in the original equations. In the numerator of x, the constants c_1 and c_2 replace a_1 and a_2. In the numerator of y, the constants c_1 and c_2 replace b_1 and b_2.

EXAMPLE 2   Solve using Cramer's rule:

$$2x + 5y = 7,$$
$$5x - 2y = -3.$$

SOLUTION   We have

$$x = \frac{\begin{vmatrix} 7 & 5 \\ -3 & -2 \end{vmatrix}}{\begin{vmatrix} 2 & 5 \\ 5 & -2 \end{vmatrix}} \qquad \text{Using Cramer's rule}$$

$$= \frac{7(-2) - (-3)5}{2(-2) - 5 \cdot 5} = -\frac{1}{29}$$

and

$$y = \frac{\begin{vmatrix} 2 & 7 \\ 5 & -3 \end{vmatrix}}{\begin{vmatrix} 2 & 5 \\ 5 & -2 \end{vmatrix}} \qquad \text{Using Cramer's rule}$$

$$= \frac{2(-3) - 5 \cdot 7}{-29} = \frac{41}{29}. \qquad \text{The denominator is the same as in the expression for } x.$$

The solution is $\left(-\frac{1}{29}, \frac{41}{29}\right)$. The check is left to the student.

Cramer's Rule: 3 × 3 Systems

Cramer's rule can be extended for systems of three linear equations. However, before doing so, we must define what a 3 × 3 determinant is.

3 × 3 Determinants   The determinant of a three-by-three matrix is defined as follows:

Subtract.   Add.

$$\begin{vmatrix} a_1 & b_1 & c_1 \\ a_2 & b_2 & c_2 \\ a_3 & b_3 & c_3 \end{vmatrix} = a_1 \begin{vmatrix} b_2 & c_2 \\ b_3 & c_3 \end{vmatrix} - a_2 \begin{vmatrix} b_1 & c_1 \\ b_3 & c_3 \end{vmatrix} + a_3 \begin{vmatrix} b_1 & c_1 \\ b_2 & c_2 \end{vmatrix}.$$

Note that the a's come from the first column. Note too that the 2 × 2 determinants above can be obtained by crossing out the row and the column in which the a occurs.

For a_1:

$$\begin{vmatrix} a_1 & b_1 & c_1 \\ a_2 & b_2 & c_2 \\ a_3 & b_3 & c_3 \end{vmatrix}$$

For a_2:

$$\begin{vmatrix} a_1 & b_1 & c_1 \\ a_2 & b_2 & c_2 \\ a_3 & b_3 & c_3 \end{vmatrix}$$

For a_3:

$$\begin{vmatrix} a_1 & b_1 & c_1 \\ a_2 & b_2 & c_2 \\ a_3 & b_3 & c_3 \end{vmatrix}$$

EXAMPLE 3   Evaluate:

$$\begin{vmatrix} -1 & 0 & 1 \\ -5 & 1 & -1 \\ 4 & 8 & 1 \end{vmatrix}.$$

SOLUTION   We have

Subtract.   Add.

$$\begin{vmatrix} -1 & 0 & 1 \\ -5 & 1 & -1 \\ 4 & 8 & 1 \end{vmatrix} = -1 \begin{vmatrix} 1 & -1 \\ 8 & 1 \end{vmatrix} - (-5) \begin{vmatrix} 0 & 1 \\ 8 & 1 \end{vmatrix} + 4 \begin{vmatrix} 0 & 1 \\ 1 & -1 \end{vmatrix}$$

$$= -1(1 + 8) + 5(0 - 8) + 4(0 - 1) \qquad \text{Evaluating the three determinants}$$

$$= -9 - 40 - 4 = -53.$$

Cramer's Rule: 3 × 3 Systems   The solution of the system

$$a_1x + b_1y + c_1z = d_1,$$
$$a_2x + b_2y + c_2z = d_2,$$
$$a_3x + b_3y + c_3z = d_3$$

can be found using the following determinants:

$$D = \begin{vmatrix} a_1 & b_1 & c_1 \\ a_2 & b_2 & c_2 \\ a_3 & b_3 & c_3 \end{vmatrix}, \qquad D_x = \begin{vmatrix} d_1 & b_1 & c_1 \\ d_2 & b_2 & c_2 \\ d_3 & b_3 & c_3 \end{vmatrix},$$

$$D_y = \begin{vmatrix} a_1 & d_1 & c_1 \\ a_2 & d_2 & c_2 \\ a_3 & d_3 & c_3 \end{vmatrix}, \qquad D_z = \begin{vmatrix} a_1 & b_1 & d_1 \\ a_2 & b_2 & d_2 \\ a_3 & b_3 & d_3 \end{vmatrix}.$$

D contains only coefficients.
In D_x, the *d*'s replace the *a*'s.
In D_y, the *d*'s replace the *b*'s.
In D_z, the *d*'s replace the *c*'s.

If a unique solution exists, it is given by

$$x = \frac{D_x}{D}, \qquad y = \frac{D_y}{D}, \qquad z = \frac{D_z}{D}.$$

EXAMPLE 4   Solve using Cramer's rule:

$$x - 3y + 7z = 13,$$
$$x + y + z = 1,$$
$$x - 2y + 3z = 4.$$

SOLUTION   We compute D, D_x, D_y, and D_z:

$$D = \begin{vmatrix} 1 & -3 & 7 \\ 1 & 1 & 1 \\ 1 & -2 & 3 \end{vmatrix} = -10; \qquad D_x = \begin{vmatrix} 13 & -3 & 7 \\ 1 & 1 & 1 \\ 4 & -2 & 3 \end{vmatrix} = 20;$$

$$D_y = \begin{vmatrix} 1 & 13 & 7 \\ 1 & 1 & 1 \\ 1 & 4 & 3 \end{vmatrix} = -6; \qquad D_z = \begin{vmatrix} 1 & -3 & 13 \\ 1 & 1 & 1 \\ 1 & -2 & 4 \end{vmatrix} = -24.$$

Then

$$x = \frac{D_x}{D} = \frac{20}{-10} = -2;$$

$$y = \frac{D_y}{D} = \frac{-6}{-10} = \frac{3}{5};$$

$$z = \frac{D_z}{D} = \frac{-24}{-10} = \frac{12}{5}.$$

The solution is $\left(-2, \frac{3}{5}, \frac{12}{5}\right)$. The check is left to the student.

In Example 4, we need not have evaluated D_z. Once *x* and *y* were found, we could have substituted them into one of the equations to find *z*.

To use Cramer's rule, we divide by D, provided $D \neq 0$. If $D = 0$ and at least one of the other determinants is not 0, then the system is inconsistent. If *all* the determinants are 0, then the equations in the system are dependent.

Determinants

Determinants can be evaluated using the DET(option of the MATRIX MATH menu. After entering the matrix, we go to the home screen and select the determinant operation. Then we enter the name of the matrix using the MATRIX NAMES menu. The graphing calculator will return the value of the determinant of the matrix. For example, if

$$\mathbf{A} = \begin{bmatrix} 1 & 6 & -1 \\ -3 & -5 & 3 \\ 0 & 4 & 2 \end{bmatrix},$$

we have

```
det([A])
            26
```
.

9.4 EXERCISE SET

FOR EXTRA HELP

MathXL · MyMathLab · InterAct Math · Tutor Center · Video Lectures on CD: Disc 5 · Student's Solutions Manual

Concept Reinforcement   *Classify each of the following as either true or false.*

1. A square matrix has the same number of rows and columns.

2. A 3×4 matrix has 3 rows and 4 columns.

3. Cramer's rule exists only for 2×2 systems.

4. Whenever Cramer's rule yields a denominator that is 0, the system has no solution.

5. Whenever Cramer's rule yields a numerator that is 0, the equations are dependent.

6. Cramer's rule allows us to solve some systems that could not be solved any other way.

Evaluate.

7. $\begin{vmatrix} 5 & 1 \\ 2 & 4 \end{vmatrix}$

8. $\begin{vmatrix} 3 & 2 \\ 2 & -3 \end{vmatrix}$

9. $\begin{vmatrix} 6 & -9 \\ 2 & 3 \end{vmatrix}$

10. $\begin{vmatrix} 3 & 2 \\ -7 & 5 \end{vmatrix}$

11. $\begin{vmatrix} 1 & 4 & 0 \\ 0 & -1 & 2 \\ 3 & -2 & 1 \end{vmatrix}$

12. $\begin{vmatrix} 3 & 0 & -2 \\ 5 & 1 & 2 \\ 2 & 0 & -1 \end{vmatrix}$

13. $\begin{vmatrix} -1 & -2 & -3 \\ 3 & 4 & 2 \\ 0 & 1 & 2 \end{vmatrix}$

14. $\begin{vmatrix} 1 & 2 & 2 \\ 2 & 1 & 0 \\ 3 & 3 & 1 \end{vmatrix}$

15. $\begin{vmatrix} -4 & -2 & 3 \\ -3 & 1 & 2 \\ 3 & 4 & -2 \end{vmatrix}$

16. $\begin{vmatrix} 2 & -1 & 1 \\ 1 & 2 & -1 \\ 3 & 4 & -3 \end{vmatrix}$

Solve using Cramer's rule.

17. $5x + 8y = 1,$
$3x + 7y = 5$

18. $3x - 4y = 6,$
$5x + 9y = 10$

19. $5x - 4y = -3,$
$7x + 2y = 6$

20. $-2x + 4y = 3,$
$3x - 7y = 1$

21. $3x - y + 2z = 1,$
$x - y + 2z = 3,$
$-2x + 3y + z = 1$

22. $3x + 2y - z = 4,$
$3x - 2y + z = 5,$
$4x - 5y - z = -1$

23. $2x - 3y + 5z = 27,$
$x + 2y - z = -4,$
$5x - y + 4z = 27$

24. $x - y + 2z = -3,$
$x + 2y + 3z = 4,$
$2x + y + z = -3$

25. $r - 2s + 3t = 6,$
$2r - s - t = -3,$
$r + s + t = 6$

26. $a \qquad - 3c = 6,$
$b + 2c = 2,$
$7a - 3b - 5c = 14$

TW 27. What is it about Cramer's rule that makes it useful?

TW 28. Which version of Cramer's rule do you find more useful: the version for 2×2 systems or the version for 3×3 systems? Why?

Focused Review

29. Solve by graphing by hand. [4.1]
$x + y = -2,$
$y = 2 - 5x$

30. Solve using a graphing calculator. [4.1]
$8y = 14 - 3x,$
$5y = 8 + x$

31. Solve using substitution. [4.2]
$x + y = 3,$
$y = 2x - 1$

32. Solve using elimination. [4.3]
$2x + 3y = 10,$
$5x - 3y = 4$

33. Solve using matrices. [9.3]
$x - 3y = 7,$
$2x + y = 8$

34. Solve using Cramer's rule. [9.4]
$2x + y = 1,$
$3x - 2y = 10$

Synthesis

TW 35. Cramer's rule states that if $a_1x + b_1y = c_1$ and $a_2x + b_2y = c_2$ are dependent, then
$$\begin{vmatrix} a_1 & b_1 \\ a_2 & b_2 \end{vmatrix} = 0.$$
Explain why this will always happen.

TW 36. Under what conditions can a 3×3 system of linear equations be consistent but unable to be solved using Cramer's rule?

Solve.

37. $\begin{vmatrix} y & -2 \\ 4 & 3 \end{vmatrix} = 44$

38. $\begin{vmatrix} 2 & x & -1 \\ -1 & 3 & 2 \\ -2 & 1 & 1 \end{vmatrix} = -12$

39. $\begin{vmatrix} m+1 & -2 \\ m-2 & 1 \end{vmatrix} = 27$

40. Show that an equation of the line through (x_1, y_1) and (x_2, y_2) can be written
$$\begin{vmatrix} x & y & 1 \\ x_1 & y_1 & 1 \\ x_2 & y_2 & 1 \end{vmatrix} = 0.$$

9.5 Business and Economics Applications

Break-Even Analysis ◼ Supply and Demand

CAUTION! Do not confuse "cost" with "price." When we discuss the *cost* of an item, we are referring to what it costs to produce the item. The *price* of an item is what a consumer pays to purchase the item and is used when calculating revenue.

Break-Even Analysis

When a company manufactures x units of a product, it spends money. This is **total cost** and can be thought of as a function C, where $C(x)$ is the total cost of producing x units. When the company sells x units of the product, it takes in money. This is **total revenue** and can be thought of as a function R, where $R(x)$ is the total revenue from the sale of x units. **Total profit** is the money taken in less the money spent, or total revenue minus total cost. Total profit from the production and sale of x units is a function P given by

$$\textbf{Profit} = \textbf{Revenue} - \textbf{Cost}, \quad \text{or} \quad P(x) = R(x) - C(x).$$

If $R(x)$ is greater than $C(x)$, there is a gain and $P(x)$ is positive. If $C(x)$ is greater than $R(x)$, there is a loss and $P(x)$ is negative. When $R(x) = C(x)$, the company breaks even.

There are two kinds of costs. First, there are costs like rent, insurance, machinery, and so on. These costs, which must be paid whether a product is produced or not, are called *fixed costs*. When a product is being produced, there are costs for labor, materials, marketing, and so on. These are called *variable costs*, because they vary according to the amount being produced. The sum of the fixed cost and the variable cost gives the *total cost* of producing a product.

EXAMPLE 1 Manufacturing Lamps. Ergs, Inc., is planning to make a new lamp. Fixed costs will be $90,000, and it will cost $15 to produce each lamp (variable costs). Each lamp sells for $26.

a) Find the total cost $C(x)$ of producing x lamps.

b) Find the total revenue $R(x)$ from the sale of x lamps.

c) Find the total profit $P(x)$ from the production and sale of x lamps.

d) What profit will the company realize from the production and sale of 3000 lamps? of 14,000 lamps?

e) Graph the total-cost, total-revenue, and total-profit functions using the same set of axes. Determine the break-even point.

SOLUTION

a) Total cost is given by

$$C(x) = (\text{Fixed costs}) \text{ plus } (\text{Variable costs}),$$
$$\text{or} \quad C(x) = \quad 90{,}000 \quad + \quad 15x,$$

where x is the number of lamps produced.

b) Total revenue is given by

$$R(x) = 26x. \qquad \text{\$26 times the number of lamps sold. We assume that}$$
$$\text{every lamp produced is sold.}$$

c) Total profit is given by

$$P(x) = R(x) - C(x) \qquad \text{Profit is revenue minus cost.}$$
$$= 26x - (90{,}000 + 15x)$$
$$= 11x - 90{,}000.$$

d) Profits will be

$$P(3000) = 11 \cdot 3000 - 90{,}000 = -\$57{,}000$$

when 3000 lamps are produced and sold, and

$$P(14{,}000) = 11 \cdot 14{,}000 - 90{,}000 = \$64{,}000$$

when 14,000 lamps are produced and sold. These values can also be found using a graphing calculator, as shown in the figure at left. Thus the company loses money if only 3000 lamps are sold, but makes money if 14,000 are sold.

e) The graphs of each of the three functions are shown below:

$$R(x) = 26x, \qquad \text{This represents the revenue function.}$$
$$C(x) = 90{,}000 + 15x, \qquad \text{This represents the cost function.}$$
$$P(x) = 11x - 90{,}000. \qquad \text{This represents the profit function.}$$

$R(x)$, $C(x)$, and $P(x)$ are all in dollars.

The revenue function has a graph that goes through the origin and has a slope of 26. The cost function has an intercept on the \$-axis of 90,000 and has a slope of 15. The profit function has an intercept on the \$-axis of $-90{,}000$ and has a slope of 11. It is shown by the dashed line. The red dashed line shows a "negative" profit, which is a loss. (That is what is known as "being in the red.") The black dashed line shows a "positive" profit, or gain. (That is what is known as "being in the black.")

$y_1 = 11x - 90000$

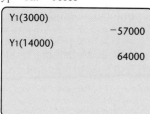

```
Y₁(3000)
                    ⁻57000
Y₁(14000)
                     64000
```

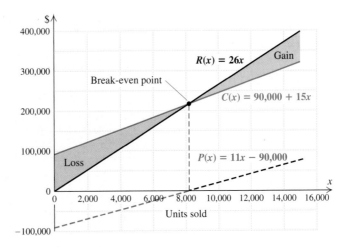

Gains occur where the revenue is greater than the cost. Losses occur where the revenue is less than the cost. The **break-even point** occurs where the graphs of R and C cross. Thus to find the break-even point, we solve a system:

$$R(x) = 26x,$$
$$C(x) = 90,000 + 15x.$$

Since both revenue and cost are in *dollars* and they are equal at the break-even point, the system can be rewritten as

$$d = 26x, \qquad (1)$$
$$d = 90,000 + 15x \qquad (2)$$

and solved using substitution:

$26x = 90,000 + 15x$ **Substituting 26x for d in equation (2)**
$11x = 90,000$
$x \approx 8181.8.$

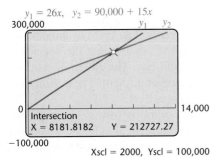

The system can also be solved using a graphing calculator, as shown in the figure at left.

The firm will break even if it produces and sells about 8182 lamps (8181 will yield a tiny loss and 8182 a tiny gain), and takes in a total of $R(8182) = 26 \cdot 8182 = \$212,732$ in revenue. Note that the x-coordinate of the break-even point can also be found by solving $P(x) = 0$. The break-even point is (8182 lamps, $212,732).

Student Notes

Supply and Demand

As the price of coffee varies, the amount sold varies. The table and graph below show that *consumers will demand less as the price goes up.*

Demand function, D

Price, p, per Kilogram	Quantity, $D(p)$ (in millions of kilograms)
$ 8.00	25
9.00	20
10.00	15
11.00	10
12.00	5

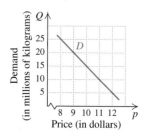

As the price of coffee varies, the amount available varies. The table and graph below show that *sellers will supply more as the price goes up.*

Supply function, S

Price, p, per Kilogram	Quantity, $S(p)$ (in millions of kilograms)
$ 9.00	5
9.50	10
10.00	15
10.50	20
11.00	25

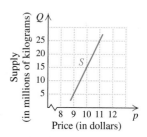

Let's look at the above graphs together. We see that as price increases, demand decreases. As price increases, supply increases. The point of intersection is called the **equilibrium point.** At that price, the amount that the seller will supply is the same amount that the consumer will buy. The situation is analogous to a buyer and a seller negotiating the price of an item. The equilibrium point is the price and quantity that they finally agree on.

Any ordered pair of coordinates from the graph is (price, quantity), because the horizontal axis is the price axis and the vertical axis is the quantity axis. If D is a demand function and S is a supply function, then the equilibrium point is where demand equals supply:

$$D(p) = S(p).$$

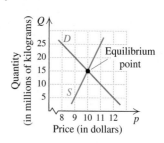

EXAMPLE 2   Find the equilibrium point for the demand and supply functions given:

$$D(p) = 1000 - 60p, \quad (1)$$
$$S(p) = 200 + 4p. \quad (2)$$

SOLUTION   Since both demand and supply are *quantities* and they are equal at the equilibrium point, we rewrite the system as

$$q = 1000 - 60p, \quad (1)$$
$$q = 200 + 4p. \quad (2)$$

We substitute $200 + 4p$ for q in equation (1) and solve:

$$200 + 4p = 1000 - 60p \qquad \text{Substituting } 200 + 4p \text{ for } q \text{ in equation (1)}$$
$$200 + 64p = 1000 \qquad \text{Adding } 60p \text{ to both sides}$$
$$64p = 800 \qquad \text{Adding } -200 \text{ to both sides}$$
$$p = \tfrac{800}{64} = 12.5.$$

Thus the equilibrium price is $12.50 per unit.

To find the equilibrium quantity, we substitute $12.50 into either $D(p)$ or $S(p)$. We use $S(p)$:

$$S(12.5) = 200 + 4(12.5) = 200 + 50 = 250.$$

Thus the equilibrium quantity is 250 units, and the equilibrium point is ($12.50, 250). The graph at left confirms the solution.   ◢

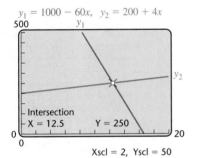

$y_1 = 1000 - 60x, \quad y_2 = 200 + 4x$

Intersection
X = 12.5   Y = 250

Xscl = 2, Yscl = 50

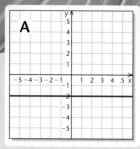

A

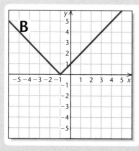

B

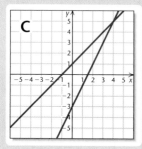

C

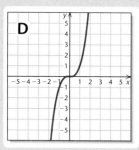

D

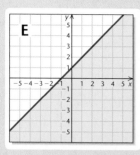

E

Visualizing the Graph

Match each equation, inequality, or set of equations or inequalities with its graph.

1. $y = -2$

2. $y = x^3$

3. $y = x^2$

4. $y = \dfrac{1}{x}$

5. $y = x + 1$

6. $y = |x + 1|$

7. $y \leq x + 1$

8. $y > 2x - 3$

9. $y = x + 1,$
   $y = 2x - 3$

10. $y \leq x + 1,$
    $y > 2x - 3$

Answers on page A-36

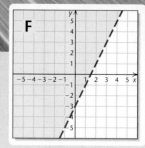

F

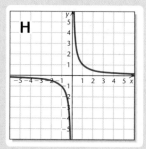

G

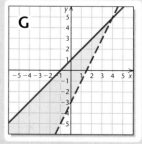

H

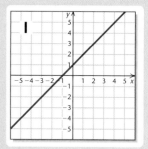

I

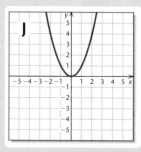

J

9.5 EXERCISE SET

FOR EXTRA HELP

MathXL

MyMathLab

InterAct
Math

AW Math
Tutor Center

Video Lectures
on CD: Disc 5

Student's
Solutions Manual

Concept Reinforcement   In each of Exercises 1–8, match the word or phrase with the most appropriate choice from the column on the right.

1. ____ Total cost

2. ____ Total revenue

3. ____ Total profit

4. ____ Fixed costs

5. ____ Variable costs

6. ____ Break-even point

7. ____ Equilibrium point

8. ____ Price

a) The amount of money that a company takes in

b) The sum of fixed costs and variable costs

c) The point at which total revenue equals total cost

d) What consumers pay per item

e) The difference between total revenue and total cost

f) What companies spend whether or not a product is produced

g) The point at which supply equals demand

h) The costs that vary according to the number of items produced

For each of the following pairs of total-cost and total-revenue functions, find (a) *the total-profit function and* (b) *the break-even point.*

9. $C(x) = 45x + 300{,}000$;
   $R(x) = 65x$

10. $C(x) = 25x + 270{,}000$;
    $R(x) = 70x$

11. $C(x) = 10x + 120{,}000$;
    $R(x) = 60x$

12. $C(x) = 30x + 49{,}500$;
    $R(x) = 85x$

13. $C(x) = 40x + 22{,}500$;
    $R(x) = 85x$

14. $C(x) = 20x + 10{,}000$;
    $R(x) = 100x$

15. $C(x) = 22x + 16{,}000$;
    $R(x) = 40x$

16. $C(x) = 15x + 75{,}000$;
    $R(x) = 55x$

Aha! 17. $C(x) = 75x + 100{,}000$;
    $R(x) = 125x$

18. $C(x) = 20x + 120{,}000$;
    $R(x) = 50x$

Find the equilibrium point for each of the following pairs of demand and supply functions.

19. $D(p) = 1000 - 10p$,
    $S(p) = 230 + p$

20. $D(p) = 2000 - 60p$,
    $S(p) = 460 + 94p$

21. $D(p) = 760 - 13p$,
    $S(p) = 430 + 2p$

22. $D(p) = 800 - 43p$,
    $S(p) = 210 + 16p$

23. $D(p) = 7500 - 25p$,
    $S(p) = 6000 + 5p$

24. $D(p) = 8800 - 30p$,
    $S(p) = 7000 + 15p$

25. $D(p) = 1600 - 53p$,
    $S(p) = 320 + 75p$

26. $D(p) = 5500 - 40p$,
    $S(p) = 1000 + 85p$

Solve.

27. *Computer Manufacturing.*   Biz.com Electronics is planning to introduce a new line of computers. The fixed costs for production are $125,300. The variable costs for producing each computer are $450. The revenue from each computer is $800. Find the following.

a) The total cost $C(x)$ of producing x computers
b) The total revenue $R(x)$ from the sale of x computers
c) The total profit $P(x)$ from the production and sale of x computers
d) The profit or loss from the production and sale of 100 computers; of 400 computers
e) The break-even point

28. *Manufacturing CD Players.*   SoundGen, Inc., is planning to manufacture a new type of CD player. The fixed costs for production are $22,500. The variable costs for producing each CD player are estimated to be $40. The revenue from each CD player is to be $85. Find the following.

a) The total cost $C(x)$ of producing x CD players
b) The total revenue $R(x)$ from the sale of x CD players
c) The total profit $P(x)$ from the production and sale of x CD players
d) The profit or loss from the production and sale of 3000 CD players; of 400 CD players
e) The break-even point

29. *Manufacturing Caps.*   Martina's Custom Printing is planning on adding painter's caps to its product line. For the first year, the fixed costs for setting up production are $16,404. The variable costs for producing a dozen caps are $6.00. The revenue on each dozen caps will be $18.00. Find the following.

a) The total cost $C(x)$ of producing x dozen caps
b) The total revenue $R(x)$ from the sale of x dozen caps
c) The total profit $P(x)$ from the production and sale of x dozen caps
d) The profit or loss from the production and sale of 3000 dozen caps; of 1000 dozen caps
e) The break-even point

30. *Sport Coat Production.*   Sarducci's is planning a new line of sport coats. For the first year, the fixed costs for setting up production are $10,000. The variable costs for producing each coat are $30.

The revenue from each coat is to be $80. Find the following.

a) The total cost $C(x)$ of producing x coats
b) The total revenue $R(x)$ from the sale of x coats
c) The total profit $P(x)$ from the production and sale of x coats
d) The profit or loss from the production and sale of 2000 coats; of 50 coats
e) The break-even point

31. *Dog Food Production.*   Great Foods will soon begin producing a new line of puppy food. The marketing department predicts that the demand function will be $D(p) = -14.97p + 987.35$ and the supply function will be $S(p) = 98.55p - 5.13$.

a) To the nearest cent, what price per unit should be charged in order to have equilibrium between supply and demand?
b) The production of the puppy food involves $5265 in fixed costs and $2.10 per unit in variable costs. If the price per unit is the value you found in part (a), how many units must be sold in order to break even?

32. *Computer Production.*   Number Solutions Computers is planning a new line of computers, each of which will sell for $970. For the first year, the fixed costs in setting up production are $1,235,580 and the variable costs for each computer are $697.

a) What is the break-even point? (Round to the nearest whole number.)
b) The marketing department at Number Solutions is not sure that $970 is the best price. Their demand function for the new computers is given by $D(p) = -304.5p + 374,580$ and their supply function is given by $S(p) = 788.7p - 576,504$. What price p would result in equilibrium between supply and demand?

33. In Example 1, the slope of the line representing Revenue is the sum of the slopes of the other two lines. This is not a coincidence. Explain why.

34. Variable costs and fixed costs are often compared to the slope and the y-intercept, respectively, of an equation for a line. Explain why you feel this analogy is or is not valid.

Skill Maintenance

Solve. [2.2]

35. $3x - 9 = 27$

36. $4x - 7 = 53$

37. $4x - 5 = 7x - 13$

38. $2x + 9 = 8x - 15$

39. $7 - 2(x - 8) = 14$

40. $6 - 4(3x - 2) = 10$

Synthesis

TW 41. Ian claims that since his fixed costs are $1000, he need sell only 20 birdbaths at $50 each in order to break even. Does this sound plausible? Why or why not?

TW 42. In this section, we examined supply and demand functions for coffee. Does it seem realistic to you for the graph of *D* to have a constant slope? Why or why not?

43. *Yo-yo Production.* Bing Boing Hobbies is willing to produce 100 yo-yo's at $2.00 each and 500 yo-yo's at $8.00 each. Research indicates that the public will buy 500 yo-yo's at $1.00 each and 100 yo-yo's at $9.00 each. Find the equilibrium point.

44. *Loudspeaker Production.* Fidelity Speakers, Inc., has fixed costs of $15,400 and variable costs of $100 for each pair of speakers produced. If the speakers sell for $250 a pair, how many pairs of speakers must be produced (and sold) in order to have enough profit to cover the fixed costs of two additional facilities? Assume that all fixed costs are identical.

45. *Peanut Butter.* The following table lists the data for supply and demand of an 18-oz jar of peanut butter at various prices.

Price	Supply (in millions)	Demand (in millions)
$1.59	23.4	22.5
1.29	19.2	24.8
1.69	26.8	22.2
1.19	18.4	29.7
1.99	30.7	19.3

a) Use linear regression to find the supply function $S(p)$ for suppliers of peanut butter at price p.

b) Use linear regression to find the demand function $D(p)$ for consumers of peanut butter at price p.

c) Find the equilibrium point.

46. *Aircraft Production.* In October 2006, due to increased cost, Airbus revised its estimated break-even point for the new A380. The break-even point rose from 270 aircraft to 420 aircraft. The following table lists the number of aircraft on order at various points in time. (*Sources: The Register; Newswire Today;* www.dw-word.de)

Number of Months After January 1, 2004	Number of Airbus A380 Orders
5	129
13	149
16	139
18	144
29	159
30	167

a) Find a linear function $A(t)$ that can be used to estimate the number of A380 aircraft on order *t* months after January 1, 2004.

b) Assuming linear growth, when will 420 aircraft be on order?

c) Airbus can produce 45 A380 aircraft each year (*Source: International Herald Tribune,* 10/19/06). If production begins in January 2007, when will Airbus produce the 420th A380?

d) Use the function from part (a) and the production rate from part (c) to estimate when A380 production will exceed the number of orders for the aircraft.

9 Chapter Summary and Review

KEY TERMS AND DEFINITIONS

SYSTEMS

System of equations, p. 681   Two or more equations that are to be solved simultaneously. A system is **consistent** if it has at least one solution. Otherwise it is **inconsistent.** The equations in a system are **dependent** if one of them can be removed without changing the solution set. Otherwise, they are **independent.**

Matrix (matrices), p. 692   A rectangular array of numbers. The **elements,** or **entries,** in a matrix are arranged in **rows** and **columns.**

APPLICATIONS

Total profit, p. 707   **Total revenue** (the amount taken in) minus **total cost** (the amount spent).

Break-even point, p. 709   The point at which total revenue equals total cost, or the point at which total profit is 0.

Equilibrium point, p. 710   The point at which **demand** equals **supply.**

IMPORTANT CONCEPTS

[Section references appear in brackets.]

Concept	Example
Solve systems of three equations.	Solve: $$x + y - z = 3,$$ $$-x + y + 2z = -5,$$ $$2x - y - 3z = 9$$ *Eliminate x using two equations:* $$\begin{aligned} x + y - z &= 3 \\ \underline{-x + y + 2z} &= \underline{-5} \\ 2y + z &= -2. \end{aligned}$$ *Solve the system of two equations for y and z:* $$\begin{aligned} 2y + z &= -2 \\ \underline{-3y - z} &= \underline{3} \\ -y &= 1 \\ y &= -1 \\ 2(-1) + z &= -2 \\ z &= 0. \end{aligned}$$ *Eliminate x again using two different equations:* $$\begin{aligned} -2x - 2y + 2z &= -6 \\ \underline{2x - y - 3z} &= \underline{9} \\ -3y - z &= 3. \end{aligned}$$ *Substitute and solve for x:* $$\begin{aligned} x + y - z &= 3 \\ \underline{x + (-1) - 0} &= \underline{3} \\ x &= 4. \end{aligned}$$ The solution is $(4, -1, 0)$. $\qquad$ [9.1]

(continued)

Solve systems of equations using matrices.	Solve: $x + 4y = 1,$ $2x - y = 3.$	*Write as a matrix in row-echelon form.* $\begin{bmatrix} 1 & 4 &\vdots& 1 \\ 2 & -1 &\vdots& 3 \end{bmatrix}$ $\begin{bmatrix} 1 & 4 &\vdots& 1 \\ 0 & -9 &\vdots& 1 \end{bmatrix}$	*Rewrite as equations and solve:* $-9y = 1$ $y = -\frac{1}{9}$ $x + 4\left(-\frac{1}{9}\right) = 1$ $x = \frac{13}{9}.$
	The solution is $\left(\frac{13}{9}, -\frac{1}{9}\right).$		[9.3]
Find the determinant of a matrix.	$\begin{bmatrix} 2 & 3 \\ -1 & 5 \end{bmatrix} = 2 \cdot 5 - (-1)(3) = 13$		[9.4]
Solve systems of equations using Cramer's rule.	Solve: $x - 3y = 7,$ $2x + 5y = 4.$	$x = \dfrac{\begin{vmatrix} 7 & -3 \\ 4 & 5 \end{vmatrix}}{\begin{vmatrix} 1 & -3 \\ 2 & 5 \end{vmatrix}}$ $y = \dfrac{\begin{vmatrix} 1 & 7 \\ 2 & 4 \end{vmatrix}}{\begin{vmatrix} 1 & -3 \\ 2 & 5 \end{vmatrix}}$ $x = \frac{47}{11}$ $y = \frac{-10}{11}$	
	The solution is $\left(\frac{47}{11}, -\frac{10}{11}\right).$		[9.4]

Review Exercises

👉 *Concept Reinforcement*   *Complete each of the following sentences.*

1. Systems of three equations in three variables are usually solved using the _____ method. [9.1]

2. If a system has at least one solution, it is said to be _____. [9.1]

3. The sum of the measures of the angles in any triangle is _____. [9.2]

4. When a matrix is in reduced row-echelon form, the first nonzero number in any nonzero row is _____. [9.3]

5. When a matrix has the same number of rows and columns, it is said to be _____. [9.4]

6. Cramer's rule is a formula in which the numerator and the denominator of each fraction is a(n) _____. [9.4]

7. Total revenue minus total cost is _____. [9.5]

8. _____ costs must be paid whether a product is produced or not. [9.5]

9. At the _____, the amount that the seller will supply is the same as the amount that the consumer will buy. [9.5]

10. At the break-even point, the value of the profit function is _____. [9.5]

Solve. If a system's equations are dependent or if there is no solution, state this. [9.1]

11. $x + 4y + 3z = 2,$
$\quad 2x + y + z = 10,$
$\quad -x + y + 2z = 8$

12. $4x + 2y - 6z = 34,$
$\quad 2x + y + 3z = 3,$
$\quad 6x + 3y - 3z = 37$

13. $2x - 5y - 2z = -4,$
$\quad 7x + 2y - 5z = -6,$
$\quad -2x + 3y + 2z = 4$

14. $2x - 3y + z = 1,$
$\quad x - y + 2z = 5,$
$\quad 3x - 4y + 3z = -2$

15. $3x + y = 2,$
$\quad x + 3y + z = 0,$
$\quad x + z = 2$

Solve.

16. In triangle *ABC*, the measure of angle *A* is four times the measure of angle *C*, and the measure of angle *B* is 45° more than the measure of angle *C*. What are the measures of the angles of the triangle? [9.2]

17. *Food Prices.* At a concession stand during a high school football game, Sara bought a hot dog, a bag of chips, and a bottle of water for $4.75. Devin bought two hot dogs and a bottle of water for $6.75. Tyla bought a bag of chips and two bottles of water for $3.25. What was the price of each item? [9.2]

Solve using matrices. Show your work. [9.3]

18. $3x + 4y = -13,$
$\quad 5x + 6y = 8$

19. $3x - y + z = -1,$
$\quad 2x + 3y + z = 4,$
$\quad 5x + 4y + 2z = 5$

Evaluate. [9.4]

20. $\begin{vmatrix} -2 & 4 \\ -3 & 5 \end{vmatrix}$

21. $\begin{vmatrix} 2 & 3 & 0 \\ 1 & 4 & -2 \\ 2 & -1 & 5 \end{vmatrix}$

Solve using Cramer's rule. Show your work. [9.4]

22. $2x + 3y = 6,$
$\quad x - 4y = 14$

23. $2x + y + z = -2,$
$\quad 2x - y + 3z = 6,$
$\quad 3x - 5y + 4z = 7$

24. Find **(a)** the total-profit function and **(b)** the break-even point for the total-cost and total-revenue functions

$$C(x) = 30x + 15{,}800$$

and

$$R(x) = 50x. \quad [9.5]$$

25. Find the equilibrium point for the demand and supply functions

$$S(p) = 60 + 7p$$

and

$$D(p) = 120 - 13p. \quad [9.5]$$

26. Auriel is beginning to produce organic honey. For the first year, the fixed costs for setting up production are $9000. The variable costs for producing each pint of honey are $0.75. The revenue from each pint of honey is $5.25. Find the following. [9.5]

 a) The total cost $C(x)$ of producing x pints of honey
 b) The total revenue $R(x)$ from the sale of x pints of honey
 c) The total profit $P(x)$ from the production and sale of x pints of honey
 d) The profit or loss from the production and sale of 1500 pints of honey; of 5000 pints of honey
 e) The break-even point

Synthesis

TW 27. How would you go about solving a problem that involves four variables? [9.2]

TW 28. Explain how a system of equations can be both dependent and inconsistent. [9.1]

29. Auriel is quitting a job that pays $27,000 a year to make honey (see Exercise 26). How many pints of honey must she produce and sell in order to make the same amount that she made in the job she left? [9.5]

30. The graph of $f(x) = ax^2 + bx + c$ contains the points $(-2, 3)$, $(1, 1)$, and $(0, 3)$. Find a, b, and c and give a formula for the function. [9.2]

Chapter Test 9

Solve. If a system's equations are dependent or if there is no solution, state this.

1. $-3x + y - 2z = 8,$
$-x + 2y - z = 5,$
$2x + y + z = -3$

2. $6x + 2y - 4z = 15,$
$-3x - 4y + 2z = -6,$
$4x - 6y + 3z = 8$

3. $2x + 2y = 0,$
$4x + 4z = 4,$
$2x + y + z = 2$

4. $3x + 3z = 0,$
$2x + 2y = 2,$
$3y + 3z = 3$

Solve using matrices. Show your work.

5. $7x - 8y = 10,$
$9x + 5y = -2$

6. $x + 3y - 3z = 12,$
$3x - y + 4z = 0,$
$-x + 2y - z = 1$

Evaluate.

7. $\begin{vmatrix} 4 & -2 \\ 3 & 7 \end{vmatrix}$

8. $\begin{vmatrix} 3 & 4 & 2 \\ 2 & -5 & 4 \\ 4 & 5 & -3 \end{vmatrix}$

9. Solve using Cramer's rule:
$8x - 3y = 5,$
$2x + 6y = 3.$

10. In triangle ABC, the measure of angle B is 5° less than three times that of angle A. The measure of angle C is 5° more than three times that of angle B. Find the angle measures.

11. An electrician, a carpenter, and a plumber are hired to work on a house. The electrician earns $21 per hour, the carpenter $19.50 per hour, and the plumber $24 per hour. The first day on the job, they worked a total of 21.5 hr and earned a total of $469.50. If the plumber worked 2 more hours than the carpenter did, how many hours did each work?

12. Find the equilibrium point for the demand and supply functions
$$D(p) = 79 - 8p \quad \text{and} \quad S(p) = 37 + 6p.$$

13. Kick Back, Inc., is producing a new hammock. For the first year, the fixed costs for setting up production are $40,000. The variable costs for producing each hammock are $25. The revenue from each hammock is $70. Find the following.
   a) The total cost $C(x)$ of producing x hammocks
   b) The total revenue $R(x)$ from the sale of x hammocks
   c) The total profit $P(x)$ from the production and sale of x hammocks
   d) The profit or loss from the production and sale of 300 hammocks; of 900 hammocks
   e) The break-even point

Synthesis

14. Solve:
$$2w - x + 5y - z = 1,$$
$$w + x - 2y + z = 2,$$
$$w + 3x - y + z = 0,$$
$$3w + x + y - z = 0.$$

15. At a county fair, an adult's ticket sold for $5.50, a senior citizen's ticket for $4.00, and a child's ticket for $1.50. On opening day, the number of adults' and senior citizens' tickets sold was 30 more than the number of children's tickets sold. The number of adults' tickets sold was 6 more than four times the number of senior citizens' tickets sold. Total receipts from the ticket sales were $11,219.50. How many of each type of ticket were sold?

1-9 Cumulative Review

1. *Meal Preparation.* In 1993, on average, adults spent 49 min per weeknight preparing dinner. This time decreased to 31 min per weeknight by 2003. (*Source*: *Parade Magazine*, 11/16/03) What was the rate of decrease? [3.4]

2. *Frequent Fliers.* The number of unredeemed frequent flier miles has increased from 5.5 trillion in 1999 to 14.2 trillion in 2005 (*Source*: WebFlyer.com). [3.7], [3.8]

   a) Find a linear function $F(t)$ that fits the data.

   b) Use the function of part (a) to predict the number of unredeemed frequent flier miles in 2007.

3. *Automobiles.* The following table lists the number of automobiles in the United States for various years. [3.7], [3.8]

Year	Registered Automobiles (in millions)
1960	67.9
1970	98.1
1980	139.8
1990	179.3
2005	220.0

Source: The Indianapolis Star, 4/28/06

   a) Use linear regression to find a function $a(t)$ that can be used to estimate the number of automobiles in the United States t years after 1960.

   b) Use the function of part (a) to estimate the number of automobiles in the United States in 2015.

4. *Fuel Economy.* The following table lists annual fuel consumption and expenditures by the fuel economy of vehicles. [7.8]

Fuel Economy (in miles per gallon)	Fuel Consumption (in billions)	Fuel Expenditures (in billions)
10.9 or less	2.3 gal	$3.0
11 to 12.9	2.0	2.5
13 to 15.9	16.3	21.7
16 to 18.9	26.6	35.5
19 to 21.9	26.1	34.9
22 to 24.9	22.1	29.4
25 to 29.9	13.6	18.1
30 or more	4.1	5.2

Source: Energy Information Administration/Household Vehicles Energy Use

   a) Use a graph to show that fuel expenditure varies directly as fuel consumption.

   b) Find an equation of variation that describes the data. Use the point (26.6, 35.5).

   c) What does the variation constant signify?

5. *College Football.* The greatest payout to a college team for a bowl game in 2005–2006 was $18.3 million. The smallest payout was $750,000. Fourteen teams received either the greatest or smallest payout, for a total of $63.15 million. How many teams received $18.3 million and how many received $750,000? [4.4]

In Exercises 6–9, match each equation with one of the following graphs.

a)
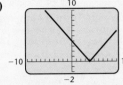

b)
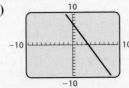

c)
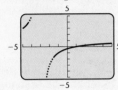

d)
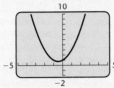

6. $y = -2x + 7$ [3.6]

7. $y = |x - 4|$ [8.2]

8. $y = x^2 + x + 1$ [6.1]

9. $y = \dfrac{x - 1}{x + 3}$ [7.1]

Graph on a plane.

10. $4x \geq 5y + 20$ [8.3]

11. $y = \frac{1}{3}x - 2$ [3.6]

12. Evaluate

$$\frac{2x - y^2}{x + y}$$

for $x = 3$ and $y = -4$. [1.8]

13. Convert to scientific notation: $5,760,000,000$. [5.2]

14. Determine the slope and the y-intercept for the line given by $7x - 4y = 12$. [3.6]

15. Find an equation for the line that passes through the points $(-1, 7)$ and $(2, -3)$. [3.7]

Perform the indicated operations and simplify.

16. $(2x^2 - 3x + 1) + (6x - 3x^3 + 7x^2 - 4)$ [5.4]

17. $(5x^3y^2)(-3xy^2)$ [5.5]

18. $(3a + b - 2c) - (-4b + 3c - 2a)$ [5.4], [5.7]

19. $(5x^2 - 2x + 1)(3x^2 + x - 2)$ [5.5]

20. $(2x^2 - y)^2$ [5.6], [5.7]

21. $(2x^2 - y)(2x^2 + y)$ [5.6], [5.7]

22. $(-5m^3n^2 - 3mn^3) +$
$(-4m^2n^2 + 4m^3n^2) - (2mn^3 - 3m^2n^2)$ [5.4], [5.7]

23. $\dfrac{y^2 - 36}{2y + 8} \cdot \dfrac{y + 4}{y + 6}$ [7.2]

24. $\dfrac{x^4 - 1}{x^2 - x - 2} \div \dfrac{x^2 + 1}{x - 2}$ [7.2]

25. $\dfrac{5ab}{a^2 - b^2} + \dfrac{a + b}{a - b}$ [7.4]

26. $\dfrac{2}{m + 1} + \dfrac{3}{m - 5} - \dfrac{m^2 - 1}{m^2 - 4m - 5}$ [7.4]

27. $y - \dfrac{2}{3y}$ [7.4]

28. Simplify: $\dfrac{\dfrac{1}{x} - \dfrac{1}{y}}{x + y}$. [7.5]

29. Divide: $(9x^3 + 5x^2 + 2) \div (x + 2)$. [5.8]

Factor.

30. $4x^3 + 18x^2$ [6.1]

31. $x^2 + 8x - 84$ [6.2]

32. $16y^2 - 81$ [6.4]

33. $64x^3 + 8$ [6.5]

34. $t^2 - 16t + 64$ [6.4]

35. $x^6 - x^2$ [6.4]

36. $0.027b^3 - 0.008c^3$ [6.5]

37. $20x^2 + 7x - 3$ [6.3]

38. $3x^2 - 17x - 28$ [6.3]

39. $x^5 - x^3y + x^2y - y^2$ [6.1]

40. If
$$f(x) = \frac{x - 2}{x - 5},$$
find the following. [7.1]

a) $f(3)$ **b)** The domain of f

41. Write the domain of f using interval notation if $f(x) = \sqrt{x - 7}$. [8.1]

42. If $f(x) = x^2 - 4$ and $g(x) = x^2 - 7x + 10$, find the domain of f/g. [6.2]

Solve.

43. $8x = 1 + 16x^2$ [6.4]

44. $625 = 49y^2$ [6.4]

45. $20 > 2 - 6x$ [2.6]

46. $x^2 - x \geq 2$ [8.4]

47. $-8 < x + 2 < 15$ [8.1]

48. $3x - 2 < -6 \text{ or } x + 3 > 9$ [8.1]

49. $|x| > 6.4$ [8.2]

50. $|4x - 1| \leq 14$ [8.2]

51. $\dfrac{2}{n} - \dfrac{7}{n} = 3$ [7.6]

52. $\dfrac{6}{x - 5} = \dfrac{2}{2x}$ [7.6]

53. $\dfrac{3x}{x-2} - \dfrac{6}{x+2} = \dfrac{24}{x^2-4}$ [7.6]

54. $\dfrac{3x^2}{x+2} + \dfrac{5x-22}{x-2} = \dfrac{-48}{x^2-4}$ [7.6]

55. $5x - 2y = -23,$
$3x + 4y = 7$ [4.3]

56. $-3x + 4y + \ z = -5,$
$\quad x - 3y - \ z = 6,$
$\quad 2x + 3y + 5z = -8$ [9.1]

57. Let $f(x) = |3x - 5|$. Find all values of x for which $f(x) = 2$. [8.2]

Solve.

58. $5m - 3n = 4m + 12$, for n [2.3]

59. $P = \dfrac{3a}{a+b}$, for a [7.8]

Find the solutions of each equation, inequality, or system from the given graph.

60. $10 - 5x = 3x + 2$ [4.5]   **61.** $f(x) \le g(x)$ [8.1]

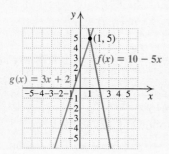

62. $y = x - 5,$
$y = 1 - 2x$ [4.2]

63. $x - 5 = 1 - 2x$
[4.5]

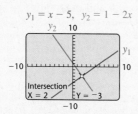

64. $x^2 - 2x + 1 = 0$ [6.1]

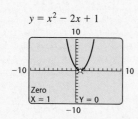

65. $|x - 3| \le 2$ [8.2]   **66.** $|x - 3| = 2$ [8.2]

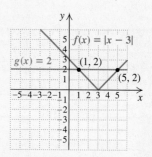

67. The sum of three numbers is 20. The first number is 3 less than twice the third number. The second number minus the third number is -7. What are the numbers? [9.2]

68. A digital data circuit can transmit a particular set of data in 4 sec. An analog phone circuit can transmit the same data in 20 sec. How long would it take, working together, for both circuits to transmit the data? [7.7]

69. The floor area of a rental trailer is rectangular. The length is 3 ft more than the width. A rug of area 54 ft² exactly fills the floor of the trailer. Find the perimeter of the trailer. [6.7]

70. The sum of the squares of three consecutive even integers is equal to 8 more than three times the square of the second number. Find the integers. [6.7]

71. The volume of wood V in a tree trunk varies jointly as the height h and the square of the girth g (girth is distance around). If the volume is 35 ft³ when the height is 20 ft and the girth is 5 ft, what is the height when the volume is 85.75 ft³ and the girth is 7 ft? [7.8]

Synthesis

72. Multiply: $(x - 4)^3$. [5.5]

73. Find all roots for $f(x) = x^4 - 34x^2 + 225$. [6.4]

Solve.

74. $4 \le |3 - x| \le 6$ [8.1], [8.2]

75. $\dfrac{18}{x-9} + \dfrac{10}{x+5} = \dfrac{28x}{x^2 - 4x - 45}$ [7.6]

76. $16x^3 = x$ [6.4]

10

Exponents and Radical Functions

In this chapter, we learn about square roots, cube roots, fourth roots, and so on. These roots are studied in connection with the manipulation of radical expressions and the solution of real-world applications. Exponents that are fractions are also studied and are used to ease some of our work with radicals. The chapter closes with an examination of the complex-number system.

APPLICATION *Firefighting.*

The number of gallons per minute discharged from a fire hose depends on the diameter of the hose and the nozzle pressure. The following table and graph show the amount of water flow for a 2-in. diameter solid bore nozzle at various nozzle pressures. Determine whether the graph indicates that a radical function can be used to model water flow.

NOZZLE PRESSURE (IN POUNDS PER SQUARE INCH, PSI)	WATER FLOW (IN GALLONS PER MINUTE, GPM)
40	752
60	921
80	1063
100	1188
120	1302
150	1455
200	1681

Source: www.firetactics.com

Xscl = 50, Yscl = 500

This problem appears as Example 14 in Section 10.1.

10.1 Radical Expressions, Functions, and Models

Square Roots and Square Root Functions   ◼   Expressions of the Form $\sqrt{a^2}$ ◼   Cube Roots   ◼   Odd and Even *n*th Roots   ◼   Radical Functions and Models

In this section, we consider roots, such as square roots and cube roots. We look at the symbolism that is used and ways in which expressions can be manipulated to get equivalent expressions. All of this will be important in problem solving.

Square Roots and Square Root Functions

When a number is multiplied by itself, we say that the number is squared. Often we need to know what number was squared in order to produce some value *a*. If such a number can be found, we call that number a *square root* of *a*.

> **Square Root**   The number *c* is a *square root* of *a* if $c^2 = a$.

For example,

  9 has -3 and 3 as square roots because $(-3)^2 = 9$ and $3^2 = 9$.
  25 has -5 and 5 as square roots because $(-5)^2 = 25$ and $5^2 = 25$.
  -4 does not have a real-number square root because there is no real number *c* such that $c^2 = -4$.

Note that every positive number has two square roots, whereas 0 has only itself as a square root. Negative numbers do not have real-number square roots, although later in this chapter we introduce the *complex-number* system in which such square roots do exist.

> **EXAMPLE 1**   Find the two square roots of 64.
>
> **SOLUTION**   The square roots of 64 are 8 and -8, because $8^2 = 64$ and $(-8)^2 = 64$.

Whenever we refer to *the* square root of a number, we mean the nonnegative square root of that number. This is often referred to as the *principal square root* of the number.

> **Principal Square Root**   The *principal square root* of a nonnegative number is its nonnegative square root. The symbol $\sqrt{}$ is called a *radical sign* and is used to indicate the principal square root of the number over which it appears.

Student Notes

It is important to remember the difference between *the* square root of 9 and *a* square root of 9. *A* square root of 9 means either 3 or -3, but *the* square root of 9, or $\sqrt{9}$, means the principal square root of 9, or 3.

EXAMPLE 2   Simplify each of the following.

a) $\sqrt{25}$ **b)** $\sqrt{\dfrac{25}{64}}$

c) $-\sqrt{64}$ **d)** $\sqrt{0.0049}$

SOLUTION

a) $\sqrt{25} = 5$ $\sqrt{}$ indicates the principal square root. Note that $\sqrt{25} \neq -5$.

b) $\sqrt{\dfrac{25}{64}} = \dfrac{5}{8}$ Since $\left(\dfrac{5}{8}\right)^2 = \dfrac{25}{64}$

c) $-\sqrt{64} = -8$ Since $\sqrt{64} = 8, -\sqrt{64} = -8$.

d) $\sqrt{0.0049} = 0.07$ $(0.07)(0.07) = 0.0049$

In addition to being read as "the principal square root of a," $\sqrt{a}$ is also read as "the square root of a," or simply "root a" or "radical a." Any expression in which a radical sign appears is called a *radical expression.* The following are radical expressions:

$$\sqrt{5}, \qquad \sqrt{a}, \qquad -\sqrt{3x}, \qquad \sqrt{\dfrac{y^2 + 7}{y}}, \qquad \sqrt{x} + 8.$$

The expression under the radical sign is called the **radicand.** In the expressions above, the radicands are 5, a, $3x$, $(y^2 + 7)/y$, and x.

All but the most basic calculators give values for square roots. For example, to calculate $\sqrt{5}$ on most graphing calculators, we press $\boxed{\checkmark}$ $\boxed{5}$ $\boxed{\;)\;}$ $\boxed{\textbf{ENTER}}$. On some calculators, $\boxed{\checkmark}$ is pressed after 5 has been entered. A calculator will display an approximation like

      2.23606798

for $\sqrt{5}$. The exact value of $\sqrt{5}$ is not given by any repeating or terminating decimal. The same is true for the square root of any whole number that is not a perfect square. We discussed such *irrational numbers* in Chapter 1.

The square-root function, given by

$$f(x) = \sqrt{x},$$

has the interval $[0, \infty)$ as its domain. We can draw its graph by selecting convenient values for x and calculating the corresponding outputs. Once these ordered pairs have been graphed, a smooth curve can be drawn.

As x increases, the output $\sqrt{x}$ increases. The range of $f(x)$ is $[0, \infty)$.

$$f(x) = \sqrt{x}$$

x	$\sqrt{x}$	$(x, f(x))$
0	0	$(0, 0)$
1	1	$(1, 1)$
4	2	$(4, 2)$
9	3	$(9, 3)$

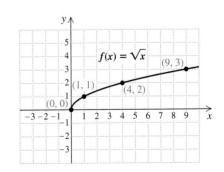

EXAMPLE 3   For each function, find the indicated function value.

a) $f(x) = \sqrt{3x - 2}$; $f(1)$

b) $g(z) = -\sqrt{6z + 4}$; $g(3)$

SOLUTION

a) $f(1) = \sqrt{3 \cdot 1 - 2}$ Substituting

$ = \sqrt{1} = 1$ Simplifying

b) $g(3) = -\sqrt{6 \cdot 3 + 4}$ Substituting

$ = -\sqrt{22}$ Simplifying. This answer is exact.

$ \approx -4.69041576$ Using a calculator to approximate $\sqrt{22}$

Expressions of the Form $\sqrt{a^2}$

It is tempting to write $\sqrt{a^2} = a$, but the next example shows that, as a rule, this is untrue.

EXAMPLE 4   Evaluate $\sqrt{x^2}$ for the following values: **(a)** 5; **(b)** 0; **(c)** -5.

SOLUTION

a) $\sqrt{5^2} = \sqrt{25} = 5$
            Same

b) $\sqrt{0^2} = \sqrt{0} = 0$
            Same

c) $\sqrt{(-5)^2} = \sqrt{25} = 5$
            Opposites   Note that $\sqrt{(-5)^2} \neq -5$.

In Example 4, we saw that $\sqrt{5^2} = 5$ and $\sqrt{(-5)^2} = 5$. Recall that $|5| = 5$ and $|-5| = 5$. Thus evaluating $\sqrt{a^2}$ is just like evaluating $|a|$.

Interactive Discovery

Use graphs or tables to determine which of the following are identities. Be sure to enclose the entire radicand in parentheses.

1. $\sqrt{x^2} = x$ **2.** $\sqrt{x^2} = -x$

3. $\sqrt{x^2} = |x|$ **4.** $\sqrt{(x + 3)^2} = x + 3$

5. $\sqrt{(x + 3)^2} = |x + 3|$ **6.** $\sqrt{x^8} = x^4$

In general, we cannot say that $\sqrt{x^2} = x$. However, we can simplify $\sqrt{x^2}$ using absolute value.

Simplifying $\sqrt{a^2}$ For any real number a,

$$\sqrt{a^2} = |a|.$$

(The principal square root of a^2 is the absolute value of a.)

When a radicand is the square of a variable expression, like $(x + 5)^2$ or $36t^2$, absolute-value signs are needed when simplifying. We use absolute-value signs unless we know that the expression being squared is nonnegative. This assures that our result is never negative.

EXAMPLE 5   Simplify each expression. Assume that the variable can represent any real number.

a) $\sqrt{(x + 1)^2}$ **b)** $\sqrt{x^2 - 8x + 16}$
c) $\sqrt{a^8}$ **d)** $\sqrt{t^6}$

SOLUTION

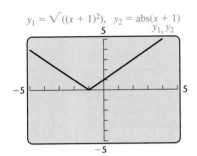

$y_1 = \sqrt{((x + 1)^2)}, \quad y_2 = \text{abs}(x + 1)$

a) $\sqrt{(x + 1)^2} = |x + 1|$ Since $x + 1$ might be negative (for example, if $x = -3$), absolute-value notation is necessary.

The graph at left confirms the identity.

b) $\sqrt{x^2 - 8x + 16} = \sqrt{(x - 4)^2} = |x - 4|$ Since $x - 4$ might be negative, absolute-value notation is necessary.

c) Note that $(a^4)^2 = a^8$ and that a^4 is never negative. Thus,

$$\sqrt{a^8} = |a^4| = a^4.$$ Absolute-value notation is unnecessary here.

d) Note that $(t^3)^2 = t^6$. Thus,

$$\sqrt{t^6} = |t^3|.$$ Since t^3 might be negative, absolute-value notation is necessary.

If we know that no radicands have been formed by raising negative quantities to even powers, we do not need absolute-value signs.

EXAMPLE 6   Simplify each expression. Assume that no radicands were formed by raising negative quantities to even powers.

a) $\sqrt{y^2}$ **b)** $\sqrt{a^{10}}$ **c)** $\sqrt{9x^2 - 6x + 1}$

SOLUTION

a) $\sqrt{y^2} = y$ We are assuming that y is nonnegative, so no absolute-value notation is necessary. When y is negative, $\sqrt{y^2} \neq y$.

b) $\sqrt{a^{10}} = a^5$ Assuming that a^5 is nonnegative. Note that $(a^5)^2 = a^{10}$.

c) $\sqrt{9x^2 - 6x + 1} = \sqrt{(3x - 1)^2} = 3x - 1$ Assuming that $3x - 1$ is nonnegative

Cube Roots

We often need to know what number was cubed in order to produce a certain value. When such a number is found, we say that we have found a *cube root*. For example,

2 is the cube root of 8 because $2^3 = 2 \cdot 2 \cdot 2 = 8$;

-4 is the cube root of -64 because $(-4)^3 = (-4)(-4)(-4) = -64$.

Cube Root   The number c is the *cube root* of a if $c^3 = a$. In symbols, we write $\sqrt[3]{a}$ to denote the cube root of a.

The cube-root function, given by

$$f(x) = \sqrt[3]{x}$$

has $\mathbb{R}$ as its domain. We can draw its graph by selecting convenient values for x and calculating the corresponding outputs. Once these ordered pairs have been graphed, a smooth curve can be drawn. Note that the range of $f(x)$ is also $\mathbb{R}$.

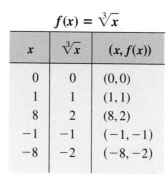

$f(x) = \sqrt[3]{x}$

x	$\sqrt[3]{x}$	$(x, f(x))$
0	0	$(0, 0)$
1	1	$(1, 1)$
8	2	$(8, 2)$
-1	-1	$(-1, -1)$
-8	-2	$(-8, -2)$

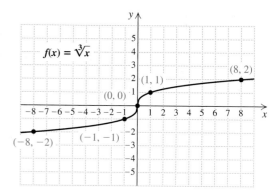

In the real-number system, every number has exactly one cube root. The cube root of a positive number is positive, the cube root of a negative number is negative, and the cube root of 0 is 0.

The following graphs illustrate that

$$\sqrt{x^2} = |x| \quad \text{and} \quad \sqrt[3]{x^3} = x.$$

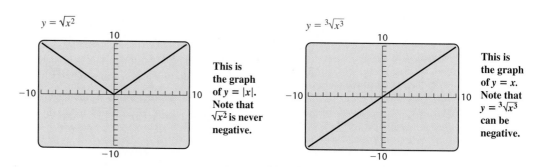

$y = \sqrt{x^2}$

This is the graph of $y = |x|$. Note that $\sqrt{x^2}$ is never negative.

$y = \sqrt[3]{x^3}$

This is the graph of $y = x$. Note that $y = \sqrt[3]{x^3}$ can be negative.

EXAMPLE 7   For each function, find the indicated function value.

a) $f(y) = \sqrt[3]{y}$; $f(125)$

b) $g(x) = \sqrt[3]{x - 3}$; $g(-24)$

SOLUTION

a) $f(125) = \sqrt[3]{125} = 5$ Since $5 \cdot 5 \cdot 5 = 125$

b) $g(-24) = \sqrt[3]{-24 - 3}$
$= \sqrt[3]{-27}$
$= -3$ Since $(-3)(-3)(-3) = -27$

EXAMPLE 8   Simplify: $\sqrt[3]{-8y^3}$.

SOLUTION

$$\sqrt[3]{-8y^3} = -2y \qquad \text{Since } (-2y)(-2y)(-2y) = -8y^3$$

Odd and Even *n*th Roots

The fourth root of a number a is the number c for which $c^4 = a$. There are also 5th roots, 6th roots, and so on. We write $\sqrt[n]{a}$ for the *n*th root. The number n is called the **index** (plural, **indices**). When the index is 2, we generally do not write it.

EXAMPLE 9   Find each of the following.

a) $\sqrt[5]{32}$ **b)** $\sqrt[5]{-32}$

c) $-\sqrt[5]{32}$ **d)** $-\sqrt[5]{-32}$

SOLUTION

a) $\sqrt[5]{32} = 2$ Since $2^5 = 32$

b) $\sqrt[5]{-32} = -2$ Since $(-2)^5 = -32$

c) $-\sqrt[5]{32} = -2$ Taking the opposite of $\sqrt[5]{32}$

d) $-\sqrt[5]{-32} = -(-2) = 2$ Taking the opposite of $\sqrt[5]{-32}$

> Every number has just one real root when n is odd. Odd roots of positive numbers are positive and odd roots of negative numbers are negative. Absolute-value signs are not used when finding odd roots.

EXAMPLE 10   Find each of the following.

a) $\sqrt[7]{x^7}$ **b)** $\sqrt[9]{(x - 1)^9}$

SOLUTION

a) $\sqrt[7]{x^7} = x$ **b)** $\sqrt[9]{(x - 1)^9} = x - 1$

When the index n is even, we say that we are taking an *even root*. Every positive real number has two real *n*th roots when n is even—one positive and one negative. Negative numbers do not have real *n*th roots when n is even.

When n is even, the notation $\sqrt[n]{a}$ indicates the nonnegative *n*th root. Thus, to write even *n*th roots, absolute-value signs are often necessary.

EXAMPLE 11   Simplify each expression, if possible. Assume that variables can represent any real number.

a) $\sqrt[4]{16}$ **b)** $-\sqrt[4]{16}$

c) $\sqrt[4]{-16}$ **d)** $\sqrt[4]{81x^4}$

e) $\sqrt[6]{(y+7)^6}$

SOLUTION

a) $\sqrt[4]{16} = 2$ Since $2^4 = 16$

b) $-\sqrt[4]{16} = -2$ Taking the opposite of $\sqrt[4]{16}$

c) $\sqrt[4]{-16}$ cannot be simplified. $\sqrt[4]{-16}$ is not a real number.

d) $\sqrt[4]{81x^4} = 3|x|$ Use absolute-value notation since x could represent a negative number.

e) $\sqrt[6]{(y+7)^6} = |y+7|$ Use absolute-value notation since $y + 7$ could be negative.

We summarize as follows.

Simplifying *n*th roots

n	*a*	$\sqrt[n]{a}$	$\sqrt[n]{a^n}$
Even	Positive	Positive	$|a|$ (or a)
	Negative	Not a real number	$|a|$ (or $-a$)
Odd	Positive	Positive	a
	Negative	Negative	a

Radical Functions and Models

A **radical function** is a function that can be described by a radical expression.

Radical Expressions and Functions

When entering radical expressions on a graphing calculator, care must be taken to place parentheses properly. For example, we enter

$$f(x) = \frac{3.5 + \sqrt{4.5 - 6x}}{5}$$

on the Y= screen as (3 . 5 + 2ND √ 4 . 5 − 6 X,T,Θ,n)) ÷ 5 . The outer parentheses enclose the numerator of the expression. The first right parenthesis) indicates the end of the radicand; the left parenthesis of the radicand is supplied by the calculator when √ is pressed.

(continued)

```
MATH NUM CPX PRB
1: ▶ Frac
2: ▶ Dec
3: 3
4: ³√(
5: ˣ√
6: fMin(
7↓fMax(
```

Cube roots can be entered using the $\sqrt[3]{}($ option in the MATH MATH menu, as shown at left. For an index other than 2 or 3, first enter the index, then choose the $\sqrt[x]{}$ option from the MATH MATH menu, and then enter the radicand. The $\sqrt[x]{}$ option may not supply the left parenthesis.

If we can determine the domain of the function algebraically, we can use that information to choose an appropriate viewing window. For example, compare the following graphs of the function $f(x) = 2\sqrt{15 - x}$. Note that the domain of f is $\{x \mid x \le 15\}$, or $(-\infty, 15]$.

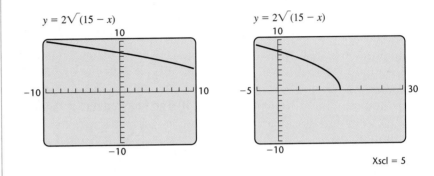

The graph on the left above uses the standard viewing window. The domain of the function is not clear from the graph, and the shape looks almost like a straight line. The viewing window for the graph on the right above is $[-5, 30, -10, 10]$. The part of the x-axis shown there includes the endpoint of the interval that gives the domain of the function. Knowing the domain of the function helps us choose Xmin and Xmax. For now, choosing appropriate values of Ymin and Ymax may require some trial and error.

If a function is given by a radical expression with an odd index, the domain is the set of all real numbers. If a function is given by a radical expression with an even index, the domain is the set of replacements for which the radicand is nonnegative.

EXAMPLE 12   Find the domain of the function given by each of the following equations. Check by graphing the function. Then, from the graph, estimate the range of the function.

a) $q(x) = \sqrt{-x}$
b) $t(x) = \sqrt{2x - 5} - 3$
c) $f(x) = \sqrt{x^2 + 1}$

SOLUTION

a) We find all values of x for which the radicand is nonnegative:

$$-x \ge 0$$
$$x \le 0. \qquad \text{Multiplying by } -1; \text{ reversing the direction of the inequality}$$

The domain is $\{x \mid x \leq 0\}$, or $(-\infty, 0]$, as indicated on the graph by the shading on the x-axis. The range appears to be $[0, \infty)$, as indicated by the shading on the y-axis. This can also be seen by examining a table of values.

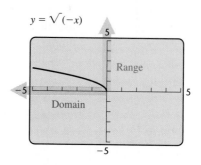

b) The function $t(x) = \sqrt{2x - 5} - 3$ is defined when $2x - 5 \geq 0$:

$$2x - 5 \geq 0$$
$$2x \geq 5 \qquad \text{Adding 5}$$
$$x \geq \tfrac{5}{2}. \qquad \text{Dividing by 2}$$

The domain is $\left\{x \mid x \geq \tfrac{5}{2}\right\}$, or $\left[\tfrac{5}{2}, \infty\right)$, as indicated on the graph by the shading on the x-axis. The range appears to be $[-3, \infty)$, as indicated by the shading on the y-axis.

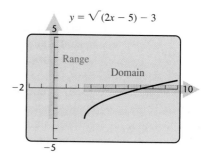

c) The radicand in $f(x) = \sqrt{x^2 + 1}$ is $x^2 + 1$. We must have

$$x^2 + 1 \geq 0$$
$$x^2 \geq -1.$$

Since x^2 is nonnegative for all real numbers x, the inequality is true for all real numbers. The domain is $(-\infty, \infty)$. The range appears to be $[1, \infty)$.

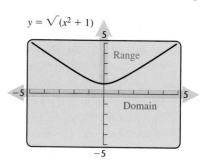

EXAMPLE 13 Determine the domain of the function given by

$$g(x) = \sqrt[6]{7 - 3x}.$$

SOLUTION Since the index is even, the radicand, $7 - 3x$, must be non-negative. We solve the inequality:

$$7 - 3x \geq 0 \qquad \text{We cannot find the 6th root of a negative number.}$$
$$-3x \geq -7$$
$$x \leq \tfrac{7}{3}. \qquad \text{Multiplying both sides by } -\tfrac{1}{3} \text{ and reversing the inequality}$$

Thus,

the domain of g is $\left\{x \,\middle|\, x \leq \tfrac{7}{3}\right\}$, or $\left(-\infty, \tfrac{7}{3}\right]$.

Some situations can be modeled using radical functions. The graphs of radical functions can have many different shapes. However, radical functions given by equations of the form

$$r(x) = \sqrt{ax + b}$$

will have the general shape of the graph of $f(x) = \sqrt{x}$.

We can determine whether a radical function might fit a set of data by plotting the points.

EXAMPLE 14 Firefighting. The number of gallons per minute discharged from a fire hose depends on the diameter of the hose and the nozzle pressure. The following table lists the amount of water flow for a 2-in. diameter solid bore nozzle at various nozzle pressures. Graph the data and determine whether a radical function can be used to model water flow.

Nozzle Pressure (in pounds per square inch, psi)	Water Flow (in gallons per minute, GPM)
40	752
60	921
80	1063
100	1188
120	1302
150	1455
200	1681

Source: www.firetactics.com

SOLUTION We graph the data, entering the nozzle pressure as L1 and the water flow as L2. The data appear to follow the pattern of the graph of a radical function. We determine that a radical function could be used to model water flow.

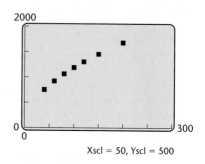

L1	L2	L3	1
40	752	------	
60	921		
80	1063		
100	1188		
120	1302		
150	1455		
200	1681		

L1(1) = 40

2000

0 ⎣_____⎦ 300
0

Xscl = 50, Yscl = 500

EXAMPLE 15 Firefighting. Water flow, as described in Example 14, depends on the diameter of the hose and the nozzle pressure. For a 2-in. diameter solid bore nozzle, the water flow is given by

$$f(x) = 118.8\sqrt{x},$$

where $f(x)$ is the water flow in gallons per minute (GPM) when the nozzle pressure is x pounds per square inch (psi) (*Source*: Houston Fire Department Continuing Education, www.houstontx.gov). Find the water flow when the nozzle pressure is 50 psi and when it is 175 psi.

SOLUTION We enter the equation and graph it along with the data in Example 14 to show that the model fits the data. Using a table with Indpnt set to Ask, we determine that when the nozzle pressure is 50 psi, the water flow is approximately 840 GPM, and at 175 psi, the water flow is approximately 1572 GPM.

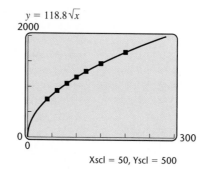

$y = 118.8\sqrt{x}$
2000

0 ⎣_____⎦ 300
0

Xscl = 50, Yscl = 500

X	Y1	
50	840.04	
175	1571.6	

X =

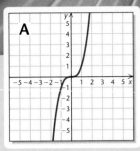

A

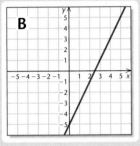

B

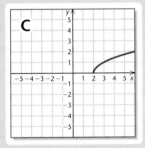

C

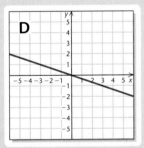

D

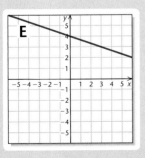

E

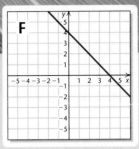

F

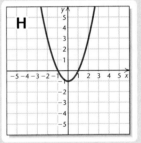

G

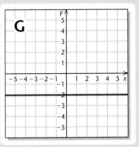

H

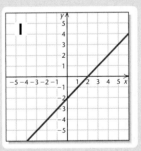

I

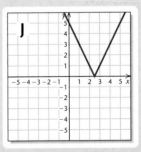

J

Visualizing the Graph

Match each function with its graph.

1. $f(x) = 2x - 5$

2. $f(x) = x^2 - 1$

3. $f(x) = \sqrt{x - 2}$

4. $f(x) = x - 2$

5. $f(x) = -\frac{1}{3}x$

6. $f(x) = x^3$

7. $f(x) = 4 - x$

8. $f(x) = |2x - 5|$

9. $f(x) = -2$

10. $f(x) = -\frac{1}{3}x + 4$

Answers on page A-38

10.1 EXERCISE SET

Concept Reinforcement *Select the appropriate word to complete each of the following.*

1. Every positive number has _____ square root(s).
negative/positive → one/two

2. The principal square root is never _____ .
negative/positive

3. For any _____ number a, we have $\sqrt{a^2} = a$.
negative/positive

4. For any _____ number a, we have $\sqrt{a^2} = -a$.
negative/positive

5. If a is a whole number that is not a perfect square, then $\sqrt{a}$ is a(n) _____ number.
irrational/rational

6. The domain of the function f given by $f(x) = \sqrt[3]{x}$ is all _____ numbers.
whole/real/positive

7. If $\sqrt[4]{x}$ is a real number, then x must be _____ .
negative/positive/nonnegative

8. If $\sqrt[3]{x}$ is negative, then x must be _____ .
negative/positive

For each number, find all of its square roots.

9. 49

10. 81

11. 144

12. 9

13. 400

14. 2500

15. 900

16. 225

Simplify.

17. $-\sqrt{\dfrac{36}{49}}$

18. $-\sqrt{\dfrac{361}{9}}$

19. $\sqrt{441}$

20. $\sqrt{196}$

21. $-\sqrt{\dfrac{16}{81}}$

22. $-\sqrt{\dfrac{81}{144}}$

23. $\sqrt{0.04}$

24. $\sqrt{0.36}$

25. $-\sqrt{0.0025}$

26. $\sqrt{0.0144}$

Identify the radicand and the index for each expression.

27. $5\sqrt{p^2} + 4$

28. $-7\sqrt{y^2} - 8$

29. $x^2 y^3 \sqrt[3]{\dfrac{x}{y+4}}$

30. $a^2 b^3 \sqrt[3]{\dfrac{a}{a^2 - b}}$

For each function, find the specified function value, if it exists. If it does not exist, state this.

31. $f(t) = \sqrt{5t - 10}$; $f(6)$, $f(2)$, $f(1)$, $f(-1)$

32. $g(x) = \sqrt{x^2 - 25}$; $g(-6)$, $g(3)$, $g(6)$, $g(13)$

33. $t(x) = -\sqrt{2x + 1}$; $t(4)$, $t(0)$, $t(-1)$, $t\left(-\frac{1}{2}\right)$

34. $p(z) = \sqrt{2z^2 - 20}$; $p(4)$, $p(3)$, $p(-5)$, $p(0)$

35. $f(t) = \sqrt{t^2 + 1}$; $f(0)$, $f(-1)$, $f(-10)$

36. $g(x) = -\sqrt{(x + 1)^2}$; $g(-3)$, $g(4)$, $g(-5)$

37. $g(x) = \sqrt{x^3 + 9}$; $g(-2)$, $g(-3)$, $g(3)$

38. $f(t) = \sqrt{t^3 - 10}$; $f(2)$, $f(3)$, $f(4)$

Simplify. Remember to use absolute-value notation when necessary. If a root cannot be simplified, state this.

39. $\sqrt{36x^2}$

40. $\sqrt{25t^2}$

41. $\sqrt{(-6b)^2}$

42. $\sqrt{(-7c)^2}$

43. $\sqrt{(8 - t)^2}$

44. $\sqrt{(a + 3)^2}$

45. $\sqrt{y^2 + 16y + 64}$

46. $\sqrt{x^2 - 4x + 4}$

47. $\sqrt{4x^2 + 28x + 49}$

48. $\sqrt{9x^2 - 30x + 25}$

49. $\sqrt[4]{256}$

50. $-\sqrt[4]{625}$

51. $\sqrt[5]{-1}$

52. $-\sqrt[5]{3^5}$

53. $\sqrt[5]{-\dfrac{32}{243}}$

54. $\sqrt[5]{-\dfrac{1}{32}}$

55. $\sqrt[6]{x^6}$

56. $\sqrt[8]{y^8}$

57. $\sqrt[4]{(6a)^4}$

58. $\sqrt[4]{(7b)^4}$

59. $\sqrt[10]{(-6)^{10}}$

60. $\sqrt[12]{(-10)^{12}}$

61. $\sqrt[414]{(a + b)^{414}}$

62. $\sqrt[1976]{(2a + b)^{1976}}$

63. $\sqrt{a^{22}}$

64. $\sqrt{x^{10}}$

65. $\sqrt{-25}$

66. $\sqrt{-16}$

Simplify. Assume that no radicands were formed by raising negative quantities to even powers.

67. $\sqrt{16x^2}$

68. $\sqrt{25t^2}$

69. $\sqrt{(3t)^2}$

70. $\sqrt{(7c)^2}$

71. $\sqrt{(a + 1)^2}$

72. $\sqrt{(5 + b)^2}$

73. $\sqrt{4x^2 + 8x + 4}$

74. $\sqrt{9x^2 + 36x + 36}$

75. $\sqrt{9t^2 - 12t + 4}$

76. $\sqrt{25t^2 - 20t + 4}$

77. $\sqrt[3]{27}$

78. $-\sqrt[3]{64}$

79. $\sqrt[4]{16x^4}$

80. $\sqrt[4]{81x^4}$

81. $\sqrt[3]{-216}$

82. $-\sqrt[5]{-100{,}000}$

83. $-\sqrt[3]{-125y^3}$

84. $-\sqrt[3]{-64x^3}$

85. $\sqrt{t^{18}}$

86. $\sqrt{a^{14}}$

87. $\sqrt{(x - 2)^8}$

88. $\sqrt{(x + 3)^{10}}$

For each function, find the specified function value, if it exists. If it does not exist, state this.

89. $f(x) = \sqrt[3]{x + 1}$; $f(7), f(26), f(-9), f(-65)$

90. $g(x) = -\sqrt[3]{2x - 1}$; $g(0), g(-62), g(-13), g(63)$

91. $g(t) = \sqrt[4]{t - 3}$; $g(19), g(-13), g(1), g(84)$

92. $f(t) = \sqrt[4]{t + 1}$; $f(0), f(15), f(-82), f(80)$

Determine the domain of each function described.

93. $f(x) = \sqrt{x - 6}$

94. $g(x) = \sqrt{x + 8}$

95. $g(t) = \sqrt[4]{t + 8}$

96. $f(x) = \sqrt[4]{x - 9}$

97. $g(x) = \sqrt[4]{2x - 10}$

98. $g(t) = \sqrt[3]{2t - 6}$

99. $f(t) = \sqrt[5]{8 - 3t}$

100. $f(t) = \sqrt[6]{4 - 3t}$

101. $h(z) = -\sqrt[6]{5z + 2}$

102. $d(x) = -\sqrt[4]{7x - 5}$

Aha! **103.** $f(t) = 7 + \sqrt[8]{t^8}$

104. $g(t) = 9 + \sqrt[6]{t^6}$

Determine algebraically the domain of each function described. Then use a graphing calculator to confirm your answer and to estimate the range.

105. $f(x) = \sqrt{5 - x}$

106. $g(x) = \sqrt{2x + 1}$

107. $f(x) = 1 - \sqrt{x + 1}$

108. $g(x) = 2 + \sqrt{3x - 5}$

109. $g(x) = 3 + \sqrt{x^2 + 4}$

110. $f(x) = 5 - \sqrt{3x^2 + 1}$

In Exercises 111–114, match each function with one of the following graphs without using a calculator.

a)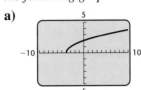

b)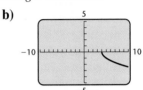

c)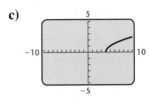

d)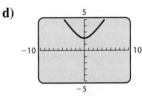

111. $f(x) = \sqrt{x - 4}$

112. $g(x) = \sqrt{x + 4}$

113. $h(x) = \sqrt{x^2 + 4}$

114. $f(x) = -\sqrt{x - 4}$

In Exercises 115–120, determine whether a radical function would be a good model of the given situation.

115. *Wind Chill.*   When the wind is blowing, the air temperature feels lower than the actual temperature. This is referred to as the wind-chill temperature. The following table lists wind-chill temperatures for various wind speeds at a thermometer reading of 15°F.

Wind Speed (in miles per hour)	Wind-Chill Temperature (in degrees Fahrenheit)
5	7
10	3
15	0
20	−2
25	−4
30	−5
35	−7
40	−8

Source: National Weather Service

116. *Farm Size.*   The following table lists the average size of United States' farms for various years from 1940 to 2002.

Year	Average Size of Farm (in acres)
1940	175
1960	303
1980	426
1997	431
2002	441

Source: U.S. Department of Agriculture

117. *Koi Growth.*   Koi, a popular fish for backyard pools, grow from $\frac{1}{40}$ cm when newly hatched to an average length of 80 cm. The following table lists the length of a koi at various ages.

Age (in months)	Length (in centimeters)
1	2.9
2	5.0
4	9.1
13	24.9
16	29.3
30	45.8
36	51.1
48	59.3
60	65.2
72	69.4

Source: www.coloradokoi.com

118. *Cancer Research.*   The following table lists the amount of federal funds allotted to the National Cancer Institute for cancer research in the United States from 2003 to 2007.

Year	Funds (in billions)
2003	$4.59
2004	4.74
2005	4.83
2006	4.79
2007*	4.75

*Requested
Source: National Cancer Institute

119. *Postage.*   The following table lists the cost to send a Media Mail Package through the U.S. Postal Service in 2006 for packages of various weights.

Weight (in pounds)	Cost
1	$1.59
2	2.07
3	2.55
4	3.03
5	3.51
6	3.99

Source: U.S. Postal Service

120. *Cable Television.*   The following table lists the number of households served by cable television from 1980 to 2002.

Year	Number of Households Served by Cable Television (in millions)
1980	17.7
1985	39.9
1990	54.9
1993	58.8
1994	60.5
1996	64.6
1998	67.6
2002	73.0

Sources: Nielsen Media Research; energycommerce.house.gov

121. *Koi Growth.*   The length, in centimeters, of a koi of age x months can be estimated using the function

$$f(x) = 0.27 + \sqrt{71.94x - 164.41}$$

(see Exercise 117). Use the function to estimate the length of a koi at 8 months and at 20 months.

122. *Cable Television.*   The number of households h, in millions, served by cable television x years after 1970 can be modeled by the function

$$h(x) = 11.67 + \sqrt{166.67x - 1659.72}$$

(see Exercise 120). Use the function to estimate the number of households served by cable television in 1992 and in 2005.

TW 123. Explain how to write the negative square root of a number using radical notation.

TW 124. Does the square root of a number's absolute value always exist? Why or why not?

Skill Maintenance

Simplify. Do not use negative exponents in your answer. [5.1], [5.2]

125. $(a^3b^2c^5)^3$

126. $(5a^7b^8)(2a^3b)$

127. $(2a^{-2}b^3c^{-4})^{-3}$

128. $(5x^{-3}y^{-1}z^2)^{-2}$

129. $\dfrac{8x^{-2}y^5}{4x^{-6}z^{-2}}$

130. $\dfrac{10a^{-6}b^{-7}}{2a^{-2}c^{-3}}$

Synthesis

TW 131. If the domain of $f = [1, \infty)$ and the range of $f = [2, \infty)$, find a possible expression for $f(x)$ and explain how such an expression is formulated.

TW 132. Kelly obtains the following graph of

$$f(x) = \sqrt{x^2 - 4x - 12}$$

and concludes that the domain of f is $(-\infty, -2]$. Is she correct? If not, what mistake is she making?

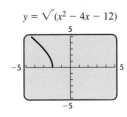

$$y = \sqrt{(x^2 - 4x - 12)}$$

TW 133. Could the following situation possibly be modeled using a radical function? Why or why not? "For each year, the yield increases. The amount of increase is smaller each year."

TW 134. Could the following situation possibly be modeled using a radical function? Why or why not? "For each year, the costs increase. The amount of increase is the same each year."

135. *Spaces in a Parking Lot.*   A parking lot has attendants to park the cars. The number N of stalls needed for waiting cars before attendants can get to them is given by the formula $N = 2.5\sqrt{A}$, where A is the number of arrivals in peak hours. Find the number of spaces needed for the given number of arrivals in peak hours: **(a)** 25; **(b)** 36; **(c)** 49; **(d)** 64.

Determine the domain of each function described. Then draw the graph of each function.

136. $g(x) = \sqrt{x} + 5$

137. $f(x) = \sqrt{x + 5}$

138. $f(x) = \sqrt{x} - 2$

139. $g(x) = \sqrt{x - 2}$

140. Find the domain of f if

$$f(x) = \frac{\sqrt{x + 3}}{\sqrt[4]{2 - x}}.$$

141. Find the domain of g if

$$g(x) = \frac{\sqrt[4]{5 - x}}{\sqrt[6]{x + 4}}.$$

142. Find the domain of F if $F(x) = \dfrac{x}{\sqrt{x^2 - 5x - 6}}.$

143. Examine the graph of the data in Exercise 118. What type of function could be used to model the data?

144. Examine the graph of the data in Exercise 119. What type of function could be used to model the data?

10.2 Rational Numbers as Exponents

Rational Exponents ▧ Negative Rational Exponents ▧
Laws of Exponents ▧ Simplifying Radical Expressions ▧

In Section 5.1, we considered the natural numbers as exponents. Our discussion of exponents was expanded to include all integers in Section 5.2. We now expand the study still further—to include all rational numbers. This will give meaning to expressions like $a^{1/3}$, $7^{-1/2}$, and $(3x)^{4/5}$. Such notation will help us simplify certain radical expressions.

Rational Exponents

Consider $a^{1/2} \cdot a^{1/2}$. To add exponents when multiplying, it must follow that $a^{1/2} \cdot a^{1/2} = a^{1/2+1/2}$, or a^1. This suggests that $a^{1/2}$ is a square root of a. Similarly, $a^{1/3} \cdot a^{1/3} \cdot a^{1/3} = a^{1/3+1/3+1/3}$, or a^1, so $a^{1/3}$ should mean $\sqrt[3]{a}$.

> $a^{1/n} = \sqrt[n]{a}$ $a^{1/n}$ means $\sqrt[n]{a}$. When a is nonnegative, n can be any natural number greater than 1. When a is negative, n must be odd.

Note that the denominator of the exponent becomes the index and the base becomes the radicand.

EXAMPLE 1   Write an equivalent expression using radical notation.

a) $x^{1/2}$ **b)** $(-8)^{1/3}$

c) $(abc)^{1/5}$ **d)** $(25x^{16})^{1/2}$

SOLUTION

a) $x^{1/2} = \sqrt{x}$

b) $(-8)^{1/3} = \sqrt[3]{-8} = -2$

c) $(abc)^{1/5} = \sqrt[5]{abc}$

d) $(25x^{16})^{1/2} = \sqrt{25x^{16}} = 5x^8$

The denominator of the exponent becomes the index. The base becomes the radicand. Recall that for square roots, the index 2 is understood without being written.

EXAMPLE 2   Write an equivalent expression using exponential notation.

a) $\sqrt[5]{9xy}$ **b)** $\sqrt[7]{\dfrac{x^3 y}{4}}$ **c)** $\sqrt{5x}$

SOLUTION   Parentheses are required to indicate the base.

a) $\sqrt[5]{9xy} = (9xy)^{1/5}$

b) $\sqrt[7]{\dfrac{x^3 y}{4}} = \left(\dfrac{x^3 y}{4}\right)^{1/7}$

The index becomes the denominator of the exponent. The radicand becomes the base.

c) $\sqrt{5x} = (5x)^{1/2}$

The index 2 is understood without being written. We assume $x \geq 0$.

> **CAUTION!**   When converting from radical notation to exponential notation, parentheses are necessary to indicate the base.
>
> $$\sqrt{5x} = (5x)^{1/2}$$

Rational Exponents

We can enter a radical expression in radical notation or by using rational exponents.

To use rational exponents, enter the radicand enclosed in parentheses, then press ⌃, and then enter the rational exponent, also enclosed in parentheses. If the rational exponent is written as a single decimal number, the parentheses around the exponent are unnecessary. If decimal notation for a rational exponent must be rounded, fraction notation should be used.

▶ **EXAMPLE 3**   Graph: $f(x) = \sqrt[4]{2x - 7}$.

SOLUTION   We can enter the equation in radical notation using $\sqrt[x]{\;}$. Alternatively, we can rewrite the radical expression using a rational exponent:

$$f(x) = (2x - 7)^{1/4}, \quad \text{or} \quad f(x) = (2x - 7)^{0.25}.$$

Then we let $y = (2x - 7) \wedge (1/4)$ or $y = (2x - 7) \wedge .25$. The screen on the left below shows all three forms of the equation; they are equivalent.

Knowing the domain of the function can help us determine an appropriate viewing window. Since the index is even, the domain is the set of all x for which the radicand is nonnegative, or $\left[\frac{7}{2}, \infty\right)$. We choose a viewing window of $[-1, 10, -1, 5]$. This will show the axes and the first quadrant. The graph is shown on the right below.

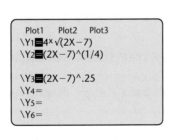

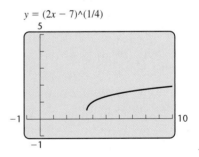

How shall we define $a^{2/3}$? If the property for multiplying exponents is to hold, we must have $a^{2/3} = (a^{1/3})^2$ and $a^{2/3} = (a^2)^{1/3}$. This would suggest that $a^{2/3} = \left(\sqrt[3]{a}\right)^2$ and $a^{2/3} = \sqrt[3]{a^2}$. We make our definition accordingly.

> **Positive Rational Exponents**   For any natural numbers m and n ($n \neq 1$) and any real number a for which $\sqrt[n]{a}$ exists,
>
> $$a^{m/n} \quad \text{means} \quad \left(\sqrt[n]{a}\right)^m, \quad \text{or} \quad \sqrt[n]{a^m}.$$

Student Notes

It is important to remember both meanings of $a^{m/n}$. When the root of the base a is known, $(\sqrt[n]{a})^m$ is generally easier to work with. When it is not known, $\sqrt[n]{a^m}$ is often more convenient.

EXAMPLE 4   Write an equivalent expression using radical notation and simplify.

a) $27^{2/3}$ **b)** $25^{3/2}$

SOLUTION

a) $27^{2/3}$ means $\left(\sqrt[3]{27}\right)^2$ or, equivalently, $\sqrt[3]{27^2}$. Let's see which is easier to simplify:

$$\left(\sqrt[3]{27}\right)^2 = 3^2 \qquad\qquad \sqrt[3]{27^2} = \sqrt[3]{729}$$
$$= 9; \qquad\qquad\qquad\qquad = 9.$$

The simplification on the left is probably easier for most people.

b) $25^{3/2}$ means $\left(\sqrt[2]{25}\right)^3$ or, equivalently, $\sqrt[2]{25^3}$ (the index 2 is normally omitted). Since $\sqrt{25}$ is more commonly known than $\sqrt{25^3}$, we use that form:

$$25^{3/2} = \left(\sqrt{25}\right)^3 = 5^3 = 125.$$

EXAMPLE 5   Write an equivalent expression using exponential notation.

a) $\sqrt[3]{9^4}$ **b)** $\left(\sqrt[4]{7xy}\right)^5$

SOLUTION

a) $\sqrt[3]{9^4} = 9^{4/3}$

b) $\left(\sqrt[4]{7xy}\right)^5 = (7xy)^{5/4}$ **The index becomes the denominator of the fraction that is the exponent.**

Rational roots of numbers can be approximated on a calculator.

EXAMPLE 6   Approximate $\sqrt[5]{(-23)^3}$. Round to the nearest thousandth.

SOLUTION   We first rewrite the expression using a rational exponent:

$$\sqrt[5]{(-23)^3} = (-23)^{3/5}.$$

Using a calculator, we have

$$(-23)^{\wedge}(3/5) \approx -6.562.$$

Negative Rational Exponents

Recall that $x^{-2} = \dfrac{1}{x^2}$. Negative rational exponents behave similarly.

Negative Rational Exponents   For any rational number m/n and any nonzero real number a for which $a^{m/n}$ exists,

$$a^{-m/n} \quad \text{means} \quad \frac{1}{a^{m/n}}.$$

CAUTION!   A negative exponent does not indicate that the expression in which it appears is negative.

EXAMPLE 7   Write an equivalent expression with positive exponents and, if possible, simplify.

a) $9^{-1/2}$ **b)** $(5xy)^{-4/5}$ **c)** $64^{-2/3}$

d) $4x^{-2/3}y^{1/5}$ **e)** $\left(\dfrac{3r}{7s}\right)^{-5/2}$

SOLUTION

a) $9^{-1/2} = \dfrac{1}{9^{1/2}}$ $9^{-1/2}$ is the reciprocal of $9^{1/2}$.

Since $9^{1/2} = \sqrt{9} = 3$, the answer simplifies to $\dfrac{1}{3}$.

b) $(5xy)^{-4/5} = \dfrac{1}{(5xy)^{4/5}}$ $(5xy)^{-4/5}$ is the reciprocal of $(5xy)^{4/5}$.

c) $64^{-2/3} = \dfrac{1}{64^{2/3}}$ $64^{-2/3}$ is the reciprocal of $64^{2/3}$.

Since $64^{2/3} = \left(\sqrt[3]{64}\right)^2 = 4^2 = 16$, the answer simplifies to $\dfrac{1}{16}$.

d) $4x^{-2/3}y^{1/5} = 4 \cdot \dfrac{1}{x^{2/3}} \cdot y^{1/5} = \dfrac{4y^{1/5}}{x^{2/3}}$

e) In Section 5.2, we found that $(a/b)^{-n} = (b/a)^n$. This property holds for *any* negative exponent:

$$\left(\dfrac{3r}{7s}\right)^{-5/2} = \left(\dfrac{7s}{3r}\right)^{5/2}.$$ Writing the reciprocal of the base and changing the sign of the exponent

Laws of Exponents

The same laws hold for rational exponents as for integer exponents.

Laws of Exponents   For any real numbers a and b and any rational exponents m and n for which a^m, a^n, and b^m are defined:

1. $a^m \cdot a^n = a^{m+n}$ In multiplying, add exponents if the bases are the same.

2. $\dfrac{a^m}{a^n} = a^{m-n}$ In dividing, subtract exponents if the bases are the same. (Assume $a \neq 0$.)

3. $(a^m)^n = a^{m \cdot n}$ To raise a power to a power, multiply the exponents.

4. $(ab)^m = a^m b^m$ To raise a product to a power, raise each factor to the power and multiply.

EXAMPLE 8   Use the laws of exponents to simplify.

a) $3^{1/5} \cdot 3^{3/5}$ **b)** $a^{1/4}/a^{1/2}$

c) $(7.2^{2/3})^{3/4}$ **d)** $(a^{-1/3}b^{2/5})^{1/2}$

SOLUTION

a) $3^{1/5} \cdot 3^{3/5} = 3^{1/5+3/5} = 3^{4/5}$ **Adding exponents**

b) $\dfrac{a^{1/4}}{a^{1/2}} = a^{1/4-1/2} = a^{1/4-2/4}$ **Subtracting exponents after finding a common denominator**

$$= a^{-1/4}, \text{ or } \frac{1}{a^{1/4}}$$ $a^{-1/4}$ **is the reciprocal of** $a^{1/4}$.

c) $(7.2^{2/3})^{3/4} = 7.2^{(2/3)(3/4)} = 7.2^{6/12}$ **Multiplying exponents**

$$= 7.2^{1/2}$$ **Using arithmetic to simplify the exponent**

d) $(a^{-1/3}b^{2/5})^{1/2} = a^{(-1/3)(1/2)} \cdot b^{(2/5)(1/2)}$ **Raising a product to a power and multiplying exponents**

$$= a^{-1/6}b^{1/5}, \text{ or } \frac{b^{1/5}}{a^{1/6}}$$

Simplifying Radical Expressions

Many radical expressions can be simplified using rational exponents.

To Simplify Radical Expressions

1. Convert radical expressions to exponential expressions.
2. Use arithmetic and the laws of exponents to simplify.
3. Convert back to radical notation as needed.

EXAMPLE 9 Use rational exponents to simplify. Do not use exponents that are fractions in the final answer.

a) $\sqrt[6]{(5x)^3}$ **b)** $\sqrt[5]{t^{20}}$

c) $\left(\sqrt[3]{ab^2c}\right)^{12}$ **d)** $\sqrt{\sqrt[3]{x}}$

$y_1 = 6 \sqrt[x]{((5x)^3)}$, $y_2 = \sqrt{(5x)}$

X	Y1	Y2
0	0	0
1	2.2361	2.2361
2	3.1623	3.1623
3	3.873	3.873
4	4.4721	4.4721
5	5	5
6	5.4772	5.4772

X = 0

SOLUTION

a) $\sqrt[6]{(5x)^3} = (5x)^{3/6}$ **Converting to exponential notation**

$$= (5x)^{1/2}$$ **Simplifying the exponent**

$$= \sqrt{5x}$$ **Returning to radical notation**

To check on a graphing calculator, we let $y_1 = \sqrt[6]{(5x)^3}$ and $y_2 = \sqrt{5x}$ and compare values in a table. If we scroll through the table (see the figure at left), we see that $y_1 = y_2$, so our simplification is probably correct.

b) $\sqrt[5]{t^{20}} = t^{20/5}$ **Converting to exponential notation**

$$= t^4$$ **Simplifying the exponent**

c) $\left(\sqrt[3]{ab^2c}\right)^{12} = (ab^2c)^{12/3}$ **Converting to exponential notation**

$$= (ab^2c)^4$$ **Simplifying the exponent**

$$= a^4b^8c^4$$ **Using the laws of exponents**

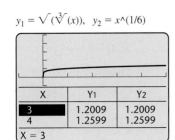

$y_1 = \sqrt{}(\sqrt[3]{}(x)), \quad y_2 = x^{\wedge}(1/6)$

X	Y1	Y2
3	1.2009	1.2009
4	1.2599	1.2599

X = 3

d) $\sqrt{\sqrt[3]{x}} = \sqrt{x^{1/3}}$ Converting the radicand to exponential notation

$\phantom{\sqrt{\sqrt[3]{x}}} = (x^{1/3})^{1/2}$ Try to go directly to this step.

$\phantom{\sqrt{\sqrt[3]{x}}} = x^{1/6}$ Using the laws of exponents

$\phantom{\sqrt{\sqrt[3]{x}}} = \sqrt[6]{x}$ Returning to radical notation

We can check by graphing $y_1 = \sqrt{\sqrt[3]{x}}$ and $y_2 = \sqrt[6]{x}$. The graphs coincide, as we also see by scrolling through the table of values shown at left.

10.2 EXERCISE SET

🔖 *Concept Reinforcement*   *In each of Exercises 1–8, match the expression with the equivalent expression from the column on the right.*

1. ____ $x^{2/5}$ **a)** $x^{3/5}$

2. ____ $x^{5/2}$ **b)** $\left(\sqrt[5]{x}\right)^4$

3. ____ $x^{-5/2}$ **c)** $\sqrt{x^5}$

4. ____ $x^{-2/5}$ **d)** $x^{1/2}$

5. ____ $x^{1/5} \cdot x^{2/5}$

6. ____ $(x^{1/5})^{5/2}$ **e)** $\dfrac{1}{\left(\sqrt{x}\right)^5}$

7. ____ $\sqrt[5]{x^4}$ **f)** $\sqrt[4]{x^5}$

8. ____ $\left(\sqrt[4]{x}\right)^5$ **g)** $\sqrt[5]{x^2}$

                        h) $\dfrac{1}{\left(\sqrt[5]{x}\right)^2}$

Note: Assume for all exercises that even roots are of nonnegative quantities and that all denominators are nonzero.

Write an equivalent expression using radical notation and, if possible, simplify.

9. $x^{1/6}$ 10. $y^{1/5}$

11. $16^{1/2}$ 12. $8^{1/3}$

13. $81^{1/4}$ 14. $64^{1/6}$

15. $9^{1/2}$ 16. $25^{1/2}$

17. $(xyz)^{1/3}$ 18. $(ab)^{1/4}$

19. $(a^2b^2)^{1/5}$ 20. $(x^3y^3)^{1/4}$

21. $t^{2/5}$ 22. $b^{3/2}$

23. $16^{3/4}$ 24. $4^{7/2}$

25. $27^{4/3}$ 26. $9^{5/2}$

27. $(81x)^{3/4}$ 28. $(125a)^{2/3}$

29. $(25x^4)^{3/2}$ 30. $(9y^6)^{3/2}$

Write an equivalent expression using exponential notation.

31. $\sqrt[3]{20}$ 32. $\sqrt[3]{19}$

33. $\sqrt{17}$ 34. $\sqrt{6}$

35. $\sqrt{x^3}$ 36. $\sqrt{a^5}$

37. $\sqrt[5]{m^2}$ 38. $\sqrt[5]{n^4}$

39. $\sqrt[4]{cd}$ 40. $\sqrt[5]{xy}$

41. $\sqrt[5]{xy^2z}$ 42. $\sqrt{x^3y^2z^2}$

43. $\left(\sqrt{3mn}\right)^3$ 44. $\left(\sqrt[3]{7xy}\right)^4$

45. $\left(\sqrt[7]{8x^2y}\right)^5$ 46. $\left(\sqrt[6]{2a^5b}\right)^7$

47. $\dfrac{2x}{\sqrt[3]{z^2}}$ 48. $\dfrac{3a}{\sqrt[5]{c^2}}$

Write an equivalent expression with positive exponents and, if possible, simplify.

49. $x^{-1/3}$ 50. $y^{-1/4}$

51. $(2rs)^{-3/4}$ 52. $(5xy)^{-5/6}$

53. $\left(\dfrac{1}{16}\right)^{-3/4}$ **54.** $\left(\dfrac{1}{8}\right)^{-2/3}$

55. $\dfrac{2c}{a^{-3/5}}$ **56.** $\dfrac{3b}{a^{-5/7}}$

57. $5x^{-2/3}y^{4/5}z$ **58.** $2a^{3/4}b^{-1/2}c^{2/3}$

59. $3^{-5/2}a^3b^{-7/3}$ **60.** $2^{-1/3}x^4y^{-2/7}$

61. $\left(\dfrac{2ab}{3c}\right)^{-5/6}$ **62.** $\left(\dfrac{7x}{8yz}\right)^{-3/5}$

63. $\dfrac{6a}{\sqrt[4]{b}}$ **64.** $\dfrac{7x}{\sqrt[3]{z}}$

Graph using a graphing calculator.

65. $f(x) = \sqrt[4]{x+7}$ **66.** $g(x) = \sqrt[5]{4-x}$

67. $r(x) = \sqrt[7]{3x-2}$ **68.** $q(x) = \sqrt[6]{2x+3}$

69. $f(x) = \sqrt[6]{x^3}$ **70.** $g(x) = \sqrt[8]{x^2}$

Approximate. Round to the nearest thousandth.

71. $\sqrt[5]{9}$ **72.** $\sqrt[6]{13}$

73. $\sqrt[4]{10}$ **74.** $\sqrt[7]{-127}$

75. $\sqrt[3]{(-3)^5}$ **76.** $\sqrt[10]{(1.5)^6}$

Use the laws of exponents to simplify. Do not use negative exponents in any answers.

77. $7^{3/4} \cdot 7^{1/8}$ **78.** $11^{2/3} \cdot 11^{1/2}$

79. $\dfrac{3^{5/8}}{3^{-1/8}}$ **80.** $\dfrac{8^{7/11}}{8^{-2/11}}$

81. $\dfrac{5.2^{-1/6}}{5.2^{-2/3}}$ **82.** $\dfrac{2.3^{-3/10}}{2.3^{-1/5}}$

83. $(10^{3/5})^{2/5}$ **84.** $(5^{5/4})^{3/7}$

85. $a^{2/3} \cdot a^{5/4}$ **86.** $x^{3/4} \cdot x^{1/3}$

Aha! **87.** $(64^{3/4})^{4/3}$ **88.** $(27^{-2/3})^{3/2}$

89. $(m^{2/3}n^{-1/4})^{1/2}$ **90.** $(x^{-1/3}y^{2/5})^{1/4}$

Use rational exponents to simplify. Do not use fraction exponents in the final answer.

91. $\sqrt[6]{x^4}$ **92.** $\sqrt[6]{a^2}$

93. $\sqrt[4]{a^{12}}$ **94.** $\sqrt[3]{x^{15}}$

95. $\sqrt[5]{a^{10}}$ **96.** $\sqrt[6]{x^{18}}$

97. $\left(\sqrt[7]{xy}\right)^{14}$ **98.** $\left(\sqrt[3]{ab}\right)^{15}$

99. $\sqrt[4]{(7a)^2}$ **100.** $\sqrt[8]{(3x)^2}$

101. $\left(\sqrt[8]{2x}\right)^6$ **102.** $\left(\sqrt[10]{3a}\right)^5$

103. $\sqrt[3]{\sqrt[6]{a}}$ **104.** $\sqrt[4]{\sqrt{x}}$

105. $\sqrt[4]{(xy)^{12}}$ **106.** $\sqrt{(ab)^6}$

107. $\left(\sqrt[5]{a^2b^4}\right)^{15}$ **108.** $\left(\sqrt[3]{x^2y^5}\right)^{12}$

109. $\sqrt[3]{\sqrt[4]{xy}}$ **110.** $\sqrt[5]{\sqrt{2a}}$

TW **111.** If $f(x) = (x+5)^{1/2}(x+7)^{-1/2}$, find the domain of f. Explain how you found your answer.

TW **112.** Explain why $\sqrt[3]{x^6} = x^2$ for any value of x, whereas $\sqrt[2]{x^6} = x^3$ only when $x \geq 0$.

Skill Maintenance

Simplify.

113. $3x(x^3 - 2x^2) + 4x^2(2x^2 + 5x)$ [5.4], [5.5]

114. $5t^3(2t^2 - 4t) - 3t^4(t^2 - 6t)$ [5.4], [5.5]

115. $(3a - 4b)(5a + 3b)$ [5.6], [5.7]

116. $(7x - y)^2$ [5.6], [5.7]

117. *Real Estate Taxes.* For homes under \$100,000, the property transfer tax in Vermont is 0.5% of the selling price. Find the selling price of a home that had a transfer tax of \$467.50. [2.4]

118. What numbers are their own squares? [6.7]

Synthesis

TW **119.** Let $f(x) = 5x^{-1/3}$. Under what condition will we have $f(x) > 0$? Why?

TW **120.** If $g(x) = x^{3/n}$, in what way does the domain of g depend on whether n is odd or even?

Use rational exponents to simplify.

121. $\sqrt{x\sqrt[3]{x^2}}$ **122.** $\sqrt[4]{\sqrt[3]{8x^3y^6}}$

123. $\sqrt[12]{p^2 + 2pq + q^2}$

*Music. The function given by $f(x) = k2^{x/12}$ can be used to determine the frequency, in cycles per second, of a musical note that is x half-steps above a note with frequency k.**

124. The frequency of concert A for a trumpet is 440 cycles per second. Find the frequency of the A that is two octaves (24 half-steps) above concert A (few trumpeters can reach this note.)

---

*This application was inspired by information provided by Dr. Homer B. Tilton of Pima Community College East.

125. Show that the G that is 7 half-steps (a "perfect fifth") above middle C (262 cycles per second) has a frequency that is about 1.5 times that of middle C.

126. Show that the C sharp that is 4 half-steps (a "major third") above concert A (see Exercise 124) has a frequency that is about 25% greater than that of concert A.

127. *Baseball.*   The statistician Bill James has found that a baseball team's winning percentage P can be approximated by

$$P = \frac{r^{1.83}}{r^{1.83} + \sigma^{1.83}},$$

where r is the total number of runs scored by that team and σ is the total number of runs scored by their opponents (*Source*: M. Bittinger, *One Man's Journey Through Mathematics*. Boston: Addison-Wesley, 2004). During a recent season, the San Francisco Giants scored 799 runs and their opponents scored 749 runs. Use James's formula to predict the Giants' winning percentage (the team actually won 55.6% of their games).

128. *Road Pavement Messages.*   In a psychological study, it was determined that the proper length L of the letters of a word printed on pavement is given by

$$L = \frac{0.000169d^{2.27}}{h},$$

where d is the distance of a car from the lettering and h is the height of the eye above the surface of the road. All units are in meters. This formula says that if a person is h meters above the surface of

the road and is to be able to recognize a message d meters away, that message will be the most recognizable if the length of the letters is L. Find L to the nearest tenth of a meter, given d and h.

a) $h = 1$ m, $d = 60$ m
b) $h = 0.9906$ m, $d = 75$ m
c) $h = 2.4$ m, $d = 80$ m
d) $h = 1.1$ m, $d = 100$ m

129. *Physics.*   The equation $m = m_0(1 - v^2c^{-2})^{-1/2}$, developed by Albert Einstein, is used to determine the mass m of an object that is moving v meters per second and has mass m_0 before the motion begins. The constant c is the speed of light, approximately 3×10^8 m/sec. Suppose that a particle with mass 8 mg is accelerated to a speed of $\frac{9}{5} \times 10^8$ m/sec. Without using a calculator, find the new mass of the particle.

130. A person's body surface area (BSA) can be approximated by the DuBois formula

$$\text{BSA} = 0.007184w^{0.425}h^{0.725},$$

where w is mass, in kilograms, h is height, in centimeters, and BSA is in square meters (*Source*: www.halls.md). What is the BSA of a child who is 122 cm tall and has a mass of 29.5 kg?

131. Using a graphing calculator, select the **MODE** SIMUL and the **FORMAT** EXPROFF. Then graph

$$y_1 = x^{1/2}, \qquad y_2 = 3x^{2/5},$$
$$y_3 = x^{4/7}, \quad \text{and} \quad y_4 = \tfrac{1}{5}x^{3/4}.$$

Looking only at coordinates, match each graph with its equation.

Collaborative Corner

Are Equivalent Fractions Equivalent Exponents?

Focus: Functions and rational exponents
Time: 10–20 minutes
Group Size: 3
Materials: Graph paper

In arithmetic, we have seen that $\frac{1}{3}, \frac{1}{6} \cdot 2$, and $2 \cdot \frac{1}{6}$ all represent the same number. Interestingly,

$$f(x) = x^{1/3},$$
$$g(x) = (x^{1/6})^2, \quad \text{and}$$
$$h(x) = (x^2)^{1/6}$$

represent three *different* functions.

ACTIVITY

1. Selecting a variety of values for x and using the definition of positive rational exponents, one group member should graph f, a second group member should graph g, and a third group member should graph h. Be sure to check whether negative x-values are in the domain of the function.

2. Compare the three graphs and check each other's work. How and why do the graphs differ?

3. Decide as a group which graph, if any, would best represent the graph of $k(x) = x^{2/6}$. Then be prepared to explain your reasoning to the entire class. (*Hint*: Study the definition of $a^{m/n}$ on p. 741 carefully.)

10.3 Multiplying Radical Expressions

Multiplying Radical Expressions ◼ Simplifying by Factoring ◼
Multiplying and Simplifying

Study Tip

Interested in Something Extra?

Students interested in extra-credit projects are often pleasantly surprised to find that their instructor will not only provide a possible project for them to pursue, but might even tailor that project to the student's interests. Remember that if you do an extra-credit project, you must still attend to the regular course work.

Multiplying Radical Expressions

Note that $\sqrt{4}\,\sqrt{25} = 2 \cdot 5 = 10$. Also $\sqrt{4 \cdot 25} = \sqrt{100} = 10$. Likewise,
$$\sqrt[3]{27}\,\sqrt[3]{8} = 3 \cdot 2 = 6 \quad \text{and} \quad \sqrt[3]{27 \cdot 8} = \sqrt[3]{216} = 6.$$

These examples suggest the following.

> **The Product Rule for Radicals** For any real numbers $\sqrt[n]{a}$ and $\sqrt[n]{b}$,
> $$\sqrt[n]{a} \cdot \sqrt[n]{b} = \sqrt[n]{a \cdot b}.$$

(The product of two nth roots is the nth root of the product of the two radicands.)

Rational exponents can be used to derive this rule:

$$\sqrt[n]{a} \cdot \sqrt[n]{b} = a^{1/n} \cdot b^{1/n} = (a \cdot b)^{1/n} = \sqrt[n]{a \cdot b}.$$

EXAMPLE 1   Multiply.

a) $\sqrt{3} \cdot \sqrt{5}$
b) $\sqrt{x + 3}\,\sqrt{x - 3}$
c) $\sqrt[3]{4} \cdot \sqrt[3]{5}$
d) $\sqrt[4]{\dfrac{y}{5}} \cdot \sqrt[4]{\dfrac{7}{x}}$

SOLUTION

a) When no index is written, roots are understood to be square roots with an unwritten index of two. We apply the product rule:

$$\sqrt{3} \cdot \sqrt{5} = \sqrt{3 \cdot 5}$$
$$= \sqrt{15}.$$

b) $\sqrt{x + 3}\,\sqrt{x - 3} = \sqrt{(x + 3)(x - 3)}$ The product of two square roots is the square root of the product.

$$= \sqrt{x^2 - 9}$$

> **CAUTION!**
> $\sqrt{x^2 - 9} \neq \sqrt{x^2} - \sqrt{9}.$

c) Both $\sqrt[3]{4}$ and $\sqrt[3]{5}$ have indices of three, so to multiply we can use the product rule:

$$\sqrt[3]{4} \cdot \sqrt[3]{5} = \sqrt[3]{4 \cdot 5} = \sqrt[3]{20}.$$

d) $\sqrt[4]{\dfrac{y}{5}} \cdot \sqrt[4]{\dfrac{7}{x}} = \sqrt[4]{\dfrac{y}{5} \cdot \dfrac{7}{x}} = \sqrt[4]{\dfrac{7y}{5x}}$ In Section 10.4, we discuss other ways to write answers like this.

> **CAUTION!**   The product rule for radicals applies only when radicals have the same index:
> $$\sqrt[n]{a} \cdot \sqrt[m]{b} \neq \sqrt[nm]{a \cdot b}.$$

Connecting the Concepts

INTERPRETING GRAPHS: DOMAINS OF RADICAL FUNCTIONS

Although, as we saw in Example 1(b), $\sqrt{x+3}\,\sqrt{x-3} = \sqrt{x^2-9}$, it is not true that $f(x) = \sqrt{x+3}\,\sqrt{x-3}$ and $g(x) = \sqrt{x^2-9}$ represent the same function. The domains of f and g are not the same.

Since f is written as the product of two rational expressions, both $\sqrt{x+3}$ and $\sqrt{x-3}$ must be defined for all inputs for f. Thus we must have

$$x + 3 \geq 0 \quad and \quad x - 3 \geq 0$$
$$x \geq -3 \quad and \quad x \geq 3.$$

The domain of f is thus $\{x \mid x \geq -3 \ and \ x \geq 3\}$, or $[3, \infty)$.

The function g is defined when $x^2 - 9 \geq 0$, or for $\{x \mid x \leq -3 \ or \ x \geq 3\}$. Thus the domain of g is $(-\infty, -3] \cup [3, \infty)$.

These domains are confirmed by the following graphs. Note that the graphs do coincide where the domains coincide. We say that the expressions are equivalent because they represent the same number for all possible replacements for *both* expressions.

$y = \sqrt{(x+3)}\sqrt{(x-3)}$

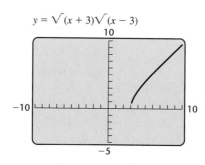

$y = \sqrt{(x^2-9)}$

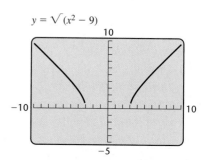

Simplifying by Factoring

An integer p is a *perfect square* if there exists a rational number q for which $q^2 = p$. We say that p is a *perfect cube* if $q^3 = p$ for some rational number q. In general, p is a *perfect nth power* if $q^n = p$ for some rational number q. The product rule allows us to simplify $\sqrt[n]{ab}$ when a or b is a perfect nth power.

Using the Product Rule to Simplify

$$\sqrt[n]{ab} = \sqrt[n]{a} \cdot \sqrt[n]{b}.$$

($\sqrt[n]{a}$ and $\sqrt[n]{b}$ must both be real numbers.)

To illustrate, suppose we wish to simplify $\sqrt{20}$. Since this is a *square* root, we check to see if there is a factor of 20 that is a perfect square. There is one, 4, so we express 20 as $4 \cdot 5$ and use the product rule.

$$\sqrt{20} = \sqrt{4 \cdot 5} \qquad \text{Factoring the radicand (4 is a perfect square)}$$
$$= \sqrt{4} \cdot \sqrt{5} \qquad \text{Factoring into two radicals}$$
$$= 2\sqrt{5} \qquad \text{Taking the square root of 4}$$

To Simplify a Radical Expression with Index *n* by Factoring

1. Express the radicand as a product in which one factor is the largest perfect *n*th power possible.
2. Take the *n*th root of each factor.
3. Simplification is complete when no radicand has a factor that is a perfect *n*th power.

It is often safe to assume that a radicand does not represent a negative number raised to an even power. We will henceforth make this assumption—unless functions are involved—and discontinue use of absolute-value notation when taking even roots.

▶ **EXAMPLE 2** Simplify by factoring: **(a)** $\sqrt{200}$; **(b)** $\sqrt{18x^2y}$; **(c)** $\sqrt[3]{72}$; **(d)** $\sqrt[4]{162x^6}$.

SOLUTION

a) $\sqrt{200} = \sqrt{100 \cdot 2}$ 100 is the largest perfect-square factor of 200.
$$= \sqrt{100} \cdot \sqrt{2} = 10\sqrt{2}$$

b) $\sqrt{18x^2y} = \sqrt{9 \cdot 2 \cdot x^2 \cdot y}$ $9x^2$ is the largest perfect-square factor of $18x^2y$.
$$= \sqrt{9x^2} \cdot \sqrt{2y} \qquad \text{Factoring into two radicals}$$
$$= 3x\sqrt{2y} \qquad \text{Taking the square root of } 9x^2$$

c) $\sqrt[3]{72} = \sqrt[3]{8 \cdot 9}$ 8 is the largest perfect-cube (third-power) factor of 72.
$$= \sqrt[3]{8} \cdot \sqrt[3]{9} = 2\sqrt[3]{9}$$

Let's look at this example another way. We write a complete factorization and look for triples of factors. Each triple of factors makes a cube:

$$\sqrt[3]{72} = \sqrt[3]{\underline{2 \cdot 2 \cdot 2} \cdot 3 \cdot 3} \qquad \text{Each triple of factors is a cube.}$$
$$= 2\sqrt[3]{3 \cdot 3} = 2\sqrt[3]{9}.$$

d) $\sqrt[4]{162x^6} = \sqrt[4]{81 \cdot 2 \cdot x^4 \cdot x^2}$ $81 \cdot x^4$ is the largest perfect fourth-power factor of $162x^6$.
$$= \sqrt[4]{81x^4} \cdot \sqrt[4]{2x^2} \qquad \text{Factoring into two radicals}$$
$$= 3x\sqrt[4]{2x^2} \qquad \text{Taking fourth roots}$$

Let's look at this example another way. We write a complete factorization and look for quadruples of factors. Each quadruple makes a perfect fourth power:

$$\sqrt[4]{162x^6} = \sqrt[4]{\underline{3 \cdot 3 \cdot 3 \cdot 3} \cdot 2 \cdot \underline{x \cdot x \cdot x \cdot x} \cdot x \cdot x} \qquad \text{Each quadruple of factors is a power of 4.}$$

$$= 3 \cdot x \cdot \sqrt[4]{2 \cdot x \cdot x} = 3x\sqrt[4]{2x^2}.$$

> **EXAMPLE 3** If $f(x) = \sqrt{3x^2 - 6x + 3}$, find a simplified form for $f(x)$.
>
> **SOLUTION**
>
> $$\begin{aligned} f(x) &= \sqrt{3x^2 - 6x + 3} \\ &= \left.\begin{aligned}&\sqrt{3(x^2 - 2x + 1)}\\&\sqrt{(x-1)^2 \cdot 3}\end{aligned}\right\} && \text{Factoring the radicand; } x^2 - 2x + 1\\ & && \text{is a perfect square.} \\ &= \sqrt{(x-1)^2} \cdot \sqrt{3} && \text{Factoring into two radicals} \\ &= |x - 1|\sqrt{3} && \text{Taking the square root of } (x-1)^2 \end{aligned}$$

We can check Example 3 by graphing $y_1 = \sqrt{3x^2 - 6x + 3}$ and $y_2 = |x - 1|\sqrt{3}$, as shown in the graph on the left below.

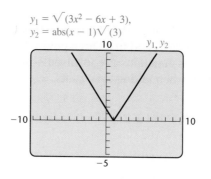

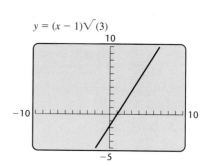

It appears that the graphs coincide, and scrolling through a table of values would show us that this is indeed the case. Note from the graph on the right above that the absolute-value sign is important since the graph of $y = (x - 1)\sqrt{3}$ is different from the graph of $y = \sqrt{3x^2 - 6x + 3}$.

> **EXAMPLE 4** Simplify: **(a)** $\sqrt{x^7 y^{11} z^9}$; **(b)** $\sqrt[3]{-16a^7 b^{14}}$.
>
> **SOLUTION**
>
> **a)** There are many ways to factor $x^7 y^{11} z^9$. Because of the square root (index of 2), we identify the largest exponents that are multiples of 2:
>
> $$\begin{aligned} \sqrt{x^7 y^{11} z^9} &= \sqrt{x^6 \cdot x \cdot y^{10} \cdot y \cdot z^8 \cdot z} && \begin{array}{l}\text{Using the largest even powers}\\\text{of } x, y, \text{ and } z\end{array} \\ &= \sqrt{x^6}\,\sqrt{y^{10}}\,\sqrt{z^8}\,\sqrt{xyz} && \begin{array}{l}\text{Factoring into several}\\\text{radicals}\end{array} \\ &= x^{6/2}\ y^{10/2}\ z^{8/2}\,\sqrt{xyz} && \begin{array}{l}\text{Converting to fraction}\\\text{exponents}\end{array} \\ &= x^3 y^5 z^4 \sqrt{xyz}. \end{aligned}$$
>
> *Check:* $$\begin{aligned} \left(x^3 y^5 z^4 \sqrt{xyz}\right)^2 &= (x^3)^2 (y^5)^2 (z^4)^2 \left(\sqrt{xyz}\right)^2 \\ &= x^6 \cdot y^{10} \cdot z^8 \cdot xyz = x^7 y^{11} z^9 \end{aligned}$$
>
> Our check shows that $x^3 y^5 z^4 \sqrt{xyz}$ is the square root of $x^7 y^{11} z^9$.
>
> **b)** Since the cube root of a negative number is negative, we can write
>
> $$\sqrt[3]{-16a^7 b^{14}} = -\sqrt[3]{16a^7 b^{14}}.$$

There are many ways to factor $16a^7b^{14}$. Because of the cube root (index of 3), we identify factors with the largest exponents that are multiples of 3:

$$-\sqrt[3]{16a^7b^{14}} = -\sqrt[3]{8 \cdot 2 \cdot a^6 \cdot a \cdot b^{12} \cdot b^2} \quad \text{Using the largest perfect-cube factors}$$

$$= -\sqrt[3]{8}\ \sqrt[3]{a^6}\ \sqrt[3]{b^{12}}\ \sqrt[3]{2ab^2} \quad \text{Factoring into several radicals}$$

$$= -2 \cdot a^{6/3} \cdot b^{12/3}\sqrt[3]{2ab^2} \quad \text{Converting to fraction exponents}$$

$$= -2a^2b^4\sqrt[3]{2ab^2}$$

As a check, let's redo the problem using a complex factorization of the radicand:

$$\sqrt[3]{-16a^7b^{14}} = -\sqrt[3]{2 \cdot 2 \cdot 2 \cdot 2 \cdot a \cdot a \cdot a \cdot a \cdot a \cdot a \cdot a \cdot b \cdot b \cdot b \cdot b \cdot b \cdot b \cdot b \cdot b \cdot b \cdot b \cdot b \cdot b \cdot b}$$

Each triple of factors makes a cube.

$$= -2 \cdot a \cdot a \cdot b \cdot b \cdot b \cdot b \cdot \sqrt[3]{2 \cdot a \cdot b \cdot b}$$
$$= -2a^2b^4\sqrt[3]{2ab^2}. \quad \text{Our answer checks.}$$

> To simplify an *n*th root, identify factors in the radicand with exponents that are multiples of *n*.

Multiplying and Simplifying

We have used the product rule for radicals to find products and also to simplify radical expressions. For some radical expressions, it is possible to do both: First find a product and then simplify.

EXAMPLE 5   Multiply and simplify.

a) $\sqrt{15}\ \sqrt{6}$ b) $3\sqrt[3]{25} \cdot 2\sqrt[3]{5}$ c) $\sqrt[4]{8x^3y^5}\ \sqrt[4]{4x^2y^3}$

SOLUTION

a) $\sqrt{15}\ \sqrt{6} = \sqrt{15 \cdot 6}$ Multiplying radicands
$= \sqrt{90} = \sqrt{9 \cdot 10}$ 9 is a perfect square.
$= 3\sqrt{10}$

b) $3\sqrt[3]{25} \cdot 2\sqrt[3]{5} = 3 \cdot 2 \cdot \sqrt[3]{25 \cdot 5}$ Using a commutative law; multiplying radicands
$= 6 \cdot \sqrt[3]{125}$ 125 is a perfect cube.
$= 6 \cdot 5$, or 30

c) $\sqrt[4]{8x^3y^5}\ \sqrt[4]{4x^2y^3} = \sqrt[4]{32x^5y^8}$ Multiplying radicands
$= \sqrt[4]{16x^4y^8 \cdot 2x}$ Identifying perfect fourth-power factors
$= \sqrt[4]{16}\ \sqrt[4]{x^4}\ \sqrt[4]{y^8}\ \sqrt[4]{2x}$ Factoring into radicals
$= 2xy^2\sqrt[4]{2x}$ Finding the fourth roots; assume $x \ge 0$.

The checks are left to the student.

10.3 EXERCISE SET

↪ *Concept Reinforcement* *Classify each of the following statements as either true or false.*

1. For any real numbers $\sqrt[n]{a}$ and $\sqrt[n]{b}$, $\sqrt[n]{a} \cdot \sqrt[n]{b} = \sqrt[n]{ab}$.

2. For any real numbers $\sqrt[n]{a}$ and $\sqrt[n]{b}$, $\sqrt[n]{a} + \sqrt[n]{b} = \sqrt[n]{a + b}$.

3. For any real numbers $\sqrt[n]{a}$ and $\sqrt[m]{b}$, $\sqrt[n]{a} \cdot \sqrt[m]{b} = \sqrt[nm]{ab}$.

4. For $x > 0$, $\sqrt{x^2 - 9} = x - 3$.

5. The expression $\sqrt[3]{X}$ is not simplified if X contains a factor that is a perfect cube.

6. It is often possible to simplify $\sqrt{A \cdot B}$ even though $\sqrt{A}$ and $\sqrt{B}$ cannot be simplified.

Multiply.

7. $\sqrt{5}\,\sqrt{7}$

8. $\sqrt{10}\,\sqrt{7}$

9. $\sqrt[3]{7}\,\sqrt[3]{2}$

10. $\sqrt[3]{2}\,\sqrt[3]{5}$

11. $\sqrt[4]{6}\,\sqrt[4]{3}$

12. $\sqrt[4]{8}\,\sqrt[4]{9}$

13. $\sqrt{2x}\,\sqrt{13y}$

14. $\sqrt{5a}\,\sqrt{6b}$

15. $\sqrt[5]{8y^3}\,\sqrt[5]{10y}$

16. $\sqrt[5]{9t^2}\,\sqrt[5]{2t}$

17. $\sqrt{y - b}\,\sqrt{y + b}$

18. $\sqrt{x - a}\,\sqrt{x + a}$

19. $\sqrt[3]{0.7y}\,\sqrt[3]{0.3y}$

20. $\sqrt[3]{0.5x}\,\sqrt[3]{0.2x}$

21. $\sqrt[5]{x - 2}\,\sqrt[5]{(x - 2)^2}$

22. $\sqrt[4]{x - 1}\,\sqrt[4]{x^2 + x + 1}$

23. $\sqrt{\dfrac{7}{t}}\,\sqrt{\dfrac{s}{11}}$

24. $\sqrt{\dfrac{x}{6}}\,\sqrt{\dfrac{7}{y}}$

25. $\sqrt[7]{\dfrac{x - 3}{4}}\,\sqrt[7]{\dfrac{5}{x + 2}}$

26. $\sqrt[6]{\dfrac{a}{b - 2}}\,\sqrt[6]{\dfrac{3}{b + 2}}$

Simplify by factoring.

27. $\sqrt{18}$

28. $\sqrt{50}$

29. $\sqrt{27}$

30. $\sqrt{45}$

31. $\sqrt{8}$

32. $\sqrt{75}$

33. $\sqrt{198}$

34. $\sqrt{325}$

35. $\sqrt{36a^4b}$

36. $\sqrt{175y^8}$

37. $\sqrt[3]{8x^3y^2}$

38. $\sqrt[3]{27ab^6}$

39. $\sqrt[3]{-16x^6}$

40. $\sqrt[3]{-32a^6}$

Find a simplified form of $f(x)$. Assume that x can be any real number.

41. $f(x) = \sqrt[3]{125x^5}$

42. $f(x) = \sqrt[3]{16x^6}$

43. $f(x) = \sqrt{49(x - 3)^2}$

44. $f(x) = \sqrt{81(x - 1)^2}$

45. $f(x) = \sqrt{5x^2 - 10x + 5}$

46. $f(x) = \sqrt{2x^2 + 8x + 8}$

Simplify. Assume that no radicands were formed by raising negative numbers to even powers.

47. $\sqrt{a^6b^7}$

48. $\sqrt{x^6y^9}$

49. $\sqrt[3]{x^5y^6z^{10}}$

50. $\sqrt[3]{a^6b^7c^{13}}$

51. $\sqrt[5]{-32a^7b^{11}}$

52. $\sqrt[4]{16x^5y^{11}}$

53. $\sqrt[5]{x^{13}y^8z^{17}}$

54. $\sqrt[5]{a^6b^8c^9}$

55. $\sqrt[3]{-80a^{14}}$

56. $\sqrt[4]{810x^9}$

Multiply and simplify.

57. $\sqrt{6}\,\sqrt{3}$

58. $\sqrt{15}\,\sqrt{5}$

59. $\sqrt{15}\,\sqrt{21}$

60. $\sqrt{10}\,\sqrt{14}$

61. $\sqrt[3]{9}\,\sqrt[3]{3}$

62. $\sqrt[3]{2}\,\sqrt[3]{4}$

Aha! 63. $\sqrt{18a^3}\,\sqrt{18a^3}$

64. $\sqrt{75x^7}\,\sqrt{75x^7}$

65. $\sqrt[3]{5a^2}\,\sqrt[3]{2a}$

66. $\sqrt[3]{7x}\,\sqrt[3]{3x^2}$

67. $\sqrt{2x^5}\,\sqrt{10x^2}$

68. $\sqrt{5a^7}\,\sqrt{15a^3}$

69. $\sqrt[3]{s^2t^4}\,\sqrt[3]{s^4t^6}$

70. $\sqrt[3]{x^2y^4}\,\sqrt[3]{x^2y^6}$

71. $\sqrt[3]{(x + 5)^2}\,\sqrt[3]{(x + 5)^4}$

72. $\sqrt[3]{(a - b)^5}\,\sqrt[3]{(a - b)^7}$

73. $\sqrt[4]{20a^3b^7}\,\sqrt[4]{4a^2b^5}$

74. $\sqrt[4]{9x^7y^2}\,\sqrt[4]{9x^2y^9}$

75. $\sqrt[5]{x^3(y+z)^6}\,\sqrt[5]{x^3(y+z)^4}$

76. $\sqrt[5]{a^3(b-c)^4}\,\sqrt[5]{a^7(b-c)^4}$

TW 77. Why do we need to know how to multiply radical expressions before learning how to simplify radical expressions?

TW 78. Why is it incorrect to say that, in general, $\sqrt{x^2} = x$?

Skill Maintenance

Perform the indicated operation and, if possible, simplify. [7.4]

79. $\dfrac{3x}{16y} + \dfrac{5y}{64x}$

80. $\dfrac{2}{a^3b^4} + \dfrac{6}{a^4b}$

81. $\dfrac{4}{x^2-9} - \dfrac{7}{2x-6}$

82. $\dfrac{8}{x^2-25} - \dfrac{3}{2x-10}$

Simplify. [5.1]

83. $\dfrac{9a^4b^7}{3a^2b^5}$

84. $\dfrac{12a^2b^7}{4ab^2}$

Synthesis

TW 85. Explain why it is true that $\sqrt[n]{ab} = \sqrt[n]{a} \cdot \sqrt[n]{b}$.

TW 86. Is the equation $\sqrt{(2x+3)^8} = (2x+3)^4$ always, sometimes, or never true? Why?

87. *Radar Range.*   The function given by

$$R(x) = \frac{1}{2}\sqrt[4]{\frac{x \cdot 3.0 \times 10^6}{\pi^2}}$$

can be used to determine the maximum range $R(x)$, in miles, of an ARSR-3 surveillance radar with a peak power of x watts (*Source*: Introduction to RADAR Techniques, Federal Aviation Administration, 1988). Determine the maximum radar range when the peak power is 5×10^4 watts.

88. *Speed of a Skidding Car.*   Police can estimate the speed at which a car was traveling by measuring its skid marks. The function given by

$$r(L) = 2\sqrt{5L}$$

can be used, where L is the length of a skid mark, in feet, and $r(L)$ is the speed, in miles per hour.

Find the exact speed and an estimate (to the nearest tenth mile per hour) for the speed of a car that left skid marks **(a)** 20 ft long; **(b)** 70 ft long; **(c)** 90 ft long. See also Exercise 102.

89. *Wind Chill Temperature.*   When the temperature is T degrees Celsius and the wind speed is v meters per second, the *wind chill temperature*, T_w, is the temperature (with no wind) that it feels like. Here is a formula for finding wind chill temperature:

$$T_w = 33 - \frac{\left(10.45 + 10\sqrt{v} - v\right)(33 - T)}{22}.$$

Estimate the wind chill temperature (to the nearest tenth of a degree) for the given actual temperatures and wind speeds.

a) $T = 7°C$, $v = 8$ m/sec
b) $T = 0°C$, $v = 12$ m/sec
c) $T = -5°C$, $v = 14$ m/sec
d) $T = -23°C$, $v = 15$ m/sec

Simplify. Assume that all variables are nonnegative.

90. $\left(\sqrt{r^3 t}\right)^7$

91. $\left(\sqrt[3]{25x^4}\right)^4$

92. $\left(\sqrt[3]{a^2b^4}\right)^5$

93. $\left(\sqrt{a^3b^5}\right)^7$

Draw and compare the graphs of each group of equations.

94. $f(x) = \sqrt{x^2 + 2x + 1}$,
$g(x) = x + 1$,
$h(x) = |x + 1|$

95. $f(x) = \sqrt{x^2 - 2x + 1}$,
$g(x) = x - 1$,
$h(x) = |x - 1|$

96. If $f(t) = \sqrt{t^2 - 3t - 4}$, what is the domain of f?

97. What is the domain of g, if $g(x) = \sqrt{x^2 - 6x + 8}$?

Solve.

98. $\sqrt[3]{5x^{k+1}}\,\sqrt[3]{25x^k} = 5x^7$, for k

99. $\sqrt[5]{4a^{3k+2}}\,\sqrt[5]{8a^{6-k}} = 2a^4$, for k

100. Use a graphing calculator to check your answers to Exercises 21 and 41.

TW 101. Rony is puzzled. When he uses a graphing calculator to graph $y = \sqrt{x} \cdot \sqrt{x}$, he gets the following screen. Explain why Rony did not get the complete line $y = x$.

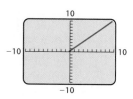

TW 102. Does a car traveling twice as fast as another car leave a skid mark that is twice as long? (See Exercise 88.) Why or why not?

10.4   Dividing Radical Expressions

Dividing and Simplifying   ◼   Rationalizing Denominators and
Numerators (Part 1)

Dividing and Simplifying

Just as the root of a product can be expressed as the product of two roots, the root of a quotient can be expressed as the quotient of two roots. For example,

$$\sqrt[3]{\dfrac{27}{8}} = \dfrac{3}{2} \quad \text{and} \quad \dfrac{\sqrt[3]{27}}{\sqrt[3]{8}} = \dfrac{3}{2}.$$

This example suggests the following.

The Quotient Rule for Radicals   For any real numbers $\sqrt[n]{a}$ and $\sqrt[n]{b}$, $b \neq 0$,

$$\sqrt[n]{\dfrac{a}{b}} = \dfrac{\sqrt[n]{a}}{\sqrt[n]{b}}.$$

Remember that an nth root is simplified when its radicand has no factors that are perfect nth powers. Recall too that we assume that no radicands represent negative quantities raised to an even power.

EXAMPLE 1   Simplify by taking the roots of the numerator and the denominator.

a) $\sqrt[3]{\dfrac{27}{125}}$

b) $\sqrt{\dfrac{25}{y^2}}$

SOLUTION

a) $\sqrt[3]{\dfrac{27}{125}} = \dfrac{\sqrt[3]{27}}{\sqrt[3]{125}} = \dfrac{3}{5}$ Taking the cube roots of the numerator and the denominator

b) $\sqrt{\dfrac{25}{y^2}} = \dfrac{\sqrt{25}}{\sqrt{y^2}} = \dfrac{5}{y}$ Taking the square roots of the numerator and the denominator. Assume $y > 0$.

As in Section 10.3, any radical expressions appearing in the answers should be simplified as much as possible.

EXAMPLE 2   Simplify: **(a)** $\sqrt{\dfrac{16x^3}{y^8}}$; **(b)** $\sqrt[3]{\dfrac{27y^{14}}{8x^3}}$.

SOLUTION

a) $\sqrt{\dfrac{16x^3}{y^8}} = \dfrac{\sqrt{16x^3}}{\sqrt{y^8}}$

$= \dfrac{\sqrt{16x^2 \cdot x}}{\sqrt{y^8}} = \dfrac{4x\sqrt{x}}{y^4}$ Simplifying the numerator and the denominator

b) $\sqrt[3]{\dfrac{27y^{14}}{8x^3}} = \dfrac{\sqrt[3]{27^{14}}}{\sqrt[3]{8x^3}}$

$= \dfrac{\sqrt[3]{27y^{12}y^2}}{\sqrt[3]{8x^3}} = \dfrac{\sqrt[3]{27y^{12}}\,\sqrt[3]{y^2}}{\sqrt[3]{8x^3}} = \dfrac{3y^4\sqrt[3]{y^2}}{2x}$ Simplifying the numerator and the denominator

If we read from right to left, the quotient rule tells us that to divide two radical expressions that have the same index, we can divide the radicands.

EXAMPLE 3   Divide and, if possible, simplify.

a) $\dfrac{\sqrt{80}}{\sqrt{5}}$ **b)** $\dfrac{5\sqrt[3]{32}}{\sqrt[3]{2}}$ **c)** $\dfrac{\sqrt{72xy}}{2\sqrt{2}}$ **d)** $\dfrac{\sqrt[4]{18a^9b^5}}{\sqrt[4]{3b}}$

SOLUTION

a) $\dfrac{\sqrt{80}}{\sqrt{5}} = \sqrt{\dfrac{80}{5}} = \sqrt{16} = 4$ **Because the indices match, we can divide the radicands.**

b) $\dfrac{5\sqrt[3]{32}}{\sqrt[3]{2}} = 5\sqrt[3]{\dfrac{32}{2}} = 5\sqrt[3]{16}$

$= 5\sqrt[3]{8 \cdot 2}$ 8 is the largest perfect-cube factor of 16.

$= 5\sqrt[3]{8}\,\sqrt[3]{2} = 5 \cdot 2\sqrt[3]{2}$

$= 10\sqrt[3]{2}$

c) $\dfrac{\sqrt{72xy}}{2\sqrt{2}} = \dfrac{1}{2}\sqrt{\dfrac{72xy}{2}}$

$\qquad\qquad = \dfrac{1}{2}\sqrt{36xy} = \dfrac{1}{2}\cdot 6\sqrt{xy} = 3\sqrt{xy}$

> Because the indices match, we can divide the radicands.

d) $\dfrac{\sqrt[4]{18a^9b^5}}{\sqrt[4]{3b}} = \sqrt[4]{\dfrac{18a^9b^5}{3b}}$

$\qquad\qquad = \sqrt[4]{6a^9b^4} = \sqrt[4]{a^8b^4}\,\sqrt[4]{6a}$

$\qquad\qquad = a^2b\sqrt[4]{6a}$

> Note that 8 is the largest power less than 9 that is a multiple of the index 4.
>
> *Partial check:* $(a^2b)^4 = a^8b^4$

Rationalizing Denominators and Numerators (Part 1)*

When a radical expression appears in a denominator, it can be useful to find an equivalent expression in which the denominator no longer contains a radical.[†] The procedure for finding such an expression is called **rationalizing the denominator.** We carry this out by multiplying by 1 in either of two ways.

One way is to multiply by 1 *under* the radical to make the denominator of the radicand a perfect power.

EXAMPLE 4   Rationalize each denominator.

a) $\sqrt{\dfrac{7}{3}}$

 b) $\sqrt[3]{\dfrac{5}{16}}$

SOLUTION

a) We multiply by 1 under the radical, using $\frac{3}{3}$. We do this so that the denominator of the radicand will be a perfect square:

$\sqrt{\dfrac{7}{3}} = \sqrt{\dfrac{7}{3}\cdot\dfrac{3}{3}}$ Multiplying by 1 under the radical

$\qquad = \sqrt{\dfrac{21}{9}}$ The denominator, 9, is now a perfect square.

$\qquad = \dfrac{\sqrt{21}}{\sqrt{9}} = \dfrac{\sqrt{21}}{3}.$

b) Note that $16 = 4^2$. Thus, to make the denominator a perfect cube, we multiply under the radical by $\frac{4}{4}$:

$\sqrt[3]{\dfrac{5}{16}} = \sqrt[3]{\dfrac{5}{4\cdot 4}\cdot\dfrac{4}{4}}$ Since the index is 3, we need 3 identical factors in the denominator.

$\qquad = \sqrt[3]{\dfrac{20}{4^3}}$ The denominator is now a perfect cube.

$\qquad = \dfrac{\sqrt[3]{20}}{\sqrt[3]{4^3}} = \dfrac{\sqrt[3]{20}}{4}.$

---

*Denominators and numerators with two terms are rationalized in Section 10.5.
[†]See Exercise 73 on p. 761.

Another way to rationalize a denominator is to multiply by 1 *outside* the radical.

> **EXAMPLE 5**   Rationalize each denominator.
>
> **a)** $\sqrt{\dfrac{4}{5b}}$
>
> **b)** $\dfrac{\sqrt[3]{a}}{\sqrt[3]{9x}}$

SOLUTION

a) We rewrite the expression as a quotient of two radicals. Then we simplify and multiply by 1:

$$\sqrt{\frac{4}{5b}} = \frac{\sqrt{4}}{\sqrt{5b}} = \frac{2}{\sqrt{5b}} \qquad \text{We assume } b > 0.$$

$$= \frac{2}{\sqrt{5b}} \cdot \frac{\sqrt{5b}}{\sqrt{5b}} \qquad \text{Multiplying by 1}$$

$$= \frac{2\sqrt{5b}}{\left(\sqrt{5b}\right)^2} \qquad \text{\textbf{Try to do this step mentally.}}$$

$$= \frac{2\sqrt{5b}}{5b}.$$

b) To rationalize the denominator $\sqrt[3]{9x}$, note that $9x$ is $3 \cdot 3 \cdot x$. In order for this radicand to be a cube, we need another factor of 3 and two more factors of x. Thus we multiply by 1, using $\sqrt[3]{3x^2}/\sqrt[3]{3x^2}$:

$$\frac{\sqrt[3]{a}}{\sqrt[3]{9x}} = \frac{\sqrt[3]{a}}{\sqrt[3]{9x}} \cdot \frac{\sqrt[3]{3x^2}}{\sqrt[3]{3x^2}} \qquad \text{Multiplying by 1}$$

$$= \frac{\sqrt[3]{3ax^2}}{\sqrt[3]{27x^3}} \quad \longleftarrow \text{\textbf{ This radicand is now a perfect cube.}}$$

$$= \frac{\sqrt[3]{3ax^2}}{3x}.$$

Sometimes in calculus it is necessary to rationalize a numerator. To do so, we multiply by 1 to make the radicand in the *numerator* a perfect power.

> **EXAMPLE 6**   Rationalize each numerator.
>
> **a)** $\sqrt{\dfrac{7}{5}}$
>
> **b)** $\dfrac{\sqrt[3]{4a^2}}{\sqrt[3]{5b}}$

SOLUTION

a) $\sqrt{\dfrac{7}{5}} = \sqrt{\dfrac{7}{5} \cdot \dfrac{7}{7}}$ **Multiplying by 1 under the radical. We also could have multiplied by $\sqrt{7}/\sqrt{7}$ outside the radical.**

$$= \sqrt{\frac{49}{35}} \qquad \text{\textbf{The numerator is now a perfect square.}}$$

$$= \frac{\sqrt{49}}{\sqrt{35}} = \frac{7}{\sqrt{35}}$$

b) $\dfrac{\sqrt[3]{4a^2}}{\sqrt[3]{5b}} = \dfrac{\sqrt[3]{4a^2}}{\sqrt[3]{5b}} \cdot \dfrac{\sqrt[3]{2a}}{\sqrt[3]{2a}}$ Multiplying by 1

$= \dfrac{\sqrt[3]{8a^3}}{\sqrt[3]{10ba}}$ ⟵ This radicand is now a perfect cube.

$= \dfrac{2a}{\sqrt[3]{10ab}}$

10.4 EXERCISE SET

FOR EXTRA HELP

MathXL   MyMathLab   InterAct Math   AW Math Tutor Center   Video Lectures on CD: Disc 5   Student's Solutions Manual

⮑ *Concept Reinforcement* In each of Exercises 1–8, match the expression with an equivalent expression from the column on the right. Assume a, b > 0.

1. ___ $\sqrt[3]{\dfrac{a^2}{b^6}}$

2. ___ $\dfrac{\sqrt[3]{a^6}}{\sqrt[3]{b^9}}$

3. ___ $\sqrt[5]{\dfrac{a^6}{b^4}}$

4. ___ $\sqrt{\dfrac{a}{b^3}}$

5. ___ $\dfrac{\sqrt[5]{a^2}}{\sqrt[5]{b^2}}$

6. ___ $\dfrac{\sqrt{5a^4}}{\sqrt{5a^3}}$

7. ___ $\dfrac{\sqrt[5]{a^2}}{\sqrt[5]{b^3}}$

8. ___ $\sqrt[4]{\dfrac{16a^6}{a^2}}$

a) $\dfrac{\sqrt[5]{a^2}\sqrt[5]{b^2}}{\sqrt[5]{b^5}}$

b) $\dfrac{a^2}{b^3}$

c) $\sqrt{\dfrac{a \cdot b}{b^3 \cdot b}}$

d) $\sqrt{a}$

e) $\dfrac{\sqrt[3]{a^2}}{b^2}$

f) $\sqrt[5]{\dfrac{a^6 b}{b^4 \cdot b}}$

g) $2a$

h) $\dfrac{\sqrt[5]{a^2 b^3}}{\sqrt[5]{b^5}}$

Simplify by taking the roots of the numerator and the denominator. Assume all variables represent positive numbers.

9. $\sqrt{\dfrac{36}{25}}$

10. $\sqrt{\dfrac{100}{81}}$

11. $\sqrt[3]{\dfrac{64}{27}}$

12. $\sqrt[3]{\dfrac{343}{1000}}$

13. $\sqrt{\dfrac{49}{y^2}}$

14. $\sqrt{\dfrac{121}{x^2}}$

15. $\sqrt{\dfrac{36y^3}{x^4}}$

16. $\sqrt{\dfrac{25a^5}{b^6}}$

17. $\sqrt[3]{\dfrac{27a^4}{8b^3}}$

18. $\sqrt[3]{\dfrac{64x^7}{216y^6}}$

19. $\sqrt[4]{\dfrac{16a^4}{b^4 c^8}}$

20. $\sqrt[4]{\dfrac{81x^4}{y^8 z^4}}$

21. $\sqrt[4]{\dfrac{a^5 b^8}{c^{10}}}$

22. $\sqrt[4]{\dfrac{x^9 y^{12}}{z^6}}$

23. $\sqrt[5]{\dfrac{32x^6}{y^{11}}}$

24. $\sqrt[5]{\dfrac{243a^9}{b^{13}}}$

25. $\sqrt[6]{\dfrac{x^6 y^8}{z^{15}}}$ **26.** $\sqrt[6]{\dfrac{a^9 b^{12}}{c^{13}}}$

Divide and, if possible, simplify. Assume all variables represent positive numbers.

27. $\dfrac{\sqrt{35x}}{\sqrt{7x}}$ **28.** $\dfrac{\sqrt{28y}}{\sqrt{4y}}$

29. $\dfrac{\sqrt[3]{270}}{\sqrt[3]{10}}$ **30.** $\dfrac{\sqrt[3]{40}}{\sqrt[3]{5}}$

31. $\dfrac{\sqrt{40xy^3}}{\sqrt{8x}}$ **32.** $\dfrac{\sqrt{56ab^3}}{\sqrt{7a}}$

33. $\dfrac{\sqrt[3]{96a^4 b^2}}{\sqrt[3]{12a^2 b}}$ **34.** $\dfrac{\sqrt[3]{189x^5 y^7}}{\sqrt[3]{7x^2 y^2}}$

35. $\dfrac{\sqrt{100ab}}{5\sqrt{2}}$ **36.** $\dfrac{\sqrt{75ab}}{3\sqrt{3}}$

37. $\dfrac{\sqrt[4]{48x^9 y^{13}}}{\sqrt[4]{3xy^{-2}}}$ **38.** $\dfrac{\sqrt[5]{64a^{11} b^{28}}}{\sqrt[5]{2ab^{-2}}}$

39. $\dfrac{\sqrt[3]{x^3 - y^3}}{\sqrt[3]{x - y}}$ ←   → **40.** $\dfrac{\sqrt[3]{r^3 + s^3}}{\sqrt[3]{r + s}}$

> *Hint*: **Factor and then simplify.**

Rationalize each denominator. Assume all variables represent positive numbers.

41. $\sqrt{\dfrac{3}{2}}$ **42.** $\sqrt{\dfrac{6}{7}}$

43. $\dfrac{2\sqrt{5}}{7\sqrt{3}}$ **44.** $\dfrac{3\sqrt{5}}{2\sqrt{2}}$

45. $\sqrt[3]{\dfrac{16}{9}}$ **46.** $\sqrt[3]{\dfrac{2}{9}}$

47. $\dfrac{\sqrt[3]{3a}}{\sqrt[3]{5c}}$ **48.** $\dfrac{\sqrt[3]{7x}}{\sqrt[3]{3y}}$

49. $\dfrac{\sqrt[3]{5y^4}}{\sqrt[3]{6x^4}}$ **50.** $\dfrac{\sqrt[3]{3a^4}}{\sqrt[3]{7b^2}}$

51. $\sqrt[3]{\dfrac{2}{x^2 y}}$ **52.** $\sqrt[3]{\dfrac{5}{ab^2}}$

53. $\sqrt{\dfrac{7a}{18}}$ **54.** $\sqrt{\dfrac{3x}{10}}$

55. $\sqrt{\dfrac{9}{20x^2 y}}$ **56.** $\sqrt{\dfrac{7}{32a^2 b}}$

Aha! **57.** $\sqrt{\dfrac{10ab^2}{72a^3 b}}$ **58.** $\sqrt{\dfrac{21x^2 y}{75xy^5}}$

Rationalize each numerator. Assume all variables represent positive numbers.

59. $\dfrac{\sqrt{5}}{\sqrt{7x}}$ **60.** $\dfrac{\sqrt{10}}{\sqrt{3x}}$

61. $\sqrt{\dfrac{14}{21}}$ **62.** $\sqrt{\dfrac{12}{15}}$

63. $\dfrac{4\sqrt{13}}{3\sqrt{7}}$ **64.** $\dfrac{5\sqrt{21}}{2\sqrt{5}}$

65. $\dfrac{\sqrt[3]{7}}{\sqrt[3]{2}}$ **66.** $\dfrac{\sqrt[3]{5}}{\sqrt[3]{4}}$

67. $\sqrt{\dfrac{7x}{3y}}$ **68.** $\sqrt{\dfrac{7a}{6b}}$

69. $\sqrt[3]{\dfrac{2a^5}{5b}}$ **70.** $\sqrt[3]{\dfrac{2a^4}{7b}}$

71. $\sqrt{\dfrac{x^3 y}{2}}$ **72.** $\sqrt{\dfrac{ab^5}{3}}$

TW 73. Explain why it is easier to approximate

$$\dfrac{\sqrt{2}}{2} \quad \text{than} \quad \dfrac{1}{\sqrt{2}}$$

if no calculator is available and $\sqrt{2} \approx 1.414213562$.

TW 74. A student *incorrectly* claims that

$$\dfrac{5 + \sqrt{2}}{\sqrt{18}} = \dfrac{5 + \sqrt{1}}{\sqrt{9}} = \dfrac{5 + 1}{3}.$$

How could you convince the student that a mistake has been made? How would you explain the correct way of rationalizing the denominator?

Skill Maintenance

Multiply. [7.2]

75. $\dfrac{3}{x - 5} \cdot \dfrac{x - 1}{x + 5}$

76. $\dfrac{7}{x + 4} \cdot \dfrac{x - 2}{x - 4}$

Simplify.

77. $\dfrac{a^2 - 8a + 7}{a^2 - 49}$ [7.1]

78. $\dfrac{t^2 + 9t - 22}{t^2 - 4}$ [7.1]

79. $(5a^3b^4)^3$ [5.1]

80. $(3x^4)^2(5xy^3)^2$ [5.1]

Synthesis

TW 81. Is the quotient of two irrational numbers always an irrational number? Why or why not?

TW 82. Is it possible to understand how to rationalize a denominator without knowing how to multiply rational expressions? Why or why not?

83. *Pendulums.*   The *period* of a pendulum is the time it takes to complete one cycle, swinging to and fro. For a pendulum that is L centimeters long, the period T is given by the formula

$$T = 2\pi\sqrt{\dfrac{L}{980}},$$

where T is in seconds. Find, to the nearest hundredth of a second, the period of a pendulum of length **(a)** 65 cm; **(b)** 98 cm; **(c)** 120 cm. Use a calculator's $\boxed{\pi}$ key if possible.

Perform the indicated operations.

84. $\dfrac{7\sqrt{a^2b}\,\sqrt{25xy}}{5\sqrt{a^{-4}b^{-1}}\,\sqrt{49x^{-1}y^{-3}}}$

85. $\dfrac{(\sqrt[3]{81mn^2})^2}{(\sqrt[3]{mn})^2}$

86. $\dfrac{\sqrt{44x^2y^9z}\,\sqrt{22y^9z^6}}{(\sqrt{11xy^8z^2})^2}$

87. $\sqrt{a^2 - 3} - \dfrac{a^2}{\sqrt{a^2 - 3}}$

88. $5\sqrt{\dfrac{x}{y}} + 4\sqrt{\dfrac{y}{x}} - \dfrac{3}{\sqrt{xy}}$

89. Provide a reason for each step in the following derivation of the quotient rule:

$$\sqrt[n]{\dfrac{a}{b}} = \left(\dfrac{a}{b}\right)^{1/n} \quad \underline{\hspace{2cm}}$$

$$= \dfrac{a^{1/n}}{b^{1/n}} \quad \underline{\hspace{2cm}}$$

$$= \dfrac{\sqrt[n]{a}}{\sqrt[n]{b}} \quad \underline{\hspace{2cm}}$$

90. Show that $\dfrac{\sqrt[n]{a}}{\sqrt[n]{b}}$ is the nth root of $\dfrac{a}{b}$ by raising it to the nth power and simplifying.

91. Let $f(x) = \sqrt{18x^3}$ and $g(x) = \sqrt{2x}$. Find $(f/g)(x)$ and specify the domain of f/g.

92. Let $f(t) = \sqrt{2t}$ and $g(t) = \sqrt{50t^3}$. Find $(f/g)(t)$ and specify the domain of f/g.

93. Let $f(x) = \sqrt{x^2 - 9}$ and $g(x) = \sqrt{x - 3}$. Find $(f/g)(x)$ and specify the domain of f/g.

10.5 Expressions Containing Several Radical Terms

Adding and Subtracting Radical Expressions ◼ Products and Quotients of Two or More Radical Terms ◼ Rationalizing Denominators and Numerators (Part 2) ◼ Terms with Differing Indices

Radical expressions like $6\sqrt{7} + 4\sqrt{7}$ or $\left(\sqrt{a} + \sqrt{b}\right)\left(\sqrt{a} - \sqrt{b}\right)$ contain more than one *radical term* and can sometimes be simplified.

Adding and Subtracting Radical Expressions

When two radical expressions have the same indices and radicands, they are said to be **like radicals.** Like radicals can be combined (added or subtracted) in much the same way that we combined like terms earlier in this text.

EXAMPLE 1 Simplify by combining like radical terms.

a) $6\sqrt{7} + 4\sqrt{7}$ b) $\sqrt[3]{2} - 7x\sqrt[3]{2} + 5\sqrt[3]{2}$

c) $6\sqrt[5]{4x} + 3\sqrt[5]{4x} - \sqrt[3]{4x}$

SOLUTION

a) $6\sqrt{7} + 4\sqrt{7} = (6 + 4)\sqrt{7}$ **Using the distributive law (factoring out $\sqrt{7}$)**

$\phantom{6\sqrt{7} + 4\sqrt{7}} = 10\sqrt{7}$ **You can think: 6 square roots of 7 plus 4 square roots of 7 results in 10 square roots of 7.**

b) $\sqrt[3]{2} - 7x\sqrt[3]{2} + 5\sqrt[3]{2} = (1 - 7x + 5)\sqrt[3]{2}$ **Factoring out $\sqrt[3]{2}$**

$\phantom{\sqrt[3]{2} - 7x\sqrt[3]{2} + 5\sqrt[3]{2}} = (6 - 7x)\sqrt[3]{2}$ **These parentheses are important!**

c) $6\sqrt[5]{4x} + 3\sqrt[5]{4x} - \sqrt[3]{4x} = (6 + 3)\sqrt[5]{4x} - \sqrt[3]{4x}$ **Try to do this step mentally.**

$\phantom{6\sqrt[5]{4x} + 3\sqrt[5]{4x} - \sqrt[3]{4x}} = 9\sqrt[5]{4x} - \sqrt[3]{4x}$ **Because the indices differ, we are done.**

Our ability to simplify radical expressions can help us to find like radicals even when, at first, it may appear that none exists.

EXAMPLE 2 Simplify by combining like radical terms, if possible.

a) $3\sqrt{8} - 5\sqrt{2}$ b) $9\sqrt{5} - 4\sqrt{3}$ c) $\sqrt[3]{2x^6y^4} + 7\sqrt[3]{2y}$

SOLUTION

a) $3\sqrt{8} - 5\sqrt{2} = 3\sqrt{4 \cdot 2} - 5\sqrt{2}$

$\phantom{3\sqrt{8} - 5\sqrt{2}} = 3\sqrt{4} \cdot \sqrt{2} - 5\sqrt{2}$ **Simplifying $\sqrt{8}$**

$\phantom{3\sqrt{8} - 5\sqrt{2}} = 3 \cdot 2 \cdot \sqrt{2} - 5\sqrt{2}$

$\phantom{3\sqrt{8} - 5\sqrt{2}} = 6\sqrt{2} - 5\sqrt{2}$

$\phantom{3\sqrt{8} - 5\sqrt{2}} = \sqrt{2}$ **Combining like radicals**

b) $9\sqrt{5} - 4\sqrt{3}$ cannot be simplified.

c) $\sqrt[3]{2x^6y^4} + 7\sqrt[3]{2y} = \sqrt[3]{x^6y^3 \cdot 2y} + 7\sqrt[3]{2y}$
$= \sqrt[3]{x^6y^3} \cdot \sqrt[3]{2y} + 7\sqrt[3]{2y}$ Simplifying $\sqrt[3]{2x^6y^4}$
$= x^2y \cdot \sqrt[3]{2y} + 7\sqrt[3]{2y}$
$= (x^2y + 7)\sqrt[3]{2y}$ Factoring to combine like radical terms

Products and Quotients of Two or More Radical Terms

Radical expressions often contain factors that have more than one term. The procedure for multiplying out such expressions is similar to finding products of polynomials. Some products will yield like radical terms, which we can now combine.

EXAMPLE 3   Multiply.

a) $\sqrt{3}(x - \sqrt{5})$
b) $\sqrt[3]{y}(\sqrt[3]{y^2} + \sqrt[3]{2})$
c) $(4\sqrt{3} + \sqrt{2})(\sqrt{3} - 5\sqrt{2})$
d) $(\sqrt{a} + \sqrt{b})(\sqrt{a} - \sqrt{b})$

SOLUTION

a) $\sqrt{3}(x - \sqrt{5}) = \sqrt{3} \cdot x - \sqrt{3} \cdot \sqrt{5}$ Using the distributive law
$= x\sqrt{3} - \sqrt{15}$ Multiplying radicals

b) $\sqrt[3]{y}(\sqrt[3]{y^2} + \sqrt[3]{2}) = \sqrt[3]{y} \cdot \sqrt[3]{y^2} + \sqrt[3]{y} \cdot \sqrt[3]{2}$ Using the distributive law
$= \sqrt[3]{y^3} + \sqrt[3]{2y}$ Multiplying radicals
$= y + \sqrt[3]{2y}$ Simplifying $\sqrt[3]{y^3}$

$$\overset{F}{} \quad \overset{O}{} \quad \overset{I}{} \quad \overset{L}{}$$
c) $(4\sqrt{3} + \sqrt{2})(\sqrt{3} - 5\sqrt{2}) = 4(\sqrt{3})^2 - 20\sqrt{3} \cdot \sqrt{2} + \sqrt{2} \cdot \sqrt{3} - 5(\sqrt{2})^2$
$= 4 \cdot 3 - 20\sqrt{6} + \sqrt{6} - 5 \cdot 2$ Multiplying radicals
$= 12 - 20\sqrt{6} + \sqrt{6} - 10$
$= 2 - 19\sqrt{6}$ Combining like terms

d) $(\sqrt{a} + \sqrt{b})(\sqrt{a} - \sqrt{b}) = (\sqrt{a})^2 - \sqrt{a}\sqrt{b} + \sqrt{a}\sqrt{b} - (\sqrt{b})^2$ Using FOIL
$= a - b$ Combining like terms

In Example 3(d) above, you may have noticed that since the outer and inner products in the multiplication are opposites, the result, $a - b$, is not itself a radical expression. Pairs of radical terms, like $\sqrt{a} + \sqrt{b}$ and $\sqrt{a} - \sqrt{b}$, are called **conjugates.**

Rationalizing Denominators and Numerators (Part 2)

The use of conjugates allows us to rationalize denominators or numerators with two terms.

EXAMPLE 4   Rationalize each denominator: **(a)** $\dfrac{4}{\sqrt{3} + x}$; **(b)** $\dfrac{4 + \sqrt{2}}{\sqrt{5} - \sqrt{2}}$.

SOLUTION

a) $\dfrac{4}{\sqrt{3} + x} = \dfrac{4}{\sqrt{3} + x} \cdot \dfrac{\sqrt{3} - x}{\sqrt{3} - x}$ 	Multiplying by 1, using the conjugate of $\sqrt{3} + x$, which is $\sqrt{3} - x$

$\qquad = \dfrac{4\left(\sqrt{3} - x\right)}{\left(\sqrt{3} + x\right)\left(\sqrt{3} - x\right)}$ 	Multiplying numerators and denominators

$\qquad = \dfrac{4\left(\sqrt{3} - x\right)}{\left(\sqrt{3}\right)^2 - x^2}$ 	Using FOIL in the denominator

$\qquad = \dfrac{4\sqrt{3} - 4x}{3 - x^2}$ 	Simplifying. No radicals remain in the denominator.

b) $\dfrac{4 + \sqrt{2}}{\sqrt{5} - \sqrt{2}} = \dfrac{4 + \sqrt{2}}{\sqrt{5} - \sqrt{2}} \cdot \dfrac{\sqrt{5} + \sqrt{2}}{\sqrt{5} + \sqrt{2}}$ 	Multiplying by 1, using the conjugate of $\sqrt{5} - \sqrt{2}$, which is $\sqrt{5} + \sqrt{2}$

$\qquad = \dfrac{\left(4 + \sqrt{2}\right)\left(\sqrt{5} + \sqrt{2}\right)}{\left(\sqrt{5} - \sqrt{2}\right)\left(\sqrt{5} + \sqrt{2}\right)}$ 	Multiplying numerators and denominators

$\qquad = \dfrac{4\sqrt{5} + 4\sqrt{2} + \sqrt{2}\,\sqrt{5} + \left(\sqrt{2}\right)^2}{\left(\sqrt{5}\right)^2 - \left(\sqrt{2}\right)^2}$ 	Using FOIL

$\qquad = \dfrac{4\sqrt{5} + 4\sqrt{2} + \sqrt{10} + 2}{5 - 2}$ 	Squaring in the denominator and the numerator

$\qquad = \dfrac{4\sqrt{5} + 4\sqrt{2} + \sqrt{10} + 2}{3}$ 	No radicals remain in the denominator.

```
(4+√ (2))/(√ (5)−√
(2))
              6.587801273
(4√ (5)+4√ (2)+√ (1
0)+2)/3
              6.587801273
```

We can check by approximating the value of the original expression and the value of the rationalized expression, as shown at left. Care must be taken to place the parentheses properly. The approximate values are the same, so we have a check.

To rationalize a numerator with two terms, we use the conjugate of the numerator.

EXAMPLE 5   Rationalize the numerator: $\dfrac{4 + \sqrt{2}}{\sqrt{5} - \sqrt{2}}$.

SOLUTION   We have

$\dfrac{4 + \sqrt{2}}{\sqrt{5} - \sqrt{2}} = \dfrac{4 + \sqrt{2}}{\sqrt{5} - \sqrt{2}} \cdot \dfrac{4 - \sqrt{2}}{4 - \sqrt{2}}$ 	Multiplying by 1, using the conjugate of $4 + \sqrt{2}$, which is $4 - \sqrt{2}$

$\qquad = \dfrac{16 - \left(\sqrt{2}\right)^2}{4\sqrt{5} - \sqrt{5}\,\sqrt{2} - 4\sqrt{2} + \left(\sqrt{2}\right)^2}$

$\qquad = \dfrac{14}{4\sqrt{5} - \sqrt{10} - 4\sqrt{2} + 2}.$

```
(4+√ (2))/(√ (5)−√
(2))
              6.587801273
14/(4√ (5)−√ (10)−
4√ (2)+2)
              6.587801273
```

We check by comparing the values of the original expression and the expression with a rationalized numerator, as shown at left.

Terms with Differing Indices

To multiply or divide radical terms with different indices, we can convert to exponential notation, use the rules for exponents, and then convert back to radical notation.

Student Notes

Expressions similar to the one in Example 6 are most easily simplified by rewriting the expression using exponents in place of radicals. After simplifying, remember to write your final result in radical notation. In general, if a problem is presented in one form, it is expected that the final result be presented in the same form.

EXAMPLE 6 Divide and, if possible, simplify: $\dfrac{\sqrt[4]{(x+y)^3}}{\sqrt{x+y}}$.

SOLUTION

$$\dfrac{\sqrt[4]{(x+y)^3}}{\sqrt{x+y}} = \dfrac{(x+y)^{3/4}}{(x+y)^{1/2}}$$ Converting to exponential notation

$$= (x+y)^{3/4-1/2}$$ Since the bases are identical, we can subtract exponents: $\frac{3}{4} - \frac{1}{2} = \frac{3}{4} - \frac{2}{4} = \frac{1}{4}$.

$$\left.\begin{array}{l} = (x+y)^{1/4} \\ = \sqrt[4]{x+y} \end{array}\right\}$$ Converting back to radical notation

---

To Simplify Products or Quotients with Differing Indices

1. Convert all radical expressions to exponential notation.
2. When the bases are identical, subtract exponents to divide and add exponents to multiply. This may require finding a common denominator.
3. Convert back to radical notation and, if possible, simplify.

EXAMPLE 7 Multiply and simplify: $\sqrt{x^3}\,\sqrt[3]{x}$.

SOLUTION

$$\sqrt{x^3}\,\sqrt[3]{x} = x^{3/2} \cdot x^{1/3}$$ Converting to exponential notation

$$= x^{11/6}$$ Adding exponents: $\frac{3}{2} + \frac{1}{3} = \frac{9}{6} + \frac{2}{6}$

$$= \sqrt[6]{x^{11}}$$ Converting back to radical notation

$$\left.\begin{array}{l} = \sqrt[6]{x^6}\,\sqrt[6]{x^5} \\ = x\sqrt[6]{x^5} \end{array}\right\}$$ Simplifying

---

EXAMPLE 8 If $f(x) = \sqrt[3]{x^2}$ and $g(x) = \sqrt{x} + \sqrt[4]{x}$, find $(f \cdot g)(x)$.

SOLUTION Recall from Section 5.9 that $(f \cdot g)(x) = f(x) \cdot g(x)$. Thus,

$$(f \cdot g)(x) = \sqrt[3]{x^2}\left(\sqrt{x} + \sqrt[4]{x}\right)$$ x is assumed to be nonnegative.

$$= x^{2/3}(x^{1/2} + x^{1/4})$$ Converting to exponential notation

$$= x^{2/3} \cdot x^{1/2} + x^{2/3} \cdot x^{1/4}$$ Using the distributive law

$$= x^{2/3+1/2} + x^{2/3+1/4}$$ Adding exponents

$$= x^{7/6} + x^{11/12}$$ $\frac{2}{3} + \frac{1}{2} = \frac{4}{6} + \frac{3}{6}$; $\frac{2}{3} + \frac{1}{4} = \frac{8}{12} + \frac{3}{12}$

$$= \sqrt[6]{x^7} + \sqrt[12]{x^{11}}$$ Converting back to radical notation

$$\left.\begin{array}{l} = \sqrt[6]{x^6}\,\sqrt[6]{x} + \sqrt[12]{x^{11}} \\ = x\sqrt[6]{x} + \sqrt[12]{x^{11}} \end{array}\right\}$$ Simplifying

If factors are raised to powers that share a common denominator, we can write the final result as a single radical expression.

EXAMPLE 9   Divide and, if possible, simplify: $\dfrac{\sqrt[3]{a^2b^4}}{\sqrt{ab}}$.

SOLUTION

$$\frac{\sqrt[3]{a^2b^4}}{\sqrt{ab}} = \frac{(a^2b^4)^{1/3}}{(ab)^{1/2}} \qquad \text{Converting to exponential notation}$$

$$= \frac{a^{2/3}b^{4/3}}{a^{1/2}b^{1/2}} \qquad \text{Using the product and power rules}$$

$$= a^{2/3-1/2}b^{4/3-1/2} \qquad \text{Subtracting exponents}$$

$$= a^{1/6}b^{5/6}$$

$$= \sqrt[6]{a}\,\sqrt[6]{b^5} \qquad \text{Converting to radical notation}$$

$$= \sqrt[6]{ab^5} \qquad \text{Using the product rule for radicals}$$

10.5 EXERCISE SET

FOR EXTRA HELP

*Math*XP			Tutor Center		
MathXL	MyMathLab	InterAct Math	AW Math Tutor Center	Video Lectures on CD: Disc 5	Student's Solutions Manual

Concept Reinforcement   *For each of Exercises 1–6, fill in the blanks by selecting from the following words (which may be used more than once).*

radicand(s)

indices

conjugate(s)

base(s)

denominator(s)

numerator(s)

1. To add radical expressions, the _____ and the _____ must be the same.

2. To multiply radical expressions, the _____ must be the same.

3. To find a product by adding exponents, the _____ must be the same.

4. To add rational expressions, the _____ must be the same.

5. To rationalize the _____ of $\dfrac{\sqrt{a+0.1}-\sqrt{a}}{0.1}$, we multiply by a form of 1, using the _____ of $\sqrt{a+0.1}-\sqrt{a}$, or $\sqrt{a+0.1}+\sqrt{a}$, to write 1.

6. To find a quotient by subtracting exponents, the _____ must be the same.

Add or subtract. Simplify by combining like radical terms, if possible. Assume that all variables and radicands represent positive real numbers.

7. $2\sqrt{5}+7\sqrt{5}$

8. $4\sqrt{7}+2\sqrt{7}$

9. $7\sqrt[3]{4}-5\sqrt[3]{4}$

10. $14\sqrt[5]{2}-6\sqrt[5]{2}$

11. $\sqrt[3]{y}+9\sqrt[3]{y}$

12. $9\sqrt[4]{t}-3\sqrt[4]{t}$

13. $8\sqrt{2} - 6\sqrt{2} + 5\sqrt{2}$

14. $\sqrt{6} + 8\sqrt{6} - 3\sqrt{6}$

15. $9\sqrt[3]{7} - \sqrt{3} + 4\sqrt[3]{7} + 2\sqrt{3}$

16. $5\sqrt{7} - 8\sqrt[4]{11} + \sqrt{7} + 9\sqrt[4]{11}$

17. $4\sqrt{27} - 3\sqrt{3}$

18. $9\sqrt{50} - 4\sqrt{2}$

19. $3\sqrt{45} + 7\sqrt{20}$

20. $5\sqrt{12} + 16\sqrt{27}$

21. $3\sqrt[3]{16} + \sqrt[3]{54}$

22. $\sqrt[3]{27} - 5\sqrt[3]{8}$

23. $\sqrt{5a} + 2\sqrt{45a^3}$

24. $4\sqrt{3x^3} - \sqrt{12x}$

25. $\sqrt[3]{6x^4} + \sqrt[3]{48x}$

26. $\sqrt[3]{54x} - \sqrt[3]{2x^4}$

27. $\sqrt{4a - 4} + \sqrt{a - 1}$

28. $\sqrt{9y + 27} + \sqrt{y + 3}$

29. $\sqrt{x^3 - x^2} + \sqrt{9x - 9}$

30. $\sqrt{4x - 4} - \sqrt{x^3 - x^2}$

Multiply. Assume all variables represent nonnegative real numbers.

31. $\sqrt{3}(4 + \sqrt{3})$

32. $\sqrt{7}(3 - \sqrt{7})$

33. $3\sqrt{5}(\sqrt{5} - \sqrt{2})$

34. $4\sqrt{2}(\sqrt{3} - \sqrt{5})$

35. $\sqrt{2}(3\sqrt{10} - 2\sqrt{2})$

36. $\sqrt{3}(2\sqrt{5} - 3\sqrt{4})$

37. $\sqrt[3]{3}(\sqrt[3]{9} - 4\sqrt[3]{21})$

38. $\sqrt[3]{2}(\sqrt[3]{4} - 2\sqrt[3]{32})$

39. $\sqrt[3]{a}(\sqrt[3]{a^2} + \sqrt[3]{24a^2})$

40. $\sqrt[3]{x}(\sqrt[3]{3x^2} - \sqrt[3]{81x^2})$

41. $(2 + \sqrt{6})(5 - \sqrt{6})$

42. $(4 - \sqrt{5})(2 + \sqrt{5})$

43. $(\sqrt{2} + \sqrt{7})(\sqrt{3} - \sqrt{7})$

44. $(\sqrt{7} - \sqrt{2})(\sqrt{5} + \sqrt{2})$

45. $(3 - \sqrt{5})(3 + \sqrt{5})$

46. $(6 - \sqrt{7})(6 + \sqrt{7})$

47. $(\sqrt{6} + \sqrt{8})(\sqrt{6} - \sqrt{8})$

48. $(\sqrt{5} + \sqrt{3})(\sqrt{5} - \sqrt{3})$

49. $(3\sqrt{7} + 2\sqrt{5})(2\sqrt{7} - 4\sqrt{5})$

50. $(4\sqrt{5} - 3\sqrt{2})(2\sqrt{5} + 4\sqrt{2})$

51. $(2 + \sqrt{3})^2$

52. $(3 + \sqrt{7})^2$

53. $(\sqrt{3} - \sqrt{2})^2$

54. $(\sqrt{5} - \sqrt{3})^2$

55. $(\sqrt{2t} + \sqrt{5})^2$

56. $(\sqrt{3x} - \sqrt{2})^2$

57. $(3 - \sqrt{x + 5})^2$

58. $(4 + \sqrt{x - 3})^2$

59. $(2\sqrt[4]{7} - \sqrt[4]{6})(3\sqrt[4]{9} + 2\sqrt[4]{5})$

60. $(4\sqrt[3]{3} + \sqrt[3]{10})(2\sqrt[3]{7} + 5\sqrt[3]{6})$

Rationalize each denominator.

61. $\dfrac{5}{4 - \sqrt{3}}$

62. $\dfrac{3}{4 - \sqrt{7}}$

63. $\dfrac{2 + \sqrt{5}}{6 + \sqrt{3}}$

64. $\dfrac{1 + \sqrt{2}}{3 + \sqrt{5}}$

65. $\dfrac{\sqrt{a}}{\sqrt{a} + \sqrt{b}}$

66. $\dfrac{\sqrt{z}}{\sqrt{x} - \sqrt{z}}$

Aha! **67.** $\dfrac{\sqrt{7} - \sqrt{3}}{\sqrt{3} - \sqrt{7}}$

68. $\dfrac{\sqrt{7} + \sqrt{5}}{\sqrt{5} + \sqrt{2}}$

69. $\dfrac{3\sqrt{2} - \sqrt{7}}{4\sqrt{2} + 2\sqrt{5}}$

70. $\dfrac{5\sqrt{3} - \sqrt{11}}{2\sqrt{3} - 5\sqrt{2}}$

Rationalize each numerator. If possible, simplify your result.

71. $\dfrac{\sqrt{7} + 2}{5}$

72. $\dfrac{\sqrt{3} + 1}{4}$

73. $\dfrac{\sqrt{6} - 2}{\sqrt{3} + 7}$

74. $\dfrac{\sqrt{10} + 4}{\sqrt{2} - 3}$

75. $\dfrac{\sqrt{x} - \sqrt{y}}{\sqrt{x} + \sqrt{y}}$

76. $\dfrac{\sqrt{a} + \sqrt{b}}{\sqrt{a} - \sqrt{b}}$

77. $\dfrac{\sqrt{a + h} - \sqrt{a}}{h}$

78. $\dfrac{\sqrt{x - h} - \sqrt{x}}{h}$

Perform the indicated operation and simplify. Assume all variables represent positive real numbers.

79. $\sqrt{a}\,\sqrt[4]{a^3}$

80. $\sqrt[3]{x^2}\,\sqrt[6]{x^5}$

81. $\sqrt[5]{b^2}\,\sqrt{b^3}$

82. $\sqrt[4]{a^3}\,\sqrt[3]{a^2}$

83. $\sqrt{xy^3}\,\sqrt[3]{x^2y}$

84. $\sqrt[5]{a^3b}\,\sqrt{ab}$

85. $\sqrt[4]{9ab^3}\,\sqrt{3a^4b}$

86. $\sqrt{2x^3y^3}\,\sqrt[3]{4xy^2}$

87. $\sqrt{a^4b^3c^4}\,\sqrt[3]{ab^2c}$

88. $\sqrt[3]{xy^2z}\,\sqrt{x^3yz^2}$

89. $\dfrac{\sqrt[3]{a^2}}{\sqrt[4]{a}}$

90. $\dfrac{\sqrt[3]{x^2}}{\sqrt[5]{x}}$

91. $\dfrac{\sqrt[4]{x^2y^3}}{\sqrt[3]{xy}}$

92. $\dfrac{\sqrt[5]{a^4b}}{\sqrt[3]{ab}}$

93. $\dfrac{\sqrt{ab^3}}{\sqrt[5]{a^2b^3}}$

94. $\dfrac{\sqrt[5]{x^3y^4}}{\sqrt{xy}}$

95. $\dfrac{\sqrt[4]{(3x-1)^3}}{\sqrt[5]{(3x-1)^3}}$

96. $\dfrac{\sqrt[3]{(2+5x)^2}}{\sqrt[4]{2+5x}}$

97. $\dfrac{\sqrt[3]{(2x+1)^2}}{\sqrt[5]{(2x+1)^2}}$

98. $\dfrac{\sqrt[4]{(5+3x)^3}}{\sqrt[3]{(5+3x)^2}}$

99. $\sqrt[3]{x^2y}\left(\sqrt{xy}-\sqrt[5]{xy^3}\right)$

100. $\sqrt[4]{a^2b}\left(\sqrt[3]{a^2b}-\sqrt[5]{a^2b^2}\right)$

101. $\left(m+\sqrt[3]{n^2}\right)\left(2m+\sqrt[4]{n}\right)$

102. $\left(r-\sqrt[4]{s^3}\right)\left(3r-\sqrt[5]{s}\right)$

In Exercises 103–106, f(x) and g(x) are as given. Find (f · g)(x). Assume all variables represent nonnegative real numbers.

103. $f(x)=\sqrt[4]{x},\ g(x)=\sqrt[4]{2x}-\sqrt[4]{x^{11}}$

104. $f(x)=\sqrt[4]{x^7}+\sqrt[4]{3x^2},\ g(x)=\sqrt[4]{x}$

105. $f(x)=x+\sqrt{7},\ g(x)=x-\sqrt{7}$

106. $f(x)=x-\sqrt{2},\ g(x)=x+\sqrt{6}$

Let f(x) = x². Find each of the following.

107. $f\left(5+\sqrt{2}\right)$

108. $f\left(7+\sqrt{3}\right)$

109. $f\left(\sqrt{3}-\sqrt{5}\right)$

110. $f\left(\sqrt{6}-\sqrt{3}\right)$

TW 111. In what way(s) is combining like radical terms similar to combining like terms that are monomials?

TW 112. Why do we need to know how to multiply radical expressions before learning how to add them?

Focused Review

Perform the indicated operation and simplify. Assume all variables represent positive real numbers.

113. $5\sqrt{3}+2\sqrt{3}$ [10.5]

114. $5\sqrt{3}\cdot2\sqrt{3}$ [10.3]

115. $\dfrac{\sqrt{200xy}}{2\sqrt{2y}}$ [10.4]

116. $\sqrt[3]{x^2y^4z}\cdot\sqrt[3]{x^4y^2z}$ [10.3]

117. $\left(5\sqrt{3}+\sqrt{x}\right)\left(5\sqrt{3}-\sqrt{x}\right)$ [10.5]

118. $\sqrt{2x}\,\sqrt[4]{8x^3}$ [10.5]

Synthesis

TW 119. Ramon *incorrectly* writes
$$\sqrt[5]{x^2}\cdot\sqrt{x^3}=x^{2/5}\cdot x^{3/2}=\sqrt[5]{x^3}.$$
What mistake do you suspect he is making?

TW 120. After examining the expression $\sqrt[4]{25xy^3}\,\sqrt{5x^4y}$, Dyan (correctly) concludes that x and y are both nonnegative. Explain how she could reach this conclusion.

Find a simplified form for f(x). Assume x ≥ 0.

121. $f(x)=\sqrt{20x^2+4x^3}-3x\sqrt{45+9x}+\sqrt{5x^2+x^3}$

122. $f(x)=\sqrt{x^3-x^2}+\sqrt{9x^3-9x^2}-\sqrt{4x^3-4x^2}$

123. $f(x)=\sqrt[4]{x^5-x^4}+3\sqrt[4]{x^9-x^8}$

124. $f(x)=\sqrt[4]{16x^4+16x^5}-2\sqrt[4]{x^8+x^9}$

Simplify.

125. $\frac12\sqrt{36a^5bc^4}-\frac12\sqrt[3]{64a^4bc^6}+\frac16\sqrt{144a^3bc^6}$

126. $7x\sqrt{(x+y)^3}-5xy\sqrt{x+y}-2y\sqrt{(x+y)^3}$

127. $\sqrt{27a^5(b+1)}\,\sqrt[3]{81a(b+1)^4}$

128. $\sqrt{8x(y+z)^5}\,\sqrt[3]{4x^2(y+z)^2}$

129. $\dfrac{\dfrac{1}{\sqrt{w}}-\sqrt{w}}{\dfrac{\sqrt{w}+1}{\sqrt{w}}}$

130. $\dfrac{1}{4+\sqrt{3}}+\dfrac{1}{\sqrt{3}}+\dfrac{1}{\sqrt{3}-4}$

Express each of the following as the product of two radical expressions.

131. $x-5$

132. $y-7$

133. $x-a$

Multiply.

134. $\sqrt{9+3\sqrt{5}}\,\sqrt{9-3\sqrt{5}}$

135. $\left(\sqrt{x+2}-\sqrt{x-2}\right)^2$

10.6 Solving Radical Equations

The Principle of Powers ■ Equations with Two Radical Terms

[Connecting the Concepts

In Sections 10.1–10.5, we learned how to manipulate radical expressions as well as expressions containing rational exponents. We performed this work to find *equivalent expressions*.

Now that we know how to work with radicals and rational exponents, we can learn how to solve a new type of equation. As in our earlier work with equations, finding *equivalent equations* will be part of our strategy. What is different, however, is that now we will use a step that does not always produce equivalent equations. Checking solutions will therefore be more important than ever.

The Principle of Powers

A **radical equation** is an equation in which the variable appears in a radicand. Examples are

$$\sqrt[3]{2x} + 1 = 5, \quad \sqrt{a} + \sqrt{a-2} = 7, \quad \text{and} \quad 4 - \sqrt{3x+1} = \sqrt{6-x}.$$

To solve such equations, we need a new principle. Suppose $a = b$ is true. If we square both sides, we get another true equation: $a^2 = b^2$. This can be generalized.

> **The Principle of Powers**   If $a = b$, then $a^n = b^n$ for any exponent n.

Note that the principle of powers is an "if–then" statement. The statement obtained by interchanging the two parts of the sentence—"if $a^n = b^n$ for some exponent n, then $a = b$"—is *not always true*. For example, "if $x = 3$, then $x^2 = 9$" is true, but the statement "if $x^2 = 9$, then $x = 3$" is *not* true when x is replaced with -3.

[Interactive Discovery

Solve each pair of equations graphically, and compare the solution sets.

1. $x = 3$; $x^2 = 9$
2. $x = -2$; $x^2 = 4$
3. $\sqrt{x} = 5$; $x = 25$
4. $\sqrt{x} = -3$; $x = 9$

We see that every solution of $x = a$ is a solution of $x^2 = a^2$, but not every solution of $x^2 = a^2$ is necessarily a solution of $x = a$. A similar statement can be made for any even exponent n; for example, the solution sets of $x^4 = 3^4$ and $x = 3$ are not the same. For this reason, when we raise both sides of an equation to an even power, it will be essential for us to check the answer in the original equation. We *will* find all the solutions; we *may* also find additional numbers that are not solutions of the original equation.

EXAMPLE 1   Solve: $\sqrt{x} - 3 = 4$.

ALGEBRAIC APPROACH

Before using the principle of powers, we need to isolate the radical term:

$$\sqrt{x} - 3 = 4$$
$$\sqrt{x} = 7 \qquad \text{Isolating the radical by adding 3 to both sides}$$
$$\left(\sqrt{x}\right)^2 = 7^2 \qquad \text{Using the principle of powers}$$
$$x = 49.$$

Check:

$$\begin{array}{c|c} \sqrt{x} - 3 = 4 & \\ \hline \sqrt{49} - 3 & 4 \\ 7 - 3 & \\ 4 \overset{?}{=} 4 & \text{TRUE} \end{array}$$

The solution is 49.

GRAPHICAL APPROACH

We let $f(x) = \sqrt{x} - 3$ and $g(x) = 4$. Since the domain of f is $[0, \infty)$, we try a viewing window of $[-1, 10, -10, 10]$.

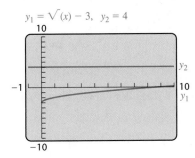

It appears as though the graphs may intersect at an x-value greater than 10. We adjust the viewing window to $[-10, 100, -5, 5]$ and find the point of intersection.

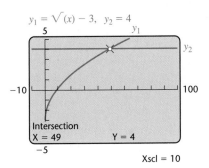

We have a solution of 49, which we can check by substituting into the original equation, as shown in the algebraic approach.

The solution is 49.

EXAMPLE 2 Solve: $\sqrt{x} - 5 = -7$.

ALGEBRAIC APPROACH

We have

$$\sqrt{x} - 5 = -7$$
$$\sqrt{x} = -2 \qquad \text{Isolating the radical by adding 5 to both sides}$$

> The equation $\sqrt{x} = -2$ has no solution because the principal square root of a number is never negative. We continue as in Example 1 for comparison.

$$\left(\sqrt{x}\right)^2 = (-2)^2 \qquad \text{Using the principle of powers (squaring)}$$
$$x = 4.$$

Check:

$$\begin{array}{c|c} \sqrt{x} - 5 \;=\; -7 & \\ \hline \sqrt{4} - 5 & -7 \\ 2 - 5 & \\ -3 \stackrel{?}{=} -7 & \text{FALSE} \end{array}$$

The number 4 does not check. Thus, $\sqrt{x} - 5 = -7$ has no solution.

GRAPHICAL APPROACH

We let $f(x) = \sqrt{x} - 5$ and $g(x) = -7$ and graph.

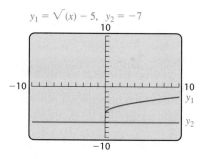

$y_1 = \sqrt{(x)} - 5, \quad y_2 = -7$

The graphs do not intersect. There is no real-number solution.

> **CAUTION!** Raising both sides of an equation to an even power may not produce an equivalent equation. In this case, a check is essential.

Note in Example 2 that $x = 4$ has solution 4 but $\sqrt{x} - 5 = -7$ has *no* solution. Thus the equations $x = 4$ and $\sqrt{x} - 5 = -7$ are *not* equivalent.

> **To Solve an Equation with a Radical Term**
>
> **1.** Isolate the radical term on one side of the equation.
> **2.** Use the principle of powers and solve the resulting equation.
> **3.** Check any possible solution in the original equation.

> **EXAMPLE 3**   Solve: $x = \sqrt{x + 7} + 5$.

ALGEBRAIC APPROACH

$$x = \sqrt{x + 7} + 5$$

$$x - 5 = \sqrt{x + 7} \qquad \text{Isolating the radical by subtracting 5 from both sides}$$

$$\left.(x - 5)^2 = \left(\sqrt{x + 7}\right)^2 \atop x^2 - 10x + 25 = x + 7 \right\} \quad \text{Using the principle of powers; squaring both sides}$$

$$x^2 - 11x + 18 = 0 \qquad \text{Adding } -x - 7 \text{ to both sides to write the quadratic equation in standard form}$$

$$(x - 9)(x - 2) = 0 \qquad \text{Factoring}$$

$$x = 9 \quad or \quad x = 2 \qquad \text{Using the principle of zero products}$$

The possible solutions are 9 and 2. Let's check.

Check: For 9:

$$\begin{array}{c|c} x = \sqrt{x + 7} + 5 \\ \hline 9 & \sqrt{9 + 7} + 5 \\ 9 \overset{?}{=} 9 \end{array} \quad \text{TRUE}$$

For 2:

$$\begin{array}{c|c} x = \sqrt{x + 7} + 5 \\ \hline 2 & \sqrt{2 + 7} + 5 \\ 2 \overset{?}{=} 8 \end{array} \quad \text{FALSE}$$

Since 9 checks but 2 does not, the solution is 9.

GRAPHICAL APPROACH

We graph $y_1 = x$ and $y_2 = \sqrt{(x + 7)} + 5$ and determine any points of intersection.

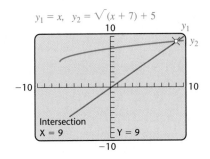

$$y_1 = x, \quad y_2 = \sqrt{(x + 7)} + 5$$

Intersection X = 9   Y = 9

We obtain a solution of 9.

It is important to isolate a radical term before using the principle of powers. Suppose in Example 3 that both sides of the equation were squared *before* isolating the radical. We then would have had the expression $\left(\sqrt{x + 7} + 5\right)^2$ or $x + 7 + 10\sqrt{x + 7} + 25$ in the third step of the algebraic approach, and the radical would have remained in the problem.

> **EXAMPLE 4**   Solve: $(2x + 1)^{1/3} + 5 = 0$.

ALGEBRAIC APPROACH

We need not use radical notation to solve:

$$(2x + 1)^{1/3} + 5 = 0$$

$$(2x + 1)^{1/3} = -5 \qquad \text{Subtracting 5 from both sides}$$

$$[(2x + 1)^{1/3}]^3 = (-5)^3 \qquad \text{Cubing both sides}$$

$$(2x + 1)^1 = (-5)^3 \qquad \text{Multiplying exponents. Try to do this mentally.}$$

$$2x + 1 = -125$$

$$2x = -126 \qquad \text{Subtracting 1 from both sides}$$

$$x = -63.$$

Because both sides were raised to an *odd* power, we need check only the accuracy of our work.

Check:
$$(2x + 1)^{1/3} + 5 = 0$$

$$
\begin{array}{c|c}
[2(-63) + 1]^{1/3} + 5 & 0 \\
{[-126 + 1]^{1/3} + 5} & \\
(-125)^{1/3} + 5 & \\
\sqrt[3]{-125} + 5 & \\
-5 + 5 & \\
0 \overset{?}{=} 0 & \text{TRUE}
\end{array}
$$

GRAPHICAL APPROACH

We let $f(x) = (2x + 1)^{1/3} + 5$ and determine any values of x for which $f(x) = 0$. We might start by graphing f using a standard viewing window, as shown on the left below.

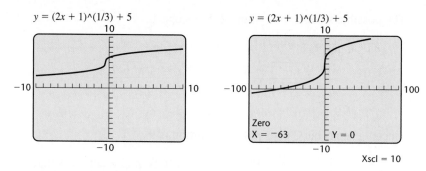

It appears that if the graph of f does have an x-intercept, it will be at a value of x less than -10. Let's try a viewing window of $[-100, 100, -10, 10]$, as shown on the right above. We find that $f(x) = 0$ when $x = -63$.

Equations with Two Radical Terms

A strategy for solving equations with two or more radical terms is as follows.

To Solve an Equation with Two or More Radical Terms

1. Isolate one of the radical terms.
2. Use the principle of powers.
3. If a radical remains, perform steps (1) and (2) again.
4. Solve the resulting equation.
5. Check possible solutions in the original equation.

EXAMPLE 5 Solve: $\sqrt{2x - 5} = 1 + \sqrt{x - 3}$.

ALGEBRAIC APPROACH

We have

$$\sqrt{2x - 5} = 1 + \sqrt{x - 3}$$
$$(\sqrt{2x - 5})^2 = (1 + \sqrt{x - 3})^2$$ One radical is already isolated. We square both sides.

> This is like squaring a binomial. We square 1, then find twice the product of 1 and $\sqrt{x - 3}$, and then the square of $\sqrt{x - 3}$.

$$2x - 5 = 1 + 2\sqrt{x - 3} + (\sqrt{x - 3})^2$$
$$2x - 5 = 1 + 2\sqrt{x - 3} + (x - 3)$$
$$x - 3 = 2\sqrt{x - 3}$$ Isolating the remaining radical term

$$(x - 3)^2 = (2\sqrt{x - 3})^2$$ Squaring both sides

$$x^2 - 6x + 9 = 4(x - 3)$$ Remember to square both the 2 and the $\sqrt{x - 3}$ on the right side.

$$x^2 - 6x + 9 = 4x - 12$$
$$x^2 - 10x + 21 = 0$$
$$(x - 7)(x - 3) = 0$$ Factoring
$$x = 7 \quad or \quad x = 3.$$ Using the principle of zero products

The possible solutions are 7 and 3. Both 7 and 3 check in the original equation and are the solutions.

GRAPHICAL APPROACH

We let $f(x) = \sqrt{2x - 5}$ and $g(x) = 1 + \sqrt{x - 3}$. The domain of f is $[\frac{5}{2}, \infty)$, and the domain of g is $[3, \infty)$. Since the intersection of their domains is $[3, \infty)$, any point of intersection will occur in that interval. We choose a viewing window of $[-1, 10, -1, 5]$.

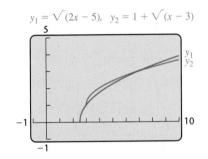

$y_1 = \sqrt{(2x - 5)}, \quad y_2 = 1 + \sqrt{(x - 3)}$

The graphs appear to intersect when x is approximately 3 and again when x is approximately 7. The points of intersection are $(3, 1)$ and $(7, 3)$. To confirm that there are no other points of intersection, we could examine the graphs using a larger viewing window.

> **CAUTION!** A common error in solving equations like
>
> $$\sqrt{2x - 5} = 1 + \sqrt{x - 3}$$
>
> is to obtain $1 + (x - 3)$ as the square of the right side. This is wrong because $(A + B)^2 \neq A^2 + B^2$. Placing parentheses around each side when squaring serves as a reminder to square the entire expression:
>
> $$(1 + \sqrt{x - 3})^2 = (1 + \sqrt{x - 3})(1 + \sqrt{x - 3})$$
> $$= 1 + 2\sqrt{x - 3} + (\sqrt{x - 3})^2.$$

EXAMPLE 6 Let $f(x) = \sqrt{x+5} - \sqrt{x-3}$. Find all x-values for which $f(x) = 2$.

ALGEBRAIC APPROACH

We must have $f(x) = 2$, or

$\sqrt{x+5} - \sqrt{x-3} = 2.$ Substituting for $f(x)$

To solve, we isolate one radical term and square both sides:

$\sqrt{x+5} = 2 + \sqrt{x-3}$ Adding $\sqrt{x-3}$. This isolates one of the radical terms.

$(\sqrt{x+5})^2 = (2 + \sqrt{x-3})^2$ Using the principle of powers (squaring both sides)

$x + 5 = 4 + 4\sqrt{x-3} + (x-3)$ Using $(A+B)^2 = A^2 + 2AB + B^2$

$5 = 1 + 4\sqrt{x-3}$ Adding $-x$ and combining like terms

$\left.\begin{array}{l} 4 = 4\sqrt{x-3} \\ 1 = \sqrt{x-3} \end{array}\right\}$ Isolating the remaining radical term

$1^2 = (\sqrt{x-3})^2$ Squaring both sides

$1 = x - 3$

$4 = x.$

GRAPHICAL APPROACH

We graph $y_1 = \sqrt{(x+5)} - \sqrt{(x-3)}$ and $y_2 = 2$ and determine the coordinates of any points of intersection.

$y_1 = \sqrt{(x+5)} - \sqrt{(x-3)}, \ y_2 = 2$

The solution is the x-coordinate of the point of intersection, which is 4.

Check: $f(4) = \sqrt{4+5} - \sqrt{4-3} = \sqrt{9} - \sqrt{1} = 3 - 1 = 2.$

We have $f(x) = 2$ when $x = 4$.

10.6 EXERCISE SET

Concept Reinforcement   *Classify each of the following as either true or false.*

1. If $x^2 = 25$, then $x = 5$.

2. If $t = 7$, then $t^2 = 49$.

3. If $\sqrt{x} = 3$, then $(\sqrt{x})^2 = 3^2$.

4. If $x^2 = 36$, then $x = 6$.

5. $\sqrt{x} - 8 = 7$ is equivalent to $\sqrt{x} = 15$.

6. $\sqrt{t} + 5 = 8$ is equivalent to $\sqrt{t} = 3$.

Solve.

7. $\sqrt{5x - 2} = 7$

8. $\sqrt{3x - 2} = 6$

9. $\sqrt{3x + 1} = 6$

10. $\sqrt{2x - 1} = 2$

11. $\sqrt{y + 1} - 5 = 8$

12. $\sqrt{x - 2} - 7 = -4$

13. $\sqrt{x - 7} + 3 = 10$

14. $\sqrt{y + 4} + 6 = 7$

15. $\sqrt[3]{x + 5} = 2$

16. $\sqrt[3]{x - 2} = 3$

17. $\sqrt[4]{y - 1} = 3$

18. $\sqrt[4]{x + 3} = 2$

19. $3\sqrt{x} = x$

20. $8\sqrt{y} = y$

21. $2y^{1/2} - 7 = 9$

22. $3x^{1/2} + 12 = 9$

23. $\sqrt[3]{x} = -3$

24. $\sqrt[3]{y} = -4$

25. $t^{1/3} - 2 = 3$

26. $x^{1/4} - 2 = 1$

Aha! **27.** $(y - 3)^{1/2} = -2$

28. $(x + 2)^{1/2} = -4$

29. $\sqrt[4]{3x + 1} - 4 = -1$

30. $\sqrt[4]{2x + 3} - 5 = -2$

31. $(x + 7)^{1/3} = 4$

32. $(y - 7)^{1/4} = 3$

33. $\sqrt[3]{3y + 6} + 7 = 8$

34. $\sqrt[3]{6x + 9} + 5 = 2$

35. $\sqrt{3t + 4} = \sqrt{4t + 3}$

36. $\sqrt{2t - 7} = \sqrt{3t - 12}$

37. $3(4 - t)^{1/4} = 6^{1/4}$

38. $2(1 - x)^{1/3} = 4^{1/3}$

39. $3 + \sqrt{5 - x} = x$

40. $x = \sqrt{x - 1} + 3$

41. $\sqrt{4x - 3} = 2 + \sqrt{2x - 5}$

42. $3 + \sqrt{z - 6} = \sqrt{z + 9}$

43. $\sqrt{20 - x} + 8 = \sqrt{9 - x} + 11$

44. $4 + \sqrt{10 - x} = 6 + \sqrt{4 - x}$

45. $\sqrt{x + 2} + \sqrt{3x + 4} = 2$

46. $\sqrt{6x + 7} - \sqrt{3x + 3} = 1$

47. If $f(x) = \sqrt{x} + \sqrt{x - 9}$, find any x for which $f(x) = 1$.

48. If $g(x) = \sqrt{x} + \sqrt{x - 5}$, find any x for which $g(x) = 5$.

49. If $f(t) = \sqrt{t - 2} - \sqrt{4t + 1}$, find any t for which $f(t) = -3$.

50. If $g(t) = \sqrt{2t + 7} - \sqrt{t + 15}$, find any t for which $g(t) = -1$.

51. If $f(x) = \sqrt{2x - 3}$ and $g(x) = \sqrt{x + 7} - 2$, find any x for which $f(x) = g(x)$.

52. If $f(x) = 2\sqrt{3x + 6}$ and $g(x) = 5 + \sqrt{4x + 9}$, find any x for which $f(x) = g(x)$.

53. If $f(t) = 4 - \sqrt{t - 3}$ and $g(t) = (t + 5)^{1/2}$, find any t for which $f(t) = g(t)$.

54. If $f(t) = 7 + \sqrt{2t - 5}$ and $g(t) = 3(t + 1)^{1/2}$, find any t for which $f(t) = g(t)$.

TW 55. Explain in your own words why it is important to check your answers when using the principle of powers.

TW 56. The principle of powers is an "if–then" statement that becomes false when the sentence parts are interchanged. Give an example of another such if–then statement from everyday life (answers will vary).

Focused Review

Solve.

57. $x^2 = x + 12$ [6.2]

58. $\frac{1}{3}x - \frac{1}{4} = 5$ [2.2]

59. $|2x - 3| > 5$ [8.2]

60. $\sqrt{x - 3} + 2 = 7$ [10.6]

61. $4 - x \leq 2x + 3$ [2.6]

62. $x^2 = 36$ [6.4]

63. $2(x - 1)^{1/3} = -4$ [10.6]

64. $\dfrac{1}{x} + \dfrac{1}{x - 1} = \dfrac{2}{x - 3}$ [7.6]

65. $2x - 3y = 8,$
     $x + 2y = 5$ [4.3]

66. $2x - 3y + z = 0,$
     $4x + 5y + z = 1,$
     $2x + y - 4z = 5$ [9.1]

Synthesis

TW 67. Describe a procedure that could be used to create radical equations that have no solution.

TW 68. Is checking essential when the principle of powers is used with an odd power n? Why or why not?

Steel Manufacturing.   *In the production of steel and other metals, the temperature of the molten metal is so great that conventional thermometers melt. Instead, sound is transmitted across the surface of the metal to a receiver on the far side and the speed of the sound is measured. The formula*

$$S(t) = 1087.7 \sqrt{\frac{9t + 2617}{2457}}$$

gives the speed of sound $S(t)$, in feet per second, at a temperature of t degrees Celsius.

69. Find the temperature of a blast furnace where sound travels 1880 ft/sec.

70. Find the temperature of a blast furnace where sound travels 1502.3 ft/sec.

71. Solve the above equation for t.

Automotive Repair.   *For an engine with a displacement of 2.8 L, the function given by*

$$d(n) = 0.75\sqrt{2.8n}$$

can be used to determine the diameter size of the carburetor's opening, in millimeters. Here n is the number of rpm's at which the engine achieves peak performance. (Source: macdizzy.com)

72. If a carburetor's opening is 81 mm, for what number of rpm's will the engine produce peak power?

73. If a carburetor's opening is 84 mm, for what number of rpm's will the engine produce peak power?

Escape Velocity.   *A formula for the escape velocity v of a satellite is*

$$v = \sqrt{2gr}\sqrt{\frac{h}{r+h}},$$

where g is the force of gravity, r is the planet or star's radius, and h is the height of the satellite above the planet or star's surface.

74. Solve for h.

75. Solve for r.

Sighting to the Horizon.   *The function $D(h) = 1.2\sqrt{h}$ can be used to approximate the distance D, in miles, that a person can see to the horizon from a height h, in feet.*

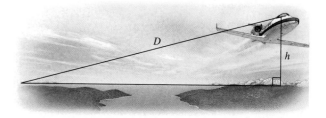

76. How far above sea level must a pilot fly in order to see a horizon that is 180 mi away?

77. How high above sea level must a sailor climb in order to see 10.2 mi out to sea?

Solve.

78. $\left(\dfrac{z}{4} - 5\right)^{2/3} = \dfrac{1}{25}$

79. $\dfrac{x + \sqrt{x+1}}{x - \sqrt{x+1}} = \dfrac{5}{11}$

80. $\sqrt{\sqrt{y} + 49} = 7$

81. $(z^2 + 17)^{3/4} = 27$

82. $x^2 - 5x - \sqrt{x^2 - 5x - 2} = 4$
(*Hint*: Let $u = x^2 - 5x - 2$.)

83. $\sqrt{8 - b} = b\sqrt{8 - b}$

Without graphing, determine the x-intercepts of the graphs given by each of the following.

84. $f(x) = \sqrt{x-2} - \sqrt{x+2} + 2$

85. $g(x) = 6x^{1/2} + 6x^{-1/2} - 37$

86. $f(x) = (x^2 + 30x)^{1/2} - x - (5x)^{1/2}$

87. Saul is trying to solve Exercise 77 using a graphing calculator. Without resorting to trial and error, how can he determine a suitable viewing window for finding the solution?

Collaborative Corner

Tailgater Alert

Focus: Radical equations and problem solving
Time: 15–25 minutes
Group size: 2–3
Materials: Calculators

The faster a car is traveling, the more distance it needs to stop. Thus it is important for drivers to allow sufficient space between their vehicle and the vehicle in front of them. Police recommend that for each 10 mph of speed, a driver allow 1 car length. Thus a driver going 30 mph should have at least 3 car lengths between his or her vehicle and the one in front.
  In Exercise Set 10.3, the function $r(L) = 2\sqrt{5L}$ was used to find the speed, in miles per hour, that a car was traveling when it left skid marks L feet long.

ACTIVITY

1. Each group member should estimate the length of a car in which he or she frequently travels. (Each should use a different length, if possible.)
2. Using a calculator as needed, each group member should complete the table at right. Column 1

gives a car's speed s, column 2 lists the minimum amount of space between cars traveling s miles per hour, as recommended by police. Column 3 is the speed that a vehicle *could* travel were it forced to stop in the distance listed in column 2, using the above function.

Column 1 s (in miles per hour)	Column 2 $L(s)$ (in feet)	Column 3 $r(L)$ (in miles per hour)
20		
30		
40		
50		
60		
70		

3. Determine whether there are any speeds at which the "1 car length per 10 mph" guideline might not suffice. On what reasoning do you base your answer? Compare tables to determine how car length affects the results. What recommendations would your group make to a new driver?

10.7 Geometric Applications

Using the Pythagorean Theorem ■ Two Special Triangles

Using the Pythagorean Theorem

There are many kinds of problems that involve powers and roots. Many also involve right triangles and the Pythagorean theorem, which we studied in Section 6.7 and restate here.

Study Tip

Making Sketches

One need not be an artist to make highly useful mathematical sketches. That said, it is important to make sure that your sketches are drawn accurately enough to represent the relative sizes within each shape. For example, if one side of a triangle is clearly the longest, make sure your drawing reflects this.

The Pythagorean Theorem* In any right triangle, if a and b are the lengths of the legs and c is the length of the hypotenuse, then

$$a^2 + b^2 = c^2.$$

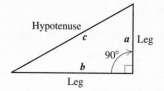

In using the Pythagorean theorem, we often make use of the following principle.

The Principle of Square Roots For any nonnegative real number n,

If $x^2 = n$, then $x = \sqrt{n}$ or $x = -\sqrt{n}$.

For most real-world applications involving length or distance, $-\sqrt{n}$ is not needed.

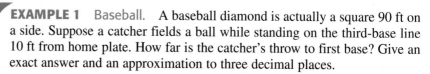

EXAMPLE 1 Baseball. A baseball diamond is actually a square 90 ft on a side. Suppose a catcher fields a ball while standing on the third-base line 10 ft from home plate. How far is the catcher's throw to first base? Give an exact answer and an approximation to three decimal places.

SOLUTION   We make a drawing and let $d =$ the distance, in feet, to first base. Note that a right triangle is formed in which the leg from home plate to first base measures 90 ft and the leg from home plate to where the catcher fields the ball measures 10 ft.

We substitute these values into the Pythagorean theorem to find d:

$$d^2 = 90^2 + 10^2$$
$$d^2 = 8100 + 100$$
$$d^2 = 8200.$$

We now use the principle of square roots: If $d^2 = 8200$, then $d = \sqrt{8200}$ or $d = -\sqrt{8200}$. Since d represents a length, it follows that d is the positive square root of 8200:

$d = \sqrt{8200}$ ft   **This is an exact answer.**

$d \approx 90.554$ ft.   **Using a calculator for an approximation**

---

*The converse of the Pythagorean theorem also holds. That is, if a, b, and c are the lengths of the sides of a triangle and $a^2 + b^2 = c^2$, then the triangle is a right triangle.

EXAMPLE 2 Guy Wires.   The base of a 40-ft-long guy wire is located 15 ft from the telephone pole that it is anchoring. How high up the pole does the guy wire reach? Give an exact answer and an approximation to three decimal places.

SOLUTION   We make a drawing and let h = the height, in feet, to which the guy wire reaches. A right triangle is formed in which one leg measures 15 ft and the hypotenuse measures 40 ft. Using the Pythagorean theorem, we have

$$h^2 + 15^2 = 40^2$$
$$h^2 + 225 = 1600$$
$$h^2 = 1375$$
$$h = \sqrt{1375}.$$

Exact answer:
$$h = \sqrt{1375} \text{ ft}$$

Approximation:
$$h \approx 37.081 \text{ ft}$$ Using a calculator

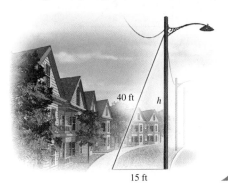

40 ft h

15 ft

Two Special Triangles

When both legs of a right triangle are the same size, we call the triangle an *isosceles right triangle*, as shown at left. If one leg of an isosceles right triangle has length a, we can find a formula for the length of the hypotenuse as follows:

$$c^2 = a^2 + b^2$$
$$c^2 = a^2 + a^2$$ Because the triangle is isosceles, both legs are the same size: $a = b$.
$$c^2 = 2a^2.$$ Combining like terms

Next, we use the principle of square roots. Because a, b, and c are lengths, there is no need to consider negative square roots or absolute values. Thus,

$$c = \sqrt{2a^2}$$ Using the principle of square roots
$$= \sqrt{a^2 \cdot 2} = a\sqrt{2}.$$

EXAMPLE 3   One leg of an isosceles right triangle measures 7 cm. Find the length of the hypotenuse. Give an exact answer and an approximation to three decimal places.

SOLUTION   We substitute:

$$c = a\sqrt{2}$$ This equation is worth memorizing.
$$= 7\sqrt{2}.$$

Exact answer: $c = 7\sqrt{2}$ cm
Approximation: $c \approx 9.899$ cm   Using a calculator

When the hypotenuse of an isosceles right triangle is known, the lengths of the legs can be found.

45°

a c

45°

a

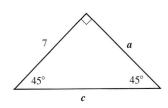

7   a

45°   45°

c

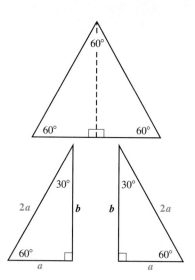

EXAMPLE 4   The hypotenuse of an isosceles right triangle is 5 ft long. Find the length of a leg. Give an exact answer and an approximation to three decimal places.

SOLUTION   We replace c with 5 and solve for a:

$$5 = a\sqrt{2} \qquad \text{Substituting 5 for } c \text{ in } c = a\sqrt{2}$$

$$\frac{5}{\sqrt{2}} = a \qquad \text{Dividing both sides by } \sqrt{2}$$

$$\frac{5\sqrt{2}}{2} = a. \qquad \text{Rationalize the denominator if desired.}$$

Exact answer: $a = \dfrac{5}{\sqrt{2}}$ ft, or $\dfrac{5\sqrt{2}}{2}$ ft

Approximation: $a \approx 3.536$ ft   **Using a calculator**

A second special triangle is known as a $30°–60°–90°$ right triangle, so named because of the measures of its angles. Note that in an equilateral triangle, all sides have the same length and all angles are $60°$. An altitude, drawn dashed in the figure, bisects, or splits in half, one angle and one side. Two $30°–60°–90°$ right triangles are thus formed.

Because of the way in which the altitude is drawn, if a represents the length of the shorter leg in a $30°–60°–90°$ right triangle, then $2a$ represents the length of the hypotenuse. We have

$$a^2 + b^2 = (2a)^2 \qquad \text{Using the Pythagorean theorem}$$
$$a^2 + b^2 = 4a^2$$
$$b^2 = 3a^2 \qquad \text{Subtracting } a^2 \text{ from both sides}$$
$$b = \sqrt{3a^2}$$
$$b = \sqrt{a^2 \cdot 3}$$
$$b = a\sqrt{3}.$$

EXAMPLE 5   The shorter leg of a $30°–60°–90°$ right triangle measures 8 in. Find the lengths of the other sides. Give exact answers and, where appropriate, an approximation to three decimal places.

SOLUTION   The hypotenuse is twice as long as the shorter leg, so we have

$$c = 2a \qquad \text{This relationship is worth memorizing.}$$
$$= 2 \cdot 8 = 16 \text{ in.} \qquad \text{This is the length of the hypotenuse.}$$

The length of the longer leg is the length of the shorter leg times $\sqrt{3}$. This gives us

$$b = a\sqrt{3} \qquad \text{This is also worth memorizing.}$$
$$= 8\sqrt{3} \text{ in.} \qquad \text{This is the length of the longer leg.}$$

Exact answer: $c = 16$ in., $b = 8\sqrt{3}$ in.

Approximation: $b \approx 13.856$ in.

EXAMPLE 6   The length of the longer leg of a $30°$–$60°$–$90°$ right triangle is 14 cm. Find the length of the hypotenuse. Give an exact answer and an approximation to three decimal places.

SOLUTION   The length of the hypotenuse is twice the length of the shorter leg. We first find a, the length of the shorter leg, by using the length of the longer leg:

$$14 = a\sqrt{3} \qquad \text{Substituting 14 for } b \text{ in } b = a\sqrt{3}$$

$$\frac{14}{\sqrt{3}} = a. \qquad \text{Dividing by } \sqrt{3}$$

Since the hypotenuse is twice as long as the shorter leg, we have

$$c = 2a$$

$$= 2 \cdot \frac{14}{\sqrt{3}} \qquad \text{Substituting}$$

$$= \frac{28}{\sqrt{3}} \text{ cm.}$$

Exact answer: $c = \dfrac{28}{\sqrt{3}}$ cm, or $\dfrac{28\sqrt{3}}{3}$ cm if the denominator is rationalized.

Approximation: $c \approx 16.166$ cm

Student Notes

Perhaps the easiest way to remember the important results listed in the adjacent box is to write out, on your own, the derivations shown on pp. 781 and 782.

Lengths Within Isosceles and 30°–60°–90° Right Triangles

The length of the hypotenuse in an isosceles right triangle is the length of a leg times $\sqrt{2}$.

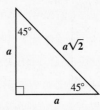

The length of the longer leg in a $30°$–$60°$–$90°$ right triangle is the length of the shorter leg times $\sqrt{3}$. The hypotenuse is twice as long as the shorter leg.

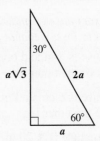

10.7 EXERCISE SET

↝ *Concept Reinforcement* *Complete each of the following sentences.*

1. In any _____ triangle, the square of the length of the _____ is the sum of the squares of the lengths of the legs.

2. The shortest side of a right triangle is always one of the two _____.

3. The principle of _____ _____ states that if $x^2 = n$, then $x = \sqrt{n}$ or $x = -\sqrt{n}$.

4. In a(n) _____ right triangle, both legs have the same length.

5. In a(n) ___–___–___ right triangle, the hypotenuse is twice as long as the shorter _____.

6. If both legs have measure a, then the hypotenuse measures _____.

In a right triangle, find the length of the side not given. Give an exact answer and, where appropriate, an approximation to three decimal places.

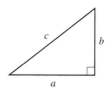

7. $a = 5, b = 3$ 8. $a = 8, b = 10$

Aha! 9. $a = 9, b = 9$ 10. $a = 10, b = 10$

11. $b = 12, c = 13$ 12. $a = 5, c = 12$

In Exercises 13–28, give an exact answer and, where appropriate, an approximation to three decimal places.

13. A right triangle's hypotenuse is 8 m and one leg is $4\sqrt{3}$ m. Find the length of the other leg.

14. A right triangle's hypotenuse is 6 cm and one leg is $\sqrt{5}$ cm. Find the length of the other leg.

15. The hypotenuse of a right triangle is $\sqrt{20}$ in. and one leg measures 1 in. Find the length of the other leg.

16. The hypotenuse of a right triangle is $\sqrt{15}$ ft and one leg measures 2 ft. Find the length of the other leg.

Aha! 17. One leg of a right triangle is 1 m and the hypotenuse measures $\sqrt{2}$ m. Find the length of the other leg.

18. One leg of a right triangle is 1 yd and the hypotenuse measures 2 yd. Find the length of the other leg.

19. *Bicycling.* Clare routinely bicycles across a rectangular parking lot on her way to class. If the lot is 200 ft long and 150 ft wide, how far does Clare travel when she rides across the lot diagonally?

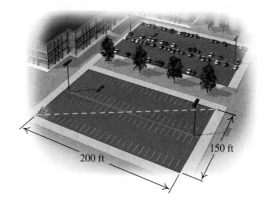

150 ft
200 ft

20. *Guy Wire.* How long is a guy wire if it reaches from the top of a 15-ft pole to a point on the ground 10 ft from the pole?

21. *Softball.* A slow-pitch softball diamond is actually a square 65 ft on a side. How far is it from home plate to second base?

22. *Baseball.* Suppose the catcher in Example 1 makes a throw to second base from the same location. How far is that throw?

23. *Television Sets.* What does it mean to refer to a 20-in. TV set or a 25-in. TV set? Such units refer to

the diagonal of the screen. A 20-in. TV set has a width of 16 in. What is its height?

20 in.

h

w

24. *Television Sets.*   A 25-in. TV set has a screen with a height of 15 in. What is its width? (See Exercise 23.)

25. *Speaker Placement.*   A stereo receiver is in a corner of a 12-ft by 14-ft room. Speaker wire will run under a rug, diagonally, to a speaker in the far corner. If 4 ft of slack is required on each end, how long a piece of wire should be purchased?

26. *Distance over Water.*   To determine the width of a pond, a surveyor locates two stakes at either end of the pond and uses instrumentation to place a third stake so that the distance across the pond is the length of a hypotenuse. If the third stake is 90 m from one stake and 70 m from the other, what is the distance across the pond?

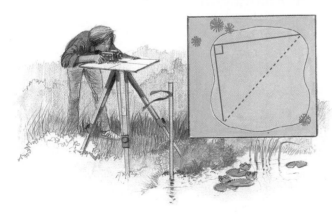

27. *Walking.*   Students at Compkin Community College have worn a path that cuts diagonally across the campus "quad." If the quad is actually a rectangle that Marissa measured to be 70 paces long and 40 paces wide, how many paces will Marissa save by using the diagonal path?

28. *Crosswalks.*   The diagonal crosswalk at the intersection of State St. and Main St. is the hypotenuse of a triangle in which the crosswalks across State St. and Main St. are the legs. If State St. is 28 ft wide and Main St. is 40 ft wide, how much shorter is the distance traveled by pedestrians using the diagonal crosswalk?

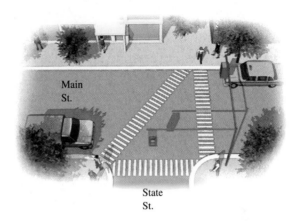

Main St.

State St.

For each triangle, find the missing length(s). Give an exact answer and, where appropriate, an approximation to three decimal places.

29.

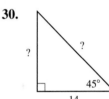

45°

?

5

?

30.

?

?

45°

14

31.

14

30°

?

?

32.

60°

18

?

?

33.

?

60°

?

15

34.

?

?

45°

8

35.

?

?

45°

13

36.

7

?

30°

?

37.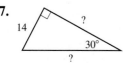

38.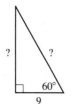

39.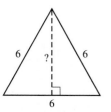

40.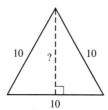

41.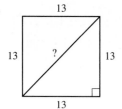

42.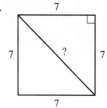

43.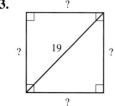

44.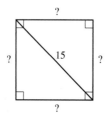

In Exercises 45–50, give an exact answer and, where appropriate, an approximation to three decimal places.

45. *Bridge Expansion.* During the summer heat, a 2-mi bridge expands 2 ft in length. If we assume that the bulge occurs straight up the middle, how high is the bulge? (The answer may surprise you. Most bridges have expansion spaces to avoid such buckling.)

46. Triangle *ABC* has sides of lengths 25 ft, 25 ft, and 30 ft. Triangle *PQR* has sides of lengths 25 ft, 25 ft, and 40 ft. Which triangle, if either, has the greater area and by how much?

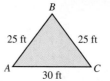 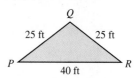

47. *Camping Tent.* The entrance to a pup tent is the shape of an equilateral triangle. If the base of the tent is 4 ft wide, how tall is the tent?

48. The length and the width of a rectangle are given by consecutive integers. The area of the rectangle is 90 cm². Find the length of a diagonal of the rectangle.

49. Find all points on the *y*-axis of a Cartesian coordinate system that are 5 units from the point (3, 0).

50. Find all points on the *x*-axis of a Cartesian coordinate system that are 5 units from the point (0, 4).

TW 51. Write a problem for a classmate to solve in which the solution is: "The height of the tepee is $5\sqrt{3}$ yd."

TW 52. Write a problem for a classmate to solve in which the solution is: "The height of the window is $15\sqrt{3}$ ft."

Skill Maintenance

Graph.

53. $f(x) = \dfrac{2}{3}x - 5$ [3.8]

54. $g(x) = 3x + 4$ [3.8]

55. $F(x) < -2x + 4$ [8.3]

56. $G(x) > -3x + 2$ [8.3]

57. $f(x) = -2$ [3.8]

58. $g(x) = |x| + 1$ [3.8]

Synthesis

TW 59. Are there any right triangles, other than those with sides measuring 3, 4, and 5, that have consecutive numbers for the lengths of the sides? Why or why not?

ᵀᵂ **60.** If a 30°–60°–90° triangle and an isosceles right triangle have the same perimeter, which will have the greater area? Why?

61. The perimeter of a regular hexagon is 72 cm. Determine the area of the shaded region shown.

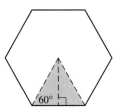

62. If the perimeter of a regular hexagon is 120 ft, what is its area? (*Hint*: See Exercise 61.)

63. Each side of a regular octagon has length s. Find a formula for the distance d between the parallel sides of the octagon.

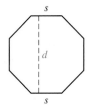

64. *Roofing.* Kit's home, which is 24 ft wide and 32 ft long, needs a new roof. By counting clapboards that are 4 in. apart, Kit determines that the peak of the roof is 6 ft higher than the sides. If one packet of shingles covers 100 ft², how many packets will the job require?

65. *Painting.* (Refer to Exercise 64.) A gallon of Benjamin Moore® exterior acrylic paint covers 450–500 ft². If Kit's house has dimensions as shown above, how many gallons of paint should be bought to paint the house? What assumption(s) is made in your answer?

66. *Contracting.* Oxford Builders has an extension cord on their generator that permits them to work, with electricity, anywhere in a circular area of 3850 ft². Find the dimensions of the largest square room they could work on without having to relocate the generator to reach each corner of the floor plan.

67. *Contracting.* Cleary Construction has a hose attached to their insulation blower that permits them to work, with electricity, anywhere in a circular area of 6160 ft². Find the dimensions of the largest square room with 12-ft ceilings in which they could reach all corners with the hose while leaving the blower centrally located. Assume that the blower sits on the floor.

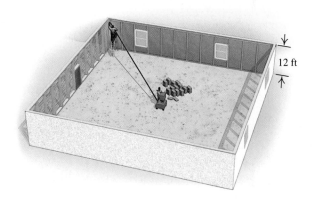

68. A cube measures 5 cm on each side. How long is the diagonal that connects two opposite corners of the cube? Give an exact answer.

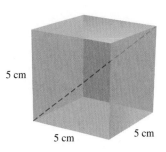

10.8   The Complex Numbers

Imaginary and Complex Numbers   ■   Addition and Subtraction   ■
Multiplication   ■   Conjugates and Division   ■   Powers of i

Imaginary and Complex Numbers

Negative numbers do not have square roots in the real-number system. However, a larger number system that contains the real-number system is designed so that negative numbers *do* have square roots. That system is called the **complex-number system,** and it will allow us to solve equations like $x^2 + 1 = 0$. The complex-number system makes use of i, a number that is, by definition, a square root of -1.

The Number i i is the unique number for which $i = \sqrt{-1}$ and $i^2 = -1$.

We can now define the square root of a negative number as follows:

$$\sqrt{-p} = \sqrt{-1}\,\sqrt{p} = i\sqrt{p} \text{ or } \sqrt{p}i, \quad \text{for any positive number } p.$$

EXAMPLE 1   Express in terms of i: **(a)** $\sqrt{-7}$; **(b)** $\sqrt{-16}$; **(c)** $-\sqrt{-13}$; **(d)** $-\sqrt{-50}$.

SOLUTION

a) $\sqrt{-7} = \sqrt{-1 \cdot 7} = \sqrt{-1} \cdot \sqrt{7} = i\sqrt{7}$, or $\sqrt{7}i$ *i is **not** under the radical.*

b) $\sqrt{-16} = \sqrt{-1 \cdot 16} = \sqrt{-1} \cdot \sqrt{16} = i \cdot 4 = 4i$

c) $-\sqrt{-13} = -\sqrt{-1 \cdot 13} = -\sqrt{-1} \cdot \sqrt{13} = -i\sqrt{13}$, or $-\sqrt{13}i$

d) $-\sqrt{-50} = -\sqrt{-1} \cdot \sqrt{25} \cdot \sqrt{2} = -i \cdot 5 \cdot \sqrt{2}$
$\phantom{-\sqrt{-50}} = -5i\sqrt{2}$, or $-5\sqrt{2}i$

Imaginary Numbers   An *imaginary number* is a number that can be written in the form $a + bi$, where a and b are real numbers and $b \neq 0$.

Don't let the name "imaginary" fool you. Imaginary numbers appear in fields such as engineering and the physical sciences. The following are examples of imaginary numbers:

$5 + 4i$,   Here $a = 5, b = 4$.

$\sqrt{3} - \pi i$,   Here $a = \sqrt{3}, b = -\pi$.

$\sqrt{7}i$ Here $a = 0, b = \sqrt{7}$.

The union of the set of all imaginary numbers and the set of all real numbers is the set of all **complex numbers.**

> **Complex Numbers**   A *complex number* is any number that can be written in the form $a + bi$, where a and b are real numbers. (Note that a and b both can be 0.)

The following are examples of complex numbers:

$7 + 3i$ (here $a \neq 0, b \neq 0$); $4i$ (here $a = 0, b \neq 0$);

8 (here $a \neq 0, b = 0$); 0 (here $a = 0, b = 0$).

Complex numbers like $17i$ or $4i$, in which $a = 0$ and $b \neq 0$, are imaginary numbers with no real part. Such numbers are called *pure imaginary numbers.*

Note that when $b = 0$, we have $a + 0i = a$, so every real number is a complex number. The relationships among various real and complex numbers are shown below.

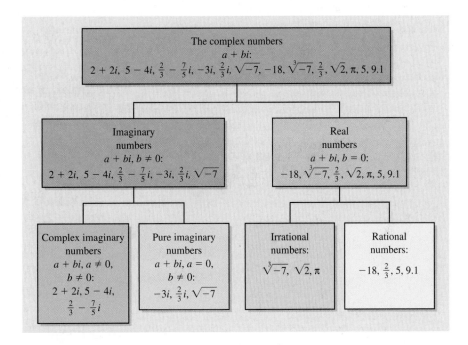

Note that although $\sqrt{-7}$ and $\sqrt[3]{-7}$ are both complex numbers, $\sqrt{-7}$ is imaginary whereas $\sqrt[3]{-7}$ is real.

Addition and Subtraction

The commutative, associative, and distributive laws hold for complex numbers. Thus we can add and subtract them as we do binomials.

EXAMPLE 2 Add or subtract and simplify.

a) $(8 + 6i) + (3 + 2i)$ **b)** $(4 + 5i) - (6 - 3i)$

SOLUTION

a) $(8 + 6i) + (3 + 2i) = (8 + 3) + (6i + 2i)$ Combining the real parts and the imaginary parts

$= 11 + (6 + 2)i = 11 + 8i$

b) $(4 + 5i) - (6 - 3i) = (4 - 6) + [5i - (-3i)]$ Note that the 6 and the $-3i$ are *both* being subtracted.

$= -2 + 8i$

Student Notes

The rule developed in Section 10.3, $\sqrt[n]{a} \cdot \sqrt[n]{b} = \sqrt[n]{a \cdot b}$, does *not* apply when n is 2 and either a or b is negative. Indeed this condition is stated on p. 748 when it is specified that $\sqrt[n]{a}$ and $\sqrt[n]{b}$ are both *real* numbers.

Multiplication

To multiply square roots of negative real numbers, we first express them in terms of i. For example,

$$\sqrt{-2} \cdot \sqrt{-5} = \sqrt{-1} \cdot \sqrt{2} \cdot \sqrt{-1} \cdot \sqrt{5}$$
$$= i \cdot \sqrt{2} \cdot i \cdot \sqrt{5}$$
$$= i^2 \cdot \sqrt{10}$$
$$= -1\sqrt{10} = -\sqrt{10} \text{ is correct!}$$

CAUTION! With complex numbers, simply multiplying radicands is *incorrect* when both radicands are negative: $\sqrt{-2} \cdot \sqrt{-5} \neq \sqrt{10}$.

With this in mind, we can now multiply complex numbers.

EXAMPLE 3 Multiply and simplify. When possible, write answers in the form $a + bi$.

a) $\sqrt{-16} \cdot \sqrt{-25}$ **b)** $\sqrt{-5} \cdot \sqrt{-7}$ **c)** $-3i \cdot 8i$
d) $-4i(3 - 5i)$ **e)** $(1 + 2i)(1 + 3i)$

SOLUTION

a) $\sqrt{-16} \cdot \sqrt{-25} = \sqrt{-1} \cdot \sqrt{16} \cdot \sqrt{-1} \cdot \sqrt{25}$
$= i \cdot 4 \cdot i \cdot 5$
$= i^2 \cdot 20$
$= -1 \cdot 20 \quad i^2 = -1$
$= -20$

b) $\sqrt{-5} \cdot \sqrt{-7} = \sqrt{-1} \cdot \sqrt{5} \cdot \sqrt{-1} \cdot \sqrt{7}$ Try to do this step mentally.
$= i \cdot \sqrt{5} \cdot i \cdot \sqrt{7}$
$= i^2 \cdot \sqrt{35}$
$= -1 \cdot \sqrt{35} \quad i^2 = -1$
$= -\sqrt{35}$

c) $-3i \cdot 8i = -24 \cdot i^2$

$\qquad\qquad = -24 \cdot (-1) \quad i^2 = -1$

$\qquad\qquad = 24$

d) $-4i(3 - 5i) = -4i \cdot 3 + (-4i)(-5i)$ Using the distributive law

$\qquad\qquad = -12i + 20i^2$

$\qquad\qquad = -12i - 20$ $\qquad\qquad i^2 = -1$

$\qquad\qquad = -20 - 12i$ $\qquad\qquad$ Writing in the form $a + bi$

e) $(1 + 2i)(1 + 3i) = 1 + 3i + 2i + 6i^2$ Multiplying each term of $1 + 3i$ by every term of $1 + 2i$ (FOIL)

$\qquad\qquad\qquad\quad = 1 + 3i + 2i - 6$ $i^2 = -1$

$\qquad\qquad\qquad\quad = -5 + 5i$ Combining like terms

Conjugates and Division

Recall that the conjugate of $4 + \sqrt{2}$ is $4 - \sqrt{2}$.

Conjugates of complex numbers are defined in a similar manner.

> **Conjugate of a Complex Number**   The *conjugate* of a complex number $a + bi$ is $a - bi$, and the *conjugate* of $a - bi$ is $a + bi$.

EXAMPLE 4   Find the conjugate of each number.

a) $-3 + 7i$ **b)** $14 - 5i$ **c)** $4i$

SOLUTION

a) $-3 + 7i$ The conjugate is $-3 - 7i$.

b) $14 - 5i$ The conjugate is $14 + 5i$.

c) $4i$ The conjugate is $-4i$. Note that $4i = 0 + 4i$.

The product of a complex number and its conjugate is a real number.

EXAMPLE 5   Multiply: $(5 + 7i)(5 - 7i)$.

SOLUTION

$$(5 + 7i)(5 - 7i) = 5^2 - (7i)^2 \quad \text{Using } (A + B)(A - B) = A^2 - B^2$$

$$= 25 - 49i^2$$

$$= 25 - 49(-1) \quad i^2 = -1$$

$$= 25 + 49 = 74$$

Conjugates are used when dividing complex numbers. The procedure is much like that used to rationalize denominators in Section 10.5.

EXAMPLE 6 Divide and simplify to the form $a + bi$.

a) $\dfrac{-5 + 9i}{1 - 2i}$ **b)** $\dfrac{7 + 3i}{5i}$

SOLUTION

a) To divide and simplify $(-5 + 9i)/(1 - 2i)$, we multiply by 1, using the conjugate of the denominator to form 1:

$$\frac{-5 + 9i}{1 - 2i} = \frac{-5 + 9i}{1 - 2i} \cdot \frac{1 + 2i}{1 + 2i}$$ Multiplying by 1 using the conjugate of the denominator in the symbol for 1

$$= \frac{-5 - 10i + 9i + 18i^2}{1^2 - 4i^2}$$ Multiplying using FOIL

$$= \frac{-5 - i - 18}{1 - 4(-1)}$$ $i^2 = -1$

$$= \frac{-23 - i}{5}$$

$$= -\frac{23}{5} - \frac{1}{5}i$$

Writing in the form $a + bi$; note that $\dfrac{X + Y}{Z} = \dfrac{X}{Z} + \dfrac{Y}{Z}$

b) The conjugate of $5i$ is $-5i$, so we *could* multiply by $-5i/(-5i)$. However, when the denominator is a pure imaginary number, it is easiest if we multiply by i/i:

$$\frac{7 + 3i}{5i} = \frac{7 + 3i}{5i} \cdot \frac{i}{i}$$ Multiplying by 1 using i/i. We can also use the conjugate of $5i$ to write $-5i/(-5i)$.

$$= \frac{7i + 3i^2}{5i^2}$$ Multiplying

$$= \frac{7i + 3(-1)}{5(-1)}$$ $i^2 = -1$

$$= \frac{7i - 3}{-5} = \frac{-3}{-5} + \frac{7}{-5}i, \text{ or } \frac{3}{5} - \frac{7}{5}i.$$ Writing in the form $a + bi$

Powers of i

Answers to problems involving complex numbers are generally written in the form $a + bi$. In the following discussion, we show why there is no need to use powers of i (other than 1) when writing answers.

Recall that -1 raised to an *even* power is 1, and -1 raised to an *odd* power is -1. Simplifying powers of i can then be done by using the fact that $i^2 = -1$ and expressing the given power of i in terms of i^2. Consider the following:

$$i^2 = -1,$$

$$i^3 = i^2 \cdot i = (-1)i = -i,$$

$$i^4 = (i^2)^2 = (-1)^2 = 1,$$

$$i^5 = i^4 \cdot i = (i^2)^2 \cdot i = (-1)^2 \cdot i = i,$$

$$i^6 = (i^2)^3 = (-1)^3 = -1. \leftarrow \text{The pattern is now repeating.}$$

The powers of i cycle themselves through the values i, -1, $-i$, and 1. Even powers of i are -1 or 1 whereas odd powers of i are i or $-i$.

EXAMPLE 7 Simplify: **(a)** i^{18}; **(b)** i^{24}.

SOLUTION

a) $i^{18} = (i^2)^9$ Using the power rule

$\qquad = (-1)^9 = -1$ Raising -1 to a power

b) $i^{24} = (i^2)^{12} = (-1)^{12} = 1$

To simplify i^n when n is odd, we rewrite i^n as $i^{n-1} \cdot i$.

EXAMPLE 8 Simplify: **(a)** i^{29}; **(b)** i^{75}.

SOLUTION

a) $i^{29} = i^{28}i^1$ Using the product rule. This is a key step when i is raised to an odd power.

$\qquad = (i^2)^{14}i$ Using the power rule

$\qquad = (-1)^{14}i = 1 \cdot i = i$

b) $i^{75} = i^{74}i^1$ Using the product rule

$\qquad = (i^2)^{37}i$ Using the power rule

$\qquad = (-1)^{37}i = -1 \cdot i = -i$

Complex Numbers

Many graphing calculators can perform operations with complex numbers. To do so, first choose the a + bi setting in the MODE screen.

```
Normal  Sci  Eng
Float   0123456789
Radian  Degree
Func    Par  Pol  Seq
Connected  Dot
Sequential  Simul
Real  a+bi  re^θi
Full  Horiz  G-T
```

```
(-1+i)(2-i)
                    -1+3i
(2-i)/(3+i)▶Frac
                    1/2-1/2i
```

Now complex numbers can be entered as they are written. Press ⌐i⌐ for the complex number i. (⌐i⌐ is often the 2nd option associated with the ⌐·⌐ key.) For example, to calculate $(-1 + i)(2 - i)$, press ⌐(⌐ ⌐(-)⌐ ⌐1⌐ ⌐+⌐ ⌐i⌐ ⌐)⌐ ⌐(⌐ ⌐2⌐ ⌐-⌐ ⌐i⌐ ⌐)⌐ ⌐ENTER⌐. As seen in the screen on the right above, the result is $-1 + 3i$. To calculate $\dfrac{2 - i}{3 + i}$ and write the result using fraction notation, press ⌐(⌐ ⌐2⌐ ⌐-⌐ ⌐i⌐ ⌐)⌐ ⌐÷⌐ ⌐(⌐ ⌐3⌐ ⌐+⌐ ⌐i⌐ ⌐)⌐ ⌐MATH⌐ ⌐1⌐ ⌐ENTER⌐. Note the need for parentheses around the numerator and the denominator of the expression. The result appears in the form $a + bi$; i is not in the denominator of the fraction. The result is $\frac{1}{2} - \frac{1}{2}i$.

10.8 EXERCISE SET

↪ *Concept Reinforcement*   *Classify each statement as either true or false.*

1. Imaginary numbers are so named because they have no real-world applications.

2. Every real number is imaginary, but not every imaginary number is real.

3. Every imaginary number is a complex number, but not every complex number is imaginary.

4. Every real number is a complex number, but not every complex number is real.

5. Addition and subtraction of complex numbers have much in common with addition and subtraction of polynomials.

6. The product of a complex number and its conjugate is always a real number.

7. The square of a complex number is always a real number.

8. The quotient of two complex numbers is always a complex number.

Express in terms of i.

9. $\sqrt{-36}$

10. $\sqrt{-25}$

11. $\sqrt{-13}$

12. $\sqrt{-19}$

13. $\sqrt{-18}$

14. $\sqrt{-98}$

15. $\sqrt{-3}$

16. $\sqrt{-4}$

17. $\sqrt{-81}$

18. $\sqrt{-27}$

19. $-\sqrt{-300}$

20. $-\sqrt{-75}$

21. $6 - \sqrt{-84}$

22. $4 - \sqrt{-60}$

23. $-\sqrt{-76} + \sqrt{-125}$

24. $\sqrt{-4} + \sqrt{-12}$

25. $\sqrt{-18} - \sqrt{-100}$

26. $\sqrt{-72} - \sqrt{-25}$

Perform the indicated operation and simplify. Write each answer in the form a + bi.

27. $(6 + 7i) + (5 + 3i)$

28. $(4 - 5i) + (3 + 9i)$

29. $(9 + 8i) - (5 + 3i)$

30. $(9 + 7i) - (2 + 4i)$

31. $(7 - 4i) - (5 - 3i)$

32. $(5 - 3i) - (9 + 2i)$

33. $(-5 - i) - (7 + 4i)$

34. $(-2 + 6i) - (-7 + i)$

35. $7i \cdot 6i$

36. $6i \cdot 9i$

37. $(-4i)(-6i)$

38. $7i \cdot (-8i)$

39. $\sqrt{-36}\sqrt{-9}$

40. $\sqrt{-49}\sqrt{-25}$

41. $\sqrt{-5}\sqrt{-2}$

42. $\sqrt{-6}\sqrt{-7}$

43. $\sqrt{-6}\sqrt{-21}$

44. $\sqrt{-15}\sqrt{-10}$

45. $5i(2 + 6i)$

46. $2i(7 + 3i)$

47. $-7i(3 - 4i)$

48. $-4i(6 - 5i)$

49. $(1 + i)(3 + 2i)$

50. $(1 + 5i)(4 + 3i)$

51. $(6 - 5i)(3 + 4i)$

52. $(5 - 6i)(2 + 5i)$

53. $(7 - 2i)(2 - 6i)$

54. $(-4 + 5i)(3 - 4i)$

55. $(-2 + 3i)(-2 + 5i)$

56. $(-3 + 6i)(-3 + 4i)$

57. $(-5 - 4i)(3 + 7i)$

58. $(2 + 9i)(-3 - 5i)$

59. $(4 - 2i)^2$

60. $(1 - 2i)^2$

61. $(2 + 3i)^2$

62. $(3 + 2i)^2$

63. $(-2 + 3i)^2$

64. $(-5 - 2i)^2$

65. $\dfrac{7}{4 + i}$

66. $\dfrac{3}{8 + i}$

67. $\dfrac{2}{3 - 2i}$

68. $\dfrac{4}{2 - 3i}$

69. $\dfrac{2i}{5 + 3i}$

70. $\dfrac{3i}{4 + 2i}$

71. $\dfrac{5}{6i}$

72. $\dfrac{4}{7i}$

73. $\dfrac{5 - 3i}{4i}$

74. $\dfrac{2 + 7i}{5i}$

Aha! **75.** $\dfrac{7i + 14}{7i}$

76. $\dfrac{6i + 3}{3i}$

77. $\dfrac{4 + 5i}{3 - 7i}$

78. $\dfrac{5 + 3i}{7 - 4i}$

79. $\dfrac{2 + 3i}{2 + 5i}$

80. $\dfrac{3 + 2i}{4 + 3i}$

81. $\dfrac{3 - 2i}{4 + 3i}$

82. $\dfrac{5 - 2i}{3 + 6i}$

Simplify.

83. i^7

84. i^{11}

85. i^{24}

86. i^{35}

87. i^{42}

88. i^{64}

89. i^9

90. $(-i)^{71}$

91. $(-i)^6$

92. $(-i)^4$

93. $(5i)^3$

94. $(-3i)^5$

95. $i^2 + i^4$

96. $5i^5 + 4i^3$

TW 97. Is the product of two imaginary numbers always an imaginary number? Why or why not?

TW 98. In what way(s) are conjugates of complex numbers similar to the conjugates used in Section 10.5?

Skill Maintenance

Factor.

99. $12x^2 - 23x + 10$ [6.3]

100. $x^3 - 3x^2 - 10x$ [6.2]

101. $20x^2y^3 + 15x^3y^2 - 35x^4y^3$ [6.1]

102. $x^2y - 3y - x^3 + 3x$ [6.1]

103. $14x^4y - 14y^5$ [6.4]

104. $27x^3 - 64y^3$ [6.5]

Synthesis

TW 105. Is the set of real numbers a subset of the set of complex numbers? Why or why not?

TW 106. Is the union of the set of imaginary numbers and the set of real numbers the set of complex numbers? Why or why not?

A function g is given by

$$g(z) = \dfrac{z^4 - z^2}{z - 1}.$$

107. Find $g(3i)$.

108. Find $g(1 + i)$.

109. Find $g(5i - 1)$.

110. Find $g(2 - 3i)$.

111. Evaluate

$$\dfrac{1}{w - w^2} \quad \text{for} \quad w = \dfrac{1 - i}{10}.$$

Simplify.

112. $\dfrac{i^5 + i^6 + i^7 + i^8}{(1 - i)^4}$

113. $(1 - i)^3(1 + i)^3$

114. $\dfrac{5 - \sqrt{5}i}{\sqrt{5}i}$

115. $\dfrac{6}{1 + \dfrac{3}{i}}$

116. $\left(\dfrac{1}{2} - \dfrac{1}{3}i\right)^2 - \left(\dfrac{1}{2} + \dfrac{1}{3}i\right)^2$

117. $\dfrac{i - i^{38}}{1 + i}$

10 Chapter Summary and Review

KEY TERMS AND DEFINITIONS

RADICALS

Square root, p. 724 c is a square root of a if $c^2 = a$. $\sqrt{a}$ means the **principal,** or nonnegative, square root of a.

Cube root, p. 728 c is a cube root of a if $c^3 = a$. $\sqrt[3]{a}$ means the cube root of a.

***n*th root,** p. 729 c is an nth root of a if $c^n = a$. The nth root of a is written

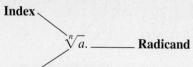

$$\sqrt[n]{a^n} = |a| \text{ when } n \text{ is even.}$$

$$\sqrt[n]{a^n} = a \text{ when } n \text{ is odd.}$$

COMPLEX NUMBERS

Complex number, p. 789   Any number that can be written in the form $a + bi$, where a and b are real numbers and $i = \sqrt{-1}$. A complex number is real when $b = 0$, is **imaginary** when $b \neq 0$, and is **pure imaginary** when $a = 0$.

APPLICATIONS

Pythagorean Theorem, p. 780   For any right triangle with legs of lengths a and b and hypotenuse of length c,

$$a^2 + b^2 = c^2.$$

Special triangles, p. 783   The length of the hypotenuse in an isosceles right triangle is the length of a leg times $\sqrt{2}$.

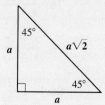

The length of the longer leg in a $30°$–$60°$–$90°$ right triangle is the length of the shorter leg times $\sqrt{3}$. The hypotenuse is twice as long as the shorter leg.

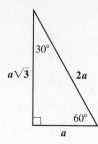

IMPORTANT CONCEPTS

[Section references appear in brackets.]

Concept	Example
The domain of a **radical function** with an even index is the set of all inputs for which the radicand is nonnegative.	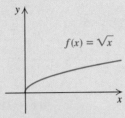 Domain of f: $[0, \infty)$ Range of f: $[0, \infty)$ **[10.1]**
Radical notation can be written using **rational exponents**. $a^{1/n}$ means $\sqrt[n]{a}$. $a^{m/n}$ means $\left(\sqrt[n]{a}\right)^m$ or $\sqrt[n]{a^m}$. When a is negative, n must be odd.	$7^{1/10} = \sqrt[10]{7}$ $x^{1/2} = \sqrt{x}$ $3^{4/5} = \sqrt[5]{3^4} = \left(\sqrt[5]{3}\right)^4$ **[10.2]**

Radical expressions can often be simplified.
 	
1. Simplify by factoring.
 	
2. Use rational exponents to simplify.
 	
3. Combine **like radical terms.**
	$\sqrt[3]{a^6 b} = \sqrt[3]{a^6}\,\sqrt[3]{b} = a^2\sqrt[3]{b}$ **[10.3]**
	
	$\sqrt[3]{p} \cdot \sqrt[4]{q^3} = p^{1/3} \cdot q^{3/4}$
	    $= p^{4/12} \cdot q^{9/12} = \sqrt[12]{p^4 q^9}$ **[10.5]**
	
	$\sqrt{8} + 3\sqrt{2} = \sqrt{4} \cdot \sqrt{2} + 3\sqrt{2}$
	    $= 2\sqrt{2} + 3\sqrt{2} = 5\sqrt{2}$ **[10.5]**

Radicands of radical expressions can be multiplied and divided if the indices are the same.
 	
The Product Rule for Radicals
For any real numbers $\sqrt[n]{a}$ and $\sqrt[n]{b}$,
    $\sqrt[n]{a}\,\sqrt[n]{b} = \sqrt[n]{a \cdot b}.$
 	
The Quotient Rule for Radicals
For any real numbers $\sqrt[n]{a}$ and $\sqrt[n]{b}$, $b \neq 0$,
    $\sqrt[n]{\dfrac{a}{b}} = \dfrac{\sqrt[n]{a}}{\sqrt[n]{b}}.$
	
	$\sqrt{15}\,\sqrt{35} = \sqrt{15 \cdot 35}$
	    $= \sqrt{3 \cdot 5 \cdot 5 \cdot 7} = 5\sqrt{21}$ **[10.3]**
	
	$\sqrt[3]{\dfrac{8x^3}{27}} = \dfrac{\sqrt[3]{8x^3}}{\sqrt[3]{27}} = \dfrac{2x}{3};$
	
	$\dfrac{\sqrt{20}}{\sqrt{2}} = \sqrt{\dfrac{20}{2}} = \sqrt{10}$ **[10.4]**

(continued)

A denominator is **rationalized** when it does not contain any irrational numbers. A denominator with two terms is rationalized using the **conjugate** of the denominator.

$$\frac{1}{\sqrt{2}} = \frac{1}{\sqrt{2}} \cdot \frac{\sqrt{2}}{\sqrt{2}} = \frac{\sqrt{2}}{2} \qquad [10.4]$$

$$\frac{3}{2 + \sqrt{5}} = \frac{3}{2 + \sqrt{5}} \cdot \frac{2 - \sqrt{5}}{2 - \sqrt{5}}$$

$$= \frac{6 - 3\sqrt{5}}{4 - 5} = \frac{6 - 3\sqrt{5}}{-1} = -6 + 3\sqrt{5} \qquad [10.5]$$

To solve a radical equation, use the principle of powers and the steps on pp. 772 and 774. Solutions must be checked in the original equation.

The Principle of Powers

If $a = b$, then $a^n = b^n$ for any exponent n.

$$x - 7 = \sqrt{x - 5} \qquad\qquad 2 + \sqrt{x} = \sqrt{x + 8}$$
$$(x - 7)^2 = (\sqrt{x - 5})^2 \qquad (2 + \sqrt{x})^2 = (\sqrt{x + 8})^2$$
$$x^2 - 14x + 49 = x - 5 \qquad 4 + 4\sqrt{x} + x = x + 8$$
$$x^2 - 15x + 54 = 0 \qquad\qquad 4\sqrt{x} = 4$$
$$(x - 6)(x - 9) = 0 \qquad\qquad \sqrt{x} = 1$$
$$x = 6 \quad or \quad x = 9 \qquad\qquad (\sqrt{x})^2 = (1)^2$$

Only 9 checks and is the solution.

$$x = 1$$

1 checks and is the solution.   [10.6]

Complex numbers can be added, subtracted, multiplied, and divided.

$$(3 + 2i) + (4 - 7i) = 7 - 5i;$$

$$(8 + 6i) - (5 + 2i) = 3 + 4i;$$

$$(2 + 3i)(4 - i) = 8 - 2i + 12i - 3i^2$$
$$= 8 + 10i - 3(-1) = 11 + 10i;$$

$$\frac{1 - 4i}{3 - 2i} = \frac{1 - 4i}{3 - 2i} \cdot \frac{3 + 2i}{3 + 2i}$$

$$= \frac{3 + 2i - 12i - 8i^2}{9 + 6i - 6i - 4i^2}$$

$$= \frac{3 - 10i - 8(-1)}{9 - 4(-1)}$$

$$= \frac{11 - 10i}{13} = \frac{11}{13} - \frac{10}{13}i. \qquad [10.8]$$

Review Exercises

👈 *Concept Reinforcement*   *Classify each of the following as either true or false.*

1. $\sqrt{ab} = \sqrt{a} \cdot \sqrt{b}$ for any real numbers $\sqrt{a}$ and $\sqrt{b}$. [10.3]

2. $\sqrt{a^2} = a$, for any real number a. [10.1]

3. $\sqrt[3]{a^3} = a$, for any real number a. [10.1]

4. $x^{2/5}$ means $\sqrt[5]{x^2}$ and $\left(\sqrt[5]{x}\right)^2$. [10.2]

5. A hypotenuse is never shorter than either leg. [10.7]

6. $i^{13} = (i^2)^6 i = i$. [10.8]

7. Some radical equations have no solution. [10.6]

8. If $f(x) = \sqrt{x - 5}$, then the domain of f is the set of all nonnegative real numbers. [10.1]

Simplify. [10.1]

9. $\sqrt{\dfrac{49}{9}}$

10. $-\sqrt{0.25}$

Let $f(x) = \sqrt{2x - 7}$. Find the following. [10.1]

11. $f(16)$

12. The domain of f

13. *Dairy Farming.*   As a calf grows, it needs more milk for nourishment. The number of pounds of milk, M, required by a calf weighing x pounds can be estimated using the formula
$$M = -5 + \sqrt{6.7x - 444}.$$
(*Source*: www.ext.vt.edu) Estimate the number of pounds of milk required by a calf of the given weight: **(a)** 300 lb; **(b)** 100 lb; **(c)** 200 lb; **(d)** 400 lb. [10.1]

Simplify. Assume that each variable can represent any real number.

14. $\sqrt{25t^2}$ [10.1]

15. $\sqrt{(c + 8)^2}$ [10.1]

16. $\sqrt{x^2 - 6x + 9}$ [10.1]

17. $\sqrt{4x^2 + 4x + 1}$ [10.1]

18. $\sqrt[5]{-32}$ [10.1]

19. $\sqrt[3]{-\dfrac{64x^6}{27}}$ [10.4]

20. $\sqrt[4]{x^{12}y^8}$ [10.3]

21. $\sqrt[6]{64x^{12}}$ [10.3]

22. Write an equivalent expression using exponential notation: $\left(\sqrt[3]{5ab}\right)^4$. [10.2]

23. Write an equivalent expression using radical notation: $(16a^6)^{3/4}$. [10.2]

Use rational exponents to simplify. Assume x, $y \geq 0$. [10.2]

24. $\sqrt{x^6 y^{10}}$

25. $\left(\sqrt[6]{x^2 y}\right)^2$

Simplify. Do not use negative exponents in the answers. [10.2]

26. $(x^{-2/3})^{3/5}$

27. $\dfrac{7^{-1/3}}{7^{-1/2}}$

28. If $f(x) = \sqrt{25(x - 6)^2}$, find a simplified form for $f(x)$. [10.3]

Perform the indicated operation and, if possible, simplify. Write all answers using radical notation.

29. $\sqrt{2x}\,\sqrt{3y}$ [10.3]

30. $\sqrt[3]{a^5 b}\,\sqrt[3]{27b}$ [10.3]

31. $\sqrt[3]{-24x^{10}y^8}\,\sqrt[3]{18x^7y^4}$ [10.3]

32. $\dfrac{\sqrt[3]{60xy^3}}{\sqrt[3]{10x}}$ [10.4]

33. $\dfrac{\sqrt{75x}}{2\sqrt{3}}$ [10.4]

34. $\sqrt[4]{\dfrac{48a^{11}}{c^8}}$ [10.4]

35. $5\sqrt[3]{x} + 2\sqrt[3]{x}$ [10.5]

36. $2\sqrt{75} - 9\sqrt{3}$ [10.5]

37. $\sqrt[3]{8x^4} + \sqrt[3]{xy^6}$ [10.5]

38. $\sqrt{50} + 2\sqrt{18} + \sqrt{32}$ [10.5]

39. $\left(\sqrt{3} - 3\sqrt{8}\right)\left(\sqrt{5} + 2\sqrt{8}\right)$ [10.5]

40. $\sqrt[4]{x}\,\sqrt{x}$ [10.5]

41. $\dfrac{\sqrt[3]{x^2}}{\sqrt[4]{x}}$ [10.5]

42. If $f(x) = x^2$, find $f(a - \sqrt{2})$. [10.5]

43. Rationalize the denominator:

$$\frac{4\sqrt{5}}{\sqrt{2} + \sqrt{3}}.\ [10.5]$$

44. Rationalize the numerator of the expression in Exercise 43. [10.5]

Solve. [10.6]

45. $\sqrt{y + 6} - 2 = 3$

46. $(x + 1)^{1/3} = -5$

47. $1 + \sqrt{x} = \sqrt{3x - 3}$

48. If $f(x) = \sqrt[4]{x + 2}$, find a such that $f(a) = 2$. [10.6]

Solve. Give an exact answer and, where appropriate, an approximation to three decimal places. [10.7]

49. The diagonal of a square has length 10 cm. Find the length of a side of the square.

50. A skate-park jump has a ramp that is 6 ft long and is 2 ft high. How long is its base?

6 ft   2 ft
?

51. Find the missing lengths. Give exact answers and, where appropriate, an approximation to three decimal places.

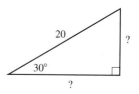

20   ?
30°
?

52. Express in terms of i and simplify: $-\sqrt{-8}$. [10.8]

53. Add: $(-4 + 3i) + (2 - 12i)$. [10.8]

54. Subtract: $(9 - 7i) - (3 - 8i)$. [10.8]

Simplify. [10.8]

55. $(2 + 5i)(2 - 5i)$ **56.** i^{18}

57. Simplify: $(6 - 3i)(2 - i)$. [10.8]

58. Divide and simplify to the form $a + bi$:

$$\frac{7 - 2i}{3 + 4i}.\ [10.8]$$

Synthesis

TW 59. What makes some complex numbers real and others imaginary? [10.8]

TW 60. Explain why $\sqrt[n]{x^n} = |x|$ when n is even, but $\sqrt[n]{x^n} = x$ when n is odd. [10.1]

61. Solve:

$$\sqrt{11x + \sqrt{6 + x}} = 6.\ [10.6]$$

62. Simplify:

$$\frac{2}{1 - 3i} - \frac{3}{4 + 2i}.\ [10.8]$$

63. Write a quotient of two imaginary numbers that is a real number (answers may vary). [10.8]

Chapter Test 10

Simplify. Assume that variables can represent any real number.

1. $\sqrt{50}$

2. $\sqrt[3]{-\dfrac{8}{x^6}}$

3. $\sqrt{81a^2}$

4. $\sqrt{x^2 - 8x + 16}$

5. $\sqrt[5]{x^{12}y^8}$

6. $\sqrt{\dfrac{25x^2}{36y^4}}$

7. $\sqrt[3]{3z}\,\sqrt[3]{5y^2}$

8. $\dfrac{\sqrt[5]{x^3y^4}}{\sqrt[5]{xy^2}}$

9. $\sqrt[4]{x^3y^2}\,\sqrt{xy}$

10. $\dfrac{\sqrt[5]{a^2}}{\sqrt[4]{a}}$

11. $8\sqrt{2} - 2\sqrt{2}$

12. $\sqrt{x^4y} + \sqrt{9y^3}$

13. $\left(7 + \sqrt{x}\right)\left(2 - 3\sqrt{x}\right)$

14. Write an equivalent expression using exponential notation: $\sqrt{7xy}$.

15. Write an equivalent expression using radical notation: $(4a^3b)^{5/6}$.

16. If $f(x) = \sqrt{2x - 10}$, determine the domain of f.

17. If $f(x) = x^2$, find $f\left(5 + \sqrt{2}\right)$.

18. Rationalize the denominator:

$$\dfrac{\sqrt{3}}{5 + \sqrt{2}}.$$

Solve.

19. $x = \sqrt{2x - 5} + 4$

20. $\sqrt{x} = \sqrt{x + 1} - 5$

■ **21.** *Falling Object.* The number of seconds t that it takes for an object to fall x meters when thrown down at a velocity of 9.5 meters per second is given by the function.

$$t(x) = -0.9694 + \sqrt{0.9397 + 0.2041x}.$$

A rock is thrown from the top of Turkey Bluff, 90 m above the Gasconade River in Missouri, at a velocity of 9.5 meters per second. After how many seconds will the rock hit the water?

Solve. Give exact answers and approximations to three decimal places.

22. One leg of a $30°$–$60°$–$90°$ right triangle is 7 cm long. Find the possible lengths of the other leg.

23. A referee jogs diagonally from one corner of a 50-ft by 90-ft basketball court to the far corner. How far does she jog?

24. Express in terms of i and simplify: $\sqrt{-50}$.

25. Subtract: $(9 + 8i) - (-3 + 6i)$.

26. Multiply: $\sqrt{-16}\,\sqrt{-36}$.

27. Multiply. Write the answer in the form $a + bi$.

$(4 - i)^2$

28. Divide and simplify to the form $a + bi$:

$$\dfrac{-2 + i}{3 - 5i}.$$

29. Simplify: i^{37}.

Synthesis

30. Solve:

$$\sqrt{2x - 2} + \sqrt{7x + 4} = \sqrt{13x + 10}.$$

31. Simplify:

$$\dfrac{1 - 4i}{4i(1 + 4i)^{-1}}.$$

32. Drake's Discount Shoe Center has two locations. The sign at the original location is shaped like an isosceles right triangle. The sign at the newer location is shaped like a $30°$–$60°$–$90°$ triangle. The hypotenuse of each sign measures 6 ft. Which sign has the greater area and by how much? (Round to three decimal places.)

11

Quadratic Functions and Equations

*I*n translating problem situations to mathematics, we often obtain a function or equation containing a second-degree polynomial in one variable. Such functions or equations are said to be *quadratic*. In this chapter, we examine a variety of equations, inequalities, and applications for which we will need to solve quadratic equations or graph quadratic functions.

APPLICATION *Sonoma Sunshine.*

The percent of days each month during which the sun shines in Sonoma, California, can be modeled by a quadratic function, as indicated by the table and graph below. Find a quadratic function that fits the data.

Month	Percent of Days with Sunshine
1 (Jan.)	46
3 (March)	75
5 (May)	90
7 (July)	98
9 (Sep.)	92
11 (Nov.)	65

Xscl = 2, Yscl = 10

This problem appears as Example 4 in Section 11.8.

11.1 Quadratic Equations

The Principle of Square Roots ▪ Completing the Square ▪ Problem Solving

In Chapter 6, we solved *quadratic equations* like $x^2 = 10 + 3x$ by graphing and by factoring. One way to solve by graphing is to rewrite the equation above, for example, as $x^2 - 3x - 10 = 0$ and then graph the *quadratic function* given by $f(x) = x^2 - 3x - 10$. The solutions of the equation, -2 and 5, are the first coordinates of the x-intercepts of the graph of f.

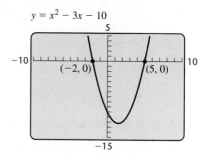

A quadratic equation will have no, one, or two real-number solutions. We can see this by examining the graphs of quadratic functions.

Interactive Discovery

Graph each function and determine the number of x-intercepts.

1. $f(x) = x^2$
2. $g(x) = -x^2$
3. $h(x) = (x - 2)^2$
4. $p(x) = 2x^2 + 1$
5. $f(x) = -1.5x^2 + x - 3$
6. $g(x) = 4x^2 - 2x - 7$
7. Describe the shape of the graph of a quadratic function.

Although we will study graphs of quadratic functions in detail in Section 11.6, we can make some general observations here. The graph of a quadratic function is a cup-shaped curve called a *parabola*. It can open upward, like the graph of $f(x) = x^2$, or downward, like the graph of $g(x) = -x^2$.

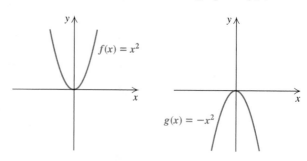

The graph of a quadratic function can have no, one, or two *x*-intercepts, as illustrated below. Thus a quadratic equation can have no, one, or two real-number roots. Quadratic equations can have imaginary-number roots. These must be found using algebraic methods.

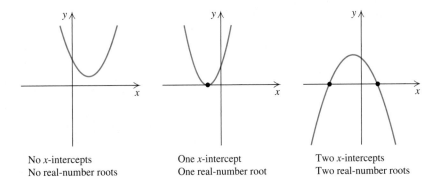

No *x*-intercepts   One *x*-intercept   Two *x*-intercepts
No real-number roots   One real-number root   Two real-number roots

To solve a quadratic equation by factoring, we write the equation in the *standard form* $ax^2 + bx + c = 0$, factor, and use the principle of zero products.

EXAMPLE 1   Solve: $3x^2 = 2 - x$.

SOLUTION   We have

$$3x^2 = 2 - x$$

$$3x^2 + x - 2 = 0 \qquad \text{Adding } -2 + x \text{ to both sides to obtain standard form}$$

$$(3x - 2)(x + 1) = 0 \qquad \text{Factoring}$$

$$3x - 2 = 0 \quad or \quad x + 1 = 0 \qquad \text{Using the principle of zero products}$$

$$3x = 2 \quad or \qquad x = -1$$

$$x = \tfrac{2}{3} \quad or \qquad x = -1.$$

We check by substituting -1 and $\tfrac{2}{3}$ into the original equation. A graphical solution (see the figure at left) provides another check.

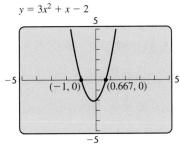

$y = 3x^2 + x - 2$

Check:

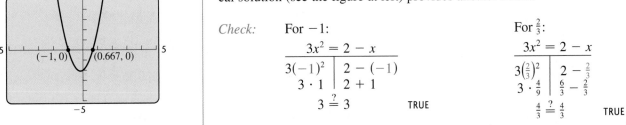

For -1:

$3x^2 = 2 - x$	
$3(-1)^2$	$2 - (-1)$
$3 \cdot 1$	$2 + 1$
$3 \overset{?}{=} 3$	TRUE

For $\tfrac{2}{3}$:

$3x^2 = 2 - x$	
$3\left(\tfrac{2}{3}\right)^2$	$2 - \tfrac{2}{3}$
$3 \cdot \tfrac{4}{9}$	$\tfrac{6}{3} - \tfrac{2}{3}$
$\tfrac{4}{3} \overset{?}{=} \tfrac{4}{3}$	TRUE

The solutions are -1 and $\tfrac{2}{3}$.

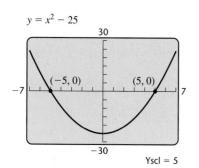

$y = x^2 - 25$

EXAMPLE 2 Solve: $x^2 = 25$.

SOLUTION We have

$$x^2 = 25$$
$$x^2 - 25 = 0 \qquad \text{Writing in standard form}$$
$$(x - 5)(x + 5) = 0 \qquad \text{Factoring}$$
$$x - 5 = 0 \quad or \quad x + 5 = 0 \qquad \text{Using the principle of zero products}$$
$$x = 5 \quad or \qquad x = -5.$$

The solutions are 5 and -5. The graph at left confirms the solutions.

The Principle of Square Roots

Consider the equation $x^2 = 25$ again. We know from Chapter 10 that the number 25 has two real-number square roots, 5 and -5, the solutions of the equation in Example 2. Thus square roots can provide a quick method for solving equations of the type $x^2 = k$.

> **The Principle of Square Roots** For any real number k, if $x^2 = k$, then $x = \sqrt{k}$ or $x = -\sqrt{k}$.

EXAMPLE 3 Solve: $3x^2 = 6$. Give exact answers and approximations to three decimal places.

SOLUTION We have

$$3x^2 = 6$$
$$x^2 = 2 \qquad \text{Isolating } x^2$$
$$x = \sqrt{2} \quad or \quad x = -\sqrt{2}. \qquad \text{Using the principle of square roots}$$

We often use the symbol $\pm\sqrt{2}$ to represent the two numbers $\sqrt{2}$ and $-\sqrt{2}$.

Check:

For $\sqrt{2}$:

$$\frac{3x^2 = 6}{\begin{array}{c|c} 3(\sqrt{2})^2 & 6 \\ 3 \cdot 2 & \\ & 6 \overset{?}{=} 6 \quad \text{TRUE} \end{array}}$$

For $-\sqrt{2}$:

$$\frac{3x^2 = 6}{\begin{array}{c|c} 3(-\sqrt{2})^2 & 6 \\ 3 \cdot 2 & \\ & 6 \overset{?}{=} 6 \quad \text{TRUE} \end{array}}$$

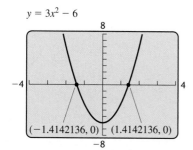

$y = 3x^2 - 6$

The solutions are $\sqrt{2}$ and $-\sqrt{2}$, or $\pm\sqrt{2}$, which round to 1.414 and -1.414. A graphical solution, shown at left, yields approximate solutions, since $\sqrt{2}$ is irrational. The solutions found graphically correspond to those found algebraically.

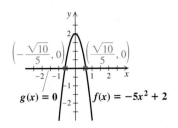

A visualization of Example 4

EXAMPLE 4 Solve: $-5x^2 + 2 = 0$.

SOLUTION We have

$$-5x^2 + 2 = 0$$

$$x^2 = \frac{2}{5} \qquad \text{Isolating } x^2$$

$$x = \sqrt{\frac{2}{5}} \quad or \quad x = -\sqrt{\frac{2}{5}}. \qquad \text{Using the principle of square roots}$$

The solutions are $\sqrt{\dfrac{2}{5}}$ and $-\sqrt{\dfrac{2}{5}}$. This can also be written as $\pm\sqrt{\dfrac{2}{5}}$, or, if we rationalize the denominator, $\pm\dfrac{\sqrt{10}}{5}$. The checks are left to the student.

Sometimes we get solutions that are imaginary numbers.

EXAMPLE 5 Solve: $4x^2 + 9 = 0$.

SOLUTION We have

$$4x^2 + 9 = 0$$

$$x^2 = -\tfrac{9}{4} \qquad \text{Isolating } x^2$$

$$x = \sqrt{-\tfrac{9}{4}} \qquad or \qquad x = -\sqrt{-\tfrac{9}{4}} \qquad \text{Using the principle of square roots}$$

$$x = \sqrt{\tfrac{9}{4}}\sqrt{-1} \qquad or \qquad x = -\sqrt{\tfrac{9}{4}}\sqrt{-1}$$

$$x = \tfrac{3}{2}i \qquad or \qquad x = -\tfrac{3}{2}i. \qquad \text{Recall that } \sqrt{-1} = i.$$

$y = 4x^2 + 9$

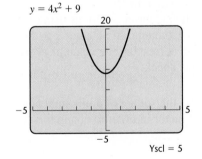

Yscl = 5

Check:

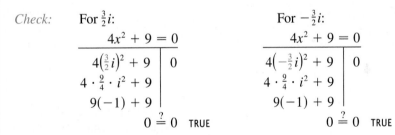

For $\tfrac{3}{2}i$:

$$\begin{array}{c|c} 4x^2 + 9 = 0 & \\ \hline 4\left(\tfrac{3}{2}i\right)^2 + 9 & 0 \\ 4 \cdot \tfrac{9}{4} \cdot i^2 + 9 & \\ 9(-1) + 9 & \\ & 0 \overset{?}{=} 0 \quad \text{TRUE} \end{array}$$

For $-\tfrac{3}{2}i$:

$$\begin{array}{c|c} 4x^2 + 9 = 0 & \\ \hline 4\left(-\tfrac{3}{2}i\right)^2 + 9 & 0 \\ 4 \cdot \tfrac{9}{4} \cdot i^2 + 9 & \\ 9(-1) + 9 & \\ & 0 \overset{?}{=} 0 \quad \text{TRUE} \end{array}$$

The solutions are $\tfrac{3}{2}i$ and $-\tfrac{3}{2}i$, or $\pm\tfrac{3}{2}i$. The graph at left confirms that there are no real-number solutions, because there are no x-intercepts.

The principle of square roots can be restated in a more general form that pertains to algebraic expressions other than just x.

The Principle of Square Roots (Generalized Form) For any real number k and any algebraic expression X,

$$\text{If} \quad X^2 = k, \quad \text{then} \quad X = \sqrt{k} \quad \text{or} \quad X = -\sqrt{k}.$$

EXAMPLE 6   Let $f(x) = (x - 2)^2$. Find all x-values for which $f(x) = 7$.

SOLUTION   We are asked to find all x-values for which

$$f(x) = 7,$$

or

$$(x - 2)^2 = 7. \qquad \text{Substituting } (x - 2)^2 \text{ for } f(x)$$

We solve both algebraically and graphically.

ALGEBRAIC APPROACH	GRAPHICAL APPROACH

ALGEBRAIC APPROACH

The generalized principle of square roots gives us

$$x - 2 = \sqrt{7} \quad or \quad x - 2 = -\sqrt{7}$$
$$x = 2 + \sqrt{7} \quad or \quad x = 2 - \sqrt{7}.$$

Check: $f\left(2 + \sqrt{7}\right) = \left(2 + \sqrt{7} - 2\right)^2 = \left(\sqrt{7}\right)^2 = 7.$

Similarly,

$$f\left(2 - \sqrt{7}\right) = \left(2 - \sqrt{7} - 2\right)^2 = \left(-\sqrt{7}\right)^2 = 7.$$

Rounded to the nearest thousandth,

$$2 + \sqrt{7} \approx 4.646 \quad \text{and} \quad 2 - \sqrt{7} \approx -0.646,$$

so the graphical solutions match the algebraic solutions.
    The solutions are $2 + \sqrt{7}$ and $2 - \sqrt{7}$, or simply $2 \pm \sqrt{7}$.

GRAPHICAL APPROACH

We graph the equations $y_1 = (x - 2)^2$ and $y_2 = 7$. If there are any real-number x-values for which $f(x) = 7$, they will be the x-coordinates of the points of intersection of the graphs.

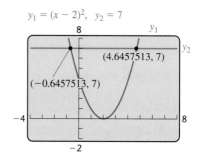

$y_1 = (x - 2)^2, \ y_2 = 7$

Rounded to the nearest thousandth, the solutions are approximately -0.646 and 4.646.
    We can check the algebraic solutions using the VALUE option of the CALC menu. If we enter $2 + \sqrt{7}$ or $2 - \sqrt{7}$ for x, the corresponding value of y_1 will be 7, so the solutions check.

In Example 6, one side of the equation is the square of a binomial and the other side is a constant. Sometimes we must factor in order to obtain this form.

EXAMPLE 7   Solve: $x^2 + 6x + 9 = 2$.

SOLUTION   We have

$$x^2 + 6x + 9 = 2 \qquad \text{The left side is the square of a binomial.}$$
$$(x + 3)^2 = 2 \qquad \text{Factoring}$$
$$x + 3 = \sqrt{2} \quad or \quad x + 3 = -\sqrt{2} \qquad \text{Using the principle of square roots}$$
$$x = -3 + \sqrt{2} \quad or \quad x = -3 - \sqrt{2}. \qquad \text{Adding } -3 \text{ to both sides}$$

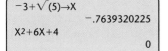

One way to check the solutions on a graphing calculator is to enter $y_1 = x^2 + 6x + 9$ and then evaluate $y_1(-3 + \sqrt{2})$ and $y_1(-3 - \sqrt{2})$, as shown at left. The second calculation can be entered quickly by copying the first entry and editing it.

The solutions are $-3 + \sqrt{2}$ and $-3 - \sqrt{2}$, or $-3 \pm \sqrt{2}$. ◢

Completing the Square

By using a method called *completing the square*, we can use the principle of square roots to solve *any* quadratic equation.

EXAMPLE 8   Solve: $x^2 + 6x + 4 = 0$.

SOLUTION   We have

$$x^2 + 6x + 4 = 0$$

$$x^2 + 6x \quad = -4 \qquad \text{Subtracting 4 from both sides}$$

$$x^2 + 6x + 9 = -4 + 9 \qquad \text{Adding 9 to both sides. We explain this shortly.}$$

$$(x + 3)^2 = 5 \qquad \text{Factoring the perfect-square trinomial}$$

$$x + 3 = \pm\sqrt{5} \qquad \text{Using the principle of square roots. Remember that } \pm\sqrt{5} \text{ represents two numbers.}$$

$$x = -3 \pm \sqrt{5}. \qquad \text{Adding } -3 \text{ to both sides}$$

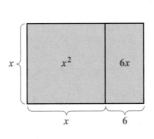

We check by storing $-3 + \sqrt{5}$ as x and evaluating $x^2 + 6x + 4$, as shown at left. We can repeat the process for $-3 - \sqrt{5}$ to check the second solution. The solutions are $-3 + \sqrt{5}$ and $-3 - \sqrt{5}$. ◢

In Example 8, we chose to add 9 to both sides because it creates a perfect-square trinomial on the left side. The 9 was determined by taking half of the coefficient of x and squaring it.

To help see why this procedure works, examine the following drawings.

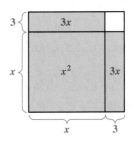

Note that the shaded areas in both figures represent the same area, $x^2 + 6x$. However, only the figure on the right, in which the $6x$ is halved, can be converted into a square with the addition of a constant term. The constant 9 is the "missing" piece that *completes* the square.

To complete the square for $x^2 + bx$, we add $\left(\dfrac{b}{2}\right)^2$.

Example 9, which follows, provides practice in finding numbers that complete the square. We will then use this skill to solve equations.

Student Notes

In problems like Examples 9(b) and (c), it is best to avoid decimal notation. Most students have an easier time recognizing $\frac{9}{64}$ as $\left(\frac{3}{8}\right)^2$ than regarding 0.140625 as 0.375^2.

EXAMPLE 9 Replace the blanks in each equation with constants to form a true equation.

a) $x^2 + 14x + \underline{\hphantom{xx}} = (x + \underline{\hphantom{xx}})^2$

b) $x^2 - 5x + \underline{\hphantom{xx}} = (x - \underline{\hphantom{xx}})^2$

c) $x^2 + \frac{3}{4}x + \underline{\hphantom{xx}} = (x + \underline{\hphantom{xx}})^2$

SOLUTION We take half of the coefficient of x and square it.

a) Half of 14 is 7, and $7^2 = 49$. Thus, $x^2 + 14x + 49$ is a perfect-square trinomial and is equivalent to $(x + 7)^2$. We have

$$x^2 + 14x + 49 = (x + 7)^2.$$

b) Half of -5 is $-\frac{5}{2}$, and $\left(-\frac{5}{2}\right)^2 = \frac{25}{4}$. Thus, $x^2 - 5x + \frac{25}{4}$ is a perfect-square trinomial and is equivalent to $\left(x - \frac{5}{2}\right)^2$. We have

$$x^2 - 5x + \frac{25}{4} = \left(x - \frac{5}{2}\right)^2.$$

c) Half of $\frac{3}{4}$ is $\frac{3}{8}$, and $\left(\frac{3}{8}\right)^2 = \frac{9}{64}$. Thus, $x^2 + \frac{3}{4}x + \frac{9}{64}$ is a perfect-square trinomial and is equivalent to $\left(x + \frac{3}{8}\right)^2$. We have

$$x^2 + \frac{3}{4}x + \frac{9}{64} = \left(x + \frac{3}{8}\right)^2.$$

We can now use the method of completing the square to solve equations similar to Example 8.

EXAMPLE 10 Solve: **(a)** $x^2 - 8x - 7 = 0$; **(b)** $x^2 + 5x - 3 = 0$.

SOLUTION

a) $x^2 - 8x - 7 = 0$

$x^2 - 8x = 7$ Adding 7 to both sides. We can now complete the square on the left side.

$x^2 - 8x + 16 = 7 + 16$ Adding 16 to both sides to complete the square: $\frac{1}{2}(-8) = -4$, and $(-4)^2 = 16$

$(x - 4)^2 = 23$ **Factoring and simplifying**

$x - 4 = \pm\sqrt{23}$ **Using the principle of square roots**

$x = 4 \pm \sqrt{23}$ **Adding 4 to both sides**

CAUTION! In Example 10(a), be sure to add 16 to *both sides* of the equation.

The solutions are $4 - \sqrt{23}$ and $4 + \sqrt{23}$, or $4 \pm \sqrt{23}$. The checks are left to the student.

b) $x^2 + 5x - 3 = 0$

$x^2 + 5x = 3$ **Adding 3 to both sides**

$x^2 + 5x + \dfrac{25}{4} = 3 + \dfrac{25}{4}$ **Completing the square: $\frac{1}{2} \cdot 5 = \frac{5}{2}$, and $\left(\frac{5}{2}\right)^2 = \frac{25}{4}$**

$\left(x + \dfrac{5}{2}\right)^2 = \dfrac{37}{4}$ **Factoring and simplifying**

$x + \dfrac{5}{2} = \pm\dfrac{\sqrt{37}}{2}$ **Using the principle of square roots and the quotient rule for radicals**

$x = -\dfrac{5}{2} \pm \dfrac{\sqrt{37}}{2}$ *or* $x = \dfrac{-5 \pm \sqrt{37}}{2}$ **Adding $-\frac{5}{2}$ to both sides**

The checks are left to the student. The solutions are $-\dfrac{5}{2} \pm \dfrac{\sqrt{37}}{2}$, or $\dfrac{-5 \pm \sqrt{37}}{2}$.

Before we complete the square, the x^2-coefficient must be 1. When it is not 1, we divide both sides of the equation by whatever that coefficient may be.

EXAMPLE 11   Find the x-intercepts of the function given by

$$f(x) = 3x^2 + 7x - 2.$$

SOLUTION   The value of $f(x)$ must be 0 at any x-intercepts. Thus,

$f(x) = 0$ **We set $f(x)$ equal to 0.**

$3x^2 + 7x - 2 = 0$ **Substituting**

$3x^2 + 7x = 2$ **Adding 2 to both sides**

$x^2 + \dfrac{7}{3}x = \dfrac{2}{3}$ **Dividing both sides by 3**

$x^2 + \dfrac{7}{3}x + \dfrac{49}{36} = \dfrac{2}{3} + \dfrac{49}{36}$ **Completing the square: $\left(\frac{1}{2} \cdot \frac{7}{3}\right)^2 = \frac{49}{36}$**

$\left(x + \dfrac{7}{6}\right)^2 = \dfrac{73}{36}$ **Factoring and simplifying**

$x + \dfrac{7}{6} = \pm\dfrac{\sqrt{73}}{6}$ **Using the principle of square roots and the quotient rule for radicals**

$x = -\dfrac{7}{6} \pm \dfrac{\sqrt{73}}{6}$,   *or* $\dfrac{-7 \pm \sqrt{73}}{6}$.   **Adding $-\frac{7}{6}$ to both sides**

The x-intercepts are

$$\left(\dfrac{-7 - \sqrt{73}}{6}, 0\right) \quad \text{and} \quad \left(\dfrac{-7 + \sqrt{73}}{6}, 0\right).$$

The checks are left to the student.

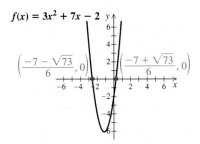

$f(x) = 3x^2 + 7x - 2$

$\left(\dfrac{-7 - \sqrt{73}}{6}, 0\right)$ $\left(\dfrac{-7 + \sqrt{73}}{6}, 0\right)$

A visualization of Example 11

The procedure used in Example 11 is important because it can be used to solve *any* quadratic equation.

To Solve a Quadratic Equation in *x* by Completing the Square

1. Isolate the terms with variables on one side of the equation, and arrange them in descending order.
2. Divide both sides by the coefficient of x^2 if that coefficient is not 1.
3. Complete the square by taking half of the coefficient of *x* and adding its square to both sides.
4. Express the trinomial as the square of a binomial (factor the trinomial) and simplify the other side.
5. Use the principle of square roots (find the square roots of both sides).
6. Solve for *x* by adding or subtracting on both sides.

Problem Solving

After one year, an amount of money *P*, invested at 4% per year, is worth 104% of *P*, or $P(1.04)$. If that amount continues to earn 4% interest per year, after the second year the investment will be worth 104% of $P(1.04)$, or $P(1.04)^2$. This is called **compounding interest** since after the first time period, interest is earned on both the initial investment *and* the interest from the first time period. Continuing the above pattern, we see that after the third year, the investment will be worth 104% of $P(1.04)^2$. Generalizing, we have the following.

The Compound-Interest Formula   If an amount of money *P* is invested at interest rate *r*, compounded annually, then in *t* years, it will grow to the amount *A* given by

$$A = P(1 + r)^t. \qquad (r \text{ is written in decimal notation.})$$

We can use quadratic equations to solve certain interest problems.

EXAMPLE 12   Investment Growth.   Rosa invested $4000 at interest rate *r*, compounded annually. In 2 yr, it grew to $4410. What was the interest rate?

SOLUTION

1. **Familiarize.** We are already familiar with the compound-interest formula. If we were not, we would need to consult an outside source.

2. **Translate.** The translation consists of substituting into the formula:

$$A = P(1 + r)^t$$
$$4410 = 4000(1 + r)^2. \qquad \text{Substituting}$$

3. **Carry out.** We solve for *r* both algebraically and graphically.

ALGEBRAIC APPROACH

We have

$$4410 = 4000(1 + r)^2$$

$$\frac{4410}{4000} = (1 + r)^2 \qquad \text{Dividing both sides by 4000}$$

$$\frac{441}{400} = (1 + r)^2 \qquad \text{Simplifying}$$

$$\pm\sqrt{\frac{441}{400}} = 1 + r \qquad \text{Using the principle of square roots}$$

$$\pm\frac{21}{20} = 1 + r \qquad \text{Simplifying}$$

$$-\frac{20}{20} \pm \frac{21}{20} = r \qquad \text{Adding } -1, \text{ or } -\frac{20}{20}, \text{ to both sides}$$

$$\frac{1}{20} = r \quad or \quad -\frac{41}{20} = r.$$

Since r represents an interest rate, we convert to decimal notation. We now have

$$r = 0.05 \quad or \quad r = -2.05.$$

GRAPHICAL APPROACH

We let $y_1 = 4410$ and $y_2 = 4000(1 + x)^2$. Using the window $[-3, 1, 0, 5000]$, we see that the graphs intersect at two points. The first coordinates of the points of intersection are -2.05 and 0.05.

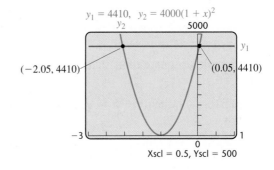

$y_1 = 4410, \quad y_2 = 4000(1 + x)^2$

$(-2.05, 4410)$ $(0.05, 4410)$

Xscl = 0.5, Yscl = 500

4. **Check.** Since the rate cannot be negative, we need only check 0.05, or 5%. If \$4000 were invested at 5% interest, compounded annually, then in 2 yr it would grow to $4000(1.05)^2$, or \$4410. The number 0.05 checks.

5. **State.** The interest rate was 5%.

EXAMPLE 13 Free-Falling Objects. The formula $s = 16t^2$ is used to approximate the distance s, in feet, that an object falls freely from rest in t seconds. Ireland's Cliffs of Moher are 702 ft tall (*Source*: Based on data from 4windstravel.com). How long will it take a stone to fall from the top? Round to the nearest tenth of a second.

SOLUTION

1. **Familiarize.** We make a drawing to help visualize the problem and agree to disregard air resistance.

2. **Translate.** We substitute into the formula:

$$s = 16t^2$$
$$702 = 16t^2.$$

3. **Carry out.** We solve for t:

$$702 = 16t^2$$

$$\frac{702}{16} = t^2$$

$$43.875 = t^2$$

$$\sqrt{43.875} = t \qquad \text{Using the principle of square roots; rejecting the negative square root since } t \text{ cannot be negative in this problem}$$

$$6.6 \approx t. \qquad \text{Using a calculator and rounding to the nearest tenth}$$

4. **Check.** Since $16(6.6)^2 \approx 697 \approx 702$, our answer checks.

5. **State.** It takes about 6.6 sec for a stone to fall freely from the Cliffs of Moher.

11.1 EXERCISE SET

Concept Reinforcement   *Complete each of the following to form true statements.*

1. The principle of square roots states that if $x^2 = k$, then $x =$ ____ or $x =$ ____.

2. If $(x + 5)^2 = 49$, then $x + 5 =$ ____ or $x + 5 =$ ____.

3. If $t^2 + 6t + 9 = 17$, then $(\_\_\_)^2 = 17$ and ____ $= \pm\sqrt{17}$.

4. The equations $x^2 + 8x +$ ____ $= 23$ and $x^2 + 8x = 7$ are equivalent.

5. The expressions $t^2 + 10t +$ ____ and $(t +$ ____$)^2$ are equivalent.

6. The expressions $x^2 - 6x +$ ____ and $(x -$ ____$)^2$ are equivalent.

Determine the number of real-number solutions of each equation from the given graph.

7. $x^2 + x - 12 = 0$

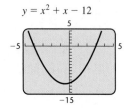
$y = x^2 + x - 12$

8. $-3x^2 - x - 7 = 0$

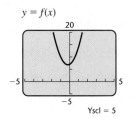
$y = -3x^2 - x - 7$

9. $4x^2 + 9 = 12x$

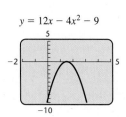
$y = 12x - 4x^2 - 9$

10. $2x^2 + 3 = 6x$

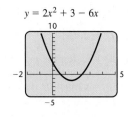
$y = 2x^2 + 3 - 6x$

11. $f(x) = 0$

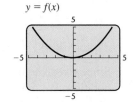
$y = f(x)$

Yscl = 5

12. $f(x) = 0$

$y = f(x)$

Solve.

13. $4x^2 = 20$

14. $7x^2 = 21$

15. $9x^2 + 16 = 0$

16. $25x^2 + 4 = 0$

17. $5t^2 - 7 = 0$

18. $3t^2 - 2 = 0$

19. $(x - 1)^2 = 49$

20. $(x + 2)^2 = 25$

21. $(a - 13)^2 = 18$

22. $(a + 5)^2 = 8$

23. $(x + 1)^2 = -9$

24. $(x - 1)^2 = -49$

25. $\left(y + \frac{3}{4}\right)^2 = \frac{17}{16}$

26. $\left(t + \frac{3}{2}\right)^2 = \frac{7}{2}$

27. $x^2 - 10x + 25 = 64$

28. $x^2 - 6x + 9 = 100$

29. Let $f(x) = (x - 5)^2$. Find x such that $f(x) = 16$.

30. Let $g(x) = (x - 2)^2$. Find x such that $g(x) = 25$.

31. Let $F(t) = (t + 4)^2$. Find t such that $F(t) = 13$.

32. Let $f(t) = (t + 6)^2$. Find t such that $f(t) = 15$.

Aha! 33. Let $g(x) = x^2 + 14x + 49$. Find x such that $g(x) = 49$.

34. Let $F(x) = x^2 + 8x + 16$. Find x such that $F(x) = 9$.

Replace the blanks in each equation with constants to complete the square and form a true equation.

35. $x^2 + 16x +$ ___ $= (x +$ ___$)^2$

36. $x^2 + 8x +$ ___ $= (x +$ ___$)^2$

37. $t^2 - 10t + \underline{\quad} = (t - \underline{\quad})^2$

38. $t^2 - 6t + \underline{\quad} = (t - \underline{\quad})^2$

39. $x^2 + 3x + \underline{\quad} = (x + \underline{\quad})^2$

40. $x^2 + 7x + \underline{\quad} = (x + \underline{\quad})^2$

41. $t^2 - 9t + \underline{\quad} = (t - \underline{\quad})^2$

42. $t^2 - 3t + \underline{\quad} = (t - \underline{\quad})^2$

43. $x^2 + \frac{2}{5}x + \underline{\quad} = (x + \underline{\quad})^2$

44. $x^2 + \frac{2}{3}x + \underline{\quad} = (x + \underline{\quad})^2$

45. $t^2 - \frac{5}{6}t + \underline{\quad} = (t - \underline{\quad})^2$

46. $t^2 - \frac{5}{3}t + \underline{\quad} = (t - \underline{\quad})^2$

Solve by completing the square. Show your work.

47. $x^2 + 6x = 7$ **48.** $x^2 + 8x = 9$

49. $t^2 - 10t = -24$ **50.** $t^2 - 10t = -21$

51. $x^2 + 10x + 9 = 0$ **52.** $x^2 + 8x + 7 = 0$

53. $t^2 + 8t - 3 = 0$ **54.** $t^2 + 6t - 5 = 0$

Complete the square to find the x-intercepts of each function given by the equation listed.

55. $f(x) = x^2 + 6x + 7$ **56.** $f(x) = x^2 + 5x + 3$

57. $g(x) = x^2 + 12x + 25$ **58.** $g(x) = x^2 + 4x + 2$

59. $f(x) = x^2 - 10x - 22$ **60.** $f(x) = x^2 - 8x - 10$

Solve by completing the square. Remember to first divide, as in Example 11, to make sure that the coefficient of x^2 is 1.

61. $9x^2 + 18x = -8$ **62.** $4x^2 + 8x = -3$

63. $3x^2 - 5x - 2 = 0$ **64.** $2x^2 - 5x - 3 = 0$

65. $5x^2 + 4x - 3 = 0$ **66.** $4x^2 + 3x - 5 = 0$

67. Find the x-intercepts of the function given by $f(x) = 4x^2 + 2x - 3$.

68. Find the x-intercepts of the function given by $f(x) = 3x^2 + x - 5$.

69. Find the x-intercepts of the function given by $g(x) = 2x^2 - 3x - 1$.

70. Find the x-intercepts of the function given by $g(x) = 3x^2 - 5x - 1$.

Interest.   Use $A = P(1 + r)^t$ to find the interest rate in Exercises 71–76. Refer to Example 12.

71. $2000 grows to $2420 in 2 yr

72. $2560 grows to $2890 in 2 yr

73. $1280 grows to $1805 in 2 yr

74. $1000 grows to $1440 in 2 yr

75. $6250 grows to $6760 in 2 yr

76. $6250 grows to $7290 in 2 yr

Free-Falling Objects.   Use $s = 16t^2$ for Exercises 77–80. Refer to Example 13 and neglect air resistance.

77. Suspended 1053 ft above the water, the bridge over Colorado's Royal Gorge is the world's highest suspension bridge (*Source: The Guinness Book of Records*). How long would it take an object to fall freely from the bridge?

78. The CN Tower in Toronto, at 1815 ft, is the world's tallest self-supporting tower on land (no guy wires) (*Source: The Guinness Book of Records*). How long would it take an object to fall freely from the top?

79. The Sears Tower in Chicago is 1454 ft tall. How long would it take an object to fall freely from the top?

80. The Gateway Arch in St. Louis is 630 ft high (*Source*: www.icivilengineer.com). How long would it take an object to fall freely from the top?

TW 81. Explain in your own words a sequence of steps that can be used to solve any quadratic equation in the quickest way.

TW 82. Describe how to write a quadratic equation that can be solved algebraically but not graphically.

Skill Maintenance

Evaluate. [1.8]

83. $at^2 - bt$, for $a = 3$, $b = 5$, and $t = 4$

84. $mn^2 - mp$, for $m = -2$, $n = 7$, and $p = 3$

Simplify. [10.3]

85. $\sqrt[3]{270}$ **86.** $\sqrt{80}$

Let $f(x) = \sqrt{3x - 5}$. Find each of the following. [10.1]

87. $f(10)$ **88.** $f(18)$

Synthesis

TW **89.** What would be better: to receive 3% interest every 6 months or to receive 6% interest every 12 months? Why?

TW **90.** Example 12 was solved with a graphing calculator by graphing each side of

$$4410 = 4000(1 + r)^2.$$

How could you determine, from a reading of the problem, a suitable viewing window?

Find b such that each trinomial is a square.

91. $x^2 + bx + 81$ **92.** $x^2 + bx + 49$

93. If $f(x) = 2x^5 - 9x^4 - 66x^3 + 45x^2 + 280x$ and $x^2 - 5$ is a factor of $f(x)$, find all a for which $f(a) = 0$.

94. If

$$f(x) = \left(x - \tfrac{1}{3}\right)(x^2 + 6)$$

and

$$g(x) = \left(x - \tfrac{1}{3}\right)\left(x^2 - \tfrac{2}{3}\right),$$

find all a for which $(f + g)(a) = 0$.

95. *Boating.* A barge and a fishing boat leave a dock at the same time, traveling at a right angle to each other. The barge travels 7 km/h slower than the fishing boat. After 4 hr, the boats are 68 km apart. Find the speed of each boat.

96. Find three consecutive integers such that the square of the first plus the product of the other two is 67.

11.2

The Quadratic Formula

Solving Using the Quadratic Formula   ■   Approximating Solutions

There are at least two reasons for learning to complete the square. One is to help graph certain equations that appear later in this chapter. Another is to develop a general formula for solving quadratic equations.

Solving Using the Quadratic Formula

Each time we solve by completing the square, the procedure is the same. In mathematics, when a procedure is repeated many times, a formula can often be developed to speed up our work.

We begin with a quadratic equation in standard form,

$$ax^2 + bx + c = 0,$$

Study Tip

Know It "By Heart"

When memorizing something like the quadratic formula, try to first understand and write out the derivation. Doing this will help you remember the formula.

with $a > 0$. For $a < 0$, a slightly different derivation is needed (see Exercise 58), but the result is the same. Let's solve by completing the square. As the steps are performed, compare them with Example 11 on p. 811.

$$ax^2 + bx = -c \qquad \text{Adding } -c \text{ to both sides}$$

$$x^2 + \frac{b}{a}x = -\frac{c}{a} \qquad \text{Dividing both sides by } a$$

Half of $\dfrac{b}{a}$ is $\dfrac{b}{2a}$ and $\left(\dfrac{b}{2a}\right)^2$ is $\dfrac{b^2}{4a^2}$. We add $\dfrac{b^2}{4a^2}$ to both sides:

$$x^2 + \frac{b}{a}x + \frac{b^2}{4a^2} = -\frac{c}{a} + \frac{b^2}{4a^2} \qquad \text{Adding } \frac{b^2}{4a^2} \text{ to complete the square}$$

$$\left(x + \frac{b}{2a}\right)^2 = -\frac{4ac}{4a^2} + \frac{b^2}{4a^2} \qquad \begin{array}{l}\text{Factoring on the left side;}\\ \text{finding a common denomi-}\\ \text{nator on the right side}\end{array}$$

$$\left(x + \frac{b}{2a}\right)^2 = \frac{b^2 - 4ac}{4a^2}$$

$$x + \frac{b}{2a} = \pm\frac{\sqrt{b^2 - 4ac}}{2a} \qquad \begin{array}{l}\text{Using the principle of square}\\ \text{roots and the quotient rule for}\\ \text{radicals; since } a > 0,\\ \sqrt{4a^2} = 2a\end{array}$$

$$x = \frac{-b \pm \sqrt{b^2 - 4ac}}{2a}. \qquad \text{Adding } -\frac{b}{2a} \text{ to both sides}$$

It is important that you remember the quadratic formula and know how to use it.

The Quadratic Formula   The solutions of $ax^2 + bx + c = 0$, $a \neq 0$, are given by

$$x = \frac{-b \pm \sqrt{b^2 - 4ac}}{2a}.$$

▶ **EXAMPLE 1**   Solve $5x^2 + 8x = -3$ using the quadratic formula.

SOLUTION   We first find standard form and determine a, b, and c:

$$5x^2 + 8x + 3 = 0; \qquad \text{Adding 3 to both sides to get 0 on one side}$$

$$a = 5, \quad b = 8, \quad c = 3.$$

Next, we use the quadratic formula:

$$x = \frac{-b \pm \sqrt{b^2 - 4ac}}{2a}$$

$$x = \frac{-8 \pm \sqrt{8^2 - 4 \cdot 5 \cdot 3}}{2 \cdot 5} \qquad \text{Substituting}$$

$$x = \frac{-8 \pm \sqrt{64 - 60}}{10}$$

> **Be sure to write the fraction bar all the way across.**

$$x = \frac{-8 \pm \sqrt{4}}{10} = \frac{-8 \pm 2}{10}$$

$$x = \frac{-8 + 2}{10} \quad or \quad x = \frac{-8 - 2}{10}$$

$$x = \frac{-6}{10} \quad or \quad x = \frac{-10}{10}$$

$$x = -\frac{3}{5} \quad or \quad x = -1.$$

The solutions are $-\frac{3}{5}$ and -1. The checks are left to the student.

Because $5x^2 + 8x + 3$ can be factored as $(5x + 3)(x + 1)$, the quadratic formula may not have been the fastest way of solving Example 1. However, because the quadratic formula works for *any* quadratic equation, we need not spend too much time struggling to solve a quadratic equation by factoring.

To Solve a Quadratic Equation

1. If the equation can be easily written in the form $ax^2 = p$ or $(x + k)^2 = d$, use the principle of square roots as in Section 11.1.
2. If step (1) does not apply, write the equation in the form $ax^2 + bx + c = 0$.
3. Try factoring and using the principle of zero products.
4. If factoring seems to be difficult or impossible, use the quadratic formula. Completing the square can also be used, but is usually slower.

The solutions of a quadratic equation can always be found using the quadratic formula. They cannot always be found by factoring.

Recall that a second-degree polynomial in one variable is said to be quadratic. Similarly, a second-degree polynomial function in one variable is said to be a **quadratic function.**

EXAMPLE 2   For the quadratic function given by $f(x) = 3x^2 - 6x - 4$, find all x for which $f(x) = 0$.

SOLUTION   We substitute and solve for x:

$$f(x) = 0$$

$$3x^2 - 6x - 4 = 0 \qquad \text{Substituting. You can try to solve this by factoring.}$$

$$a = 3, \quad b = -6, \quad c = -4.$$

We then substitute into the quadratic formula:

$$x = \frac{-(-6) \pm \sqrt{(-6)^2 - 4 \cdot 3 \cdot (-4)}}{2 \cdot 3}$$

$$= \frac{6 \pm \sqrt{36 + 48}}{6}$$

$$= \frac{6 \pm \sqrt{84}}{6}$$

$$= \frac{6}{6} \pm \frac{\sqrt{84}}{6} \qquad \text{Note that 4 is a perfect-square factor of 84.}$$

$$= 1 \pm \frac{\sqrt{4}\sqrt{21}}{6} \qquad 84 = 4 \cdot 21$$

$$\left.\begin{array}{l} = 1 \pm \dfrac{2\sqrt{21}}{6} \\[2ex] = 1 \pm \dfrac{\sqrt{21}}{3}. \end{array}\right\} \quad \textbf{Simplifying}$$

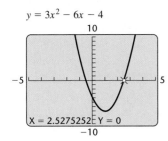

$y = 3x^2 - 6x - 4$

X = 2.5275252   Y = 0

To check, we graph $y_1 = 3x^2 - 6x - 4$, and press ⟨TRACE⟩. When we enter $1 + \sqrt{}$ (21)/3, a rational approximation and the y-value 0 appear, as shown in the graph at left. The number $1 - \sqrt{21}/3$ also checks.

The solutions are $1 - \dfrac{\sqrt{21}}{3}$ and $1 + \dfrac{\sqrt{21}}{3}$. ◢

Some quadratic equations have solutions that are imaginary numbers.

EXAMPLE 3   Solve: $x^2 + 2 = -x$.

SOLUTION   We first find standard form:

$$x^2 + x + 2 = 0. \qquad \textbf{Adding } x \textbf{ to both sides}$$

Since we cannot factor $x^2 + x + 2$, we use the quadratic formula with $a = 1$, $b = 1$, and $c = 2$:

$$x = \frac{-1 \pm \sqrt{1^2 - 4 \cdot 1 \cdot 2}}{2 \cdot 1} \qquad \textbf{Substituting}$$

$$= \frac{-1 \pm \sqrt{1 - 8}}{2} = \frac{-1 \pm \sqrt{-7}}{2}$$

$$= \frac{-1 \pm i\sqrt{7}}{2}, \text{ or } -\frac{1}{2} \pm \frac{\sqrt{7}}{2}i$$

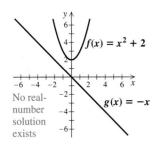

$f(x) = x^2 + 2$

$g(x) = -x$

No real-number solution exists

A visualization of Example 3

The solutions are $-\dfrac{1}{2} - \dfrac{\sqrt{7}}{2}i$ and $-\dfrac{1}{2} + \dfrac{\sqrt{7}}{2}i$. The checks are left to the student.

Note from the graph at left that the graphs of $f(x) = x^2 + 2$ and $g(x) = -x$ do not intersect. Thus there are no real-number solutions of the equation. ◢

The quadratic formula can be used to solve certain rational equations.

> **EXAMPLE 4** If $f(x) = 2 + \dfrac{7}{x}$ and $g(x) = \dfrac{4}{x^2}$, find all x for which $f(x) = g(x)$.
>
> **SOLUTION** We set $f(x)$ equal to $g(x)$ and solve:
>
> $$f(x) = g(x)$$
> $$2 + \frac{7}{x} = \frac{4}{x^2}. \qquad \text{Substituting. Note that } x \neq 0.$$

This is a rational equation similar to those in Section 7.6. To solve, we multiply both sides by the LCD, x^2:

$$x^2 \left(2 + \frac{7}{x} \right) = x^2 \cdot \frac{4}{x^2}$$
$$2x^2 + 7x = 4 \qquad \text{Simplifying}$$
$$2x^2 + 7x - 4 = 0. \qquad \text{Subtracting 4 from both sides}$$

We have

$$a = 2, \quad b = 7, \quad \text{and} \quad c = -4.$$

Substituting then gives us

$$x = \frac{-7 \pm \sqrt{7^2 - 4 \cdot 2 \cdot (-4)}}{2 \cdot 2}$$
$$= \frac{-7 \pm \sqrt{49 + 32}}{4} = \frac{-7 \pm \sqrt{81}}{4} = \frac{-7 \pm 9}{4}$$
$$x = \frac{-7 + 9}{4} = \frac{1}{2} \quad \text{or} \quad x = \frac{-7 - 9}{4} = -4. \qquad \text{\textbf{Both answers should check since } } x \neq 0.$$

You can confirm that $f\left(\frac{1}{2}\right) = g\left(\frac{1}{2}\right)$ and $f(-4) = g(-4)$. The solutions are $\frac{1}{2}$ and -4.

Approximating Solutions

When the solution of an equation is irrational, a rational-number approximation is often useful. This is often the case in real-world applications similar to those found in Section 11.3.

> **EXAMPLE 5** Use a calculator to approximate the solutions of Example 2.
>
> **SOLUTION** On most graphing calculators, the following sequence of keystrokes can be used to approximate $1 + \sqrt{21}/3$:
>
> (1) (+) (√) (2) (1) (**)**) (÷) (3) (**ENTER**).
>
> Similar keystrokes can be used to approximate $1 - \sqrt{21}/3$.
>   The solutions are approximately 2.527525232 and -0.5275252317.

11.2 EXERCISE SET

FOR EXTRA HELP

         Tutor Center

MathXL   MyMathLab   InterAct Math   AW Math Tutor Center   Video Lectures on CD: Disc 6   Student's Solutions Manual

☙ *Concept Reinforcement*   *Classify each of the following as either true or false.*

1. The quadratic formula can be used to solve *any* quadratic equation.

2. The quadratic formula does not work if solutions are imaginary numbers.

3. Solving by factoring is always slower than using the quadratic formula.

4. The steps used to derive the quadratic formula are the same as those used when solving by completing the square.

5. A quadratic equation can have as many as four solutions.

6. It is possible for a quadratic equation to have no real-number solutions.

Solve.

7. $x^2 + 7x - 3 = 0$

8. $x^2 - 7x + 4 = 0$

9. $3p^2 = 18p - 6$

10. $3u^2 = 8u - 5$

11. $x^2 + x + 1 = 0$

12. $x^2 + x + 2 = 0$

13. $x^2 + 13 = 4x$

14. $x^2 + 13 = 6x$

15. $h^2 + 4 = 6h$

16. $r^2 + 3r = 8$

17. $\dfrac{1}{x^2} - 3 = \dfrac{8}{x}$

18. $\dfrac{9}{x} - 2 = \dfrac{5}{x^2}$

19. $3x + x(x - 2) = 4$

20. $4x + x(x - 3) = 5$

21. $12t^2 + 9t = 1$

22. $15t^2 + 7t = 2$

23. $25x^2 - 20x + 4 = 0$

24. $36x^2 + 84x + 49 = 0$

25. $7x(x + 2) + 5 = 3x(x + 1)$

26. $5x(x - 1) - 7 = 4x(x - 2)$

27. $14(x - 4) - (x + 2) = (x + 2)(x - 4)$

28. $11(x - 2) + (x - 5) = (x + 2)(x - 6)$

29. $5x^2 = 13x + 17$

30. $25x = 3x^2 + 28$

31. $x^2 + 9 = 4x$

32. $x^2 + 7 = 3x$

33. $x^3 - 8 = 0$ (*Hint*: Factor the difference of cubes. Then use the quadratic formula.)

34. $x^3 + 1 = 0$

35. Let $f(x) = 3x^2 - 5x + 2$. Find x such that $f(x) = 0$.

36. Let $g(x) = 4x^2 - 2x - 3$. Find x such that $g(x) = 0$.

37. Let
$$f(x) = \frac{7}{x} + \frac{7}{x + 4}.$$
Find all x for which $f(x) = 1$.

38. Let
$$g(x) = \frac{2}{x} + \frac{2}{x + 3}.$$
Find all x for which $g(x) = 1$.

39. Let
$$F(x) = \frac{x + 3}{x} \quad \text{and} \quad G(x) = \frac{x - 4}{3}.$$
Find all x for which $F(x) = G(x)$.

40. Let
$$f(x) = \frac{3 - x}{4} \quad \text{and} \quad g(x) = \frac{1}{4x}.$$
Find all x for which $f(x) = g(x)$.

41. Let
$$f(x) = \frac{15 - 2x}{6} \quad \text{and} \quad g(x) = \frac{3}{x}.$$
Find all x for which $f(x) = g(x)$.

42. Let

$$f(x) = x + 5 \quad \text{and} \quad g(x) = \frac{3}{x - 5}.$$

Find all x for which $f(x) = g(x)$.

Solve. Use a calculator to approximate, as precisely as possible, the solutions as rational numbers.

43. $x^2 + 4x - 7 = 0$

44. $x^2 + 6x + 4 = 0$

45. $x^2 - 6x + 4 = 0$

46. $x^2 - 4x + 1 = 0$

47. $2x^2 - 3x - 7 = 0$

48. $3x^2 - 3x - 2 = 0$

TW 49. Are there any equations that can be solved by the quadratic formula but not by completing the square? Why or why not?

TW 50. If you had to choose between remembering the method of completing the square and remembering the quadratic formula, which would you choose? Why?

Skill Maintenance

Simplify.

51. $\dfrac{x^2 + xy}{2x}$ [7.1]

52. $\dfrac{a^3 - ab^2}{ab}$ [7.1]

53. $\sqrt{27a^2b^5} \cdot \sqrt{6a^3b}$ [10.3]

54. $\sqrt{8a^3b} \cdot \sqrt{12ab^5}$ [10.3]

55. $\dfrac{\dfrac{3}{x - 1}}{\dfrac{1}{x + 1} + \dfrac{2}{x - 1}}$ [7.5]

56. $\dfrac{\dfrac{4}{a^2b}}{\dfrac{3}{a} - \dfrac{4}{b^2}}$ [7.5]

Synthesis

TW 57. Suppose you had a large number of quadratic equations to solve and none of the equations had a constant term. Would you use factoring or the quadratic formula to solve these equations? Why?

TW 58. If $a < 0$ and $ax^2 + bx + c = 0$, then $-a$ is positive and the equivalent equation, $-ax^2 - bx - c = 0$, can be solved using the quadratic formula.

a) Find this solution, replacing a, b, and c in the formula with $-a$, $-b$, and $-c$ from the equation.

b) How does the result of part (a) indicate that the quadratic formula "works" regardless of the sign of a?

For Exercises 59–61, let

$$f(x) = \frac{x^2}{x - 2} + 1 \quad \text{and} \quad g(x) = \frac{4x - 2}{x - 2} + \frac{x + 4}{2}.$$

59. Find the x-intercepts of the graph of f.

60. Find the x-intercepts of the graph of g.

61. Find all x for which $f(x) = g(x)$.

Solve.

62. $x^2 - 0.75x - 0.5 = 0$

63. $z^2 + 0.84z - 0.4 = 0$

64. $(1 + \sqrt{3})x^2 - (3 + 2\sqrt{3})x + 3 = 0$

65. $\sqrt{2}x^2 + 5x + \sqrt{2} = 0$

66. $ix^2 - 2x + 1 = 0$

67. One solution of $kx^2 + 3x - k = 0$ is -2. Find the other.

TW 68. Can a graph be used to solve *any* quadratic equation? Why or why not?

TW 69. Solve Example 2 graphically and compare with the algebraic solution. Which method is faster? Which method is more precise?

TW 70. Solve Example 4 graphically and compare with the algebraic solution. Which method is faster? Which method is more precise?

11.3 Applications Involving Quadratic Equations

Solving Problems   ∎   Solving Formulas

Solving Problems

As we found in Section 7.7, some problems translate to rational equations. The solution of such rational equations can involve quadratic equations.

EXAMPLE 1   Motorcycle Travel.   Madison rode her motorcycle 300 mi at a certain average speed. Had she averaged 10 mph more, the trip would have taken 1 hr less. Find the average speed of the motorcycle.

SOLUTION

1. **Familiarize.** We make a drawing, labeling it with the information provided. As in Section 6.5, we can create a table. We let r represent the rate, in miles per hour, and t the time, in hours, for Madison's trip.

Time t 300 miles   Speed r

Time $t - 1$ 300 miles   Speed $r + 10$

Distance	Speed	Time
300	r	t
300	$r + 10$	$t - 1$

$$r = \frac{300}{t}$$

$$r + 10 = \frac{300}{t - 1}$$

Recall that the definition of speed, $r = d/t$, relates the three quantities.

2. **Translate.** From the first two lines of the table, we obtain

$$r = \frac{300}{t} \quad \text{and} \quad r + 10 = \frac{300}{t - 1}.$$

3. Carry out. A system of equations has been formed. We substitute for r from the first equation into the second and solve the resulting equation:

$$\frac{300}{t} + 10 = \frac{300}{t-1}$$

Substituting $300/t$ for r

$$t(t-1) \cdot \left[\frac{300}{t} + 10\right] = t(t-1) \cdot \frac{300}{t-1}$$

Multiplying by the LCD

$$\not{t}(t-1) \cdot \frac{300}{\not{t}} + t(t-1) \cdot 10 = t\not{(t-1)} \cdot \frac{300}{\not{t-1}}$$

Using the distributive law and removing factors that equal 1: $\frac{t}{t}=1; \frac{t-1}{t-1}=1$

$$300(t-1) + 10(t^2 - t) = 300t$$
$$300t - 300 + 10t^2 - 10t = 300t$$
$$10t^2 - 10t - 300 = 0$$

Rewriting in standard form

$$t^2 - t - 30 = 0$$

Multiplying by $\frac{1}{10}$ or dividing by 10

$$(t-6)(t+5) = 0$$

Factoring

$$t = 6 \quad or \quad t = -5.$$

Principle of zero products

4. Check. As a partial check, we graph $y_1 = 300/x + 10$ and $y_2 = 300/(x-1)$. The graph at left shows that the solutions are -5 and 6.

Note that we have solved for t, not r as required. Since negative time has no meaning here, we disregard the -5 and use 6 hr to find r:

$$r = \frac{300 \text{ mi}}{6 \text{ hr}} = 50 \text{ mph}.$$

> **CAUTION!** Always make sure that you find the quantity asked for in the problem.

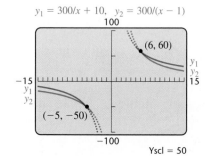

$y_1 = 300/x + 10, \quad y_2 = 300/(x-1)$

To see if 50 mph checks, we increase the speed 10 mph to 60 mph and see how long the trip would have taken at that speed:

$$t = \frac{d}{r} = \frac{300 \text{ mi}}{60 \text{ mph}} = 5 \text{ hr.}$$

Note that mi/mph $=$ mi $\div \frac{\text{mi}}{\text{hr}} =$ mi $\cdot \frac{\text{hr}}{\text{mi}} =$ hr.

This is 1 hr less than the trip actually took, so the answer checks.

5. State. Madison's motorcycle traveled at an average speed of 50 mph.

Solving Formulas

Recall that to solve a formula for a certain letter, we use the principles for solving equations to get that letter alone on one side.

EXAMPLE 2   Period of a Pendulum.   The time T required for a pendulum of length l to swing back and forth (complete one period) is given by the formula $T = 2\pi\sqrt{l/g}$, where g is the earth's gravitational constant. Solve for l.

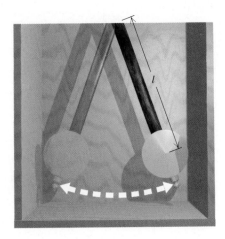

SOLUTION   We have

$$T = 2\pi\sqrt{\dfrac{l}{g}}$$ **This is a radical equation (see Section 10.6).**

$$T^2 = \left(2\pi\sqrt{\dfrac{l}{g}}\right)^2$$ **Principle of powers (squaring both sides)**

$$T^2 = 2^2\pi^2\dfrac{l}{g}$$

$$gT^2 = 4\pi^2 l$$ **Multiplying both sides by g to clear fractions**

$$\dfrac{gT^2}{4\pi^2} = l.$$ **Dividing both sides by $4\pi^2$**

We now have l alone on one side and l does not appear on the other side, so the formula is solved for l.

In formulas for which variables represent nonnegative numbers, there is no need for absolute-value signs when taking square roots.

EXAMPLE 3   Hang Time.*   An athlete's *hang time* is the amount of time that the athlete can remain airborne when jumping. A formula relating an athlete's vertical leap V, in inches, to hang time T, in seconds, is $V = 48T^2$. Solve for T.

---

*This formula is taken from an article by Peter Brancazio, "The Mechanics of a Slam Dunk," *Popular Mechanics*, November 1991. Courtesy of Professor Peter Brancazio, Brooklyn College.

SOLUTION

$$48T^2 = V$$

$$T^2 = \frac{V}{48} \qquad \text{Dividing by 48 to get } T^2 \text{ alone}$$

$$T = \frac{\sqrt{V}}{\sqrt{48}} \qquad \begin{array}{l}\text{Using the principle of square roots and the}\\ \text{quotient rule for radicals. We assume } V,\, T \geq 0.\end{array}$$

$$\left.\begin{array}{l} = \dfrac{\sqrt{V}}{\sqrt{16}\,\sqrt{3}} = \dfrac{\sqrt{V}}{4\sqrt{3}} \\[2ex] = \dfrac{\sqrt{V}}{4\sqrt{3}} \cdot \dfrac{\sqrt{3}}{\sqrt{3}} = \dfrac{\sqrt{3V}}{12} \end{array}\right\} \quad \text{Rationalizing the denominator}$$

EXAMPLE 4   Falling Distance.   An object tossed downward with an initial speed (velocity) of v_0 will travel a distance of s meters, where $s = 4.9t^2 + v_0 t$ and t is measured in seconds. Solve for t.

SOLUTION   Since t is squared in one term and raised to the first power in the other term, the equation is quadratic in t.

$$4.9t^2 + v_0 t = s$$

$$4.9t^2 + v_0 t - s = 0 \qquad \text{Writing standard form}$$

$$a = 4.9, \quad b = v_0, \quad c = -s$$

$$t = \frac{-v_0 \pm \sqrt{v_0^2 - 4(4.9)(-s)}}{2(4.9)} \qquad \text{Using the quadratic formula}$$

Since the negative square root would yield a negative value for t, we use only the positive root:

$$t = \frac{-v_0 + \sqrt{v_0^2 + 19.6s}}{9.8}.$$

The following list of steps should help you when solving formulas for a given letter. Try to remember that when solving a formula, you use the same approach that you would to solve an equation.

Student Notes

After identifying which numbers to use as *a*, *b*, and *c*, be careful to replace only the *letters* in the quadratic formula.

To Solve a Formula for a Letter—Say, *b*

1. Clear fractions and use the principle of powers, as needed. Perform these steps until radicals containing *b* are gone and *b* is not in any denominator.
2. Combine all like terms.
3. If the only power of *b* is b^1, the equation can be solved as in Sections 2.3 and 7.8.
4. If b^2 appears but *b* does not, solve for b^2 and use the principle of square roots to solve for *b*.
5. If there are terms containing both *b* and b^2, put the equation in standard form and use the quadratic formula.

11.3 EXERCISE SET

Solve.

1. *Car Trips.* During the first part of a trip, Raena's Honda traveled 120 mi at a certain speed. Raena then drove another 100 mi at a speed that was 10 mph slower. If the total time of Raena's trip was 4 hr, what was her speed on each part of the trip?

2. *Canoeing.* During the first part of a canoe trip, Alex covered 60 km at a certain speed. He then traveled 24 km at a speed that was 4 km/h slower. If the total time for the trip was 8 hr, what was the speed on each part of the trip?

3. *Car Trips.* Petra's Plymouth travels 200 mi averaging a certain speed. If the car had gone 10 mph faster, the trip would have taken 1 hr less. Find Petra's average speed.

4. *Car Trips.* Sandi's Subaru travels 280 mi averaging a certain speed. If the car had gone 5 mph faster, the trip would have taken 1 hr less. Find Sandi's average speed.

5. *Air Travel.* A Cessna flies 600 mi at a certain speed. A Beechcraft flies 1000 mi at a speed that is 50 mph faster, but takes 1 hr longer. Find the speed of each plane.

6. *Air Travel.* A turbo-jet flies 50 mph faster than a super-prop plane. If a turbo-jet goes 2000 mi in 3 hr less time than it takes the super-prop to go 2800 mi, find the speed of each plane.

7. *Bicycling.* Naoki bikes the 40 mi to Hillsboro averaging a certain speed. The return trip is made at a speed that is 6 mph slower. Total time for the round trip is 14 hr. Find Naoki's average speed on each part of the trip.

8. *Car Speed.* On a sales trip, Gail drives the 600 mi to Richmond averaging a certain speed. The return trip is made at an average speed that is 10 mph slower. Total time for the round trip is 22 hr. Find Gail's average speed on each part of the trip.

9. *Navigation.* The Hudson River flows at a rate of 3 mph. A patrol boat travels 60 mi upriver and returns in a total time of 9 hr. What is the speed of the boat in still water?

10. *Navigation.* The current in a typical Mississippi River shipping route flows at a rate of 4 mph. In order for a barge to travel 24 mi upriver and then return in a total of 5 hr, approximately how fast must the barge be able to travel in still water?

11. *Filling a Pool.* A well and a spring are filling a swimming pool. Together, they can fill the pool in 4 hr. The well, working alone, can fill the pool in 6 hr less time than the spring. How long would the spring take, working alone, to fill the pool?

12. *Filling a Tank.* Two pipes are connected to the same tank. Working together, they can fill the tank in 2 hr. The larger pipe, working alone, can fill the tank in 3 hr less time than the smaller one. How long would the smaller one take, working alone, to fill the tank?

13. *Paddleboats.* Ellen paddles 1 mi upstream and 1 mi back in a total time of 1 hr. The speed of the river is 2 mph. Find the speed of Ellen's paddleboat in still water.

14. *Rowing.* Dan rows 10 km upstream and 10 km back in a total time of 3 hr. The speed of the river is 5 km/h. Find Dan's speed in still water.

Solve each formula for the indicated letter. Assume that all variables represent nonnegative numbers.

15. $A = 4\pi r^2$, for r
(Surface area of a sphere of radius r)

16. $A = 6s^2$, for s
(Surface area of a cube with sides of length s)

17. $A = 2\pi r^2 + 2\pi rh$, for r
(Surface area of a right cylindrical solid with radius r and height h)

18. $F = \dfrac{Gm_1m_2}{r^2}$, for r
(Law of gravity)

19. $N = \dfrac{kQ_1Q_2}{s^2}$, for s
(Number of phone calls between two cities)

20. $A = \pi r^2$, for r
(Area of a circle)

21. $T = 2\pi\sqrt{\dfrac{l}{g}}$, for g
(A pendulum formula)

22. $a^2 + b^2 = c^2$, for b
(Pythagorean formula in two dimensions)

23. $a^2 + b^2 + c^2 = d^2$, for c
(Pythagorean formula in three dimensions)

24. $N = \dfrac{k^2 - 3k}{2}$, for k
(Number of diagonals of a polygon with k sides)

25. $s = v_0 t + \dfrac{gt^2}{2}$, for t
(A motion formula)

26. $A = \pi r^2 + \pi rs$, for r
(Surface area of a cone)

27. $N = \frac{1}{2}(n^2 - n)$, for n
(Number of games if n teams play each other once)

28. $A = A_0(1 - r)^2$, for r
(A business formula)

29. $V = 3.5\sqrt{h}$, for h
(Distance to horizon from a height)

30. $W = \sqrt{\dfrac{1}{LC}}$, for L
(An electricity formula)

Aha! 31. $at^2 + bt + c = 0$, for t
(An algebraic formula)

32. $A = P_1(1 + r)^2 + P_2(1 + r)$, for r
(Amount in an account when P_1 is invested for 2 yr and P_2 for 1 yr at interest rate r)

Solve.

33. *Falling Distance.* (Use $4.9t^2 + v_0 t = s$.)
   a) A bolt falls off an airplane at an altitude of 500 m. Approximately how long does it take the bolt to reach the ground?
   b) A ball is thrown downward at a speed of 30 m/sec from an altitude of 500 m. Approximately how long does it take the ball to reach the ground?
   c) Approximately how far will an object fall in 5 sec, when thrown downward at an initial velocity of 30 m/sec from a plane?

34. *Falling Distance.* (Use $4.9t^2 + v_0 t = s$.)
   a) A ring is dropped from a helicopter at an altitude of 75 m. Approximately how long does it take the ring to reach the ground?
   b) A coin is tossed downward with an initial velocity of 30 m/sec from an altitude of 75 m. Approximately how long does it take the coin to reach the ground?
   c) Approximately how far will an object fall in 2 sec, if thrown downward at an initial velocity of 20 m/sec from a helicopter?

35. *Bungee Jumping.*   Jesse is tied to one end of a 40-m elasticized (bungee) cord. The other end of the cord is tied to the middle of a bridge. If Jesse jumps off the bridge, for how long will he fall before the cord begins to stretch? (Use $4.9t^2 = s$.)

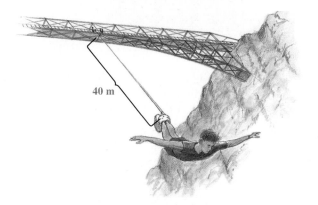

40 m

36. *Bungee Jumping.*   Sheila is tied to a bungee cord (see Exercise 35) and falls for 2.5 sec before her cord begins to stretch. How long is the bungee cord?

37. *Hang Time.*   The NBA's Steve Francis reportedly has a vertical leap of 45 in. (*Source*: www.maximonline.com). What is his hang time? (Use $V = 48T^2$.)

38. *League Schedules.*   In a bowling league, each team plays each of the other teams once. If a total of 66 games is played, how many teams are in the league? (See Exercise 27.)

For Exercises 39 and 40, use $4.9t^2 + v_0 t = s$.

39. *Downward Speed.*   An object thrown downward from a 100-m cliff travels 51.6 m in 3 sec. What was the initial velocity of the object?

40. *Downward Speed.*   An object thrown downward from a 200-m cliff travels 91.2 m in 4 sec. What was the initial velocity of the object?

For Exercises 41 and 42, use $A = P_1(1 + r)^2 + P_2(1 + r)$. (See Exercise 32.)

41. *Compound Interest.*   A firm invests $3000 in a savings account for 2 yr. At the beginning of the second year, an additional $1700 is invested. If a total of $5253.70 is in the account at the end of the second year, what is the annual interest rate?

42. *Compound Interest.*   A business invests $10,000 in a savings account for 2 yr. At the beginning of the second year, an additional $3500 is invested. If a total of $15,569.75 is in the account at the end of the second year, what is the annual interest rate?

43. Marti is tied to a bungee cord that is twice as long as the cord tied to Pedro. Will Marti's fall take twice as long as Pedro's before their cords begin to stretch? Why or why not? (See Exercises 35 and 36.)

44. Under what circumstances would a negative value for *t*, time, have meaning?

Focused Review

Solve.

45. *Credits.*   Tivon transferred to Oak College with 32 credits, and is taking 16 credits a semester. Alexis transferred with 38 credits, and is taking 14 credits a semester. Determine, in terms of an inequality, when Tivon will have more credits than Alexis. [2.7]

46. *Donuts.*   South Street Bakers charges $1.10 for a cream-filled donut and 85¢ for a glazed donut. On a recent Sunday, a total of 90 glazed and cream-filled donuts were sold for $88.00. How many of each type were sold? [4.4]

47. *Picture Frames.*   A rectangular picture frame measures 10 in. by 13 in., and 88 in² of picture shows. Find the width of the frame. [6.7]

48. *Fast Food.*   Jaime can process customer orders twice as fast as Cheri, a newly hired employee. Working together, they process one customer's order in 2 min. How long would it take each of them, working alone, to process the order? [7.7]

49. *File Download.*   Ethan has downloaded 254 kilobytes of a file. He has twice as much left to download. How big is the file he is downloading? [2.5]

50. *Commuting.*   Jennifer commutes 90 mi to her work in Washington D.C., averaging a certain speed. If she could average 15 mph faster, the trip would take $\frac{1}{2}$ hr less. Find Jennifer's average speed. [11.3]

Synthesis

TW 51. Write a problem for a classmate to solve. Devise the problem so that **(a)** the solution is found after solving a rational equation and **(b)** the solution is "The express train travels 90 mph."

TW 52. In what ways do the motion problems of this section (like Example 1) differ from the motion problems in Chapter 7 (see p. 587)?

53. *Biochemistry.*   The equation

$$A = 6.5 - \frac{20.4t}{t^2 + 36}$$

is used to calculate the acid level A in a person's blood t minutes after sugar is consumed. Solve for t.

54. *Special Relativity.*   Einstein found that an object with initial mass m_0 and traveling velocity v has mass

$$m = \frac{m_0}{\sqrt{1 - \dfrac{v^2}{c^2}}},$$

where c is the speed of light. Solve the formula for c.

55. Find a number for which the reciprocal of 1 less than the number is the same as 1 more than the number.

56. *Purchasing.*   A discount store bought a quantity of beach towels for \$250 and sold all but 15 at a profit of \$3.50 per towel. With the total amount received, the manager could buy 4 more than twice as many as were bought before. Find the cost per towel.

57. *Art and Aesthetics.*   For over 2000 yr, artists, sculptors, and architects have regarded the proportions of a "golden" rectangle as visually appealing. A rectangle of width w and length l is considered "golden" if

$$\frac{w}{l} = \frac{l}{w + l}.$$

Solve for l.

58. *Diagonal of a Cube.*   Find a formula that expresses the length of the three-dimensional diagonal of a cube as a function of the cube's surface area.

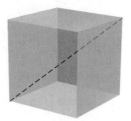

59. Solve for n:

$$mn^4 - r^2pm^3 - r^2n^2 + p = 0.$$

60. *Surface Area.*   Find a formula that expresses the diameter of a right cylindrical solid as a function of its surface area and its height. (See Exercise 17.)

61. A sphere is inscribed in a cube as shown in the figure below. Express the surface area of the sphere as a function of the surface area S of the cube. (See Exercise 15.)

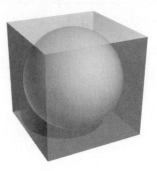

11.4 Studying Solutions of Quadratic Equations

The Discriminant ■ Writing Equations from Solutions

The Discriminant

It is sometimes enough to know what *type* of number a solution will be, without actually solving the equation. Suppose we want to know if $4x^2 + 7x - 15 = 0$ has rational solutions (and thus can be solved by factoring). Using the quadratic formula, we would have

$$x = \frac{-b \pm \sqrt{b^2 - 4ac}}{2a}$$
$$= \frac{-7 \pm \sqrt{7^2 - 4 \cdot 4 \cdot (-15)}}{2 \cdot 4}.$$

Note that the radicand, $7^2 - 4 \cdot 4 \cdot (-15)$, determines what type of number the solutions will be. Since $7^2 - 4 \cdot 4 \cdot (-15) = 49 - 16(-15) = 289$, and since 289 is a perfect square $\left(\sqrt{289} = 17\right)$, we know that the solutions of the equation will be two rational numbers. This means that $4x^2 + 7x - 15 = 0$ *can* be solved by factoring.

It is $b^2 - 4ac$, known as the **discriminant,** that determines what type of number the solutions of a quadratic equation are. If a, b, and c are rational, then:

- When $b^2 - 4ac$ simplifies to 0, it doesn't matter if we use $+\sqrt{b^2 - 4ac}$ or $-\sqrt{b^2 - 4ac}$; we get the same solution twice. Thus, when the discriminant is 0, there is one *repeated* solution and it is rational.

  Example: $9x^2 + 6x + 1 = 0 \rightarrow b^2 - 4ac = 6^2 - 4 \cdot 9 \cdot 1 = 0.$

- When $b^2 - 4ac$ is positive, there are two different real-number solutions: If $b^2 - 4ac$ is a perfect square, these solutions are rational numbers.

  Example: $6x^2 + 5x + 1 = 0 \rightarrow b^2 - 4ac = 5^2 - 4 \cdot 6 \cdot 1 = 1.$

- When $b^2 - 4ac$ is positive, but not a perfect square, there are two irrational solutions and they are conjugates of each other (see p. 570).

  Example: $2x^2 + 4x + 1 = 0 \rightarrow b^2 - 4ac = 4^2 - 4 \cdot 2 \cdot 1 = 8.$

- When $b^2 - 4ac$ is negative, there are two imaginary-number solutions and they are complex conjugates of each other.

  Example: $3x^2 + 2x + 1 = 0 \rightarrow b^2 - 4ac = 2^2 - 4 \cdot 3 \cdot 1 = -8.$

Note that any equation for which $b^2 - 4ac$ is a perfect square can be solved by factoring.

Study Tip

When Something Seems Easy

Every so often, you may encounter a lesson that you remember from a previous math course. When this occurs, make sure that *all* of that lesson is understood and review any tricky spots.

Discriminant $b^2 - 4ac$	Nature of Solutions
0	One solution; a rational number

Positive
   Perfect square
   Not a perfect square
	Two different real-number solutions
	   Solutions are rational.
	   Solutions are irrational conjugates.

Negative	Two different imaginary-number solutions (complex conjugates)

Note that all quadratic equations have one or two solutions. They can always be found algebraically; only real-number solutions can be found graphically.

▶ **EXAMPLE 1** For each equation, determine what type of number the solutions are and how many solutions exist.

a) $9x^2 - 12x + 4 = 0$ **b)** $x^2 + 5x + 8 = 0$

c) $2x^2 + 7x - 3 = 0$

SOLUTION

a) For $9x^2 - 12x + 4 = 0$, we have

$$a = 9, \quad b = -12, \quad c = 4.$$

We substitute and compute the discriminant:

$$b^2 - 4ac = (-12)^2 - 4 \cdot 9 \cdot 4$$
$$= 144 - 144 = 0.$$

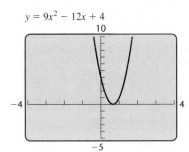

$y = 9x^2 - 12x + 4$

There is exactly one solution, and it is rational. This indicates that $9x^2 - 12x + 4 = 0$ can be solved by factoring. The graph at left confirms that $9x^2 - 12x + 4 = 0$ has just one solution.

b) For $x^2 + 5x + 8 = 0$, we have

$$a = 1, \quad b = 5, \quad c = 8.$$

We substitute and compute the discriminant:

$$b^2 - 4ac = 5^2 - 4 \cdot 1 \cdot 8$$
$$= 25 - 32 = -7.$$

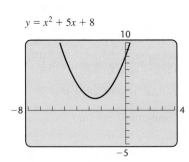

$y = x^2 + 5x + 8$

Since the discriminant is negative, there are two imaginary-number solutions that are complex conjugates of each other. As the graph at left shows, there are no x-intercepts of the graph of $y = x^2 + 5x + 8$, and thus no real solutions of the equation $x^2 + 5x + 8 = 0$.

c) For $2x^2 + 7x - 3 = 0$, we have

$$a = 2, \quad b = 7, \quad c = -3;$$
$$b^2 - 4ac = 7^2 - 4 \cdot 2(-3)$$
$$= 49 - (-24) = 73.$$

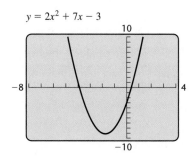

$y = 2x^2 + 7x - 3$

The discriminant is a positive number that is not a perfect square. Thus there are two irrational solutions that are conjugates of each other. From the graph at left, we see that there are two solutions of the equation. We cannot tell from simply observing the graph whether the solutions are rational or irrational.

As we saw in Example 1, discriminants can be used to determine the number of real-number solutions of $ax^2 + bx + c = 0$. This can be used as an aid in graphing.

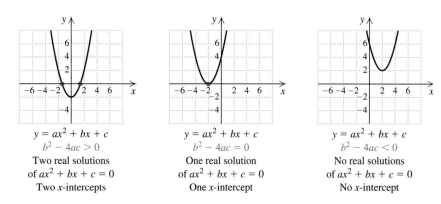

$y = ax^2 + bx + c$	$y = ax^2 + bx + c$	$y = ax^2 + bx + c$
$b^2 - 4ac > 0$	$b^2 - 4ac = 0$	$b^2 - 4ac < 0$
Two real solutions	One real solution	No real solutions
of $ax^2 + bx + c = 0$	of $ax^2 + bx + c = 0$	of $ax^2 + bx + c = 0$
Two x-intercepts	One x-intercept	No x-intercept

Writing Equations from Solutions

We know by the principle of zero products that $(x - 2)(x + 3) = 0$ has solutions 2 and -3. If we know the solutions of an equation, we can write an equation, using the principle in reverse.

> **EXAMPLE 2**   Find an equation for which the given numbers are solutions.
>
> **a)** 3 and $-\frac{2}{5}$ **b)** $2i$ and $-2i$
>
> **c)** $5\sqrt{7}$ and $-5\sqrt{7}$ **d)** $-4, 0,$ and 1
>
> **SOLUTION**
>
> **a)** $x = 3$ *or* $x = -\frac{2}{5}$
>
> $x - 3 = 0$ *or* $x + \frac{2}{5} = 0$ **Getting 0's on one side**
>
> $(x - 3)\left(x + \frac{2}{5}\right) = 0$ **Using the principle of zero products (multiplying)**
>
> $x^2 + \frac{2}{5}x - 3x - 3 \cdot \frac{2}{5} = 0$ **Multiplying**
>
> $x^2 - \frac{13}{5}x - \frac{6}{5} = 0$ **Combining like terms**
>
> $5x^2 - 13x - 6 = 0$ **Multiplying both sides by 5 to clear fractions**

Note that multiplying both sides by the LCD, 5, clears the equations of fractions. Had we preferred, we could have multiplied $x + \frac{2}{5} = 0$ by 5, thus clearing fractions *before* using the principle of zero products.

b)
$$x = 2i \quad or \quad x = -2i$$
$$x - 2i = 0 \quad or \quad x + 2i = 0 \qquad \text{Getting 0's on one side}$$
$$(x - 2i)(x + 2i) = 0 \qquad \text{Using the principle of zero products (multiplying)}$$
$$x^2 - (2i)^2 = 0 \qquad \text{Finding the product of a sum and difference}$$
$$x^2 - 4i^2 = 0$$
$$x^2 + 4 = 0 \qquad i^2 = -1$$

c)
$$x = 5\sqrt{7} \quad or \quad x = -5\sqrt{7}$$
$$x - 5\sqrt{7} = 0 \quad or \quad x + 5\sqrt{7} = 0 \qquad \text{Getting 0's on one side}$$
$$\left(x - 5\sqrt{7}\right)\left(x + 5\sqrt{7}\right) = 0 \qquad \text{Using the principle of zero products}$$
$$x^2 - \left(5\sqrt{7}\right)^2 = 0 \qquad \text{Finding the product of a sum and difference}$$
$$x^2 - 25 \cdot 7 = 0$$
$$x^2 - 175 = 0$$

d)
$$x = -4 \quad or \quad x = 0 \quad or \quad x = 1$$
$$x + 4 = 0 \quad or \quad x = 0 \quad or \quad x - 1 = 0 \qquad \text{Getting 0's on one side}$$
$$(x + 4)x(x - 1) = 0 \qquad \text{Using the principle of zero products}$$
$$x(x^2 + 3x - 4) = 0 \qquad \text{Multiplying}$$
$$x^3 + 3x^2 - 4x = 0$$

11.4 EXERCISE SET

FOR EXTRA HELP

MathXP   MathXL   MyMathLab   InterAct Math   AW Math Tutor Center   Video Lectures on CD: Disc 6   Student's Solutions Manual

Concept Reinforcement   *Complete each of the following.*

1. In the quadratic formula, the expression $b^2 - 4ac$ is called the _____.

2. When $b^2 - 4ac$ is 0, there is/are _____ solution(s).

3. When $b^2 - 4ac$ is positive, there is/are _____ solution(s).

4. When $b^2 - 4ac$ is negative, there is/are _____ solution(s).

5. When $b^2 - 4ac$ is a perfect square, the answers are _____ numbers.

6. When $b^2 - 4ac$ is negative, the answer(s) is/are _____ numbers.

For each equation, determine what type of number the solutions are and how many solutions exist.

7. $x^2 - 7x + 5 = 0$ **8.** $x^2 - 5x + 3 = 0$

9. $x^2 + 3 = 0$ **10.** $x^2 + 5 = 0$

11. $x^2 - 5 = 0$ **12.** $x^2 - 3 = 0$

13. $4x^2 + 8x - 5 = 0$

14. $4x^2 - 12x + 9 = 0$

15. $x^2 + 4x + 6 = 0$

16. $x^2 - 2x + 4 = 0$

17. $9t^2 - 48t + 64 = 0$

18. $6t^2 - 19t - 20 = 0$

19. $10x^2 - x - 2 = 0$

20. $6x^2 + 5x - 4 = 0$

Aha! **21.** $9t^2 - 3t = 0$

22. $4m^2 + 7m = 0$

23. $x^2 + 4x = 8$

24. $x^2 + 5x = 9$

25. $2a^2 - 3a = -5$

26. $3a^2 + 5 = 7a$

27. $y^2 + \frac{9}{4} = 4y$

28. $x^2 = \frac{1}{2}x - \frac{3}{5}$

Write a quadratic equation having the given numbers as solutions.

29. $-7, 3$

30. $-6, 4$

31. 3, only solution
(*Hint*: It must be a repeated solution.)

32. -5, only solution

33. $-1, -3$

34. $-2, -5$

35. $5, \frac{3}{4}$

36. $4, \frac{2}{3}$

37. $-\frac{1}{4}, -\frac{1}{2}$

38. $\frac{1}{2}, \frac{1}{3}$

39. $2.4, -0.4$

40. $-0.6, 1.4$

41. $-\sqrt{3}, \sqrt{3}$

42. $-\sqrt{7}, \sqrt{7}$

43. $2\sqrt{5}, -2\sqrt{5}$

44. $3\sqrt{2}, -3\sqrt{2}$

45. $4i, -4i$

46. $3i, -3i$

47. $2 - 7i, 2 + 7i$

48. $5 - 2i, 5 + 2i$

49. $3 - \sqrt{14}, 3 + \sqrt{14}$

50. $2 - \sqrt{10}, 2 + \sqrt{10}$

51. $1 - \dfrac{\sqrt{21}}{3}, 1 + \dfrac{\sqrt{21}}{3}$

52. $\dfrac{5}{4} - \dfrac{\sqrt{33}}{4}, \dfrac{5}{4} + \dfrac{\sqrt{33}}{4}$

Write a third-degree equation having the given numbers as solutions.

53. $-2, 1, 5$

54. $-5, 0, 2$

55. $-1, 0, 3$

56. $-2, 2, 3$

TW **57.** Under what condition(s) is the discriminant *not* the fastest way to determine how many and what type of solutions exist?

TW **58.** While solving a quadratic equation of the form $ax^2 + bx + c = 0$ with a graphing calculator, Shawn-Marie gets the following screen. How

could the sign of the discriminant help her check the graph?

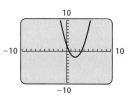

Skill Maintenance

Simplify. [5.1]

59. $(3a^2)^4$

60. $(4x^3)^2$

Find the x-intercepts of the graph of f. [6.2]

61. $f(x) = x^2 - 7x - 8$

62. $f(x) = x^2 - 6x + 8$

63. During a one-hour television show, there were 12 commercials. Some of the commercials were 30 sec long and the others were 60 sec long. The amount of time for 30-sec commercials was 6 min less than the total number of minutes of commercial time during the show. How many 30-sec commercials were used? [4.4]

64. Graph: $y = -\frac{3}{7}x + 4$. [3.6]

Synthesis

TW **65.** If we assume that a quadratic equation has integers for coefficients, will the product of the solutions always be a real number? Why or why not?

TW **66.** Can a fourth-degree equation have exactly three irrational solutions? Why or why not?

67. The graph of an equation of the form

$$y = ax^2 + bx + c$$

is a curve similar to the one shown below. Determine *a*, *b*, and *c* from the information given.

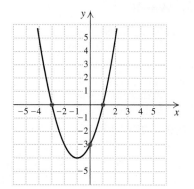

68. Show that the product of the solutions of $ax^2 + bx + c = 0$ is c/a.

*For each equation under the given condition, **(a)** find k and **(b)** find the other solution.*

69. $kx^2 - 2x + k = 0$; one solution is -3

70. $x^2 - kx + 2 = 0$; one solution is $1 + i$

71. $x^2 - (6 + 3i)x + k = 0$; one solution is 3

72. Show that the sum of the solutions of $ax^2 + bx + c = 0$ is $-b/a$.

73. Show that whenever there is just one solution of $ax^2 + bx + c = 0$, that solution is of the form $-b/(2a)$.

74. Find h and k, where $3x^2 - hx + 4k = 0$, the sum of the solutions is -12, and the product of the solutions is 20. (*Hint*: See Exercises 68 and 72.)

75. Suppose that $f(x) = ax^2 + bx + c$, with $f(-3) = 0, f\left(\frac{1}{2}\right) = 0$, and $f(0) = -12$. Find a, b, and c.

76. Find an equation for which $2 - \sqrt{3}, 2 + \sqrt{3}$, $5 - 2i$, and $5 + 2i$ are solutions.

Aha! **77.** Find an equation with integer coefficients for which $1 - \sqrt{5}$ and $3 + 2i$ are two of the solutions.

TW **78.** A discriminant that is a perfect square indicates that factoring can be used to solve the quadratic equation. Why?

11.5 Equations Reducible to Quadratic

Recognizing Equations in Quadratic Form ◾
Radical and Rational Equations

Recognizing Equations in Quadratic Form

Certain equations that are not really quadratic can be thought of in such a way that they can be solved as quadratic. For example, because the square of x^2 is x^4, the equation $x^4 - 9x^2 + 8 = 0$ is said to be "quadratic in x^2":

$$x^4 \;-\; 9x^2 \;+\; 8 = 0$$

$$(x^2)^2 - 9(x^2) + 8 = 0 \qquad \text{Thinking of } x^4 \text{ as } (x^2)^2$$

$$u^2 \;-\; 9u \;+\; 8 = 0. \qquad \text{To make this clearer, write } u \text{ instead of } x^2.$$

The equation $u^2 - 9u + 8 = 0$ can be solved by factoring or by the quadratic formula. Then, remembering that $u = x^2$, we can solve for x. Equations that can be solved like this are *reducible to quadratic*, or *in quadratic form*.

EXAMPLE 1   Solve: $x^4 - 9x^2 + 8 = 0$.

SOLUTION   We solve both algebraically and graphically.

ALGEBRAIC APPROACH

Let $u = x^2$. Then we solve by substituting u for x^2 and u^2 for x^4:

$$u^2 - 9u + 8 = 0$$
$$(u - 8)(u - 1) = 0 \qquad \text{Factoring}$$
$$u - 8 = 0 \quad or \quad u - 1 = 0 \qquad \text{Principle of zero products}$$
$$u = 8 \quad or \qquad u = 1.$$

We replace u with x^2 and solve these equations:

$$x^2 = 8 \qquad or \quad x^2 = 1$$
$$x = \pm\sqrt{8} \quad or \quad x = \pm 1$$
$$x = \pm 2\sqrt{2} \quad or \quad x = \pm 1.$$

To check, note that for both $x = 2\sqrt{2}$ and $-2\sqrt{2}$, we have $x^2 = 8$ and $x^4 = 64$. Similarly, for both $x = 1$ and -1, we have $x^2 = 1$ and $x^4 = 1$. Thus instead of making four checks, we need make only two.

Check:

For $\pm 2\sqrt{2}$:

$$\frac{x^4 - 9x^2 + 8 = 0}{\left(\pm 2\sqrt{2}\right)^4 - 9\left(\pm 2\sqrt{2}\right)^2 + 8 \;\Big|\; 0}$$
$$64 - 9 \cdot 8 + 8$$
$$0 \overset{?}{=} 0 \quad \text{TRUE}$$

For ± 1:

$$\frac{x^4 - 9x^2 + 8 = 0}{(\pm 1)^4 - 9(\pm 1)^2 + 8 \;\Big|\; 0}$$
$$1 - 9 + 8$$
$$0 \overset{?}{=} 0 \quad \text{TRUE}$$

The solutions are 1, -1, $2\sqrt{2}$, and $-2\sqrt{2}$.

GRAPHICAL APPROACH

We let $f(x) = x^4 - 9x^2 + 8$ and determine the zeros of the function.

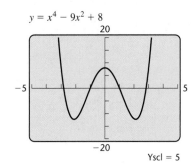

It appears as though the graph has 4 x-intercepts. Recall from Chapter 6 that a fourth-degree polynomial function has at most 4 zeros, so we know that we have not missed any zeros. Using ZERO four times, we obtain the solutions -2.8284271, -1, 1, and 2.8284271. The solutions are ± 1 and, approximately, ± 2.8284271. Note that since $2\sqrt{2} \approx 2.8284271$, the solutions found graphically are the same as those found algebraically.

Student Notes

To recognize an equation in quadratic form, note that there will be two terms with variables. The exponent of the first term is twice the exponent of the second term. It is the variable expression in the *second* term that is written as u.

Example 1 can be solved directly by factoring:

$$x^4 - 9x^2 + 8 = 0$$
$$(x^2 - 1)(x^2 - 8) = 0$$
$$x^2 - 1 = 0 \quad or \quad x^2 - 8 = 0$$
$$x^2 = 1 \qquad or \qquad x^2 = 8$$
$$x = \pm 1 \quad or \qquad x = \pm 2\sqrt{2}.$$

There is nothing wrong with this approach. However, in the examples that follow, you will note that it becomes increasingly difficult to solve the equation without first making a substitution.

Radical and Rational Equations

Sometimes rational equations, radical equations, or equations containing exponents that are fractions are reducible to quadratic. It is especially important that answers to these equations be checked in the original equation.

EXAMPLE 2 Solve: $x - 3\sqrt{x} - 4 = 0$.

SOLUTION This radical equation could be solved using the method discussed in Section 10.6. However, if we note that the square of $\sqrt{x}$ is x, we can regard the equation as "quadratic in $\sqrt{x}$."

We let $u = \sqrt{x}$ and consequently $u^2 = x$:

$$x - 3\sqrt{x} - 4 = 0$$
$$u^2 - 3u - 4 = 0 \qquad \text{Substituting}$$
$$(u - 4)(u + 1) = 0$$
$$u = 4 \quad or \quad u = -1. \qquad \text{Using the principle of zero products}$$

> **CAUTION!** A common error is to solve for u but forget to solve for x. Remember to solve for the *original* variable!

Next, we replace u with $\sqrt{x}$ and solve these equations:

$$\sqrt{x} = 4 \quad or \quad \sqrt{x} = -1.$$

Squaring gives us $x = 16$ or $x = 1$ and also makes checking essential.

Check:

For 16:

$$\frac{x - 3\sqrt{x} - 4 = 0}{16 - 3\sqrt{16} - 4 \mid 0}$$
$$16 - 3 \cdot 4 - 4 \mid$$
$$0 \overset{?}{=} 0 \quad \text{TRUE}$$

For 1:

$$\frac{x - 3\sqrt{x} - 4 = 0}{1 - 3\sqrt{1} - 4 \mid 0}$$
$$1 - 3 \cdot 1 - 4 \mid$$
$$-6 \overset{?}{=} 0 \quad \text{FALSE}$$

The number 16 checks, but 1 does not. Had we noticed that $\sqrt{x} = -1$ has no solution (since principal roots are never negative), we could have solved only the equation $\sqrt{x} = 4$. The graph at left provides further evidence that although 16 is a solution, -1 is not. The solution is 16. ◢

$y = x - 3\sqrt{(x)} - 4$

Xscl = 5

EXAMPLE 3 Find the x-intercepts of the graph of

$$f(x) = (x^2 - 1)^2 - (x^2 - 1) - 2.$$

SOLUTION The x-intercepts occur where $f(x) = 0$ so we must have

$$(x^2 - 1)^2 - (x^2 - 1) - 2 = 0. \qquad \text{Setting } f(x) \text{ equal to 0}$$

The equation is quadratic in $x^2 - 1$, so we let $u = x^2 - 1$ and $u^2 = (x^2 - 1)^2$:

$$u^2 - u - 2 = 0 \qquad \text{Substituting in } (x^2 - 1)^2 - (x^2 - 1) - 2 = 0$$
$$(u - 2)(u + 1) = 0$$
$$u = 2 \quad or \quad u = -1. \qquad \text{Using the principle of zero products}$$

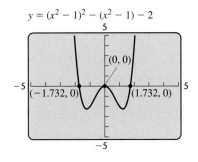

$$y = (x^2 - 1)^2 - (x^2 - 1) - 2$$

Next, we replace u with $x^2 - 1$ and solve these equations:

$$x^2 - 1 = 2 \quad or \quad x^2 - 1 = -1$$
$$x^2 = 3 \quad or \quad x^2 = 0 \qquad \text{Adding 1 to both sides}$$
$$x = \pm\sqrt{3} \quad or \quad x = 0. \qquad \text{Using the principle of square roots}$$

The x-intercepts occur at $\left(-\sqrt{3}, 0\right)$, $(0, 0)$, and $\left(\sqrt{3}, 0\right)$. These solutions are confirmed by the graph at left.

Sometimes great care must be taken in deciding what substitution to make.

EXAMPLE 4   Solve: $m^{-2} - 6m^{-1} + 4 = 0$.

SOLUTION   Note that the square of m^{-1} is $(m^{-1})^2$, or m^{-2}. This allows us to regard the equation as quadratic in m^{-1}.
We let $u = m^{-1}$ and $u^2 = m^{-2}$:

$$u^2 - 6u + 4 = 0 \qquad \text{Substituting}$$
$$u = \frac{-(-6) \pm \sqrt{(-6)^2 - 4 \cdot 1 \cdot 4}}{2 \cdot 1} \qquad \text{Using the quadratic formula}$$
$$\left. \begin{array}{l} u = \dfrac{6 \pm \sqrt{20}}{2} = \dfrac{2 \cdot 3 \pm 2\sqrt{5}}{2} \\[2mm] u = 3 \pm \sqrt{5}. \end{array} \right\} \qquad \text{Simplifying}$$

Next, we replace u with m^{-1} and solve:

$$m^{-1} = 3 \pm \sqrt{5}$$
$$\frac{1}{m} = 3 \pm \sqrt{5} \qquad \text{Recall that } m^{-1} = \frac{1}{m}.$$
$$1 = m\left(3 \pm \sqrt{5}\right) \qquad \text{Multiplying both sides by } m$$
$$\frac{1}{3 \pm \sqrt{5}} = m. \qquad \text{Dividing both sides by } 3 \pm \sqrt{5}$$

We can check both solutions as follows.

Check:

For $1/\left(3 - \sqrt{5}\right)$:

$$\begin{array}{r|l} m^{-2} - 6m^{-1} + 4 = 0 & \\ \hline \left(\dfrac{1}{3 - \sqrt{5}}\right)^{-2} - 6\left(\dfrac{1}{3 - \sqrt{5}}\right)^{-1} + 4 & 0 \\ \left(3 - \sqrt{5}\right)^2 - 6\left(3 - \sqrt{5}\right) + 4 & \\ 9 - 6\sqrt{5} + 5 - 18 + 6\sqrt{5} + 4 & \\ & 0 \overset{?}{=} 0 \quad \text{TRUE} \end{array}$$

For $1/\left(3 + \sqrt{5}\right)$:

$$\begin{array}{r|l} m^{-2} - 6m^{-1} + 4 = 0 & \\ \hline \left(\dfrac{1}{3 + \sqrt{5}}\right)^{-2} - 6\left(\dfrac{1}{3 + \sqrt{5}}\right)^{-1} + 4 & 0 \\ \left(3 + \sqrt{5}\right)^2 - 6\left(3 + \sqrt{5}\right) + 4 & \\ 9 + 6\sqrt{5} + 5 - 18 - 6\sqrt{5} + 4 & \\ & 0 \overset{?}{=} 0 \quad \text{TRUE} \end{array}$$

Student Notes

When it is difficult to decide what substitution to make, try letting u represent one variable expression in the equation. If u^2 is the other variable expression, then you have found an appropriate substitution.

In Example 4, you might try two substitutions:

$$u = m^{-2} \Rightarrow u^2 = m^{-4};$$
$$u = m^{-1} \Rightarrow u^2 = m^{-2}.$$

Only for $u = m^{-1}$ are both u and u^2 in the equation.

We could also check the solutions using a calculator by entering $y_1 = x^{-2} - 6x^{-1} + 4$ and evaluating y_1 for both numbers, as shown on the left below.

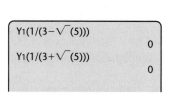

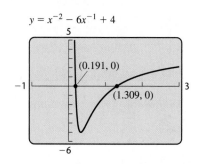

As another check, we solve the equation graphically. The first coordinates of the x-intercepts of the graph on the right above are approximate solutions.

Both numbers check. The solutions are $1/(3 - \sqrt{5})$ and $1/(3 + \sqrt{5})$, or approximately 1.309 and 0.191.

EXAMPLE 5 Solve: $t^{2/5} - t^{1/5} - 2 = 0$.

SOLUTION Note that the square of $t^{1/5}$ is $(t^{1/5})^2$, or $t^{2/5}$. The equation is therefore quadratic in $t^{1/5}$, so we let $u = t^{1/5}$ and $u^2 = t^{2/5}$:

$$u^2 - u - 2 = 0 \qquad \text{Substituting}$$
$$(u - 2)(u + 1) = 0$$
$$u = 2 \quad or \quad u = -1. \qquad \text{Using the principle of zero products}$$

Now we replace u with $t^{1/5}$ and solve:

$$t^{1/5} = 2 \quad or \quad t^{1/5} = -1$$
$$t = 32 \quad or \quad t = -1. \qquad \text{Principle of powers; raising to the 5th power}$$

Check:

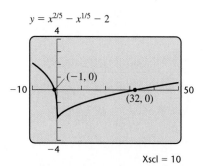

For 32:

$$
\begin{array}{c|c}
t^{2/5} - t^{1/5} - 2 = 0 & \\
\hline
32^{2/5} - 32^{1/5} - 2 & 0 \\
(32^{1/5})^2 - 32^{1/5} - 2 & \\
2^2 - 2 - 2 & \\
& 0 \stackrel{?}{=} 0 \quad \text{TRUE}
\end{array}
$$

For -1:

$$
\begin{array}{c|c}
t^{2/5} - t^{1/5} - 2 = 0 & \\
\hline
(-1)^{2/5} - (-1)^{1/5} - 2 & 0 \\
[(-1)^{1/5}]^2 - (-1)^{1/5} - 2 & \\
(-1)^2 - (-1) - 2 & \\
& 0 \stackrel{?}{=} 0 \quad \text{TRUE}
\end{array}
$$

Both numbers check and are confirmed by the graph at left. The solutions are 32 and -1.

The following tips may prove useful.

> ### To Solve an Equation That Is Reducible to Quadratic
>
> 1. The equation is quadratic in form if the variable factor in one term is the square of the variable factor in the other variable term.
> 2. Write down any substitutions that you are making.
> 3. Whenever you make a substitution, be sure to solve for the variable that is used in the original equation.
> 4. Check possible answers in the original equation.

11.5 EXERCISE SET

 Concept Reinforcement *In each of Exercises 1–8, match the equation with an appropriate substitution from the column on the right that could be used to reduce the equation to quadratic form.*

1. ____ $4x^6 - 2x^3 + 1 = 0$ **a)** $u = x^{-1/3}$

2. ____ $3x^4 + 4x^2 - 7 = 0$ **b)** $u = x^{1/3}$

3. ____ $5x^8 + 2x^4 - 3 = 0$ **c)** $u = x^{-2}$

4. ____ $2x^{2/3} - 5x^{1/3} + 4 = 0$ **d)** $u = x^2$

5. ____ $3x^{4/3} + 4x^{2/3} - 7 = 0$ **e)** $u = x^{-2/3}$

6. ____ $2x^{-2/3} + x^{-1/3} + 6 = 0$ **f)** $u = x^3$

7. ____ $4x^{-4/3} - 2x^{-2/3} + 3 = 0$ **g)** $u = x^{2/3}$

8. ____ $3x^{-4} + 4x^{-2} - 2 = 0$ **h)** $u = x^4$

Solve.

9. $x^4 - 5x^2 + 4 = 0$

10. $x^4 - 10x^2 + 9 = 0$

11. $x^4 - 9x^2 + 20 = 0$

12. $x^4 - 12x^2 + 27 = 0$

13. $4t^4 - 19t^2 + 12 = 0$

14. $9t^4 - 14t^2 + 5 = 0$

15. $r - 2\sqrt{r} - 6 = 0$

16. $s - 4\sqrt{s} - 1 = 0$

17. $(x^2 - 7)^2 - 3(x^2 - 7) + 2 = 0$

18. $(x^2 - 1)^2 - 5(x^2 - 1) + 6 = 0$

19. $(1 + \sqrt{x})^2 + 5(1 + \sqrt{x}) + 6 = 0$

20. $(3 + \sqrt{x})^2 + 3(3 + \sqrt{x}) - 10 = 0$

21. $x^{-2} - x^{-1} - 6 = 0$

22. $2x^{-2} - x^{-1} - 1 = 0$

23. $4x^{-2} + x^{-1} - 5 = 0$

24. $m^{-2} + 9m^{-1} - 10 = 0$

25. $t^{2/3} + t^{1/3} - 6 = 0$

26. $w^{2/3} - 2w^{1/3} - 8 = 0$

27. $y^{1/3} - y^{1/6} - 6 = 0$

28. $t^{1/2} + 3t^{1/4} + 2 = 0$

29. $t^{1/3} + 2t^{1/6} = 3$

30. $m^{1/2} + 6 = 5m^{1/4}$

31. $(3 - \sqrt{x})^2 - 10(3 - \sqrt{x}) + 23 = 0$

32. $(5 + \sqrt{x})^2 - 12(5 + \sqrt{x}) + 33 = 0$

33. $16\left(\dfrac{x-1}{x-8}\right)^2 + 8\left(\dfrac{x-1}{x-8}\right) + 1 = 0$

34. $9\left(\dfrac{x+2}{x+3}\right)^2 - 6\left(\dfrac{x+2}{x+3}\right) + 1 = 0$

Find all x-intercepts of the given function f. If none exists, state this.

35. $f(x) = 5x + 13\sqrt{x} - 6$

36. $f(x) = 3x + 10\sqrt{x} - 8$

37. $f(x) = (x^2 - 3x)^2 - 10(x^2 - 3x) + 24$

38. $f(x) = (x^2 - 6x)^2 - 2(x^2 - 6x) - 35$

39. $f(x) = x^{2/5} + x^{1/5} - 6$

40. $f(x) = x^{1/2} - x^{1/4} - 6$

Aha! **41.** $f(x) = \left(\dfrac{x^2+2}{x}\right)^4 + 7\left(\dfrac{x^2+2}{x}\right)^2 + 5$

42. $f(x) = \left(\dfrac{x^2+1}{x}\right)^4 + 4\left(\dfrac{x^2+1}{x}\right)^2 + 12$

TW **43.** To solve $25x^6 - 10x^3 + 1 = 0$, Don lets $u = 5x^3$ and Robin lets $u = x^3$. Can they both be correct? Why or why not?

TW **44.** While trying to solve $0.05x^4 - 0.8 = 0$ with a graphing calculator, Murray gets the following screen. Can Murray solve this equation with a graphing calculator? Why or why not?

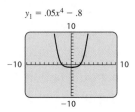

$y_1 = .05x^4 - .8$

Focused Review

Solve.

45. $x^2 + x = 6$ [6.2]

46. $2x^2 - 3x + 4 = 0$ [11.2]

47. $(x - 7)^2 = 5$ [11.1]

48. $4x^2 + x = 6$ [11.2]

49. $x^4 - 5x^2 + 6$ [11.5]

50. $2(3 + \sqrt{x})^2 + 5(3 + \sqrt{x}) + 2 = 0$ [11.5]

Synthesis

TW **51.** Describe a procedure that could be used to solve any equation of the form $ax^4 + bx^2 + c = 0$.

TW **52.** Describe a procedure that could be used to write an equation that is quadratic in $3x^2 - 1$. Then explain how the procedure could be adjusted to write equations that are quadratic in $3x^2 - 1$ and have no real-number solution.

Solve.

53. $5x^4 - 7x^2 + 1 = 0$

54. $3x^4 + 5x^2 - 1 = 0$

55. $(x^2 - 4x - 2)^2 - 13(x^2 - 4x - 2) + 30 = 0$

56. $(x^2 - 5x - 1)^2 - 18(x^2 - 5x - 1) + 65 = 0$

57. $\dfrac{x}{x-1} - 6\sqrt{\dfrac{x}{x-1}} - 40 = 0$

58. $\left(\sqrt{\dfrac{x}{x-3}}\right)^2 - 24 = 10\sqrt{\dfrac{x}{x-3}}$

59. $a^5(a^2 - 25) + 13a^3(25 - a^2) + 36a(a^2 - 25) = 0$

60. $a^3 - 26a^{3/2} - 27 = 0$

61. $x^6 - 28x^3 + 27 = 0$

62. $x^6 + 7x^3 - 8 = 0$

11.6 Quadratic Functions and Their Graphs

The Graph of $f(x) = ax^2$ ▣   The Graph of $f(x) = a(x - h)^2$ ▣
The Graph of $f(x) = a(x - h)^2 + k$

We have already used quadratic functions when we solved equations earlier in this chapter. In this section and the next, we learn to graph such functions.

The Graph of $f(x) = ax^2$

The most basic quadratic function is $f(x) = x^2$.

EXAMPLE 1   Graph: $f(x) = x^2$.

SOLUTION   We choose some values for x and compute $f(x)$ for each. Then we plot the ordered pairs and connect them with a smooth curve.

x	$f(x) = x^2$	$(x, f(x))$
-3	9	$(-3, 9)$
-2	4	$(-2, 4)$
-1	1	$(-1, 1)$
0	0	$(0, 0)$
1	1	$(1, 1)$
2	4	$(2, 4)$
3	9	$(3, 9)$

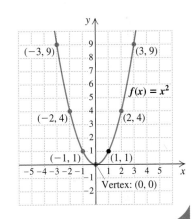

All quadratic functions have graphs similar to the one in Example 1. Such curves are called *parabolas*. They are U-shaped and symmetric with respect to a vertical line known as the parabola's *axis of symmetry*. For the graph of $f(x) = x^2$, the y-axis (or the line $x = 0$) is the axis of symmetry. Were the paper folded on this line, the two halves of the curve would match. The point $(0, 0)$ is known as the *vertex* of this parabola.

Since x^2 is a polynomial, we know that the domain of $f(x) = x^2$ is the set of all real numbers, or $(-\infty, \infty)$. We see from the graph and the table in Example 1 that the range of f is $\{y \mid y \geq 0\}$, or $[0, \infty)$. The function has a *minimum value* of 0. This minimum value occurs when $x = 0$.

Now let's consider a quadratic function of the form $g(x) = ax^2$. How does the constant a affect the graph of the function?

Student Notes

By paying attention to the symmetry of each parabola and the location of the vertex, you save yourself considerable work. Note too that if the x^2-coefficient is a, the x-values 1 unit to the right or left of the vertex are paired with the y-value a units above the vertex. Thus the graph of $y = \frac{3}{2}x^2$ includes the points $\left(-1, \frac{3}{2}\right)$ and $\left(1, \frac{3}{2}\right)$.

Interactive Discovery

Graph each of the following equations along with $y_1 = x^2$ in a $[-5, 5, -10, 10]$ window. If your calculator has a Transfrm application, run that application and graph $y_2 = Ax^2$. Then enter the values for A indicated in Exercises 1–6. For each equation, answer the following questions.

a) What is the vertex of the graph of y_2?

b) What is the axis of symmetry of the graph of y_2?

c) Does the graph of y_2 open upward or downward?

d) Is the parabola narrower or wider than the parabola given by $y_1 = x^2$?

1. $y_2 = 3x^2$	**2.** $y_2 = \frac{1}{3}x^2$	**3.** $y_2 = 0.2x^2$
4. $y_2 = -x^2$	**5.** $y_2 = -4x^2$	**6.** $y_2 = -\frac{2}{3}x^2$

7. Describe the effect of multiplying x^2 by a when $a > 1$ and when $0 < a < 1$.

8. Describe the effect of multiplying x^2 by a when $a < -1$ and when $-1 < a < 0$.

You may have noticed that for $f(x) = ax^2$, the constant a, when $|a| \neq 1$, made the graph wider or narrower than the graph of $g(x) = x^2$. When a was negative, the graph opened downward.

Note that if a parabola opens upward $(a > 0)$, the function value, or y-value, at the vertex is a least, or *minimum,* value. That is, it is less than the y-value at any other point on the graph. If the parabola opens downward $(a < 0)$, the function value at the vertex is a greatest, or *maximum,* value.

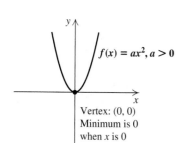

$f(x) = ax^2, a > 0$

Vertex: $(0, 0)$
Minimum is 0
when x is 0

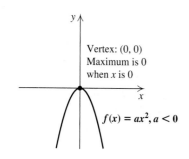

Vertex: $(0, 0)$
Maximum is 0
when x is 0

$f(x) = ax^2, a < 0$

Graphing $f(x) = ax^2$

The graph of $f(x) = ax^2$ is a parabola with $x = 0$ as its axis of symmetry. Its vertex is the origin. The domain of f is $(-\infty, \infty)$.

For $a > 0$, the parabola opens upward. The range of the function is $[0, \infty)$. A minimum function value of 0 occurs when $x = 0$.

For $a < 0$, the parabola opens downward. The range of the function is $(-\infty, 0]$. A maximum function value of 0 occurs when $x = 0$.

If $|a|$ is greater than 1, the parabola is narrower than $y = x^2$.

If $|a|$ is between 0 and 1, the parabola is wider than $y = x^2$.

EXAMPLE 2 Graph: $g(x) = \frac{1}{2}x^2$.

SOLUTION The function is in the form $g(x) = ax^2$, where $a = \frac{1}{2}$. The vertex is $(0,0)$ and the axis of symmetry is $x = 0$. Since $\frac{1}{2} > 0$, the parabola opens upward. The parabola is wider than $y = x^2$ since $|\frac{1}{2}|$ is between 0 and 1.

To graph the function, we calculate and plot points. Because we know the general shape of the graph, we can sketch the graph using only a few points.

x	$g(x) = \frac{1}{2}x^2$
-3	$\frac{9}{2}$
-2	2
-1	$\frac{1}{2}$
0	0
1	$\frac{1}{2}$
2	2
3	$\frac{9}{2}$

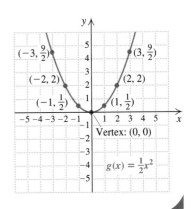

Note in Example 2 that since a parabola is symmetric, we can calculate function values for inputs to one side of the vertex and then plot the mirror images of those points on the other "half" of the graph.

The Graph of $f(x) = a(x - h)^2$

We could now consider graphs of

$$f(x) = ax^2 + bx + c,$$

where b and c are not both 0. In effect, we will do so, but in a different form. It turns out to be convenient to first graph $f(x) = a(x - h)^2$, where h is some constant. This allows us to observe similarities to the graphs drawn above.

Interactive Discovery

Graph each of the following pairs of equations in a $[-5, 5, -10, 10]$ window. For each pair, answer the following questions.

a) What is the vertex of the graph of y_2?

b) What is the axis of symmetry of the graph of y_2?

c) Is the graph of y_2 narrower, wider, or the same shape as the graph of y_1?

1. $y_1 = x^2$; $y_2 = (x - 3)^2$
2. $y_1 = 2x^2$; $y_2 = 2(x + 1)^2$
3. $y_1 = -\frac{1}{2}x^2$; $y_2 = -\frac{1}{2}\left(x - \frac{3}{2}\right)^2$
4. $y_1 = -3x^2$; $y_2 = -3(x + 2)^2$
5. Describe the effect of h on the graph of $g(x) = a(x - h)^2$.

The graph of $g(x) = a(x - h)^2$ looks just like the graph of $f(x) = ax^2$, except that it is moved, or *translated*, $|h|$ units horizontally.

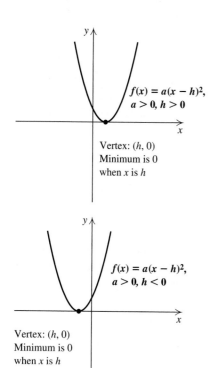

Vertex: $(h, 0)$
Minimum is 0
when x is h

$f(x) = a(x - h)^2,$
$a > 0, h > 0$

$f(x) = a(x - h)^2,$
$a > 0, h < 0$

Vertex: $(h, 0)$
Minimum is 0
when x is h

Graphing $f(x) = a(x - h)^2$

The graph of $f(x) = a(x - h)^2$ has the same shape as the graph of $y = ax^2$. The domain of f is $(-\infty, \infty)$.

If h is positive, the graph of $y = ax^2$ is shifted h units to the right.

If h is negative, the graph of $y = ax^2$ is shifted $|h|$ units to the left.

The vertex is $(h, 0)$, and the axis of symmetry is $x = h$.

For $a > 0$, the range of f is $[0, \infty)$. A minimum function value of 0 occurs when $x = h$.

For $a < 0$, the range of f is $(-\infty, 0]$. A maximum function value of 0 occurs when $x = h$.

EXAMPLE 3 Graph $g(x) = -2(x + 3)^2$, and determine the maximum value of $g(x)$ and where it occurs.

SOLUTION We rewrite the equation as $g(x) = -2[x - (-3)]^2$. In this case, $a = -2$ and $h = -3$, so the graph looks like that of $y = 2x^2$ translated 3 units to the left and, since $-2 < 0$, it opens downward. The vertex is $(-3, 0)$, and the axis of symmetry is $x = -3$. Plotting points as needed, we obtain the graph shown below. We say that g is a *reflection* of the graph of $f(x) = 2(x + 3)^2$ across the x-axis.

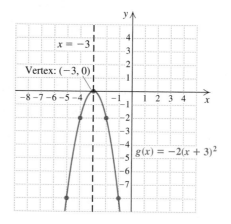

$x = -3$

Vertex: $(-3, 0)$

$g(x) = -2(x + 3)^2$

The maximum value of $g(x)$ is 0 when $x = -3$.

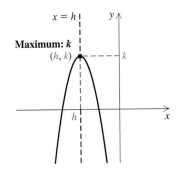

Maximum: *k*
(*h*, *k*)

$f(x) = a(x - h)^2 + k,$
$a < 0$

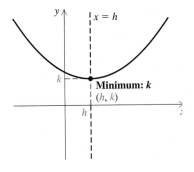

Minimum: *k*
(*h*, *k*)

$f(x) = a(x - h)^2 + k,$
$a < 0$

The Graph of $f(x) = a(x - h)^2 + k$

If we add a positive constant k to the graph of $f(x) = a(x - h)^2$, the graph is moved up. If we add a negative constant k, the graph is moved down. The axis of symmetry for the parabola remains at $x = h$, but the vertex will be at (h, k).

Graphing $f(x) = a(x - h)^2 + k$

The graph of $f(x) = a(x - h)^2 + k$ has the same shape as the graph of $y = a(x - h)^2$.

If k is positive, the graph of $y = a(x - h)^2$ is shifted k units up.

If k is negative, the graph of $y = a(x - h)^2$ is shifted $|k|$ units down.

The vertex is (h, k), and the axis of symmetry is $x = h$.

The domain of f is $(-\infty, \infty)$.

For $a > 0$, the range of f is $[k, \infty)$. The minimum function value is k, which occurs when $x = h$.

For $a < 0$, the range of f is $(-\infty, k]$. The maximum function value is k, which occurs when $x = h$.

EXAMPLE 4 Graph $g(x) = (x - 3)^2 - 5$, and find the minimum function value and the range of g.

SOLUTION The graph will look like that of $f(x) = (x - 3)^2$ but shifted 5 units down. You can confirm this by plotting some points.

The vertex is now $(3, -5)$, and the minimum function value is -5. The range of g is $[-5, \infty)$.

x	$g(x) = (x - 3)^2 - 5$
0	4
1	−1
2	−4
3	−5	←Vertex
4	−4
5	−1
6	4

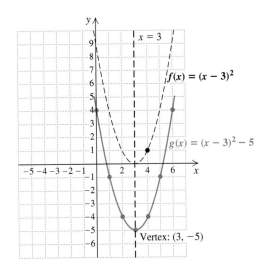

▶ **EXAMPLE 5** Graph $h(x) = \frac{1}{2}(x-3)^2 + 5$, and find the minimum function value and the range of h.

SOLUTION The graph looks just like that of $f(x) = \frac{1}{2}x^2$ but moved 3 units to the right and 5 units up. The vertex is $(3,5)$, and the axis of symmetry is $x = 3$. We draw $f(x) = \frac{1}{2}x^2$ and then shift the curve over and up.

By plotting some points, we have a check. The minimum function value is 5, and the range is $[5, \infty)$.

x	$h(x) = \frac{1}{2}(x-3)^2 + 5$	
0	$9\frac{1}{2}$	
1	7	
3	5	← Vertex
5	7	
6	$9\frac{1}{2}$	

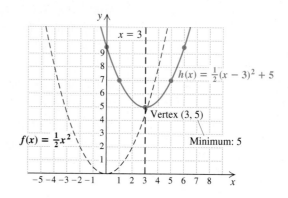

▶ **EXAMPLE 6** Graph $f(x) = -2(x+3)^2 + 5$. Find the vertex, the axis of symmetry, the maximum or minimum function value, and the range of f.

SOLUTION We first express the equation in the equivalent form

$$f(x) = -2[x - (-3)]^2 + 5.$$

The graph looks like that of $y = -2x^2$ translated 3 units to the left and 5 units up. The vertex is $(-3, 5)$, and the axis of symmetry is $x = -3$. Since -2 is negative, we know that 5, the second coordinate of the vertex, is the maximum function value. The range of f is $(-\infty, 5]$.

We compute a few points as needed, selecting convenient x-values on either side of the vertex. The graph is shown here.

x	$f(x) = -2(x+3)^2 + 5$	
-4	3	
-3	5	← Vertex
-2	3	

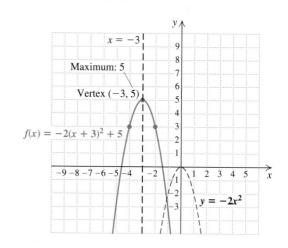

Because we can read the vertex and determine the orientation directly from the equation, it is often faster to graph $y = a(x - h)^2 + k$ by hand than by using a graphing calculator.

11.6 EXERCISE SET

🖐 *Concept Reinforcement*   *In each of Exercises 1–8, match the equation with the corresponding graph from those shown.*

1. _____ $f(x) = 2(x - 1)^2 + 3$

2. _____ $f(x) = -2(x - 1)^2 + 3$

3. _____ $f(x) = 2(x + 1)^2 + 3$

4. _____ $f(x) = 2(x - 1)^2 - 3$

5. _____ $f(x) = -2(x + 1)^2 + 3$

6. _____ $f(x) = -2(x + 1)^2 - 3$

7. _____ $f(x) = 2(x + 1)^2 - 3$

8. _____ $f(x) = -2(x - 1)^2 - 3$

a)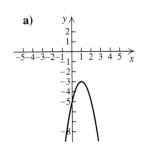

b)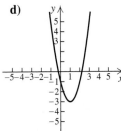

c)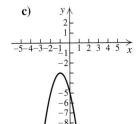

d)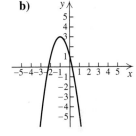

e)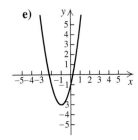

f)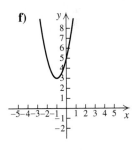

g)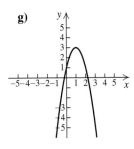

h)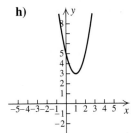

For each graph of a quadratic function

$$f(x) = a(x - h)^2 + k$$

in Exercises 9–14:

a) Tell whether a is positive or negative.
b) Determine the vertex.
c) Determine the axis of symmetry.
d) Determine the range.

9.

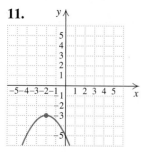

10.

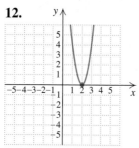

11.

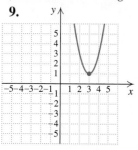

12.

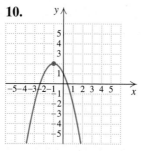

13.

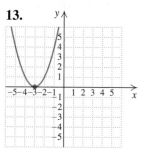

14.

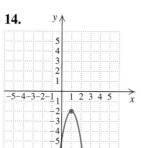

Graph.

15. $f(x) = x^2$

16. $f(x) = -x^2$

17. $f(x) = -2x^2$

18. $f(x) = -3x^2$

19. $g(x) = \frac{1}{3}x^2$

20. $g(x) = \frac{1}{4}x^2$

Aha! **21.** $h(x) = -\frac{1}{3}x^2$

22. $h(x) = -\frac{1}{4}x^2$

23. $f(x) = \frac{5}{2}x^2$

24. $f(x) = \frac{3}{2}x^2$

For each of the following, graph the function, label the vertex, and draw the axis of symmetry.

25. $g(x) = (x + 1)^2$

26. $g(x) = (x + 4)^2$

27. $f(x) = (x - 2)^2$

28. $f(x) = (x - 1)^2$

29. $h(x) = (x - 3)^2$

30. $h(x) = (x - 4)^2$

31. $f(x) = -(x + 1)^2$

32. $f(x) = -(x - 1)^2$

33. $g(x) = -(x - 2)^2$

34. $g(x) = -(x + 4)^2$

35. $f(x) = 2(x + 1)^2$

36. $f(x) = 2(x + 4)^2$

37. $h(x) = -\frac{1}{2}(x - 4)^2$

38. $h(x) = -\frac{3}{2}(x - 2)^2$

39. $f(x) = \frac{1}{2}(x - 1)^2$

40. $f(x) = \frac{1}{3}(x + 2)^2$

41. $f(x) = -2(x + 5)^2$

42. $f(x) = 2(x + 7)^2$

43. $h(x) = -3\left(x - \frac{1}{2}\right)^2$

44. $h(x) = -2\left(x + \frac{1}{2}\right)^2$

For each of the following, graph the function and find the vertex, the axis of symmetry, and the maximum value or the minimum value.

45. $f(x) = (x - 5)^2 + 2$

46. $f(x) = (x + 3)^2 - 2$

47. $f(x) = (x + 1)^2 - 3$

48. $f(x) = (x - 1)^2 + 2$

49. $g(x) = (x + 4)^2 + 1$

50. $g(x) = -(x - 2)^2 - 4$

51. $h(x) = -2(x - 1)^2 - 3$

52. $h(x) = -2(x + 1)^2 + 4$

53. $f(x) = 2(x + 4)^2 + 1$

54. $f(x) = 2(x - 5)^2 - 3$

55. $g(x) = -\frac{3}{2}(x - 1)^2 + 4$

56. $g(x) = \frac{3}{2}(x + 2)^2 - 3$

Without graphing, find the vertex, the axis of symmetry, and the maximum value or the minimum value.

57. $f(x) = 6(x - 8)^2 + 7$

58. $f(x) = 4(x + 5)^2 - 6$

59. $h(x) = -\frac{2}{7}(x + 6)^2 + 11$

60. $h(x) = -\frac{3}{11}(x - 7)^2 - 9$

61. $f(x) = 7\left(x + \frac{1}{4}\right)^2 - 13$

62. $f(x) = 6\left(x - \frac{1}{4}\right)^2 + 15$

63. $f(x) = \sqrt{2}(x + 4.58)^2 + 65\pi$

64. $f(x) = 4\pi(x - 38.2)^2 - \sqrt{34}$

TW 65. While trying to graph $y = -\frac{1}{2}x^2 + 3x + 1$, Omar gets the following screen. How can Omar tell at a glance that a mistake has been made?

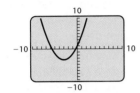

TW 66. Explain, without plotting points, why the graph of $y = (x + 2)^2$ looks like the graph of $y = x^2$ translated 2 units to the left.

Skill Maintenance

Graph using intercepts. [3.3]

67. $2x - 7y = 28$ **68.** $6x - 3y = 36$

Solve each system. [4.3]

69. $3x + 4y = -19,$ **70.** $5x + 7y = 9,$
$7x - 6y = -29$ $3x - 4y = -11$

Replace the blanks with constants to form a true equation. [11.1]

71. $x^2 + 5x + \underline{\quad} = (x + \underline{\quad})^2$

72. $x^2 - 9x + \underline{\quad} = (x - \underline{\quad})^2$

Synthesis

TW 73. Before graphing a quadratic function, Sophie always plots five points. First, she calculates and plots the coordinates of the vertex. Then she plots *four* more points after calculating *two* more ordered pairs. How is this possible?

TW 74. If the graphs of $f(x) = a_1(x - h_1)^2 + k_1$ and $g(x) = a_2(x - h_2)^2 + k_2$ have the same shape, what, if anything, can you conclude about the a's, the h's, and the k's? Why?

Write an equation for a function having a graph with the same shape as the graph of $f(x) = \frac{3}{5}x^2$, but with the given point as the vertex.

75. $(4, 1)$ **76.** $(2, 6)$ **77.** $(3, -1)$

78. $(5, -6)$ **79.** $(-2, -5)$ **80.** $(-4, -2)$

For each of the following, write the equation of the parabola that has the shape of $f(x) = 2x^2$ or $g(x) = -2x^2$ and has a maximum or minimum value at the specified point.

81. Minimum: $(2, 0)$ **82.** Minimum: $(-4, 0)$

83. Maximum: $(0, 3)$ **84.** Maximum: $(3, 8)$

Find an equation for a quadratic function F that satisfies the following conditions.

85. The graph of F is the same shape as the graph of f, where $f(x) = 3(x + 2)^2 + 7$, and $F(x)$ is a minimum at the same point that $g(x) = -2(x - 5)^2 + 1$ is a maximum.

86. The graph of F is the same shape as the graph of f, where $f(x) = -\frac{1}{3}(x - 2)^2 + 7$, and $F(x)$ is a maximum at the same point that $g(x) = 2(x + 4)^2 - 6$ is a minimum.

Functions other than parabolas can be translated. When calculating $f(x)$, if we replace x with $x - h$, where h is a constant, the graph will be moved horizontally. If we replace $f(x)$ with $f(x) + k$, the graph will be moved vertically. Use the graph below for Exercises 87–92.

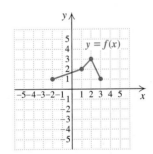

Draw a graph of each of the following.

87. $y = f(x - 1)$ **88.** $y = f(x + 2)$

89. $y = f(x) + 2$ **90.** $y = f(x) - 3$

91. $y = f(x + 3) - 2$ **92.** $y = f(x - 3) + 1$

Collaborative Corner

Match the Graph

Focus: Graphing quadratic functions
Time: 15–20 minutes
Group Size: 6
Materials: Index cards

ACTIVITY

1. On each of six index cards, write one of the following equations:

$$y = \tfrac{1}{2}(x - 3)^2 + 1; \qquad y = \tfrac{1}{2}(x - 1)^2 + 3;$$
$$y = \tfrac{1}{2}(x + 1)^2 - 3; \qquad y = \tfrac{1}{2}(x + 3)^2 + 1;$$
$$y = \tfrac{1}{2}(x + 3)^2 - 1; \qquad y = \tfrac{1}{2}(x + 1)^2 + 3.$$

2. Fold each index card and mix up the six cards in a hat or bag. Then, one by one, each group member should select one of the equations. Do not let anyone see your equation.

3. Each group member should carefully graph the equation selected. Make the graph large enough so that when it is finished, it can be easily viewed by the rest of the group. Be sure to scale the axes and label the vertex, but **do not label the graph with the equation used.**

4. When all group members have drawn a graph, place the graphs in a pile. The group should then match and agree on the correct equation for each graph *with no help from the person who drew the graph.* If a mistake has been made and a graph has no match, determine what its equation *should* be.

5. Compare your group's labeled graphs with those of other groups to reach consensus within the class on the correct label for each graph.

11.7 More About Graphing Quadratic Functions

Finding the Vertex ■ Finding Intercepts

Finding the Vertex

By *completing the square* (see Section 11.1), we can rewrite any polynomial $ax^2 + bx + c$ in the form $a(x - h)^2 + k$. Once that has been done, the procedures discussed in Section 11.6 will enable us to graph any quadratic function.

> **EXAMPLE 1** Graph: $g(x) = x^2 - 6x + 4$.
>
> **SOLUTION** We have
>
> $$g(x) = x^2 - 6x + 4$$
> $$= (x^2 - 6x) + 4.$$

Study Tip

Use What You Know

An excellent and common strategy for solving any new type of problem is to rewrite the problem in an equivalent form that we already know how to solve. Although this is not always feasible, when it is—as in most of the problems in this section—it can make a new topic much easier to learn.

To complete the square inside the parentheses, we take half the x-coefficient, $\frac{1}{2} \cdot (-6) = -3$, and square it to get $(-3)^2 = 9$. Then we add $9 - 9$ inside the parentheses:

$$g(x) = (x^2 - 6x + 9 - 9) + 4 \qquad \text{The effect is of adding 0.}$$
$$= (x^2 - 6x + 9) + (-9 + 4) \qquad \text{Using the associative law of addition to regroup}$$
$$= (x - 3)^2 - 5. \qquad \text{Factoring and simplifying}$$

This equation appeared as Example 4 of Section 11.6. The graph is that of $f(x) = x^2$ translated 3 units to the right and 5 units down. The vertex is $(3, -5)$.

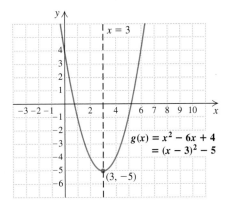

When the leading coefficient is not 1, we factor out that number from the first two terms. Then we complete the square and use the distributive law.

EXAMPLE 2   Graph: $f(x) = 3x^2 + 12x + 13$.

SOLUTION   Since the coefficient of x^2 is not 1, we need to factor out that number—in this case, 3—from the first two terms. Remember that we want the form $f(x) = a(x - h)^2 + k$:

$$f(x) = 3x^2 + 12x + 13 = 3(x^2 + 4x) + 13.$$

Now we complete the square as before. We take half of the x-coefficient, $\frac{1}{2} \cdot 4 = 2$, and square it: $2^2 = 4$. Then we add $4 - 4$ inside the parentheses:

$$f(x) = 3(x^2 + 4x + 4 - 4) + 13. \qquad \text{Adding 4 − 4, or 0, inside the parentheses}$$

The distributive law allows us to separate the -4 from the perfect-square trinomial so long as it is multiplied by 3:

The -4 was added inside the parentheses. To separate it, we *must* multiply it by 3.

$$f(x) = 3(x^2 + 4x + 4) + 3(-4) + 13 \qquad \text{This leaves a perfect-square trinomial inside the parentheses.}$$

$$= 3(x + 2)^2 + 1. \qquad \text{Factoring and simplifying}$$

The vertex is $(-2, 1)$, and the axis of symmetry is $x = -2$. The coefficient of x^2 is 3, so the graph is narrow and opens upward. We choose a few *x*-values on either side of the vertex, compute *y*-values, and then graph the parabola.

x	$f(x) = 3(x + 2)^2 + 1$
-2	1
-3	4
-1	4

←Vertex

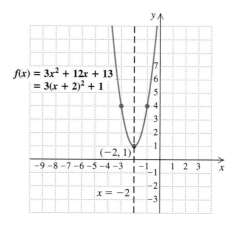

EXAMPLE 3   Graph: $f(x) = -2x^2 + 10x - 7$.

SOLUTION   We first find the vertex by completing the square. To do so, we factor out -2 from the first two terms of the expression. This makes the coefficient of x^2 inside the parentheses 1:

$$f(x) = -2x^2 + 10x - 7$$
$$= -2(x^2 - 5x) - 7.$$

Now we complete the square as before. We take half of the *x*-coefficient and square it to get $\frac{25}{4}$. Then we add $\frac{25}{4} - \frac{25}{4}$ inside the parentheses:

$$f(x) = -2\left(x^2 - 5x + \tfrac{25}{4} - \tfrac{25}{4}\right) - 7$$
$$= -2\left(x^2 - 5x + \tfrac{25}{4}\right) + (-2)\left(-\tfrac{25}{4}\right) - 7 \qquad \text{Multiplying by } -2, \text{ using the distributive law, and regrouping}$$
$$= -2\left(x - \tfrac{5}{2}\right)^2 + \tfrac{11}{2}. \qquad \text{Factoring and simplifying}$$

The vertex is $\left(\frac{5}{2}, \frac{11}{2}\right)$, and the axis of symmetry is $x = \frac{5}{2}$. The coefficient of x^2, -2, is negative, so the graph opens downward. We plot a few points on either side of the vertex, including the *y*-intercept, $f(0)$, and graph the parabola.

x	$f(x)$	
$\frac{5}{2}$	$\frac{11}{2}$	← Vertex
0	-7	← *y*-intercept
1	1	
4	1	

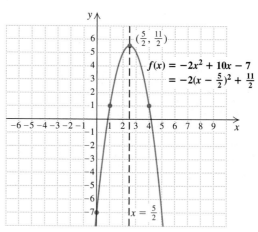

The method used in Examples 1–3 can be generalized to find a formula for locating the vertex. We complete the square as follows:

$$f(x) = ax^2 + bx + c$$

$$= a\left(x^2 + \frac{b}{a}x\right) + c. \qquad \text{Factoring } a \text{ out of the first two terms. Check by multiplying.}$$

Half of the x-coefficient, $\dfrac{b}{a}$, is $\dfrac{b}{2a}$. We square it to get $\dfrac{b^2}{4a^2}$ and add $\dfrac{b^2}{4a^2} - \dfrac{b^2}{4a^2}$ inside the parentheses. Then we distribute the a and regroup terms:

$$f(x) = a\left(x^2 + \frac{b}{a}x + \frac{b^2}{4a^2} - \frac{b^2}{4a^2}\right) + c$$

$$= a\left(x^2 + \frac{b}{a}x + \frac{b^2}{4a^2}\right) + a\left(-\frac{b^2}{4a^2}\right) + c \qquad \text{Using the distributive law}$$

$$= a\left(x + \frac{b}{2a}\right)^2 + \frac{-b^2}{4a} + \frac{4ac}{4a} \qquad \begin{array}{l}\text{Factoring and} \\ \text{finding a common} \\ \text{denominator}\end{array}$$

$$= a\left[x - \left(-\frac{b}{2a}\right)\right]^2 + \frac{4ac - b^2}{4a}.$$

Thus we have the following.

The Vertex of a Parabola   The vertex of the parabola given by $f(x) = ax^2 + bx + c$ is

$$\left(-\frac{b}{2a}, f\left(-\frac{b}{2a}\right)\right), \quad \text{or} \quad \left(-\frac{b}{2a}, \frac{4ac - b^2}{4a}\right).$$

The x-coordinate of the vertex is $-b/(2a)$. The axis of symmetry is $x = -b/(2a)$. The second coordinate of the vertex is most commonly found by computing $f\left(-\dfrac{b}{2a}\right)$.

Student Notes

An excellent way to remember a formula is to understand its derivation. Check with your instructor to determine what, if any, formulas you will be expected to remember.

Let's reexamine Example 3 to see how we could have found the vertex directly. From the formula above,

$$\text{the } x\text{-coordinate of the vertex is } -\frac{b}{2a} = -\frac{10}{2(-2)} = \frac{5}{2}.$$

Substituting $\frac{5}{2}$ into $f(x) = -2x^2 + 10x - 7$, we find the second coordinate of the vertex:

$$f\left(\tfrac{5}{2}\right) = -2\left(\tfrac{5}{2}\right)^2 + 10\left(\tfrac{5}{2}\right) - 7$$

$$= -2\left(\tfrac{25}{4}\right) + 25 - 7$$

$$= -\tfrac{25}{2} + 18$$

$$= -\tfrac{25}{2} + \tfrac{36}{2} = \tfrac{11}{2}.$$

The vertex is $\left(\tfrac{5}{2}, \tfrac{11}{2}\right)$. The axis of symmetry is $x = \tfrac{5}{2}$.

Note that a quadratic function has a maximum or a minimum value at the vertex of its graph. Thus determining the maximum or minimum value of the function allows us to find the vertex of the graph.

Maximums and Minimums

On most graphing calculators, we can find a maximum or minimum function value for any given interval. This MAXIMUM or MINIMUM feature is often found in the CALC menu. (CALC is the 2nd option associated with the TRACE key.)

To find a maximum or a minimum, enter and graph the function, choosing a viewing window that will show the vertex. Next, press ⬭CALC⬭ and choose either the MAXIMUM or MINIMUM option in the menu. The graphing calculator will find the maximum or minimum function value over a specified interval, so the left and right endpoints, or bounds, of the interval must be entered as well as a guess near where the maximum or minimum occurs. The calculator will return the coordinates of the point for which the function value is a maximum or minimum within the interval.

EXAMPLE 4 Use a graphing calculator to determine the vertex of the graph of the function given by $f(x) = -2x^2 + 10x - 7$.

SOLUTION The coefficient of x^2 is negative, so the parabola opens downward and the function has a maximum value. We enter and graph the function, choosing a window that will show the vertex.

We choose the MAXIMUM option from the CALC menu. The calculator will prompt us first for a left bound. After visually locating the vertex, we either move the cursor to a point left of the vertex using the left and right arrow keys or type a value of x that is less than the x-value at the vertex. (See the graph on the left below.) After pressing **ENTER**, we choose a right bound in a similar manner (as shown in the graph on the right), and press **ENTER** again. Finally, we enter a guess that is close to the vertex (as in the graph on the left at the top of the next page), and press **ENTER**. (Many calculators will use the right bound as a guess if we simply press **ENTER**.) The calculator indicates that the coordinates of the vertex are $(2.5, 5.5)$. (See the graph on the right at the top of the next page.)

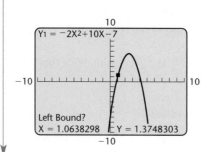
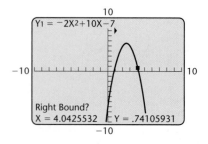

Student Notes

It is easy to press a wrong key when using a calculator. Always check to see if your answer is reasonable. For example, if the MINIMUM option were chosen in Example 4 instead of MAXIMUM, one endpoint of the interval would have been returned instead of the maximum. In this case, noting that the highest point on the graph is marked is a quick check that we did indeed find the maximum.

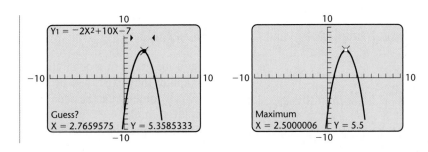

In Example 3, we found the coordinates of the vertex of the graph of $f(x) = -2x^2 + 10x - 7$, $\left(\frac{5}{2}, \frac{11}{2}\right)$, by completing the square. On p. 855, we also found these coordinates using the formula for the vertex of a parabola. The coordinates found using a calculator in Example 4 were in decimal notation. Because of the calculator's method used to find the maximum function value, the coordinates may not be exact and can indeed vary slightly for choices of windows. Since $2.5000006 \approx \frac{5}{2}$, and since $5.5 = \frac{11}{2}$, the coordinates check.

We have actually developed three methods for finding the vertex. One is by completing the square, the second is by using a formula, and the third is by using a graphing calculator. You should check with your instructor about which method to use.

Finding Intercepts

For any function, the y-intercept occurs at $f(0)$. Therefore, for $f(x) = ax^2 + bx + c$, the y-intercept is simply $(0, c)$. To find x-intercepts, we look for points where $y = 0$ or $f(x) = 0$. Thus, for $f(x) = ax^2 + bx + c$, the x-intercepts occur at those x-values for which $ax^2 + bx + c = 0$.

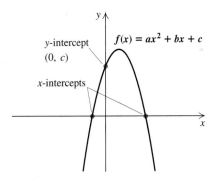

EXAMPLE 5 Find the x- and y-intercepts of the graph of

$$f(x) = x^2 - 2x - 2.$$

SOLUTION The y-intercept is simply $(0, f(0))$, or $(0, -2)$. To find the x-intercepts, we solve the equation

$$0 = x^2 - 2x - 2.$$

We are unable to factor $x^2 - 2x - 2$, so we use the quadratic formula and get $x = 1 \pm \sqrt{3}$. Thus the x-intercepts are $\left(1 - \sqrt{3}, 0\right)$ and $\left(1 + \sqrt{3}, 0\right)$.

If graphing, we would approximate, to get $(-0.7, 0)$ and $(2.7, 0)$.

Connecting the Concepts

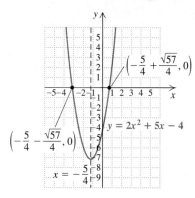

$$\left(-\frac{5}{4}+\frac{\sqrt{57}}{4},0\right)$$

$$y = 2x^2 + 5x - 4$$

$$\left(-\frac{5}{4}-\frac{\sqrt{57}}{4},0\right)$$

$$x = -\frac{5}{4}$$

Because the graph of a quadratic equation is symmetric, the x-intercepts of the graph, if they exist, will be symmetric with respect to the axis of symmetry. This symmetry can be seen directly if the x-intercepts are found using the quadratic formula.

For example, the x-intercepts of the graph of $y = 2x^2 + 5x - 4$ are

$$\left(-\frac{5}{4}+\frac{\sqrt{57}}{4},0\right) \quad \text{and} \quad \left(-\frac{5}{4}-\frac{\sqrt{57}}{4},0\right).$$

For this equation, the axis of symmetry is $x = -\dfrac{5}{4}$ and the x-intercepts

are $\dfrac{\sqrt{57}}{4}$ units to the left and right of $-\dfrac{5}{4}$ on the x-axis.

The Graph of a Quadratic Function Given by $f(x) = ax^2 + bx + c$ or $f(x) = a(x - h)^2 + k$

The graph is a parabola.

The vertex is (h, k) or $\left(-\dfrac{b}{2a}, f\left(-\dfrac{b}{2a}\right)\right)$.

The axis of symmetry is $x = h$.

The y-intercept of the graph is $(0, c)$.

The x-intercepts can be found by solving $ax^2 + bx + c = 0$.

    If $b^2 - 4ac > 0$, there are two x-intercepts.
    If $b^2 - 4ac = 0$, there is one x-intercept.
    If $b^2 - 4ac < 0$, there are no x-intercepts.

The domain of the function is $(-\infty, \infty)$.

If a is positive: The graph opens upward.
                     The function has a minimum value, given by k.
                        This occurs when $x = h$.
                     The range of the function is $[k, \infty)$.

If a is negative: The graph opens downward.
                     The function has a maximum value, given by k.
                        This occurs when $x = h$.
                     The range of the function is $(-\infty, k]$.

A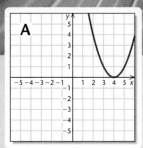

F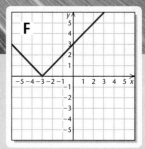

Visualizing the Graph

Match each function with its graph.

1. $f(x) = 3x^2$

2. $f(x) = x^2 - 4$

3. $f(x) = (x - 4)^2$

4. $f(x) = x - 4$

5. $f(x) = -2x^2$

6. $f(x) = x + 3$

7. $f(x) = |x + 3|$

8. $f(x) = (x + 3)^2$

9. $f(x) = \sqrt{x + 3}$

10. $f(x) = (x + 3)^2 - 4$

Answers on page A-45

B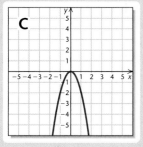

C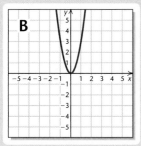

D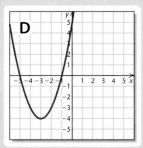

E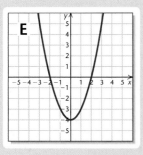

G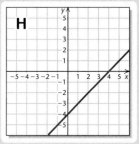

H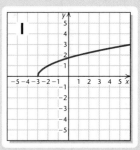

I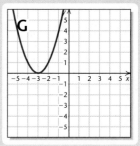

J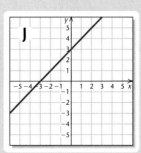

11.7 EXERCISE SET

Concept Reinforcement *Complete each of the following.*

1. The expressions $x^2 + 6x + 5$ and $(x^2 + 6x + 9) - \underline{} + 5$ are equivalent.

2. The expressions $x^2 + 8x + 3$ and $(x^2 + 8x + \underline{}) - 16 + 3$ are equivalent.

3. The functions given by $f(x) = 2x^2 + 12x - 7$ and $f(x) = 2(x^2 + 6x + \underline{}) - 18 - 7$ are equivalent.

4. The functions given by $g(x) = 3x^2 - 6x - 5$ and $g(x) = 3(x^2 - 2x + \underline{}) - 3 - 5$ are equivalent.

5. The graph of $f(x) = -2(x - \underline{})^2 + 7$ has its vertex at $(3, 7)$.

6. The graph of $g(x) = -3(x - 4)^2 + \underline{}$ has its vertex at $(4, 1)$.

7. The graph of $f(x) = 2\left(x - \underline{}\right)^2 + \underline{}$ has its axis of symmetry at $x = \frac{5}{2}$ and has its vertex at $\left(\frac{5}{2}, -4\right)$.

8. The graph of $g(x) = \frac{1}{2}(x - \underline{})^2 + \frac{7}{2}$ has $x = -2$ as its axis of symmetry and $\underline{}$ as its vertex.

*For each quadratic function, **(a)** write the function in the form $f(x) = a(x - h)^2 + k$ and **(b)** find the vertex and the axis of symmetry.*

9. $f(x) = x^2 - 4x + 3$

10. $f(x) = x^2 + 6x + 13$

11. $f(x) = -x^2 + 3x - 10$

12. $f(x) = x^2 + 5x + 6$

13. $f(x) = 2x^2 - 7x + 1$

14. $f(x) = -2x^2 + 5x - 1$

*For each quadratic function, **(a)** find the vertex and the axis of symmetry and **(b)** graph the function.*

15. $f(x) = x^2 + 4x + 5$ **16.** $f(x) = x^2 + 2x - 5$

17. $g(x) = x^2 - 6x + 13$ **18.** $g(x) = x^2 - 4x + 5$

19. $f(x) = x^2 + 8x + 20$

20. $f(x) = x^2 - 10x + 21$

21. $h(x) = 2x^2 - 16x + 25$

22. $h(x) = 2x^2 + 16x + 23$

23. $f(x) = -x^2 + 2x + 5$

24. $f(x) = -x^2 - 2x + 7$

25. $g(x) = x^2 + 3x - 10$

26. $g(x) = x^2 + 5x + 4$

27. $f(x) = 3x^2 - 24x + 50$

28. $f(x) = 4x^2 + 8x - 3$

29. $h(x) = x^2 + 7x$

30. $h(x) = x^2 - 5x$

31. $f(x) = -2x^2 - 4x - 6$

32. $f(x) = -3x^2 + 6x + 2$

33. $g(x) = 2x^2 - 8x + 3$

34. $g(x) = 2x^2 + 5x - 1$

35. $f(x) = -3x^2 + 5x - 2$

36. $f(x) = -3x^2 - 7x + 2$

37. $h(x) = \frac{1}{2}x^2 + 4x + \frac{19}{3}$

38. $h(x) = \frac{1}{2}x^2 - 3x + 2$

Use a graphing calculator to find the vertex of the graph of each function.

39. $f(x) = x^2 + x - 6$

40. $f(x) = x^2 + 2x - 5$

41. $f(x) = 5x^2 - x + 1$

42. $f(x) = -4x^2 - 3x + 7$

43. $f(x) = -0.2x^2 + 1.4x - 6.7$

44. $f(x) = 0.5x^2 + 2.4x + 3.2$

Find the x- and y-intercepts. If no x-intercepts exist, state this.

45. $f(x) = x^2 - 6x + 3$

46. $f(x) = x^2 + 5x + 2$

47. $g(x) = -x^2 + 2x + 3$

48. $g(x) = x^2 - 6x + 9$

aha! **49.** $f(x) = x^2 - 9x$

50. $f(x) = x^2 - 7x$

51. $h(x) = -x^2 + 4x - 4$

52. $h(x) = 4x^2 - 12x + 3$

53. $f(x) = 2x^2 - 4x + 6$

54. $f(x) = x^2 - x + 2$

For each quadratic function, find **(a)** *the maximum or minimum value and* **(b)** *the x- and y-intercepts. Round to the nearest hundredth.*

55. $f(x) = 2.31x^2 - 3.135x - 5.89$

56. $f(x) = -18.8x^2 + 7.92x + 6.18$

57. $g(x) = -1.25x^2 + 3.42x - 2.79$

58. $g(x) = 0.45x^2 - 1.72x + 12.92$

TW 59. Does the graph of every quadratic function have a y-intercept? Why or why not?

TW 60. Is it possible for the graph of a quadratic function to have only one x-intercept if the vertex is off the x-axis? Why or why not?

Focused Review

Graph.

61. $f(x) = \frac{3}{2}x$ [3.8]

62. $f(x) = -\frac{2}{3}x$ [3.8]

63. $f(x) = \frac{2}{x}$ [7.1]

64. $f(x) = \frac{3}{x}$ [7.1]

65. $f(x) = 3x - 1$ [3.8]

66. $f(x) = (x - 2)^2 - 1$ [11.6]

67. $f(x) = x^2 - x - 6$ [11.7]

68. $f(x) = 2x^2 + x + 1$ [11.7]

Synthesis

TW 69. If the graphs of two quadratic functions have the same x-intercepts, will they also have the same vertex? Why or why not?

TW 70. Suppose that the graph of $f(x) = ax^2 + bx + c$ has $(x_1, 0)$ and $(x_2, 0)$ as x-intercepts. Explain why the graph of $g(x) = -ax^2 - bx - c$ will also have $(x_1, 0)$ and $(x_2, 0)$ as x-intercepts.

71. Graph the function

$$f(x) = x^2 - x - 6.$$

Then use the graph to approximate solutions to each of the following equations.

a) $x^2 - x - 6 = 2$

b) $x^2 - x - 6 = -3$

72. Graph the function

$$f(x) = \frac{x^2}{2} + x - \frac{3}{2}.$$

Then use the graph to approximate solutions to each of the following equations.

a) $\frac{x^2}{2} + x - \frac{3}{2} = 0$

b) $\frac{x^2}{2} + x - \frac{3}{2} = 1$

c) $\frac{x^2}{2} + x - \frac{3}{2} = 2$

Find an equivalent equation of the type

$$f(x) = a(x - h)^2 + k.$$

73. $f(x) = mx^2 - nx + p$

74. $f(x) = 3x^2 + mx + m^2$

75. A quadratic function has $(-1, 0)$ as one of its intercepts and $(3, -5)$ as its vertex. Find an equation for the function.

76. A quadratic function has $(4, 0)$ as one of its intercepts and $(-1, 7)$ as its vertex. Find an equation for the function.

Graph.

77. $f(x) = |x^2 - 1|$

78. $f(x) = |x^2 - 3x - 4|$

79. $f(x) = |2(x - 3)^2 - 5|$

11.8 Problem Solving and Quadratic Functions

Maximum and Minimum Problems   ■   Fitting Quadratic Functions to Data

Let's look now at some of the many situations in which quadratic functions are used for problem solving.

Maximum and Minimum Problems

We have seen that for any quadratic function f, the value of $f(x)$ at the vertex is either a maximum or a minimum. Thus problems in which a quantity must be maximized or minimized can often be solved by finding the coordinates of a vertex, assuming the problem can be modeled with a quadratic function.

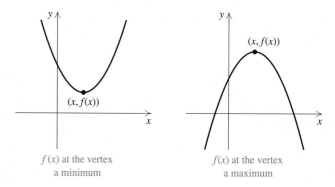

$f(x)$ at the vertex
a minimum

$f(x)$ at the vertex
a maximum

▸ **EXAMPLE 1**   Newborn Calves.   The number of pounds of milk per day recommended for a calf that is x weeks old can be approximated by $p(x)$, where $p(x) = -0.2x^2 + 1.3x + 6.2$ (*Source*: C. Chaloux, University of Vermont, 1998). When is a calf's milk consumption greatest and how much milk does it consume at that time?

SOLUTION

1., 2. Familiarize and **Translate.** We are given the function for milk consumption by a calf. Note that it is a quadratic function of x, the calf's age in weeks. Since the coefficient of x^2 is negative, the calf's consumption increases and then decreases. The graph at left confirms this.

3. Carry out. We can either complete the square,

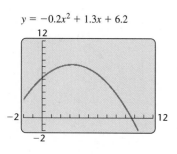

$y = -0.2x^2 + 1.3x + 6.2$

$$
\begin{aligned}
p(x) &= -0.2x^2 + 1.3x + 6.2 \\
&= -0.2(x^2 - 6.5x) + 6.2 \\
&= -0.2(x^2 - 6.5x + 3.25^2 - 3.25^2) + 6.2 \qquad \text{\small Completing the square; } \tfrac{6.5}{2} = 3.25 \\
&= -0.2(x^2 - 6.5x + 3.25^2) + (-0.2)(-3.25^2) + 6.2 \\
&= -0.2(x - 3.25)^2 + 8.3125, \qquad \text{\small Factoring and simplifying}
\end{aligned}
$$

or we can use $-b/(2a) = -1.3/(-0.4) = 3.25$. Using a calculator, we find that

$$p(3.25) = -0.2(3.25)^2 + 1.3(3.25) + 6.2 = 8.3125.$$

4. **Check.** Both of the approaches in step (3) indicate that a maximum occurs when $x = 3.25$, or $3\frac{1}{4}$. We could also use the MAXIMUM feature on a graphing calculator as a check.

5. **State.** A calf's milk consumption is greatest when the calf is $3\frac{1}{4}$ weeks old. At that time, it drinks about 8.3 lb of milk per day.

EXAMPLE 2   Swimming Area.   A lifeguard has 100 m of roped-together flotation devices with which to cordon off a rectangular swimming area at Lakeside Beach. If the shoreline forms one side of the rectangle, what dimensions will maximize the size of the area for swimming?

SOLUTION

1. **Familiarize.** We make a drawing and label it, letting $w =$ the width of the rectangle, in meters, and $l =$ the length of the rectangle, in meters.

Recall that Area $= l \cdot w$ and Perimeter $= 2w + 2l$. Since the beach forms one length of the rectangle, the flotation devices comprise three sides. Thus,

$$2w + l = 100.$$

To get a better feel for the problem, we can look at some possible dimensions for a rectangular area that can be enclosed with 100 m of flotation devices. All possibilities are chosen so that $2w + l = 100$.

We can make a table by hand or use a graphing calculator. To use a calculator, we let x represent the width of the swimming area. Solving $2w + l = 100$ for l, we have $l = 100 - 2w$. Thus, if $y_1 = 100 - 2x$, then the area is $y_2 = x \cdot y_1$.

$y_1 = 100 - 2x, y_2 = x \cdot y_1$

X	Y1	Y2
20	60	1200
22	56	1232
24	52	1248
26	48	1248
28	44	1232
30	40	1200
32	36	1152
X = 20		

What choice of X will maximize Y2?

l	w	Rope Length	Area
40 m	30 m	100 m	1200 m²
30 m	35 m	100 m	1050 m²
20 m	40 m	100 m	800 m²
.	.	.	.
.	.	.	.
.	.	.	.

What choice of l and w will maximize A?

The table helps us understand the problem, but the maximum area may or may not be one of the values shown. To find the maximum area, we use algebra.

2. **Translate.** We have two equations: One guarantees that all 100 m of flotation devices are used; the other expresses area in terms of length and width.

$$2w + l = 100,$$
$$A = l \cdot w$$

3. **Carry out.** We solve the system of equations both algebraically and graphically.

ALGEBRAIC APPROACH

We need to express A as a function of l or w but not both. To do so, we solve for l in the first equation to obtain $l = 100 - 2w$. Substituting for l in the second equation, we get a quadratic function:

$$A = (100 - 2w)w \quad \text{Substituting for } l$$
$$= 100w - 2w^2. \quad \text{This represents a parabola opening downward, so a maximum exists.}$$

Factoring and completing the square, we get

$$A = -2(w^2 - 50w + 625 - 625) \quad \text{We could also use the vertex formula.}$$
$$= -2(w - 25)^2 + 1250. \quad \text{This suggests a maximum of 1250 m}^2 \text{ when } w = 25 \text{ m.}$$

The maximum area, 1250 m², occurs when $w = 25$ m and $l = 100 - 2(25)$, or 50 m.

GRAPHICAL APPROACH

As in the *Familiarize* step, we let x represent the width of the area. Then the length is given by $y_1 = 100 - 2x$ and the area is $y_2 = x \cdot y_1$, or $y = 100x - 2x^2$. The maximum area is the second coordinate of the vertex of the graph of $y = 100x - 2x^2$.

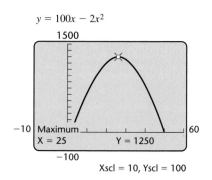

The maximum is 1250 m² when the width, x, is 25 m and the length, $100 - 2x$, is 50 m.

4. **Check.** Note that 1250 m² is greater than any of the values for A found in the *Familiarize* step. To be more certain, we could check values other than those used in that step. For example, if $w = 26$ m, then $l = 100 - 2 \cdot 26 = 48$ m, and $A = 26 \cdot 48 = 1248$ m². Since 1250 m² is greater than 1248 m², it appears that we have a maximum.

5. **State.** The largest rectangular area for swimming that can be enclosed is 25 m by 50 m.

Fitting Quadratic Functions to Data

We can now model some real-world situations using quadratic functions. As always, before attempting to fit an equation to data, we should graph the data and compare the result to the shape of the graphs of different types of functions and ask what type of function seems appropriate.

Connecting the Concepts

The general shapes of the graphs of many functions that we have studied are shown below.

Linear function:
$f(x) = mx + b$

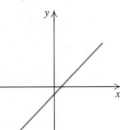

Absolute-value function:
$f(x) = |x|$

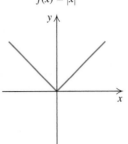

Rational function:
$f(x) = \dfrac{1}{x}$

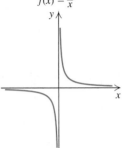

Radical function:
$f(x) = \sqrt{x}$

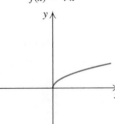

Quadratic function:
$f(x) = ax^2 + bx + c,\, a > 0$

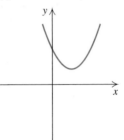

Quadratic function:
$f(x) = ax^2 + bx + c,\, a < 0$

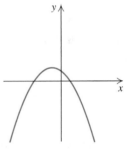

We see that in order for a quadratic function to fit a set of data, the graph of the data must approximate the shape of a parabola. Data that resemble one half of a parabola might also be modeled using a quadratic function with a restricted domain, as illustrated in the figure below.

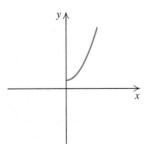

Quadratic function:
$f(x) = ax^2 + bx + c,$
$a > 0,\ x \geqslant 0$

Quadratic function:
$f(x) = ax^2 + bx + c,$
$a < 0,\ x \geqslant 0$

EXAMPLE 3 Determine whether a quadratic function can be used to model each of the following situations.

a) Sonoma Sunshine.   The percent of days in each month that the sun shines in Sonoma, California.

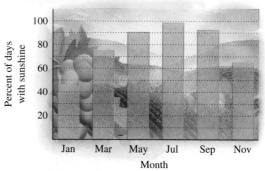

Source: www.city-data.com

b) Sonoma Precipitation.   The amount of rainfall each month in Sonoma, California.

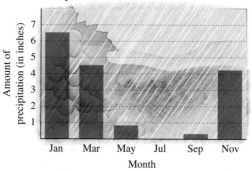

Source: www.city-data.com

c) Vehicles.   The number of vehicles registered in the United States.

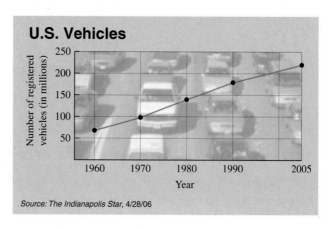

d) Alternative-Fueled Vehicles.   The number of hybrid-electric passenger vehicles sold in the United States.

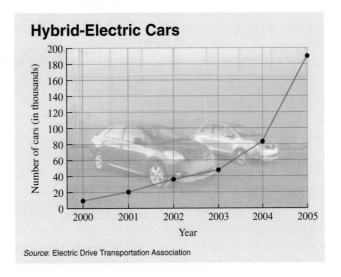

Hybrid-Electric Cars

Source: Electric Drive Transportation Association

e) Olympic Volunteers.   The number of volunteers at the Winter Olympics.

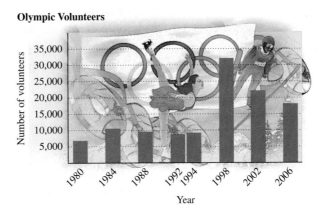

Olympic Volunteers

SOLUTION

a) The data rise and then fall, resembling a parabola that opens downward. The situation could be modeled by a quadratic function $f(x) = ax^2 + bx + c$, $a < 0$.

b) The data fall and then rise, resembling a parabola that opens upward. A quadratic function $f(x) = ax^2 + bx + c$, $a > 0$, might be used as a model for this situation.

c) The data appear nearly linear. A linear function is a better model for this situation than a quadratic function.

d) The data resemble the right half of a parabola that opens upward. We could use a quadratic function $f(x) = ax^2 + bx + c$, $a > 0$, $x \geq 0$, to model the situation.

e) The data do not resemble a parabola. A quadratic function would not be a good model for this situation.

EXAMPLE 4 Sonoma Sunshine. The percent of days each month during which the sun shines in Sonoma, California, can be modeled by a quadratic function, as seen in Example 3(a). To do this, we let x = the number of the month, where 1 represents January, 2 represents February, and so on.

Month	Percent of Days with Sunshine
1 (Jan.)	46
3 (March)	75
5 (May)	90
7 (July)	98
9 (Sep.)	92
11 (Nov.)	65

a) Use the data points (1, 46), (5, 90), and (11, 65) to fit a quadratic function $f(x)$ to the data.

b) Use the function from part (a) to estimate the percent of days in April that the sun shines in Sonoma.

c) Use the REGRESSION feature of a graphing calculator to fit a quadratic function $g(x)$ to the data.

d) Use the function from part (c) to estimate the percent of days in April that the sun shines in Sonoma, and compare the estimate with the estimate from part (b).

SOLUTION

a) Only one parabola will go through any three noncollinear points. The quadratic function found will depend on the choice of those points. We are looking for a function of the form $f(x) = ax^2 + bx + c$, where $f(x)$ is the percent of days with sunshine in month x. To solve for a, b, and c, we can use any of the methods discussed in Chapter 9 for solving a system of three equations. Here we use elimination, and substitute the given values of x and $f(x)$ from the three data points listed:

$$46 = a(1)^2 + b(1) + c, \qquad \text{Using the data point (1, 46)}$$
$$90 = a(5)^2 + b(5) + c, \qquad \text{Using the data point (5, 90)}$$
$$65 = a(11)^2 + b(11) + c. \qquad \text{Using the data point (11, 65)}$$

After simplifying, we have the system of three equations

$$46 = \quad a + \quad b + c, \qquad (1)$$
$$90 = 25a + 5b + c, \qquad (2)$$
$$65 = 121a + 11b + c. \qquad (3)$$

We multiply equation (1) by -25 and add that to equation (2) to eliminate a. We also multiply equation (1) by -121 and add that to equation (3) to eliminate a again.

$$-1150 = -25a - 25b - 25c \qquad -5566 = -121a - 121b - 121c$$
$$\underline{90 = \quad 25a + \ 5b + \quad c} \qquad \underline{65 = \quad 121a + \ 11b + \quad c}$$
$$-1060 = \qquad -20b - 24c \qquad -5501 = \qquad -110b - 120c$$

We obtain the system of equations

$$-1060 = -20b - 24c,$$
$$-5501 = -110b - 120c.$$

We can divide both sides of the first equation by -4, and both sides of the second equation by -1, to obtain

$$265 = \quad 5b + \quad 6c, \qquad (4)$$
$$5501 = 110b + 120c. \qquad (5)$$

To solve, we multiply equation (4) by -22 and add:

$$-5830 = -110b - 132c$$
$$\underline{5501 = \quad 110b + 120c}$$
$$-329 = \qquad -12c$$
$$\tfrac{329}{12} = c.$$

Next, we solve for b, using equation (4) above:

$265 = 5b + 6\left(\tfrac{329}{12}\right)$	**Substituting**
$265 = 5b + \tfrac{329}{2}$	**Multiplying**
$530 = 10b + 329$	**Clearing fractions**
$201 = 10b$	
$\tfrac{201}{10} = b.$	

Finally, we use equation (1) to solve for a:

$46 = a + \tfrac{201}{10} + \tfrac{329}{12}$	**Substituting**
$\tfrac{2760}{60} = a + \tfrac{1206}{60} + \tfrac{1645}{60}$	**The LCD is 60.**
$-\tfrac{91}{60} = a.$	**Solving for a**

Thus the function

$$f(x) = -\tfrac{91}{60}x^2 + \tfrac{201}{10}x + \tfrac{329}{12}$$

fits the three points given. As a partial check, we graph the function along with the data, as shown at left. The function goes through the three given points.

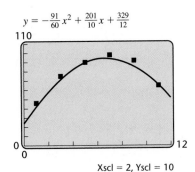

$y = -\tfrac{91}{60}x^2 + \tfrac{201}{10}x + \tfrac{329}{12}$

Xscl = 2, Yscl = 10

b) Since April is month 4, $f(4)$ will give an estimate of the percent of days of sunshine in April:

$$f(x) = -\tfrac{91}{60}(4)^2 + \tfrac{201}{10}(4) + \tfrac{329}{12} \approx 84.$$

According to this model, the sun shines on about 84% of April days in Sonoma.

QuadReg
$y = ax^2 + bx + c$
$a = -1.625$
$b = 21.7$
$c = 24.925$

c) Using regression allows us to consider all the data given in the problem, not just three points. After entering the data, we choose the QUADREG option in the STAT CALC menu.

From the first screen at left, we obtain the quadratic function

$$g(x) = -1.625x^2 + 21.7x + 24.925.$$

On the second screen at left, we graph $g(x)$, along with the data points.

$y = -1.625x^2 + 21.7x + 24.925$

110

0
0   12

Xscl = 2, Yscl = 10

d) To estimate the percent of days of sunshine in April, we evaluate $g(4)$ and obtain $g(4) \approx 86$. According to this model, the sun will shine about 86% of April days in Sonoma. This estimate is slightly higher than the 84% found in part (b). The second model, which took into consideration all the data, is a better fit for the data. Without further information, it is difficult to tell which model more accurately describes the sunshine pattern in Sonoma.

11.8 EXERCISE SET

↪ *Concept Reinforcement*   *In each of Exercises 1–6, match the description with the graph that displays that characteristic.*

1. ___ A minimum value of $f(x)$ exists.

2. ___ A maximum value of $f(x)$ exists.

3. ___ No maximum or minimum value of $f(x)$ exists.

4. ___ The data points appear to suggest a linear model.

5. ___ The data points appear to suggest a quadratic model with a maximum.

6. ___ The data points appear to suggest a quadratic model with a minimum.

a)

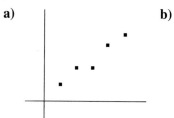

b) $f(x)$

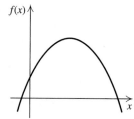

c) $f(x)$

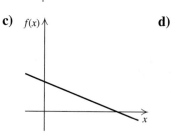

d)

e) $f(x)$

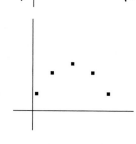

f)

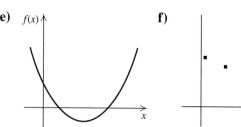

Solve.

7. *Ticket Sales.* The number of tickets sold each day for an upcoming Los Lobos show can be approximated by

$$N(x) = -0.4x^2 + 9x + 11,$$

where x is the number of days since the concert was first announced. When will daily ticket sales peak and how many tickets will be sold that day?

8. *Stock Prices.* The value of a share of I. J. Solar can be represented by $V(x) = x^2 - 6x + 13$, where x is the number of months after January 2004. What is the lowest value $V(x)$ will reach, and when did that occur?

9. *Minimizing Cost.* Sweet Harmony Crafts has determined that when x hundred Dobros are built, the average cost per Dobro can be estimated by

$$C(x) = 0.1x^2 - 0.7x + 2.425,$$

where $C(x)$ is in hundreds of dollars. What is the minimum average cost per Dobro and how many Dobros should be built to achieve that minimum?

10. *Maximizing Profit.* Recall that total profit P is the difference between total revenue R and total cost C. Given $R(x) = 1000x - x^2$ and $C(x) = 3000 + 20x$, find the total profit, the maximum value of the total profit, and the value of x at which it occurs.

11. *Furniture Design.* A furniture builder is designing a rectangular end table with a perimeter of 128 in. What dimensions will yield the maximum area?

12. *Architecture.* An architect is designing an atrium for a hotel. The atrium is to be rectangular with a perimeter of 720 ft of brass piping. What dimensions will maximize the area of the atrium?

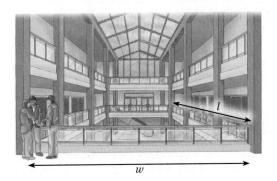

13. *Patio Design.* A stone mason has enough stones to enclose a rectangular patio with 60 ft of perimeter, assuming that the attached house forms one side of the rectangle. What is the maximum area that the mason can enclose? What should the dimensions of the patio be in order to yield this area?

14. *Garden Design.* Ginger is fencing in a rectangular garden, using the side of her house as one side of the rectangle. What is the maximum area that she can enclose with 40 ft of fence? What should the dimensions of the garden be in order to yield this area?

15. *Molding Plastics.* Economite Plastics plans to produce a one-compartment vertical file by bending the long side of an 8-in. by 14-in. sheet of plastic along two lines to form a U shape. How tall should the file be in order to maximize the volume that the file can hold?

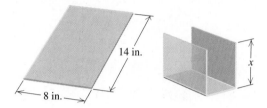

16. *Composting.* A rectangular compost container is to be formed in a corner of a fenced yard, with 8 ft of chicken wire completing the other two sides of the rectangle. If the chicken wire is 3 ft high, what dimensions of the base will maximize the container's volume?

17. What is the maximum product of two numbers that add to 18? What numbers yield this product?

18. What is the maximum product of two numbers that add to 26? What numbers yield this product?

19. What is the minimum product of two numbers that differ by 8? What are the numbers?

20. What is the minimum product of two numbers that differ by 7? What are the numbers?

Aha! **21.** What is the maximum product of two numbers that add to −10? What numbers yield this product?

22. What is the maximum product of two numbers that add to −12? What numbers yield this product?

Choosing Models. *For the scatterplots and graphs in Exercises 23–32, determine which, if any, of the following functions might be used as a model for the data: Linear, with $f(x) = mx + b$; quadratic, with $f(x) = ax^2 + bx + c, a > 0$; quadratic, with $f(x) = ax^2 + bx + c, a < 0$; neither quadratic nor linear.*

23.

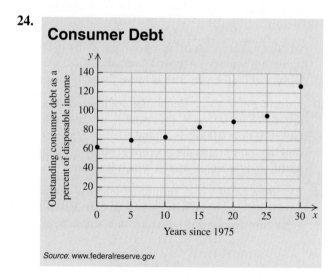

*Estimated
Source: Statistical Abstract of the United States, 2006*

24.

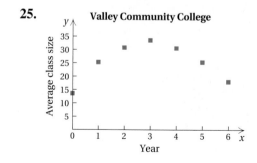

Source: www.federalreserve.gov

25.

26.

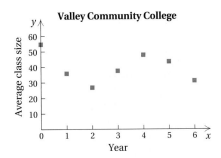

Valley Community College

27.

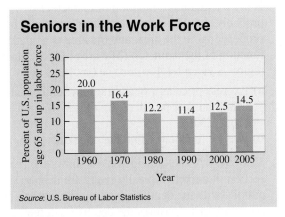

Seniors in the Work Force

Source: U.S. Bureau of Labor Statistics

28.

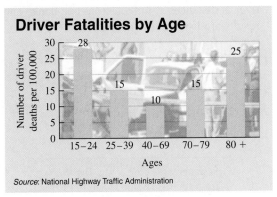

Driver Fatalities by Age

Source: National Highway Traffic Administration

29.

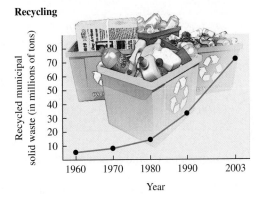

Recycling

Source: Environmental Protection Agency

30.

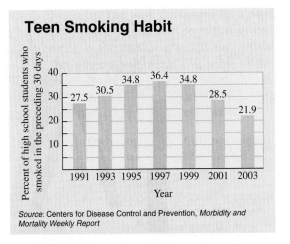

Teen Smoking Habit

Source: Centers for Disease Control and Prevention, *Morbidity and Mortality Weekly Report*

31.

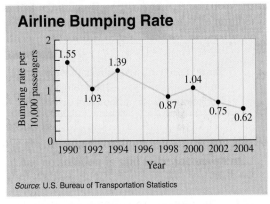

Airline Bumping Rate

Source: U.S. Bureau of Transportation Statistics

32.

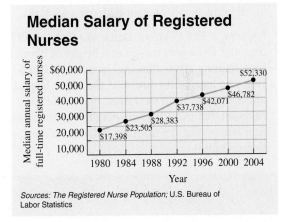

Median Salary of Registered Nurses

Sources: The Registered Nurse Population; U.S. Bureau of Labor Statistics

Find a quadratic function that fits the set of data points.

33. $(1, 4), (-1, -2), (2, 13)$

34. $(1, 4), (-1, 6), (-2, 16)$

35. $(2, 0), (4, 3), (12, -5)$

36. $(-3, -30), (3, 0), (6, 6)$

37. a) Find a quadratic function that fits the following data.

Travel Speed (in kilometers per hour)	Number of Nighttime Accidents (for every 200 million kilometers driven)
60	400
80	250
100	250

b) Use the function to estimate the number of nighttime accidents that occur at 50 km/h.

38. a) Find a quadratic function that fits the following data.

Travel Speed (in kilometers per hour)	Number of Daytime Accidents (for every 200 million kilometers driven)
60	100
80	130
100	200

b) Use the function to estimate the number of daytime accidents that occur at 50 km/h.

39. *Archery.* The Olympic flame tower at the 1992 Summer Olympics was lit at a height of about 27 m by a flaming arrow that was launched about 63 m from the base of the tower. If the arrow landed about 63 m beyond the tower, find a quadratic function that expresses the height h of the arrow as a function of the distance d that it traveled horizontally.

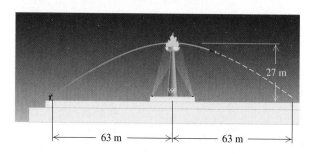

40. *Pizza Prices.* Pizza Unlimited has the following prices for pizzas.

Diameter	Price
8 in.	$ 6.00
12 in.	$ 8.50
16 in.	$11.50

Is price a quadratic function of diameter? It probably should be, because the price should be proportional to the area, and the area is a quadratic function of the diameter. (The area of a circular region is given by $A = \pi r^2$ or $(\pi/4) \cdot d^2$.)

a) Express price as a quadratic function of diameter using the data points (8, 6), (12, 8.50), and (16, 11.50).

b) Use the function to find the price of a 14-in. pizza.

41. *Hydrology.* The drawing below shows the cross section of a river. Typically rivers are deepest in the middle, with the depth decreasing to 0 at the edges. A hydrologist measures the depths D, in feet, of a river at distances x, in feet, from one bank. The results are listed in the table below.

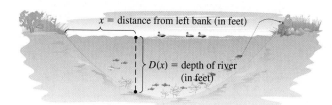

Distance x, from the Left Bank (in feet)	Depth, D, of the River (in feet)
0	0
15	10.2
25	17
50	20
90	7.2
100	0

a) Use regression to find a quadratic function that fits the data.

b) Use the function to estimate the depth of the river 70 ft from the left bank.

42. *Work Force.*   The graph in Exercise 27 indicates that the percent of the U.S. population age 65 and older that is in the work force is increasing after several decades of decrease.

  a) Use regression to find a quadratic function that can be used to estimate the percent p of the population 65 and older that is in the work force x years after 1960.

  b) Use the function found in part (a) to predict the percent of those 65 and older who will be in the work force in 2010.

43. *Alternative-Fueled Vehicles.*   The number of hybrid-electric passenger vehicles sold in the United States during several years is shown in the table below. As we saw in Example 3, a quadratic function can be used to model this data.

Year	Number of Hybrid-Electric Passenger Vehicles
2000	9,367
2001	20,287
2002	35,961
2003	47,525
2004	83,153
2005	189,916

Source: Electric Drive Transportation Association

  a) Use regression to find a quadratic function that can be used to estimate the number of hybrid-electric vehicles v sold x years after 2000.

  b) Use the function found in part (a) to predict the number of hybrid-electric vehicles sold in 2008.

44. *Teen Smoking.*   The graph in Exercise 30 indicates that the percent of high school students who smoke is decreasing after several years of increase.

  a) Use regression to find a quadratic function that can be used to estimate the percent t of high school students who smoked in the preceding 30 days x years after 1990.

  b) Use the function found in part (a) to estimate the percent of high school students who smoked in the preceding 30 days in 2005.

45. Does every nonlinear function have a minimum or a maximum value? Why or why not?

46. Explain how the leading coefficient of a quadratic function can be used to determine if a maximum or a minimum function value exists.

Skill Maintenance

Simplify.

47. $\dfrac{x}{x^2 + 17x + 72} - \dfrac{8}{x^2 + 15x + 56}$ [7.4]

48. $\dfrac{x^2 - 9}{x^2 - 8x + 7} \div \dfrac{x^2 + 6x + 9}{x^2 - 1}$ [7.2]

Solve. [2.6]

49. $5x - 9 < 31$ **50.** $3x - 8 \geq 22$

51. *Registered Nurses' Salary.*   The average annual salary of full-time registered nurses is shown in Exercise 32. Use the points (4, 23,505) and (20, 46,782) to find a linear function that can be used to estimate the average annual salary r of registered nurses x years after 1980. [3.7], [3.8]

52. Use the data in Exercise 32 and regression to find a linear function that can be used to predict the average annual salary r of registered nurses x years after 1980. [3.7], [3.8]

Synthesis

53. Write a problem for a classmate to solve. Design the problem so that its solution requires finding a minimum or maximum function value.

54. Explain what restrictions should be placed on the quadratic functions developed in Exercises 37 and 42 and why such restrictions are needed.

55. *Bridge Design.*   The cables supporting a straight-line suspension bridge are nearly parabolic in shape. Suppose that a suspension bridge is being designed with concrete supports 160 ft apart and with vertical cables 30 ft above road level at the midpoint of the bridge and 80 ft above road level at a point 50 ft from the midpoint of the bridge. How long are the longest vertical cables?

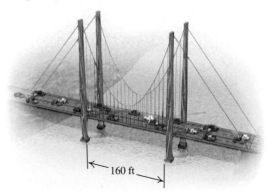

160 ft

56. *Trajectory of a Launched Object.*   The height above the ground of a launched object is a quadratic function of the time that it is in the air. Suppose that a flare is launched from a cliff 64 ft above sea level. If 3 sec after being launched the flare is again level with the cliff, and if 2 sec after that it lands in the sea, what is the maximum height that the flare will reach?

57. *Norman Window.*   A *Norman window* is a rectangle with a semicircle on top. Big Sky Windows is designing a Norman window that will require 24 ft of trim. What dimensions will allow the maximum amount of light to enter a house?

58. *Crop Yield.*   An orange grower finds that she gets an average yield of 40 bushels (bu) per tree when she plants 20 trees on an acre of ground. Each time she adds a tree to an acre, the yield per tree decreases by 1 bu, due to congestion. How many trees per acre should she plant for maximum yield?

59. *Cover Charges.*   When the owner of Sweet Sounds charges a $10 cover charge, an average of 80 people will attend a show. For each 25¢ increase in admission price, the average number attending decreases by 1. What should the owner charge in order to make the most money?

60. *Minimizing Area.*   A 36-in. piece of string is cut into two pieces. One piece is used to form a circle while the other is used to form a square. How should the string be cut so that the sum of the areas is a minimum?

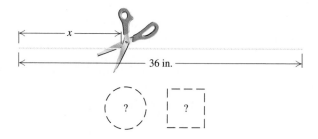

11 Chapter Summary and Review

KEY TERMS AND DEFINITIONS

QUADRATIC EQUATIONS

Quadratic equation, p. 804   Any equation that can be written in the form $ax^2 + bx + c = 0$, with a, b, and c constant.

Discriminant, p. 831   The radicand, $b^2 - 4ac$, in the quadratic formula.

$b^2 - 4ac = 0 \rightarrow$ One solution, a rational number

$b^2 - 4ac > 0 \rightarrow$ Two real solutions, both rational if $b^2 - 4ac$ is a perfect square

$b^2 - 4ac < 0 \rightarrow$ Two imaginary-number solutions

Vertex of a parabola, $f(x) = ax^2 + bx + c$, p. 855

$$\left(-\frac{b}{2a}, f\left(-\frac{b}{2a}\right)\right), \quad \text{or} \quad \left(-\frac{b}{2a}, \frac{4ac - b^2}{4a}\right)$$

APPLICATIONS

Compound interest

$$A = P(1 + r)^t$$

Free-fall distance (in feet)

$$s = 16t^2$$

IMPORTANT CONCEPTS

[Section references appear in brackets.]

Concept	Example
Graphs of quadratic equations are **parabolas**.	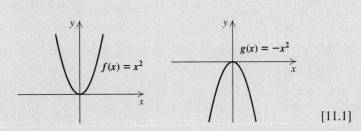 [11.1]
The **principle of square roots** can be used to solve a quadratic equation if a perfect-square trinomial is on one side and a constant is on the other side. For any real number k, if $x^2 = k$, then $x = \sqrt{k}$ or $x = -\sqrt{k}$.	$x^2 - 8x + 16 = 25$ $(x - 4)^2 = 25$ $x - 4 = -5 \quad or \quad x - 4 = 5$ $x = -1 \quad or \qquad x = 9$ [11.1]

(continued)

Factoring and using the **principle of zero products** is the easiest method to use to solve a quadratic equation if you can factor the polynomial.	$x^2 - 3x - 10 = 0$ $(x + 2)(x - 5) = 0$ $x + 2 = 0 \quad or \quad x - 5 = 0$ $\quad x = -2 \quad or \qquad x = 5$ [11.1]
Completing the square can be used to solve any quadratic equation. Calculations can be lengthy.	$x^2 + 6x = 1$ $x^2 + 6x + \left(\frac{6}{2}\right)^2 = 1 + \left(\frac{6}{2}\right)^2$ $x^2 + 6x + 9 = 1 + 9$ $(x + 3)^2 = 10$ $x + 3 = \pm\sqrt{10}$ $x = -3 \pm \sqrt{10}$ [11.1]
The **quadratic formula** can be used to solve any quadratic equation. If $ax^2 + bx + c = 0$, then $x = \dfrac{-b \pm \sqrt{b^2 - 4ac}}{2a}$.	$x^2 - 2x - 5 = 0$ $x = \dfrac{-(-2) \pm \sqrt{(-2)^2 - 4(1)(-5)}}{2 \cdot 1}$ $x = \dfrac{2 \pm \sqrt{4 + 20}}{2}$ $x = \dfrac{2}{2} \pm \dfrac{2\sqrt{6}}{2}$ $x = 1 \pm \sqrt{6}$ [11.2]
Equations that are **reducible to quadratic** or **quadratic in form** can be solved by making an appropriate substitution, solving a quadratic equation, and then solving for the original variable.	$x^4 - 10x^2 + 9 = 0$ Let $u = x^2$. Then $u^2 = x^4$. $u^2 - 10u + 9 = 0$ $(u - 9)(u - 1) = 0$ $u - 9 = 0 \quad or \quad u - 1 = 0$ $\quad u = 9 \quad or \qquad u = 1$ $\quad x^2 = 9 \quad or \qquad x^2 = 1$ $\quad x = \pm 3 \quad or \qquad x = \pm 1.$ [11.5]

(continued)

The graph of $f(x) = ax^2 + bx + c = a(x - h)^2 + k$ is a parabola.

The vertex is (h, k).

The axis of symmetry is $x = h$.

The domain of f is $(-\infty, \infty)$.

If $a > 0$:

   The parabola opens up.

   The minimum function value is k.

   The range of f is $[k, \infty)$.

If $a < 0$:

   The parabola opens down.

   The maximum function value is k.

   The range of f is $(-\infty, k]$.

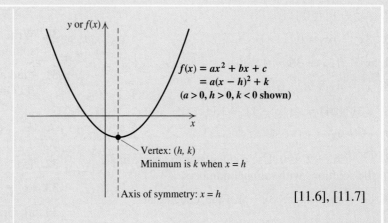

y or $f(x)$

$f(x) = ax^2 + bx + c$
$= a(x - h)^2 + k$
$(a > 0, h > 0, k < 0$ shown$)$

x

Vertex: (h, k)
Minimum is k when $x = h$

Axis of symmetry: $x = h$

[11.6], [11.7]

Review Exercises

Classify each statement as either true or false.

1. Every quadratic equation has two different solutions. [11.4]

2. Every quadratic equation has at least one solution. [11.4]

3. If an equation cannot be solved by completing the square, it cannot be solved by the quadratic formula. [11.2]

4. A negative discriminant indicates two imaginary-number solutions of a quadratic equation. [11.4]

5. Certain radical or rational equations can be written in quadratic form. [11.5]

6. The graph of $f(x) = 2(x + 3)^2 - 4$ has its vertex at $(3, -4)$. [11.6]

7. The graph of $g(x) = 5x^2$ has $x = 0$ as its axis of symmetry. [11.6]

8. The graph of $f(x) = -2x^2 + 1$ has no minimum value. [11.6]

9. The zeros of $g(x) = x^2 - 9$ are -3 and 3. [11.6]

10. The graph of every quadratic function has at least one x-intercept. [11.7]

11. Given the following graph of $f(x) = ax^2 + bx + c$.

   a) State the number of real-number solutions of $ax^2 + bx + c = 0$. [11.1]

   b) State whether a is positive or negative. [11.6]

   c) Determine the minimum value of f. [11.6]

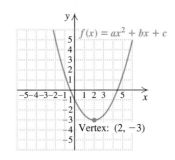

y

$f(x) = ax^2 + bx + c$

x

Vertex: $(2, -3)$

Solve.

12. $4x^2 - 9 = 0$ [11.1]

13. $8x^2 + 6x = 0$ [11.1]

14. $x^2 - 12x + 36 = 9$ [11.1]

15. $x^2 - 4x + 8 = 0$ [11.2]

16. $x(3x + 4) = 4x(x - 1) + 15$ [11.2]

17. $x^2 + 9x = 1$ [11.2]

18. $x^2 - 5x - 2 = 0$. Use a calculator to approximate the solutions with rational numbers. [11.2]

19. Let $f(x) = 4x^2 - 3x - 1$. Find x such that $f(x) = 0$. [11.2]

Replace the blanks with constants to form a true equation. [11.1]

20. $x^2 - 12x +$ ___ $= (x -$ ___$)^2$

21. $x^2 + \frac{3}{5}x +$ ___ $= \left(x +$ ___$\right)^2$

22. Solve by completing the square. Show your work.
$$x^2 - 6x + 1 = 0 \text{ [11.1]}$$

23. $2500 grows to $3025 in 2 yr. Use the formula $A = P(1 + r)^t$ to find the interest rate. [11.1]

24. The Peachtree Center Plaza in Atlanta, Georgia, is 723 ft tall. Use $s = 16t^2$ to approximate how long it would take an object to fall from the top. [11.1]

Solve. [11.3]

25. A corporate pilot must fly from company head-quarters to a manufacturing plant and back in 4 hr. The distance between headquarters and the plant is 300 mi. If there is a 20-mph headwind going and a 20-mph tailwind returning, how fast must the plane be able to travel in still air?

26. Working together, Erica and Shawna can answer a day's worth of customer-service e-mails in 4 hr. Working alone, Erica takes 6 hr longer than Shawna. How long would it take Shawna to answer the e-mails alone?

For each equation, determine whether the solutions are real or imaginary. If they are real, specify whether they are rational or irrational. [11.4]

27. $x^2 + 3x - 6 = 0$ **28.** $x^2 + 2x + 5 = 0$

29. Write a quadratic equation having the solutions $\sqrt{5}$ and $-\sqrt{5}$. [11.4]

30. Write a quadratic equation having -4 as its only solution. [11.4]

31. Find all x-intercepts of the graph of $f(x) = x^4 - 13x^2 + 36$. [11.5]

Solve. [11.5]

32. $15x^{-2} - 2x^{-1} - 1 = 0$

33. $(x^2 - 4)^2 - (x^2 - 4) - 6 = 0$

34. **a)** Graph: $f(x) = -3(x + 2)^2 + 4$. [11.6]
  b) Label the vertex.
  c) Draw the axis of symmetry.
  d) Find the maximum or the minimum value.

35. For the function given by $f(x) = 2x^2 - 12x + 23$: [11.7]
  a) find the vertex and the axis of symmetry;
  b) graph the function.

36. Find the x- and y-intercepts of
$$f(x) = x^2 - 9x + 14. \text{ [11.7]}$$

37. Solve $N = 3\pi\sqrt{1/p}$ for p. [11.3]

38. Solve $2A + T = 3T^2$ for T. [11.3]

Determine which, if any, of the following functions might be used as a model for the data: Linear, with $f(x) = mx + b$; quadratic, with $f(x) = ax^2 + bx + c$, $a > 0$; quadratic, with $f(x) = ax^2 + bx + c$, $a < 0$; neither quadratic nor linear. [11.8]

39.

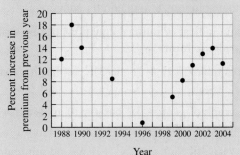

40.

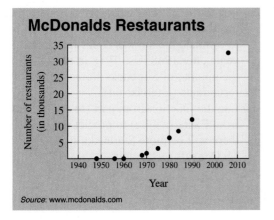

McDonalds Restaurants

Source: www.mcdonalds.com

41.

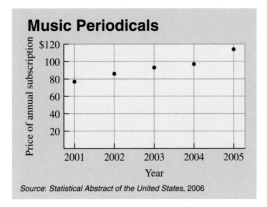

Music Periodicals

Source: Statistical Abstract of the United States, 2006

42. Eastgate Consignments wants to build a rectangular area in a corner for children to play in while their parents shop. They have 30 ft of low fencing. What is the maximum area they can enclose? What dimensions will yield this area? [11.8]

43. *McDonalds.* The following table lists the number of McDonalds restaurants for various years (see Exercise 40). [11.8]

Year	Number of McDonalds Restaurants
1948	1
1956	14
1960	228
1968	1,000
1970	1,600
1975	3,076
1980	6,263
1984	8,300
1990	11,800
2006	32,000

a) Use the data points (0, 1), (20, 1000), and (36, 8300) to find a quadratic function that can be used to estimate the number M of McDonalds restaurants x years after 1948.

b) Use the function to estimate the number of McDonalds restaurants in 2010.

44. a) Use the REGRESSION feature of a graphing calculator and all the data in Exercise 43 to find a quadratic function that can be used to estimate the number M of McDonalds restaurants x years after 1948. [11.8]

b) Use the function to estimate the number of McDonalds restaurants in 2010. [11.8]

Synthesis

TW 45. Discuss two ways in which completing the square was used in this chapter. [11.1], [11.2], [11.7]

TW 46. Compare the results of Exercises 43(b) and 44(b). Which model do you think gives a better prediction? [11.8]

47. A quadratic function has x-intercepts at -3 and 5. If the y-intercept is at -7, find an equation for the function. [11.7]

48. Find h and k if, for $3x^2 - hx + 4k = 0$, the sum of the solutions is 20 and the product of the solutions is 80. [11.4]

49. The average of two positive integers is 171. One of the numbers is the square root of the other. Find the integers. [11.5]

50. *Starbucks.* The following table lists the number of Starbucks stores for various years. [11.8]

Year	Number of Starbucks Stores
2001	3475
2002	4242
2003	5201
2004	6132
2005	7950

Source: www.starbucks.com

a) Use the REGRESSION feature of a graphing calculator to find a quadratic function that can be used to estimate the number S of Starbucks stores x years after 1948. Use 1948 in order to compare the number of Starbucks stores and McDonalds restaurants (Exercise 44).

b) Use the function from part (a) and the function from Exercise 44(a) to predict the years in which there will be more Starbucks stores than McDonalds restaurants.

Chapter Test 11

1. Given the following graph of $f(x) = ax^2 + bx + c$.
  a) State the number of real-number solutions of $ax^2 + bx + c = 0$.
  b) State whether a is positive or negative.
  c) Determine the maximum function value of f.

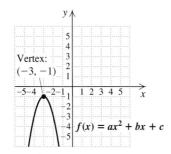

Solve.

2. $3x^2 - 16 = 0$

3. $4x(x - 2) - 3x(x + 1) = -18$

4. $x^2 + x + 1 = 0$

5. $2x + 5 = x^2$

6. $x^{-2} - x^{-1} = \frac{3}{4}$

7. $x^2 + 3x = 5$. Use a calculator to approximate the solutions with rational numbers.

8. Let $f(x) = 12x^2 - 19x - 21$. Find x such that $f(x) = 0$.

Replace the blanks with constants to form a true equation.

9. $x^2 - 16x + \underline{\quad} = (x - \underline{\quad})^2$

10. $x^2 + \frac{2}{7}x + \underline{\quad} = \left(x + \underline{\quad}\right)^2$

11. Solve by completing the square. Show your work.
$$x^2 + 10x + 15 = 0$$

Solve.

12. The Connecticut River flows at a rate of 4 km/h for the length of a popular scenic route. In order for a cruiser to travel 60 km upriver and then return in a total of 8 hr, how fast must the boat be able to travel in still water?

13. Brock and Ian can assemble a swing set in $1\frac{1}{2}$ hr. Working alone, it takes Ian 4 hr longer than Brock to assemble the swing set. How long would it take Brock, working alone, to assemble the swing set?

14. Determine the type of number that the solutions of $x^2 + 5x + 17 = 0$ will be.

15. Write a quadratic equation having solutions -2 and $\frac{1}{3}$.

16. Find all x-intercepts of the graph of
$$f(x) = (x^2 + 4x)^2 + 2(x^2 + 4x) - 3.$$

17. **a)** Graph: $f(x) = 4(x - 3)^2 + 5$.
  b) Label the vertex.
  c) Draw the axis of symmetry.
  d) Find the maximum or the minimum function value.

18. For the function $f(x) = 2x^2 + 4x - 6$:
  a) find the vertex and the axis of symmetry;
  b) graph the function.

19. Find the x- and y-intercepts of
$$f(x) = x^2 - x - 6.$$

20. Solve $V = \frac{1}{3}\pi(R^2 + r^2)$ for r. Assume all variables are positive.

21. State whether the graph appears to represent a linear function, a quadratic function, or neither.

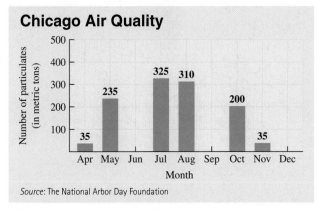

Chicago Air Quality

Source: The National Arbor Day Foundation

22. Jay's Custom Pickups has determined that when x hundred truck caps are built, the average cost per cap is given by
$$C(x) = 0.2x^2 - 1.3x + 3.4025,$$

where $C(x)$ is in hundreds of dollars. What is the minimum cost per truck cap and how many caps should be built to achieve that minimum?

23. Trees improve air quality in part by retaining airborne particles, called particulates, from the air. The following table shows the number of metric tons of particulates retained by trees in Chicago during the spring, summer, and autumn months. (See Exercise 21.)

Month		Amount of Particulates Retained (in metric tons)
April	(0)	35
May	(1)	235
July	(3)	325
August	(4)	310
October	(6)	200
November	(7)	35

Source: The National Arbor Day Foundation

Use the data points $(0, 35)$, $(4, 310)$, and $(6, 200)$ to find a quadratic function that can be used to estimate the number of metric tons p of particulates retained x months after April.

24. Use regression to find a quadratic function that fits the data in Exercise 23.

Synthesis

25. One solution of $kx^2 + 3x - k = 0$ is -2. Find the other solution.

26. Find a fourth-degree polynomial equation, with integer coefficients, for which $2 - \sqrt{3}$ and $5 - i$ are solutions.

27. Find a polynomial equation, with integer coefficients, for which 5 is a repeated root and $\sqrt{2}$ and $\sqrt{3}$ are solutions.

12

Exponential and Logarithmic Functions

The functions that we consider in this chapter are interesting not only from a purely intellectual point of view, but also for their rich applications to many fields. We will look at applications such as compound interest and population growth, to name just two.

The theory centers on functions with variable exponents (*exponential functions*). We study these functions, their inverses, and their properties.

APPLICATION *Cruise-Ship Passengers.*

In 1970, cruise lines carried approximately 500,000 passengers. This number was projected to increase to 15 million in 2006. (*Sources*: Cruise Lines International; *Annual Cruise Review* 2004) Use the data in the following table to estimate the number of cruise-ship passengers in 2012.

YEAR	NUMBER OF PASSENGERS (IN MILLIONS)
1970	0.5
1980	1.4
1990	3.6
2000	6.5
2004	13.4
2006	15 (proj.)

Xscl = 5, Yscl = 5

This problem appears as Example 9 in Section 12.7.

12.1

Composite and Inverse Functions

Composite Functions ■ Inverses and One-to-One Functions ■
Finding Formulas for Inverses ■ Graphing Functions and
Their Inverses ■ Inverse Functions and Composition

Composite Functions

In the real world, functions frequently occur in which some quantity depends on a variable that, in turn, depends on another variable. For instance, a firm's profits may depend on the number of items the firm produces, which may in turn depend on the number of employees hired. Functions like this are called **composite functions.**

For example, the function g that gives a correspondence between women's shoe sizes in the United States and those in Italy is given by $g(x) = 2x + 24$, where x is the U.S. size and $g(x)$ is the Italian size. Thus a U.S. size 4 corresponds to a shoe size of $g(4) = 2 \cdot 4 + 24$, or 32, in Italy.

A different function gives a correspondence between women's shoe sizes in Italy and those in Britain. This particular function is given by $f(x) = \frac{1}{2}x - 14$, where x is the Italian size and $f(x)$ is the corresponding British size. Thus an Italian size 32 corresponds to a British size $f(32) = \frac{1}{2} \cdot 32 - 14$, or 2.

It seems reasonable to conclude that a shoe size of 4 in the United States corresponds to a size of 2 in Britain and that some function h describes this correspondence. Can we find a formula for h? If we look at the following tables, we might guess that such a formula is $h(x) = x - 2$, and that is indeed correct. But, for more complicated formulas, we would need to use algebra.

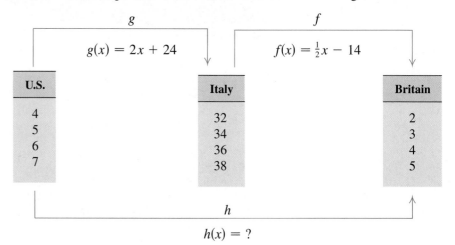

Size x shoes in the United States correspond to size $g(x)$ shoes in Italy, where

$$g(x) = 2x + 24.$$

Size n shoes in Italy correspond to size $f(n)$ shoes in Britain. Thus size $g(x)$ shoes in Italy correspond to size $f(g(x))$ shoes in Britain. Since the x in the expression $f(g(x))$ represents a U.S. shoe size, we can find the British shoe size that corresponds to a U.S. size x as follows:

$$f(g(x)) = f(2x + 24) = \tfrac{1}{2} \cdot (2x + 24) - 14 \quad \text{Using } g(x) \text{ as an input}$$
$$= x + 12 - 14 = x - 2.$$

This gives a formula for h: $h(x) = x - 2$. Thus U.S. size 4 corresponds to British size $h(4) = 4 - 2$, or 2. The function h is the *composition* of f and g and is denoted $f \circ g$ (read "the composition of f and g," "f composed with g," or "f circle g").

> **Composition of Functions** The *composite function $f \circ g$*, the *composition* of f and g, is defined as
>
> $$(f \circ g)(x) = f(g(x)).$$

We can visualize the composition of functions as follows.

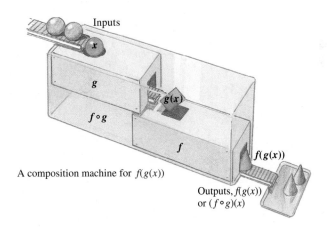

A composition machine for $f(g(x))$

EXAMPLE 1   Given $f(x) = 3x$ and $g(x) = 1 + x^2$:

a) Find $(f \circ g)(5)$ and $(g \circ f)(5)$.

b) Find $(f \circ g)(x)$ and $(g \circ f)(x)$.

SOLUTION   Consider each function separately:

$$f(x) = 3x \qquad \textbf{This function multiplies each input by 3.}$$

and $\quad g(x) = 1 + x^2. \quad \textbf{This function adds 1 to the square of each input.}$

a) To find $(f \circ g)(5)$, we first find $g(5)$ and then use that as an input for f:

$$(f \circ g)(5) = f(g(5)) = f(1 + 5^2) \qquad \text{Using } g(x) = 1 + x^2$$
$$= f(26) = 3 \cdot 26 = 78. \qquad \text{Using } f(x) = 3x$$

To find $(g \circ f)(5)$, we first find $f(5)$ and then use that as an input for g:

$$(g \circ f)(5) = g(f(5)) = g(3 \cdot 5) \qquad \text{Note that } f(5) = 3 \cdot 5 = 15.$$
$$= g(15) = 1 + 15^2 = 1 + 225 = 226.$$

b) We find $(f \circ g)(x)$ by substituting $g(x)$ for x in the equation for $f(x)$:

$$(f \circ g)(x) = f(g(x)) = f(1 + x^2) \qquad \text{Using } g(x) = 1 + x^2$$
$$= 3 \cdot (1 + x^2) = 3 + 3x^2. \qquad \text{Using } f(x) = 3x. \textit{ These}$$

parentheses indicate multiplication.

To find $(g \circ f)(x)$, we substitute $f(x)$ for x in the equation for $g(x)$:

$$(g \circ f)(x) = g(f(x)) = g(3x) \qquad \text{Substituting } 3x \text{ for } f(x)$$
$$= 1 + (3x)^2 = 1 + 9x^2.$$

As a check, note that $(g \circ f)(5) = 1 + 9 \cdot 5^2 = 1 + 9 \cdot 25 = 226$, as expected from part (a) above.

Example 1 shows that, in general, $(f \circ g)(5) \neq (g \circ f)(5)$ and $(f \circ g)(x) \neq (g \circ f)(x)$.

EXAMPLE 2 Given $f(x) = \sqrt{x}$ and $g(x) = x - 1$, find $(f \circ g)(x)$ and $(g \circ f)(x)$.

SOLUTION

$$(f \circ g)(x) = f(g(x)) = f(x - 1) = \sqrt{x - 1} \qquad \text{Using } g(x) = x - 1$$
$$(g \circ f)(x) = g(f(x)) = g(\sqrt{x}) = \sqrt{x} - 1 \qquad \text{Using } f(x) = \sqrt{x}$$

To check using a graphing calculator, let

$$y_1 = f(x) = \sqrt{}(x),$$
$$y_2 = g(x) = x - 1,$$
$$y_3 = \sqrt{}(x - 1),$$

and $\quad y_4 = y_1(y_2(x)).$

If our work is correct, y_4 will be equivalent to y_3. We form a table of values, selecting only y_3 and y_4. The table on the left below indicates that the functions are probably equivalent.

X	Y₃	Y₄
1	0	0
1.5	.70711	.70711
2	1	1
2.5	1.2247	1.2247
3	1.4142	1.4142
3.5	1.5811	1.5811
4	1.7321	1.7321
X = 1		

X	Y₅	Y₆
1	0	0
1.5	.22474	.22474
2	.41421	.41421
2.5	.58114	.58114
3	.73205	.73205
3.5	.87083	.87083
4	1	1
X = 1		

Next, we let $y_5 = \sqrt{}(x) - 1$ and $y_6 = y_2(y_1(x))$ and select only y_5 and y_6. The table on the right above indicates that the functions are probably equivalent. Thus we conclude that

$$(f \circ g)(x) = \sqrt{x - 1} \quad \text{and} \quad (g \circ f)(x) = \sqrt{x} - 1.$$

Functions that are not defined by an equation can still be composed when the output of the first function is in the domain of the second function.

EXAMPLE 3 Use the following table to find each value, if possible.

a) $(y_2 \circ y_1)(3)$ **b)** $(y_1 \circ y_2)(3)$

X	Y1	Y2
0	−2	3
1	2	5
2	−1	7
3	1	9
4	0	11
5	0	13
6	−3	15

X = 0

SOLUTION

a) Since $(y_2 \circ y_1)(3) = y_2(y_1(3))$, we first find $y_1(3)$. We locate 3 in the x-column and then move across to the y_1-column to find $y_1(3) = 1$.

X	Y1	Y2
0	−2	3
1	2	5
2	−1	7
3	1	9
4	0	11
5	0	13
6	−3	15

Y2 = 5

The output 1 now becomes an input for the function y_2:

$$y_2(y_1(3)) = y_2(1).$$

To find $y_2(1)$, we locate 1 in the x-column and then move across to the y_2-column to find $y_2(1) = 5$. Thus, $(y_2 \circ y_1)(3) = 5$.

b) Since $(y_1 \circ y_2)(3) = y_1(y_2(3))$, we find $y_2(3)$ by locating 3 in the x-column and moving across to the y_2-column. We have $y_2(3) = 9$, so

$$(y_1 \circ y_2)(3) = y_1(y_2(3)) = y_1(9).$$

However, y_1 is not defined for $x = 9$, so $(y_1 \circ y_2)(3)$ is undefined. ◢

In fields ranging from chemistry to geology and economics, one needs to recognize how a function can be regarded as the composition of two "simpler" functions. This is sometimes called *de*composition.

EXAMPLE 4 If $h(x) = (7x + 3)^2$, find f and g such that $h(x) = (f \circ g)(x)$.

SOLUTION To find $h(x)$, we can think of first forming $7x + 3$ and then squaring. This suggests that $g(x) = 7x + 3$ and $f(x) = x^2$. We check by forming the composition:

$$(f \circ g)(x) = f(g(x)) = f(7x + 3) = (7x + 3)^2 = h(x), \text{ as desired.}$$

Student Notes

Throughout this chapter, keep in mind that functions are defined using dummy variables that represent an arbitrary member of the domain. Thus, $f(x) = x^2 - 3x$ and $f(t) = t^2 - 3t$ describe the same function f.

There are other less "obvious" answers. For example, if

$$f(x) = (x - 1)^2 \quad \text{and} \quad g(x) = 7x + 4,$$

then

$$(f \circ g)(x) = f(g(x)) = f(7x + 4)$$
$$= (7x + 4 - 1)^2 = (7x + 3)^2 = h(x).$$

Inverses and One-to-One Functions

Let's consider the following two functions. We think of them as relations, or correspondences.

Countries and their capitals

Domain (Set of Inputs)	Range (Set of Outputs)
Australia	→ Canberra
China	→ Beijing
Cyprus	→ Nicosia
Egypt	→ Cairo
Haiti	→ Port-au-Prince
Peru	→ Lima

U.S. Senators and their states

Domain (Set of Inputs)	Range (Set of Outputs)
Bunning, McConnell	→ Kentucky
Kerry, Kennedy	→ Massachusetts
Smith, Wyden	→ Oregon

Suppose we reverse the arrows. We obtain what is called the **inverse relation.** Are these inverse relations functions?

Countries and their capitals

Range (Set of Outputs)	Domain (Set of Inputs)
Australia ←	Canberra
China ←	Beijing
Cyprus ←	Nicosia
Egypt ←	Cairo
Haiti ←	Port-au-Prince
Peru ←	Lima

U.S. Senators and their states

Range (Set of Outputs)	Domain (Set of Inputs)
Bunning ←, McConnell	Kentucky
Kerry ←, Kennedy ←	Massachusetts
Smith ←, Wyden ←	Oregon

Recall that for each input, a function provides exactly one output. However, it is possible for different inputs to correspond to the same output. Only when this possibility is *excluded* will the inverse be a function. For the functions listed above, this means the inverse of the "Capital" correspondence is a function, but the inverse of the "U.S. Senator" correspondence is not.

In the Capital function, different inputs have different outputs, so it is a **one-to-one function.** In the U.S. Senator function, *Kerry* and *Kennedy* are both paired with *Massachusetts.* Thus the U.S. Senator function is not one-to-one.

One-To-One Function   A function f is *one-to-one* if different inputs have different outputs. That is, if for a and b in the domain of f with $a \neq b$, we have $f(a) \neq f(b)$, then the function f is one-to-one. If a function is one-to-one, then its inverse correspondence is also a function.

How can we tell graphically whether a function is one-to-one?

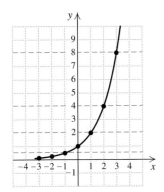

EXAMPLE 5   Shown at left is the graph of a function similar to those we will study in Section 12.2. Determine whether the function is one-to-one and thus has an inverse that is a function.

SOLUTION   A function is one-to-one if different inputs have different outputs—that is, if no two x-values have the same y-value. For this function, we cannot find two x-values that have the same y-value. Note that this means that no horizontal line can be drawn so that it crosses the graph more than once. The function is one-to-one so its inverse is a function.   ◢

The graph of every function must pass the vertical-line test. In order for a function to have an inverse that is a function, it must pass the *horizontal-line test* as well.

The Horizontal-Line Test   If it is impossible to draw a horizontal line that intersects a function's graph more than once, then the function is one-to-one. For every one-to-one function, an inverse function exists.

EXAMPLE 6   Determine whether the function $f(x) = x^2$ is one-to-one and thus has an inverse that is a function.

SOLUTION   The graph of $f(x) = x^2$ is shown here. Many horizontal lines cross the graph more than once. For example, the line $y = 4$ crosses where the first coordinates are -2 and 2. Although these are different inputs, they have the same output. That is, $-2 \neq 2$, but $f(-2) = f(2) = 4$.

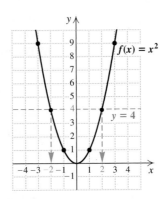

Thus the function is not one-to-one and no inverse function exists.   ◢

Finding Formulas for Inverses

When the inverse of f is also a function, it is denoted f^{-1} (read "f-inverse").

> **CAUTION!**   The -1 in f^{-1} is *not* an exponent!

Suppose a function is described by a formula. If its inverse is a function, how do we find a formula for that inverse? For any equation in two variables, if we interchange the variables, we form an equation of the inverse correspondence. If it is a function, we proceed as follows to find a formula for f^{-1}.

> **To Find a Formula for f^{-1}**
>
> First make sure that f is one-to-one. Then:
>
> **1.** Replace $f(x)$ with y.
> **2.** Interchange x and y. (This gives the inverse function.)
> **3.** Solve for y.
> **4.** Replace y with $f^{-1}(x)$. (This is inverse function notation.)

> **EXAMPLE 7**   Determine whether each function is one-to-one and if it is, find a formula for $f^{-1}(x)$.
>
> **a)** $f(x) = x + 2$ **b)** $f(x) = 2x - 3$

SOLUTION

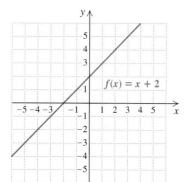

a) The graph of $f(x) = x + 2$ is shown at left. It passes the horizontal-line test, so it is one-to-one. Thus its inverse is a function.

1. Replace $f(x)$ with y: $y = x + 2.$

2. Interchange x and y: $x = y + 2.$ This gives the inverse function.

3. Solve for y: $x - 2 = y.$

4. Replace y with $f^{-1}(x)$: $f^{-1}(x) = x - 2.$ We also "reversed" the equation.

In this case, the function f adds 2 to all inputs. Thus, to "undo" f, the function f^{-1} must subtract 2 from its inputs.

b) The function $f(x) = 2x - 3$ is also linear. Any linear function that is not constant will pass the horizontal-line test. Thus, f is one-to-one.

1. Replace $f(x)$ with y: $y = 2x - 3.$

2. Interchange x and y: $x = 2y - 3.$

3. Solve for y: $x + 3 = 2y$

$$\frac{x + 3}{2} = y.$$

4. Replace y with $f^{-1}(x)$: $f^{-1}(x) = \dfrac{x + 3}{2}.$

In this case, the function f doubles all inputs and then subtracts 3. Thus, to "undo" f, the function f^{-1} adds 3 to each input and then divides by 2.

Graphing Functions and Their Inverses

How do the graphs of a function and its inverse compare?

EXAMPLE 8   Graph $f(x) = 2x - 3$ and $f^{-1}(x) = (x + 3)/2$ on the same set of axes. Then compare.

SOLUTION   The graph of each function follows. Note that the graph of f^{-1} can be drawn by reflecting the graph of f across the line $y = x$. That is, if we graph $f(x) = 2x - 3$ in wet ink and fold the paper along the line $y = x$, the graph of $f^{-1}(x) = (x + 3)/2$ will appear as the impression made by f.

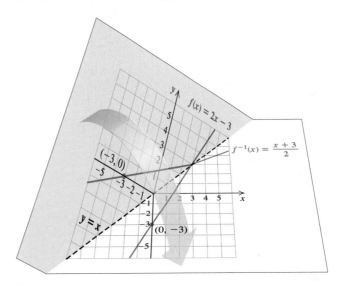

When x and y are interchanged to find a formula for the inverse, we are, in effect, reflecting or flipping the graph of $f(x) = 2x - 3$ across the line $y = x$. For example, when the coordinates of the y-intercept of the graph of f, $(0, -3)$, are reversed, we get $(-3, 0)$, the x-intercept of the graph of f^{-1}.

Visualizing Inverses

The graph of f^{-1} is a reflection of the graph of f across the line $y = x$.

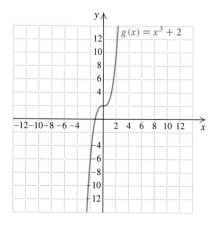

Student Notes

Since we interchange x and y to find an equation for an inverse, the domain of the function is the range of the inverse, and the range of the function is the domain of the inverse.

EXAMPLE 9 Consider $g(x) = x^3 + 2$.

a) Determine whether the function is one-to-one.

b) If it is one-to-one, find a formula for its inverse.

c) Graph the inverse, if it exists.

SOLUTION

a) The graph of $g(x) = x^3 + 2$ is shown at left. It passes the horizontal-line test and thus has an inverse.

b) **1.** Replace $g(x)$ with y: $\qquad\qquad y = x^3 + 2.$ $\qquad$ Using $g(x) = x^3 + 2$

$\quad$ **2.** Interchange x and y: $\qquad\qquad x = y^3 + 2.$

$\quad$ **3.** Solve for y: $\qquad\qquad\qquad x - 2 = y^3$

$\qquad\qquad\qquad\qquad\qquad\qquad \sqrt[3]{x - 2} = y.$ $\quad$ Each number has only one cube root, so we can solve for y.

$\quad$ **4.** Replace y with $g^{-1}(x)$: $\quad g^{-1}(x) = \sqrt[3]{x - 2}.$

c) To find the graph, we reflect the graph of $g(x) = x^3 + 2$ across the line $y = x$, as we did in Example 8. We can also substitute into $g^{-1}(x) = \sqrt[3]{x - 2}$ and plot points. Note that $(2, 10)$ is on the graph of g, whereas $(10, 2)$ is on the graph of g^{-1}. The graphs of g and g^{-1} are shown together below.

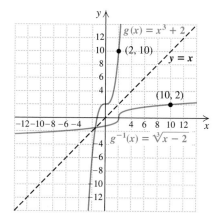

Inverse Functions and Composition

Let's consider inverses of functions in terms of function machines. Suppose that a one-to-one function f is programmed into a machine. If the machine has a reverse switch, when the switch is thrown, the machine performs the inverse function f^{-1}. Inputs then enter at the opposite end, and the entire process is reversed.

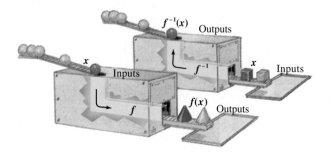

Consider $g(x) = x^3 + 2$ and $g^{-1}(x) = \sqrt[3]{x - 2}$ from Example 9. For the input 3,

$$g(3) = 3^3 + 2 = 27 + 2 = 29.$$

The output is 29. Now we use 29 for the input in the inverse:

$$g^{-1}(29) = \sqrt[3]{29 - 2} = \sqrt[3]{27} = 3.$$

The function g takes 3 to 29. The inverse function g^{-1} takes the number 29 back to 3.

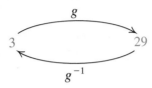

In general, for any output $f(x)$, the function f^{-1} takes that output back to x. Similarly, for any output $f^{-1}(x)$, the function f takes that output back to x.

Composition and Inverses   If a function f is one-to-one, then f^{-1} is the unique function for which

$$(f^{-1} \circ f)(x) = f^{-1}(f(x)) = x \quad \text{and}$$
$$(f \circ f^{-1})(x) = f(f^{-1}(x)) = x.$$

EXAMPLE 10   Let $f(x) = 2x + 1$. Show that

$$f^{-1}(x) = \frac{x - 1}{2}.$$

SOLUTION   We find $(f^{-1} \circ f)(x)$ and $(f \circ f^{-1})(x)$ and check to see that each is x.

$$(f^{-1} \circ f)(x) = f^{-1}(f(x)) = f^{-1}(2x + 1)$$
$$= \frac{(2x + 1) - 1}{2}$$
$$= \frac{2x}{2} = x$$

$$(f \circ f^{-1})(x) = f(f^{-1}(x)) = f\left(\frac{x - 1}{2}\right)$$
$$= 2 \cdot \frac{x - 1}{2} + 1$$
$$= x - 1 + 1 = x$$

Inverse Functions: Graphing and Composition

In Example 9, we found that the inverse of $y_1 = x^3 + 2$ is $y_2 = \sqrt[3]{x - 2}$. There are several ways to check this result using a graphing calculator.

1. We can check visually by graphing both functions, along with the line $y = x$, on a "squared" set of axes. The graph of y_2 *appears* to be the reflection of the graph of y_1 across the line $y = x$. This is only a partial check since we are comparing graphs visually.

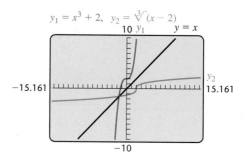

2. A second, more precise, visual check begins with graphing $y_1 = x^3 + 2$ using a squared viewing window. We then press DRAW 8 VARS) 1 1 ENTER to select the DrawInv option of the DRAW menu and graph the inverse. The resulting graph should coincide with the graph of y_2.

3. For a third check, note that if y_2 is the inverse of y_1, then $(y_2 \circ y_1)(x) = x$ and $(y_1 \circ y_2)(x) = x$. We enter $y_3 = y_2(y_1(x))$ and $y_4 = y_1(y_2(x))$, and form a table to compare x, y_3, and y_4. Note that for the values shown, $y_3 = y_4 = x$.

TblStart = −3, ΔTbl = 1

X	Y3	Y4
−3	−3	−3
−2	−2	−2
−1	−1	−1
0	0	0
1	1	1
2	2	2
3	3	3

X = −3

12.1 EXERCISE SET

Concept Reinforcement   *Classify each statement as either true or false.*

1. The composition of two functions f and g is written $f \circ g$.

2. The notation $(f \circ g)(x)$ means $f(g(x))$.

3. If $f(x) = x^2$ and $g(x) = x + 3$, then $(g \circ f)(x) = (x + 3)^2$.

4. If $f(2) = 15$ and $g(15) = 25$, then $(g \circ f)(2) = 25$.

5. The function f is one-to-one if $f(1) = 1$.

6. For f^{-1} to be a function, f cannot be one-to-one.

7. The function f is the inverse of f^{-1}.

8. If g and h are inverses of each other, then $(g \circ h)(x) = x$.

Find $(f \circ g)(1)$, $(g \circ f)(1)$, $(f \circ g)(x)$, *and* $(g \circ f)(x)$.

9. $f(x) = x^2 + 1$; $g(x) = 2x - 3$

10. $f(x) = 2x + 1$; $g(x) = x^2 - 5$

11. $f(x) = x - 3$; $g(x) = 2x^2 - 7$

12. $f(x) = 3x^2 + 4$; $g(x) = 4x - 1$

13. $f(x) = x + 7$; $g(x) = 1/x^2$

14. $f(x) = 1/x^2$; $g(x) = x + 2$

15. $f(x) = \sqrt{x}$; $g(x) = x + 3$

16. $f(x) = 10 - x$; $g(x) = \sqrt{x}$

17. $f(x) = \sqrt{4x}$; $g(x) = 1/x$

18. $f(x) = \sqrt{x + 3}$; $g(x) = 13/x$

19. $f(x) = x^2 + 4$; $g(x) = \sqrt{x - 1}$

20. $f(x) = x^2 + 8$; $g(x) = \sqrt{x + 17}$

Use the following table to find each value, if possible.

X	Y₁	Y₂
-3	-4	1
-2	-1	-2
-1	2	-3
0	5	-2
1	8	1
2	11	6
3	14	11
X =		

21. $(y_1 \circ y_2)(-3)$

22. $(y_2 \circ y_1)(-3)$

23. $(y_1 \circ y_2)(-1)$

24. $(y_2 \circ y_1)(-1)$

25. $(y_2 \circ y_1)(1)$

26. $(y_1 \circ y_2)(1)$

Use the following table to find each value, if possible.

x	$f(x)$	$g(x)$
1	0	1
2	3	5
3	2	8
4	6	5
5	4	1

27. $(f \circ g)(2)$

28. $(g \circ f)(4)$

29. $f(g(3))$

30. $g(f(5))$

Find $f(x)$ and $g(x)$ such that $h(x) = (f \circ g)(x)$. Answers may vary.

31. $h(x) = (7 + 5x)^2$

32. $h(x) = (3x - 1)^2$

33. $h(x) = \sqrt{2x + 7}$

34. $h(x) = \sqrt{5x + 2}$

35. $h(x) = \dfrac{2}{x - 3}$

36. $h(x) = \dfrac{3}{x} + 4$

37. $h(x) = \dfrac{1}{\sqrt{7x + 2}}$

38. $h(x) = \sqrt{x - 7} - 3$

39. $h(x) = \dfrac{1}{\sqrt{3x}} + \sqrt{3x}$

40. $h(x) = \dfrac{1}{\sqrt{2x}} - \sqrt{2x}$

Determine whether each function is one-to-one.

Aha! **41.** $f(x) = x - 5$ **42.** $f(x) = 5 - 2x$

43. $f(x) = x^2 + 1$ **44.** $f(x) = 1 - x^2$

45. 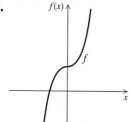   **46.**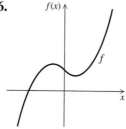

47. 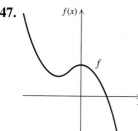   **48.**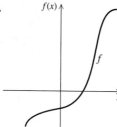

For each function, (a) determine whether it is one-to-one; (b) if it is one-to-one, find a formula for the inverse.

49. $f(x) = x + 4$ **50.** $f(x) = x + 2$

51. $f(x) = 2x$ **52.** $f(x) = 3x$

53. $g(x) = 3x - 1$ **54.** $g(x) = 2x - 5$

55. $f(x) = \frac{1}{2}x + 1$ **56.** $f(x) = \frac{1}{3}x + 2$

57. $g(x) = x^2 + 5$ **58.** $g(x) = x^2 - 4$

59. $h(x) = -2x + 4$ **60.** $h(x) = -3x + 1$

Aha! **61.** $f(x) = \dfrac{1}{x}$ **62.** $f(x) = \dfrac{3}{x}$

63. $G(x) = 4$ **64.** $H(x) = 2$

65. $f(x) = \dfrac{2x + 1}{3}$ **66.** $f(x) = \dfrac{3x + 2}{5}$

67. $f(x) = x^3 - 5$ **68.** $f(x) = x^3 + 7$

69. $g(x) = (x - 2)^3$ **70.** $g(x) = (x + 7)^3$

71. $f(x) = \sqrt{x}$ **72.** $f(x) = \sqrt{x - 1}$

Graph each function and its inverse using the same set of axes.

73. $f(x) = \frac{2}{3}x + 4$ **74.** $g(x) = \frac{1}{4}x + 2$

75. $f(x) = x^3 + 1$ **76.** $f(x) = x^3 - 1$

77. $g(x) = \frac{1}{2}x^3$ **78.** $g(x) = \frac{1}{3}x^3$

79. $F(x) = -\sqrt{x}$ **80.** $f(x) = \sqrt{x}$

81. $f(x) = -x^2, x \geq 0$ **82.** $f(x) = x^2 - 1, x \leq 0$

83. Let $f(x) = \sqrt[3]{x - 4}$. Use composition to show that $f^{-1}(x) = x^3 + 4$.

84. Let $f(x) = 3/(x + 2)$. Use composition to show that
$$f^{-1}(x) = \frac{3}{x} - 2.$$

85. Let $f(x) = (1 - x)/x$. Use composition to show that
$$f^{-1}(x) = \frac{1}{x + 1}.$$

86. Let $f(x) = x^3 - 5$. Use composition to show that $f^{-1}(x) = \sqrt[3]{x + 5}$.

Use a graphing calculator to help determine whether or not the given pairs of functions are inverses of each other.

87. $f(x) = 0.75x^2 + 2$; $g(x) = \sqrt{\dfrac{4(x - 2)}{3}}$

88. $f(x) = 1.4x^3 + 3.2$; $g(x) = \sqrt[3]{\dfrac{x - 3.2}{1.4}}$

89. $f(x) = \sqrt{2.5x + 9.25}$;
$g(x) = 0.4x^2 - 3.7, x \geq 0$

90. $f(x) = 0.8x^{1/2} + 5.23$;
$g(x) = 1.25(x^2 - 5.23), x \geq 0$

In Exercises 91 and 92 here and on the following page, match the graph of each function in Column A with the graph of its inverse in Column B.

91. Column A   Column B

(1)   A.

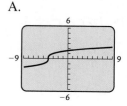

(2)

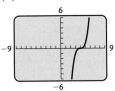

(3)

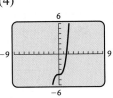

(4)

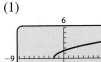

B.

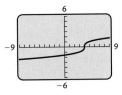

C.

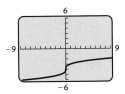

D.

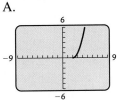

92. Column A

(1)

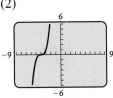

(2)

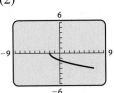

(3)

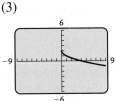

(4)

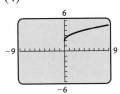

Column B

A.

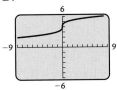

B.

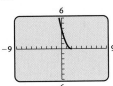

C.

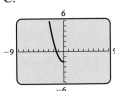

D.

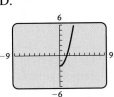

93. *Dress Sizes in the United States and France.*
A size-6 dress in the United States is size 38 in France. A function that converts dress sizes in the United States to those in France is

$$f(x) = x + 32.$$

a) Find the dress sizes in France that correspond to sizes 8, 10, 14, and 18 in the United States.

b) Determine whether this function has an inverse that is a function. If so, find a formula for the inverse.

c) Use the inverse function to find dress sizes in the United States that correspond to sizes 40, 42, 46, and 50 in France.

94. *Dress Sizes in the United States and Italy.*   A size-6 dress in the United States is size 36 in Italy. A function that converts dress sizes in the United States to those in Italy is

$$f(x) = 2(x + 12).$$

a) Find the dress sizes in Italy that correspond to sizes 8, 10, 14, and 18 in the United States.

b) Determine whether this function has an inverse that is a function. If so, find a formula for the inverse.

c) Use the inverse function to find dress sizes in the United States that correspond to sizes 40, 44, 52, and 60 in Italy.

TW **95.** Is there a one-to-one relationship between the numbers and letters on the keypad of a telephone? Why or why not?

TW **96.** Mathematicians usually try to select "logical" words when forming definitions. Does the term "one-to-one" seem logical? Why or why not?

Focused Review

Let $f(x) = 2x + \sqrt{x}$ and $g(x) = x^2$. Find each of the following.

97. $f(3)$ [3.8]

98. $f(-3)$ [10.1]

99. $(f + g)(x)$ [5.9]

100. $(f/g)(x)$ [5.9]

101. The domain of f [10.1]

102. The domain of g [11.1]

103. The domain of $f + g$ [5.9], [10.1]

104. The domain of f/g [5.9], [10.1]

105. $(f \circ g)(x)$ [12.1]

106. $(g \circ f)(x)$ [12.1]

Synthesis

TW 107. The function $V(t) = 750(1.2)^t$ is used to predict the value, $V(t)$, of a certain rare stamp t years from 2005. Do not calculate $V^{-1}(t)$, but explain how V^{-1} could be used.

TW 108. An organization determines that the cost per person of chartering a bus is given by the function

$$C(x) = \frac{100 + 5x}{x},$$

where x is the number of people in the group and $C(x)$ is in dollars. Determine $C^{-1}(x)$ and explain how this inverse function could be used.

For Exercises 109 and 110, graph the inverse of f.

109.

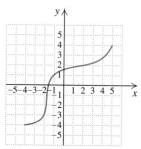

110.

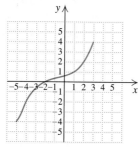

111. *Dress Sizes in France and Italy.* Use the information in Exercises 93 and 94 to find a function for the French dress size that corresponds to a size x dress in Italy.

112. *Dress Sizes in Italy and France.* Use the information in Exercises 93 and 94 to find a function for the Italian dress size that corresponds to a size x dress in France.

TW 113. What relationship exists between the answers to Exercises 111 and 112? Explain how you determined this.

114. Show that function composition is associative by showing that $((f \circ g) \circ h)(x) = (f \circ (g \circ h))(x)$.

115. Show that if $h(x) = (f \circ g)(x)$, then $h^{-1}(x) = (g^{-1} \circ f^{-1})(x)$. (*Hint*: Use Exercise 114.)

116. Match each function in Column A with its inverse from Column B.

Column A

Column B

(1) $y = 5x^3 + 10$ **A.** $y = \dfrac{\sqrt[3]{x} - 10}{5}$

(2) $y = (5x + 10)^3$ **B.** $y = \sqrt[3]{\dfrac{x}{5} - 10}$

(3) $y = 5(x + 10)^3$ **C.** $y = \sqrt[3]{\dfrac{x - 10}{5}}$

(4) $y = (5x)^3 + 10$ **D.** $y = \dfrac{\sqrt[3]{x - 10}}{5}$

TW 117. Examine the following table. Is it possible that f and g could be inverses of each other? Why or why not?

x	$f(x)$	$g(x)$
6	6	6
7	6.5	8
8	7	10
9	7.5	12
10	8	14
11	8.5	16
12	9	18

118. Assume in Exercise 117 that f and g are both linear functions. Find equations for $f(x)$ and $g(x)$. Are f and g inverses of each other?

119. Let $c(w)$ represent the cost of mailing a package that weighs w pounds. Let $f(n)$ represent the weight, in pounds, of n copies of a certain book. Explain what $(c \circ f)(n)$ represents.

120. Let $g(a)$ represent the number of gallons of sealant needed to seal a bamboo floor with area a. Let $c(s)$ represent the cost of s gallons of sealant. Which composition makes sense: $(c \circ g)(a)$ or $(g \circ c)(s)$? What does it represent?

*The following graphs show the rate of flow R, in liters per minute, of blood from the heart in a man who bicycles for 20 min, and the pressure P, in millimeters of mercury, in the artery leading to the lungs for a rate of blood flow R from the heart.**

Blood Flow

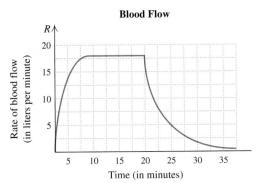

Artery Pressure

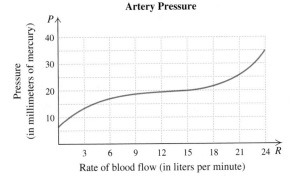

121. Estimate $P(R(10))$.

122. Explain what $P(R(10))$ represents.

123. Estimate $P^{-1}(20)$.

124. Explain what $P^{-1}(20)$ represents.

125. *Real Estate.*   The following table lists the number of existing homes sold and the Federal Funds rate for several months in 2005 and in 2006 (*Source:* Mortgage Banking, April 2006)

Month	Existing Homes Sold (in thousands of units)	Federal Funds Rate
Sept. 2005	6290	3.62%
Oct. 2005	6180	3.78
Nov. 2005	6150	4.00
Dec. 2005	5860	4.16
Jan. 2006	5770	4.29

a) Use linear regression to find a linear function that can be used to predict the number of existing homes h sold as a function of the number of months t after September 2005. Let $y_1 = h(t)$.

b) Use linear regression to find a linear function that can be used to predict the Federal Funds rate r as a function of the number of months t after September 2005. Let $y_2 = r(t)$.

c) Use linear regression to find a linear function that can be used to predict the number of existing homes H sold as a function of the Federal Funds rate r. Let $y_3 = H(r)$.

d) Let $y_4 = y_3(y_2(x))$. Use a table of values to compare the functions y_1 and y_4. Use composition of functions to write the relationship between the functions h, r, and H.

*This problem was suggested by Kandace Kling, Portland Community College, Sylvania, Oregon.

12.2 Exponential Functions

Graphing Exponential Functions ▪ Equations with *x* and *y* Interchanged ▪ Applications of Exponential Functions

Composite and inverse functions, as shown in Section 12.1, are included in this chapter because they are needed in order to understand the logarithmic functions that appear in Section 12.3. Here in Section 12.2, we introduce a new type of function, the *exponential function*, so that we can study both it and its inverse in Sections 12.3–12.7.

Consider the graph below. The rapidly rising curve approximates the graph of an *exponential function*. We now consider such functions and some of their applications.

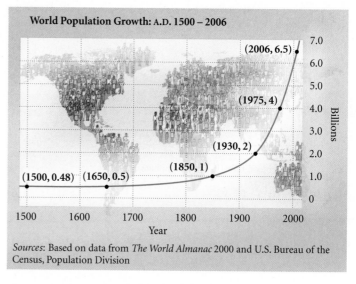

World Population Growth: A.D. 1500 – 2006

(2006, 6.5)
(1975, 4)
(1930, 2)
(1850, 1)
(1500, 0.48) (1650, 0.5)

Sources: Based on data from *The World Almanac* 2000 and U.S. Bureau of the Census, Population Division

Graphing Exponential Functions

In Chapter 10, we studied exponential expressions with rational-number exponents, such as

$$5^{1/4}, \qquad 3^{-3/4}, \qquad 7^{2.34}, \qquad 5^{1.73}.$$

For example, $5^{1.73}$, or $5^{173/100}$, represents the 100th root of 5 raised to the 173rd power. What about expressions with irrational exponents, such as $5^{\sqrt{3}}$ or $7^{-\pi}$? To attach meaning to $5^{\sqrt{3}}$, consider a rational approximation, r, of $\sqrt{3}$. As r gets closer to $\sqrt{3}$, the value of 5^r gets closer to some real number p.

r closes in on $\sqrt{3}$.	5^r closes in on some real number p.
$1.7 < r < 1.8$	$15.426 \approx 5^{1.7} < p < 5^{1.8} \approx 18.119$
$1.73 < r < 1.74$	$16.189 \approx 5^{1.73} < p < 5^{1.74} \approx 16.452$
$1.732 < r < 1.733$	$16.241 \approx 5^{1.732} < p < 5^{1.733} \approx 16.267$

We define $5^{\sqrt{3}}$ to be the number p. To eight decimal places,

$$5^{\sqrt{3}} \approx 16.24245082.$$

Any positive irrational exponent can be interpreted in a similar way. Negative irrational exponents are then defined using reciprocals. Thus, so long as a is positive, a^x has meaning for *any* real number x. All of the laws of exponents still hold, but we will not prove that here. We now define an *exponential function*.

Exponential Function   The function $f(x) = a^x$, where a is a positive constant, $a \neq 1$, is called the *exponential function*, base a.

We require the base a to be positive to avoid imaginary numbers that would result from taking even roots of negative numbers. The restriction $a \neq 1$ is made to exclude the constant function $f(x) = 1^x$, or $f(x) = 1$.

The following are examples of exponential functions:

$$f(x) = 2^x, \qquad f(x) = \left(\tfrac{1}{3}\right)^x, \qquad f(x) = 5^{-3x}. \qquad \text{Note that } 5^{-3x} = (5^{-3})^x.$$

Like polynomial functions, the domain of an exponential function is the set of all real numbers. Unlike polynomial functions, exponential functions have a variable exponent. Because of this, graphs of exponential functions either rise or fall dramatically.

EXAMPLE 1   Graph the exponential function given by $y = f(x) = 2^x$.

SOLUTION   We compute some function values, thinking of y as $f(x)$, and list the results in a table. It is a good idea to start by letting $x = 0$.

$$f(0) = 2^0 = 1; \qquad\qquad f(-1) = 2^{-1} = \frac{1}{2^1} = \frac{1}{2};$$
$$f(1) = 2^1 = 2;$$
$$f(2) = 2^2 = 4; \qquad\qquad f(-2) = 2^{-2} = \frac{1}{2^2} = \frac{1}{4};$$
$$f(3) = 2^3 = 8;$$
$$f(-3) = 2^{-3} = \frac{1}{2^3} = \frac{1}{8}$$

Next, we plot these points and connect them with a smooth curve.

x	y, or $f(x)$
0	1
1	2
2	4
3	8
-1	$\frac{1}{2}$
-2	$\frac{1}{4}$
-3	$\frac{1}{8}$

The curve comes very close to the x-axis, but does not touch or cross it.

$$y = f(x) = 2^x$$

Be sure to plot enough points to determine how steeply the curve rises.

Note that as x increases, the function values increase without bound. As x decreases, the function values decrease, getting very close to 0. The x-axis, or the line $y = 0$, is a horizontal *asymptote*, meaning that the curve gets closer and closer to this line the further we move to the left.

EXAMPLE 2 Graph: $y = f(x) = \left(\frac{1}{2}\right)^x$.

SOLUTION We compute some function values, thinking of y as $f(x)$, and list the results in a table. Before we do this, note that

$$y = f(x) = \left(\tfrac{1}{2}\right)^x = (2^{-1})^x = 2^{-x}.$$

Then we have

$$f(0) = 2^{-0} = 1;$$

$$f(1) = 2^{-1} = \frac{1}{2^1} = \frac{1}{2};$$

$$f(2) = 2^{-2} = \frac{1}{2^2} = \frac{1}{4};$$

$$f(3) = 2^{-3} = \frac{1}{2^3} = \frac{1}{8};$$

$$f(-1) = 2^{-(-1)} = 2^1 = 2;$$

$$f(-2) = 2^{-(-2)} = 2^2 = 4;$$

$$f(-3) = 2^{-(-3)} = 2^3 = 8.$$

x	y, or $f(x)$
0	1
1	$\frac{1}{2}$
2	$\frac{1}{4}$
3	$\frac{1}{8}$
-1	2
-2	4
-3	8

Next, we plot these points and connect them with a smooth curve. This curve is a mirror image, or *reflection*, of the graph of $y = 2^x$ (see Example 1) across the y-axis. The line $y = 0$ is again an asymptote.

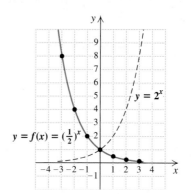

If the values of $f(x)$ increase as x increases, we say that f is an *increasing function*. The function in Example 1, $f(x) = 2^x$, is increasing. If the values of $f(x)$ decrease as x increases, we say that f is a *decreasing function*. The function in Example 2, $f(x) = \left(\frac{1}{2}\right)^x$, is decreasing.

1. Graph each of the following functions, and determine whether the graph increases or decreases from left to right.

    a) $f(x) = 3^x$ **b)** $g(x) = 4^x$

    c) $h(x) = 1.5^x$ **d)** $r(x) = \left(\frac{1}{3}\right)^x$

    e) $t(x) = 0.75^x$

2. How can you tell from the number a in $f(x) = a^x$ whether the graph of f increases or decreases?

3. Compare the graphs of f, g, and h. Which curve is steepest?

4. Compare the graphs of r and t. Which curve is steeper?

We can make the following observations.

A. For $a > 1$, the graph of $f(x) = a^x$ increases from left to right. The greater the value of a, the steeper the curve. (See the figure on the left below.)

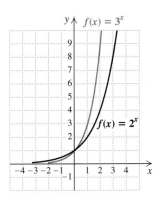
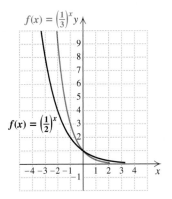

B. For $0 < a < 1$, the graph of $f(x) = a^x$ decreases from left to right. For smaller values of a, the curve becomes steeper. (See the figure on the right above.)

C. All graphs of $f(x) = a^x$ go through the y-intercept $(0, 1)$.

D. All graphs of $f(x) = a^x$ have the x-axis as the asymptote.

E. If $f(x) = a^x$, with $a > 0$, $a \neq 1$, the domain of f is all real numbers, and the range of f is all positive real numbers.

F. For $a > 0$, $a \neq 1$, the function given by $f(x) = a^x$ is one-to-one. Its graph passes the horizontal-line test.

Student Notes

When using translations, make sure that you are shifting in the correct direction. When in doubt, substitute a value for x and make some calculations.

EXAMPLE 3   Graph: $y = f(x) = 2^{x-2}$.

SOLUTION   We construct a table of values. Then we plot the points and connect them with a smooth curve. Here $x - 2$ is the *exponent*.

$$f(0) = 2^{0-2} = 2^{-2} = \frac{1}{4}; \qquad f(-1) = 2^{-1-2} = 2^{-3} = \frac{1}{8};$$

$$f(1) = 2^{1-2} = 2^{-1} = \frac{1}{2}; \qquad f(-2) = 2^{-2-2} = 2^{-4} = \frac{1}{16}$$

$$f(2) = 2^{2-2} = 2^0 = 1;$$

$$f(3) = 2^{3-2} = 2^1 = 2;$$

$$f(4) = 2^{4-2} = 2^2 = 4;$$

x	y, or $f(x)$
0	$\frac{1}{4}$
1	$\frac{1}{2}$
2	1
3	2
4	4
-1	$\frac{1}{8}$
-2	$\frac{1}{16}$

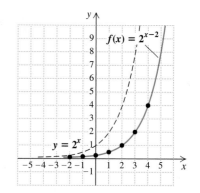

The graph looks just like the graph of $y = 2^x$, but it is translated 2 units to the right. The y-intercept of $y = 2^x$ is $(0, 1)$. The y-intercept of $y = 2^{x-2}$ is $\left(0, \frac{1}{4}\right)$. The line $y = 0$ is again the asymptote.

As we saw in Example 3, the graph of $f(x) = 2^{x-2}$ looks like the graph of $y = 2^x$ translated 2 units to the right. In general, the graph of $f(x) = 2^{x-h}$ will look like the graph of $y = 2^x$ translated right or left. Similarly, the graph of $f(x) = 2^x + k$ will look like the graph of 2^x translated up or down.

This observation is true in general.

The graph of $f(x) = a^{x-h} + k$ looks like the graph of $y = a^x$ translated $|h|$ units left or right and $|k|$ units up or down.

- If $h > 0$, $y = a^x$ is translated h units right.
- If $h < 0$, $y = a^x$ is translated $|h|$ units left.
- If $k > 0$, $y = a^x$ is translated k units up.
- If $k < 0$, $y = a^x$ is translated $|k|$ units down.

Equations with *x* and *y* Interchanged

It will be helpful in later work to be able to graph an equation in which the x and the y in $y = a^x$ are interchanged.

EXAMPLE 4   Graph: $x = 2^y$.

SOLUTION   Note that x is alone on one side of the equation. To find ordered pairs that are solutions, we choose values for y and then compute values for x:

For $y = 0$, $x = 2^0 = 1$.

For $y = 1$, $x = 2^1 = 2$.

For $y = 2$, $x = 2^2 = 4$.

For $y = 3$, $x = 2^3 = 8$.

For $y = -1$, $x = 2^{-1} = \dfrac{1}{2}$.

For $y = -2$, $x = 2^{-2} = \dfrac{1}{4}$.

For $y = -3$, $x = 2^{-3} = \dfrac{1}{8}$.

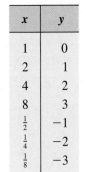

x	y
1	0
2	1
4	2
8	3
$\frac{1}{2}$	−1
$\frac{1}{4}$	−2
$\frac{1}{8}$	−3

(1) Choose values for y.

(2) Compute values for x.

We plot the points and connect them with a smooth curve.

The curve does not touch or cross the y-axis, which serves as a vertical asymptote.

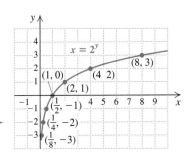

Note too that this curve looks just like the graph of $y = 2^x$, except that it is reflected across the line $y = x$, as shown here.

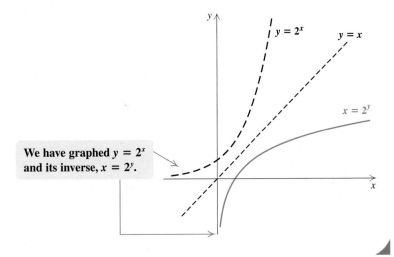

We have graphed $y = 2^x$ and its inverse, $x = 2^y$.

Applications of Exponential Functions

▶ **EXAMPLE 5** Interest Compounded Annually. The amount of money A that a principal P will be worth after t years at interest rate i, compounded annually, is given by the formula

$$A = P(1 + i)^t.$$ **You might review Example 12 in Section 11.1.**

Suppose that $100,000 is invested at 8% interest, compounded annually.

a) Find a function for the amount in the account after t years.

b) Find the amount of money in the account at $t = 0$, $t = 4$, $t = 8$, and $t = 10$.

c) Graph the function.

SOLUTION

a) If $P = \$100,000$ and $i = 8\% = 0.08$, we can substitute these values and form the following function:

$$A(t) = \$100,000(1 + 0.08)^t \quad \text{Using } A = P(1 + i)^t$$
$$= \$100,000(1.08)^t.$$

b) An efficient way to find the function values using a graphing calculator is to let $y_1 = 100,000(1.08)^{\wedge} x$ and use the TABLE feature with Indpnt set to Ask, as shown at left. We highlight a table entry if we want to view the function value to more decimal places than is shown in the table. Note that large values in a table will be written in scientific notation.

Alternatively, we can use $y_1(\ )$ notation to evaluate the function for each value. Using either procedure gives us

$$A(0) = \$100,000,$$
$$A(4) \approx \$136,048.90,$$
$$A(8) \approx \$185,093.02,$$
$$\text{and} \quad A(10) \approx \$215,892.50.$$

c) We can use the function values computed in part (b), and others if we wish, to draw the graph by hand. Whether we are graphing by hand or using a graphing calculator, the axes will be scaled differently because of the large function values.

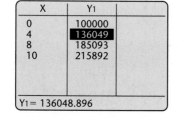

X	Y₁
0	100000
4	136049
8	185093
10	215892

Y₁ = 136048.896

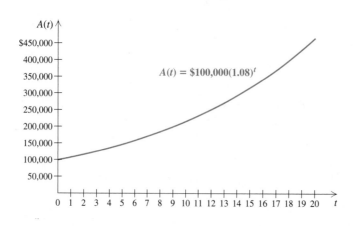

$A(t) = \$100,000(1.08)^t$

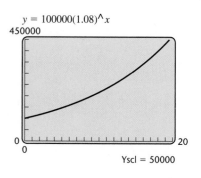

$y = 100000(1.08)^{\wedge}x$
450000

Yscl = 50000

Connecting the Concepts

INTERPRETING GRAPHS: LINEAR AND EXPONENTIAL GROWTH

When the graph of a quantity over time approximates the graph of an exponential function, as in the World Population graph on p. 902, we say that there is *exponential* growth. This is different from *linear* growth, which we examined in Chapter 3.

Linear functions increase (or decrease) by a constant amount. The rate of change (slope) of a linear graph is constant. The rate of change of an exponential graph is not constant.

To illustrate the difference between linear and

exponential functions, compare the salaries for two hypothetical jobs, shown in the following table and graph. Note that the starting salaries are the same, but the salary for job A increases by a constant amount (linearly) and the salary for job B increases by a constant percent (exponentially).

Note that the salary for job A is larger for the first few years only. The rate of change of the salary for job B increases each year, since the increase is based on a larger salary each year.

	Job A	Job B
Starting salary	$50,000	$50,000
Guaranteed raise	$5000 per year	8% per year
Salary function	$f(x) = 50,000 + 5000x$	$g(x) = 50,000(1.08)^x$

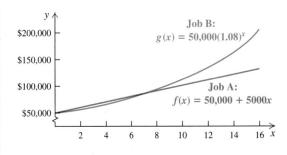

12.2 EXERCISE SET

FOR EXTRA HELP

 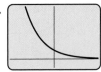

MathXL   MyMathLab   InterAct Math   AW Math Tutor Center   Video Lectures on CD: Disc 6   Student's Solutions Manual

Concept Reinforcement *Classify each of the following as either true or false.*

1. The graph of $f(x) = a^x$ always passes through the point (0, 1).

2. The graph of $g(x) = \left(\frac{1}{2}\right)^x$ gets closer and closer to the x-axis as x gets larger and larger.

3. The graph of $f(x) = 2^{x-3}$ looks just like the graph of $y = 2^x$, but it is translated 3 units to the right.

4. The graph of $g(x) = 2^x - 3$ looks just like the graph of $y = 2^x$, but it is translated 3 units up.

5. The graph of $y = 3^x$ gets close to, but never touches, the y-axis.

6. The graph of $x = 3^y$ gets close to, but never touches, the y-axis.

In each of Exercises 7–10 is the graph of a function $f(x) = a^x$. Determine from the graph whether $a > 1$ or $0 < a < 1$.

7.

8.

9.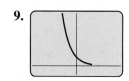

10.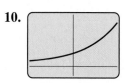

Graph.

11. $y = f(x) = 2^x$

12. $y = f(x) = 3^x$

13. $y = 5^x$

14. $y = 6^x$

15. $y = 2^x + 3$

16. $y = 2^x + 1$

17. $y = 3^x - 1$

18. $y = 3^x - 2$

19. $y = 2^{x-1}$

20. $y = 2^{x-2}$

21. $y = 2^{x+3}$

22. $y = 2^{x+1}$

23. $y = \left(\frac{1}{5}\right)^x$

24. $y = \left(\frac{1}{4}\right)^x$

25. $y = \left(\frac{1}{2}\right)^x$

26. $y = \left(\frac{1}{3}\right)^x$

27. $y = 2^{x-3} - 1$

28. $y = 2^{x+1} - 3$

29. $y = 1.7^x$

30. $y = 4.8^x$

31. $y = 0.15^x$

32. $y = 0.98^x$

33. $x = 3^y$

34. $x = 6^y$

35. $x = 2^{-y}$

36. $x = 3^{-y}$

37. $x = 5^y$

38. $x = 4^y$

39. $x = \left(\frac{3}{2}\right)^y$

40. $x = \left(\frac{4}{3}\right)^y$

Graph each pair of equations using the same set of axes.

41. $y = 3^x$, $x = 3^y$

42. $y = 2^x$, $x = 2^y$

43. $y = \left(\frac{1}{2}\right)^x$, $x = \left(\frac{1}{2}\right)^y$

44. $y = \left(\frac{1}{4}\right)^x$, $x = \left(\frac{1}{4}\right)^y$

Aha! *In Exercises 45–50, match each equation with one of the following graphs.*

a)

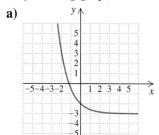

b)

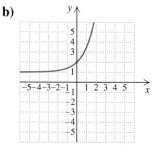

c)

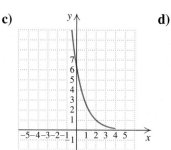

d)

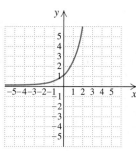

e)

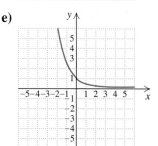

f)

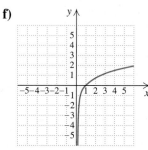

45. $y = \left(\frac{5}{2}\right)^x$

46. $y = \left(\frac{2}{5}\right)^x$

47. $x = \left(\frac{5}{2}\right)^y$

48. $y = \left(\frac{2}{5}\right)^x - 3$

49. $y = \left(\frac{2}{5}\right)^{x-2}$

50. $y = \left(\frac{5}{2}\right)^x + 1$

Solve.

51. *Population Growth.* The world population $P(t)$, in billions, t years after 1980, can be approximated by

$$P(t) = 4.495(1.015)^t$$

(*Source*: Based on data from U.S. Bureau of the Census, International Data Base).

a) Predict the world population in 2008, in 2012, and in 2016.

b) Graph the function.

52. *Growth of Bacteria.* The bacteria *Escherichia coli* are commonly found in the human bladder. Suppose that 3000 of the bacteria are present at time $t = 0$. Then t minutes later, the number of bacteria present can be approximated by

$$N(t) = 3000(2)^{t/20}.$$

a) How many bacteria will be present after 10 min? 20 min? 30 min? 40 min? 60 min?

b) Graph the function.

53. *Smoking Cessation.* The percentage of smokers P who, with telephone counseling to quit smoking, are still successful t months later can be approximated by

$$P(t) = 21.4(0.914)^t$$

(*Sources*: New England Journal of Medicine; data from California's Smokers' Hotline).

a) Estimate the percentage of smokers receiving telephone counseling who are successful in quitting for 1 month, 3 months, and 1 year.

b) Graph the function.

54. *Smoking Cessation.*   The percentage of smokers P who, without telephone counseling, have successfully quit smoking for t months (see Exercise 53) can be approximated by

$$P(t) = 9.02(0.93)^t$$

(*Sources*: New England Journal of Medicine; data from California's Smokers' Hotline).

a) Estimate the percentage of smokers not receiving telephone counseling who are successful in quitting for 1 month, 3 months, and 1 year.

b) Graph the function.

55. *Marine Biology.*   Due to excessive whaling prior to the mid 1970s, the humpback whale is considered an endangered species. The worldwide population of humpbacks, $P(t)$, in thousands, t years after 1900 ($t < 70$) can be approximated by*

$$P(t) = 150(0.960)^t.$$

a) How many humpback whales were alive in 1930? in 1960?

b) Graph the function.

56. *Salvage Value.*   A photocopier is purchased for $5200. Its value each year is about 80% of the value of the preceding year. Its value, in dollars, after t years is given by the exponential function

$$V(t) = 5200(0.8)^t.$$

a) Find the value of the machine after 0 yr, 1 yr, 2 yr, 5 yr, and 10 yr.

b) Graph the function.

57. *Marine Biology.*   As a result of preservation efforts in most countries in which whaling was common, the humpback whale population has grown since the 1970s. The worldwide population of humpbacks, $P(t)$, in thousands, t years after 1982 can be approximated by*

$$P(t) = 5.5(1.08)^t.$$

a) How many humpback whales were alive in 1992? in 2006?

b) Graph the function.

58. *Recycling Aluminum Cans.*   About $\frac{1}{2}$ of all aluminum cans will be recycled. A beverage company distributes 250,000 cans. The number in use after t years is given by the exponential function

$$N(t) = 250{,}000\left(\tfrac{1}{2}\right)^t$$

(*Source*: The Aluminum Association, Inc., May 2004).

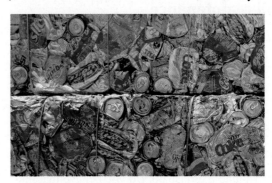

a) How many cans are still in use after 0 yr? 1 yr? 4 yr? 10 yr?

b) Graph the function.

59. *Spread of Zebra Mussels.*   Beginning in 1988, infestations of zebra mussels began to threaten water treatment facilities, power plants, and entire ecosystems in North America.[†] The function

$$A(t) = 10 \cdot 34^t$$

can be used to estimate the number of square centimeters of lake bottom that will be covered with mussels t years after an infestation covering 10 cm^2 first occurs.

a) How many square centimeters of lake bottom will be covered with mussels 5 yr after an infestation covering 10 cm^2 first appears? 7 yr after the infestation first appears?

b) Graph the function.

60. *Cell Phones.*   The number N of cell phones in use in the United States, in millions, can be estimated by

$$N(t) = 7.12(1.3)^t,$$

where t is the number of years after 1990 (*Source*: Cellular Telecommunications and Internet Association).

a) Estimate the number of cell phones in use in 1995, in 2005, and in 2010.

b) Graph the function.

---

*Based on information from the American Cetacean Society, 2006, and the ASK Archive, 1998.

[†]Many thanks to Dr. Gerald Mackie of the Department of Zoology at the University of Guelph in Ontario for the background information for this exercise.

61. Without using a calculator, explain why 2^π must be greater than 8 but less than 16.

62. Suppose that $1000 is invested for 5 yr at 7% interest, compounded annually. In what year will the most interest be earned? Why?

Skill Maintenance

Simplify.

63. 5^{-2} [5.2]

64. 2^{-5} [5.2]

65. $1000^{2/3}$ [10.2]

66. $25^{-3/2}$ [10.2]

67. $\dfrac{10a^8b^7}{2a^2b^4}$ [5.1]

68. $\dfrac{24x^6y^4}{4x^2y^3}$ [5.1]

Synthesis

69. Examine Exercise 60. Do you believe that the equation for the number of cell phones in use in the United States will be accurate 20 yr from now? Why or why not?

70. Why was it necessary to discuss irrational exponents before graphing exponential functions?

Determine which of the two given numbers is larger. Do not use a calculator.

71. $\pi^{1.3}$ or $\pi^{2.4}$

72. $\sqrt{8^3}$ or $8^{\sqrt{3}}$

Graph.

73. $y = 2^x + 2^{-x}$

74. $y = \left|\left(\frac{1}{2}\right)^x - 1\right|$

75. $y = |2^x - 2|$

76. $y = 2^{-(x-1)^2}$

77. $y = \left|2^{x^2} - 1\right|$

78. $y = 3^x + 3^{-x}$

Graph both equations using the same set of axes.

79. $y = 3^{-(x-1)}$, $x = 3^{-(y-1)}$

80. $y = 1^x$, $x = 1^y$

81. *Sales of MP3 Players.* Sales of MP3 players have grown from $80 million in 2000 to $205 million in 2002 to $1204 million in 2004 (*Source*: Consumer Electronics Association). Use the ExpReg option in the STAT CALC menu to find an exponential function that can be used to estimate sales of MP3 players x years after 2000. Then use that function to predict sales in 2008.

Collaborative Corner

The True Cost of a New Car

Focus: Car loans and exponential functions
Time: 30 minutes
Group Size: 2
Materials: Calculators with exponentiation keys

The formula

$$M = \frac{Pr}{1 - (1 + r)^{-n}}$$

is used to determine the payment size, M, when a loan of P dollars is to be repaid in n equally sized monthly payments. Here r represents the monthly interest rate. Loans repaid in this fashion are said to be *amortized* (spread out equally) over a period of n months.

ACTIVITY

1. Suppose one group member is selling the other a car for $2600, financed at 1% interest per month for 24 months. What should be the size of each monthly payment?

2. Suppose both group members are shopping for the same model new car. Each group member visits a different dealer. One dealer offers the car for $13,000 at 10.5% interest (0.00875 monthly interest) for 60 months (no down payment). The other dealer offers the same car for $12,000, but at 12% interest (0.01 monthly interest) for 48 months (no down payment).

    a) Determine the monthly payment size for each offer. Then determine the total amount paid for the car under each offer. How much of each total is interest?

    b) Work together to find the annual interest rate for which the total cost of 60 monthly payments for the $13,000 car would equal the total amount paid for the $12,000 car (as found in part a above).

12.3 Logarithmic Functions

Graphs of Logarithmic Functions ■ Common Logarithms ■
Equivalent Equations ■ Solving Certain Logarithmic Equations

We are now ready to study inverses of exponential functions. These functions have many applications and are referred to as *logarithm*, or *logarithmic*, *functions*.

Graphs of Logarithmic Functions

Consider the exponential function $f(x) = 2^x$. Like all exponential functions, f is one-to-one. Can a formula for f^{-1} be found?

To answer this, we use the method of Section 12.1:

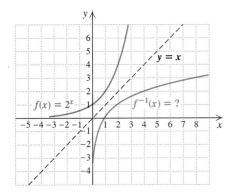

1. Replace $f(x)$ with y: $y = 2^x$.
2. Interchange x and y: $x = 2^y$.
3. Solve for y: $y = $ the power to which we raise 2 to get x.
4. Replace y with $f^{-1}(x)$: $f^{-1}(x) = $ the power to which we raise 2 to get x.

We now define a new symbol to replace the words "the power to which we raise 2 to get x":

**$\log_2 x$, read "the logarithm, base 2, of x," or "log, base 2, of x,"
means "the exponent to which we raise 2 to get x."**

Thus if $f(x) = 2^x$, then $f^{-1}(x) = \log_2 x$. Note that $f^{-1}(8) = \log_2 8 = 3$, because 3 is *the exponent to which we raise 2 to get* 8.

EXAMPLE 1 Simplify: **(a)** $\log_2 32$; **(b)** $\log_2 1$; **(c)** $\log_2 \frac{1}{8}$.

SOLUTION

a) Think of $\log_2 32$ as the exponent to which we raise 2 to get 32. That exponent is 5. Therefore, $\log_2 32 = 5$.

b) We ask ourselves: "To what exponent do we raise 2 in order to get 1?" That exponent is 0 (recall that $2^0 = 1$). Thus, $\log_2 1 = 0$.

c) To what exponent do we raise 2 in order to get $\frac{1}{8}$? Since $2^{-3} = \frac{1}{8}$, we have $\log_2 \frac{1}{8} = -3$.

Although numbers like $\log_2 13$ can be only approximated, we must remember that $\log_2 13$ represents *the exponent to which we raise 2 to get 13*. That is, $2^{\log_2 13} = 13$. A calculator can be used to show that $\log_2 13 \approx 3.7$ and $2^{3.7} \approx 13$. Later in this chapter, we will use a calculator to find such approximations.

For any exponential function $f(x) = a^x$, the inverse is called a **logarithmic function, base** *a*. The graph of the inverse can be drawn by reflecting the graph of $f(x) = a^x$ across the line $y = x$. It will be helpful to remember that the inverse of $f(x) = a^x$ is given by $f^{-1}(x) = \log_a x$.

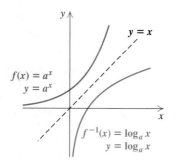

Student Notes

As an aid in remembering what $\log_a x$ means, note that *a* is called the *base*, just as it is the base in $a^y = x$.

The Meaning of log*ₐ* x

For $x > 0$ and *a* a positive constant other than 1, $\log_a x$ is the exponent to which *a* must be raised in order to get *x*. Thus,

$$\log_a x = m \text{ means } a^m = x$$

or equivalently,

$$\log_a x \text{ is that unique exponent for which } a^{\log_a x} = x.$$

It is important to remember that *a logarithm is an exponent*. It might help to repeat several times: "The logarithm, base *a*, of a number *x* is the exponent to which *a* must be raised in order to get *x*."

EXAMPLE 2 Simplify: $7^{\log_7 85}$.

SOLUTION Remember that $\log_7 85$ is the exponent to which 7 is raised to get 85. Raising 7 to that exponent, we have

$$7^{\log_7 85} = 85.$$

Because logarithmic and exponential functions are inverses of each other, the result in Example 2 should come as no surprise: If $f(x) = \log_7 x$, then

$$\text{for} \quad f(x) = \log_7 x, \text{ we have } f^{-1}(x) = 7^x$$
$$\text{and} \quad f^{-1}(f(x)) = f^{-1}(\log_7 x) = 7^{\log_7 x} = x.$$

Thus, $f^{-1}(f(85)) = 7^{\log_7 85} = 85$.

The following is a comparison of exponential and logarithmic functions.

Exponential Function	Logarithmic Function
$y = a^x$	$x = a^y$
$f(x) = a^x$	$g(x) = \log_a x$
$a > 0, a \neq 1$	$a > 0, a \neq 1$
The domain is $\mathbb{R}$.	The range is $\mathbb{R}$.
The range is $(0, \infty)$.	The domain is $(0, \infty)$.
$f^{-1}(x) = \log_a x$	$g^{-1}(x) = a^x$
The graph of $f(x)$ contains the points $(0, 1)$ and $(1, a)$.	The graph of $g(x)$ contains the points $(1, 0)$ and $(a, 1)$.

Student Notes

Since $g(x) = \log_a x$ is the inverse of $f(x) = a^x$, the domain of f is the range of g, and the range of f is the domain of g.

EXAMPLE 3   Graph: $y = f(x) = \log_5 x$.

SOLUTION   If $y = \log_5 x$, then $5^y = x$. We can find ordered pairs that are solutions by choosing values for y and computing the x-values.

For $y = 0$, $x = 5^0 = 1$.
For $y = 1$, $x = 5^1 = 5$.
For $y = 2$, $x = 5^2 = 25$.
For $y = -1$, $x = 5^{-1} = \frac{1}{5}$.
For $y = -2$, $x = 5^{-2} = \frac{1}{25}$.

(1) Select y.
(2) Compute x.

This table shows the following:

$\left. \begin{array}{l} \log_5 1 = 0; \\ \log_5 5 = 1; \\ \log_5 25 = 2; \\ \log_5 \frac{1}{5} = -1; \\ \log_5 \frac{1}{25} = -2. \end{array} \right\}$ **These can all be checked using the equations above.**

x, or 5^y	y
1	0
5	1
25	2
$\frac{1}{5}$	-1
$\frac{1}{25}$	-2

We plot the set of ordered pairs and connect the points with a smooth curve. The graphs of $y = 5^x$ and $y = x$ are shown only for reference.

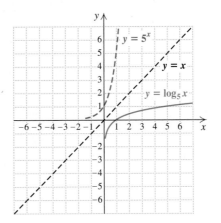

Common Logarithms

Any positive number other than 1 can serve as the base of a logarithmic function. However, some numbers are easier to use than others.

Base-10 logarithms, called **common logarithms,** are useful because they have the same base as our "commonly" used decimal system. Before calculators became so widely available, common logarithms helped with tedious calculations. In fact, that is why logarithms were developed. Before the advent of calculators, tables were developed to list common logarithms. Today we find common logarithms using calculators.

The abbreviation **log,** with no base written, is generally understood to mean logarithm base 10, or a common logarithm. Thus,

$$\log 17 \quad \text{means} \quad \log_{10} 17.$$ It is important to remember this abbreviation.

Common Logarithms

To find the common logarithm of a number on most graphing calculators, press **LOG**, the number, and then **ENTER**. Some calculators automatically supply the left parenthesis when **LOG** is pressed. Of these, some require that a right parenthesis be entered and some assume one is present at the end of the expression. Even if a calculator does not require an ending parenthesis, it is a good idea to include one.

For example, if we are using a calculator that supplies the left parenthesis, log 5 can be found by pressing **LOG** **5** **)** **ENTER**. Note from the screen on the left below that log 5 ≈ 0.6990.

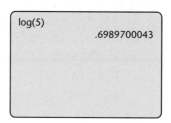

log(5)	log(250/.03)
.6989700043	3.920818754
	log(250)/log(.03
	)
	-1.574609939

Parentheses must be placed carefully when evaluating expressions containing logarithms. For example, to evaluate

$$\log \frac{250}{0.03},$$

press **LOG** **2** **5** **0** **÷** **.** **0** **3** **)** **ENTER**. To evaluate

$$\frac{\log 250}{\log 0.03},$$

press **LOG** **2** **5** **0** **)** **÷** **LOG** **.** **0** **3** **)** **ENTER**. Note from the screen on the right above that $\log \dfrac{250}{0.03} \approx 3.9208$ and that $\dfrac{\log 250}{\log 0.03} \approx -1.5746.$

(continued)

The inverse of a logarithmic function is an exponential function. Because of this, on many calculators the ⟨**LOG**⟩ key serves as the ⟨**10ˣ**⟩ key after the ⟨**2ND**⟩ key is pressed. As with logarithms, some calculators automatically supply a left parenthesis before the exponent. Using such a calculator, we find $10^{3.417}$ by pressing ⟨**10ˣ**⟩ ⟨**3**⟩ ⟨**.**⟩ ⟨**4**⟩ ⟨**1**⟩ ⟨**7**⟩ ⟨**)**⟩ ⟨**ENTER**⟩. The result is approximately 2612.1614. If we then press ⟨**LOG**⟩ ⟨**ANS**⟩ ⟨**)**⟩ ⟨**ENTER**⟩, the result is 3.417. This illustrates that $f(x) = \log x$ and $g(x) = 10^x$ are inverse functions.

```
10^(3.417)
                  2612.161354
log(Ans)
                         3.417
```

A graphing calculator can quickly draw graphs of logarithmic functions, base 10. It is important to remember to place parentheses properly when entering logarithmic expressions.

EXAMPLE 4   Graph: $f(x) = \log \dfrac{x}{5} + 1$.

SOLUTION   We enter $y = \log (x/5) + 1$. Since logarithms of negative numbers are not defined, we set the window accordingly. The graph is shown below. Note that although on the calculator the graph appears to touch the y-axis, a table of values or TRACE would show that the y-axis is indeed an asymptote.

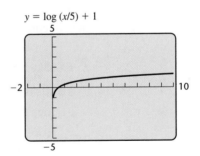

$y = \log (x/5) + 1$

Equivalent Equations

We use the definition of logarithm to rewrite a *logarithmic equation* as an equivalent *exponential equation* or the other way around:

$$m = \log_a x \quad \text{is equivalent to} \quad a^m = x.$$

> **CAUTION!**   **Do not forget this relationship!** It is probably the most important definition in the chapter. Many times this definition will be used to justify a property we are considering.

EXAMPLE 5   Rewrite each as an equivalent exponential equation: **(a)** $y = \log_3 5$; **(b)** $-2 = \log_a 7$; **(c)** $a = \log_b d$.

SOLUTION

a) $y = \log_3 5$ is equivalent to $3^y = 5$ The logarithm is the exponent.

The base remains the base.

b) $-2 = \log_a 7$ is equivalent to $a^{-2} = 7$
c) $a = \log_b d$ is equivalent to $b^a = d$

EXAMPLE 6   Rewrite each as an equivalent logarithmic equation: **(a)** $8 = 2^x$; **(b)** $y^{-1} = 4$; **(c)** $a^b = c$.

SOLUTION

a) $8 = 2^x$ is equivalent to $x = \log_2 8$ The exponent is the logarithm.

The base remains the base.

b) $y^{-1} = 4$ is equivalent to $-1 = \log_y 4$
c) $a^b = c$ is equivalent to $b = \log_a c$

Solving Certain Logarithmic Equations

Logarithmic equations are often solved by rewriting them as equivalent exponential equations.

EXAMPLE 7   Solve: **(a)** $\log_2 x = -3$; **(b)** $\log_x 16 = 2$.

SOLUTION

a) $\log_2 x = -3$

    $2^{-3} = x$ Rewriting as an exponential equation

    $\frac{1}{8} = x$ Computing 2^{-3}

Check: $\log_2 \frac{1}{8}$ is the exponent to which 2 is raised to get $\frac{1}{8}$. Since that exponent is -3, we have a check. The solution is $\frac{1}{8}$.

b) $\log_x 16 = 2$

    $x^2 = 16$ Rewriting as an exponential equation

    $x = 4$ *or* $x = -4$ Principle of square roots

Check: $\log_4 16 = 2$ because $4^2 = 16$. Thus, 4 is a solution of $\log_x 16 = 2$. Because all logarithmic bases must be positive, -4 cannot be a solution. Logarithmic bases must be positive because logarithms are defined using exponential functions that require positive bases. The solution is 4.

One method for solving certain logarithmic and exponential equations relies on the following property, which results from the fact that exponential functions are one-to-one.

The Principle of Exponential Equality   For any real number b, where $b \neq -1, 0$, or 1,

$$b^x = b^y \quad \text{is equivalent to} \quad x = y.$$

(Powers of the same base are equal if and only if the exponents are equal.)

EXAMPLE 8   Solve: **(a)** $\log_{10} 1000 = x$; **(b)** $\log_4 1 = t$.

SOLUTION

a) We rewrite $\log_{10} 1000 = x$ in exponential form and solve:

$$10^x = 1000 \qquad \text{Rewriting as an exponential equation}$$
$$10^x = 10^3 \qquad \text{Writing 1000 as a power of 10}$$
$$x = 3. \qquad \text{Equating exponents}$$

Check: This equation can also be solved directly by determining the exponent to which we raise 10 in order to get 1000. In both cases, we find that $\log_{10} 1000 = 3$, so we have a check. The solution is 3.

b) We rewrite $\log_4 1 = t$ in exponential form and solve:

$$4^t = 1 \qquad \text{Rewriting as an exponential equation}$$
$$4^t = 4^0 \qquad \text{Writing 1 as a power of 4. This can be done mentally.}$$
$$t = 0. \qquad \text{Equating exponents}$$

Check: As in part (a), this equation can be solved directly by determining the exponent to which we raise 4 in order to get 1. In both cases, we find that $\log_4 1 = 0$, so we have a check. The solution is 0.

Example 8 illustrates an important property of logarithms.

$\log_a 1$

The logarithm, base a, of 1 is always 0: $\log_a 1 = 0$.

This follows from the fact that $a^0 = 1$ is equivalent to the logarithmic equation $\log_a 1 = 0$. Thus, $\log_{10} 1 = 0$, $\log_7 1 = 0$, and so on.

Another property results from the fact that $a^1 = a$. This is equivalent to the equation $\log_a a = 1$.

> **$\log_a a$**
>
> The logarithm, base a, of a is always 1: $\log_a a = 1$.

Thus, $\log_{10} 10 = 1$, $\log_8 8 = 1$, and so on.

12.3 EXERCISE SET

FOR EXTRA HELP

MathXL   MyMathLab   InterAct Math   AW Math Tutor Center   Video Lectures on CD: Disc 6   Student's Solutions Manual

Concept Reinforcement   *In each of Exercises 1–8, match the expression or equation with an equivalent expression or equation from the column on the right.*

1. _____ $\log_5 25$ **a)** 1

2. _____ $2^5 = x$ **b)** x

3. _____ $\log_5 5$ **c)** $x^5 = 27$

4. _____ $\log_2 1$ **d)** $\log_2 x = 5$

5. _____ $\log_5 5^x$ **e)** $\log_2 8 = x$

6. _____ $\log_x 27 = 5$ **f)** $\log_x 5 = -2$

7. _____ $8 = 2^x$ **g)** 2

8. _____ $x^{-2} = 5$ **h)** 0

Simplify.

9. $\log_{10} 1000$ **10.** $\log_{10} 100$

11. $\log_2 16$ **12.** $\log_2 8$

13. $\log_3 81$ **14.** $\log_3 27$

15. $\log_4 \frac{1}{16}$ **16.** $\log_4 \frac{1}{4}$

17. $\log_7 \frac{1}{7}$ **18.** $\log_7 \frac{1}{49}$

19. $\log_5 625$ **20.** $\log_5 125$

21. $\log_8 8$ **22.** $\log_7 1$

23. $\log_8 1$ **24.** $\log_8 8$

Aha! **25.** $\log_9 9^5$ **26.** $\log_9 9^{10}$

27. $\log_{10} 0.01$ **28.** $\log_{10} 0.1$

29. $\log_9 3$ **30.** $\log_{16} 4$

31. $\log_9 27$ **32.** $\log_{16} 64$

33. $\log_{1000} 100$ **34.** $\log_{27} 9$

35. $5^{\log_5 7}$ **36.** $6^{\log_6 13}$

Graph by hand.

37. $y = \log_{10} x$ **38.** $y = \log_2 x$

39. $y = \log_3 x$ **40.** $y = \log_7 x$

41. $f(x) = \log_6 x$ **42.** $f(x) = \log_4 x$

43. $f(x) = \log_{2.5} x$ **44.** $f(x) = \log_{1/2} x$

Graph both functions using the same set of axes.

45. $f(x) = 3^x$, $f^{-1}(x) = \log_3 x$

46. $f(x) = 4^x$, $f^{-1}(x) = \log_4 x$

Use a calculator to find each of the following rounded to four decimal places.

47. $\log 4$ **48.** $\log 5$

49. $\log 13{,}400$ **50.** $\log 93{,}100$

51. $\log 0.527$ **52.** $\log 0.493$

Use a calculator to find each of the following rounded to four decimal places.

53. $10^{2.3}$ **54.** $10^{0.173}$

55. $10^{-2.9523}$ **56.** $10^{4.8982}$

57. $10^{0.0012}$ **58.** $10^{-3.89}$

Graph using a graphing calculator.

59. $\log (x + 2)$

60. $\log (x - 5)$

61. $\log (1 - 2x)$

62. $\log (3x + 2.7)$

63. $\log (x^2)$

64. $\log (x^2 + 1)$

Rewrite each of the following as an equivalent exponential equation. Do not solve.

65. $t = \log_5 9$

66. $h = \log_7 10$

67. $\log_5 25 = 2$

68. $\log_6 6 = 1$

69. $\log_{10} 0.1 = -1$

70. $\log_{10} 0.01 = -2$

71. $\log_{10} 7 = 0.845$

72. $\log_{10} 3 = 0.4771$

73. $\log_c m = 8$

74. $\log_b n = 23$

75. $\log_t Q = r$

76. $\log_m P = a$

77. $\log_e 0.25 = -1.3863$

78. $\log_e 0.989 = -0.0111$

79. $\log_r T = -x$

80. $\log_c M = -w$

Rewrite each of the following as an equivalent logarithmic equation. Do not solve.

81. $10^2 = 100$

82. $10^4 = 10,000$

83. $4^{-5} = \frac{1}{1024}$

84. $5^{-3} = \frac{1}{125}$

85. $16^{3/4} = 8$

86. $8^{1/3} = 2$

87. $10^{0.4771} = 3$

88. $10^{0.3010} = 2$

89. $z^m = 6$

90. $m^n = r$

91. $p^m = V$

92. $Q^t = x$

93. $e^3 = 20.0855$

94. $e^2 = 7.3891$

95. $e^{-4} = 0.0183$

96. $e^{-2} = 0.1353$

Solve.

97. $\log_3 x = 2$

98. $\log_4 x = 3$

99. $\log_5 125 = x$

100. $\log_4 64 = x$

101. $\log_2 16 = x$

102. $\log_3 27 = x$

103. $\log_x 7 = 1$

104. $\log_x 8 = 1$

105. $\log_3 x = -2$

106. $\log_2 x = -1$

107. $\log_{32} x = \frac{2}{5}$

108. $\log_8 x = \frac{2}{3}$

109. Express in words what number is represented by $\log_b c$.

110. Is it true that $2 = b^{\log_b 2}$? Why or why not?

Skill Maintenance

Simplify.

111. $\frac{x^{12}}{x^4}$ [5.1]

112. $\frac{a^{15}}{a^3}$ [5.1]

113. $(a^4 b^6)(a^3 b^2)$ [5.1]

114. $(x^3 y^5)(x^2 y^7)$ [5.1]

115. $\dfrac{\frac{3}{x} - \frac{2}{xy}}{\frac{2}{x^2} + \frac{1}{xy}}$ [7.5]

116. $\dfrac{\frac{4 + x}{x^2 + 2x + 1}}{\frac{3}{x+1} - \frac{2}{x+2}}$ [7.5]

Synthesis

117. Would a manufacturer be pleased or unhappy if sales of a product grew logarithmically? Why?

118. Explain why the number $\log_2 13$ must be between 3 and 4.

119. Graph both equations using the same set of axes:
$$y = \left(\tfrac{3}{2}\right)^x, \qquad y = \log_{3/2} x.$$

Graph by hand.

120. $y = \log_2(x - 1)$

121. $y = \log_3 |x + 1|$

Solve.

122. $|\log_3 x| = 2$

123. $\log_4 (3x - 2) = 2$

124. $\log_8 (2x + 1) = -1$

125. $\log_{10} (x^2 + 21x) = 2$

Simplify.

126. $\log_{1/4} \frac{1}{64}$

127. $\log_{1/5} 25$

128. $\log_{81} 3 \cdot \log_3 81$

129. $\log_{10} (\log_4 (\log_3 81))$

130. $\log_2 (\log_2 (\log_4 256))$

131. Show that $b^x = b^y$ is *not* equivalent to $x = y$ for $b = 0$ or $b = 1$.

132. If $\log_b a = x$, does it follow that $\log_a b = 1/x$? Why or why not?

12.4 Properties of Logarithmic Functions

Logarithms of Products ▪ Logarithms of Powers ▪
Logarithms of Quotients ▪ Using the Properties Together

Logarithmic functions are important in many applications and in more advanced mathematics. We now establish some basic properties that are useful in manipulating expressions involving logarithms.

Interactive Discovery

For each of the following, use a calculator to determine which is the equivalent logarithmic expression.

1. $\log (20 \cdot 5)$
   a) $\log (20) \cdot \log (5)$
   b) $5 \cdot \log (20)$
   c) $\log (20) + \log (5)$

2. $\log (20^5)$
   a) $\log (20) \cdot \log (5)$
   b) $5 \cdot \log (20)$
   c) $\log (20) - \log (5)$

3. $\log \left(\dfrac{20}{5} \right)$
   a) $\log (20) \cdot \log (5)$
   b) $\log (20)/\log (5)$
   c) $\log (20) - \log (5)$

4. $\log \left(\dfrac{1}{20} \right)$
   a) $-\log (20)$
   b) $\log (1)/\log (20)$
   c) $\log (1) \cdot \log (20)$

In this section, we will state and prove the results you may have observed. As their proofs reveal, the properties of logarithms are related to the properties of exponents.

Logarithms of Products

The first property we discuss is related to the product rule for exponents: $a^m \cdot a^n = a^{m+n}$. Its proof appears immediately after Example 2.

> **The Product Rule for Logarithms**   For any positive numbers M, N, and a $(a \neq 1)$,
>
> $$\log_a (MN) = \log_a M + \log_a N.$$
>
> (The logarithm of a product is the sum of the logarithms of the factors.)

EXAMPLE 1 Express as an equivalent expression that is a sum of logarithms: $\log_2 (4 \cdot 16)$.

SOLUTION We have

$$\log_2 (4 \cdot 16) = \log_2 4 + \log_2 16. \qquad \text{Using the product rule for logarithms}$$

As a check, note that

$$\log_2 (4 \cdot 16) = \log_2 64 = 6 \qquad 2^6 = 64$$

and that

$$\log_2 4 + \log_2 16 = 2 + 4 = 6. \qquad 2^2 = 4 \text{ and } 2^4 = 16$$

EXAMPLE 2 Express as an equivalent expression that is a single logarithm: $\log_b 7 + \log_b 5$.

SOLUTION We have

$$\log_b 7 + \log_b 5 = \log_b (7 \cdot 5) \qquad \text{Using the product rule for logarithms}$$
$$= \log_b 35.$$

A PROOF OF THE PRODUCT RULE. Let $\log_a M = x$ and $\log_a N = y$. Converting to exponential equations, we have $a^x = M$ and $a^y = N$.

Now we multiply the left sides and the right sides of the equations to obtain

$$MN = a^x \cdot a^y, \quad \text{or} \quad MN = a^{x+y}.$$

Converting back to a logarithmic equation, we get

$$\log_a (MN) = x + y.$$

Recalling what x and y represent, we get

$$\log_a (MN) = \log_a M + \log_a N.$$

Study Tip

Go as Slow as Necessary

When new material seems challenging to you, do not pay attention to how quickly or slowly a classmate absorbs the concepts. We each move at our own speed when it comes to digesting new material.

Logarithms of Powers

The second basic property is related to the power rule for exponents: $(a^m)^n = a^{mn}$. Its proof follows Example 3.

The Power Rule for Logarithms For any positive numbers M and a ($a \neq 1$), and any real number p,

$$\log_a M^p = p \cdot \log_a M.$$

(The logarithm of a power of M is the exponent times the logarithm of M.)

To better understand the power rule, note that

$$\log_a M^3 = \log_a (M \cdot M \cdot M) = \log_a M + \log_a M + \log_a M = 3 \log_a M.$$

EXAMPLE 3 Use the power rule for logarithms to write an equivalent expression that is a product: **(a)** $\log_a 9^{-5}$; **(b)** $\log_7 \sqrt[3]{x}$.

SOLUTION

a) $\log_a 9^{-5} = -5 \log_a 9$ — Using the power rule for logarithms

b) $\log_7 \sqrt[3]{x} = \log_7 x^{1/3}$ — Writing exponential notation

$\quad\quad\quad = \frac{1}{3} \log_7 x$ — Using the power rule for logarithms

A Proof of the Power Rule. Let $x = \log_a M$. The equivalent exponential equation is $a^x = M$. Raising both sides to the pth power, we get

$$(a^x)^p = M^p, \quad \text{or} \quad a^{xp} = M^p. \quad \text{Multiplying exponents}$$

Converting back to a logarithmic equation gives us

$$\log_a M^p = xp.$$

But $x = \log_a M$, so substituting, we have

$$\log_a M^p = (\log_a M)p = p \cdot \log_a M. \quad \blacksquare$$

Logarithms of Quotients

The third property that we study is similar to the quotient rule for exponents: $\frac{a^m}{a^n} = a^{m-n}$. Its proof follows Example 5.

The Quotient Rule for Logarithms For any positive numbers M, N, and a ($a \neq 1$),

$$\log_a \frac{M}{N} = \log_a M - \log_a N.$$

(The logarithm of a quotient is the logarithm of the dividend minus the logarithm of the divisor.)

To better understand the quotient rule, note that

$$\log_a\left(\frac{b^5}{b^3}\right) = \log_a b^2 = 2 \log_a b = 5 \log_a b - 3 \log_a b$$

$$= \log_a b^5 - \log_a b^3.$$

EXAMPLE 4 Express as an equivalent expression that is a difference of logarithms: $\log_t (6/U)$.

SOLUTION

$$\log_t \frac{6}{U} = \log_t 6 - \log_t U \quad \text{Using the quotient rule for logarithms}$$

EXAMPLE 5 Express as an equivalent expression that is a single logarithm: $\log_b 17 - \log_b 27$.

SOLUTION

$$\log_b 17 - \log_b 27 = \log_b \frac{17}{27}$$ Using the quotient rule for logarithms "in reverse"

A PROOF OF THE QUOTIENT RULE. Our proof uses both the product and power rules:

$$\log_a \frac{M}{N} = \log_a MN^{-1}$$ Rewriting $\frac{M}{N}$ as MN^{-1}

$$= \log_a M + \log_a N^{-1}$$ Using the product rule for logarithms

$$= \log_a M + (-1)\log_a N$$ Using the power rule for logarithms

$$= \log_a M - \log_a N.$$ ■

Using the Properties Together

EXAMPLE 6 Express as an equivalent expression, using the individual logarithms of x, y, and z.

a) $\log_b \dfrac{x^3}{yz}$ **b)** $\log_a \sqrt[4]{\dfrac{xy}{z^3}}$

SOLUTION

a) $\log_b \dfrac{x^3}{yz} = \log_b x^3 - \log_b yz$ Using the quotient rule for logarithms

$$= 3 \log_b x - \log_b yz$$ Using the power rule for logarithms

$$= 3 \log_b x - (\log_b y + \log_b z)$$ Using the product rule for logarithms. Because of the subtraction, parentheses are essential.

$$= 3 \log_b x - \log_b y - \log_b z$$ Using the distributive law

b) $\log_a \sqrt[4]{\dfrac{xy}{z^3}} = \log_a \left(\dfrac{xy}{z^3}\right)^{1/4}$ Writing exponential notation

$$= \frac{1}{4} \cdot \log_a \frac{xy}{z^3}$$ Using the power rule for logarithms

$$= \frac{1}{4}\left(\log_a xy - \log_a z^3\right)$$ Using the quotient rule for logarithms. Parentheses are important.

$$= \frac{1}{4}\left(\log_a x + \log_a y - 3 \log_a z\right)$$ Using the product and power rules for logarithms

> **CAUTION!** Because the product and quotient rules replace one term with two, parentheses are often necessary, as in Example 6.

EXAMPLE 7 Express as an equivalent expression that is a single logarithm.

a) $\dfrac{1}{2} \log_a x - 7 \log_a y + \log_a z$ **b)** $\log_a \dfrac{b}{\sqrt{x}} + \log_a \sqrt{bx}$

SOLUTION

a) $\dfrac{1}{2} \log_a x - 7 \log_a y + \log_a z$

$= \log_a x^{1/2} - \log_a y^7 + \log_a z$ **Using the power rule for logarithms**

$= \left(\log_a \sqrt{x} - \log_a y^7\right) + \log_a z$ **Using parentheses to emphasize the order of operations; $x^{1/2} = \sqrt{x}$**

$= \log_a \dfrac{\sqrt{x}}{y^7} + \log_a z$ **Using the quotient rule for logarithms**

$= \log_a \dfrac{z\sqrt{x}}{y^7}$ **Using the product rule for logarithms**

b) $\log_a \dfrac{b}{\sqrt{x}} + \log_a \sqrt{bx} = \log_a \dfrac{b \cdot \sqrt{bx}}{\sqrt{x}}$ **Using the product rule for logarithms**

$= \log_a b\sqrt{b}$ **Removing a factor equal to 1: $\dfrac{\sqrt{x}}{\sqrt{x}} = 1$**

$= \log_a b^{3/2},\ \text{or}\ \dfrac{3}{2} \log_a b$ **Since $b\sqrt{b} = b^1 \cdot b^{1/2}$**

If we know the logarithms of two different numbers (to the same base), the properties allow us to calculate other logarithms.

EXAMPLE 8   Given $\log_a 2 = 0.431$ and $\log_a 3 = 0.683$, calculate a numerical value for each of the following, if possible.

a) $\log_a 6$ **b)** $\log_a \frac{2}{3}$ **c)** $\log_a 81$

d) $\log_a \frac{1}{3}$ **e)** $\log_a 2a$ **f)** $\log_a 5$

SOLUTION

a) $\log_a 6 = \log_a (2 \cdot 3) = \log_a 2 + \log_a 3$ **Using the product rule for logarithms**

$= 0.431 + 0.683 = 1.114$

Check: $a^{1.114} = a^{0.431} \cdot a^{0.683} = 2 \cdot 3 = 6$

b) $\log_a \frac{2}{3} = \log_a 2 - \log_a 3$ **Using the quotient rule for logarithms**

$= 0.431 - 0.683 = -0.252$

c) $\log_a 81 = \log_a 3^4 = 4 \log_a 3$ **Using the power rule for logarithms**

$= 4(0.683) = 2.732$

d) $\log_a \frac{1}{3} = \log_a 1 - \log_a 3$ **Using the quotient rule for logarithms**

$= 0 - 0.683 = -0.683$

e) $\log_a 2a = \log_a 2 + \log_a a$ **Using the product rule for logarithms**

$= 0.431 + 1 = 1.431$

f) $\log_a 5$ *cannot be found using these properties.* ($\log_a 5 \neq \log_a 2 + \log_a 3$)

A final property follows from the product rule: Since $\log_a a^k = k \log_a a$, and $\log_a a = 1$, we have $\log_a a^k = k$.

The Logarithm of the Base to an Exponent   For any base a,

$$\log_a a^k = k.$$

(The logarithm, base a, of a to an exponent is the exponent.)

This property also follows from the definition of logarithm: k is the exponent to which you raise a in order to get a^k.

EXAMPLE 9   Simplify: **(a)** $\log_3 3^7$; **(b)** $\log_{10} 10^{-5.2}$.

SOLUTION

a) $\log_3 3^7 = 7$ **7 is the exponent to which you raise 3 in order to get 3^7.**

b) $\log_{10} 10^{-5.2} = -5.2$

We summarize the properties covered in this section as follows.

For any positive numbers M, N, and a ($a \neq 1$):

$$\log_a (MN) = \log_a M + \log_a N; \qquad \log_a M^p = p \cdot \log_a M;$$

$$\log_a \frac{M}{N} = \log_a M - \log_a N; \qquad \log_a a^k = k.$$

12.4 EXERCISE SET

FOR EXTRA HELP

*Math*XP   MathXL   MyMathLab   InterAct Math   Tutor Center   AW Math Tutor Center   Video Lectures on CD: Disc 6   Student's Solutions Manual

↪ *Concept Reinforcement*   *In each of Exercises 1–6, match the expression with an equivalent expression from the column on the right.*

1. ____ $\log_7 20$

2. ____ $\log_7 5^4$

3. ____ $\log_7 \frac{5}{4}$

4. ____ $\log_7 7$

5. ____ $\log_7 1$

6. ____ $\log_7 5 + \log_7 6$

a) $\log_7 5 - \log_7 4$

b) 1

c) 0

d) $\log_7 30$

e) $\log_7 5 + \log_7 4$

f) $4 \log_7 5$

Express as an equivalent expression that is a sum of logarithms.

7. $\log_3 (81 \cdot 27)$

8. $\log_2 (16 \cdot 32)$

9. $\log_4 (64 \cdot 16)$

10. $\log_5 (25 \cdot 125)$

11. $\log_c (rst)$

12. $\log_t (3ab)$

Express as an equivalent expression that is a single logarithm.

13. $\log_a 5 + \log_a 14$

14. $\log_b 65 + \log_b 2$

15. $\log_c t + \log_c y$

16. $\log_t H + \log_t M$

Express as an equivalent expression that is a product.

17. $\log_a r^8$ **18.** $\log_b t^5$

19. $\log_c y^6$ **20.** $\log_{10} y^7$

21. $\log_b C^{-3}$ **22.** $\log_c M^{-5}$

Express as an equivalent expression that is a difference of two logarithms.

23. $\log_2 \frac{25}{13}$ **24.** $\log_3 \frac{23}{9}$

25. $\log_b \frac{m}{n}$ **26.** $\log_a \frac{y}{x}$

Express as an equivalent expression that is a single logarithm.

27. $\log_a 17 - \log_a 6$ **28.** $\log_b 32 - \log_b 7$

29. $\log_b 36 - \log_b 4$ **30.** $\log_a 26 - \log_a 2$

31. $\log_a 7 - \log_a 18$ **32.** $\log_b 5 - \log_b 13$

Express as an equivalent expression, using the individual logarithms of w, x, y, and z.

33. $\log_a (xyz)$ **34.** $\log_a (wxy)$

35. $\log_a (x^3 z^4)$ **36.** $\log_a (x^2 y^5)$

37. $\log_a (x^2 y^{-2} z)$ **38.** $\log_a (xy^2 z^{-3})$

39. $\log_a \frac{x^4}{y^3 z}$ **40.** $\log_a \frac{x^4}{yz^2}$

41. $\log_b \frac{xy^2}{wz^3}$ **42.** $\log_b \frac{w^2 x}{y^3 z}$

43. $\log_a \sqrt{\frac{x^7}{y^5 z^8}}$ **44.** $\log_c \sqrt[3]{\frac{x^4}{y^3 z^2}}$

45. $\log_a \sqrt[3]{\frac{x^6 y^3}{a^2 z^7}}$ **46.** $\log_a \sqrt[4]{\frac{x^8 y^{12}}{a^3 z^5}}$

Express as an equivalent expression that is a single logarithm and, if possible, simplify.

47. $8 \log_a x + 3 \log_a z$

48. $2 \log_b m + \frac{1}{2} \log_b n$

49. $\log_a x^2 - 2 \log_a \sqrt{x}$

50. $\log_a \frac{a}{\sqrt{x}} - \log_a \sqrt{ax}$

51. $\frac{1}{2} \log_a x + 5 \log_a y - 2 \log_a x$

52. $\log_a 2x + 3(\log_a x - \log_a y)$

53. $\log_a (x^2 - 4) - \log_a (x + 2)$

54. $\log_a (2x + 10) - \log_a (x^2 - 25)$

Given $\log_b 3 = 0.792$ and $\log_b 5 = 1.161$. If possible, calculate numerical values for each of the following.

55. $\log_b 15$ **56.** $\log_b \frac{5}{3}$

57. $\log_b \frac{3}{5}$ **58.** $\log_b \frac{1}{3}$

59. $\log_b \frac{1}{5}$ **60.** $\log_b \sqrt{b}$

61. $\log_b \sqrt{b^3}$ **62.** $\log_b 3b$

63. $\log_b 8$ **64.** $\log_b 45$

Simplify.

Aha! **65.** $\log_t t^7$ **66.** $\log_p p^4$

67. $\log_e e^m$ **68.** $\log_Q Q^{-2}$

Find each of the following.

Aha! **69.** $\log_5 (125 \cdot 625)$ **70.** $\log_3 (9 \cdot 81)$

71. $\log_2 \left(\frac{128}{16} \right)$ **72.** $\log_3 \left(\frac{243}{27} \right)$

TW **73.** A student *incorrectly* reasons that

$$\log_b \frac{1}{x} = \log_b \frac{x}{xx}$$

$$= \log_b x - \log_b x + \log_b x = \log_b x.$$

What mistake has the student made?

TW **74.** How could you convince someone that

$$\log_a c \neq \log_c a?$$

Skill Maintenance

Graph. [10.1]

75. $f(x) = \sqrt{x} - 3$ **76.** $g(x) = \sqrt{x} + 2$

77. $g(x) = \sqrt[3]{x} + 1$ **78.** $f(x) = \sqrt[3]{x} - 1$

Simplify. [5.1]

79. $(a^3 b^2)^5 (a^2 b^7)$ **80.** $(x^5 y^3 z^2)(x^2 yz^2)^3$

Synthesis

TW **81.** Is it possible to express $\log_b \frac{x}{5}$ as an equivalent expression that is a difference of two logarithms without using the quotient rule? Why or why not?

TW **82.** Is it true that $\log_a x + \log_b x = \log_{ab} x$? Why or why not?

Express as an equivalent expression that is a single logarithm and, if possible, simplify.

83. $\log_a (x^8 - y^8) - \log_a (x^2 + y^2)$

84. $\log_a (x + y) + \log_a (x^2 - xy + y^2)$

Express as an equivalent expression that is a sum or difference of logarithms and, if possible, simplify.

85. $\log_a \sqrt{1 - s^2}$

86. $\log_a \dfrac{c - d}{\sqrt{c^2 - d^2}}$

87. If $\log_a x = 2$, $\log_a y = 3$, and $\log_a z = 4$, what is

$$\log_a \frac{\sqrt[3]{x^2 z}}{\sqrt[3]{y^2 z^{-2}}}?$$

88. If $\log_a x = 2$, what is $\log_a (1/x)$?

89. If $\log_a x = 2$, what is $\log_{1/a} x$?

Classify each of the following as true or false. Assume a, x, P, and Q > 0, a ≠ 1.

90. $\log_a \left(\dfrac{P}{Q}\right)^x = x \log_a P - \log_a Q$

91. $\log_a (Q + Q^2) = \log_a Q + \log_a (Q + 1)$

92. Use graphs to show that

$$\log x^2 \neq \log x \cdot \log x.$$

(*Note:* log means $\log_{10}$.)

12.5
Natural Logarithms and Changing Bases

The Base *e* and Natural Logarithms ▪ Changing Logarithmic Bases ▪ Graphs of Exponential and Logarithmic Functions, Base *e*

There are logarithm bases that fit into certain applications more naturally than others. We have already looked at logarithms with one such base, base-10 logarithms, or common logarithms. Another logarithm base widely used today is an irrational number named *e*.

The Base *e* and Natural Logarithms

When interest is computed *n* times a year, the compound interest formula is

$$A = P\left(1 + \frac{r}{n}\right)^{nt},$$

where A is the amount that an initial investment P will be worth after t years at interest rate r. Suppose that $1 is invested at 100% interest for 1 year (no bank would pay this). The preceding formula becomes a function A defined in terms of the number of compounding periods n:

$$A(n) = \left(1 + \frac{1}{n}\right)^n.$$

Interactive Discovery

1. What happens to the function values of $A(n) = \left(1 + \dfrac{1}{n}\right)^n$ as n gets larger? To find out, use the TABLE feature with Indpnt set to Ask to fill in the following table. Round each entry to six decimal places.

n	$A(n) = \left(1 + \dfrac{1}{n}\right)^n$
1 (compounded annually)	$\left(1 + \dfrac{1}{1}\right)^1$, or \$2.00
2 (compounded semiannually)	$\left(1 + \dfrac{1}{2}\right)^2$, or \$☐
3	
4 (compounded quarterly)	
12 (compounded monthly)	
52 (compounded weekly)	
365 (compounded daily)	
8760 (compounded hourly)	

2. Which of the following statements appears to be true?

   a) $A(n)$ gets very large as n gets very large.

   b) $A(n)$ gets very small as n gets very large.

   c) $A(n)$ approaches a certain number as n gets very large.

The numbers in the table approach a very important number in mathematics, called e. Because e is irrational, its decimal representation does not terminate or repeat.

The Number e $e \approx 2.7182818284\ldots$

Logarithms base e are called **natural logarithms,** or **Napierian logarithms,** in honor of John Napier (1550–1617), who first "discovered" logarithms.
 The abbreviation "ln" is generally used with natural logarithms. Thus,

$\ln 53$ means $\log_e 53$.   **Remember:** $\log x$ **means** $\log_{10} x$, **and** $\ln x$ **means** $\log_e x$.

Study Tip

Is Your Answer Reasonable?

It is always a good idea—especially when using a calculator—to check that your answer is reasonable. It is easy for an incorrect calculation or keystroke to result in an answer that is clearly too big or too small.

Natural Logarithms

Natural logarithms are entered on a graphing calculator much like common logarithms are (see pp. 916–917). For example, if we are using a calculator that supplies the left parenthesis, $\ln 8$ can be found by pressing $\boxed{LN}$ $\boxed{8}$ $\boxed{)}$ $\boxed{ENTER}$. Note from the screen below that $\ln 8 \approx 2.0794$.

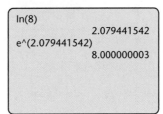

```
ln(8)
              2.079441542
e^(2.079441542)
              8.000000003
```

On many calculators, the $\boxed{LN}$ key serves as the $\boxed{e^x}$ key after the $\boxed{2ND}$ key is pressed. Some calculators automatically supply a left parenthesis before the exponent. If such a calculator is used,

$$e^{2.079441542}$$

is found by pressing $\boxed{e^x}$ $\boxed{2}$ $\boxed{.}$ $\boxed{0}$ $\boxed{7}$ $\boxed{9}$ $\boxed{4}$ $\boxed{4}$ $\boxed{1}$ $\boxed{5}$ $\boxed{4}$ $\boxed{2}$ $\boxed{)}$ $\boxed{ENTER}$. As shown on the screen above, $e^{2.079441542} \approx 8$, as we would expect, since $f(x) = e^x$ is the inverse function of $g(x) = \ln x$.

Changing Logarithmic Bases

Most calculators can find both common logarithms and natural logarithms. To find a logarithm with some other base, a conversion formula is usually needed.

The Change-of-Base Formula   For any logarithmic bases a and b, and any positive number M,

$$\log_b M = \frac{\log_a M}{\log_a b}.$$

(To find the log, base b, we typically compute $\log M/\log b$ or $\ln M/\ln b$.)

PROOF.   Let $x = \log_b M$. Then,

$b^x = M$	$\log_b M = x$ is equivalent to $b^x = M$.
$\log_a b^x = \log_a M$	Taking the logarithm, base a, on both sides
$x \log_a b = \log_a M$	Using the power rule for logarithms
$x = \dfrac{\log_a M}{\log_a b}.$	Dividing both sides by $\log_a b$

But at the outset we stated that $x = \log_b M$. Thus, by substitution, we have

$$\log_b M = \frac{\log_a M}{\log_a b}.$$ This is the change-of-base formula.   ∎

EXAMPLE 1 Find $\log_5 8$ using the change-of-base formula.

SOLUTION We use the change-of-base formula with $a = 10$, $b = 5$, and $M = 8$:

$$\log_5 8 = \frac{\log_{10} 8}{\log_{10} 5}$$ Substituting into $\log_b M = \dfrac{\log_a M}{\log_a b}$

$$\approx \frac{0.903089987}{0.6989700043}$$ Using **LOG** twice

$$\approx 1.2920.$$ When using a calculator, it is best not to round before dividing.

The figure below shows the computation using a graphing calculator.

```
log(8)/log(5)
              1.292029674
```

To check, note that $\ln 8/\ln 5 \approx 1.2920$. We can also use a calculator to verify that $5^{1.2920} \approx 8$.

EXAMPLE 2 Find $\log_4 31$.

SOLUTION As indicated in the check of Example 1, base e can also be used.

$$\log_4 31 = \frac{\log_e 31}{\log_e 4}$$ Substituting into $\log_b M = \dfrac{\log_a M}{\log_a b}$

$$= \frac{\ln 31}{\ln 4}$$ Using **LN** twice

$$\approx 2.4771.$$ *Check:* $4^{2.4771} \approx 31$

```
ln(31)/ln(4)
              2.477098155
log(31)/log(4)
              2.477098155
4^Ans
                       31
```

The screen at left illustrates that we find the same solution using either natural or common logarithms.

Graphs of Exponential and Logarithmic Functions, Base e

EXAMPLE 3 Graph $f(x) = e^x$ and $g(x) = e^{-x}$ and state the domain and the range of f and g.

SOLUTION We use a calculator with an $\boxed{e^x}$ key to find approximate values of e^x and e^{-x}. Using these values, we can graph the functions.

x	e^x	e^{-x}
0	1	1
1	2.7	0.4
2	7.4	0.1
−1	0.4	2.7
−2	0.1	7.4

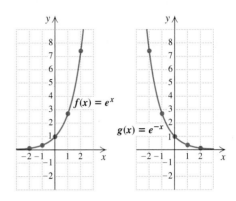

The domain of each function is $\mathbb{R}$ and the range of each function is $(0, \infty)$.

EXAMPLE 4 Graph $f(x) = e^{-x} + 2$ and state the domain and the range of f.

SOLUTION To graph by hand, we find some solutions with a calculator, plot them, and then draw the graph. For example, $f(2) = e^{-2} + 2 \approx 2.1$. The graph is exactly like the graph of $g(x) = e^{-x}$, but is translated 2 units up.

To graph using a graphing calculator, we enter $y = e \,{}^\wedge\, (-x) + 2$ and choose a viewing window that shows more of the y-axis than the x-axis, since the function is exponential.

x	$e^{-x} + 2$
0	3
1	2.4
2	2.1
−1	4.7
−2	9.4

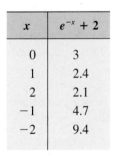
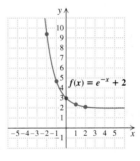
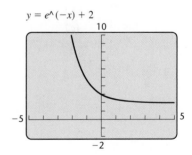

The domain of f is $\mathbb{R}$ and the range is $(2, \infty)$.

EXAMPLE 5 Graph and state the domain and the range of each function.

a) $g(x) = \ln x$ **b)** $f(x) = \ln (x + 3)$

SOLUTION

a) To graph by hand, we find some solutions with a calculator and then draw the graph. As expected, the graph is a reflection across the line $y = x$ of the graph of $y = e^x$. We can also graph $y = \ln (x)$ using a graphing calculator. Since a logarithmic function is the inverse of an exponential function, we choose a viewing window that shows more of the x-axis than the y-axis.

x	$\ln x$
1	0
4	1.4
7	1.9
0.5	−0.7

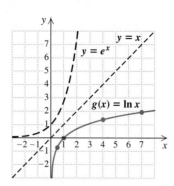

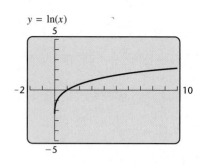

The domain of g is $(0, \infty)$ and the range is $\mathbb{R}$.

b) We find some solutions with a calculator, plot them, and draw the graph by hand. Note that the graph of $y = \ln(x + 3)$ is the graph of $y = \ln x$ translated 3 units to the left. To graph using a graphing calculator, we enter $y = \ln(x + 3)$.

x	$\ln(x + 3)$
0	1.1
1	1.4
2	1.6
3	1.8
4	1.9
−1	0.7
−2	0
−2.5	−0.7

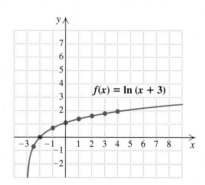

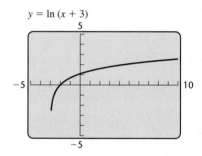

Since $x + 3$ must be positive, the domain of f is $(-3, \infty)$ and the range is $\mathbb{R}$.

Logarithmic functions with bases other than 10 or e can be graphed on a graphing calculator using the change-of-base formula.

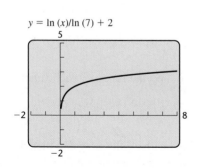

EXAMPLE 6   Graph: $f(x) = \log_7 x + 2$.

SOLUTION   We use the change-of-base formula with natural logarithms. (We would get the same graph if we used common logarithms.)

$$f(x) = \log_7 x + 2 \qquad \text{Note that this is not } \log_7(x + 2).$$

$$= \frac{\ln x}{\ln 7} + 2 \qquad \text{Using the change-of-base formula for } \log_7 x$$

We graph $y = \ln(x)/\ln(7) + 2$.

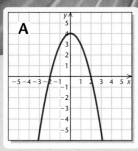

A

F

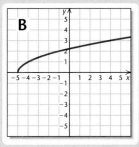

B

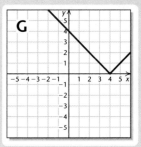

G

Visualizing the Graph

Match each function with its graph.

1. $f(x) = 2x - 3$

2. $f(x) = 2x^2 + 1$

3. $f(x) = \sqrt{x + 5}$

4. $f(x) = |4 - x|$

5. $f(x) = \ln x$

6. $f(x) = 2^{-x}$

7. $f(x) = -4$

8. $f(x) = \log x + 3$

9. $f(x) = 2^x$

10. $f(x) = 4 - x^2$

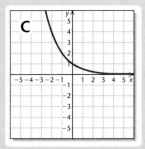

C

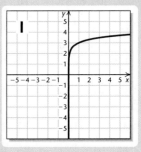

H

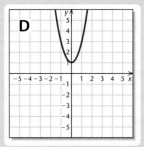

D

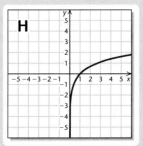

I

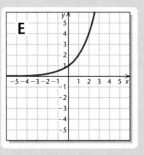

E

Answers on page A-52

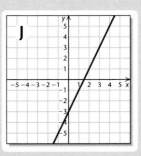

J

12.5 EXERCISE SET

Concept Reinforcement *Classify each of the following as either true or false.*

1. The expression log 23 means $\log_{10} 23$.

2. The expression ln 7 means $\log_e 7$.

3. The number e is approximately 2.7.

4. The expressions log 9 and log 18/log 2 are equivalent.

5. The expressions $\log_2 9$ and $\ln 9/\ln 2$ are equivalent.

6. The domain of the function given by $f(x) = \ln(x + 2)$ is $(-2, \infty)$.

7. The range of the function given by $g(x) = e^x$ is $(0, \infty)$.

8. The range of the function given by $f(x) = \ln x$ is $(-\infty, \infty)$.

Use a calculator to find each of the following to the nearest ten-thousandth.

9. log 6   **10.** log 5   **11.** log 72.8

12. log 73.9   *Aha!* **13.** log 1000   **14.** log 100

15. log 0.527   **16.** log 0.493   **17.** $\dfrac{\log 8200}{\log 150}$

18. $\dfrac{\log 5700}{\log 90}$ **19.** $10^{2.3}$ **20.** $10^{3.4}$

21. $10^{0.173}$ **22.** $10^{0.247}$ **23.** $10^{-2.9523}$

24. $10^{-3.2046}$ **25.** ln 5   **26.** ln 2

27. ln 57   **28.** ln 30   **29.** ln 0.0062

30. ln 0.00073   **31.** $\dfrac{\ln 2300}{0.08}$ **32.** $\dfrac{\ln 1900}{0.07}$

33. $e^{2.71}$ **34.** $e^{3.06}$ **35.** $e^{-3.49}$

36. $e^{-2.64}$ **37.** $e^{4.7}$ **38.** $e^{1.23}$

Find each of the following logarithms using the change-of-base formula. Round answers to the nearest ten-thousandth.

39. $\log_6 92$ **40.** $\log_3 78$

41. $\log_2 100$ **42.** $\log_7 100$

43. $\log_7 65$ **44.** $\log_5 42$

45. $\log_{0.5} 5$ **46.** $\log_{0.1} 3$

47. $\log_2 0.2$ **48.** $\log_2 0.08$

49. $\log_\pi 58$ **50.** $\log_\pi 200$

Graph by hand or using a graphing calculator and state the domain and the range of each function.

51. $f(x) = e^x$ **52.** $f(x) = e^{-x}$

53. $f(x) = e^x + 3$ **54.** $f(x) = e^x + 2$

55. $f(x) = e^x - 2$ **56.** $f(x) = e^x - 3$

57. $f(x) = 0.5e^x$ **58.** $f(x) = 2e^x$

59. $f(x) = 0.5e^{2x}$ **60.** $f(x) = 2e^{-0.5x}$

61. $f(x) = e^{x-3}$ **62.** $f(x) = e^{x-2}$

63. $f(x) = e^{x+2}$ **64.** $f(x) = e^{x+3}$

65. $f(x) = -e^x$ **66.** $f(x) = -e^{-x}$

67. $g(x) = \ln x + 1$ **68.** $g(x) = \ln x + 3$

69. $g(x) = \ln x - 2$ **70.** $g(x) = \ln x - 1$

71. $g(x) = 2\ln x$ **72.** $g(x) = 3\ln x$

73. $g(x) = -2\ln x$ **74.** $g(x) = -\ln x$

75. $g(x) = \ln(x + 2)$ **76.** $g(x) = \ln(x + 1)$

77. $g(x) = \ln(x - 1)$ **78.** $g(x) = \ln(x - 3)$

Write an equivalent expression for the function that could be graphed using a graphing calculator. Then graph the function.

79. $f(x) = \log_5 x$

80. $f(x) = \log_3 x$

81. $f(x) = \log_2 (x - 5)$

82. $f(x) = \log_5 (2x + 1)$

83. $f(x) = \log_3 x + x$

84. $f(x) = \log_2 x - x + 1$

TW 85. Using a calculator, Zeno *incorrectly* says that log 79 is between 4 and 5. How could you convince him, without using a calculator, that he is mistaken?

TW 86. Examine Exercise 85. What mistake do you believe Zeno made?

Skill Maintenance

87. Find an equation of variation if y varies directly as x, and $y = 7.2$ when $x = 0.8$. [7.8]

88. Find an equation of variation if y varies inversely as x, and $y = 3.5$ when $x = 6.1$. [7.8]

Solve. [11.3]

89. $T = 2\pi\sqrt{L/32}$, for L

90. $E = mc^2$, for c
(Assume $E, m, c > 0$.)

91. Joni can key in a musical score in 2 hr. Miles takes 3 hr to key in the same score. How long would it take them, working together, to key in the score? [7.7]

TW 92. The side exit at the Flynn Theater can empty a capacity crowd in 25 min. The main exit can empty a capacity crowd in 15 min. How long will it take to empty a capacity crowd when both exits are in use? [7.7]

Synthesis

TW 93. In an attempt to solve $\ln x = 1.5$, Emma gets the following graph. How can Emma tell at a glance that she has made a mistake?

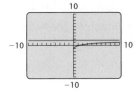

TW 94. Explain how the graph of $f(x) = \ln x$ could be used to graph the function given by $g(x) = e^{x-1}$.

Knowing only that log 2 ≈ 0.301 *and* log 3 ≈ 0.477, *find each of the following.*

95. $\log_6 81$

96. $\log_9 16$

97. $\log_{12} 36$

98. Find a formula for converting common logarithms to natural logarithms.

99. Find a formula for converting natural logarithms to common logarithms.

Solve for x.

100. $\log (275x^2) = 38$

101. $\log (492x) = 5.728$

102. $\dfrac{3.01}{\ln x} = \dfrac{28}{4.31}$

103. $\log 692 + \log x = \log 3450$

For each function given below, **(a)** *determine the domain and the range,* **(b)** *set an appropriate window, and* **(c)** *draw the graph.*

104. $f(x) = 7.4e^x \ln x$

105. $f(x) = 3.4 \ln x - 0.25e^x$

106. $f(x) = x \ln (x - 2.1)$

107. $f(x) = 2x^3 \ln x$

12.6 Solving Exponential and Logarithmic Equations

Solving Exponential Equations ◼ Solving Logarithmic Equations

Solving Exponential Equations

Equations with variables in exponents, such as $5^x = 12$ and $2^{7x} = 64$, are called **exponential equations.** In Section 12.3, we solved certain logarithmic equations by using the principle of exponential equality. We restate that principle below.

> **The Principle of Exponential Equality** For any real number b, where $b \neq -1, 0$, or 1,
>
> $b^x = b^y$ is equivalent to $x = y$.
>
> (Powers of the same base are equal if and only if the exponents are equal.)

EXAMPLE 1 Solve: $4^{3x-5} = 16$.

SOLUTION Note that $16 = 4^2$. Thus we can write each side as a power of the same number:

$$4^{3x-5} = 4^2.$$

Since the base is the same, 4, the exponents must be the same. Thus,

$$3x - 5 = 2 \qquad \textbf{Equating exponents}$$
$$3x = 7$$
$$x = \tfrac{7}{3}.$$

Check:

$$\begin{array}{c|c} \multicolumn{2}{c}{4^{3x-5} = 16} \\ \hline 4^{3(7/3)-5} & 16 \\ 4^{7-5} & \\ 4^2 \overset{?}{=} 16 & \text{TRUE} \end{array}$$

The solution is $\tfrac{7}{3}$.

When it seems impossible to write both sides of an equation as powers of the same base, we use the following principle and write an equivalent logarithmic equation.

> **The Principle of Logarithmic Equality**   For any logarithmic base a, and for x, $y > 0$,
>
> $$x = y \quad \text{is equivalent to} \quad \log_a x = \log_a y.$$
>
> (Two expressions are equal if and only if the logarithms of those expressions are equal.)

Because calculators can generally find only common or natural logarithms (without resorting to the change-of-base formula), we usually take the common or natural logarithm on both sides of the equation.

The principle of logarithmic equality, used together with the power rule for logarithms, allows us to solve equations with a variable in an exponent.

▸ **EXAMPLE 2**   Solve: $7^{x-2} = 60$.

ALGEBRAIC APPROACH

We have

$$7^{x-2} = 60$$

$$\log 7^{x-2} = \log 60$$ Using the principle of logarithmic equality to take the common logarithm on both sides. Natural logarithms also would work.

$$(x - 2) \log 7 = \log 60$$ Using the power rule for logarithms

$$x - 2 = \frac{\log 60}{\log 7}$$ **CAUTION!** This is *not* $\log 60 - \log 7$.

$$x = \frac{\log 60}{\log 7} + 2$$ Adding 2 to both sides

$$x \approx 4.1041.$$ Using a calculator and rounding to four decimal places

Since $7^{4.1041-2} \approx 60.0027$, we have a check. The solution is $\dfrac{\log 60}{\log 7} + 2$, or approximately 4.1041.

GRAPHICAL APPROACH

We graph $y_1 = 7^\wedge (x - 2)$ and $y_2 = 60$.

$y_1 = 7^{x-2}, \quad y_2 = 60$

Intersection
X = 4.1040769   Y = 60
Yscl = 10

Rounded to four decimal places, the x-coordinate of the point of intersection is 4.1041. The solution is 4.1041.

EXAMPLE 3 Solve: $e^{0.06t} = 1500$.

ALGEBRAIC APPROACH

Since one side is a power of e, we take the *natural logarithm* on both sides:

$\ln e^{0.06t} = \ln 1500$	Taking the natural logarithm on both sides
$0.06t = \ln 1500$	Finding the logarithm of the base to a power: $\log_a a^k = k$
$t = \dfrac{\ln 1500}{0.06}$	Dividing both sides by 0.06
$\approx 121.887.$	Using a calculator and rounding to three decimal places

The solution is 121.887.

GRAPHICAL APPROACH

We graph $y_1 = e^{\wedge}(0.06x)$ and $y_2 = 1500$. Since $y_2 = 1500$, we choose a value for Ymax that is greater than 1500. It may require trial and error to choose appropriate units for the x-axis.

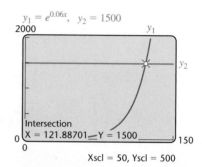

Rounded to three decimal places, the solution is 121.887.

To Solve an Equation of the Form $a^t = b$ for t

1. Take the logarithm (either natural or common) of both sides.
2. Use the power rule for exponents so that the variable is no longer written as an exponent.
3. Divide both sides by the coefficient of the variable to isolate the variable.
4. If appropriate, use a calculator to find an approximate solution in decimal form.

Some equations, like the one in Example 3, are readily solved algebraically. There are other exponential equations for which we do not have the tools to solve algebraically, but we can nevertheless solve graphically.

EXAMPLE 4   Solve: $xe^{3x-1} = 5$.

SOLUTION   We graph $y_1 = xe\char`\^(3x - 1)$ and $y_2 = 5$ and determine the coordinates of any points of intersection.

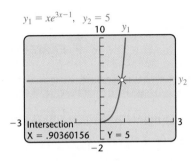

$$y_1 = xe^{3x-1}, \quad y_2 = 5$$

The *x*-coordinate of the point of intersection is approximately 0.90360156. Thus the solution is approximately 0.904.

Solving Logarithmic Equations

Equations containing logarithmic expressions are called **logarithmic equations.** We saw in Section 12.3 that certain logarithmic equations can be solved by writing an equivalent exponential equation.

EXAMPLE 5   Solve: **(a)** $\log_4 (8x - 6) = 3$; **(b)** $\ln 5x = 27$.

SOLUTION

a) $\log_4 (8x - 6) = 3$

$$4^3 = 8x - 6 \qquad \text{Writing the equivalent exponential equation}$$

$$64 = 8x - 6$$

$$70 = 8x \qquad \text{Adding 6 to both sides}$$

$$x = \tfrac{70}{8}, \text{ or } \tfrac{35}{4}.$$

Check:

$$
\begin{array}{c|c}
\log_4 (8x - 6) = 3 & \\
\hline
\log_4 \left(8 \cdot \tfrac{35}{4} - 6 \right) & 3 \\
\log_4 (2 \cdot 35 - 6) & \\
\log_4 64 & \\
3 \overset{?}{=} 3 & \text{TRUE}
\end{array}
$$

The solution is $\tfrac{35}{4}$.

b) $\ln 5x = 27$ $\qquad$ Remember: $\ln 5x$ means $\log_e 5x$.

$$e^{27} = 5x \qquad \text{Writing the equivalent exponential equation}$$

$$\frac{e^{27}}{5} = x \qquad \text{This is a very large number.}$$

The solution is $\dfrac{e^{27}}{5}$. The check is left to the student.

Often the properties for logarithms are needed. The goal is to first write an equivalent equation in which the variable appears in just one logarithmic expression. We then isolate that expression and solve as in Example 5. Since logarithms of negative numbers are not defined, all possible solutions must be checked.

EXAMPLE 6   Solve: $\log x + \log (x - 3) = 1$.

ALGEBRAIC APPROACH

To increase understanding, we write in the base, 10.

$$\log_{10} x + \log_{10} (x - 3) = 1$$

$$\log_{10} [x(x - 3)] = 1$$

Using the product rule for logarithms to obtain a single logarithm

$$x(x - 3) = 10^1$$

Writing an equivalent exponential equation

$$x^2 - 3x = 10$$

$$x^2 - 3x - 10 = 0$$

$$(x + 2)(x - 5) = 0$$ Factoring

$$x + 2 = 0 \quad or \quad x - 5 = 0$$ Using the principle of zero products

$$x = -2 \quad or \quad x = 5$$

The possible solutions are -2 and 5.

GRAPHICAL APPROACH

We graph $y_1 = \log (x) + \log (x - 3)$ and $y_2 = 1$ and determine the coordinates of any points of intersection.

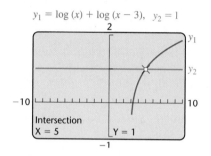

$y_1 = \log (x) + \log (x - 3), \ y_2 = 1$

There is one point of intersection, $(5, 1)$. The solution is the first coordinate of that point, or 5.

Note that the algebraic approach resulted in two possible solutions and the graphical approach in only one. To check, we substitute both -2 and 5 in the original equation.

Student Notes

It is essential that you remember the properties of logarithms from Section 12.4. Consider reviewing the properties before attempting to solve equations similar to those in Examples 6–8.

Check:

For -2:

$$\frac{\log x + \log (x - 3) = 1}{\log (-2) + \log (-2 - 3) \ \overset{?}{=} \ 1} \quad \text{FALSE}$$

For 5:

$$\begin{array}{c|c} \log x + \log (x - 3) = 1 \\ \hline \log 5 + \log (5 - 3) & 1 \\ \log 5 + \log 2 & \\ \log 10 & \\ & 1 \overset{?}{=} 1 \quad \text{TRUE} \end{array}$$

The number -2 *does not check* because the logarithm of a negative number is undefined. The solution is 5.

EXAMPLE 7   Solve: $\log_2 (x + 7) - \log_2 (x - 7) = 3$.

ALGEBRAIC APPROACH

We have

$$\log_2 (x + 7) - \log_2 (x - 7) = 3$$

$$\log_2 \frac{x + 7}{x - 7} = 3 \qquad \text{Using the quotient rule for logarithms to obtain a single logarithm}$$

$$\frac{x + 7}{x - 7} = 2^3 \qquad \text{Writing an equivalent exponential equation}$$

$$\frac{x + 7}{x - 7} = 8$$

$$x + 7 = 8(x - 7) \qquad \text{Multiplying by the LCD, } x - 7$$

$$x + 7 = 8x - 56 \qquad \text{Using the distributive law}$$

$$63 = 7x$$

$$9 = x. \qquad \text{Dividing by 7}$$

GRAPHICAL APPROACH

We first use the change-of-base formula to write the base-2 logarithms using common logarithms. Then we graph and determine the coordinates of any points of intersection.

$y_1 = \log (x + 7)/\log (2) - \log (x - 7)/\log (2),$
$y_2 = 3$

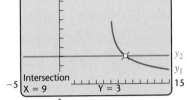

The graphs intersect at $(9, 3)$. The solution is 9.

Check:

$$\frac{\log_2 (x + 7) - \log_2 (x - 7) = 3}{\log_2 (9 + 7) - \log_2 (9 - 7) \ \Big| \ 3}$$
$$\log_2 16 - \log_2 2$$
$$4 - 1$$
$$3 \overset{?}{=} 3 \quad \text{TRUE}$$

The solution is 9.

EXAMPLE 8   Solve: $\log_7 (x + 1) + \log_7 (x - 1) = \log_7 8$.

ALGEBRAIC APPROACH

We have

$$\log_7 (x + 1) + \log_7 (x - 1) = \log_7 8$$

$$\log_7 [(x + 1)(x - 1)] = \log_7 8 \qquad \text{Using the product rule for logarithms}$$

$$\log_7 (x^2 - 1) = \log_7 8 \qquad \text{Multiplying. Note that both sides are base-7 logarithms.}$$

$$x^2 - 1 = 8 \qquad \text{Using the principle of logarithmic equality. Study this step carefully.}$$

$$x^2 - 9 = 0$$

$$(x - 3)(x + 3) = 0 \qquad \text{Solving the quadratic equation}$$

$$x = 3 \quad or \quad x = -3.$$

The student should confirm that 3 checks but -3 does not. The solution is 3.

GRAPHICAL APPROACH

Using the change-of-base formula, we graph

$$y_1 = \log(x+1)/\log(7) + \log(x-1)/\log(7)$$

and

$$y_2 = \log(8)/\log(7).$$

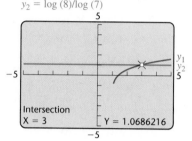

$y_1 = \log(x+1)/\log(7) + \log(x-1)/\log(7),$
$y_2 = \log(8)/\log(7)$

Intersection
X = 3   Y = 1.0686216

The graphs intersect at (3, 1.0686216). The solution is 3.

12.6 EXERCISE SET

FOR EXTRA HELP

MathXL   MyMathLab   InterAct Math   AW Math Tutor Center   Video Lectures on CD: Disc 6   Student's Solutions Manual

Concept Reinforcement   In each of Exercises 1–8, match the equation with an equivalent equation from the column on the right that could be the next step in the solution process.

1. _____ $5^x = 3$

2. _____ $e^{5x} = 3$

3. _____ $\ln x = 3$

4. _____ $\log_x 5 = 3$

5. _____ $\log_5 x + \log_5(x-2) = 3$

6. _____ $\log_5 x - \log_5(x-2) = 3$

7. _____ $\ln x - \ln(x-2) = 3$

8. _____ $\log x + \log(x-2) = 3$

a) $\ln e^{5x} = \ln 3$

b) $\log_5(x^2 - 2x) = 3$

c) $\log(x^2 - 2x) = 3$

d) $\log_5 \dfrac{x}{x-2} = 3$

e) $\log 5^x = \log 3$

f) $e^3 = x$

g) $\ln \dfrac{x}{x-2} = 3$

h) $x^3 = 5$

Solve. Where appropriate, include approximations to the nearest thousandth.

9. $2^x = 19$

10. $2^x = 15$

11. $8^{x-1} = 17$

12. $4^{x+1} = 13$

13. $e^t = 1000$

14. $e^t = 100$

15. $e^{0.03t} + 2 = 7$

16. $e^{-0.07t} + 3 = 3.08$

17. $5 = 3^{x+1}$

18. $7 = 3^{x-1}$

Aha! 19. $2^{x+3} = 16$

20. $4^{x+1} = 64$

21. $4.9^x - 87 = 0$

22. $7.2^x - 65 = 0$

23. $19 = 2e^{4x}$

24. $29 = 3e^{2x}$

25. $7 + 3e^{5x} = 13$

26. $4 + 5e^{4x} = 9$

Aha! 27. $\log_3 x = 4$

28. $\log_2 x = 6$

29. $\log_2 x = -3$

30. $\log_5 x = 3$

31. $\ln x = 5$

32. $\ln x = 4$

33. $\log_8 x = \frac{1}{3}$

34. $\log_4 x = \frac{1}{2}$

35. $\ln 4x = 3$

36. $\ln 3x = 2$

37. $\log x = 2.5$

38. $\log x = 0.5$

39. $\ln (2x + 1) = 4$

40. $\ln (4x - 2) = 3$

Aha! 41. $\ln x = 1$

42. $\log x = 1$

43. $5 \ln x = -15$

44. $3 \ln x = -3$

45. $\log_2 (8 - 6x) = 5$

46. $\log_5 (2x - 7) = 3$

47. $\log (x - 9) + \log x = 1$

48. $\log (x + 9) + \log x = 1$

49. $\log x - \log (x + 3) = 1$

50. $\log x - \log (x + 7) = -1$

51. $\log_4 (x + 3) = 2 + \log_4 (x - 5)$

52. $\log_2 (x + 3) = 4 + \log_2 (x - 3)$

53. $\log_7 (x + 1) + \log_7 (x + 2) = \log_7 6$

54. $\log_6 (x + 3) + \log_6 (x + 2) = \log_6 20$

55. $\log_5 (x + 4) + \log_5 (x - 4) = \log_5 20$

56. $\log_4 (x + 2) + \log_4 (x - 7) = \log_4 10$

57. $\ln (x + 5) + \ln (x + 1) = \ln 12$

58. $\ln (x - 6) + \ln (x + 3) = \ln 22$

59. $\log_2 (x - 3) + \log_2 (x + 3) = 4$

60. $\log_3 (x - 4) + \log_3 (x + 4) = 2$

61. $\log_{12} (x + 5) - \log_{12} (x - 4) = \log_{12} 3$

62. $\log_6 (x + 7) - \log_6 (x - 2) = \log_6 5$

63. $\log_2 (x - 2) + \log_2 x = 3$

64. $\log_4 (x + 6) - \log_4 x = 2$

65. $e^{0.5x} - 7 = 2x + 6$

66. $e^{-x} - 3 = x^2$

67. $\ln (3x) = 3x - 8$

68. $\ln (x^2) = -x^2$

69. Find the value of x for which the natural logarithm is the same as the common logarithm.

70. Find all values of x for which the common logarithm of the square of x is the same as the square of the common logarithm of x.

71. Could Example 2 have been solved by taking the natural logarithm on both sides? Why or why not?

72. Christina finds that the solution of $\log_3 (x + 4) = 1$ is -1, but rejects -1 as an answer because it is negative. What mistake is she making?

Focused Review

Solve.

73. $4x^2 - 25 = 0$ [6.4]

74. $5x^2 = 7x$ [6.1]

75. $|8 - 3x| = 4$ [8.2]

76. $9 - 13x = 3x + 9$ [2.2]

77. $x^{1/2} - 6x^{1/4} + 8 = 0$ [11.5]

78. $\sqrt{4x - 4} = \sqrt{x + 4} + 1$ [10.6]

79. $x^2 + 3x + 5 = 0$ [11.2]

80. $2^{x+5} = 8^{4x}$ [12.6]

81. $e^x = 1.5$ [12.6]

82. $\log x + \log (x - 3) = 1$ [12.6]

83. $\frac{4}{x} = 3 + x$ [7.6]

84. $\frac{x + 1}{x + 2} = \frac{x + 3}{x + 4}$ [7.6]

85. $5x - 3y = 16,$
$4x + 2y = 4$ [4.3]

86. $4a - 5b + c = 3,$
$3a - 4b + 2c = 3,$
$a + b - 7c = -2$ [9.1]

Synthesis

TW **87.** Can the principle of logarithmic equality be expanded to include all functions? That is, is the statement "$m = n$ is equivalent to $f(m) = f(n)$" true for any function f? Why or why not?

TW **88.** Explain how Exercises 37 and 38 could be solved using the graph of $f(x) = \log x$.

Solve. If no solution exists, state this.

89. $27^x = 81^{2x-3}$

90. $8^x = 16^{3x+9}$

91. $\log_x (\log_3 27) = 3$

92. $\log_6 (\log_2 x) = 0$

93. $x \log \frac{1}{8} = \log 8$

94. $\log_5 \sqrt{x^2 - 9} = 1$

95. $2^{x^2+4x} = \frac{1}{8}$

96. $\log (\log x) = 5$

97. $\log_5 |x| = 4$

98. $\log x^2 = (\log x)^2$

99. $\log \sqrt{2x} = \sqrt{\log 2x}$

100. $1000^{2x+1} = 100^{3x}$

101. $3^{x^2} \cdot 3^{4x} = \frac{1}{27}$

102. $3^{3x} \cdot 3^{x^2} = 81$

103. $\log x^{\log x} = 25$

104. $3^{2x} - 8 \cdot 3^x + 15 = 0$

105. $(81^{x-2})(27^{x+1}) = 9^{2x-3}$

106. $3^{2x} - 3^{2x-1} = 18$

107. Given that $2^y = 16^{x-3}$ and $3^{y+2} = 27^x$, find the value of $x + y$.

108. If $x = (\log_{125} 5)^{\log_5 125}$, what is the value of $\log_3 x$?

12.7 Applications of Exponential and Logarithmic Functions

Applications of Logarithmic Functions   ■   Applications of Exponential Functions

We now consider applications of exponential and logarithmic functions.

Applications of Logarithmic Functions

EXAMPLE 1   Sound Levels.   To measure the volume, or "loudness," of a sound, the *decibel* scale is used. The loudness L, in decibels (dB), of a sound is given by

$$L = 10 \cdot \log \frac{I}{I_0},$$

where I is the intensity of the sound, in watts per square meter (W/m²), and $I_0 = 10^{-12}$ W/m². (I_0 is approximately the intensity of the softest sound that can be heard by the human ear.)

a) The intensity of sound inside a racecar reaches 10^1 W/m². How loud, in decibels, is the sound level?

b) The Occupational Safety and Health Administration (OSHA) considers sound levels of 85 dB and above unsafe. What is the intensity of such sounds?

Danica Patrick, the first woman to lead an Indianapolis 500

SOLUTION

a) To find the loudness, in decibels, we use the above formula:

$$L = 10 \cdot \log \frac{I}{I_0}$$

$$= 10 \cdot \log \frac{10^1}{10^{-12}} \quad \text{Substituting}$$

$$= 10 \cdot \log 10^{13} \quad \text{Subtracting exponents}$$

$$= 10 \cdot 13 \quad \log 10^a = a$$

$$= 130.$$

The sound inside a racecar reaches 130 dB.

b) We substitute and solve for I:

$$L = 10 \cdot \log \frac{I}{I_0}$$

$$85 = 10 \cdot \log \frac{I}{10^{-12}} \quad \text{Substituting}$$

$$8.5 = \log \frac{I}{10^{-12}} \quad \text{Dividing both sides by 10}$$

$$8.5 = \log I - \log 10^{-12} \quad \text{Using the quotient rule for logarithms}$$

$$8.5 = \log I - (-12) \quad \log 10^a = a$$

$$-3.5 = \log I \quad \text{Adding } -12 \text{ to both sides}$$

$$10^{-3.5} = I. \quad \text{Converting to an exponential equation}$$

Earplugs would be recommended for sounds with intensities exceeding $10^{-3.5}$ W/m^2.

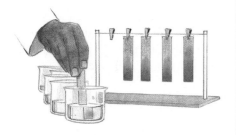

EXAMPLE 2 Chemistry: pH of Liquids. In chemistry, the pH of a liquid is a measure of its acidity. We calculate pH as follows:

$$pH = -\log [H^+],$$

where $[H^+]$ is the hydrogen ion concentration in moles per liter.

a) The hydrogen ion concentration of human blood is normally about 3.98×10^{-8} moles per liter. Find the pH.

b) The pH of seawater is about 8.3. Find the hydrogen ion concentration.

SOLUTION

a) To find the pH of blood, we use the above formula:

$$pH = -\log [H^+]$$

$$= -\log [3.98 \times 10^{-8}]$$

$$\approx -(-7.400117) \quad \text{Using a calculator}$$

$$\approx 7.4.$$

The pH of human blood is normally about 7.4.

b) We substitute and solve for $[H^+]$:

$$8.3 = -\log[H^+] \qquad \text{Using pH} = -\log[H^+]$$
$$-8.3 = \log[H^+] \qquad \text{Dividing both sides by } -1$$
$$10^{-8.3} = [H^+] \qquad \text{Converting to an exponential equation}$$
$$5.01 \times 10^{-9} \approx [H^+]. \qquad \text{Using a calculator; writing scientific notation}$$

The hydrogen ion concentration of seawater is about 5.01×10^{-9} moles per liter.

Applications of Exponential Functions

EXAMPLE 3 Interest Compounded Annually.   Suppose that \$25,000 is invested at 4% interest, compounded annually. (See Example 5 in Section 12.2.) In t years, it will grow to the amount A given by the function

$$A(t) = 25,000(1.04)^t.$$

a) How long will it take to accumulate \$80,000 in the account?

b) Find the amount of time it takes for the \$25,000 to double itself.

SOLUTION

a) We set $A(t) = 80,000$ and solve for t:

$$80,000 = 25,000(1.04)^t$$
$$\frac{80,000}{25,000} = 1.04^t \qquad \text{Dividing both sides by 25,000}$$
$$3.2 = 1.04^t$$
$$\log 3.2 = \log 1.04^t \qquad \text{Taking the common logarithm on both sides}$$
$$\log 3.2 = t \log 1.04 \qquad \text{Using the power rule for logarithms}$$
$$\frac{\log 3.2}{\log 1.04} = t \qquad \text{Dividing both sides by } \log 1.04$$
$$29.7 \approx t. \qquad \text{Using a calculator}$$

Remember that when doing a calculation like this on a calculator, it is best to wait until the end to round off. We check by solving graphically, as shown at left. At an interest rate of 4% per year, it will take about 29.7 yr for \$25,000 to grow to \$80,000.

b) To find the *doubling time*, we replace $A(t)$ with 50,000 and solve for t:

$$50,000 = 25,000(1.04)^t$$
$$2 = (1.04)^t \qquad \text{Dividing both sides by 25,000}$$
$$\log 2 = \log(1.04)^t \qquad \text{Taking the common logarithm on both sides}$$
$$\log 2 = t \log 1.04 \qquad \text{Using the power rule for logarithms}$$
$$t = \frac{\log 2}{\log 1.04} \approx 17.7. \qquad \text{Dividing both sides by } \log 1.04 \text{ and using a calculator}$$

At an interest rate of 4% per year, the doubling time is about 17.7 yr.

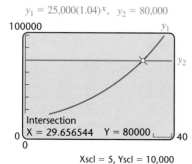

$y_1 = 25,000(1.04)^x, \quad y_2 = 80,000$

Intersection
X = 29.656544   Y = 80000

Xscl = 5, Yscl = 10,000

Student Notes

Study the different steps in the solution of Example 3(b). Note that if 50,000 and 25,000 are replaced with 8000 and 4000, the doubling time is unchanged.

Like investments, populations often grow exponentially.

Exponential Growth

An **exponential growth model** is a function of the form

$$P(t) = P_0 e^{kt}, \quad k > 0,$$

where P_0 is the population at time 0, $P(t)$ is the population at time t, and k is the **exponential growth rate** for the situation. The **doubling time** is the amount of time necessary for the population to double in size.

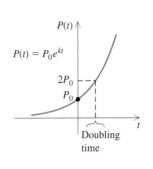

The exponential growth rate is the rate of growth of a population at any *instant* in time. Since the population is continually growing, the percent of total growth after one year will exceed the exponential growth rate.

EXAMPLE 4   Growth of Zebra Mussel Populations.   Zebra mussels, inadvertently imported from Europe, began fouling North American waters in 1988. These mussels are so prolific that lake and river bottoms, as well as water intake pipes, can become blanketed with them, altering an entire ecosystem. In 2000, a portion of the Hudson River contained an average of 10 zebra mussels per square mile. The exponential growth rate was 340% per year.

a) Find the exponential growth function that models the data.

b) Predict the number of mussels per square mile in 2007.

SOLUTION

a) In 2000, at $t = 0$, the population was $10/\text{mi}^2$. We substitute 10 for P_0 and 340%, or 3.4, for k. This gives the exponential growth function

$$P(t) = 10e^{3.4t}.$$

b) In 2007, we have $t = 7$ (since 7 yr have passed since 2000). To find the population in 2007, we compute $P(7)$:

$$P(7) = 10e^{3.4(7)} \qquad \text{Using } P(t) = 10e^{3.4t} \text{ from part (a)}$$
$$= 10e^{23.8}$$
$$\approx 217{,}000{,}000{,}000. \qquad \text{Using a calculator}$$

The population of zebra mussels in the specified portion of the Hudson River will reach approximately 217,000,000,000 per square mile in 2007.

EXAMPLE 5   Spread of a Computer Virus.   The number of computers infected by a virus t hours after it first appears usually increases exponentially. In 2004, the "MyDoom" worm spread from 100 computers to about 100,000 computers in 24 hr (*Source*: Based on data from IDG News Service).

a) Find the exponential growth rate and the exponential growth function.

b) Assuming exponential growth, estimate how long it took the MyDoom worm to infect 9000 computers.

SOLUTION

a) We use $N(t) = N_0 e^{kt}$, where t is the number of hours since the first 100 computers were infected. Substituting 100 for N_0 gives

$$N(t) = 100e^{kt}.$$

To find the exponential growth rate, k, note that after 24 hr, 100,000 computers had been infected:

$$\left.\begin{array}{l} N(24) = 100e^{k \cdot 24} \\ 100{,}000 = 100e^{24k} \end{array}\right\} \quad \text{\textbf{Substituting}}$$

$$1000 = e^{24k} \qquad \text{\textbf{Dividing both sides by 100}}$$

$$\ln 1000 = \ln e^{24k} \qquad \text{\textbf{Taking the natural logarithm on both sides}}$$

$$\ln 1000 = 24k \qquad \text{\textbf{ln } } e^a = a$$

$$\frac{\ln 1000}{24} = k \qquad \text{\textbf{Dividing both sides by 24}}$$

$$0.288 \approx k. \qquad \text{\textbf{Using a calculator and rounding}}$$

The exponential growth rate is 28.8% and the exponential growth function is given by $N(t) = 100e^{0.288t}$.

b) To estimate how long it took for 9000 computers to be infected, we replace $N(t)$ with 9000 and solve for t:

$$9000 = 100e^{0.288t}$$

$$90 = e^{0.288t} \qquad \text{\textbf{Dividing both sides by 100}}$$

$$\ln 90 = \ln e^{0.288t} \qquad \text{\textbf{Taking the natural logarithm on both sides}}$$

$$\ln 90 = 0.288t \qquad \text{\textbf{ln } } e^a = a$$

$$\frac{\ln 90}{0.288} = t \qquad \text{\textbf{Dividing both sides by 0.288}}$$

$$15.6 \approx t. \qquad \text{\textbf{Using a calculator}}$$

Rounding up to 16, we see that, according to this model, it took about 16 hr for 9000 computers to be infected.

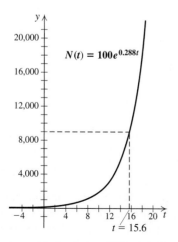

A graphical solution of Example 5

EXAMPLE 6   Interest Compounded Continuously.   When an amount of money P_0 is invested at interest rate k, compounded *continuously*, interest is computed every "instant" and added to the original amount. The balance $P(t)$, after t years, is given by the exponential growth model

$$P(t) = P_0 e^{kt}.$$

a) Suppose that \$30,000 is invested and grows to \$44,754.75 in 5 yr. Find the exponential growth function.

b) What is the doubling time?

SOLUTION

a) We have $P(0) = 30,000$. Thus the exponential growth function is

$$P(t) = 30,000e^{kt}, \quad \text{where } k \text{ must still be determined.}$$

Knowing that for $t = 5$ we have $P(5) = 44,754.75$, it is possible to solve for k:

$$44,754.75 = 30,000e^{k(5)} = 30,000e^{5k}$$

$$\frac{44,754.75}{30,000} = e^{5k} \qquad \textbf{Dividing both sides by 30,000}$$

$$1.491825 = e^{5k}$$

$$\ln 1.491825 = \ln e^{5k} \qquad \textbf{Taking the natural logarithm on both sides}$$

$$\ln 1.491825 = 5k \qquad \textbf{ln } e^a = a$$

$$\frac{\ln 1.491825}{5} = k \qquad \textbf{Dividing both sides by 5}$$

$$0.08 \approx k. \qquad \textbf{Using a calculator and rounding}$$

The interest rate is about 0.08, or 8%, compounded continuously. Because interest is being compounded continuously, the yearly interest rate is a bit more than 8%. The exponential growth function is

$$P(t) = 30,000e^{0.08t}.$$

b) To find the doubling time T, we replace $P(T)$ with 60,000 and solve for T:

$$60,000 = 30,000e^{0.08T}$$

$$2 = e^{0.08T} \qquad \textbf{Dividing both sides by 30,000}$$

$$\ln 2 = \ln e^{0.08T} \qquad \textbf{Taking the natural logarithm on both sides}$$

$$\ln 2 = 0.08T \qquad \textbf{ln } e^a = a$$

$$\frac{\ln 2}{0.08} = T \qquad \textbf{Dividing both sides by 0.08}$$

$$8.7 \approx T. \qquad \textbf{Using a calculator and rounding}$$

Thus the original investment of \$30,000 will double in about 8.7 yr. ◢

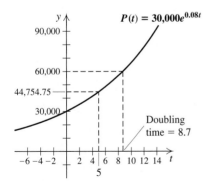

A graphical solution of Example 6

For any specified interest rate, continuous compounding gives the highest yield and the shortest doubling time.

In some real-life situations, a quantity or population is *decreasing* or *decaying* exponentially.

Exponential Decay

An **exponential decay model** is a function of the form

$$P(t) = P_0 e^{-kt}, \quad k > 0,$$

where P_0 is the quantity present at time 0, $P(t)$ is the amount present at time t, and k is the **decay rate.** The **half-life** is the amount of time necessary for half of the quantity to decay.

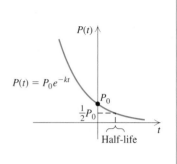

$P(t) = P_0 e^{-kt}$

$P(t)$

P_0

$\frac{1}{2}P_0$

Half-life

t

Chaco Canyon, New Mexico

EXAMPLE 7 Carbon Dating.   The radioactive element carbon-14 has a half-life of 5750 yr. The percentage of carbon-14 present in the remains of organic matter can be used to determine the age of that organic matter. Recently, while digging in Chaco Canyon, New Mexico, archaeologists found corn pollen that had lost 38.1% of its carbon-14. The age of this corn pollen was evidence that Indians had been cultivating crops in the Southwest centuries earlier than scientists had thought. What was the age of the pollen? (*Source*: *American Anthropologist*)

SOLUTION   We first find k. To do so, we use the concept of half-life. When $t = 5750$ (the half-life), $P(t)$ will be half of P_0. Then

$$0.5P_0 = P_0 e^{-k(5750)} \qquad \text{Substituting in } P(t) = P_0 e^{-kt}$$

$$0.5 = e^{-5750k} \qquad \text{Dividing both sides by } P_0$$

$$\ln 0.5 = \ln e^{-5750k} \qquad \text{Taking the natural logarithm on both sides}$$

$$\ln 0.5 = -5750k \qquad \ln e^a = a$$

$$\frac{\ln 0.5}{-5750} = k \qquad \text{Dividing}$$

$$0.00012 \approx k. \qquad \text{Using a calculator and rounding}$$

Now we have a function for the decay of carbon-14:

$$P(t) = P_0 e^{-0.00012t}. \longleftarrow \text{This completes the first part of our solution.}$$

If the corn pollen has lost 38.1% of its carbon-14 from an initial amount P_0, then $100\% - 38.1\%$, or 61.9%, of P_0 is still present. To find the age t of the pollen, we solve this equation for t:

$$0.619P_0 = P_0 e^{-0.00012t} \qquad \text{We want to find } t \text{ for which } P(t) = 0.619P_0.$$

$$0.619 = e^{-0.00012t} \qquad \text{Dividing both sides by } P_0$$

$$\ln 0.619 = \ln e^{-0.00012t} \qquad \text{Taking the natural logarithm on both sides}$$

$$\ln 0.619 = -0.00012t \qquad \ln e^a = a$$

$$\frac{\ln 0.619}{-0.00012} = t \qquad \text{Dividing}$$

$$4000 \approx t. \qquad \text{Using a calculator}$$

The pollen is about 4000 yr old.

The equation

$$P(t) = P_0 e^{-0.00012t}$$

can be used for any subsequent carbon-dating problem.

By looking at the graph of a set of data, we can tell whether a population or other quantity is growing or decaying exponentially.

EXAMPLE 8   For each of the following graphs, determine whether an exponential function might fit the data.

a)

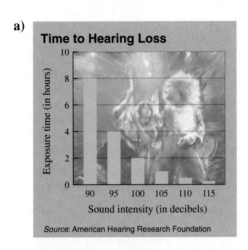

Time to Hearing Loss

Source: American Hearing Research Foundation

b)

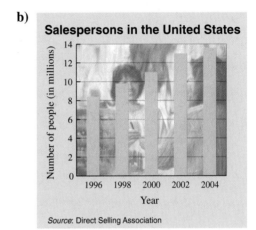

Salespersons in the United States

Source: Direct Selling Association

c)

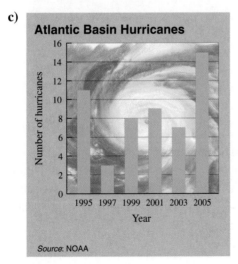

Atlantic Basin Hurricanes

Source: NOAA

d)

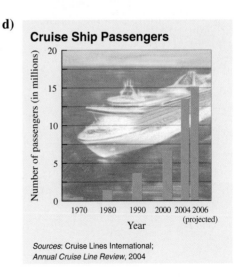

Cruise Ship Passengers

Sources: Cruise Lines International;
Annual Cruise Line Review, 2004

Student Notes

Like linear functions, two points determine an exponential function. It is important to determine whether data follow a linear or an exponential model (if either) before fitting a function to the data.

SOLUTION

a) As the sound intensity increases, the safe exposure time decreases. The amount of decrease gets smaller as the intensity increases. It appears that an exponential decay function might fit the data.

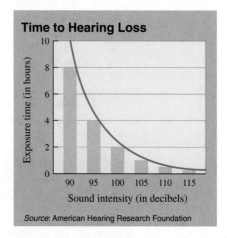

Time to Hearing Loss

Source: American Hearing Research Foundation

b) The number of salespersons increased between 1996 and 2004 at approximately the same rate each year. It does not appear that an exponential function models the data; instead, a linear function might be appropriate.

Salespersons in the United States

Source: Direct Selling Association

c) The number of hurricanes first fell, then rose, then fell again, then rose again. This does not fit an exponential model.

d) The number of cruise-ship passengers increased from 1970 to 2006, and the amount of yearly growth also increased during that time. This suggests that an exponential growth function could be used to model this situation.

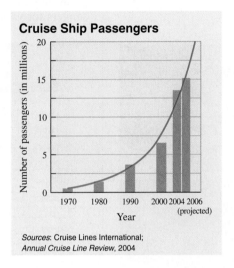

Cruise Ship Passengers

Sources: Cruise Lines International;
Annual Cruise Line Review, 2004

Exponential Regression

To fit an exponential function to a set of data, we use the ExpReg option in the STAT CALC menu. After entering the data, with the independent variable as L1 and the dependent variable as L2, choose the ExpReg option. For the data shown on the left below, the calculator will return a screen like that shown on the right below.

L1	L2	L3	1
1	12	-------	
2	60		
3	300		
4	1500		
5	7500		
6	37500		
-------	-------		
L1(1)=1			

```
ExpReg
 y=a*b^x
 a=2.4
 b=5
```

The exponential function found is of the form $f(x) = ab^x$.

We can use the function in this form or convert it to an exponential function of the form $f(x) = ae^{kx}$. The number a remains unchanged. To find k, we solve the equation $b^x = e^{kx}$ for k:

$$b^x = e^{kx}$$
$$b^x = (e^k)^x \qquad \text{Writing both sides as a power of } x$$
$$b = e^k \qquad \text{The bases must be equal.}$$
$$\ln b = \ln e^k \qquad \text{Taking the natural logarithm on both sides}$$
$$\ln b = k. \qquad \ln e^k = k$$

Thus to write $f(x) = 2.4(5)^x$ in the form $f(x) = ae^{kx}$, we can write $f(x) = 2.4e^{(\ln 5)x}$, or

$$f(x) \approx 2.4e^{1.609x}.$$

EXAMPLE 9 Cruise-Ship Passengers. In 1970, cruise lines carried approximately 500,000 passengers. This number was projected to increase to 15 million in 2006. The following table and graph show the number of passengers for various years.

Year	Number of Passengers (in millions)
1970	0.5
1980	1.4
1990	3.6
2000	6.5
2004	13.4
2006	15 (proj.)

Sources: Cruise Lines International; *Annual Cruise Review* 2004

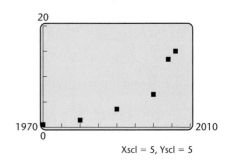

Xscl = 5, Yscl = 5

a) Use regression to fit an exponential function to the data and graph the function.

b) Determine the exponential growth rate.

c) The cruise industry projects that in 2012 there will be 20 million cruise-ship passengers. Use the exponential function from part (a) to estimate the number of passengers in 2012. How does this estimate correspond to the cruise-industry projection?

SOLUTION

a) We let x represent the number of years since 1970, and enter the data, as shown on the left below. Then we use regression to find the function and copy it as y_1:

$$f(x) = 0.5240654694(1.096501437)^x.$$

The graph of the function, along with the data points, is shown on the right below.

L1	L2	L3	1
0	.5	-----	
10	1.4		
20	3.6		
30	6.5		
34	13.4		
36	15		
-----	-----		

L1(1) = 0

$y = 0.5240654694(1.096501437)^x$

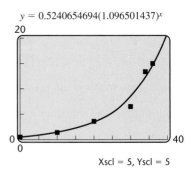

Xscl = 5, Yscl = 5

b) The exponential growth rate is the number k in the equation $f(x) = ae^{kx}$. To write $f(x) = 0.5240654694(1.096501437)^x$ in the form $f(x) = ae^{kx}$, we find k:

$$k = \ln b \approx \ln 1.096501437 \approx 0.0921245994.$$

Thus the exponential growth rate is approximately 0.092, or 9.2%.

c) Since the year 2012 is 42 yr after 1970, we find $f(42)$ using a table of values or $Y_1(42)$:

$$f(42) \approx 25.$$

Using the function, we estimate that there will be 25 million cruise-ship passengers in 2012. This is 5 million more than the cruise-industry projection for that year.

Connecting the Concepts

We can now add the exponential functions $f(t) = P_0e^{kt}$ and $f(t) = P_0e^{-kt}$, $k > 0$, and the logarithmic function $f(x) = \log_b x$, $b > 1$, to our library of functions.

Linear function:
$f(x) = mx + b$

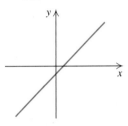

Absolute-value function:
$f(x) = |x|$

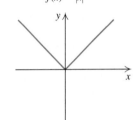

Rational function:
$f(x) = \dfrac{1}{x}$

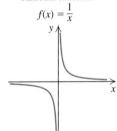

Radical function:
$f(x) = \sqrt{x}$

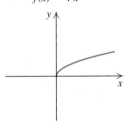

Quadratic function:
$f(x) = ax^2 + bx + c, a > 0$

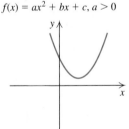

Quadratic function:
$f(x) = ax^2 + bx + c, a < 0$

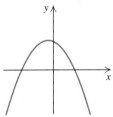

Exponential growth function:
$f(t) = P_0e^{kt}, k > 0$

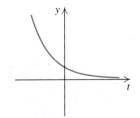

Exponential decay function:
$f(t) = P_0e^{-kt}, k > 0$

Logarithmic function:
$f(x) = \log_b x, b > 1$

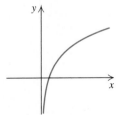

Solve.

1. *Digital Marketing.* The amount spent by U.S. companies on digital marketing, in billions of dollars, t years after 2000 can be estimated by

$$d(t) = 1.47(1.56)^t$$

(*Source*: Based on data from Jupiter Media Matrix).

a) Determine the year in which companies first spent $10 billion on digital marketing.

b) What is the doubling time for digital marketing spending?

2. *Cell Phones.* The number of cell phones in use in the United States, in millions, t years after 1995 can be estimated by

$$N(t) = 36.9(1.2)^t$$

(*Source*: Cellular Telecommunications and Internet Association).

a) In what year did the number of cell phones in use first reach 200 million?

b) What is the doubling time for the number of cell phones in use?

3. *Skateboarding.* The number of skateboarders of age x, in thousands, can be approximated by

$$N(x) = 1089(0.9)^x, \quad 7 \le x \le 75$$

(*Sources*: Based on figures from the National Sporting Goods Association and *Statistical Abstract of the United States*, 2006).

a) Estimate the number of 21-year-old skateboarders.

b) At what age are there only 6300 skateboarders?

4. *Recycling Aluminum Cans.* Approximately one-half of all aluminum cans distributed will be recycled each year. A beverage company distributes 250,000 cans. The number still in use after t years is given by the function

$$N(t) = 250{,}000\left(\tfrac{1}{2}\right)^t$$

(*Source*: The Aluminum Association, Inc., May 2004).

a) After how many years will 60,000 cans still be in use?

b) After what amount of time will only 1000 cans still be in use?

5. *Student Loan Repayment.* A college loan of $29,000 is made at 3% interest, compounded annually. After t years, the amount due, A, is given by the function

$$A(t) = 29{,}000(1.03)^t.$$

a) After what amount of time will the amount due reach $35,000?

b) Find the doubling time.

6. *Spread of a Rumor.* The number of people who have heard a rumor increases exponentially. If all who hear a rumor repeat it to two people a day, and if 20 people start the rumor, the number of people N who have heard the rumor after t days is given by

$$N(t) = 20(3)^t.$$

a) After what amount of time will 1000 people have heard the rumor?

b) What is the doubling time for the number of people who have heard the rumor?

7. *Telephone Lines.* As more Americans make cell phones their *only* phones, the percent of phone lines that are land lines has been shrinking. The percent of U.S. phone lines that are land lines $P(t)$, in use t years after 2000, can be estimated by

$$P(t) = 63.03(0.95)^t$$

(*Sources*: Based on data from Federal Communications Commission; Cellular Telecommunications and Internet Association).

a) In what year did the percentage of phones that are land lines drop below 50%?

b) In what year will the percentage of phones that are land lines drop below 25%?

8. *Smoking.*   The percentage of smokers who received telephone counseling and had successfully quit smoking for t months is given by

$$P(t) = 21.4(0.914)^t.$$

(*Sources: New England Journal of Medicine*; data from California's Smoker's Hotline).

a) In what month will 15% of those who quit and used telephone counseling still be smoke-free?

b) In what month will 5% of those who quit and used phone counseling still be smoke-free?

9. *Marine Biology.*   As a result of preservation efforts in countries in which whaling was once common, the humpback whale population has grown since the 1970s. The worldwide population $P(t)$, in thousands, t years after 1982 can be estimated by

$$P(t) = 5.5(1.08)^t.$$

a) In what year will the humpback whale population reach 40,000?

b) Find the doubling time.

10. *World Population.*   The world population $P(t)$, in billions, t years after 1980 can be approximated by

$$P(t) = 4.495(1.015)^t$$

(*Sources*: Based on data from U.S. Bureau of the Census; International Data Base).

a) In what year will the world population reach 7 billion?

b) Find the doubling time.

Use $pH = -\log[H^+]$ *for Exercises 11–14.*

11. *Chemistry.*   The hydrogen ion concentration of fresh-brewed coffee is about 1.3×10^{-5} moles per liter. Find the pH.

12. *Chemistry.*   The hydrogen ion concentration of milk is about 1.6×10^{-7} moles per liter. Find the pH.

13. *Medicine.*   When the pH of a patient's blood drops below 7.4, a condition called *acidosis* sets in. Acidosis can be deadly when the patient's pH reaches 7.0. What would the hydrogen ion concentration of the patient's blood be at that point?

14. *Medicine.*   When the pH of a patient's blood rises above 7.4, a condition called *alkalosis* sets in. Alkalosis can be deadly when the patient's pH reaches 7.8. What would the hydrogen ion concentration of the patient's blood be at that point?

Use $L = 10 \cdot \log \dfrac{I}{I_0}$ *for Exercises 15–18, where* $I_0 = 10^{-12}$ W/m^2.

15. *Audiology.*   The intensity of sound in normal conversation is about 3.2×10^{-6} W/m^2. How loud in decibels is this sound level?

16. *Audiology.*   The intensity of a riveter at work is about 3.2×10^{-3} W/m^2. How loud in decibels is this sound level?

17. *Music.*   The band U2 recently performed and sound measurements of 105 dB were recorded. What is the intensity of such sounds?

18. *Music.*   The band Strange Folk performed in Burlington, VT, and reached sound levels of 111 dB (*Source*: Melissa Garrido, *Burlington Free Press*). What is the intensity of such sounds?

19. *Stellar Magnitude.*   The apparent stellar magnitude *m* of a star with received intensity *I* is given by

$$m(I) = -(19 + 2.5 \cdot \log I),$$

where *I* is in W/m² (*Source:* The Columbus Optical SETI Observatory). The smaller the apparent stellar magnitude, the brighter the star appears.

a) The intensity of light received from the sun is 1390 W/m². What is the apparent stellar magnitude of the sun?

b) The 5-m diameter Hale telescope on Mt. Palomar can detect a star with magnitude +23. What is the received intensity of light from such a star?

20. *Richter Scale.*   The Richter scale, developed in 1935, has been used for years to measure earthquake magnitude. The Richter magnitude of an earthquake *m* is given by the formula

$$m(A) = \log \frac{A}{A_0},$$

where *A* is the maximum amplitude of the earthquake and A_0 is a constant. What is the magnitude on the Richter scale of an earthquake with an amplitude that is a million times A_0?

Use $P(t) = P_0 e^{kt}$ for Exercises 21 and 22.

21. *Interest Compounded Continuously.*   Suppose that P_0 is invested in a savings account where interest is compounded continuously at 2.5% per year.

a) Express $P(t)$ in terms of P_0 and 0.025.

b) Suppose that $5000 is invested. What is the balance after 1 yr? after 2 yr?

c) When will an investment of $5000 double itself?

22. *Interest Compounded Continuously.*   Suppose that P_0 is invested in a savings account where interest is compounded continuously at 3.1% per year.

a) Express $P(t)$ in terms of P_0 and 0.031.

b) Suppose that $1000 is invested. What is the balance after 1 yr? after 2 yr?

c) When will an investment of $1000 double itself?

23. *Population Growth.*   In 2006, the population of the United States was 300 million and the exponential growth rate was 0.9% per year (*Source:* U.S. Bureau of the Census).

a) Find the exponential growth function.

b) Predict the U.S. population in 2010.

c) When will the U.S. population reach 325 million?

24. *World Population Growth.*   In 2006, the world population was 6.5 billion and the exponential growth rate was 1.1% per year (*Source:* U.S. Bureau of the Census).

a) Find the exponential growth function.

b) Predict the world population in 2010.

c) When will the world population be 8.0 billion?

25. *iPod Sales.*   The number of iPods sold since January 1, 2003, has grown at an exponential growth rate of 11.2% per month (*Source:* Based on data from iLounge.com and the register.co.uk). What is the doubling time for iPod sales?

26. *Population Growth.*   The exponential growth rate of the population of Somalia is 3.4% per year (one of the highest in the world) (*Source: Time Almanac* 2006). What is the doubling time?

27. *World Population.*   The function

$$Y(x) = 67.17 \ln \frac{x}{4.5}$$

can be used to estimate the number of years $Y(x)$ after 1980 required for the world population to reach *x* billion people (*Sources:* Based on data from U.S. Bureau of the Census; International Data Base).

a) In what year will the world population reach 7 billion?

b) In what year will the world population reach 8 billion?

c) Graph the function.

28. *Marine Biology.* The function

$$Y(x) = 13 \ln \frac{x}{5.5}$$

can be used to estimate the number of years $Y(x)$ after 1982 required for the world's humpback whale population to reach x thousand whales.

a) In what year will the whale population reach 40,000?

b) In what year will the whale population reach 50,000?

c) Graph the function.

29. *Forgetting.* Students in an English class took a final exam. They took equivalent forms of the exam at monthly intervals thereafter. The average score $S(t)$, in percent, after t months was found to be given by

$$S(t) = 68 - 20 \log (t + 1), \quad t \geq 0.$$

a) What was the average score when they initially took the test, $t = 0$?

b) What was the average score after 4 months? after 24 months?

c) Graph the function.

d) After what time t was the average score 50%?

30. *Advertising.* A model for advertising response is given by

$$N(a) = 2000 + 500 \log a, \quad a \geq 1,$$

where $N(a)$ is the number of units sold and a is the amount spent on advertising, in thousands of dollars.

a) How many units were sold after spending $1000 ($a = 1$) on advertising?

b) How many units were sold after spending $8000?

c) Graph the function.

d) How much would have to be spent in order to sell 5000 units?

31. *MP3 Sales.* Sales of MP3 players have grown exponentially since 2002, when 1.7 million units were sold. This increased to 11.0 million units in 2005. (*Source:* NPD Group)

a) Find an exponential growth function that fits the data.

b) Predict the number of units sold in 2007.

32. *iPod Sales.* Sales of iPods have grown exponentially since January 1, 2003, at which time 656,000 units had been sold. Approximately 34 months later, in November 2005, the 30-millionth unit was sold. (*Sources:* Based on data from iLounge.com and the register.co.uk)

a) Find an exponential growth function that fits the data.

b) Predict the number of units sold as of June 2006.

33. *Decline in Farmland.* The number of acres of farmland in the United States has decreased from 987 million acres in 1990 to 938 million acres in 2002 (*Source: Statistical Abstract of the United States*, 2006). Assume the number of acres of farmland is decreasing exponentially.

a) Find the value k, and write an equation for an exponential function that can be used to predict the number of acres of U.S. farmland t years after 1990.

b) Predict the number of acres of farmland in 2008.

c) In what year will there be only 800 million acres of U.S. farmland remaining?

34. *Decline in Cases of Mumps.* The number of cases of mumps has dropped exponentially from 900 in 1995 to 300 in 2001 (*Source:* U.S. Centers for Disease Control and Prevention).

a) Find the value k, and write an exponential function that can be used to estimate the number of cases t years after 1995.

b) Estimate the number of cases of mumps in 2008.

c) In what year (theoretically) will there be only 1 case of mumps?

35. *Archaeology.* When archaeologists found the Dead Sea Scrolls, they determined that the linen wrapping had lost 22.3% of its carbon-14. How old is the linen wrapping? (See Example 7.)

36. *Archaeology.* In 1996, researchers found an ivory tusk that had lost 18% of its carbon-14. How old was the tusk? (See Example 7.)

37. *Chemistry.* The exponential decay rate of iodine-131 is 9.6% per day. What is its half-life?

38. *Chemistry.* The decay rate of krypton-85 is 6.3% per year. What is its half-life?

39. *Home Construction.* The chemical urea formaldehyde was found in some insulation used in houses built during the mid to late 1960s. Unknown at the time was the fact that urea formaldehyde emitted toxic fumes as it decayed. The half-life of urea formaldehyde is 1 yr. What is its decay rate?

40. *Plumbing.* Lead pipes and solder are often found in older buildings. Unfortunately, as lead decays, toxic chemicals can get in the water resting in the pipes. The half-life of lead is 22 yr. What is its decay rate?

41. *Value of a Sports Card.* Legend has it that because he objected to smoking, and because his first baseball card was issued in cigarette packs, the great shortstop Honus Wagner halted production of his card before many were produced. One of these cards was purchased in 1991 by hockey great Wayne Gretzky (and a partner) for $451,000. The same card was sold in 2000 for $1.1 million. For the following questions, assume that the card's value increases exponentially, as it has for many years.

WAGNER, PITTSBURG

a) Find the exponential growth rate k, and determine an exponential function V that can be used to estimate the dollar value, $V(t)$, of the card t years after 1991.
b) Estimate the value of the card in 2006.
c) What is the doubling time for the value of the card?
d) In what year will the value of the card first exceed $3,000,000?

42. *Art Masterpieces.* In August 2004, a collector paid $104,168,000, for Pablo Picasso's "Garçon à la Pipe." The same painting sold for $30,000 in 1950. (*Source*: BBC News, 5/6/04)

a) Find the exponential growth rate k, and determine the exponential growth function V, for which $V(t)$ is the painting's value, in millions of dollars, t years after 1950.
b) Estimate the value of the painting in 2009.
c) What is the doubling time for the value of the painting?
d) How long after 1950 will the value of the painting be $1 billion?

In Exercises 43–46, determine whether an exponential function might fit the data.

43.

Motor Vehicle Accidents

Source: National Safety Council

44.

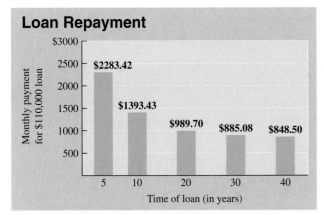

Loan Repayment

45. Boston Red Sox infield roof box seats

Year	Season Price
1980	$ 8
1991	16
2000	45
2003	65
2006	90

Sources: harvard magazine.com, boston.com, and redsox.mlb.com

46. World automobile production

Year	Number of Automobiles Produced (in millions)
1950	8
1960	13
1970	23
1980	29
1990	36
2002	41

Sources: ibike.org and mindfully.org

47. *Baseball.* The price of an infield roof box seat for Boston Red Sox fans has been growing exponentially. The table in Exercise 45 shows the price per game for season ticket holders for various years.

a) Use regression to find an exponential function that can be used to estimate the price p of an infield roof box seat x years after 1980.
b) Determine the exponential growth rate.
c) Predict the price of an infield roof box seat in 2010.

48. *World Population.* The population of the world is growing exponentially, as shown by the graph on p. 716.

a) Use the four most recent data points and regression to fit an exponential function of the form $P(t) = P_0 e^{kt}$ to the data, where t is the year and P is in billions.
b) Predict the world population in 2050.

49. *Hearing Loss.* Prolonged exposure to high noise levels can lead to hearing loss. The following table lists the safe exposure time, in hours, for different sound intensities.

Sound Intensity (in decibels)	Safe Exposure Time (in hours)
90	8
100	2
105	1
110	0.5
115	0.25

Source: American Hearing Research Foundation

a) Use regression to fit an exponential function of the form $f(x) = ab^x$ to the data.
b) Estimate the safe exposure time for a sound intensity of 95 dB.

50. *Loan Repayment.* The size of a monthly loan repayment decreases exponentially as the time of the loan increases. The table in Exercise 44 shows the size of monthly loan payments for a $110,000 loan at a fixed interest rate for various lengths of loans.

a) Use regression to fit an exponential function of the form $f(x) = ab^x$ to the data.
b) Estimate the monthly loan payment for a 15-yr $110,000 loan.

51. Will the model used to predict the number of cell phones in Exercise 2 still be realistic in 2020? Why or why not?

TW 52. Examine the restriction on t in Exercise 29.

  a) What upper limit might be placed on t?

  b) In practice, would this upper limit ever be enforced? Why or why not?

Skill Maintenance

Graph. [11.7]

53. $y = x^2 - 8x$

54. $y = x^2 - 5x - 6$

55. $f(x) = 3x^2 - 5x - 1$

56. $g(x) = 2x^2 - 6x + 3$

Solve by completing the square. [11.1]

57. $x^2 - 8x = 7$

58. $x^2 + 10x = 6$

Synthesis

TW 59. *Atmospheric Pressure.* Atmospheric pressure P at altitude a is given by

$$P(a) = P_0 e^{-0.00005a},$$

where P_0 is the pressure at sea level $\approx 14.7 \text{ lb/in}^2$ (pounds per square inch). Explain how a barometer, or some other device for measuring atmospheric pressure, can be used to find the height of a skyscraper.

TW 60. Write a problem for a classmate to solve in which information is provided and the classmate is asked to find an exponential growth function. Make the problem as realistic as possible.

61. *Sports Salaries.* As part of Alex Rodriguez's 10-yr $252-million contract with the New York Yankees, he will receive $24 million in 2010 (part from the Yankees and part from his former team, the Texas Rangers) (*Source: The San Francisco Chronicle*). How much money would need to be invested in 2004, at 4% interest compounded continuously, in order to have $24 million for Rodriguez in 2010? (This is much like finding what $24 million in 2010 is worth in 2004 dollars.)

62. *Supply and Demand.* The supply and demand for the sale of stereos by Sound Ideas are given by

$$S(x) = e^x \quad \text{and} \quad D(x) = 162{,}755e^{-x},$$

where $S(x)$ is the price at which the company is willing to supply x stereos and $D(x)$ is the demand price for a quantity of x stereos. Find the equilibrium point. (For reference, see Section 3.8.)

63. Use Exercise 7 to form a model for the percentage of U.S. phone lines that are cellular t years after 2000.

TW 64. Use the model developed in Exercise 63 to predict the percentage of U.S. phone lines that will be cellular in 2020. Does your prediction seem plausible? Why or why not?

65. *Nuclear Energy.* Plutonium-239 (Pu-239) is used in nuclear energy plants. The half-life of Pu-239 is 24,360 yr. (*Source: Microsoft Encarta 97 Encyclopedia*) How long will it take for a fuel rod of Pu-239 to lose 90% of its radioactivity?

66. *Growth of Bacteria.* The bacteria *Escherichia coli* (*E. coli*) are commonly found in the human bladder. Suppose that 3000 of the bacteria are present at time $t = 0$. Then t minutes later, the number of bacteria present is

$$N(t) = 3000(2)^{t/20}.$$

If 100,000,000 bacteria accumulate, a bladder infection can occur. If, at 11:00 A.M., a patient's bladder contains 25,000 *E. coli* bacteria, at what time can infection occur?

67. Show that for exponential growth at rate k, the doubling time T is given by $T = \dfrac{\ln 2}{k}$.

68. Show that for exponential decay at rate k, the half-life T is given by $T = \dfrac{\ln 2}{k}$.

69. *Sports Medicine.* The temperature of the ankle surface begins to increase logarithmically a few minutes after ice is applied (*Source: Journal of Athletic Training*, Vol 41, Number 2, June 2006). After 10 min, the temperature is 15.5°C. After 20 min, the temperature is 21.1°C. If the ice is then removed, in 10 more min the temperature will be 23.6°C, and in 10 more min, the temperature will be 24.9°C.

  a) Let $t = 0$ represent the time at which the ice is applied. Enter the data into a table and graph the data.

  b) Use the LnReg option of the STAT CALC menu to find a logarithmic function that could be used to estimate the ankle surface temperature t min after the ice is applied.

  c) The surface temperature before ice is applied was 27.6°C. How many minutes after the ice was applied will the surface temperature be back to 27.6°C?

Logistic Curves.   *Realistically, most quantities that are growing exponentially eventually level off. The quantity may continue to increase, but at a decreasing rate. This pattern of growth can be modeled by a logistic function*

$$f(x) = \frac{c}{1 + ae^{-bx}}.$$

The general shape of this family of functions is shown by the following graph.

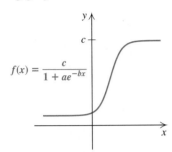

Many graphing calculators can fit a logistic function to a set of data.

70. *Internet Access.*   The percentage of U.S. households with Internet access for various years is listed in the following table.

Year	Percentage of Households with Internet Access
1997	18.6
1998	26.2
2000	41.5
2001	50.3
2003	54.6

Source: www.ntia.doc.gov

a) Find a logistic function

$$f(x) = \frac{c}{1 + ae^{-bx}}$$

that could be used to estimate the percentage of U.S. households with Internet access x years after 1997.

b) Use the function found in part (a) to predict the percentage of U.S. households with Internet access in 2010.

 71. *Heart Transplants.*   In 1967, Dr. Christiaan Barnard of South Africa stunned the world by performing the first heart transplant. Since that time, the operation's popularity has both grown and declined, as shown in the table below.

Year	Number of Heart Transplants Worldwide
1982	189
1983	318
1984	669
1985	1189
1986	2167
1987	2720
1988	3157
1989	3378
1990	4016
1991	4186
1992	4199
1993	4346
1994	4402
1995	4314
1996	4128
1997	4039
1998	3744
1999	3419
2000	3246
2001	3122
2002	3265
2003	3020

Source: International Society for Heart & Lung Transplantation

a) Using 1982 as $t = 0$, graph the data.

b) Does it appear that an exponential function might have ever served as an appropriate model for these data? If so, for what years would this have been the case?

c) Considering *all* the data points on your graph, which would be the most appropriate model: a linear, a quadratic, or an exponential function? Why?

12 Chapter Summary and Review

KEY TERMS AND DEFINITIONS

FUNCTIONS

One-to-one function, p. 891 A function for which different inputs have different outputs; a function whose graph passes the **horizontal-line test.**

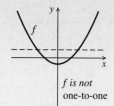

f is not one-to-one

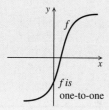

f is one-to-one

Inverse function, p. 891 A function formed by interchanging the members of the domain and the members of the range of a function. A one-to-one function has an inverse.

EXPONENTIAL AND LOGARITHMIC FUNCTIONS

Logarithm, p. 914 $\log_a x$ is the exponent to which a must be raised in order to get x. $\log_a x = m$ means $a^m = x$. *A logarithm is an exponent.*

Exponential function, p. 903

$$f(x) = a^x$$

Domain: $\mathbb{R}$

Range: $(0, \infty)$

Logarithmic function, p. 914

$$g(x) = \log_a x$$

Domain: $(0, \infty)$

Range: $\mathbb{R}$

Common logarithm, p. 916 A logarithm, base 10.
Natural logarithm, p. 930 A logarithm, base e.
e, p. 930 An irrational number; $e \approx 2.718281828$.

APPLICATIONS

Interest compounded annually

$$A = P(1 + i)^t$$

Interest compounded continuously

$$P(t) = P_0 e^{kt}$$

Loudness of sound

$$L = 10 \cdot \log \frac{I}{I_0}$$

pH

$$\text{pH} = -\log[\text{H}^+]$$

Carbon dating

$$P(t) = P_0 e^{-0.00012t}$$

IMPORTANT CONCEPTS

[Section references appear in brackets.]

Concept	Example
The **composition** of f and g is defined as $$(f \circ g)(x) = f(g(x)).$$	If $f(x) = x^2 + 1$ and $g(x) = x - 5$, then $$\begin{aligned}(f \circ g)(x) &= f(g(x)) \\ &= f(x - 5) \\ &= (x - 5)^2 + 1 \\ &= x^2 - 10x + 25 + 1 \\ &= x^2 - 10x + 26.\end{aligned}$$ [12.1]
If f is one-to-one, it is possible to find its inverse: **1.** Replace $f(x)$ with y. **2.** Interchange x and y. **3.** Solve for y. **4.** Replace y with $f^{-1}(x)$.	If $f(x) = 2x - 3$, find $f^{-1}(x)$. **1.** $y = 2x - 3$ **2.** $x = 2y - 3$ **3.** $x + 3 = 2y$ $\dfrac{x + 3}{2} = y$ **4.** $\dfrac{x + 3}{2} = f^{-1}(x)$ [12.1]
The graph of $f(x) = a^x$ contains the points $(0, 1)$ and $(1, a)$. It increases if $a > 1$ and decreases if $0 < a < 1$. The graph of $g(x) = \log_a x$ contains the points $(1, 0)$ and $(a, 1)$.	[12.2], [12.3], [12.5]

(continued)

Properties of Logarithms

$\log_a(MN) = \log_a M + \log_a N$	$\log_7 10 = \log_7 5 + \log_7 2$
$\log_a \dfrac{M}{N} = \log_a M - \log_a N$	$\log_5 \dfrac{14}{3} = \log_5 14 - \log_5 3$
$\log_a M^p = p \cdot \log_a M$	$\log_8 5^{12} = 12\log_8 5$
$\log_a 1 = 0$	$\log_9 1 = 0$
$\log_a a = 1$	$\log_4 4 = 1$
$\log_a a^k = k$	$\log_3 3^8 = 8$
$\log M = \log_{10} M$	$\log 43 = \log_{10} 43$
$\ln M = \log_e M$	$\ln 37 = \log_e 37$
$\log_b M = \dfrac{\log_a M}{\log_a b}$	$\log_6 31 = \dfrac{\log 31}{\log 6} = \dfrac{\ln 31}{\ln 6}$ [12.3], [12.4], [12.5]

The Principle of Exponential Equality
For any real number b, $b \neq -1, 0,$ or 1:

$$b^x = b^y \text{ is equivalent to } x = y.$$

$25 = 5^x$

$5^2 = 5^x$

$2 = x$ [12.6]

The Principle of Logarithmic Equality
For any logarithm base a, and for $x, y > 0$:

$$x = y \text{ is equivalent to } \log_a x = \log_a y.$$

$83 = 7^x$

$\log 83 = \log 7^x$

$\log 83 = x \log 7$

$\dfrac{\log 83}{\log 7} = x$ [12.6]

Exponential Growth Model

$$P(t) = P_0 e^{kt},\ k > 0$$

P_0 is the population at time 0.

$P(t)$ is the population at time t.

k is the **exponential growth rate.**

The **doubling time** is the amount of time necessary for the population to double in size.

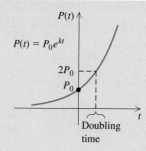

     [12.7]

Exponential Decay Model

$$P(t) = P_0 e^{-kt},\ k > 0$$

P_0 is the quantity present at time 0.

$P(t)$ is the amount present at time t.

k is the **exponential decay rate.**

The **half-life** is the amount of time necessary for half of the quantity to decay.

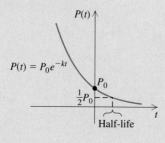

     [12.7]

Review Exercises

Concept Reinforcement　In each of Exercises 1–10, classify the statement as either true or false.

1. The functions given by $f(x) = e^x$ and $g(x) = \ln x$ are inverses of each other. [12.5]

2. A function's doubling time is the amount of time t for which $f(t) = 2f(0)$. [12.7]

3. A radioactive isotope's half-life is the amount of time t for which $f(t) = \frac{1}{2} f(0)$. [12.7]

4. $\ln (ab) = \ln a - \ln b$ [12.4]

5. $\log x^a = x \ln a$ [12.4]

6. $\log_a \dfrac{m}{n} = \log_a m - \log_a n$ [12.4]

7. For $f(x) = 3^x$, the domain of f is $[0, \infty)$. [12.2]

8. For $g(x) = \log_2 x$, the domain of g is $[0, \infty)$. [12.3]

9. The function F is not one-to-one if $F(-2) = F(5)$. [12.1]

10. The function g is one-to-one if it passes the vertical-line test. [12.1]

11. Find $(f \circ g)(x)$ and $(g \circ f)(x)$ if $f(x) = x^2 + 1$ and $g(x) = 2x - 3$. [12.1]

12. If $h(x) = \sqrt{3 - x}$, find $f(x)$ and $g(x)$ such that $h(x) = (f \circ g)(x)$. Answers may vary. [12.1]

13. Determine whether $f(x) = 4 - x^2$ is one-to-one. [12.1]

Find a formula for the inverse of each function. [12.1]

14. $f(x) = x - 8$

15. $g(x) = \dfrac{3x + 1}{2}$

16. $f(x) = 27x^3$

Graph by hand.

17. $f(x) = 3^x + 1$ [12.2]　　18. $x = \left(\frac{1}{4}\right)^y$ [12.2]

19. $y = \log_5 x$ [12.3]

Simplify. [12.3]

20. $\log_3 9$

21. $\log_{10} \frac{1}{100}$

22. $\log_5 5^7$

23. $\log_9 3$

Rewrite as an equivalent logarithmic equation. [12.3]

24. $10^{-2} = \frac{1}{100}$

25. $25^{1/2} = 5$

Rewrite as an equivalent exponential equation. [12.3]

26. $\log_4 16 = x$

27. $\log_8 1 = 0$

Express as an equivalent expression using the individual logarithms of x, y, and z. [12.4]

28. $\log_a x^4 y^2 z^3$

29. $\log_a \dfrac{x^5}{yz^2}$

30. $\log \sqrt[4]{\dfrac{z^2}{x^3 y}}$

Express as an equivalent expression that is a single logarithm and, if possible, simplify. [12.4]

31. $\log_a 7 + \log_a 8$

32. $\log_a 72 - \log_a 12$

33. $\frac{1}{2} \log a - \log b - 2 \log c$

34. $\frac{1}{3}[\log_a x - 2 \log_a y]$

Simplify. [12.4]

35. $\log_m m$

36. $\log_m 1$

37. $\log_m m^{17}$

Given $\log_a 2 = 1.8301$ and $\log_a 7 = 5.0999$, find each of the following. [12.4]

38. $\log_a 14$

39. $\log_a \frac{2}{7}$

40. $\log_a 28$

41. $\log_a 3.5$

42. $\log_a \sqrt{7}$

43. $\log_a \frac{1}{4}$

Use a calculator to find each of the following to the nearest ten-thousandth. [12.3], [12.5]

44. $\log 75$

45. $10^{1.789}$

46. $\ln 0.05$

47. $e^{-0.98}$

Find each of the following logarithms using the change-of-base formula. Round answers to the nearest ten-thousandth. [12.5]

48. $\log_5 2$

49. $\log_{12} 70$

Graph and state the domain and the range of each function. [12.5]

50. $f(x) = e^x - 1$

51. $g(x) = 0.6 \ln x$

Solve. Where appropriate, include approximations to the nearest ten-thousandth. [12.6]

52. $2^x = 32$

53. $3^x = \frac{1}{9}$

54. $\log_3 x = -4$

55. $\log_x 16 = 4$

56. $\log x = -3$

57. $3 \ln x = -6$

58. $4^{2x-5} = 19$

59. $2^{x^2} \cdot 2^{4x} = 32$

60. $4^x = 8.3$

61. $e^{-0.1t} = 0.03$

62. $3e^{2x} = 6$

63. $\log_3 (2x - 5) = 1$

64. $\log_4 x + \log_4 (x - 6) = 2$

65. $\log x + \log (x - 15) = 2$

66. $\log_3 (x - 4) = 3 - \log_3 (x + 4)$

67. In a business class, students were tested at the end of the course with a final exam. They were then tested again 6 months later. The forgetting formula was determined to be

$$S(t) = 82 - 18 \log (t + 1),$$

where t is the time, in months, after taking the final exam. [12.7]

a) Determine the average score when they first took the exam (when $t = 0$).

b) What was the average score after 6 months?

c) After what time was the average score 54?

68. A color photocopier is purchased for $5200. Its value each year is about 80% of its value in the preceding year. Its value in dollars after t years is given by the exponential function

$$V(t) = 5200(0.8)^t. \ [12.7]$$

a) After what amount of time will the copier's value be $1200?

b) After what amount of time will the copier's value be half the original value?

69. *Cell Phones.* In 1997, there were 107.8 million cell phones sold, worldwide. This number grew exponentially to 816.6 million in 2005. (*Source:* Gartner Dataquest) [12.7]

a) Find the value k, and write an exponential function that describes the number of cell phones sold t years after 1997.

b) Predict the number of cell phones that will be sold in 2009.

c) In what year will 1.5 billion cell phones be sold?

70. *MLB Salaries.* The average salary of a Major League baseball player has grown exponentially. Average salaries for various years are listed in the following table. [12.7]

Year	Average Salary
1970	$ 29,803
1980	143,756
1990	578,930
2000	1,998,034
2005	2,632,655

Source: Major League Baseball

a) Use regression to find an exponential function of the form $f(x) = ab^x$ that can be used to estimate the average salary of a Major League baseball player x years after 1970.

b) Use the function of part (a) to estimate the average salary of a Major League baseball player in 2006.

c) The average salary of a Major League baseball player in 2006 was $2,866,544. By what percent was the prediction of part (b) off?

71. The value of Jose's stock market portfolio doubled in 3 yr. What was the exponential growth rate? [12.7]

72. How long will it take $7600 to double itself if it is invested at 4.2%, compounded continuously? [12.7]

73. How old is a skull that has lost 34% of its carbon-14? (Use $P(t) = P_0 e^{-0.00012t}$.) [12.7]

74. What is the pH of a substance if its hydrogen ion concentration is 2.3×10^{-7} moles per liter? (Use pH $= -\log [H^+]$.) [12.7]

75. The intensity of the sound of water at the foot of the Niagara Falls is about 10^{-3} W/m^2.* How loud in decibels is this sound level? [12.7]

$$\left(\text{Use } L = 10 \cdot \log \frac{I}{10^{-12}}. \right)$$

Synthesis

76. Explain why negative numbers do not have logarithms. [12.3]

77. Explain why taking the natural or common logarithm on each side of an equation produces an equivalent equation. [12.6]

Sound and Hearing, Life Science Library. (New York: Time Incorporated, 1965), p. 173.

Solve. [12.6]

78. $\ln (\ln x) = 3$

79. $2^{x^2+4x} = \frac{1}{8}$

80. Solve the system:
$$5^{x+y} = 25,$$
$$2^{2x-y} = 64. \ [12.6]$$

81. *Blogs.* The doubling time for the number of blogs is 6 months (*Source*: Technorati, *State of the Blogosphere*, April 2006). At the beginning of 2003, there were about 0.6 million blogs. Find an exponential function that can be used to predict the number of blogs t months after January 2003. [12.7]

Chapter Test 12

1. Find $(f \circ g)(x)$ and $(g \circ f)(x)$ if $f(x) = x + x^2$ and $g(x) = 2x + 1$.

2. If
$$h(x) = \frac{1}{2x^2 + 1},$$
find $f(x)$ and $g(x)$ such that $h(x) = (f \circ g)(x)$. Answers may vary.

3. Determine whether $f(x) = |x - 3|$ is one-to-one.

Find a formula for the inverse of each function.

4. $f(x) = 3x + 4$

5. $g(x) = (x + 1)^3$

Graph by hand.

6. $f(x) = 2^x - 3$

7. $g(x) = \log_7 x$

Simplify.

8. $\log_5 125$

9. $\log_{100} 10$

10. $3^{\log_3 18}$

Rewrite as an equivalent logarithmic equation.

11. $4^{-3} = \frac{1}{64}$

12. $256^{1/2} = 16$

Rewrite as an equivalent exponential equation.

13. $m = \log_7 49$

14. $\log_3 81 = 4$

15. Express as an equivalent expression using the individual logarithms of a, b, and c:
$$\log \frac{a^3 b^{1/2}}{c^2}.$$

16. Express as an equivalent expression that is a single logarithm:
$$\frac{1}{3} \log_a x + 2 \log_a z.$$

Simplify.

17. $\log_p p$

18. $\log_t t^{23}$

19. $\log_c 1$

Given $\log_a 2 = 0.301$, $\log_a 6 = 0.778$, and $\log_a 7 = 0.845$, find each of the following.

20. $\log_a 14$

21. $\log_a 3$

22. $\log_a 16$

Use a calculator to find each of the following to the nearest ten-thousandth.

23. log 12.3

24. $10^{-0.8}$

25. ln 0.035

26. $e^{4.8}$

27. Find $\log_3 14$ using the change-of-base formula. Round to the nearest ten-thousandth.

Graph and state the domain and the range of each function.

28. $f(x) = e^x + 3$

29. $g(x) = \ln (x - 4)$

Solve. Where appropriate, include approximations to the nearest ten-thousandth.

30. $2^x = \frac{1}{32}$

31. $\log_x 25 = 2$

32. $\log_4 x = \frac{1}{2}$

33. $\log x = 4$

34. $5^{4-3x} = 87$

35. $7^x = 1.2$

36. $\ln x = \frac{1}{4}$

37. $\log (x - 3) + \log (x + 1) = \log 5$

38. The average walking speed R of people living in a city of population P, in thousands, is given by $R = 0.37 \ln P + 0.05$, where R is in feet per second.

   a) The population of Tucson, Arizona, is 512,000. Find the average walking speed.

   b) Philadelphia, Pennsylvania, has an average walking speed of about 2.76 ft/sec. Find the population.

39. The population of Nigeria was about 128.8 million in 2005, and the exponential growth rate was 2.4% per year.

   a) Write an exponential function describing the population of Nigeria.

   b) What will the population be in 2007? in 2012?

   c) When will the population be 175 million?

   d) What is the doubling time?

40. The average cost of a year at a private four-year college grew exponentially from $18,039 in 1997 to $23,503 in 2003 (*Source*: National Center for Education Statistics, Digest of Education Statistics, 2003).

   a) Find the value k, and write an exponential function that approximates the cost of a year of college t years after 1997.

   b) Predict the cost of a year of college in 2010.

   c) In what year will the average cost of college be $50,000?

41. *Online Advertising.* The amount of money spent in online advertising grew exponentially from 2000 through 2006, as listed in the following table.

Year	Online Advertising Spending (in billions)
2000	$ 5.4
2001	5.7
2002	6.8
2003	8.6
2004	10.6
2005	12.9
2006	15.4

Source: Jupiter Media Metrix

   a) Use regression to find an exponential function of the form $f(x) = ab^x$ that can be used to predict online advertising spending t years after 2000.

   b) Use the function to estimate online advertising spending in 2008.

42. An investment with interest compounded continuously doubled itself in 15 yr. What is the interest rate?

43. How old is an animal bone that has lost 43% of its carbon-14? (Use $P(t) = P_0 e^{-0.00012t}$.)

44. The sound of traffic at a busy intersection averages 75 dB. What is the intensity of such a sound?

$$\left(\text{Use } L = 10 \cdot \log \frac{I}{I_0}. \right)$$

45. The hydrogen ion concentration of water is 1.0×10^{-7} moles per liter. What is the pH? (Use pH $= -\log [\text{H}^+]$.)

Synthesis

46. Solve: $\log_5 |2x - 7| = 4$.

47. If $\log_a x = 2$, $\log_a y = 3$, and $\log_a z = 4$, find
$$\log_a \frac{\sqrt[3]{x^2 z}}{\sqrt[3]{y^2 z^{-1}}}.$$

1–12 Cumulative Review

Travel. *Prices for a massage at Passport Travel Spa in an airport waiting area are shown in the following graph.*

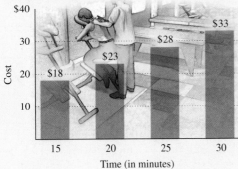

Airport Massage

Source: Passport Travel Spa, Indianapolis International Airport

1. Determine whether the data can be modeled by a linear, quadratic, or exponential function. [3.7]

2. Find the rate of change, in dollars per minute. [3.4]

3. Find a linear function $f(x) = mx + b$ that fits the data, where f is the cost of the massage and x is the length, in minutes. [3.7], [3.8]

4. Use the function in Exercise 3 to estimate the cost of a 10-min massage. [3.8]

5. For the function in Exercise 3, what do the values of m and b signify? [3.6]

Life Insurance. *The monthly premium for a term life insurance policy increases as the age of the insured person increases. The following graph shows some premiums for a $250,000 policy for a male.*

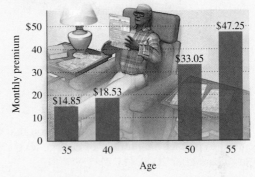

Life Insurance

Source: Insurance company advertisement

6. Determine whether the data can be modeled by a linear, quadratic, or exponential function. [11.8], [12.7]

7. Use regression to find an exponential function of the form $f(x) = ab^x$ that can be used to estimate the monthly insurance premium m for a male who is x years old. [12.7]

8. Use the function in Exercise 7 to estimate the monthly premium for a 45-year-old male. [12.7]

College Texts.   *The number of college textbooks sold each year from 2000–2005 is listed in the following graph.*

College Texts

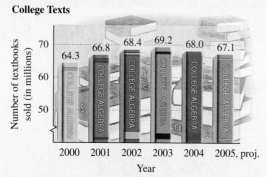

Source: Book Industry Trends, 2005

9. Determine whether the data can be modeled by a linear, quadratic, or exponential function. [11.8]

10. Use regression to find a quadratic function that can be used to predict the number of college texts c sold t years after 2000. [11.8]

11. Use the function in Exercise 10 to estimate the number of college texts sold in 2006. [11.8]

12. The perimeter of a rectangular garden is 112 m. The length is 16 m more than the width. Find the length and the width. [2.5]

13. In triangle ABC, the measure of angle B is three times the measure of angle A. The measure of angle C is 105° greater than the measure of angle A. Find the angle measures. [2.5]

14. Good's Candies of Indiana makes all their chocolates by hand. It takes Anne 10 min to coat a tray of candies in chocolate. It takes Clay 12 min to coat a tray of candies. How long would it take Anne and Clay, working together, to coat the candies? [7.7]

15. Joe's Thick and Tasty salad dressing gets 45% of its calories from fat. The Light and Lean dressing gets 20% of its calories from fat. How many ounces of each should be mixed in order to get 15 oz of dressing that gets 30% of its calories from fat? [4.4]

16. A cruise ship can move at a speed of 21 mph in still water. The ship travels 113 mi down the Amazon River in the same time that it takes to travel 55 mi upriver. What is the speed of the Amazon? [7.7]

17. What is the minimum product of two numbers whose difference is 14? What are the numbers that yield this product? [11.8]

In Exercises 18–21, match each function with one of the following graphs.

a)
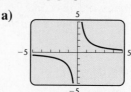

b)
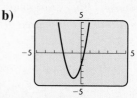

c)
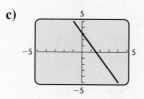

d)
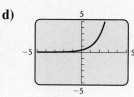

18. $f(x) = -2x + 3.1$ [3.6], [3.8]

19. $p(x) = 3x^2 + 5x - 2$ [11.6]

20. $h(x) = \dfrac{2.3}{x}$ [7.1]

21. $g(x) = e^{x-1}$ [12.5]

22. Find the domain of the function f given by
$$f(x) = \frac{-4}{3x^2 - 5x - 2}. \quad [6.3]$$

23. For the function described by
$$h(x) = -3x^2 + 4x + 8,$$
find $h(-2)$. [3.8]

24. Find the inverse of f if $f(x) = 9 - 2x$. [12.1]

25. Find a linear function with a graph that contains the points $(0, -8)$ and $(-1, 2)$. [3.7], [3.8]

26. Find an equation of the line whose graph has a y-intercept of $(0, 7)$ and is perpendicular to the line given by $2x + y = 6$. [3.6]

Graph by hand.

27. $5x = 15 + 3y$ [3.3]

28. $y = 2x^2 - 4x - 1$ [11.7]

29. $y = \log_3 x$ [12.3]

30. $y = 3^x$ [12.2]

31. $-2x - 3y \le 12$ [8.3]

32. Graph: $f(x) = 2(x + 3)^2 + 1$. [11.6]

    a) Label the vertex.
    b) Draw the axis of symmetry.
    c) Find the maximum or minimum value.

33. Graph $f(x) = 2e^x$ and determine the domain and the range. [12.5]

34. Evaluate $\dfrac{x^0 + y}{-z}$ for $x = 6$, $y = 9$, and $z = -5$.
[1.8], [5.1]

Simplify.

35. $\left| -\frac{5}{2} + \left(-\frac{7}{2} \right) \right|$ [1.5]

36. $(-2x^2y^{-3})^{-4}$ [5.2]

37. $(-5x^4y^{-3}z^2)(-4x^2y^2)$ [5.2]

38. $\dfrac{3x^4y^6z^{-2}}{-9x^4y^2z^3}$ [5.2]

39. $4x - 3 - 2[5 - 3(2 - x)]$ [1.8]

40. $3^3 + 2^2 - (32 \div 4 - 16 \div 8)$ [1.8]

Perform the indicated operations and simplify.

41. $(5p^2q^3 + 6pq - p^2 + p) +$
$(2p^2q^3 + p^2 - 5pq - 9)$ [5.4], [5.7]

42. $(11x^2 - 6x - 3) - (3x^2 + 5x - 2)$ [5.4]

43. $(3x^2 - 2y)^2$ [5.6], [5.7]

44. $(5a + 3b)(2a - 3b)$ [5.6], [5.7]

45. $\dfrac{x^2 + 8x + 16}{2x + 6} \div \dfrac{x^2 + 3x - 4}{x^2 - 9}$ [7.2]

46. $\dfrac{1 + \dfrac{3}{x}}{x - 1 - \dfrac{12}{x}}$ [7.5]

47. $\dfrac{a^2 - a - 6}{a^3 - 27} \cdot \dfrac{a^2 + 3a + 9}{6}$ [7.2]

48. $\dfrac{3}{x + 6} - \dfrac{2}{x^2 - 36} + \dfrac{4}{x - 6}$ [7.4]

Factor.

49. $xy + 2xz - xw$ [6.1]

50. $8 - 125x^3$ [6.5]

51. $6x^2 + 8xy - 8y^2$ [6.3]

52. $x^4 - 4x^3 + 7x - 28$ [6.1]

53. $2m^2 + 12mn + 18n^2$ [6.4]

54. $x^4 - 16y^4$ [6.4]

55. Divide: $(x^4 - 5x^3 + 2x^2 - 6) \div (x - 3)$. [5.8]

56. Multiply $(5.2 \times 10^4)(3.5 \times 10^{-6})$. Write scientific notation for the answer. [5.2]

For the radical expressions that follow, assume that all variables represent positive numbers.

57. Divide and simplify:
$\dfrac{\sqrt[3]{40xy^8}}{\sqrt[3]{5xy}}$. [10.4]

58. Multiply and simplify: $\sqrt{7xy^3} \cdot \sqrt{28x^2y}$. [10.3]

59. Write as an equivalent expression without rational exponents: $(27a^6b)^{4/3}$. [10.4]

60. Rationalize the denominator:
$\dfrac{3 - \sqrt{y}}{2 - \sqrt{y}}$. [10.5]

61. Divide and simplify:
$\dfrac{\sqrt{x + 5}}{\sqrt[5]{x + 5}}$. [10.5]

62. Multiply these complex numbers:
$(2 - i\sqrt{3})(6 + 2i\sqrt{3})$. [10.8]

63. Add: $(8 + 2i) + (5 - 3i)$. [10.8]

64. Express in terms of logarithms of a, b, and c:
$\log\left(\dfrac{a^2c^3}{b} \right)$. [12.4]

65. Express as a single logarithm:
$3 \log x - \frac{1}{2} \log y - 2 \log z$. [12.4]

66. Convert to an exponential equation: $\log_a 5 = x$. [12.3]

67. Convert to a logarithmic equation: $x^3 = t$. [12.3]

Find each of the following using a calculator. Round to the nearest ten-thousandth. [12.3], [12.5]

68. $\log 0.05566$ **69.** $10^{2.89}$

70. $\ln 12.78$ **71.** $e^{-1.4}$

Solve.

72. $5(2x - 3) = 9 - 5(2 - x)$ [2.2]

73. $4x - 3y = 15$,
$3x + 5y = 4$ [4.3]

74. $x + y - 3z = -1,$
$2x - y + z = 4,$
$-x - y + z = 1$ [8.1]

75. $x(x - 3) = 10$ [6.2]

76. $\dfrac{7}{x^2 - 5x} - \dfrac{2}{x - 5} = \dfrac{4}{x}$ [7.6]

77. $\dfrac{8}{x + 1} + \dfrac{11}{x^2 - x + 1} = \dfrac{24}{x^3 + 1}$ [7.6], [11.2]

78. $\sqrt{4 - 5x} = 2x - 1$ [10.6]

79. $\sqrt[3]{2x} = 1$ [10.6]

80. $3x^2 + 75 = 0$ [11.1]

81. $x - 8\sqrt{x} + 15 = 0$ [11.5]

82. $x^4 - 13x^2 + 36 = 0$ [11.5]

83. $\log_7 x = 1$ [12.3]

84. $\log_x 36 = 2$ [12.3]

85. $9^x = 27$ [12.6]

86. $3^{5x} = 7$ [12.6]

87. $\ln x - \ln(x - 8) = 1$ [12.6]

88. $x^2 + 4x > 5$ [8.4]

89. If $f(x) = x^2 + 6x$, find a such that $f(a) = 11$. [11.2]

90. If $f(x) = |2x - 3|$, find all x for which $f(x) \geq 7$. [8.2]

Solve.

91. $D = \dfrac{ab}{b + a}$, for a [7.8]

92. $\dfrac{1}{p} + \dfrac{1}{q} = \dfrac{1}{f}$, for q [7.8]

93. $M = \dfrac{2}{3}(A + B)$, for B [2.3]

Students in a biology class just took a final exam. A formula for determining what the average exam grade on a similar test will be t months later is

$$S(t) = 78 - 15 \log(t + 1).$$

94. The average score when the students first took the exam occurs when $t = 0$. Find the students' average score on the final exam. [12.7]

95. What would the average score be on a retest after 4 months? [12.7]

The population of Kenya was 33.8 million in 2005, and the exponential growth rate was 2.6% per year.

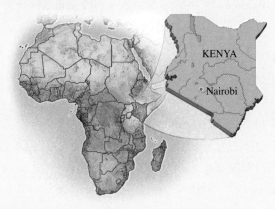

96. Write an exponential function describing the growth of the population of Kenya. [12.7]

97. Predict what the population will be in 2008 and in 2014. [12.7]

98. What is the doubling time of the population? [12.7]

99. y varies directly as the square of x and inversely as z, and $y = 2$ when $x = 5$ and $z = 100$. What is y when $x = 3$ and $z = 4$? [7.8]

Synthesis

Solve.

100. $\dfrac{5}{3x - 3} + \dfrac{10}{3x + 6} = \dfrac{5x}{x^2 + x - 2}$ [7.6]

101. $\log \sqrt{3x} = \sqrt{\log 3x}$ [12.6]

102. A train travels 280 mi at a certain speed. If the speed had been increased by 5 mph, the trip could have been made in 1 hr less time. Find the actual speed. [11.3]

R Elementary Algebra Review

This chapter is a review of the first six chapters of this text. Each section corresponds to a chapter of the text. For further explanation of the topics in this chapter, refer to the section or pages referenced in the margin.

R.1 Introduction to Algebraic Expressions

The Real Numbers ■ Operations on Real Numbers ■ Algebraic Expressions

The Real Numbers

Sets of real numbers, Section 1.4

Sets of Numbers

Natural numbers: $\{1, 2, 3, \ldots\}$
Whole numbers: $\{0, 1, 2, 3, \ldots\}$
Integers: $\{\ldots, -3, -2, -1, 0, 1, 2, 3, \ldots\}$
Rational numbers: $\left\{ \dfrac{a}{b} \,\middle|\, a \text{ and } b \text{ are integers and } b \neq 0 \right\}$

Rational numbers can always be written as **terminating** or **repeating** decimals. **Irrational numbers,** like $\sqrt{2}$ or π, can be thought of as nonterminating and nonrepeating decimals. The set of **real numbers** consists of all rational and irrational numbers, taken together.

Real numbers can be represented by points on the number line.

We can compare, or **order,** real numbers by their graphs on the number line. For any two numbers, the one to the left is less than the one to the right.

Sentences like $\frac{1}{4} = 0.25$, containing an equals sign, are called **equations.** An **inequality** is a sentence containing $>$ (is greater than), $<$ (is less than), $\geq$ (is greater than or equal to), or $\leq$ (is less than or equal to). Equations and inequalities can be true or false.

EXAMPLE 1   Write true or false for each equation or inequality.

a) $-2\frac{1}{3} = -\frac{7}{3}$ **b)** $1 = -1$ **c)** $-5 < -2$

d) $-3 \geq 2$ **e)** $1.1 \leq 1.1$

SOLUTION

a) $-2\frac{1}{3} = -\frac{7}{3}$ is *true* because $-2\frac{1}{3}$ and $-\frac{7}{3}$ represent the same number.

b) $1 = -1$ is a *false* equation.

c) $-5 < -2$ is *true* because -5 is to the left of -2 on the number line.

d) $-3 \geq 2$ is *false* because neither $-3 > 2$ nor $-3 = 2$ is true.

e) $1.1 \leq 1.1$ is *true* because $1.1 = 1.1$ is true.

The distance of a number from 0 is called the **absolute value** of the number. The notation $|-4|$ represents the absolute value of -4. The absolute value of a number is never negative.

EXAMPLE 2   Find the absolute value: **(a)** $|-4|$; **(b)** $\left|\frac{11}{3}\right|$; **(c)** $|0|$.

SOLUTION

a) $|-4| = 4$ since -4 is 4 units from 0.

b) $\left|\frac{11}{3}\right| = \frac{11}{3}$ since $\frac{11}{3}$ is $\frac{11}{3}$ units from 0.

c) $|0| = 0$ since 0 is 0 units from itself.

Operations on Real Numbers

Rules for Addition of Real Numbers

1. *Positive numbers*: Add as usual. The answer is positive.
2. *Negative numbers*: Add absolute values and make the answer negative.
3. *A positive and a negative number*: Subtract absolute values. Then:
   a) If the positive number has the greater absolute value, the answer is positive.
   b) If the negative number has the greater absolute value, the answer is negative.
   c) If the numbers have the same absolute value, the answer is 0.
4. *One number is zero*: The sum is the other number.

Opposite, p. 48

Every real number has an **opposite.** The opposite of -6 is 6, the opposite of 3.7 is -3.7, and the opposite of 0 is itself. When opposites are added, the result is 0. Finding the opposite of a number is often called "changing its sign."

Subtraction of real numbers is defined in terms of addition and opposites.

Subtraction, Section 1.6

Subtraction of Real Numbers   To subtract, add the opposite of the number being subtracted.

The rules for multiplication and division can be stated together.

Multiplication and division, Section 1.7

Rules for Multiplication and Division   To multiply or divide two real numbers:

1. Using the absolute values, multiply or divide, as indicated.
2. If the signs are the same, the answer is positive.
3. If the signs are different, the answer is negative.

EXAMPLE 3   Perform the indicated operations: **(a)** $-13 + (-9)$; **(b)** $-\frac{4}{5} + \frac{1}{10}$; **(c)** $-6 - (-7.3)$; **(d)** $3(-1.5)$; **(e)** $\left(-\frac{4}{9}\right) \div \left(-\frac{2}{5}\right)$.

SOLUTION

a) $-13 + (-9) = -22$

Two negatives. *Think*: Add the absolute values, 13 and 9, to get 22. Make the answer *negative*, -22.

b) $-\frac{4}{5} + \frac{1}{10} = -\frac{8}{10} + \frac{1}{10} = -\frac{7}{10}$

A negative and a positive. *Think*: The difference of absolute values is $\frac{8}{10} - \frac{1}{10}$, or $\frac{7}{10}$. The negative number has the larger absolute value, so the answer is *negative*, $-\frac{7}{10}$.

c) $-6 - (-7.3) = -6 + 7.3 = 1.3$ Change the subtraction to addition and add the opposite.

d) $3(-1.5) = -4.5$ *Think*: $3(1.5) = 4.5$. The signs are different, so the answer is negative.

e) $\left(-\frac{4}{9}\right) \div \left(-\frac{2}{5}\right) = \left(-\frac{4}{9}\right) \cdot \left(-\frac{5}{2}\right)$ Multiplying by the reciprocal. The answer is positive.
$= \frac{20}{18} = \frac{10}{9} \cdot \frac{2}{2} = \frac{10}{9}$

Division by 0, p. 61

Exponential notation, p. 65

Addition, subtraction, and multiplication are defined for all real numbers, but we cannot **divide by 0**. For example, $\frac{0}{3} = 0 \div 3 = 0$, but $\frac{3}{0} = 3 \div 0$ is **undefined**.

A product like $2 \cdot 2 \cdot 2 \cdot 2$, in which the factors are the same, is called a **power**. Powers are often written using **exponential notation**:

$$2 \cdot 2 \cdot 2 \cdot 2 = 2^4. \longleftarrow \text{There are 4 factors; 4 is the } \textit{exponent.}$$
$$\text{2 is the } \textit{base.}$$

A number raised to the exponent of 1 is the number itself; for example, $3^1 = 3$.

An expression containing a series of operations is not necessarily evaluated from left to right. Instead, we perform the operations according to the following rules.

Rules for Order of Operations

1. Calculate within the innermost grouping symbols, ( ), [ ], { }, | |, and above or below fraction bars.
2. Simplify all exponential expressions.
3. Perform all multiplication and division, working from left to right.
4. Perform all addition and subtraction, working from left to right.

▶ **EXAMPLE 4**   Simplify: $3 - [(4 \times 5) + 12 \div 2^3 \times 6] + 5$.

SOLUTION

$3 - [(4 \times 5) + 12 \div 2^3 \times 6] + 5$

$= 3 - [20 + 12 \div 2^3 \times 6] + 5$ Doing the calculations in the innermost parentheses first

$= 3 - [20 + 12 \div 8 \times 6] + 5$ Working inside the brackets; evaluating 2^3

$= 3 - [20 + 1.5 \times 6] + 5$ $12 \div 8$ is the first multiplication or division working from left to right.

$= 3 - [20 + 9] + 5$ Multiplying

$= 3 - 29 + 5$ Completing the calculations within the brackets

$\left.\begin{array}{l} = -26 + 5 \\ = -21 \end{array}\right\}$ Adding and subtracting from left to right

Algebraic Expressions

In an **algebraic expression** like $2xt^3$, the number 2 is a **constant** and x and t are **variables.** When numbers are **substituted** for x and t, we say that we are **evaluating** the expression.

Algebraic expressions containing variables can be evaluated by substituting a number for each variable in the expression and following the rules for order of operations.

EXAMPLE 5   The perimeter P of a rectangle of length l and width w is given by the formula $P = 2l + 2w$. Find the perimeter when l is 16 in. and w is 7.5 in.

SOLUTION   We evaluate, substituting 16 in. for l and 7.5 in. for w and carrying out the operations:

$$P = 2l + 2w$$
$$= 2 \cdot 16 + 2 \cdot 7.5$$
$$= 32 + 15$$
$$= 47 \text{ in.}$$

Expressions that represent the same number are said to be **equivalent.** The laws that follow provide methods for writing equivalent expressions.

Laws and Properties of Real Numbers

Commutative laws:	$a + b = b + a;\ ab = ba$
Associative laws:	$a + (b + c) = (a + b) + c;$ $a(bc) = (ab)c$
Distributive law:	$a(b + c) = ab + ac$
Identity property of 1:	$1 \cdot a = a \cdot 1 = a$
Identity property of 0:	$a + 0 = 0 + a = a$
Law of opposites:	$a + (-a) = 0$
Multiplicative property of 0:	$0 \cdot a = a \cdot 0 = 0$
Property of -1:	$-1 \cdot a = -a$
Opposite of a sum:	$-(a + b) = -a + (-b)$
$\dfrac{-a}{b} = \dfrac{a}{-b} = -\dfrac{a}{b},$	$\dfrac{-a}{-b} = \dfrac{a}{b}$

The distributive law can be used to multiply and to **factor** expressions. We factor an expression when we write an equivalent expression that is a product.

EXAMPLE 6   Write an equivalent expression as indicated.

a) Multiply: $-2(5x - 3)$.   **b)** Factor: $5x + 10y + 5$.

SOLUTION

$$
\begin{aligned}
\textbf{a) } -2(5x - 3) &= -2(5x + (-3)) && \text{Adding the opposite}\\
&= -2 \cdot 5x + (-2) \cdot (-3) && \text{Using the distributive law}\\
&= (-2 \cdot 5)x + 6 && \text{Using the associative law for multiplication}\\
&= -10x + 6
\end{aligned}
$$

$$
\begin{aligned}
\textbf{b) } 5x + 10y + 5 &= 5 \cdot x + 5 \cdot 2y + 5 \cdot 1 && \text{The common factor is 5.}\\
&= 5(x + 2y + 1) && \text{Using the distributive law}
\end{aligned}
$$

Factoring can be checked by multiplying:

$$5(x + 2y + 1) = 5 \cdot x + 5 \cdot 2y + 5 \cdot 1 = 5x + 10y + 5.$$

Terms, p. 18

Combine like terms, pp. 44, 68

The **terms** of an algebraic expression are separated by plus signs. When two terms have variable factors that are exactly the same, the terms are called **like, or similar, terms.** The distributive law enables us to **combine, or collect, like terms.**

EXAMPLE 7   Combine like terms: $-5m + 3n - 4n + 10m$.

SOLUTION

$$
\begin{aligned}
-5m &+ 3n - 4n + 10m\\
&= -5m + 3n + (-4n) + 10m && \text{Rewriting as addition}\\
&= -5m + 10m + 3n + (-4n) && \text{Using the commutative law of addition}\\
&= (-5 + 10)m + (3 + (-4))n && \text{Using the distributive law}\\
&= 5m + (-n)\\
&= 5m - n && \text{Rewriting as subtraction}
\end{aligned}
$$

We can also use the distributive law to help simplify algebraic expressions containing parentheses.

EXAMPLE 8   Simplify: **(a)** $4x - (y - 2x)$; **(b)** $3(t + 2) - 6(t - 1)$.

SOLUTION

$$
\begin{aligned}
\textbf{a) } 4x - (y - 2x) &= 4x - y + 2x && \text{Removing parentheses and changing the sign of every term}\\
&= 6x - y && \text{Combining like terms}
\end{aligned}
$$

$$
\begin{aligned}
\textbf{b) } 3(t + 2) - 6(t - 1) &= 3t + 6 - 6t + 6 && \text{Multiplying each term of } t + 2 \text{ by 3 and each term of } t - 1 \text{ by } -6\\
&= -3t + 12 && \text{Combining like terms}
\end{aligned}
$$

We have seen that algebraic expressions can be evaluated for specified numbers. If the expressions on each side of an equation have the same **value** for a given number, then that number is a **solution** of the equation.

EXAMPLE 9   Determine whether each number is a solution of $x - 2 = -5$.

a) 3   **b)** -3

SOLUTION

a) We have:

$$x - 2 = -5 \qquad \text{Writing the equation}$$
$$\overline{3 - 2 \ \big| \ -5} \qquad \text{Substituting 3 for } x$$
$$1 \overset{?}{=} -5 \qquad 1 = -5 \text{ is FALSE}$$

Since $3 - 2 = -5$ is false, 3 is not a solution of $x - 2 = -5$.

b) We have

$$x - 2 = -5$$
$$\overline{-3 - 2 \ \big| \ -5}$$
$$-5 \overset{?}{=} -5 \quad \text{TRUE}$$

Since $-3 - 2 = -5$ is true, -3 is a solution of $x - 2 = -5$.   ◢

Certain word phrases can be translated to algebraic expressions. These in turn can often be used to translate problems to equations.

EXAMPLE 10   Energy Use.   Translate this problem to an equation. Do not solve.

On average, a home spa costs about $192 a year to operate. This is 16 times as much as the average annual operating cost of a home computer. (*Source*: U.S. Department of Energy) How much does it cost to operate a home computer for a year?

SOLUTION   We let c represent the annual cost of operating a home computer. We then reword the problem to make the translation more direct.

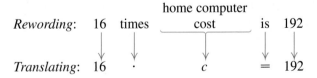

Rewording: 16   times   home computer cost   is   192

Translating: 16   · c =   192   ◢

R.1 EXERCISE SET

Classify each equation or inequality as true or false.

1. $2.3 = 2.31$

2. $-3 \geq -3$

3. $-10 < -1$

4. $0 \leq -1$

5. $5 > 0$

6. $\frac{1}{10} = 0.1$

Find each absolute value.

7. $|4|$

8. $\left|\frac{11}{4}\right|$

9. $|-1.3|$

10. $|-105|$

Simplify.

11. $(-13) + (-12)$

12. $3 - (-2)$

13. $-\frac{1}{3} - \frac{2}{5}$

14. $\frac{3}{8} \div \frac{3}{5}$

15. $4.2 - 10.7$

16. $(-1.3)(2.8)$

17. $-15 + 0$

18. $\left(-\frac{1}{2}\right) + \frac{1}{8}$

19. $0 \div (-10)$

20. $0 - 32$

21. $\left(-\frac{3}{10}\right) + \left(-\frac{1}{5}\right)$

22. $\left(-\frac{4}{7}\right)\left(\frac{7}{4}\right)$

23. $-3.8 + 9.6$

24. $-0.01 + 1$

25. $(-12) \div 4$

26. $(-3)(30)$

27. $32 - (-7)$

28. $-100 + 35$

29. $(-10)(-17.5)$

30. $-10 - 2.68$

31. $(-68) + 36$

32. $175 \div (-25)$

33. $2 + (-3) + 7 + 10$

34. $-5 + (-15) + 13 + (-1)$

35. $3 \cdot (-2) \cdot (-1) \cdot (-1)$

36. $(-6) \cdot (-5) \cdot (-4) \cdot (-3) \cdot (-2) \cdot (-1)$

37. $(-1)^4 + 2^3$

38. $(-1)^5 + 2^4$

39. $2 \cdot 6 - 3 \cdot 5$

40. $12 \div 4 + 15 \div 3$

41. $3 - (2 \cdot 4 + 11)$

42. $3 - 2 \cdot 4 + 11$

43. $4 \cdot 5^2$

44. $7 \cdot 2^3$

45. $25 - 8 \cdot 3 + 1$

46. $12 - 16 \cdot 5 + 4$

47. $2 - (3^3 + 16 \div (-2)^3)$

48. $-7 - (8 + 10 \cdot 2^2)$

49. $|6(-3)| + |(-2)(-9)|$

50. $3 - |2 - 7 + 4|$

51. $\dfrac{7000 + (-10)^3}{10^2 \cdot (2 + 4)}$

52. $\dfrac{3 - 2 \cdot 6 - 5}{2(3 + 7)^2}$

53. $2 + 8 \div 2 \cdot 2$

54. $2 + 8 \div (2 \cdot 2)$

Evaluate.

55. $x - y$, for $x = 10$ and $y = 3$

56. $2m - n$, for $m = 6$ and $n = 11$

57. $-3 - x^2 + 12x$, for $x = 5$

58. $14 + (y - 5)^2 - 12 \div y$, for $y = -2$

59. The area of a parallelogram with base b and height h is bh. Find the area of the parallelogram when the height is 3.5 cm and the base is 8 cm.

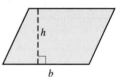

60. The area of a triangle with base b and height h is $\frac{1}{2}bh$. Find the area of the triangle when the height is 2 in. and the base is 6.2 in.

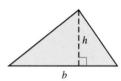

Multiply.

61. $4(2x + 7)$

62. $3(5y + 1)$

63. $5(x - 10)$

64. $4(3x - 2)$

65. $-2(15 - 3x)$

66. $-7(3x - 5)$

67. $2(4a + 6b - 3c)$

68. $5(8p + q - 5r)$

Factor.

69. $8x + 6y$

70. $7p + 14q$

71. $3 + 3w$

72. $4x + 4y$

73. $10x + 50y + 100$

74. $81p + 27q + 36$

Combine like terms.

75. $3p - 2p$

76. $4x + 3x$

77. $4m + 10 - 5m + 12$

78. $3a - 4b - 5b - 6a$

79. $-6x + 7 + 9x$

80. $16r + (-7r) + 3s$

Remove parentheses and simplify.

81. $2p - (7 - 4p)$

82. $4r - (3r + 5)$

83. $2x + 5y - 7(x - y)$

84. $14m - 6(2n - 3m) + n$

85. $6[2a + 4(a - 2b)]$

86. $2[2a + 1 - (3a - 6)]$

87. $3 - 2[5(x - 10y) - (3 + 2y)]$

88. $7 - 4[2(3 - 2x) - 5(4x - 3)]$

Determine whether the given number is a solution of the given equation.

89. $4;\ 3x - 2 = 10$

90. $12;\ 100 = 4x + 50$

91. $-3;\ 4 - x = 1$

92. $-1;\ 2 = 5 + 3x$

93. $4.6;\ \dfrac{x}{2} = 2.3$

94. $144;\ \dfrac{x}{9} = 16$

Translate each problem to an equation. Do not solve.

95. Three times what number is 348?

96. What number added to 256 is 113?

97. *Fast-Food Calories.* A McDonald's Big Mac® contains 500 calories. This is 69 more calories than a Taco Bell Beef Burrito® provides. How many calories are in a Taco Bell Beef Burrito?

98. *Coca-Cola® Consumption.* The average U.S. citizen consumes 296 servings of Cola-Cola each year. This is 7.4 times the international average. What is the international average per capita consumption of Coke?

99. *Vegetable Production.* It takes 42 gal of water to produce 1 lb of broccoli. This is twice the amount of water used to produce 1 lb of lettuce. How many gallons of water does it take to produce 1 lb of lettuce?

100. *Sports Costs.* The average annual cost for scuba diving is $470. This is $458 more than the average annual cost to play badminton. What is the average annual cost to play badminton?

R.2 Equations, Inequalities, and Problem Solving

Solving Equations and Formulas ◾ Solving Inequalities ◾
Problem Solving

Solving Equations and Formulas

Any replacement for the variable in an equation that makes the equation true is called a *solution* of the equation. To **solve** an equation means to find all of its solutions.

We use the following principles to write **equivalent equations,** or equations with the same solutions.

Equivalent equations, p. 84

The Addition and Multiplication Principles for Equations

The Addition Principle

For any real numbers a, b, and c,

$$a = b \quad \text{is equivalent to} \quad a + c = b + c.$$

The Multiplication Principle

For any real numbers a, b, and c, with $c \neq 0$,

$$a = b \quad \text{is equivalent to} \quad a \cdot c = b \cdot c.$$

To solve $x + a = b$ for x, we add $-a$ to (or subtract a from) both sides. To solve $ax = b$ for x, we multiply both sides by $\frac{1}{a}$ (or divide both sides by a).

To solve an equation like $-3x - 10 = 14$, we first isolate the variable term, $-3x$, using the addition principle. Then we use the multiplication principle to get the variable by itself.

EXAMPLE 1   Solve: $-3x - 10 = 14$.

SOLUTION

$$-3x - 10 = 14$$

$$-3x - 10 + 10 = 14 + 10 \qquad \text{Using the addition principle: Adding 10 to both sides}$$

First isolate the x-term.
$$-3x = 24 \qquad \text{Simplifying}$$

$$\frac{-3x}{-3} = \frac{24}{-3} \qquad \text{Dividing both sides by } -3$$

Then isolate x.
$$x = -8 \qquad \text{Simplifying}$$

Check:
$$\begin{array}{c|c} -3x - 10 = 14 \\ \hline -3(-8) - 10 & 14 \\ 24 - 10 & \\ 14 \overset{?}{=} 14 & \text{TRUE} \end{array}$$

The solution is -8.

Equations are generally easier to solve when they do not contain fractions. The easiest way to clear an equation of fractions is to multiply *every term on both sides* of the equation by the least common denominator.

EXAMPLE 2   Solve: $\frac{5}{2} - \frac{1}{6}t = \frac{2}{3}$.

SOLUTION   The number 6 is the least common denominator, so we multiply both sides by 6.

$$6\left(\frac{5}{2} - \frac{1}{6}t\right) = 6 \cdot \frac{2}{3}$$ Multiplying both sides by 6

$$6 \cdot \frac{5}{2} - 6 \cdot \frac{1}{6}t = 6 \cdot \frac{2}{3}$$ Using the distributive law. Be sure to multiply every term by 6.

$$15 - t = 4$$ The fractions are cleared.

$$15 - t - 15 = 4 - 15$$ Subtracting 15 from both sides

$$-t = -11$$

$$(-1)(-t) = (-1)(-11)$$ Multiplying both sides by -1 to change the sign

$$t = 11$$

Check:
$$\frac{5}{2} - \frac{1}{6}t = \frac{2}{3}$$

$$\begin{array}{c|c} \frac{5}{2} - \frac{1}{6}(11) & \frac{2}{3} \\ \frac{5}{2} - \frac{11}{6} & \\ \frac{15}{6} - \frac{11}{6} & \\ & \frac{2}{3} \overset{?}{=} \frac{2}{3} \quad \text{TRUE} \end{array}$$

The solution is 11.

To solve equations that contain parentheses, we can use the distributive law to first remove the parentheses. If like terms appear in an equation, we combine them and then solve.

EXAMPLE 3   Solve: $1 - 3(4 - x) = 2(x + 5) - 3x$.

SOLUTION

$$1 - 3(4 - x) = 2(x + 5) - 3x$$

$$1 - 12 + 3x = 2x + 10 - 3x$$ Using the distributive law

$$-11 + 3x = -x + 10$$ Combining like terms; $1 - 12 = -11$ and $2x - 3x = -x$

$$-11 + 3x + x = 10$$ Adding x to both sides to get all x-terms on one side

$$-11 + 4x = 10$$ Combining like terms

$$4x = 10 + 11$$ Adding 11 to both sides to isolate the x-term

$$4x = 21$$ Simplifying

$$x = \frac{21}{4}$$ Dividing both sides by 4

Check:
$$1 - 3(4 - x) = 2(x + 5) - 3x$$

$$\begin{array}{c|c} 1 - 3\left(4 - \frac{21}{4}\right) & 2\left(\frac{21}{4} + 5\right) - 3\left(\frac{21}{4}\right) \\ 1 - 3\left(-\frac{5}{4}\right) & 2\left(\frac{41}{4}\right) - \frac{63}{4} \\ 1 + \frac{15}{4} & \frac{82}{4} - \frac{63}{4} \\ \frac{19}{4} \overset{?}{=} \frac{19}{4} & \quad \text{TRUE} \end{array}$$

We can also check using a calculator, as shown at left. The solution is $\frac{21}{4}$.

```
1−3(4−21/4)
                    4.75
2(21/4+5)−3*21/4
                    4.75
```

The equations in Examples 1–3 are **conditional**—that is, they are true for some values of x and false for other values of x. An **identity** is an equation like $x + 1 = x + 1$ that is true for all values of x. A **contradiction** is an equation like $x + 1 = x + 2$ that is never true.

A **formula** is an equation using two or more letters that represents a relationship between two or more quantities. A formula can be solved for a specified letter using the principles for solving equations.

EXAMPLE 4   The formula

$$A = \frac{a + b + c + d}{4}$$

gives the average A of four test scores a, b, c, and d. Solve for d.

SOLUTION   We have

$$A = \frac{a + b + c + d}{4}$$ We want the letter d alone.

$$4A = a + b + c + d$$ Multiplying by 4 to clear the fraction

$$4A - a - b - c = d.$$ Subtracting $a + b + c$ from (or adding $-a - b - c$ to) both sides. The letter d is now isolated.

We can also write this as $d = 4A - a - b - c$. This formula can be used to determine the test score needed to obtain a specified average if three tests have already been taken.

Solving Inequalities

A **solution of an inequality** is a replacement of the variable that makes the inequality true. The solutions of an inequality in one variable can be **graphed,** or represented by a drawing, on the number line. All points that are solutions are shaded. A parenthesis indicates an endpoint that is not a solution and a bracket indicates an endpoint that is a solution.

EXAMPLE 5   Graph each inequality: **(a)** $m \leq 2$; **(b)** $-1 \leq x < 4$.

SOLUTION

a) The solutions of $m \leq 2$ are shown on the number line by shading points to the left of 2 as well as the point at 2. The bracket at 2 indicates that 2 is a part of the graph (that is, it is a solution of $m \leq 2$).

b) In order to be a solution of the inequality $-1 \leq x < 4$, a number must be a solution of both $-1 \leq x$ and $x < 4$. The solutions are shaded on the number line, with a parenthesis indicating that 4 is not a solution and a bracket indicating that -1 is a solution.

Set-builder notation, p. 134

In Example 5, note that $m \leq 2$ and $-1 \leq x < 4$ are inequalities that describe a solution set. Since it is impossible to list all the solutions, we use **set-builder notation** or **interval notation** to write such sets.

Using set-builder notation, we write the solution set of Example 5(a) as

$$\{m \mid m \leq 2\},$$

read

"the set of all m such that m is less than or equal to 2."

Interval notation uses parentheses, ( ), and brackets, [ ], to describe a set of real numbers.

Interval notation, p. 134

Interval Notation	Set-builder Notation	Graph
(a, b) open interval	$\{x \mid a < x < b\}$	(a, b)
$[a, b]$ closed interval	$\{x \mid a \leq x \leq b\}$	$[a, b]$
$(a, b]$ half-open interval	$\{x \mid a < x \leq b\}$	$(a, b]$
$[a, b)$ half-open interval	$\{x \mid a \leq x < b\}$	$[a, b)$
(a, ∞)	$\{x \mid x > a\}$	
$[a, \infty)$	$\{x \mid x \geq a\}$	
$(-\infty, a)$	$\{x \mid x < a\}$	
$(-\infty, a]$	$\{x \mid x \leq a\}$	

Equivalent inequalities, p. 136

Thus, for example, the solution set of Example 5(a), in interval notation, is $(-\infty, 2]$.

As with equations, our goal when solving inequalities is to isolate the variable on one side. We use principles that enable us to write **equivalent inequalities**—inequalities having the same solution set. The addition principle is similar to the addition principle for equations; the multiplication principle contains an important difference.

The addition principle for inequalities, p. 136

The multiplication principle for inequalities, p. 138

The Addition and Multiplication Principles for Inequalities

The Addition Principle

For any real numbers a, b, and c,

$$a < b \quad \text{is equivalent to} \quad a + c < b + c, \qquad \text{and}$$
$$a > b \quad \text{is equivalent to} \quad a + c > b + c.$$

The Multiplication Principle

For any real numbers a and b, and for any *positive* number c,

$$a < b \quad \text{is equivalent to} \quad ac < bc, \qquad \text{and}$$
$$a > b \quad \text{is equivalent to} \quad ac > bc.$$

For any real numbers a and b, and for any *negative* number c,

$$a < b \quad \text{is equivalent to} \quad ac > bc, \qquad \text{and}$$
$$a > b \quad \text{is equivalent to} \quad ac < bc.$$

Similar statements hold for $\leq$ and $\geq$.

Note that when we multiply both sides of an inequality by a negative number, we must reverse the direction of the inequality symbol in order to have an equivalent inequality.

EXAMPLE 6   Solve $-2x \geq 5$ and then graph the solution.

SOLUTION   We have

$$-2x \geq 5$$

$$\frac{-2x}{-2} \leq \frac{5}{-2} \qquad \text{Multiplying by } -\frac{1}{2} \text{ or dividing by } -2$$
$$\text{The symbol must be reversed!}$$

$$x \leq -\frac{5}{2}.$$

Any number less than or equal to $-\frac{5}{2}$ is a solution. The graph is as follows:

The solution set is $\left\{x \,\middle|\, x \leq -\frac{5}{2}\right\}$, or $\left(-\infty, -\frac{5}{2}\right]$.

We can use the addition and multiplication principles together to solve inequalities. We can also combine like terms, remove parentheses, and clear fractions and decimals.

▸**EXAMPLE 7**   Solve: $2 - 3(x + 5) > 4 - 6(x - 1)$.

SOLUTION   We have

$$2 - 3(x + 5) > 4 - 6(x - 1)$$

$2 - 3x - 15 > 4 - 6x + 6$	Using the distributive law to remove parentheses
$-3x - 13 > -6x + 10$	Simplifying
$-3x + 6x > 10 + 13$	Adding $6x$ and 13, to get all x-terms on one side and the remaining terms on the other side
$3x > 23$	Combining like terms
$x > \frac{23}{3}.$	Multiplying by $\frac{1}{3}$. The inequality symbol stays the same because $\frac{1}{3}$ is positive.

The solution set is $\left\{x \mid x > \frac{23}{3}\right\}$, or $\left(\frac{23}{3}, \infty\right)$.   ◢

Problem Solving

One of the most important uses of algebra is as a tool for problem solving. The following five steps can be used to help solve problems of many types.

Problem solving, Section 2.5

Five Steps for Problem Solving in Algebra

1. *Familiarize* yourself with the problem.
2. *Translate* to mathematical language. (This often means write an equation.)
3. *Carry out* some mathematical manipulation. (This often means *solve* an equation.)
4. *Check* your possible answer in the original problem.
5. *State* the answer clearly.

▸**EXAMPLE 8**   Kitchen Cabinets.   Cherry kitchen cabinets cost 10% more than oak cabinets. Shelby Custom Cabinets designs a kitchen using $7480 worth of cherry cabinets. How much would the same kitchen cost using oak cabinets?

SOLUTION

Familiarization step, p. 119

Percent, Section 2.4

1. **Familiarize.** The *Familiarize* step is often the most important of the five steps, and may require a significant amount of time. Sometimes it helps to make a drawing or a table, make a guess and check it, or look up further information. For this problem, we could review percent notation. We could also make a guess. Let's suppose that the oak cabinets cost $6500. Then the cherry cabinets would cost 10% more, or an additional $(0.10) \times (\$6500) = \$650$. Altogether the cherry cabinets would cost $\$6500 + \$650 = \$7150$. Since $\$7150 \neq \$7480$, our guess is incorrect, but we see that 10% of the price of the oak cabinets must be added to the price of the oak cabinets to get the price of the cherry cabinets. We let c = the cost of the oak cabinets.

2. **Translate.** What we learned in the *Familiarize* step leads to the translation of the problem to an equation.

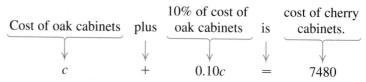

	Cost of oak cabinets	plus	10% of cost of oak cabinets	is	cost of cherry cabinets.
	c	$+$	$0.10c$	$=$	7480

3. **Carry out.** We solve the equation:

$$c + 0.10c = 7480$$
$$1c + 0.10c = 7480 \qquad \text{Writing } c \text{ as } 1c \text{ before combining terms}$$
$$1.10c = 7480 \qquad \text{Combining like terms}$$
$$c = \frac{7480}{1.10} \qquad \text{Dividing by 1.10}$$
$$c = 6800.$$

4. **Check.** We check in the wording of the stated problem: Cherry cabinets cost 10% more, so the additional cost is

$$10\% \text{ of } \$6800 = 0.10(\$6800) = \$680.$$

The total cost of the cherry cabinets is then

$$\$6800 + \$680 = \$7480,$$

which is the amount stated in the problem.

5. **State.** The oak cabinets would cost $6800.

Sometimes the translation of a problem is an inequality.

EXAMPLE 9   Long-Distance Telephone Usage.   Elyse pays a flat rate of 15¢ per minute for long-distance telephone calls. The monthly charge for her local calls is $21.50. How many minutes can she spend calling long distance in a month and not exceed her telephone budget of $50?

SOLUTION

1. **Familiarize.** Suppose that Elyse spends 5 hr, or 300 min, making long-distance calls one month. Then her bill would be the local service charge plus the long-distance charges, or

$$\$21.50 + \$0.15(300) = \$66.50.$$

This exceeds $50, so we know that the number of long-distance minutes must be less than 300. We let $m =$ the number of minutes of long-distance calls in a month.

2. **Translate.** The *Familiarize* step helps us reword and translate.

Rewording:	The local service charge	plus	the long-distance charges	cannot exceed	$50.
Translating:	21.50	$+$	$0.15m$	$\leq$	50

Solving applications with inequalities, Section 2.7

3. **Carry out.** We solve the inequality:

$$21.50 + 0.15m \leq 50$$

$$0.15m \leq 28.50 \qquad \text{Subtracting 21.50 from both sides}$$

$$m \leq 190. \qquad \text{Dividing by 0.15. The inequality symbol stays the same.}$$

4. **Check.** As a partial check, note that the telephone bill for 190 min of long-distance charges is

$$\$21.50 + \$0.15(190) = \$50.$$

Since this does not exceed the $50 budget, and fewer minutes will cost even less, our answer checks. We also note that 190 is less than 300 min, as noted in the *Familiarize* step.

5. **State.** Elyse will not exceed her budget if she talks long distance for no more than 190 min.

R.2 EXERCISE SET

FOR EXTRA HELP

MathXL   MyMathLab   InterAct Math   AW Math Tutor Center   Student's Solutions Manual

Solve.

1. $y + 5 = 13$

2. $y + 7 = -3$

3. $-3 + m = 9$

4. $-11 = 4 + x$

5. $t + \frac{1}{3} = \frac{1}{4}$

6. $-\frac{2}{3} + p = \frac{1}{6}$

7. $-1.9 = x - 1.1$

8. $x + 4.6 = 1.7$

9. $3y = 13$

10. $2x = 18$

11. $-x = \frac{5}{3}$

12. $-y = -\frac{2}{5}$

13. $-\frac{2}{7}x = -12$

14. $-\frac{1}{4}x = 3$

15. $\dfrac{-t}{5} = 1$

16. $\dfrac{2}{3} = -\dfrac{z}{8}$

17. $3x + 7 = 13$

18. $4x + 3 = -1$

19. $3y - 10 = 15$

20. $12 = 5y - 18$

21. $4x + 7 = 3 - 5x$

22. $2x = 5 + 7x$

23. $2x - 7 = 5x + 1 - x$

24. $a + 7 - 2a = 14 + 7a - 10$

25. $\frac{2}{5} + \frac{1}{3}t = 5$

26. $-\frac{5}{6} + t = \frac{1}{2}$

27. $x + 0.45 = 2.6x$

28. $1.8x + 0.16 = 4.2 - 0.05x$

29. $8(3 - m) + 7 = 47$

30. $2(5 - m) = 5(6 + m)$

31. $4 - (6 + x) = 13$

32. $18 = 9 - (3 - x)$

33. $2 + 3(4 + c) = 1 - 5(6 - c)$

34. $b + (b + 5) - 2(b - 5) = 18 + b$

35. $0.1(a - 0.2) = 1.2 + 2.4a$

36. $\frac{2}{3}\left(\frac{1}{2} - x\right) + \frac{5}{6} = \frac{3}{2}\left(\frac{2}{3}x + 1\right)$

37. $A = lw$, for l

38. $A = lw$, for w

39. $p = 30q$, for q

40. $d = 20t$, for t

41. $I = \dfrac{P}{V}$, for P

42. $b = \dfrac{A}{h}$, for A

43. $q = \dfrac{p + r}{2}$, for p

44. $q = \dfrac{p - r}{2}$, for r

45. $A = \pi r^2 + \pi r^2 h$, for π

46. $ax + by = c$, for a

Determine whether each number is a solution of the given inequality.

47. $x \le -5$

    a) 5   **b)** -5

    c) 0   **d)** -10

48. $y > 0$

    a) -1 **b)** 1

    c) 0   **d)** 100

Solve and graph. Write each answer in set-builder notation and in interval notation.

49. $x + 3 \le 15$ **50.** $y + 7 < -10$

51. $m - 17 > -5$ **52.** $x + 9 \ge -8$

53. $2x \ge -3$ **54.** $-\frac{1}{2}n \le 4$

55. $-5t > 15$ **56.** $3x > 10$

Solve. Write each answer in set-builder notation and in interval notation.

57. $2y - 7 > 13$ **58.** $2 - 6y \le 18$

59. $6 - 5a \le a$ **60.** $4b + 7 > 2 - b$

61. $2(3 + 5x) \ge 7(10 - x)$ **62.** $2(x + 5) < 8 - 3x$

63. $\frac{2}{3}(6 - x) < \frac{1}{4}(x + 3)$ **64.** $\frac{2}{3}t + \frac{8}{9} \ge \frac{4}{6} - \frac{1}{4}t$

65. $0.7(2 + x) \ge 1.1x + 5.75$

66. $0.4x + 5.7 \le 2.6 - 3(1.2x - 7)$

Solve. Use the five-step problem-solving process.

67. Three less than the sum of 2 and some number is 6. What is the number?

68. Five times some number is 10 less than the number. What is the number?

69. The sum of two consecutive even integers is 34. Find the numbers.

70. The sum of three consecutive integers is 195. Find the numbers.

71. *Reading.* Leisa is reading a 500-page book. She has twice as many pages to read as she has already finished. How many pages has she already read?

72. *Mowing.* It takes Caleb 50 min to mow his lawn. It will take him three times as many minutes to finish as he has already spent mowing. How long has he already spent mowing?

73. *Triangles.* The second angle of a triangle is one third as large as the first. The third angle is 5° more than the first. Find the measure of the second angle.

74. *Perimeter of a Rectangle.* The perimeter of a rectangle is 28 cm. The width is 5 cm less than the length. Find the width and the length.

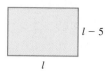

75. *Water Usage.* Rural Water Company charges a monthly service fee of $9.70 plus a volume charge of $2.60 for every hundred cubic feet of water used. How much water was used if the monthly bill is $33.10?

76. *Telephone Bills.* Brandon pays $4.95 a month for a long-distance telephone service that offers a flat rate of 7¢ per minute. One month his total long-distance telephone bill was $10.69. How many minutes of long-distance telephone calls were made that month?

77. *Sale Prices.* A can of tomatoes is on sale at 20% off for 64¢. What is the normal selling price of the tomatoes?

78. *Plywood.* The price of a piece of plywood rose 5% to $42. What was the original price of the plywood?

79. *Practice.* Dierdre's basketball coach requires each team member to average at least 15 min a day shooting baskets. One week Dierdre spent 10 min, 20 min, 5 min, 0 min, 25 min, and 15 min shooting baskets. How long must she practice shooting baskets on the seventh day if she is to meet the requirement?

80. *Perimeter of a Garden.* The perimeter of Garry's rectangular garden cannot exceed the 100 ft of fencing that he purchased. He wants the length to be twice the width. What widths of the garden will meet these conditions?

81. *Meeting Costs.* The Winds charges a $75 cleaning fee plus $45 an hour for the use of its meeting room. Complete Consultants has budgeted $200 to rent a room for a seminar. For how many hours can they rent the meeting room at The Winds?

82. *Meeting Costs.* Spring Haven charges a $15 setup fee, a $30 cleanup fee, and $50 an hour for the use of its meeting room. For what lengths of time will Spring Haven's room be less expensive than the room at The Winds (see Exercise 81)?

R.3 Introduction to Graphing and Functions

Points and Ordered Pairs ■ Graphs and Slope ■
Linear Equations ■ Functions

Points and Ordered Pairs

We can represent, or graph, pairs of numbers such as $(2, -5)$ on a plane. To do so, we use two perpendicular number lines called **axes.** The axes cross at a point called the **origin.** Arrows on the axes show the positive directions.

The order of the **coordinates,** or numbers in a pair, is important. The **first coordinate** indicates horizontal position and the **second coordinate** indicates vertical position. Such pairs of numbers are called **ordered pairs.** Thus the ordered pairs $(1, -2)$ and $(-2, 1)$ correspond to different points, as shown in the figure below.

The axes divide the plane into four regions, or **quadrants,** as indicated by Roman numerals in the figure below. Points on the axes are not considered to be in any quadrant. The horizontal axis is often labeled the x-axis, and the vertical axis the y-axis.

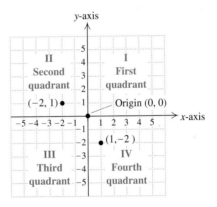

Graphs and Slope

When an equation contains two variables, solutions must be ordered pairs. Unless stated otherwise, the first number in each pair replaces the variable that occurs first alphabetically.

EXAMPLE 1 Determine whether $(1, 4)$ is a solution of $y - x = 3$.

SOLUTION We substitute 1 for x and 4 for y since x occurs first alphabetically:

$$\frac{y - x = 3}{4 - 1 \mid 3}$$
$$3 \overset{?}{=} 3 \quad \text{TRUE}$$

Since $3 = 3$ is true, the pair $(1, 4)$ *is* a solution.

A curve or line that represents all the solutions of an equation is called its **graph.**

▶**EXAMPLE 2**   Graph: $y = -2x + 1$.

SOLUTION   We select a value for x, calculate the corresponding value of y, and form an ordered pair.

If $x = 0$, then $y = -2 \cdot 0 + 1 = 1$, and $(0, 1)$ is a solution. Repeating this step, we find other ordered pairs and list the results in a table. We then plot the points corresponding to the pairs. They appear to form a straight line, so we draw a line through the points.

$$y = -2x + 1$$

x	y	(x, y)
0	1	$(0, 1)$
-1	3	$(-1, 3)$
3	-5	$(3, -5)$
1	-1	$(1, -1)$

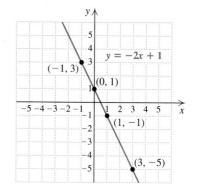

To graph using a graphing calculator, we let $y_1 = -2x + 1$, as shown at left. ◢

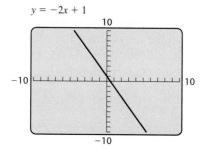

$y = -2x + 1$

The graph in Example 2 is a straight line. An equation whose graph is a straight line is a **linear equation.** The *rate of change* of y with respect to x is called the **slope** of a graph. A linear graph has constant slope. It can be found using any two points on a line.

Slope   The *slope* of the line containing points (x_1, y_1) and (x_2, y_2) is given by

$$m = \frac{\text{change in } y}{\text{change in } x} = \frac{\text{rise}}{\text{run}} = \frac{y_2 - y_1}{x_2 - x_1}.$$

▶**EXAMPLE 3**   Find the slope of the line containing the points $(-2, 1)$ and $(3, -4)$.

SOLUTION   From $(-2, 1)$ to $(3, -4)$, the change in y, or the rise, is $-4 - 1$, or -5. The change in x, or the run, is $3 - (-2)$, or 5. Thus

$$\text{Slope} = \frac{\text{change in } y}{\text{change in } x} = \frac{\text{rise}}{\text{run}} = \frac{-4 - 1}{3 - (-2)} = \frac{-5}{5} = -1. \quad ◢$$

The slope of a line indicates the direction and steepness of its slant. The larger the absolute value of the slope, the steeper the line. The direction of the slant is indicated by the sign of the slope, as shown in the figures below.

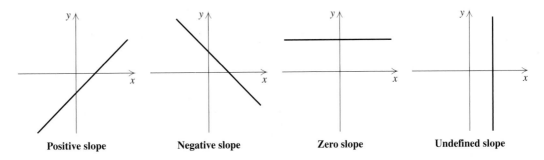

| Positive slope | Negative slope | Zero slope | Undefined slope |

x-intercept, p. 188

y-intercept, p. 188

The **x-intercept** of a line, if it exists, is the point at which the graph crosses the *x*-axis. To find an *x*-intercept, we replace *y* with 0 and calculate *x*.

The **y-intercept** of a line, if it exists, is the point at which the graph crosses the *y*-axis. To find a *y*-intercept, we replace *x* with 0 and calculate *y*.

Linear Equations

Any equation that can be written in the **standard form** $Ax + By = C$ is linear. Linear equations can also be written in other forms.

Forms of Linear Equations

Standard form: $Ax + By = C$

Slope–intercept form: $y = mx + b$

Point–slope form: $y - y_1 = m(x - x_1)$

The slope and *y*-intercept of a line can be read from the slope–intercept form of the line's equation.

Slope and *y*-intercept

For the graph of any equation $y = mx + b$,

- the slope is *m*, and
- the *y*-intercept is $(0, b)$.

EXAMPLE 4   Find the slope and the *y*-intercept of the line given by the equation $4x - 3y = 9$.

SOLUTION   We write the equation in slope–intercept form $y = mx + b$:

$$4x - 3y = 9 \qquad \text{We must solve for } y.$$
$$-3y = -4x + 9 \qquad \text{Adding } -4x \text{ to both sides}$$
$$y = \tfrac{4}{3}x - 3. \qquad \text{Dividing both sides by } -3$$

The slope is $\tfrac{4}{3}$ and the *y*-intercept is $(0, -3)$.

If we know that the graph of an equation is a straight line, we can plot two points on the line and draw the line through those points. The intercepts are often convenient points to use.

EXAMPLE 5   Graph $2x - 5y = 10$ using intercepts.

SOLUTION   To find the *x*-intercept, we let $y = 0$ and solve for *x*:

$$2x - 5 \cdot 0 = 10 \qquad \text{Replacing } y \text{ with } 0$$
$$2x = 10$$
$$x = 5.$$

To find the *y*-intercept, we let $x = 0$ and solve for *y*:

$$2 \cdot 0 - 5y = 10 \qquad \text{Replacing } x \text{ with } 0$$
$$-5y = 10$$
$$y = -2.$$

Thus the *x*-intercept is $(5, 0)$ and the *y*-intercept is $(0, -2)$. The graph is a line, since $2x - 5y = 10$ is in the form $Ax + By = C$. It passes through these two points.

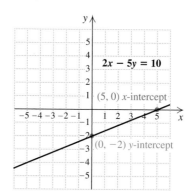

Alternatively, if we know a point on the line and its slope, we can plot the point and "count off" its slope to locate another point on the line.

▶ **EXAMPLE 6**   Graph: $y = -\dfrac{1}{2}x + 3$.

SOLUTION   The equation is in slope–intercept form, so we can read the slope and y-intercept directly from the equation.

Slope: $-\dfrac{1}{2}$

y-intercept: $(0, 3)$

We plot the y-intercept and use the slope to find another point.

Another way to write the slope is $\dfrac{-1}{2}$. This means for a run of 2 units, there is a negative rise, or a fall, of 1 unit. Starting at $(0, 3)$, we move 2 units in the positive horizontal direction and then 1 unit down, to locate the point $(2, 2)$. Then we draw the graph. A third point can be calculated and plotted as a check.

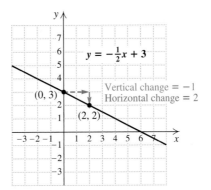

Horizontal and vertical lines intersect only one axis.

Horizontal and Vertical Lines

Horizontal Line	*Vertical Line*
$y = b$	$x = a$
y-intercept $(0, b)$	x-intercept $(a, 0)$
Slope is 0	Undefined slope
Example: $y = -3$	Example: $x = 2$

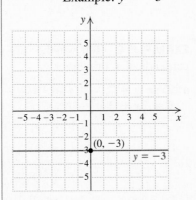

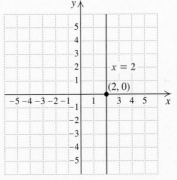

If we know the slope of a line and the coordinates of a point on the line, we can find an equation of the line, using either the slope–intercept equation $y = mx + b$, or the point–slope equation $y - y_1 = m(x - x_1)$.

EXAMPLE 7   Find the slope–intercept equation of a line given the following:

a) The slope is 2, and the *y*-intercept is $(0, -5)$.

b) The graph contains the points $(-2, 1)$ and $(3, -4)$.

SOLUTION

a) Since the slope and the *y*-intercept are given, we use the slope–intercept equation:

$$y = mx + b$$
$$y = 2x - 5. \qquad \text{Substituting 2 for } m \text{ and } -5 \text{ for } b$$

b) To use the point–slope equation, we need a point on the line and its slope. The slope can be found from the points given:

$$m = \frac{1 - (-4)}{-2 - 3} = \frac{5}{-5} = -1.$$

Either point can be used for (x_1, y_1). Using $(-2, 1)$, we have

$$y - y_1 = m(x - x_1)$$
$$y - 1 = -1(x - (-2)) \qquad \begin{array}{l}\text{Substituting } -2 \text{ for } x_1, \\ 1 \text{ for } y_1, \text{ and } -1 \text{ for } m\end{array}$$
$$y - 1 = -(x + 2)$$
$$y - 1 = -x - 2$$
$$y = -x - 1. \qquad \text{This is in slope–intercept form.}$$

We can tell from the slopes of two lines whether they are parallel or perpendicular.

Parallel lines, p. 227

Perpendicular lines, p. 228

Parallel and Perpendicular Lines

Two lines are parallel if they have the same slope.
Two lines are perpendicular if the product of the slopes is -1.

EXAMPLE 8   Tell whether the graphs of each pair of lines are parallel, perpendicular, or neither.

a) $2x - y = 7,$
      $y = 2x + 3$

b) $4x - y = 8,$
      $x + 4y = 8$

SOLUTION

a) The slope of $y = 2x + 3$ is 2. To find the slope of $2x - y = 7$, we solve for y:

$$2x - y = 7$$
$$-y = -2x + 7$$
$$y = 2x - 7.$$

The slope of $2x - y = 7$ is also 2. Since the slopes are equal, the lines are parallel.

b) We solve both equations for y in order to determine the slopes of the lines:

$$4x - y = 8$$
$$-y = -4x + 8$$
$$y = 4x - 8.$$

The slope of $4x - y = 8$ is 4.
For the second line, we have

$$x + 4y = 8$$
$$4y = -x + 8$$
$$y = -\tfrac{1}{4}x + 2.$$

The slope of $x + 4y = 8$ is $-\tfrac{1}{4}$. Since $4 \cdot \left(-\tfrac{1}{4}\right) = -1$, the lines are perpendicular.

Functions

Correspondence, p. 254

A **correspondence** from one set to another is a pairing of one element of the first set with one element of a second set. Such a match can be written as an ordered pair. One type of correspondence used extensively in mathematics is called a **function.**

Function, p. 255

Domain, p. 254

Range, p. 254

> **Function**   A *function* is a correspondence between a first set, called the *domain,* and a second set, called the *range,* such that each member of the domain corresponds to *exactly one* member of the range.

If one element of the domain is paired with more than one element of the range, the correspondence is not a function. So, for example, a correspondence containing the ordered pairs $(1, 3)$ and $(1, 5)$ is not a function.

Function notation, p. 260

Output, p. 260

Input, p. 260

A function is often labeled with a letter, such as f or g. The notation $f(x)$—read "f of x," "f at x," or "the value of f at x"—represents the **output** that corresponds to the **input** x. For example, if a function f contains the ordered pair $(0, -5)$, we can write

$$f(0) = -5.$$

Many functions are described by equations. To find a function value, we substitute a given number or expression for the variable in the equation.

▼**EXAMPLE 9**   For $f(x) = 2x^2 - x + 4$, find $f(-1)$.

SOLUTION   We have

$$f(-1) = 2(-1)^2 - (-1) + 4$$
$$= 2(1) + 1 + 4$$
$$= 7.$$

When the domain and the range of a function are sets of numbers, we can graph the function. If a graph contains two or more points with the same first coordinate, that graph cannot represent a function.

Vertical-line test, p. 259

The Vertical-Line Test   A graph represents a function if it is not possible to draw a vertical line that intersects the graph more than once.

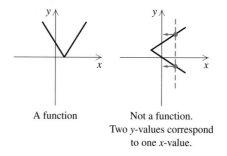

A function   Not a function.
                  Two *y*-values correspond
                  to one *x*-value.

Restrictions on domain, p. 264

The domains of some functions are described along with the equation for the function. Sometimes the domain of a function used in an application is limited by the application. If a domain is not specified, we assume that the domain is the set of all real numbers for which the function value can be computed.

▼**EXAMPLE 10**   Determine the domain of each of the following functions.

a) $f(x) = 2x^2 + x - 10$ **b)** $g(x) = \dfrac{1}{x + 2}$

c) $h(x) = \sqrt{7 - x}$

SOLUTION   We look for any restrictions on the domain—that is, any numbers for which the function value cannot be computed.

a) We can compute $2x^2 + x - 10$ for any number that replaces x. Therefore, the domain of f is $(-\infty, \infty)$, or $\{x \mid x$ is a real number$\}$, or $\mathbb{R}$. We can confirm this by examining a graph of $f(x) = 2x^2 + x - 10$, as shown at left.

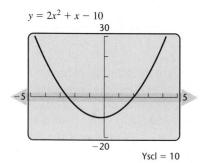

$y = 2x^2 + x - 10$

Yscl = 10

b) The expression $1/(x + 2)$ cannot be computed when the denominator, $x + 2$, is 0. We set $x + 2$ equal to 0 and solve:

$$x + 2 = 0$$
$$x = -2.$$

Thus, -2 is *not* in the domain of g, but all other real numbers are. The domain of g is $\{x \mid x \text{ is a real number } and \; x \neq -2\}$. This is confirmed by the graph shown below.

$y = 1/(x + 2)$

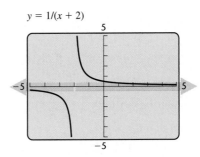

c) Since the square root of a negative number is not a real number, only inputs for which $7 - x$ is nonnegative are in the domain of the function. We solve an inequality:

$$7 - x \geq 0$$
$$7 \geq x.$$

We see that all numbers less than or equal to 7 will be in the domain. The domain of h is $(-\infty, 7]$, or $\{x \mid x \leq 7\}$. The graph of $h(x) = \sqrt{7 - x}$, as shown below, confirms this result.

$y = \sqrt{(7 - x)}$

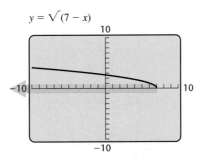

R.3 EXERCISE SET

1. Plot these points.

$(2, -3), (5, 1), (0, 2), (-1, 0),$
$(0, 0), (-2, -5), (-1, 1), (1, -1)$

2. Plot these points.

$(0, -4), (-4, 0), (5, -2), (2, 5),$
$(3, 3), (-3, -1), (-1, 4), (0, 1)$

In which quadrant is each point located?

3. $(2, 1)$ **4.** $(-2, 5)$

5. $(3, -2.6)$ **6.** $(-1.7, -5.9)$

7. First coordinates are positive in quadrants
_____ and _____.

8. Second coordinates are negative in quadrants
_____ and _____.

Determine whether each equation has the given ordered pair as a solution.

9. $y = 2x - 5;\ (1, 3)$

10. $4x + 3y = 8;\ (-1, 4)$

11. $a - 5b = -3;\ (2, 1)$

12. $c = d + 1;\ (1, 2)$

Graph by hand.

13. $y = x + 5$ **14.** $y = 3 - x$

15. $y = -\frac{1}{2}x + 3$ **16.** $y = 2x + 5$

 Graph using a graphing calculator.

17. $y = x^2 - 7$ **18.** $y = 2x^2 - x + 3$

19. $y = |x + 3|$ **20.** $y = \sqrt{3 - x}$

21. $y = \sqrt{x^2 + 1}$ **22.** $y = |x^2 - x - 1|$

Find the slope of the line containing each given pair of points.

23. $(3, 6)$ and $(2, 5)$

24. $(-1, 7)$ and $(-3, 4)$

25. $(0, 3)$ and $(1, -2)$

26. $(-3, 8)$ and $(2, -7)$

27. $\left(-2, -\frac{1}{2}\right)$ and $\left(5, -\frac{1}{2}\right)$

28. $(6.8, 7.5)$ and $(6.8, -3.2)$

Find the slope and the y-intercept of each equation.

29. $y = 2x - 5$ **30.** $y = 4x - 8$

31. $2x + y = 1$ **32.** $x - 2y = 3$

Find the intercepts. Then graph.

33. $3 - y = 2x$ **34.** $2x + 5y = 10$

35. $y = 3x + 5$ **36.** $y = -x + 7$

37. $3x - 2y = 6$ **38.** $2y + 1 = x$

Determine the coordinates of the y-intercept of each equation. Then graph the equation.

39. $y = 2x - 5$ **40.** $y = -\frac{5}{4}x - 3$

41. $2y + 4x = 6$ **42.** $3y + x = 4$

Find the slope of each line, and graph.

43. $y = 4$ **44.** $x = -5$

45. $x = 3$ **46.** $y = -1$

Find the slope–intercept equation of a line given the conditions.

47. The slope is $\frac{1}{3}$ and the y-intercept is $(0, 1)$.

48. The slope is -1 and the y-intercept is $(0, -5)$.

49. The graph contains the points $(0, 3)$ and $(-1, 4)$.

50. The graph contains the points $(5, 1)$ and $(8, 0)$.

Determine whether each pair of lines is parallel, perpendicular, or neither.

51. $x + y = 5,$
$\quad x - y = 1$

52. $2x + y = 3,$
$\quad y = 4 - 2x$

53. $2x + 3y = 1,$
$\quad 2x - 3y = 5$

54. $y = \frac{1}{3}x - 7,$
$\quad y + 3x = 1$

Find the function values.

55. $g(x) = \frac{1}{3}x + 7$
    a) $g(-3)$ **b)** $g(4)$ **c)** $g(a+1)$

56. $h(x) = \dfrac{1}{x+2}$
    a) $h(0)$ **b)** $h(2)$ **c)** $h(-2)$

57. $f(x) = 5x^2 + 2x + 3$
    a) $f(-1)$ **b)** $f(0)$ **c)** $f(2a)$

58. $f(t) = 6 - t^2$
    a) $f(-1)$ **b)** $f(1)$ **c)** $f(3a)$

Determine whether each of the following is the graph of a function.

59.
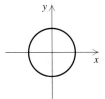

60.
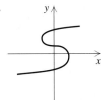

61.
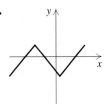

62.
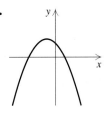

Determine the domain of each function.

63. $f(x) = \dfrac{x}{x-3}$

64. $g(x) = \sqrt{x}$

65. $g(x) = 2x + 3$

66. $f(x) = |x + 3|$

67. $f(x) = \sqrt{\frac{1}{2}x + 3}$

68. $g(x) = \dfrac{1}{2x+1}$

R.4   Systems of Equations in Two Variables

Solving Systems Graphically   ■   Solving Systems Algebraically   ■
Applications   ■   Solving Equations by Graphing

A **system of equations** is a set of two or more equations that are to be solved simultaneously. An ordered pair is a solution of a system of equations in two variables if it makes *all* of the equations in the system true. We can solve systems of equations in two variables graphically or algebraically.

Solving Systems Graphically

Intersect, p. 287

The solution of a system of equations is given by the point of intersection of the graphs of the equations.

EXAMPLE 1   Solve by graphing:

$$2x + y = 1,$$
$$x + y = -2.$$

SOLUTION

BY HAND

We graph each equation. The point of intersection has coordinates that make *both* equations true. Apparently, $(3, -5)$ is the solution. Because solving by graphing is not always accurate, we must check. As shown below, $(3, -5)$ is indeed the solution.

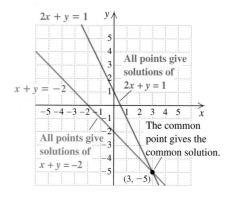

Check:

$$\begin{array}{c} 2x + y = 1 \\ \hline 2 \cdot 3 + (-5) \mid 1 \\ 1 \stackrel{?}{=} 1 \quad \text{TRUE} \end{array} \qquad \begin{array}{c} x + y = -2 \\ \hline 3 + (-5) \mid -2 \\ -2 \stackrel{?}{=} -2 \quad \text{TRUE} \end{array}$$

WITH A GRAPHING CALCULATOR

We first solve each equation for y:

$$\begin{aligned} 2x + y &= 1, & x + y &= -2 \\ y &= 1 - 2x; & y &= -2 - x. \end{aligned}$$

Then we enter and graph both equations using the same viewing window.

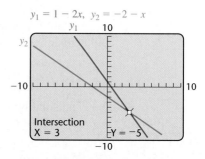

The point of intersection is $(3, -5)$. This ordered pair makes both equations true, so it is the solution of the system.

The graphs of two equations can intersect at one point, can be parallel with no points of intersection, or can have the same graph. Thus systems of two equations can have one solution, no solution, or an infinite number of solutions.

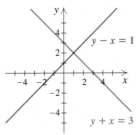

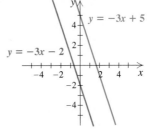

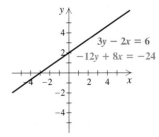

Graphs intersect at one point.
The system is *consistent* and has one solution. Since neither equation is a multiple of the other, they are *independent*. The solution is (1, 2).

Graphs are parallel.
The system is *inconsistent* because there is no solution. Since the equations are not equivalent, they are *independent*. The solution set is ∅.

Equations have the same graph.
The system is *consistent* and has an infinite number of solutions. The equations are *dependent* since they are equivalent. The solution set is $\{(x, y) | 3y - 2x = 6\}$.

Solving Systems Algebraically

Solving by substitution, Section 4.2

Algebraic (nongraphical) methods for solving systems yield exact answers. The **substitution method** relies on having a variable isolated.

▶ **EXAMPLE 2**   Solve the system

$$x + y = 4, \qquad (1)$$
$$x - y = 1. \qquad (2)$$

We have numbered the equations for easy reference.

SOLUTION   First, we select an equation and solve for one variable. Here we choose to solve for x in equation (2):

$$x - y = 1 \qquad (2)$$
$$x = y + 1. \qquad (3) \qquad \text{Adding } y \text{ to both sides}$$

Equation (3) says that x and $y + 1$ name the same number. Thus we can substitute $y + 1$ for x in equation (1):

$$x + y = 4 \qquad \text{Equation (1)}$$
$$(y + 1) + y = 4. \qquad \text{Substituting } y + 1 \text{ for } x$$

We solve this last equation, using methods learned earlier:

$$(y + 1) + y = 4$$
$$2y + 1 = 4 \qquad \text{Removing parentheses and combining like terms}$$
$$2y = 3 \qquad \text{Subtracting 1 from both sides}$$
$$y = \tfrac{3}{2}. \qquad \text{Dividing by 2}$$

We now return to the original pair of equations and substitute $\frac{3}{2}$ for y in either equation so that we can solve for x. For this problem, calculations are slightly easier if we use equation (3):

$$x = y + 1 \qquad \text{Equation (3)}$$
$$= \tfrac{3}{2} + 1 \qquad \text{Substituting } \tfrac{3}{2} \text{ for } y$$
$$= \tfrac{3}{2} + \tfrac{2}{2} = \tfrac{5}{2}.$$

We obtain the ordered pair $\left(\frac{5}{2}, \frac{3}{2}\right)$. A check ensures that it is a solution:

Check:

$x + y = 4$		$x - y = 1$	
$\tfrac{5}{2} + \tfrac{3}{2}$	4	$\tfrac{5}{2} - \tfrac{3}{2}$	1
$\tfrac{8}{2}$		$\tfrac{2}{2}$	
$4 \overset{?}{=} 4$ TRUE		$1 \overset{?}{=} 1$ TRUE	

Since $\left(\frac{5}{2}, \frac{3}{2}\right)$ checks, it is the solution.

Solving by elimination, Section 4.3

A second algebraic method, the **elimination method,** makes use of the addition principle: If $a = b$, then $a + c = b + c$. To use the elimination method, we multiply, if necessary, so that the coefficients of one variable are opposites. Then we add the equations, eliminating that variable.

EXAMPLE 3 Solve the system

$$2x + 3y = 17, \qquad (1)$$
$$5x + 7y = 29. \qquad (2)$$

SOLUTION We multiply so that the x-terms are eliminated.

$2x + 3y = 17, \rightarrow$ Multiplying both sides by 5 $\rightarrow$ $10x + 15y = 85$ The coefficients of x, 10

$5x + 7y = 29 \rightarrow$ Multiplying both sides by -2 $\rightarrow$ $\underline{-10x - 14y = -58}$ and -10, are opposites.

$\qquad\qquad\qquad\qquad\qquad\qquad\qquad\qquad 0 + \ y = \ 27$ Adding

$$\qquad\qquad\qquad\qquad\qquad\qquad\qquad\qquad\qquad y = \ 27$$

Next, we substitute to find x:

$$2x + 3 \cdot 27 = 17 \qquad \text{Substituting 27 for } y \text{ in equation (1)}$$
$$2x + 81 = 17$$
$$\left.\begin{array}{l} 2x = -64 \\ x = -32. \end{array}\right\} \quad \text{Solving for } x$$

Check:

$2x + 3y = 17$		$5x + 7y = 29$	
$2(-32) + 3(27)$	17	$5(-32) + 7(27)$	29
$-64 + 81$		$-160 + 189$	
$17 \overset{?}{=} 17$ TRUE		$29 \overset{?}{=} 29$ TRUE	

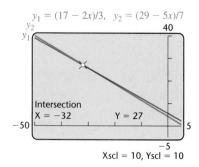

$y_1 = (17 - 2x)/3, \ y_2 = (29 - 5x)/7$

Intersection
X = −32   Y = 27

Xscl = 10, Yscl = 10

We can also check using a graphing calculator, as shown in the figure at left. We obtain $(-32, 27)$, or $x = -32$, $y = 27$, as the solution.

When solving algebraically, we can tell whether a system has no solutions or an infinite number of solutions.

> **Rules for Special Cases**   When solving a system of two linear equations in two variables:
>
> 1. If an identity is obtained, such as $0 = 0$, then the system has an infinite number of solutions. The equations are dependent and, since a solution exists, the system is consistent.
> 2. If a contradiction is obtained, such as $0 = 7$, then the system has no solution. The system is inconsistent.

Applications

Many problems that can be translated using two variables can be solved using systems of equations.

Mixture problems, p. 321

EXAMPLE 4   Blending Coffees.   The Java Joint wants to mix organic Kenyan beans that sell for $8.25 per pound with organic Venezuelan beans that sell for $9.50 per pound to form a 50-lb batch of Morning Blend that sells for $9.00 per pound. How many pounds of each type of bean should go into the blend?

SOLUTION

1. **Familiarize.** This problem is an example of a **mixture problem.** We know the price per pound of two kinds of coffee and the weight and price per pound of the batch of Morning Blend being made. Note that we can easily calculate the value of the batch of Morning Blend by multiplying 50 lb times $9.00 per pound. We let $k =$ the number of pounds of organic Kenyan coffee used and $v =$ the number of pounds of organic Venezuelan coffee used.

2. **Translate.** Since a 50-lb batch is being made, we must have

$$k + v = 50.$$

To find a second equation, we consider the total value of the 50-lb batch. That value must be the same as the value of the Kenyan beans and the value of the Venezuelan beans that go into the blend.

Rewording: The value of the   plus   the value of the   is   the value of the
                Kenyan beans   Venezuelan beans   Morning Blend.

Translating: $k \cdot 8.25$ $+$ $v \cdot 9.50$ $=$ $50 \cdot 9.00$

This information can be presented in a table.

	Kenyan	**Venezuelan**	**Morning Blend**
Number of Pounds	k	v	50
Price per Pound	$8.25	$9.50	$9.00
Value of Beans	$8.25k$	$9.50v$	$50 \cdot 9$, or 450

$k + v = 50$

$8.25k + 9.50v = 450$

We have translated to a system of equations:

$$k + v = 50, \qquad (1)$$
$$8.25k + 9.50v = 450. \qquad (2)$$

3. **Carry out.** When equation (1) is solved for k, we have $k = 50 - v$. We then substitute $50 - v$ for k in equation (2):

$8.25(50 - v) + 9.50v = 450$	Solving by substitution
$412.50 - 8.25v + 9.50v = 450$	Using the distributive law
$1.25v = 37.50$	Combining like terms; subtracting 412.50 from both sides
$v = 30.$	Dividing both sides by 1.25

If $v = 30$, we see from equation (1) that $k = 20$.

4. **Check.** If 20 lb of organic Kenyan beans and 30 lb of organic Venezuelan beans are mixed, a 50-lb blend will result. The value of 20 lb of Kenyan beans is $20(\$8.25)$, or \$165. The value of 30 lb of Venezuelan beans is $30(\$9.50)$, or \$285, so the value of the blend is $\$165 + \$285 = \$450$. A 50-lb blend priced at \$9.00 a pound is also worth \$450, so our answer checks.

5. **State.** The Morning Blend should be made by combining 20 lb of organic Kenyan beans with 30 lb of organic Venezuelan beans.

When a problem deals with distance, speed (rate), and time, we use the following.

Distance, Rate, and Time Equations If r represents rate, t represents time, and d represents distance, then

$$d = rt, \qquad r = \frac{d}{t}, \quad \text{and} \quad t = \frac{d}{r}.$$

EXAMPLE 5 Marine Travel.   A Coast-Guard patrol boat travels 4 hr on a trip downstream with a 6-mph current. The return trip against the same current takes 5 hr. Find the speed of the boat in still water.

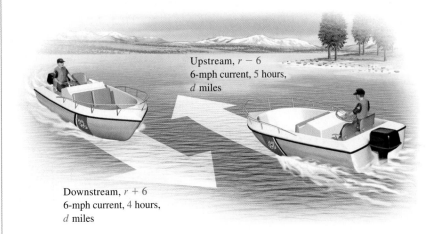

Upstream, $r - 6$
6-mph current, 5 hours,
d miles

Downstream, $r + 6$
6-mph current, 4 hours,
d miles

1. **Familiarize.** We first make a drawing and note that the distances are the same. We let $d =$ the distance, in miles, and $r =$ the speed of the boat in still water, in miles per hour. Then, when the boat is traveling downstream, its speed is $r + 6$ (the current helps the boat along). When it is traveling upstream, its speed is $r - 6$ (the current holds the boat back). We can organize the information in a table. We use the formula $d = rt$.

$$d \;\; = \;\; r \;\; \cdot \;\; t$$

	Distance	Rate	Time	
Downstream	d	$r + 6$	4	$\longrightarrow d = (r + 6)4$
Upstream	d	$r - 6$	5	$\longrightarrow d = (r - 6)5$

2. **Translate.** From each row of the table, we get an equation, $d = rt$:

$$d = 4r + 24, \qquad (1)$$
$$d = 5r - 30. \qquad (2)$$

3. **Carry out.** We solve the system by the substitution method:

$4r + 24 = 5r - 30$ **Substituting $4r + 24$ for d in equation (2)**

$\left. \begin{array}{l} 24 = r - 30 \\ 54 = r. \end{array} \right\}$ **Solving for r**

4. **Check.** If $r = 54$, then $r + 6 = 60$; and $60 \cdot 4 = 240$, the distance traveled downstream. If $r = 54$, then $r - 6 = 48$; and $48 \cdot 5 = 240$, the distance traveled upstream. The speed of the boat in still water, 54 mph, checks.

5. **State.** The speed of the boat in still water is 54 mph.

Solving Equations by Graphing

Solving equations graphically, Section 4.5

When two graphs intersect, that point of intersection represents a solution of *both* equations. We can use this fact to solve equations graphically.

EXAMPLE 6   Solve by graphing: $\frac{1}{2}x - 3 = 2(x + 7)$.

SOLUTION   We graph two equations, each corresponding to one side of the equation we wish to solve. For this equation, we graph $y_1 = \frac{1}{2}x - 3$ and $y_2 = 2(x + 7)$ using the same viewing window. We may need to adjust the dimensions of the viewing window in order to see the point of intersection.

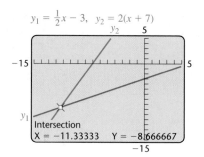

Using INTERSECT, we see that the point of intersection is approximately $(-11.33333, -8.666667)$. The x-coordinate of the point of intersection is the solution of the original equation. Note that this solution is approximate. To find the exact solution, we can solve algebraically. The solution is $-\frac{34}{3}$, or approximately -11.33333.

Intersect method, p. 335

Zero method, p. 338

Zero of a function, p. 338

We solved the equation in Example 6 using the **Intersect method.** To use the **Zero method,** we find the **zeros of a function,** or inputs that make an output zero.

EXAMPLE 7   Find any zeros of the function given by $f(x) = 3 - 2x$.

SOLUTION   A zero is an input x such that $f(x) = 0$:

$$f(x) = 0 \qquad \text{The output } f(x) \text{ must be 0.}$$
$$3 - 2x = 0 \qquad \text{Substituting}$$
$$\left.\begin{array}{l} 3 = 2x \\ \frac{3}{2} = x. \end{array}\right\} \quad \text{Solving for } x$$

Since $f\left(\frac{3}{2}\right) = 3 - 2 \cdot \frac{3}{2} = 0$, $\frac{3}{2}$ is the zero of f.

EXAMPLE 8   Solve by graphing: $3x + 1 = 5x - 7$.

SOLUTION   We use the Zero method, first getting 0 on one side of the equation:

$$3x + 1 = 5x - 7$$
$$-2x + 1 = -7 \qquad \text{Subtracting } 5x \text{ from both sides}$$
$$-2x + 8 = 0. \qquad \text{Adding 7 to both sides}$$

We let $f(x) = -2x + 8$, and find the **zero** of the function by locating the x-intercept of its graph.

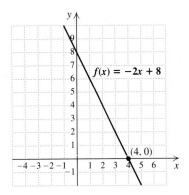

Since $f(4) = 0$, the zero of the function is 4. We check 4 in the original equation.

Check:
$$3x + 1 = 5x - 7$$

$3 \cdot 4 + 1$	$5 \cdot 4 - 7$
$12 + 1$	$20 - 7$

$$13 \overset{?}{=} 13 \qquad \text{TRUE}$$

The solution is 4.

R.4 EXERCISE SET

Solve each system using the indicated method. Be sure to check your solution. If a system has an infinite number of solutions, use set-builder notation to write the solution set, as illustrated in the third graph after Example 1. If a system has no solution, state this.

Solve each system graphically.

1. $x + y = 7,$
  $x - y = 1$

2. $x - y = 1,$
  $x + y = 3$

3. $y = -2x + 5,$
  $x + y = 4$

4. $y = 2x - 5,$
  $x + y = 4$

5. $y = x - 3,$
  $y = -2x + 3$

6. $y = -3x + 2,$
  $y = x - 2$

7. $4x - 20 = 5y,$
  $8x - 10y = 12$

8. $6x + 12 = 2y,$
  $6 - y = -3x$

9. $x = 6,$
  $y = -1$

10. $x = -2,$
  $y = 5$

11. $y = \frac{1}{5}x + 4,$
  $2y = \frac{2}{5}x + 8$

12. $y = \frac{1}{3}x + 2,$
  $y = \frac{1}{3}x - 7$

Solve each system using the substitution method.

13. $y = 5 - 4x,$
  $2x - 3y = 13$

14. $2y + x = 9,$
  $x = 3y - 3$

15. $3x + 5y = 3,$
  $x = 8 - 4y$

16. $9x - 2y = 3,$
  $3x - 6 = y$

17. $3s - 4t = 14,$
  $5s + t = 8$

18. $m - 2n = 16,$
  $4m + n = 1$

19. $4x - 2y = 6,$
  $2x - 3 = y$

20. $t = 4 - 2s,$
  $t + 2s = 6$

21. $x - 4y = 3,$
  $5x + 3y = 4$

22. $2a + 2b = 5,$
  $3a - b = 7$

23. $2x - 3 = y,$
  $y - 2x = 1$

24. $a - 2b = 3,$
  $3a = 6b + 9$

Solve each system using the elimination method.

25. $x + 3y = 7,$
  $-x + 4y = 7$

26. $2x + y = 6,$
  $x - y = 3$

27. $2x - y = -3,$
  $x + y = 9$

28. $x - 2y = 6,$
  $-x + 3y = -4$

29. $9x + 3y = -3,$
  $2x - 3y = -8$

30. $6x - 3y = 18,$
  $6x + 3y = -12$

31. $5x + 3y = 19,$
  $2x - 5y = 11$

32. $3x + 2y = 3,$
  $9x - 8y = -2$

33. $5r - 3s = 24,$
  $3r + 5s = 28$

34. $5x - 7y = -16,$
  $2x + 8y = 26$

35. $6s + 9t = 12,$
  $4s + 6t = 5$

36. $10a + 6b = 8,$
  $5a + 3b = 2$

37. $12x - 6y = -15,$
  $-4x + 2y = 5$

38. $8s + 12t = 16,$
  $6s + 9t = 12$

Solve using a system of equations.

39. The sum of two numbers is 89. One number is 3 more than the other. Find the numbers.

40. Find two numbers for which the sum is 108 and the difference is 50.

41. *Supplementary Angles.* Supplementary angles are angles for which the sum of their measures is 180°. Two angles are supplementary. One angle is 33° more than twice the other. Find the measure of each angle.

42. *Complementary Angles.* Complementary angles are angles for which the sum of their measures is 90°. Two angles are complementary. Their difference is 12°. Find the measure of each angle.

43. *Basketball Scoring.* In winning the Western Conference finals, the San Antonio Spurs once scored 76 of their points on a combination of 36 two- and three-pointers (*Source*: National Basketball Association). How many shots of each type did they make?

44. *Zoo Admissions.* From October 27 through April 2, the Bronx Zoo charges $8 for adults and $6 for children and seniors (*Source*: Bronx Zoo). One December day, a total of $2610 was collected from 394 admissions. How many adult admissions were there?

45. *Dance Lessons.* Jean charges $20 for a private tap lesson and $12 for a group class. One Wednesday, Jean earned $216 from 14 students. How many students of each type did Jean teach?

46. *Music and Movies.* Freedom Tunes sells used CDs for $5.99 and used videos for $7.99. Holly recently purchased a total of 14 CDs and videos for $93.86 (before tax). How many CDs and how many videos did she buy?

47. *Mixed Nuts.* A grocer wishes to mix peanuts worth $2.52 per pound with Brazil nuts worth $3.80 per pound to make 480 lb of a mixture worth $3.44 per pound. How much of each should be used?

48. *Mixed Nuts.* The Nuthouse has 10 kg of mixed cashews and pecans worth $8.40 per kilogram. Cashews alone sell for $8.00 per kilogram, and pecans sell for $9.00 per kilogram. How many kilograms of each are in the mixture?

49. *Production.* Streakfree window cleaner is 12% alcohol and Sunstream window cleaner is 30% alcohol. How much of each should be used to make 90 oz of a cleaner that is 20% alcohol? Complete the following table as part of the *Translate* step.

Type of Solution	Streakfree	Sunstream	Mixture
Amount of Solution	x	y	
Percent Alcohol	12%		20%
Amount of Alcohol in Solution		$0.3y$	

50. *Chemistry.* E-Chem Testing has a solution that is 80% base and another that is 30% base. A technician needs 200 L of a solution that is 62% base. The 200 L will be prepared by mixing the two solutions on hand. How much of each should be used?

51. *Car Travel.* Two cars leave Denver traveling in opposite directions. One car travels at a speed of 75 km/h and the other at 88 km/h. In how many hours will they be 489 km apart?

52. *Canoeing.* Darren paddled for 4 hr with a 6-km/h current to reach a campsite. The return trip against the same current took 16 hr. Find the speed of Darren's canoe in still water.

53. *Air Travel.* Rod is a pilot for Crossland Airways. He computes his flight time against a headwind for a trip of 2900 mi at 5 hr. The flight would take 4 hr and 50 min if the headwind were half as great. Find the headwind and the plane's air speed.

54. *Motorcycle Travel.* Sally and Rocky travel on motorcycles toward each other from Chicago and Indianapolis, which are about 350 km apart, and they are biking at rates of 110 km/h and 90 km/h. They started at the same time. In how many hours will they meet?

Find the zero of each function.

55. $f(x) = 7 - x$

56. $f(x) = \frac{1}{2}x + 2$

57. $f(x) = 2x + 9$

58. $f(x) = 3 - 4x$

Solve by graphing.

59. $2x - 5 = 3x + 1$

60. $x - 7 = 2x + 5$

61. $3x = x + 6$

62. $4x - 7 = x + 2$

63. $5x + 1 = 11x - 3$

64. $2x - 5 = 8x$

R.5 Polynomials

Exponents

We know that x^4 means $x \cdot x \cdot x \cdot x$ and that x^1 means x. Exponential notation is also defined for zero and negative exponents.

Zero and Negative Exponents For any real number a, $a \neq 0$,

$$a^0 = 1 \quad \text{and} \quad a^{-n} = \frac{1}{a^n}.$$

EXAMPLE 1 Simplify: **(a)** $(-36)^0$; **(b)** $(-2x)^0$.

SOLUTION

The exponent zero, p. 354

a) $(-36)^0 = 1$ since any number (other than 0 itself) raised to the 0 power is 1.

b) $(-2x)^0 = 1$ for any $x \neq 0$.

EXAMPLE 2 Write an equivalent expression using positive exponents.

a) x^{-2} **b)** $\dfrac{1}{x^{-2}}$ **c)** $7y^{-1}$

SOLUTION

Negative exponents, p. 361

a) $x^{-2} = \dfrac{1}{x^2}$ x^{-2} is the reciprocal of x^2.

b) $\dfrac{1}{x^{-2}} = x^{-(-2)} = x^2$ The reciprocal of x^{-2} is $x^{-(-2)}$, or x^2.

c) $7y^{-1} = 7\left(\dfrac{1}{y^1}\right) = \dfrac{7}{y}$ y^{-1} is the reciprocal of y^1.

The following properties hold for any integers m and n and any real numbers a and b, provided no denominators are 0 and 0^0 is not considered.

Properties of Exponents

The Product Rule: $\qquad\qquad a^m \cdot a^n = a^{m+n}$

The Quotient Rule: $\qquad\qquad \dfrac{a^m}{a^n} = a^{m-n}$

The Power Rule: $\qquad\qquad (a^m)^n = a^{mn}$

Raising a product to a power: $\;\;(ab)^n = a^n b^n$

Raising a quotient to a power: $\;\;\left(\dfrac{a}{b}\right)^n = \dfrac{a^n}{b^n}$

These properties are often used to simplify exponential expressions.

EXAMPLE 3   Simplify.

a) $(x^2y^{-1})(xy^{-3})$ 　　　**b)** $\dfrac{(3p)^3}{(3p)^{-2}}$ 　　　**c)** $\left(\dfrac{ab^2}{3c^3}\right)^{-4}$

The product rule, p. 352

The quotient rule, p. 353

The power rule, p. 355

Raising a product to a power, p. 356

Raising a quotient to a power, p. 357

SOLUTION

a)
$$(x^2y^{-1})(xy^{-3}) = x^2y^{-1}xy^{-3} \qquad \text{Using an associative law}$$
$$= x^2x^1y^{-1}y^{-3} \qquad \text{Using a commutative law; } x = x^1$$
$$= x^{2+1}y^{-1+(-3)} \qquad \text{Using the product rule: Adding exponents}$$
$$= x^3y^{-4}, \text{ or } \dfrac{x^3}{y^4}$$

b)
$$\dfrac{(3p)^3}{(3p)^{-2}} = (3p)^{3-(-2)} \qquad \text{Using the quotient rule: Subtracting exponents}$$
$$= (3p)^5$$
$$= 3^5p^5 \qquad \text{Raising each factor to the fifth power}$$
$$= 243p^5$$

c)
$$\left(\dfrac{ab^2}{3c^3}\right)^{-4} = \dfrac{(ab^2)^{-4}}{(3c^3)^{-4}} \qquad \text{Raising the numerator and the denominator to the } -4 \text{ power}$$
$$= \dfrac{a^{-4}(b^2)^{-4}}{3^{-4}(c^3)^{-4}} \qquad \text{Raising each factor to the } -4 \text{ power}$$
$$= \dfrac{a^{-4}b^{-8}}{3^{-4}c^{-12}} \qquad \text{Multiplying exponents}$$
$$= \dfrac{3^4c^{12}}{a^4b^8}, \text{ or } \dfrac{81c^{12}}{a^4b^8} \qquad \text{Rewriting without negative exponents}$$

Polynomials

Algebraic expressions like

$$2x^3 + 3x - 5, \qquad 4x, \qquad -7, \quad \text{and} \quad 2a^3b^2 + ab^3$$

are all examples of **polynomials.** All variables in a polynomial are raised to whole-number powers, and there are no variables in a denominator. The **terms** of a polynomial are separated by addition signs. The **degree of a term** is the number of variable factors in that term. The **leading term** of a polynomial is the term of highest degree. The **degree of a polynomial** is the degree of the leading term. A polynomial is written in *descending order* when the leading term appears first, followed by the term of next highest degree, and so on.

The number -2 in the term $-2y^3$ is called the **coefficient** of that term. The coefficient of the leading term is the **leading coefficient** of the polynomial. To illustrate this terminology, consider the polynomial

$$4y^2 - 8y^5 + y^3 - 6y + 7.$$

The *terms* are $4y^2$, $-8y^5$, y^3, $-6y$, and 7.

The *coefficients* are 4, -8, 1, -6, and 7.

The *degree of each term* is 2, 5, 3, 1, and 0.

The *leading term* is $-8y^5$ and the *leading coefficient* is -8.

The *degree of the polynomial* is 5.

Polynomials are classified by the number of terms and by degree.

A **monomial** has one term.	*Example*: $-2x^3y$
A **binomial** has two terms.	*Example*: $1.4x^2 - 10$
A **trinomial** has three terms.	*Example*: $x^2 - 3x - 6$
A **constant** polynomial has degree 0.	*Example*: 7
A **linear** polynomial has degree 1.	*Example*: $3x + 5$
A **quadratic** polynomial has degree 2.	*Example*: $5x^2 - x$
A **cubic** polynomial has degree 3.	*Example*: $x^3 + 2x^2 - \frac{1}{3}$
A **quartic** polynomial has degree 4.	*Example*: $-6x^4 - 2x^2 + 19$

Like, or *similar*, *terms* are either constant terms or terms containing the same variable(s) raised to the same power(s). Polynomials containing like terms can be simplified by *combining* those terms.

EXAMPLE 4   Combine like terms: $4x^2y + 2xy - x^2y + xy^2$.

SOLUTION   The like terms are $4x^2y$ and $-x^2y$. Thus we have

$$\begin{aligned} 4x^2y + 2xy - x^2y + xy^2 &= 4x^2y - x^2y + 2xy + xy^2 \\ &= 3x^2y + 2xy + xy^2. \end{aligned}$$

A polynomial can be evaluated by replacing the variable or variables with a number or numbers. We use parentheses when substituting negative numbers.

Evaluating a polynomial, p. 375

EXAMPLE 5   Evaluate $-a^2 + 2ab + 5b^2$ for $a = -1$ and $b = 3$.

SOLUTION   We replace a with -1 and b with 3 and calculate the value using the rules for order of operations:

$$-a^2 + 2ab + 5b^2 = -(-1)^2 + 2 \cdot (-1) \cdot 3 + 5 \cdot 3^2$$
$$= -1 - 6 + 45 = 38.$$

A function described by a polynomial is called a **polynomial function.**

EXAMPLE 6   Find $P(-2)$ for the polynomial function given by $P(x) = -x^3 + 3x - 7$.

SOLUTION   We substitute -2 for x and carry out the operations using the rules for order of operations:

$$P(-2) = -(-2)^3 + 3(-2) - 7$$
$$= -(-8) + (-6) - 7 \qquad \text{We cube the input before taking the opposite.}$$
$$= 8 - 6 - 7$$
$$= -5.$$

Polynomials can be added, subtracted, multiplied, and divided.

Addition and Subtraction of Polynomials

Addition of polynomials, Section 5.4

To add two polynomials, we write a plus sign between them and combine like terms.

EXAMPLE 7   Add: $(4x^3 + 3x^2 + 2x - 7) + (-5x^2 + x - 10)$.

SOLUTION

$$(4x^3 + 3x^2 + 2x - 7) + (-5x^2 + x - 10)$$
$$= 4x^3 + (3 - 5)x^2 + (2 + 1)x + (-7 - 10)$$
$$= 4x^3 - 2x^2 + 3x - 17$$

Opposite of a polynomial, p. 391

To find the **opposite of a polynomial,** we replace each term with its opposite. This process is also called *changing the sign* of each term. For example, the opposite of

$$3y^4 - 7y^2 - \tfrac{1}{3}y + 17$$

is

$$-(3y^4 - 7y^2 - \tfrac{1}{3}y + 17) = -3y^4 + 7y^2 + \tfrac{1}{3}y - 17.$$

Subtraction of polynomials, Section 5.4

To subtract polynomials, we add the opposite of the polynomial being subtracted.

EXAMPLE 8 Subtract: $(3a^4 - 2a + 7) - (-a^3 + 5a - 1)$.

SOLUTION

$$(3a^4 - 2a + 7) - (-a^3 + 5a - 1)$$
$$= 3a^4 - 2a + 7 + a^3 - 5a + 1 \qquad \text{Adding the opposite}$$
$$= 3a^4 + a^3 - 7a + 8 \qquad \text{Combining like terms}$$

Multiplication of Polynomials

Multiplication of polynomials, Section 5.5

To multiply two monomials, we multiply coefficients and then multiply variables using the product rule for exponents. To multiply a monomial and a polynomial, we multiply each term of the polynomial by the monomial, using the distributive property.

EXAMPLE 9 Multiply: $4x^3(3x^4 - 2x^3 + 7x - 5)$.

SOLUTION

$$\textit{Think:} \quad \underbrace{4x^3 \cdot 3x^4}_{} - \underbrace{4x^3 \cdot 2x^3}_{} + \underbrace{4x^3 \cdot 7x}_{} - \underbrace{4x^3 \cdot 5}_{}$$
$$4x^3(3x^4 - 2x^3 + 7x - 5) = 12x^7 \quad - \quad 8x^6 \quad + \quad 28x^4 \quad - \quad 20x^3$$

To multiply any two polynomials P and Q, we select one of the polynomials—say, P. We then multiply each term of P by every term of Q and combine like terms.

EXAMPLE 10 Multiply: $(2a^3 + 3a - 1)(a^2 - 4a)$.

SOLUTION It is often helpful to use columns for a long multiplication. We multiply each term at the top by every term at the bottom, write like terms in columns, and add the results.

$$
\begin{array}{r}
2a^3 \quad + 3a - 1 \\
a^2 - 4a \\
\hline
-8a^4 \qquad - 12a^2 + 4a \\
2a^5 \qquad + 3a^3 \quad - a^2 \\
\hline
2a^5 - 8a^4 + 3a^3 - 13a^2 + 4a
\end{array}
$$

Multiplying the top row by $-4a$

Multiplying the top row by a^2

Combining like terms. Be sure that like terms are lined up in columns.

We could multiply two binomials in the same manner in which we multiplied the polynomials in Example 10. However, by noting the pattern of the products formed, we can develop a method of multiplying two binomials more efficiently.

The FOIL Method   To multiply two binomials, $A + B$ and $C + D$, multiply the First terms AC, the Outer terms AD, the Inner terms BC, and then the Last terms BD. Then combine like terms, if possible.

$$(A + B)(C + D) = AC + AD + BC + BD$$

1. Multiply First terms: AC.
2. Multiply Outer terms: AD.
3. Multiply Inner terms: BC.
4. Multiply Last terms: BD.

$$\downarrow$$

FOIL

EXAMPLE 11   Multiply: $(3x + 4)(x - 2)$.

SOLUTION

$$(3x + 4)(x - 2) = 3x^2 - 6x + 4x - 8$$
$$= 3x^2 - 2x - 8 \quad \text{Combining like terms}$$

Special products occur so often that specific formulas or methods for computing them have been developed.

FOIL, p. 405

Multiplying sums and differences of two terms, p. 406

Squaring binomials, p. 408

Special Products   The product of a sum and a difference of the same two terms:

$$(A + B)(A - B) = \underbrace{A^2 - B^2}$$

This is called a *difference of squares*.

The square of a binomial:

$$(A + B)^2 = A^2 + 2AB + B^2$$
$$(A - B)^2 = A^2 - 2AB + B^2$$

EXAMPLE 12 Multiply: **(a)** $(x + 3y)(x - 3y)$; **(b)** $(x^3 + 2)^2$.

SOLUTION

$$(A + B)\ (A - B) = A^2 - B^2$$

a) $(x + 3y)(x - 3y) = x^2 - (3y)^2$ $A = x$ and $B = 3y$
$$= x^2 - 9y^2$$

$$(A + B)^2 = A^2 + 2 \cdot A \cdot B + B^2$$

b) $(x^3 + 2)^2 = (x^3)^2 + 2 \cdot x^3 \cdot 2 + 2^2$ $A = x^3$ and $B = 2$
$$= x^6 + 4x^3 + 4$$

Division of Polynomials

Division of polynomials, Section 5.8

Polynomial division is similar to division in arithmetic. To divide a polynomial by a monomial, we divide each term by the monomial.

EXAMPLE 13 Divide: $(3x^5 + 8x^3 - 12x) \div (4x)$.

SOLUTION This division can be written

$$\frac{3x^5 + 8x^3 - 12x}{4x} = \frac{3x^5}{4x} + \frac{8x^3}{4x} - \frac{12x}{4x} \qquad \text{Dividing each term by } 4x$$

$$= \frac{3}{4}x^{5-1} + \frac{8}{4}x^{3-1} - \frac{12}{4}x^{1-1} \qquad \begin{array}{l}\text{Dividing} \\ \text{coefficients and} \\ \text{subtracting} \\ \text{exponents}\end{array}$$

$$= \frac{3}{4}x^4 + 2x^2 - 3.$$

To check, we multiply the quotient by $4x$:

$$\left(\tfrac{3}{4}x^4 + 2x^2 - 3\right)4x = 3x^5 + 8x^3 - 12x. \qquad \textbf{The answer checks.}$$

To use long division, we write polynomials in descending order, including terms with 0 coefficients for missing terms. As shown below in Example 14, the procedure ends when the degree of the remainder is less than the degree of the divisor.

EXAMPLE 14 Divide: $(4x^3 - 7x + 1) \div (2x + 1)$.

SOLUTION The polynomials are already written in descending order, but there is no x^2-term in the dividend. We fill in $0x^2$ for that term.

$$2x + 1\overline{)4x^3 + 0x^2 - 7x + 1} \qquad \begin{array}{l}\textbf{Divide the first term of the dividend,} \\ 4x^3, \textbf{ by the first term in the divisor,} \\ 2x: 4x^3/(2x) = 2x^2.\end{array}$$

with $2x^2$ above, $\underline{4x^3 + 2x^2}$ — **Multiply $2x^2$ by the divisor, $2x + 1$.**

$-2x^2$ — **Subtract:** $(4x^3 + 0x^2) - (4x^3 + 2x^2) = -2x^2$.

Then we bring down the next term of the dividend, $-7x$.

$$\begin{array}{r} 2x^2 - x \\ 2x+1\overline{\smash{\big)}\ 4x^3 + 0x^2 - 7x + 1} \\ \underline{4x^3 + 2x^2} \\ -2x^2 - 7x \\ \underline{-2x^2 - x} \\ -6x \end{array}$$

— Divide the first term of $-2x^2 - 7x$ by the first term in the divisor: $-2x^2/(2x) = -x$.

— The $-7x$ has been "brought down."

— Multiply $-x$ by the divisor, $2x + 1$.

— Subtract: $(-2x^2 - 7x) - (-2x^2 - x) = -6x$.

Since the degree of the remainder, $-6x$, is *not* less than the degree of the divisor, we must continue dividing.

$$\begin{array}{r} 2x^2 - x - 3 \\ 2x+1\overline{\smash{\big)}\ 4x^3 + 0x^2 - 7x + 1} \\ \underline{4x^3 + 2x^2} \\ -2x^2 - 7x \\ \underline{-2x^2 - x} \\ -6x + 1 \\ \underline{-6x - 3} \\ 4 \end{array}$$

— Divide the first term of $-6x + 1$ by the first term in the divisor: $-6x/(2x) = -3$.

— The 1 has been "brought down."

— Multiply -3 by $2x + 1$.

— Subtract.

The answer is $2x^2 - x - 3$ with R 4, or

$$\text{Quotient} \longrightarrow 2x^2 - x - 3 + \frac{4}{2x+1}.$$

— Remainder
— Divisor

To check, we can multiply by the divisor and add the remainder:

$$(2x+1)(2x^2 - x - 3) + 4 = 4x^3 - 7x - 3 + 4$$
$$= 4x^3 - 7x + 1.$$

R.5 EXERCISE SET

FOR EXTRA HELP

MathXL   MyMathLab   InterAct Math   AW Math Tutor Center   Student's Solutions Manual

Simplify.

1. a^0, for $a = -25$

2. y^0, for $y = 6.97$

3. $4^0 - 4^1$

4. $8^1 - 8^0$

Write an equivalent expression using positive exponents. Then, if possible, simplify.

5. 8^{-2}

6. 2^{-5}

7. $(-2)^{-3}$

8. $(-3)^{-2}$

9. $(ab)^{-2}$

10. ab^{-2}

11. $\dfrac{1}{y^{-10}}$

12. $\dfrac{1}{x^{-t}}$

Write an equivalent expression using negative exponents.

13. $\dfrac{1}{y^4}$

14. $\dfrac{1}{a^2 b^3}$

15. $\dfrac{1}{x^t}$

16. $\dfrac{1}{n}$

Simplify. Write answers with positive exponents.

17. $x^5 \cdot x^8$

18. $a^4 \cdot a^{-2}$

19. $\dfrac{a}{a^{-5}}$

20. $\dfrac{p^{-3}}{p^{-8}}$

21. $\dfrac{(4x)^{10}}{(4x)^2}$

22. $\dfrac{a^2 b^9}{a^9 b^2}$

23. $(7^8)^5$

24. $(x^3)^{-7}$

25. $(x^{-2} y^{-3})^{-4}$

26. $(-2a^2)^3$

27. $\left(\dfrac{y^2}{4}\right)^3$

28. $\left(\dfrac{ab^2}{c^3}\right)^4$

29. $\left(\dfrac{2p^3}{3q^4}\right)^{-2}$

30. $\left(\dfrac{2}{x}\right)^{-5}$

Identify the terms of each polynomial.

31. $8x^3 - 6x^2 + x - 7$

32. $-a^2 b + 4a^2 - 8b + 17$

Determine the coefficient and the degree of each term in each polynomial. Then find the degree of each polynomial.

33. $18x^3 + 36x^9 - 7x + 3$ **34.** $-8y^7 + y + 19$

35. $-x^2 y + 4y^3 - 2xy$ **36.** $8 - x^2 y^4 + y^7$

Determine the leading term and the leading coefficient of each polynomial.

37. $-p^2 + 4 + 8p^4 - 7p$ **38.** $13 + 20t - 30t^2 - t^3$

Combine like terms. Write each answer in descending order.

39. $3x^3 - x^2 + x^4 + x^2$ **40.** $5t - 8t^2 + 4t^2$

41. $3 - 2t^2 + 8t - 3t - 5t^2 + 7$

42. $8x^5 - \frac{1}{3} + \frac{4}{5}x + 1 - \frac{1}{2}x$

Evaluate each polynomial for the given replacements of the variables.

43. $3x^2 - 7x + 10$, for $x = -2$

44. $-y + 3y^2 + 2y^3$, for $y = 3$

45. $a^2 b^3 + 2b^2 - 6a$, for $a = 2$ and $b = -1$

46. $2pq^3 - 5q^2 + 8p$, for $p = -4$ and $q = -2$

The distance s, in feet, traveled by a body falling freely from rest in t seconds is approximated by
$$s = 16t^2.$$

47. A pebble is dropped into a well and takes 3 sec to hit the water. How far down is the surface of the water?

48. An acorn falls from the top of an oak tree and takes 2 sec to hit the ground. How high is the tree?

Add or subtract, as indicated.

49. $(3x^3 + 2x^2 + 8x) + (x^3 - 5x^2 + 7)$

50. $(-6x^4 + 3x^2 - 16) + (4x^2 + 4x - 7)$

51. $(8y^2 - 2y - 3) - (9y^2 - 7y - 1)$

52. $(4t^2 + 6t - 7) - (t + 5)$

53. $(-x^2 y + 2y^2 + y) - (3y^2 + 2x^2 y - 7y)$

54. $(ab + x^2 y^2) + (2ab - x^2 y^2)$

Multiply.

55. $4x^2(3x^3 - 7x + 7)$

56. $a^2 b(a^3 + b^2 - ab - 2b)$

57. $(2a + y)(4a + b)$

58. $(x + 7y)(y - 3x)$

59. $(x + 7)(x^2 - 3x + 1)$

60. $(2x - 3)(x^2 - x - 1)$

61. $(x + 7)(x - 7)$

62. $(2x + 1)^2$

63. $(x + y)^2$

64. $(xy + 1)(xy - 1)$

65. $(2x^2 + 7)(3x^2 - 2)$

66. $(x^2 + 2)^2$

67. $(a - 3b)^2$

68. $(1.1x^2 + 5)(0.1x^2 - 2)$

69. $(6a - 5y)(7a + 3y)$

70. $(3p^2 - q^3)^2$

Divide and check.

71. $(3t^5 + 9t^3 - 6t^2 + 15t) \div (-3t)$

72. $(4x^5 + 10x^4 - 16x^2) \div (4x^2)$

73. $(15x^2 - 16x - 15) \div (3x - 5)$

74. $(x^3 - 2x^2 - 14x + 1) \div (x - 5)$

75. $(2x^3 - x^2 + 1) \div (x + 1)$

76. $(2x^3 + 3x^2 - 50) \div (2x - 5)$

77. $(5x^3 + 3x^2 - 5x) \div (x^2 - 1)$

78. $(2x^3 + 3x^2 + 6x + 10) \div (x^2 + 3)$

<table>
<tr><td>**R.6**</td><td>**Polynomial Factorizations and Equations**</td></tr>
</table>

Common Factors and Factoring by Grouping   ■   Factoring Trinomials   ■
Factoring Special Forms   ■   Solving Polynomial Equations

To *factor* a polynomial is to find an equivalent expression that is a product. To factor a monomial, we find two monomials whose product is equivalent to the original monomial. For example, three factorizations of $50x^6$ are $5 \cdot 10x^6$, $5x^3 \cdot 10x^3$, and $2x \cdot 25x^5$.

Common Factors and Factoring by Grouping

Factor, p. 453

If all the terms in a polynomial share a common factor, that factor can be "factored out" of the polynomial. Whenever you are factoring a polynomial with two or more terms, try to first find the largest common factor of the terms, if one exists.

Common factor, p. 453

EXAMPLE 1   Factor: $3x^6 + 15x^4 - 9x^3$.

SOLUTION   The largest factor common to 3, 15, and -9 is 3. The largest power of x common to x^6, x^4, and x^3 is x^3. Thus the largest common factor of the terms of the polynomial is $3x^3$. We factor as follows:

$$3x^6 + 15x^4 - 9x^3 = 3x^3 \cdot x^3 + 3x^3 \cdot 5x - 3x^3 \cdot 3 \qquad \text{Factoring each term}$$
$$= 3x^3(x^3 + 5x - 3). \qquad \text{Factoring out } 3x^3$$

Factorizations can always be checked by multiplying:

$$3x^3(x^3 + 5x - 3) = 3x^6 + 15x^4 - 9x^3.$$

A polynomial with two or more terms can be a common factor.

EXAMPLE 2   Factor: $3x^2(x - 2) + 5(x - 2)$.

SOLUTION   The binomial $x - 2$ is a factor of both $3x^2(x - 2)$ and $5(x - 2)$. Thus we have

$$3x^2(x - 2) + 5(x - 2) = (x - 2)(3x^2 + 5). \qquad \text{Factoring out the common factor, } x - 2$$

If a polynomial with four terms can be split into two groups of terms, and both groups share a common binomial factor, the polynomial can be factored. This method is known as **factoring by grouping.**

> **EXAMPLE 3**   Factor by grouping: $2x^3 + 6x^2 - x - 3$.
>
> **SOLUTION**   First, we consider the polynomial as two groups of terms, $2x^3 + 6x^2$ and $-x - 3$. Then we factor each group separately:
>
> $$2x^3 + 6x^2 - x - 3 = 2x^2(x + 3) - 1(x + 3)$$
>
> Factoring out $2x^2$ and -1 to give the common binomial factor, $x + 3$
>
> $$= (x + 3)(2x^2 - 1).$$
>
> The check is left to the student.

Not every polynomial with four terms is factorable by grouping. A polynomial that is not factorable is said to be **prime.**

Factoring Trinomials

Many trinomials that have no common factor can be written as the product of two binomials. We look first at trinomials of the form

$$x^2 + bx + c,$$

for which the leading coefficient is 1.

Factoring trinomials involves a trial-and-error process. In order for the product of two binomials to be $x^2 + bx + c$, the binomials must look like

$$(x + p)(x + q),$$

where p and q are constants that must be determined. We look for two numbers p and q whose product is c and whose sum is b.

> **EXAMPLE 4**   Factor.
>
> **a)** $x^2 + 10x + 16$ **b)** $x^2 - 8x + 15$
> **c)** $x^2 - 2x - 24$ **d)** $3t^2 - 33st + 84s^2$
>
> **SOLUTION**
>
> **a)** The factorization is of the form
>
> $$(x +\quad)(x +\quad).$$
>
> To find the constant terms, we need a pair of factors whose product is 16 and whose sum is 10. Since 16 is positive, its factors will have the same sign as 10—that is, we need consider only positive factors of 16.

We list the possible factorizations in a table and calculate the sum of each pair of factors.

Pairs of Factors of 16	Sums of Factors
1, 16	17
2, 8	10 ←
4, 4	8

The numbers we seek are 2 and 8.

The factorization of $x^2 + 10x + 16$ is $(x + 2)(x + 8)$. To check, we multiply.

Check: $(x + 2)(x + 8) = x^2 + 8x + 2x + 16 = x^2 + 10x + 16.$

b) For $x^2 - 8x + 15$, c is positive and b is negative. Therefore, the factors of 15 will be negative. Again, we list the possible factorizations in a table.

Pairs of Factors of 15	Sums of Factors
−1, −15	−16
−3, −5	−8 ←

The numbers we need are −3 and −5.

The factorization is $(x - 3)(x - 5)$. The check is left to the student.

Constant term negative, p. 466

c) For $x^2 - 2x - 24$, c is negative, so one factor of −24 will be negative and one will be positive. Since b is also negative, the negative factor must have the larger absolute value.

Pairs of Factors of −24	Sums of Factors
1, −24	−23
2, −12	−10
3, −8	−5
4, −6	−2 ←

The numbers we need are 4 and −6.

The factorization is $(x + 4)(x - 6)$.

Check: $(x + 4)(x - 6) = x^2 - 6x + 4x - 24 = x^2 - 2x - 24.$

d) Always look first for a common factor. There is a common factor, 3, which we factor out first:

$$3t^2 - 33st + 84s^2 = 3(t^2 - 11st + 28s^2).$$

Now we consider $t^2 - 11st + 28s^2$. Think of $28s^2$ as the "constant" term c and $-11s$ as the "coefficient" b of the middle term. We try to express $28s^2$ as the product of two factors whose sum is $-11s$. These factors are $-4s$ and $-7s$. Thus the factorization of $t^2 - 11st + 28s^2$ is

$$(t - 4s)(t - 7s). \qquad \text{This is not the entire factorization of } 3t^2 - 33st + 84s^2.$$

We now include the common factor, 3, and write

$$3t^2 - 33st + 84s^2 = 3(t - 4s)(t - 7s). \qquad \text{This is the factorization.}$$

Check: $\quad 3(t - 4s)(t - 7s) = 3(t^2 - 11st + 28s^2)$
$$= 3t^2 - 33st + 84s^2.$$

Factoring trinomials of the type $ax^2 + bx + c$, Section 6.3

When the leading coefficient of a trinomial is not 1, the number of trials needed to find a factorization can increase dramatically. We will consider two methods for factoring trinomials of the type $ax^2 + bx + c$: factoring by reversing FOIL and the *ac*-method.

To Factor $ax^2 + bx + c$ by Reversing FOIL

1. Factor out the largest common factor, if one exists. Here we assume none does.
2. Find two First terms whose product is ax^2:

$$(\blacksquare x + \quad)(\blacksquare x + \quad) = ax^2 + bx + c.$$
$$\text{FOIL}$$

3. Find two Last terms whose product is c:

$$(\quad x + \blacksquare)(\quad x + \blacksquare) = ax^2 + bx + c.$$
$$\text{FOIL}$$

4. Repeat steps (2) and (3) until a combination is found for which the sum of the Outer and Inner products is bx:

$$(\blacksquare x + \blacksquare)(\blacksquare x + \blacksquare) = ax^2 + bx + c.$$
$$\text{I} \qquad \text{FOIL}$$
$$\text{O}$$

5. Always check by multiplying.

Factoring with FOIL, p. 473

EXAMPLE 5   Factor: $20x^3 - 22x^2 - 12x$.

SOLUTION

1) First, we factor out the largest common factor, $2x$:

$$20x^3 - 22x^2 - 12x = 2x(10x^2 - 11x - 6).$$

2) Next, in order to factor the trinomial $10x^2 - 11x - 6$, we search for two terms whose product is $10x^2$. The possibilities are

$$(x + \quad)(10x + \quad) \quad \text{or} \quad (2x + \quad)(5x + \quad).$$

3) There are four pairs of factors of -6. Since the first terms of the binomials are different, the order of the factors is important. So there are eight possibilities for the last terms:

$$1, -6 \qquad -1, 6 \qquad 2, -3 \qquad -2, 3$$

and

$$-6, 1 \qquad 6, -1 \qquad -3, 2 \qquad 3, -2.$$

4) Since each of the eight possibilities from step (3) could be used in either of the two possibilities from step (2), there are $2 \cdot 8$, or 16, possible factorizations. We check the possibilities systematically until we find one that gives the correct factorization. Let's first try factors with $(2x + \quad)(5x + \quad)$.

Trial	Product	
$(2x + 1)(5x - 6)$	$10x^2 - 7x - 6$	←Wrong middle term
$(2x - 1)(5x + 6)$	$10x^2 + 7x - 6$	←Wrong middle term. Note that changing the signs in the binomials changed the sign of the middle term in the product.
$(2x + 2)(5x - 3)$	$10x^2 + 4x - 6$	←Wrong middle term. We need not consider $(2x - 2)(5x + 3)$.
$(2x - 6)(5x + 1)$	$10x^2 - 28x - 6$	←Wrong middle term. We need not consider $(2x + 6)(5x - 1)$.
$(2x - 3)(5x + 2)$	$10x^2 - 11x - 6$	←Correct middle term

We can stop when we find a correct factorization. Including the common factor $2x$, we now have

$$20x^3 - 22x^2 - 12x = 2x(2x - 3)(5x + 2).$$

This can be checked by multiplying.

With practice, some of the trials can be skipped or performed mentally.

The second method of factoring trinomials of the type $ax^2 + bx + c$ involves factoring by grouping.

Student Note

Both the third and the fourth trial factorizations in Example 5 contain a binomial with a common factor: $2x + 2 = 2(x + 1)$ and $2x - 6 = 2(x - 3)$. If all common factors are factored out as a first step, you can immediately eliminate any trial factorization containing a binomial with a common factor.

The *ac*-method, p. 476

To Factor $ax^2 + bx + c$ Using the *ac*-Method

1. Factor out the largest common factor, if one exists.
2. Multiply the leading coefficient a and the constant c.
3. Find a pair of factors of ac whose sum is b.
4. Rewrite the middle term, bx, as a sum or a difference using the factors found in step (3).
5. Factor by grouping.
6. Always check by multiplying.

R-54   CHAPTER R   Elementary Algebra Review

EXAMPLE 6   Factor: $7x^2 + 31x + 12$.

SOLUTION

1) There is no common factor (other than 1 or -1).

2) We multiply the leading coefficient, 7, and the constant, 12:

$$7 \cdot 12 = 84.$$

3) We look for a pair of factors of 84 whose sum is 31. Since both 84 and 31 are positive, we need consider only positive factors.

Pairs of Factors of 84	Sums of Factors
1, 84	85
2, 42	44
3, 28	31 ←————— $3 + 28 = 31$

4) Next, we rewrite $31x$ using the factors 3 and 28:

$$31x = 3x + 28x.$$

5) We now factor by grouping:

$$7x^2 + 31x + 12 = 7x^2 + 3x + 28x + 12 \qquad \textbf{Substituting } 3x + 28x \textbf{ for } 31x$$

$$= x(7x + 3) + 4(7x + 3)$$

$$= (7x + 3)(x + 4). \qquad \textbf{Factoring out the common factor, } 7x + 3$$

6) *Check:* $(7x + 3)(x + 4) = 7x^2 + 31x + 12$.

Factoring Special Forms

We can factor certain types of polynomials directly, without using trial and error.

Factoring Formulas

Perfect-square trinomial: $A^2 + 2AB + B^2 = (A + B)^2,$
$\qquad\qquad\qquad\qquad\qquad A^2 - 2AB + B^2 = (A - B)^2$

Difference of squares: $A^2 - B^2 = (A + B)(A - B)$

Sum of cubes: $A^3 + B^3 = (A + B)(A^2 - AB + B^2)$

Difference of cubes: $A^3 - B^3 = (A - B)(A^2 + AB + B^2)$

Before using the factoring formulas, it is important to check carefully that the expression being factored is indeed in one of the forms listed. Note that there is no factoring formula for the sum of two squares.

EXAMPLE 7 Factor: **(a)** $2x^2 - 2$; **(b)** $x^2y^2 + 20xy + 100$; **(c)** $p^3 - 64$; **(d)** $3y^2 + 27$.

SOLUTION

a) We first factor out a common factor, 2:

$$2x^2 - 2 = 2(x^2 - 1).$$

Looking at $x^2 - 1$, we see that it is a difference of squares, with $A = x$ and $B = 1$. The factorization is thus

$$2x^2 - 2 = 2(x^2 - 1) = 2(x + 1)(x - 1).$$
$$\underset{A^2 - B^2}{\uparrow\quad\uparrow}\qquad\underset{(A + B)\,(A - B)}{\uparrow\quad\uparrow\quad\uparrow\quad\uparrow}$$

b) First, we check for a common factor; there is none. The polynomial is a perfect-square trinomial, since x^2y^2 and 100 are squares; there is no minus sign before either square; and $20xy$ is $2 \cdot xy \cdot 10$, where xy and 10 are square roots of x^2y^2 and 100, respectively. The factorization is thus

$$x^2y^2 + 20xy + 100 = (xy)^2 + 2 \cdot xy \cdot 10 + 10^2 = (xy + 10)^2.$$
$$\underset{A^2\ +\ 2\ \cdot\ A\ \cdot\ B\ +\ B^2\ =\ (A\ +\ B)^2}{\uparrow\qquad\quad\uparrow\quad\uparrow\qquad\uparrow\qquad\quad\uparrow\qquad\uparrow}$$

c) This is a difference of cubes, with $A = p$ and $B = 4$:

$$p^3 - 64 = (p)^3 - (4)^3$$
$$= (p - 4)(p^2 + 4p + 16).$$

d) We factor out the common factor, 3:

$$3y^2 + 27 = 3(y^2 + 9).$$

Since $y^2 + 9$ is a sum of squares, no further factorization is possible.

A polynomial is said to be *factored completely* when no factor can be factored further.

EXAMPLE 8 Factor completely: $x^4 - 1$.

SOLUTION

$$x^4 - 1 = (x^2 + 1)(x^2 - 1) \qquad \text{Factoring a difference of squares}$$
$$= (x^2 + 1)(x + 1)(x - 1) \qquad \text{The factor } x^2 - 1 \text{ is itself a difference of squares.}$$

Solving Polynomial Equations

Whenever two polynomials are set equal to each other, we have a **polynomial equation.** We can solve polynomial equations graphically using the Intersect or Zero method.

▶ **EXAMPLE 9**   Solve: $3x^3 = 5x^2 + 2x$.

INTERSECT METHOD

We graph $y_1 = 3x^3$ and $y_2 = 5x^2 + 2x$ and look for any points of intersection. The x-coordinates of those points will be the solutions of the equation.

From the first graph, we see that one point of intersection is $(2, 24)$. In the second graph, we enlarge the section of the graph near the origin and see that $(-0.33333, -0.11111)$ and $(0, 0)$ are also points of intersection.

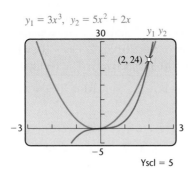

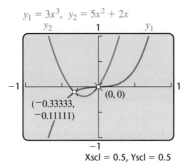

The solutions of the equation are -0.33333, 0, and 2.

ZERO METHOD

We rewrite the equation as $3x^3 - 5x^2 - 2x = 0$. We graph $f(x) = 3x^3 - 5x^2 - 2x$ and find the zeros of the function. Those zeros will be the solutions of the equation.

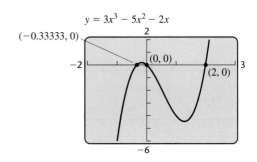

We see that $f(x) = 0$ when $x = -0.33333$, 0, or 2. Thus the solutions of the equation are -0.33333, 0, and 2.

◢

Note in Example 9 that one of the solutions found is an approximation. If we can factor the polynomial that describes the function, we can find exact solutions using the following principle.

> **The Principle of Zero Products**   An equation $ab = 0$ is true if and only if $a = 0$ or $b = 0$, or both. (A product is 0 if and only if at least one factor is 0.)

If we can write an equation as a product that equals 0, we can try to use the principle of zero products to solve the equation.

EXAMPLE 10   Solve.

a) $x^2 - 11x = 12$ **b)** $5x^2 + 10x + 5 = 0$

c) $9x^2 = 1$

SOLUTION

a) We must have 0 on one side of the equation before using the principle of zero products:

$$x^2 - 11x = 12$$
$$x^2 - 11x - 12 = 0 \qquad \text{Subtracting 12 from both sides}$$
$$(x - 12)(x + 1) = 0 \qquad \text{Factoring}$$
$$x - 12 = 0 \quad or \quad x + 1 = 0 \qquad \text{Using the principle of zero products}$$
$$x = 12 \quad or \qquad x = -1.$$

The solutions are 12 and -1. The check is left to the student.

b) We have

$$5x^2 + 10x + 5 = 0$$
$$5(x^2 + 2x + 1) = 0 \qquad \text{Factoring out a common factor}$$
$$5(x + 1)(x + 1) = 0 \qquad \text{Factoring completely}$$
$$x + 1 = 0 \quad or \quad x + 1 = 0 \qquad \text{Using the principle of zero products}$$
$$x = -1 \quad or \qquad x = -1.$$

There is only one solution, -1. The check is left to the student.

c) We have

$$9x^2 = 1$$
$$9x^2 - 1 = 0 \qquad \text{Subtracting 1 from both sides to get 0 on one side}$$
$$(3x + 1)(3x - 1) = 0 \qquad \text{Factoring a difference of squares}$$
$$3x + 1 = 0 \quad or \quad 3x - 1 = 0 \qquad \text{Using the principle of zero products}$$
$$3x = -1 \quad or \qquad 3x = 1$$
$$x = -\tfrac{1}{3} \quad or \qquad x = \tfrac{1}{3}.$$

The solutions are $\frac{1}{3}$ and $-\frac{1}{3}$. The check is left to the student.

Quadratic equations can be used to solve problems. One important result that uses squared quantities is the Pythagorean theorem. It relates the lengths of the sides of a **right triangle,** that is, a triangle with a 90° angle. The side opposite the 90° angle is called the **hypotenuse,** and the other sides are called the **legs.**

The Pythagorean Theorem   The sum of the squares of the legs of a right triangle is equal to the square of the hypotenuse:

$$a^2 + b^2 = c^2.$$

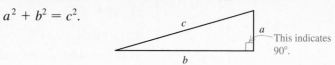

This indicates 90°.

EXAMPLE 11   Swing Sets.   The length of a slide on a swing set is 5 ft. The distance from the base of the ladder to the base of the slide is 1 ft more than the height of the ladder. Find the height of the ladder.

SOLUTION

1. **Familiarize.** We first make a drawing and let x = the height of the ladder, in feet. We know then that the other leg of the triangle is $x + 1$, since it is 1 ft longer than the ladder. The hypotenuse has length 5 ft.

2. **Translate.** Applying the Pythagorean theorem gives us

$$a^2 + b^2 = c^2$$
$$x^2 + (x + 1)^2 = 5^2. \qquad \text{Substituting}$$

3. **Carry out.** We solve the equation:

$$\begin{aligned}
x^2 + (x + 1)^2 &= 5^2 \\
x^2 + x^2 + 2x + 1 &= 25 && \text{Squaring } x + 1 \text{; squaring 5} \\
2x^2 + 2x + 1 &= 25 && \text{Combining like terms} \\
2x^2 + 2x - 24 &= 0 && \text{Getting 0 on one side} \\
2(x^2 + x - 12) &= 0 && \text{Factoring out a common factor} \\
2(x + 4)(x - 3) &= 0 && \text{Factoring a trinomial} \\
x + 4 = 0 \quad or \quad x - 3 &= 0 && \text{Using the principle of zero products} \\
x = -4 \quad or \qquad x &= 3.
\end{aligned}$$

4. **Check.** We know that the integer -4 is not a solution because the height of the ladder cannot be negative. When $x = 3$, the distance from the base of the ladder to the base of the slide is $x + 1 = 4$, and $3^2 + 4^2 = 5^2$. So the solution 3 checks.

5. **State.** The ladder is 3 ft high.

R.6 EXERCISE SET

Factor completely. If a polynomial is prime, state this.

1. $3x^3 + 6x^2 - 9x$

2. $x^2y^4 - 2xy^5 + 3x^3y^6$

3. $y^2 - 6y + 9$

4. $4z^2 - 25$

5. $2p^3(p + 2) + (p + 2)$

6. $6y^2 + y - 1$

7. $16x^2 + 25$

8. $y^3 - 1$

9. $8t^3 + 27$

10. $a^2b^2 + 24ab + 144$

11. $m^2 + 13m + 42$

12. $2x^3 - 6x^2 + x - 3$

13. $x^4 - 81$

14. $x^2 + x + 1$

15. $8x^2 + 22x + 15$

16. $4x^2 - 40x + 100$

17. $x^3 + 2x^2 - x - 2$

18. $(x + 2y)(x - 1) + (x + 2y)(x - 2)$

19. $0.001t^6 - 0.008$

20. $x^2 - 20 - x$

21. $-\frac{1}{16} + x^4$

22. $5x^8 - 5z^{16}$

23. $a^2 + 6a + 9 - y^2$

24. $t^6 - p^6$

25. $5mn + m^2 - 150n^2$

26. $\frac{1}{27} + x^3$

27. $24x^2y - 6y - 10xy$

28. $-3y^2 - 12y - 12$

29. $y^2 + 121 - 22y$

30. $p^2 - m^2 - 2mn - n^2$

Solve.

31. $(x - 2)(x + 7) = 0$

32. $(3x - 5)(7 - 4x) = 0$

33. $8x(4.7 - x) = 0$

34. $(x - 3)(x + 1)(2x - 9) = 0$

35. $x^2 = 100$

36. $8x^2 = 5x$

37. $4x^2 - 18x = 70$

38. $x^2 + 2x + 1 = 0$

39. $2x^3 - 10x^2 = 0$

40. $100x^2 = 81$

41. $(a + 1)(a - 5) = 7$

42. $d(d - 3) = 40$

43. $x^2 + 6x - 55 = 0$

44. $x^2 + 7x - 60 = 0$

45. $\frac{1}{2}x^2 + 5x + \frac{25}{2} = 0$

46. $3 + 10x^2 = 11x$

47. *Landscaping.*   A triangular flower garden is 3 ft longer than it is wide. The area of the garden is 20 ft². What are the dimensions of the garden?

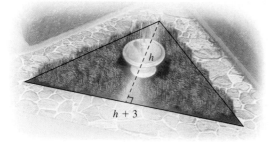

48. *Page Numbers.*   The product of the page numbers on two facing pages of a book is 156. Find the page numbers.

49. *Right Triangles.*   The hypotenuse of a right triangle is 17 ft. One leg is 1 ft shorter than twice the length of the other leg. Find the length of the legs.

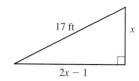

50. *Hiking.*   Jenna hiked 500 ft up a steep incline. Her global positioning unit indicated that her horizontal position had changed by 100 ft more than her vertical position had changed. What was the change in altitude?

Appendixes

A Working with Units

Calculating with Dimension Symbols ■ Making Unit Changes ■
Dimension Analysis

Quantities such as lengths, weights, and time are described using standard units such as centimeters, pounds, and seconds, respectively. It is important to know how to work with units when combining quantities, using formulas, and changing units. Analyzing the units used can even serve as a partial check of an answer to a problem.

Calculating with Dimension Symbols

Inside calculations, units can often be treated somewhat like variables. We can add, subtract, multiply, and divide quantities.

EXAMPLE 1 Rondel worked 8 hr on Friday and 9 hr on Saturday. How many hours did he work in all?

SOLUTION We can think of this addition as

$$8 \text{ hr} + 9 \text{ hr} = (8 + 9) \text{ hr} = 17 \text{ hr}.$$

Rondel worked 17 hr.

The addition in Example 1 is similar to the addition

$$8x + 9x = (8 + 9)x = 17x.$$

The 8 hr and 9 hr are, in this situation, "like terms." We could not have combined quantities with different units—say, 8 hr and 9 min—in this way.

When we substitute quantities into a formula, handling dimension symbols correctly should give us the correct unit for the answer.

EXAMPLE 2   Find the area of this square.

5 ft

5 ft

SOLUTION   The formula for the area of a square is

$$A = s^2.$$

We treat the symbol "ft" as though it were a variable and obtain the correct unit:

$$A = s^2$$
$$= (5 \text{ ft})^2 \quad \textbf{Substituting 5 ft for } s$$
$$= 5 \text{ ft} \cdot 5 \text{ ft}$$
$$= (5 \cdot 5)(\text{ft} \cdot \text{ft})$$
$$= 25 \text{ ft}^2.$$

EXAMPLE 3   Lorrie drove 150 km in 2 hr. What was her average speed?

SOLUTION   Since speed is given by

$$r = \frac{d}{t},$$

the average speed r is

$$r = \frac{d}{t} = \frac{150 \text{ km}}{2 \text{ hr}} = \frac{150}{2} \cdot \frac{\text{km}}{\text{hr}} = 75\frac{\text{km}}{\text{hr}}. \qquad d = 150 \text{ km}; t = 2 \text{ hr}$$

Lorrie's average speed was 75 km/h.

EXAMPLE 4   The amount of work done is given by the formula

$$work = force \times distance.$$

How much work is done when a 100-lb rock is lifted 10 ft?

SOLUTION   We substitute and calculate:

$$work = force \times distance = 100 \text{ lb} \times 10 \text{ ft} \qquad \textbf{Substituting}$$
$$= 100 \times 10 \times \text{lb} \times \text{ft}$$
$$= 1000 \text{ foot-pounds}. \qquad \textbf{A unit used to}$$
$$\textbf{represent work}$$
$$\textbf{is foot-pounds.}$$

EXAMPLE 5 The formula for the period T of a pendulum is $T = 2\pi\sqrt{\dfrac{l}{g}}$, where l is the length of the pendulum and g is the gravitational constant. Find the period of a pendulum of length 4 m.

SOLUTION Since l is given in meters, we use the value for g that is given in m/sec²: $g \approx 9.8$ m/sec². Substituting, we have

$$T = 2\pi\sqrt{\frac{l}{g}}$$
Recall that π is a constant with no unit.

$$\approx 2\pi\sqrt{\frac{4\text{ m}}{\dfrac{9.8\text{ m}}{\text{sec}^2}}}$$
Substituting

$$= 2\pi\sqrt{4\text{ m}\cdot\frac{\text{sec}^2}{9.8\text{ m}}}$$
Dividing by multiplying by the reciprocal

$$= 2\pi\sqrt{\frac{4}{9.8}}\cdot\sqrt{\frac{\text{m}\cdot\text{sec}^2}{\text{m}}}$$
$\sqrt{ab} = \sqrt{a}\,\sqrt{b}$

$$= 2\pi\sqrt{\frac{4}{9.8}}\cdot\sqrt{\text{sec}^2}$$
Removing a factor equal to 1: $\dfrac{\text{m}}{\text{m}} = 1$

$$\approx 4.0\text{ sec.}$$
$\sqrt{\text{sec}^2} = \text{sec}$

Making Unit Changes

It is often necessary to write a quantity using a different unit. Some unit equivalencies are given in the following tables.

American Units of Length	
1 ft	12 in.
1 yd	3 ft
1 mi	5280 ft

Metric Units of Length	
1 cm	10 mm
1 m	100 cm
1 km	1000 m

We can change units using substitution or multiplication by 1.

EXAMPLE 6 Convert 3 feet to inches.

SOLUTION

Substitution: We know that 1 ft = 12 in., so we substitute:

$$
\begin{aligned}
3\text{ ft} &= 3 \cdot 1\text{ ft} && \text{Rewriting to make the substitution clear}\\
&= 3 \cdot 12\text{ in.} && \text{Substituting 12 in. for 1 ft}\\
&= 36\text{ in.}
\end{aligned}
$$

Multiplication by 1: Two symbols for 1 can be written using the equivalency 1 ft = 12 in.:

$$\frac{1 \text{ ft}}{12 \text{ in.}} = 1 \quad \text{and} \quad \frac{12 \text{ in.}}{1 \text{ ft}} = 1.$$

We are converting from "ft" to "in." Since we want to eliminate the unit "ft" and end up with the unit "in.," we use the symbol for 1 with "ft" in the denominator and "in." in the numerator.

$$3 \text{ ft} = 3 \text{ ft} \cdot 1$$

$$= \frac{3 \text{ ft}}{1} \cdot \frac{12 \text{ in.}}{1 \text{ ft}} \qquad \text{"ft" in the numerator and "ft" in the denominator will cancel.}$$

$$= \frac{3 \cdot 12}{1 \cdot 1} \cdot \frac{\cancel{\text{ft}} \cdot \text{in.}}{\cancel{\text{ft}}} \qquad \frac{\text{ft}}{\text{ft}} = 1$$

$$= 36 \text{ in.}$$

We can multiply by 1 several times to make successive conversions.

EXAMPLE 7 The speed limit on route I-35 in Oklahoma City is $60\frac{\text{mi}}{\text{hr}}$. Convert this to $\frac{\text{ft}}{\text{sec}}$.

SOLUTION We write successive conversions, but do not carry out the operations until the last step.

$$60\frac{\text{mi}}{\text{hr}} = 60\frac{\text{mi}}{\text{hr}} \cdot \frac{5280 \text{ ft}}{1 \text{ mi}} \qquad \text{Converting mi to ft. The unit is now ft/hr.}$$

$$= 60\frac{\text{mi}}{\text{hr}} \cdot \frac{5280 \text{ ft}}{1 \text{ mi}} \cdot \frac{1 \text{ hr}}{60 \text{ min}} \qquad \text{Converting hr to min. The unit is now ft/min.}$$

$$= 60\frac{\text{mi}}{\text{hr}} \cdot \frac{5280 \text{ ft}}{1 \text{ mi}} \cdot \frac{1 \text{ hr}}{60 \text{ min}} \cdot \frac{1 \text{ min}}{60 \text{ sec}} \qquad \text{Converting min to sec. The unit is now ft/sec.}$$

$$= \frac{60 \cdot 5280 \cdot 1 \cdot 1}{1 \cdot 60 \cdot 60} \cdot \frac{\cancel{\text{mi}} \cdot \text{ft} \cdot \cancel{\text{hr}} \cdot \cancel{\text{min}}}{\cancel{\text{hr}} \cdot \cancel{\text{mi}} \cdot \cancel{\text{min}} \cdot \text{sec}} \qquad \frac{\text{mi} \cdot \text{hr} \cdot \text{min}}{\text{mi} \cdot \text{hr} \cdot \text{min}} = 1$$

$$= 88\frac{\text{ft}}{\text{sec}} \qquad \text{Carrying out the calculations}$$

The speed limit is $88\frac{\text{ft}}{\text{sec}}$.

Dimension Analysis

By noting the units used in calculations, we can help assure the accuracy of our work. Some quantities may need to be written with different units in order to answer a problem correctly.

EXAMPLE 8 Alanna is driving at a speed of 65 mph. How far will she travel in 10 min?

SOLUTION We use the formula $d = rt$. Since the speed is given in miles per *hour*, we must write the time traveled, 10 min, in terms of hours.

$$d = rt = \frac{65 \text{ mi}}{\text{hr}} \cdot 10 \text{ min} \qquad \textbf{Substituting}$$

$$= \frac{65 \text{ mi}}{\text{hr}} \cdot \frac{10 \text{ min}}{1} \cdot \frac{1 \text{hr}}{60 \text{ min}} \qquad \textbf{Converting min to hr}$$

$$= \frac{650}{60} \text{ mi} = 10\frac{5}{6} \text{ mi} \qquad \frac{\text{min} \cdot \text{hr}}{\text{min} \cdot \text{hr}} = 1$$

Alanna will travel $10\dfrac{5}{6}$ mi.

Some formulas are derived using specific units. These units must be used in the calculations.

EXAMPLE 9 Body mass index (BMI), one measure of appropriate weight to height, is given by

$$\text{BMI} = \frac{w}{h^2} \times 703,$$

where w is weight, in pounds, and h is height, in inches. The index is rounded to the nearest whole number and does not have a unit. Jolene is 5 ft 3 in. tall and weighs 130 lb. What is her BMI?

SOLUTION We begin by examining the units used in the formula. Weights must be in pounds and height in inches. Jolene's weight is given in pounds, but we must convert her height to inches before substituting in the formula:

$$5 \text{ ft 3 in.} = 5 \text{ ft} + 3 \text{ in.} \qquad \textbf{Writing as an addition}$$

$$= 5(12 \text{ in.}) + 3 \text{ in.} \qquad \textbf{Substituting 12 in. for 1 ft}$$

$$= 63 \text{ in.}$$

We substitute 130 for w and 63 for h and calculate:

$$\text{BMI} = \frac{w}{h^2} \times 703 = \frac{130}{63^2} \times 703 \approx 23.$$

Jolene's BMI is 23.

Sometimes we know that the result should be stated in certain units. For example, measurements of lengths are given in linear units, such as cm. Areas are measured in square units, such as cm^2, and volumes are measured in cubic units, such as cm^3. If the answer we get has the correct units, this provides a partial check of our calculations. Analysis of units used can even help us check the formulas we use.

EXAMPLE 10   Chad needs to calculate the area of a circular flower garden with a radius of 5 ft. He is certain that one of two formulas he remembers from geometry gives the area of a circle:

$$X = 2\pi r \quad \text{and} \quad X = \pi r^2.$$

Which could be the formula for the area of a circle?

SOLUTION   Area is always expressed in square units. Chad can substitute 5 ft for r in both formulas and check the units.

Substituting in $X = 2\pi r$: $X = 2\pi(5 \text{ ft}) = 10\pi \text{ ft}$ This is a measure of length.

Substituting in $X = \pi r^2$: $X = \pi(5 \text{ ft})^2 = 25\pi \text{ ft}^2$ This is a measure of area.

Since the second formula gives the answer in square feet, $X = \pi r^2$ could be the formula for the area of a circle.

Working with Units

- Write all length quantities with the same units, all time quantities with the same units, and so on.
- Use units specified by any formulas.
- Check the answer units to see if they are appropriate.

A EXERCISE SET

Perform the indicated operations.

1. $12 \text{ ft} + 8 \text{ ft}$
2. $350 \text{ mph} + 20 \text{ mph}$
3. $19 \text{ g} - 8 \text{ g}$
4. $14.2 \text{ lb} - 16 \text{ lb}$
5. $16 \text{ ft}^3 + 8 \text{ ft}^3$
6. $9 \text{ m}^2 - 3 \text{ m}^2$
7. $(4 \text{ cm})(3 \text{ cm})$
8. $(2 \text{ in.})(5 \text{ in.})$
9. $36 \text{ ft} \cdot \dfrac{1 \text{ yd}}{3 \text{ ft}}$
10. $55\dfrac{\text{mi}}{\text{hr}} \cdot 4 \text{ hr}$
11. $\dfrac{3 \text{ ft} \cdot 8 \text{ lb}}{6 \text{ sec}}$
12. $\dfrac{60 \text{ men} \cdot 8 \text{ hr}}{20 \text{ day}}$

Find the average speeds, given total distance and total time.

13. 160 km, 8 hr
14. 90 mi, 6 hr
15. 8.8 cm, 2 days
16. 1038 ft, 0.1 sec

Evaluate each formula using the given values.

17. $A = l \cdot w$; $l = 16 \text{ cm}, w = 3 \text{ cm}$ (Area of a rectangle)
18. $V = s^3$; $s = 10 \text{ ft}$ (Volume of a cube)

19. $C = 2\pi r$; $r = 4$ in.
(Circumference of a circle)

20. $A = \pi r^2$; $r = 4$ in.
(Area of a circle)

21. $s = \frac{1}{2}at^2$; $a = 5$ m/sec^2, $t = 10$ sec
(Distance traveled)

22. $\lambda = \dfrac{v}{f}$; $v = 400$ m/sec, $f = 250$/sec
(Wavelength)
(*Note*: f is in cycles/sec, but cycles is not used in the formula.)

23. $S = \dfrac{I}{Mr}$; $I = 324$ kg-cm^2, $M = 0.340$ kg, $r = 60$ cm
(Sweet spot of tennis racquet)

24. $EBV = w(ABV)$; $w = 50$ kg, $ABV = 6.5$ mL/kg
(Estimated blood volume)

25. $KE = \frac{1}{2}mv^2$; $m = 50$ kg, $v = 15$ m/sec
(Kinetic energy)

26. $v = \sqrt{\dfrac{2 \cdot KE}{m}}$; $KE = 980$ kg-m^2/sec^2, $m = 50$ kg
(Velocity)

Make the unit changes.

27. 6.2 km to m

28. 82 cm to m

29. 35 mi/hr to ft/min

30. $9 per hour to ¢/min

31. 1 billion sec to yr (Use 365 days = 1 yr.)

32. 2 days to sec

33. 216 in^2 to ft^2

34. 10 m^3 to cm^3

35. The speed of light is 186,000 mi/sec. What is the speed of light in mi/yr?

36. The American woodcock can fly as slowly as 5 mph. What is the speed in ft/sec?

Solve.

37. A piece of lumber is 12 ft long and 8 in. wide. How many square feet of wood is in the piece of lumber?

38. A *board foot* is an amount of wood equivalent to 1 ft square and 1 in. thick. How many board feet of wood is in an old barn timber that is 4 in. thick, 8 in. wide, and 14 ft long?

39. The number of calories needed daily for a woman of weight w, in kilograms, height h, in centimeters, and age a, in years, can be estimated by the formula

$$C = 655 + 9.6w + 1.8h - 4.7a.$$

If Tawny weighs 54.7 kg, is 1.6 m tall, and is 28 years old, what is her daily caloric requirement?

40. Dr. Thomas orders a Dopamine drip of 20 mcg/kg/min. Charmagne uses an 800-mg vial of Dopamine in 500 mL of solution for a patient weighing 70 kg. What should the drip rate of the solution be, in mL/hr? (*Hint*: 1 mg = 1000 mcg.)

For Exercises 41–44, determine which formula could be the correct one to use to find the given measure.

41. Volume of a cylinder
 A. $X = \pi r^2 h$ **B.** $X = 2\pi rh + 2\pi r^2$

42. Surface area of a sphere
 A. $X = \frac{4}{3}\pi r^3$ **B.** $X = 4\pi r^2$

43. Surface area of a cone
 A. $X = \frac{1}{3}\pi r^2 h$ **B.** $X = \pi r^2 + \pi rs$

44. Slant height of a cone
 A. $X = \pi r^2 + \pi rs$ **B.** $X = \sqrt{r^2 + h^2}$

B   Distance, Midpoints, and Circles

The Distance and Midpoint Formulas   ■   Circles

The Distance and Midpoint Formulas

If two points are on a horizontal line, they have the same second coordinate. We can find the distance between them by subtracting their first coordinates. This difference may be negative, depending on the order in which we subtract. So, to make sure we get a positive number, we take the absolute value of this difference. The distance between the points (x_1, y_1) and (x_2, y_1) on a horizontal line is thus $|x_2 - x_1|$. Similarly, the distance between the points (x_2, y_1) and (x_2, y_2) on a vertical line is $|y_2 - y_1|$.

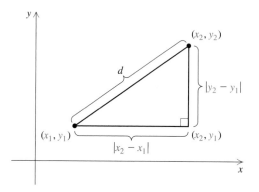

Now consider *any* two points (x_1, y_1) and (x_2, y_2). If $x_1 \neq x_2$ and $y_1 \neq y_2$, these points, along with the point (x_2, y_1), describe a right triangle. The lengths of the legs are $|x_2 - x_1|$ and $|y_2 - y_1|$. We find d, the length of the hypotenuse, by using the Pythagorean theorem:

$$d^2 = |x_2 - x_1|^2 + |y_2 - y_1|^2.$$

Since the square of a number is the same as the square of its opposite, we can replace the absolute-value signs with parentheses:

$$d^2 = (x_2 - x_1)^2 + (y_2 - y_1)^2.$$

Taking the principal square root, we have a formula for distance.

The Distance Formula   The distance d between any two points (x_1, y_1) and (x_2, y_2) is given by

$$d = \sqrt{(x_2 - x_1)^2 + (y_2 - y_1)^2}.$$

EXAMPLE 1   Find the distance between $(5, -1)$ and $(-4, 6)$. Find an exact answer and an approximation to three decimal places.

SOLUTION   We substitute into the distance formula:

$$d = \sqrt{(-4 - 5)^2 + [6 - (-1)]^2} \quad \text{Substituting}$$
$$= \sqrt{(-9)^2 + 7^2}$$
$$= \sqrt{130} \qquad\qquad \text{This is exact.}$$
$$\approx 11.402. \qquad\qquad \text{Using a calculator for an approximation}$$

The distance formula is needed to develop the formula for a circle, which follows. It is also needed to verify a formula for the coordinates of the *midpoint* of a segment connecting two points.

Student Notes

To help remember the formulas correctly, note that the distance formula (a variation on the Pythagorean theorem) involves both subtraction and addition, whereas the midpoint formula does not include any subtraction.

The Midpoint Formula   If the endpoints of a segment are (x_1, y_1) and (x_2, y_2), then the coordinates of the midpoint are

$$\left(\frac{x_1 + x_2}{2}, \frac{y_1 + y_2}{2} \right).$$

(To locate the midpoint, average the x-coordinates and average the y-coordinates.)

EXAMPLE 2   Find the midpoint of the segment with endpoints $(-2, 3)$ and $(4, -6)$.

SOLUTION   Using the midpoint formula, we obtain

$$\left(\frac{-2 + 4}{2}, \frac{3 + (-6)}{2} \right), \quad \text{or} \quad \left(\frac{2}{2}, \frac{-3}{2} \right), \quad \text{or} \quad \left(1, -\frac{3}{2} \right).$$

Circles

A **circle** is a set of points in a plane that are a fixed distance r, called the **radius** (plural, **radii**), from a fixed point (h, k), called the **center.** Note that the word radius can mean either any segment connecting a point on a circle to the center or the length of such a segment. If (x, y) is on the circle, then by the definition of a circle and the distance formula, it follows that

$$r = \sqrt{(x - h)^2 + (y - k)^2}.$$

Squaring both sides gives the equation of a circle in standard form.

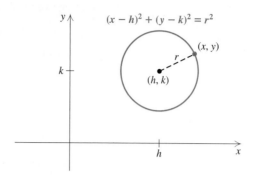

Equation of a Circle (Standard Form)

The equation of a circle, centered at (h, k), with radius r, is given by

$$(x - h)^2 + (y - k)^2 = r^2.$$

Note that for $h = 0$ and $k = 0$, the circle is centered at the origin. Otherwise, the circle is translated $|h|$ units horizontally and $|k|$ units vertically.

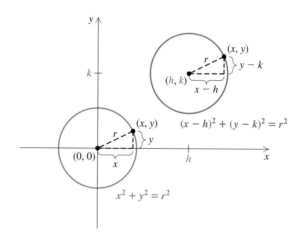

EXAMPLE 3   Find an equation of the circle having center (4, 5) and radius 6.

SOLUTION   Using the standard form, we obtain

$$(x - 4)^2 + (y - 5)^2 = 6^2, \quad \text{Using } (x - h)^2 + (y - k)^2 = r^2$$

or

$$(x - 4)^2 + (y - 5)^2 = 36.$$

EXAMPLE 4   Find the center and the radius and then graph each circle.

a) $(x - 2)^2 + (y + 3)^2 = 4^2$

b) $x^2 + y^2 + 8x - 2y + 15 = 0$

SOLUTION

a) We write standard form:

$$(x - 2)^2 + [y - (-3)]^2 = 4^2.$$

The center is $(2, -3)$ and the radius is 4. To graph, we plot the points $(2, 1)$, $(2, -7)$, $(-2, -3)$, and $(6, -3)$, which are, respectively, 4 units above, below, left, and right of $(2, -3)$. We then either sketch a circle by hand or use a compass.

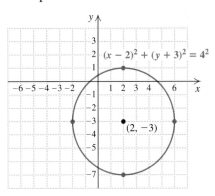

b) To write the equation $x^2 + y^2 + 8x - 2y + 15 = 0$ in standard form, we complete the square twice, once with $x^2 + 8x$ and once with $y^2 - 2y$:

$$x^2 + y^2 + 8x - 2y + 15 = 0$$

$$x^2 + 8x \qquad + y^2 - 2y \qquad = -15$$

Grouping the x-terms and the y-terms; adding -15 to both sides

$$x^2 + 8x + 16 + y^2 - 2y + 1 = -15 + 16 + 1$$

Adding $\left(\frac{8}{2}\right)^2$, or 16, and $\left(-\frac{2}{2}\right)^2$ or 1, to both sides to get standard form

$$(x + 4)^2 + (y - 1)^2 = 2$$

Factoring

$$[x - (-4)]^2 + (y - 1)^2 = \left(\sqrt{2}\right)^2.$$

Writing standard form

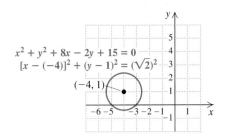

The center is $(-4, 1)$ and the radius is $\sqrt{2}$.

B EXERCISE SET

Find the distance between each pair of points. Where appropriate, find an approximation to three decimal places.

1. (1, 6) and (5, 9)   **2.** (1, 10) and (7, 2)

3. (0, −7) and (3, −4)   **4.** (6, 2) and (6, −8)

5. (−4, 4) and (6, −6)   **6.** (5, 21) and (−3, 1)

Aha! **7.** (8.6, −3.4) and (−9.2, −3.4)

8. (5.9, 2) and (3.7, −7.7)

9. $\left(\frac{5}{7}, \frac{1}{14}\right)$ and $\left(\frac{1}{7}, \frac{11}{14}\right)$

10. $\left(0, \sqrt{7}\right)$ and $\left(\sqrt{6}, 0\right)$

11. $\left(-\sqrt{6}, \sqrt{2}\right)$ and (0, 0)

12. $\left(\sqrt{5}, -\sqrt{3}\right)$ and (0, 0)

13. (−4, −2) and (−7, −11)

14. (−3, −7) and (−1, −5)

Find the midpoint of each segment with the given endpoints.

15. (−7, 6) and (9, 2)   **16.** (6, 7) and (7, −9)

17. (2, −1) and (5, 8)   **18.** (−1, 2) and (1, −3)

19. (−8, −5) and (6, −1)   **20.** (8, −2) and (−3, 4)

21. (−3.4, 8.1) and (2.9, −8.7)

22. (4.1, 6.9) and (5.2, −6.9)

23. $\left(\frac{1}{6}, -\frac{3}{4}\right)$ and $\left(-\frac{1}{3}, \frac{5}{6}\right)$

24. $\left(-\frac{4}{5}, -\frac{2}{3}\right)$ and $\left(\frac{1}{8}, \frac{3}{4}\right)$

25. $\left(\sqrt{2}, -1\right)$ and $\left(\sqrt{3}, 4\right)$

26. $\left(9, 2\sqrt{3}\right)$ and $\left(-4, 5\sqrt{3}\right)$

Find an equation of the circle satisfying the given conditions.

27. Center (0, 0), radius 6

28. Center (0, 0), radius 5

29. Center (7, 3), radius $\sqrt{5}$

30. Center (5, 6), radius $\sqrt{2}$

31. Center (−4, 3), radius $4\sqrt{3}$

32. Center (−2, 7), radius $2\sqrt{5}$

33. Center (−7, −2), radius $5\sqrt{2}$

34. Center (−5, −8), radius $3\sqrt{2}$

35. Center (0, 0), passing through (−3, 4)

36. Center (3, −2), passing through (11, −2)

37. Center (−4, 1), passing through (−2, 5)

38. Center (−1, −3), passing through (−4, 2)

Find the center and the radius of each circle. Then graph the circle.

39. $x^2 + y^2 = 64$ **40.** $x^2 + y^2 = 36$

41. $(x + 1)^2 + (y + 3)^2 = 36$

42. $(x - 2)^2 + (y + 3)^2 = 4$

43. $(x - 4)^2 + (y + 3)^2 = 10$

44. $(x + 5)^2 + (y - 1)^2 = 15$

45. $x^2 + y^2 = 10$ **46.** $x^2 + y^2 = 7$

47. $(x - 5)^2 + y^2 = \frac{1}{4}$ **48.** $x^2 + (y - 1)^2 = \frac{1}{25}$

49. $x^2 + y^2 + 8x - 6y - 15 = 0$

50. $x^2 + y^2 + 6x - 4y - 15 = 0$

51. $x^2 + y^2 - 8x + 2y + 13 = 0$

52. $x^2 + y^2 + 6x + 4y + 12 = 0$

53. $x^2 + y^2 + 10y - 75 = 0$

54. $x^2 + y^2 - 8x - 84 = 0$

55. $x^2 + y^2 + 7x - 3y - 10 = 0$

56. $x^2 + y^2 - 21x - 33y + 17 = 0$

57. $36x^2 + 36y^2 = 1$

58. $4x^2 + 4y^2 = 1$

Glossary

Absolute value [1.4] The number of units that a number is from 0 on the number line.

***ac*-method** [6.3] A method for factoring a trinomial that uses factoring by grouping.

Additive identity [1.5] The number 0.

Additive inverse [1.6] A number's opposite. Two numbers are additive inverses of each other if their sum is zero.

Algebraic expression [1.1] An expression consisting of a number or a variable or a collection of numbers, variables, operation signs, and grouping symbols.

Arithmetic sequence [14.2] A sequence in which the difference between any two successive terms is constant.

Arithmetic series [14.2] A series for which the associated sequence is arithmetic.

Ascending order [5.4] A polynomial in one variable written with the terms arranged according to degree, from least to greatest.

Associative law for addition [1.2] The statement that when three numbers are added, regrouping the addends gives the same sum.

Associative law for multiplication [1.2] The statement that when three numbers are multiplied, regrouping the factors gives the same product.

Asymptote [13.3] A line that a graph approaches more and more closely as x increases or as x decreases.

Average [2.7] Most commonly, the mean of a set of numbers.

Axes [3.1] Two perpendicular number lines used to identify points in a plane.

Axis of symmetry [11.6] A line that can be drawn through a graph such that the part of the graph on one side of the line is an exact reflection of the part on the opposite side.

Bar graph [3.1] A graphic display of data using bars proportional in length to the numbers represented.

Base [1.8] In exponential notation, the number being raised to an exponent.

Binomial [5.3] A polynomial composed of two terms.

Branches [13.3] The two curves that comprise a hyperbola.

Break-even point [9.5] In business, the point of intersection of the revenue function and the cost function.

Circle [13.1] A set of points in a plane that are a fixed distance r, called the radius, from a fixed point (h, k), called the center.

Circle graph [3.1] A graphic display of data using sectors of a circle to represent percents.

Circumference [2.3] The distance around a circle.

Closed interval $[a, b]$ [2.6] The set of all numbers x for which $a \le x \le b$. Thus, $[a, b] = \{x \,|\, a \le x \le b\}$.

Coefficient [2.1] The numerical multiplier of a variable.

Columns of a matrix [9.3] Vertically-aligned elements of a matrix.

Combined variation [7.8] A mathematical relationship in which a variable varies directly and/or inversely, at the same time, with more than one other variable.

Common logarithm [12.3] A logarithm with base 10.

Commutative law for addition [1.2] The statement that when two real numbers are added, the order in which the numbers are added does not affect the sum.

Commutative law for multiplication [1.2] The statement that when two real numbers are multiplied, the order in which the numbers are multiplied does not affect the product.

Completing the square [11.1] Adding a particular constant to an expression so that the resulting sum is a perfect square.

Complex number [10.8] Any number that can be written $a + bi$, where a and b are real numbers.

Complex rational expression [7.5] A rational expression that has one or more rational expressions within its numerator and/or denominator.

Complex-number system [10.8] A number system that contains the real-number system and is designed so that negative numbers have square roots.

Composite function [12.1] A function in which a quantity depends on a variable that, in turn, depends on another variable.

Composite number [1.3] A natural number, other than 1, that is not prime.

Compound inequality [8.1] A statement in which two or more inequalities are combined using the word *and* or the word *or*.

Compound interest [11.1] Interest computed on the sum of an original principal and the interest previously accrued by that principal.

Conditional equation [2.2] An equation that is true for some replacements and false for others.

Conic section [13.1] A curve formed by the intersection of a plane and a cone.

Conjugate of a complex number [10.8] The conjugate of a complex number $a + bi$ is $a - bi$.

Conjugates [10.5] Pairs of radical terms, like $\sqrt{a} + \sqrt{b}$ and $\sqrt{a} - \sqrt{b}$, for which the product does not have a radical term.

Conjunction [8.1] A sentence in which two statements are joined by the word *and*.

Consecutive even integers [2.5] Integers that are even and two units apart.

Consecutive integers [2.5] Integers that are one unit apart.

Consecutive odd integers [2.5] Integers that are odd and two units apart.

Consistent system of equations [4.1] A system of equations that has at least one solution.

Constant [1.1] A specific number that never changes.

Constant function [3.8] A function given by an equation of the form $f(x) = b$, where b is a real number.

Constant of proportionality [7.8] The constant, k, in an equation of direct or inverse variation.

Contradiction [2.2] An equation that is never true.

Coordinates [3.1] The numbers in an ordered pair.

Cube root [10.1] The number c is called the cube root of a if $c^3 = a$.

Cubic polynomial [5.3] A polynomial in one variable of degree 3.

Curve fitting [3.7] The process of analyzing data to determine if it has a recognized pattern and if it does, fitting an equation to the data.

Degree of a polynomial [5.3] The degree of the term of highest degree in a polynomial.

Degree of a term [5.3] The number of variable factors in a term.

Demand function [9.5] A function modeling the relationship between the price of a good and the quantity of that good demanded.

Denominator [1.3] The number below the fraction bar in a fraction.

Dependent equations [4.1] The equations in a system are dependent if one equation can be removed without changing the solution set.

Dependent variable [2.3] In an equation with two variables, the variable that *depends* on the value of the other.

Descending order [5.3] A polynomial in one variable written with the terms arranged according to degree, from greatest to least.

Determinant [9.4] The determinant of a two-by-two matrix $\begin{bmatrix} a & c \\ b & d \end{bmatrix}$ is denoted by $\begin{vmatrix} a & c \\ b & d \end{vmatrix}$ and represents $ad - bc$.

Difference of squares [6.4] An expression that can be written in the form $A^2 - B^2$.

Direct variation [7.8] A situation that translates to an equation of the form $y = kx$, where k is a nonzero constant.

Discriminant [11.4] The expression $b^2 - 4ac$ from the quadratic formula.

Disjunction [8.1] A sentence in which two statements are joined by the word *or*.

Distributive law [1.2] The statement that multiplying a factor by the sum of two numbers gives the same result as multiplying the factor by each of the two numbers and then adding.

Domain [3.8] The set of all first coordinates of the ordered pairs in a function.

Double root [6.4] A repeated root that appears twice.

Doubling time [12.7] The time necessary for a population to double in size.

Elements of a matrix [9.3] The individual numbers in a matrix.

Elimination method [4.3] An algebraic method that uses the addition principle to solve a system of equations.

Ellipse [13.2] The set of all points in a plane for which the sum of the distances from two fixed points F_1 and F_2 is constant.

Empty set [2.2] The set containing no elements, denoted $\varnothing$ or $\{\ \}$.

Equation [1.1] A number sentence formed by placing an equals sign between two expressions.

Equation of variation [7.8] An equation used to represent direct, inverse, or combined variation.

Equilibrium point [9.5] The point of intersection between the demand function and the supply function.

Equivalent equations [2.1] Equations that have the same solutions.

Equivalent expressions [1.2] Expressions that have the same value for all allowable replacements.

Equivalent inequalities [2.6] Inequalities that have the same solution set.

Evaluate [1.1] To substitute a value for each occurrence of a variable in an expression and carry out the operations.

Exponent [1.8] In expressions of the form a^n, the number n is an exponent. For n a natural number, a^n represents n factors of a.

Exponential decay [12.7] A decrease in quantity over time that can be modeled by an exponential equation of the form $P(t) = P_0 e^{-kt}, k > 0$.

Exponential equation [12.6] An equation in which a variable appears as an exponent.

Exponential function [12.2] A function that can be described by an exponential equation.

Exponential growth [12.7] An increase in quantity over time that can be modeled by an exponential function of the form $P(t) = P_0 e^{kt}, k > 0$.

Exponential notation [1.8] A representation of a number using a base raised to a power.

Extrapolation [3.7] The process of predicting a future value on the basis of given data.

Factor [1.2] *Verb*: to write an equivalent expression that is a product. *Noun*: a multiplier.

Factoring [1.2] The process of rewriting a sum or a difference as a product.

Finite sequence [14.1] A function having for its domain a set of natural numbers: $\{1, 2, 3, 4, 5, \ldots, n\}$, for some natural number n.

Fixed costs [9.5] In business, costs that are incurred whether or not a product is produced.

Focus [13.2] One of two fixed points that determine the points of an ellipse.

FOIL method [5.6] To multiply two binomials by multiplying the First terms, the Outside terms, the Inside terms, and then the Last terms.

Formula [2.3] An equation that uses two or more letters to represent a relationship between two or more quantities.

Fraction notation [1.3] A number written using a numerator and a denominator.

Function [3.8] A correspondence between a first set, called the *domain*, and a second set, called the *range*, such that each member of the domain corresponds to *exactly one* member of the range.

General term of a sequence [14.1] The nth term, denoted a_n.

Geometric sequence [14.3] A sequence in which the ratio of every pair of successive terms is constant.

Geometric series [14.3] A series for which the associated sequence is geometric.

Grade [3.5] The ratio of the vertical distance a road rises over the horizontal distance it runs, expressed as a percent.

Graph [2.6] A picture or diagram of the data in a table. A line, curve, or collection of points that represents all the solutions of an equation.

Greatest common factor [6.1] The common factor of the terms of a polynomial with the largest possible coefficient and the largest possible exponent(s).

Half-life [12.7] The amount of time necessary for half of a quantity to decay.

Half-open interval [2.6] An interval that includes exactly one of two endpoints.

Half-plane [8.3] The graph of a linear inequality.

Horizontal-line test [12.1] If it is impossible to draw a horizontal line that intersects the graph of a function more than once, then that function is one-to-one.

Hyperbola [13.3] The set of all points P in the plane such that the difference of the distance from P to two fixed points is constant.

Hypotenuse [6.7] In a right triangle, the side opposite the right angle.

Identity [2.2] An equation that is true for all replacements.

Identity property of 0 [1.5] The statement that the sum of a number and 0 is always the original number.

Identity property of 1 [1.3] The statement that the product of a number and 1 is always the original number.

Imaginary number [10.8] A number that can be written in the form $a + bi$, where a and b are real numbers and $b \neq 0$.

Imaginary number i [10.8] The square root of -1. That is, $i = \sqrt{-1}$ and $i^2 = -1$.

Inconsistent system of equations [4.1] A system of equations for which there is no solution.

Independent equations [4.1] Equations that are not dependent.

Independent variable [2.3] In an equation with two variables, the variable that does *not depend* on the other.

Index [10.1] In the radical $\sqrt[n]{a}$, the number n is called the index.

Inequality [1.4] A mathematical sentence formed when an inequality symbol, $<$, $>$, $\leq$, $\geq$, or $\neq$, is placed between two algebraic expressions.

Infinite geometric series [14.3] The sum of the terms of an infinite geometric sequence.

Infinite sequence [14.1] A function having for its domain the set of natural numbers: $\{1, 2, 3, 4, 5, \ldots\}$.

Input [3.8] An element of the domain of a function.

Integers [1.4] The whole numbers and their opposites.

Interpolation [3.7] The process of estimating a value between given values.

Intersection of A and B [8.1] The set of all elements that are common to *both* A and B.

Interval notation [2.6] The use of a pair of numbers inside parentheses and brackets to represent the set of all numbers between those two numbers. *See also Closed interval* and *Open interval.*

Inverse relation [12.1] The relation formed by interchanging the members of the domain and the range of a relation.

Inverse variation [7.8] A situation that translates to an equation of the form $y = k/x$, where k is a nonzero constant.

Irrational number [1.4] A real number that when written as a decimal, neither terminates nor repeats.

Isosceles right triangle [10.7] A right triangle in which both legs have the same length.

Joint variation [7.8] A situation that translates to an equation of the form $y = kxz$, where k is a nonzero constant.

Leading coefficient [5.3] The coefficient of the term of highest degree in a polynomial.

Leading term [5.3] The term of highest degree in a polynomial.

Least common denominator [7.3] The least common multiple of the denominators.

Legs [6.7] In a right triangle, the two sides that form the right angle.

Like radicals [10.5] Radical expressions that have the same index and radicand.

Like terms [1.5] Terms that have exactly the same variable factors.

Line graph [3.1] A graph in which quantities are represented as points connected by straight-line segments.

Linear equation [2.2] Any equation whose graph is a straight line, and can be written in the form $y = mx + b$, or $Ax + By = C$, where x and y are variables.

Linear equation in three variables [9.1] An equation in the form $Ax + By + Cz = D$, where A, B, C, and D are real numbers.

Linear function [3.8] A function whose graph is a straight line; $f(x) = mx + b$.

Linear inequality [8.3] An inequality whose related equation is a linear equation.

Linear regression [3.7] A method of finding a "best" line to fit a set of data.

Logarithmic equation [12.6] An equation containing a logarithmic expression.

Logarithmic function, base *a* [12.3] The inverse of an exponential function with base *a*.

Mathematical model [1.1] A representation, using mathematics, of a real-world situation.

Matrix [9.3] A rectangular array of numbers.

Maximum value [11.6] The largest function value (output) achieved by a function.

Mean [2.7] The sum of a set of numbers divided by the number of addends.

Minimum value [11.6] The smallest function value (output) achieved by a function.

Monomial [5.3] A constant, a variable, or a product of a constant and one or more variables.

Motion problem [7.7] A problem that deals with distance, speed (rate), and time.

Multiplicative identity [1.3] The number 1.

Multiplicative inverses [1.3] Reciprocals; two numbers whose product is 1.

Multiplicative property of zero [1.7] The statement that the product of 0 and any real number is 0.

Natural logarithm [12.5] A logarithm with base *e*.

Natural numbers [1.3] The counting numbers: $\{1, 2, 3, 4, 5, \ldots\}$.

Nonlinear equation [3.2] An equation whose graph is not a straight line.

***n*th root** [10.1] A number c is called the *n*th root of a if $c^n = a$. In the case of n an even number, c is called *the* *n*th root of a (or the *principal square root of a*) if c is an *n*th root and c is nonnegative.

Numerator [1.3] The number above the fraction bar in a fraction.

One-to-one function [12.1] A function for which different inputs have different outputs.

Open interval (a, b) [2.6] The set of all numbers x for which $a < x < b$. Thus, $(a, b) = \{x \mid a < x < b\}$.

Opposite [1.6] The opposite, or additive inverse, of a number a is written $-a$. Opposites are the same distance from 0 on the number line but on different sides of 0.

Opposite of a polynomial [5.4] The *opposite* of a polynomial P can be written $-P$, or equivalently, by replacing each term in P with its opposite.

Ordered pair [3.1] A pair of numbers of the form (a, b) for which the order in which the numbers are listed is important.

Origin [3.1] The point, $(0, 0)$, on a graph where the two axes intersect.

Output [3.8] An element of the range of a function.

Parabola [11.6] A graph of a quadratic function.

Parallel lines [3.6] Lines that extend indefinitely without intersecting.

Pascal's triangle [14.4] A triangular array of coefficients of the expansion $(a + b)^n$ for $n = 0, 1, 2, \ldots$.

Perfect-square trinomial [6.4] A trinomial that is the square of a binomial.

Perpendicular lines [3.6] Lines that form a right angle.

Point–slope form of an equation [3.7] An equation of the type $y - y_1 = m(x - x_1)$, where x and y are variables, m is the slope, and (x_1, y_1) is a point through which the line passes.

Polynomial [5.3] A monomial or a sum of monomials.

Polynomial equation [6.1] An equation in which two polynomials are set equal to each other.

Polynomial function [5.3] A function in which outputs are determined by evaluating a polynomial.

Polynomial inequality [8.4] An inequality that is equivalent to an inequality with a polynomial as one side and 0 as the other.

Prime factorization [1.3] The factorization of a whole number into a product of its prime factors.

Prime number [1.3] A natural number that has exactly two different factors: the number itself and 1.

Prime polynomial [6.1] A polynomial that cannot be factored.

Principal square root [10.1] The nonnegative square root of a number.

Proportion [7.7] An equation stating that two ratios are equal.

Pure imaginary number [10.8] A complex number of the form $a + bi$, with $a = 0$ and $b \neq 0$.

Pythagorean theorem [6.7] In any right triangle, if a and b are the lengths of the legs and c is the length of the hypotenuse, then $a^2 + b^2 = c^2$.

Quadrants [3.1] The four regions into which the axes divide a plane.

Quadratic equation [11.1] An equation equivalent to one of the form $ax^2 + bx + c = 0$, where $a \neq 0$.

Quadratic formula [11.2] The solutions of $ax^2 + bx + c = 0$, $a \neq 0$, are given by the equation $x = \left(-b \pm \sqrt{b^2 - 4ac}\right)/(2a)$.

Quadratic inequality [8.4] A second-degree polynomial inequality in one variable.

Quadratic polynomial [11.2] A second-degree polynomial function in one variable.

Quartic polynomial [5.3] A polynomial in one variable of degree 4.

Radical equation [10.6] An equation in which a variable appears in a radicand.

Radical expression [10.1] An algebraic expression in which a radical sign appears.

Radical function [10.1] A function that is described by a radical expression.

Radical sign [10.1] The symbol $\sqrt{\ }$.

Radical term [10.5] A term in which a radical sign appears.

Radicand [10.1] The expression under the radical sign.

Radius [13.1] The distance from the center of a circle to a point on the circle. Also, a segment connecting the center to a point on the circle.

Range [3.8] The set of all second coordinates of the ordered pairs in a function.

Rate [3.4] A ratio that indicates how two quantities change with respect to each other.

Ratio [7.7] The ratio of a to b is a/b, also written $a:b$.

Rational equation [7.6] An equation containing one or more rational expressions.

Rational expression [7.1] An expression that consists of a polynomial divided by a nonzero polynomial.

Rational function [7.1] A function described by a rational expression.

Rational inequality [8.4] An inequality containing a rational expression.

Rational number [1.4] A number that can be written in the form a/b, where a and b are integers and $b \neq 0$.

Rationalizing the denominator [10.4] A procedure for finding an equivalent expression without a radical in the denominator.

Rationalizing the numerator [10.4] A procedure for finding an equivalent expression without a radical in the numerator.

Real number [1.4] Any number that is either rational or irrational.

Reciprocal [1.3] A multiplicative inverse. Two numbers are reciprocals if their product is 1.

Reflection [11.6] The mirror image of a graph.

Relation [3.8] A correspondence between a set called the domain and a set called the range such that each member of the domain corresponds to *at least one* member of the range.

Repeated root [6.4] The number c is a repeated root of an equation of the form $P(x) = 0$ if $P(x)$ has $(x - c)$ as a factor more than once.

Repeating decimal [1.4] A decimal in which a number pattern repeats indefinitely.

Right triangle [6.7] A triangle that includes a right angle.

Root of an equation [6.1] Any solution of an equation in one variable.

Root of multiplicity two [6.4] A repeated root that appears twice.

Row-equivalent operations [9.3] Operations used to produce equivalent systems of equations.

Rows of a matrix [9.3] Horizontally-aligned elements of a matrix.

Scientific notation [5.2] A number written in the form $N \times 10^m$, where m is an integer, $1 \leq N < 10$, and N is expressed in decimal notation.

Sequence [14.1] A function for which the domain is a set of consecutive positive integers beginning with 1.

Series [14.1] The sum of specified terms in a sequence.

Set [1.4] A collection of objects.

Set-builder notation [2.6] The naming of a set by describing specific conditions under which an element is in the set.

Sigma notation [14.1] The naming of a sum using the Greek letter Σ (sigma) as part of an abbreviated form.

Significant digits [5.2] When working with decimals in science, significant digits are used to determine how accurate a measurement is.

Similar triangles [7.7] Triangles in which corresponding sides are proportional and corresponding angles have the same measurement.

Simplify [1.3] To rewrite an expression in an equivalent, abbreviated, form.

Slope [3.5] The ratio of the rise to the run for any two points on a line.

Slope–intercept form of an equation [3.6] An equation of the form $y = mx + b$, where x and y are variables, m is the slope, and $(0, b)$ is the y-intercept.

Solution [1.1] A replacement or substitution that makes an equation or inequality true.

Solution set [2.2] The set of all solutions of an equation, an inequality, or a system of equations or inequalities.

Solve [2.1] To find all solutions of an equation, an inequality, or a system of equations or inequalities; to find the solution(s) of a problem.

Speed [3.4] The ratio of distance traveled to the time required to travel that distance.

Square matrix [9.4] A matrix with the same number of rows and columns.

Square root [10.1] The number c is a square root of a if $c^2 = a$.

Standard form of a linear equation [3.3] An equation of the form $Ax + By = C$, where A, B, and C are real numbers and A and B are not both 0.

Substitute [1.1] To replace a variable with a number.

Substitution method [4.2] An algebraic method for solving systems of equations.

Supply function [9.5] A function modeling the relationship between the price of a good and the quantity of that good supplied.

Synthetic division [5.8] A method used to divide a polynomial by a binomial of the form $x - a$.

System of equations [4.1] A set of two or more equations, in two or more variables, that are to be solved simultaneously.

Term [1.2] A number, a variable, or a product or a quotient of numbers and/or variables.

Terminating decimal [1.4] A decimal that can be written using a finite number of decimal places.

Total cost [9.5] The amount spent to produce a product.

Total profit [9.5] The amount taken in less the money spent, or total revenue minus total cost.

Total revenue [9.5] The amount taken in from the sale of a product.

Trinomial [5.3] A polynomial that is composed of three terms.

Undefined [1.7] An expression that has no meaning attached to it.

Union of A and B [8.1] The set of all elements belonging to either A or B.

Value [1.1] The numerical result after a number has been substituted into an expression.

Variable [1.1] A letter that represents an unknown number.

Variable costs [9.5] In business, costs that vary according to the amount of products produced.

Variable expression [1.1] An expression containing a variable.

Vertex [11.7] The point at which the graph of a quadratic equation crosses its axis of symmetry.

Vertical asymptote [7.1] When graphing a rational function, one or more lines that the graph approaches, but never touches.

Vertical-line test [3.8] The statement that a graph represents a function if it is impossible to draw a vertical line that intersects the graph more than once.

Whole numbers [1.4] The set of natural numbers and 0: $\{0, 1, 2, 3, 4, 5, \ldots\}$.

x-axis [3.1] The horizontal axis in a coordinate plane.

x-intercept [3.3] The point at which a graph crosses the x-axis.

y-axis [3.1] The vertical axis in a coordinate plane.

y-intercept [3.3] The point at which a graph crosses the y-axis.

Zeros [6.1] The x-values for which a function $f(x)$ is 0.

Photo Credits

Answers

Chapter 1

1. Expression   **2.** Equation   **3.** Equation
4. Expression   **5.** Equation   **6.** Equation
7. Expression   **8.** Equation   **9.** Equation
10. Expression   **11.** Expression   **12.** Expression
13. 27   **15.** 8   **17.** 4   **19.** 7   **21.** 3   **23.** 3438
25. 24 ft^2 **27.** 15 cm^2 **29.** 0.368   **31.** Let r represent Ron's age; $r + 5$, or $5 + r$ **33.** $b + 6$, or $6 + b$
35. $c - 9$ **37.** $6 + q$, or $q + 6$ **39.** Let p represent

Phil's speed; $9p$, or $p \cdot 9$ **41.** $y - x$ **43.** $x \div w$, or $\dfrac{x}{w}$

45. $n - m$ **47.** Let l represent the length of the box and h represent the height; $l + h$, or $h + l$ **49.** $9 \cdot 2m$, or $2m \cdot 9$
51. Let y represent "some number"; $\frac{1}{4}y$, or $\frac{y}{4}$ **53.** Let w represent the number of women attending; 64% of w, or
$0.64w$ **55.** Yes   **57.** No   **59.** Yes   **61.** Yes
63. Let x represent the unknown number; $73 + x = 201$
65. Let x represent the unknown number; $42x = 2352$
67. Let s represent the number of unoccupied squares;
$s + 19 = 64$ **69.** Let w represent the amount of solid waste generated, in millions of tons; 28.5% of $w = 110.4$, or
$0.285w = 110.4$ **71.** (f)   **73.** (d)   **75.** (g)   **77.** (e)
79. $p = r + 19$ **81.** $f = a + 5$ **83.** $v = 10,000d$
85. TW   **87.** TW   **89.** $337.50   **91.** 2   **93.** 6
95. $w + 4$ **97.** $l + w + l + w$, or $2l + 2w$ **99.** $t + 8$
101. TW

1. Commutative   **2.** Associative   **3.** Associative
4. Commutative   **5.** Distributive   **6.** Associative
7. Associative   **8.** Commutative   **9.** Commutative
10. Distributive   **11.** $x + 7$ **13.** $c + ab$ **15.** $3y + 9x$
17. $5(1 + a)$ **19.** $a \cdot 2$ **21.** ts **23.** $5 + ba$
25. $(a + 1)5$ **27.** $a + (5 + b)$ **29.** $(r + t) + 7$
31. $ab + (c + d)$ **33.** $8(xy)$ **35.** $(2a)b$
37. $(3 \cdot 2)(a + b)$ **39.** $(r + t) + 6$; $(t + 6) + r$
41. $17(ab)$; $b(17a)$

43. $(5 + x) + 2 = (x + 5) + 2$ Commutative law
$= x + (5 + 2)$ Associative law
$= x + 7$ Simplifying
45. $(m \cdot 3)7 = m(3 \cdot 7)$ Associative law
$= m \cdot 21$ Simplifying
$= 21m$ Commutative law
47. $4a + 12$ **49.** $6 + 6x$ **51.** $3x + 3$ **53.** $24 + 8y$
55. $18x + 54$ **57.** $5r + 10 + 15t$ **59.** $2a + 2b$

61. $5x + 5y + 10$ **63.** $x, xyz, 19$ **65.** $2a, \dfrac{a}{b}, 5b$

67. $2(a + b)$ **69.** $7(1 + y)$ **71.** $3(6x + 1)$
73. $5(x + 2 + 3y)$ **75.** $3(4x + 3)$ **77.** $3(a + 3b)$
79. $11(4x + y + 2z)$ **81.** s, t **83.** $3, (x + y)$
85. $7, a$ **87.** $(a - b), (x - y)$ **89.** TW   **91.** Let k

represent Kara's salary; $2k$ **92.** $\dfrac{1}{2} \cdot m$, or $\dfrac{m}{2}$ **93.** TW

95. Yes; distributive law   **97.** Yes; distributive law and commutative law of multiplication   **99.** No; for example, let $x = 1$ and $y = 2$. Then $30 \cdot 2 + 1 \cdot 15 = 60 + 15 = 75$ and $5[2(1 + 3 \cdot 2)] = 5[2(7)] = 5 \cdot 14 = 70$.   **101.** TW

1. Composite   **2.** Composite   **3.** Prime   **4.** Composite
5. Composite   **6.** Prime   **7.** Prime   **8.** Neither
9. Neither   **10.** Composite   **11.** (b)   **12.** (c)
13. (d)   **14.** (a)   **15.** $2 \cdot 25$; $5 \cdot 10$; 1, 2, 5, 10, 25, 50
17. $3 \cdot 14$; $6 \cdot 7$; 1, 2, 3, 6, 7, 14, 21, 42   **19.** $2 \cdot 13$
21. $2 \cdot 3 \cdot 5$ **23.** $3 \cdot 3 \cdot 3$ **25.** $2 \cdot 3 \cdot 3$
27. $2 \cdot 2 \cdot 2 \cdot 5$ **29.** Prime   **31.** $2 \cdot 3 \cdot 5 \cdot 7$
33. $5 \cdot 23$ **35.** $\frac{2}{3}$ **37.** $\frac{2}{7}$ **39.** $\frac{1}{8}$ **41.** 7   **43.** $\frac{1}{4}$
45. 6   **47.** $\frac{21}{25}$ **49.** $\frac{60}{41}$ **51.** $\frac{15}{7}$ **53.** $\frac{3}{14}$ **55.** $\frac{27}{8}$
57. $\frac{1}{2}$ **59.** $\frac{7}{6}$ **61.** $\dfrac{3b}{7a}$ **63.** $\dfrac{7}{a}$ **65.** $\frac{5}{6}$ **67.** 1
69. $\frac{5}{18}$ **71.** 0   **73.** $\frac{35}{18}$ **75.** $\frac{10}{3}$ **77.** 28   **79.** 1
81. $\frac{6}{35}$ **83.** 18   **85.** TW   **87.** $5(3 + x)$; answers may vary   **88.** $7 + (b + a)$, or $(a + b) + 7$ **89.** TW
91. Row 1: 7, 2, 36, 14, 8, 8; row 2: 9, 18, 2, 10, 12, 21

93. $\frac{2}{5}$ **95.** $\frac{3q}{t}$ **97.** $\frac{6}{25}$ **99.** $\frac{5ap}{2cm}$ **101.** $\frac{23r}{18t}$

103. $\frac{28}{45}$ m^2 **105.** $14\frac{2}{9}$ m   **107.** $27\frac{3}{5}$ cm

Exercise Set 1.4, pp. 39–41

1. Repeating   **2.** Terminating   **3.** Integer   **4.** Whole number   **5.** Rational number   **6.** Irrational number
7. Natural number   **8.** Absolute value   **9.** -1395, 29,035
11. 950,000,000, -460 **13.** 2, -12 **15.** 750, -125
17. Jets: -34; Strikers: 34   **19.**

21.   **23.**

25. 0.875   **27.** -0.75 **29.** $1.1\overline{6}$ **31.** $0.\overline{6}$ **33.** -0.5
35. $0.1\overline{3}$ **37.** $<$ **39.** $>$ **41.** $<$ **43.** $<$ **45.** $>$
47. $<$ **49.** $<$ **51.** $x < -7$ **53.** $y \geq -10$
55. True   **57.** False   **59.** True   **61.** 58   **63.** 17
65. 5.6   **67.** 329   **69.** $\frac{9}{7}$ **71.** 0   **73.** 8
75. $-83, -4.7, 0, \frac{5}{9}, 8.31, 62$ **77.** $-83, 0, 62$
79. $-83, -4.7, 0, \frac{5}{9}, \pi, \sqrt{17}, 8.31, 62$ **81.** TW   **83.** 42
84. $ba + 5$, or $5 + ab$ **85.** TW   **87.** TW
89. $-23, -17, 0, 4$ **91.** $-\frac{4}{3}, \frac{4}{9}, \frac{4}{8}, \frac{4}{6}, \frac{4}{5}, \frac{4}{3}, \frac{4}{2}$ **93.** $<$
95. $=$ **97.** $-7, 7$ **99.** $-4, -3, 3, 4$ **101.** $\frac{3}{3}$
103. $\frac{70}{9}$ **105.** TW

Exercise Set 1.5, pp. 45–47

1. (f)   **2.** (d)   **3.** (e)   **4.** (a)   **5.** (b)   **6.** (c)
7. -3 **9.** 4   **11.** 0   **13.** -8 **15.** -27 **17.** -8
19. 0   **21.** -41 **23.** 0   **25.** 7   **27.** -2 **29.** 11
31. -33 **33.** 0   **35.** 18   **37.** -45 **39.** 0   **41.** 20
43. -1.7 **45.** -9.1 **47.** $\frac{1}{5}$ **49.** $\frac{-6}{7}$ **51.** $-\frac{1}{15}$
53. $\frac{2}{9}$ **55.** -3 **57.** 0   **59.** The price rose 9¢.
61. Her new balance was \$95.   **63.** The total gain was 22 yd.   **65.** Lyle owes \$85.   **67.** The elevation of the peak is 13,796 ft.   **69.** $14a$ **71.** $9x$ **73.** $13t$
75. $-2m$ **77.** $-7a$ **79.** $1 - 2x$ **81.** $12x + 17$
83. $7r + 8t + 16$ **85.** $18n + 16$ **87.** TW
89. $21z + 7y + 14$ **90.** $\frac{28}{3}$ **91.** TW   **93.** \$65.25
95. $-5y$ **97.** $-7m$ **99.** $-7t, -23$ **101.** 1 under par

Exercise Set 1.6, pp. 53–55

1. (d)   **2.** (g)   **3.** (f)   **4.** (h)   **5.** (a)   **6.** (c)
7. (b)   **8.** (e)   **9.** Four minus ten   **11.** Two minus negative nine   **13.** Nine minus the opposite of t
15. The opposite of x minus y **17.** Negative three minus the opposite of n **19.** -39 **21.** 9   **23.** 3.14

25. -23 **27.** $\frac{14}{3}$ **29.** -0.101 **31.** 72   **33.** $-\frac{2}{5}$
35. 1   **37.** -7 **39.** -2 **41.** -5 **43.** -6
45. -10 **47.** -6 **49.** 0   **51.** -5 **53.** -10
55. 2   **57.** 0   **59.** 0   **61.** 8   **63.** -11 **65.** 16
67. -19 **69.** -1 **71.** 17   **73.** -5 **75.** -3
77. -21 **79.** 5   **81.** -8 **83.** 10   **85.** -23
87. -68 **89.** -58 **91.** -5.5 **93.** -0.928
95. $-\frac{7}{11}$ **97.** $-\frac{4}{5}$ **99.** $\frac{5}{17}$ **101.** $3.8 - (-5.2)$; 9
103. $114 - (-79)$; 193   **105.** -58 **107.** 34   **109.** 41
111. -62 **113.** -139 **115.** 0   **117.** $-7x, -4y$
119. $9, -5t, -3st$ **121.** $-3x$ **123.** $-5a + 4$
125. $-7n - 9$ **127.** $-6x + 5$ **129.** $-8t - 7$
131. $-12x + 3y + 9$ **133.** $8x + 66$ **135.** 9.24 points
137. 30,345 ft   **139.** 116 m   **141.** TW   **143.** 432 ft^2
144. $2 \cdot 2 \cdot 2 \cdot 2 \cdot 2 \cdot 3 \cdot 3 \cdot 3$ **145.** TW
147. 11:00 P.M., August 14   **149.** False. For example, let $m = -3$ and $n = -5$. Then $-3 > -5$, but $-3 + (-5) = -8 \not> 0$.   **151.** True. For example, for $m = 4$ and $n = -4$, $4 = -(-4)$ and $4 + (-4) = 0$; for $m = -3$ and $n = 3$, $-3 = -3$ and $-3 + 3 = 0$.
153. (-) 9 − (-) 7 ENTER

Exercise Set 1.7, pp. 61–63

1. 1   **2.** 0   **3.** 0   **4.** 1   **5.** 0   **6.** 1   **7.** 1   **8.** 0
9. 1   **10.** 0   **11.** -24 **13.** -56 **15.** -24
17. -72 **19.** 42   **21.** 45   **23.** 190   **25.** -144
27. 1200   **29.** 98   **31.** -78 **33.** 21.7   **35.** $-\frac{2}{5}$
37. $\frac{1}{12}$ **39.** -11.13 **41.** $-\frac{5}{12}$ **43.** 252   **45.** 0
47. $\frac{1}{28}$ **49.** 150   **51.** 0   **53.** -720 **55.** $-30,240$
57. -7 **59.** -4 **61.** -7 **63.** 4   **65.** -8 **67.** 2
69. -12 **71.** -8 **73.** Undefined   **75.** -4 **77.** 0
79. 0   **81.** $-\frac{8}{3}; \frac{8}{-3}$ **83.** $-\frac{29}{35}; \frac{-29}{35}$ **85.** $\frac{-7}{3}; \frac{7}{-3}$
87. $-\frac{x}{2}; \frac{x}{-2}$ **89.** $-\frac{5}{4}$ **91.** $-\frac{13}{47}$ **93.** $-\frac{1}{10}$
95. $\frac{1}{4.3}$, or $\frac{10}{43}$ **97.** $-\frac{4}{9}$ **99.** Does not exist   **101.** $\frac{21}{20}$
103. $\frac{12}{55}$ **105.** -1 **107.** 1   **109.** $-\frac{9}{11}$ **111.** $-\frac{7}{4}$
113. -12 **115.** -3 **117.** 1   **119.** 7   **121.** $-\frac{2}{9}$
123. $\frac{1}{10}$ **125.** $-\frac{7}{6}$ **127.** $\frac{6}{7}$ **129.** $-\frac{14}{15}$ **131.** TW
133. $\frac{22}{39}$ **134.** No   **135.** TW   **137.** For 2 and 3, the reciprocal of the sum is $1/(2 + 3)$, or $1/5$. But $1/5 \neq 1/2 + 1/3$.   **139.** Negative   **141.** Negative
143. Negative   **145.** (a) m and n have different signs; (b) either m or n is zero; (c) m and n have the same sign.
147. TW

Interactive Discovery, p. 66

1. Multiplication   **2.** $4 + (2 \times 5)$

Exercise Set 1.8, pp. 71–73

1. (a) Division; **(b)** subtraction; **(c)** addition;
(d) multiplication; **(e)** subtraction; **(f)** multiplication
2. (a) Multiplication; **(b)** subtraction; **(c)** addition;
(d) subtraction; **(e)** division; **(f)** multiplication
3. 2^3 **5.** x^7 **7.** $(3t)^5$ **9.** 9 **11.** 16 **13.** -16
15. 64 **17.** 625 **19.** 7 **21.** $81t^4$ **23.** $-343x^3$
25. 26 **27.** 86 **29.** 7 **31.** 5 **33.** 1 **35.** 298
37. 11 **39.** -36 **41.** 1291 **43.** 14 **45.** 152
47. 36 **49.** 1 **51.** -26 **53.** -2 **55.** $-\frac{9}{2}$ **57.** (a)
59. (d) **61.** -1.026 **63.** 110.342 **65.** -12.86
67. -11 **69.** -3 **71.** -15 **73.** 9 **75.** 30
77. 6 **79.** -17 **81.** 15 **83.** 25.125 **85.** 13,778
87. $-9x - 1$ **89.** $-5 + 6x$ **91.** $-4a + 3b - 7c$
93. $-3x^2 - 5x + 1$ **95.** $2x - 7$ **97.** $-3a + 9$
99. $5x - 6$ **101.** $-3t - 11r$ **103.** $9y - 25z$
105. $x^2 + 2$ **107.** $-t^3 - 2t$ **109.** $37a^2 - 23ab + 35b^2$
111. $-22t^3 - t^2 + 9t$ **113.** $2x - 25$ **115.** TW
117. Let n represent the number; $2n + 9$, or $9 + 2n$
118. Let m and n represent the two numbers; $\frac{1}{2}(m + n)$
119. TW **121.** $-6r - 5t + 21$ **123.** $-2x - f$
125. TW **127.** True **129.** False **131.** 0
133. 39,000 **135.** $44x^3$

Review Exercises: Chapter 1, pp. 77–79

1. True **2.** True **3.** False **4.** True **5.** False
6. False **7.** True **8.** False **9.** False **10.** True
11. 15 **12.** -7 **13.** -5 **14.** $z - 7$ **15.** xz
16. Let m and n represent the numbers; $mn + 1$, or $1 + mn$
17. No **18.** Let d represent the number of digital prints
made in 2006, in billions; $14.1 = d + 3.2$ **19.** $c = 300t$
20. $t \cdot 3 + 5$ **21.** $2x + (y + z)$ **22.** $(4x)y, 4(yx), (4y)x$;
answers may vary **23.** $18x + 30y$ **24.** $40x + 24y + 16$
25. $3(7x + 5y)$ **26.** $7(5x + 2 + y)$ **27.** $2 \cdot 2 \cdot 13$
28. $\frac{5}{12}$ **29.** $\frac{9}{4}$ **30.** $\frac{31}{36}$ **31.** $\frac{3}{16}$ **32.** $\frac{3}{5}$ **33.** $\frac{72}{25}$
34. $-45, 72$ **35.**
$$\overset{\frac{-1}{3}}{\xleftarrow{\hspace{1em}}\!\!\!\underset{-5\ -4\ -3\ -2\ -1\ \ 0\ \ 1\ \ 2\ \ 3\ \ 4\ \ 5}{|\ |\ |\ |\ |\ \bullet\ |\ |\ |\ |\ |}\!\!\!\xrightarrow{\hspace{1em}}}$$
36. $x > -3$ **37.** True **38.** False **39.** -0.875
40. 1 **41.** -9 **42.** -3 **43.** $-\frac{7}{12}$ **44.** 0 **45.** -5
46. 5 **47.** $-\frac{7}{5}$ **48.** -7.9 **49.** 54 **50.** -9.18
51. $-\frac{2}{7}$ **52.** -140 **53.** -7 **54.** -3 **55.** $\frac{3}{4}$
56. 92 **57.** 62 **58.** 48 **59.** 168 **60.** $\frac{21}{8}$ **61.** $\frac{103}{17}$
62. $7a - 3b$ **63.** $-2x + 5y$ **64.** 7 **65.** $-\frac{1}{7}$
66. $(2x)^4$ **67.** $-125x^3$ **68.** $-3a + 9$ **69.** $-2b + 21$
70. $-3x + 9$ **71.** $12y - 34$ **72.** $5x + 24$
73. TW The value of a constant never varies. A variable can
represent a variety of numbers. **74.** TW A term is one of
the parts of an expression that is separated from the other
parts by plus signs. A factor is part of a product.

75. TW The distributive law is used in factoring algebraic
expressions, multiplying algebraic expressions, combining
like terms, finding the opposite of a sum, and subtracting
algebraic expressions. **76.** TW A negative number raised
to an even exponent is positive; a negative number raised to an
odd exponent is negative. **77.** 25,281
78. (a) $\frac{3}{11}$; **(b)** $\frac{10}{11}$ **79.** $-\frac{5}{8}$ **80.** -2.1 **81.** (i) **82.** (j)
83. (a) **84.** (h) **85.** (k) **86.** (b) **87.** (c) **88.** (e)
89. (d) **90.** (f) **91.** (g)

Test: Chapter 1, p. 80

1. [1.1] 4 **2.** [1.1] Let x represent the number; $x - 9$
3. [1.1] 240 ft^2 **4.** [1.2] $q + 3p$ **5.** [1.2] $(x \cdot 4) \cdot y$
6. [1.1] No **7.** [1.1] Let p represent the maximum
production capability; $p - 282 = 2518$ **8.** [1.2] $35 - 7x$
9. [1.7] $-5y + 10$ **10.** [1.2] $11(1 - 4x)$
11. [1.2] $7(x + 3 + 2y)$ **12.** [1.3] $2 \cdot 2 \cdot 3 \cdot 5 \cdot 5$
13. [1.3] $\frac{2}{7}$ **14.** [1.4] $<$ **15.** [1.4] $>$ **16.** [1.4] $\frac{9}{4}$
17. [1.4] 2.7 **18.** [1.6] $-\frac{2}{3}$ **19.** [1.7] $-\frac{7}{4}$ **20.** [1.6] 8
21. [1.4] $-2 \geq x$ **22.** [1.6] 7.8 **23.** [1.5] -8
24. [1.6] -2.5 **25.** [1.6] $\frac{7}{8}$ **26.** [1.7] -48 **27.** [1.7] $\frac{3}{16}$
28. [1.7] -6 **29.** [1.7] $\frac{3}{4}$ **30.** [1.7] -9.728
31. [1.8] -173 **32.** [1.6] 15 **33.** [1.8] -4
34. [1.8] 448 **35.** [1.6] $21a + 22y$ **36.** [1.8] $16x^4$
37. [1.8] $x + 7$ **38.** [1.8] $9a - 12b - 7$
39. [1.8] $68y - 8$ **40.** [1.1] 5
41. [1.8] $9 - (3 - 4) + 5 = 15$ **42.** [1.8] 15
43. [1.8] $4a$ **44.** [1.8] False

Chapter 2

Exercise Set 2.1, pp. 90–91

1. (f) **2.** (c) **3.** (a) **4.** (b) **5.** (d) **6.** (e)
7. 17 **9.** -11 **11.** -21 **13.** -31 **15.** 13
17. 19 **19.** -4 **21.** $\frac{7}{3}$ **23.** $-\frac{13}{10}$ **25.** $\frac{41}{24}$ **27.** $-\frac{1}{20}$
29. 1.5 **31.** -5 **33.** 14 **35.** 4 **37.** 12 **39.** -23
41. 8 **43.** -7 **45.** 8 **47.** -88 **49.** 20 **51.** -54
53. $\frac{5}{9}$ **55.** 1 **57.** $\frac{9}{2}$ **59.** -7.6 **61.** -2.5 **63.** -15
65. 18 **67.** -6 **69.** -128 **71.** $-\frac{1}{2}$ **73.** -15
75. 12 **77.** 310.756 **79.** TW **81.** -34 **82.** 41
83. 1 **84.** -16 **85.** TW **87.** 9.4 **89.** 2
91. $-13, 13$ **93.** 9000 **95.** 250

Exercise Set 2.2, pp. 98–99

1. (c) **2.** (e) **3.** (a) **4.** (f) **5.** (b) **6.** (d)
7. 8 **9.** 7 **11.** 5 **13.** 14 **15.** -7 **17.** -11
19. -24 **21.** 19 **23.** $\frac{10}{9}$ **25.** 3 **27.** 15 **29.** -4

31. $-\frac{28}{3}$ **33.** All real numbers; identity   **35.** -3
37. 5   **39.** 2   **41.** 0   **43.** 8   **45.** 0   **47.** 10
49. 4   **51.** 0   **53.** 2   **55.** -8 **57.** 2
59. No solution; contradiction   **61.** $-\frac{2}{5}$ **63.** $\frac{64}{3}$ **65.** $\frac{2}{5}$
67. 3   **69.** -4 **71.** $1.\overline{6}$ **73.** $-\frac{40}{37}$ **75.** 11   **77.** 6
79. $\frac{16}{15}$ **81.** $-\frac{51}{31}$ **83.** 2   **85.** TW   **87.** -7 **88.** 15
89. -15 **90.** -28 **91.** TW   **93.** $\dfrac{1136}{909}$, or $1.\overline{2497}$

95. No solution; contradiction   **97.** No solution;
contradiction   **99.** $\frac{2}{3}$ **101.** 0   **103.** 0   **105.** -2

Exercise Set 2.3, pp. 106–109

1. 2 mi   **3.** 1423 students   **5.** 54,000 Btu's   **7.** 255 mg
9. $b = \dfrac{A}{h}$ **11.** $r = \dfrac{d}{t}$ **13.** $P = \dfrac{I}{rt}$ **15.** $m = 65 - H$
17. $l = \dfrac{P - 2w}{2}$, or $l = \dfrac{P}{2} - w$ **19.** $\pi = \dfrac{A}{r^2}$ **21.** $h = \dfrac{2A}{b}$
23. $m = \dfrac{E}{c^2}$ **25.** $d = 2Q - c$ **27.** $b = 3A - a - c$
29. $A = Ms$ **31.** $C = \frac{5}{9}(F - 32)$ **33.** $t = \dfrac{A}{a + b}$
35. $h = \dfrac{2A}{a + b}$ **37.** $L = W - \dfrac{N(R - r)}{400}$, or
$L = \dfrac{400W - NR + Nr}{400}$ **39.** TW   **41.** 0   **42.** 9.18
43. -13 **44.** 65   **45.** TW   **47.** 35 yr   **49.** 27 in³
51. $a = \dfrac{w}{c} \cdot d$ **53.** $c = \dfrac{d}{a - b}$ **55.** $a = \dfrac{c}{3 + b + d}$
57. $K = 917 + 13.2276w + 2.3622h - 6a$

Exercise Set 2.4, pp. 114–118

1. (d)   **2.** (c)   **3.** (e)   **4.** (b)   **5.** (c)   **6.** (d)
7. (f)   **8.** (a)   **9.** (b)   **10.** (e)   **11.** 0.49   **13.** 0.02
15. 0.77   **17.** 0.2   **19.** 0.6258   **21.** 0.007   **23.** 1.25
25. 64%   **27.** 10.6%   **29.** 42%   **31.** 90%
33. 0.49%   **35.** 108%   **37.** 230%   **39.** 80%
41. 32%   **43.** 25%   **45.** 24%   **47.** $46\frac{2}{3}$, or $\frac{140}{3}$
49. 2.5   **51.** 84   **53.** 125%   **55.** 0.8   **57.** 50%
59. $198   **61.** $1584   **63.** $528   **65.** 75 credits
67. About 626 at-bats   **69.** (a) 16%; (b) $29
71. About 47%   **73.** $280   **75.** 285 women
77. $18/hr   **79.** 150%   **81.** $36   **83.** $148.50
85. About 31.5 lb   **87.** 7410 brochures
89. About 165 calories   **91.** TW   **93.** Let n represent the
number; $n + 5$ **94.** Let t represent Tino's weight; $t - 4$
95. $8 \cdot 2a$ **96.** Let x and y represent the two numbers;
$xy + 1$ **97.** TW   **99.** 18,500 people
101. About 5 ft 6 in.   **103.** About 1.4%; 2004: 14.3 per
1000 women; 2005: 14.5 per 1000 women   **105.** TW

Exercise Set 2.5, pp. 127–132

1. 8   **3.** 11   **5.** $85   **7.** $85   **9.** Approximately
$62\frac{2}{3}$ mi   **11.** 290 mi   **13.** 1204 and 1205   **15.** 396 and
398   **17.** 19, 20, 21   **19.** Bride: 102 yr; groom: 83 yr
21. Bathrooms: $11\frac{2}{3}$ billion; kitchens: $23\frac{1}{3}$ billion
23. 140 and 141   **25.** Width: 100 ft; length: 160 ft; area:
16,000 ft²   **27.** Width: 50 ft; length: 84 ft   **29.** $1\frac{3}{4}$ in. by
$3\frac{1}{2}$ in.   **31.** 30°, 90°, 60°   **33.** 95°   **35.** Bottom: 144 ft;
middle: 72 ft; top: 24 ft   **37.** $8\frac{3}{4}$ mi   **39.** $128\frac{1}{3}$ mi
41. 65°, 25°   **43.** Length: 27.9 cm; width: 21.6 cm
45. $6600   **47.** 1049 points   **49.** 160 chirps per minute
51. 32 scores   **53.** August 19   **55.** TW   **57.** <
58. <   **59.** <   **60.** >   **61.** TW   **63.** $37   **65.** 20
67. Half-dollars: 5; quarters: 10; dimes: 20; nickels: 60
69. 120 apples   **71.** 30 games   **73.** 76   **75.** TW
77. Width: 23.31 cm; length: 27.56 cm

Exercise Set 2.6, pp. 141–143

1. $\geq$ **2.** $\leq$ **3.** <   **4.** >   **5.** Equivalent
6. Equivalent   **7.** Equivalent   **8.** Not equivalent
9. (a) Yes; (b) yes; (c) yes; (d) no; (e) yes
11. (a) No; (b) no; (c) yes; (d) yes; (e) no
13. $x \leq 7$
15. $t > -2$
17. $1 \leq m$
19. $-3 < x \leq 5$
21. $0 < x < 3$
23. $\{x \mid x > -4\}$, or $(-4, \infty)$ **25.** $\{x \mid x \leq 2\}$, or $(-\infty, 2]$
27. $\{x \mid x < -1\}$, or $(-\infty, -1)$ **29.** $\{x \mid x \geq 0\}$, or $[0, \infty)$
31. $\{y \mid y > 7\}$, or $(7, \infty)$,
33. $\{x \mid x \leq -18\}$, or $(-\infty, -18]$,
35. $\{x \mid x < 10\}$, or $(-\infty, 10)$,
37. $\{t \mid t \geq -3\}$, or $[-3, \infty)$,
39. $\{y \mid y > -5\}$, or $(-5, \infty)$,
41. $\{x \mid x \leq 5\}$, or $(-\infty, 5]$,
43. $\{y \mid y \leq \frac{1}{2}\}$, or $\left(-\infty, \frac{1}{2}\right]$ **45.** $\{t \mid t > \frac{5}{8}\}$, or $\left(\frac{5}{8}, \infty\right)$
47. $\{x \mid x < 0\}$, or $(-\infty, 0)$ **49.** $\{t \mid t < 23\}$, or $(-\infty, 23)$
51. $\{x \mid x < 7\}$, or $(-\infty, 7)$,
53. $\{x \mid x > -\frac{13}{7}\}$, or $\left(-\frac{13}{7}, \infty\right)$,
55. $\{t \mid t < -3\}$, or $(-\infty, -3)$,

57. $\left\{y \mid y \geq -\frac{2}{7}\right\}$, or $\left[-\frac{2}{7}, \infty\right)$ **59.** $\left\{y \mid y \geq -\frac{1}{10}\right\}$, or $\left[-\frac{1}{10}, \infty\right)$
61. $\left\{x \mid x > \frac{4}{5}\right\}$, or $\left(\frac{4}{5}, \infty\right)$ **63.** $\{x \mid x < 9\}$, or $(-\infty, 9)$
65. $\{y \mid y \geq 4\}$, or $[4, \infty)$ **67.** $\{t \mid t \leq 7\}$, or $(-\infty, 7]$
69. $\{y \mid y < -4\}$, or $(-\infty, -4)$ **71.** $\{x \mid x > -4\}$, or
$(-4, \infty)$ **73.** $\left\{y \mid y < -\frac{10}{3}\right\}$, or $\left(-\infty, -\frac{10}{3}\right)$
75. $\{x \mid x > -10\}$, or $(-10, \infty)$ **77.** $\{y \mid y < 2\}$, or $(-\infty, 2)$
79. $\{y \mid y \geq 3\}$, or $[3, \infty)$ **81.** $\{x \mid x > -4\}$, or $(-4, \infty)$
83. $\{n \mid n \geq 70\}$, or $[70, \infty)$ **85.** $\{x \mid x \leq 15\}$, or $(-\infty, 15]$
87. $\{t \mid t < 14\}$, or $(-\infty, 14)$ **89.** $\{y \mid y < 6\}$, or $(-\infty, 6)$
91. $\{t \mid t \leq -4\}$, or $(-\infty, -4]$ **93.** $\{r \mid r > -3\}$, or $(-3, \infty)$
95. $\{x \mid x \geq 8\}$, or $[8, \infty)$ **97.** $\left\{x \mid x < \frac{11}{18}\right\}$, or $\left(-\infty, \frac{11}{18}\right)$
99. TW   **101.** 1   **102.** $\{x \mid x > 1\}$, or $(1, \infty)$ **103.** $\frac{19}{5}$
104. $\left\{x \mid x \geq \frac{19}{5}\right\}$, or $\left[\frac{19}{5}, \infty\right)$ **105.** TW
107. $\{x \mid x$ is a real number$\}$, or $(-\infty, \infty)$
109. $\left\{x \mid x \leq \frac{5}{6}\right\}$, or $\left(-\infty, \frac{5}{6}\right]$ **111.** $\{x \mid x > 7\}$, or $(7, \infty)$
113. $\left\{x \mid x < \dfrac{y-b}{a}\right\}$, or $\left(-\infty, \dfrac{y-b}{a}\right)$
115. $\{x \mid x$ is a real number$\}$, or $(-\infty, \infty)$

Exercise Set 2.7, pp. 148–153

1. $b \leq a$ **2.** $b < a$ **3.** $a \leq b$ **4.** $a < b$ **5.** $b \leq a$
6. $a \leq b$ **7.** $b < a$ **8.** $a < b$ **9.** Let n represent the
number; $n \geq 8$ **11.** Let t represent the temperature; $t \leq -3$
13. Let p represent the price of Pat's PT Cruiser; $p > 21{,}900$
15. Let d represent the distance to Normandale Community
College; $d \leq 15$ **17.** Let n represent the number; $n > -2$
19. Let p represent the number of people attending the
Million Man March; $400{,}000 < p < 1{,}200{,}000$ **21.** More
than 2.5 hr   **23.** More than 18 one-way trips per month
25. Scores greater than or equal to 97   **27.** 8 credits or
more   **29.** 21 calls or more   **31.** Lengths greater than
6 cm   **33.** Depths less than 437.5 ft   **35.** Blue-book value
is greater than or equal to $10,625   **37.** Lengths greater
than or equal to 5 in.   **39.** Temperatures greater than 37°C
41. Heights at least 4 ft   **43.** A serving contains at least
16 g of fat.   **45.** Dates after September 16   **47.** 14 or
fewer copies   **49.** Years after 1995   **51.** Mileages less
than or equal to 193   **53.** For 363 min or less   **55.** More
than 25 checks   **57.** Gross sales greater than $7000
59. About 6.8 gal or less   **61.** TW   **63.** 2   **64.** $\frac{1}{2}$
65. $-\frac{10}{3}$ **66.** $-\frac{1}{5}$ **67.** TW   **69.** Parties of more than
80 guests   **71.** More than 6 hr   **73.** Lengths less than or
equal to 8 cm   **75.** They contain at least 7.5 g of fat per
serving.   **77.** At least $42   **79.** TW

Review Exercises: Chapter 2, pp. 156–157

1. True   **2.** False   **3.** True   **4.** True   **5.** True
6. False   **7.** True   **8.** True   **9.** -25 **10.** 7
11. -65 **12.** 1   **13.** -5 **14.** 1.11   **15.** $\frac{1}{2}$

16. $-\frac{15}{64}$ **17.** $\frac{38}{5}$ **18.** -8 **19.** -5 **20.** $-\frac{1}{3}$
21. 4   **22.** 3   **23.** 4   **24.** 16   **25.** 7   **26.** $-\frac{7}{5}$
27. 12   **28.** No solution; contradiction   **29.** $d = \dfrac{C}{\pi}$
30. $B = \dfrac{3V}{h}$ **31.** $b = 2A - a$ **32.** 0.009   **33.** 44%
34. 70%   **35.** 140   **36.** Yes   **37.** No   **38.** Yes
39.
$$5x - 6 < 2x + 3$$
40.
$$-2 < x \leq 5$$
41.
$$t > 0$$
42. $\left\{t \mid t \geq -\frac{1}{2}\right\}$, or $\left[-\frac{1}{2}, \infty\right)$
43. $\{x \mid x \geq 7\}$, or $[7, \infty)$ **44.** $\{y \mid y > 3\}$, or $(3, \infty)$
45. $\{y \mid y \leq -4\}$, or $(-\infty, -4]$ **46.** $\{x \mid x < -11\}$, or
$(-\infty, -11)$ **47.** $\{y \mid y > -7\}$, or $(-7, \infty)$
48. $\{x \mid x > -6\}$, or $(-6, \infty)$ **49.** $\left\{x \mid x > -\frac{9}{11}\right\}$, or $\left(-\frac{9}{11}, \infty\right)$
50. $\{t \mid t \leq -12\}$, or $(-\infty, -12]$ **51.** $\{x \mid x \leq -8\}$, or
$(-\infty, -8]$ **52.** 20 bottles   **53.** 15 ft, 17 ft
54. $240 billion   **55.** 57, 59   **56.** Width: 11 cm;
length: 17 cm   **57.** $160   **58.** $102 million
59. 35°, 85°, 60°   **60.** 7 subscriptions   **61.** $105 or less
62. Widths greater than 17 cm   **63.** TW Multiplying both
sides of an equation by *any* nonzero number results in an
equivalent equation. When multiplying on both sides of an
inequality, the sign of the number being multiplied by must
be considered. If the number is positive, the direction of the
inequality symbol remains unchanged; if the number is
negative, the direction of the inequality symbol must be
reversed to produce an equivalent inequality.
64. TW The solutions of an equation can usually each be
checked. The solutions of an inequality are normally too
numerous to check. Checking a few numbers from the
solution set found cannot guarantee that the answer is correct,
although if any number does not check, the answer found is
incorrect.   **65.** 25 hr 39 min   **66.** Nile: 6690 km;
Amazon: 6296 km   **67.** $18,600   **68.** $-23, 23$
69. $-20, 20$ **70.** $a = \dfrac{y-3}{2-b}$

Test: Chapter 2, p. 158

1. [2.1] 9   **2.** [2.1] 15   **3.** [2.1] -3 **4.** [2.1] 49
5. [2.1] -12 **6.** [2.2] 2   **7.** [2.1] -8 **8.** [2.1] $-\frac{7}{20}$
9. [2.2] 7   **10.** [2.2] $-\frac{5}{3}$ **11.** [2.2] $\frac{23}{3}$ **12.** [2.2] All real
numbers; identity   **13.** [2.6] $\{x \mid x > -5\}$, or $(-5, \infty)$
14. [2.6] $\{x \mid x > -13\}$, or $(-13, \infty)$ **15.** [2.6] $\left\{x \mid x < \frac{21}{8}\right\}$,
or $\left(-\infty, \frac{21}{8}\right)$ **16.** [2.6] $\{y \mid y \geq -13\}$, or $[-13, \infty)$
17. [2.6] $\{y \mid y \leq -8\}$, or $(-\infty, -8]$ **18.** [2.6] $\left\{x \mid x \leq -\frac{1}{20}\right\}$,
or $\left(-\infty, -\frac{1}{20}\right]$ **19.** [2.6] $\{x \mid x < -6\}$, or $(-\infty, -6)$
20. [2.6] $\{x \mid x \leq -1\}$, or $(-\infty, -1]$ **21.** [2.3] $r = \dfrac{A}{2\pi h}$

22. [2.3] $l = 2w - P$ **23.** [2.4] 2.3   **24.** [2.4] 5.4%
25. [2.4] 16   **26.** [2.4] 44%
27. [2.6]   **28.** [2.6]

$y < 4$
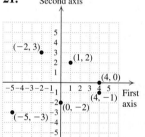

$-2 \le x \le 2$
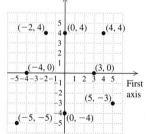

29. [2.5] Width: 7 cm; length: 11 cm   **30.** [2.5] 60 mi
31. [2.5] 81 mm, 83 mm, 85 mm   **32.** [2.4] $65
33. [2.7] Mileages less than or equal to 525.8 mi

34. [2.3] $d = \dfrac{a}{3}$ **35.** [1.4], [2.2] $-15, 15$

36. [2.5] 60 tickets

Chapter 3

Exercise Set 3.1, pp. 170–173

1. (a)   **2.** (c)   **3.** (b)   **4.** (d)   **5.** 2 drinks
7. The person weighs more than 200 lb.   **9.** About 3,000,000
11. About 1,500,000   **13.** About 26.7 million tons
15. About 2.8 million tons   **17.** 120,000,000 phones
19. 2004

21.

23.
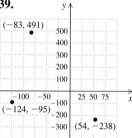

25.
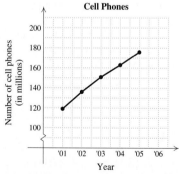

27. $A(-4, 5)$; $B(-3, -3)$; $C(0, 4)$; $D(3, 4)$; $E(3, -4)$
29. $A(4, 1)$; $B(0, -5)$; $C(-4, 0)$; $D(-3, -2)$; $E(3, 0)$

31.
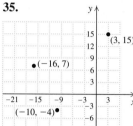

33.
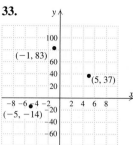

35.
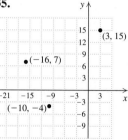

37.
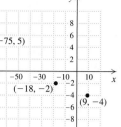

39.
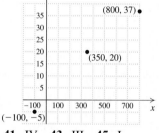

41. IV   **43.** III   **45.** I

47. II   **49.** I and IV   **51.** I and III   **53.** TW   **55.** -18
56. -31 **57.** 6   **58.** 1   **59.** $y = \frac{3}{2}x - 3$
60. $y = \frac{7}{4}x - \frac{7}{2}$ **61.** TW   **63.** II or IV   **65.** $(-1, -5)$

67.
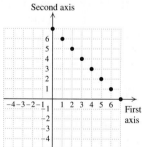

69. $\dfrac{65}{2}$ sq units

71. Latitude 27° North; longitude 81° West   **73.** TW

Exercise Set 3.2, pp. 184–186

1. No   **2.** Yes   **3.** No   **4.** No   **5.** Yes   **6.** Yes

7.

$$\begin{array}{c|c} y = x + 3 \\ \hline 2 & -1 + 3 \\ 2 \overset{?}{=} 2 \end{array} \quad \text{True}$$

$$\begin{array}{c|c} y = x + 3 \\ \hline 7 & 4 + 3 \\ 7 \overset{?}{=} 7 \end{array} \quad \text{True}$$

$(2, 5)$; answers may vary

9.

$$\begin{array}{c|c} y = \frac{1}{2}x + 3 \\ \hline 5 & \frac{1}{2} \cdot 4 + 3 \\ & 2 + 3 \\ 5 \overset{?}{=} 5 \end{array} \quad \text{True}$$

$$\begin{array}{c|c} y = \frac{1}{2}x + 3 \\ \hline 2 & \frac{1}{2}(-2) + 3 \\ & -1 + 3 \\ 2 \overset{?}{=} 2 \end{array} \quad \text{True}$$

$(0, 3)$; answers may vary

11.

$$\begin{array}{c|c} y + 3x = 7 \\ \hline 1 + 3 \cdot 2 & 7 \\ 1 + 6 & \\ 7 \overset{?}{=} 7 \quad \text{True} \end{array}$$

$$\begin{array}{c|c} y + 3x = 7 \\ \hline -5 + 3 \cdot 4 & 7 \\ -5 + 12 & \\ 7 \overset{?}{=} 7 \quad \text{True} \end{array}$$

$(1, 4)$; answers may vary

13.

$$\begin{array}{c|c} 4x - 2y = 10 \\ \hline 4 \cdot 0 - 2(-5) & 10 \\ 0 + 10 & \\ 10 \overset{?}{=} 10 \quad \text{True} \end{array}$$

$$\begin{array}{c|c} 4x - 2y = 10 \\ \hline 4 \cdot 4 - 2 \cdot 3 & 10 \\ 16 - 6 & \\ 10 \overset{?}{=} 10 \quad \text{True} \end{array}$$

$(2, -1)$; answers may vary

15.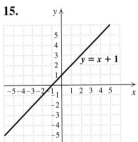

17.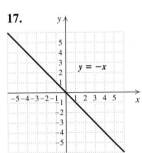

19.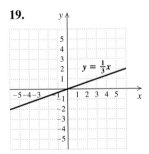

21.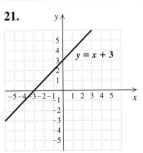

23.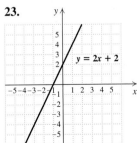

25.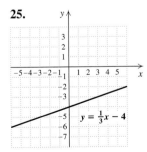

27.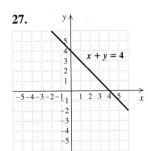

29.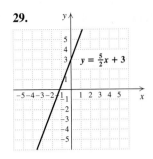

31.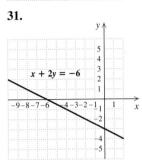

33.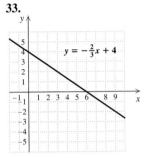

35.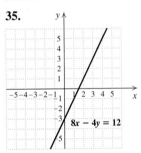

37.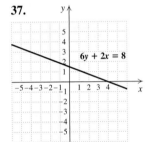

39.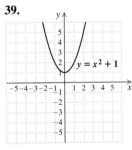

41.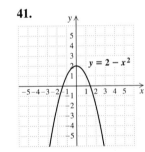

43. (a)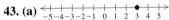

(b)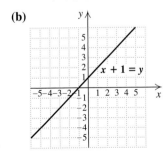

45. (a)

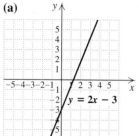

(b)

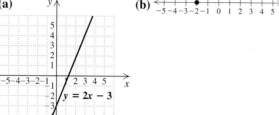

47. $y = (-3/2)x + 1$

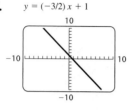

49. $4y - 3x = 1$, or $y = (3x + 1)/4$

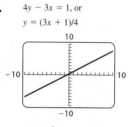

51. $y = -2$

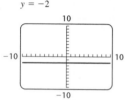

53. $y = -x^2$

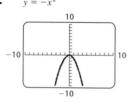

55. $x^2 - y = 3$, or $y = x^2 - 3$

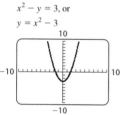

57. $y = x \wedge 3$

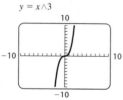

59. (b)   61. (a)   63. (b)

65.

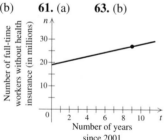

27 million workers

67.

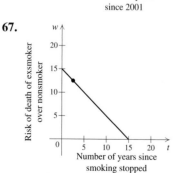

$12\frac{1}{2}$ times

69.

$300

71.

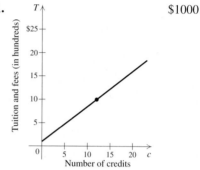

$1000

73.

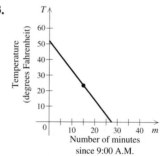

24°F   **75. TW**

77. $\frac{12}{5}$ **78.** $\frac{9}{2}$ **79.** $-\frac{5}{2}$ **80.** $p = \dfrac{w}{q+1}$

81. $y = \dfrac{C - Ax}{B}$ **82.** $Q = 2A - T$ **83. TW**

85. $s + n = 18$

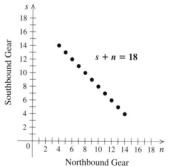

87. $x + y = 2$, or $y = -x + 2$
89. $5x - 3y = 15$, or $y = \frac{5}{3}x - 5$

91.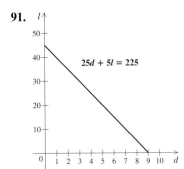

Answers may vary. 1 dinner, 40 lunches; 5 dinners, 20 lunches; 8 dinners, 5 lunches

93. 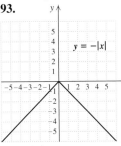 **95.**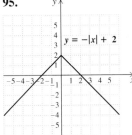

97. TW

Exercise Set 3.3, pp. 193–195

1. (f)   **2.** (e)   **3.** (d)   **4.** (c)   **5.** (b)   **6.** (a)
7. Linear   **9.** Linear   **11.** Not linear   **13.** Linear
15. Not linear   **17. (a)** $(0, 5)$; **(b)** $(2, 0)$ **19. (a)** $(0, -4)$;
(b) $(3, 0)$ **21. (a)** $(0, -2)$; **(b)** $(-3, 0), (3, 0)$
23. (a) $(0, 4)$; **(b)** $(-3, 0), (3, 0), (5, 0)$ **25. (a)** $(0, 5)$;
(b) $(3, 0)$ **27. (a)** $(0, -14)$; **(b)** $(4, 0)$ **29. (a)** $(0, 50)$;
(b) $\left(-\frac{75}{2}, 0\right)$ **31. (a)** $(0, 9)$; **(b)** none   **33. (a)** None;
(b) $(-7, 0)$

35. 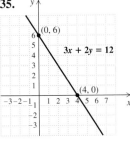 **37.**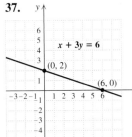

39. 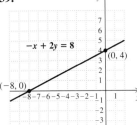 **41.**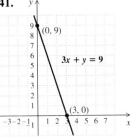

43. 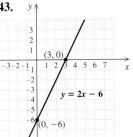 **45.**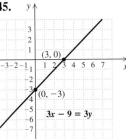

47. 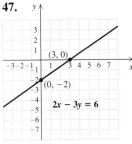 **49.**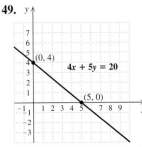

51. 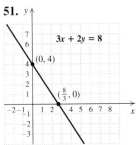 **53.**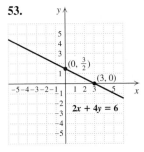

55. 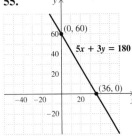 **57.**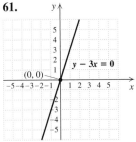

59. 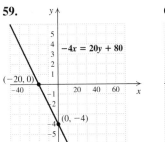 **61.**

63.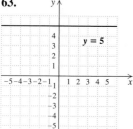

65.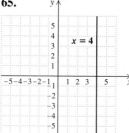

67.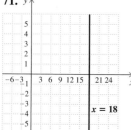

69.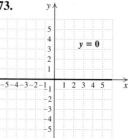

71.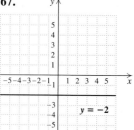

73.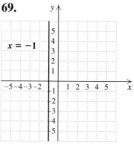

75.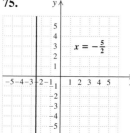

77.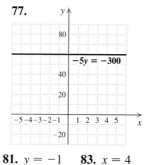

79.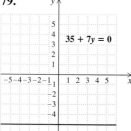

81. $y = -1$ **83.** $x = 4$

85. $y = 0$ **87.** $(5, 0), (0, 20)$; (c)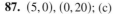

89. $(200, 0), (0, 7000)$; (d)   **91.** $y = -0.72x - 15$

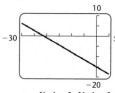

Xscl = 5, Yscl = 5

93. $5x + 6y = 84$, or
$y = (84 - 5x)/6$

95. $19x - 17y = 200$, or
$y = (19x - 200)/17$

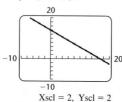

Xscl = 2, Yscl = 2

Xscl = 5, Yscl = 5

97. TW   **99.** $d - 7$ **100.** $5 + w$, or $w + 5$
101. Let n represent the number; $2 + n$ **102.** Let n
represent the number; $3n$ **103.** Let x and y represent the
numbers; $2(x + y)$ **104.** Let a and b represent the numbers;
$\frac{1}{2}(a + b)$ **105.** TW   **107.** $y = 0$ **109.** $x = -2$
111. $(-3, -3)$ **113.** $-5x + 3y = 15$, or $y = \frac{5}{3}x + 5$
115. -24

Exercise Set 3.4, pp. 200–204

1. (a) 21 mpg; **(b)** \$39.33/day; **(c)** 91 mi/day; **(d)** 43¢/mile
3. (a) 6 mph; **(b)** \$4/hr; **(c)** \$0.67/mi   **5. (a)** \$16/hr;
(b) 4.5 pages/hr; **(c)** \$3.56/page   **7.** \$145/yr
9. (a) 14.5 floors/min; **(b)** 4.14 sec/floor
11. (a) 3.71 ft/min; **(b)** 0.27 min/ft
13.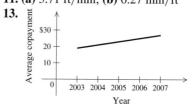

15.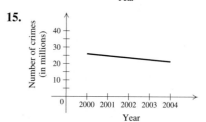

17.

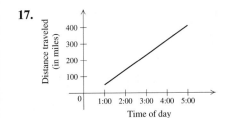

19.

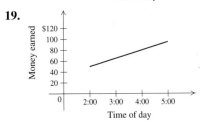

21.

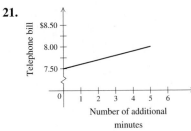

23. 2 haircuts/hr   **25.** 75 mi/hr   **27.** 7¢/min
29. −$500/yr   **31.** 0.04 gal/mi   **33.** (e)   **35.** (d)
37. (b)   **39.** TW   **41.** 5   **42.** −6   **43.** −1   **44.** −$\frac{4}{3}$
45. −$\frac{4}{3}$ **46.** −$\frac{4}{5}$ **47.** TW   **49.**

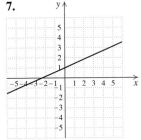

51.   **53.** 13 ft/sec

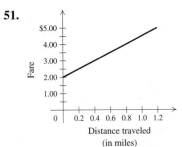

55. About 41.7 min   **57.** 4:20 P.M.

Exercise Set 3.5, pp. 213–220

1. Positive   **2.** Negative   **3.** Negative   **4.** Positive
5. Positive   **6.** Negative   **7.** Zero   **8.** Positive
9. Negative   **10.** Zero   **11.** 3 million people/yr

13. 1.3%/yr   **15.** $\frac{3}{4}$ point/$1000 income
17. −2.1°/min   **19.** $\frac{3}{4}$ **21.** $\frac{3}{2}$ **23.** $\frac{1}{3}$ **25.** −1   **27.** 0
29. −$\frac{1}{3}$ **31.** Undefined   **33.** −$\frac{1}{4}$ **35.** $\frac{3}{2}$ **37.** 0
39. −3   **41.** $\frac{3}{2}$ **43.** −$\frac{4}{5}$ **45.** $\frac{7}{9}$ **47.** −$\frac{2}{3}$ **49.** −$\frac{1}{2}$
51. 0   **53.** −$\frac{11}{6}$ **55.** Undefined   **57.** Undefined   **59.** 0
61. Undefined   **63.** 0   **65.** 8%   **67.** 8.$\overline{3}$%   **69.** $\frac{29}{98}$, or
about 30%   **71.** About 29%   **73.** (a) II; (b) IV; (c) I;
(d) III   **75.** TW   **77.** $y = \dfrac{c - ax}{b}$ **78.** $r = \dfrac{p + mn}{x}$
79. $y = \dfrac{ax - c}{b}$ **80.** $t = \dfrac{q - rs}{n}$ **81.** 3   **82.** 2
83. TW   **85.** $\{m \mid -\frac{7}{4} \le m \le 0\}$ **87.** $\dfrac{18 - x}{x}$ **89.** $\frac{1}{4}$
91. 0.364, or 36.4%

Interactive Discovery, p. 222

1. Yes   **2.** The origin, or $(0,0)$ **3.** The origin, or $(0,0)$

Interactive Discovery, p. 223

1. The value of y_2 is 5 more than that of y_1 for the same value
of x. The value of y_3 is 7 less than that of y_1.   **2.** The graph
of y_4 will look like the graph of y_1, shifted down 3.2 units.
3. The graph is shifted up or down, depending on the sign of b.

Interactive Discovery, p. 224

1. Yes   **2.** No   **3.** No   **4.** Yes   **5.** m

Exercise Set 3.6, pp. 230–233

1. (c)   **2.** (e)   **3.** (f)   **4.** (b)   **5.** (d)   **6.** (a)
7.   **9.**

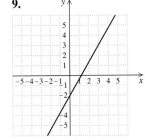

11.   **13.**

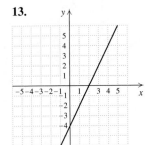

15.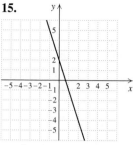

17. $-\frac{2}{7}; (0,5)$ **19.** $\frac{5}{8}; (0,3)$

21. $\frac{9}{5}; (0,-4)$ **23.** $3; (0,7)$ **25.** $-\frac{5}{2}; (0,4)$

27. $0; (0,4)$ **29.** $\frac{2}{5}; \left(0,\frac{8}{5}\right)$ **31. (a)** II; **(b)** IV; **(c)** III; **(d)** I

33. $y = 3x + 7$ **35.** $y = \frac{7}{8}x - 1$ **37.** $y = -\frac{5}{3}x - 8$

39. $y = 3$ **41.** $y = \frac{8}{9}x + 9$, where y is the number of gallons per person and x is the number of years since 1990

43. $y = 15x + 250$, where y is the number of jobs, in thousands, and x is the number of years since 1998

45.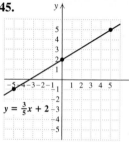

47.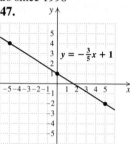

49.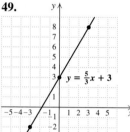

51.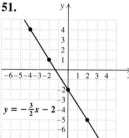

53.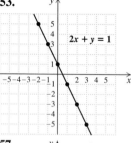

55.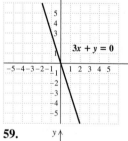

57.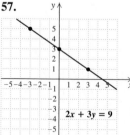

59.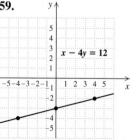

61. Yes **63.** No **65.** Yes **67.** Yes **69.** No

71. Yes **73.** $y = 5x + 11$ **75.** $y = \frac{1}{2}x$ **77.** $y = x + 3$

79. $y = x - 4$ **81.** TW

83.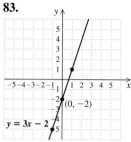

84.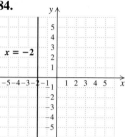

85.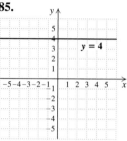

86.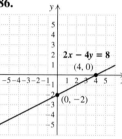

87.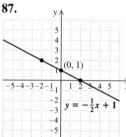

88.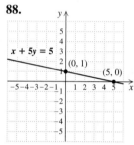

89. TW

91. When $x = 0$, $y = b$, so $(0, b)$ is on the line. When $x = 1$, $y = m + b$, so $(1, m + b)$ is on the line. Then

$$\text{slope} = \frac{(m + b) - b}{1 - 0} = m.$$

93. $y = 1.5x + 16$ **95.**

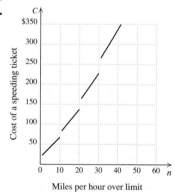

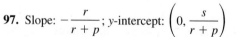

97. Slope: $-\dfrac{r}{r + p}$; y-intercept: $\left(0, \dfrac{s}{r + p}\right)$ **99.** TW

Visualizing the Graph, p. 247

1. C **2.** G **3.** F **4.** B **5.** D **6.** A **7.** I
8. H **9.** J **10.** E

Exercise Set 3.7, pp. 248–253

1. (g) **2.** (d) **3.** (e) **4.** (a) **5.** (b) **6.** (h)
7. (f) **8.** (c) **9.** (c) **11.** (d) **13.** $y - 2 = 5(x - 6)$
15. $y - 1 = -4(x - 3)$ **17.** $y - (-4) = \frac{3}{2}(x - 5)$
19. $y - 6 = \frac{5}{4}(x - (-2))$ **21.** $y - (-1) = -2(x - (-4))$
23. $y - 8 = 1(x - (-2))$ **25.** $\frac{2}{7}; (8, 9)$ **27.** $-5; (7, -2)$
29. $-\frac{5}{3}; (-2, 4)$ **31.** $\frac{4}{7}; (0, 0)$ **33.** $y = 2x - 3$
35. $y = \frac{7}{4}x - 9$ **37.** $y = -3x + 3$ **39.** $y = -4x - 9$
41. $y = -\frac{5}{6}x + 4$ **43.** $y = -\frac{1}{2}x + 9$ **45.** $y = 2x - 7$
47. $y = \frac{5}{3}x - \frac{28}{3}$ **49.** $y = 2x - 8$ **51.** $y = -\frac{5}{3}x - \frac{41}{3}$
53. $y = -\frac{1}{2}x + 6$ **55.** $y = -x + 6$ **57.** $y = \frac{2}{3}x + 3$
59. $y = \frac{2}{5}x - 2$ **61.** $y = \frac{3}{4}x - \frac{5}{2}$
63.
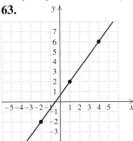
65.
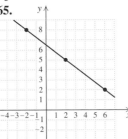
67.
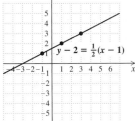
69.
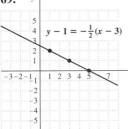
71.
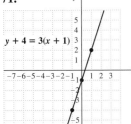
73.
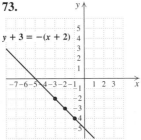

75.
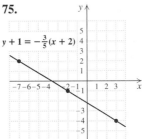

77. **(a)** $y = -1.62x + 63.72$; **(b)** 49.14 births per 1000 females; **(c)** 34.56 births per 1000 females
79. **(a)** $A = 0.2t + 78.2$; **(b)** 78.8 million acres; **(c)** 80.2 million acres **81.** **(a)** $C = \frac{168}{29}t + 2450$; **(b)** 2664 calories; **(c)** approximately 2023
83. **(a)** $E = 0.13t + 78.41$; **(b)** 81.01
85. **(a)** $N = 12.09t + 189.6$; **(b)** \$358.86 million
87. Linear **89.** Not linear **91.** Linear
93. **(a)** $W = 0.1673799884x + 63.2346443$; **(b)** 81.65 yr; this estimate is higher **95.** **(a)** $B = 885.3333333x + 81693$; **(b)** 93,202 financial institutions **97.** TW **99.** 4
100. Undefined **101.** 0 **102.** -6 **103.** $y = 5x - 1$
104. $y = -2$ **105.** $y = -2x + 3$ **106.** $y = \frac{1}{2}x + 3$
107. $y = -2x - 1$ **108.** $y = -\frac{3}{2}x + \frac{17}{2}$ **109.** TW
111.
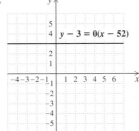
113. $y = 2x - 9$

115. $y = -\frac{4}{3}x + \frac{23}{3}$ **117.** $y = \frac{1}{5}x - 2$ **119.** $y = \frac{1}{2}x + 1$
121. 21.1°C **123.** \$11,000

Exercise Set 3.8, pp. 267–272

1. Domain **2.** Range **3.** Exactly **4.** More
5. Horizontal **6.** Vertical **7.** "f of 3," "f at 3," or "the value of f at 3" **8.** Function **9.** Yes **11.** Yes
13. Yes **15.** No **17.** Function **19.** Function
21. **(a)** -2; **(b)** $\{x \mid -2 \le x \le 5\}$, or $[-2, 5]$; **(c)** 4; **(d)** $\{y \mid -3 \le y \le 4\}$, or $[-3, 4]$
23. **(a)** 3; **(b)** $\{x \mid -1 \le x \le 4\}$, or $[-1, 4]$; **(c)** 3; **(d)** $\{y \mid 1 \le y \le 4\}$, or $[1, 4]$
25. **(a)** -2; **(b)** $\{x \mid -4 \le x \le 2\}$, or $[-4, 2]$; **(c)** -2; **(d)** $\{y \mid -3 \le y \le 3\}$, or $[-3, 3]$
27. **(a)** 3; **(b)** $\{x \mid -4 \le x \le 3\}$, or $[-4, 3]$; **(c)** -3; **(d)** $\{y \mid -2 \le y \le 5\}$, or $[-2, 5]$
29. **(a)** 1; **(b)** $\{-3, -1, 1, 3, 5\}$; **(c)** 3; **(d)** $\{-1, 0, 1, 2, 3\}$

31. (a) 4; (b) $\{x\,|-3 \le x \le 4\}$, or $[-3, 4]$; (c) $-1, 3$;
(d) $\{y\,|-4 \le y \le 5\}$, or $[-4, 5]$ **33.** (a) 1;
(b) $\{x\,|-4 < x \le 5\}$, or $(-4, 5]$; (c) $\{x\,|2 < x \le 5\}$, or $(2, 5]$;
(d) $\{-1, 1, 2\}$ **35.** Domain: $\mathbb{R}$; range: $\mathbb{R}$ **37.** Domain: $\mathbb{R}$;
range: $\{4\}$ **39.** Domain: $\mathbb{R}$; range: $\{y\,|y \ge 1\}$, or $[1, \infty)$
41. Domain: $\{x\,|x$ is a real number $and\ x \ne -2\}$;
range: $\{y\,|y$ is a real number $and\ y \ne -4\}$
43. Domain: $\{x\,|x \ge 0\}$, or $[0, \infty)$; range: $\{y\,|y \ge 0\}$, or $[0, \infty)$
45. Yes **47.** Yes **49.** No **51.** No **53.** (a) 3;
(b) -5; (c) -11; (d) 19; (e) $2a + 7$; (f) $2a + 5$ **55.** (a) 0;
(b) 1; (c) 57; (d) $5t^2 + 4t$; (e) $20a^2 + 8a$; (f) $10a^2 + 8a$
57. (a) $\frac{3}{5}$; (b) $\frac{1}{3}$; (c) $\frac{4}{7}$; (d) 0; (e) $\dfrac{x - 1}{2x - 1}$
59. $4\sqrt{3}$ cm$^2 \approx 6.93$ cm^2 **61.** 36π in$^2 \approx 113.10$ in^2
63. $1\frac{20}{33}$ atm; $1\frac{10}{11}$ atm; $4\frac{1}{33}$ atm **65.** 11 **67.** 0 **69.** $-\frac{21}{2}$
71. $\frac{25}{6}$ **73.** -3 **75.** -25
77. $\{x\,|x$ is a real number $and\ x \ne 3\}$
79. $\{x\,|x$ is a real number $and\ x \ne \frac{1}{2}\}$ **81.** $\mathbb{R}$ **83.** $\mathbb{R}$
85. $\{x\,|x$ is a real number $and\ x \ne 9\}$ **87.** $\mathbb{R}$ **89.** $\mathbb{R}$
91. (a) -5; (b) 1; (c) 21 **93.** (a) 0; (b) 2; (c) 7
95. (a) 100; (b) 100; (c) 131 **97.** 25 signifies that the cost
per person is \$25; 75 signifies that the setup cost for the party
is \$75. **99.** $\frac{1}{2}$ signifies that Ty's hair grows $\frac{1}{2}$ in. per month;
1 signifies that his hair is 1 in. long when cut. **101.** $\frac{1}{7}$ signifies that the life expectancy of American women increases $\frac{1}{7}$ yr
per year, for years after 1970; 75.5 signifies that the life expectancy in 1970 was 75.5 yr. **103.** 0.227 signifies that the
price increases \$0.227 per year, for years since 1995; 4.29 sig-
nifies that the average cost of a movie ticket in 1995 was
\$4.29. **105.** 2 signifies that the cost per mile of a taxi ride
is \$2; 2.5 signifies that the minimum cost of a taxi ride is
\$2.50. **107.** (a) -5000 signifies that the depreciation is
\$5000 per year; 90,000 signifies that the original value of the
truck was \$90,000; (b) 18 yr; (c) $\{t\,|0 \le t \le 18\}$
109. (a) -150 signifies that the depreciation is \$150 per win-
ter of use; 900 signifies that the original value of the snow-
blower was \$900; (b) after 4 winters of use; (c) $\{n\,|0 \le n \le 6\}$
111. TW
113.

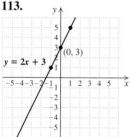
$y = 2x + 3$

114.

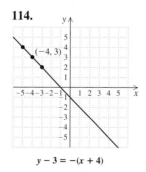
$y - 3 = -(x + 4)$

115.

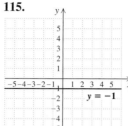
$y = -1$

116.

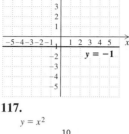
$f(x) = \frac{1}{2}x - 2$
$(0, -2)$

117.

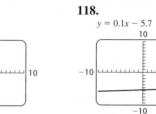
$y = x^2$

118.

$y = 0.1x - 5.7$

119.

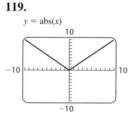
$y = \text{abs}(x)$

120.

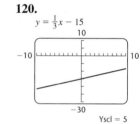
$y = \frac{1}{3}x - 15$
Yscl = 5

121. TW **123.** 26; 99 **125.** Worm
127. About 2 min 50 sec **129.** 1 every 3 min
131. $g(x) = \frac{15}{4}x - \frac{13}{4}$ **133.** False **135.** False

Review Exercises: Chapter 3, pp. 276–279

1. True **2.** True **3.** False **4.** False **5.** True
6. True **7.** True **8.** False **9.** True **10.** True
11. \$54,000 **12.** \$269.50
13.–15.

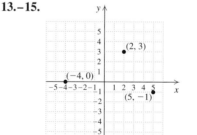
$(2, 3)$
$(-4, 0)$
$(5, -1)$

16. IV **17.** III
18. II **19.** $(-5, -1)$ **20.** $(-2, 5)$ **21.** $(3, 0)$
22. **23.** (a) Yes; (b) no

$(25, 7)$
$(-10, 6)$
$(-65, -2)$

24.

$$
\begin{array}{c|c}
2x - y = 3 \\
\hline
2 \cdot 0 - (-3) & 3 \\
0 + 3 & \\
3 \stackrel{?}{=} 3 \quad \text{True}
\end{array}
\qquad
\begin{array}{c|c}
2x - y = 3 \\
\hline
2 \cdot 2 - 1 & 3 \\
4 - 1 & \\
3 \stackrel{?}{=} 3 \quad \text{True}
\end{array}
$$

$(-1, -5)$; answers may vary

25.
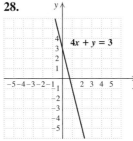

26.

27.
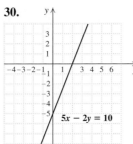

28.

29.
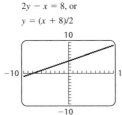

30.

31. $y = x^2 + 1$
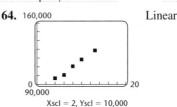

32. $2y - x = 8$, or $y = (x + 8)/2$
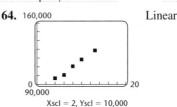

33.
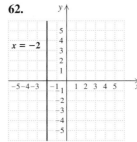

Percent of households engaged in flower planting — Number of years since 2000

28%

34. No **35.** Yes **36.** **(a)** $\frac{4}{9}$ meal/min; **(b)** $2\frac{1}{4}$ min/meal
37. $\frac{1}{12}$ gal/mi **38.** **(a)** $(0, -2)$; **(b)** none; **(c)** 0
39. **(a)** $(0, 2)$; **(b)** $(4, 0)$; **(c)** $-\frac{1}{2}$ **40.** **(a)** $(0, -3)$; **(b)** $(2, 0)$;
(c) $\frac{3}{2}$ **41.** $\frac{3}{2}$ **42.** 0 **43.** Undefined **44.** 2
45. 28% **46.** $-\frac{7}{10}$ **47.** 0 **48.** Undefined **49.** $\frac{3}{2}$
50. x-intercept: $(6, 0)$; y-intercept: $(0, 9)$ **51.** $-\frac{1}{2}$; $(0, 5)$
52. Perpendicular **53.** Parallel **54.** $y = -\frac{3}{4}x + 6$
55. $y - 6 = -\frac{1}{2}(x - 3)$ **56.** $y = \frac{5}{4}x - \frac{13}{4}$
57. $y = -\frac{5}{3}x - \frac{5}{3}$ **58.** **(a)** $r = 4.2t + 56$; **(b)** \$68.6 billion;
(c) \$89.6 billion **59.**
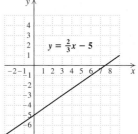

60.
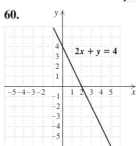

61.
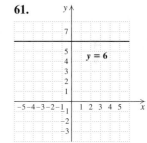

62.
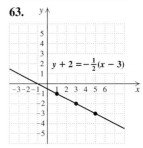

63.

64.
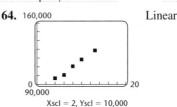

Linear

Xscl = 2, Yscl = 10,000

65. $B = 3692.377049t + 80935.5082$
66. 151,091 twin births **67.** **(a)** 3; **(b)** $\{x \mid -2 \le x \le 4\}$, or
$[-2, 4]$; **(c)** -1; **(d)** $\{y \mid 1 \le y \le 5\}$, or $[1, 5]$ **68.** $\frac{3}{2}$
69. $4a^2 + 4a - 3$ **70.** **(a)** 10.53 yr; **(b)** 0.233 signifies that
the median age of cars increases 0.233 yr per year; 5.87 signifies that the median age of cars was 5.87 in 1990
71. **(a)** Yes; **(b)** domain: $\mathbb{R}$; range: $\{y \mid y \ge 0\}$, or $[0, \infty)$

72. (a) No **73. (a)** No **74. (a)** Yes; **(b)** domain: $\mathbb{R}$; range: $\{-2\}$ **75.** $\mathbb{R}$ **76.** $\{x \mid x \text{ is a real number } and \ x \neq 1\}$ **77. (a)** 5; **(b)** 4; **(c)** 16; **(d)** 35 **78.** 🅣🅦 Two perpendicular lines share the same y-intercept if their point of intersection is on the y-axis. **79.** 🅣🅦 Two functions that have the same domain and range are not necessarily identical. For example, the functions $f \colon \{(-2, 1), (-3, 2)\}$ and $g \colon \{(-2, 2), (-3, 1)\}$ have the same domain and range but are different functions. **80.** -1 **81.** 19 **82.** Area: 45 sq units; perimeter: 28 units **83.** $(0, 4), (1, 3), (-1, 3)$; answers may vary **84.** Domain: $\{x \mid x \geq -4 \text{ and } x \neq 2\}$; range: $\{y \mid y \geq 0 \text{ and } y \neq 3\}$

Test: Chapter 3, pp. 279–281

1. [3.1] \$205.20 **2.** [3.1] \$638 **3.** [3.1] II **4.** [3.1] III **5.** [3.1] $(3, 4)$ **6.** [3.1] $(0, -4)$ **7.** [3.1] $(-5, 2)$ **8.** [3.2]

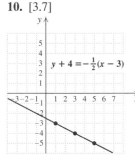

9. [3.3]

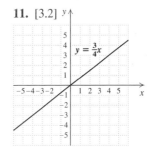

$2x - 4y = -8$

10. [3.7]

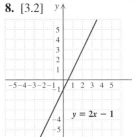

$y + 4 = -\frac{1}{2}(x - 3)$

11. [3.2]

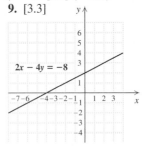

$y = \frac{3}{4}x$

12. [3.3]

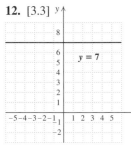

$y = 7$

13. [3.2]

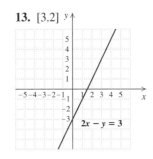

$2x - y = 3$

14. [3.3]

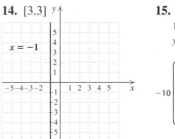

$x = -1$

15. [3.2]

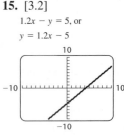

$1.2x - y = 5$, or $y = 1.2x - 5$

16. [3.3] x-intercept: $(6, 0)$; y-intercept: $(0, -10)$ **17.** [3.5] $\frac{9}{2}$ **18.** [3.4] $\frac{1}{3}$ km/min **19.** [3.5] 31.5% **20.** [3.6] 3; $(0, 7)$ **21.** [3.6] Parallel **22.** [3.6] Perpendicular **23.** [3.7] $y - 8 = -3(x - 6)$ **24.** [3.7] $y = -x + 10$ **25.** [3.7] **(a)** $c = \frac{31}{21}t + 16$; **(b)** approximately 39.6 hr; **(c)** approximately 57.3 hr **26.** [3.7] $d = 355.7t - 159.7$ **27.** [3.7] Approximately 2330 million DVDs **28.** [3.8] **(a)** No **29.** [3.8] **(a)** Yes; **(b)** domain: $\{x \mid -4 \leq x \leq 5\}$, or $[-4, 5]$; range: $\{y \mid -2 \leq y \leq 4\}$, or $[-2, 4]$ **30.** [3.8] **(a)** 0; **(b)** -4 **31.** [3.8] **(a)** $-\frac{5}{3}$; **(b)** $\{x \mid x \text{ is a real number } and \ x \neq -\frac{1}{2}\}$ **32.** [3.8] **(a)** 3; **(b)** 10; **(c)** 0 **33.** [3.6] $y = \frac{2}{5}x + 9$ **34.** [3.1] Area: 25 sq units; perimeter: 20 units **35.** [3.2], [3.7] $(0, 12), (-3, 15), (5, 7)$

Cumulative Review: Chapters 1–3, pp. 281–283

1. 0.5 million bicycles per year **2. (a)** $c = \frac{85}{3}r + 70$; **(b)** approximately 353 calories per hour **3. (a)** $c = 2.64w + 8.\overline{6}$; **(b)** approximately 365 calories per hour **4.** \$210 billion **5.** \$50,919 **6.** 15.6 million Americans **7.** \$120 **8.** 50 m, 53 m, 40 m **9.** No more than 8 hr **10.** \$40/person **11.** 7 **12.** $12x - 15y + 21$ **13.** $3(5x - 3y + 1)$ **14.** $2 \cdot 3 \cdot 7$ **15.** 0.45 **16.** 4 **17.** $\frac{1}{4}$ **18.** -4 **19.** $-x - y$ **20.** 0.785 **21.** $\frac{11}{60}$ **22.** 2.6 **23.** 7.28 **24.** $-\frac{5}{12}$ **25.** -3 **26.** 27 **27.** $-2y - 7$ **28.** $5x + 11$ **29.** -1.2 **30.** -21 **31.** 9 **32.** $-\frac{20}{3}$ **33.** 2 **34.** $\frac{13}{8}$ **35.** $-\frac{17}{21}$ **36.** -17 **37.** 2 **38.** $\{x \mid x < 16\}$ **39.** $\{x \mid x \leq -\frac{11}{8}\}$ **40.** $h = \dfrac{A - \pi r^2}{2\pi r}$ **41.** IV

42.

$-1 < x \leq 2$

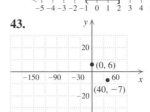

43.

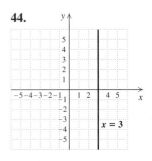

44.

45. 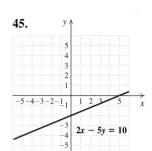   **46.**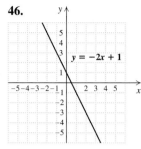

47. 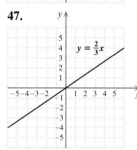   **48.**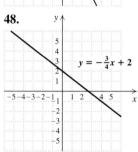

49.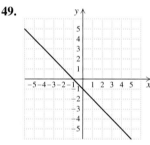

$y - 1 = -(x + 2)$

50. $(10.5, 0); (0, -3)$

51. $(-1.25, 0); (0, 5)$ **52.** $-\frac{1}{3}$ **53.** $y = \frac{2}{7}x - 4$
54. $-\frac{1}{3}; (0, 3)$ **55.** $y = -\frac{1}{2}x + 5$
56. $\{-5, -3, -1, 1, 3\}; \{-3, -2, 1, 4, 5\}, -2; 3$
57. (a) -7; (b) $\{x | x \text{ is a real number } and \ x \neq \frac{1}{2}\}$
58. \$25,000   **59.** $-4, 4$ **60.** 2   **61.** -5 **62.** 3
63. No solution; contradiction   **64.** $Q = \dfrac{2 - pm}{p}$
65. $y = -\frac{7}{3}x + 7; y = -\frac{7}{3}x - 7; y = \frac{7}{3}x - 7; y = \frac{7}{3}x + 7$
66. $m = -\frac{5}{9}, b = -\frac{2}{9}$

Chapter 4

Visualizing the Graph, p. 295

1. C   **2.** H   **3.** J   **4.** G   **5.** D   **6.** I   **7.** A   **8.** F
9. E   **10.** B

Exercise Set 4.1, pp. 296–299

1. True   **2.** False   **3.** True   **4.** True   **5.** True
6. False   **7.** False   **8.** True   **9.** Yes   **11.** No   **13.** Yes
15. Yes   **17.** $(4, 1)$ **19.** $(2, -1)$ **21.** $(4, 3)$
23. $(-3, -2)$ **25.** $(-3, 2)$ **27.** $(3, -7)$ **29.** $(7, 2)$

31. $(4, 0)$ **33.** No solution   **35.** $\{(x, y) | y = 3 - x\}$
37. Approximately $(1.53, 2.58)$ **39.** No solution
41. Approximately $(-6.37, -18.77)$ **43.** All except
33 and 39   **45.** 35   **47.** Full-time faculty:
$y = 8.198033080x + 436.7572642$; part-time faculty:
$y = 13.14975414x + 197.344658$, where x is the number of
years after 1980; approximately 2028   **49.** Waste generated:
$y = 0.7588447653x + 75.05631769$; waste recycled:
$y = 1.521480144x + 28.96534296$, where x is the number of
years after 1990; approximately 2050   **51.** TW   **53.** 15
54. $\frac{19}{12}$ **55.** $\frac{9}{20}$ **56.** $\frac{13}{3}$ **57.** $y = -\frac{3}{4}x + \frac{7}{4}$
58. $y = \frac{2}{5}x - \frac{9}{5}$ **59.** TW   **61.** Answers may vary.
(a) $x + y = 6, x - y = 4$; (b) $x + y = 1, 2x + 2y = 3$;
(c) $x + y = 1, 2x + 2y = 2$ **63.** $A = -\frac{17}{4}, B = -\frac{12}{5}$
65. $(0, 0), (1, 1)$ **67.** (c)   **69.** (b)
71. (a) $n = -1.1t + 33.1; n = 2.9t + 2.8$; (b) 2008

Exercise Set 4.2, pp. 304–307

1. False   **2.** True   **3.** True   **4.** True   **5.** $(2, 5)$
7. $(2, 1)$ **9.** $(4, 3)$ **11.** $(2, -4)$ **13.** $(-2, 4)$
15. No solution   **17.** $(-1, -3)$ **19.** $\left(\frac{17}{3}, \frac{2}{3}\right)$
21. $\{(x, y) | x - 2y = 7\}$ **23.** $\{(x, y) | y = 2x + 5\}$
25. $\left(\frac{25}{8}, -\frac{11}{4}\right)$ **27.** $(-3, 0)$ **29.** $(6, 3)$ **31.** No solution
33. $(-3, -4)$ **35.** No solution   **37.** 39, 44   **39.** 42, 51
41. 12, 28   **43.** $55°, 125°$ **45.** $36°, 54°$ **47.** $1\frac{3}{4}$ in. by
$3\frac{1}{2}$ in.   **49.** Length: 380 mi; width: 270 mi   **51.** Length:
90 yd; width: 50 yd   **53.** Height: 20 ft; width: 5 ft
55. TW   **57.** $-5x - 3y$ **58.** $-11x$ **59.** $-11y$
60. $-6y - 25$ **61.** $-11y$ **62.** $23x$ **63.** TW
65. $(7, -1)$ **67.** $(4.38, 4.33)$ **69.** 34 yr
71. $(30, 50, 100)$ **73.** TW

Exercise Set 4.3, pp. 313–315

1. False   **2.** True   **3.** True   **4.** True   **5.** $(9, 3)$
7. $(5, 1)$ **9.** $(2, 7)$ **11.** $(-1, 3)$ **13.** $\left(-1, \frac{1}{5}\right)$
15. $\{(a, b) | 3a - 6b = 8\}$ **17.** $(-3, -5)$ **19.** $(4, 5)$
21. $(4, 1)$ **23.** No solution   **25.** $(1, -1)$
27. $(-3, -1)$ **29.** No solution   **31.** $(50, 18)$
33. $(-2, 2)$ **35.** $(2, -1)$ **37.** $\left(\frac{231}{202}, \frac{117}{202}\right)$ **39.** 50 mi
41. $26°, 64°$ **43.** 165 min   **45.** $37°, 143°$
47. White: 92 loaves; whole-wheat: 83 loaves
49. Length: 6 ft; width: 3 ft   **51.** TW   **53.** $(4, 3)$
54. $(3, 0)$ **55.** $(-19, -15)$ **56.** $\left(-\frac{1}{3}, -\frac{5}{3}\right)$ **57.** $(1, 1)$
58. $(1, 2)$ **59.** No solution   **60.** $\{(x, y) | y + 4 = 2x\}$
61. $\left(2, -\frac{2}{5}\right)$ **62.** $\left(\frac{10}{3}, -\frac{11}{3}\right)$ **63.** TW   **65.** $(2, 5)$
67. $\left(\frac{1}{2}, -\frac{1}{2}\right)$ **69.** $(0, 3)$ **71.** $x = \dfrac{c - b}{a - 1}; y = \dfrac{ac - b}{a - 1}$
73. Rabbits: 12; pheasants: 23
75. Man: 45 yr; his daughter: 10 yr

Exercise Set 4.4, pp. 329–333

1. Verbal: 508; math: 520   **3.** Simon: 122 properties;
DeBartolo: 61 properties   **5.** 3-credit courses: 19;
4-credit courses: 8   **7.** 5-cent bottles or cans: 336;
10-cent bottles or cans: 94   **9.** 2225 motorcycles
11. Nonrecycled sheets: 38; recycled sheets: 112
13. General Electric bulbs: 60; SLi bulbs: 140
15. HP cartridges: 15; Epson cartridges: 35
17. Kenyan: 8 lb; Sumatran: 12 lb
19. Sunflower seed: 30 lb; rolled oats: 20 lb
21. 50%-acid solution: 80 mL; 80%-acid solution: 120 mL

Type of Solution	50%-acid	80%-acid	68%-acid mix
Amount of Solution	x	y	200
Percent Acid	50%	80%	68%
Amount of Acid in Solution	$0.5x$	$0.8y$	136

23. Deep Thought: 12 lb; Oat Dream: 8 lb
25. $7500 at 6%; $4500 at 9%
27. Arctic Antifreeze: 12.5 L; Frost No-More: 7.5 L
29. 87-octane: 4 gal; 93-octane: 8 gal
31. Whole milk: $169\frac{3}{13}$ lb; cream: $30\frac{10}{13}$ lb
33. Foul shots: 28; two-pointers: 36   **35.** 375 km
37. 24 mph   **39.** About 1489 mi
41. 30-sec commercials: 4; 60-sec commercials: 8
43. Quarters: 17; fifty-cent pieces: 13   **45.** 🇹🇼
47. About 29 million cases   **48.** About $62 billion
49. $13.5 billion   **50.** 11 named storms
51. Length: 76 m; width: 19 m
52. $8.50-brushes: 32; $9.75-brushes: 13   **53.** 🇹🇼
55. 0%: 20 reams; 30%: 40 reams   **57.** 1.8 L
59. 180 members   **61.** Brown: 0.8 gal; neutral: 0.2 gal
63. City: 261 mi; highway: 204 mi

Exercise Set 4.5, pp. 342–345

1. (e)   **2.** (d)   **3.** (f)   **4.** (b)   **5.** (a)   **6.** (c)
7. -2 **9.** -4 **11.** 8   **13.** 0   **15.** 5   **17.** -2
19. None   **21.** $-2, 2$ **23.** 5   **25.** -20 **27.** 2.7
29. $-\frac{7}{3}$ **31.** 7   **33.** 3   **35.** 9   **37.** 1   **39.** 6
41. $1\frac{1}{3}$ **43.** $10,500 over $100   **45.** 5 months
47. 2 hr 15 min   **49.** 150 lb   **51.** 🇹🇼   **53.** $-4.9x + 6.66$
54. 32   **55.** $\frac{333}{245}$ **56.** $\{x \mid x \le -12\}$, or $(-\infty, -12]$
57. $\left(\frac{13}{2}, -\frac{7}{2}\right)$ **58.** $\left(\frac{4}{3}, -\frac{1}{3}\right)$ **59.** 🇹🇼   **61.** $-2, 2$
63. 1   **65.** $-6, 2$ **67.** $-1, 2$

69.

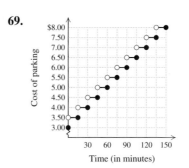

Review Exercises: Chapter 4, pp. 348–349

1. Substitution   **2.** Elimination   **3.** Approximate
4. Dependent   **5.** Inconsistent   **6.** Infinite   **7.** Parallel
8. Zero   **9.** Alphabetical   **10.** x-coordinate
11. $(-2, 1)$ **12.** $(2, 1)$ **13.** $\left(-\frac{11}{15}, -\frac{43}{30}\right)$
14. No solution   **15.** $\left(-\frac{4}{5}, \frac{2}{5}\right)$ **16.** $\left(\frac{37}{19}, \frac{53}{19}\right)$
17. $\left(\frac{76}{17}, -\frac{2}{119}\right)$ **18.** $(2, 2)$ **19.** $\{(x, y) \mid 3x + 4y = 6\}$
20. Length: 265 ft; width: 165 ft   **21.** DVD: $17; CD: $14
22. Private lessons: 7 students; group lessons: 5 students
23. 61°; 119°   **24.** 4 hr   **25.** 8% juice: 10 L;
15% juice: 4 L   **26.** 1300-word pages: 7;
1850-word pages: 5   **27.** -2 **28.** 1   **29.** $\frac{4}{7}$ **30.** 3
31. 🇹🇼 A solution of a system of two equations is an ordered
pair that makes both equations true. The graph of an equation
represents all ordered pairs that make that equation true. So in
order for an ordered pair to make *both* equations true, it must
be on both graphs.   **32.** 🇹🇼 Both methods involve finding
the coordinates of the point of intersection of two graphs. The
solution of a system of equations is the ordered pair at the
point of intersection; the solution of an equation is the
x-coordinate of the point of intersection.   **33.** $(0, 2), (1, 3)$
34. $C = 1, D = 3$ **35.** 24   **36.** $33,600

Test: Chapter 4, p. 350

1. [4.1] $(2, 4)$ **2.** [4.1] No solution   **3.** [4.2] $\left(3, -\frac{11}{3}\right)$
4. [4.2] $\left(\frac{15}{7}, -\frac{18}{7}\right)$ **5.** [4.2] $\{(x, y) \mid x = 5y - 10\}$
6. [4.3] $\left(-\frac{3}{2}, -\frac{3}{2}\right)$ **7.** [4.3] No solution   **8.** [4.3] $(0, 1)$
9. [4.3] $(12, -6)$ **10.** [4.2] Length: 94 ft; width: 50 ft
11. [4.3] 38°; 52°   **12.** [4.4] Hardbacks: 11; paperbacks: 12
13. [4.4] Pepperidge Farm Goldfish: 120 g;
Rold Gold Pretzels: 500 g   **14.** [4.4] 20 mph
15. [4.4] Nickels: 10; quarters: 3   **16.** [4.5] -2
17. [4.5] 10   **18.** [4.2] $(2, 0)$ **19.** [4.1] $C = -\frac{19}{2}, D = \frac{14}{3}$
20. [4.4] 9 people   **21.** [3.8], [4.3] $m = 7, b = 10$

Chapter 5

Exercise Set 5.1, pp. 358–360

1. (b) **2.** (f) **3.** (e) **4.** (h) **5.** (g) **6.** (a)
7. (c) **8.** (d) **9.** r^{10} **11.** 9^8 **13.** a^7 **15.** 8^{11}
17. $(3y)^{12}$ **19.** $(5t)^7$ **21.** a^5b^9 **23.** $(x+1)^{12}$
25. r^{10} **27.** x^4y^7 **29.** 7^3 **31.** x^{12} **33.** t^4 **35.** $5a$
37. 1 **39.** $\frac{3}{4}m^3$ **41.** $4a^7b^6$ **43.** m^9n^4 **45.** 1 **47.** 5
49. 2 **51.** -4 **53.** x^{28} **55.** 5^{16} **57.** m^{35} **59.** t^{80}
61. $49x^2$ **63.** $-8a^3$ **65.** $16m^6$ **67.** $a^{14}b^7$ **69.** x^8y^7
71. $24x^{19}$ **73.** $\dfrac{a^3}{64}$ **75.** $\dfrac{49}{25a^2}$ **77.** $\dfrac{a^{20}}{b^{15}}$ **79.** $\dfrac{y^6}{4}$
81. $\dfrac{x^8y^4}{z^{12}}$ **83.** $\dfrac{a^{12}}{16b^{20}}$ **85.** $\dfrac{125x^{21}y^3}{8z^{12}}$ **87.** 1 **89.** ᵀᵂ
91. $3(s-r+t)$ **92.** $-7(x-y+z)$ **93.** $8x$
94. $-3a-6b$
95.

96.

97. ᵀᵂ **99.** ᵀᵂ **101.** Let $a=1$; then $(a+5)^2=36$,
but $a^2+5^2=26$. **103.** Let $a=0$; then $\dfrac{a+7}{7}=1$, but
$a=0$. **105.** a^{8k} **107.** $\frac{16}{375}$ **109.** 13 **111.** $<$
113. $<$ **115.** $>$ **117.** $4{,}000{,}000$; $4{,}194{,}304$; $194{,}304$
119. $2{,}000{,}000{,}000$; $2{,}147{,}483{,}648$; $147{,}483{,}648$
121. $1{,}536{,}000$ bytes, or approximately $1{,}500{,}000$ bytes

Exercise Set 5.2, pp. 368–371

1. Positive power of 10 **2.** Negative power of 10
3. Negative power of 10 **4.** Positive power of 10
5. Positive power of 10 **6.** Negative power of 10
7. $\dfrac{1}{7^2}$, or $\dfrac{1}{49}$ **9.** $\dfrac{1}{(-2)^6}$, or $\dfrac{1}{64}$ **11.** $\dfrac{1}{a^3}$ **13.** 5^3, or 125
15. $\dfrac{1}{7}$ **17.** $\dfrac{8}{x^3}$ **19.** $\dfrac{3a^8}{b^6}$ **21.** $\dfrac{1}{3x^5z^4}$ **23.** $\dfrac{y^7z^4}{x^2}$
25. $\left(\dfrac{2}{a}\right)^3$, or $\dfrac{8}{a^3}$ **27.** 8^{-4} **29.** $(-5)^{-6}$ **31.** x^{-1}
33. $\dfrac{1}{x^{-5}}$ **35.** $\dfrac{4}{x^{-3}}$ **37.** $\dfrac{y^{-3}}{5}$ **39.** 8^{-6}, or $\dfrac{1}{8^6}$
41. b^{-3}, or $\dfrac{1}{b^3}$ **43.** a^2 **45.** $10a^{-6}b^{-2}$, or $\dfrac{10}{a^6b^2}$
47. 10^{-9}, or $\dfrac{1}{10^9}$ **49.** 2^{-2}, or $\dfrac{1}{2^2}$, or $\dfrac{1}{4}$ **51.** y^9

53. $-3ab^{-1}$, or $\dfrac{-3a}{b}$ **55.** $-\dfrac{1}{4}x^3y^{-2}z^{11}$, or $-\dfrac{x^3z^{11}}{4y^2}$
57. 9^{-12}, or $\dfrac{1}{9^{12}}$ **59.** t^{40} **61.** $x^{-18}y^6$, or $\dfrac{y^6}{x^{18}}$ **63.** x^{-8}, or
$\dfrac{1}{x^8}$ **65.** $32a^{-4}$, or $\dfrac{32}{a^4}$ **67.** 1 **69.** $9a^{-8}$, or $\dfrac{9}{a^8}$
71. $\dfrac{m^{-3}}{n^{-12}}$, or $\dfrac{n^{12}}{m^3}$ **73.** $\dfrac{5^4}{(-4)^4}x^{-20}y^{24}$, or $\dfrac{5^4y^{24}}{4^4x^{20}}$ **75.** 1
77. -4096 **79.** 0.0625 **81.** 0.648 **83.** 0.0004
85. $673{,}000{,}000$ **87.** 0.0000000008923
89. $90{,}300{,}000{,}000$ **91.** 4.7×10^{10} **93.** 1.6×10^{-8}
95. 4.07×10^{11} **97.** 6.03×10^{-7} **99.** 5.02×10^{18}
101. -3.05×10^{-10} **103.** 9.7×10^{-5} **105.** 1.3×10^{-11}
107. 6.0 **109.** 1.5×10^3 **111.** 3.0×10^{-5}
113. 1.79×10^{20} **115.** 1.2×10^{24}
117. 2.0×10^9 neutrinos **119.** 5.8×10^8 mi
121. 1×10^5 light years **123.** $4.2\times10^1\,\text{m}^3$
125. 7.9×10^7 bacteria **127.** Approximately $5\times10^2\,\text{in}^3$,
or $3\times10^{-1}\,\text{ft}^3$ **129.** ᵀᵂ **131.** $-x-8$ **132.** $-x+2$
133. -8 **134.** 2 **135.** $\{x\,|\,x\ge-8\}$, or $[-8,\infty)$
136. $\{x\,|\,x<2\}$, or $(-\infty,2)$ **137.** ᵀᵂ **139.** ᵀᵂ
141. 5 **143.** 7.0×10^{23} **145.** 3^{11} **147.** 1.25×10^{22}
149. 8×10^{18} grains

Interactive Discovery, p. 378

Answers may include: The graph of a polynomial function has
no sharp corners; there are no holes or breaks; the domain is
the set of all real numbers.

Visualizing the Graph, p. 381

1. C **2.** B **3.** A **4.** D **5.** J **6.** E **7.** I
8. F **9.** G **10.** H

Exercise Set 5.3, pp. 382–387

1. (b) **2.** (h) **3.** (d) **4.** (f) **5.** (g) **6.** (a)
7. (c) **8.** (e) **9.** Yes **11.** No **13.** Yes
15. $7x^4,\ x^3,\ -5x,\ 8$ **17.** $-t^4,\ 7t^3,\ -3t^2,\ 6$
19. Coefficients: 4, 7; degrees: 5, 1
21. Coefficients: 9, -3, 4; degrees: 2, 1, 0
23. Coefficients: 1, -1, 4, -3; degrees: 4, 3, 1, 0
25. (a) 3, 5, 2; (b) $7a^5$, 7; (c) 5
27. (a) 4, 2, 7, 0; (b) x^7, 1; (c) 7
29. (a) 1, 4, 0, 3; (b) $-a^4$, -1; (c) 4

31.

Term	Coefficient	Degree of the Term	Degree of the Polynomial
$8x^5$	8	5	
$-\frac{1}{2}x^4$	$-\frac{1}{2}$	4	
$-4x^3$	-4	3	5
$7x^2$	7	2	
6	6	0	

33. Trinomial   **35.** None of these   **37.** Binomial
39. Monomial   **41.** $11x^2 + 3x$ **43.** $4a^4$
45. $6x^2 - 3x$ **47.** $5x^3 - x + 5$ **49.** $-x^4 - x^3$
51. $\frac{1}{15}x^4 + 10$ **53.** $3.4x^2 + 1.3x + 5.5$
55. $10t^4 - 12t^3 + 17t$ **57.** $-17; 25$ **59.** $16; 34$
61. $-67; 17$ **63.** $-27; 81$ **65.** $-117; -63$ **67.** $47; 7$
69. $-45; -8\frac{19}{27}$ **71.** 9   **73.** $2.17 trillion   **75.** 1112 ft
77. 62.8 cm   **79.** 153.86 m²   **81.** About 25.5 mph
83. About 29.6 ft   **85.** 26.4 million   **87.** 6.4 million
89. 14; 55 oranges   **91.** About 2.3 mcg/mL   **93.** $[0, 10]$
95. $(-\infty, 3]$ **97.** $(-\infty, \infty)$ **99.** $[-4, \infty)$
101. $[-65, \infty)$ **103.** $[0, \infty)$ **105.** $(-\infty, 5]$
107. $(-\infty, \infty)$ **109.** $[-6.7, \infty)$ **111.** ᵀᵂ   **113.** 5
114. -9 **115.** $5(x + 3)$ **116.** $7(a - 3)$
117. 6.25¢/mi   **118.** 274 and 275   **119.** ᵀᵂ
121. $2x^5 + 4x^4 + 6x^2 + 8$; answers may vary   **123.** 10
125. $5x^9 + 4x^8 + x^2 + 5x$ **127.** $x^3 - 2x^2 - 6x + 3$
129. 85.0
131.

t	$-t^2 + 10t - 18$
3	3
4	6
5	7
6	6
7	3

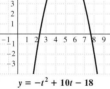

$y = -t^2 + 10t - 18$

133.

d	$-0.0064d^2 + 0.8d + 2$
0	2
30	20.24
60	26.96
90	22.16
120	5.84

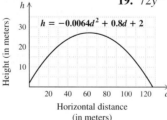

Height (in meters)

$h = -0.0064d^2 + 0.8d + 2$

Horizontal distance (in meters)

Interactive Discovery, p. 390

1. No. The graphs of $y_1 = (x^2 - 4x) + (x^2 - 3x)$ and $y_2 = 2x^2 - x$ are not the same; note that $-4x + (-3x) = -7x$, not $-x$.   **2.** Yes. The graph of $y = (2x^2 + 5) + (x^2 - 8) - (3x^2 - 3)$ is the x-axis.   **3.** No. The graphs of $y_1 = (x^2 - 5x) + (3x^2 - 6)$ and $y_2 = 4x^2 - 11x$ are not the same, and the values in the table for y_1 and y_2 are not the same for a given value of x; note that $-5x - 6 \neq -11x$.

Exercise Set 5.4, pp. 394–397

1. $-2x + 6$ **3.** $x^2 - 5x - 1$ **5.** $9t^2 + 5t - 3$
7. $6m^3 + 3m^2 - 3m - 9$ **9.** $7 + 13a + 6a^2 + 14a^3$
11. $9x^8 + 8x^7 - 3x^4 + 2x^2 - 2x + 5$
13. $-\frac{1}{2}x^4 + \frac{2}{3}x^3 + x^2$ **15.** $4.2t^3 + 3.5t^2 - 6.4t - 1.8$
17. $-3x^4 + 3x^2 + 4x$
19. $1.05x^4 + 0.36x^3 + 14.22x^2 + x + 0.97$
21. $-(-t^3 + 4t^2 - 9); t^3 - 4t^2 + 9$
23. $-(12x^4 - 3x^3 + 3); -12x^4 + 3x^3 - 3$ **25.** $-8x + 9$
27. $-3a^4 + 5a^2 - 9$ **29.** $4x^4 - 6x^2 - \frac{3}{4}x + 8$
31. $9x + 3$ **33.** $-t^2 - 8t + 7$
35. $6x^4 + 3x^3 - 4x^2 + 3x - 4$
37. $4.6x^3 + 9.2x^2 - 3.8x - 23$ **39.** 0
41. $1 + 2a + 7a^2 - 3a^3$ **43.** $\frac{3}{4}x^3 - \frac{1}{2}x$
45. $0.05t^3 - 0.07t^2 + 0.01t + 1$ **47.** $3x + 5$
49. $11x^4 + 12x^3 - 10x^2$ **51.** (a) $5x^2 + 4x$; (b) 145; 273
53. $14y + 25$ **55.** $(r + 11)(r + 9); 9r + 99 + 11r + r^2$
57. $(x + 3)^2; x^2 + 3x + 9 + 3x$ **59.** $\pi r^2 - 25\pi$
61. $18z - 64$ **63.** $(x^2 - 12)$ ft²   **65.** $(z^2 - 16\pi)$ ft²
67. $\left(144 - \dfrac{d^2}{4}\pi\right)$ m²   **69.** Not correct   **71.** Correct
73. Not correct   **75.** ᵀᵂ   **77.** 0   **78.** 0   **79.** $13t + 14$
80. $14t - 15$ **81.** $\{x \mid x < \frac{14}{3}\}$, or $\left(-\infty, \frac{14}{3}\right)$
82. $\{x \mid x \leq 12\}$, or $(-\infty, 12]$ **83.** ᵀᵂ
85. $9t^2 - 20t + 11$ **87.** $-10y^2 - 2y - 10$
89. $-3y^4 - y^3 + 5y - 2$ **91.** $250.591x^3 + 2.812x$
93. $20w + 42$ **95.** $2x^2 + 20x$ **97.** $8x + 36$ **99.** ᵀᵂ

Exercise Set 5.5, pp. 402–404

1. (b)   **2.** (c)   **3.** (d)   **4.** (a)   **5.** $36x^3$ **7.** x^3
9. $-x^8$ **11.** $28t^8$ **13.** $-0.02x^{10}$ **15.** $\frac{1}{15}x^4$ **17.** 0
19. $72y^{10}$ **21.** $-3x^2 + 15x$ **23.** $4x^2 + 4x$

25. $4a^2 - 28a$ **27.** $x^5 + x^2$ **29.** $6x^3 - 18x^2 + 3x$
31. $15t^3 + 30t^2$ **33.** $-6x^4 - 6x^3$ **35.** $4a^9 - 8a^7 - \frac{5}{12}a^4$
37. $x^2 + 7x + 6$ **39.** $x^2 + 3x - 10$ **41.** $a^2 - 13a + 42$
43. $x^2 - 9$ **45.** $25 - 15x + 2x^2$ **47.** $t^2 + \frac{17}{6}t + 2$
49. $\frac{3}{16}a^2 + \frac{5}{4}a - 2$ **51.**

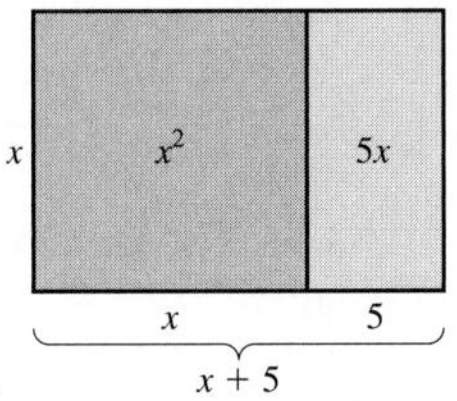

53.

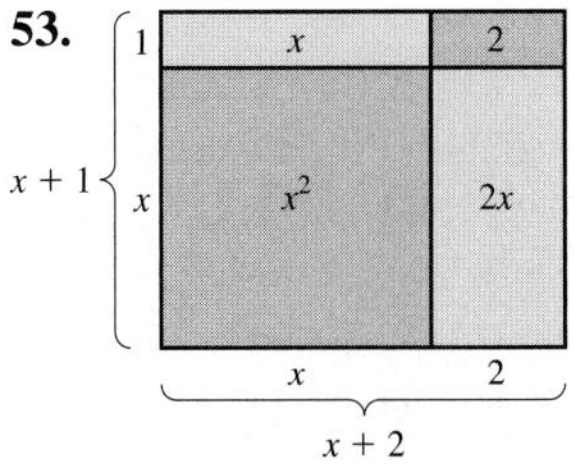

55.

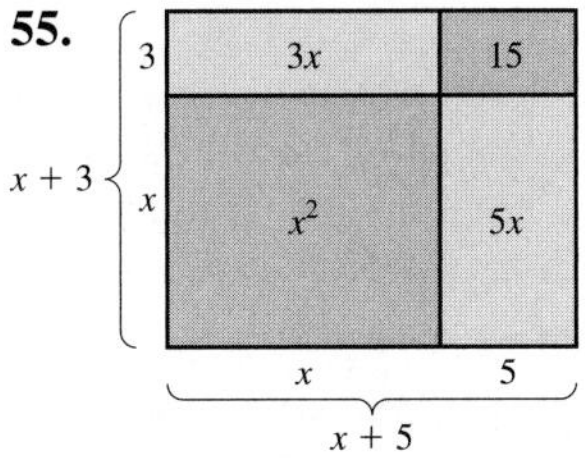

57. $x^3 + 4x + 5$ **59.** $2a^3 - a^2 - 11a + 10$
61. $2y^5 - 13y^3 + y^2 - 7y - 7$
63. $27x^2 + 39x + 14$ **65.** $x^4 - 2x^3 - x + 2$
67. $6t^4 - 17t^3 - 5t^2 - t - 4$
69. $x^4 + 8x^3 + 12x^2 + 9x + 4$ **71.** TW
73. $4x^2 - 3x - 5$ **74.** $-4x^2 + 9x + 9$
75. $12x^3 - 10x^2 - 33x - 14$
76. $-12x^3 + 10x^2 + 33x + 14$ **77.** TW
79. $75y^2 - 45y$ **81.** 5 **83.** $V = (4x^3 - 48x^2 + 144x)$ in^3; $S = (-4x^2 + 144)$ in^2 **85.** $(x^3 - 5x^2 + 8x - 4)$ cm^3
87. $(x^3 + 2x^2 - 210)$ m^3 **89.** 0 **91.** 0

Interactive Discovery, p. 406

1. Identity **2.** Not an identity **3.** Not an identity
4. Not an identity

Interactive Discovery, p. 407

1. Not an identity **2.** Identity **3.** Not an identity
4. Not an identity **5.** Identity

Exercise Set 5.6, pp. 410–412

1. True **2.** False **3.** False **4.** True
5. $x^3 + 3x^2 + 5x + 15$ **7.** $x^4 + 2x^3 + 6x + 12$
9. $y^2 - y - 6$ **11.** $9x^2 + 21x + 10$ **13.** $5x^2 + 4x - 12$
15. $2 + 3t - 9t^2$ **17.** $4x^2 - 49$ **19.** $p^2 - \frac{1}{16}$
21. $x^2 + 0.2x + 0.01$ **23.** $-2x^2 - 11x + 6$
25. $a^2 + 18a + 81$ **27.** $1 - 2t - 15t^2$
29. $x^5 + 3x^3 - x^2 - 3$ **31.** $3x^6 - 2x^4 - 6x^2 + 4$
33. $4t^6 + 16t^3 + 15$ **35.** $8x^5 + 16x^3 + 5x^2 + 10$
37. $4x^3 - 12x^2 + 3x - 9$ **39.** $x^2 - 49$ **41.** $4x^2 - 1$
43. $25m^2 - 4$ **45.** $9x^8 - 1$ **47.** $x^8 - 49$ **49.** $t^2 - \frac{9}{16}$
51. $x^2 + 4x + 4$ **53.** $9x^{10} + 6x^5 + 1$ **55.** $a^2 - \frac{4}{5}a + \frac{4}{25}$
57. $x^4 - x^3 + 3x^2 - x + 2$ **59.** $4 - 12x^4 + 9x^8$
61. $25 + 60t^2 + 36t^4$ **63.** $49x^2 - 4.2x + 0.09$
65. $10a^5 - 5a^3$ **67.** $a^3 - a^2 - 10a + 12$
69. $9 - 12x^3 + 4x^6$ **71.** $4x^3 + 24x^2 - 12x$
73. $t^6 - 2t^3 + 1$ **75.** $15t^5 - 3t^4 + 3t^3$
77. $36x^8 - 36x^4 + 9$ **79.** $12x^3 + 8x^2 + 15x + 10$
81. $25 - 60x^4 + 36x^8$ **83.** $a^3 + 1$ **85.** $a^2 + 2a + 1$
87. $x^2 + 7x + 10$ **89.** $x^2 + 14x + 49$
91. $t^2 + 10t + 24$ **93.** $t^2 + 13t + 36$
95. $9x^2 + 24x + 16$
97.

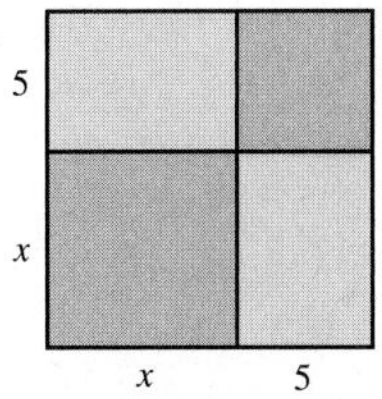

99.

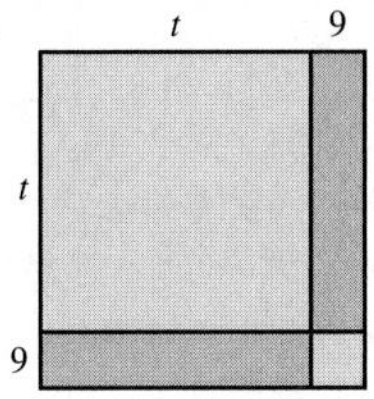

101.

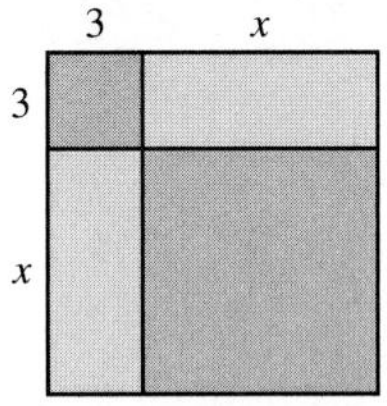

103. TW

105. Lamps: 500 watts; air conditioner: 2000 watts; television: 50 watts **106.** II
107. $y = \dfrac{8}{5x}$ **108.** $a = \dfrac{c}{3b}$ **109.** $x = \dfrac{b + c}{a}$
110. $t = \dfrac{u - r}{s}$ **111.** TW **113.** $16x^4 - 81$
115. $81t^4 - 72t^2 + 16$ **117.** $t^{24} - 4t^{18} + 6t^{12} - 4t^6 + 1$
119. 396 **121.** -7 **123.** $y^2 - 4y + 4$ **125.** $l^3 - l$
127. $Q(Q - 14) - 5(Q - 14)$, $(Q - 5)(Q - 14)$; other equivalent expressions are possible.
129. $(y + 1)(y - 1)$, $y(y + 1) - y - 1$; other equivalent expressions are possible.

Exercise Set 5.7, pp. 418–421

1. (a) **2.** (b) **3.** (b) **4.** (a) **5.** (c) **6.** (c)
7. (a) **8.** (a) **9.** -7 **11.** -92 **13.** 2.97 L
15. About 2494 calories **17.** 20.60625 in^2 **19.** 66.4 m
21. Coefficients: 1, −2, 3, −5; degrees: 4, 2, 2, 0; 4
23. Coefficients: 17, −3, −7; degrees: 5, 5, 0; 5
25. $3a - 2b$ **27.** $3x^2y - 2xy^2 + x^2 + 5x$
29. $8u^2v - 5uv^2 + 7u^2$ **31.** $6a^2c - 7ab^2 + a^2b$
33. $3x^2 - 4xy + 3y^2$ **35.** $-6a^4 - 8ab + 7ab^2$
37. $-6r^2 - 5rt - t^2$ **39.** $3x^3 - x^2y + xy^2 - 3y^3$
41. $10y^4x^2 - 8y^3x$ **43.** $-8x + 8y$ **45.** $6z^2 + 7uz - 3u^2$
47. $x^2y^2 + 3xy - 28$ **49.** $4a^2 - b^2$ **51.** $15r^2t^2 - rt - 2$
53. $m^6n^2 + 2m^3n - 48$ **55.** $30x^2 - 28xy + 6y^2$
57. $0.01 - p^2q^2$ **59.** $x^2 + 2xh + h^2$
61. $16a^2 + 40ab + 25b^2$ **63.** $c^4 - d^2$ **65.** $a^2b^2 - c^2d^4$
67. $a^2 + 2ab + b^2 - c^2$ **69.** $a^2 - b^2 - 2bc - c^2$

71. $a^2 + ab + ac + bc$ **73.** $x^2 - z^2$
75. $x^2 + y^2 + z^2 + 2xy + 2xz + 2yz$ **77.** $\frac{1}{2}x^2 + \frac{1}{2}xy - y^2$

79. We draw a rectangle with   **81.**
dimensions $r + s$ by $u + v$.

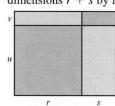

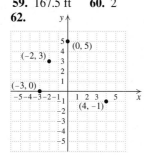

83. (a) $t^2 - 2t + 6$; (b) $2ah + h^2$; (c) $2ah - h^2$ **85.** TW
87. 12   **88.** 5   **89.** 27   **90.** 36   **91.** 7   **92.** 5
93. TW   **95.** $2\pi ab - \pi b^2$ **97.** $a^2 - 4b^2$
99. $x^3 + 2y^3 + x^2y + xy^2$
101. $2\pi nh + 2\pi mh + 2\pi n^2 - 2\pi m^2$ **103.** TW
105. $P + 2Pr + Pr^2$ **107.** \$15,638.03

Exercise Set 5.8, pp. 428–429

1. $4x^5 - 3x$ **3.** $1 - 2u + u^6$ **5.** $5t^2 - 8t + 2$
7. $6t^2 - 10t + \frac{3}{2}$ **9.** $-5x^5 + 7x^2 + 1$
11. $x^4 + \frac{3}{2}x^2 + \frac{1}{x}$ **13.** $-3rs - r + 2s$
15. $1 - ab^2 - a^3b^4$ **17.** $x + 3$ **19.** $a - 12 + \frac{32}{a + 4}$
21. $x - 4 + \frac{1}{x - 5}$ **23.** $y - 5$ **25.** $a^2 - 2a + 4$
27. $t + 4 + \frac{3}{t - 4}$ **29.** $t^2 - 3t + 1$
31. $3a + 1 + \frac{3}{2a + 5}$ **33.** $t^2 - 2t + 3 + \frac{-4}{t + 1}$
35. $t^2 - 1 + \frac{3t - 1}{t^2 + 5}$ **37.** $2x^2 + 1 + \frac{-x}{2x^2 - 3}$
39. $x^2 - 3x + 5 + \frac{-12}{x + 1}$ **41.** $a + 5 + \frac{-4}{a + 3}$
43. $x^2 - 5x - 23 + \frac{-43}{x - 2}$ **45.** $3x^2 - 2x + 2 + \frac{-3}{x + 3}$
47. $y^2 + 2y + 1 + \frac{12}{y - 2}$ **49.** $x^4 + 2x^3 + 4x^2 + 8x + 16$
51. $3x^2 + 6x - 3 + \frac{2}{x + \frac{1}{3}}$ **53.** TW   **55.** -17
56. -23 **57.** -2 **58.** 5   **59.** 167.5 ft   **60.** 2
61.

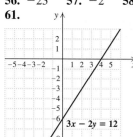

62.

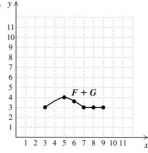

63. TW   **65.** $5x^{6k} - 16x^{3k} + 14$ **67.** $3t^{2h} + 2t^h - 5$
69. $a + 3 + \frac{5}{5a^2 - 7a - 2}$ **71.** $2x^2 + x - 3$ **73.** 3
75. -1 **77.** TW

Interactive Discovery, p. 433

1. 0   **2.** -1 **3.** $-1, 0$ **4.** Yes   **5.** Yes   **6.** No;
3 is not in the domain of f/g.

Exercise Set 5.9, pp. 436–439

1. Domain   **2.** Subtract   **3.** Evaluate   **4.** Domains
5. Excluding   **6.** Sum   **7.** 1   **9.** -41 **11.** 12
13. $\frac{13}{18}$ **15.** 5   **17.** $x^2 - 3x + 3$
19. $-3x^3 + x^2 - 6x + 2$ **21.** $x^2 - x + 3$ **23.** 23
25. 5   **27.** 56   **29.** $\frac{x^2 - 2}{5 - x}, x \neq 5$ **31.** $x^2 + x - 7$
33. $\frac{2}{7}$ **35.** 4%   **37.** $1.3 + 2.2 = 3.5$ million
39. About 50 million; the number of passengers using
Newark Liberty and LaGuardia in 1998   **41.** About 81 million; the number of passengers using the three airports in 2002
43. About 51 million; the number of passengers using
LaGuardia and Newark Liberty in 2002
45. $\mathbb{R}$ **47.** $\{x \mid x$ is a real number *and* $x \neq 3\}$
49. $\{x \mid x$ is a real number *and* $x \neq 0\}$
51. $\{x \mid x$ is a real number *and* $x \neq 1\}$
53. $\{x \mid x$ is a real number *and* $x \neq 2$ *and* $x \neq 4\}$
55. $\{x \mid x$ is a real number *and* $x \neq 3\}$
57. $\{x \mid x$ is a real number *and* $x \neq 4\}$
59. $\{x \mid x$ is a real number *and* $x \neq 4$ *and* $x \neq 5\}$
61. $\left\{x \mid x$ is a real number *and* $x \neq -1$ *and* $x \neq -\frac{5}{2}\right\}$
63. $4x^2 - 6x + 9, x \neq -\frac{3}{2}$ **65.** $2x - 5, x \neq -\frac{2}{3}$
67. 4; 3   **69.** 5; -1 **71.** $\{x \mid 0 \leq x \leq 9\}$;
$\{x \mid 3 \leq x \leq 10\}$; $\{x \mid 3 \leq x \leq 9\}$
73.

75. TW   **77.** $x = \frac{7}{4}y + 2$ **78.** $y = \frac{3}{8}x - \frac{5}{8}$
79. $y = -\frac{5}{2}x - \frac{3}{2}$ **80.** $x = -\frac{5}{6}y - \frac{1}{3}$ **81.** TW
83. $\left\{x \mid x$ is a real number *and* $x \neq -\frac{5}{2}$ *and* $x \neq -3$ *and* $x \neq 1$
and $x \neq -1\right\}$

85. Answers may vary.

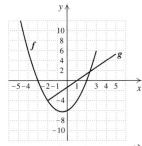

87. $\{x \,|\, x \text{ is a real number } and -1 < x < 5 \text{ and } x \neq \frac{3}{2}\}$

89. Answers may vary. $f(x) = \dfrac{1}{x+2}$, $g(x) = \dfrac{1}{x-5}$

91. The domain of $y_3 = \dfrac{2.5x + 1.5}{x - 3}$ is $\{x \,|\, x \text{ is a real number}$ *and* $x \neq 3\}$. The CONNECTED mode graph contains the line $x = 3$, whereas the DOT mode graph contains no points having 3 as the first coordinate. Thus the DOT mode graph represents y_3 more accurately.

Review Exercises: Chapter 5, pp. 443–445

1. False   **2.** True   **3.** True   **4.** False   **5.** True
6. False   **7.** True   **8.** True   **9.** y^{11} **10.** $(3x)^{14}$

11. t^8 **12.** 4^3, or 64   **13.** 1   **14.** $\dfrac{9t^8}{4s^6}$ **15.** $-8x^3y^6$

16. $18x^5$ **17.** a^7b^6 **18.** $\dfrac{1}{m^7}$ **19.** t^{-8} **20.** $\dfrac{1}{7^2}$, or $\dfrac{1}{49}$

21. $\dfrac{1}{a^{13}b^7}$ **22.** $\dfrac{1}{x^{12}}$ **23.** $\dfrac{x^6}{4y^2}$ **24.** $\dfrac{y^3}{8x^3}$ **25.** 8,300,000

26. 3.28×10^{-5} **27.** 2.1×10^4 **28.** 5.1×10^{-5}
29. 2.28×10^{11} platelets   **30.** $3x^2$, $6x$, $\frac{1}{2}$ **31.** $-4y^5$, $7y^2$, $-3y$, -2 **32.** 7, -1, 7   **33.** 4, 6, -5, $\frac{5}{3}$
34. (a) 2, 0, 5; **(b)** $15t^5$, 15; **(c)** 5   **35. (a)** 5, 4, 2, 1;
(b) $-2x^5$, -2; **(c)** 5   **36.** Binomial   **37.** None of these
38. Monomial   **39.** $-x^2 + 9x$ **40.** $-\frac{1}{4}x^3 + 4x^2 + 7$
41. $-3x^5 + 25$ **42.** $-2x^2 - 3x + 2$
43. $10x^4 - 7x^2 - x - \frac{1}{2}$ **44.** 10   **45.** 28   **46.** $(-\infty, 2]$
47. About 340 mg   **48.** About 185 mg   **49.** $[0, 345]$
50. $[0, 6]$ **51.** $x^5 + 3x^4 + 6x^3 - 2x - 9$
52. $2x^2 - 4x - 6$ **53.** $x^5 - 3x^3 - 2x^2 + 8$
54. $\frac{3}{4}x^4 + \frac{1}{4}x^3 - \frac{1}{3}x^2 - \frac{7}{4}x + \frac{3}{8}$
55. $-x^5 + x^4 - 5x^3 - 2x^2 + 2x$ **56. (a)** $4w + 6$;
(b) $w^2 + 3w$ **57.** $-12x^3$ **58.** $49x^2 + 14x + 1$
59. $a^2 - 3a - 28$ **60.** $m^2 - 25$
61. $12x^3 - 23x^2 + 13x - 2$ **62.** $x^2 - 18x + 81$
63. $15t^5 - 6t^4 + 12t^3$ **64.** $a^2 - 49$
65. $x^2 - 1.05x + 0.225$
66. $x^7 + x^5 - 3x^4 + 3x^3 - 2x^2 + 5x - 3$
67. $9x^2 - 30x + 25$ **68.** $2t^4 - 11t^2 - 21$
69. $a^2 + \frac{1}{6}a - \frac{1}{3}$ **70.** $9x^4 - 16$ **71.** $4 - x^2$
72. $2x^2 - 7xy - 15y^2$ **73.** 49
74. Coefficients: 1, -7, 9, -8; degrees: 6, 2, 2, 0; 6

75. Coefficients: 1, -1, 1; degrees: 16, 40, 23; 40
76. $-y + 9w - 5$ **77.** $6m^3 + 4m^2n - mn^2$
78. $-x^2 - 10xy$ **79.** $11x^3y^2 - 8x^2y - 6x^2 - 6x + 6$
80. $p^3 - q^3$ **81.** $9a^8 - 2a^4b^3 + \frac{1}{9}b^6$ **82.** $\frac{1}{2}x^2 - \frac{1}{2}y^2$
83. $5x^2 - \frac{1}{2}x + 3$ **84.** $3x^2 - 7x + 4 + \dfrac{1}{2x+3}$
85. $t^3 + 2t - 3$ **86.** 102   **87.** -17 **88.** $-\frac{9}{2}$
89. $x^2 + 3x - 5$ **90.** $3x^3 - 6x^2 + 3x - 6$
91. $\dfrac{x^2 + 1}{3x - 6}$, $x \neq 2$ **92.** $\mathbb{R}$
93. $\{x \,|\, x \text{ is a real number } and x \neq 2\}$
94. 🅣🅦 In the expression $5x^3$, the exponent refers only to the x. In the expression $(5x)^3$, the entire expression within the parentheses is cubed.   **95.** 🅣🅦 The sum of two polynomials of degree n will also have degree n, since only the coefficients are added and the variables remain unchanged. An exception to this occurs when the leading terms of the two polynomials are opposites. The sum of those terms is then zero and the sum of the polynomials will have a degree less than n.
96. (a) 3; **(b)** 2   **97.** $-28x^8$ **98.** $8x^4 + 4x^3 + 5x - 2$
99. $-16x^6 + x^2 - 10x + 25$ **100.** $\frac{94}{13}$

Test: Chapter 5, p. 446

1. [5.1] t^8 **2.** [5.1] $(x+3)^{11}$ **3.** [5.1] 3^3, or 27
4. [5.1] 1   **5.** [5.1] x^6 **6.** [5.1] $-27y^6$ **7.** [5.1] $-24x^{17}$
8. [5.1] a^6b^5 **9.** [5.2] $\dfrac{1}{5^3}$ **10.** [5.2] y^{-8} **11.** [5.2] $\dfrac{1}{t^6}$
12. [5.2] $\dfrac{y^5}{x^5}$ **13.** [5.2] $\dfrac{b^4}{16a^{12}}$ **14.** [5.2] $\dfrac{c^3}{a^3b^3}$
15. [5.2] 3.9×10^9 **16.** [5.2] 0.00000005
17. [5.2] 1.8×10^{17} **18.** [5.2] 1.3×10^{22}
19. [5.2] 1.5×10^4 files   **20.** [5.3] Binomial   **21.** [5.3] $\frac{1}{3}$, -1, 7   **22.** [5.3] Degrees of terms: 3, 1, 5, 0; leading term: $7t^5$; leading coefficient: 7; degree of polynomial: 5
23. [5.3] -7 **24.** [5.3] $5a^2 - 6$ **25.** [5.3] $\frac{7}{4}y^2 - 4y$
26. [5.3] $x^5 + 2x^3 + 4x^2 - 8x + 3$ **27.** [5.3] $(-\infty, \infty)$
28. [5.4] $5x^4 + 5x^2 + x + 5$
29. [5.4] $-4x^4 + x^3 - 8x - 3$
30. [5.4] $-x^5 + 1.3x^3 - 0.8x^2 - 3$
31. [5.5] $-12x^4 + 9x^3 + 15x^2$ **32.** [5.6] $x^2 - \frac{2}{3}x + \frac{1}{9}$
33. [5.6] $25t^2 - 49$ **34.** [5.6] $3b^2 - 4b - 15$
35. [5.6] $x^{14} - 4x^8 + 4x^6 - 16$ **36.** [5.6] $48 + 34y - 5y^2$
37. [5.5] $6x^3 - 7x^2 - 11x - 3$ **38.** [5.6] $64a^2 + 48a + 9$
39. [5.7] $-5x^3y - x^2y^2 + xy^3 - y^3 + 19$
40. [5.7] $8a^2b^2 + 6ab + 6ab^2 + ab^3 - 4b^3$
41. [5.7] $9x^{10} - 16y^{10}$ **42.** [5.8] $4x^2 + 3x - 5$
43. [5.8] $6x^2 - 20x + 26 + \dfrac{-39}{x+2}$ **44.** [5.9] -130
45. [5.9] $-3x^3 - 4x^2 - 3x - 4$
46. [5.9] $\{x \,|\, x \text{ is a real number } and x \neq -\frac{4}{3}\}$
47. [5.5], [5.6] $V = l(l-2)(l-1) = l^3 - 3l^2 + 2l$
48. [2.2], [5.6] $\frac{100}{21}$

Chapter 6

Exercise Set 6.1, pp. 459–463

1. False **2.** False **3.** True **4.** True **5.** True
6. True **7.** True **8.** False **9.** $-3, 5$ **11.** $-2, 0$
13. $-3, 1$ **15.** $-4, 2$ **17.** $0, 5$ **19.** $1, 3$ **21.** $10, 15$
23. $0, 1, 2$ **25.** $-15, 6, 12$ **27.** $-0.42857, 0.33333$
29. $-5, 9$ **31.** $-0.5, 7$ **33.** $-1, 0, 3$ **35.** III **37.** I
39. Expression **41.** Equation **43.** Equation
45. $2t(t + 4)$ **47.** $y^2(y + 9)$ **49.** $5x^2(3 - x^2)$
51. $4xy(x - 3y)$ **53.** $3(y^2 - y - 3)$
55. $2a(3b - 2d + 6c)$ **57.** $3x^2y^4z^2(3xy^2 - 4x^2z^2 + 5yz)$
59. $-5(x - 7)$ **61.** $-6(y + 12)$ **63.** $-2(x^2 - 2x + 6)$
65. $-3(-y + 8x)$, or $-3(8x - y)$ **67.** $-7(-s + 2t)$, or
$-7(2t - s)$ **69.** $-(x^2 - 5x + 9)$
71. $-a(a^3 - 2a^2 + 13)$ **73.** $(b - 5)(a + c)$
75. $(x + 7)(2x - 3)$ **77.** $(x - y)(a^2 - 5)$
79. $(c + d)(a + b)$ **81.** $(b - 1)(b^2 + 2)$
83. $(a - 3)(a^2 - 2)$ **85.** $12x(6x^2 - 3x + 2)$
87. $x^3(x - 1)(x^2 + 1)$ **89.** $(y^2 + 3)(2y^2 + 5)$
91. (a) $h(t) = -8t(2t - 9)$; **(b)** $h(1) = 56$ ft
93. $R(n) = n(n - 1)$ **95.** $P(x) = x(x - 3)$
97. $R(x) = 0.4x(700 - x)$ **99.** $N(x) = \frac{1}{6}(x^3 + 3x^2 + 2x)$
101. $H(n) = \frac{1}{2}n(n - 1)$ **103.** $-1, 0$ **105.** $0, 3$
107. $-3, 0$ **109.** $-\frac{1}{3}, 0$ **111.** TW **113.** -26
114. -1 **115.** -30 **116.** -19 **117.** $56, 58, 60$
118. 18,245 shopping centers **119.** TW
121. $(x + 4)(x - 2)$ **123.** $x^5y^4 + x^4y^6 = x^3y(x^2y^3 + xy^5)$
125. $(x^2 - x + 5)(r + s)$
127. $(x^4 + x^2 + 5)(a^4 + a^2 + 5)$ **129.** $x^{-9}(x^3 + 1 + x^6)$
131. $x^{1/3}(1 - 5x^{1/6} + 3x^{5/12})$ **133.** $x^{-5/2}(1 + x)$
135. $x^{-7/5}(x^{3/5} - 1 + x^{16/15})$ **137.** $3a^n(a + 2 - 5a^2)$
139. $y^{a+b}(7y^a - 5 + 3y^b)$

Exercise Set 6.2, pp. 471–473

1. True **2.** True **3.** False **4.** True **5.** False
6. False **7.** True **8.** True **9.** $(x + 2)(x + 6)$
11. $(t + 3)(t + 5)$ **13.** $(x - 9)(x + 3)$
15. $2(n - 5)(n - 5)$, or $2(n - 5)^2$ **17.** $a(a + 8)(a - 9)$
19. $(x + 9)(x + 5)$ **21.** $(x + 5)(x - 2)$
23. $3(x + 2)(x + 3)$ **25.** $-(x - 8)(x + 7)$, or
$(-x + 8)(x + 7)$, or $(8 - x)(7 + x)$
27. $-y(y - 8)(y + 4)$, or $y(-y + 8)(y + 4)$, or
$y(8 - y)(4 + y)$ **29.** $x^2(x + 16)(x - 5)$ **31.** Prime
33. $(p - 8q)(p + 3q)$ **35.** $(y + 4z)(y + 4z)$, or
$(y + 4z)^2$ **37.** $p^2(p + 1)(p + 79)$ **39.** $-6, -2$
41. 5 **43.** $-8, 0, 9$ **45.** $-5, 1$ **47.** $-3, 2$
49. $-5, 9$ **51.** $-1, 0, 3$ **53.** $-9, 5$ **55.** $0, 9$
57. $-5, 0, 8$ **59.** $-4, 5$ **61.** $-7, 5$
63. $(x + 22)(x - 12)$ **65.** $(x + 24)(x + 16)$
67. $(x + 64)(x - 38)$ **69.** $f(x) = (x + 1)(x - 2)$, or

$f(x) = x^2 - x - 2$ **71.** $f(x) = (x + 7)(x + 10)$, or
$f(x) = x^2 + 17x + 70$ **73.** $f(x) = x(x - 1)(x - 2)$, or
$f(x) = x^3 - 3x^2 + 2x$ **75.** TW **77.** $3x^2(3x^3 - 8x + 1)$
78. $x^2(x + 1)(x + 2)$ **79.** $(x + 2)(y + 3)$
80. $(x - 7)(2x - 3)$ **81.** $15x^2y^3(3xy + 2y^2 + 4x^2)$
82. $-2(t^3 + 2t - 4)$ **83.** $(w - z)(x - y)$
84. $(x + 5)(x + 10)$ **85.** TW **87.** $\{-1, 3\}; (-2, 4)$, or
$\{x \,|\, -2 < x < 4\}$ **89.** $f(x) = 5x^3 - 20x^2 + 5x + 30$;
answers may vary **91.** 6.90 **93.** 3.48
95. $ab^2(2a^3b^4 - 3ab - 20)$ **97.** $\left(x + \frac{4}{5}\right)\left(x - \frac{1}{5}\right)$
99. $(y - 0.1)(y + 0.5)$ **101.** $(x + a)(x + b)$
103. $a^2(p^a + 2)(p^a - 1)$
105. $76, -76, 28, -28, 20, -20$ **107.** $x - 365$

Exercise Set 6.3, pp. 480–482

1. (f) **2.** (c) **3.** (e) **4.** (a) **5.** (g) **6.** (d)
7. (h) **8.** (b) **9.** $(3x + 5)(2x - 5)$
11. $y(2y - 3)(5y + 4)$ **13.** $2(4a - 1)(3a - 1)$
15. $(7y + 4)(5y + 2)$ **17.** $2(5t - 3)(t + 1)$
19. $4(x - 4)(2x + 1)$ **21.** $x^2(2x - 3)(7x + 1)$
23. $4(3a - 4)(a + 1)$ **25.** $(3x + 4)(3x + 1)$
27. $(4x + 3)(x + 3)$ **29.** $-2(2t - 3)(2t + 5)$
31. $(4 + 3z)(2 - 3z)$ **33.** $xy(6y + 5)(3y - 2)$
35. $(x - 2)(24x + 1)$ **37.** $3x(7x + 3)(3x + 4)$
39. $2x^2(4x - 3)(6x + 5)$ **41.** $(4a - 3b)(3a - 2b)$
43. $(2x - 3y)(x + 2y)$ **45.** $(3x - 4y)(2x - 7y)$
47. $(3x - 5y)(3x - 5y)$, or $(3x - 5y)^2$
49. $(9xy - 4)(xy + 1)$ **51.** $-\frac{5}{3}, \frac{5}{2}$ **53.** $-\frac{4}{3}, \frac{2}{3}$
55. $-\frac{4}{3}, -\frac{3}{7}, 0$ **57.** $\frac{2}{3}, 2$ **59.** $-2, -\frac{3}{4}, 0$ **61.** $-\frac{1}{3}, \frac{5}{2}$
63. $-\frac{7}{4}, \frac{4}{3}$ **65.** $-\frac{1}{2}, 7$ **67.** $-8, -4$ **69.** $-4, \frac{3}{2}$
71. $-9, -3$ **73.** $\{x \,|\, x$ is a real number $and \, x \neq 5$ and
$x \neq -1\}$ **75.** $\{x \,|\, x$ is a real number $and \, x \neq 0$ and $x \neq \frac{1}{2}\}$
77. $\{x \,|\, x$ is a real number $and \, x \neq 0$ and $x \neq 2$ and $x \neq 5\}$
79. TW **81.** $(x + 1)(x + 3)$ **82.** $(2x - 5)(x + 1)$
83. $y(x + y)(x + 2)$ **84.** $x(3x - 1)(5x + 3)$
85. $(4x - 3)(3x + 1)$ **86.** $2(x - 7)(x + 2)$
87. $(a - b)(c - d)$ **88.** $4x^2y^3(3xy - 6y^2 + 1)$
89. $3a(a - 2)$ **90.** $-8(x^2 - x - 1)$ **91.** TW
93. $(2x + 15)(2x + 45)$ **95.** $3x(x + 68)(x - 18)$
97. $-\frac{11}{8}, -\frac{1}{4}, \frac{2}{3}$ **99.** 1 **101.** $(3ab + 2)(6ab - 5)$
103. Prime **105.** $(5t^5 - 1)^2$ **107.** $(10x^n + 3)(2x^n + 1)$
109. $[7(t - 3)^n - 2][(t - 3)^n + 1]$
111. $(2a^2b^3 + 5)(a^2b^3 - 4)$ **113.** $(2x^a - 3)(2x^a + 1)$
115. Since $ax^2 + bx + c = (mx + r)(nx + s)$, from FOIL
we know that $a = mn$, $c = rs$, and $b = ms + rn$. If $P = ms$
and $Q = rn$, then $b = P + Q$. Since $ac = mnrs = msrn$, we
have $ac = PQ$.

Exercise Set 6.4, pp. 489–490

1. Difference of two squares   **2.** Perfect-square trinomial
3. Perfect-square trinomial   **4.** Difference of two squares
5. None of these   **6.** Polynomial having a common factor
7. Polynomial having a common factor   **8.** None of these
9. Perfect-square trinomial   **10.** Polynomial having a
common factor   **11.** $(t + 3)^2$ **13.** $(a - 7)^2$
15. $4(a - 2)^2$ **17.** $(y + 6)^2$ **19.** $y(y - 9)^2$
21. $2(x - 10)^2$ **23.** $(1 - 4d)^2$ **25.** $y(y + 4)^2$
27. $(0.5x + 0.3)^2$ **29.** $(p^2 + q^2)^2$ **31.** $(5a + 3b)^2$
33. $5(a - b)^2$ **35.** $(y + 10)(y - 10)$
37. $(m + 8)(m - 8)$ **39.** $(pq + 5)(pq - 5)$
41. $8(x + y)(x - y)$ **43.** $7x(y^2 + z^2)(y + z)(y - z)$
45. $a(2a + 7)(2a - 7)$
47. $3(x^4 + y^4)(x^2 + y^2)(x + y)(x - y)$
49. $a^2(3a + 5b^2)(3a - 5b^2)$ **51.** $\left(\frac{1}{7} + x\right)\left(\frac{1}{7} - x\right)$
53. $(a + b + 3)(a + b - 3)$
55. $(x - 3 + y)(x - 3 - y)$ **57.** $(t + 8)(t + 1)(t - 1)$
59. $(r - 3)^2(r + 3)$ **61.** $(m - n + 5)(m - n - 5)$
63. $(6 + x + y)(6 - x - y)$
65. $(r - 1 + 2s)(r - 1 - 2s)$
67. $(4 + a + b)(4 - a - b)$
69. $(x + 5)(x + 2)(x - 2)$ **71.** $(a - 2)(a + b)(a - b)$
73. 1   **75.** 6   **77.** $-3, 3$ **79.** $-\frac{1}{5}, \frac{1}{5}$ **81.** $-\frac{1}{2}, \frac{1}{2}$
83. $-1, 1, 3$ **85.** $-1.541, 4.541$ **87.** $-3.871, -0.129$
89. $-2.414, -1, 0.414$ **91.** 6   **93.** $-4, 4$ **95.** -1
97. $-1, 1, 2$ **99.** TW   **101.** $(3x - 1)(2x - 5)$
102. $(4x + 1)(4x - 1)$ **103.** $(4x + 1)^2$
104. $3x(x + 1)(x - 2)$ **105.** $(x^2 - 2)(x + 3)$
106. $(2x + 5)(2x - 3)$ **107.** $\left(\frac{1}{8} + x\right)\left(\frac{1}{8} - x\right)$
108. $(10 + t + s)(10 - t - s)$ **109.** $12xy(x - y)^2$
110. $-3x(x^3 - 6x^2 - 1)$ **111.** TW   **113.** $-\frac{1}{54}(4r + 3s)^2$
115. $(0.3x^4 + 0.8)^2$, or $\frac{1}{100}(3x^4 + 8)^2$
117. $(r + s + 1)(r - s - 9)$ **119.** $(x^{2a} + y^b)(x^{2a} - y^b)$
121. $(5y^a + x^b - 1)(5y^a - x^b + 1)$ **123.** $3(x + 3)^2$
125. $(3x^n - 1)^2$ **127.** $(t + 1 - 3)(t + 1 - 1)$, or $t(t - 2)$
129. $(m + 2n)(m + 2n + 5)$ **131.** $h(2a + h)$
133. (a) $\pi h(R + r)(R - r)$; **(b)** 3,014,400 cm^3

Exercise Set 6.5, pp. 495–496

1. Difference of two cubes   **2.** Sum of two cubes
3. Difference of two squares   **4.** Prime polynomial
5. Sum of two cubes   **6.** Difference of two cubes
7. None of these   **8.** Both a difference of two cubes *and* a
difference of two squares   **9.** Both a difference of two
cubes *and* a difference of two squares   **10.** None of these
11. $(x + 4)(x^2 - 4x + 16)$ **13.** $(z - 1)(z^2 + z + 1)$
15. $(x - 3)(x^2 + 3x + 9)$ **17.** $(3x + 1)(9x^2 - 3x + 1)$
19. $(4 - 5x)(16 + 20x + 25x^2)$
21. $(3y + 4)(9y^2 - 12y + 16)$
23. $(x - y)(x^2 + xy + y^2)$
25. $\left(a + \frac{1}{2}\right)\left(a^2 - \frac{1}{2}a + \frac{1}{4}\right)$ **27.** $8(t - 1)(t^2 + t + 1)$

29. $2(3x + 1)(9x^2 - 3x + 1)$
31. $a(b + 5)(b^2 - 5b + 25)$
33. $5(x - 2z)(x^2 + 2xz + 4z^2)$
35. $(x + 0.1)(x^2 - 0.1x + 0.01)$
37. $8(2x^2 - t^2)(4x^4 + 2x^2t^2 + t^4)$
39. $2y(y - 4)(y^2 + 4y + 16)$
41. $(z + 1)(z^2 - z + 1)(z - 1)(z^2 + z + 1)$
43. $(t^2 + 4y^2)(t^4 - 4t^2y^2 + 16y^4)$
45. $(x^4 - yz^4)(x^8 + x^4yz^4 + y^2z^8)$ **47.** -1 **49.** $\frac{3}{2}$
51. 10   **53.** TW   **55.** $5(x - z)(x^2 + xz + z^2)$
56. $(x - 3)(x + 7)$ **57.** $-8xz(x^2 - 3xz + 2z^2 + 1)$
58. $(1 + 10x)(1 - 10x + 100x^2)$
59. $3(x^2 + z^2)(x + z)(x - z)$ **60.** $(2x - 5)^2$
61. $(x + 1)(x^2 - x + 1)(x - 1)(x^2 + x + 1)$
62. $(x^2 + 1)(x + 1)$ **63.** $3x(x + 10)^2$
64. $x^2y^3(x + 2)(x + 6)$ **65.** $(4x + 3)(3x + 4)$
66. $(x + 1 + y)(x + 1 - y)$ **67.** TW
69. $(x^{2a} - y^b)(x^{4a} + x^{2a}y^b + y^{2b})$ **71.** $2x(x^2 + 75)$
73. $5\left(xy^2 - \frac{1}{2}\right)\left(x^2y^4 + \frac{1}{2}xy^2 + \frac{1}{4}\right)$ **75.** $-(3x^{4a} + 3x^{2a} + 1)$
77. $(t - 8)(t - 1)(t^2 + t + 1)$
79. $h(2a + h)(a^2 + ah + h^2)(3a^2 + 3ah + h^2)$

Exercise Set 6.6, pp. 501–502

1. Common factor   **2.** Perfect-square trinomial
3. Grouping   **4.** Multiplying   **5.** $10(a + 8)(a - 8)$
7. $(y - 7)^2$ **9.** $(2t + 3)(t + 4)$ **11.** $x(x - 9)^2$
13. $(x - 5)^2(x + 5)$ **15.** $3t(3t + 1)(3t - 1)$
17. $3x(3x - 5)(x + 3)$ **19.** Prime
21. $3(x + 3)(2x - 5)$ **23.** $-2a^4(a - 2)^2$
25. $5x(x^2 + 4)(x + 2)(x - 2)$ **27.** $(t^2 + 3)(t^2 - 3)$
29. $-x^4(x^2 - 2x + 7)$ **31.** $(x - y)(x^2 + xy + y^2)$
33. $a(x^2 + y^2)$ **35.** $9mn(4 - mn)$ **37.** $2\pi r(h + r)$
39. $(a + b)(5a + 3b)$ **41.** $(x + 1)(x + y)$
43. $(n - 5 + 3m)(n - 5 - 3m)$ **45.** $(3x - 2y)(x + 5y)$
47. $(a - 2b)^2$ **49.** $(4x + 3y)^2$ **51.** Prime
53. $(2t + 1)(4t^2 - 2t + 1)(2t - 1)(4t^2 + 2t + 1)$
55. $p(4p^2 + 4pq - q^2)$ **57.** $(3b - a)(b + 6a)$
59. $-1(xy + 2)(xy + 6)$, or $-(xy + 2)(xy + 6)$
61. $(t^4 + s^5 + 6)(t^4 - s^5 - 6)$
63. $2a(3a + 2b)(9a^2 - 6ab + 4b^2)$
65. $x^4(x + 2y)(x - y)$ **67.** $\left(6a - \frac{5}{4}\right)^2$ **69.** $\left(\frac{1}{9}x - \frac{4}{3}\right)^2$
71. $(1 + 4x^6y^6)(1 + 2x^3y^3)(1 - 2x^3y^3)$
73. $(2ab + 3)^2$ **75.** $a(a^2 + 8)(a + 8)$ **77.** TW
79. For $(-1, 11)$:

$y = -4x + 7$	
11	$-4(-1) + 7$
	$4 + 7$
$11 \overset{?}{=} 11$	True

For $(0, 7)$:

$y = -4x + 7$	
7	$-4 \cdot 0 + 7$
	$0 + 7$
$7 \overset{?}{=} 7$	True

For $(3, -5)$:

$y = -4x + 7$	
-5	$-4 \cdot 3 + 7$
	$-12 + 7$
$-5 \overset{?}{=} -5$	True

80. $\frac{4}{5}$ **81.** $-\frac{7}{3}$ **82.** $-\frac{9}{2}$ **83.** $\frac{9}{4}$
84.   **85.** TW

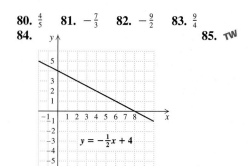

$y = -\frac{1}{2}x + 4$

87. $-x(x^2 + 9)(x^2 - 2)$
89. $-3(a + 1)(a - 1)(a + 2)(a - 2)$
91. $(y + 1)(y - 7)(y + 3)$
93. $(2x - 2 + 3y)(3x - 3 - y)$
95. $(a + 3)^2(2a + b + 4)(a - b + 5)$

Visualizing the Graph, p. 511

1. D   **2.** J   **3.** A   **4.** B   **5.** E   **6.** C   **7.** I   **8.** F
9. G   **10.** H

Exercise Set 6.7 pp. 512–516

1. $-12, 11$ **3.** Length: 12 cm; width: 7 cm   **5.** 3 m
7. 3 cm   **9.** 10 ft   **11.** 16, 18, 20   **13.** Height: 6 ft;
base: 4 ft   **15.** Height: 16 m; base: 7 m   **17.** 41 ft
19. Length: 100 m; width: 75 m   **21.** 2 sets   **23.** 2 sec
25. 5 sec   **27.** $1\frac{1}{2}$ yr, 6 yr
29. (a) $C(x) = 1.703571429x^2 - 11.75357143x + 32.2$;
(b) \$47.2 million; (c) about 6.7 yr after 2000, or
approximately 2007
31. (a) $L(x) = 0.0078989899x^3 - 1.040519481x^2 +$
$46.47265512x - 678.4848485$; (b) about \$58; (c) about 64
33. (a) $B(x) = 9.289417614x^4 - 263.4086174x^3 +$
$2056.113163x^2 - 939.209145x + 37459.76515$; (b) about
78,667 degrees; (c) approximately 1996, 2001, and 2004
35. TW   **37.** $\frac{14}{3}$ **38.** $\{x \mid x > \frac{10}{3}\}$, or $\left(\frac{10}{3}, \infty\right)$ **39.** $-\frac{1}{3}, 2$
40. $(-2, -9)$ **41.** 2   **42.** -1 **43.** $\left(\frac{18}{7}, -\frac{1}{7}\right)$
44. $\left(6, \frac{4}{3}\right)$ **45.** $-\frac{1}{3}, \frac{1}{3}$ **46.** $-1, 6$ **47.** TW
49. Length: 28 cm; width: 14 cm   **51.** About 5.7 sec

Review Exercises: Chapter 6, pp. 519–520

1. False   **2.** True   **3.** True   **4.** False   **5.** True
6. False   **7.** True   **8.** False   **9.** $x(7x + 6)$
10. $3y^2(3y^2 - 1)$ **11.** $3x(5x^3 - 6x^2 + 7x - 3)$
12. $(a - 9)(a - 3)$ **13.** $(3m + 2)(m + 4)$
14. $(5x + 2)^2$ **15.** $4(y + 2)(y - 2)$
16. $x(x - 2)(x + 7)$ **17.** $(a + 2b)(x - y)$
18. $(y + 2)(3y^2 - 5)$ **19.** $(a^2 + 9)(a + 3)(a - 3)$
20. $4(x^4 + x^2 + 5)$ **21.** $(3x - 2)(9x^2 + 6x + 4)$
22. $(0.4b - 0.5c)(0.16b^2 + 0.2bc + 0.25c^2)$

23. $y(y^4 + 1)$ **24.** $2z^6(z^2 - 8)$
25. $2y(3x^2 - 1)(9x^4 + 3x^2 + 1)$ **26.** $4(3x - 5)^2$
27. $(3t + p)(2t + 5p)$ **28.** $(x + 3)(x - 3)(x + 2)$
29. $(a - b + 2t)(a - b - 2t)$ **30.** $-2, 1, 5$ **31.** $4, 7$
32. 10   **33.** $\frac{2}{3}, \frac{3}{2}$ **34.** $0, \frac{7}{4}$ **35.** $-4, 4$ **36.** $-3, 0, 7$
37. $-1, 6$ **38.** $-4, 4, 5$ **39.** $12, 15$
40. $-1.646, 3.646$ **41.** $-4, 11$
42. $\{x \mid x$ is a real number $and\ x \neq -7\ and\ x \neq \frac{2}{3}\}$
43. 5   **44.** $3, 5, 7; -7, -5, -3$
45. Length: 8 in.; width: 5 in.   **46.** 17 ft
47. (a) $P(x) = 0.0256348234x^2 - 1.01295601x +$
24.99905124; (b) 25.5; (c) approximately 1971 and 2009
48. TW The roots of a polynomial function are the
x-coordinates of the points at which the graph of the function
crosses or touches the x-axis.   **49.** TW The principle of zero
products states that if a product is equal to 0, at least one of
the factors must be 0. If a product is nonzero, we cannot
conclude that any one of the factors is a particular value.
50. $2(2x - y)(4x^2 + 2xy + y^2)(2x + y)(4x^2 - 2xy + y^2)$
51. $-2(3x^2 + 1)$ **52.** $a^3 - b^3 + 3b^2 - 3b + 1$
53. $-1, -\frac{1}{2}$

Test: Chapter 6, pp. 521–522

1. [6.1] $5x^2(3 - x^2)$ **2.** [6.4] $(y + 5)(y + 2)(y - 2)$
3. [6.2] $(p - 14)(p + 2)$ **4.** [6.3] $(6m + 1)(2m + 3)$
5. [6.4] $(3y + 5)(3y - 5)$ **6.** [6.5] $3(r - 1)(r^2 + r + 1)$
7. [6.4] $(3x - 5)^2$
8. [6.4] $(x^4 + y^4)(x^2 + y^2)(x + y)(x - y)$
9. [6.4] $(y + 4 + 10t)(y + 4 - 10t)$
10. [6.4] $5(2a - b)(2a + b)$ **11.** [6.3] $2(4x - 1)(3x - 5)$
12. [6.5] $2ab(2a^2 + 3b^2)(4a^4 - 6a^2b^2 + 9b^4)$
13. [6.1] $4xy(y^3 + 9x + 2x^2y - 4)$ **14.** [6.1] $-3, -1, 2, 4$
15. [6.3] $-\frac{5}{2}, 8$ **16.** [6.2] $-3, 6$ **17.** [6.4] $-5, 5$
18. [6.3] $-7, -\frac{3}{2}$ **19.** [6.1] $-\frac{1}{3}, 0$ **20.** [6.4] 9
21. [6.2] $-3.372, 0, 2.372$ **22.** [6.3] $0, 5$
23. [6.3] $\{x \mid x$ is a real number $and\ x \neq -1\}$
24. [6.7] Length: 8 cm; width: 5 cm   **25.** [6.7] $4\frac{1}{2}$ sec
26. [6.7] (a) $E(x) = 0.7x^2 - 0.6x + 33.28$;
(b) about \$84.6 billion; (c) about 10.2 yr after 2000, or
approximately 2010   **27.** (a) [5.5] $x^5 + x + 1$;
(b) [5.4], [6.5] $(x^2 + x + 1)(x^3 - x^2 + 1)$
28. [6.3] $(3x^n + 4)(2x^n - 5)$

Cumulative Review: Chapters 1–6, pp. 522–524

1. Approximately \$57,662   **2.** 172 invitations
3. 36 two-point shots; 28 foul shots   **4.** 21 Humalog vials;
29 Novolog vials   **5.** 14 ft   **6.** $-278, -276$
7. Emily: \$840; Andrew: \$420   **8.** Not linear   **9.** Linear
10. (a) $y = -0.1x + 13.1$; (b) 7.1 miles per gallon
11. $f(x) = 0.0821428571x + 1.330952381$, where f is in
trillions of dollars and x is the number of years after 1995

12. 17 **13.** -30.7 **14.** $-8x + 52$ **15.** $-x - 7$
16. $-\frac{43}{8}$ **17.** 1 **18.** 10 **19.** $-7, 7$ **20.** $\{x | x > \frac{9}{2}\}$,
or $(\frac{9}{2}, \infty)$ **21.** -2 **22.** $-10, -1$ **23.** -8
24. $(\frac{17}{11}, \frac{2}{11})$ **25.** $-4, \frac{1}{2}$ **26.** $\{x | x > 43\}$, or $(43, \infty)$
27. $(\frac{17}{28}, -\frac{2}{7})$ **28.** $-\frac{7}{2}, 5$ **29.** $b = 3a - c + 9$
30. $y = \frac{2}{3}z - \frac{1}{2}x$ **31.** $\frac{3}{2}x + 2y - 3z$
32. $-4x^3 - \frac{1}{7}x^2 - 2$
33.

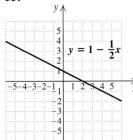

$y = 1 - \frac{1}{2}x$

34.

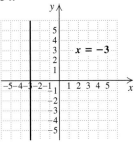

$x = -3$

35.

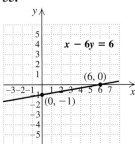

$x - 6y = 6$
$(6, 0)$
$(0, -1)$

36.

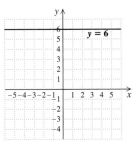

$y = 6$

37.

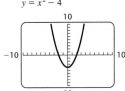

$y = x^2 - 4$

38. -2

39. Slope: $-\frac{3}{4}$; y-intercept: $(0, 2)$ **40.** $y = -\frac{3}{2}x + 4$
41. $y = \frac{1}{2}x - 7$ **42.** $y = -x + 3$ **43.** x^{-2}, or $\frac{1}{x^2}$
44. y^{-8}, or $\frac{1}{y^8}$ **45.** $-4a^4b^{14}$ **46.** $-y^3 - 2y^2 - 2y + 7$
47. $15a - 10b + 5c$ **48.** $2x^5 + x^3 - 6x^2 - x + 3$
49. $36x^2 - 60xy + 25y^2$ **50.** $2x^2 - x - 21$
51. $4x^6 - 1$ **52.** $5x^2 - 4x + 2 + \frac{2}{3x} + \frac{6}{x^2}$
53. $x^3 - x^2 + 9x - 25 + \frac{93}{x + 3}$ **54.** $2x(3 - x - 12x^3)$
55. $(4x + 9)(4x - 9)$ **56.** $(t - 4)(t - 6)$
57. $(4x + 3)(2x + 1)$ **58.** $2(3x - 2)(x - 4)$
59. $2(x + 5)(x^2 - 5x + 25)$ **60.** $(4x + 5)^2$
61. $(3x - 2)(x + 4)$ **62.** $(x + 2)(x^3 - 3)$
63. Domain: $\mathbb{R}$; range: $\{3\}$ **64.** Domain: $\mathbb{R}$; range: $\mathbb{R}$

65. Domain: $\mathbb{R}$; range: $\{y | y \geq -3\}$, or $[-3, \infty)$
66. Domain: $\mathbb{R}$; range: $\{y | y \geq -4\}$, or $[-4, \infty)$ **67.** 12
68. $16y^6 - y^4 + 6y^2 - 9$
69. $2(a^{16} + 81b^{20})(a^8 + 9b^{10})(a^4 + 3b^5)(a^4 - 3b^5)$
70. $-7, 4, 12$ **71.** -7 **72.** $-5, -3, 3, 5$

Chapter 7

Interactive Discovery, p. 533

1. One vertical asymptote: $x = -\frac{1}{2}$ **2.** $x = -\frac{1}{2}$

Visualizing the Graph, p. 535

1. I **2.** B **3.** G **4.** A **5.** D **6.** C **7.** E
8. J **9.** H **10.** F

Exercise Set 7.1, pp. 536–539

1. (e) **2.** (j) **3.** (g) **4.** (c) **5.** (i) **6.** (h)
7. (a) **8.** (f) **9.** (d) **10.** (b) **11.** $\frac{40}{13}$ hr, or $3\frac{1}{13}$ hr
13. $\frac{2}{3}$; 28; $\frac{163}{10}$ **15.** $-\frac{9}{4}$; does not exist; $-\frac{11}{9}$ **17.** 0
19. -8 **21.** 4 **23.** $-4, 7$ **25.** $-5, 5$ **27.** $\frac{3}{x}$
29. $\frac{2}{3t^4}$ **31.** $a - 5$ **33.** $\frac{x - 4}{x + 5}$ **35.** $\frac{3(2a - 1)}{7(a - 1)}$
37. $-\frac{1}{2}$ **39.** $\frac{a + 4}{2(a - 4)}$ **41.** $\frac{x + 4}{x - 4}$ **43.** $t - 1$
45. $\frac{y^2 + 4}{y + 2}$ **47.** $\frac{1}{2}$ **49.** -1 **51.** $-\frac{1}{1 + 2t}$
53. $\frac{a - 5}{a + 5}$ **55.** -1 **57.** $\frac{x^2 + x + 1}{x + 1}$ **59.** $3(y + 2)$
61. $f(x) = \frac{3}{x}, x \neq -7, 0$ **63.** $g(x) = \frac{x - 3}{5}, x \neq -3$
65. $h(x) = -\frac{1}{5}, x \neq 4$ **67.** $f(t) = \frac{t + 4}{t - 4}, t \neq 4$
69. $g(t) = -\frac{7}{3}, t \neq 3$ **71.** $h(t) = \frac{t + 4}{t - 9}, t \neq -1, 9$
73. $f(x) = 3x + 2, x \neq \frac{2}{3}$ **75.** $g(t) = \frac{4 + t}{4 - t}, t \neq 4$
77. $x = -5$ **79.** No vertical asymptotes **81.** $x = -3$
83. $x = 2, x = 4$ **85.** (b) **87.** (f) **89.** (a) **91.** TW
93. $-\frac{7}{30}$ **94.** $-\frac{13}{40}$ **95.** $\frac{5}{14}$ **96.** $\frac{1}{35}$
97. $4x^3 - 7x^2 + 9x - 5$ **98.** $-2t^4 + 11t^3 - t^2 + 10t - 3$
99. TW **101.** $2a + h$ **103.** $\frac{(x^2 + y^2)(x + y)}{(x - y)^3}$
105. $\frac{1}{x - 1}$ **107.** $\frac{a^2 + 2}{a^2 - 3}$ **109.** 1
111. Domain: $\{x | x \text{ is a real number } and \, x \neq -2 \text{ and } x \neq 1\}$;
range: $\{y | y \text{ is a real number } and \, x \neq 2 \text{ and } x \neq 3\}$

113. Domain: $\{x \mid x \text{ is a real number } and \ x \neq -1 \text{ and } x \neq 1\}$; range: $\{y \mid y \leq -1 \text{ or } y > 0\}$ **115.** TW

Exercise Set 7.2, pp. 543–545

1. $\dfrac{7x(x-5)}{5(2x+1)}$ **3.** $\dfrac{(a-4)(a+2)}{(a+6)^2}$ **5.** $\dfrac{(2x+3)(x+1)}{4(x-5)}$

7. $\dfrac{5a^2}{3}$ **9.** $\dfrac{4}{c^2d}$ **11.** $\dfrac{y+4}{4}$ **13.** $\dfrac{x+2}{x-2}$

15. $\dfrac{(a^2+25)(a-5)}{(a-3)(a-1)(a+5)}$ **17.** $\dfrac{5(a+3)}{a(a+4)}$

19. $\dfrac{2a}{a-2}$ **21.** $\dfrac{t-5}{t+5}$ **23.** $\dfrac{5(a+6)}{a-1}$ **25.** 1

27. $\dfrac{t+2}{t+4}$ **29.** $-\dfrac{a+1}{2+a}$ **31.** 1   **33.** $c(c-2)$

35. $\dfrac{a^2+ab+b^2}{3(a+2b)}$ **37.** $\dfrac{7}{3x}$ **39.** $\dfrac{1}{a^3-8a}$ **41.** $\frac{35}{24}$

43. $\dfrac{x^2}{20}$ **45.** $6x^4y^7$ **47.** $\dfrac{y+5}{2y}$ **49.** $\dfrac{5}{x^5}$ **51.** $4(y-2)$

53. $-\dfrac{a}{b}$ **55.** $\dfrac{(y+3)(y^2+1)}{y+1}$ **57.** $\frac{15}{16}$ **59.** $\dfrac{a-5}{3(a-1)}$

61. $-\dfrac{5x+2}{x-3}$ **63.** $\dfrac{(2x-1)(2x+1)}{x-5}$

65. $\dfrac{(a-5)(a+3)}{(a+4)^2}$ **67.** $\dfrac{1}{(c-5)(5c-3)}$

69. $\dfrac{x^2+4x+16}{(x+4)^2}$ **71.** $\dfrac{(2a+b)^2}{2(a+b)}$ **73.** TW

75. $\frac{19}{12}$ **76.** $\frac{41}{24}$ **77.** $\frac{1}{18}$ **78.** $-\frac{1}{6}$ **79.** $-\frac{37}{20}$ **80.** $\frac{49}{45}$

81. TW   **83.** 1   **85.** 1   **87.** $\dfrac{a^2-2b}{a^2+3b}$ **89.** $\dfrac{2s}{r+2s}$

91. $\dfrac{6t^2-26t+30}{8t^2-15t-21}$

93. (a) $\dfrac{16(x+1)}{(x-1)^2(x^2+x+1)}, x \neq 1, x \neq -1$;

(b) $\dfrac{x^2+x+1}{(x+1)^3}, x \neq 1, x \neq -1$;

(c) $\dfrac{(x+1)^3}{x^2+x+1}, x \neq 1, x \neq -1$

Exercise Set 7.3, pp. 553–555

1. Numerators; denominator   **2.** Grouping; term
3. Least common denominator; LCD   **4.** Factorizations;
denominators   **5.** $\dfrac{10}{x}$ **7.** $\dfrac{3x+5}{12}$ **9.** $\dfrac{9}{a+3}$

11. $\dfrac{6}{a+2}$ **13.** $\dfrac{2y+7}{2y}$ **15.** 11   **17.** $\dfrac{3x+5}{x+1}$

19. $a+5$ **21.** $x-4$ **23.** 0   **25.** $\dfrac{1}{x+2}$

27. $\dfrac{(3a-7)(a-2)}{(a+6)(a-1)}$ **29.** $\dfrac{t-4}{t+3}$ **31.** $\dfrac{x+5}{x-6}$

33. $-\dfrac{5}{x-4}$, or $\dfrac{5}{4-x}$ **35.** $-\dfrac{1}{x-1}$, or $\dfrac{1}{1-x}$ **37.** 135

39. 72   **41.** 126   **43.** $12x^3$ **45.** $30a^4b^8$

47. $6(y-3)$ **49.** $(x-2)(x+2)(x+3)$

51. $t(t-4)(t+2)^2$ **53.** $30x^2y^2z^3$ **55.** $(a+1)(a-1)^2$

57. $(m-3)(m-2)^2$ **59.** $(t^2-9)^2$

61. $12x^3(x-5)(x-3)(x-1)$

63. $(t+1)(t-1)(t^2+t+1)$ **65.** $\dfrac{10}{12x^5}, \dfrac{x^2y}{12x^5}$

67. $\dfrac{12b}{8a^2b^2}, \dfrac{7a}{8a^2b^2}$ **69.** $\dfrac{2x(x+3)}{(x-2)(x+2)(x+3)}$,

$\dfrac{4x(x-2)}{(x-2)(x+2)(x+3)}$ **71.** TW   **73.** $-\dfrac{7}{9}, \dfrac{-7}{9}$

74. $\dfrac{-3}{2}, \dfrac{3}{-2}$ **75.** $-\frac{11}{36}$ **76.** $-\frac{7}{60}$ **77.** $x^2-9x+18$

78. $s^2 - \pi r^2$ **79.** TW   **81.** $\dfrac{18x+5}{x-1}$ **83.** $\dfrac{x}{3x+1}$

85. 30 strands   **87.** 60 strands   **89.** $120(x-1)^2(x+1)$

91. 70 min   **93.** 7:55 A.M.   **95.** TW

Exercise Set 7.4, pp. 563–565

1. LCD   **2.** Missing; denominator   **3.** Numerators; LCD

4. Simplify   **5.** $\dfrac{4x+9}{x^2}$ **7.** $-\dfrac{5}{24r}$ **9.** $\dfrac{2d^2+7c}{c^2d^3}$

11. $\dfrac{-2xy-18}{3x^2y^3}$ **13.** $\dfrac{5x+7}{18}$ **15.** $\dfrac{a+8}{4}$

17. $\dfrac{5a^2+7a-3}{9a^2}$ **19.** $\dfrac{-7x-13}{4x}$ **21.** $\dfrac{c^2+3cd-d^2}{c^2d^2}$

23. $\dfrac{3y^2-3xy-6x^2}{2x^2y^2}$ **25.** $\dfrac{10x}{(x-1)(x+1)}$

27. $\dfrac{2z+6}{(z-1)(z+1)}$ **29.** $\dfrac{11x+15}{4x(x+5)}$ **31.** $\dfrac{16-9t}{6t(t-5)}$

33. $\dfrac{x^2-x}{(x-5)(x+5)}$ **35.** $\dfrac{4t-5}{4(t-3)}$ **37.** $\dfrac{2x+10}{(x+3)^2}$

39. $\dfrac{3x-14}{(x-2)(x+2)}$ **41.** $\dfrac{9a}{4(a-5)}$ **43.** 0

45. $\dfrac{12a-11}{(a-3)(a-1)(a+2)}$ **47.** $\dfrac{x-5}{(x+5)(x+3)}$

49. $\dfrac{3z^2+19z-20}{(z-2)^2(z+3)}$ **51.** $\dfrac{-5}{x^2+17x+16}$ **53.** $\dfrac{5x-3}{5}$

55. $y+3$ **57.** $\dfrac{1}{t^2+4}$ **59.** $\dfrac{1}{m^2+mn+n^2}$

61. $\dfrac{y^2+10y+11}{(y-7)(y+7)}$ **63.** $\dfrac{3x^2-7x-4}{3(x-2)(x+2)}$

65. $\dfrac{a-2}{(a-3)(a+3)}$ **67.** $\dfrac{2x-3}{2-x}$ **69.** 2   **71.** 0

73. $f(x) = \dfrac{3(x + 4)}{x + 3}, x \neq \pm 3$

75. $f(x) = \dfrac{(x - 7)(2x - 1)}{(x - 4)(x - 1)(x + 3)}, x \neq -4, -3, 1, 4$

77. $f(x) = \dfrac{-2}{(x + 1)(x + 2)}, x \neq -3, -2, -1$ **79. TW**

81. $\dfrac{3y^6z^7}{7x^5}$ **82.** $\dfrac{7b^{11}c^7}{9a^2}$ **83.** $\dfrac{7}{9}$ **84.** $-\dfrac{5}{6}$

85.

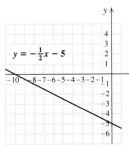

$y = -\frac{1}{2}x - 5$

86.

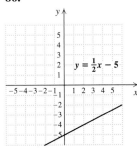

$y = \frac{1}{2}x - 5$

87. TW   **89.** Perimeter: $\dfrac{10x - 14}{(x - 5)(x + 4)}$;

area: $\dfrac{6}{(x - 5)(x + 4)}$ **91.** $\dfrac{30}{(x - 3)(x + 4)}$

93. $\dfrac{x^4 + 4x^3 - 5x^2 - 126x - 441}{(x + 2)^2(x + 7)^2}$ **95.** $\dfrac{-x^2 - 3}{(2x - 3)(x - 3)}$

97. $\dfrac{9x^2 + 28x + 15}{(x - 3)(x + 3)^2}$ **99.** $\dfrac{a}{a - b} + \dfrac{3b}{b - a}$; Answers may

vary.   **101.** $\dfrac{x^4 + 6x^3 + 2x^2}{(x + 2)(x - 2)(x + 5)}$

103. $\dfrac{x^5}{(x^2 - 4)(x^2 + 3x - 10)}$

105. $\{x \mid x \text{ is a real number } and \ x \neq -5 \ and \ x \neq -2$ $and \ x \neq 2\}$

107. Domain: $\{x \mid x \text{ is a real number } and \ x \neq -1\}$; range: $\{y \mid y \text{ is a real number } and \ y \neq 3\}$

109. Domain: $\{x \mid x \text{ is a real number } and \ x \neq 0 \ and \ x \neq 1\}$; range: $\{y \mid y > 0\}$

Exercise Set 7.5, pp. 572–575

1. (e)   **2.** (d)   **3.** (f)   **4.** (c)   **5.** (b)   **6.** (a)

7. $\dfrac{(x + 2)(x - 3)}{(x - 1)(x + 4)}$ **9.** $\dfrac{5b - 4a}{2b + 3a}$ **11.** $\dfrac{2z + 3y}{4y - z}$

13. $\dfrac{a + b}{a}$ **15.** $\dfrac{3}{3x + 2}$ **17.** $\dfrac{1}{x - y}$ **19.** $\dfrac{1}{a(a - h)}$

21. $\dfrac{(a - 2)(a - 7)}{(a + 1)(a - 6)}$ **23.** $\dfrac{x + 2}{x + 3}$ **25.** $\dfrac{1 + 2y}{1 - 3y}$

27. $\dfrac{y^2 + 1}{y^2 - 1}$ **29.** $\dfrac{4x - 7}{7x - 9}$ **31.** $\dfrac{a^2 - 3a - 6}{a^2 - 2a - 3}$

33. $\dfrac{a + 1}{2a + 5}$ **35.** $\dfrac{-1 - 3x}{8 - 2x}$, or $\dfrac{3x + 1}{2x - 8}$ **37.** $\dfrac{1}{y + 5}$

39. $-y$ **41.** $\dfrac{6a^2 + 30a + 60}{3a^2 + 2a + 4}$ **43.** $\dfrac{(2x + 1)(x + 2)}{2x(x - 1)}$

45. $\dfrac{(2a - 3)(a + 5)}{2(a - 3)(a + 2)}$ **47.** -1 **49.** $\dfrac{2x^2 - 11x - 27}{2x^2 + 21x + 13}$

51. TW   **53.** $\dfrac{15}{4}$ **54.** $\dfrac{4t - 3}{4(t - 3)}$

55. $\dfrac{a^2 + 9a - 21}{(a - 3)(a - 1)(a + 2)}$ **56.** $\dfrac{t - 5}{t + 5}$ **57.** 1

58. $\dfrac{x - 1}{x - 4}$ **59. TW**   **61.** $\dfrac{5(y + x)}{3(y - x)}$ **63.** $\dfrac{8c}{17}$

65. $\dfrac{-3}{x(x + h)}$ **67.** $\dfrac{2}{(1 + x + h)(1 + x)}$

69. $\{x \mid x \text{ is a real number } and \ x \neq \pm 1 \ and \ x \neq \pm 4$ $and \ x \neq \pm 5\}$ **71.** $\dfrac{2 + a}{3 + a}, a \neq -2, -3$

73. $\dfrac{x^4}{81}$; $\{x \mid x \text{ is a real number } and \ x \neq 3\}$

Exercise Set 7.6, pp. 580–582

1. Equation   **2.** Expression   **3.** Expression
4. Equation   **5.** Equation   **6.** Equation   **7.** Equation
8. Expression   **9.** Expression   **10.** Equation   **11.** $\frac{42}{5}$
13. -8 **15.** $-\frac{11}{9}$ **17.** 5   **19.** -3 **21.** No solution
23. $-4, -1$ **25.** $-2, 6$ **27.** -13 **29.** -5
31. $-\frac{10}{3}$ **33.** -1 **35.** 2, 3   **37.** -145
39. No solution   **41.** -1 **43.** 4   **45.** $-6, 5$
47. $-\frac{3}{2}, 5$ **49.** 14   **51.** $\frac{3}{4}$ **53.** $\frac{5}{14}$ **55.** $\frac{3}{5}$ **57. TW**
59. 38   **60.** $\{x \mid x > \frac{7}{10}\}$, or $\left(\frac{7}{10}, \infty\right)$ **61.** $-1, 1$
62. $-\frac{1}{2}, 8$ **63.** $(-3, -14)$ **64.** $(1, -1)$ **65.** $\frac{2}{5}$
66. $-2, 3$ **67.** $-4, 5$ **68.** $-5, 0, \frac{1}{2}$ **69. TW**   **71.** $\frac{1}{5}$
73. $-\frac{7}{2}$ **75.** 0.0854697   **77.** Yes

Exercise Set 7.7, pp. 593–597

1. 2   **3.** $-3, -2$ **5.** 6 and 7, -7 and -6 **7.** $3\frac{3}{14}$ hr
9. $8\frac{4}{7}$ hr   **11.** $19\frac{4}{5}$ min   **13.** Canon: 36 min; HP: 72 min
15. Blueair 402: 30 min; Panasonic F-P20HU1: 40 min
17. Erickson Air-Crane: 10 hr; S-58T: 40 hr
19. Zsuzanna: $1\frac{1}{3}$ hr; Stan: 4 hr   **21.** 8 hr   **23.** 7 mph
25. 4.3 ft/sec   **27.** AMTRAK: 80 km/h; B&M: 66 km/h
29. Express: 45 mph; local: 38 mph   **31.** 9 km/h
33. 2 km/h   **35.** 25 mph   **37.** 20 mph   **39.** 10.5
41. $\frac{8}{3}$ **43.** $3\frac{3}{4}$ in.   **45.** 20 ft   **47.** $2\frac{2}{3}$ ft   **49.** 12.6
51. 105 steps per minute   **53.** 702 photos   **55.** $32,340
57. 20 duds   **59.** 184   **61.** (a) 1.92 T; (b) 28.8 lb
63. TW   **65.** $5a^4b^6$ **66.** $5x^6y^4$ **67.** $4s^{10}t^8$
68. $-2x^4 - 7x^2 + 11x$ **69.** $-8x^3 + 28x^2 - 25x + 19$
70. $11x^4 + 7x^3 - 2x^2 - 4x - 10$ **71. TW**   **73.** $49\frac{1}{2}$ hr
75. 11.25 min   **77.** 2250 people per hour   **79.** $14\frac{7}{8}$ mi
81. Page 278   **83.** $8\frac{2}{11}$ min after 10:30   **85.** $51\frac{3}{7}$ mph

Exercise Set 7.8, pp. 607–612

1. LCD   **2.** Same   **3.** Factor   **4.** Constant; proportionality   **5.** Inverse   **6.** Direct   **7.** Direct
8. Inverse   **9.** Inverse   **10.** Direct   **11.** $d = \dfrac{L}{f}$
13. $v_1 = \dfrac{2s}{t} - v_2$, or $\dfrac{2s - tv_2}{t}$ **15.** $b = \dfrac{at}{a - t}$
17. $R = \dfrac{2V}{I} - 2r$, or $\dfrac{2V - 2Ir}{I}$ **19.** $g = \dfrac{Rs}{s - R}$
21. $n = \dfrac{IR}{E - Ir}$ **23.** $q = \dfrac{pf}{p - f}$ **25.** $t_1 = \dfrac{H}{Sm} + t_2$, or $\dfrac{H + Smt_2}{Sm}$ **27.** $r = \dfrac{Re}{E - e}$ **29.** $r = 1 - \dfrac{a}{S}$, or $\dfrac{S - a}{S}$
31. $a + b = \dfrac{f}{c^2}$ **33.** $r = \dfrac{A}{P} - 1$, or $\dfrac{A - P}{P}$
35. $t_2 = \dfrac{d_2 - d_1}{v} + t_1$, or $\dfrac{d_2 - d_1 + t_1 v}{v}$ **37.** $b^2 = \dfrac{a^2 y^2}{a^2 - x^2}$
39. $Q = \dfrac{2Tt - 2AT}{A - q}$ **41.** $k = 7; y = 7x$
43. $k = 1.7; y = 1.7x$ **45.** $k = 6; y = 6x$
47. $k = 60; y = \dfrac{60}{x}$ **49.** $k = 112; y = \dfrac{112}{x}$
51. $k = 9; y = \dfrac{9}{x}$ **53.** 241,920,000 cans   **55.** 6 amperes
57. 3.5 hr   **59.** 32 kg   **61.** 50 min   **63.** 20 min
65. 122,269,231 tons   **67.** $y = \tfrac{2}{3}x^2$ **69.** $y = \dfrac{54}{x^2}$
71. $y = 0.3xz^2$ **73.** $y = \dfrac{4wx^2}{z}$ **75.** About 2.9 sec
77. 308 cm^3 **79.** About 57.42 mph   **81.** (a) Inverse;
(b) $y = \dfrac{300}{x}$; **(c)** 100 min   **83.** (a) Directly; (b) $y \approx 1.08x$;
(c) 8.64 million people   **85.** TW   **87.** ℝ   **88.** ℝ
89. $\{x \mid x \text{ is a real number } and \ x \neq 3\}$
90. $8a^3 - 2a$ **91.** $(t + 2b)(t^2 - 2bt + 4b^2)$
92. $-\tfrac{5}{3}, \tfrac{7}{2}$ **93.** TW   **95.** 567 mi   **97.** Ratio is $\dfrac{a + 12}{a + 6}$;
percent increase is $\dfrac{6}{a + 6} \cdot 100\%$, or $\dfrac{600}{a + 6}\%$
99. $t_1 = t_2 + \dfrac{(d_2 - d_1)(t_4 - t_3)}{a(t_4 - t_2)(t_4 - t_3) + d_3 - d_4}$
101. The intensity is halved.   **103.** About 1.697 m
105. $d(s) = \dfrac{28}{s}$; 70 yd

Review Exercises: Chapter 7, pp. 615–617

1. False   **2.** True   **3.** False   **4.** True   **5.** True
6. False   **7.** False   **8.** False   **9.** True   **10.** True
11. (a) $-\tfrac{2}{9}$; (b) $-\tfrac{3}{4}$; (c) 0   **12.** 0   **13.** 4   **14.** $-6, 6$
15. $-6, 5$ **16.** $\dfrac{x - 2}{x + 1}$ **17.** $\dfrac{y - 5}{y + 5}$ **18.** $-5(x + 2y)$
19. $\dfrac{a - 6}{5}$ **20.** $\dfrac{8(t + 1)}{(2t - 1)(t - 1)}$ **21.** $-32t$
22. $\dfrac{2x(x - 1)}{x + 1}$ **23.** $\dfrac{(x^2 + 1)(2x + 1)}{(x - 2)(x + 1)}$ **24.** $\dfrac{(t + 4)^2}{t + 1}$
25. $48x^3$ **26.** $(x + 5)(x - 2)(x - 4)$ **27.** $\dfrac{15 - 3x}{x + 3}$
28. -1 **29.** $\dfrac{4}{x - 4}$ **30.** $\dfrac{x + 5}{2x}$ **31.** $\dfrac{2x + 5}{x - 2}$
32. $\dfrac{2a}{a - 1}$ **33.** $d + c$ **34.** $\dfrac{-x^2 + x + 26}{(x + 1)(x - 5)(x + 5)}$
35. $\dfrac{19x + 8}{10x(x + 2)}$ **36.** $f(x) = \dfrac{7x + 3}{x - 3}, x \neq \tfrac{1}{2}, x \neq 3$
37. $f(x) = \dfrac{2(x - 2)}{x + 2}, x \neq -2, x \neq 2$ **38.** $\tfrac{4}{9}$
39. $\dfrac{a^2 b^2}{2(b^2 - ba + a^2)}$ **40.** $\dfrac{(y + 11)(y + 5)}{(y - 5)(y + 2)}$
41. $\dfrac{(14 - 3x)(x + 3)}{2x^2 + 16x + 6}$ **42.** 2   **43.** 6   **44.** No solution
45. 0   **46.** $-1, 4$ **47.** $5\tfrac{1}{7}$ hr   **48.** Celeron: 45 sec;
Pentium 4: 30 sec   **49.** 24 mph   **50.** Motorcycle: 62 mph;
car: 70 mph   **51.** 60   **52.** 6   **53.** $s = \dfrac{Rg}{g - R}$
54. $m = \dfrac{H}{S(t_1 - t_2)}$ **55.** $c = \dfrac{b + 3a}{2}$ **56.** $t_1 = \dfrac{-A}{vT} + t_2$,
or $\dfrac{-A + vTt_2}{vT}$ **57.** About 22 lb   **58.** 64 L   **59.** $y = \dfrac{\tfrac{3}{4}}{x}$
60. (a) Inverse; (b) $y = \dfrac{24}{x}$; (c) 3 oz
61. TW   The least common denominator was used to add and subtract rational expressions, to simplify complex rational expressions, and to solve rational equations.
62. TW   A rational *expression* is a quotient of two polynomials. Expressions can be simplified, multiplied, or added, but they cannot be solved for a variable. A rational *equation* is an equation containing rational expressions. In a rational equation, we often can solve for a variable.
63. All real numbers except 0 and 13   **64.** 45
65. $9\tfrac{9}{19}$ sec

Test: Chapter 7, pp. 617–618

1. [7.1] 0   **2.** [7.1] 1, 2

3. [7.3] $(y - 3)(y + 3)(y + 7)$ **4.** [7.2] $\dfrac{-2(a + 5)}{3}$

5. [7.2] $\dfrac{(5y + 1)(y + 1)}{3y(y + 2)}$

6. [7.2] $\dfrac{(2x + 1)(2x - 1)(x^2 + 1)}{(x - 1)^2(x - 2)}$

7. [7.2] $(x + 3)(x - 3)$ **8.** [7.3] $\dfrac{-3x + 9}{x^3}$

9. [7.3] $\dfrac{-2t + 8}{t^2 + 1}$ **10.** [7.4] $\dfrac{3}{3 - x}$ **11.** [7.4] $\dfrac{2x - 5}{x - 3}$

12. [7.4] $\dfrac{11t - 8}{t(t - 2)}$ **13.** [7.4] $\dfrac{-x^2 - 7x - 15}{(x + 4)(x - 4)(x + 1)}$

14. [7.4] $\dfrac{-2(2x^2 + 5x + 20)}{(x - 4)(x + 4)(x^2 + 4x + 16)}$ **15.** [7.1] $\dfrac{3x + 7}{x + 3}$

16. [7.4] $f(x) = \dfrac{x - 4}{(x + 3)(x - 2)}, x \neq -3, 2$

17. [7.5] $\dfrac{a(2b + 3a)}{5a + b}$ **18.** [7.5] $\dfrac{(x - 9)(x - 6)}{(x + 6)(x - 3)}$

19. [7.5] $\dfrac{2x^2 - 7x + 1}{3x^2 + 10x - 17}$ **20.** [7.6] $-\dfrac{21}{4}$ **21.** [7.6] 15

22. [7.1] 5; 0   **23.** [7.6] $\frac{5}{3}$ **24.** [7.8] $b_1 = \dfrac{2A}{h} - b_2$, or

$\dfrac{2A - b_2 h}{h}$ **25.** [7.7] 5 and 6; −6 and −5

26. [7.7] $3\frac{3}{11}$ mph   **27.** [7.7] $1\frac{31}{32}$ hr   **28.** [7.7] $2\frac{1}{7}$ c

29. [7.8] 30 workers   **30.** [7.8] 637 in²   **31.** [7.6] $-\frac{19}{3}$

32. [7.6] $\{x \mid x \text{ is a real number } and\ x \neq 0\ and\ x \neq 15\}$

33. [7.5], [7.6] x-intercept: $(11, 0)$; y-intercept: $\left(0, -\frac{33}{5}\right)$

34. [7.7] Hans: 56 lawns; Franz: 42 lawns

Chapter 8

Interactive Discovery, p. 629

1. Domain of $f = (-\infty, 3]$; domain of $g = [-1, \infty)$
2. Domain of $f + g =$ domain of $f - g =$ domain of $f \cdot g = [-1, 3]$ **3.** By finding the intersection of the domains of f and g

Exercise Set 8.1, pp. 630–636

1. (h)   **2.** (j)   **3.** (f)   **4.** (a)   **5.** (e)   **6.** (d)
7. (b)   **8.** (g)   **9.** (c)   **10.** (i)
11. $\{x \mid x \geq 2\}$, or $[2, \infty)$ **13.** $\{x \mid x < 3\}$, or $(-\infty, 3)$
15. (a) $\{x \mid x \leq -2\}$, or $(-\infty, -2]$; (b) $\{x \mid x > \frac{8}{3}\}$ or $\left(\frac{8}{3}, \infty\right)$;
(c) $\{x \mid x > \frac{4}{5}\}$, or $\left(\frac{4}{5}, \infty\right)$ **17.** $\{x \mid x < 7\}$, or $(-\infty, 7)$
19. $\{x \mid x \geq 2\}$, or $[2, \infty)$ **21.** $\{x \mid x < 8\}$, or $(-\infty, 8)$

23. $\{x \mid x \leq 2\}$, or $(-\infty, 2]$ **25.** $\{x \mid x \leq -\frac{4}{9}\}$, or $\left(-\infty, -\frac{4}{9}\right]$
27. At least 625 people
29. $n(x) = 1.274096386x + 58.45783133$; years after 2021
31. $n(x) = -2.05x + 178.6444444$,
$v(x) = 4.566666667x + 55.4$; years after 2018
33. $f(x) = 0.0056862962x + 1.891866531$; years after 2006
35. $\{9, 11\}$ **37.** $\{0, 5, 10, 15, 20\}$ **39.** $\{b, d, f\}$
41. $\{r, s, t, u, v\}$ **43.** $\varnothing$ **45.** $\{3, 5, 7\}$
47. $(3, 7)$
49. $[-6, -2]$
51. $(-\infty, -1) \cup (4, \infty)$
53. $(-\infty, -2] \cup (1, \infty)$
55. $(-2, 4]$
57. $(-2, 4)$
59. $(-\infty, 5) \cup (7, \infty)$
61. $(-\infty, -4] \cup [5, \infty)$
63. $[-3, 7)$
65. $[3, 7)$
67. $(-\infty, 5)$
69. $\{t \mid -3 < t < 7\}$, or $(-3, 7)$
71. $\{x \mid -1 < x \leq 4\}$, or $(-1, 4]$
73. $\{a \mid -2 \leq a < 2\}$, or $[-2, 2)$
75. $\mathbb{R}$, or $(-\infty, \infty)$
77. $\{x \mid 7 < x < 23\}$, or $(7, 23)$
79. $\{x \mid -3 \leq x \leq 2\}$, or $[-3, 2]$
81. $\{x \mid 1 \leq x \leq 3\}$, or $[1, 3]$
83. $\{x \mid -\frac{7}{2} < x \leq 7\}$, or $\left(-\frac{7}{2}, 7\right]$
85. $\{x \mid x \leq 1\ or\ x \geq 3\}$, or $(-\infty, 1] \cup [3, \infty)$
87. $\{x \mid x < 2\ or\ x > 6\}$, or $(-\infty, 2) \cup (6, \infty)$
89. $\{a \mid a < \frac{7}{2}\}$, or $\left(-\infty, \frac{7}{2}\right)$
91. $\{a \mid a < -5\}$, or $(-\infty, -5)$
93. $\mathbb{R}$, or $(-\infty, \infty)$

95. $\{t \mid t \le 6\}$, or $(-\infty, 6]$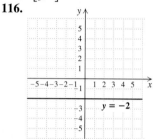
97. $(-1, 6)$ **99.** $(-\infty, -8) \cup (-8, \infty)$ **101.** $[6, \infty)$
103. $(-\infty, 4) \cup (4, 5) \cup (5, \infty)$ **105.** $\left[-\frac{7}{2}, \infty\right)$
107. $(-\infty, 4]$ **109.** $[5, \infty)$ **111.** $\left[\frac{2}{3}, 3\right]$ **113.** TW
115.   **116.**

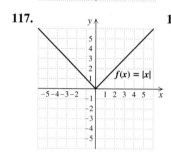

117.   **118.**

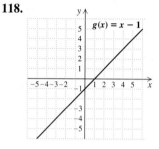

119. $(8, 5)$ **120.** $(-5, -3)$ **121.** TW
123. From 2000 to 2025   **125.** Sizes between 6 and 13
127. $1965 \le y \le 1981$
129. $\left\{m \mid m < \frac{6}{5}\right\}$, or $\left(-\infty, \frac{6}{5}\right)$

131. $\left\{x \mid -\frac{1}{8} < x < \frac{1}{2}\right\}$, or $\left(-\frac{1}{8}, \frac{1}{2}\right)$

133. True   **135.** False   **137.** $(-\infty, -7) \cup \left(-7, \frac{3}{4}\right]$

Exercise Set 8.2, pp. 645–647

1. True   **2.** True   **3.** False   **4.** True   **5.** True
6. True   **7.** False   **8.** False   **9.** $\{-5, 1\}$ **11.** $(-5, 1)$
13. $(-\infty, -5] \cup [1, \infty)$ **15.** $\{-7, 7\}$ **17.** $\varnothing$ **19.** $\{0\}$
21. $\{-5.5, 5.5\}$ **23.** $\left\{-\frac{1}{2}, \frac{7}{2}\right\}$ **25.** $\varnothing$ **27.** $\{-4, 8\}$
29. $\{2, 8\}$ **31.** $\{-2, 16\}$ **33.** $\{-8, 8\}$ **35.** $\left\{-\frac{11}{7}, \frac{11}{7}\right\}$
37. $\{-7, 8\}$ **39.** $\{-12, 2\}$ **41.** $\left\{-\frac{1}{3}, 3\right\}$ **43.** $\{-7, 1\}$
45. $\{-8.7, 8.7\}$ **47.** $\left\{-\frac{8}{3}, 4\right\}$ **49.** $\{1, 11\}$ **51.** $\left\{-\frac{1}{2}\right\}$
53. $\left\{-\frac{3}{5}, 5\right\}$ **55.** $\mathbb{R}$ **57.** $\{1\}$ **59.** $\left\{32, \frac{8}{3}\right\}$
61. $\{a \mid -9 \le a \le 9\}$, or $[-9, 9]$
63. $\{x \mid x < -8 \text{ or } x > 8\}$, or $(-\infty, -8) \cup (8, \infty)$

65. $\{t \mid t < 0 \text{ or } t > 0\}$, or $(-\infty, 0) \cup (0, \infty)$

67. $\{x \mid -3 < x < 5\}$, or $(-3, 5)$
69. $\{x \mid -8 \le x \le 4\}$, or $[-8, 4]$
71. $\{x \mid x < -2 \text{ or } x > 8\}$, or $(-\infty, -2) \cup (8, \infty)$

73. $\mathbb{R}$, or $(-\infty, \infty)$

75. $\left\{a \mid a \le -\frac{2}{3} \text{ or } a \ge \frac{10}{3}\right\}$, or $\left(-\infty, -\frac{2}{3}\right] \cup \left[\frac{10}{3}, \infty\right)$

77. $\{y \mid -9 < y < 15\}$, or $(-9, 15)$

79. $\{x \mid x \le -8 \text{ or } x \ge 0\}$, or $(-\infty, -8] \cup [0, \infty)$

81. $\left\{y \mid y < -\frac{4}{3} \text{ or } y > 4\right\}$, or $\left(-\infty, -\frac{4}{3}\right) \cup (4, \infty)$

83. $\varnothing$ **85.** $\left\{x \mid x \le -\frac{2}{15} \text{ or } x \ge \frac{14}{15}\right\}$, or $\left(-\infty, -\frac{2}{15}\right] \cup \left[\frac{14}{15}, \infty\right)$

87. $\{m \mid -9 \le m \le 3\}$, or $[-9, 3]$
89. $\{a \mid -6 < a < 0\}$, or $(-6, 0)$
91. $\left\{x \mid -\frac{1}{2} \le x \le \frac{7}{2}\right\}$, or $\left[-\frac{1}{2}, \frac{7}{2}\right]$

93. $\left\{x \mid x \le -\frac{7}{3} \text{ or } x \ge 5\right\}$, or $\left(-\infty, -\frac{7}{3}\right] \cup [5, \infty)$

95. $\{x \mid -4 < x < 5\}$, or $(-4, 5)$
97. TW   **99.** 6   **100.** $\{x \mid x \ge 6\}$, or $[6, \infty)$ **101.** $(6, 21)$
102. $\left\{x \mid x \le -\frac{1}{3}\right\}$, or $\left(-\infty, -\frac{1}{3}\right]$
103. $\{x \mid -11 < x < 1\}$, or $(-11, 1)$
104. $\{-11, 1\}$ **105.** TW   **107.** $\left\{t \mid t \ge \frac{5}{3}\right\}$, or $\left[\frac{5}{3}, \infty\right)$
109. $\mathbb{R}$, or $(-\infty, \infty)$ **111.** $\left\{-\frac{1}{7}, \frac{7}{3}\right\}$ **113.** $|x| < 3$
115. $|x| \ge 6$ **117.** $|x + 3| > 5$
119. $|x - 7| < 2$, or $|7 - x| < 2$ **121.** $|x - 3| \le 4$
123. $|x + 4| < 3$ **125.** Between 80 ft and 100 ft

Visualizing the Graph, p. 657

1. B   **2.** F   **3.** J   **4.** A   **5.** E   **6.** G   **7.** C   **8.** D
9. I   **10.** H

Exercise Set 8.3, pp. 658–660

1. (e)   **2.** (c)   **3.** (d)   **4.** (a)   **5.** (b)   **6.** (f)

7. Yes   **9.** No

11.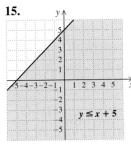

13.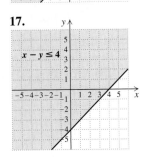

15.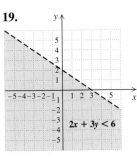

17.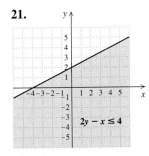

19.

21.

23.

25.

27.

29.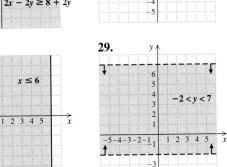

31.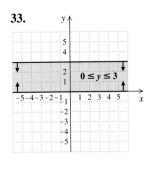

$-4 \leq x \leq 2$

33.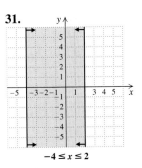

$0 \leq y \leq 3$

35. $y > x + 3.5$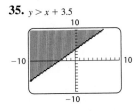

37. $8x - 2y < 11$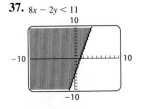

39.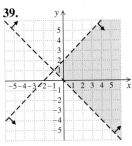

41.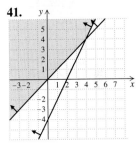

43.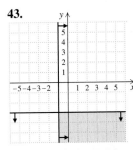

45.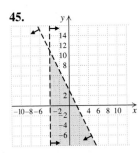

47.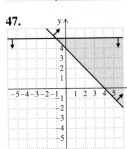

49.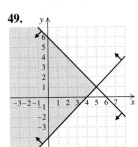

51.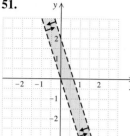

53.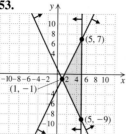

55.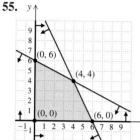

57.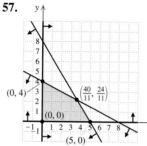

59.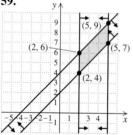

61. **TW**   **63.** Peanuts: $6\frac{2}{3}$ lb; fancy nuts: $3\frac{1}{3}$ lb
64. Hendersons: 10 bags; Savickis: 4 bags   **65.** Activity-card holders: 128; noncard holders: 75   **66.** Students: 70;
adults: 130   **67.** 25 ft   **68.** 2.75%   **69.** **TW**

71.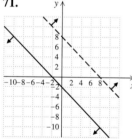

73.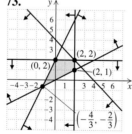

75. $w > 0,$
$h > 0,$
$w + h + 30 \le 62,$ or
$w + h \le 32,$
$2w + 2h + 30 \le 130,$ or
$w + h \le 50$

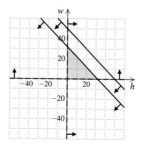

77. $35c + 75a > 1000,$
$c \ge 0,$
$a \ge 0$

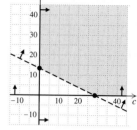

79. $y \le x,$
$y \le 2$

81. $y \le x + 2,$
$y \le -x + 4,$
$y \ge 0$

Exercise Set 8.4, pp. 666–669

1. True   **2.** False   **3.** True   **4.** True   **5.** True
6. True   **7.** False   **8.** False   **9.** $\left[-4, \frac{3}{2}\right]$
11. $(-\infty, -2) \cup (0, 2) \cup (3, \infty)$ **13.** $\left(-\infty, -\frac{7}{2}\right) \cup (-2, \infty)$
15. $(-4, 3),$ or $\{x \mid -4 < x < 3\}$
17. $(-\infty, -7] \cup [2, \infty),$ or $\{x \mid x \le -7 \text{ or } x \ge 2\}$
19. $(-\infty, -1) \cup (2, \infty),$ or $\{x \mid x < -1 \text{ or } x > 2\}$ **21.** $\varnothing$
23. $(-2, 6),$ or $\{x \mid -2 < x < 6\}$
25. $(-\infty, -2) \cup (0, 2),$ or $\{x \mid x < -2 \text{ or } 0 < x < 2\}$
27. $[-2, 1] \cup [4, \infty),$ or $\{x \mid -2 \le x \le 1 \text{ or } x \ge 4\}$
29. $[-0.78, 1.59],$ or $\{x \mid -0.78 \le x \le 1.59\}$
31. $(-\infty, -2) \cup (1, 3),$ or $\{x \mid x < -2 \text{ or } 1 < x < 3\}$
33. $[-2, 2],$ or $\{x \mid -2 \le x \le 2\}$ **35.** $(-1, 2) \cup (3, \infty),$ or $\{x \mid -1 < x < 2 \text{ or } x > 3\}$ **37.** $(-\infty, 0] \cup [2, 5],$ or $\{x \mid x \le 0 \text{ or } 2 \le x \le 5\}$ **39.** $(-\infty, -5),$ or $\{x \mid x < -5\}$
41. $(-\infty, -1] \cup (3, \infty),$ or $\{x \mid x \le -1 \text{ or } x > 3\}$
43. $(-\infty, -6),$ or $\{x \mid x < -6\}$ **45.** $(-\infty, -1] \cup [2, 5),$ or $\{x \mid x \le -1 \text{ or } 2 \le x < 5\}$ **47.** $(-\infty, -3) \cup [0, \infty),$ or $\{x \mid x < -3 \text{ or } x \ge 0\}$ **49.** $(0, \infty),$ or $\{x \mid x > 0\}$
51. $(-\infty, -4) \cup [1, 3),$ or $\{x \mid x < -4 \text{ or } 1 \le x < 3\}$
53. $\left(-\frac{3}{4}, \frac{5}{2}\right],$ or $\{x \mid -\frac{3}{4} < x \le \frac{5}{2}\}$ **55.** $(-\infty, 2) \cup [3, \infty),$ or $\{x \mid x < 2 \text{ or } x \ge 3\}$ **57.** **TW**   **59.** $[3, \infty),$ or $\{x \mid x \ge 3\}$
60. $\left(-\infty, -\frac{11}{2}\right) \cup \left(\frac{5}{2}, \infty\right),$ or $\{x \mid x < -\frac{11}{2} \text{ or } x > \frac{5}{2}\}$
61. $\left[-\frac{8}{5}, \frac{12}{5}\right],$ or $\{x \mid -\frac{8}{5} \le x \le \frac{12}{5}\}$ **62.** $\varnothing$
63. $(-1, 7),$ or $\{x \mid -1 < x < 7\}$
64. $(-\infty, 0) \cup [1, \infty),$ or $\{x \mid x < 0 \text{ or } x \ge 1\}$ **65.** **TW**
67. $\varnothing$ **69.** $\{0\}$ **71.** (a) $(10, 200),$ or $\{x \mid 10 < x < 200\};$
(b) $[0, 10) \cup (200, \infty),$ or $\{x \mid 0 \le x < 10 \text{ or } x > 200\}$
73. $\{n \mid n \text{ is an integer } and\ 12 \le n \le 25\}$ **75.** $f(x)$ has no
zeros; $f(x) < 0$ for $(-\infty, 0),$ or $\{x \mid x < 0\}; f(x) > 0$ for
$(0, \infty),$ or $\{x \mid x > 0\}$ **77.** $f(x) = 0$ for $x = -1, 0; f(x) < 0$
for $(-\infty, -3) \cup (-1, 0),$ or $\{x \mid x < -3 \text{ or } -1 < x < 0\};$
$f(x) > 0$ for $(-3, -1) \cup (0, 2) \cup (2, \infty),$ or
$\{x \mid -3 < x < -1 \text{ or } 0 < x < 2 \text{ or } x > 2\}$
79. **(a)** $h(x) = -0.3553571429x^2 + 4.185x + 16.77142857;$
(b) more than 11 yr after 1994, or after 2005

Review Exercises: Chapter 8, pp. 671–672

1. True **2.** False **3.** True **4.** True **5.** False
6. True **7.** True **8.** False **9.** False **10.** True
11. $\{x | x > -4\}$, or $(-4, \infty)$ **12.** $\{x | x \geq -6\}$, or $[-6, \infty)$
13. $\{1, 5, 9\}$ **14.** $\{1, 2, 3, 5, 6, 9\}$
15. 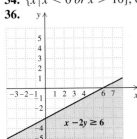 $(-5, 3]$

16. ⟵+++++++++++⟶ $(-\infty, \infty)$
　　　$-5\ -4\ -3\ -2\ -1\ \ 0\ \ 1\ \ 2\ \ 3\ \ 4\ \ 5$

17. $\{x | -12 < x \leq -3\}$, or $(-12, -3]$
⟵+++++++++++⟶
$-12\,\text{-}11\text{-}10\text{-}9\text{-}8\text{-}7\text{-}6\text{-}5\text{-}4\text{-}3\text{-}2$

18. $\{x | -\frac{5}{4} < x < \frac{5}{2}\}$, or $\left(-\frac{5}{4}, \frac{5}{2}\right)$　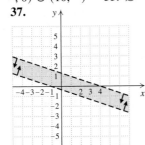

19. $\{x | x < -3 \text{ or } x > 1\}$, or $(-\infty, -3) \cup (1, \infty)$
⟵++++++++++⟶
$-5\ -4\ -3\ -2\ -1\ \ 0\ \ 1\ \ 2\ \ 3\ \ 4\ \ 5$

20. $\{x | x < -11 \text{ or } x \geq -6\}$, or $(-\infty, -11) \cup [-6, \infty)$
⟵+++++++++⟶
$-12\text{-}10\text{-}8\ \text{-}6\ \text{-}4\ \text{-}2\ \ 0\ \ 2\ \ 4\ \ 6\ \ 8$

21. $\{x | x \leq -6 \text{ or } x \geq 8\}$, or $(-\infty, -6] \cup [8, \infty)$
⟵+++++++++⟶
$-10\text{-}8\ \text{-}6\ \text{-}4\ \text{-}2\ \ 0\ \ 2\ \ 4\ \ 6\ \ 8\ \ 10$

22. $\{x | x < -\frac{2}{5} \text{ or } x > \frac{8}{5}\}$, or $\left(-\infty, -\frac{2}{5}\right) \cup \left(\frac{8}{5}, \infty\right)$

　　　$-\frac{2}{5}$　$\frac{8}{5}$
⟵++++++++⟶
$-5\ -4\ -3\ -2\ -1\ \ 0\ \ 1\ \ 2\ \ 3\ \ 4\ \ 5$

23. $(-\infty, 8) \cup (8, \infty)$ **24.** $[-5, \infty)$ **25.** $\left(-\infty, \frac{8}{3}\right]$
26. $\{-5, 5\}$ **27.** $\{t | t \leq -3.5 \text{ or } t \geq 3.5\}$, or
$(-\infty, -3.5] \cup [3.5, \infty)$ **28.** $\{-4, 10\}$
29. $\{x | -\frac{17}{2} < x < \frac{7}{2}\}$, or $\left(-\frac{17}{2}, \frac{7}{2}\right)$
30. $\{x | x \leq -\frac{11}{3} \text{ or } x \geq \frac{19}{3}\}$, or $\left(-\infty, -\frac{11}{3}\right] \cup \left[\frac{19}{3}, \infty\right)$
31. $\{-14, \frac{4}{3}\}$ **32.** $\varnothing$ **33.** $\{x | -16 \leq x \leq 8\}$, or $[-16, 8]$
34. $\{x | x < 0 \text{ or } x > 10\}$, or $(-\infty, 0) \cup (10, \infty)$ **35.** $\varnothing$
36.

37.

$x - 2y \geq 6$

38.

$(21, 6)$

$(-9, 6)$

$(6, 1)$

39. $(-1, 0) \cup (3, \infty)$, or $\{x | -1 < x < 0 \text{ or } x > 3\}$
40. $(-3, 5]$, or $\{x | -3 < x \leq 5\}$
41. **TW** The equation $|X| = p$ has two solutions when p is
positive because X can be either p or $-p$. The same equation
has no solution when p is negative because no number has a
negative absolute value. **42.** **TW** The solution set of a
system of inequalities is all ordered pairs that make *all* the
individual inequalities true. This consists of ordered pairs
that are common to all the individual solution sets, or the
intersection of the graphs.
43. $\{x | -\frac{8}{3} \leq x \leq -2\}$, or $\left[-\frac{8}{3}, -2\right]$
44. $|d - 1.1| \leq 0.03$ **45.** $|t - 21.5| \leq 3.5$

Test: Chapter 8, p. 673

1. [8.1] $\{x | x \leq 3\}$, or $(-\infty, 3]$
2. [8.1] $\{x | x > -3\}$, or $(-3, \infty)$ **3.** [8.1] $\{3, 5\}$
4. [8.1] $\{1, 3, 5, 7, 9, 11, 13\}$
5. [8.1] $(-\infty, -3) \cup (-3, 1) \cup (1, \infty)$ **6.** [8.1] $(-\infty, 4]$
7. [8.1] $\{x | 1 < x < 8\}$, or $(1, 8)$ ⟵+======+⟶
　　　　　　　　　　　　　　　　　　$0\ 1$　　　　　　　8

8. [8.1] $\{t | -\frac{2}{5} < t \leq \frac{9}{5}\}$, or $\left(-\frac{2}{5}, \frac{9}{5}\right]$
⟵++++++⟶
　$-\frac{2}{5}\ 0$　　$\frac{9}{5}$

9. [8.1] $\{x | x < 3 \text{ or } x > 6\}$, or $(-\infty, 3) \cup (6, \infty)$
⟵+===+⟶
　　0　　3　　6

10. [8.1] $\{x | x < -4 \text{ or } x > -\frac{5}{2}\}$, or $(-\infty, -4) \cup \left(-\frac{5}{2}, \infty\right)$
⟵+===+⟶
$-4\ \ -\frac{5}{2}\ \ \ 0$

11. [8.1] $\{x | 4 \leq x < \frac{15}{2}\}$, or $\left[4, \frac{15}{2}\right)$
⟵++====+⟶
　0　　4　　$\frac{15}{2}$

12. [8.2] $\{-13, 13\}$ ⟵●++++●⟶
　　　　　　　　　　-13　0　13
13. [8.2] $\{a | a < -7 \text{ or } a > 7\}$, or $(-\infty, -7) \cup (7, \infty)$
⟵+====+⟶
-7　0　7

14. [8.2] $\{x | -2 < x < \frac{8}{3}\}$, or $\left(-2, \frac{8}{3}\right)$
⟵+====+⟶
-2　0　$2\ \frac{8}{3}$

15. [8.2] $\{t | t \leq -\frac{13}{5} \text{ or } t \geq \frac{7}{5}\}$, or $\left(-\infty, -\frac{13}{5}\right] \cup \left[\frac{7}{5}, \infty\right)$
⟵+====+⟶
$-\frac{13}{5}$　0　$\frac{7}{5}$

16. [8.2] $\varnothing$ **17.** [8.1] $\{x | x < \frac{1}{2} \text{ or } x > \frac{7}{2}\}$, or
$\left(-\infty, \frac{1}{2}\right) \cup \left(\frac{7}{2}, \infty\right)$ ⟵+===+⟶
　　　　　　　　　　　　　　　　　　$0\ \frac{1}{2}$　　　$\frac{7}{2}$

18. [8.2] $\{1\}$

19. [8.3]

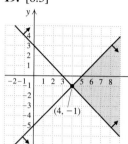

20. [8.3]

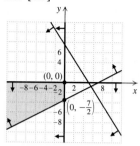

21. [8.4] $[-6, 1]$, or $\{x \mid -6 \le x \le 1\}$
22. [8.4] $(-1, 0) \cup (1, \infty)$, or $\{x \mid -1 < x < 0 \text{ or } x > 1\}$
23. [8.2] $[-1, 0] \cup [4, 6]$ **24.** [8.1] $\left(\frac{1}{5}, \frac{4}{5}\right)$
25. [8.2] $|x + 3| \le 5$

Chapter 9

Exercise Set 9.1, pp. 683–684

1. True   **2.** False   **3.** False   **4.** True   **5.** True
6. False   **7.** No   **9.** $(4, 0, 2)$ **11.** $(2, -2, 2)$
13. $(3, -2, 1)$ **15.** No solution   **17.** $(2, 1, 3)$
19. $(2, -5, 6)$ **21.** The equations are dependent.
23. $\left(\frac{1}{2}, 4, -6\right)$ **25.** $\left(\frac{1}{2}, \frac{1}{3}, \frac{1}{6}\right)$ **27.** $\left(\frac{1}{2}, \frac{2}{3}, -\frac{5}{6}\right)$
29. $(15, 33, 9)$ **31.** $(3, 4, -1)$ **33.** $(10, 23, 50)$
35. No solution   **37.** The equations are dependent.
39. TW   **41.** $(9, -4)$ **42.** $(-12, -29)$ **43.** $\left(\frac{7}{4}, \frac{31}{8}\right)$
44. $\left(\frac{17}{7}, \frac{22}{7}\right)$ **45.** $(5, 1, 2)$ **46.** $(4, 5)$ **47.** TW
49. $(1, -1, 2)$ **51.** $(-3, -1, 0, 4)$ **53.** $\left(-\frac{1}{2}, -1, -\frac{1}{3}\right)$
55. 14   **57.** $z = 8 - 2x - 4y$

Exercise Set 9.2, pp. 689–692

1. 16, 19, 22   **3.** 8, 21, -3 **5.** 32°, 96°, 52°
7. Individual adult: $64; spouse: $57; child: $43   **9.** Bran
muffin: 1.5 g; banana: 3 g; 1 cup Wheaties: 3 g
11. $27,415; 4WD: $1970; sunroof: $800   **13.** Elrod: 20
ft/hr; Dot: 24 ft/hr; Wendy: 30 ft/hr   **15.** 12-oz cups: 17;
16-oz cups: 25; 20-oz cups: 13   **17.** Small: 10; medium: 25;
large: 5   **19.** Roast beef: 2 servings; baked potato: 1 serving;
broccoli: 2 servings   **21.** Asia: 4.8 billion; Africa: 1.8 billion;
rest of world: 2.5 billion   **23.** Two-point field goals: 32;
three-point field goals: 5; foul shots: 13   **25.** TW   **27.** -8
28. 33   **29.** -55 **30.** -71 **31.** $-14x + 21y - 35z$
32. $-24a - 42b + 54c$ **33.** $-5a$ **34.** $11x$ **35.** TW
37. Applicant: $102; spouse: $58; first child: $43; second
child: $40   **39.** 20 yr   **41.** 35 tickets

Exercise Set 9.3, pp. 699–700

1. Horizontal; columns   **2.** Equation   **3.** Entry
4. Matrices   **5.** Multiple   **6.** First   **7.** $\left(-\frac{1}{3}, -4\right)$

9. $(-4, 3)$ **11.** $\left(\frac{3}{2}, \frac{5}{2}\right)$ **13.** $\left(2, \frac{1}{2}, -2\right)$ **15.** $(2, -2, 1)$
17. $\left(4, \frac{1}{2}, -\frac{1}{2}\right)$ **19.** $(1, -3, -2, -1)$ **21.** Dimes: 18;
nickels: 24   **23.** $4.05-granola: 5 lb; $2.70-granola: 10 lb
25. $400 at 7%; $500 at 8%; $1600 at 9%   **27.** TW
29. 13   **30.** -22 **31.** 37   **32.** 422   **33.** TW
35. 1324

Exercise Set 9.4, pp. 705–706

1. True   **2.** True   **3.** False   **4.** False   **5.** False
6. False   **7.** 18   **9.** 36   **11.** 27   **13.** -3 **15.** -5
17. $(-3, 2)$ **19.** $\left(\frac{9}{19}, \frac{51}{38}\right)$ **21.** $\left(-1, -\frac{6}{7}, \frac{11}{7}\right)$
23. $(2, -1, 4)$ **25.** $(1, 2, 3)$ **27.** TW   **29.** $(1, -3)$
30. Approximately $(0.26, 1.65)$, or $\left(\frac{6}{23}, \frac{38}{23}\right)$ **31.** $\left(\frac{4}{3}, \frac{5}{3}\right)$
32. $(2, 2)$ **33.** $\left(\frac{31}{7}, -\frac{6}{7}\right)$ **34.** $\left(\frac{12}{7}, -\frac{17}{7}\right)$ **35.** TW
37. 12   **39.** 10

Visualizing the Graph, p. 712

1. A   **2.** D   **3.** J   **4.** H   **5.** I   **6.** B   **7.** E
8. F   **9.** C   **10.** G

Exercise Set 9.5, pp. 713–715

1. (b)   **2.** (a)   **3.** (e)   **4.** (f)   **5.** (h)   **6.** (c)
7. (g)   **8.** (d)   **9.** (a) $P(x) = 20x - 300,000$;
(b) (15,000 units, $975,000)
11. (a) $P(x) = 50x - 120,000$; **(b)** (2400 units, $144,000)
13. (a) $P(x) = 45x - 22,500$; **(b)** (500 units, $42,500)
15. (a) $P(x) = 18x - 16,000$; **(b)** (889 units, $35,560)
17. (a) $P(x) = 50x - 100,000$; **(b)** (2000 units, $250,000)
19. ($70, 300)   **21.** ($22, 474)   **23.** ($50, 6250)
25. ($10, 1070)   **27. (a)** $C(x) = 125,300 + 450x$;
(b) $R(x) = 800x$; **(c)** $P(x) = 350x - 125,300$; **(d)** $90,300
loss, $14,700 profit; **(e)** (358 computers, $286,400)
29. (a) $C(x) = 16,404 + 6x$; **(b)** $R(x) = 18x$;
(c) $P(x) = 12x - 16,404$; **(d)** $19,596 profit, $4404 loss;
(e) (1367 dozen caps, $24,606)   **31. (a)** $8.74; **(b)** 793 units
33. TW   **35.** 12   **36.** 15   **37.** $\frac{8}{3}$ **38.** 4   **39.** $\frac{9}{2}$
40. $\frac{1}{3}$ **41.** TW   **43.** ($5, 300 yo-yo's)
45. (a) $S(p) = 15.97p - 1.05$;
(b) $D(p) = -11.26p + 41.16$; **(c)** ($1.55, 23.7 million jars)

Review Exercises: Chapter 9, pp. 717–718

1. Elimination   **2.** Consistent   **3.** 180°   **4.** 1
5. Square   **6.** Determinant   **7.** Total profit   **8.** Fixed
9. Equilibrium point   **10.** Zero   **11.** $(4, -8, 10)$
12. The equations are dependent.   **13.** $(2, 0, 4)$ **14.** No
solution   **15.** $\left(\frac{8}{9}, -\frac{2}{3}, \frac{10}{9}\right)$ **16.** A: 90°; B: 67.5°; C: 22.5°
17. Hot dog: $2.75; chips: $0.75; water: $1.25

18. $\left(55, -\frac{89}{2}\right)$ **19.** $(-1, 1, 3)$ **20.** 2   **21.** 9
22. $(6, -2)$ **23.** $(-3, 0, 4)$
24. **(a)** $P(x) = 20x - 15{,}800$; **(b)** (790 units, $39,500)
25. ($3, 81)   **26.** **(a)** $C(x) = 0.75x + 9000$;
(b) $R(x) = 5.25x$; **(c)** $P(x) = 4.5x - 9000$; **(d)** $2250 loss,
$13,500 profit; **(e)** (2000 pints of honey, $10,500)
27. 🅣🅦 To solve a problem involving four variables, go
through the *Familiarize* and *Translate* steps as usual. The
resulting system of equations can be solved using the elimination method just as for three variables but likely with more
steps.   **28.** 🅣🅦 A system of equations can be both dependent and inconsistent if it is equivalent to a system with fewer
equations that has no solution. An example is a system of
three equations in three unknowns in which two of the equations represent the same plane, and the third represents a parallel plane.   **29.** 8000 pints
30. $a = -\frac{2}{3}, b = -\frac{4}{3}, c = 3; f(x) = -\frac{2}{3}x^2 - \frac{4}{3}x + 3$

Test: Chapter 9, p. 719

1. [9.1] The equations are dependent.   **2.** [9.1] $\left(2, -\frac{1}{2}, -1\right)$
3. [9.1] No solution   **4.** [9.1] $(0, 1, 0)$
5. [9.3] $\left(\frac{34}{107}, -\frac{104}{107}\right)$ **6.** [9.3] $(3, 1, -2)$
7. [9.4] 34   **8.** [9.4] 133   **9.** [9.4] $\left(\frac{13}{18}, \frac{7}{27}\right)$
10. [9.2] A: 15°; B: 40°; C: 125°
11. [9.2] Electrician: 3.5 hr; carpenter: 8 hr; plumber: 10 hr
12. [9.5] ($3, 55)   **13.** [9.5] **(a)** $C(x) = 25x + 40{,}000$;
(b) $R(x) = 70x$; **(c)** $P(x) = 45x - 40{,}000$; **(d)** $26,500 loss,
$500 profit; **(e)** (889 hammocks, $62,230)
14. [9.1] $(1, -1, 0, 2)$ **15.** [9.2] Adults' tickets: 1346;
senior citizens' tickets: 335; children's tickets: 1651

Cumulative Review: Chapters 1–9, pp. 720–722

1. 1.8 min/yr   **2.** **(a)** $F(t) = 1.45t + 5.5$, where t is the
number of years since 1999 and F is in trillions of miles;
(b) 17.1 trillion miles   **3.** **(a)** $a(t) = 3.482704918t +$
67.88319672; **(b)** approximately 259 million automobiles
4. **(a)**

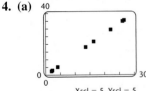

$\text{Xscl} = 5, \text{Yscl} = 5$
(b) $y \approx 1.33x$; **(c)** The fuel cost is approximately $1.33/gal.
5. 3 teams received $18.3 million, and 11 teams received
$750,000   **6.** (b)   **7.** (a)   **8.** (d)   **9.** (c)

10.

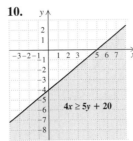

$4x \geq 5y + 20$

11.

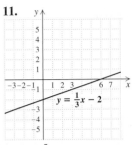

$y = \frac{1}{3}x - 2$

12. 10   **13.** 5.76×10^9 **14.** Slope: $\frac{7}{4}$; y-intercept: $(0, -3)$
15. $y = -\frac{10}{3}x + \frac{11}{3}$ **16.** $-3x^3 + 9x^2 + 3x - 3$
17. $-15x^4y^4$ **18.** $5a + 5b - 5c$
19. $15x^4 - x^3 - 9x^2 + 5x - 2$ **20.** $4x^4 - 4x^2y + y^2$
21. $4x^4 - y^2$ **22.** $-m^3n^2 - m^2n^2 - 5mn^3$
23. $\dfrac{y - 6}{2}$ **24.** $x - 1$ **25.** $\dfrac{a^2 + 7ab + b^2}{(a - b)(a + b)}$
26. $\dfrac{-m^2 + 5m - 6}{(m + 1)(m - 5)}$ **27.** $\dfrac{3y^2 - 2}{3y}$ **28.** $\dfrac{y - x}{xy(x + y)}$
29. $9x^2 - 13x + 26 + \dfrac{-50}{x + 2}$ **30.** $2x^2(2x + 9)$
31. $(x - 6)(x + 14)$ **32.** $(4y - 9)(4y + 9)$
33. $8(2x + 1)(4x^2 - 2x + 1)$ **34.** $(t - 8)^2$
35. $x^2(x - 1)(x + 1)(x^2 + 1)$
36. $(0.3b - 0.2c)(0.09b^2 + 0.06bc + 0.04c^2)$
37. $(4x - 1)(5x + 3)$ **38.** $(3x + 4)(x - 7)$
39. $(x^2 - y)(x^3 + y)$ **40.** **(a)** $-\frac{1}{2}$;
(b) $\{x \mid x \text{ is a real number } and\ x \neq 5\}$, or $(-\infty, 5) \cup (5, \infty)$
41. $[7, \infty)$ **42.** $\{x \mid x \text{ is a real number } and\ x \neq 2 \text{ and } x \neq 5\}$,
or $(-\infty, 2) \cup (2, 5) \cup (5, \infty)$ **43.** $\frac{1}{4}$ **44.** $-\frac{25}{7}, \frac{25}{7}$
45. $\{x \mid x > -3\}$, or $(-3, \infty)$ **46.** $\{x \mid x \leq -1 \text{ or } x \geq 2\}$, or
$(-\infty, -1] \cup [2, \infty)$ **47.** $\{x \mid -10 < x < 13\}$, or $(-10, 13)$
48. $\{x \mid x < -\frac{4}{3} \text{ or } x > 6\}$, or $\left(-\infty, -\frac{4}{3}\right) \cup (6, \infty)$
49. $\{x \mid x < -6.4 \text{ or } x > 6.4\}$, or $(-\infty, -6.4) \cup (6.4, \infty)$
50. $\{x \mid -\frac{13}{4} \leq x \leq \frac{15}{4}\}$, or $\left[-\frac{13}{4}, \frac{15}{4}\right]$ **51.** $-\frac{5}{3}$ **52.** -1
53. No solution   **54.** $\frac{1}{3}$ **55.** $(-3, 4)$ **56.** $(-2, -3, 1)$
57. $1, \frac{7}{3}$ **58.** $n = \dfrac{m - 12}{3}$ **59.** $a = \dfrac{Pb}{3 - P}$ **60.** 1
61. $\{x \mid x \geq 1\}$, or $[1, \infty)$ **62.** $(2, -3)$ **63.** 2   **64.** 1
65. $\{x \mid 1 \leq x \leq 5\}$, or $[1, 5]$ **66.** $\{1, 5\}$ **67.** $12, \frac{1}{2}, 7\frac{1}{2}$
68. $3\frac{1}{3}$ sec   **69.** 30 ft   **70.** All such sets of even integers
satisfy this condition.   **71.** 25 ft
72. $x^3 - 12x^2 + 48x - 64$ **73.** $-3, 3, -5, 5$
74. $\{x \mid -3 \leq x \leq -1 \text{ or } 7 \leq x \leq 9\}$, or $[-3, -1] \cup [7, 9]$
75. All real numbers except 9 and -5 **76.** $-\frac{1}{4}, 0, \frac{1}{4}$

Chapter 10

Interactive Discovery, p. 726

1. Not an identity   **2.** Not an identity   **3.** Identity
4. Not an identity   **5.** Identity   **6.** Identity

Visualizing the Graph, p. 735

1. B   **2.** H   **3.** C   **4.** I   **5.** D   **6.** A   **7.** F
8. J   **9.** G   **10.** E

Exercise Set 10.1, pp. 736–739

1. Two   **2.** Negative   **3.** Positive   **4.** Negative
5. Irrational   **6.** Real   **7.** Nonnegative   **8.** Negative
9. $7, -7$ **11.** $12, -12$ **13.** $20, -20$ **15.** $30, -30$
17. $-\frac{6}{7}$ **19.** 21 **21.** $-\frac{4}{9}$ **23.** 0.2 **25.** -0.05
27. $p^2; 2$ **29.** $\frac{x}{y+4}; 3$ **31.** $\sqrt{20}; 0$; does not exist;
does not exist   **33.** $-3; -1$; does not exist; 0
35. $1; \sqrt{2}; \sqrt{101}$ **37.** 1; does not exist; 6 **39.** $6|x|$
41. $6|b|$ **43.** $|8 - t|$ **45.** $|y + 8|$ **47.** $|2x + 7|$
49. 4 **51.** -1 **53.** $-\frac{2}{3}$ **55.** $|x|$ **57.** $6|a|$ **59.** 6
61. $|a + b|$ **63.** $|a^{11}|$ **65.** Cannot be simplified
67. $4x$ **69.** $3t$ **71.** $a + 1$ **73.** $2(x + 1)$, or $2x + 2$
75. $3t - 2$ **77.** 3 **79.** $2x$ **81.** -6 **83.** $5y$ **85.** t^9
87. $(x - 2)^4$ **89.** $2; 3; -2; -4$ **91.** 2; does not exist;
does not exist; 3 **93.** $\{x \mid x \geq 6\}$, or $[6, \infty)$
95. $\{t \mid t \geq -8\}$, or $[-8, \infty)$ **97.** $\{x \mid x \geq 5\}$, or $[5, \infty)$
99. $\mathbb{R}$ **101.** $\{z \mid z \geq -\frac{2}{5}\}$, or $[-\frac{2}{5}, \infty)$ **103.** $\mathbb{R}$
105. Domain: $\{x \mid x \leq 5\}$, or $(-\infty, 5]$;
range: $\{y \mid y \geq 0\}$, or $[0, \infty)$
107. Domain: $\{x \mid x \geq -1\}$, or $[-1, \infty)$;
range: $\{y \mid y \leq 1\}$, or $(-\infty, 1]$
109. Domain: $\mathbb{R}$; range: $\{y \mid y \geq 5\}$, or $[5, \infty)$
111. (c)   **113.** (d)   **115.** Yes   **117.** Yes   **119.** No
121. 20.5 cm; 36.0 cm   **123.** TW   **125.** $a^9 b^6 c^{15}$
126. $10a^{10} b^9$ **127.** $\frac{a^6 c^{12}}{8b^9}$ **128.** $\frac{x^6 y^2}{25z^4}$ **129.** $2x^4 y^5 z^2$
130. $\frac{5c^3}{a^4 b^7}$ **131.** TW   **133.** TW
135. (a) 13; (b) 15; (c) 18; (d) 20
137. $\{x \mid x \geq -5\}$, or $[-5, \infty)$ **139.** $\{x \mid x \geq 0\}$, or $[0, \infty)$

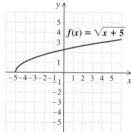

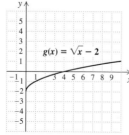

141. $\{x \mid -4 < x \leq 5\}$, or $(-4, 5]$ **143.** Cubic

Exercise Set 10.2, pp. 745–747

1. (g)   **2.** (c)   **3.** (e)   **4.** (h)   **5.** (a)   **6.** (d)
7. (b)   **8.** (f)   **9.** $\sqrt[6]{x}$ **11.** 4 **13.** 3 **15.** 3
17. $\sqrt[3]{xyz}$ **19.** $\sqrt[5]{a^2 b^2}$ **21.** $\sqrt[5]{t^2}$ **23.** 8 **25.** 81
27. $27\sqrt[4]{x^3}$ **29.** $125x^6$ **31.** $20^{1/3}$ **33.** $17^{1/2}$
35. $x^{3/2}$ **37.** $m^{2/5}$ **39.** $(cd)^{1/4}$ **41.** $(xy^2 z)^{1/5}$
43. $(3mn)^{3/2}$ **45.** $(8x^2 y)^{5/7}$ **47.** $\frac{2x}{z^{2/3}}$ **49.** $\frac{1}{x^{1/3}}$
51. $\frac{1}{(2rs)^{3/4}}$ **53.** 8 **55.** $2a^{3/5} c$ **57.** $\frac{5y^{4/5} z}{x^{2/3}}$
59. $\frac{a^3}{3^{5/2} b^{7/3}}$ **61.** $\left(\frac{3c}{2ab}\right)^{5/6}$ **63.** $\frac{6a}{b^{1/4}}$
65. $y = (x + 7) \wedge (1/4)$ **67.** $y = (3x - 2) \wedge (1/7)$

69. $y = x \wedge (3/6)$ **71.** 1.552 **73.** 1.778

75. -6.240 **77.** $7^{7/8}$ **79.** $3^{3/4}$ **81.** $5.2^{1/2}$ **83.** $10^{6/25}$
85. $a^{23/12}$ **87.** 64 **89.** $\frac{m^{1/3}}{n^{1/8}}$ **91.** $\sqrt[3]{x^2}$ **93.** a^3
95. a^2 **97.** $x^2 y^2$ **99.** $\sqrt{7a}$ **101.** $\sqrt[4]{8x^3}$ **103.** $\sqrt[18]{a}$
105. $x^3 y^3$ **107.** $a^6 b^{12}$ **109.** $\sqrt[12]{xy}$ **111.** TW
113. $11x^4 + 14x^3$ **114.** $-3t^6 + 28t^5 - 20t^4$
115. $15a^2 - 11ab - 12b^2$ **116.** $49x^2 - 14xy + y^2$
117. $\$93,500$ **118.** $0, 1$ **119.** TW   **121.** $\sqrt[6]{x^5}$
123. $\sqrt[6]{p + q}$ **125.** $2^{7/12} \approx 1.498 \approx 1.5$ **127.** 53.0%
129. 10 mg   **131.**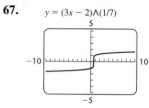

Exercise Set 10.3, pp. 754–756

1. True   **2.** False   **3.** False   **4.** False   **5.** True
6. True   **7.** $\sqrt{35}$ **9.** $\sqrt[3]{14}$ **11.** $\sqrt[4]{18}$ **13.** $\sqrt{26xy}$
15. $\sqrt[5]{80y^4}$ **17.** $\sqrt{y^2 - b^2}$ **19.** $\sqrt[3]{0.21y^2}$
21. $\sqrt[5]{(x - 2)^3}$ **23.** $\sqrt{\frac{7s}{11t}}$ **25.** $\sqrt[7]{\frac{5x - 15}{4x + 8}}$
27. $3\sqrt{2}$ **29.** $3\sqrt{3}$ **31.** $2\sqrt{2}$ **33.** $3\sqrt{22}$
35. $6a^2 \sqrt{b}$ **37.** $2x\sqrt[3]{y^2}$ **39.** $-2x^2 \sqrt[3]{2}$
41. $f(x) = 5x^3 \sqrt[3]{x^2}$ **43.** $f(x) = |7(x - 3)|$, or $7|x - 3|$
45. $f(x) = |x - 1|\sqrt{5}$ **47.** $a^3 b^3 \sqrt{b}$ **49.** $xy^2 z^3 \sqrt[3]{x^2 z}$
51. $-2ab^2 \sqrt[5]{a^2 b}$ **53.** $x^2 yz^3 \sqrt[5]{x^3 y^3 z^2}$ **55.** $-2a^4 \sqrt[3]{10a^2}$
57. $3\sqrt{2}$ **59.** $3\sqrt{35}$ **61.** 3 **63.** $18a^3$ **65.** $a\sqrt[3]{10}$

67. $2x^3\sqrt{5x}$ **69.** $s^2t^3\sqrt[3]{t}$ **71.** $(x+5)^2$ **73.** $2ab^3\sqrt[4]{5a}$
75. $x(y+z)^2\sqrt[5]{x}$ **77. TW**   **79.** $\dfrac{12x^2+5y^2}{64xy}$
80. $\dfrac{2a+6b^3}{a^4b^4}$ **81.** $\dfrac{-7x-13}{2(x-3)(x+3)}$
82. $\dfrac{-3x+1}{2(x-5)(x+5)}$ **83.** $3a^2b^2$ **84.** $3ab^5$ **85. TW**
87. 175.6 mi   **89. (a)** $-3.3°C$; **(b)** $-16.6°C$; **(c)** $-25.5°C$;
(d) $-54.0°C$ **91.** $25x^5\sqrt[3]{25x}$ **93.** $a^{10}b^{17}\sqrt{ab}$
95.

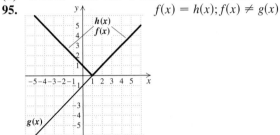

$f(x)=h(x);f(x)\ne g(x)$

97. $\{x\,|\,x\le 2 \ or \ x\ge 4\}$, or $(-\infty,2]\cup[4,\infty)$ **99.** 6
101. , **TW**

Exercise Set 10.4, pp. 760–762

1. (e)   **2. (b)**   **3. (f)**   **4. (c)**   **5. (h)**   **6. (d)**
7. (a)   **8. (g)**   **9.** $\frac{6}{5}$ **11.** $\frac{4}{3}$ **13.** $\dfrac{7}{y}$ **15.** $\dfrac{6y\sqrt{y}}{x^2}$
17. $\dfrac{3a\sqrt[3]{a}}{2b}$ **19.** $\dfrac{2a}{bc^2}$ **21.** $\dfrac{ab^2}{c^2}\sqrt[4]{\dfrac{a}{c^2}}$ **23.** $\dfrac{2x}{y^2}\sqrt[5]{\dfrac{x}{y}}$
25. $\dfrac{xy}{z^2}\sqrt[6]{\dfrac{y^2}{z^3}}$ **27.** $\sqrt{5}$ **29.** 3   **31.** $y\sqrt{5y}$
33. $2\sqrt[3]{a^2b}$ **35.** $\sqrt{2ab}$ **37.** $2x^2y^3\sqrt[4]{y^3}$
39. $\sqrt[3]{x^2+xy+y^2}$ **41.** $\dfrac{\sqrt{6}}{2}$ **43.** $\dfrac{2\sqrt{15}}{21}$ **45.** $\dfrac{2\sqrt[3]{6}}{3}$
47. $\dfrac{\sqrt[3]{75ac^2}}{5c}$ **49.** $\dfrac{y\sqrt[3]{180x^2y}}{6x^2}$ **51.** $\dfrac{\sqrt[3]{2xy^2}}{xy}$ **53.** $\dfrac{\sqrt{14a}}{6}$
55. $\dfrac{3\sqrt{5y}}{10xy}$ **57.** $\dfrac{\sqrt{5b}}{6a}$ **59.** $\dfrac{5}{\sqrt{35x}}$ **61.** $\dfrac{2}{\sqrt{6}}$ **63.** $\dfrac{52}{3\sqrt{91}}$
65. $\dfrac{7}{\sqrt[3]{98}}$ **67.** $\dfrac{7x}{\sqrt{21xy}}$ **69.** $\dfrac{2a^2}{\sqrt[3]{20ab}}$ **71.** $\dfrac{x^2y}{\sqrt{2xy}}$
73. TW   **75.** $\dfrac{3(x-1)}{(x-5)(x+5)}$ **76.** $\dfrac{7(x-2)}{(x+4)(x-4)}$
77. $\dfrac{a-1}{a+7}$ **78.** $\dfrac{t+11}{t+2}$ **79.** $125a^9b^{12}$ **80.** $225x^{10}y^6$
81. TW   **83. (a)** 1.62 sec; **(b)** 1.99 sec; **(c)** 2.20 sec
85. $9\sqrt[3]{9n^2}$ **87.** $\dfrac{-3\sqrt{a^2-3}}{a^2-3}$, or $\dfrac{-3}{\sqrt{a^2-3}}$
89. Step 1: $\sqrt[n]{a}=a^{1/n}$, by definition; Step 2: $\left(\dfrac{a}{b}\right)^n=\dfrac{a^n}{b^n}$,
raising a quotient to a power; Step 3: $a^{1/n}=\sqrt[n]{a}$, by definition

91. $(f/g)(x)=3x$, where x is a real number and $x>0$
93. $(f/g)(x)=\sqrt{x+3}$, where x is a real number and $x>3$

Exercise Set 10.5, pp. 767–769

1. Radicands; indices   **2.** Indices   **3.** Bases
4. Denominators   **5.** Numerator; conjugate   **6.** Bases
7. $9\sqrt{5}$ **9.** $2\sqrt[3]{4}$ **11.** $10\sqrt[3]{y}$ **13.** $7\sqrt{2}$
15. $13\sqrt[3]{7}+\sqrt{3}$ **17.** $9\sqrt{3}$ **19.** $23\sqrt{5}$ **21.** $9\sqrt[3]{2}$
23. $(1+6a)\sqrt{5a}$ **25.** $(x+2)\sqrt[3]{6x}$ **27.** $3\sqrt{a-1}$
29. $(x+3)\sqrt{x-1}$ **31.** $4\sqrt{3}+3$ **33.** $15-3\sqrt{10}$
35. $6\sqrt{5}-4$ **37.** $3-4\sqrt[3]{63}$ **39.** $a+2a\sqrt[3]{3}$
41. $4+3\sqrt{6}$ **43.** $\sqrt{6}-\sqrt{14}+\sqrt{21}-7$ **45.** 4
47. -2 **49.** $2-8\sqrt{35}$ **51.** $7+4\sqrt{3}$ **53.** $5-2\sqrt{6}$
55. $2t+5+2\sqrt{10t}$ **57.** $14+x-6\sqrt{x+5}$
59. $6\sqrt[4]{63}+4\sqrt[4]{35}-3\sqrt[4]{54}-2\sqrt[4]{30}$ **61.** $\dfrac{20+5\sqrt{3}}{13}$
63. $\dfrac{12-2\sqrt{3}+6\sqrt{5}-\sqrt{15}}{33}$ **65.** $\dfrac{a-\sqrt{ab}}{a-b}$ **67.** -1
69. $\dfrac{12-3\sqrt{10}-2\sqrt{14}+\sqrt{35}}{6}$ **71.** $\dfrac{3}{5\sqrt{7}-10}$
73. $\dfrac{2}{14+2\sqrt{3}+3\sqrt{2}+7\sqrt{6}}$ **75.** $\dfrac{x-y}{x+2\sqrt{xy}+y}$
77. $\dfrac{1}{\sqrt{a+h}+\sqrt{a}}$ **79.** $a\sqrt[4]{a}$ **81.** $b\sqrt[10]{b^9}$
83. $xy\sqrt[6]{xy^5}$ **85.** $3a^2b\sqrt[4]{ab}$ **87.** $a^2b^2c^2\sqrt[6]{a^2bc^2}$
89. $\sqrt[12]{a^5}$ **91.** $\sqrt[12]{x^2y^5}$ **93.** $\sqrt[10]{ab^9}$ **95.** $\sqrt[20]{(3x-1)^3}$
97. $\sqrt[15]{(2x+1)^4}$ **99.** $x^6\sqrt[15]{xy^5}-\sqrt[15]{x^{13}y^{14}}$
101. $2m^2+m\sqrt[4]{n}+2m\sqrt[3]{n^2}+\sqrt[12]{n^{11}}$ **103.** $\sqrt[4]{2x^2}-x^3$
105. x^2-7 **107.** $27+10\sqrt{2}$ **109.** $8-2\sqrt{15}$
111. TW   **113.** $7\sqrt{3}$ **114.** 30   **115.** $5\sqrt{x}$
116. $x^2y^2\sqrt[3]{z^2}$ **117.** $75-x$ **118.** $2x\sqrt[4]{2x}$ **119. TW**
121. $f(x)=-6x\sqrt{5+x}$
123. $f(x)=(x+3x^2)\sqrt[4]{x-1}$
125. $ac^2\left[(3a+2c)\sqrt{ab}-2\sqrt[3]{ab}\right]$
127. $9a^2(b+1)\sqrt[6]{243a^5(b+1)^5}$ **129.** $1-\sqrt{w}$
131. $\left(\sqrt{x}+\sqrt{5}\right)\left(\sqrt{x}-\sqrt{5}\right)$
133. $\left(\sqrt{x}+\sqrt{a}\right)\left(\sqrt{x}-\sqrt{a}\right)$ **135.** $2x-2\sqrt{x^2-4}$

Interactive Discovery, p. 770

1. $\{3\};\{-3,3\}$ **2.** $\{-2\};\{-2,2\}$ **3.** $\{25\};\{25\}$
4. $\varnothing;\{9\}$

Exercise Set 10.6, pp. 776–778

1. False   **2.** True   **3.** True   **4.** False   **5.** True
6. True   **7.** $\frac{51}{5}$ **9.** $\frac{25}{2}$ **11.** 168   **13.** 56   **15.** 3
17. 82   **19.** 0, 9   **21.** 64   **23.** -27 **25.** 125
27. No solution   **29.** $\frac{80}{3}$ **31.** 57   **33.** $-\frac{5}{3}$ **35.** 1

37. $\frac{106}{27}$ **39.** 4   **41.** 3, 7   **43.** $\frac{80}{9}$ **45.** -1
47. No solution   **49.** 2, 6   **51.** 2   **53.** 4   **55.** TW
57. $-3, 4$ **58.** $\frac{63}{4}$ **59.** $\{x \mid x < -1 \; or \; x > 4\}$, or
$(-\infty, -1) \cup (4, \infty)$ **60.** 28   **61.** $\left\{x \mid x \geq \frac{1}{3}\right\}$, or $\left[\frac{1}{3}, \infty\right)$
62. $-6, 6$ **63.** -7 **64.** $\frac{3}{5}$ **65.** $\left(\frac{31}{7}, \frac{2}{7}\right)$ **66.** $\left(\frac{1}{2}, 0, -1\right)$
67. TW   **69.** 524.8°C   **71.** $t = \frac{1}{9}\left(\frac{S^2 \cdot 2457}{1087.7^2} - 2617\right)$

73. 4480 rpm   **75.** $r = \frac{v^2 h}{2gh - v^2}$ **77.** 72.25 ft   **79.** $-\frac{8}{9}$
81. $-8, 8$ **83.** 1, 8   **85.** $\left(\frac{1}{36}, 0\right), (36, 0)$ **87.** 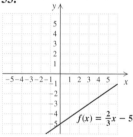, TW

Exercise Set 10.7, pp. 784–787

1. Right; hypotenuse   **2.** Legs   **3.** Square roots
4. Isosceles   **5.** 30°, 60°, 90°; leg   **6.** $a\sqrt{2}$
7. $\sqrt{34}$; 5.831   **9.** $9\sqrt{2}$; 12.728   **11.** 5   **13.** 4 m
15. $\sqrt{19}$ in.; 4.359 in.   **17.** 1 m   **19.** 250 ft
21. $\sqrt{8450}$, or $65\sqrt{2}$ ft; 91.924 ft   **23.** 12 in.
25. $\left(\sqrt{340} + 8\right)$ ft; 26.439 ft   **27.** $\left(110 - \sqrt{6500}\right)$ paces;
29.377 paces   **29.** Leg = 5; hypotenuse = $5\sqrt{2} \approx 7.071$
31. Shorter leg = 7; longer leg = $7\sqrt{3} \approx 12.124$
33. Leg = $5\sqrt{3} \approx 8.660$; hypotenuse = $10\sqrt{3} \approx 17.321$
35. Both legs = $\frac{13\sqrt{2}}{2} \approx 9.192$

37. Leg = $14\sqrt{3} \approx 24.249$; hypotenuse = 28
39. $3\sqrt{3} \approx 5.196$ **41.** $13\sqrt{2} \approx 18.385$
43. $\frac{19\sqrt{2}}{2} \approx 13.435$ **45.** $\sqrt{10,561}$ ft ≈ 102.767 ft
47. $h = 2\sqrt{3}$ ft ≈ 3.464 ft   **49.** $(0, -4), (0, 4)$ **51.** TW
53.
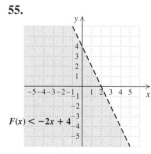
$f(x) = \frac{2}{3}x - 5$

54.
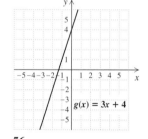
$g(x) = 3x + 4$

55.
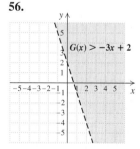
$F(x) < -2x + 4$

56.
$G(x) > -3x + 2$

57.
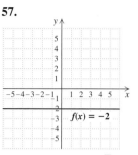
$f(x) = -2$

58.
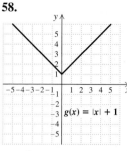
$g(x) = |x| + 1$

59. TW   **61.** $36\sqrt{3}$ cm²; 62.354 cm²   **63.** $d = s + s\sqrt{2}$
65. 5 gal. The total area of the doors and windows is
134 ft² or more.   **67.** 60.28 ft by 60.28 ft

Exercise Set 10.8, pp. 794–795

1. False   **2.** False   **3.** True   **4.** True   **5.** True
6. True   **7.** False   **8.** True   **9.** $6i$ **11.** $i\sqrt{13}$, or
$\sqrt{13}i$ **13.** $3i\sqrt{2}$, or $3\sqrt{2}i$ **15.** $i\sqrt{3}$, or $\sqrt{3}i$ **17.** $9i$
19. $-10i\sqrt{3}$, or $-10\sqrt{3}i$ **21.** $6 - 2i\sqrt{21}$, or $6 - 2\sqrt{21}i$
23. $\left(-2\sqrt{19} + 5\sqrt{5}\right)i$ **25.** $\left(3\sqrt{2} - 10\right)i$ **27.** $11 + 10i$
29. $4 + 5i$ **31.** $2 - i$ **33.** $-12 - 5i$ **35.** -42
37. -24 **39.** -18 **41.** $-\sqrt{10}$ **43.** $-3\sqrt{14}$
45. $-30 + 10i$ **47.** $-28 - 21i$ **49.** $1 + 5i$
51. $38 + 9i$ **53.** $2 - 46i$ **55.** $-11 - 16i$
57. $13 - 47i$ **59.** $12 - 16i$ **61.** $-5 + 12i$
63. $-5 - 12i$ **65.** $\frac{28}{17} - \frac{7}{17}i$ **67.** $\frac{6}{13} + \frac{4}{13}i$ **69.** $\frac{3}{17} + \frac{5}{17}i$
71. $-\frac{5}{6}i$ **73.** $-\frac{3}{4} - \frac{5}{4}i$ **75.** $1 - 2i$ **77.** $-\frac{23}{58} + \frac{43}{58}i$
79. $\frac{19}{29} - \frac{4}{29}i$ **81.** $\frac{6}{25} - \frac{17}{25}i$ **83.** $-i$ **85.** 1   **87.** -1
89. i **91.** -1 **93.** $-125i$ **95.** 0   **97.** TW
99. $(3x - 2)(4x - 5)$ **100.** $x(x + 2)(x - 5)$
101. $5x^2y^2(4y + 3x - 7x^2y)$ **102.** $(x^2 - 3)(y - x)$
103. $14y(x^2 + y^2)(x + y)(x - y)$
104. $(3x - 4y)(9x^2 + 12xy + 16y^2)$ **105.** TW
107. $-9 - 27i$ **109.** $50 - 120i$ **111.** $\frac{250}{41} + \frac{200}{41}i$
113. 8   **115.** $\frac{3}{5} + \frac{9}{5}i$ **117.** 1

Review Exercises: Chapter 10, pp. 799–800

1. True   **2.** False   **3.** True   **4.** True   **5.** True
6. True   **7.** True   **8.** False   **9.** $\frac{7}{3}$ **10.** -0.5 **11.** 5
12. $\left\{x \mid x \geq \frac{7}{2}\right\}$, or $\left[\frac{7}{2}, \infty\right)$ **13.** (a) 34.6 lb; (b) 10.0 lb;
(c) 24.9 lb; (d) 42.3 lb   **14.** $5|t|$ **15.** $|c + 8|$
16. $|x - 3|$ **17.** $|2x + 1|$ **18.** -2 **19.** $-\frac{4x^2}{3}$
20. $|x^3y^2|$, or $|x^3|y^2$ **21.** $2x^2$ **22.** $(5ab)^{4/3}$
23. $8a^4\sqrt{a}$ **24.** x^3y^5 **25.** $\sqrt[3]{x^2y}$ **26.** $\frac{1}{x^{2/5}}$ **27.** $7^{1/6}$
28. $f(x) = 5|x - 6|$ **29.** $\sqrt{6xy}$ **30.** $3a\sqrt[3]{a^2b^2}$

31. $-6x^5y^4\sqrt[3]{2x^2}$ **32.** $y\sqrt[3]{6}$ **33.** $\dfrac{5\sqrt{x}}{2}$ **34.** $\dfrac{2a^2\sqrt[4]{3a^3}}{c^2}$
35. $7\sqrt[3]{x}$ **36.** $\sqrt{3}$ **37.** $(2x+y^2)\sqrt[3]{x}$ **38.** $15\sqrt{2}$
39. $\sqrt{15}+4\sqrt{6}-6\sqrt{10}-48$ **40.** $\sqrt[4]{x^3}$ **41.** $\sqrt[12]{x^5}$
42. $a^2-2a\sqrt{2}+2$ **43.** $-4\sqrt{10}+4\sqrt{15}$
44. $\dfrac{20}{\sqrt{10}+\sqrt{15}}$ **45.** 19 **46.** -126 **47.** 4 **48.** 14
49. $5\sqrt{2}$ cm; 7.071 cm **50.** $\sqrt{32}$ ft; 5.657 ft
51. Short leg = 10; long leg = $10\sqrt{3}\approx17.321$
52. $-2i\sqrt{2}$, or $-2\sqrt{2}i$ **53.** $-2-9i$ **54.** $6+i$
55. 29 **56.** -1 **57.** $9-12i$ **58.** $\frac{13}{25}-\frac{34}{25}i$
59. TW A complex number $a+bi$ is real when $b=0$. It is imaginary when $b\neq0$.
60. TW An absolute-value sign must be used to simplify $\sqrt[n]{x^n}$ when n is even, since x may be negative. If x is negative while n is even, the radical expression cannot be simplified to x, since $\sqrt[n]{x^n}$ represents the principal, or positive, root. When n is odd, there is only one root, and it will be positive or negative depending on the sign of x. Thus there is no absolute-value sign when n is odd.
61. 3 **62.** $-\frac{2}{5}+\frac{9}{10}i$ **63.** $\dfrac{2i}{3i}$; answers may vary

Test: Chapter 10, p. 801

1. [10.3] $5\sqrt{2}$ **2.** [10.4] $-\dfrac{2}{x^2}$ **3.** [10.1] $9|a|$
4. [10.1] $|x-4|$ **5.** [10.3] $x^2y\sqrt[5]{x^2y^3}$ **6.** [10.4] $\left|\dfrac{5x}{6y^2}\right|$, or $\dfrac{5|x|}{6y^2}$ **7.** [10.3] $\sqrt[3]{15y^2z}$ **8.** [10.4] $\sqrt[5]{x^2y^2}$
9. [10.5] $xy\sqrt[4]{x}$ **10.** [10.5] $\sqrt[20]{a^3}$ **11.** [10.5] $6\sqrt{2}$
12. [10.5] $(x^2+3y)\sqrt{y}$ **13.** [10.5] $14-19\sqrt{x}-3x$
14. [10.2] $(7xy)^{1/2}$ **15.** [10.2] $\sqrt[6]{(4a^3b)^5}$
16. [10.1] $\{x\,|\,x\geq5\}$, or $[5,\infty)$ **17.** [10.5] $27+10\sqrt{2}$
18. [10.5] $\dfrac{5\sqrt{3}-\sqrt{6}}{23}$ **19.** [10.6] 7 **20.** [10.6] No
solution **21.** [10.1] 3.4 sec **22.** [10.7] $7\sqrt{3}$ cm or
12.124 cm (if the shorter leg is 7 cm); $\dfrac{7}{\sqrt{3}}$ cm or 4.041 cm
(if the longer leg is 7 cm) **23.** [10.7] $\sqrt{10{,}600}$ ft; 102.956 ft
24. [10.8] $5i\sqrt{2}$, or $5\sqrt{2}i$ **25.** [10.8] $12+2i$
26. [10.8] -24 **27.** [10.8] $15-8i$
28. [10.8] $-\frac{11}{34}-\frac{7}{34}i$ **29.** [10.8] i
30. [10.6] 3 **31.** [10.8] $-\frac{17}{4}i$
32. [10.7] The isosceles right triangle is larger by 1.206 ft^2.

Chapter 11

Interactive Discovery, p. 804

1. 1 **2.** 1 **3.** 1 **4.** 0 **5.** 0 **6.** 2 **7.** A cup-shaped curve opening up or down

Exercise Set 11.1, pp. 814–816

1. $\sqrt{k}$; $-\sqrt{k}$ **2.** 7; -7 **3.** $t+3$; $t+3$ **4.** 16
5. 25; 5 **6.** 9; 3 **7.** 2 **9.** 1 **11.** 0 **13.** $\pm\sqrt{5}$
15. $\pm\frac{4}{3}i$ **17.** $\pm\sqrt{\dfrac{7}{5}}$, or $\pm\dfrac{\sqrt{35}}{5}$ **19.** $-6,8$
21. $13\pm3\sqrt{2}$ **23.** $-1\pm3i$ **25.** $-\dfrac{3}{4}\pm\dfrac{\sqrt{17}}{4}$, or $\dfrac{-3\pm\sqrt{17}}{4}$ **27.** $-3,13$ **29.** $1,9$ **31.** $-4\pm\sqrt{13}$
33. $-14,0$ **35.** $x^2+16x+64=(x+8)^2$
37. $t^2-10t+25=(t-5)^2$ **39.** $x^2+3x+\frac{9}{4}=\left(x+\frac{3}{2}\right)^2$
41. $t^2-9t+\frac{81}{4}=\left(t-\frac{9}{2}\right)^2$ **43.** $x^2+\frac{2}{5}x+\frac{1}{25}=\left(x+\frac{1}{5}\right)^2$
45. $t^2-\frac{5}{6}t+\frac{25}{144}=\left(t-\frac{5}{12}\right)^2$ **47.** $-7,1$ **49.** $4,6$
51. $-9,-1$ **53.** $-4\pm\sqrt{19}$
55. $\left(-3-\sqrt{2},0\right),\left(-3+\sqrt{2},0\right)$
57. $\left(-6-\sqrt{11},0\right),\left(-6+\sqrt{11},0\right)$
59. $\left(5-\sqrt{47},0\right),\left(5+\sqrt{47},0\right)$ **61.** $-\frac{4}{3},-\frac{2}{3}$
63. $-\frac{1}{3},2$ **65.** $-\dfrac{2}{5}\pm\dfrac{\sqrt{19}}{5}$, or $\dfrac{-2\pm\sqrt{19}}{5}$
67. $\left(-\dfrac{1}{4}-\dfrac{\sqrt{13}}{4},0\right),\left(-\dfrac{1}{4}+\dfrac{\sqrt{13}}{4},0\right)$, or $\left(\dfrac{-1-\sqrt{13}}{4},0\right),\left(\dfrac{-1+\sqrt{13}}{4},0\right)$
69. $\left(\dfrac{3}{4}-\dfrac{\sqrt{17}}{4},0\right),\left(\dfrac{3}{4}+\dfrac{\sqrt{17}}{4},0\right)$, or $\left(\dfrac{3-\sqrt{17}}{4},0\right),\left(\dfrac{3+\sqrt{17}}{4},0\right)$ **71.** 10% **73.** 18.75%
75. 4% **77.** About 8.1 sec **79.** About 9.5 sec
81. TW **83.** 28 **84.** -92 **85.** $3\sqrt[3]{10}$ **86.** $4\sqrt{5}$
87. 5 **88.** 7 **89.** TW **91.** ±18 **93.** $-\frac{7}{2},-\sqrt{5},0,$
$\sqrt{5},8$ **95.** Barge: 8 km/h; fishing boat: 15 km/h

Exercise Set 11.2, pp. 821–822

1. True **2.** False **3.** False **4.** True **5.** False
6. True **7.** $-\dfrac{7}{2}\pm\dfrac{\sqrt{61}}{2}$ **9.** $3\pm\sqrt{7}$
11. $-\dfrac{1}{2}\pm\dfrac{\sqrt{3}}{2}i$ **13.** $2\pm3i$ **15.** $3\pm\sqrt{5}$

17. $-\dfrac{4}{3} \pm \dfrac{\sqrt{19}}{3}$ **19.** $-\dfrac{1}{2} \pm \dfrac{\sqrt{17}}{2}$ **21.** $-\dfrac{3}{8} \pm \dfrac{\sqrt{129}}{24}$

23. $\frac{2}{5}$ **25.** $-\dfrac{11}{8} \pm \dfrac{\sqrt{41}}{8}$ **27.** 5, 10   **29.** $\dfrac{13}{10} \pm \dfrac{\sqrt{509}}{10}$

31. $2 \pm \sqrt{5}i$ **33.** $2, -1 \pm \sqrt{3}i$ **35.** $\frac{2}{3}$, 1

37. $5 \pm \sqrt{53}$ **39.** $\dfrac{7}{2} \pm \dfrac{\sqrt{85}}{2}$ **41.** $\frac{3}{2}$, 6

43. -5.31662479, 1.31662479 **45.** 0.7639320225, 5.236067978 **47.** -1.265564437, 2.765564437 **49.** ™

51. $\dfrac{x + y}{2}$ **52.** $\dfrac{a^2 - b^2}{b}$ **53.** $9a^2 b^3 \sqrt{2a}$

54. $4a^2 b^3 \sqrt{6}$ **55.** $\dfrac{3(x + 1)}{3x + 1}$ **56.** $\dfrac{4b}{3ab^2 - 4a^2}$

57. ™   **59.** $(-2, 0), (1, 0)$ **61.** $4 - 2\sqrt{2}, 4 + 2\sqrt{2}$

63. -1.179210116, 0.3392101158 **65.** $\dfrac{-5\sqrt{2} \pm \sqrt{34}}{4}$

67. $\frac{1}{2}$ **69.** ▲, ™

Exercise Set 11.3, pp. 827–830

1. First part: 60 mph; second part: 50 mph   **3.** 40 mph
5. Cessna: 150 mph, Beechcraft: 200 mph; or
Cessna: 200 mph, Beechcraft: 250 mph
7. To Hillsboro: 10 mph; return trip: 4 mph
9. About 14 mph   **11.** 12 hr   **13.** About 3.24 mph

15. $r = \dfrac{1}{2}\sqrt{\dfrac{A}{\pi}}$, or $\dfrac{\sqrt{A\pi}}{2\pi}$

17. $r = \dfrac{-\pi h + \sqrt{\pi^2 h^2 + 2\pi A}}{2\pi}$

19. $s = \sqrt{\dfrac{kQ_1 Q_2}{N}}$, or $\dfrac{\sqrt{kQ_1 Q_2 N}}{N}$ **21.** $g = \dfrac{4\pi^2 l}{T^2}$

23. $c = \sqrt{d^2 - a^2 - b^2}$ **25.** $t = \dfrac{-v_0 + \sqrt{v_0^2 + 2gs}}{g}$

27. $n = \dfrac{1 + \sqrt{1 + 8N}}{2}$ **29.** $h = \dfrac{V^2}{12.25}$

31. $t = \dfrac{-b \pm \sqrt{b^2 - 4ac}}{2a}$ **33. (a)** 10.1 sec; **(b)** 7.49 sec;
(c) 272.5 m   **35.** 2.9 sec   **37.** 0.968 sec   **39.** 2.5 m/sec
41. 7%   **43.** ™   **45.** $\{x \,|\, x > 3\}$, where x is the number
of semesters   **46.** Cream-filled: 46; glazed: 44   **47.** 1 in.
48. Jaime: 3 min; Cheri: 6 min   **49.** 762 kilobytes
50. 45 mph   **51.** ™

53. $t = \dfrac{-10.2 + 6\sqrt{-A^2 + 13A - 39.36}}{A - 6.5}$

55. $\pm\sqrt{2}$ **57.** $l = \dfrac{w + w\sqrt{5}}{2}$

59. $n = \pm\sqrt{\dfrac{r^2 \pm \sqrt{r^4 + 4m^4 r^2 p - 4mp}}{2m}}$ **61.** $A(S) = \dfrac{\pi S}{6}$

Exercise Set 11.4, pp. 834–836

1. Discriminant   **2.** One   **3.** Two   **4.** Two
5. Rational   **6.** Imaginary   **7.** Two irrational
9. Two imaginary   **11.** Two irrational   **13.** Two rational
15. Two imaginary   **17.** One rational   **19.** Two rational
21. Two rational   **23.** Two irrational   **25.** Two imaginary
27. Two irrational   **29.** $x^2 + 4x - 21 = 0$
31. $x^2 - 6x + 9 = 0$ **33.** $x^2 + 4x + 3 = 0$
35. $4x^2 - 23x + 15 = 0$ **37.** $8x^2 + 6x + 1 = 0$
39. $x^2 - 2x - 0.96 = 0$ **41.** $x^2 - 3 = 0$
43. $x^2 - 20 = 0$ **45.** $x^2 + 16 = 0$
47. $x^2 - 4x + 53 = 0$ **49.** $x^2 - 6x - 5 = 0$
51. $3x^2 - 6x - 4 = 0$ **53.** $x^3 - 4x^2 - 7x + 10 = 0$
55. $x^3 - 2x^2 - 3x = 0$ **57.** ™   **59.** $81a^8$ **60.** $16x^6$
61. $(-1, 0), (8, 0)$ **62.** $(2, 0), (4, 0)$ **63.** 6 commercials
64.

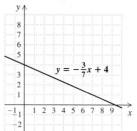

65. ™   **67.** $a = 1, b = 2, c = -3$ **69. (a)** $-\frac{3}{5}$; **(b)** $-\frac{1}{3}$
71. (a) $9 + 9i$; **(b)** $3 + 3i$
73. The solutions of $ax^2 + bx + c = 0$ are
$x = \dfrac{-b \pm \sqrt{b^2 - 4ac}}{2a}$. When there is just one solution,
$b^2 - 4ac$ must be 0, so $x = \dfrac{-b \pm 0}{2a} = \dfrac{-b}{2a}$.
75. $a = 8, b = 20, c = -12$
77. $x^4 - 8x^3 + 21x^2 - 2x - 52 = 0$

Exercise Set 11.5, pp. 841–842

1. (f)   **2.** (d)   **3.** (h)   **4.** (b)   **5.** (g)   **6.** (a)
7. (e)   **8.** (c)   **9.** $\pm 1, \pm 2$ **11.** $\pm\sqrt{5}, \pm 2$

13. $\pm\dfrac{\sqrt{3}}{2}, \pm 2$ **15.** $8 + 2\sqrt{7}$ **17.** $\pm 2\sqrt{2}, \pm 3$

19. No solution   **21.** $-\frac{1}{2}, \frac{1}{3}$ **23.** $-\frac{4}{5}, 1$ **25.** $-27, 8$
27. 729   **29.** 1   **31.** No solution   **33.** $\frac{12}{5}$ **35.** $\left(\frac{4}{25}, 0\right)$

37. $\left(\dfrac{3}{2} + \dfrac{\sqrt{33}}{2}, 0\right), \left(\dfrac{3}{2} - \dfrac{\sqrt{33}}{2}, 0\right), (4, 0), (-1, 0)$

39. $(-243, 0), (32, 0)$ **41.** No x-intercepts   **43.** ™

45. $-3, 2$ **46.** $\dfrac{3}{4} \pm \dfrac{\sqrt{23}}{4}i$ **47.** $7 \pm \sqrt{5}$

48. $-\dfrac{1}{8} \pm \dfrac{\sqrt{97}}{8}$, or $\dfrac{-1 \pm \sqrt{97}}{8}$ **49.** $\pm\sqrt{2}, \pm\sqrt{3}$

50. No solution   **51.** TW   **53.** $\pm\sqrt{\dfrac{7 \pm \sqrt{29}}{10}}$

55. $-2, -1, 5, 6$ **57.** $\frac{100}{99}$ **59.** $-5, -3, -2, 0, 2, 3, 5$

61. $1, 3, -\dfrac{1}{2} + \dfrac{\sqrt{3}}{2}i, -\dfrac{1}{2} - \dfrac{\sqrt{3}}{2}i, -\dfrac{3}{2} + \dfrac{3\sqrt{3}}{2}i,$

$-\dfrac{3}{2} - \dfrac{3\sqrt{3}}{2}i$

Interactive Discovery, p. 844

1. (a) $(0, 0)$; **(b)** $x = 0$; **(c)** upward; **(d)** narrower
2. (a) $(0, 0)$; **(b)** $x = 0$; **(c)** upward; **(d)** wider
3. (a) $(0, 0)$; **(b)** $x = 0$; **(c)** upward; **(d)** wider
4. (a) $(0, 0)$; **(b)** $x = 0$; **(c)** downward; **(d)** neither narrower
nor wider   **5. (a)** $(0, 0)$; **(b)** $x = 0$; **(c)** downward;
(d) narrower   **6. (a)** $(0, 0)$; **(b)** $x = 0$; **(c)** downward;
(d) wider   **7.** When $a > 1$, the graph of $y = ax^2$ is narrower
than the graph of $y = x^2$. When $0 < a < 1$, the graph of
$y = ax^2$ is wider than the graph of $y = x^2$.
8. When $a < -1$, the graph of $y = ax^2$ is narrower than the
graph of $y = x^2$ and the graph opens downward. When
$-1 < a < 0$, the graph of $y = ax^2$ is wider than the graph of
$y = x^2$ and the graph opens downward.

Interactive Discovery, p. 845

1. (a) $(3, 0)$; **(b)** $x = 3$; **(c)** same shape   **2. (a)** $(-1, 0)$;
(b) $x = -1$; **(c)** same shape   **3. (a)** $\left(\frac{3}{2}, 0\right)$; **(b)** $x = \frac{3}{2}$;
(c) same shape   **4. (a)** $(-2, 0)$; **(b)** $x = -2$; **(c)** same shape
5. The graph of $g(x) = a(x - h)^2$ looks like the graph of
$f(x) = ax^2$, except that it is moved left or right.

Exercise Set 11.6, pp. 849–851

1. (h)   **2.** (g)   **3.** (f)   **4.** (d)   **5.** (b)   **6.** (c)
7. (e)   **8.** (a)   **9. (a)** Positive; **(b)** $(3, 1)$;
(c) $x = 3$; **(d)** $[1, \infty)$ **11. (a)** Negative; **(b)** $(-2, -3)$;
(c) $x = -2$; **(d)** $(-\infty, -3]$ **13. (a)** Positive; **(b)** $(-3, 0)$;
(c) $x = -3$; **(d)** $[0, \infty)$

15.

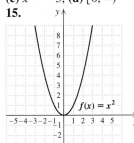

17.

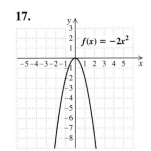

19.

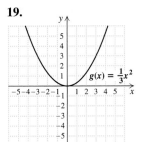

21.

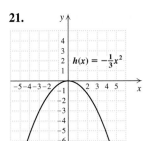

23.

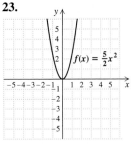

25. Vertex: $(-1, 0)$;
axis of symmetry: $x = -1$

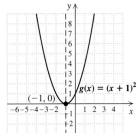

27. Vertex $(2, 0)$;
axis of symmetry: $x = 2$

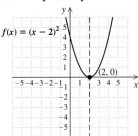

29. Vertex: $(3, 0)$;
axis of symmetry: $x = 3$

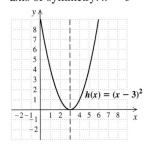

31. Vertex: $(-1, 0)$;
axis of symmetry: $x = -1$

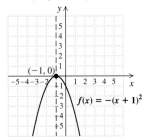

33. Vertex: $(2, 0)$;
axis of symmetry: $x = 2$

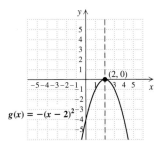

35. Vertex: $(-1, 0)$;
axis of symmetry: $x = -1$

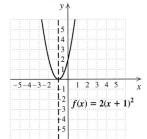

37. Vertex: (4, 0); axis of symmetry: $x = 4$

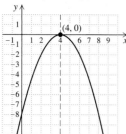

$$h(x) = -\tfrac{1}{2}(x - 4)^2$$

39. Vertex: (1, 0); axis of symmetry: $x = 1$

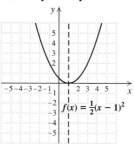

$$f(x) = \tfrac{1}{2}(x - 1)^2$$

41. Vertex: (−5, 0); axis of symmetry: $x = -5$

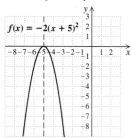

$$f(x) = -2(x + 5)^2$$

43. Vertex: $\left(\tfrac{1}{2}, 0\right)$; axis of symmetry: $x = \tfrac{1}{2}$

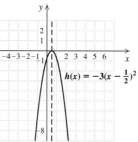

$$h(x) = -3\left(x - \tfrac{1}{2}\right)^2$$

45. Vertex: (5, 2); axis of symmetry: $x = 5$; minimum: 2

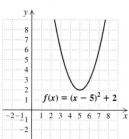

$$f(x) = (x - 5)^2 + 2$$

47. Vertex: (−1, −3); axis of symmetry: $x = -1$; minimum: −3

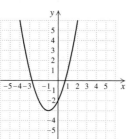

$$f(x) = (x + 1)^2 - 3$$

49. Vertex: (−4, 1); axis of symmetry: $x = -4$; minimum: 1

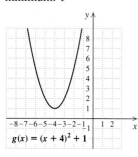

$$g(x) = (x + 4)^2 + 1$$

51. Vertex: (1, −3); axis of symmetry: $x = 1$; maximum: −3

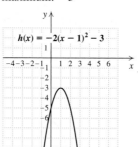

$$h(x) = -2(x - 1)^2 - 3$$

53. Vertex: (−4, 1); axis of symmetry: $x = -4$; minimum: 1

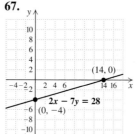

$$f(x) = 2(x + 4)^2 + 1$$

55. Vertex: (1, 4); axis of symmetry: $x = 1$; maximum: 4

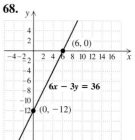

$$g(x) = -\tfrac{3}{2}(x - 1)^2 + 4$$

57. Vertex: (8, 7); axis of symmetry: $x = 8$; minimum: 7
59. Vertex: (−6, 11); axis of symmetry: $x = -6$; maximum: 11   **61.** Vertex: $\left(-\tfrac{1}{4}, -13\right)$; axis of symmetry: $x = -\tfrac{1}{4}$; minimum: −13
63. Vertex: (−4.58, 65π); axis of symmetry: $x = -4.58$; minimum: 65π **65.** TW

67.

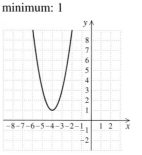

68.

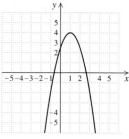

69. (−5, −1)   **70.** (−1, 2)   **71.** $x^2 + 5x + \tfrac{25}{4} = \left(x + \tfrac{5}{2}\right)^2$
72. $x^2 - 9x + \tfrac{81}{4} = \left(x - \tfrac{9}{2}\right)^2$ **73.** TW
75. $f(x) = \tfrac{3}{5}(x - 4)^2 + 1$ **77.** $f(x) = \tfrac{3}{5}(x - 3)^2 - 1$
79. $f(x) = \tfrac{3}{5}(x + 2)^2 - 5$ **81.** $f(x) = 2(x - 2)^2$
83. $g(x) = -2x^2 + 3$ **85.** $F(x) = 3(x - 5)^2 + 1$
87.

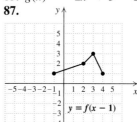

$$y = f(x - 1)$$

89.

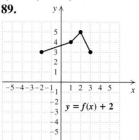

$$y = f(x) + 2$$

91.

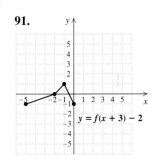

$y = f(x + 3) - 2$

Visualizing the Graph, p. 859

1. B   **2.** E   **3.** A   **4.** H   **5.** C   **6.** J   **7.** F
8. G   **9.** I   **10.** D

Exercise Set 11.7, pp. 860–861

1. 9   **2.** 16   **3.** 9   **4.** 1   **5.** 3   **6.** 1   **7.** $\frac{5}{2}$; (-4)
8. (-2); $\left(-2, \frac{7}{2}\right)$ **9. (a)** $f(x) = (x - 2)^2 - 1$;
(b) vertex: $(2, -1)$; axis of symmetry: $x = 2$
11. (a) $f(x) = -\left(x - \frac{3}{2}\right)^2 - \frac{31}{4}$; **(b)** vertex: $\left(\frac{3}{2}, -\frac{31}{4}\right)$; axis of
symmetry: $x = \frac{3}{2}$ **13. (a)** $f(x) = 2\left(x - \frac{7}{4}\right)^2 - \frac{41}{8}$;
(b) vertex: $\left(\frac{7}{4}, -\frac{41}{8}\right)$; axis of symmetry: $x = \frac{7}{4}$
15. (a) Vertex: $(-2, 1)$; axis of symmetry: $x = -2$;
(b)

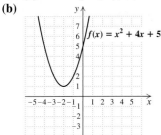

$f(x) = x^2 + 4x + 5$

17. (a) Vertex: $(3, 4)$; axis of symmetry: $x = 3$;
(b)

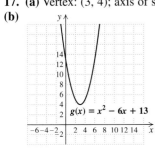

$g(x) = x^2 - 6x + 13$

19. (a) Vertex: $(-4, 4)$; axis of symmetry: $x = -4$;
(b)

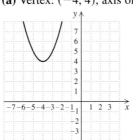

$f(x) = x^2 + 8x + 20$

21. (a) Vertex: $(4, -7)$; axis of symmetry: $x = 4$;
(b)

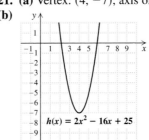

$h(x) = 2x^2 - 16x + 25$

23. (a) Vertex: $(1, 6)$; axis of symmetry: $x = 1$;
(b)

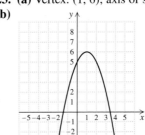

$f(x) = -x^2 + 2x + 5$

25. (a) Vertex: $\left(-\frac{3}{2}, -\frac{49}{4}\right)$; axis of symmetry: $x = -\frac{3}{2}$;
(b)

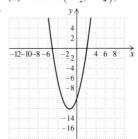

$g(x) = x^2 + 3x - 10$

27. (a) Vertex: (4, 2); axis of symmetry: $x = 4$;
(b)

$$f(x) = 3x^2 - 24x + 50$$

29. (a) Vertex: $\left(-\frac{7}{2}, -\frac{49}{4}\right)$; axis of symmetry: $x = -\frac{7}{2}$;
(b)

$$h(x) = x^2 + 7x$$

31. (a) Vertex: $(-1, -4)$; axis of symmetry: $x = -1$;
(b)

$$f(x) = -2x^2 - 4x - 6$$

33. (a) Vertex: $(2, -5)$; axis of symmetry: $x = 2$;
(b)

$$g(x) = 2x^2 - 8x + 3$$

35. (a) Vertex: $\left(\frac{5}{6}, \frac{1}{12}\right)$; axis of symmetry: $x = \frac{5}{6}$;
(b)

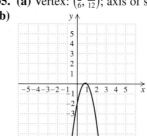

$$f(x) = -3x^2 + 5x - 2$$

37. (a) Vertex: $\left(-4, -\frac{5}{3}\right)$; axis of symmetry: $x = -4$;
(b)

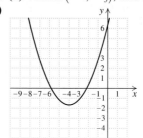

$$h(x) = \frac{1}{2}x^2 + 4x + \frac{19}{3}$$

39. $(-0.5, -6.25)$ **41.** $(0.1, 0.95)$ **43.** $(3.5, -4.25)$
45. $\left(3 - \sqrt{6}, 0\right), \left(3 + \sqrt{6}, 0\right)$; $(0, 3)$ **47.** $(-1, 0), (3, 0)$;
$(0, 3)$ **49.** $(0, 0), (9, 0)$; $(0, 0)$ **51.** $(2, 0)$; $(0, -4)$
53. No x-intercept; $(0, 6)$ **55. (a)** Minimum: -6.95;
(b) $(-1.06, 0), (2.41, 0)$; $(0, -5.89)$
57. (a) Maximum: -0.45; **(b)** no x-intercept; $(0, -2.79)$
59. TW
61.

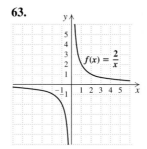

$$f(x) = \frac{3}{2}x$$

62.

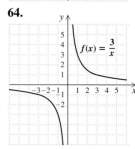

$$f(x) = -\frac{2}{3}x$$

63.

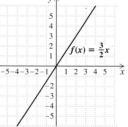

$$f(x) = \frac{2}{x}$$

64.

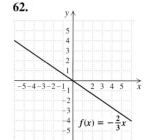

$$f(x) = \frac{3}{x}$$

65.

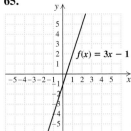

$f(x) = 3x - 1$

66.

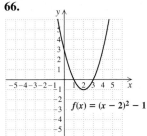

$f(x) = (x - 2)^2 - 1$

67.

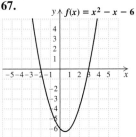

$f(x) = x^2 - x - 6$

68.

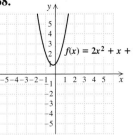

$f(x) = 2x^2 + x + 1$

69. ᵀᵂ   **71. (a)** $-2.4, 3.4$; **(b)** $-1.3, 2.3$

73. $f(x) = m\left(x - \dfrac{n}{2m}\right)^2 + \dfrac{4mp - n^2}{4m}$

75. $f(x) = \frac{5}{16}x^2 - \frac{15}{8}x - \frac{35}{16}$, or $f(x) = \frac{5}{16}(x - 3)^2 - 5$

77.

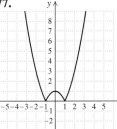

$f(x) = |x^2 - 1|$

79.

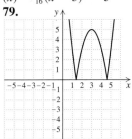

$f(x) = |2(x - 3)^2 - 5|$

Exercise Set 11.8, pp. 870–876

1. (e)   **2.** (b)   **3.** (c)   **4.** (a)   **5.** (d)   **6.** (f)
7. 11 days after the concert was announced; about 62 tickets
9. \$120/Dobro; 350 Dobros   **11.** 32 in. by 32 in.
13. 450 ft²; 15 ft by 30 ft (The house serves as a 30-ft side.)
15. 3.5 in.   **17.** 81; 9 and 9   **19.** -16; 4 and -4
21. 25; -5 and -5 **23.** $f(x) = mx + b$
25. $f(x) = ax^2 + bx + c, a < 0$
27. $f(x) = ax^2 + bx + c, a > 0$
29. $f(x) = ax^2 + bx + c, a > 0$
31. Neither quadratic nor linear
33. $f(x) = 2x^2 + 3x - 1$ **35.** $f(x) = -\frac{1}{4}x^2 + 3x - 5$
37. (a) $A(s) = \frac{3}{16}s^2 - \frac{135}{4}s + 1750$; **(b)** about 531 accidents
39. $h(d) = -0.0068d^2 + 0.8571d$
41. (a) $D(x) = -0.0082833093x^2 + 0.8242996891x + 0.2121786608$; **(b)** 17.325 ft

43. (a) $v(x) = 9982.696429x^2 - 18{,}401.85357x + 18{,}864.75$; **(b)** 510,542 vehicles   **45.** ᵀᵂ
47. $\dfrac{x - 9}{(x + 9)(x + 7)}$ **48.** $\dfrac{(x - 3)(x + 1)}{(x - 7)(x + 3)}$
49. $\{x \mid x < 8\}$, or $(-\infty, 8)$ **50.** $\{x \mid x \geq 10\}$, or $[10, \infty)$
51. $r(x) = 1454.8125x + 17{,}685.75$
52. $r(x) = 1473.553571x + 17{,}775.5$ **53.** ᵀᵂ
55. 158 ft   **57.** The radius of the circular portion of the window and the height of the rectangular portion should each be $\dfrac{24}{\pi + 4}$ ft.   **59.** \$15

Review Exercises: Chapter 11, pp. 879–882

1. False   **2.** True   **3.** True   **4.** True   **5.** True
6. False   **7.** True   **8.** True   **9.** True   **10.** False
11. (a) 2; **(b)** positive; **(c)** -3 **12.** $\pm\frac{3}{2}$ **13.** $0, -\frac{3}{4}$
14. 3, 9   **15.** $2 \pm 2i$ **16.** 3, 5   **17.** $-\dfrac{9}{2} \pm \dfrac{\sqrt{85}}{2}$
18. $-0.3722813233, 5.3722813233$ **19.** $-\frac{1}{4}, 1$
20. $x^2 - 12x + 36 = (x - 6)^2$
21. $x^2 + \frac{3}{5}x + \frac{9}{100} = \left(x + \frac{3}{10}\right)^2$ **22.** $3 \pm 2\sqrt{2}$
23. 10%   **24.** 6.7 sec   **25.** About 153 mph   **26.** 6 hr
27. Two irrational   **28.** Two imaginary   **29.** $x^2 - 5 = 0$
30. $x^2 + 8x + 16 = 0$ **31.** $(-3, 0), (-2, 0), (2, 0), (3, 0)$
32. $-5, 3$ **33.** $\pm\sqrt{2}, \pm\sqrt{7}$
34.

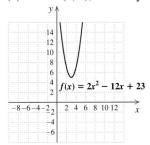

$f(x) = -3(x + 2)^2 + 4$
Maximum: 4

35. (a) Vertex: $(3, 5)$; axis of symmetry: $x = 3$;
(b)

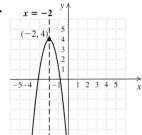

$f(x) = 2x^2 - 12x + 23$

36. $(2, 0), (7, 0); (0, 14)$ **37.** $p = \dfrac{9\pi^2}{N^2}$
38. $T = \dfrac{1 \pm \sqrt{1 + 24A}}{6}$ **39.** Neither quadratic nor linear
40. $f(x) = ax^2 + bx + c, a > 0$ **41.** $f(x) = mx + b$

42. 225 ft²; 15 ft by 15 ft
43. (a) $M(x) = \frac{4063}{360}x^2 - \frac{31,639}{180}x + 1$; (b) 32,487 restaurants
44. (a) $M(x) = 14.14783742x^2 - 298.8611333x + 995.3006199$; (b) 36,850 restaurants
45. 🆃🆆 Completing the square was used to solve quadratic equations and to graph quadratic functions by rewriting the function in the form $f(x) = a(x - h)^2 + k$.
46. 🆃🆆 The model found in Exercise 44 predicts 4363 more restaurants in 2010 than the model from Exercise 43. The higher prediction seems to fit the pattern better. Since the function in Exercise 44 considers all the data, we would expect it to be a better model.
47. $f(x) = \frac{7}{15}x^2 - \frac{14}{15}x - 7$ **48.** $h = 60, k = 60$
49. 18, 324   **50.** (a) $S(x) = 148.1428571x^2 - 15,211.71429x + 393,615.8571$; (b) more than 68 yr after 1948, or for years after 2016

Test: Chapter 11, pp. 882–883

1. (a) [11.1] 0; (b) [11.6] negative; (c) [11.6] −1
2. [11.1] $\pm\frac{4\sqrt{3}}{3}$ **3.** [11.2] 2, 9   **4.** [11.2] $\frac{-1 \pm i\sqrt{3}}{2}$
5. [11.2] $1 \pm \sqrt{6}$ **6.** [11.5] $-2, \frac{2}{3}$
7. [11.2] $-4.192582404, 1.192582404$ **8.** [11.2] $-\frac{3}{4}, \frac{7}{3}$
9. [11.1] $x^2 - 16x + 64 = (x - 8)^2$
10. [11.1] $x^2 + \frac{2}{7}x + \frac{1}{49} = \left(x + \frac{1}{7}\right)^2$ **11.** [11.1] $-5 \pm \sqrt{10}$
12. [11.3] 16 km/h   **13.** [11.3] 2 hr
14. [11.4] Two imaginary   **15.** [11.4] $3x^2 + 5x - 2 = 0$
16. [11.5] $(-3, 0), (-1, 0), \left(-2 - \sqrt{5}, 0\right), \left(-2 + \sqrt{5}, 0\right)$
17. [11.6]

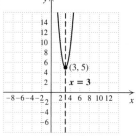

$f(x) = 4(x - 3)^2 + 5$
Minimum: 5
18. [11.7] (a) $(-1, -8), x = -1$;
(b)

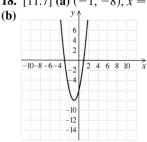

$f(x) = 2x^2 + 4x - 6$

19. [11.7] $(-2, 0), (3, 0); (0, -6)$
20. [11.3] $r = \sqrt{\dfrac{3V}{\pi} - R^2}$ **21.** [11.8] Quadratic; the data approximate a parabola opening downward.
22. [11.8] Minimum \$129/cap when 325 caps are built
23. [11.8] $p(x) = -\frac{165}{8}x^2 + \frac{605}{4}x + 35$
24. [11.8] $p(x) = -23.54166667x^2 + 162.2583333x + 57.61666667$ **25.** [11.4] $\frac{1}{2}$
26. [11.4] $x^4 - 14x^3 + 67x^2 - 114x + 26 = 0$; answers may vary.   **27.** [11.4] $x^6 - 10x^5 + 20x^4 + 50x^3 - 119x^2 - 60x + 150 = 0$; answers may vary

Chapter 12

Exercise Set 12.1, pp. 897–901

1. True   **2.** True   **3.** False   **4.** True   **5.** False
6. False   **7.** True   **8.** True
9. $(f \circ g)(1) = 2; (g \circ f)(1) = 1$;
$(f \circ g)(x) = 4x^2 - 12x + 10; (g \circ f)(x) = 2x^2 - 1$
11. $(f \circ g)(1) = -8; (g \circ f)(1) = 1$;
$(f \circ g)(x) = 2x^2 - 10; (g \circ f)(x) = 2x^2 - 12x + 11$
13. $(f \circ g)(1) = 8; (g \circ f)(1) = \frac{1}{64}$;
$(f \circ g)(x) = \frac{1}{x^2} + 7; (g \circ f)(x) = \frac{1}{(x + 7)^2}$
15. $(f \circ g)(1) = 2; (g \circ f)(1) = 4$;
$(f \circ g)(x) = \sqrt{x + 3}; (g \circ f)(x) = \sqrt{x} + 3$
17. $(f \circ g)(1) = 2; (g \circ f)(1) = \frac{1}{2}; (f \circ g)(x) = \sqrt{\dfrac{4}{x}}$;
$(g \circ f)(x) = \dfrac{1}{\sqrt{4x}}$ **19.** $(f \circ g)(1) = 4; (g \circ f)(1) = 2$;
$(f \circ g)(x) = x + 3; (g \circ f)(x) = \sqrt{x^2 + 3}$ **21.** 8
23. -4 **25.** Not defined   **27.** 4   **29.** Not defined
31. $f(x) = x^2; g(x) = 7 + 5x$ **33.** $f(x) = \sqrt{x}$;
$g(x) = 2x + 7$ **35.** $f(x) = \dfrac{2}{x}; g(x) = x - 3$
37. $f(x) = \dfrac{1}{\sqrt{x}}; g(x) = 7x + 2$ **39.** $f(x) = \dfrac{1}{x} + x$;
$g(x) = \sqrt{3x}$ **41.** Yes   **43.** No   **45.** Yes   **47.** No
49. (a) Yes; (b) $f^{-1}(x) = x - 4$ **51.** (a) Yes;
(b) $f^{-1}(x) = \dfrac{x}{2}$ **53.** (a) Yes; (b) $g^{-1}(x) = \dfrac{x + 1}{3}$
55. (a) Yes; (b) $f^{-1}(x) = 2x - 2$ **57.** (a) No
59. (a) Yes; (b) $h^{-1}(x) = \dfrac{x - 4}{-2}$
61. (a) Yes; (b) $f^{-1}(x) = \dfrac{1}{x}$ **63.** (a) No
65. (a) Yes; (b) $f^{-1}(x) = \dfrac{3x - 1}{2}$

67. (a) Yes; **(b)** $f^{-1}(x) = \sqrt[3]{x+5}$ **69. (a)** Yes;
(b) $g^{-1}(x) = \sqrt[3]{x} + 2$ **71. (a)** Yes; **(b)** $f^{-1}(x) = x^2, x \ge 0$
73.   **75.**

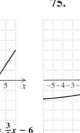

77.   **79.**

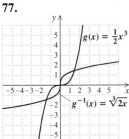
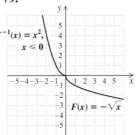

81.

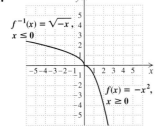

83. (1) $(f^{-1} \circ f)(x) = f^{-1}(f(x))$
$= f^{-1}(\sqrt[3]{x-4}) = (\sqrt[3]{x-4})^3 + 4$
$= x - 4 + 4 = x;$
(2) $(f \circ f^{-1})(x) = f(f^{-1}(x))$
$= f(x^3 + 4) = \sqrt[3]{x^3 + 4 - 4}$
$= \sqrt[3]{x^3} = x$
85. (1) $(f^{-1} \circ f)(x) = f^{-1}(f(x)) = f^{-1}\left(\dfrac{1-x}{x}\right)$
$= \dfrac{1}{\left(\dfrac{1-x}{x}\right) + 1}$
$= \dfrac{1}{\dfrac{1-x+x}{x}}$
$= x;$

(2) $(f \circ f^{-1})(x) = f(f^{-1}(x)) = f\left(\dfrac{1}{x+1}\right)$

$= \dfrac{1 - \left(\dfrac{1}{x+1}\right)}{\left(\dfrac{1}{x+1}\right)}$

$= \dfrac{\dfrac{x+1-1}{x+1}}{\dfrac{1}{x+1}} = x$

87. No   **89.** Yes   **91.** (1) C; (2) D; (3) B; (4) A
93. (a) 40, 42, 46, 50; **(b)** yes; $f^{-1}(x) = x - 32$;
(c) 8, 10, 14, 18   **95. TW**   **97.** $6 + \sqrt{3}$
98. Not defined   **99.** $x^2 + 2x + \sqrt{x}$ **100.** $\dfrac{2x + \sqrt{x}}{x^2}$
101. $\{x \mid x \ge 0\}$, or $[0, \infty)$ **102.** $\mathbb{R}$ **103.** $\{x \mid x \ge 0\}$, or
$[0, \infty)$ **104.** $\{x \mid x > 0\}$, or $(0, \infty)$ **105.** $2x^2 + |x|$
106. $4x^2 + 4x\sqrt{x} + x$ **107. TW**

109.   **111.** $g(x) = \dfrac{x}{2} + 20$

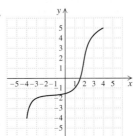

113. TW
115. Suppose that $h(x) = (f \circ g)(x)$. First, note that for
$I(x) = x$, $(f \circ I)(x) = f(I(x)) = f(x)$ for any function f.
(i) $((g^{-1} \circ f^{-1}) \circ h)(x) = ((g^{-1} \circ f^{-1}) \circ (f \circ g))(x)$
$= ((g^{-1} \circ (f^{-1} \circ f)) \circ g)(x)$
$= ((g^{-1} \circ I) \circ g)(x)$
$= (g^{-1} \circ g)(x) = x$
(ii) $(h \circ (g^{-1} \circ f^{-1}))(x) = ((f \circ g) \circ (g^{-1} \circ f^{-1}))(x)$
$= ((f \circ (g \circ g^{-1})) \circ f^{-1})(x)$
$= ((f \circ I) \circ f^{-1})(x)$
$= (f \circ f^{-1})(x) = x.$
Therefore, $(g^{-1} \circ f^{-1})(x) = h^{-1}(x)$.
117. TW   **119.** The cost of mailing n copies of the book
121. 22 mm   **123.** 15 L/min
125. (a) $h(t) = -136t + 6322$; **(b)** $r(t) = 0.172t + 3.626$;
(c) $H(r) = -776.5100671r + 9132.744966$; **(d)** $y_1 \approx y_4$;
$h(t) = (H \circ r)(t)$

Interactive Discovery, p. 905

1. (a) Increases; **(b)** increases; **(c)** increases; **(d)** decreases;
(e) decreases   **2.** If $a > 1$, the graph increases; if
$0 < a < 1$, the graph decreases.   **3.** g **4.** r

Exercise Set 12.2, pp. 909–912

1. True **2.** True **3.** True **4.** False **5.** False
6. True **7.** $a > 1$ **9.** $0 < a < 1$
11.

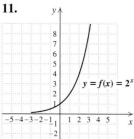

13.

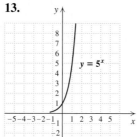

15.

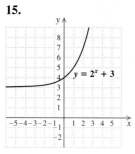

17.

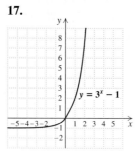

19.

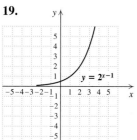

21.

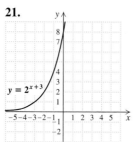

23.

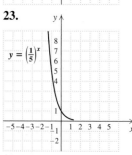

25.

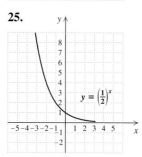

27.

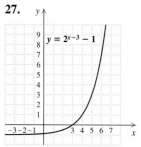

29.

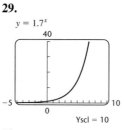

31.

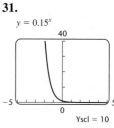

33.

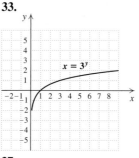

35.

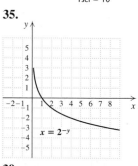

37.

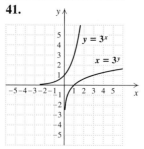

39.

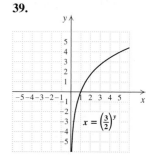

41.

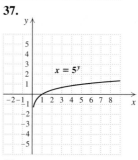

43.

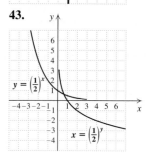

45. (d) **47.** (f) **49.** (c)
51. (a) About 6.8 billion; about 7.2 billion; about 7.7 billion;
(b)

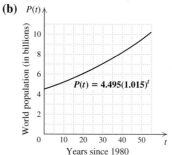

53. (a) 19.6%; 16.3%; 7.3%;
(b)

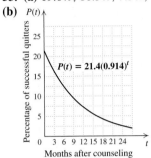

55. (a) About 44,079 whales; about 12,953 whales;
(b)

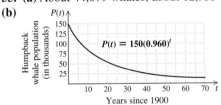

57. (a) About 11,900 whales; about 35,000 whales;
(b)

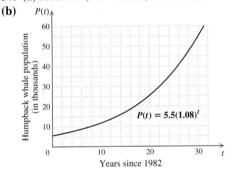

59. (a) 454,354,240 cm²; 525,233,501,440 cm²;
(b)

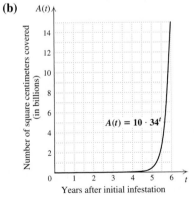

61. TW   **63.** $\frac{1}{25}$ **64.** $\frac{1}{32}$ **65.** 100   **66.** $\frac{1}{125}$
67. $5a^6b^3$ **68.** $6x^4y$ **69.** TW   **71.** $\pi^{2.4}$

73.

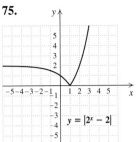

75.

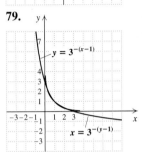

77.

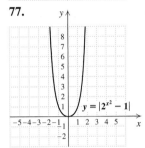

79.

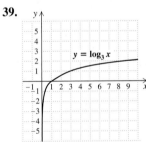

81. $m(x) = 69.67150549(1.969627617)^x$; \$15,781 million

Exercise Set 12.3, pp. 920–921

1. (g)   **2.** (d)   **3.** (a)   **4.** (h)   **5.** (b)   **6.** (c)
7. (e)   **8.** (f)   **9.** 3   **11.** 4   **13.** 4   **15.** −2
17. −1   **19.** 4   **21.** 1   **23.** 0   **25.** 5   **27.** −2
29. $\frac{1}{2}$ **31.** $\frac{3}{2}$ **33.** $\frac{2}{3}$ **35.** 7
37.

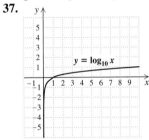

39.

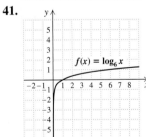

41.

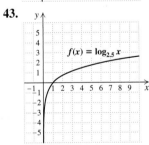

43.

45.

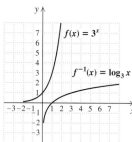

47. 0.6021   **49.** 4.1271

51. -0.2782 **53.** 199.5262   **55.** 0.0011   **57.** 1.0028
59. $y = \log(x + 2)$

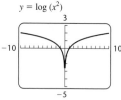

61. $y = \log(1 - 2x)$

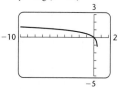

63. $y = \log(x^2)$

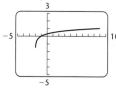

65. $5^t = 9$ **67.** $5^2 = 25$

69. $10^{-1} = 0.1$ **71.** $10^{0.845} = 7$ **73.** $c^8 = m$
75. $t^r = Q$ **77.** $e^{-1.3863} = 0.25$ **79.** $r^{-x} = T$
81. $2 = \log_{10} 100$ **83.** $-5 = \log_4 \frac{1}{1024}$ **85.** $\frac{3}{4} = \log_{16} 8$
87. $0.4771 = \log_{10} 3$ **89.** $m = \log_z 6$ **91.** $m = \log_p V$
93. $3 = \log_e 20.0855$ **95.** $-4 = \log_e 0.0183$ **97.** 9
99. 3   **101.** 4   **103.** 7   **105.** $\frac{1}{9}$ **107.** 4   **109.** TW
111. x^8 **112.** a^{12} **113.** $a^7 b^8$ **114.** $x^5 y^{12}$
115. $\dfrac{x(3y - 2)}{2y + x}$ **116.** $\dfrac{x + 2}{x + 1}$ **117.** TW
119.

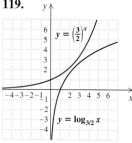

121.

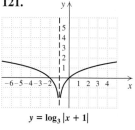

123. 6   **125.** $-25, 4$ **127.** -2 **129.** 0
131. Let $b = 0$, and suppose that $x_1 = 1$ and $x_2 = 2$. Then
$0^1 = 0^2$, but $1 \neq 2$. Then let $b = 1$, and suppose that $x_1 = 1$
and $x_2 = 2$. Then $1^1 = 1^2$, but $1 \neq 2$.

Interactive Discovery, p. 922

1. (c)   **2.** (b)   **3.** (c)   **4.** (a)

Exercise Set 12.4, pp. 927–929

1. (e)   **2.** (f)   **3.** (a)   **4.** (b)   **5.** (c)   **6.** (d)
7. $\log_3 81 + \log_3 27$ **9.** $\log_4 64 + \log_4 16$
11. $\log_c r + \log_c s + \log_c t$ **13.** $\log_a (5 \cdot 14)$, or $\log_a 70$
15. $\log_c (t \cdot y)$ **17.** $8 \log_a r$ **19.** $6 \log_c y$
21. $-3 \log_b C$ **23.** $\log_2 25 - \log_2 13$
25. $\log_b m - \log_b n$ **27.** $\log_a \frac{17}{6}$ **29.** $\log_b \frac{36}{4}$, or $\log_b 9$
31. $\log_a \frac{7}{18}$ **33.** $\log_a x + \log_a y + \log_a z$
35. $3 \log_a x + 4 \log_a z$ **37.** $2 \log_a x - 2 \log_a y + \log_a z$
39. $4 \log_a x - 3 \log_a y - \log_a z$
41. $\log_b x + 2 \log_b y - \log_b w - 3 \log_b z$
43. $\frac{1}{2}(7 \log_a x - 5 \log_a y - 8 \log_a z)$
45. $\frac{1}{3}(6 \log_a x + 3 \log_a y - 2 - 7 \log_a z)$ **47.** $\log_a (x^8 z^3)$
49. $\log_a x$ **51.** $\log_a \dfrac{y^5}{x^{3/2}}$ **53.** $\log_a (x - 2)$ **55.** 1.953
57. -0.369 **59.** -1.161 **61.** $\frac{3}{2}$ **63.** Cannot be found
65. 7   **67.** m **69.** 7   **71.** 3   **73.** TW
75.

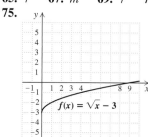

76.

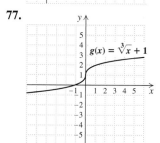

77.

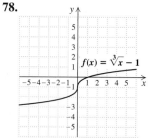

78.

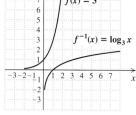

79. $a^{17} b^{17}$ **80.** $x^{11} y^6 z^8$ **81.** TW
83. $\log_a (x^6 - x^4 y^2 + x^2 y^4 - y^6)$
85. $\frac{1}{2} \log_a (1 - s) + \frac{1}{2} \log_a (1 + s)$ **87.** $\frac{10}{3}$ **89.** -2
91. True

Interactive Discovery, p. 930

1. $\$2.25$; $\$2.370370$; $\$2.441406$; $\$2.613035$; $\$2.692597$;
$\$2.714567$; $\$2.718127$ **2.** (c)

Visualizing the Graph, p. 935

1. J   **2.** D   **3.** B   **4.** G   **5.** H   **6.** C   **7.** F
8. I   **9.** E   **10.** A

Exercise Set 12.5, pp. 936–937

1. True   **2.** True   **3.** True   **4.** False   **5.** True
6. True   **7.** True   **8.** True   **9.** 0.7782   **11.** 1.8621
13. 3   **15.** −0.2782   **17.** 1.7986   **19.** 199.5262
21. 1.4894   **23.** 0.0011   **25.** 1.6094   **27.** 4.0431
29. −5.0832   **31.** 96.7583   **33.** 15.0293   **35.** 0.0305
37. 109.9472   **39.** 2.5237   **41.** 6.6439   **43.** 2.1452
45. −2.3219   **47.** −2.3219   **49.** 3.5471

51. Domain: $\mathbb{R}$; range: $(0, \infty)$

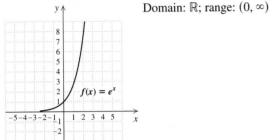

53. Domain: $\mathbb{R}$; range: $(3, \infty)$

55. Domain: $\mathbb{R}$; range: $(-2, \infty)$

57. Domain: $\mathbb{R}$; range: $(0, \infty)$

59. Domain: $\mathbb{R}$; range: $(0, \infty)$

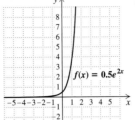

61. Domain: $\mathbb{R}$; range: $(0, \infty)$

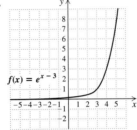

63. Domain: $\mathbb{R}$; range: $(0, \infty)$

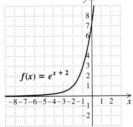

65. Domain: $\mathbb{R}$; range: $(-\infty, 0)$

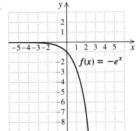

67. Domain: $(0, \infty)$; range: $\mathbb{R}$

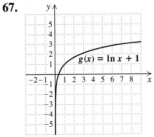

69. 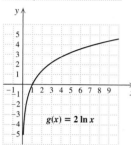 Domain: $(0, \infty)$; range: $\mathbb{R}$

71. 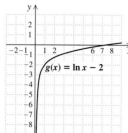 Domain: $(0, \infty)$; range: $\mathbb{R}$

73. Domain: $(0, \infty)$; range: $\mathbb{R}$

75. Domain: $(-2, \infty)$; range: $\mathbb{R}$

77. Domain: $(1, \infty)$; range: $\mathbb{R}$

79. $f(x) = \log(x)/\log(5)$, or $f(x) = \ln(x)/\ln(5)$

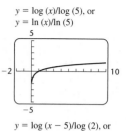

$y = \log(x)/\log(5)$, or $y = \ln(x)/\ln(5)$

81. $f(x) = \log(x-5)/\log(2)$, or $f(x) = \ln(x-5)/\ln(2)$

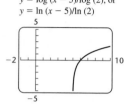

$y = \log(x-5)/\log(2)$, or $y = \ln(x-5)/\ln(2)$

83. $f(x) = \log(x)/\log(3) + x$, or $f(x) = \ln(x)/\ln(3) + x$

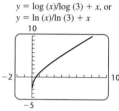

$y = \log(x)/\log(3) + x$, or $y = \ln(x)/\ln(3) + x$

85. TW   **87.** $y = 9x$ **88.** $y = \dfrac{21.35}{x}$ **89.** $L = \dfrac{8T^2}{\pi^2}$

90. $c = \sqrt{\dfrac{E}{m}}$ **91.** $1\frac{1}{5}$ hr

92. $9\frac{3}{8}$ min   **93.** TW

95. 2.452   **97.** 1.442   **99.** $\log M = \dfrac{\ln M}{\ln 10}$

101. 1086.5129   **103.** 4.9855

105. (a) Domain: $\{x \mid x > 0\}$, or $(0, \infty)$; range: $\{y \mid y < 0.5135\}$, or $(-\infty, 0.5135)$; **(b)** $[-1, 5, -10, 5]$;

(c) $y = 3.4 \ln x - 0.25e^x$

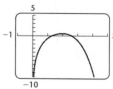

107. (a) Domain: $\{x \mid x > 0\}$, or $(0, \infty)$; range: $\{y \mid y > -0.2453\}$, or $(-0.2453, \infty)$; **(b)** $[-1, 5, -1, 10]$;

(c) $y = 2x^3 \ln x$

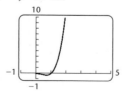

Exercise Set 12.6, pp. 944–946

1. (e)　**2.** (a)　**3.** (f)　**4.** (h)　**5.** (b)　**6.** (d)

7. (g)　**8.** (c)　**9.** $\dfrac{\log 19}{\log 2} \approx 4.248$

11. $\dfrac{\log 17}{\log 8} + 1 \approx 2.362$　**13.** $\ln 1000 \approx 6.908$

15. $\dfrac{\ln 5}{0.03} \approx 53.648$　**17.** $\dfrac{\log 5}{\log 3} - 1 \approx 0.465$　**19.** 1

21. $\dfrac{\log 87}{\log 4.9} \approx 2.810$　**23.** $\dfrac{\ln\left(\frac{19}{2}\right)}{4} \approx 0.563$

25. $\dfrac{\ln 2}{5} \approx 0.139$　**27.** 81　**29.** $\frac{1}{8}$　**31.** $e^5 \approx 148.413$

33. 2　**35.** $\dfrac{e^3}{4} \approx 5.021$　**37.** $10^{2.5} \approx 316.228$

39. $\dfrac{e^4 - 1}{2} \approx 26.799$　**41.** $e \approx 2.718$　**43.** $e^{-3} \approx 0.050$

45. -4　**47.** 10　**49.** No solution　**51.** $\frac{83}{15}$　**53.** 1
55. 6　**57.** 1　**59.** 5　**61.** $\frac{17}{2}$　**63.** 4
65. $-6.480, 6.519$　**67.** $0.000112, 3.445$　**69.** 1
71. 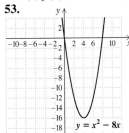　**73.** $-\frac{5}{2}, \frac{5}{2}$　**74.** $0, \frac{7}{5}$　**75.** $\frac{4}{3}, 4$　**76.** 0

77. 16, 256　**78.** 5　**79.** $\dfrac{-3}{2} \pm \dfrac{\sqrt{11}}{2} i$　**80.** $\frac{5}{11}$

81. $\ln 1.5 \approx 0.405$　**82.** 5　**83.** $-4, 1$　**84.** No solution
85. $(2, -2)$　**86.** $(3, 2, 1)$　**87.** 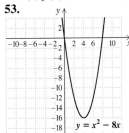　**89.** $\frac{12}{5}$　**91.** $\sqrt[3]{3}$
93. -1　**95.** $-3, -1$　**97.** $-625, 625$　**99.** $\frac{1}{2}, 5000$
101. $-3, -1$　**103.** $\frac{1}{100,000}, 100,000$　**105.** $-\frac{1}{3}$　**107.** 38

Exercise Set 12.7, pp. 958–965

1. (a) About 2004; (b) 1.6 yr　**3.** (a) 119,157 skateboarders;
(b) 49　**5.** (a) 6.4 yr; (b) 23.4 yr　**7.** (a) About 2005;
(b) about 2018　**9.** (a) About 2008; (b) 9.0 yr　**11.** 4.9
13. 10^{-7} moles per liter　**15.** 65 dB　**17.** $10^{-1.5}$ W/m²
19. (a) -26.9; (b) 1.58×10^{-17} W/m²
21. (a) $P(t) = P_0 e^{0.025t}$; (b) \$5126.58; \$5256.36; (c) 27.7 yr
23. (a) $P(t) = 300 e^{0.009t}$, where t is the number of years after
2006 and $P(t)$ is in millions; (b) 311 million;
(c) about 2015　**25.** 6.2 months
27. (a) About 2010; (b) about 2019;
(c)

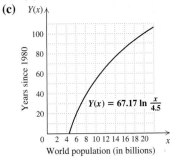

29. (a) 68%; (b) 54%, 40%

(c)

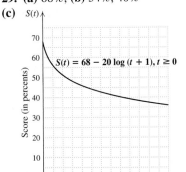

(d) 6.9 months

31. (a) $P(t) = 1.7e^{0.622t}$, where t is the number of years since
2002 and $P(t)$ is in millions; (b) 38.1 million units
33. (a) $k \approx 0.004$; $P(t) = 987e^{-0.004t}$, where t is the number
of years after 1990 and $P(t)$ is in millions;
(b) 918 million acres; (c) about 2043　**35.** About 2103 yr
37. About 7.2 days　**39.** 69.3% per year
41. (a) $k \approx 0.099$; $V(t) = 451{,}000 e^{0.099t}$, where t is the
number of years after 1991; (b) about \$1.99 million;
(c) about 7.0 yr; (d) about 2010　**43.** No　**45.** Yes
47. (a) $p(x) = 7.019404696(1.099514502)^x$; (b) 0.095, or
9.5%; (c) about \$121
49. (a) $f(x) = 2097152(0.8705505633)^x$; (b) 4 hr　**51.**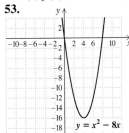
53.

54.

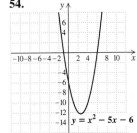

55.

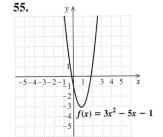

56.

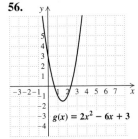

57. $4 \pm \sqrt{23}$　**58.** $-5 \pm \sqrt{31}$　**59.**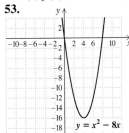
61. \$18.9 million　**63.** $P(t) = 100 - 63.03(0.95)^t$
65. About 80,922 yr, or with rounding of k, about 80,792 yr

67. Consider an exponential growth function $P(t) = P_0 e^{kt}$. At time T, $P(T) = 2P_0$.
Solve for T:
$$2P_0 = P_0 e^{kT}$$
$$2 = e^{kT}$$
$$\ln 2 = kT$$
$$\frac{\ln 2}{k} = T.$$

69. (a)

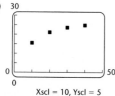

Xscl = 10, Yscl = 5

(b) $s(t) = -0.0300167067 + 6.879024467 \ln t$;
(c) about 55.5 min

71. (a)

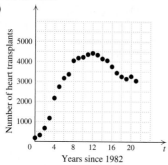

Review Exercises: Chapter 12, pp. 969–971

1. True   **2.** True   **3.** True   **4.** False   **5.** False
6. True   **7.** False   **8.** False   **9.** True   **10.** False
11. $(f \circ g)(x) = 4x^2 - 12x + 10$; $(g \circ f)(x) = 2x^2 - 1$
12. $f(x) = \sqrt{x}$; $g(x) = 3 - x$ **13.** No
14. $f^{-1}(x) = x + 8$ **15.** $g^{-1}(x) = \dfrac{2x - 1}{3}$
16. $f^{-1}(x) = \dfrac{\sqrt[3]{x}}{3}$

17.

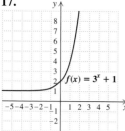

$f(x) = 3^x + 1$

18.

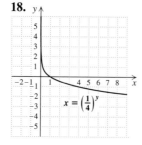

$x = \left(\dfrac{1}{4}\right)^y$

19.

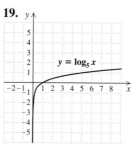

$y = \log_5 x$

20. 2   **21.** -2 **22.** 7   **23.** $\frac{1}{2}$ **24.** $\log_{10} \frac{1}{100} = -2$
25. $\log_{25} 5 = \frac{1}{2}$ **26.** $16 = 4^x$ **27.** $1 = 8^0$
28. $4 \log_a x + 2 \log_a y + 3 \log_a z$
29. $5 \log_a x - (\log_a y + 2 \log_a z)$, or
$5 \log_a x - \log_a y - 2 \log_a z$
30. $\frac{1}{4}(2 \log z - 3 \log x - \log y)$
31. $\log_a (7 \cdot 8)$, or $\log_a 56$ **32.** $\log_a \frac{72}{12}$, or $\log_a 6$
33. $\log \dfrac{a^{1/2}}{bc^2}$ **34.** $\log_a \sqrt[3]{\dfrac{x}{y^2}}$ **35.** 1   **36.** 0
37. 17   **38.** 6.93   **39.** -3.2698 **40.** 8.7601
41. 3.2698   **42.** 2.54995   **43.** -3.6602 **44.** 1.8751
45. 61.5177   **46.** -2.9957 **47.** 0.3753   **48.** 0.4307
49. 1.7097
50.

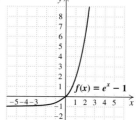

$f(x) = e^x - 1$

Domain: $\mathbb{R}$; range: $(-1, \infty)$

51.

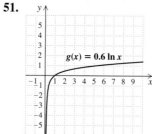

$g(x) = 0.6 \ln x$

Domain: $(0, \infty)$; range: $\mathbb{R}$

52. 5   **53.** -2 **54.** $\frac{1}{81}$ **55.** 2   **56.** $\frac{1}{1000}$
57. $e^{-2} \approx 0.1353$ **58.** $\dfrac{1}{2}\left(\dfrac{\log 19}{\log 4} + 5\right) \approx 3.5620$
59. $-5, 1$ **60.** $\dfrac{\log 8.3}{\log 4} \approx 1.5266$ **61.** $\dfrac{\ln 0.03}{-0.1} \approx 35.0656$
62. $\dfrac{\ln 2}{2} \approx 0.3466$ **63.** 4   **64.** 8   **65.** 20   **66.** $\sqrt{43}$
67. **(a)** 82; **(b)** 66.8; **(c)** 35 months   **68.** **(a)** 6.6 yr;
(b) 3.1 yr   **69.** **(a)** $k \approx 0.253$, $P(t) = 107.8 e^{0.253t}$;
(b) about 2.2 billion phones; **(c)** about 2007

70. (a) $f(x) = 35380.82167(1.139052882)^x$;
(b) about \$3,840,147; **(c)** about 34%   **71.** 23.105% per year
72. 16.5 yr   **73.** 3463 yr   **74.** 6.6   **75.** 90 dB
76. **TW** Negative numbers do not have logarithms because logarithm bases are positive, and there is no exponent to which a positive number can be raised to yield a negative number.   **77.** **TW** Taking the logarithm on each side of an equation produces an equivalent equation because the logarithm function is one-to-one. If two quantities are equal, their logarithms must be equal, and if the logarithms of two quantities are equal, the quantities must be the same.
78. e^{e^3} **79.** $-3, -1$ **80.** $\left(\frac{8}{3}, -\frac{2}{3}\right)$ **81.** $P(t) = 0.6e^{0.116t}$, where P is the number of blogs, in millions, t months after January 2003

Test: Chapter 12, pp. 971–972

1. [12.1] $(f \circ g)(x) = 2 + 6x + 4x^2$;
$(g \circ f)(x) = 2x^2 + 2x + 1$ **2.** [12.1] $f(x) = \frac{1}{x}$;
$g(x) = 2x^2 + 1$ **3.** [12.1] No   **4.** [12.1] $f^{-1}(x) = \frac{x-4}{3}$
5. [12.1] $g^{-1}(x) = \sqrt[3]{x} - 1$
6. [12.2]   **7.** [12.3]

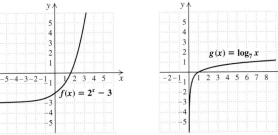

8. [12.3] 3   **9.** [12.3] $\frac{1}{2}$ **10.** [12.3] 18
11. [12.3] $\log_4 \frac{1}{64} = -3$ **12.** [12.3] $\log_{256} 16 = \frac{1}{2}$
13. [12.3] $49 = 7^m$ **14.** [12.3] $81 = 3^4$
15. [12.4] $3 \log a + \frac{1}{2} \log b - 2 \log c$
16. [12.4] $\log_a (z^2 \sqrt[3]{x})$ **17.** [12.4] 1   **18.** [12.4] 23
19. [12.4] 0   **20.** [12.4] 1.146   **21.** [12.4] 0.477
22. [12.4] 1.204   **23.** [12.3] 1.0899   **24.** [12.3] 0.1585
25. [12.5] -3.3524 **26.** [12.5] 121.5104
27. [12.5] 2.4022
28. [12.5]

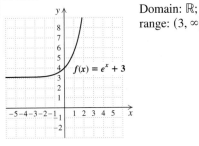

Domain: $\mathbb{R}$; range: $(3, \infty)$

29. [12.5]

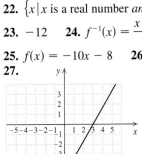

Domain: $(4, \infty)$; range: $\mathbb{R}$

30. [12.6] -5 **31.** [12.6] 5   **32.** [12.6] 2
33. [12.6] 10,000   **34.** [12.6] $-\frac{1}{3}\left(\frac{\log 87}{\log 5} - 4\right) \approx 0.4084$
35. [12.6] $\frac{\log 1.2}{\log 7} \approx 0.0937$ **36.** [12.6] $e^{1/4} \approx 1.2840$
37. [12.6] 4   **38.** [12.7] **(a)** 2.36 ft/sec; **(b)** about 1,517,000
39. [12.7] **(a)** $P(t) = 128.8e^{0.024t}$, where t is the number of years after 2005 and $P(t)$ is in millions; **(b)** 135.1 million; 152.4 million; **(c)** 2018; **(d)** 28.9 yr
40. [12.7] **(a)** $k \approx 0.044$; $C(t) = 18,039e^{0.044t}$, where t is the number of years after 1997; **(b)** \$31,962; **(c)** 2020
41. [12.7] **(a)** $a(t) = 4.975711955(1.20499705)^t$;
(b) about \$22.1 billion   **42.** [12.7] 4.6%   **43.** [12.7] About 4684 yr   **44.** [12.7] $10^{-4.5}$ W/m²   **45.** [12.7] 7.0
46. [12.6] $-309, 316$ **47.** [12.4] 2

Cumulative Review: Chapters 1–12, pp. 973–976

1. Linear   **2.** \$1/min   **3.** $f(x) = x + 3$ **4.** \$13
5. $m = 1$ signifies the cost per minute; $b = 3$ signifies the startup cost of each massage   **6.** Exponential, or perhaps quadratic   **7.** $m(x) = 1.893702275(1.059578007)^x$
8. \$25.60   **9.** Quadratic   **10.** $c(t) = -0.5035714286x^2 + 3.043571429x + 64.30714286$ **11.** 64.4 million texts
12. Length: 36 m; width: 20 m   **13.** $A: 15°; B: 45°; C: 120°$
14. $5\frac{5}{11}$ min   **15.** Thick and Tasty: 6 oz; Light and Lean: 9 oz
16. 7.25 mph   **17.** $-49; -7$ and 7   **18.** (c)   **19.** (b)
20. (a)   **21.** (d)
22. $\left\{x \mid x \text{ is a real number } and\ x \neq -\frac{1}{3} and\ x \neq 2\right\}$
23. -12 **24.** $f^{-1}(x) = \frac{x-9}{-2}$, or $f^{-1}(x) = \frac{9-x}{2}$
25. $f(x) = -10x - 8$ **26.** $y = \frac{1}{2}x + 7$
27.   **28.**

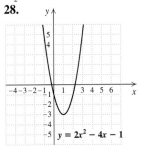

29.
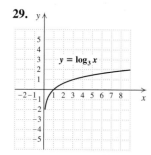

30.
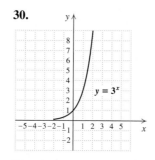

31.
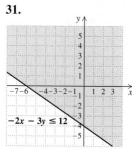

$-2x - 3y \le 12$

32.
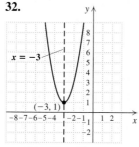

$f(x) = 2(x + 3)^2 + 1$
Minimum: 1

33.
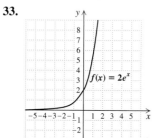

Domain: $\mathbb{R}$; range: $(0, \infty)$

34. 2   **35.** 6   **36.** $\dfrac{y^{12}}{16x^8}$ **37.** $\dfrac{20x^6z^2}{y}$ **38.** $\dfrac{-y^4}{3z^5}$

39. $-2x - 1$ **40.** 25   **41.** $7p^2q^3 + pq + p - 9$
42. $8x^2 - 11x - 1$ **43.** $9x^4 - 12x^2y + 4y^2$

44. $10a^2 - 9ab - 9b^2$ **45.** $\dfrac{(x + 4)(x - 3)}{2(x - 1)}$ **46.** $\dfrac{1}{x - 4}$

47. $\dfrac{a + 2}{6}$ **48.** $\dfrac{7x + 4}{(x + 6)(x - 6)}$ **49.** $x(y + 2z - w)$
50. $(2 - 5x)(4 + 10x + 25x^2)$ **51.** $2(3x - 2y)(x + 2y)$
52. $(x^3 + 7)(x - 4)$ **53.** $2(m + 3n)^2$
54. $(x - 2y)(x + 2y)(x^2 + 4y^2)$

55. $x^3 - 2x^2 - 4x - 12 + \dfrac{-42}{x - 3}$ **56.** 1.8×10^{-1}

57. $2y^2\sqrt[3]{y}$ **58.** $14xy^2\sqrt{x}$ **59.** $81a^8b^3\sqrt{b}$

60. $\dfrac{6 + \sqrt{y} - y}{4 - y}$ **61.** $\sqrt[10]{(x + 5)^3}$ **62.** $18 - 2\sqrt{3}i$

63. $13 - i$ **64.** $2 \log a + 3 \log c - \log b$

65. $\log\left(\dfrac{x^3}{y^{1/2}z^2}\right)$ **66.** $a^x = 5$ **67.** $\log_x t = 3$

68. -1.2545 **69.** 776.2471 **70.** 2.5479 **71.** 0.2466
72. $\frac{14}{5}$ **73.** $(3, -1)$ **74.** $(1, -2, 0)$ **75.** $-2, 5$

76. $\frac{9}{2}$ **77.** $\frac{5}{8}$ **78.** $\frac{3}{4}$ **79.** $\frac{1}{2}$ **80.** $\pm 5i$ **81.** 9, 25
82. $\pm 2, \pm 3$ **83.** 7   **84.** 6   **85.** $\frac{3}{2}$

86. $\dfrac{\log 7}{5 \log 3} \approx 0.3542$ **87.** $\dfrac{8e}{e - 1} \approx 12.6558$

88. $(-\infty, -5) \cup (1, \infty)$, or $\{x \,|\, x < -5 \ or \ x > 1\}$
89. $-3 \pm 2\sqrt{5}$ **90.** $\{x \,|\, x \le -2 \ or \ x \ge 5\}$, or

$(-\infty, -2] \cup [5, \infty)$ **91.** $a = \dfrac{Db}{b - D}$ **92.** $q = \dfrac{pf}{p - f}$

93. $B = \dfrac{3M - 2A}{2}$, or $B = \frac{3}{2}M - A$ **94.** 78   **95.** 67.5

96. $P(t) = 33.8e^{0.026t}$, where t is the number of years after
2005   **97.** 36.5 million; 42.7 million   **98.** 26.7 yr
99. 18   **100.** All real numbers except 1 and -2
101. $\frac{1}{3}, \frac{10{,}000}{3}$ **102.** 35 mph

Chapter 13

Exercise Set 13.1, pp. 986–990

1. (f)   **2.** (e)   **3.** (g)   **4.** (h)   **5.** (c)   **6.** (b)
7. (d)   **8.** (a)

9.
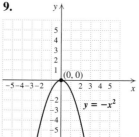

11.
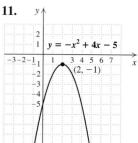

13.
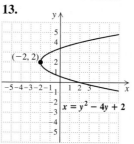

15.
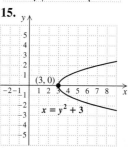

17.
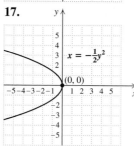

19.
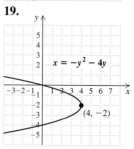

21.

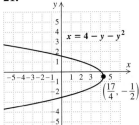

23.

75. $(5, 0); \frac{1}{2}$

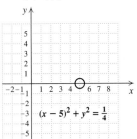

77. $(-4, 3); \sqrt{40}$, or $2\sqrt{10}$

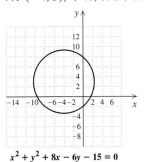

25.

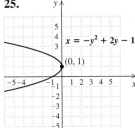

27.

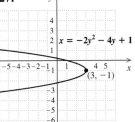

79. $(4, -1); 2$

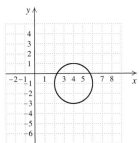

81. $(0, -5); 10$

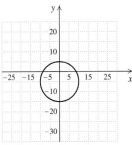

29. 5   **31.** $\sqrt{18} \approx 4.243$ **33.** $\sqrt{200} \approx 14.142$
35. 17.8   **37.** $\dfrac{\sqrt{41}}{7} \approx 0.915$ **39.** $\sqrt{8} \approx 2.828$
41. $\sqrt{90} \approx 9.487$ **43.** $(1, 4)$ **45.** $\left(\frac{7}{2}, \frac{7}{2}\right)$
47. $(-1, -3)$ **49.** $(-0.25, -0.3)$ **51.** $\left(-\frac{1}{12}, \frac{1}{24}\right)$
53. $\left(\dfrac{\sqrt{2} + \sqrt{3}}{2}, \dfrac{3}{2}\right)$ **55.** $x^2 + y^2 = 36$
57. $(x - 7)^2 + (y - 3)^2 = 5$
59. $(x + 4)^2 + (y - 3)^2 = 48$
61. $(x + 7)^2 + (y + 2)^2 = 50$ **63.** $x^2 + y^2 = 25$
65. $(x + 4)^2 + (y - 1)^2 = 20$
67. $(0, 0); 8$

83. $\left(-\dfrac{7}{2}, \dfrac{3}{2}\right); \sqrt{\dfrac{98}{4}}$, or $\dfrac{7\sqrt{2}}{2}$ **85.** $(0, 0); \frac{1}{6}$

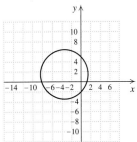

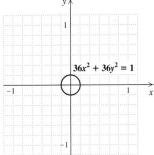

69. $(-1, -3); 6$

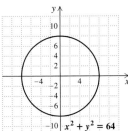

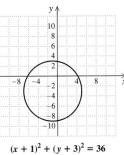

87. $x^2 + y^2 - 16 = 0$

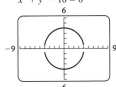

89. $x^2 + y^2 + 14x - 16y + 54 = 0$

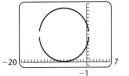

91. ™   **93.** $-\frac{2}{3}$ **94.** $\frac{25}{6}$ **95.** 4 in.   **96.** 2640 mi
97. $\left(\frac{35}{17}, \frac{5}{34}\right)$ **98.** $\left(0, -\frac{9}{5}\right)$ **99.** ™
101. $(x - 3)^2 + (y + 5)^2 = 9$ **103.** $(x - 3)^2 + y^2 = 25$
105. $(0, 4)$ **107.** $\frac{17}{4}\pi$ m², or approximately 13.4 m²
109. 7700 mm   **111. (a)** $(0, -3)$; **(b)** 5 ft
113. $x^2 + (y - 30.6)^2 = 590.49$

71. $(4, -3); \sqrt{10}$

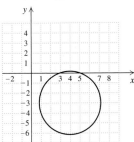

73. $(0, 0); \sqrt{10}$

115.

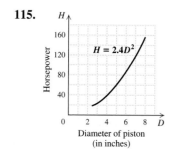

117. (a) $y = -1 \pm \sqrt{-x^2 + 6x + 7}$; **(b)**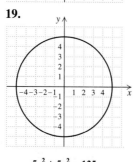

119. TW

Exercise Set 13.2, pp. 994–997

1. True   **2.** True   **3.** True   **4.** False   **5.** False
6. False   **7.** True   **8.** True
9.

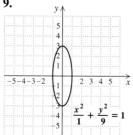

$\dfrac{x^2}{1} + \dfrac{y^2}{9} = 1$

11.

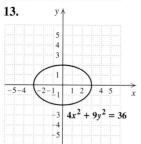

$\dfrac{x^2}{25} + \dfrac{y^2}{9} = 1$

13.

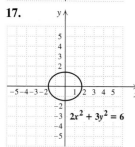

$4x^2 + 9y^2 = 36$

15.

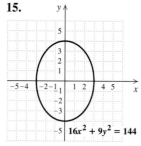

$16x^2 + 9y^2 = 144$

17.

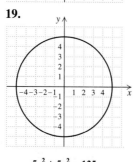

$2x^2 + 3y^2 = 6$

19.

$5x^2 + 5y^2 = 125$

21.

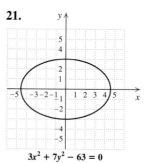

$3x^2 + 7y^2 - 63 = 0$

23.

$8x^2 = 96 - 3y^2$

25.

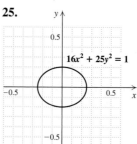

$16x^2 + 25y^2 = 1$

27.

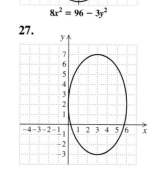

$\dfrac{(x-3)^2}{9} + \dfrac{(y-2)^2}{25} = 1$

29.

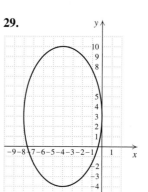

$\dfrac{(x+4)^2}{16} + \dfrac{(y-3)^2}{49} = 1$

31.

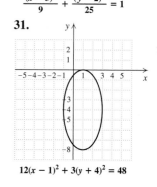

$12(x-1)^2 + 3(y+4)^2 = 48$

33.

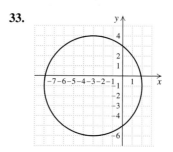

$4(x+3)^2 + 4(y+1)^2 - 10 = 90$

35. TW   **37.** $\frac{16}{9}$ **38.** $-\frac{19}{8}$ **39.** $5 \pm \sqrt{3}$ **40.** $3 \pm \sqrt{7}$
41. $\frac{3}{2}$ **42.** No solution   **43.** TW   **45.** $\dfrac{x^2}{81} + \dfrac{y^2}{121} = 1$

47. $\dfrac{(x-2)^2}{16} + \dfrac{(y+1)^2}{9} = 1$ **49.** $\dfrac{x^2}{9} + \dfrac{y^2}{25} = 1$

51. (a) Let $F_1 = (-c, 0)$ and $F_2 = (c, 0)$. Then the sum of the distances from the foci to P is $2a$. By the distance formula,

$$\sqrt{(x+c)^2 + y^2} + \sqrt{(x-c)^2 + y^2} = 2a, \text{ or}$$
$$\sqrt{(x+c)^2 + y^2} = 2a - \sqrt{(x-c)^2 + y^2}.$$

Squaring, we get

$$(x+c)^2 + y^2 = 4a^2 - 4a\sqrt{(x-c)^2 + y^2} + (x-c)^2 + y^2,$$

or

$$x^2 + 2cx + c^2 + y^2 = 4a^2 - 4a\sqrt{(x-c)^2 + y^2}$$
$$+ x^2 - 2cx + c^2 + y^2.$$

Thus

$$-4a^2 + 4cx = -4a\sqrt{(x-c)^2 + y^2}$$
$$a^2 - cx = a\sqrt{(x-c)^2 + y^2}.$$

Squaring again, we get

$$a^4 - 2a^2cx + c^2x^2 = a^2(x^2 - 2cx + c^2 + y^2)$$
$$a^4 - 2a^2cx + c^2x^2 = a^2x^2 - 2a^2cx + a^2c^2 + a^2y^2,$$

or

$$x^2(a^2 - c^2) + a^2y^2 = a^2(a^2 - c^2)$$
$$\dfrac{x^2}{a^2} + \dfrac{y^2}{a^2 - c^2} = 1.$$

(b) When P is at $(0, b)$, it follows that $b^2 = a^2 - c^2$. Substituting, we have

$$\dfrac{x^2}{a^2} + \dfrac{y^2}{b^2} = 1.$$

53. 5.66 ft

55.

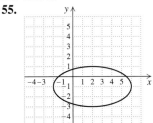

$$\dfrac{(x-2)^2}{16} + \dfrac{(y+1)^2}{4} = 1$$

57. 152,100,000 km

Visualizing the Graph, p. 1007

1. C   **2.** A   **3.** F   **4.** B   **5.** J   **6.** D   **7.** H
8. I   **9.** G   **10.** E

Exercise Set 13.3, pp. 1008–1009

1. (d)   **2.** (f)   **3.** (h)   **4.** (a)   **5.** (g)   **6.** (b)
7. (c)   **8.** (e)

9.

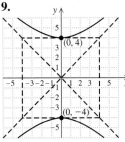

$$\dfrac{y^2}{16} - \dfrac{x^2}{16} = 1$$

11.

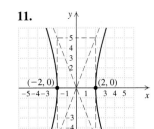

$$\dfrac{x^2}{4} - \dfrac{y^2}{25} = 1$$

13.

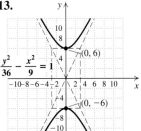

$$\dfrac{y^2}{36} - \dfrac{x^2}{9} = 1$$

15.

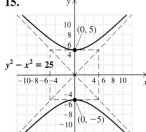

$$y^2 - x^2 = 25$$

17.

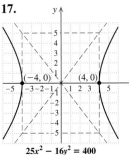

$$25x^2 - 16y^2 = 400$$

19.

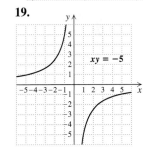

$$xy = -5$$

21.

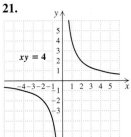

$$xy = 4$$

23.

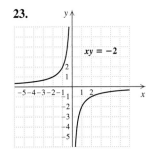

$$xy = -2$$

25.

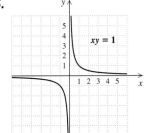

$$xy = 1$$

27. Circle   **29.** Ellipse   **31.** Hyperbola   **33.** Circle
35. Ellipse   **37.** Hyperbola   **39.** Parabola

41. Hyperbola **43.** Circle **45.** Ellipse **47.** ᴛᴡ
49.

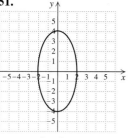

$(x - 1)^2 + (y + 2)^2 = 9$

50.

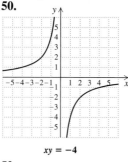

$xy = -4$

51.

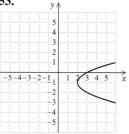

$4x^2 + y^2 = 16$

52.

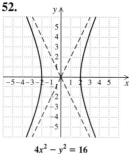

$4x^2 - y^2 = 16$

53.

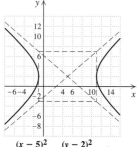

$x = y^2 + 2y + 3$

54.

$y = x^2 + 2x + 3$

55. ᴛᴡ **57.** $\dfrac{y^2}{36} - \dfrac{x^2}{4} = 1$ **59.** $C: (5, 2); V: (-1, 2),$
$(11, 2);$ asymptotes: $y - 2 = \frac{5}{6}(x - 5), y - 2 = -\frac{5}{6}(x - 5)$

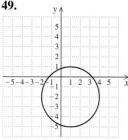

$\dfrac{(x - 5)^2}{36} - \dfrac{(y - 2)^2}{25} = 1$

61. $\dfrac{(y + 3)^2}{4} - \dfrac{(x - 4)^2}{16} = 1; C: (4, -3); V: (4, -5), (4, -1);$
asymptotes: $y + 3 = \frac{1}{2}(x - 4), y + 3 = -\frac{1}{2}(x - 4)$

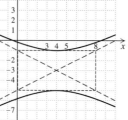

$8(y + 3)^2 - 2(x - 4)^2 = 32$

63. $\dfrac{(x + 3)^2}{1} - \dfrac{(y - 2)^2}{4} = 1; C: (-3, 2); V: (-4, 2), (-2, 2);$
asymptotes: $y - 2 = 2(x + 3), y - 2 = -2(x + 3)$

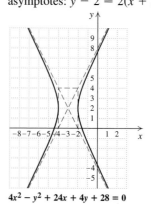

$4x^2 - y^2 + 24x + 4y + 28 = 0$

Exercise Set 13.4, pp. 1016–1019

1. True **2.** True **3.** False **4.** False **5.** True
6. True **7.** $(-4, -3), (3, 4)$ **9.** $(0, 2), (3, 0)$
11. $(-2, 1)$ **13.** $\left(\dfrac{5 + \sqrt{70}}{3}, \dfrac{-1 + \sqrt{70}}{3}\right),$
$\left(\dfrac{5 - \sqrt{70}}{3}, \dfrac{-1 - \sqrt{70}}{3}\right)$ **15.** $\left(4, \frac{3}{2}\right), (3, 2)$
17. $\left(\frac{7}{3}, \frac{1}{3}\right), (1, -1)$ **19.** $\left(\frac{11}{4}, -\frac{5}{4}\right), (1, 4)$
21. $\left(\dfrac{7 - \sqrt{33}}{2}, \dfrac{7 + \sqrt{33}}{2}\right), \left(\dfrac{7 + \sqrt{33}}{2}, \dfrac{7 - \sqrt{33}}{2}\right)$
23. $(3, -5), (-1, 3)$ **25.** $(-5, -8), (8, 5)$
27. $(0, 0), (1, 1), \left(-\dfrac{1}{2} + \dfrac{\sqrt{3}}{2}i, -\dfrac{1}{2} - \dfrac{\sqrt{3}}{2}i\right),$
$\left(-\dfrac{1}{2} - \dfrac{\sqrt{3}}{2}i, -\dfrac{1}{2} + \dfrac{\sqrt{3}}{2}i\right)$ **29.** $(-3, 0), (3, 0)$
31. $(-4, -3), (-3, -4), (3, 4), (4, 3)$

33. $\left(\dfrac{16}{3}, \dfrac{5\sqrt{7}}{3}i\right), \left(\dfrac{16}{3}, -\dfrac{5\sqrt{7}}{3}i\right), \left(-\dfrac{16}{3}, \dfrac{5\sqrt{7}}{3}i\right),$

$\left(-\dfrac{16}{3}, -\dfrac{5\sqrt{7}}{3}i\right)$ **35.** $(-3, -\sqrt{5}), (-3, \sqrt{5}), (3, -\sqrt{5}),$

$(3, \sqrt{5})$ **37.** $(4, 2), (-4, -2), (2, 4), (-2, -4)$

39. $(4, 1), (-4, -1), (2, 2), (-2, -2)$ **41.** $(2, 1), (-2, -1)$

43. $\left(2, -\dfrac{4}{5}\right), \left(-2, -\dfrac{4}{5}\right), (5, 2), (-5, 2)$

45. $\left(-\sqrt{2}, \sqrt{2}\right), \left(\sqrt{2}, -\sqrt{2}\right)$

47. Length: 8 cm; width: 6 cm

49. Length: 5 in.; width: 4 in.

51. Length: 12 ft; width: 5 ft   **53.** 6 and 10; -6 and -10

55. 24 ft, 16 ft   **57.** 13 and 12   **59.** TW   **61.** -16

62. -32 **63.** 1   **64.** $-\dfrac{1}{4}$ **65.** 44   **66.** 28   **67.** TW

69. $(x + 2)^2 + (y - 1)^2 = 4$

71. $(-2, 3), (2, -3), (-3, 2), (3, -2)$ **73.** Length: 55 ft;

width: 45 ft   **75.** 10 in. by 7 in. by 5 in.

77. Length: 61.02 in.; height: 34.32 in.

79. $(-1.50, -1.17); (3.50, 0.50)$

Review Exercises: Chapter 13, pp. 1022–1023

1. True   **2.** False   **3.** False   **4.** False   **5.** True

6. True   **7.** False   **8.** True   **9.** 4   **10.** 5

11. $\sqrt{90.1} \approx 9.492$ **12.** $\sqrt{9 + 4a^2}$ **13.** $\left(\dfrac{9}{2}, -1\right)$

14. $(-3, 7)$ **15.** $\left(\dfrac{3}{4}, \dfrac{\sqrt{3} - \sqrt{2}}{2}\right)$ **16.** $\left(\dfrac{1}{2}, 2a\right)$

17. $(-3, 2), \sqrt{7}$ **18.** $(5, 0), 7$ **19.** $(3, 1), 3$

20. $(-4, 3), \sqrt{35}$ **21.** $(x + 4)^2 + (y - 3)^2 = 48$

22. $(x - 7)^2 + (y + 2)^2 = 20$

23. Circle   **24.** Ellipse

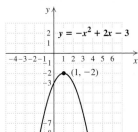

$5x^2 + 5y^2 = 80$

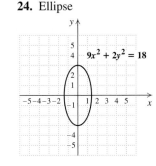

$9x^2 + 2y^2 = 18$

25. Parabola   **26.** Hyperbola

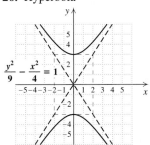

$y = -x^2 + 2x - 3$ $(1, -2)$

$\dfrac{y^2}{9} - \dfrac{x^2}{4} = 1$

27. Hyperbola

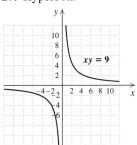

$xy = 9$

28. Parabola

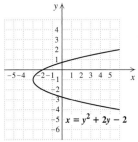

$x = y^2 + 2y - 2$

29. Ellipse

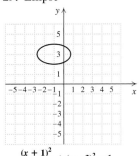

$\dfrac{(x + 1)^2}{3} + (y - 3)^2 = 1$

30. Circle

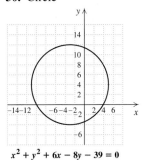

$x^2 + y^2 + 6x - 8y - 39 = 0$

31. $(7, 4)$ **32.** $(2, 2), \left(\dfrac{32}{9}, -\dfrac{10}{9}\right)$ **33.** $(0, -3), (2, 1)$

34. $(4, 3), (4, -3), (-4, 3), (-4, -3)$

35. $(2, 1), \left(\sqrt{3}, 0\right), (-2, 1), \left(-\sqrt{3}, 0\right)$

36. $(3, -3), \left(-\dfrac{3}{5}, \dfrac{21}{5}\right)$

37. $(6, 8), (6, -8), (-6, 8), (-6, -8)$

38. $(2, 2), (-2, -2), \left(2\sqrt{2}, \sqrt{2}\right), \left(-2\sqrt{2}, -\sqrt{2}\right)$

39. Length: 12 m; width: 7 m   **40.** Length: 12 in.;

width: 9 in.   **41.** 32 cm, 20 cm   **42.** 3 ft, 11 ft

43. TW   The graph of a parabola has one branch whereas the

graph of a hyperbola has two branches. A hyperbola has

asymptotes, but a parabola does not.

44. TW   Function notation rarely appears in this chapter

because many of the relations are not functions. Function

notation could be used for vertical parabolas and for

hyperbolas that have the axes as asymptotes.

45. $\left(-5, -4\sqrt{2}\right), \left(-5, 4\sqrt{2}\right), \left(3, -2\sqrt{2}\right), \left(3, 2\sqrt{2}\right)$

46. $(0, 6), (0, -6)$ **47.** $(x - 2)^2 + (y + 1)^2 = 25$

48. $\dfrac{x^2}{81} + \dfrac{y^2}{25} = 1$ **49.** $\left(\dfrac{9}{4}, 0\right)$

Test: Chapter 13, pp. 1023–1024

1. [13.1] $9\sqrt{2} \approx 12.728$ **2.** [13.1] $2\sqrt{9 + a^2}$

3. [13.1] $\left(-\dfrac{1}{2}, \dfrac{7}{2}\right)$ **4.** [13.1] $(0, 0)$ **5.** [13.1] $(-5, 1), 9$

6. [13.1] $(-2, 3), 3$

7. [13.1], [13.3] Parabola

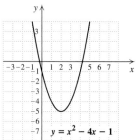

$y = x^2 - 4x - 1$

8. [13.1], [13.3] Circle

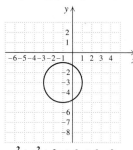

$x^2 + y^2 + 2x + 6y + 6 = 0$

9. [13.3] Hyperbola

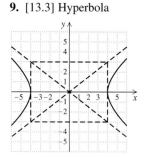

$\dfrac{x^2}{16} - \dfrac{y^2}{9} = 1$

10. [13.2], [13.3] Ellipse

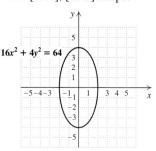

$16x^2 + 4y^2 = 64$

11. [13.3] Hyperbola

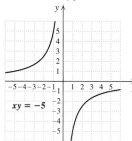

$xy = -5$

12. [13.1], [13.3] Parabola

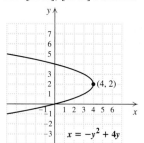

$(4, 2)$

$x = -y^2 + 4y$

13. [13.4] $(0, 3)$, $\left(\frac{8}{5}, \frac{9}{5}\right)$ **14.** [13.4] $(4, 0)$, $(-4, 0)$

15. [13.4] $(3, 2)$, $(-3, -2)$, $\left(2\sqrt{2}i, -\dfrac{3\sqrt{2}}{2}i\right)$,

$\left(-2\sqrt{2}i, \dfrac{3\sqrt{2}}{2}i\right)$ **16.** [13.4] $\left(\sqrt{6}, 2\right)$, $\left(\sqrt{6}, -2\right)$,

$\left(-\sqrt{6}, 2\right)$, $\left(-\sqrt{6}, -2\right)$ **17.** [13.4] 2 by 11

18. [13.4] $\sqrt{5}$ m, $\sqrt{3}$ m   **19.** [13.4] Length: 32 ft;

width: 24 ft   **20.** [13.4] $1200, 6%

21. [13.2] $\dfrac{(x-6)^2}{25} + \dfrac{(y-3)^2}{9} = 1$ **22.** [13.1] $\left(0, -\frac{31}{4}\right)$

23. [13.4] 9   **24.** [13.2] $\dfrac{x^2}{16} + \dfrac{y^2}{49} = 1$

Chapter 14

Exercise Set 14.1, pp. 1033–1035

1. (f)   **2.** (a)   **3.** (d)   **4.** (b)   **5.** (c)   **6.** (e)
7. 5, 7, 9, 11; 23; 33   **9.** 3, 6, 11, 18; 102; 227
11. $0, \frac{3}{5}, \frac{4}{5}, \frac{15}{17}; \frac{99}{101}; \frac{112}{113}$ **13.** $1, -\frac{1}{2}, \frac{1}{4}, -\frac{1}{8}; -\frac{1}{512}; \frac{1}{16,384}$
15. $-4, 5, -6, 7; 13; -18$ **17.** $0, 7, -26, 63; 999; -3374$
19. 13   **21.** 364   **23.** -23.5 **25.** -363 **27.** $\dfrac{441}{400}$
29. $2n$ **31.** $(-1)^{n+1}$ **33.** $(-1)^n \cdot n$ **35.** $2n + 1$
37. $(-1)^n \cdot 2 \cdot (3)^{n-1}$ **39.** $\dfrac{n}{n+1}$ **41.** 5^n **43.** $(-1)^n \cdot n^2$
45. 4   **47.** 30   **49.** $\dfrac{1}{2} + \dfrac{1}{4} + \dfrac{1}{6} + \dfrac{1}{8} + \dfrac{1}{10} = \dfrac{137}{120}$
51. $3^0 + 3^1 + 3^2 + 3^3 + 3^4 = 121$
53. $2 + \dfrac{3}{2} + \dfrac{4}{3} + \dfrac{5}{4} + \dfrac{6}{5} + \dfrac{7}{6} + \dfrac{8}{7} = \dfrac{1343}{140}$
55. $(-1)^2 2^1 + (-1)^3 2^2 + (-1)^4 2^3 + (-1)^5 2^4 + (-1)^6 2^5 +$
$(-1)^7 2^6 + (-1)^8 2^7 + (-1)^9 2^8 = -170$
57. $(0^2 - 2 \cdot 0 + 3) + (1^2 - 2 \cdot 1 + 3) +$
$(2^2 - 2 \cdot 2 + 3) + (3^2 - 2 \cdot 3 + 3) + (4^2 - 2 \cdot 4 + 3) +$
$(5^2 - 2 \cdot 5 + 3) = 43$
59. $\dfrac{(-1)^3}{3 \cdot 4} + \dfrac{(-1)^4}{4 \cdot 5} + \dfrac{(-1)^5}{5 \cdot 6} = -\dfrac{1}{15}$ **61.** $\displaystyle\sum_{k=1}^{5} \dfrac{k+1}{k+2}$
63. $\displaystyle\sum_{k=1}^{6} k^2$ **65.** $\displaystyle\sum_{k=2}^{n} (-1)^k k^2$ **67.** $\displaystyle\sum_{k=1}^{\infty} 5k$
69. $\displaystyle\sum_{k=1}^{\infty} \dfrac{1}{k(k+1)}$ **71.** TW   **73.** 77   **74.** 23
75. $x^3 + 3x^2 y + 3xy^2 + y^3$ **76.** $a^3 - 3a^2 b + 3ab^2 - b^3$
77. $8a^3 - 12a^2 b + 6ab^2 - b^3$
78. $8x^3 + 12x^2 y + 6xy^2 + y^3$ **79.** TW
81. 1, 3, 13, 63, 313, 1563   **83.** $5200, $3900, $2925,
$2193.75, $1645.31, $1233.98, $925.49, $694.12, $520.59,
$390.44   **85.** $S_{100} = 0$; $S_{101} = -1$ **87.** $i, -1, -i, 1, i; i$
89. 11th term

Interactive Discovery, p. 1037

1. The points lie on a straight line with positive slope.
2. The points lie on a straight line with positive slope.
3. The points lie on a straight line with negative slope.
4. The points lie on a straight line with negative slope.

Exercise Set 14.2, pp. 1042–1045

1. True   **2.** True   **3.** False   **4.** False   **5.** True
6. True   **7.** False   **8.** False   **9.** $a_1 = 2, d = 4$
11. $a_1 = 7, d = -4$ **13.** $a_1 = \frac{3}{2}, d = \frac{3}{4}$
15. $a_1 = $5.12, d = $0.12   **17.** 49   **19.** -94
21. $-$1628.16   **23.** 26th   **25.** 57th   **27.** 82   **29.** 5
31. 28   **33.** $a_1 = 8; d = -3; 8, 5, 2, -1, -4$

35. $a_1 = 1; d = 1$ **37.** 780   **39.** 31,375   **41.** 2550
43. 918   **45.** 1030   **47.** 35 marchers; 315 marchers
49. 180 stones   **51.** \$49.60   **53.** 722 seats   **55.** ᵀᵂ
57. $\dfrac{13}{30x}$ **58.** $\dfrac{23}{36t}$ **59.** $a^k = P$ **60.** $e^a = t$
61. $x^2 + y^2 = 81$ **62.** $(x + 2)^2 + (y - 5)^2 = 18$
63. ᵀᵂ   **65.** 33 jumps   **67.** \$8760, \$7961.77, \$7163.54,
\$6365.31, \$5567.08; \$4768.85, \$3970.62, \$3172.39, \$2374.16,
\$1575.93   **69.** Let $d =$ the common difference. Since p, m,
and q form an arithmetic sequence, $m = p + d$ and
$q = p + 2d$. Then $\dfrac{p + q}{2} = \dfrac{p + (p + 2d)}{2} = p + d = m$.
71. 156,375   **73.** Arithmetic; $a_n = -0.75n + 150.75$,
where $n = 1$ corresponds to age 20, $n = 2$ to age 21, and so
on   **75.** Not arithmetic

Interactive Discovery, p. 1047

1. The points lie on an exponential curve with $a > 1$.
2. The points lie on an exponential curve with $a > 1$.
3. The points lie on an exponential curve with $0 < a < 1$.
4. The points lie on an exponential curve with $0 < a < 1$.

Interactive Discovery, p. 1050

1. **(a)** $\frac{1}{3}$; **(b)** 1, 1.33333, 1.44444, 1.48148, 1.49383, 1.49794,
1.49931, 1.49977, 1.49992, 1.49997 (sums are rounded to
5 decimal places); **(c)** 1.5
2. **(a)** $-\frac{7}{2}$; **(b)** 1, -2.5, 9.75, -33.125, 116.9375,
-408.28125, 1429.984375, -5003.945313, 17514.80859,
-61300.83008; **(c)** does not exist
3. **(a)** 2; **(b)** 4, 12, 28, 60, 124, 252, 508, 1020, 2044, 4092;
(c) does not exist
4. **(a)** -0.4; **(b)** 1.3, 0.78, 0.988, 0.9048, 0.93808, 0.92477,
0.93009, 0.92796, 0.92881, 0.92847 (sums are rounded to
5 decimal places); **(c)** 0.928   **5.** S_∞ does not exist if $|r| > 1$.

Visualizing the Graph, p. 1054

1. J   **2.** G   **3.** A   **4.** H   **5.** I   **6.** B   **7.** E
8. D   **9.** F   **10.** C

Exercise Set 14.3, pp. 1055–1058

1. Geometric sequence   **2.** Arithmetic sequence
3. Arithmetic sequence   **4.** Geometric sequence
5. Geometric series   **6.** Arithmetic series
7. Geometric series   **8.** None of these   **9.** 2
11. -0.1 **13.** $-\frac{1}{2}$ **15.** $\frac{1}{5}$ **17.** $\dfrac{6}{m}$ **19.** 192
21. $112\sqrt{2}$ **23.** 52,488   **25.** \$2331.64   **27.** $a_n = 5^{n-1}$

29. $a_n = (-1)^{n-1}$, or $a_n = (-1)^{n+1}$
31. $a_n = \dfrac{1}{x^n}$, or $a_n = x^{-n}$ **33.** 3066   **35.** $\frac{547}{18}$ **37.** $\dfrac{1 - x^8}{1 - x}$,
or $(1 + x)(1 + x^2)(1 + x^4)$ **39.** \$5134.51   **41.** $\frac{64}{3}$
43. $\frac{49}{4}$ **45.** No   **47.** No   **49.** $\frac{43}{99}$ **51.** \$25,000
53. $\frac{7}{9}$ **55.** $\frac{830}{99}$ **57.** $\frac{5}{33}$ **59.** $\frac{5}{1024}$ ft   **61.** 155,797
63. 2710 flies   **65.** 10,723,491 apartments and houses
67. 3100.35 ft   **69.** 20.48 in.   **71.** Arithmetic
73. Geometric   **75.** Geometric   **77.** ᵀᵂ
79. $a_1 = 2, d = 2$ **80.** $a_1 = 2, r = 2$ **81.** -19
82. $-\frac{1}{16}$ **83.** 210   **84.** 5,592,405   **85.** ᵀᵂ   **87.** 54
89. $\dfrac{x^2[1 - (-x)^n]}{1 + x}$ **91.** 512 cm²

Exercise Set 14.4, pp. 1066–1067

1. 2^5, or 32   **2.** 8   **3.** 9   **4.** 4!   **5.** $\begin{pmatrix} 8 \\ 5 \end{pmatrix}$ **6.** $\begin{pmatrix} 10 \\ 2 \end{pmatrix}$,
or 45   **7.** x^7y^2 **8.** 10 choose 4   **9.** 362,880
11. 39,916,800   **13.** 56   **15.** 3024   **17.** 35   **19.** 126
21. 4060   **23.** 780   **25.** $a^4 - 4a^3b + 6a^2b^2 - 4ab^3 + b^4$
27. $p^7 + 7p^6q + 21p^5q^2 + 35p^4q^3 + 35p^3q^4 + 21p^2q^5 + 7pq^6 + q^7$
29. $2187c^7 - 5103c^6d + 5103c^5d^2 - 2835c^4d^3 + 945c^3d^4 - 189c^2d^5 + 21cd^6 - d^7$
31. $t^{-12} + 12t^{-10} + 60t^{-8} + 160t^{-6} + 240t^{-4} + 192t^{-2} + 64$
33. $x^5 - 5x^4y + 10x^3y^2 - 10x^2y^3 + 5xy^4 - y^5$
35. $19{,}683s^9 + \dfrac{59{,}049s^8}{t} + \dfrac{78{,}732s^7}{t^2} + \dfrac{61{,}236s^6}{t^3} + \dfrac{30{,}618s^5}{t^4} + \dfrac{10{,}206s^4}{t^5} + \dfrac{2268s^3}{t^6} + \dfrac{324s^2}{t^7} + \dfrac{27s}{t^8} + \dfrac{1}{t^9}$
37. $x^{15} - 10x^{12}y + 40x^9y^2 - 80x^6y^3 + 80x^3y^4 - 32y^5$
39. $125 + 150\sqrt{5}t + 375t^2 + 100\sqrt{5}t^3 + 75t^4 + 6\sqrt{5}t^5 + t^6$
41. $x^{-3} - 6x^{-2} + 15x^{-1} - 20 + 15x - 6x^2 + x^3$
43. $15a^4b^2$ **45.** $-64{,}481{,}508a^3$ **47.** $1120x^{12}y^2$
49. $1{,}959{,}552u^5v^{10}$ **51.** y^8 **53.** ᵀᵂ   **55.** 4   **56.** $\frac{5}{2}$
57. 5.6348   **58.** ±5 **59.** ᵀᵂ
61. List all the subsets of size 3: $\{a, b, c\}, \{a, b, d\}, \{a, b, e\}$,
$\{a, c, d\}, \{a, c, e\}, \{a, d, e\}, \{b, c, d\}, \{b, c, e\}, \{b, d, e\}$,
$\{c, d, e\}$.

There are exactly 10 subsets of size 3 and $\begin{pmatrix} 5 \\ 3 \end{pmatrix} = 10$, so

there are exactly $\begin{pmatrix} 5 \\ 3 \end{pmatrix}$ ways of forming a subset of size 3

from $\{a, b, c, d, e\}$.

63. $\begin{pmatrix} 8 \\ 5 \end{pmatrix}(0.15)^3(0.85)^5 \approx 0.084$

65. $\binom{8}{6}(0.15)^2(0.85)^6 + \binom{8}{7}(0.15)(0.85)^7 +$
$\binom{8}{8}(0.85)^8 \approx 0.89$

67. $\binom{n}{n-r} = \dfrac{n!}{[n-(n-r)]!\,(n-r)!}$
$= \dfrac{n!}{r!\,(n-r)!} = \binom{n}{r}$

69. $\dfrac{-\sqrt[3]{q}}{2p}$ **71.** $x^7 + 7x^6y + 21x^5y^2 + 35x^4y^3 +$
$35x^3y^4 + 21x^2y^5 + 7xy^6 + y^7$

Review Exercises: Chapter 14, pp. 1070–1071

1. False   **2.** True   **3.** True   **4.** False   **5.** True
6. True   **7.** False   **8.** False   **9.** 1, 5, 9, 13; 29; 45
10. $0, \frac{1}{5}, \frac{1}{5}, \frac{3}{17}, \frac{7}{65}, \frac{11}{145}$ **11.** $a_n = 7n$
12. $a_n = (-1)^n(2n-1)$
13. $-2 + 4 + (-8) + 16 + (-32) = -22$
14. $-3 + (-5) + (-7) + (-9) + (-11) + (-13) = -48$
15. $\sum\limits_{k=1}^{5} 4k$ **16.** $\sum\limits_{k=1}^{5}\dfrac{1}{(-2)^k}$ **17.** 85   **18.** $\frac{8}{3}$
19. $a_1 = \frac{45}{4}, d = \frac{5}{4}$ **20.** -544 **21.** 8580   **22.** $1024\sqrt{2}$
23. $\frac{2}{3}$ **24.** $a_n = 2(-1)^n$ **25.** $a_n = 3\left(\dfrac{x}{4}\right)^{n-1}$ **26.** 4095
27. $-4095x$ **28.** 12   **29.** $\frac{49}{11}$ **30.** No   **31.** No
32. \$40,000   **33.** $\frac{5}{9}$ **34.** $\frac{46}{33}$ **35.** \$24.30   **36.** 903 poles
37. \$15,791.18   **38.** 6 m   **39.** 5040   **40.** 56
41. $190a^{18}b^2$ **42.** $x^4 - 8x^3y + 24x^2y^2 - 32xy^3 + 16y^4$
43. 🅣🅦   For a geometric sequence with $|r| < 1$, as n gets larger, the absolute value of the terms gets smaller, since $|r^n|$ gets smaller.
44. 🅣🅦   The first form of the binomial theorem draws the coefficients from Pascal's triangle; the second form uses factorial notation. The second form avoids the need to compute all preceding rows of Pascal's triangle, and is generally easier to use when only one term of an expansion is needed. When several terms of an expansion are needed and n is not large (say, $n \le 8$), it is often easier to use Pascal's triangle.
45. $\dfrac{1 - (-x)^n}{x + 1}$
46. $x^{-15} + 5x^{-9} + 10x^{-3} + 10x^3 + 5x^9 + x^{15}$

Test: Chapter 14, pp. 1071–1072

1. [14.1] 1, 7, 13, 19, 25; 67   **2.** [14.1] $a_n = 4\left(\frac{1}{3}\right)^n$
3. [14.1] $-1 + (-5) + (-13) + (-29) + (-61) = -109$
4. [14.1] $\sum\limits_{k=1}^{5}(-1)^{k+1}k^3$ **5.** [14.2] -51 **6.** [14.2] $\frac{3}{8}$
7. [14.2] $a_1 = 31.2; d = -3.8$ **8.** [14.2] 2508

9. [14.3] $\frac{9}{128}$ **10.** [14.3] $\frac{2}{3}$ **11.** [14.3] 3^n
12. [14.3] $511 + 511x$ **13.** [14.3] 1   **14.** [14.3] No
15. [14.3] $\frac{\$25,000}{23} \approx \$1086.96$ **16.** [14.3] $\frac{85}{99}$
17. [14.2] 63 seats   **18.** [14.2] \$17,100
19. [14.3] \$8981.05   **20.** [14.3] 36 m   **21.** [14.4] 220
22. [14.4] $x^{10} - 15x^8y + 90x^6y^2 - 270x^4y^3 + 405x^2y^4 - 243y^5$
23. [14.4] $220a^9x^3$ **24.** [14.2] $n(n+1)$
25. [14.3] $\dfrac{1 - \left(\dfrac{1}{x}\right)^n}{1 - \dfrac{1}{x}}$, or $\dfrac{x^n - 1}{x^{n-1}(x-1)}$

Cumulative Review: Chapters 1–14, pp. 1072–1076

1. $1\frac{1}{4}$ in.   **2.** $8\frac{2}{5}$ hr, or 8 hr 24 min   **3.** 69
4. More than 4 purchases   **5.** $11\frac{3}{7}$ **6.** \$2.68 herb: 10 oz; \$4.60 herb: 14 oz   **7.** 350 mph   **8.** 20   **9.** 5000 ft^2
10. 5 ft by 12 ft   **11.** \$652.39   **12.** (a) $f(t) = 0.48t + 21.3$, where $f(t)$ is in millions; (b) 27.06 million households
13. (a) $k \approx 0.242$; $P(t) = 0.12e^{0.242t}$, where $P(t)$ is in millions; (b) about 50.9 million computers   **14.** $y = 3x - 8$

15.
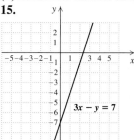
$3x - y = 7$

16.
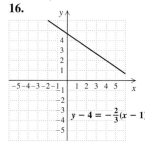
$y - 4 = -\frac{2}{3}(x - 1)$

17.
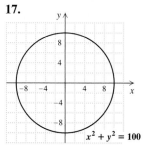
$x^2 + y^2 = 100$

18.
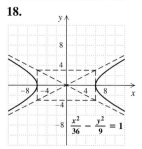
$\dfrac{x^2}{36} - \dfrac{y^2}{9} = 1$

19.
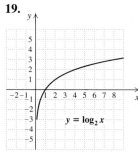
$y = \log_2 x$

20.
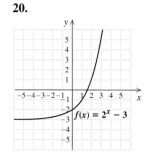
$f(x) = 2^x - 3$

21.

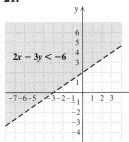

$2x - 3y < -6$

22.

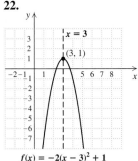

$x = 3$

$(3, 1)$

$f(x) = -2(x - 3)^2 + 1$
Maximum: 1

23.

$y = \sqrt{(4 - x)}$

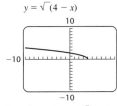

Domain: $(-\infty, 4]$;
range: $[0, \infty)$

24. 20   **25.** $\{x \mid x \leq \frac{5}{3}\}$, or $\left(-\infty, \frac{5}{3}\right]$

26. $\{x \mid x$ is a real number *and* $x \neq 1\}$, or $(-\infty, 1) \cup (1, \infty)$

27. (a) Linear; **(b)** 2   **28. (a)** Logarithmic; **(b)** -1

29. (a) Quadratic; **(b)** $-1, 4$ **30. (a)** Exponential;

(b) no real zeros   **31.** $\frac{3}{5}$ **32.** $-\frac{6}{5}, 4$ **33.** $\mathbb{R}$, or $(-\infty, \infty)$

34. $(-1, 1)$ **35.** $(2, -1, 1)$ **36.** 2   **37.** $\pm 2, \pm 5$

38. $\left(\sqrt{5}, \sqrt{3}\right), \left(\sqrt{5}, -\sqrt{3}\right), \left(-\sqrt{5}, \sqrt{3}\right), \left(-\sqrt{5}, -\sqrt{3}\right)$

39. 1.4037   **40.** 1005   **41.** $\frac{1}{25}$ **42.** $-\frac{1}{2}$

43. $\{x \mid -2 \leq x \leq 3\}$ or $[-2, 3]$ **44.** $\pm i\sqrt{2}$

45. $-2 \pm \sqrt{7}$ **46.** $\{y \mid y < -5 \ or \ y > 2\}$, or

$(-\infty, -5) \cup (2, \infty)$ **47.** $-6, 8$ **48.** 3

49. $r = \dfrac{V - P}{-Pt}$, or $\dfrac{P - V}{Pt}$ **50.** $R = \dfrac{Ir}{1 - I}$

51. $-35x^6 y^{-4}$, or $\dfrac{-35x^6}{y^4}$ **52.** 6.3   **53.** $-4y + 17$

54. 280   **55.** $\frac{7}{6}$ **56.** $3a^2 - 8ab - 15b^2$

57. $13x^3 - 7x^2 - 6x + 6$ **58.** $6a^2 + 7a - 5$

59. $9a^4 - 30a^2 y + 25y^2$ **60.** $\dfrac{4}{x + 2}$ **61.** $\dfrac{x - 4}{4(x + 2)}$

62. $\dfrac{(x + y)(x^2 + xy + y^2)}{x^2 + y^2}$ **63.** $x - a$ **64.** $(2x - 3)^2$

65. $(3a - 2)(9a^2 + 6a + 4)$ **66.** $(a + 3)(a^2 - b)$

67. $3(y^2 + 3)(5y^2 - 4)$

68. $7x^3 + 9x^2 + 19x + 38 + \dfrac{72}{x - 2}$ **69.** 6.8×10^{-12}

70. $8x^2 \sqrt{y}$ **71.** $125x^2 y^{3/4}$ **72.** $\dfrac{\sqrt[3]{5xy}}{y}$

73. $\dfrac{1 - 2\sqrt{x} + x}{1 - x}$ **74.** $26 - 13i$ **75.** $x^2 - 50 = 0$

76. $(2, -3)$; 6   **77.** $\log_a \dfrac{\sqrt[3]{x^2} \cdot z^5}{\sqrt{y}}$ **78.** $a^5 = c$

79. 3.7541   **80.** 0.0003   **81.** 8.6442   **82.** 0.0277

83. 5   **84.** -121 **85.** 875   **86.** $16\left(\frac{1}{4}\right)^{n-1}$

87. $13,440a^4 b^6$ **88.** $74.88671875x$ **89. (a)** Quadratic;

(b) $A(t) = 0.1930762006t^2 - 3.682999186t + 39.4379563$

90. (a) Exponential; **(b)** $I(t) = 15.58543135(1.082414612)^t$

91. (a) Linear; **(b)** $S(t) = 1.889081456t + 14.48700173$

92. More than 23 yr after 1985, or years after 2008

93. All real numbers except 0 and -12 **94.** 81

95. y gets divided by 8   **96.** $-\dfrac{7}{13} + \dfrac{2\sqrt{30}}{13}i$ **97.** 84 yr

Chapter R

Exercise Set R.1, pp. R-8–R-9

1. False   **3.** True   **5.** True   **7.** 4   **9.** 1.3   **11.** -25

13. $-\frac{11}{15}$ **15.** -6.5 **17.** -15 **19.** 0   **21.** $-\frac{1}{2}$

23. 5.8   **25.** -3 **27.** 39   **29.** 175   **31.** -32

33. 16   **35.** -6 **37.** 9   **39.** -3 **41.** -16

43. 100   **45.** 2   **47.** -23 **49.** 36   **51.** 10   **53.** 10

55. 7   **57.** 32   **59.** 28 cm²   **61.** $8x + 28$

63. $5x - 50$ **65.** $-30 + 6x$ **67.** $8a + 12b - 6c$

69. $2(4x + 3y)$ **71.** $3(1 + w)$ **73.** $10(x + 5y + 10)$

75. p **77.** $-m + 22$ **79.** $3x + 7$ **81.** $6p - 7$

83. $-5x + 12y$ **85.** $36a - 48b$ **87.** $-10x + 104y + 9$

89. Yes   **91.** No   **93.** Yes   **95.** Let n represent the

number; $3n = 348$ **97.** Let c represent the number of

calories in a Taco Bell Beef Burrito; $c + 69 = 500$

99. Let l represent the amount of water used to produce 1 lb

of lettuce; $42 = 2l$

Exercise Set R.2, pp. R-17–R-18

1. 8   **3.** 12   **5.** $-\frac{1}{12}$ **7.** -0.8 **9.** $\frac{13}{3}$ **11.** $-\frac{5}{3}$

13. 42   **15.** -5 **17.** 2   **19.** $\frac{25}{3}$ **21.** $-\frac{4}{9}$ **23.** -4

25. $\frac{69}{5}$ **27.** $\frac{9}{32}$ **29.** -2 **31.** -15 **33.** $\frac{43}{2}$ **35.** $-\frac{61}{115}$

37. $l = \dfrac{A}{w}$ **39.** $q = \dfrac{p}{30}$ **41.** $P = IV$ **43.** $p = 2q - r$

45. $\pi = \dfrac{A}{r^2 + r^2 h}$ **47. (a)** No; **(b)** yes; **(c)** no; **(d)** yes

49. $\{x \mid x \leq 12\}$, or $(-\infty, 12]$

51. $\{m \mid m > 12\}$, or $(12, \infty)$

53. $\{x \mid x \geq -\frac{3}{2}\}$, or $\left[-\frac{3}{2}, \infty\right)$

55. $\{t \mid t < -3\}$, or $(-\infty, -3)$

57. $\{y \mid y > 10\}$, or $(10, \infty)$ **59.** $\{a \mid a \geq 1\}$, or $[1, \infty)$

61. $\{x \mid x \geq \frac{64}{17}\}$, or $\left[\frac{64}{17}, \infty\right)$ **63.** $\{x \mid x > \frac{39}{11}\}$, or $\left(\frac{39}{11}, \infty\right)$

65. $\{x \mid x \leq -10.875\}$, or $(-\infty, -10.875]$ **67.** 7

69. 16, 18   **71.** $166\frac{2}{3}$ pages   **73.** $25°$

75. 900 cubic feet   **77.** 80¢   **79.** 30 min or more

81. For $2\frac{7}{9}$ hr or less

Exercise Set R.3, pp. R-28–R-29

1.

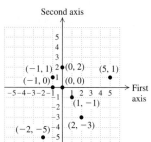

3. I **5.** IV
7. I, IV **9.** No
11. Yes

13.

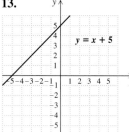

15.

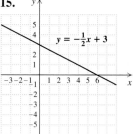

17.

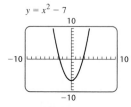

19.

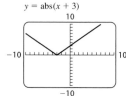

21.

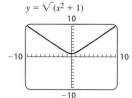

23. 1 **25.** -5 **27.** 0

29. Slope: 2; y-intercept: $(0, -5)$
31. Slope: -2; y-intercept: $(0, 1)$
33.

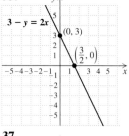

35.

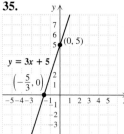

37.

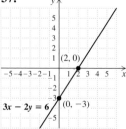

39.

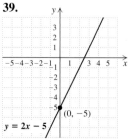

41.

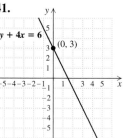

43. 0

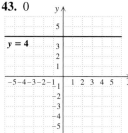

45. Undefined

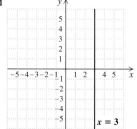

47. $y = \frac{1}{3}x + 1$ **49.** $y = -x + 3$ **51.** Perpendicular
53. Neither **55.** (a) 6; (b) $8\frac{1}{3}$; (c) $\dfrac{a}{3} + \dfrac{22}{3}$
57. (a) 6; (b) 3; (c) $20a^2 + 4a + 3$ **59.** No **61.** Yes
63. $\{x \mid x$ is a real number $and\ x \neq 3\}$ **65.** $\mathbb{R}$
67. $\{x \mid x \geq -6\}$, or $[-6, \infty)$

Exercise Set R.4, pp. R-38–R-39

1. $(4, 3)$ **3.** $(1, 3)$ **5.** $(2, -1)$ **7.** No solution
9. $(6, -1)$ **11.** $\{(x, y) \mid y = \frac{1}{5}x + 4\}$ **13.** $(2, -3)$
15. $(-4, 3)$ **17.** $(2, -2)$ **19.** $\{(x, y) \mid 2x - 3 = y\}$
21. $\left(\frac{25}{23}, -\frac{11}{23}\right)$ **23.** No solution **25.** $(1, 2)$ **27.** $(2, 7)$
29. $(-1, 2)$ **31.** $\left(\frac{128}{31}, -\frac{17}{31}\right)$ **33.** $(6, 2)$
35. No solution **37.** $\{(x, y) \mid -4x + 2y = 5\}$ **39.** 43, 46
41. 49°, 131° **43.** Two-pointers: 32; three-pointers: 4
45. Private lessons: 6 students; group lessons: 8 students
47. Peanuts: 135 lb; Brazil nuts: 345 lb
49.

x	y	90
12%	30%	20%
$0.12x$	$0.3y$	$0.2(90)$

Streakfree: 50 oz;
Sunstream: 40 oz
51. 3 hr **53.** Headwind: 40 mph; plane: 620 mph
55. 7 **57.** $-\frac{9}{2}$ **59.** -6 **61.** 3 **63.** 0.666667, or $\frac{2}{3}$

Exercise Set R.5, pp. R-47–R-48

1. 1 **3.** -3 **5.** $\dfrac{1}{8^2} = \dfrac{1}{64}$ **7.** $\dfrac{1}{(-2)^3} = -\dfrac{1}{8}$ **9.** $\dfrac{1}{(ab)^2}$
11. y^{10} **13.** y^{-4} **15.** x^{-t} **17.** x^{13} **19.** a^6
21. $(4x)^8$ **23.** 7^{40} **25.** x^8y^{12} **27.** $\dfrac{y^6}{64}$ **29.** $\dfrac{9q^8}{4p^6}$

31. $8x^3, -6x^2, x, -7$ **33.** $18, 36, -7, 3; 3, 9, 1, 0; 9$
35. $-1, 4, -2; 3, 3, 2; 3$ **37.** $8p^4; 8$ **39.** $x^4 + 3x^3$
41. $-7t^2 + 5t + 10$ **43.** 36 **45.** -14 **47.** 144 ft
49. $4x^3 - 3x^2 + 8x + 7$ **51.** $-y^2 + 5y - 2$
53. $-3x^2y - y^2 + 8y$ **55.** $12x^5 - 28x^3 + 28x^2$
57. $8a^2 + 2ab + 4ay + by$ **59.** $x^3 + 4x^2 - 20x + 7$
61. $x^2 - 49$ **63.** $x^2 + 2xy + y^2$ **65.** $6x^4 + 17x^2 - 14$
67. $a^2 - 6ab + 9b^2$ **69.** $42a^2 - 17ay - 15y^2$
71. $-t^4 - 3t^2 + 2t - 5$ **73.** $5x + 3$

75. $2x^2 - 3x + 3 + \dfrac{-2}{x + 1}$ **77.** $5x + 3 + \dfrac{3}{x^2 - 1}$

Exercise Set R.6, pp. R-59–R-60

1. $3x(x - 1)(x + 3)$ **3.** $(y - 3)^2$ **5.** $(p + 2)(2p^3 + 1)$
7. Prime **9.** $(2t + 3)(4t^2 - 6t + 9)$
11. $(m + 6)(m + 7)$ **13.** $(x^2 + 9)(x + 3)(x - 3)$
15. $(2x + 3)(4x + 5)$ **17.** $(x + 2)(x + 1)(x - 1)$
19. $(0.1t^2 - 0.2)(0.01t^4 + 0.02t^2 + 0.04)$
21. $\left(x^2 + \frac{1}{4}\right)\left(x + \frac{1}{2}\right)\left(x - \frac{1}{2}\right)$ **23.** $(a + 3 + y)(a + 3 - y)$
25. $(m + 15n)(m - 10n)$ **27.** $2y(3x + 1)(4x - 3)$
29. $(y - 11)^2$ **31.** $-7, 2$ **33.** $0, 4.7$ **35.** $-10, 10$
37. $-\frac{5}{2}, 7$ **39.** $0, 5$ **41.** $-2, 6$ **43.** $-11, 5$
45. -5 **47.** Base: 8 ft; height: 5 ft **49.** 8 ft, 15 ft

Appendixes

Exercise Set A, pp. AP-6–AP-7

1. 20 ft **3.** 11 g **5.** 24 ft^3 **7.** 12 cm^2 **9.** 12 yd
11. 4 ft-lb/sec **13.** 20 km/hr **15.** 4.4 cm/day
17. 48 cm^2 **19.** 25.1 in. **21.** 250 m
23. Approximately 15.9 cm **25.** 5625 kg-m^2/sec^2, or
5625 Joules **27.** 6.2 km = 6200 m
29. 35 mi/hr = 3080 ft/min **31.** 1 billion sec $\approx$ 31.7 yr
33. 216 in^2 = 1.5 ft^2 **35.** 5,865,696,000,000 mi/yr
37. 8 ft^2 **39.** Approximately 1337 calories **41.** A
43. B

Exercise Set B, p. AP-12

1. 5 **3.** $\sqrt{18} \approx 4.243$ **5.** $\sqrt{200} \approx 14.142$
7. 17.8 **9.** $\dfrac{\sqrt{41}}{7} \approx 0.915$ **11.** $\sqrt{8} \approx 2.828$
13. $\sqrt{90} \approx 9.487$ **15.** $(1, 4)$ **17.** $\left(\frac{7}{2}, \frac{7}{2}\right)$
19. $(-1, -3)$ **21.** $(-0.25, -0.3)$ **23.** $\left(-\frac{1}{12}, \frac{1}{24}\right)$
25. $\left(\dfrac{\sqrt{2} + \sqrt{3}}{2}, \dfrac{3}{2}\right)$ **27.** $x^2 + y^2 = 36$

29. $(x - 7)^2 + (y - 3)^2 = 5$
31. $(x + 4)^2 + (y - 3)^2 = 48$
33. $(x + 7)^2 + (y + 2)^2 = 50$ **35.** $x^2 + y^2 = 25$
37. $(x + 4)^2 + (y - 1)^2 = 20$
39. $(0, 0); 8$ **41.** $(-1, -3); 6$

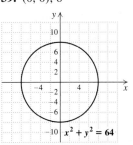
$x^2 + y^2 = 64$

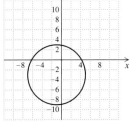
$(x + 1)^2 + (y + 3)^2 = 36$

43. $(4, -3); \sqrt{10}$ **45.** $(0, 0); \sqrt{10}$

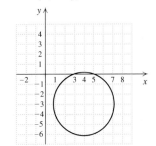
$(x - 4)^2 + (y + 3)^2 = 10$

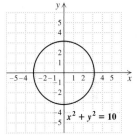
$x^2 + y^2 = 10$

47. $(5, 0); \frac{1}{2}$ **49.** $(-4, 3); \sqrt{40}$, or $2\sqrt{10}$

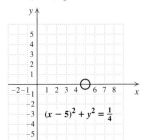
$(x - 5)^2 + y^2 = \frac{1}{4}$

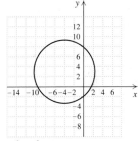
$x^2 + y^2 + 8x - 6y - 15 = 0$

51. $(4, -1); 2$ **53.** $(0, -5); 10$

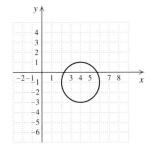
$x^2 + y^2 - 8x + 2y + 13 = 0$

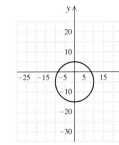
$x^2 + y^2 + 10y - 75 = 0$

55. $\left(-\dfrac{7}{2}, \dfrac{3}{2}\right)$; $\sqrt{\dfrac{98}{4}}$, or $\dfrac{7\sqrt{2}}{2}$ **57.** $(0, 0)$; $\frac{1}{6}$

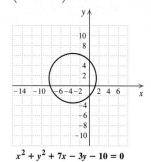

$x^2 + y^2 + 7x - 3y - 10 = 0$

$36x^2 + 36y^2 = 1$

Index*

*Page numbers followed by [CD] pertain to Chapters 13 and 14, found on the CD enclosed at the back of the book.**

Index of Applications*

*Page numbers followed by [CD] pertain to Chapters 13 and 14, found on the CD enclosed at the back of the book.